P9-ASM-666

Modern Carpentry

Building Construction Details in Easy-to-Understand Form

Willis H. Wagner
Professor Emeritus,
Industrial Technology
University of Northern Iowa, Cedar Falls

Howard Bud Smith
Technical Author and Editor
Lee Howard Associates
Bayfield, WI

Publisher
The Goodheart-Willcox Company, Inc.
Tinley Park, Illinois

Library of Congress Catalog Card Number 95-12170
International Standard Book Number 1-56637-198-8

4567890-96-01 00 99 98

Library of Congress Cataloging-in-Publication Data

Wagner, Willis H.
Modern carpentry: building construction details in easy-to-understand form / by Willis H. Wagner, (Howard Sylvester Smith).
p. cm.
Includes index.
ISBN 1-56637-198-8
1. Carpentry. I. Smith, Howard Sylvester. II. Title.
TH 5606. W34 1996
694—dc20 95-12170
CIP

Introduction

Modern Carpentry is a colorful, easy-to-understand source of authoritative and up-to-date information on building materials and construction methods. It provides detailed coverage of all aspects of light frame construction. Included are site clearing, site layout, foundations, framing, insulating, sheathing, roofing, windows and doors, exterior finish, interior finish, and mechanical systems. Units are arranged in a logical sequence—similar to the order in which the various phases of construction are performed. Special emphasis is placed on safety and the use of modern tools, materials, and prefabricated components.

Information about building materials includes size and grade descriptions and also basic technical information that covers their physical properties and other important characteristics. Scientific and technical discoveries have led to the development of many new materials. The proper use and application of these materials depend on a craftperson who has considerable knowledge of the material and how it will function in a completed structure.

Modern Carpentry includes basic information covering stair construction, chimney and fireplaces, systems-built structures, solar construction, remodeling, cabinetmaking, painting, and decorating.

Modern Carpentry also serves as an introduction to other building trades. Information about electrical wiring, plumbing systems, and heating, ventilation, and air conditioning (HVAC) has been included to provide more exposure to related building trades.

Modern Carpentry contains more than 1600 carefully selected photos and drawings. Illustrations are accurately coordinated with written instructions and descriptions that are easy to read.

Modern Carpentry is designed to provide basic instruction for students in high school, vocational-technical schools, college classes, and apprentice training programs. It can also serve as a valuable reference for students in architectural drafting classes and for journeymen carpenters and construction supervisors. **Modern Carpentry** will enable do-it-yourselfers to handle many construction jobs that they would otherwise be reluctant to undertake.

Willis H. Wagner
Howard Bud Smith

New Additions to this Edition

Carpenters work closely with members of other building trades. A carpenter should have a general knowledge of what these other workers are doing and the relationship between the workers' tasks. Four new units have been added that discuss other trades:

- Unit 20—Painting, Finishing, and Decorating
- Unit 26—Electrical Wiring
- Unit 27—Plumbing Systems
- Unit 28—Heating, Ventilation, and Air Conditioning

Safety on the worksite is of paramount importance. Carpentry trainees must be taught to recognize and correct unsafe conditions and practices. Unit 2, General Safety Rules, has been added in this edition to provide this training.

An understanding of basic mathematics is a tool a carpenter will find as useful as any saw, drill, or hammer. Math is used to ensure that roofs slope properly, stairs rise evenly, and studs fit correctly. Appendix A, Carpentry Math Review, has been added to provide a "refresher course" of basic mathematics.

Special Features

Throughout the text, the following special features are used.

Safety

Safety-related items have been printed with red text to make them more noticeable. Read these items carefully—they are the most important lessons you will learn from this book.

Step-By-Step Procedures

Tasks that require a series of steps to complete are set off from the text with a drawing of a sawhorse and the title of the task, such as:

Installing a Window

An ordered list of the procedures follows the title. When using the book for reference, these step-by-step procedures will be easy to find.

Tips

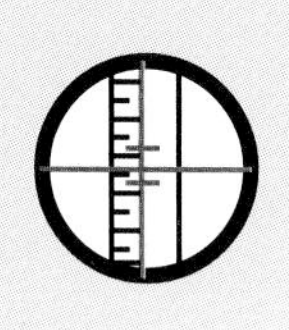

Brief suggestions and safety items appear in this form. A small boxed sketch, related to the topic, appears to the left of the text. Be sure to pay special attention to these tips.

Acknowledgments

The authors wish to thank the individuals and organizations listed below for the valuable information, photographs, and line illustrations they so willingly provided.

ABTco, Incorporated, Troy, MI
Acoustical and Board Products Association, Palatine, IL
ACE Hardware, Sister Bay, WI
Acme Brick Company, Fort Worth, TX
Agricultural Extension Service, University of Minnesota
Ahnen, Gene, Bayfield, WI
Alcoa Building Products, Inc., Pittsburgh, PA
Alum-A-Pole Corporation, Scranton, PA
American Building Components, Omaha, NE
American Chemwood Corporation, Aurora, IL
American Institute of Timber Construction, Englewood, CO
American Olean Tile Company, Lansdale, PA
American Plywood Association, Tacoma, WA
American Standard, Incorporated, Chicago, IL
Amerock Corporation, Rockford, IL
Amoco Foam Products Company, Atlanta, GA
Andersen Corporation, Bayport, MN
Architectural Woodwork Institute, Arlington, VA
Arcways, Inc., Neenah, WI
Ark-Seal, Incorporated, Denver, CO
Armstrong World Industries, Incorporated, Lancaster, PA
Asphalt Roofing Manufacturers Association, New York, NY
Baldwin Hardware Corporation, Reading, PA
Bayfield Lumber Company, Bayfield, WI
Beecham Home Improvement Products, Dayton, OH
Benjamin Obdyke, Incorporated, Warminster, PA
Binks Manufacturing Company, Franklin Park, IL
Bird & Son, East Walpole, MA
Harry Black, Anamosa, IA
Black & Decker, Incorporated, Hunt Valley, MD
Blandin Wood Products Company, Grand Rapids, MI
Boise-Cascade Corporation, Boise, ID
Borden, Incorporated, New York, NY
Brammer Manufacturing Company, Chicago, IL
Bruce Hardwood Floors, Dallas, TX
Bullard Haven Technical School, Bridgeport, CT
The Burke Company, San Mateo, CA
C-E Morgan Building Products, Oshkosh, WI
Calculated Industries, Incorporated, Yorba Linda, CA
Canadian Plywood Association, North Vancouver, B.C., Canada
Cardinal Industries, Incorporated, Columbus, OH
Carrier International Corporation, New York, NY
Cascade Precast Concrete Products, Cascade, IA
Cella Barr Associates, Tucson, AZ
Cemco, Incorporated, Louisville, KY
CertainTeed Corporation, Valley Forge, PA
Citation Homes, Spirit Lake, IA
Clearfield Conveyors, Clearfield, UT
Colonial Stair and Woodwork Company, Jeffersonville, OH
Columns, Incorporated, Pearland, TX
Conestoga Wood Specialties, Incorporated, East Earl, PA
Construction Training School, St. Louis, MO
Council of Forrest Industries, North Vancouver, B.C., Canada
Crown Aluminum Industries, Corporation, Pittsburgh, PA
Dal-Tile Corporation, Dallas, TX
Daniel Woodhead Company, Northbrook, IL
David White Instruments, Division of Realist, Inc., Menomonee Falls, WI
Deft, Incorporated, Irvine, CA
Del Webb's Sun City, Tucson, AZ
Delta International Machine, Corporation, Pittsburgh, PA
Des Champs Laboratories, Incorporated, Natural Bridge, VA
Des Moines, Iowa Public Schools, Des Moines, IA
DeVilbiss Company, Toledo, OH
Dexter Industries, Windsor Locks, CT
Dickinson Homes, Incorporated, Kingsford, MI
District 1 Technical Institute, Eau Claire, WI
Duo-Fast Corporation, Franklin Park, IL
Dutcher Glass & Paint, Cedar Falls, IA
E.I. Dupont de Nemours and Company, Wilmington, DE
Eaton, Yale & Towne, Incorporated, White Plains, NY
Ekco Building Products Company, Canton, OH
Enercept, Incorporated, Watertown, SD
Flintkote Company, New York, NY
Foley Manufacturing Company, Minneapolis, MN
Folkers Construction, Dike, IA
Forest Products Laboratory, Madison, WI
Forestry Suppliers, Incorporated, Jackson, MS
Formica Corporation, Cincinnati, OH
Forslund Building Supply, Incorporated, Ashland, WI

Frank Paxton Lumber Company, Des Moines, IA
GAF Building Materials Corporation, Wayne, NJ
GB Electrical, Incorporated, Milwaukee, WI
Gamble Brothers, Incorporated, Louisville, KY
Gang-Nail Systems, Incorporated, Miami, FL
Garlinghouse Company, Incorporated, Topeka, KS
General Products Company, Incorporated, Fredericksburg, VA
Georgia-Pacific Corporation, Atlanta, GA
Gold Bond Building Products, Charlotte, NC
Goldblatt Tool Company, Kansas City, KS
Al Gordon Plumbing and Heating, Waterloo, IA
Gory Associated Industries, Incorporated, North Miami, FL
Greco Painting, Tucson, AZ
Joe Griffith Construction, Cedar Falls, IA
Grimes-Port Jones-Schwerdtfeger Architects, Incorporated, Waterloo, IA
Grosse Steel Company, Incorporated, Cedar Falls, IA
Gypsum Association, Evanston, IL
H.L. Stud Corporation, Columbus, OH
Haas Cabinet Company, Incorporated, Sellersburg, IN
Hadley-Hobley Construction, Kansas City, MO
John Hall Laminate Work, Cedar Falls, IA
Harrington, Los Angeles, CA
Headlee Roofing, Phoenix, AZ
Homewood Scavenger Services, Incorporated, East Hazelcrest, IL
Honeywell, Incorporated, Minneapolis, MN
Hoxan America, Incorporated, Piscataway, NY
HUD, Washington, DC
I-XL Furniture Company, Goshen, IN
Ideal Company, Waco, TX
Independent Nail and Packing Company, Bridgewater, MA
Insulspan, Nashville, TN
Iowa Energy Policy Council, Des Moines, IA
ITW Paslode, An Illinois Tool Works Co., Lincolnshire, IL
J. Rouleau and Associates, Hanover, NH
Jacuzzi Whirlpool Bath, Incorporated, Walnut Creek, CA
James Hardie Building Products, Fontana, CA
John Dutcher Glass and Paint, Cedar Falls, IA
John S. Tilley Ladders Company, Incorporated, Davenport, IA
Johnson Manley Lumber Company, Tucson, AZ
Journal of Light Construction, Richmond, VT
Kasten-Weiler Construction, Fish Creek, WI
KCPL
Kentron Division, North American Reiss, Belle Mead, NJ
Kitchen Kompact, Incorporated, Jeffersonville, IN
Kohler Company, Kohler, WI
KraftMaid Cabinetry, Incorporated, Cleveland, OH
Kunkle Valve Company, Fort Wayne, IN
L.J. Smith, Incorporated, Bowerston, OH
L.S. Starrett, Company, Athol, MA
Lennox Industries, Incorporated, Dallas, TX
Libbey-Owens, Ford Glass Company, Toledo, OH
LiteForm, Incorporated, Sioux City, IA
Louisiana-Pacific Corporation, Portland, OR
LTL Home Products, Incorporated, Schuylkill, PA
Luxaire Heating and Air Conditioning, York, PA
Macklanburg-Duncan, Oklahoma City, OK
Majestic Company, Incorporated, Huntington, IN
Makita USA, Incorporated, La Mirada, CA
Malm Fireplaces, Incorporated, Santa Rosa, CA
Manville Building Materials Corporation, Denver, CO
Marquart Block Company, Waterloo, IA
Marshfield Homes, Incorporated, Marshfield, WI
Marvin Windows and Doors, Warroad, MN
Masonite Corporation, Chicago, IL
McDaniels Construction Company, Columbus, OH
McRae True Value Hardware
Mellin Well Service, Ashland, WI
Memphis Hardwood Flooring Company, Memphis, TN
Merrilat Industries, Incorporated, Adrian, MI
MET-TILE, Incorporated, Ontario, Canada
Milwaukee Electrical Tool Corporation, Brookfield, WI
Monier Roof Tile Company, Irvine, CA
Montachusett Regional Vo-Tech School, Fitchburg, MA
National Building Code, Washington, D.C.
National Decorating Products Association, St. Louis, MO
National Forest Products Association, Washington, DC
National Gypsum Company, Dallas, TX
National Oak Flooring Manufacturers Association, Memphis, TN
National Solar Heating and Cooling Center, Rockville, MD
North Bennet Street School, Boston, MA
Oak Flooring Institute, Memphis, TN
Omni Products, Addison, IL
Orem Research, Hinsdale, IL
Osmose, Buffalo, NY
Owens-Corning Fiberglass, Corporation, Toledo, OH
Owner/Builder Directory, Incorporated, Berkeley, CA
The Panel Clip Company, Farmington, MI
Pass & Seymour, Incorporated, Syracuse, NY
Patent Scaffolding Company, Long Island City, NY
Pease Industries, Incorporated, Fairfield, OH
Perlite Institute, Incorporated, New York, NY
Peters Construction Company, Waterloo, IA
Pierce Custom Homes, Limited, Green Valley, AZ
Pittsburgh Corning Corporation, Pittsburgh, PA
Pittsburgh Plate Glass Company, Pittsburgh, PA
Porter-Cable Corporation, Jackson, TN
Portland Cement Association, Skokie, IL
Preway Incorporated, Wisconsin Rapids, WI
RECON–Reconstruction Unlimited, Incorporated, Columbus, OH
Red Cedar Shingle and Handsplit Shake Bureau, Seattle, WA
Redman Industries, Incorporated, Dallas, TX
Reed Manufacturing Company, Erie, PA
Riviera Cabinets, Incorporated, Chesapeake, VA

Robbins/Sykes, Warren, AR
Robert Bosch Power Tool Corporation, New Bern, NC
Rock Island Millwork, Waterloo, IA
Rokes Building and Supply, Waterloo, IA
Rolscreen Company, Pella, IA
Ronthor, Plastics Division, US Manufacturing Corporation, New York, NY
Santa Rita High School, Tucson, AZ
Senco Products, Incorporation, Cincinnati, OH
Shakertown Corporation, Cleveland, OH
Sherwin-Williams Company, Cleveland, OH
Ship and Shore General Store, Dauphin Island, AL
Simplex Products, Adrian, MI
Simpson Strong Tie Company, Incorporated, San Leandro, CA
Slant/Fin Corporation, Greenvale, NY
Southern Forest Products Association, New Orleans, LA
Spectra-Physics Laserplane, Incorporated, Dayton, OH
Speed Cut, Incorporated, Corvallis, OR
St. Paul Technical College, St. Paul, MN
Stan Greer Millwork, Sierra Vista, AZ
Stanley Door Systems, Farmington, CT
Stanley Tools, Covington, GA
Stanley Works, New Britain, CT
Sterling, Rolling Meadows, IL
Superior Fireplace Company, Fullerton, CA
T.W. Lewis Construction Company, Tempe, AZ
Tapco International, Plymouth, MI
Technology Systems
TECO/Lumberlok, Hayward, CA
Therma-Tru Corporation, Bowling Green, OH
Tibbias Flooring Company, Oneida, TN
Timber Engineering Company, Washington, DC
Timberpeg, Claremont, NH
Tony's Construction, Incorporated, Tucson, AZ
Trane Company, La Crosse, WI
Trudeau Construction Company, Bayfield, WI
Truss Plate Institute, Madison, WI
Trussworks, Incorporated, Hayward, WI
TrusWal Systems, Incorporated, Troy, MI
U.S. Department of Agriculture, Washington, D.C.
Ungrodt Hardware Company, Washburn, WI
United Brotherhood of Carpenters & Joiners of America, Washington, DC
United States Gypsum Company, Chicago, IL
United States Steel Corporation, Pittsburgh, PA
United Technologies Carrier, Indianapolis, IN
Universal Form Clamp Company, Chicago, IL
USG Corporation, Chicago, IL
Vermiculite Institute, Minneapolis, MN
Vermont American Tool Company, Lincolnton, NC
VICA (Vocational Industrial Clubs of America), Leesburg, VA
Village of Flossmoor, Flossmoor, IL
Visador Company, Jacksonville, FL
Wageman Construction, Cedar Falls, IA
Wallace Murray Corporation, Nampa, ID
Waterloo-Cedar Falls Iowa Daily Courier, Waterloo, IA
Wausau Homes, Incorporated, Wausau, WI
Wellborn Cabinet, Incorporated, Ashland, AL
Weller, Division of Cooper Tools
Western Wood Products Association, Portland, OR
Weyerhaeuser Company, Tacoma, WA
Whirlpool Corporation, Benton Harbor, MI
William Powell Company
Wiss
Wolmanized Wood Producers
Wood Conversion Company, St. Paul, MN
Yankee Barn Homes, Incorporated, Granthan, NH
Zimmerman Builders, Waterloo, IA

Contents

Section 1
Preparing to Build

Section 2
Footings, Foundations, and Framing

Section 3
Closing In

Section 4 Finishing

Section 5
Special Construction

Section 6
Mechanical Systems

Section 7
Scaffolds and Careers

Preparing to Build

Unit 1 Building Materials

Unit 2 General Safety Rules

Unit 3 Hand Tools

Unit 4 Power Tools

Unit 5 Leveling Instruments

Unit 6 Plans, Specifications, and Codes

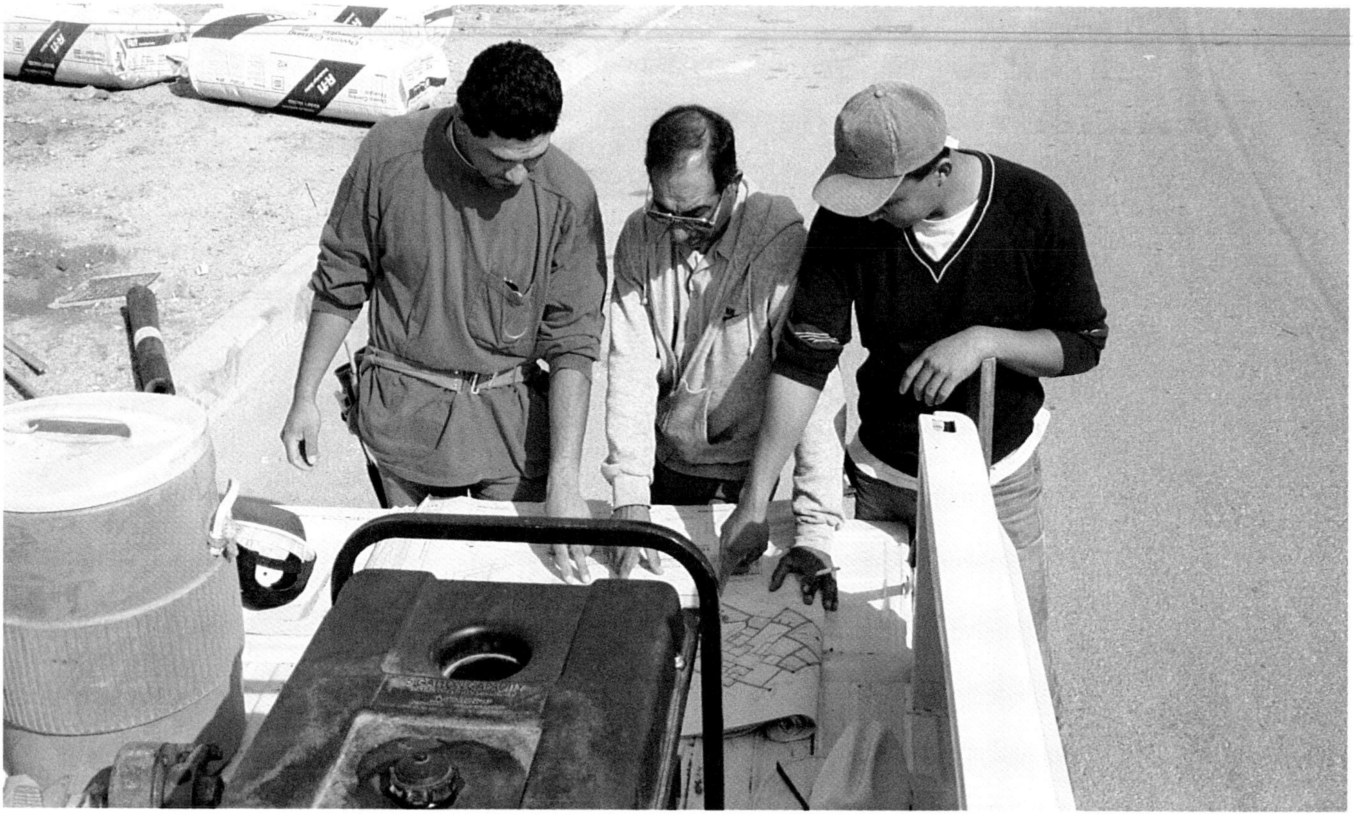

Construction drawings must be used as a guide throughout the construction process.

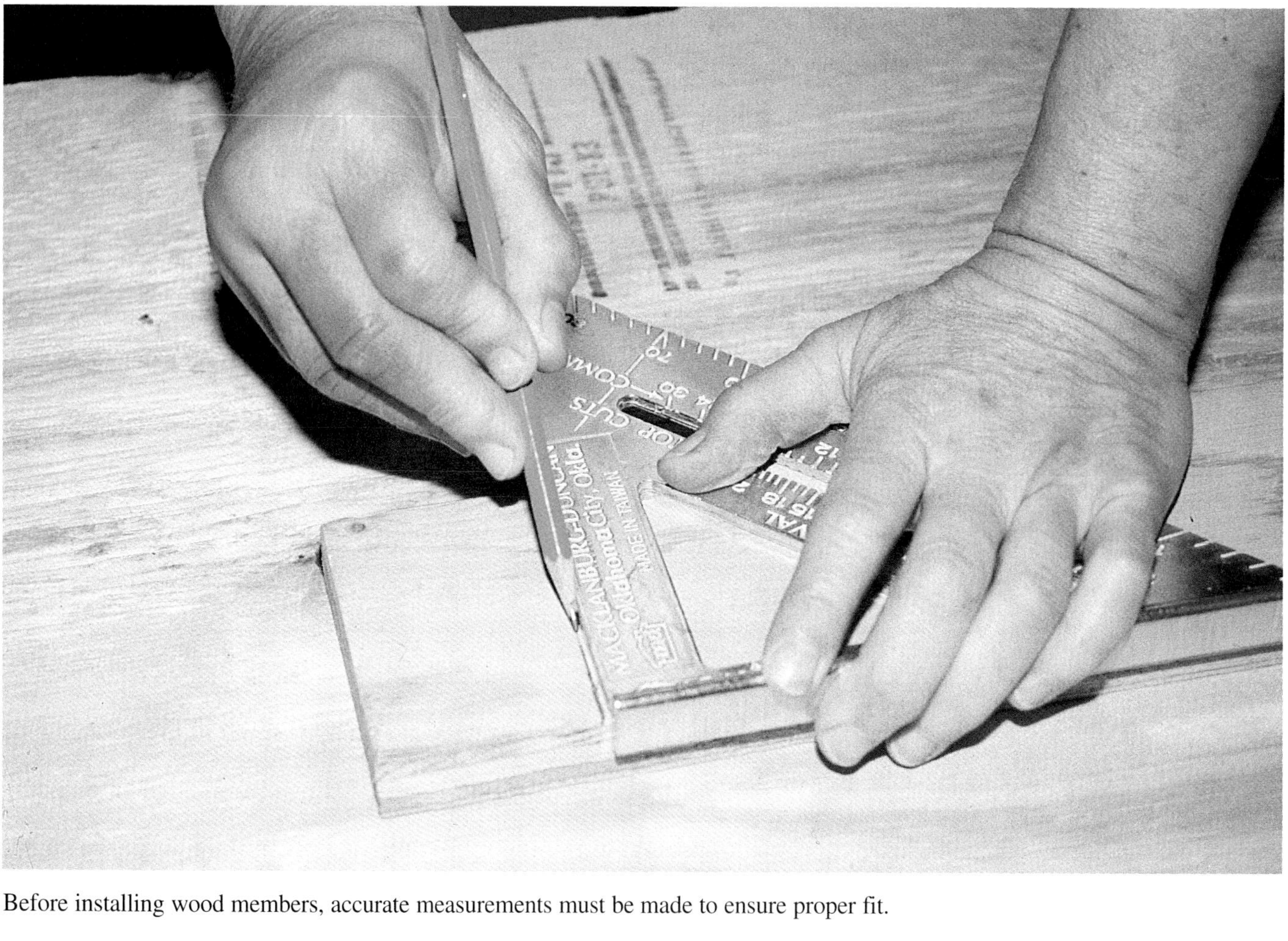

Before installing wood members, accurate measurements must be made to ensure proper fit.

1

Building Materials

Many different types of materials go into the construction of a modern residence, Figure 1-1. A carpenter should be familiar with all of these materials—each has special properties that make it suited to certain building applications.

Before being used in construction, materials are treated to improve their quality. Some are composites of different materials which are designed to do a certain job better than a natural material. Construction materials include:

- Sawed lumber.
- Engineered lumber.
- Plywood.
- Particleboard, hardboard, and waferboard.
- Wood and nonwood materials for shingles and flooring.
- Steel and aluminum.
- Concrete.
- Adhesives and sealers.
- Gypsum board and fibrous manufactured ceiling tiles.

Lumber

Wood is one of our greatest natural resources. When cut into pieces uniform in thickness, width, and length, it

Figure 1-1. About 80% of our nation's homes have a structural framework built of wood. Other products used include asphalt, aluminum, particleboard or other manufactured board, concrete, and steel.

becomes lumber. This material has always been widely used for residential construction.

Lumber is the name given to natural or engineered products of the sawmill. Lumber includes:

- Boards used for flooring, sheathing, paneling, and trim.
- Dimension lumber used for sills, plates, studs, joists, rafters, and other framing members.
- Timbers used for posts, beams, and heavy stringers.
- Numerous specialty items.

Carpenters must have a good working knowledge of lumber. They must be familiar with kinds, grades, sizes, and other details to properly select and use lumber. To understand how to handle and treat wood, carpenters should also know something about its growth, structure, and characteristics.

Figure 1-2. Leaf silhouettes of several species of hardwoods and softwoods. (Gamble Brothers, Inc.)

Wood Structure and Growth

Wood is made up of long narrow tubes or cells called fibers or *tracheids*. The cells are no larger around than human hair. Their length varies from about 1/25" for hardwoods to approximately 1/8" for softwoods. Tiny strands of cellulose make up the cell walls. The cells are held together with a natural cement called *lignin*. This cellular structure makes it possible to drive nails and screws into the wood. It also accounts for the light weight, low heat transmission, and sound absorption qualities of wood.

The growing parts of a tree are:

- The tips of the roots.
- The leaves.
- A layer of cells just inside the bark called the *cambium*.

Different types of leaves are shown in Figure 1-2. Water absorbed by the roots travels through the sapwood to the leaves, where it is combined with carbon dioxide from the air. Through the chemical process of *photosynthesis*, sunlight changes these elements to a food called carbohydrates. The sap carries this food back to the various parts of the tree.

Figure 1-3 shows the location of the *cambium layer*, where new cells are formed. The inside area of the cambium layer is called *xylem*. It develops new wood cells. The outside area, known as *phloem*, develops cells that form the bark.

Annular Rings

Growth in the cambium layer takes place in the spring and summer. Separate layers form each season. These layers are called *annular rings*, Figure 1-4. Each ring is composed of two layers: springwood and summerwood.

In the spring, trees grow rapidly and the cells produced are large and thin-walled. As growth slows down

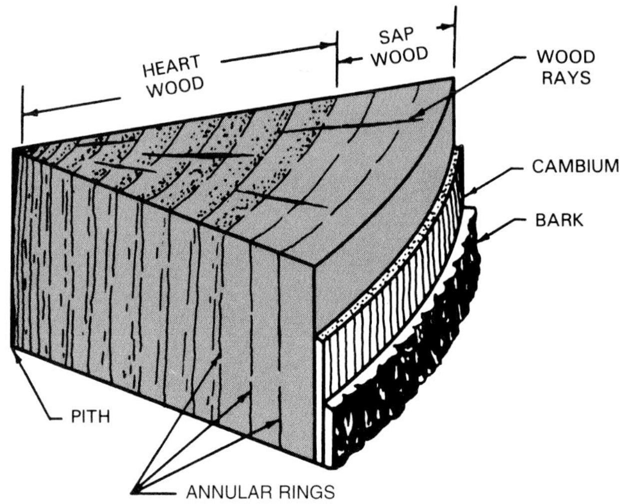

Figure 1-3. A tree is made up of many different layers.

Figure 1-4. Annular rings are formed each year and indicate the age of the tree. Drought, disease, or insects can interrupt growth, causing an extra or false ring to form. (Forest Products Lab.)

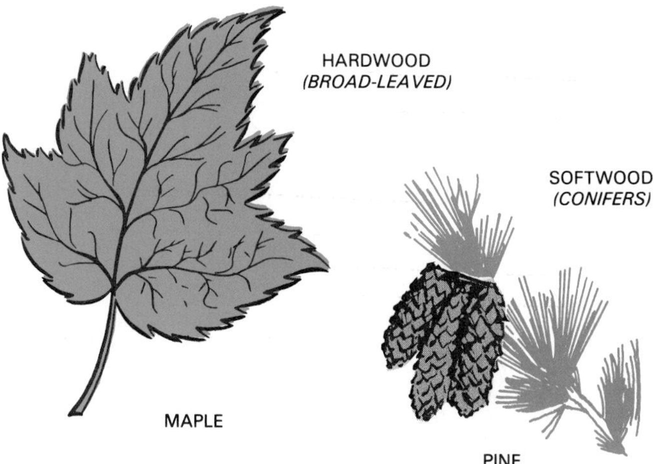

Figure 1-5. Woods are classified as either hardwoods or softwoods. Hardwoods have broad leaves; softwoods have needles.

during summer months, the cells produced are smaller, thick-walled, and darker in color. These annual growth rings are largely responsible for the grain patterns that are seen in the surface of boards.

Sapwood is located inside the cambium layer. It contains living cells and may be several inches thick. Sapwood carries sap to the leaves. The *heartwood* of the tree is formed as the sapwood becomes inactive. Usually, it turns darker in color because of the presence of gums and resins. In some woods such as hemlock, spruce, and basswood, there is little or no difference in the appearance of sapwood and heartwood. Sapwood is as strong and heavy as heartwood but is not as durable when exposed to weather.

SOFTWOODS	HARDWOODS
Douglas Fir Southern Pine Western Larch	Basswood Willow American Elm
Hemlock White Fir Spruce	*Mahogany Sweet Gum *White Ash
Ponderosa Pine Western Red Cedar Redwood	Beech Birch Cherry
Cypress White Pine Sugar Pine	Maple *Oak *Walnut

*Open grained wood

Figure 1-6. List of popular woods for residential use. Usually, softwoods are used as framing lumber.

Kinds of Wood

Lumber is either softwood or hardwood. Softwoods come from the evergreen, or needle-bearing, trees. These are called *conifers* because many of them bear cones. See Figure 1-5. Hardwoods come from broadleaf (*deciduous*) trees that shed their leaves at the end of the growing season. This classification is somewhat confusing because many of the hardwood trees produce a softer wood than some of the so-called softwood trees. Several of the more common kinds of commercial softwoods and hardwoods are listed in Figure 1-6.

A number of hardwoods have large pores in the cellular structure and are called *open grain woods*. They require additional operations during finishing. Different kinds of wood will also vary in weight, strength, workability, color, texture, grain pattern, and odor.

Study the full-color wood samples at the end of this unit as a first step in wood identification. To further develop ability to identify woods, study actual specimens. Several of the softwoods used in construction work are similar in appearance. Considerable experience is required to make accurate identification.

Most of the samples shown in this text were cut from plain-sawed or flat-grained stock. Edge-grained views would look different.

Availability of different species (kinds) of lumber varies from one part of the country to another. This is especially true of framing lumber, which is expensive to transport long distances. It is usually more economical to select building materials found in the area.

Cutting Methods

Most lumber is cut so that the annular rings form an angle of less than 45° with the surface of the board. This produces lumber called *flat-grained* if it is softwood, or *plain-sawed* if it is hardwood. This method produces the least waste and more desirable grain patterns.

Lumber can also be cut so the annular rings form an angle of more than 45° with the surface of the board, Figure 1-7. This method produces lumber called *edge-grained* if it is softwood, and *quarter-sawed* if it is hardwood. It is more difficult and expensive to use this method. However, it produces lumber that swells and shrinks less across its width and is not as likely to warp.

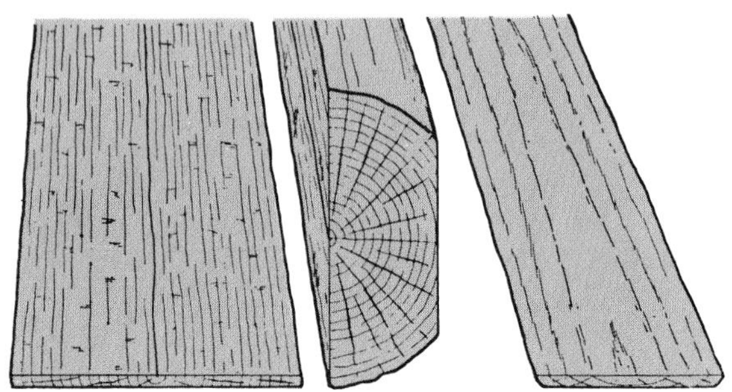

Figure 1-7. Lumber may be sawed in different ways. Left—Board is edge-grain or quarter-sawed. Saw cut was made roughly parallel to a line running through center of log. Right—Flat-grained or plain-sawed board.

Moisture Content and Shrinkage

Before wood can be used commercially, a large part of the moisture (sap) must be removed. When a living tree is cut, more than half of its weight may be moisture.

Lumber used for framing and outside finish should be dried to a moisture content of 15%. Cabinet and furniture woods are dried to a moisture content of 7 to 10%.

The amount of moisture or *moisture content* (M.C.) in wood is given as a percent of the oven-dry weight. To determine the moisture content, a sample is first weighed. It is then placed in an oven and dried at a temperature of about 212°F (100°C). The drying is continued until the wood no longer loses weight. The sample is weighed again and this oven-dry weight is subtracted from the initial weight. The difference is then divided by the oven-dry weight, Figure 1-8.

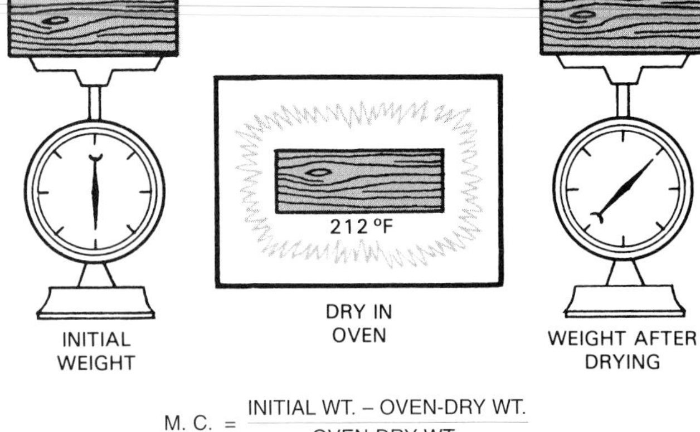

Figure 1-8. Moisture content of wood can be determined by weighing it before and after it is dried.

Moisture contained in the cell cavities is called free water. Moisture contained in the cell walls is called bound water. As the wood is dried, moisture first leaves the cell cavities. When the cells are empty but the cell walls are still full of moisture, the wood has reached a condition called the *fiber saturation point*. For most woods this is about 30%, Figure 1-9.

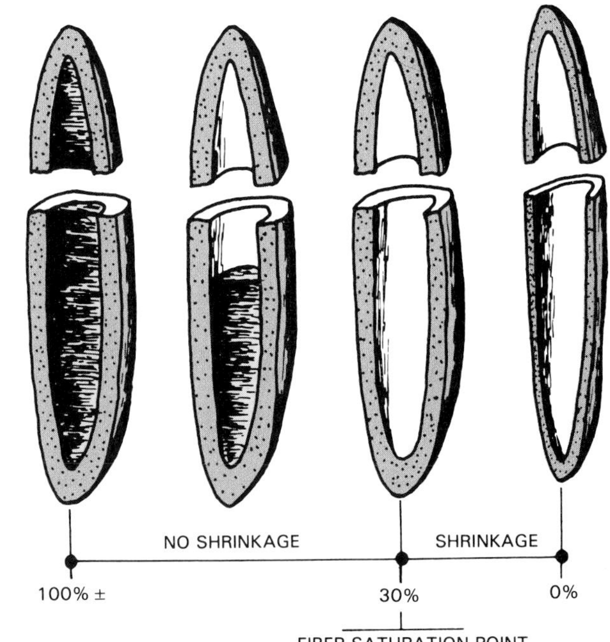

Figure 1-9. How a wood cell dries. First the free water in the cell cavity is removed. Then the cell wall dries and shrinks.

The fiber saturation point is important because wood does not start to shrink until this point is reached. As the M.C. drops below 30%, moisture is removed from the cell walls and they shrink. Figure 1-10 shows the actual shrinkage in a 2 x 10 joist.

Wood shrinks most along the direction of the annual rings (tangentially) and about one-half as much across

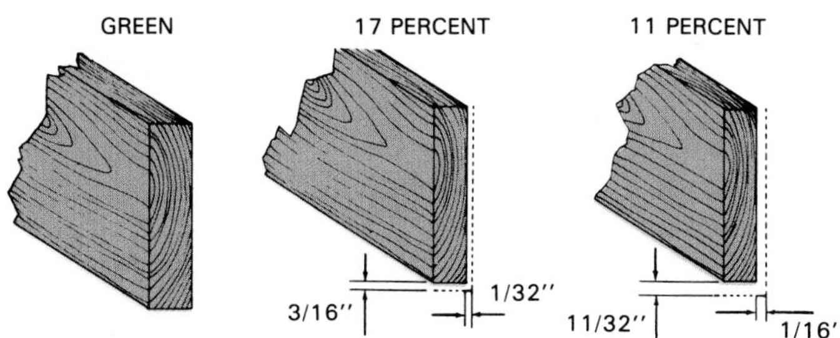

Figure 1-10. A 2 x 10 joist may shrink 1/16" across its shortest dimension.

these rings. There is little shrinkage in the length. How this shrinkage affects lumber cut from different parts of a log is shown in Figure 1-11. As wood takes on moisture, it swells in the same proportion as the shrinkage that took place.

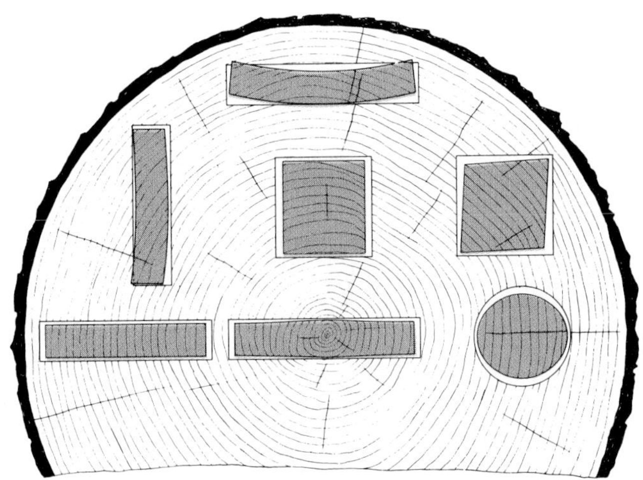

Figure 1-11. Shrinkage and distortion of flat, square, and round pieces is affected by the direction of the annual rings.

Equilibrium Moisture Content

A piece of wood will give off or take on moisture from the air around it until the moisture in the wood is balanced with that in the air. At this point the wood is said to be at *equilibrium moisture content* (E.M.C.).

Since wood is exposed to daily and seasonal changes in the relative humidity of the air, its moisture content is always changing. Therefore, its dimensions are also changing. This is the reason doors and drawers often stick during humid weather.

Ideally, a wood structure should be framed with lumber at an M.C. equal to that of the environment it is located in. This is not practical. Lumber with such a low moisture content is seldom available and would likely gain moisture during construction. Standard practice is to use lumber with a moisture content in the range of 15% to 19%. In heated structures, it will eventually reach a level of about 8%. However, this will vary in different geographical areas, Figure 1-12.

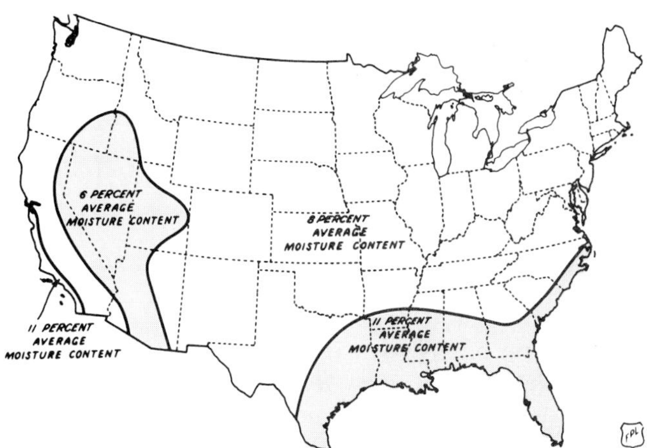

Figure 1-12. Map shows average moisture content of interior woodwork for various regions of the United States.

Carpenters understand that shrinkage is inevitable. They make allowances where needed to reduce the effects of shrinkage. The first, and by far the greatest, change in moisture content occurs during the first year after construction, particularly during the first heating season.

When "green" lumber (more than 20% M.C.) is used, shrinkage will be excessive. Warping, plaster cracks, nail pops, squeaky floors, and other difficulties are almost impossible to prevent.

Seasoning Lumber

Seasoning is reducing the moisture content of lumber to the required level specified for its grade and use. In air drying, the lumber is simply exposed to the outside air. It is carefully stacked with stickers (wood strips) between layers so air can circulate through the pile. Boards are also spaced within the layers so air can reach edges. Air drying is a slower process than kiln drying. It often creates additional defects in the wood.

Lumber is *kiln dried* by placing it in huge ovens where the temperature and humidity can be carefully controlled. When the green lumber is first placed in the kiln, steam is used to keep the humidity high. The temperature, meanwhile, is kept at a low level. Gradually the temperature is raised while the humidity is reduced. Fans keep the air in constant circulation. See Figure 1-13.

Bundles of lumber may carry a stamp to indicate that they have been kiln dried. The letters "k.d." mean "kiln dried," while "p.k.d." stands for "partly kiln dried."

Figure 1-13. Huge kilns are used to season lumber at a modern sawmill. (Forest Products Lab.)

Moisture Meters

The moisture content of wood can be determined by:

- Oven drying a sample, as previously described.
- Using an electric moisture meter.

Although the oven drying method is more accurate, meters are often used because readings can be secured rapidly and conveniently. The meters are usually calibrated to cover a range from 7% to 25%. Accuracy is plus or minus 1% of the moisture content.

There are two basic types of moisture meters. One determines the moisture content by measuring the electrical resistance between two pin-type electrodes that are driven into the wood. The other type measures the capacity of a condenser in a high-frequency circuit where the wood serves as the dielectric (nonconducting) material of the condenser. See Figure 1-14.

Lumber Defects

A defect is an irregularity occurring in or on wood that reduces its strength, durability, or usefulness. It may also detract from or improve the appearance. For example, knots, commonly considered a defect, may add to the appearance of pine paneling. An imperfection that impairs only the appearance of wood is called a blemish. Some of the common defects include:

- *Knots*—See Figure 1-15. Caused by an embedded branch or limb, knots reduce the strength of the wood. The extent of damage depends upon the type of knot, its size, and its location. See Figure 1-16.
- *Splits and checks*—Separations of the wood fibers which run along the grain and across the annular growth rings, splits usually occur at the ends of lumber that has been unevenly seasoned.
- *Shakes*—Separations along the grain and between the annular growth rings, shakes are likely to occur only in species with abrupt change from spring to summer growth.

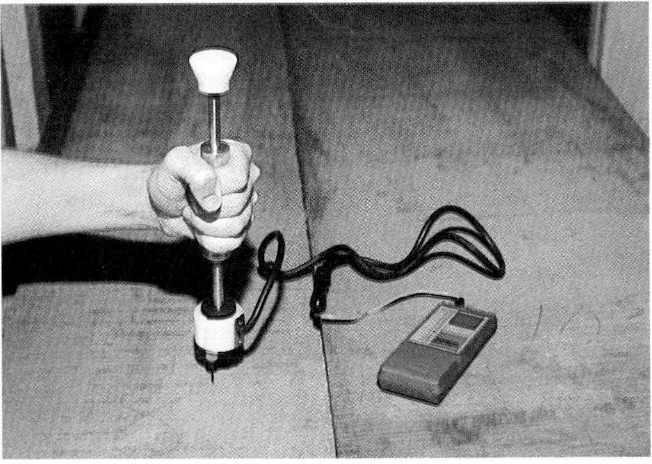

Figure 1-14. Several moisture meters. Top—Modern moisture meter. LEDs indicate correct moisture content. Middle—Probes inserted into end cut. Bottom—Hammer probe provides readings up to 1" below surface. (Forestry Suppliers Inc.)

- *Pitch pockets*—Pitch pockets are cavities that contain or have contained pitch in solid or liquid form.
- *Honeycombing*—Separation of the wood fibers inside the tree, honeycombing may not be visible on the board's surface.

Figure 1-15. Knots are the most frequently occurring defect in lumber. They affect the strength, stiffness, and appearance of lumber. These are the most common types of knots. (Western Wood Products Assoc.)

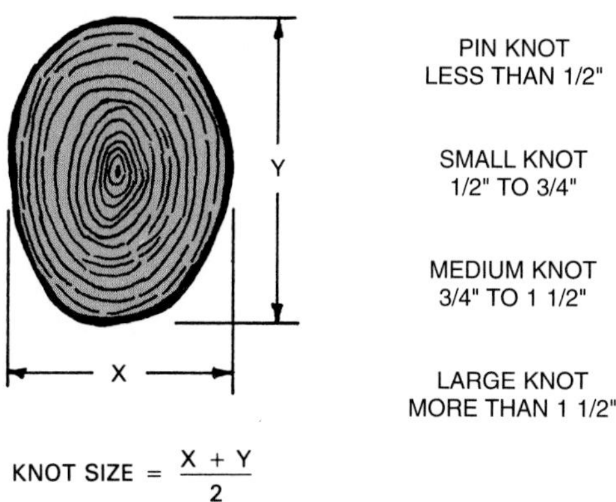

Figure 1-16. Knot size is determined by adding the width and length of the knot and dividing by two.

- *Wane*—The presence of bark or the absence of wood along the edge of a board, wane forms a bevel and/or reduces width.
- *Blue stain*—Discoloration caused by a moldlike fungus, blue stain has little or no effect on the strength of the wood. However, it is objectionable for appearance in some grades of lumber.
- *Decay*—Disintegration of wood fibers due to fungi, decay is often difficult to recognize in its early stages. In advanced stages, wood is soft, spongy, and crumbles easily.
- *Holes*—Caused by handling equipment or boring insects and worms, holes will lower the lumber grade.
- *Warp*—Any variation from true or plane surface, warp may include any one or combinations of the following: cup, bow, crook, and twist (or wind). See Figure 1-17.

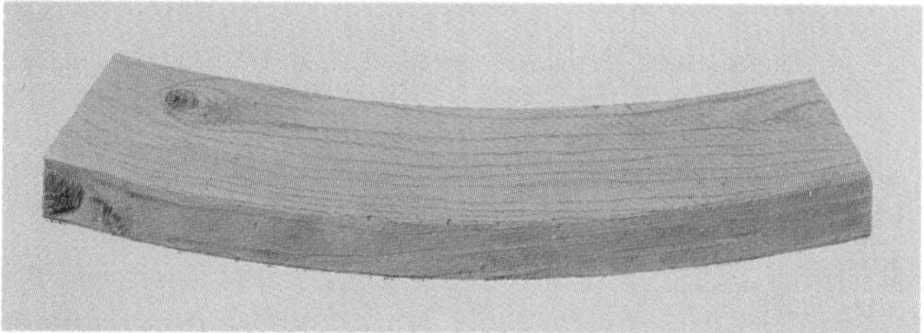

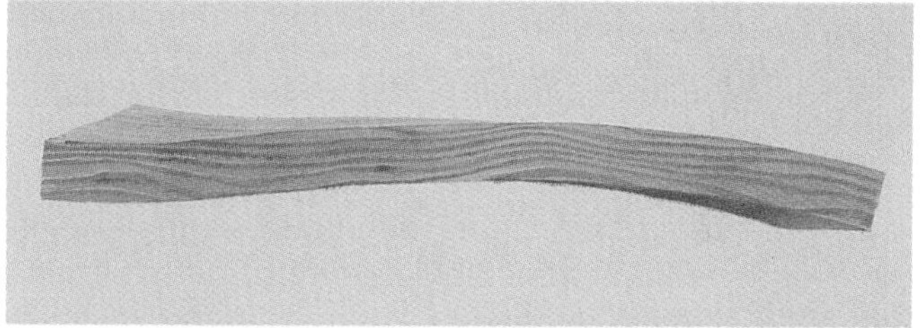

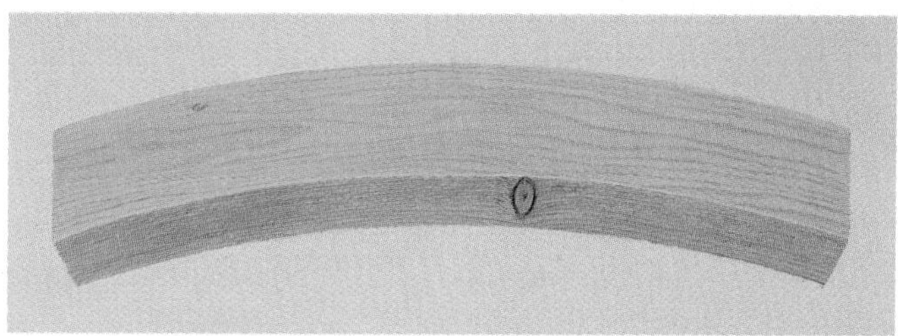

Figure 1-17. Warping and other wood defects will affect quality, strength, and appearance.

Softwood Grades

Basic principles of grading lumber, set down by the American Lumber Standards Committee, are published by the U.S. Department of Commerce. Detailed rules are developed and applied by the various associations of lumber producers—Western Wood Products Association, Southern Forest Products Association, California Redwood Association, and similar groups. These agencies publish grading rules for the species of lumber produced in their regions. They also have qualified personnel who supervise grading standards at sawmills.

Basic classifications of softwood grading include boards, dimension, and timbers. The grades within these classifications are shown in Figure 1-18.

Another classification called *factory and shop lumber* is graded primarily for remanufacturing purposes. It is used by millwork plants in the fabrication of windows, doors, moldings, and other trim items.

boards

APPEARANCE GRADES	SELECTS	B & BETTER (IWP—SUPREME) C SELECT (IWP—CHOICE) D SELECT (IWP—QUALITY)		
	FINISH	SUPERIOR PRIME E		
	PANELING	CLEAR (ANY SELECT OR FINISH GRADE) NO. 2 COMMON SELECTED FOR KNOTTY PANELING NO. 3 COMMON SELECTED FOR KNOTTY PANELING		
	SIDING (BEVEL, BUNGALOW)	SUPERIOR PRIME		
			ALTERNATE BOARD GRADES	
BOARDS SHEATHING		NO. 1 COMMON (IWP—COLONIAL) NO. 2 COMMON (IWP—STERLING) NO. 3 COMMON (IWP—STANDARD) NO. 4 COMMON (IWP—UTILITY)	SELECT MERCHANTABLE CONSTRUCTION STANDARD UTILITY	

SPECIFICATION CHECK LIST

- ☐ Grades listed in order of quality.
- ☐ Include all species suited to project.
- ☐ For economy, specify lowest grade that will satisfy job requirement.
- ☐ Specify surface texture desired.
- ☐ Specify moisture content suited to project.
- ☐ Specify (WWP) grade stamp. For finish and exposed pieces, specify stamp on back or ends.

WESTERN RED CEDAR

FINISH PANELING AND CEILING	CLEAR HEART A B
BEVEL SIDING	CLEAR — V.G. HEART A — BEVEL SIDING B — BEVEL SIDING C — BEVEL SIDING

dimension

LIGHT FRAMING 2" to 4" Thick 2" to 4" Wide	CONSTRUCTION STANDARD UTILITY ECONOMY	This category for use where high strength values are **NOT** required; such as studs, plates, sills, cripples, blocking, etc.
	STUD ECONOMY STUD	An optional all-purpose grade limited to 10 feet and shorter. Characteristics affecting strength and stiffness values are limited so that the "Stud" grade is suitable for all stud uses, including load bearing walls.
STRUCTURAL LIGHT FRAMING 2" to 4" Thick 2" to 4" Wide	SELECT STRUCTURAL NO. 1 NO. 2 NO. 3 ECONOMY	These grades are designed to fit those engineering applications where higher bending strength ratios are needed in light framing sizes. Typical uses would be for trusses, concrete pier wall forms, etc.
STRUCTURAL JOISTS & PLANKS 2" to 4" Thick 6" and Wider	SELECT STRUCTURAL NO. 1 NO. 2 NO. 3 ECONOMY	These grades are designed especially to fit in engineering applications for lumber six inches and wider, such as joists, rafters and general framing uses.

timbers

BEAMS & STRINGERS	SELECT STRUCTURAL NO. 1 NO. 2 (NO. 1 MINING) NO. 3 (NO. 2 MINING)	**POSTS & TIMBERS**	SELECT STRUCTURAL NO. 1 NO. 2 (NO. 1 MINING NO. 3 (NO. 2 MINING)

Figure 1-18. Softwood lumber classifications and grades. Names of grades and their specifications will vary among lumber manufacturers' associations and among regions producing lumber.

The carpenter must understand that quality construction does not require that all lumber be of the best grade. Today, lumber is graded for specific uses. In a given structure, several grades may be appropriate. The key to good economical construction is matching the proper grade with its function.

Hardwood Grades

Grades for hardwood lumber are established by the National Hardwood Lumber Association. FAS (firsts and seconds) is the best grade. It specifies that pieces be no less than 6" wide by 8' long and yield at least 83 1/3% clear cuttings. The next lower grade, *selects,* permits pieces 4" wide by 6' long. A still lower grade is *No. 1 common.* Lumber in this group is expected to yield 66 2/3% clear cuttings.

Lumber Stress Values

In softwood lumber, all dimension and timber grades except Economy and Mining are assigned stress values. Slope of grain, knot sizes, and knot locations are critical considerations.

There are two methods of assigning stress values:

- Visual.
- Machine rated.

In the machine rated method, lumber is fed into a special machine and subjected to bending forces. The stiffness of each piece (modulus of elasticity, E) is measured and marked on each piece. Machine stress-rated lumber (MSR) must also meet certain visual requirements.

Lumber Sizes

When listing and calculating the size and amount of lumber, the nominal dimension is always used. Figure 1-19 illustrates the nominal and dressed sizes for various classifications of lumber. Note that nominal sizes are sometimes listed in quarters. For example: 1 1/4" material is given as 5/4. This nominal dimension is its rough unfinished measurement, Figure 1-20. The dressed size is less than the nominal size as a result of seasoning and surfacing. Dressed sizes of lumber, established by the American Lumber Standards, are applied consistently throughout the industry.

Calculating Board Footage

The unit of measure for lumber is the board foot. This is a piece 1" thick and 12" square or its equivalent (144 cu. in.).

Standard size pieces can be quickly calculated by visualizing the board feet included. For example: a board 1 x 12 and 10' long will contain 10 bd. ft. If it were only 6" wide, it would be 5 bd. ft. If the original board had been 2" thick, it would have contained 20 bd. ft.

The following formula can be applied to calculate board feet. *T* represents the thickness of the board (in.), *W* is the width (in.), and *L* is the length (ft.):

$$\text{Bd. ft.} = \frac{\text{No. pcs.} \times T \times W \times L}{12}$$

An example of the application of the formula follows: find the number of board feet in six pieces of lumber that measure 1 x 8 x 14':

$$\text{Bd. ft.} = \frac{6 \times 1 \times 8 \times 14}{12}$$

$$\text{Bd. ft.} = \frac{\overset{1}{\cancel{6}} \times 1 \times 8 \times 14}{\underset{2}{\cancel{12}}}$$

$$\text{Bd. ft.} = \frac{1 \times 1 \times \overset{4}{\cancel{8}} \times 14}{\underset{1}{\cancel{2}}}$$

$$\text{Bd. ft.} = \frac{1 \times 1 \times 4 \times 14}{1}$$

$$= 56 \text{ Bd ft.}$$

Stock less than 1" thick is figured as though it were 1". When the stock is thicker than 1", the nominal size is used. When this size contains a mixed fraction, such as 1 1/4, change it to an improper fraction (5/4) and place the numerator above the formula line and the denominator below. For example: find the board footage in two pieces of lumber that measure 1 1/4" x 10" x 8':

$$\text{Bd. ft.} = \frac{2 \times 5 \times 10 \times 8}{4 \times 12}$$

$$= \frac{2 \times 5 \times \overset{5}{\cancel{10}} \times \overset{2}{\cancel{8}}}{\underset{1}{\cancel{4}} \times \underset{6}{\cancel{12}}}$$

$$= \frac{\overset{1}{\cancel{2}} \times 5 \times 5 \times 2}{1 \times \underset{3}{\cancel{6}}}$$

$$= \frac{1 \times 5 \times 5 \times 2}{1 \times 3}$$

$$= \frac{50}{3}$$

$$= 16\ 2/3 \text{ Bd ft.}$$

Use the nominal size of the material when figuring the footage. Items such as moldings, furring strips, and rounds are priced and sold by the lineal foot. Thickness and width are disregarded.

Metric Lumber Measure

Metric sized lumber gives thickness and width in millimeters (mm) and length in meters. There is little difference between metric and conventional dimensions for common sizes of lumber. For example, the common 1 x 4 board is 25 mm x 100 mm. Visually, they would appear to be about the same size. Metric lumber lengths start at 1.8 m (about 6') and increase in steps of 300 mm (about a foot) to 6.3 m. This is a little more than 20'. See Figure 1-21 for a chart of standard sizes. Metric lumber is sold by the cubic meter (m^3). See the Technical Information Section for more information.

BOARD MEASURE

The term "board measure" indicates that a board foot is the unit for measuring lumber. A board foot is one inch thick and 12 inches square.

The number of board feet in a piece is obtained by multiplying the nominal thickness in inches by the nominal width in inches by the length in feet and dividing by 12: $\frac{(T \times W \times L)}{12}$.

Lumber less than one inch in thickness is figured as one-inch.

product classification

	thickness in.	width in.		thickness in.	width in.
board lumber	1″	2″ or more	**beams & stringers**	5″ and thicker	more than 2″ greater than thickness
light framing	2″ to 4″	2″ to 4″	**posts & timbers**	5″ x 5″ and larger	not more than 2″ greater than thickness
studs	2″ to 4″	2″ to 4″ 10′ and shorter	**decking**	2″ to 4″	4″ to 12″ wide
structural light framing	2″ to 4″	2″ to 4″	**siding**	thickness expressed by dimension of butt edge	
joists & planks	2″ to 4″	6″ and wider	**mouldings**	size at thickest and widest points	

Lengths of lumber generally are 6 feet and longer in multiples of 2′

Standard Lumber Sizes / Nominal, Dressed, Based on WWPA Rules

Product	Description	Nominal Size: Thickness In.	Nominal Size: Width In.	Dressed Dimensions: Thicknesses and Widths In.: Surfaced Dry	Dressed Dimensions: Thicknesses and Widths In.: Surfaced Unseasoned	Lengths Ft.
FRAMING	S4S	2 3 4	2 3 4 6 8 10 12 Over 12	1-1/2 2-1/2 3-1/2 5-1/2 7-1/4 9-1/4 11-1/4 Off 3/4	1-9/16 2-9/16 3-9/16 5-5/8 7-1/2 9-1/2 11-1/2 Off 1/2	6 ft. and longer in multiples of 1′
				Thickness In.	**Width In.**	
TIMBERS	Rough or S4S	5 and Larger		1/2 Off Nominal		Same
		Nominal Size		**Dressed Dimensions Surfaced Dry**		
		Thickness In.	**Width In.**	**Thickness In.**	**Width In.**	**Lengths Ft.**
DECKING Decking is usually surfaced to single T&G in 2″ thickness and double T&G in 3″ and 4″ thicknesses	2″ Single T&G	2	6 8 10 12	1-1/2	5 6-3/4 8-3/4 10-3/4	6 ft. and longer in multiples of 1′
	3″ and 4″ Double T&G	3 4	6	2-1/2 3-1/2	5-1/4	
FLOORING	(D & M), (S2S & CM)............	3/8 1/2 5/8 1 1-1/4 1-1/2	2 3 4 5 6	5/16 7/16 9/16 3/4 1 1-1/4	1-1/8 2-1/8 3-1/8 4-1/8 5-1/8	4 ft. and longer in multiples of 1′
CEILING AND PARTITION	(S2S & CM)	3/8 1/2 5/8 3/4	3 4 5 6	5/16 7/16 9/16 11/16	2-1/8 3-1/8 4-1/8 5-1/8	4 ft. and longer in multiples of 1′
FACTORY AND SHOP LUMBER	S2S	1 (4/4) 1-1/4 (5/4) 1-1/2 (6/4) 1-3/4 (7/4) 2 (8/4) 2-1/2 (10/4) 3 (12/4) 4 (16/4)	5 and wider (4″ and wider in 4/4 No. 1 Shop and 4/4 No. 2 Shop)	25/32 (4/4) 1-5/32 (5/4) 1-13/32 (6/4) 1-19/32 (7/4) 1-13/16 (8/4) 2-3/8 (10/4) 2-3/4 (12/4) 3-3/4 (16/4)	Usually sold random width	4 ft. and longer in multiples of 1′

ABBREVIATIONS
Abbreviated descriptions appearing in the size table are explained below.
S1S — Surfaced one side.
S2S — Surfaced two sides.
S4S — Surfaced four sides.
S1S1E — Surfaced one side, one edge.
S1S2E — Surfaced one side, two edges
CM — Center matched.
D & M — Dressed and matched.
T & G — Tongue and grooved.
EV1S — Edge vee on one side.
S1E — Surfaced one edge.

Figure 1-19. Standard lumber sizes are set by government agencies and lumber associations. (Western Wood Products Assoc.)

Panel Materials

Wood panels for construction are manufactured in several different ways:

- As plywood, where thin sheets are laminated to various thicknesses.
- As composite plywood, where veneer faces are bonded to different kinds of wood cores.
- As nonveneered panels, including waferboard, particleboard, and oriented strand board.

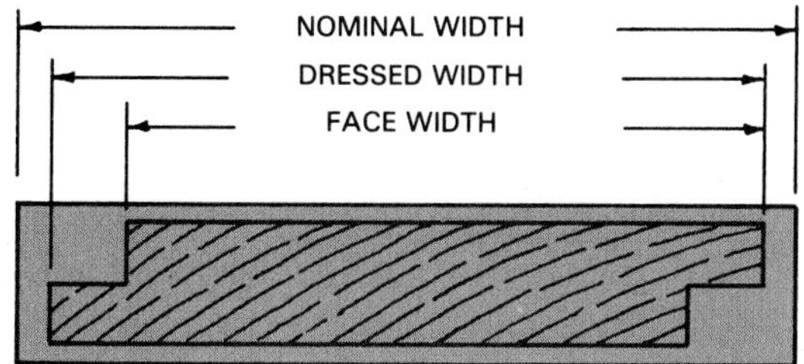

Figure 1-20. Nominal size is greater than dressed size.

Plywood

Plywood is constructed by gluing together a number of layers (plies) of wood with the grain direction turned at right angles in each successive layer. An odd number (3, 5, 7) of plies are used so they will be balanced on either side of a center core and so the grain of the outside layers will run in the same direction. The outer plies are called *faces* or face and back. The next layers under these are called *cross-bands* and the other inside layer or layers are called the *core.* See Figure 1-22. A thin plywood panel made of three layers consists of two faces and a core.

There are two basic types of plywood:

- *Exterior plywood* is bonded with waterproof glues. It can be used for siding, concrete forms, and other

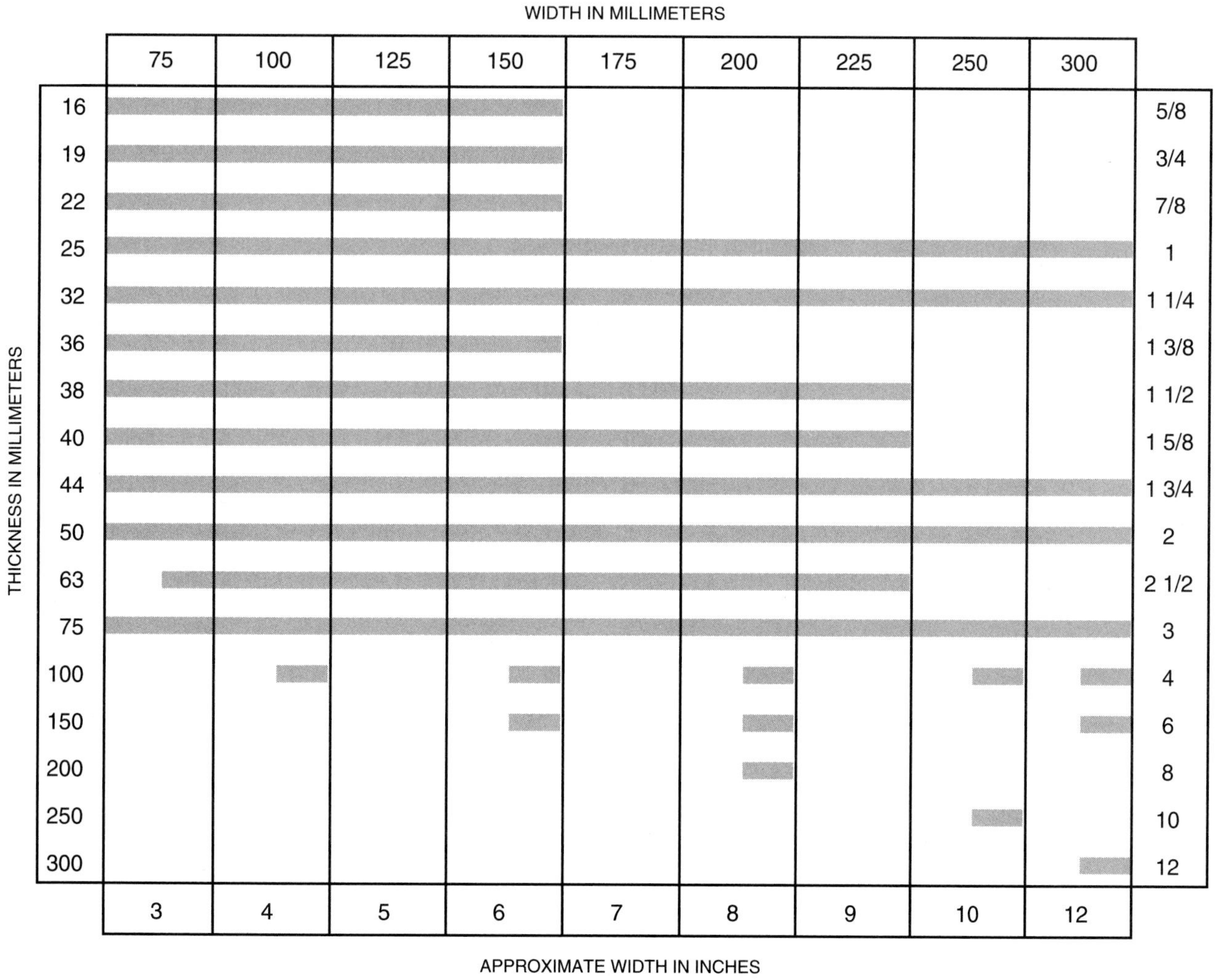

WIDTH IN MILLIMETERS

THICKNESS IN MILLIMETERS	75	100	125	150	175	200	225	250	300	APPROXIMATE THICKNESS IN INCHES
16	X	X	X	X						5/8
19	X	X	X	X						3/4
22	X	X	X	X						7/8
25	X	X	X	X	X	X	X	X	X	1
32	X	X	X	X	X	X	X	X	X	1 1/4
36	X	X	X	X						1 3/8
38	X	X	X	X	X	X	X			1 1/2
40	X	X	X	X	X	X	X			1 5/8
44	X	X	X	X	X	X	X	X	X	1 3/4
50	X	X	X	X	X	X	X	X	X	2
63	X	X	X	X	X	X	X			2 1/2
75	X	X	X	X	X	X	X	X	X	3
100		X		X		X		X	X	4
150				X		X			X	6
200						X				8
250								X		10
300									X	12
	3	4	5	6	7	8	9	10	12	

APPROXIMATE WIDTH IN INCHES

Figure 1-21. Dimensions of metric lumber are given in millimeters. Lengths are always given in meters and range from 1.8 m to 6.3 m in 0.3 m (about 1') increments.

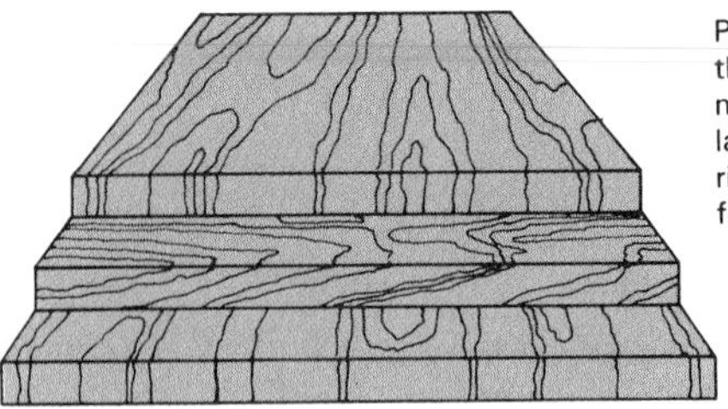

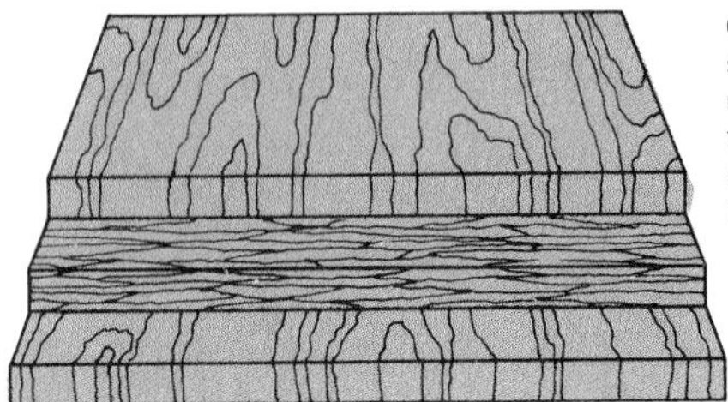

Figure 1-22. Panel construction. Top—Standard plywood construction. Bottom—Composite plywood construction. (Georgia-Pacific)

constructions where it will be exposed to the weather or excessive moisture.

- *Interior plywood* may be manufactured with either interior or exterior glue. Today, exterior glue is generally used. The grade of the panel is determined mainly by the quality of the layers. Plywood is made in thicknesses from 1/8" to more than 1" with common sizes being 1/4", 3/8", 1/2", and 3/4". Thicknesses of 15/32", 19/32", and 23/32" are also available. A standard panel size is 4' wide by 8' long.

Metric plywood panels are slightly smaller than the standard 4' x 8' panel. Figure 1-23 compares the two.

Softwood Plywood Grades

Softwood plywood for general construction is manufactured in accordance with U.S. Product Standard PS 1-83 for Construction and Industrial Plywood. This provides for designating species, strength, type of glue, and appearance.

Many species of softwood are used in making plywood. There are five separate plywood groups based on stiffness and strength. Group 1 includes the stiffest and strongest, Figure 1-24.

Grade-Trademark Stamp

Construction and industrial panels are marked for quality in two different ways:

- A grade lettering system may be used to indicate the quality of the veneer used on the face and back of the panel. The letters and their meanings are given in Figure 1-25.
- A name indicating the panel's intended use or "performance rating."

The APA (American Plywood Association) has a rigid testing program based upon PS 1-83. Mills that are members of the association may use the official grade-trademark. It is stamped on each piece of plywood.

A typical stamp for an engineered grade of structural panel is shown in Figure 1-26. The span rating shows a pair of numbers separated by a slash mark (/). The number on the left indicates the maximum recommended span in inches when the plywood is used as roof decking (sheathing). The number on the right indicates the maximum recommended span when the plywood is used as subflooring. The rating applies only when the sheet is placed with its long dimension spanning across three or more supports. Generally, the larger the span rating, the greater the panel's stiffness.

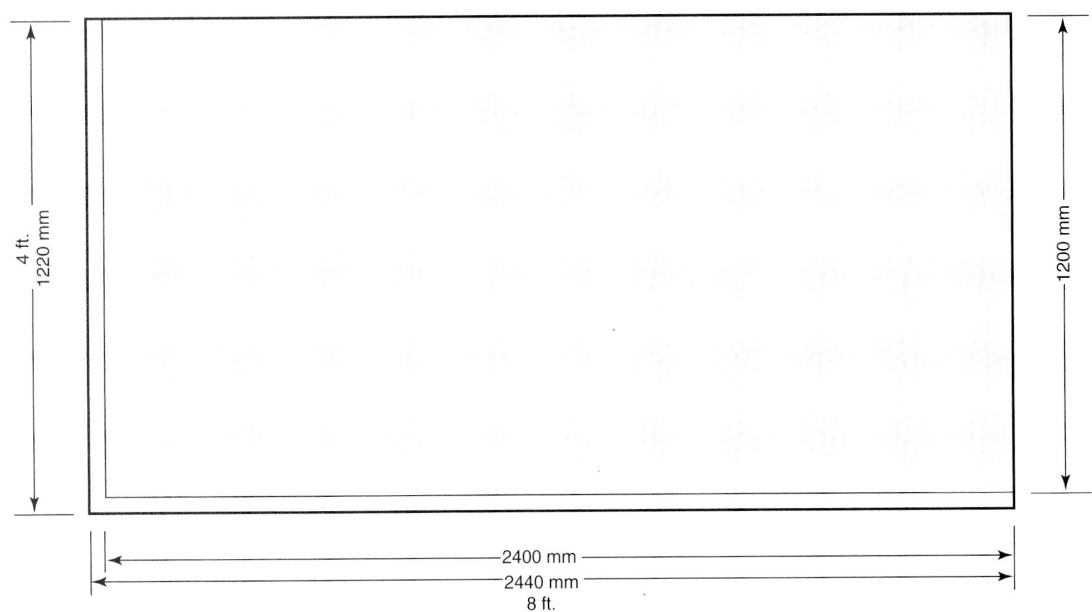

Figure 1-23. Comparison of metric and conventional size plywood panels. Metric size is slightly smaller each way.

GROUP 1	GROUP 2	GROUP 3	GROUP 4	GROUP 5
Apitong	Cedar, Port	Alder, Red	Aspen	Basswood
Beech,	Orford	Birch, Paper	Bigtooth	Poplar,
American	Cypress	Cedar, Alaska	Quaking	Balsam
Birch	Douglas	Fir,	Cativo	
Sweet	Fir 2	Subalpine	Cedar	
Yellow	Fir	Hemlock,	Incense	
Douglas	California	Eastern	Western	
Fir 1[a]	Red	Maple,	Red	
Kapur	Grand	Bigleaf	Cottonwood	
Keruing	Noble	Pine	Eastern	
Larch,	Pacific	Jack	Black	
Western	Silver	Lodgepole	(Western	
Maple, Sugar	White	Ponderosa	Poplar)	
Pine	Hemlock,	Spruce	Pine	
Caribbean	Western	Redwood	Eastern	
Ocote	Lauan	Spruce	White	
Pine, South.	Almon	Engelmann	Sugar	
Loblolly	Bagtikan	White		
Longleaf	Mayapis			
Shortleaf	Red Lauan			
Slash	Tangile			
Tanoak	White Lauan			
	Maple, Black			
	Mengkulang			
	Meranti, Red			
	Mersawa			
	Pine			
	Pond			
	Red			
	Virginia			
	Western			
	White			
	Spruce			
	Red			
	Sitka			
	Sweetgum			
	Tamarack			
	Yellow-			
	poplar			

Figure 1-24. Classification of softwood plywood rates species for strength and stiffness. Group 1 represents strongest woods. (American Plywood Assoc.)

Grade	Description
N	Smooth surface "natural finish" veneer. Select, all heartwood or all sapwood. Free of open defects. Allows not more than 6 repairs, wood only, per 4x8 panel, made parallel to grain and well matched for grain and color.
A	Smooth paintable. Not more than 18 neatly made repairs, boat, sled, or router type, and parallel to grain, permitted. May be used for natural finish in less demanding application. Synthetic or wood repairs permitted.
B	Solid surface. Shims, circular repair plugs and tight knots to 1 inch across grain permitted. Some minor splits permitted. Synthetic or wood repairs permitted.
C Plugged	Improved C veneer with splits limited to 1/8 inch width and knotholes and borer holes limited to 1/4 x 1/2 inch. Allows some broken grain. Synthetic or wood repairs permitted.
C	Tight knots to 1-1/2 inch. Knotholes to 1 inch across grain and some to 1-1/2 inch if total width of knots and knotholes is within specified limits. Synthetic or wood repairs. Discoloration and sanding defects that do not impair strength permitted. Limited splits allowed. Stitching permitted.
D	Knots and knotholes to 2-1/2 inch width across grain and 1/2 inch larger within specified limits. Limited splits are permitted. Stitching permitted. Limited to Exposure 1 or 2 or Interior panels.

Figure 1-25. Description of softwood plywood veneer grades. (American Plywood Assoc.)

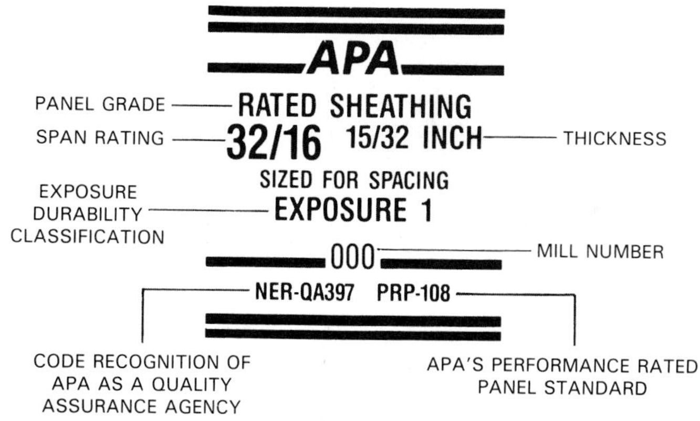

Figure 1-26. Typical grade-trademark that is stamped on all panels manufactured in compliance with national plywood standard PS 1-83. (American Plywood Assoc.)

Figure 1-27 lists some typical engineered grades of plywood. Included are descriptions and the most common uses.

Exposure Ratings

The grade-trademark stamp gives an "exposure durability" classification to plywood. There are two basic types:

- Exterior type, which has 100% waterproof glueline.
- Interior type, with moisture resistant glueline.

However, panels can be manufactured in three exposure durability classifications:

- Exterior.
- Exposure 1.
- Exposure 2.

Panels marked "Exterior" can be used outdoors and may be continually exposed to weather and moisture. Panels marked "Exposure 1" can withstand moisture during extended periods but should only be used indoors. Panels designated as "Exposure 2" should be used in protected locations only. They may be subjected to some water leakage or high humidity but, generally, should be protected from weather.

Most plywood is manufactured with waterproof exterior glue. However, interior panels may be manufactured with intermediate or interior glue.

Span Ratings of Plywood

In 1980, the APA issued APA PRP-108, Performance Standards and Policies for Structural Use Panels. This standard was a response to a need to conserve lumber without affecting the strength of structural plywood. At that time, APA also began placing its trademark on panel thicknesses of 15/32", 19/32", and 23/32". Today, the trademark is placed on sheathing, floor underlayments, and siding panels with numbers called *span ratings*. This rating is the

	Grade Designation	Descriptions & Common Uses	Typical Trademarks
PROTECTED EXTERIOR USE	**APA RATED SHEATHING EXTERIOR**	For use where appearance of only one side is important in exterior applications, such as soffits, fences, structural uses, boxcar and truck linings, farm buildings, tanks, trays, commercial refrigerators, etc. EXPOSURE DURABILITY CLASSIFICATION: Exterior. COMMON THICKNESSES: 1/4, 11/32, 3/8, 15/32, 1/2, 19/32, 5/8, 23/32, 3/4.	APA A-C GROUP 1 EXTERIOR 000 PS 1-83
	APA RATED UTILITY EXTERIOR	Utility panel for farm service and work buildings, boxcar and truck linings, containers, tanks, agricultural equipment, as a base for exterior coatings and other exterior uses or applications subject to high or continuous moisture. EXPOSURE DURABILITY CLASSIFICATION: Exterior. COMMON THICKNESSES: 1/4, 11/32, 3/8, 15/32, 1/2, 19/32, 5/8, 23/32, 3/4.	APA B-C GROUP 1 EXTERIOR 000 PS 1-83
PROTECTED INTERIOR USE	**APA STRUCTURAL I & II RATED SHEATHING EXP 1**	For use where appearance of only one side is important in interior applications, such as paneling, built-ins, shelving, partitions, flow racks, etc. EXPOSURE DURABILITY CLASSIFICATION: Interior, Exposure 1. COMMON THICKNESSES: 1/4, 11/32, 3/8, 15/32, 1/2, 19/32, 5/8, 23/32, 3/4.	APA A-D GROUP 1 EXPOSURE 1 000 PS 1-83
	APA RATED EXP 1	Utility panel for backing, sides of built-ins, industry shelving, slip sheets, separator boards, bins and other interior or protected applications. EXPOSURE DURABILITY CLASSIFICATION: Interior, Exposure 1. COMMON THICKNESSES: 1/4, 11/32, 3/8, 15/32, 1/2, 19/32, 5/8, 23/32, 3/4.	APA B-D GROUP 2 EXPOSURE 1 000 PS 1-83
	APA RATED DECORATIVE INTERIOR	Rough-sawn, brushed, grooved, or striated faces. For paneling, interior accent walls, built-ins, counter facing, exhibit displays. Can also be made by some manufacturers in Exterior for exterior siding, gable ends, fences and other exterior applications. Use recommendations for Exterior panels vary with the particular product. Check with the manufacturer. EXPOSURE DURABILITY CLASSIFICATION: Interior, Exposure 1, Exterior. COMMON THICKNESSES: 5/16, 3/8, 1/2, 5/8.	APA DECORATIVE GROUP 2 INTERIOR 000 PS 1-83
	APA STRUCTURAL UNDERLAYMENT EXP 1	For application over structural subfloor. Provides smooth surface for application of carpet and processes high concentrated and impact load resistance. EXPOSURE DURABILITY CLASSIFICATION: Interior, Exposure 1. COMMON THICKNESSES(4): 3/8, 1/2, 19/32, 5/8, 23/32, 3/4.	APA UNDERLAYMENT GROUP 1 EXPOSURE 1 000 PS 1-83
	APA RATED II EXP 1	For built-ins, cable reels, separator boards and other interior or protected applications. Not a substitute for Underlayment or APA Rated Sturd-I-Floor as it lacks their puncture resistance. EXPOSURE DURABILITY CLASSIFICATION: Interior, Exposure 1. COMMON THICKNESSES: 3/8, 1/2, 19/32, 5/8, 23/32, 3/4.	APA C-D PLUGGED GROUP 2 EXPOSURE 1 000 PS 1-83

(1) Specific grades, thicknesses, constructions and exposure durability classifications may be in limited supply in some areas. Check with your supplier before specifying.
(2) Sanded exterior plywood panels, C-C Plugged, C-D Plugged and Underlayment grades can also be manufactured in Structural I (all plies limited to Group 1 species).
(3) Some manufacturers also produce plywood panels with premium N-grade veneer on one or both faces. Available only by special order. Check with the manufacturer.
(4) Panels 1/2 inch and thicker are span rated and do not contain species group number in trademark.

Figure 1-27. Selected list of engineered grades of softwood plywood. (American Plywood Assoc.)

maximum recommended center-to-center distance in inches between supports when the long dimension of the panel is at right angles to the supports.

On APA rated sheathing, this span rating appears on the trademark as two numbers separated by a slash. The exception is panel products intended only as wall sheathing; in this case only one number is used.

APA Rated Sheathing is typically manufactured with span ratings of 24/0, 24/16, 40/20, or 48/24. Typical Sturd-I-Floor ratings are 16", 20", 24", 32", and 48". This single number is the recommended center-to-center spacing of floor joists when panels are installed with the long dimension at right angles to the supports.

Siding is manufactured with ratings of 16" and 24". Panel and lap siding rated at 16" may be applied directly to studs spaced at 16" O.C. Panel and lap siding can be applied to studs spaced 24" O.C. When applied over nailable structural sheathing, the span rating indicates the maximum recommended spacing of vertical rows of nails.

The left-hand number is the maximum recommended spacing of supports when the panel is used for roof sheathing placed with the panel's long dimension spanning at least three supports. The right-hand number indicates the maximum recommended spacing of supports when the panel is used for subflooring with the long dimension across three or more supports.

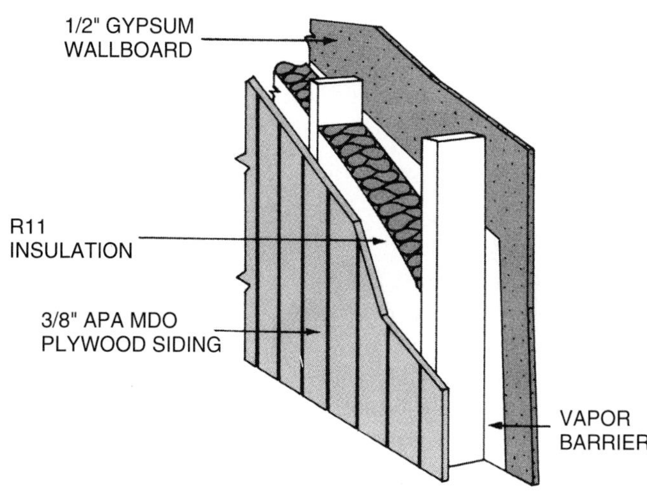

Figure 1-28. Sketch shows the make-up of APA Sturd-I-Wall which uses MDO plywood as an exterior finish. This wall has an R value of approximately 12.77. (American Plywood Assoc.)

HDO and MDO Plywood

HDO and MDO plywood combine the properties of exterior type plywood and plywood with the superior wearability of overlaid surfaces. HDO (high density overlay) plywood is manufactured with a thermosetting resin-impregnated fiber surface that is bonded under heat and pressure to both sides of the panel. It is more rugged than MDO and is suited for such punishing applications as concrete forming. The resin overlay requires no additional finish and resists abrasion, moisture penetration, and damage from common chemicals and solvents.

Like HDO, MDO (medium density overlay) is an exterior-type panel manufactured with 100% waterproof adhesive. It accepts paint readily and is suited for structural siding, exterior color accented panels, soffits, and other applications where long-lasting paint or coating performance is required. See Figure 1-28.

MDO panels that are to be used outside should be edge-sealed as soon as possible. One or two coats of high quality exterior house paint primer formulated for wood should be used. Edges are more easily sealed while panels are in a stack.

Hardwood Plywood Grades

The Hardwood Plywood Institute uses a numbering system for grading the faces and backs of panels. A grading specification of 1-2 would indicate a good face with grain carefully matched and a good back but without grain carefully matched. A No. 3 back would permit noticeable defects and patching but would be generally sound. A special or *premium* grade of hardwood is known as "architectural" or "sequence-matched." This usually requires an order to a plywood mill for a series of matched plywood panels.

For either softwood or hardwood plywood, it is common practice to designate the grade by a symbol. G2S means good two sides. G1S means good one side.

In addition to the various types and grades, hardwood plywood is made with different core constructions. The two most common are the veneer core and the lumber core. See Figure 1-29. *Veneer cores* are the least expensive,

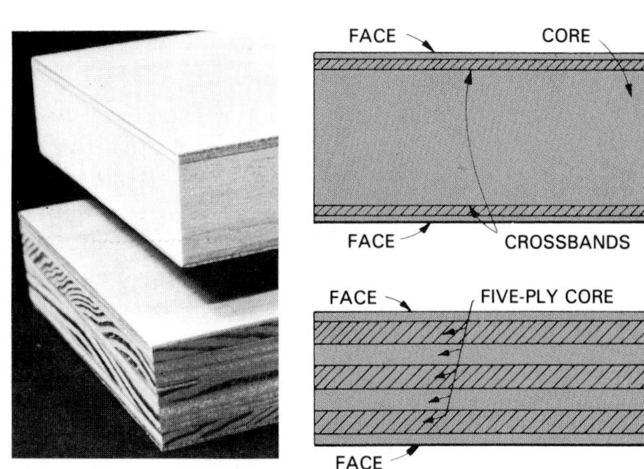

Figure 1-29. Hardwood plywood is of two types: Top—Lumber core. Bottom—Veneer core.

fairly stable, and warp resistant. *Lumber cores* are easier to cut, the edges are better for shaping and finishing, and they hold nails and screws better. Plywood is also manufactured with a particleboard core. It is made by gluing veneers directly to a particleboard surface.

Composite Board

Panels made up of a core of reconstituted wood with a thin veneer on either side are called *composite board* or *composite panels.* These materials are widely used in modern construction. They are good as sheathing, subflooring, siding, and interior wall surfaces.

In cabinetwork, hardboard and particleboard serve as appropriate materials for drawer bottoms and concealed panels in cases, cabinets, and chests. They are manufactured by many different companies and sold under various trade names.

Hardboard

Hardboard is made of refined wood fibers pressed together to form a hard, dense material (50–80 lb. per cu. ft.). There are two types: standard and tempered.

Tempered hardboard is impregnated (filled) with oils and resins. These materials make it harder, slightly heavier, more water resistant, and darker in appearance. Hardboard is manufactured with one side smooth (S1S) or both sides smooth (S2S). It is available in thicknesses from 1/12" to 5/16". The most common thicknesses are 1/8", 3/16", and 1/4". Panels are 4' wide and come in standard lengths of 8', 10', 12', and 16'.

Particleboard

Particleboard is made of wood flake, chips, and shavings bonded together with resins or adhesives. It is not as heavy as hardboard (about 40 lb. per cu. ft.) and is available in thicker panels. Particleboard may be constructed of layers made of different size wood particles. Large ones in the center provide strength. Fine ones at the surface provide smoothness.

Particleboard is used as a base for veneers and laminates. It is important in the construction of countertops, cabinets, drawers and shelving, many types of folding and sliding doors, room dividers, and a variety of built-ins.

It is popular because it has a smooth, grain-free surface and is stable. Its surface qualities make it a popular choice as a base for laminates. Particleboard doors do not warp and require little adjustment after installation.

Particleboard is available in thicknesses ranging from 1/4" to 17/16". The most common panel size is 4' x 8'.

Waferboard

Waferboard, Figure 1-30, is produced from high quality flakes of wood that are about 1 1/2" square. These flakes are bonded together under heat and pressure with phenolic resin, a waterproof adhesive. Both sides of waferboard have the same textured surface, Figure 1-31. This surface has a natural slickness that can be minimized by special treatments. The density of waferboard is about 40 lb. per cu. ft. Standard panel size is 4' x 8'. Thicknesses range from 1/4" to 1 1/8".

Figure 1-30. Waferboard is widely used for sheathing. It is made from flakes of wood that are bonded together with waterproof adhesives.

Oriented Strand Board

Somewhat like waferboard in appearance, *oriented strand board* is also made up of wood fibers adhered to each other with suitable resins and glues. The fibers are put down in successive layers arranged at right angles to one another.

A relatively new product, oriented strand board is approved by all model building codes in the United States and Canada. Its applications are many:

- A combination subfloor and underlayment.
- Sheathing for exterior walls. Oriented strand board provides an excellent nailing base for siding. The large size of the panels also reduces air infiltration through the building's walls.
- APA rated siding. When special surface treatments are used, oriented strand board is suitable as an exterior finish.

Wood Treatments

Wood and wood products that will be exposed to high levels of moisture should be protected from attack by fungi, insects, and borers. Millwork plants employ extensive treatment processes in the manufacture of such items as door frames and window units.

Structures that are continually exposed to weather—outside stairs, fences, decks, and furniture—should be fab-

A

B

Figure 1-31. Wood construction materials. A—Modern pressure-treated lumber includes a preservative and water repellent. (Wolmanized Wood Producers) B—Top. Oriented strand board. Center. Waferwood or waferboard. These panels can be used as sheathing, subflooring, or as interior wall finish. (Georgia-Pacific)

ricated from pressure-treated lumber that has long-lasting resistance to termites and fungal decay. In addition, the lumber should be pressure-treated with a liquid repellent that slows the rate at which the moisture is absorbed and released. See Figure 1-31(A).

The Environmental Protection Agency (EPA) has produced a Consumer Information Sheet that outlines uses and handling precautions for treated wood products. Copies of the sheet are available at home centers and lumber outlets. Treated woods should only be used where protection against decay and insects is important. Treated wood should never be used where the waterborne arsenical preservatives in the treatment may become a component of food, animal feed, or drinking water.

When wood is treated under pressure and with controlled conditions, the treatment penetrates deeply into the wood's cellular structure. Thus treated, the wood resists rot, decay, and termites. Even when exposed to severe conditions, the wood provides excellent service.

There are three major types of liquid preservatives:

- Waterborne, which are used for residential, commercial, recreational, marine, agricultural, and industrial applications. Of the waterborne preservatives, chromated copper arsenate (CCA), acid copper chromate (ACC) and chromated zinc chloride (CZC) are used. Waterborne treatments leave the wood clear, odorless, and paintable.
- Oilborne. The best known oilborne preservative is pentachlorophenol, which can be mixed with various solvents. It is highly toxic to insects and fungi. However, treated wood may become discolored. It is used on farms, around utility poles, and in industry.
- Creosote. This is a mixture of creosote and coal tar in heavy oil. It is suitable for treatment of pilings, utility poles, and railroad ties. Creosoted wood cannot be painted.

Often, treated lumber carries an identifying end tag or stamp, Figure 1-32. This tag or stamp indicates the type of preservative, proper exposure conditions, and other information. More information on treated lumber can be obtained by contacting:

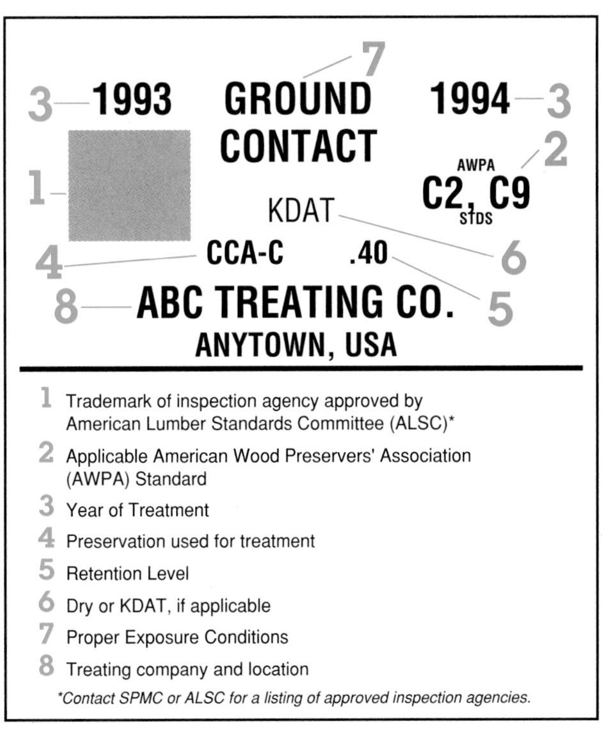

Figure 1-32. An end tag identifying treated lumber for ground-contact use. Other tags denote lumber suitable for above-ground use. (Southern Forest Products Assoc.)

American Wood Preservers Bureau
P.O. Box 5283
Springfield, VA 22150

Southern Pine Inspection Bureau
4709 Scenic Highway
Pensacola, FL 32504

Handling and Storing

Building materials are expensive and should be maintained in the best condition practical. After they are delivered to the construction site, this becomes the responsibility of the carpenter.

Piles of framing lumber and sheathing should be laid on level skids raised at least 6" above the ground. Be sure all pieces are supported and lying straight, Figure 1-33.

Figure 1-33. When lumber is delivered to the construction site, it should be carefully stacked on level skids. The skids should provide at least 6" of clearance above the ground.

Cover the material with canvas or waterproof paper. Polyethylene film also provides a watertight covering, Figure 1-34.

If moisture absorption is likely, cut steel banding on panel materials to prevent edge damage when fibers expand. Keep coverings open and away from the sides and bottom of lumber stacks to promote good ventilation. Tight coverings promote mold growth.

Exterior finish materials, door frames, and window units should not be delivered until the structure is partially enclosed and the roof surfaced. In cold weather, the entire structure should be enclosed and heated before interior finish and cabinetwork are delivered and stored.

APA structural wood panels delivered on site should be properly stored and handled. Protect ends and edges from damage. Whenever possible, store panels under a cover. Keep sanded panels and other appearance grades away from high traffic areas. Weight down the top panel to prevent warpage. Cut the steel band on bundles if rain or heavy moisture is expected.

Panels stored outdoors should be stacked on a level surface atop 4 x 4 stringers or other blocking. Use at least three stringers. Never allow panels to touch the ground. Keep panels well ventilated with stickers to prevent mildew growth.

When finish lumber is received at a different moisture content than it will attain in the structure, it should be open-stacked with wood strips so air can circulate around each piece. Plywood, especially the fine hardwoods, must be handled with care. Sanded faces become soiled and scarred if not protected. While stored, the panels should lie flat.

Engineered Lumber

Builders have relied on traditional wood products for residential and light commercial framing. Modern tree-farming technology has assured that this material continues to be available, keeping up with the increased demand for wood structures. It once took more than 80 years to grow a tree to sawmill size; tree farming has shortened the span to 27 years. In the process, unfortunately, the tree's annular rings, which determine its tensile and comprehensive strength, are much farther apart, resulting in weaker dimension lumber. The industry has found a way to strengthen lumber through the development of *engineered lumber*.

Engineered lumber includes those wood structural units that have been altered through manufacturing processes to make them strong, straight, and dimensionally stable. Components are glued together in different configurations, some solid, some shaped like steel I-beams, and some in truss form. These products include:

- Laminated-veneer lumber (LVL), including I-beams.
- Glue-laminated beams, or glue-lams.
- Wood I-beams.
- Open-web trusses.
- Parallel-strand lumber, or PSL.
- Laminated-strand lumber, or LSL.

Figure 1-35 shows laminated products. Figure 1-36 shows open truss joists.

Laminated-Veneer Lumber

LVL is produced under various trade names (Micro-lam™, Paral-lam™, and Gang-Lam™, to mention three). It is produced much like plywood. First a log is debarked and cut into an 8' wide by 1/8" sheet of 1/10" veneer. Once dried, defects are removed and the veneer panels are coated with phenolic glue and stacked with their ends randomly staggered. A heat press cures the layup, which then is ready to be cut to specified lengths and widths.

Figure 1-34. During inclement weather, building materials that are stored outside should be covered with a tarpaulin, polyethylene film, or other waterproof material.

Another widely used process cuts the veneer into 1/2" wide strips. These strips are coated with waterproof phenolic glue and fed into a series of automated machines. The resulting billet comes out measuring 12' x 17" x 66'. The billet can be cut into sizes commonly used by the construction industry, where it is used as headers, beams, and columns.

LVL products are dimensionally stable, uniform in size throughout, and can be nailed, drilled, and cut with ordinary construction tools. LVLs take stain easily and can be left exposed, where they become a natural part of the interior decor.

Although a cross section of LVL looks like plywood, there is an important difference—grain orientation. In conventional plywood, each veneer panel is attached at a 90° angle to the previous layer to change the grain direction. In LVL, all of the veneer panels have their grain running in the same direction. Called *parallel lamination*, this provides greater strength, allowing the member to carry the heavy loads required of beams, headers, and truss components. Laminated beams are 1 1/2" or 1 3/4" thick with depths of 5 1/2", 7 1/4", 9 1/4", 11 1/4", 11 7/8", 14", 16", and 18".

Glue-Laminated Beams

Glue-laminated beams are made by gluing and then applying heavy pressure to a stack of 2 x 4 or 2 x 6 lumber to form a beam that can be up to 30" deep. Glue-lams are normally used for rafters, floor-support beams, and stair stringers. They can be ordered in curved shapes for supporting arched roofs.

Wood I-Beams

Wood I-beam flanges consist of solid or laminated-veneer lumber while webs are plywood or oriented strand board 3/8" thick. The web is fastened into grooves in the flanges with waterproof glue. No nails or other type of fastener are used. Wood I-beams, being light, straight, and strong, are commonly used as joists and rafters. They come in lengths up to 60' and depths from 9 1/2" to 32". The most common sizes are: 9 1/2", 11 1/2", 14", and 16".

Open-Web Trusses

Open-web trusses, Figure 1-36, are often used in place of floor joists, especially when dealing with long spans. These trusses are fabricated in factories from solid 2 x 4 lumber. Sometimes, steel webs are used instead of wood. Depths of 14" and 16" are most common. The open webs of the trusses allow installation of pipes, wiring, drains, and other mechanical systems without having to cut openings, as in other types of joists.

Figure 1-35. LVL engineered lumber. A—Laminated-veneer beam is made up of many layers of veneered lumber and is stronger than solid lumber. B—A wooden I-beam is made up of top and bottom flanges of solid or veneered 2 x 3 wood and a web of layered veneer or oriented strand board. They are bundled as shown for easy transporting.

Parallel-Strand Lumber

Parallel-strand lumber (PSL) is made by laying up 8' strands of veneer that have been soaked with adhesive. Bonded under pressure, it forms blocks up to 66' long. It is used as exposed posts and beams. When pressure treated, it is suitable for outdoor use in porches, decks, and gazebos.

Laminated-Strand Lumber

Laminated-strand lumber (LSL) is layed up from 1/32 x 1 x 12" strands of wood bonded with polyurethane adhesive. It is available in two thicknesses: 1 1/4" and 3 1/2". Depths vary up to a maximum 16" and lengths vary up to 35'. Uses include the following: door and window headers, rim joists in floors, and core stock for flush doors with veneer overlays.

Nonwood Materials

The carpenter works with a number of materials other than lumber and wood-based products. Some of the more common items include:

- Metal framing members, especially joists and studs.
- Gypsum and metal lath.
- Wallboard and sheathing.
- Insulating boards, batts, and loose insulation.
- Shingles of asphalt, metal, fiberglass, tile, and concrete, Figure 1-37.
- Metal flashing material.
- Caulking materials, Figure 1-38.
- Resilient flooring materials and carpeting.

Metal Structural Materials

Though not widely used at present for residential construction, steel framing is often found in light commercial buildings. For reasons of economy, it is finding its way into home building.

Steel framing members are manufactured in various widths and gages and are used as studs and joists. See Figure 1-39 and Figure 1-40. Most manufacturers use a color code to prevent the different gages from being mixed at the construction site. Prices are based on thousands of lineal feet.

Studs, joists, and track are manufactured by brake-forming and punching galvanized coil and sheet stock. Originally designed for commercial and institutional construction, their use is now being extended to residential and light commercial structures.

The studs are welded or attached to base and ceiling channels with screws or clips. Often wood is used for sole plates and wall plates, Figure 1-41. A typical stud consists of a metal channel with openings through which electrical and plumbing lines can be installed. See Figure 1-42.

Wall surface material, such as drywall or paneling, is attached to the metal stud with self-tapping drywall screws. Some web-type studs have a special metal edge with a gap into which nails can be driven. Metal stud systems are often used for nonload-bearing walls and partitions.

Figure 1-36. Open-web trusses as shipped from the factory. Webs and chords are fastened with metal plate connectors. Typically, such trusses are stronger than solid joists of the same depth. The arrow points to offsets designed to receive a 1 x 4 ribbon that ties joists together and holds them upright.

Figure 1-37. Various types of roofing shingles. Left—In southwest United States, cement shingles are popular. Right—Asphalt and fiberglass shingles are important roofing materials for residential construction. (CertainTeed)

Framing with steel requires some special tools. These include variable speed drill and screw gun, hearing protectors, vise clamps, metal snips, metal punch, metal cut-off blade for portable saw, magnetic level, metal cutoff saw, and a right-angle drill. In some cases, welding equipment is needed.

Advantages of Steel Framing

In general, steel framing is more affordable than wood framing. Steel members are normally straight and consistent, unlike wood, which warps, swells, and may have knots and other imperfections. There is also little limitation

Figure 1-38. Caulking and sealing materials are designed to resist moisture and prevent air infiltration through cracks and joints. Be sure to follow the manufacturer's recommendations for a specific application. (Beecham Home Improvement Products, Inc.)

Table 1:
Sizes of Steel Framing

Studs
Widths: $1\frac{5}{8}$"*, $2\frac{1}{2}$", $3\frac{5}{8}$", 4", 6"
Gauges: 25*, 20, 18, 16, 14, 12

Joists
Widths: 4", 6", 8", $9\frac{1}{4}$", 10", 12", 14"
Gauges: 18, 16, 14, 12

** Non-loadbearing uses only*
Gauge equivalents: 25 = .019", 20 =.0346", 18 = 0.451", 16 = .0566", 14 = .0713", 12 = .1017"

Figure 1-40. Steel framing for studs and joists comes in several dimensions. (Journal of Light Construction)

Figure 1-39. Steel framing units as delivered to a building site. Due to the rising costs of lumber, steel members are finding greater use in modern residential and light commercial buildings.

Figure 1-41. Some steel framing, as shown here, is designed to be fastened to wood plates using hammers, nailers, or screw guns. (HL Stud Corp.)

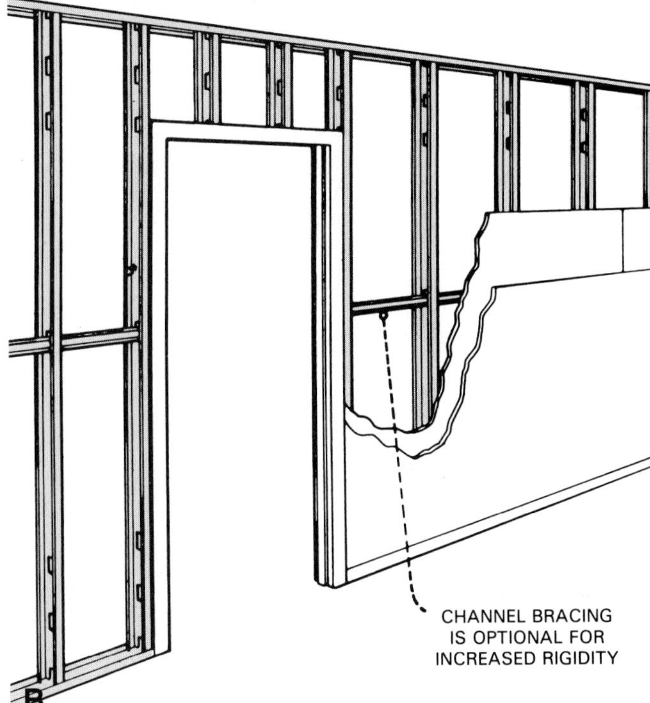

Figure 1-42. Metal studs are stamped and prepunched to accommodate installation of mechanical systems such as electrical wiring and plumbing. (National Gypsum Co.)

in the length of steel framing. Joists can be manufactured in lengths up to 40'.

Disadvantages of Steel Framing

Steel has some disadvantages. The standard stud for nonload-bearing walls, 25 gage, is flimsy. Care must be taken when handling them. Steel edges are sharp and may cause injury, even if hemmed track components are used, Figure 1-43.

Some carpenters do not use steel studs on outside walls because they are poor insulators and can reduce R values by as much as 50%. Finally, installing blocking, sheathing, and siding takes longer with steel studs than with wooden studs.

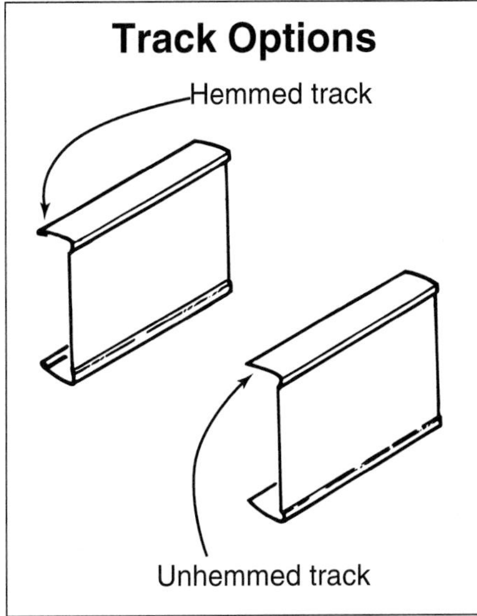

Figure 1-43. Steel framing with a hemmed track reduces injuries from sharp steel edges. (Journal of Light Construction)

Metal Framing Connectors

At one time, all wood-to-wood connections in a wood frame were made with nails. While this practice still continues and produces adequate strength for the structure, metal connectors are faster to install and improve uniformity in strength.

Basically, metal framing connectors are stamped brackets or strapping designed to make wood-to-wood, wood-to-masonry, or wood-to-concrete connections. Unless the connector is designed to be exposed and decorative, it is made from galvanized metal in various gages (thicknesses).

Strapping or Ties

Strapping or ties, Figure 1-44, are designed to hold parts of a frame together and are often used to "quake proof" structures, tying frames to foundations, and roofs to walls. Straps are perforated so that they can be fastened with nails without first drilling holes.

Hangers

Hangers connect the end of one framing member to another. Hangers are used where floor or ceiling joists intersect another framing member, such as a beam. Designs are available for either solid lumber or laminated lumber. See Figure 1-45.

Other Connector Types

Other connectors are designed for special purposes. *Tension bridging* can be used instead of solid wood or wood braces to transfer loads from joist to joist. They can be used on either solid lumber joists or on wood I-beams.

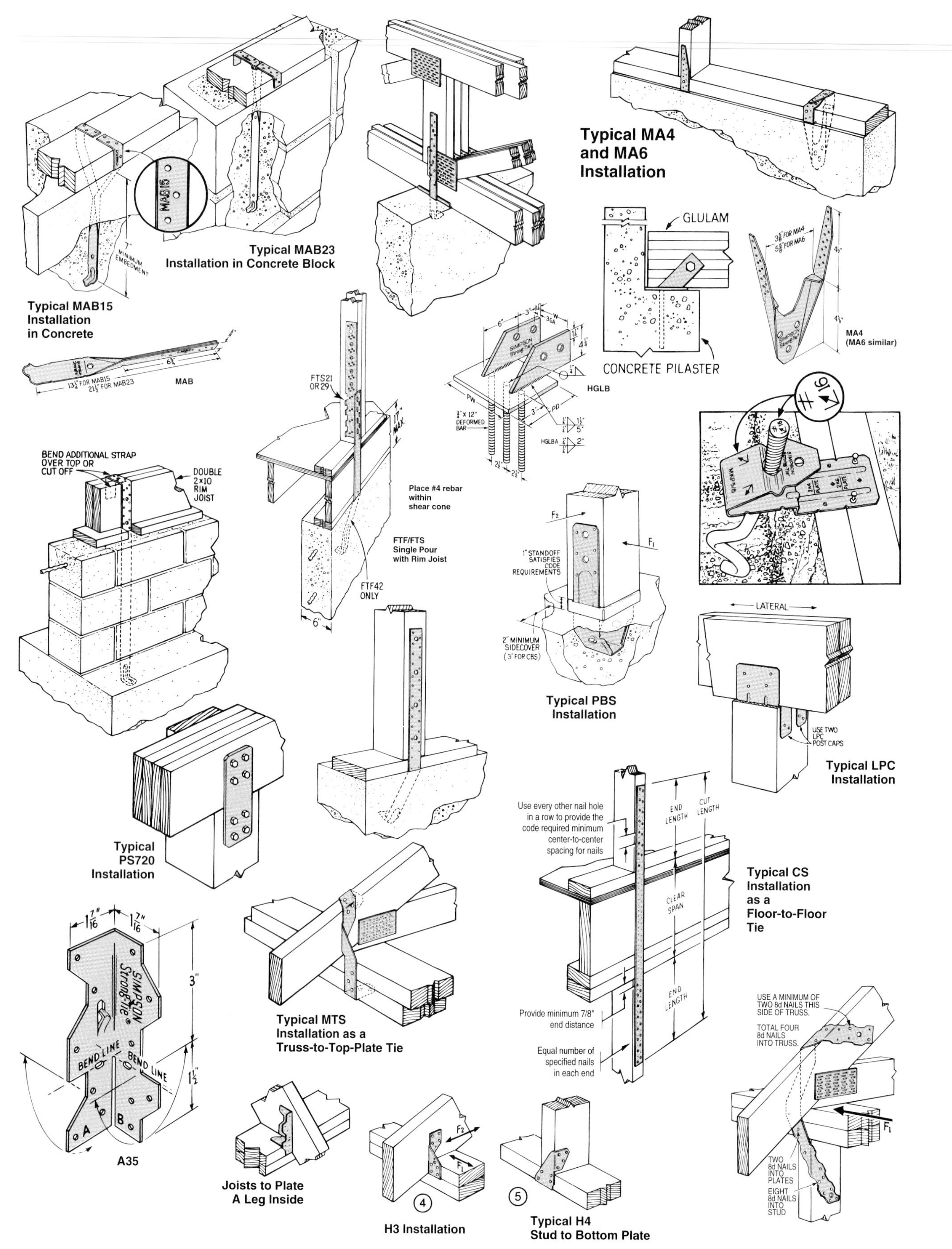

Figure 1-44. These framing connectors and ties are made in many different configurations to secure walls to foundations and roofs to walls. (© Simpson Strong-Tie Company, Inc.)

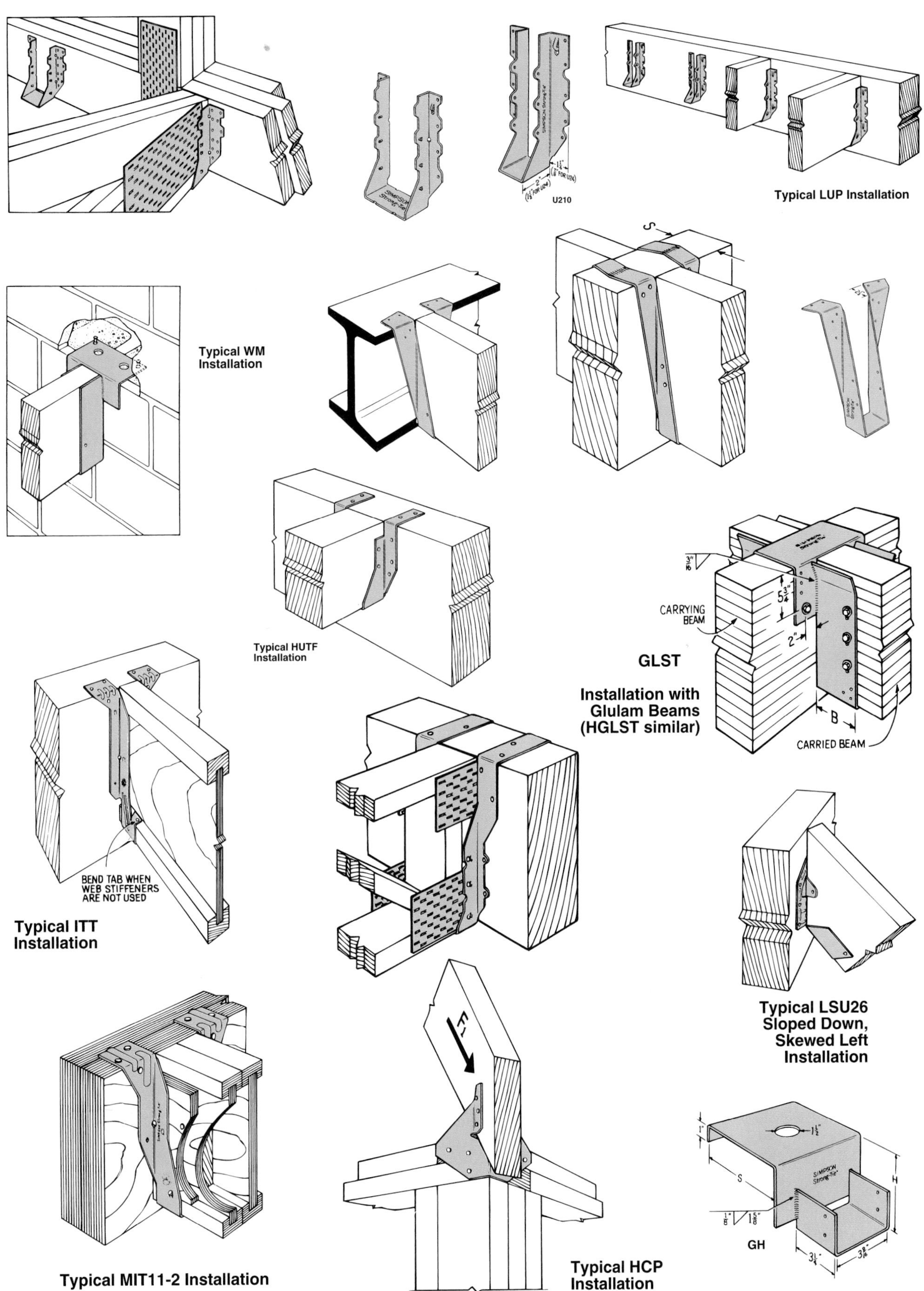

Figure 1-45. Wood-framing hangers can be used at various framing joints in residential construction to provide support of loads and stresses placed on joists and beams. (© Simpson Strong-Tie Company, Inc.)

Metal corner braces are another type of tension connector. See Figure 1-46.

Fasteners for Connectors

Fasteners used for metal connectors should be able to withstand shear stresses placed on them by the connectors. Nails or screws can be used. Some builders secure the fasteners with drywall screws, which penetrate wood quickly and cannot be easily withdrawn. However, drywall screws have little shear strength and, are not recommended. Generally, only nails are to be used, Figure 1-47. Bolts may sometimes be used. Nail length varies with the type of connector. Manufacturer's literature should be consulted. Also, see Figure 1-47(B).

Metal Lath

Where stucco is the exterior wall finish, metal lath is attached to the sheathing as a base for the stucco plaster, Figure 1-48. The lath comes in rolls and is attached with staples.

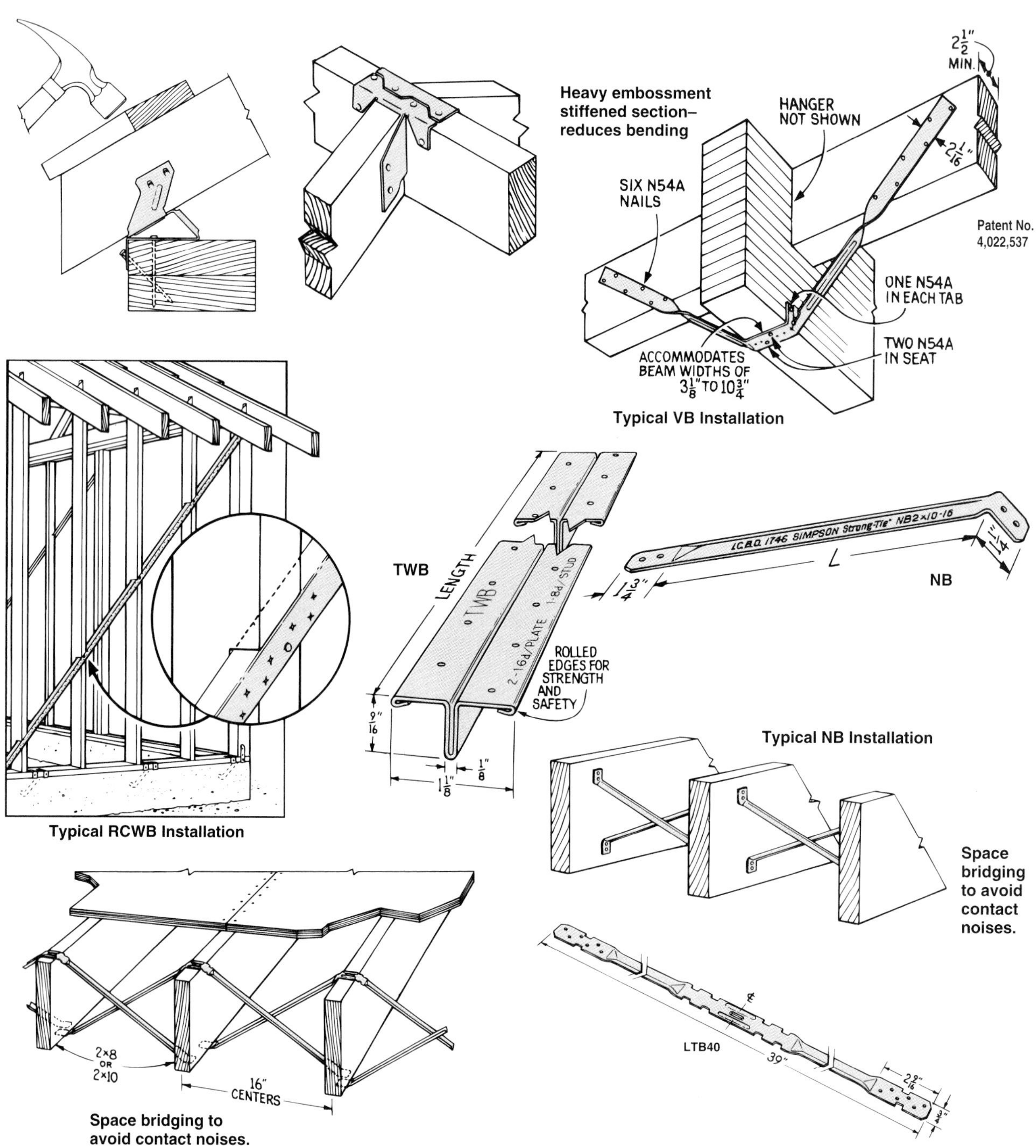

Figure 1-46. Some strapping and ties are designed to work under tension to prevent joists from tipping or building frames from racking. (© Simpson Strong-Tie Company, Inc.)

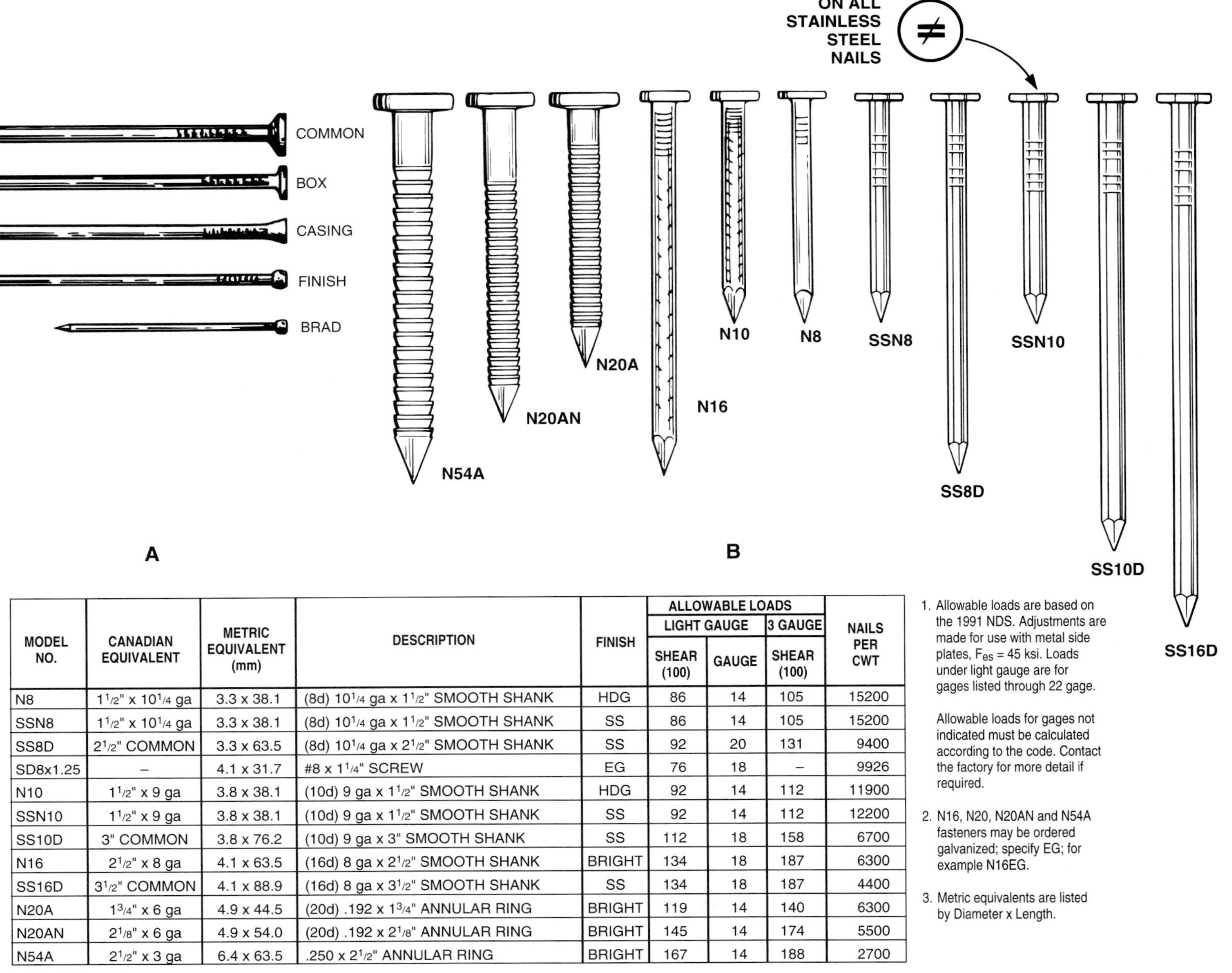

MODEL NO.	CANADIAN EQUIVALENT	METRIC EQUIVALENT (mm)	DESCRIPTION	FINISH	ALLOWABLE LOADS: LIGHT GAUGE SHEAR (100)	ALLOWABLE LOADS: LIGHT GAUGE GAUGE	ALLOWABLE LOADS: 3 GAUGE SHEAR (100)	NAILS PER CWT
N8	1 1/2" x 10 1/4 ga	3.3 x 38.1	(8d) 10 1/4 ga x 1 1/2" SMOOTH SHANK	HDG	86	14	105	15200
SSN8	1 1/2" x 10 1/4 ga	3.3 x 38.1	(8d) 10 1/4 ga x 1 1/2" SMOOTH SHANK	SS	86	14	105	15200
SS8D	2 1/2" COMMON	3.3 x 63.5	(8d) 10 1/4 ga x 2 1/2" SMOOTH SHANK	SS	92	20	131	9400
SD8x1.25	–	4.1 x 31.7	#8 x 1 1/4" SCREW	EG	76	18	–	9926
N10	1 1/2" x 9 ga	3.8 x 38.1	(10d) 9 ga x 1 1/2" SMOOTH SHANK	HDG	92	14	112	11900
SSN10	1 1/2" x 9 ga	3.8 x 38.1	(10d) 9 ga x 1 1/2" SMOOTH SHANK	SS	92	14	112	12200
SS10D	3" COMMON	3.8 x 76.2	(10d) 9 ga x 3" SMOOTH SHANK	SS	112	18	158	6700
N16	2 1/2" x 8 ga	4.1 x 63.5	(16d) 8 ga x 2 1/2" SMOOTH SHANK	BRIGHT	134	18	187	6300
SS16D	3 1/2" COMMON	4.1 x 88.9	(16d) 8 ga x 3 1/2" SMOOTH SHANK	SS	134	18	187	4400
N20A	1 3/4" x 6 ga	4.9 x 44.5	(20d) .192 x 1 3/4" ANNULAR RING	BRIGHT	119	14	140	6300
N20AN	2 1/8" x 6 ga	4.9 x 54.0	(20d) .192 x 2 1/8" ANNULAR RING	BRIGHT	145	14	174	5500
N54A	2 1/2" x 3 ga	6.4 x 63.5	.250 x 2 1/2" ANNULAR RING	BRIGHT	167	14	188	2700

Figure 1-47. The most common fastener is nails. A—Carpenters generally use these five basic types of nails. B—Nail types and specifications for securing connectors and hangers. (© Simpson Strong-Tie Company, Inc.)

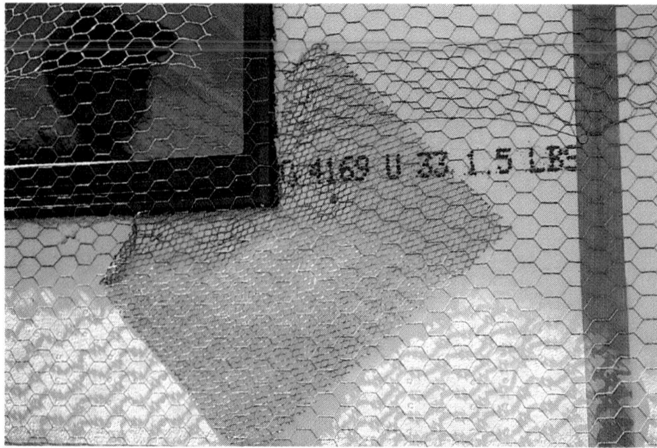

Figure 1-48. Metal lath has been attached preparing for applying stucco to the walls of a new home. (Pierce Construction Limited)

Metal Fasteners

Nails, the metal fasteners commonly used by carpenters, are available in a wide range of types and sizes. Basic kinds are illustrated in Figure 1-49.

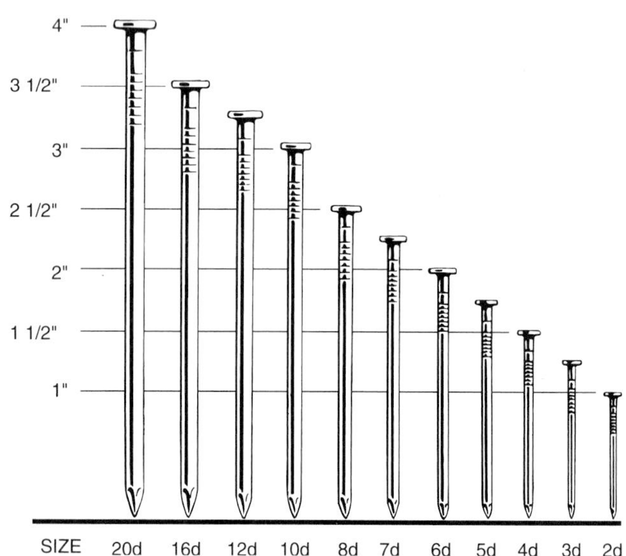

Figure 1-49. Nail sizes are given in a unit called the "penny." It is written as "d." (United States Steel)

The common nail has a heavy cross section and is designed for rough framing. The thinner box nail is used for toe nailing in frame construction and light work. The casing nail is the same weight as nails used in finish carpentry work to attach door and window casings and other wood trim. Finishing nails and brads are quite similar and have the thinnest cross section and the smallest head.

The nail size unit is called a "penny" and is abbreviated with the lower case letter d. It indicates the length of the nail. A 2d (2 penny) nail is 1" long. A 6d (6 penny) nail is 2" long. See Figure 1-50. This measurement applies to common, box casing, and finish nails. Brads and small box nails are specified by their actual length and gage number.

		COMMON		BOX	
Size	Length"	Diam."	No./Lb.	Diam."	No./Lb.
4d	1 1/2	.102	316	.083	473
5d	1 3/4	.102	271	.083	406
6d	2	.115	181	.102	236
7d	2 1/4	.115	161	.102	210
8d	2 1/2	.131	106	.115	145
10d	3	.148	69	.127	94
12d	3 1/4	.148	63	.127	88
16d	3 1/2	.165	49	.134	71
20d	4	.203	31	.148	52
30d	4 1/2	.220	24	.148	46
40d	5	.238	18	.165	35

Figure 1-50. This chart shows nail sizes and approximate number in a pound. (Georgia-Pacific)

Nails for power nailing are shown in Figure 1-51. Figure 1-52 shows a few of the many specialized nails. Each is designed for a special purpose. Annular or spiral threads greatly increase holding power. Some nails have special coatings of zinc, cement, or resin. Coating or threading increases nail holding power. Nails are made from such material as iron, steel, copper, bronze, aluminum, and stainless steel.

Wood screws have greater holding power than nails and are often used for interior construction and cabinetwork. Their size is determined by the length and diameter (gage number). Screws are classified according to the shape of head, surface finish, and the material from which they are made. See Figure 1-53.

Wood screws are available in lengths from 1/4" to 6", and in gage numbers from 0 to 24. The gage number can vary for a given length of screw. For example, a 3/4" screw is available in gage numbers of 4 through 12. The No. 4 is a thin screw, while the No. 12 has a large diameter. From one gage number to the next, the size of the wood screw changes by 13 thousandths (.013) of an inch.

Most wood screws used today are made of mild steel with a zinc chromate finish. They are labeled as F.H., which stands for flat head. Nickel and chromium plated screws, also screws made of brass, are available for special work. Wood screws are usually priced and sold by the box.

Additional useful fasteners include lag screws, hanger bolts, carriage bolts (specially designed for woodwork), corrugated fasteners, and metal splines. Specialized metal fasteners are described in other sections of this book.

Figure 1-51. Various sizes and kinds of nails and staples for power application. Flat blanks are called clips and round units are called coils. Most of the fasteners have a special thermoplastic polymer coating that makes them easier to drive. Once the nail is driven, the coating hardens like glue to increase the nail's holding power. (ITW Paslode)

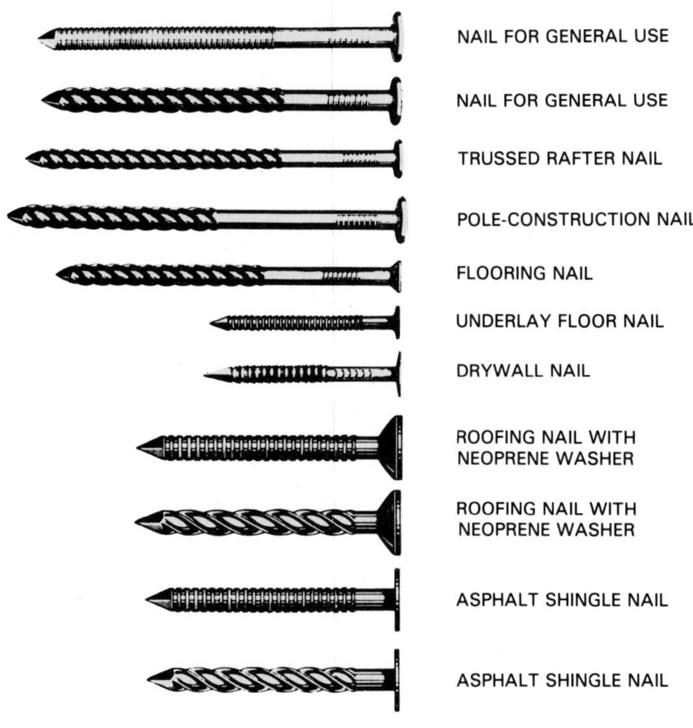

Figure 1-52. Annular and spiral threaded nails are designed for special purposes. (Independent Nail and Packing Co.)

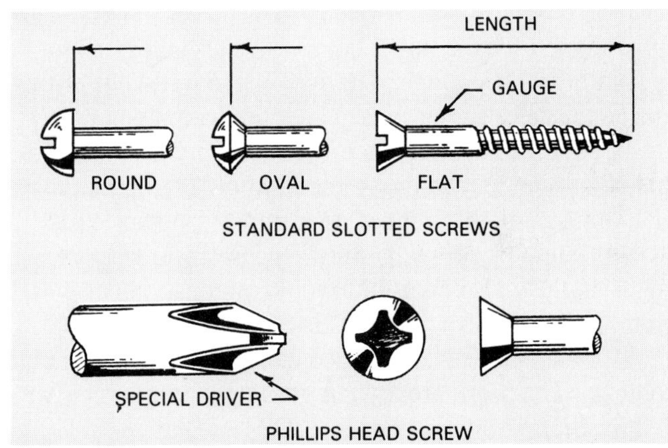

Figure 1-53. Kinds of heads found on common wood screws.

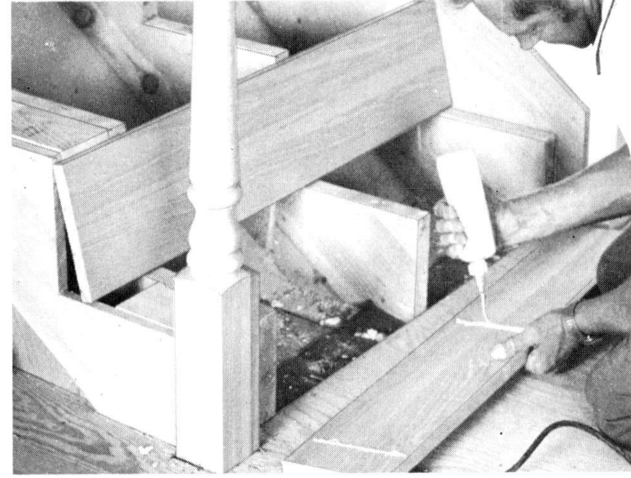

Figure 1-54. A polyvinyl resin emulsion glue is being used to assemble stairs.

Adhesives

The adhesives that carpenters use may be classified as glue and mastics. Research and development have created many new products in this area. Some are highly specialized, being designed for a specific material and/or application. Brief descriptions of several of the commonly used glues follows:

Polyvinyl resin emulsion glue (generally called polyvinyl or white glue) is excellent for interior construction. It comes ready to use in plastic squeeze bottles, Figure 1-54, and is easily applied. This glue sets up rapidly, does not stain the wood or dull tools, and holds wood parts securely.

Polyvinyl glue hardens when its moisture is removed through absorption by the wood or through evaporation. It is not waterproof and, therefore, is not appropriate for assemblies that will be subjected to high humidity or moisture. The vinyl-acetate materials used in the glue are thermoplastic. Under heat they will soften. They should not be used in constructions where the temperature may rise above 165°F (74°C).

Urea-formaldehyde resin glue (usually called urea resin) is available in a dry powder form that contains the hardening agent, or catalyst. It is mixed with water to a creamy consistency before use.

Urea resin is moisture resistant, dries to a light brown color, and holds wood securely. It hardens through chemical action when water is added and sets at room temperature in 4 to 8 hours.

Contact cement is applied to each surface and allowed to dry until a piece of paper will not stick to the film. The cemented surfaces are then pressed firmly together and bonding takes place immediately. The pieces must be carefully aligned for the initial contact because they cannot be moved after they touch. The bonding time is not critical and can usually be performed anytime within one hour.

Contact cement is made with a neoprene rubber base and is an excellent adhesive for applying plastic laminates or joining parts that cannot be clamped together easily. It works well for applying thin veneer strips to plywood edges and can also be used to join combinations of wood, cloth, leather, rubber, and plastics.

Contact cement usually contains volatile, flammable solvents. The work area where it is applied must be well ventilated.

Casein glue is made from milk curd, hydrated lime, and sodium hydroxide. It is supplied in powder form and is mixed with cold water for use. After mixing, it should set for about 15 minutes before it is applied. It is classified as a water resistant glue.

Casein glue is used for structural laminating and works well with wood that has a high moisture content. It has good joint filling qualities and, therefore, is often used on materials that have not been carefully surfaced. Casein is used for gluing oily woods such as teak, padouk, and lemon wood. Its main disadvantages are that it stains the wood, especially such species as oak, maple, and redwood, and has an abrasive effect on tool edges.

Mastics

Mastics are a heavy, pasty type of adhesive that have revolutionized the methods used in the application of wallboards, wood paneling, and some types of floors. They vary in their characteristics and application methods, and are usually designed for a specific type of material. Some are waterproof, others must be used where there is no excessive moisture.

One application method consists of placing several gobs on the surface of the material and then pressing the unit firmly in place. This causes the mastic to spread over a wider area. Some mastics are spread over the surface with a notched trowel. Still others are designed for caulking gun application, Figure 1-55.

Mastics are usually packaged in metal containers or gun cartridges ready for application. Always follow the directions of the manufacturer.

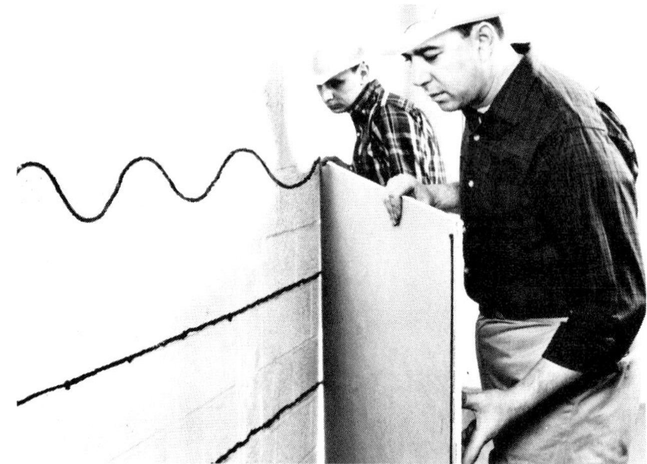

Figure 1-55. Mounting gypsum wallboard on concrete block wall. Beads of adhesive or mastic were applied with a caulking gun. (National Gypsum Co.)

Important Terms

Annular rings
Cambium
Casein glue
Composite board
Conifers
Contact cement
Core
Cross-bands
Edge grain
Engineered lumber
Equilibrium moisture content
Exterior plywood
Faces
Factory and shop lumber
Fiber saturation point
Flat-grained
Glue-laminated beams
Hardboard
Heartwood
Interior plywood
Kiln-dried
Laminated-veneer lumber
Lignin
Lumber cores
Mastics
Moisture content
No. 1 common
Open grain woods
Oriented strand board
Parallel laminations
Parallel-strand lumber
Particleboard
Phloem
Plain sawed
Polyvinyl resin emulsion glue
Quarter-sawed
Sapwood
Selects
Span ratings
Steel framing members
Tension bridging
Tracheids
Urea-formaldehyde resin glue
Veneer cores
Waferboard
Xylem

Test Your Knowledge

1. The natural cement that holds wood cells together is called _____________.
2. New wood cells are formed in which layer?
3. Which of the following kinds of wood are classified as a hardwood? Hemlock, redwood, willow, spruce.
4. When a softwood log is cut so the annular rings form an angle greater than 45° with the surface of the boards, the lumber is called _____________.
5. What is the moisture content of a board if a test sample that originally weighed 11.5 oz. was found to weigh 10 oz. after oven drying?
6. The fiber saturation point is about _____________ M.C. for nearly all kinds of wood.
7. The letters E.M.C. are an abbreviation for the term _____________ moisture content.
8. A large knot is defined as one that is over _____________ inches in size.
9. Where should a plywood panel marked "Exposure 1" be used?
10. The best grade of "selects" softwood lumber is _____________.
11. The best available grade of hardwood lumber is _____________.
12. How many board feet of lumber are contained in a pile of 24 pieces of 2 x 4 x 8'?
13. What is a span rating for plywood and what is the purpose of the rating?
14. Name the two principal types of engineered wood used as dimension lumber.
15. What is the thickness in inches of a 25 gage steel joist?

Outside Assignments

1. Prepare a visual aid that shows various metal fasteners used in carpentry. Include nails, brads, screws, carriage bolts, staples (such as are used in power staplers), and other items. Include spiral, ring groove, and coated nails. Label each item, giving the correct name, size, and other information.
2. Obtain a group of softwood samples that will be representative of the species of lumber used in your locality. Instead of writing the proper name on each piece, use a number that corresponds with your master list. This will permit you to give a wood identification quiz to members of your class. As the samples are passed around, the students can record the numbers and their answers on a sheet of paper.
3. Visit a local building supply center and secure information and literature concerning the various grades and species of lumber normally carried in stock. Prepare written descriptions of the defects permitted in several of the grades commonly selected by builders in the area. Obtain list prices of these grades to gain some understanding of the savings that can be gained by using a lower classification. Make a summary report to your class on your findings and conclusions. Work out the difference in cost between two grades of 2 x 4s if you were purchasing 50 eight-foot boards.
4. After a study of reference materials, prepare a paper on stress-rated lumber. Include information on grade stamps and their interpretation. Also define the f (fiber stress in bending) and the corresponding E (stiffness) rating. Try to include a description of modern equipment and/or new techniques used in grading.

A wood deck, when carefully designed and constructed, can add to the usefulness and beauty of a home. It is important to use pressure-treated wood, and zinc-coated or stainless steel fasteners. Be sure to follow the manufacturer's recommendations when working with treated materials. (Wolmanized Wood Producers)

Wood Identification

A key element in woodworking and in carpentry is the proper identification of the wood.

These samples are intended as a guide and an aid to the student in learning to identify various woods. They show typical color and grain characteristics for 16 different species.

CYPRESS. Light in weight, soft, and easily worked. Fairly coarse texture with annual growth rings clearly defined (sample shows edge grain). Source: Southeastern Coast of the United States. Noted for its durability against decay. Used for exterior construction and interior wall paneling.

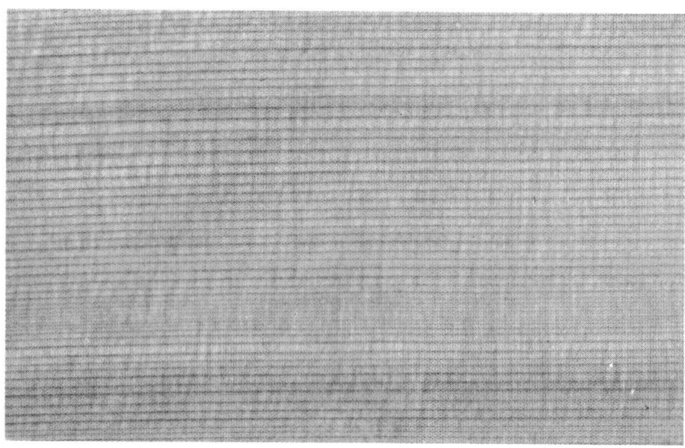

CEDAR, RED, WESTERN. A softwood, light in weight (23 lb. per cu. ft.). Similar to redwood except for cedar-like odor. Pronounced transition from spring to summer growth (see edge-grain sample). Source: Western coast of North America, especially Washington. Used for shingles, siding, structural timbers, and utility poles.

FIR, DOUGLAS. A strong, moderately heavy (34 lb. per cu. ft.) softwood. Straight close grain with heavy contrast between spring and summer growth. Splinters easily. Used for wall and roof framing and other structural work. Vast amounts are used for plywood. Machines and sands poorly. Seldom used for finish.

BIRCH. A hard, strong, wood (47 lb. per cu. ft.). Works well with machines and has excellent finishing characteristics. Heartwood, reddish-brown with white sapwood. Fine grain and texture. Used extensively for quality furniture, cabinetwork, doors, interior trim, and plywood. Also used for dowels, spools, toothpicks, and clothespins.

GUM, SWEET. Also called Red Gum. Fairly hard and strong (36 lb. per cu. ft.). A close-grained wood that machines well but has a tendency to warp. Heartwood is reddish-brown and may be highly figured. Used extensively in furniture and cabinetmaking. Stains well, often used in combination with more expensive woods.

MAHOGANY, PHILIPPINE. Medium density and hardness (37 lb. per cu. ft.). Open grain and coarse texture. Works fairly well with hand or machine tools. Varies in color from dark red (Tanguile) to light tan (Lauan). Used for medium price furniture, fixtures, trim, wall paneling. Also, boat building and core stock in plywood.

OAK, RED. Heavy (45 lb. per cu. ft.) and hard with the same general characteristics as White Oak. Heartwood is reddish-brown in color. No tyloses in wood pores. Used for flooring, millwork, and inside trim. Difficult to work with hand tools.

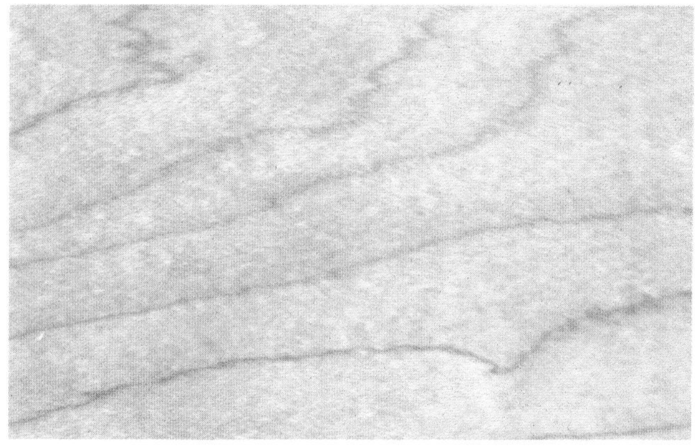

MAPLE, SUGAR. Also called Hard Maple. It is hard, strong, and heavy (44 lb. per cu. ft.). Fine texture and grain pattern. Light tan color, with occasional dark streaks. Hard to work with hand tools but machines easily. Is an excellent turning wood. Used for floors, bowling alleys, woodenware, handles, and quality furniture.

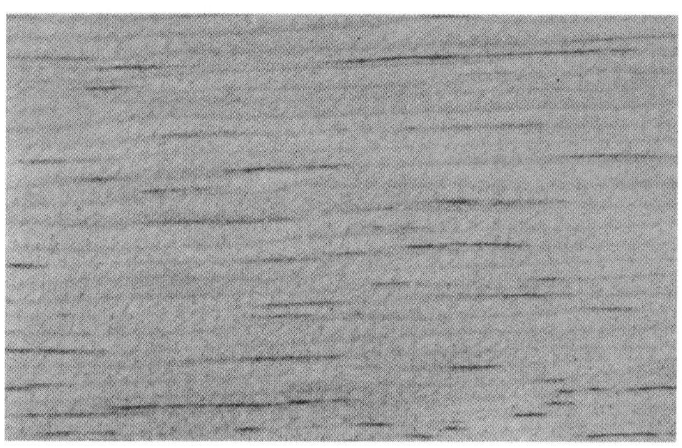

PINE, SUGAR. Lightweight (26 lb. per cu. ft.) soft, and uniform texture. Heartwood, light brown with many tiny resin canals that appear as brown flecks. Straight grained and warp resistant. Cuts and works very easily with hand tools. Used for foundry patterns, sash and door construction, and quality millwork.

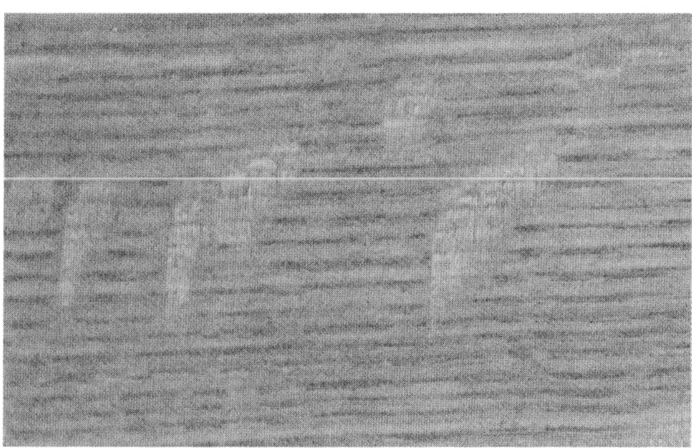

OAK, QUARTERED. Sawing or slicing oak in a radial direction results in a striking pattern as shown. The "flakes" are formed by large wood rays that reflect light. Used where dramatic wood grain effects are desired.

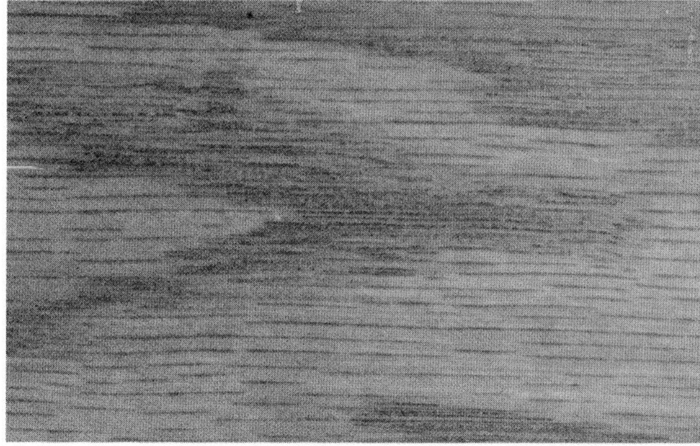

OAK, WHITE. Heavy (47 lb. per cu. ft.), very hard, durable, and strong. Works best with power tools. Heartwood is greyish-brown with open pores that are distinct and plugged with a hair-like growth called tyloses. Used for high quality millwork, interior finish, furniture, carvings, boat structures, barrels, and kegs.

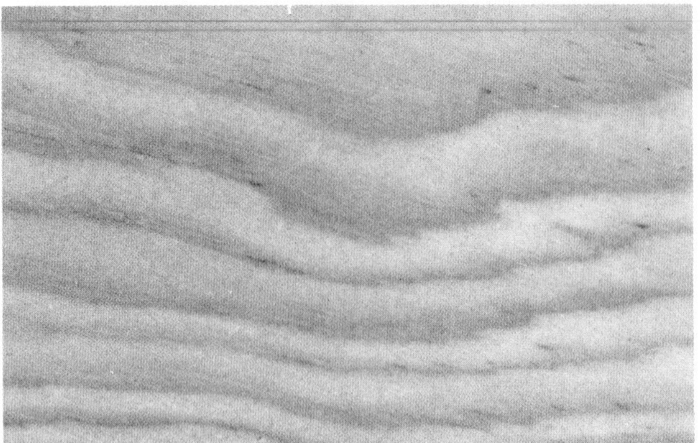

PINE, PONDEROSA. Lightweight (28 lb. per cu. ft.) and soft. Straight grained and uniform texture. Not a strong wood but works easily and has little tendency to warp. Heartwood is a light reddish brown. Change from springwood to summerwood is abrupt. Used for window and door frames, moldings, and other millwork, toys, and models.

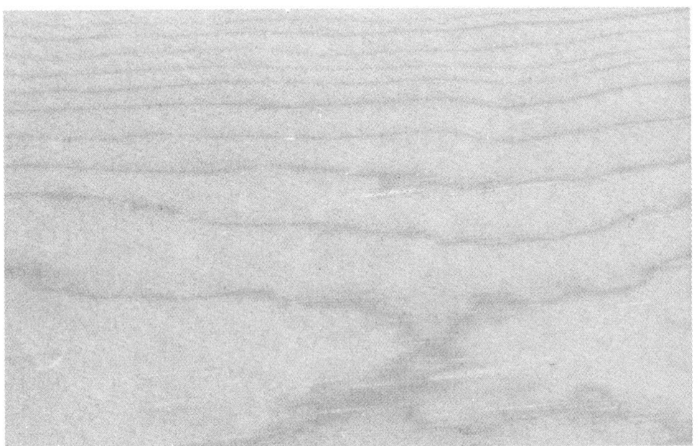

PINE, WHITE. Soft, light (28 lb. per cu. ft.), and even texture. Cream colored with some resin canals but not as prevalent as Sugar Pine. Used for interior and exterior trim and millwork item. Knotty grades often used for wall paneling. Works easily with hand or machine tools.

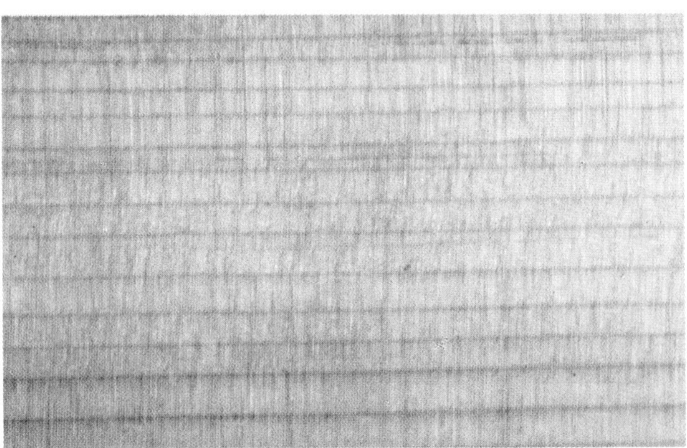

SPRUCE. A softwood, light in weight (24 lb. per cu. ft.). Transition from spring to summer growth is gradual (see edge grain sample). There are several species: Sitka, Englemen, and a general classification called Eastern. Source: various parts of the United States and Canada. Used for pulpwood, light construction, and carpentry.

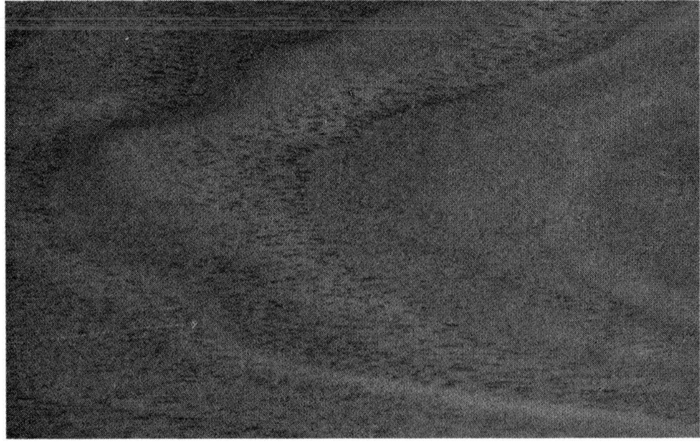

WALNUT, BLACK. Fairly dense and hard. Very strong in comparison to its weight (38 lb. per cu. ft.). Excellent machining and finishing properties. A fine textured open grain wood with beautiful grain patterns. Heartwood is a chocolate brown with sapwood near white. Used on quality furniture, gun stocks, fine cabinetwork, etc.

REDWOOD. Soft and light in weight (28 lb. per cu. ft.). Texture varies but is usually fine and even grained. Easy to work and durable. Heartwood is reddish-brown. Used for structures, outside finish, and sometimes for interior paneling. Its durability makes it especially valuable for products exposed to water and moisture.

2

General Safety Rules

Good carpenters recognize that safety is an important part of the job. They know that accidents can occur easily in building construction and that they often result in partial or total disability. Even minor cuts and bruises can be painful.

Safety is based on knowledge, skill, and an attitude of care and concern. Carpenters should know correct and proper procedures for performing the work. They should also be familiar with the potential hazards—how they can be minimized or eliminated.

Good attitudes toward safety are important. This includes belief in the importance of safety and willingness to give time and effort to a continuous study of the safest ways to perform work. It means working carefully and following the rules.

Clothing

Wear clothing appropriate for the work and weather conditions. Trousers or overalls should fit properly and have legs without cuffs. Keep shirts and jackets buttoned. Sleeves should also be buttoned or rolled up. Never wear loose or ragged clothing, especially around moving machinery.

Shoes should be sturdy, with thick soles to protect feet from protruding nails. Tennis or lightweight canvas shoes are not satisfactory. Never wear shoes with leather soles. They will not provide satisfactory traction on smooth wood surfaces—roofs in particular.

Construction work will require a *hard hat* (to prevent head injury from falling objects). Headgear should provide the necessary protection, be comfortable, permit good visibility, and shade your eyes. All clothing should be maintained in a satisfactory state of repair and not be permitted to become badly soiled.

Personal Protective Equipment

Safety glasses should be worn whenever work involves even the slightest hazard to your eyes. Standard specifications state that a safety lens must withstand the blow of a 1/8" diameter steel ball dropped from a height of 50".

Safety boots and shoes are required on heavy construction jobs. They consist of special reinforced toes that will withstand a load of 2500 lb. Hard hats should be worn whenever you are exposed to any possibility of falling objects. Standard specifications require that such hats withstand a certain degree of denting. They must be able to resist breaking when struck with an 80 lb. ball dropped from a 5' height.

Wear gloves of an appropriate type when handling rough materials. Use a respirator when working in dusty areas, while installing insulation, or where finishing materials are being sprayed.

Hand Tools

Always select the correct type and size of tool for your work. Be sure it is sharp and properly adjusted. Guard against using any tool if the handle is loose or in poor condition. Dull tools are hazardous to use because additional force must be applied to make them cut. Oil or dirt on a tool may cause it to slip, resulting in injury.

When using tools, hold them correctly. Most edge tools should be held in both hands with the cutting action away from yourself. Be careful when using your hand or fingers as a guide to start a cut.

Handle and carry tools carefully. Keep edged and pointed tools turned downward. Carry only a few tools at one time unless they are mounted in a special holder. Do not carry sharp tools in pockets of your clothing. When not in use, tools should be kept in special boxes, chests, or cabinets.

Power Tools

Before operating any power tool or machine, you must be thoroughly familiar with the way it works and the correct procedures to follow. In general, when you learn to use equipment the correct way, you also learn to use it the safe way, Figure 2-1.

There are a number of general safety rules that apply to power equipment. In addition, special safety rules must be observed in the operation of each individual tool or machine. Those that apply directly to the power tools commonly used for modern carpentry are listed in Unit 4. Study and follow them carefully.

Figure 2-1. Using a table saw the safe, correct way. Always wear safety glasses and use push sticks to move work through the blade. (Photo © Des Moines, Iowa Public Schools)

Good Housekeeping

This refers to the neatness and good order of the construction site. Maintaining a clean site contributes to the efficiency of the worker and is an important factor in the prevention of accidents.

Place building materials and supplies in neat piles. Locate them to allow adequate aisles and walkways. Rubbish and scrap should be placed in containers until disposal can be made, Figure 2-2. Do not permit blocks of wood, nails, bolts, empty cans, or pieces of wire to accumulate. They interfere with your work and constitute a tripping hazard.

Keep tools and equipment not being used in panels or chests. This will provide protection for the tools as well as the workers. In addition to improving efficiency and safety, good housekeeping helps maintain a better appearance at the construction project. This, in turn, will contribute to the morale of all workers.

Figure 2-2. An on-site dumpster provides safe disposal for construction scrap.

Decks and Floors

To perform an operation safely, either with hand or power tools, the carpenter should stand on a firm, solid base. The surface should be smooth but not slippery. Do not attempt to work over rough piles of earth or on stacks of material that are unstable. Stay well away from floor openings, floor edges, and excavations as much as possible. Where this cannot be done, install adequate guardrails or barricades. In cold weather, remove ice or cover it with sand or calcium chloride (salt).

Excavations

Shoring and adequate bracing must be placed across the face of any excavation where the ground is cracked or caving is likely to occur. Inspect the excavation and shoring daily, especially after rain. Follow state and local regulations. Never climb into an open trench until proper reinforcement against cave-in has been installed or until the sides have been sloped to the *angle of repose* of the material being excavated.

Before beginning excavations determine whether there are underground utilities in the area. If so, locate and arrange protection for them during excavation operations.

Excavated soil and rock must be stored at least 2' away from the edge of an excavation. Use ladders or steps to enter trenches which are more than 4' deep.

Scaffolds and Ladders

Scaffolds should have a minimum *safety factor* of four. This means that the scaffold will carry a load four times greater than the maximum load it will be required to support. All scaffolding should be constructed under the direction of an experienced carpenter. Inspections should be made daily before use.

Ladders should be checked at frequent and regular intervals. Their use should be limited to climbing from one level to another. Working while being supported on a ladder is hazardous and should be kept to a minimum. There are many safety rules that must be observed in the use of scaffolds and ladders. These are covered in Unit 29.

Falling Objects

When working on upper levels of a structure, you should be especially cautious in handling tools and materials so there is no chance of them falling on workers below. Do not place tools on the edge of scaffolds, stepladders, window sills, or on any other surface where they might be knocked off.

If long pieces of lumber must be placed temporarily on end and leaned against the side of the structure, be sure they will not fall sideways. When moving through a building under construction, be aware of overhead work, and, wherever possible, avoid passing directly underneath. Stay clear of materials being hoisted. Wear an approved hard hat whenever there is a possibility of falling objects. See Figure 2-3.

Figure 2-3. Hard hats are required wherever there is danger from falling objects.

Handling Hazardous Materials

Pressure-treated wood, because of the chemical preservatives used, requires special care in handling for safety of the construction worker. Avoid prolonged breathing in of sawdust particles. Sawing should be done outdoors while wearing a *dust mask*. Wear safety goggles when power sawing or machining.

Before eating any food, carefully wash any skin that has come into contact with treated wood. Clothing soiled by contact with the wood should be laundered before reuse. The clothing should be washed separately from other clothing.

When handled in this way, treated wood does not pose a health hazard.

Never burn scraps of treated wood. Rather, bury them or place them in an ordinary trash collection bin or dumpster placed on site during construction.

Use great care when spraying paints. Use an approved respirator and protect exposed skin by covering it with clothing, Figure 2-4.

Figure 2-4. Avoid breathing in fumes from paint spraying. Use a mask or respirator and cover exposed skin when necessary. (Greco Painting)

Lifting and Carrying

Injuries may be caused by improper lifting and carrying heavy objects. When lifting, stand close to the load, bend your knees, and grasp the object firmly. Then lift by straightening your legs and keeping your body as close to vertical as possible. To lower the object, reverse the procedure.

When carrying a heavy load, do not turn or twist your body but make adjustments in position by shifting your feet. If the load is heavy or bulky, have others help. Never underestimate the weight to be moved or overestimate your own ability. Always have assistance when carrying long pieces of lumber.

Fire Protection

Carpenters should have a good understanding of fire hazards. They should know the causes of fires and methods of controlling fires.

Class A fires result from burning wood and debris. Class B fires involve highly volatile materials such as gasoline, oil, paints, and oil soaked rags. Class C fires are caused by electrical wiring and equipment. Any of these fires can occur on a typical construction site.

Approved fire prevention practices should be followed throughout the construction project. Good housekeeping is an important aspect. Special precautions should be taken during the final stages of construction when heating and wiring are being installed and when highly flammable surface finishes are being applied. Always keep containers of volatile materials closed when not in use. Dispose of oily rags and combustible materials promptly.

Fire extinguishers should be available on the construction site. Be sure to use the proper kind for each type of fire. Study and follow local regulations.

First Aid

A knowledge of first aid is important. You should understand approved procedures and be able to exercise good judgment in applying them. Remember that an accident victim may receive additional injury from unskilled treatment. Information of this nature can be secured from your local Red Cross.

As a preventative measure against infection, keep an approved first aid kit on the job site. Because of the nature of the material being handled and the dirty conditions of the work area, even superficial wounds should be treated promptly. Clean, sterilize, and bandage all cuts and nicks.

Important Terms

Angle of repose
Dust mask
Hard hat
Pressure-treated wood
Safety factor
Safety glasses

Test Your Knowledge

1. Which of the following is considered safe clothing?
   A. Unbuttoned sleeves.
   B. Trousers or overalls without cuffs.
   C. Sandals.
   D. Necktie.
   E. All of the above.
2. When should safety glasses be worn?
3. When should a hard hat be worn?
   A. After you get a bad haircut.
   B. When the sun is bright.
   C. Wherever there is a danger of falling objects.
   D. All of the above.
   E. None of the above.
4. Safety shoes are required to hold a load of ______________.
   A. 50 lb.
   B. 1500 lb.
   C. 2500 lb.
   D. 3500 lb.
   E. 5000 lb.
5. When not being used, tools should be ______________.
   A. Left lying on the ground.
   B. Stored on a shelf, out of reach.
   C. Thrown in the back of a truck.
   D. Stored in panels or chests.
   E. None of the above.
6. True or False? Burning wood is considered a Class C fire.

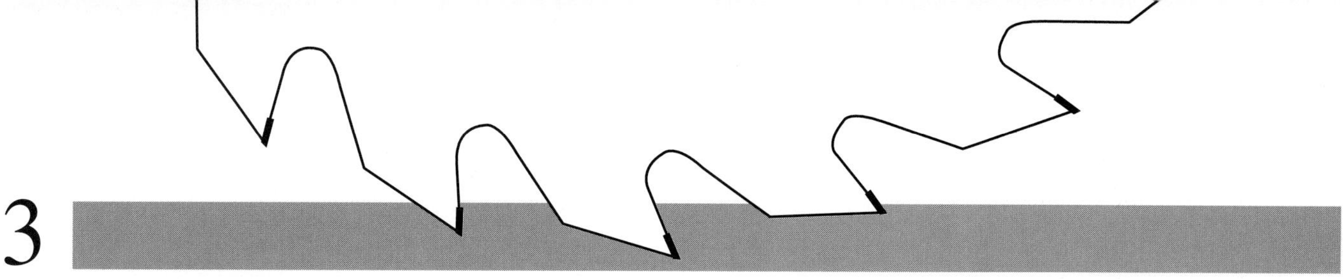

3 Hand Tools

Hand tools are essential to every aspect of carpentry work. A great variety of tools is required because residential and light commercial construction covers a broad range of activities.

The carpenter, a skilled worker, carefully selects the kind, type, and size of tools that best suit personal requirements. Tools become an important part of the carpenter's life, helping him or her to perform with speed and accuracy the various tasks of the trade.

Although the basic design of common woodworking tools has changed little in many years, modern technology and industrial "know-how" have brought numerous improvements. Special tools have been developed to do specific jobs. Experienced carpenters appreciate the importance of having good tools and select those that are accurately made from quality materials. They know that such tools last longer and will enable them to do better work.

A detailed study of the selection, care, and use of all the hand tools available is not practical for this text. Only a general description of the tools most commonly used by the carpenter will be included.

Measuring and Layout Tools

Measuring tools must be handled with considerable care and kept clean. Only then can a high level of accuracy be assured.

A *folding wood rule* or *measuring tape* is indispensable. Standard length for a rule is 6'. Some are equipped with a metal slide to aid in taking inside measurements, see Figure 3-1. Common lengths of 6' to 30' are common for measuring tapes. Tapes with a 1" blade (width) are stiffer than thin-bladed tapes. This makes the wide-bladed tapes easier to use. Long tapes are available in lengths of 50' and longer.

Squares

The *framing square*, also called a *rafter square*, is specially designed for the carpenter. Its uses are many and varied. A description of the square and how it is used in framing is included in Unit 10. It is available in steel, aluminum, or steel with a copper-clad or blued finish.

Try squares are available with blades 6" to 12" long. Handles are made of wood or metal. These are used to check the squareness of surfaces and edges and to lay out lines perpendicular to an edge, Figure 3-2.

The *combination square* serves a similar purpose and is also used to lay out miter joints. The adjustable sliding blade allows it to be conveniently used as a gauging tool.

The *T-bevel* has an adjustable blade making it possible to transfer an angle from one place to another. It is useful in laying out cuts for hip and valley rafters. Figure 3-3 illustrates how the framing square is used to set the T-bevel at a specified angle of 45°.

Wing dividers with locking legs are available in several sizes and serve a number of purposes. Dimensions can be stepped off along a layout or transferred from one position to another. Circles and arcs can also be scribed on surfaces, Figure 3-4.

Marking Gauge

A *marking gauge* is used to lay out parallel lines along the edges of material. It may also be used to:

- Transfer a dimension from one place to another.
- Check sizes of material.

The butt gauge can be used for the same purposes. However, it is normally used to lay out the gain (recess) for hinges.

A combination square has several uses. It can be used as a marking gauge to make parallel lines and as a square for 90° and 45° angles. Refer to Figure 3-5.

In layout work a *scratch awl* is used to scribe lines on the surfaces of materials. It is also used to mark points and to form starter holes for small screws or nails.

Using a *chalk line* is an easy way to mark long, straight lines. The line, which is covered with chalk, is held tight and close to the surface. Then it is snapped, Figure 3-6.This action drives the chalk onto the surface forming a distinct mark. A special reel rechalks the line each time it is wound back into the case.

The *level*, Figure 3-7, and *plumb bob* are important devices for laying out vertical and horizontal lines. A standard level is 24" long. The body may be made of wood, aluminum, or special lightweight alloys. It is often used in connection with a straightedge when the span or height of the work is greater than the length of the level. (A straightedge is a straight strip of wood usually laminated for greater stability.) Some carpenters prefer a 4' to 6' long level. It eliminates the need for a straightedge on such work as installing door and window frames.

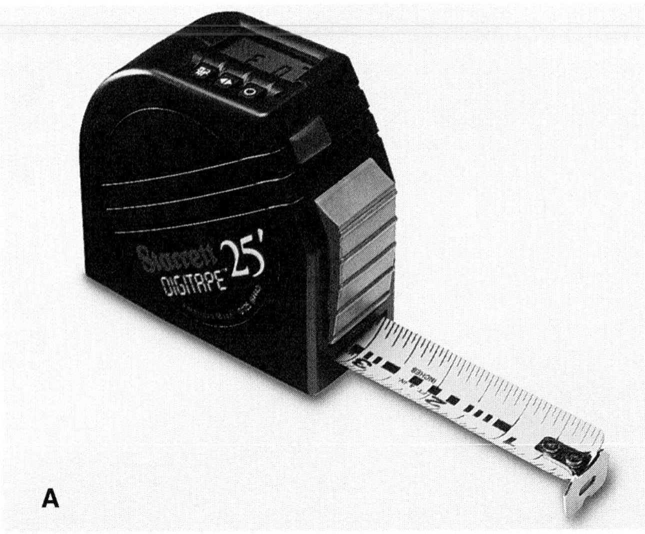

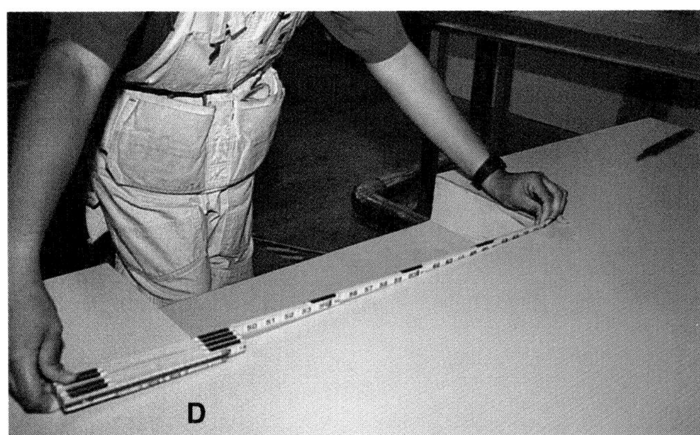

Figure 3-1. Steel tape rules are available in lengths from 6' to 25'. A—Newer models have direct digital readouts. (L.S. Starrett Co.) B—Rules are equipped with a hook that grips the end of the workpiece as a measurement is taken. (Santa Rita High School, Tucson) C—Being flexible, the tape easily bends around corners. (McDaniels Construction Co., Inc.) D—Using a folding ruler to take a measurement. (Bullard-Haven Technical School)

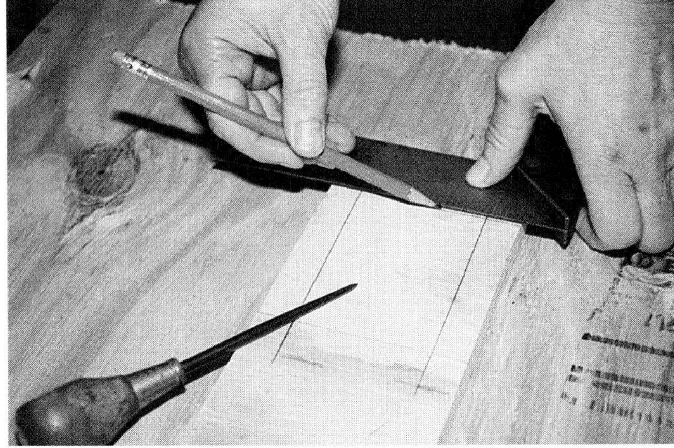

Figure 3-2. Try square is handy for laying out lines perpendicular to an edge. Scratch awl is sometimes used to mark lines when precision work is done.

Figure 3-3. Framing square can be used to set a T-bevel at a 45° angle.

Figure 3-4. Dividers may be used to draw arcs and circles or to step off short distances.

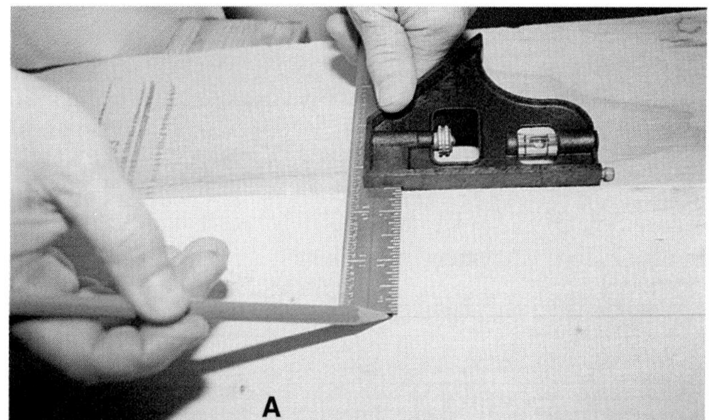

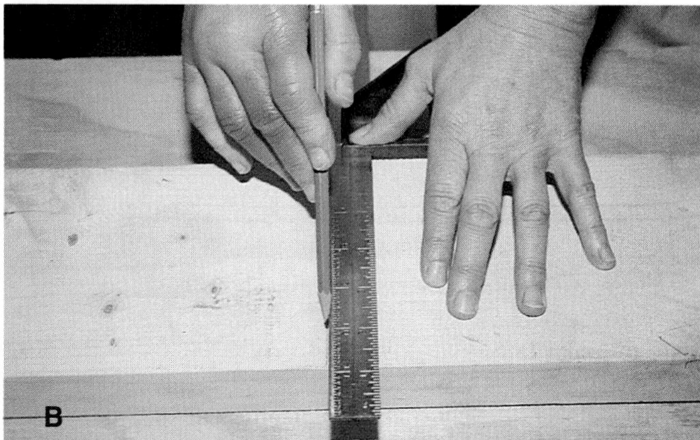

Figure 3-5. A combination square has several uses. A—Drawing a line parallel to the edge of a board. B—Marking a square cut. C—Marking a 45° cut.

Figure 3-6. Snapping a chalk line is a fast, simple way to lay down straight lines over long distances on a building. (American Plywood Assoc.)

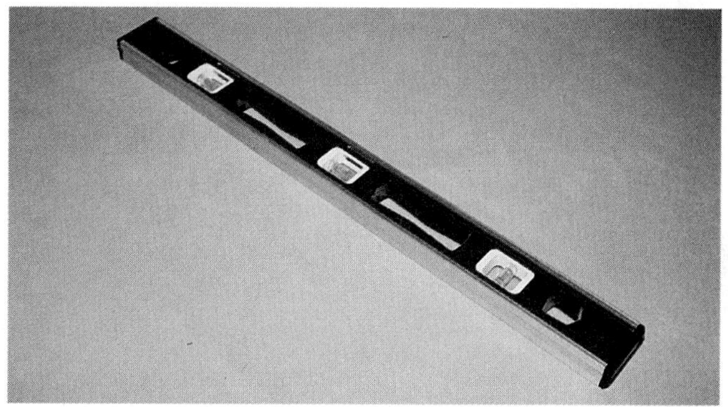

Figure 3-7. Levels and plumbs are needed to make sure materials are horizontal or vertical. Top—A level can be used to lay out level lines as well as check building frames for level and plumb by centering a bubble. Bottom—Checking a wall frame for plumb. Over great distances, a straightedge is used in connection with the level. (The Bennett Street School)

The plumb bob establishes a vertical line when attached to a line and suspended. Its weight pulls the line in a true vertical position for lay out and checking. The point of the plumb bob is always directly below the point from which it hangs. Figure 3-8 shows the tools used for measuring and layout.

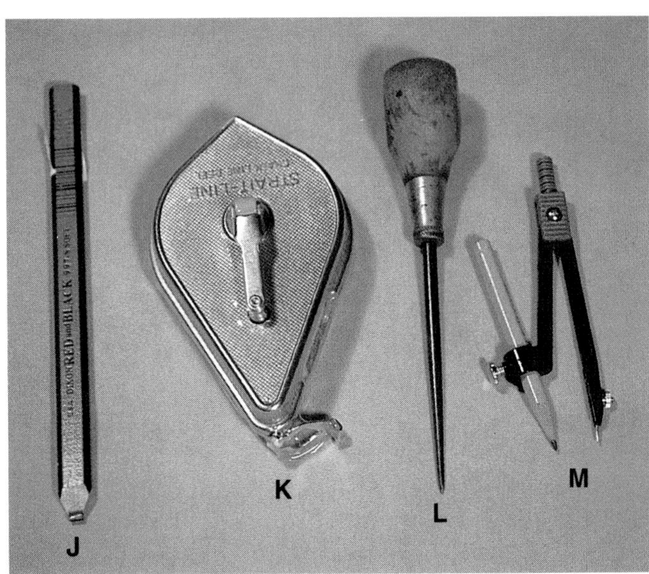

Figure 3-8. These measurement and layout tools are often used by carpenters. A—Rafter square. B—Try square. C—Long tape measure. D—Layout tape rule. E—Combination square. F—Steel rule. G—"Quick" or super square. H—Folding wood rule. I—4' steel rule. J—Carpenter's pencil. K—Chalk line. L—Awl. M—Dividers.

Many carpenters now use a framing tool called a *quick square*, or *super square*. This tool can be used to mark any angle for rafter cuts by aligning degree marks on the tool with the edge of the rafter being cut. See Figure 3-9.

Figure 3-9. Quick square can be used for marking square cuts and angle cuts. (Construction Training School, St. Louis)

Saws

The principal types of saws used by the carpenter are illustrated in Figure 3-10. These are available in several different lengths as well as various tooth sizes (given as teeth per inch, or points per inch).

Crosscut saws, as the name implies, are designed to cut across the wood grain. Their teeth are pointed, Figure 3-11(A). Using the thumb as a guide will assure getting a crosscut started on the line, Figure 3-12.

Rip saws have chisel-shaped teeth which cut best along the grain. Refer to Figure 3-11(B). Note that the teeth are set (bent) alternately from side to side so the *kerf* (cut in the wood) will be large enough for the blade to run freely. Some saws have a taper, ground from the toothed edge to the back edge. It eliminates the need for a large amount of set on the teeth.

Every carpenter needs a good crosscut saw with a tooth size ranging from 8 to 11 points. A crosscut saw for general use has 8 teeth per inch. A finishing saw used for fine cutting has 10 or 11 teeth per inch.

Most ripping operations are now performed with power saws, making the rip saw an optional tool in the carpenter's arsenal.

Another saw occasionally used for cutting curves is the *coping saw*, Figure 3-13. This saw has a thin, flexible blade that is pulled tight by the saw. With a blade that has 15 teeth per inch, the coping saw can make very fine cuts.

The *backsaw* has a thin blade reinforced with a steel strip along the back edge. The teeth are small (14–16

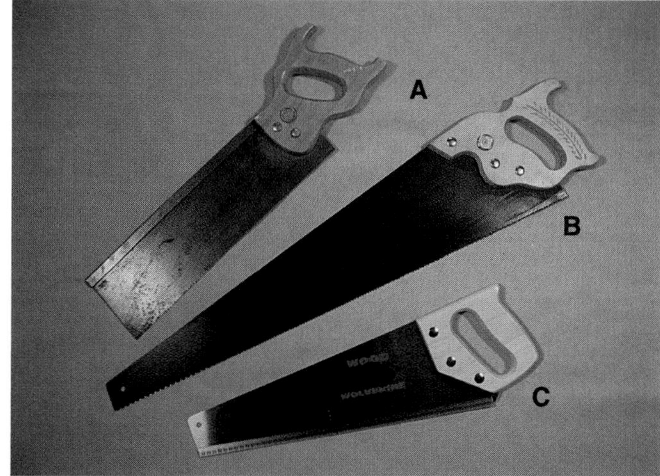

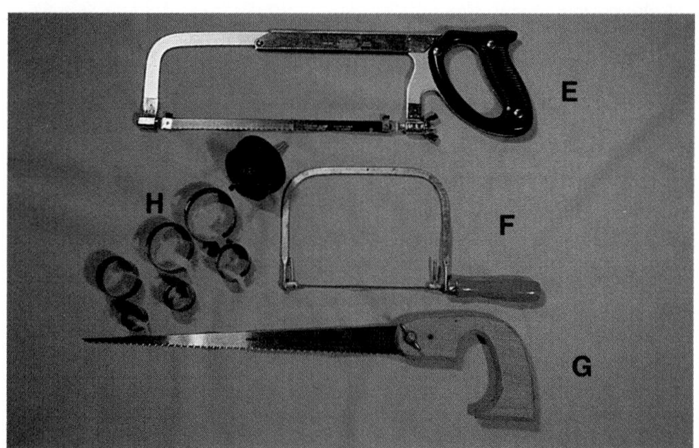

Figure 3-10. Carpenters use many different kinds of saws in the course of their work. These are used most often. A—Backsaw. B—Crosscut saw. C—Drywall saw. D—Rip saw. E—Hacksaw. F—Coping saw. G—Keyhole saw. H—Hole saws (used with electric drills). (McRae True Value Hardware)

points). Thus, the cuts produced are fine. It is used mostly for interior finish work. A similar type of saw is used in the *miter box,* Figure 3-14. Slots or holders accurately guide the blade when forming miters and other types of joints.

Many compass and keyhole saws have a quick-change feature that permits the use of various sizes and kinds of blades. Although originally designed to cut keyholes, they now serve as general purpose saws for irregular cuts or work where space is limited.

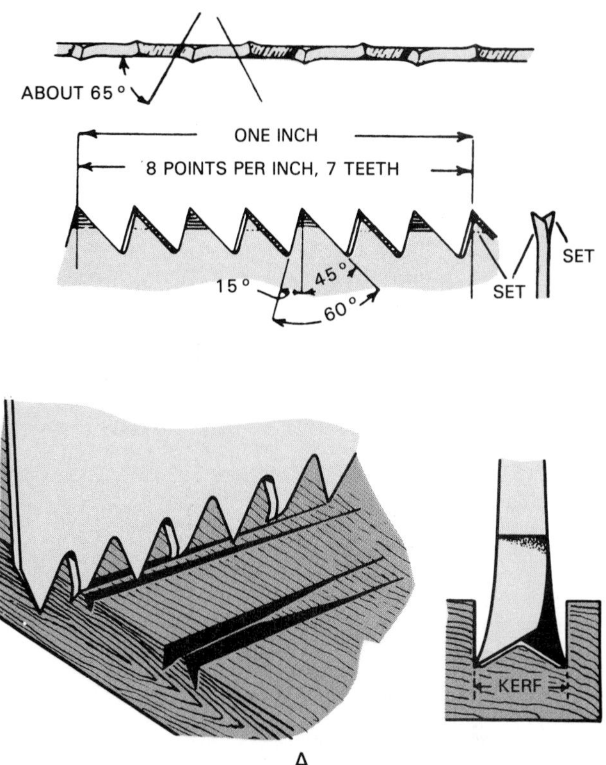

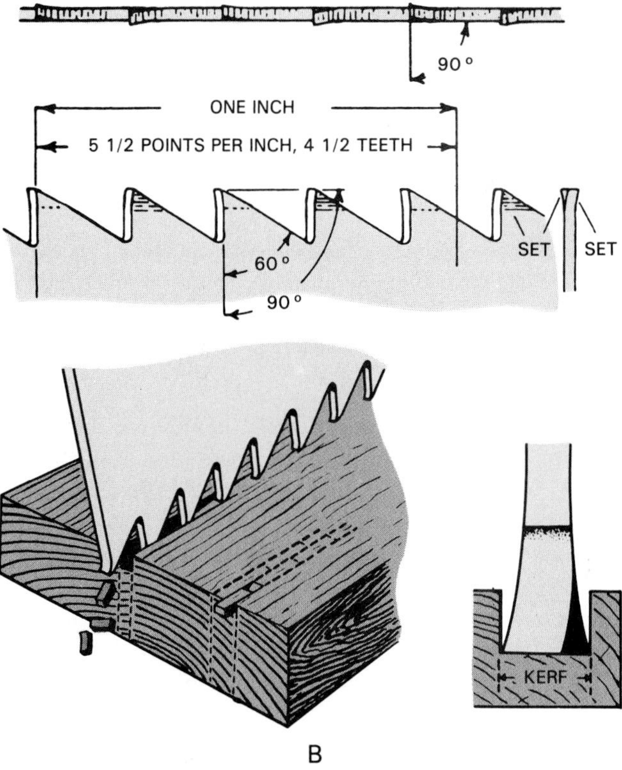

Figure 3-11. The teeth of crosscut and rip saws are markedly different. A—Crosscut teeth cut like a knife. Top shows shape and angle of teeth. Bottom shows how teeth cut. B—Rip saw teeth cut like a chisel. Top. Angle of 90° is often increased to give negative rake. Bottom. How rip teeth cut.

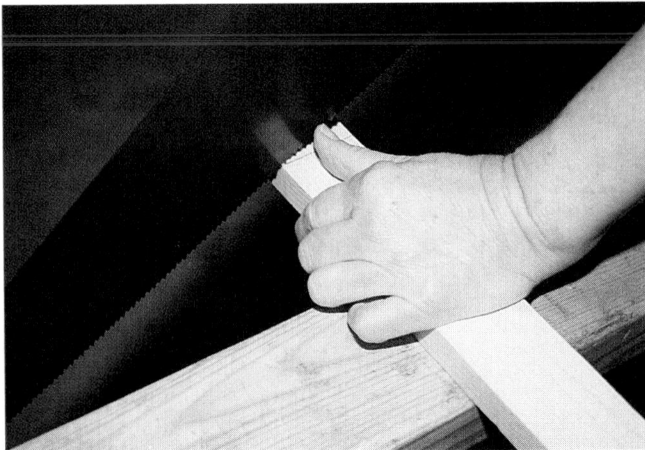

Figure 3-12. Use the thumb against the saw blade when starting a cut with the hand saw.

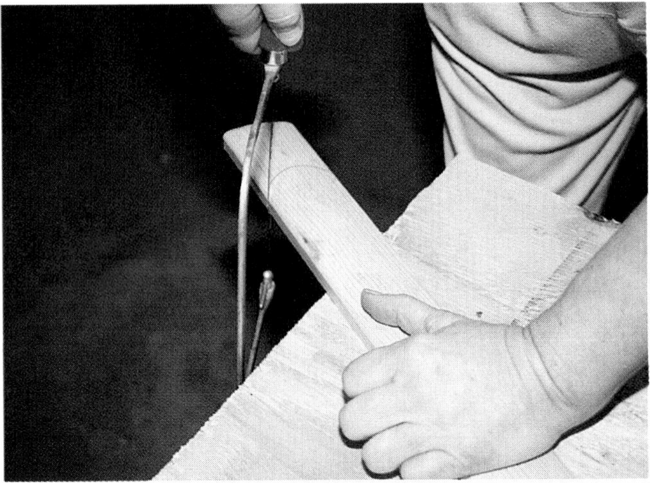

Figure 3-13. A coping saw is designed to cut along curves. Thin, flexible blade is supported under tension from the saw's frame. Blade has about 15 teeth per inch.

The *sheetrock-drywall saw* has large, specially designed teeth for cutting through paper facings, backings, and the gypsum core. Gullets are rounded to prevent their clogging from the gypsum material.

Due to the wide range of work that the carpenter must be prepared to handle, a *hacksaw* should be included in the hand tool assortment. This will be needed to cut nails, bolts, other metal fasteners, and metal trim. Most hacksaws have an adjustable frame, permitting the use of several sizes of blades.

Hacksaw Blades

Hacksaw blades are made of high speed steel, tungsten alloy steel, molybdenum steel, and other special alloys. *All-hard blades* are heat-treated. This makes them very brittle and easily broken if misused. *Flexible-back blades* are hardened only around the teeth. They are usually preferred for use on metals.

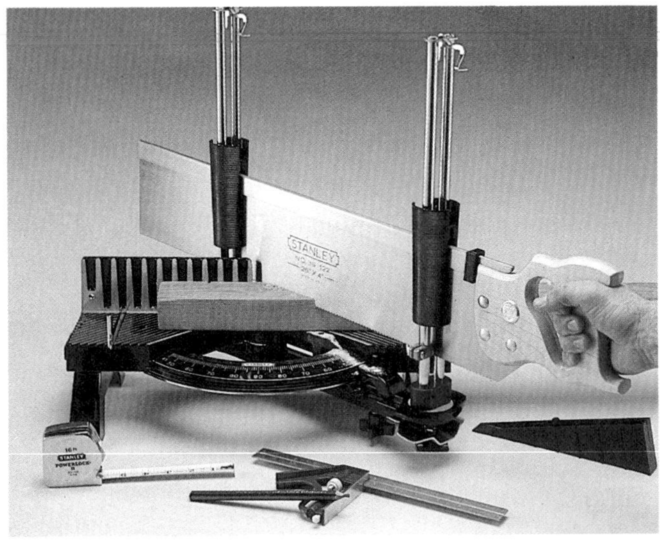

Figure 3-14. Miter box and backsaw are useful for making fine, accurate cuts. (Stanley Tools)

The blade's cutting edge may have anywhere from 14 to 32 teeth per inch. As a rule, you should use a blade with 14 teeth per inch for brass, aluminum, cast iron, and soft iron. For drill rod, mild steel, tool steel, and general work, 18 teeth per inch is recommended. For tubing and pipe, 24 teeth per inch is best. Generally, the thinner the metal being cut, the finer the blade should be. At least two or three teeth should rest on the metal; otherwise the sawing motion may shear off teeth.

Like wood-cutting saws, hacksaw blades have a set. This provides clearance for the blade to slide through the cut. Edge views showing types of set are illustrated in Figure 3-15.

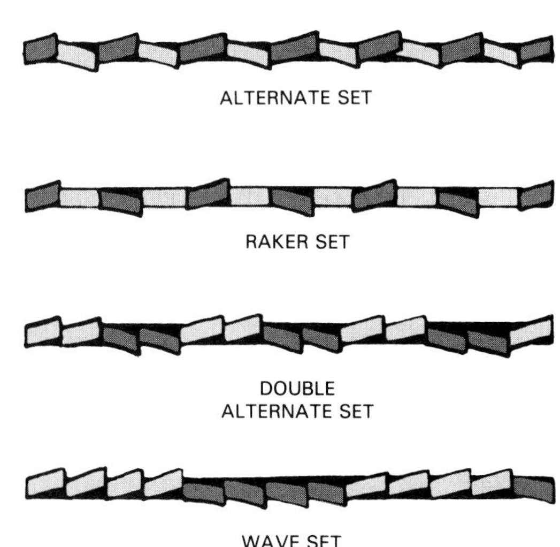

Figure 3-15. Enlarged view of hacksaw blades shows different types of set.

Blades should be installed with teeth pointing forward, away from the handle. Use both hands to operate the saw. Apply enough pressure on forward strokes to allow each tooth to remove a small amount of metal. Remove pressure on the return stroke to reduce wear on the blade. Saw with long steady strokes paced at 40 to 50 strokes per minute.

Planing, Smoothing, and Shaping Tools

The most important tools in this group are the planes and chisels. Several other tools that depend on a cutting edge to perform the work are included in the illustration, Figure 3-16.

Standard surfacing planes include the *smooth plane* (8–9" long), *jack plane* (14" long), and the *fore and jointer plane* (18–24" long). Figure 3-17 illustrates the parts and how they fit together. The jack plane is commonly selected for general purpose work.

Another surfacing plane, the *block plane*, is especially useful. It is small (6–7" long) and can be used with one hand. The blade is mounted at a low angle and the bevel of the cutter is turned up. This plane produces a fine, smooth cut, making it suitable for fitting and trimming work.

Router and *rabbet planes* are designed to form *dados*, grooves, and *rabbets*. The need for these specialized hand planes has diminished in recent years due to the availability of power driven carpentry equipment.

Wood chisels are used to trim and cut away wood or composition materials to form joints or recesses, Figure 3-18. They are also helpful in paring and smoothing small, interior surfaces that are inaccessible for other edge tools.

Width sizes range from 1/8" to 2". For general work, the carpenter usually selects 3/8", 1/2", 3/4", and 1 1/4" sizes. A soft-face hammer or mallet should always be used to drive the chisel when making deep cuts.

Tin snips are used to cut asphalt shingles and light sheet metals, such as flashing. Utility knives have very sharp blades and are useful for trimming wood, cutting veneer, hardboard, particle board, vapor barrier, tar paper,

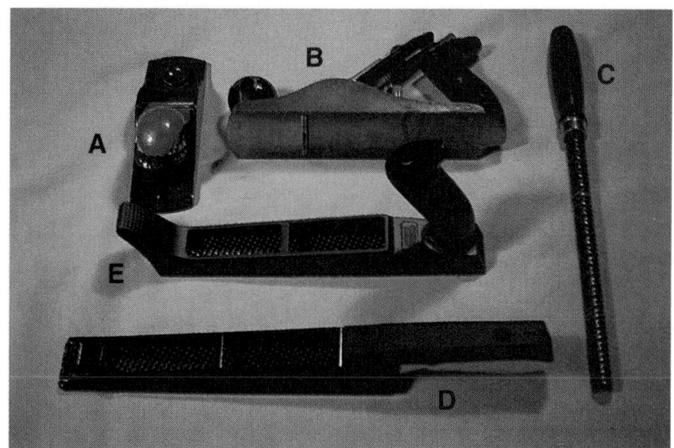

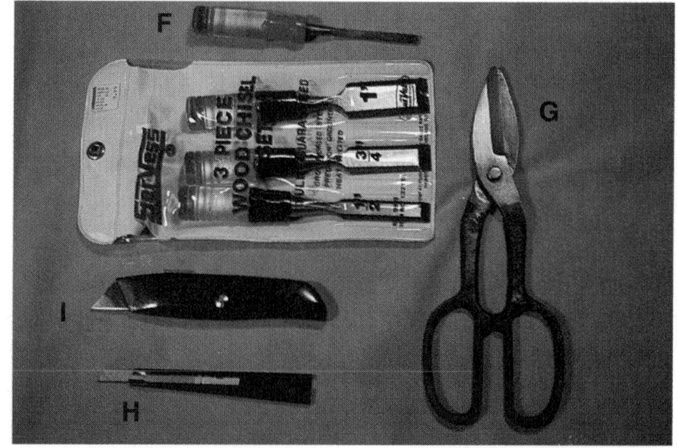

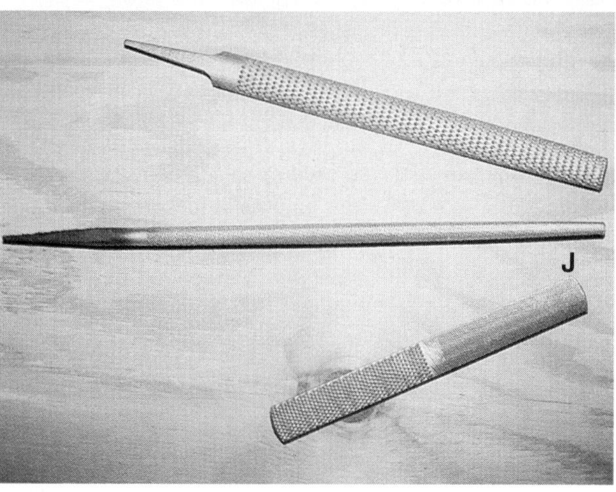

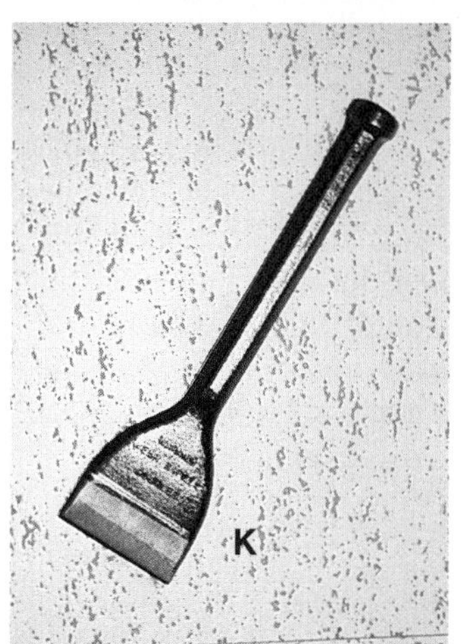

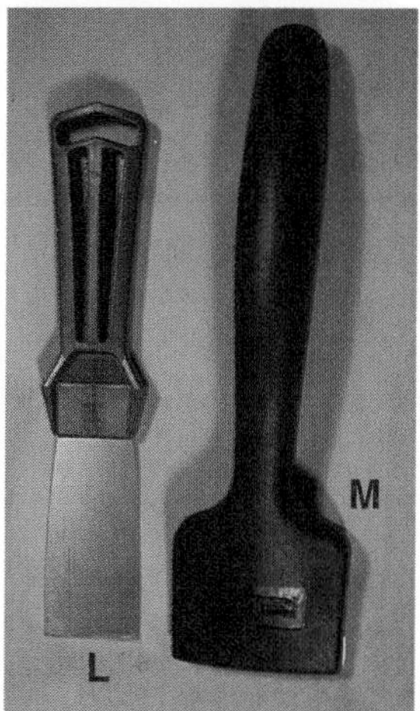

Figure 3-16. Edge tools are used in carpentry for planing, smoothing, and shaping wood. A—Block plane. B—Jack plane. C—Surform (round). D—Surform file. E—Surform, plane type. F—Wood chisels. G—Tin snips. H—Utility knife. I—Utility knife. J—Rasps. K—Flooring chisel. L—Putty knife. M—Scraper. (MacRae True Value Hardware)

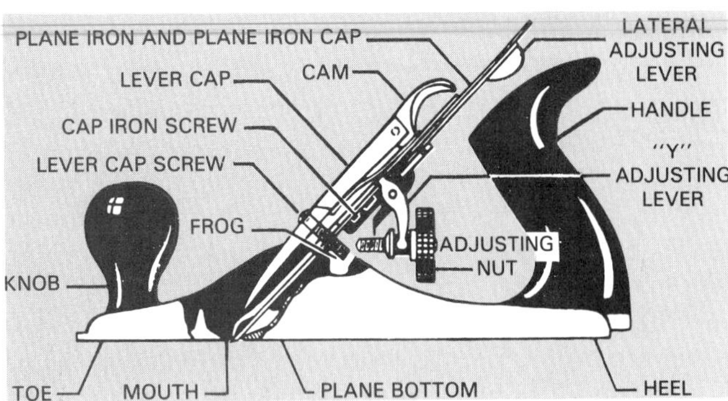

Figure 3-17. A standard plane is made up of many parts.

Figure 3-18. Cutting a recess with a wood chisel. Handles should be able to withstand light hammer or mallet blows. (Note that saw cuts form the sides of the recess.)

and house wrap. They are also used to cut bat insulation and for accurate layout.

Rasps are used for trimming, shaping, and smoothing wood. A round rasp is useful for forming inside curves and shaping or enlarging holes. Refer again to Figure 3-16(J).

A *scraper's* edge is hooked, Figure 3-16(M). It produces thin shaving-like cuttings. The edge is usually filed, and then a tool called a burnisher is used to form the hook.

There are a number of different types of scrapers. The carpenter generally prefers one with a blade that can be quickly replaced.

Two tools used for smoothing and shaping are the tungsten carbide coated file and the multiblade forming tool (Surform). Both cut rapidly and will do a considerable amount of work before becoming dull. The surface produced is rough and requires sanding to remove the tool marks, Figure 3-19. The cornering tool, shown with the planes and chisels, is used to remove sharp corners from exposed wooden parts.

Figure 3-19. A surform tool is made up of many tiny planing surfaces. In the view above, a drywall worker is trimming the edges of gypsum wallboard.

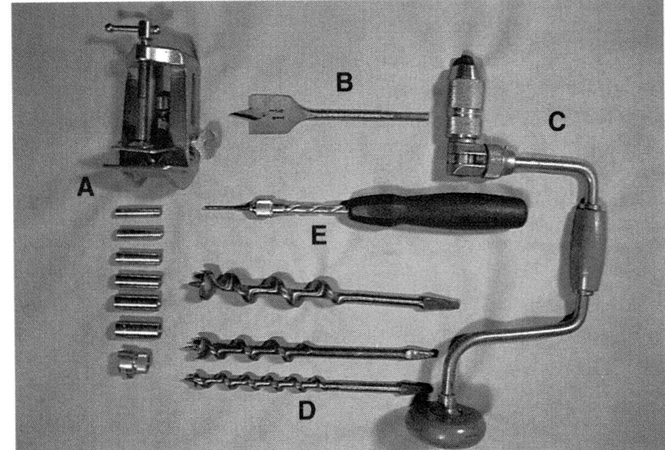

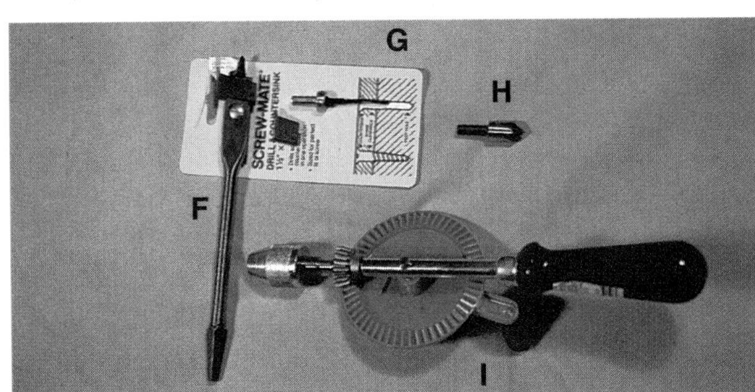

Figure 3-20. Various tools are used by the carpenter for drilling and boring holes. A—Doweling jig with guides. B—Spade bit (usually used with a power driver, but also used in a hand drill). C—Brace (used with augers). D—Augers. E—Push drill. F—Expansive bit. G—Screw mate (used with either a hand drill or power drill). H—Countersink. I—Hand drill.

Drilling and Boring Tools

Holes larger than 1/4" are made with auger bits or adjustable *expansive bits*. The operation is called *boring*.

Small holes are formed with *hand* or *push drills*. Tools in this group are illustrated in Figure 3-20.

Auger bits vary in the shape and design of the twist. The size (diameter) of standard bits ranges from 3/16" to 2". The common range is from 1/4" to 1". The size is stamped on the tang, or shank. Often, the size is expressed as the number of sixteenth of the diameter (e.g. a number 7 bit would have a diameter of 7/16").

Boring bits are mounted in a brace that holds and turns them into the wood. Figure 3-21 shows the proper method for boring with a brace and bit. First, locate the point where the hole is to be bored. Place the point of the auger on the spot. Hold the brace and bit exactly vertical and turn the crank with your free hand. Use your body weight to apply downward pressure. If there is danger of splintering, clamp a piece of scrap wood to the underside of the workpiece. Other tools designed for the brace include:

- The countersink, which forms a recess for screw heads.
- The screwdriver bit, which sets regular wood screws.

An expansive bit is adjustable and can be used to bore large holes. When boring all the way through the stock, it is best to back up the work with a scrap piece as shown in Figure 3-22.

The size of a hand drill is determined by the capacity of its chuck. Usually, the chuck capacity is 1/4" or 3/8". Regular twist drill bits are used in the hand drill. A practical set for the carpenter should range from 1/16" to 1/4" with increments (steps) of 1/32".

Push drills are designed to form small holes quickly. When the handle is pushed down, the drill chuck revolves. A spring inside the handle forces it back out to its original position when pressure is released. Push drills use special fluted bits with sizes from 1/16" to 3/16". Carpenters frequently use this type of drill to make holes for nails and screws. It can be operated with one hand. See Figure 3-23 for operation of a push drill.

Figure 3-22. An expansive bit can be used to bore large holes through wood. Backing up the work with scrap lumber prevents splintering as the bit exits the work.

Figure 3-21. Proper method for using the brace and bit. Use your body to apply downward pressure.

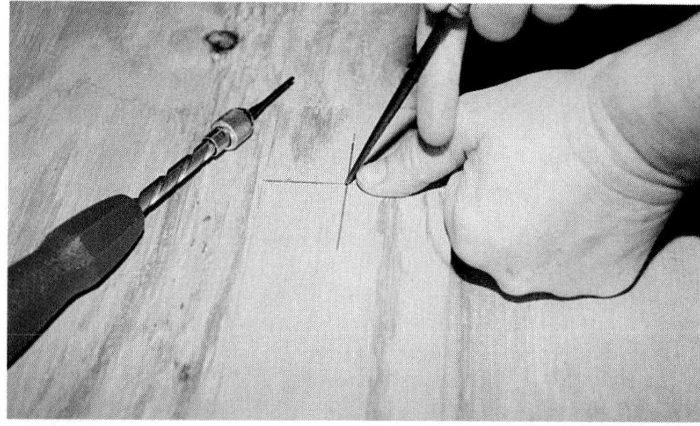

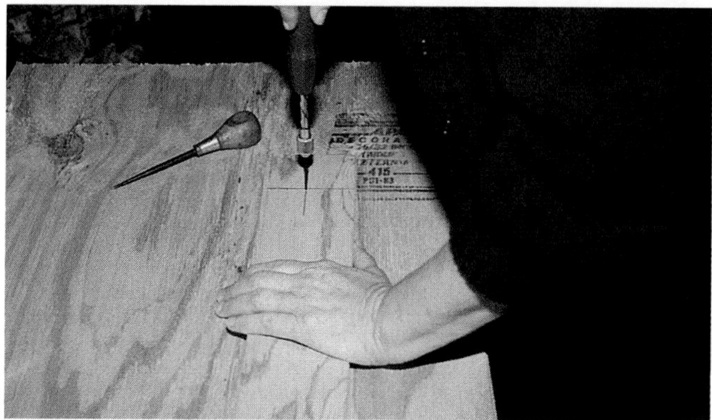

Figure 3-23. Using a push drill. Top—Use an awl to make a pilot hole. Carefully locate the point of the awl on the work, hold awl vertical and make a starting hole. Bottom—Drill the hole by alternately pushing and releasing the pressure on the drill.

Fastening Tools

Much of the carpentry work consists of fastening parts together. Nails, screws, bolts, and other types of connectors are used. Figure 3-24 shows a group of hand tools commonly used to perform operations in this area.

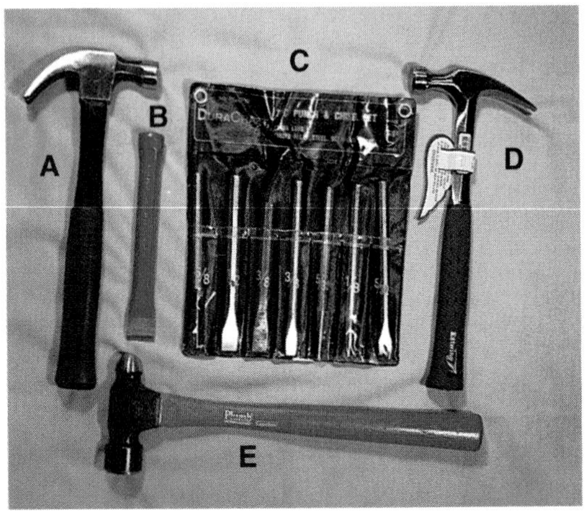

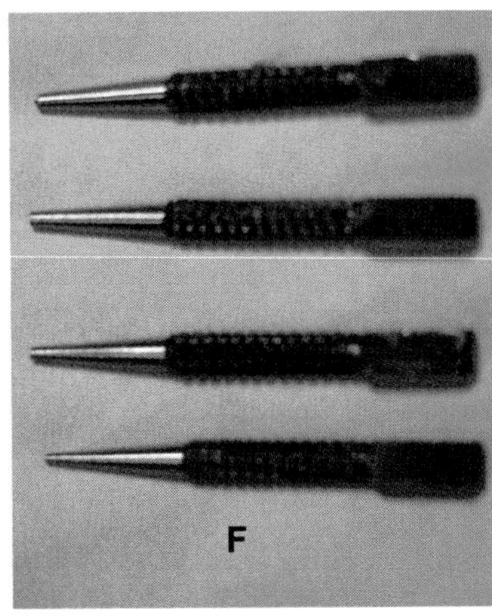

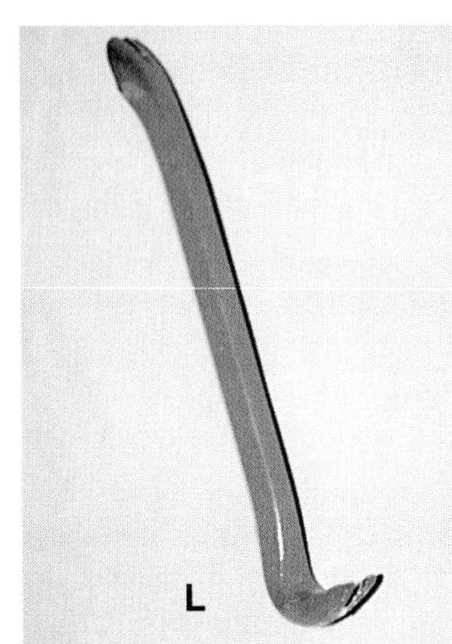

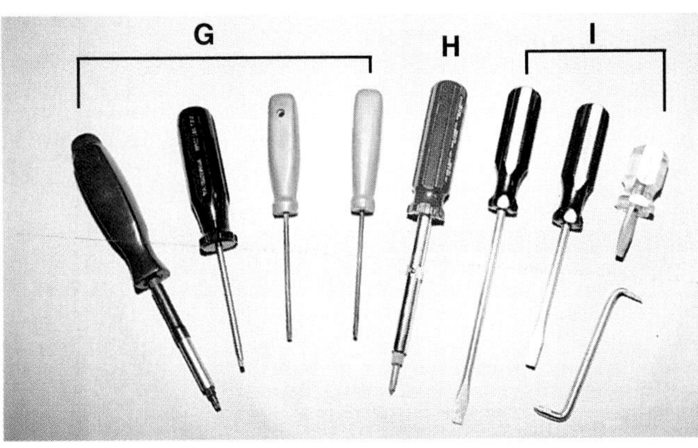

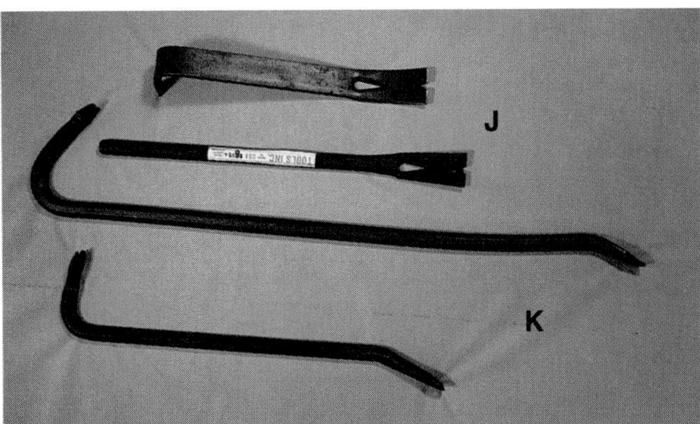

Figure 3-24. These tools are used for fastening and prying apart wood building components. A—Curved claw hammer. B—Large cold chisel. C—Cold chisels in various sizes. D—Straight or ripping claw hammer. E—Ball peen hammer. F—Nail sets. G—Torx screwdrivers. H—Phillips screwdriver. I—Slotted or flat blade screwdriver. J—Ripping chisels. K—Pry bars or crow bars. L—Nail claw. (McRae True Value Hardware)

Of the tools shown, the *claw hammer* is used most often. Carpenters usually carry a hammer in a holster attached to their belts. Two shapes of hammer heads are in common use:

- The curved claw.
- The ripping (straight) claw.

The curved claw is the most common and is best suited for pulling nails. The ripping claw, which may be driven between fastened pieces, is used somewhat like a chisel to pry them apart.

Parts of a ripping claw hammer are shown in Figure 3-25. The face can be either flat or slightly rounded (bell-faced). The bell face is most often used. It will drive nails flush with the surface without leaving hammer marks on the wood. The hammer head is forged from high quality steel and is heat-treated to give the poll and face extra hardness.

The size of a claw hammer is determined by the weight of its head. They are available from 7 to 20 oz. The 13 oz. size is the most popular for general purpose

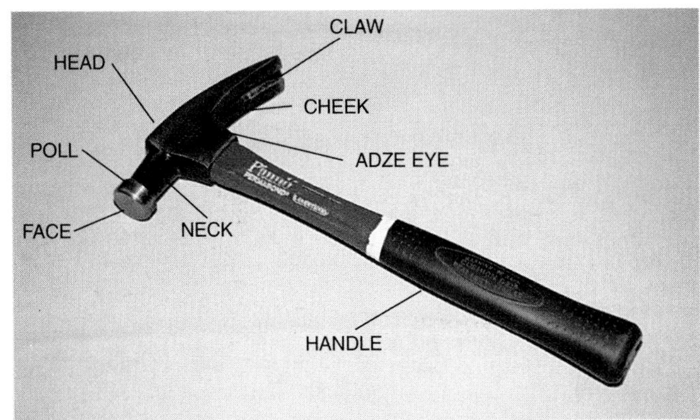

Figure 3-25. Parts of a ripping claw hammer. (McRae True Value Hardware)

work. Carpenters generally use either the 16 or 20 oz. size for rough framing.

A hammer should be given good care. It is especially important to keep the handle tight and the face clean. If a wooden handle becomes loose, it can be tightened by driving the wedges deeper or by installing new ones. Figure 3-26 illustrates the procedure for pulling nails to relieve strain on the handle and protect the work.

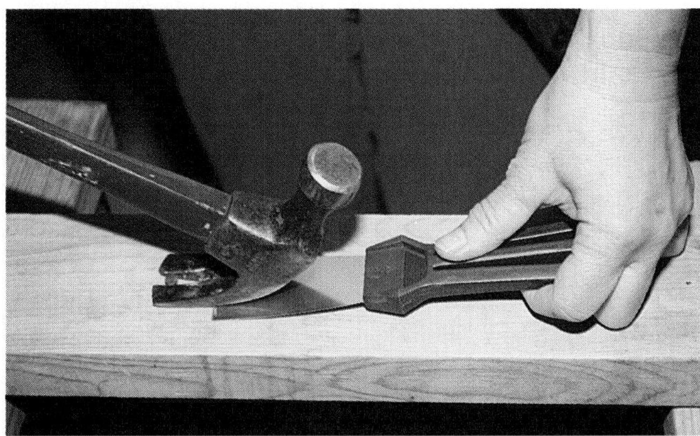

Figure 3-26. Proper way to pull a nail and separate two wood pieces. Top—Place a putty knife under the hammer to protect the wood surface. Middle—Use a block of wood to increase leverage and protect the surface. Bottom—A rip claw hammer works best to separate two nailed surfaces. Rip claw can also be used to split short boards.

Hatchets are used for rough work on such jobs as making stakes and building concrete forms. Some are designed for wood shingle work, especially wood shakes, Figure 3-27. See also Unit 11.

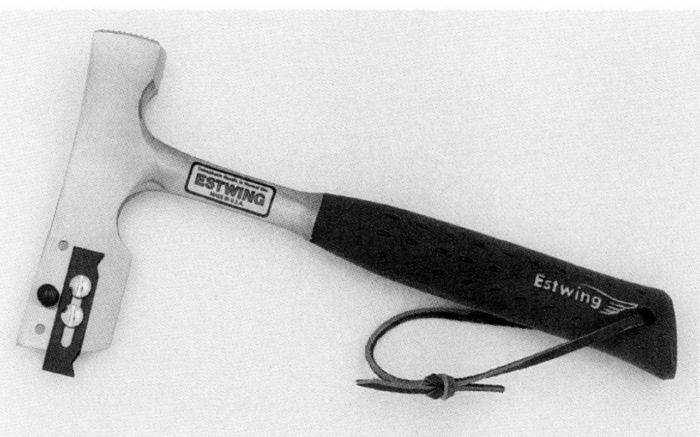

Figure 3-27. A shingling hatchet is designed for installing wood shingles and shakes. The stops are for measuring the exposure on shingles. (Exposure is the amount of shingle surface exposed to the weather.) (Estwing Mfg. Co.)

Ripping bars, also called *wrecking bars*, vary in length from 12" to 36". They are used to strip concrete forms, disassemble scaffolding, and other rough work involving prying, scraping, and nail pulling.

The *nail set* is designed to drive the heads of casing and finishing nails below the surface of the wood. Tips range in diameter from 1/32" to 5/32" by increments of 1/32". Overall length is usually 4".

A number of screwdriver sizes and styles are available. Sizes are specified by giving the length of the blade, measuring from the ferrule to the tip. The most common sizes for the carpenter are 3", 4", 6", and 8".

The size of a Phillips screwdriver is given as a point number ranging from a No. 0 (the smallest), to a No. 4 (the largest). Size numbers 1, 2, and 3 will fit most of the screws used in carpentry work.

Tips of screwdrivers must be carefully picked for the job. For a slotted screw, tips must be square, have the correct width, and fit snugly into the screw. The width of the tip should be equal to the length of the bottom of the screw slot. The sides of the screwdriver tip should be carefully ground to an included angle of not more than 8° and to a thickness that will fit the screw slot.

Today, mechanical *tackers, staplers*, and *nailers* perform a variety of operations formerly accomplished by hand nailing. See Figure 3-28. They provide an efficient method of attaching insulation, roofing material, underlayment, ceiling tile, and many other products. An advantage of the regular stapler is that it leaves one hand free to hold the material. Figure 3-29 illustrates a heavy duty nailer-tacker that is operated by striking it with a mallet.

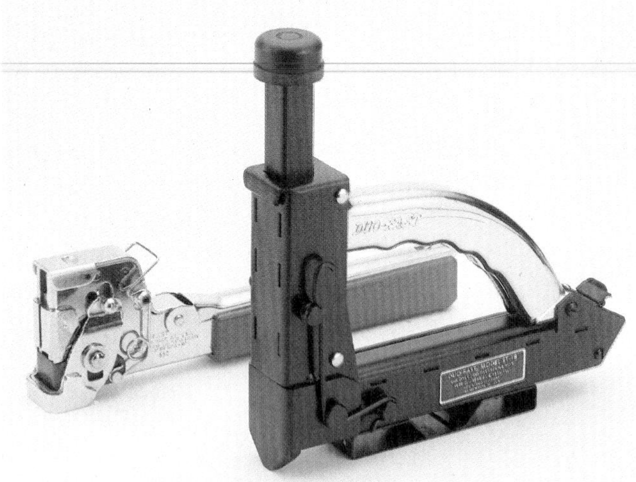

Figure 3-28. Tackers are useful for many tacking and nailing jobs. Left—Hammer tacker allows carpenter to hold work with one hand while tacking with the other. Right—Mallet-operated tacker.

Figure 3-29. This nailer-tacker is being used to fasten underlayment. Note that the mallet has a soft head. (Duo-Fast Corp.)

Prying Tools

Sometimes, particularly when remodeling, it is necessary to take apart nailed structural members. The same is true when forms or temporary braces are removed from a building under construction. Pry bars and ripping chisels were shown in Figure 3-24. Lengths range from 12" for ripping chisels to 30" for pry bars.

Ripping chisels are easily recognized by the small tear-drop slot at one end. Its purpose is to grip and pull nails while in confined spaces where a larger tool cannot be used. The ripping chisel may also be used for light prying.

Pry bars are designed for removing larger nails or spikes from lumber of larger dimensions. The 30" bar is mostly used because it fits into a standard carpentry tool box.

A *nail claw*, which is much smaller than a pry bar, is only used to pull nails above the surface of the wood, where it can be gripped easily. A hammer is used to drive its sharpened claw under the nail, raising it sufficiently to be grasped by the hammer or pry bar.

A *nail puller* is a specialty tool with a fixed jaw and a movable jaw and lever. A sliding driver on the end of the handle drives the jaws under the nail. Leverage tightens the jaw's grip on the nail, raising it above the wood's surface. See Figure 3-30 for information on the use of prying tools.

Safety with Prying Tools

There are certain safety precautions that should be observed while using ripping bars. Be sure to have good balance before applying force to the bar. Grip the bar in such a way that fingers are not bruised by coming into contact with any part of the building structure as the fastener or piece comes free.

Gripping and Clamping Tools

Carpenters find that certain gripping and clamping tools are helpful, Figure 3-31. *Wood clamps* and *C-clamps* are especially useful and adapt to a wide range of assemblies where parts need to be held together:

- While metal fasteners are being attached.
- While adhesives are setting.
- While wood parts are being worked on, Figure 3-32.

Sometimes clamps are used to hold jigs or fixtures to machines for some special setup. C-clamps are available in sizes from 1" to 12". The clamp sizes represent the tool's largest opening.

Hand screws, shown in Figure 3-33, are ideal for woodworking because the jaws are broad and distribute the pressure over a wide area. Sizes (length of the jaw) range from 4" to 24".

Bar or pipe clamps, Figure 3-34, are useful for holding larger pieces while they are receiving glue or mechanical fasteners. Such clamps are adjustable at both jaws.

Figure 3-35 shows a *wood vise* that can be fastened to a sawhorse or a workbench. Some types attach with a clamp while others are fastened with screws.

Vise-grip pliers, Figure 3-36, are an adjustable, all-purpose tool with a lever action that locks the jaws when the handles are fully closed. The pliers can be substituted for a vise or can be used in place of a pipe wrench, open-end wrench, or adjustable wrench.

Figure 3-37 shows various types of wrenches that the carpenter may use to service and adjust equipment. Tool maintenance is an important responsibility. Well-maintained tools are safe tools.

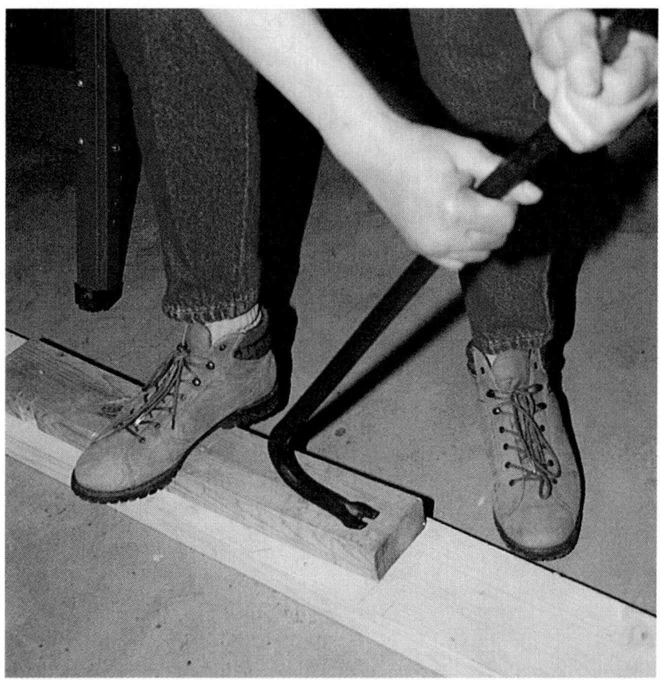

Figure 3-30. Using prying tools. Top—The pry bar is used to remove larger fasteners. Long handle provides considerable leverage. Bottom—Nail puller claws are first driven under the nail with a weight that slides along the handle. A small fulcrum attached to one claw tightens the puller's grip on the nail.

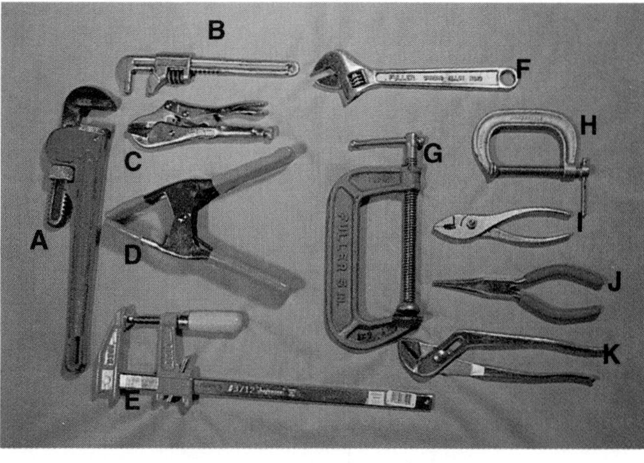

Figure 3-31. Clamping and gripping tools are helpful in carpentry work. A—Straight pipe wrench. B—Monkey wrench. C—Vise-grip pliers. D—Spring clamp. E—Bar clamp. F—Adjustable wrench. G—Large C-clamp. H—Small C-clamp. I—Slip joint pliers. J—Long-nosed pliers. K—Tongue-and-groove pliers. (McRae True Value Hardware)

Figure 3-32. Clamps can be used to provide temporary pressure on parts being shaped or glued. Top—A carriage clamp, a type of C-clamp. Bottom—Hand screw, or wood clamp. (Montachusett Regional Vo-Tech School)

Many carpenters use tool belts and tool holders that hang from the waist. These items keep tools handy and ready for use while freeing hands for other tasks when the tools are not in use. See Figure 3-38.

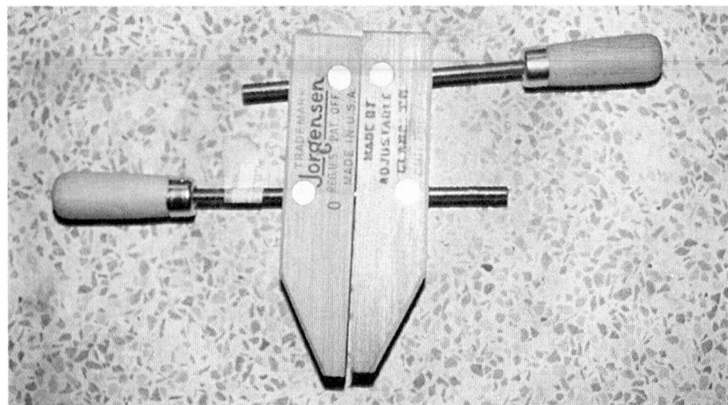

Figure 3-33. Hand screw. Top—Typical hand screw or wood clamp. Bottom—Wood clamps are ideal for clamping wood because the pressure is distributed over a broad area. (Ace Hardware, Sister Bay, WI)

Figure 3-34. Bar clamps are being used here to hold a 2 x 4 assembly too large for wood screws or C-clamps. (Montachusett Regional Vo-Tech School)

Figure 3-35. A wood vise is useful for holding small work. (McRae True Value Hardware)

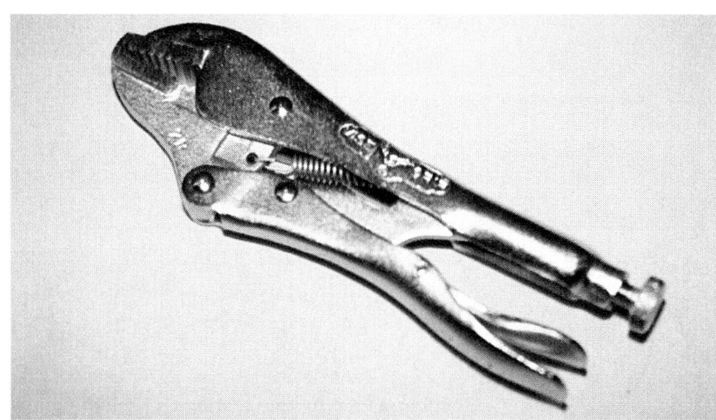

Figure 3-36. Vise-grip pliers can be used like standard pliers or they can be locked and used like a vise.

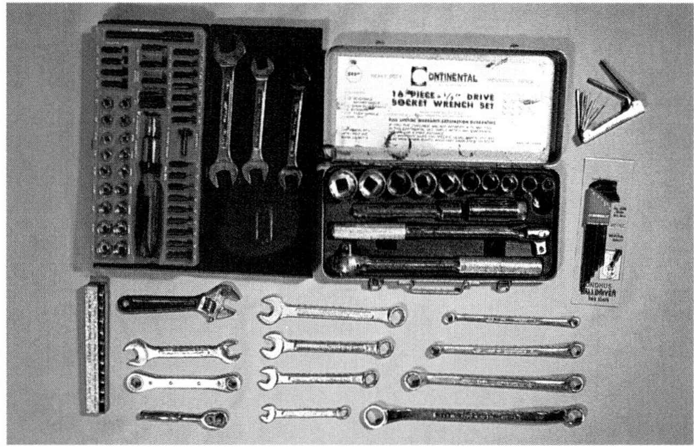

Figure 3-37. A variety of metalworking tools are useful to the carpenter in maintaining his or her tools and other equipment.

Tool Storage

Some type of chest or cabinet is needed to store and transport tools. In addition to being portable, the chest or cabinet should protect the tools from weather damage, loss, and theft.

Folding tool panels provide a practical solution to transportation and storage. Figure 3-39 shows a design that

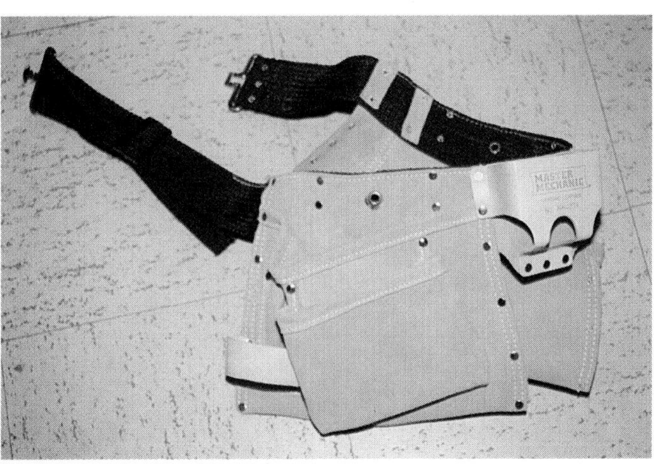

Figure 3-38. Toolholders free up carpenter's hands when tools are not in use. (McRae True Value Hardware)

Figure 3-39. Folding tool panel may be closed and carried like a huge suitcase. (American Plywood Assoc.)

can be carried like a large suitcase. When opened, it is a freestanding tool panel. Toolholders, especially designed to hold a given item, should be used to secure the tools, rather than nails, screws, or hooks. Attach the holder to a small subpanel and then mount the unit on a main panel as shown in Figure 3-40. Special locking devices may be required to hold the tools while the case is being moved.

One of the important advantages of an organized tool panel is the ease with which tools can be checked. At the end of the day, you can determine if any tools are missing by checking to see if there are empty holders. This will help you to keep from losing tools.

Figure 3-40. Custom-made toolholders mounted on a vertical panel.

Care and Maintenance of Tools

Experienced carpenters take pride in their tools and keep them in good working condition. They know that even high quality tools will not perform satisfactorily if they are dull or out of adjustment.

Tools should be wiped clean after being used. Occasionally wet the wiping cloth lightly with oil. Some carpenters use a lemon oil furniture polish. It has a slight cleaning action and also leaves a thin oil film that protects metal surfaces from rust.

Keep handles on all tools tight. When handles and fittings are broken they should be replaced.

It is a simple matter to hone edge tools on an *oilstone*, Figure 3-41. For tools with single-bevel edges like planes and chisels, place the tool on the stone with the bevel flat on the surface. Raise the back edge of the tool a few degrees so only the cutting edge is in contact. Then move the tool back and forth until a fine wire edge can be detected by pulling the finger over the edge. Now place the back of the tool flat on the oilstone and stroke lightly several times. Turn the tool over and again stroke the beveled side lightly. Repeat this total operation several times until the wire edge has disappeared from the cutting edge.

If not damaged, a cutting edge can be honed serveral times before grinding is required. When the bevel becomes blunt, reshape it by grinding, Figure 3-42. An edge can be ground many times.

Figure 3-41. Honing a chisel on an oilstone. Place a few drops of oil on the stone first.

The grinding angle will vary somewhat depending on the work the tool is used for. Figure 3-43 shows grinding and honing angles recommended for a plane iron.

Some tools may be sharpened with a file. For auger bits, it is best to use a special auger bit file, Figure 3-44. Sharpen the lips, or cutters, by stroking up through the throat. Do not file the underside. File the inside of the spurs as shown, being sure to keep them the same length.

Saws require filing as well as setting. Before filing, they often require *jointing*. In this operation the height of the teeth are struck off evenly. Filing a saw is a tedious

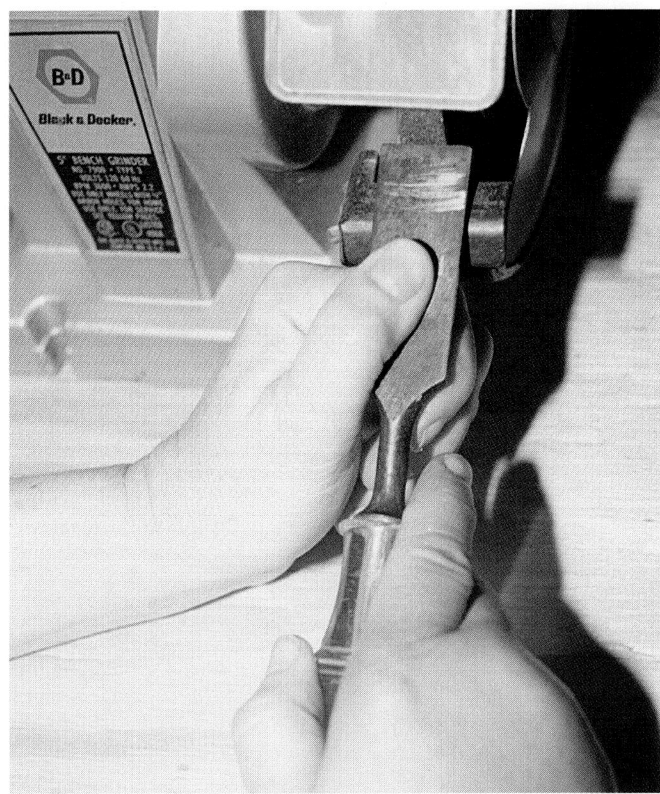

Figure 3-42. Grinding a wood chisel. Some grinders have special attachments making it possible to get a more accurate bevel.

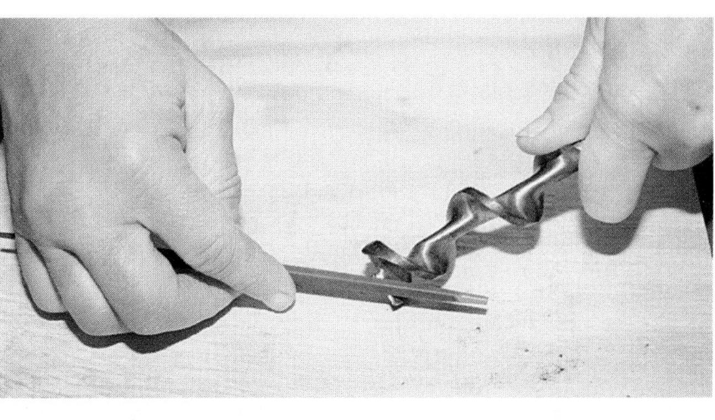

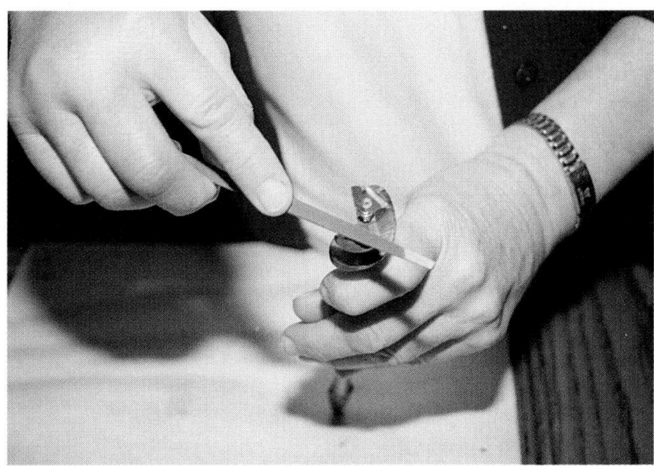

Figure 3-44. Sharpening an auger bit requires a small file. Top—Filing the cutting lip. Bottom—Filing the spur (inside surface only).

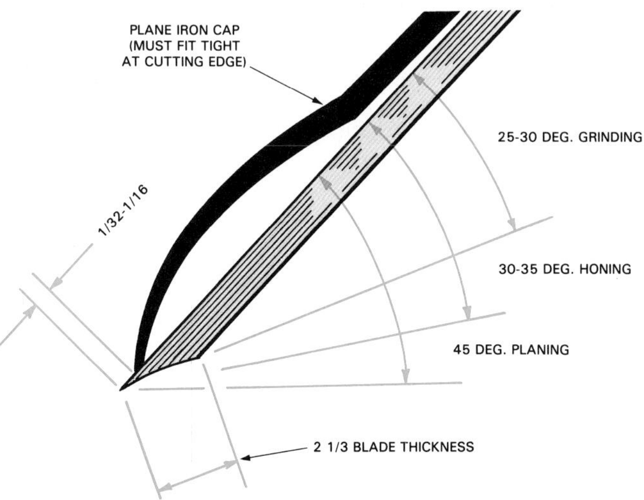

Figure 3-43. Follow these grinding and honing angles for a plane iron.

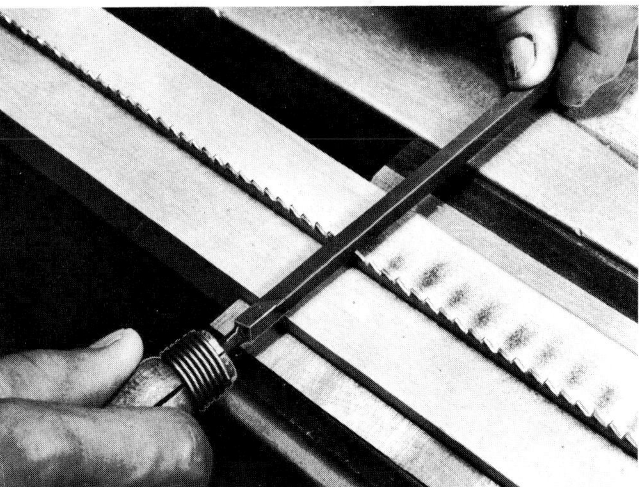

Figure 3-45. Filing a rip saw. The file should contact the back of a tooth set away from the operator.

operation. Most carpenters prefer to send their saws to a shop where they can be machine sharpened by an expert.

When a saw is only slightly dull, the teeth can often be sharpened with a few file strokes, as illustrated in Figure 3-45. Use a triangular saw file. Be sure to match the original angle of the teeth. File the back of one tooth and the front of an adjacent tooth in a single stroke. Figure 3-46 shows a saw set in operation. Figure 3-47 shows how correctly filed teeth should appear.

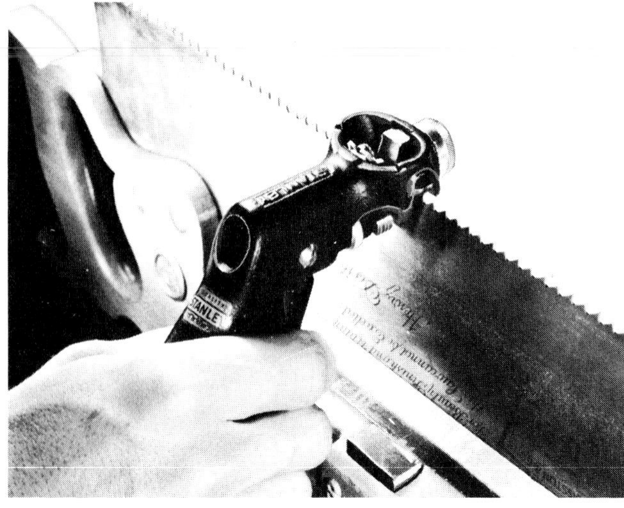

Figure 3-46. Setting the teeth. In the position shown, the saw set will bend the tooth away from the operator.

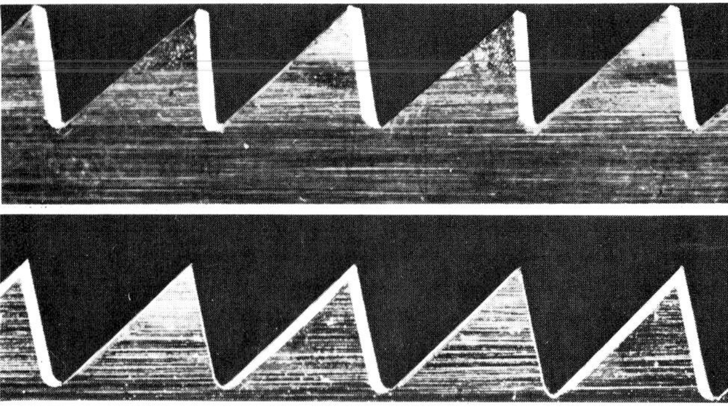

Figure 3-47. Correctly filed saw teeth. Top—Rip teeth. Bottom—Crosscut teeth.

Important Terms

All-hard blades
Backsaw
Bar or pipe clamps
Block plane
Boring
C-clamp
Chalk line
Claw hammer
Combination square
Coping saw
Crosscut saw
Dados
Expansive bits
Flexible-back blades
Folding wood rule
Fore and jointer plane
Framing square
Hacksaw
Hand drill
Hand screws
Jack plane
Jointing
Kerf
Level
Marking gauge
Measuring tape
Miter box
Nail claw
Nailers
Nail puller
Nail set
Oilstone
Plumb bob
Pry bar
Push drill
Quick square
Rabbet
Rabbet plane
Rafter square
Rasps
Ripping bar
Rip saw
Router plane
Scraper
Scratch awl
Sheetrock-drywall saw
Smooth plane
Staplers
Super square
Tackers
T-bevel
Tin snips
Try square
Wing dividers
Wood chisel
Wood clamp
Wood vise
Wrecking bar

Test Your Knowledge

1. A standard folding wood rule is _____________ feet long.
2. The blade of a framing square is 24" long. How long is the tongue (other leg)?
   A. 10"
   B. 12"
   C. 16"
   D. 20"
3. When checking structural members to see if they are horizontal or vertical, what tool should be used?
4. A 10 point saw will have 12 teeth per inch. True or False?
5. A backsaw is used for fine work. It usually has _____________.
   A. 8 teeth per inch.
   B. 8–10 teeth per inch.
   C. 12–14 teeth per inch.
   D. 14–16 teeth per inch.
6. The bevel of a block plane blade is turned _____________.
   A. up
   B. down
7. An auger bit with the number 10 stamped on the tang (shank) would bore a hole with a diameter of _____________ inches.
8. Never strike a wood chisel with a mallet. True or False?
9. The size of a claw hammer is determined by the _____________.
   A. length of the handle
   B. length of the head
   C. weight of entire hammer
   D. weight of the head
10. Name the different types of claw hammers and explain the purpose of each.
11. Where would a carpenter use tin snips?
12. How are sizes of slotted screwdrivers and Phillips screwdrivers determined?
13. Spurs of an auger bit are filed on the _____________.
    A. inside
    B. outside
14. The operation of filing off the points of saw teeth until all are level is called _____________.

Outside Assignments

1. After a study of reference books and manufacturers' catalogs, prepare a list of tools you believe the carpenter will need for rough framing and exterior finish of a typical residential structure.
2. Prepare a report on the historical development of woodworking tools used by the carpenter. Reference books on woodworking and encyclopedias will contain helpful material. If you make a report to your class, use an overhead projector or other visual aids to show students pictures and drawings of early tools. If possible, secure actual specimens from a collector.
3. Make a list of hand tools described in this chapter. At a hardware store or home improvement center, determine the cost of each tool. Add up the cost and report to the class the cost of equipping a carpenter with a complete tool kit.

Carpentry apprentice using 16 oz. hammer, nails temporary bracing to a wall frame. The hammer is the most used of all fastening tools. (Waterloo Iowa Daily Courier)

Tool belts increase a carpenter's efficiency. (Pierce Custom Homes)

A laser plane is a useful tool for laying out grade level on a building site. (Spectra-Physics Laserplane Inc.)

A hand-held calculator solves construction math problems in seconds. Included are rafter and stair calculations. Answers can be given in feet and inches, fractions, decimal feet and inches, and in metric units. Special function buttons also handle problems in square and cubic measure. (Calculated Industries, Inc.).

4

Power Tools

Modern power tools greatly reduce the time required to perform many of the operations in carpentry work. Heavy sawing, planing, routing, and boring can be accomplished in less time with less work. Moreover, when proper tools are used in the correct way, high levels of accuracy can be easily achieved.

There are two general types of power tools:

- Portable (can be carried to the work).
- Stationary (remain at one spot on the job site and workpiece is brought to them).

Portable tools are lightweight and intended to be carried around by the carpenter. They may be used anywhere on the construction site or on any part of the structure.

Stationary tools, usually called machines, are heavy equipment mounted on benches or stands. Benches and stands rest firmly on the floor or ground. The machines must be sitting level when used on the job site.

Space permits only a brief description of the kinds of power tools most commonly associated with carpentry work. This should be supplemented with woodworking textbooks and reference books devoted to power tool operation. Manufacturer's bulletins and operator's manuals are also a good source of information.

Figure 4-1. When operating a power tool, give full attention to the work. (Photo © Des Moines, Iowa Public Schools)

Power Tool Safety

Safety must be practiced continually. Before operating any power tool, you must become thoroughly familiar with:

- The way it works.
- The correct way to use it.

You must be wide awake and alert. Never operate a power tool when tired or ill. Think through the operation before performing it. Know what you are going to do and what the tool will do. Make all adjustments before turning on the power. Be sure blades and cutters are sharp and are of the correct type for the work.

While operating a power tool, do not allow yourself to be distracted. See Figure 4-1. Do not distract the attention of others while they are operating power tools. Keep all safety guards in place and wear safety glasses. See Figure 4-2.

Figure 4-2. When working with power equipment, always wear safety glasses and make sure all guards are in place. Note – worker is using a push board for added safety while using the planer. (Bullard-Haven Technical School)

Feed the work carefully and only as fast as the tool will cut it easily. Overloading is hazardous to the operator and will likely damage the tool or work. When the operation is complete, turn off the power and wait until the moving parts have stopped before leaving the machine.

Electrical Safety

Always make sure that the source of electric power is the correct voltage and that the tool switch is in the "off" position before it is plugged into an electrical outlet. Power tools will operate on either 120 V or 240 V electric power. This information is usually shown on a plate attached to the power tool's case. If in doubt about the power source voltage, examine the receptacle, Figure 4-3.

Stationary power tools are factory equipped with *magnetic starters*. These are safety devices which will automatically turn the switch to the "off" position in case of power failure. This feature is important because personal injury or damage to the equipment could result if power is reestablished with the switch "on." Make sure these switches are in good working order.

The electrical cord and plug must be in good condition and must provide a ground for the tool. This means that extension cords should be the three-wire type. Make sure that the conducting wire is large enough to prevent excessive voltage drop.

Be careful in stringing electrical extension cords around the work site. Place them where they will not be damaged or interfere with other workers. Make certain that extension cords maintain ground continuity. (This means that the third conductor (wire) in a cord is providing an unbroken path for current back to a grounded terminal. Thus, in case of a short in a tool case, current will bleed off harmlessly to ground.) Continuity can be checked with a neon tester or a *continuity monitor*, Figure 4-4.

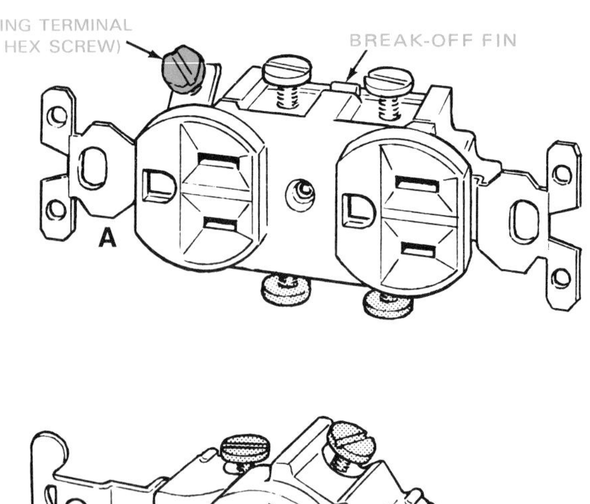

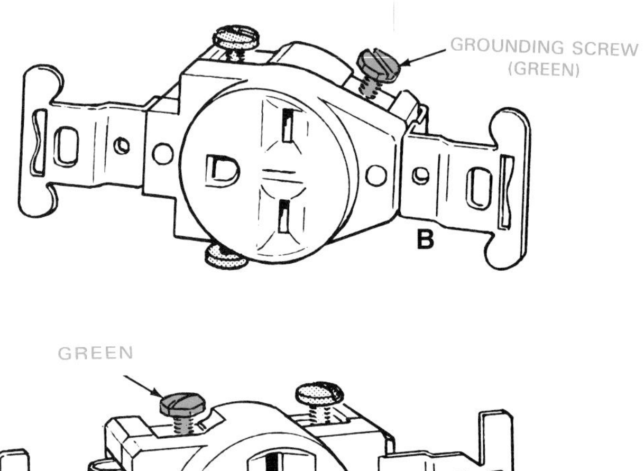

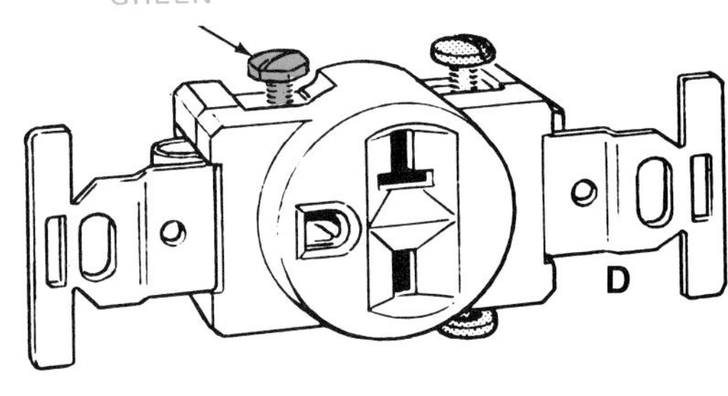

Figure 4-3. Different types of receptacles. A–Receptacles for 120 V power source always have this configuration. B–240 V receptacle with tandem blades and U-shaped ground. C–240 V receptacle for three-bladed plug. D–240 V receptacle for horizontal and vertical blades and U-shaped ground.

Shock Protection

Electrical shock is one of the potential hazards of working with power tools. Always be sure that proper grounding is provided. Receptacles should be of the concealed contact type with a grounding terminal for continuous ground. Plugs and cords should be an approved type.

Portable power tools should be double-insulated or otherwise grounded to protect the worker from dangerous electrical shock. Even though the circuit may be grounded, an operator of a portable power tool could be electrocuted should a bare conductor ground on a metal tool case. A *ground fault circuit interrupter* (GFCI) should be used on all construction job sites. These units can be installed in a circuit or can be plugged into an outlet that is grounded. These units "sense" when a short has occurred and will turn off power to the tool. See Figure 4-5.

Portable Circular Saws

This power tool is also called an electric hand saw or builders' saw. Its size is determined by the diameter of the largest blade it will take. Most carpenters prefer a 7" or 8" saw. The depth of cut is adjusted by raising or lowering the base or shoe. On many saws it is possible to make bevel cuts by tilting the shoe. Refer to Figure 4-6.

Portable saws are often guided along the layout line "free-hand." Therefore, extra clearance in the saw kerf is required. To provide this clearance, teeth usually have a

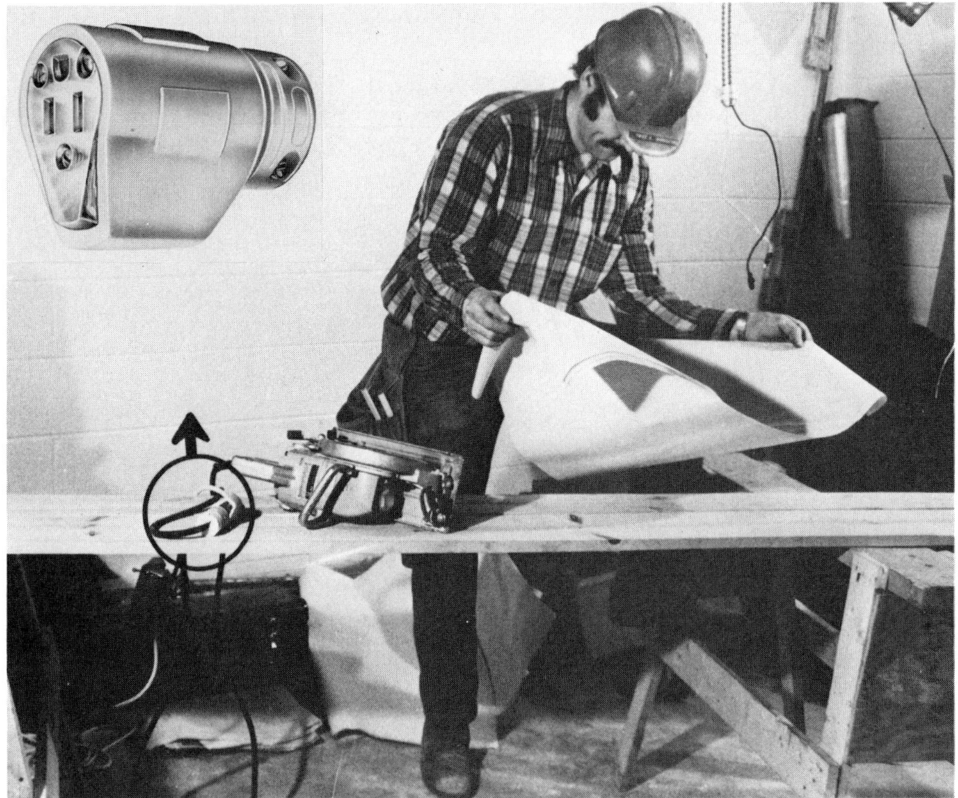

Figure 4-4. For electrical safety on construction job sites, a ground continuity monitor constantly checks that an extension cord is properly wired and grounded. Monitor (inset) is wired into the cord. A light will glow when the cord is safe. (Daniel Woodhead Co.)

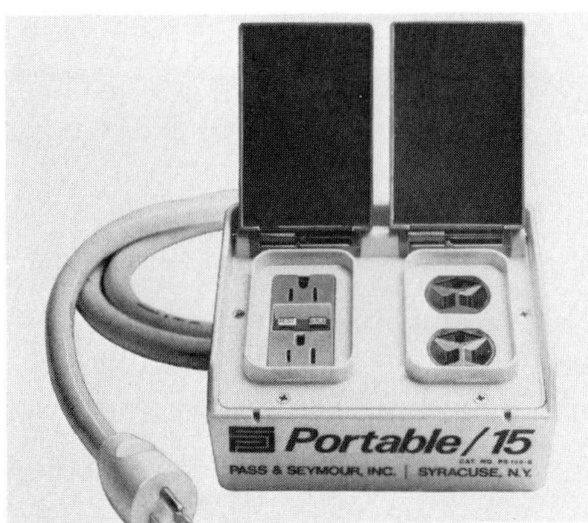

Figure 4-5. A portable ground fault interrupter is used where permanent ground fault protection is not available. Power tools are plugged into it to protect the operator. Should a ground occur in the tool, the GFI will trip at around 5 milliamperes, turning off power to the grounded tool. (Pass & Seymour, Inc.)

wide set. Figure 4-7 shows standard types of blades. The rough-cut combination blade is popular because it is suitable for both ripping and crosscutting. Some carpenters prefer carbide-tipped teeth because they usually stay sharp longer than teeth of a standard blade, Figure 4-8.

Making a Cut

To use a portable saw, grasp the handle firmly in one hand with the forefinger ready to operate the trigger switch. The other hand should be placed on the stock, well away from the cutting line. Some saws require both hands on the machine.

Rest the base on the work and align the guide mark with the layout line. Turn on the switch, allow the motor to reach full speed, and then feed it smoothly into the stock as shown in Figure 4-9. Release the switch as soon as the cut is finished. Hold the saw until the blade stops.

The portable saw may be used to make cuts in assembled work. For example, flooring and roofing boards are often nailed into place before ends are trimmed.

Safety for Portable Circular Saws

- **Unplug saw to replace saw blade. Some saws have a control that locks the arbor during this operation, Figure 4-10.**

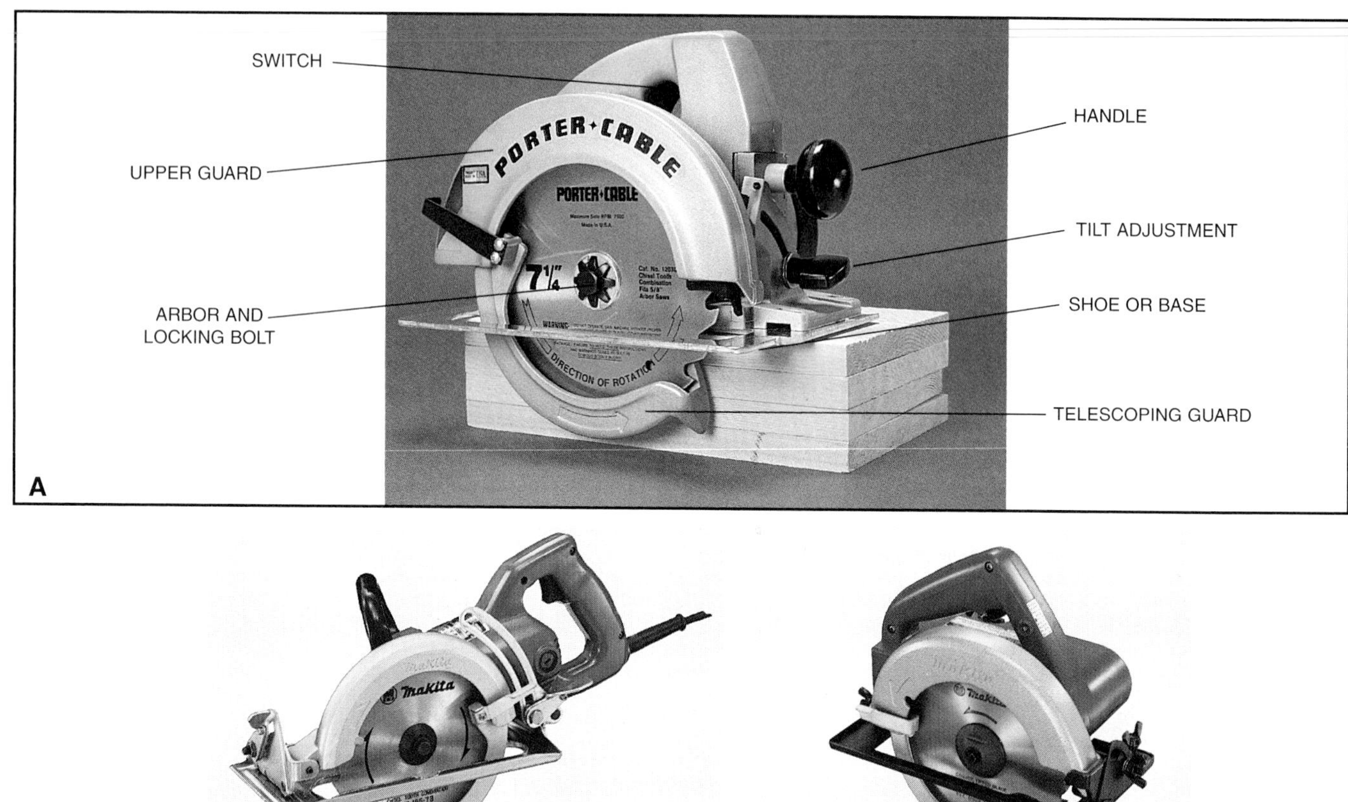

Figure 4-6. A—Parts of a portable circular saw. During use, a telescoping guard is pushed back by the stock. A spring returns the guard when the cut is completed. This type saw is also known as a "sidewinder." (Porter-Cable) B—This style of portable circular saw is known as a hypoid or worm-drive circular saw. (Makita USA, Inc.) C—A cordless circular saw is practical where electric service is not available.

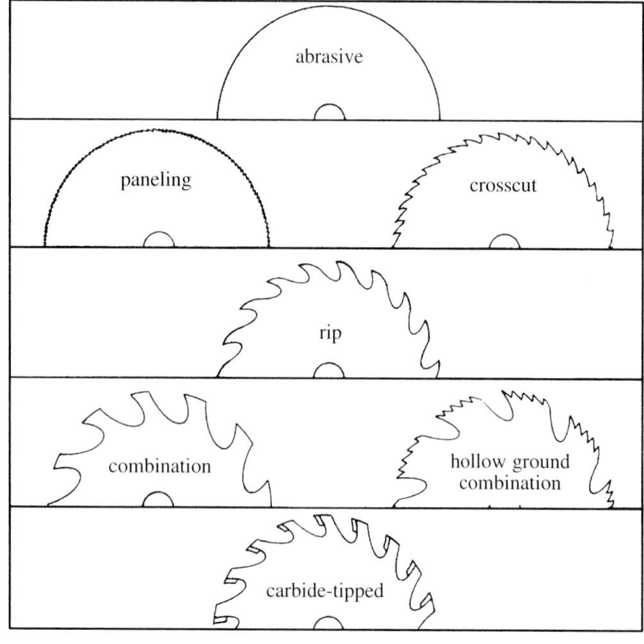

Figure 4-7. Standard blade types. Carbide-tipped blades are popular because they stay sharp longer.(Owner/Builder Directory)

Figure 4-8. Handle carbide-tipped blades carefully to prevent damage to points. (Black & Decker)

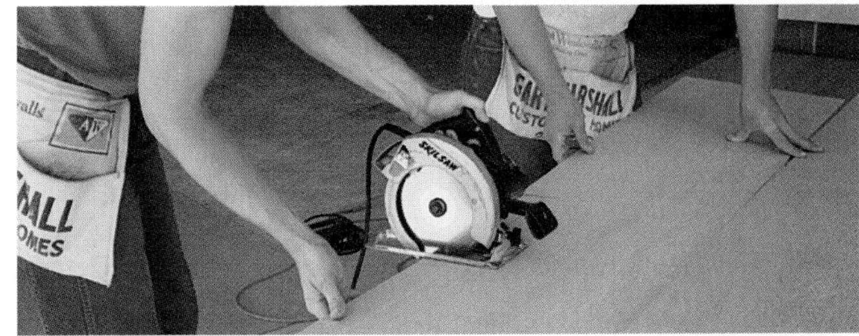

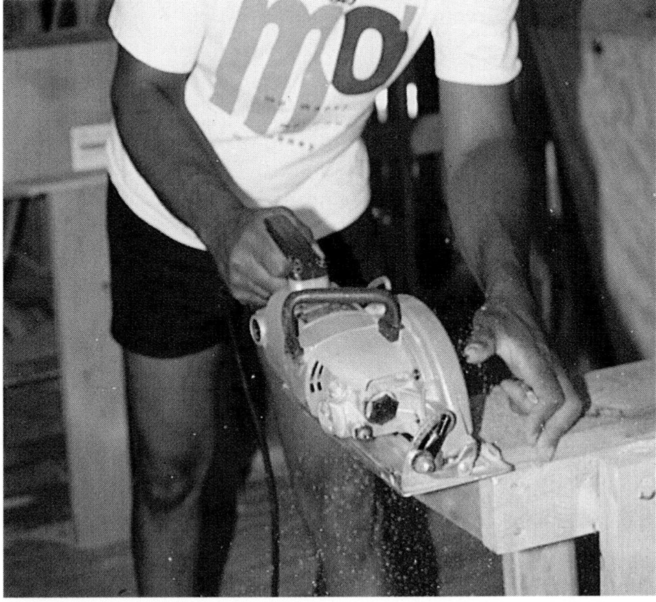

Figure 4-9. Top–Using a portable circular saw to cut a piece of light plywood. (Good side of the plywood is placed down because the saw blade rotates upward.) Note the workpiece is well-supported during cutting so saw will not bind. Bottom–Cutting an inch board on a job site using a worm-drive circular saw. (Hadley-Hobley Construction)

Figure 4-10. Always disconnect saw from power source while adjusting or changing blades. Pressing plunger locks arbor during blade change. (Black & Decker)

- **If the saw has two handles, keep both hands on them during the cut. See Figure 4-11.**
- **Support stock being cut in such a way that the kerf will not close up and bind the blade during or at the end of the cut.**
- **Support thin materials near the cut. Small pieces should be clamped to a bench top or saw horse.**
- **Be careful not to cut into the sawhorse or other supporting device. See Figure 4-12.**
- **Before starting the saw, adjust the depth of cut to the thickness of the stock, plus about 1/8".**
- **Check the base and angle adjustments to be sure they are tight. Plug the cord into a grounded outlet and be sure it will not become tangled in the work.**

Figure 4-11. If a portable saw has two handles, grasp the saw firmly with both hands. (Black & Decker)

Figure 4-12. If sawhorses are not available, support the workpiece on scrap lumber before making a cut. Shoe rests on the lumber as cut is made. (Johnson-Manley Lumber Co.)

- **Always place the saw base on the stock with the blade clear before turning on the switch.**
- **During the cut, stand to one side of the cutting line. Never reach under the workpiece during the cut.**
- **If saw strays off the cutting line, stop the saw, pull back, and start the cut over. Otherwise, the saw may bind and cause an injury.**
- **Keep hands clear of the cutting line.**
- **Have the saw at full power and up to speed before beginning a cut.**
- **Always use a sharp blade of the proper type for the material and cut.**
- **Never force the saw through the material.**
- **Being light and portable, hand power saws can be carried anywhere on the job site. Use care. Dropping the saw or allowing it to fall can damage or destroy it. Never use a damaged saw.**

Saber Saws

The *saber saw* is also called a portable jig saw. It is useful for a wide range of light work. Carpenters, cabinetmakers, electricians, and home craftspeople use it.

A standard model is shown in Figure 4-13. The stroke of the blade is about 1/2". The saw operates at a speed of approximately 2500 strokes per minute.

Blades for wood cutting have from 6 to 12 teeth per inch, Figure 4-14. For general purpose work, a blade with 10 teeth per inch is satisfactory. Always select a blade that will have at least two teeth in contact with the edge being cut.

Saws will vary in the way the blade is mounted in the chuck. Follow directions in the manufacturer's manual. Also, follow the lubrication schedule specified in this manual.

Figure 4-13. Saber saws cut with an action much like a hand saw. This particular model has variable speed control and four-position orbital action. (Black & Decker)

The saber saw can be used to make straight or bevel cuts as shown in Figure 4-15. Curves are usually cut by guiding the saw along a layout line. However, circular cuts may be made more accurately with a special guide or attachment.

Figure 4-14. Saber saw blades are designed for cutting various kinds of material.

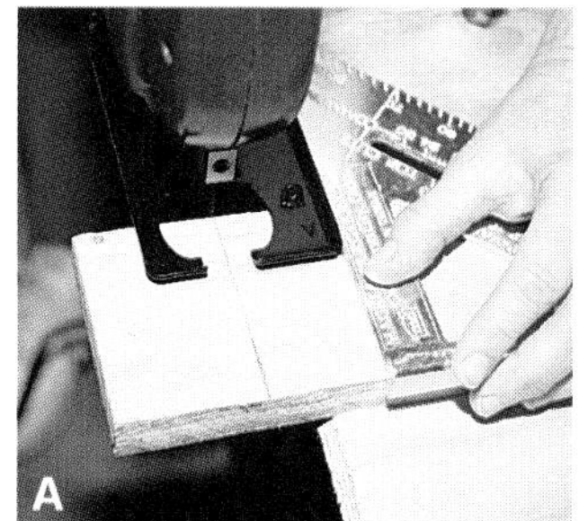

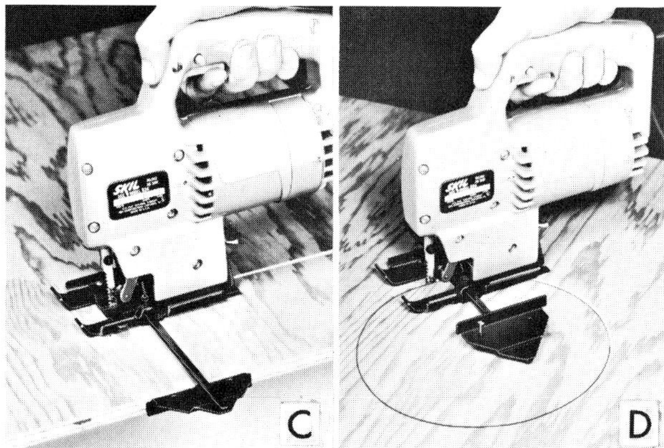

Figure 4-15. How to use a saber saw. A–Straight cutting, using quick square as a guide. B–Angle cutting. C–Using a fence. D–Using a circle cutting guide.

Since the blade cuts on the upstroke, splintering will take place on the top side of the work. This must be considered when making finished cuts, especially in fine hardwood plywood. Always hold the base of the saw firmly against the surface of the material being cut.

When cutting internal openings, a starting hole can be drilled in the waste stock, or the saw can be held on end so the blade will cut its own opening. See Figure 4-16. This is called plunge cutting and must be undertaken with considerable care. Rest the toe of the base firmly on the work and turn on the motor. Then slowly lower the blade into the stock.

Another portable power tool is the *reciprocating saw*, shown in Figure 4-17. Operation is similar to the saber saw. Like the saber saw, its cutting action is up and down, not circular. It is useful under conditions where a circular saw would not be safe or practical. Carpenters use it in remodeling work where sections of framing, sheathing, or inside walls must be removed. Blades are available to cut through metal, plaster, fiberglass, and all kinds of metals.

Safety Rules for Saber Saws

- **Select the correct blade for your work and be sure it is properly mounted in the saw.**
- **Disconnect the saw from power to change blades or to make adjustments.**
- **Make certain the saw is properly grounded through the electrical cord. The tool's switch must be in the "off" position before it is connected to the power source.**
- **Place the base of the saw firmly on the stock before starting the cut. Blade should be clear of the wood before tool is up to cutting speed.**
- **Turn on the motor before the blade contacts the work.**

Figure 4-16. On internal cuts a saber saw will make its own opening. Be sure to rest the base on the workpiece.

A

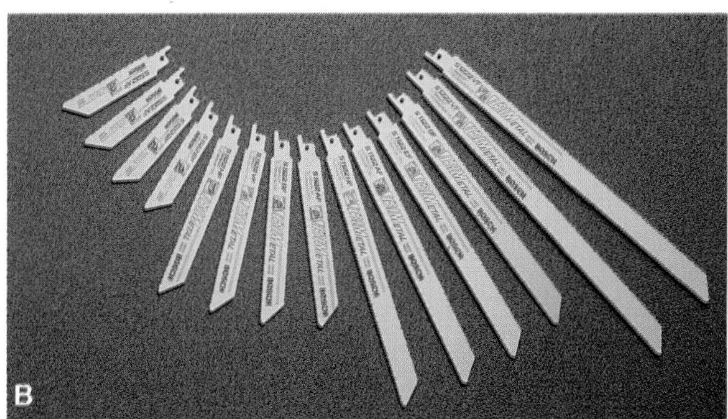

B

C

Figure 4-17. A–A reciprocating saw, so called because its blade moves up and down in a stroke about 1 1/8", can be fitted with a variety of blades to suit all purposes. (Milwaukee Electric Tool Corp.) B–Typical reciprocating saw blades. (Bosch) C–Reciprocating saw is used in situations where space does not permit use of a circular saw. (Black & Decker)

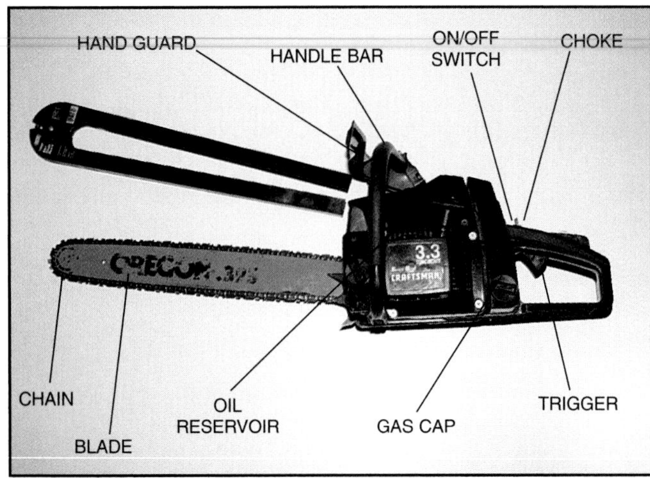

Figure 4-18. A chain saw may be used by the carpenter to cut heavy timbers. Left–Parts of a chain saw. Right–Use chain saw in a safe manner. Wear safety goggles and stand to one side of the chain.

- **Do not attempt to cut curves so sharp that the blade will be twisted.**
- **Make certain the work is well supported. Check clearance so that you do not cut into sawhorses or other supports.**

Chain Saws

Occasionally, a *chain saw* may be needed to cut heavy timbers, posts, and pilings. Both gas and electric models are available, but gasoline powered types are more powerful. Blade lengths vary from about 10" to 20". Chain saws are also useful for demolition during remodeling jobs. An 18" blade is recommended for the carpenter. See Figure 4-18.

Safety Rules For Chain Saws

- **Beware of kickback; never let the moving chain contact any object at the tip of the guide bar.**
- **Keep the chain sharp and at proper tension.**
- **To avoid danger of kickback, never cut at less than full throttle.**
- **Cut only one log or workpiece at a time.**
- **Never attempt a plunge cut.**
- **Use both hands during any cut.**
- **Always stand slightly to the left side of the working saw, never in direct line with the cutting chain.**
- **Stand with weight evenly balanced on both feet.**
- **Do not overreach or cut above shoulder height.**
- **Wear protective clothing and goggles or face screen. Ear plugs will protect from excessive noise.**
- **Stop the engine before setting the saw down.**
- **Never use a chain saw to cut nonwood materials.**

Portable Electric Drills

Portable *electric drills* come in a wide range of types and sizes. The size is determined by the chuck capacity; 1/4" and 3/8" generally being selected by the carpenter. Figure 4-19 illustrates the basic parts and reduction gears of a typical model. Speeds of about 1000 rpm are best for woodworking. Bits designed for use in a portable electric drill are shown in Figure 4-20.

A variable speed drill is shown in Figure 4-21. The trigger switch has an adjusting knob for presetting desired speed.

Safety Rules for Portable Drills

- **Select the correct drill or bit for your work and mount it securely in the chuck.**

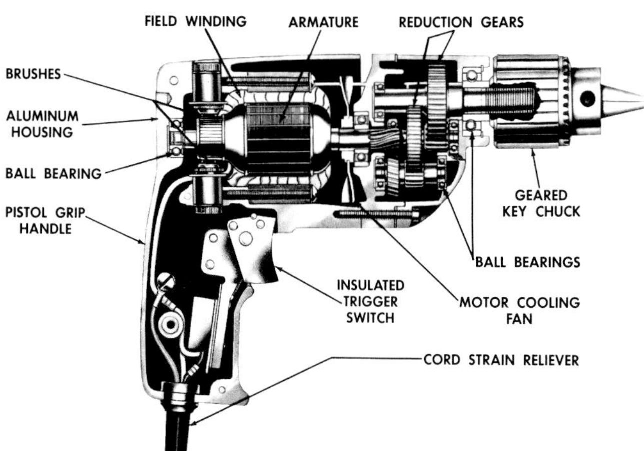

Figure 4-19. A portable electric drill is made up of many parts. Housing may be aluminum or plastic. Many are cordless.

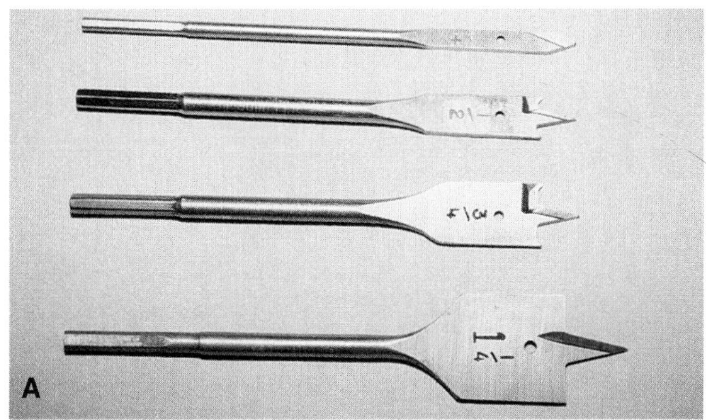

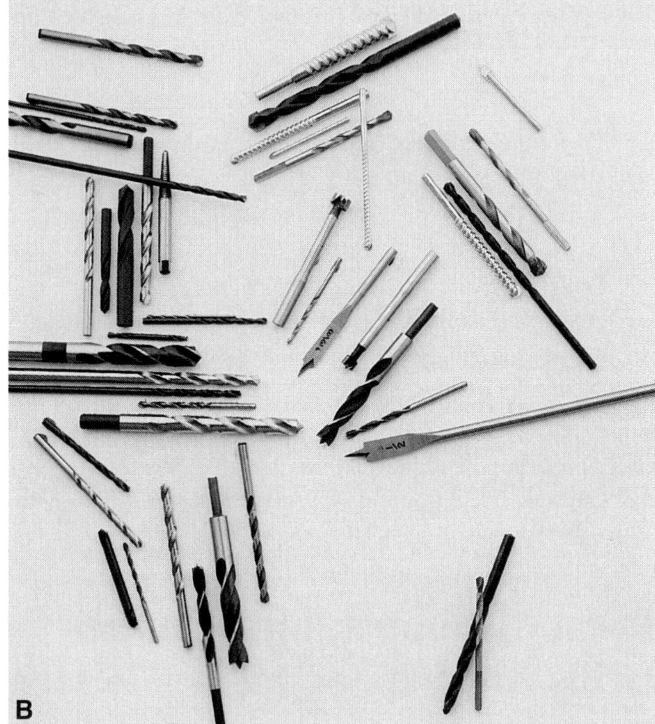

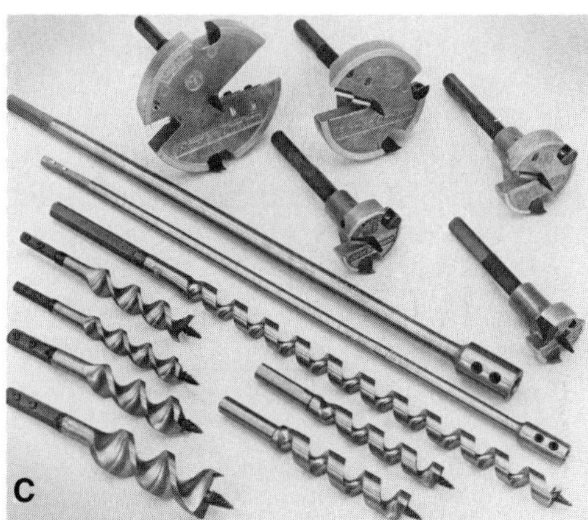

Figure 4-20. Bits for portable electric drills come in several types. A–Spade bits for light duty. B–High-speed twist drills. (Black & Decker) C–From top to bottom: self feeding, large-hole boring bits; 18" extenders; double twist bits (left); ship augers (right). (Black & Decker)

Figure 4-21. Top–Variable speed electric drill. Direction of rotation can also be reversed. (Bosch Power Tool Corp.) Bottom–Electric drills have all but replaced hand-powered drills, being faster and more accurate. (Santa Rita High School)

- **Stock must be held so it will not move during the operation.**
- **Connect the drill to a properly grounded outlet with switch in the "off" position.**
- **Turn on the switch for a moment to see if the bit is properly centered and running true.**
- **With the switch off, place the point of the bit in the punched layout hole.**
- **Hold the drill firmly in one or both hands and at the correct drilling angle.**
- **Turn on the switch and feed the drill into the work. The pressure required will vary with the size of the drill and the kind of wood being drilled. See Figure 4-22.**

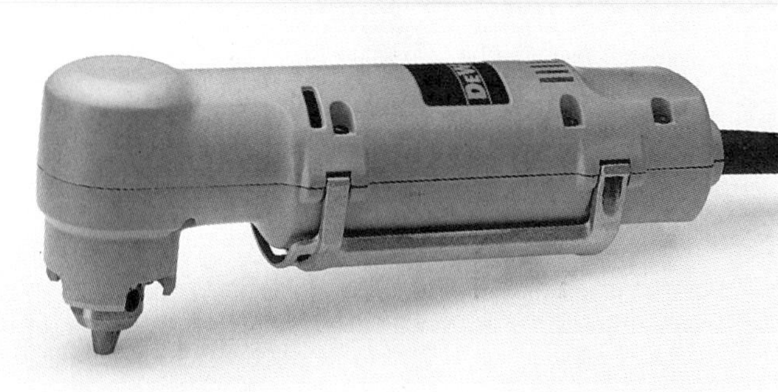

Figure 4-22. Right-angle drills are useful for working where space is limited. Pressure required will vary with drill size, type of wood, and diameter of the drill bit. (DeWALT Industrial Tool Co.)

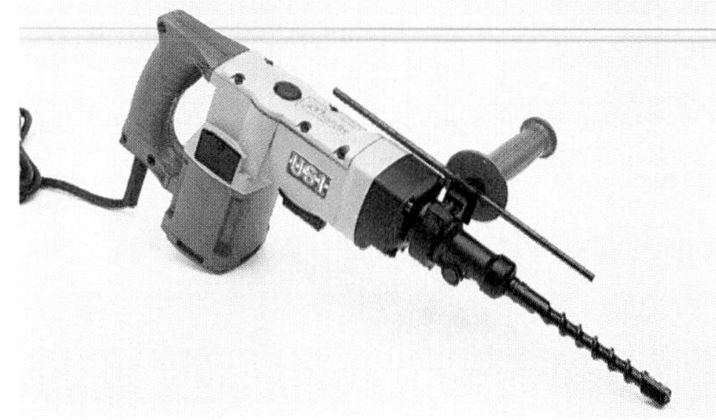

Figure 4-24. A 1 1/2" rotary hammer fitted with a spine shank four-cutter bit for fast rotary cutting. Depth rod allows drilling holes to a preset depth. A lever disengages rotary action when hammer action only is needed.
(Milwaukee Electric Tool Corp.)

Figure 4-23. Top–Cordless drills are useful where power is not close enough for power cords to be strung. (Makita U.S.A.) Bottom–Drilling a hole in a masonry wall with a cordless drill. (Porter-Cable Corp.)

- **During operation, keep the drill aligned with the direction of the hole.**
- **When drilling deep holes, especially with a twist drill, withdraw the drill several times to clear the cuttings.**
- **Always remove the bit from the drill as soon as you have completed your work.**

Cordless portable drills, Figure 4-23, are handy for many jobs. Power is supplied by a small nickel-cadmium battery that can be recharged. Such drills are used for general maintenance work and on production jobs where there are no power lines.

Rotary Hammer and Hammer Drills

Rotary hammer drills are used to drill holes in concrete and other masonry materials. Anchor devices are then placed in the holes to receive bolts or screws that fasten materials to floors, walls, or ceilings. Various type bits are available. Wood bits include flat boring bit, auger bit, ship auger bit, and hole saw. Concrete bits include carbide tip bit and screw fastener. Percussion carbide-tip bits are also available for faster cutting and greater durability in drilling concrete and masonry. Twist drills are also offered for use in drilling steel.

Rotary hammers are heavy duty tools for drilling large holes. Depending on the type of bit used, it can drill holes from 3/16" to 6" in diameter. See Figure 4-24.

Safety Rules for Rotary Hammer and Hammer Drills

- **Maintain good balance and a tight grip on the handles. Bits can bind in the hole, tearing the tool from the operator's grasp and possibly knocking him or her off balance.**

- **Push steadily into the workpiece, pulling back frequently to clear away chips.**
- **Keep bits sharp.**
- **Disconnect drill from power while installing or removing bits.**
- **Never lock the trigger in the "on" position.**
- **Check that switch is in the "off" position before plugging it in.**

Power Planes

The *power plane* produces finished wood surfaces with speed and accuracy, Figure 4-25. The motor, which operates at a speed of about 20,000 rpm, drives a spiral cutter. The depth of cut is adjusted by raising or lowering the front shoe. The rear shoe (main bed) must be kept level with the cutting edge of the cutterhead.

The power plane is equipped with a fence that is adjustable for planing bevels and chamfers. For surfacing operations, it is removed.

Hold and operate the power plane in about the same manner as a hand plane. The work should be rigidly supported in a position that will permit the operation to be easily performed. Start the cut with the front shoe resting firmly on the work and the cutterhead slightly behind the surface. Refer to Figure 4-26. Be sure that the electric cord is kept clear. Start the motor before the cutter head engages the stock. Move the plane forward with smooth, even pressure on the work. When finishing the cut, apply heavier pressure on the rear shoe.

Safety Rules for Power Planes

- **Study the manufacturer's instructions for adjustment and operation.**
- **Be sure the machine is properly grounded.**
- **Hold the standard power plane in both hands before you pull the trigger switch. Continue to hold it in both hands until the motor stops after releasing the switch.**
- **Always clamp the work securely in the best position to perform the operation.**
- **Do not attempt to operate a power plane that was designed for two hands with one hand.**
- **Disconnect the electric cord before making adjustments or changing cutters.**

The *power block plane*, Figure 4-27, can be used on small surfaces. It has about the same features and adjustments as the regular power plane. Being small, it is designed to be operated with one hand. When using this tool, the work should be securely held or clamped in place.

In planing small stock, kickbacks may occur. Be sure the hand not holding the plane is kept well out of the way.

Figure 4-25. Portable power planes are replacing hand planes because they are faster and more accurate. (Makita U.S.A.)

Figure 4-26. A portable power plane is being used to place a chamfer on the work. Note arrow. (Bosch Power Tool Corp.)

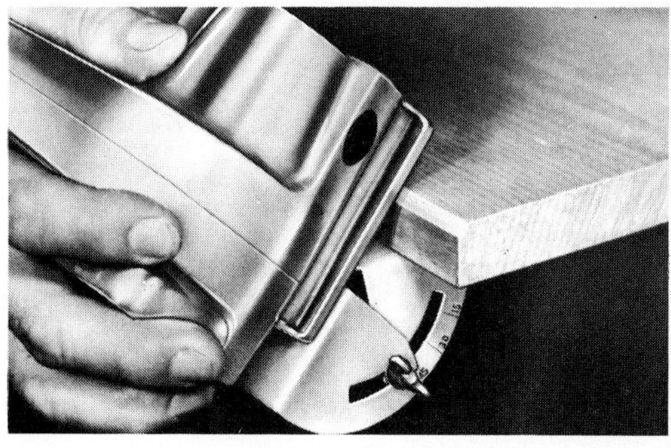

Figure 4-27. The power block plane's small size makes it useful for jobs such as cutting a chamfer on a small workpiece.

Portable Routers

Routers are used to cut irregular shapes and to form various contours on edges, Figure 4-28. When equipped with special guides, they can be used to cut dados, grooves, mortises, and dovetail joints. Important uses in carpentry include the cutting of gains for hinges when hanging passage doors and routing housed stringers for stair construction. See Units 17 and 18.

When changing bits on the router, it may be necessary to remove the base. However, it is usually possible to do this through openings in the base. See Figure 4-29.

Figure 4-28. A portable router has a motor mounted in an adjustable base. Motor revolves in a clockwise direction when viewed from above. Base rests on workpiece. Always use two hands when routing. (Montachusett Regional Vo-Tech School)

Changing Router Bits

1. Disconnect the router cord from power.
2. Lock the shaft or hold it with a wrench, depending on the kind and size of the router.
3. Loosen the chuck with a wrench, Figure 4-29(A).
4. Remove the bit, Figure 4-29(B).
5. Select new bit, Figure 4-29(C). (Some routers have a sleeve that is part of the bit assembly.)
6. Install bit on the arbor, Figure 4-29(D).
7. Tighten the chuck with a wrench, Figure 4-29(E).

When mounting bits in the router, the base is usually removed.

Straight bits are used when cutting dados and. grooves. See Figure 4-30.

Some bits for shaping and forming edges have a pilot tip that guides the router. The router motor revolves in a

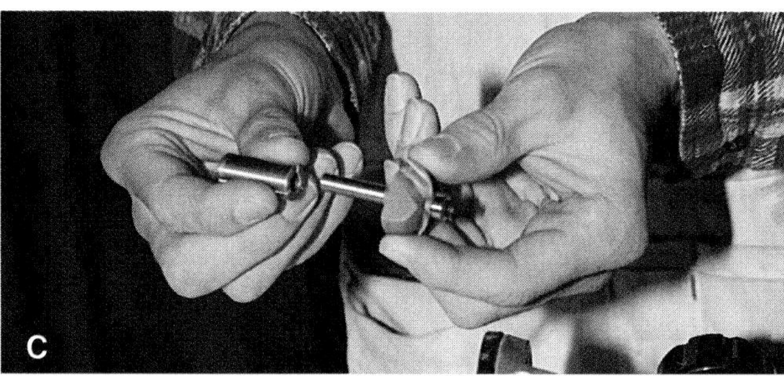

Figure 4-29. Procedure for changing router bit. A–Use a wrench or wrenches to loosen the chuck. B–Remove bit being replaced. C–Select new bit. D–Slide new bit on arbor. E–Tighten chuck and adjust depth. (Bullard-Haven Vo-Tech School)

clockwise direction (when viewed from above) and should be fed from left to right when making a cut along an edge, as illustrated in Figure 4-31. When cutting around the outside of oblong or circular pieces, always move in a counterclockwise direction.

Fixtures and templates are available that will guide the router through various decorative or blanking cuts. Figure 4-32 shows several router attachments that improve both the quality and efficiency of work.

Figure 4-30. Routers perform many edge-shaping operations. Left–Making a dado with a series of cuts. Right–Cutting a groove in the edge of a board.

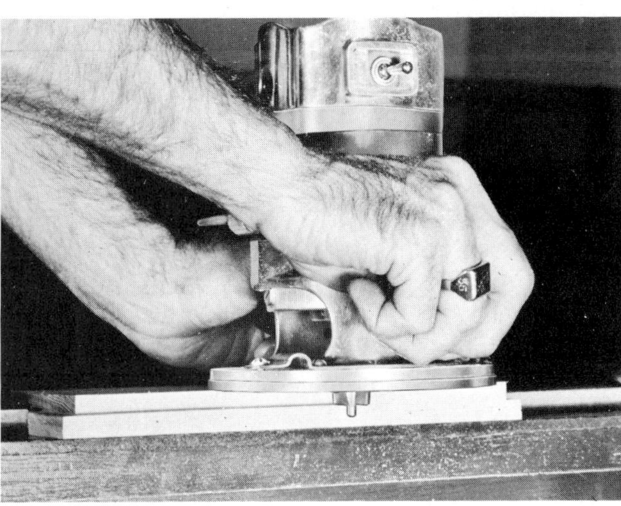

Figure 4-31. The pilot tip on cutter controls the cut. Apply light pressure between tip and material. Too much pressure will cause burn marks.

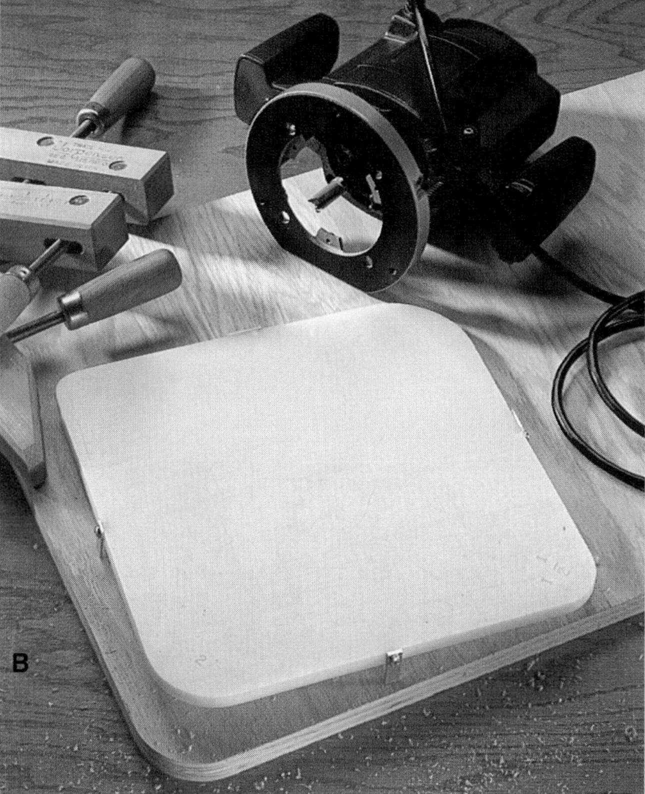

Figure 4-32. Router attachments. A–Circle cutting attachment produces circles from 8" to 48" in diameter. B–Radius edge guide is used to cut smooth rounded corners. C–Converter allows the router to act as a biscuit joiner. D–Clear offset base allows clear visibility of the cutting zone in free-hand edging and engraving. (Vermont America)

Safety Rules for Portable Routers

- **The bit must be securely mounted in the chuck and the base must be tight.**
- **Be sure the motor is properly grounded.**
- **Wear eye protection.**
- **Be certain the work is securely clamped so it will remain stationary during the routing operation.**
- **Place the router base on the work, template, or guide—with the bit clear of the wood—before turning on the power. Hold it firmly when turning on the motor. Starting torque could wrench the tool from your grasp.**
- **Hold the router with both hands and feed it smoothly through the cut in the correct direction.**
- **When the cut is complete, turn off the motor. Do not lift the machine from the work until the motor has stopped.**
- **Always unplug the motor when mounting bits or making adjustments.**

Portable Sanders

Portable sanders include three basic types:

- Belt.
- Disc.
- Finish.

They vary widely in size and design. Manufacturer's instructions should be followed carefully in the mounting of abrasive belts, discs, and sheets, Figure 4-33. Be sure to follow the manufacturer's lubrication schedule, as well.

The *belt sander's* size is determined by the width of the belt. Using the sander takes some skill. Support stock firmly. Switch must always be in the "off" position before plugging in the electric cord. Like all portable power tools, the sander should be properly grounded. Check the belt and make sure it is tracking properly.

Hold the sander over the work. Start the motor. Then, lower the sander carefully and evenly onto the surface. When using belt and finish sanders make sure to travel with the grain. Move it forward and backward over the surface in even strokes. At the end of each stroke, shift it sideways about one-half the width of the belt.

Continue over the entire surface, holding the sander level and sanding each area the same amount. Do not press down on the sander — its weight is sufficient to provide the proper pressure for the cutting action. When work is complete, raise the machine from the surface and allow the motor to stop.

Finishing sanders, Figure 4-34, are used for final

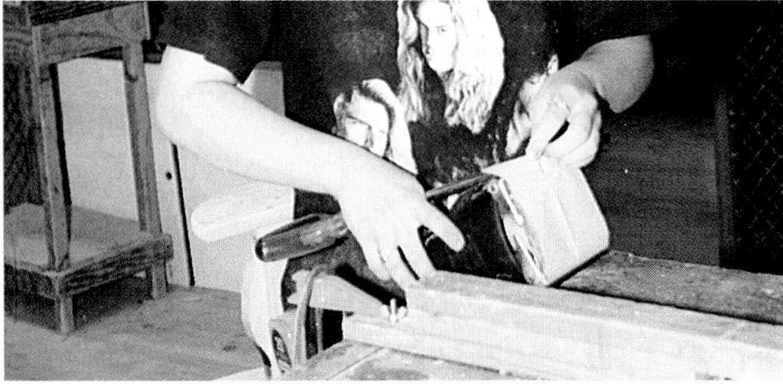

Figure 4-33. Follow the manufacturer's instructions for changing sandpaper on a power sander. Disconnect from power before attempting to replace sandpaper. (Montachusett Regional Vo-Tech School)

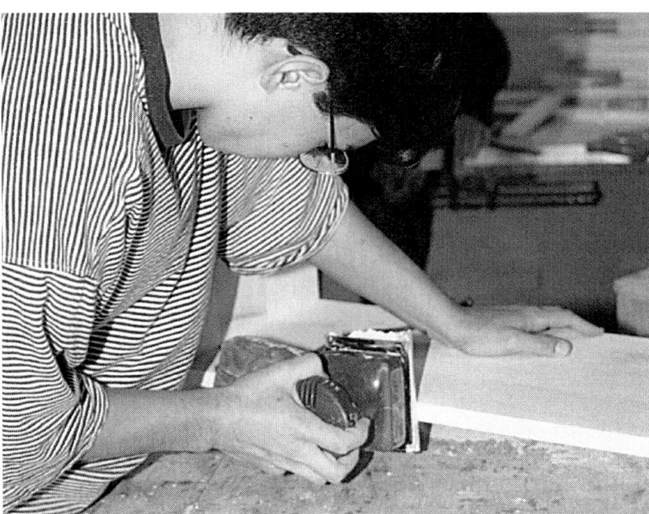

Figure 4-34. Finishing sanders. Top–Some orbital finishing sanders are equipped with a dust bag. The unit shown operates at a speed of 10,000 orbits per minute. (Bosch Power Tool Corp.) Bottom–Using an orbital sander to finish sanding cabinetwork. (Montachusett Regional Vo-Tech School)

sanding where only a small amount of material needs to be removed. They are also used for cutting down and rubbing finishing coats. There are two general types:

- Orbital sander.
- Oscillating sander.

Staplers and Nailers

A wide variety of power *staplers* and *nailers* is available. Most of them are air (pneumatic) powered. Those that are electrically operated should be properly grounded. Figure 4-35 shows pneumatic and electrically powered tools that drive nails, staples, and screws.

Using nailers and staplers speeds up work and allows the carpenter to work in tight places where swinging a hammer might be difficult or impossible. Staplers offer an alternative to nails, especially for securing sheathing and shingles. Nailer types include:

- Strip-fed. These hold a strip of nails in a long magazine located under the handle
- Coil-fed. This type carries nails in a round canister usually located ahead of a trigger handle.

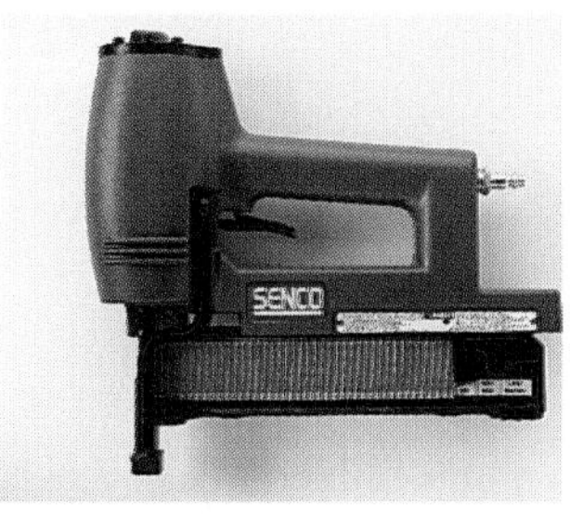

A

B

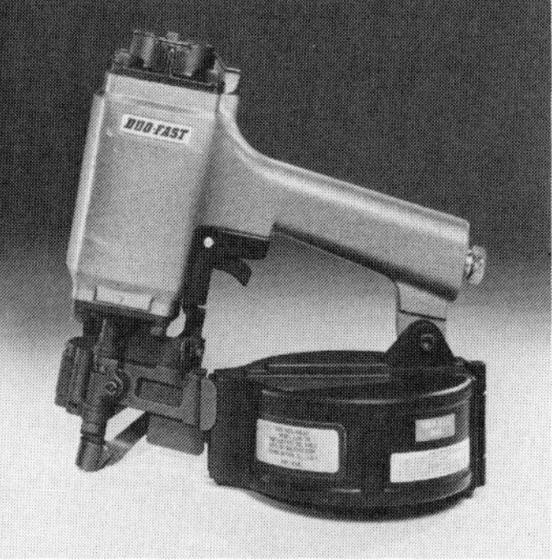

C

D

Figure 4-35. Pneumatic and electrically powered fastening tools. A–Pneumatic stapler drives 15 gauge staples from 1 1/4" to 2 1/2" long. (Senco Fastening Systems) B–Roofing stapler installs staples up to 1 1/4" long. (Paslode Co.) C–Coil nailer can drive 25 different nails from 1" to 2" long. D–Automatic screw fastener is designed for driving screws into drywall to wood or metal studs. (Duo-Fast Corp.)

To operate a nailer, first press the nose against the work to disarm the safety device that would otherwise keep it from driving the fastener. Then press the trigger to drive the nail as in Figure 4-36.

Most compressors used on job sites are portable, Figure 4-37. They may be driven by either electric motors or gas engines. Manufacturers of pneumatic tools will assist carpenters in selecting a compressor large enough to handle air needs on the job.

Safety Rules for Power Staplers and Nailers

- **Study the manufacturer's operating directions and follow them carefully.**
- **Use the correct type and size of fastener recommended by the manufacturer.**
- **For air-powered nailers, always use the correct pressure (seldom over 90 psi). Be sure the compressed air is free of dust and excessive moisture.**
- **Always keep the nose of the stapler or nailer pointed toward the work. Never aim it toward yourself or other workers.**
- **Check all safety features and be sure they are working. Make a test by driving the staples or nails into a block of wood.**
- **During use on the job, hold the nose firmly against the surface being stapled or nailed.**
- **Always disconnect the power tool from the air or electrical power supply when it is not being used.**
- **Always wear eye protection while using a power nailer or stapler. Hearing protection is also recommended.**
- **Disconnect tool from power before attempting to clear a jam.**
- **Remove finger from trigger when carrying a nailer or stapler. An accidental discharge could cause injury.**
- **Never use bottled gas to operate a pneumatic nailer. Sparks could cause a fire and many gases are bottled at extremely high pressures.**
- **Keep hoses in good condition.**
- **Keep your free hand safely out of the way of the tool.**
- **Disconnect air hose from a pneumatic tool before attempting to clear a jam or make adjustments.**
- **Never operate power nailers or staplers around flammable materials. Sparks may cause a fire.**
- **Always move forward when nailing or stapling on a roof. This will avoid backing off the roof.**
- **Secure hose of pneumatic tool with rope or other tie when working on scaffolding; this will prevent the tool from falling on other workers.**

Radial Arm Saws

The motor and blade of the *radial arm saw* are carried by an overhead arm. The stock is supported on a stationary table. The arm is attached to a column at the back of the

Figure 4-36. Pneumatic nailers speed work and are useful in close quarters where using a hammer may be difficult. Top–This finish nailer requires no lubrication and drives 1-2" brads. Middle–Framing nailer drives 2" through 3 1/2" smooth shank nails, 2" through 3" screw-shank nails, and 2" and 2 3/8" ring-shanked nails. (Hadley-Hobley Construction) Bottom–Pneumatic nailer is being used to fabricate truss rafters. (Trussworks, Inc.)

Figure 4-37. Air compressors suitable for operating pneumatic tools must deliver up to 110 pounds per square inch (psi) of pressure. Top–Typical portable compressor. Bottom–Generator and compressor on a job site power electrical as well as pneumatic tools. (Johnson-Manley Lumber Co.)

table. The depth of cut is controlled by raising or lowering the overhead arm. Figure 4-38 shows the parts of a typical radial arm saw.

The motor is mounted in a yoke and may be tilted for angle cuts. The yoke is suspended from the arm on a pivot which permits the motor to be rotated in a horizontal plane. Adjustments make it possible to perform many sawing operations.

When crosscutting, mitering, beveling, and dadoing, the work is held firmly on the table and the saw is pulled through the cut, Figure 4-39. For ripping and grooving, the blade is turned parallel to the table and locked into this position. Stock is then fed into the blade in somewhat the same manner as feeding a table saw, Figure 4-40.

For regular crosscuts and miters, first be sure the saw is against the column. Then place your work on the table and align the cut. Hold the stock firmly against the table fence with your hand at least 6" away from the path of the saw blade. Turn on the motor. Grasp the saw handle, pulling the saw firmly and slowly through the cut. See

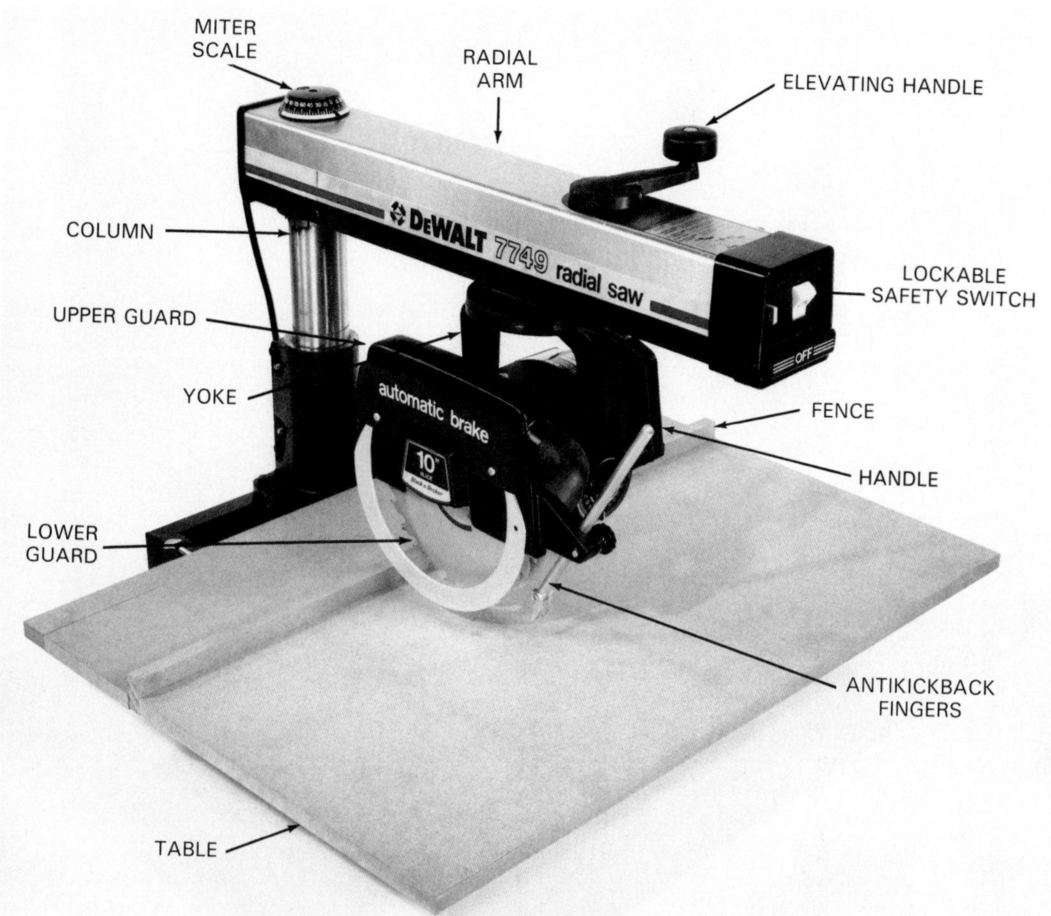

Figure 4-38. Radial arm saws make most cuts with material held stationary. (Black & Decker Co.)

Figure 4-39. Crosscutting on a radial arm saw. Lower guard is in place. (Montachusett Regional Vo-Tech School)

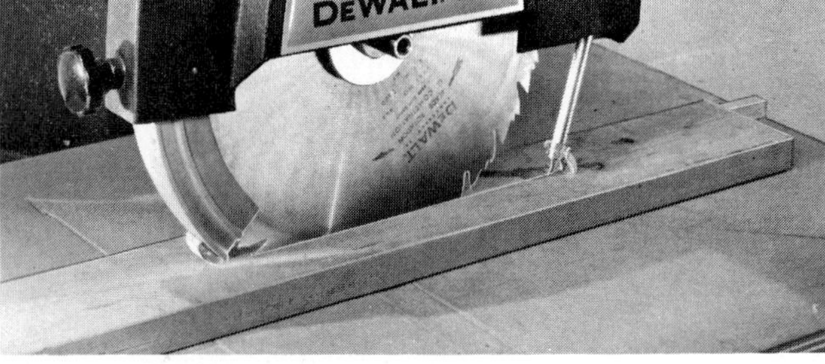

Figure 4-40. Ripping operation. Fence (not visible) is set in the table to guide the work.

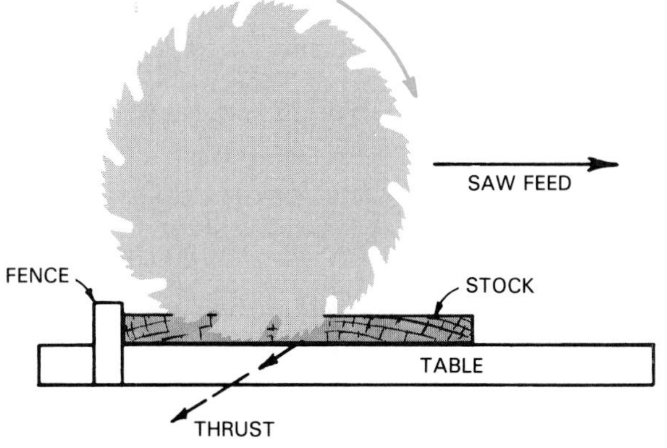

Figure 4-41. Blade rotation is in direction of the saw feed on a radial arm saw.

Figure 4-41. The radial arm saw may tend to "feed itself." You must control the rate of feed. When the cut is completed, return the saw to the rear of the table and shut off the motor.

The radial arm saw is especially useful in cutting compound miters. It is also a good tool for cutting larger dimension lumber which is difficult to slide across a saw table. See Figure 4-42. The proper saw setup for cutting a large sheet of plywood is shown in Figure 4-43.

Figure 4-42. Using a radial arm saw to cut dimension lumber. Materials of this size would be difficult to slide through a table saw. (Delta International Machinery Corp.)

Figure 4-43. Setup for cutting large sheet material on radial saw. Blade is turned at right angle to arm and material is passed through.

Safety Rules for Radial Arm Saws

- **Stock must be held firmly on the table and against the fence for all crosscutting operations. The ends of long boards must be supported level with the table.**
- **Before turning on the motor, be sure clamps and locking devices are tight. Check depth of cut and table slope. It must be slightly lower at back than front to prevent blade from "running" forward.**
- **Keep the guard and anti-kickback device in position.**
- **Always return the saw to the rear of the table after completing a crosscut or miter cut. Never remove stock from the table until the saw has been returned.**
- **Maintain a 6" margin of safety. Keep your hands this distance away from the path of the saw blade.**
- **Shut off the motor and wait for the blade to stop before making any adjustments.**
- **Do not leave the machine before the blade has stopped.**
- **Keep the table clean and free of scrap pieces and excessive amounts of sawdust. Do not attempt to clean off the table while the saw is running.**
- **In crosscutting, always pull blade toward you.**
- **Stock to be ripped must be flat with one straight edge to guide it along the fence.**
- **When ripping, always feed stock into the blade so that the bottom teeth are turning toward you. This will be the side opposite the anti-kickback fingers.**

Table Saws

Table saws are basic machines used in cabinetmaking. They are frequently used by carpenters on projects which include on-the-job built cabinets and built-ins. Carpenters use them, to some extent, for cutting and fitting moldings and other inside trim work.

When used for carpentry, the smaller sizes (4" to 6" jointers and 8" to 10" table saws) are usually selected because they can be easily moved from one job to another.

Space in this book does not permit more than a brief introduction to this equipment. Woodworking textbooks provide a complete description of the wide variety of work they will do along with instruction on how to use them.

The table saw, also called a circular saw, is used for ripping stock to width and cutting it to length. It also will cut bevels, chamfers, and tapers. Properly set up, the table saw can be used to produce grooves, dados, rabbets, and other forms basic to a wide variety of joints. The size of the saw is determined by the largest blade it will take. Figure 4-44 shows a typical model with the parts identified.

Stock to be ripped must have at least one flat face to rest on the table and one straight edge to run along the fence. Figure 4-45 shows correct procedure for making a rip cut. Be sure to follow the safety rules for sawing.

Figure 4-46 shows a standard crosscutting operation. A line is squared across the stock to show where the cut is to be located. For accurate work, make a check mark on the side of the line where the saw kerf will be located. The guard tends to hide the blade; it is helpful, when aligning the cut, to use a line scribed in the table surface. Since most of the work will be located to the left, it should extend back from the left side of the blade as shown in Figure 4-47.

Figure 4-48 shows a compound angle being sawed on a table saw. To make these cuts both the miter gage and the saw blade must be set at an angle. Tables of compound angles are available to determine the proper angles.

Safety Rules for Table Saws

- **Be certain the blade is sharp and right for the job at hand.**
- **Make sure the saw is equipped with a guard and use it.**
- **Set the blade so it extends about 1/4" above the stock to be cut.**
- **Stand to one side of the operating blade and do not reach across it.**
- **Maintain a 4" margin of safety. (Do not let your hands come closer than 4" to the operating blade even though the guard is in position.)**
- **Stock should be surfaced and at least one edge jointed before being cut on the saw.**
- **Use the fence or miter gauge to control the stock. Do not cut stock free hand.**
- **Always use push sticks when ripping short, narrow pieces.**

Figure 4-44. Table saws are most useful for finish work such as cutting moldings and components for built-ins.

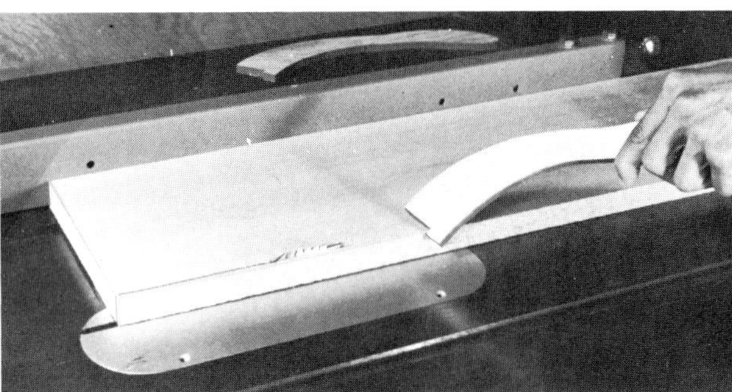

Figure 4-45. Ripping stock with a table saw. Guard should always be in place for safe operation. Top–Guard in position. Bottom–Guard removed to show operation. (Des Moines Public Schools)

- **Stop the saw before making adjustments.**
- **Do not let small scrap cuttings accumulate around the saw blade. Use a push stick to move them away.**
- **Resawing setups and other special setups must be carefully made and checked before the power is turned on.**

Figure 4-47. Line scribed on saw table will help you align stock for more accurate cutting to a line. This line should align with left side of saw kerf.

Figure 4-48. Cutting a compound angle on 2 x 4 stock. To make these cuts, both blade and miter gauge are set at an angle.

Figure 4-46. Crosscutting. Left–Squaring stock to a marked line. Right–Guard has been removed to show operation.

- **Remove the dado head or any special blades after use.**
- **Other workers, helping to "tail-off" the saw, should not push or pull on the stock but only support it. The operator must control the feed and direction of the cut.**
- **As work is completed, turn off the machine and remain until the blade has stopped. Clear the saw table and place waste in a scrap box.**

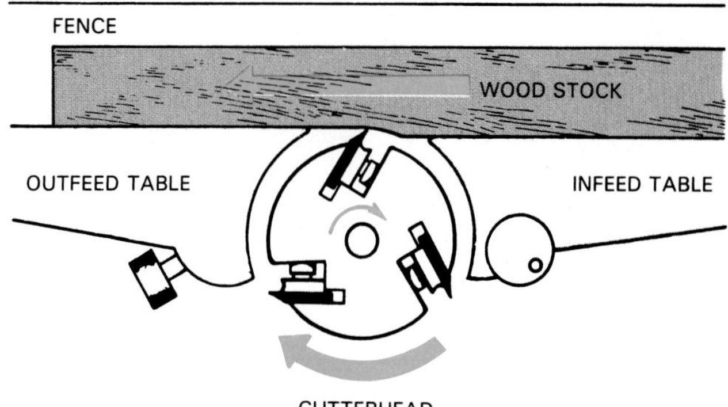

Figure 4-50. How a jointer works. Note direction of the wood grain. Avoid working against the grain.

Jointers

Jointers are commonly used to dress the edges and ends of boards. They may also be used for planing a face, cutting a rabbet, bevel, chamfer, or taper. The cutter head is cylindrical and usually has three or four knives. As the cylinder rotates at high speed, the knives cut away small chips. This cutting action produces a smooth surface on the wooden workpiece.

Principal parts of a jointer are shown in Figure 4-49. The cutter head revolves at a speed of about 4500 rpm. The size of the jointer is determined by the length of the knives.

The three main adjustable parts are:

- The infeed table.
- The outfeed table.
- The fence.

The outfeed table must be the same height as the knife edges at their highest point of rotation. This is a critical adjustment. See Figure 4-50. If the table is too high the stock will be gradually raised out of the cut and a slight taper will be formed. If it is too low, the tail end of the stock will drop as it leaves the infeed table and cause a "bite" in the surface or edge.

The fence guides the stock over the table and knives. When jointing an edge square with a face, it should be perpendicular to the table surface, Figure 4-51. The fence is tilted when cutting chamfers or bevels.

Safety Rules for Jointers

- **Before turning on the machine, make adjustments for depth of cut and position of fence.**
- **Be sure the guard is in place and is operating properly.**
- **The maximum cut for jointing on a small jointer is 1/8" for an edge and 1/16" for a flat surface.**
- **Stock must be at least 12" long. Stock to be surfaced must be at least 3/8" thick unless a special feather board is used.**
- **Feed the work so the knives will cut "with the grain." Use stock that is free from knots, splits, and checks.**
- **Keep your hands away from the cutter head even though the guard is in position. Maintain at least a 4" margin of safety.**
- **Use a push block when planing a flat surface. Do not apply pressure directly over the knives with your hand.**

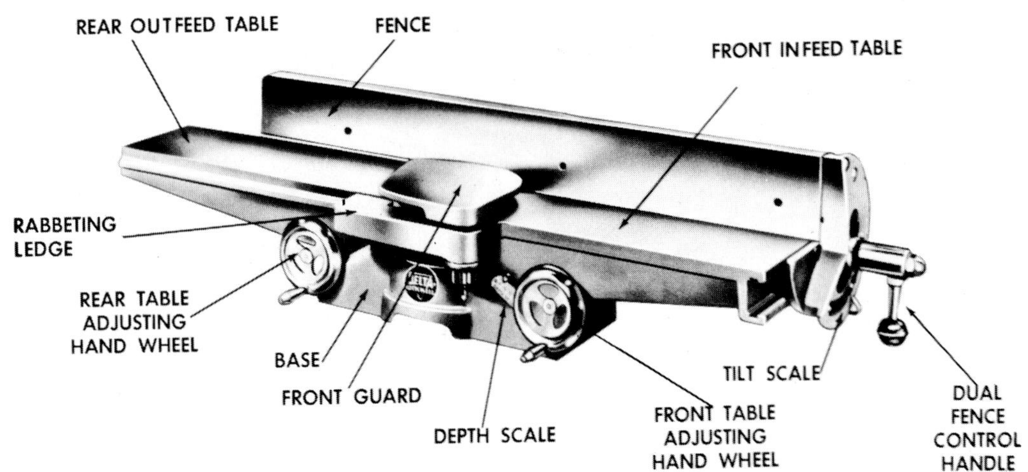

Figure 4-49. Small 4" to 6" jointers can be moved easily from one job site to another.

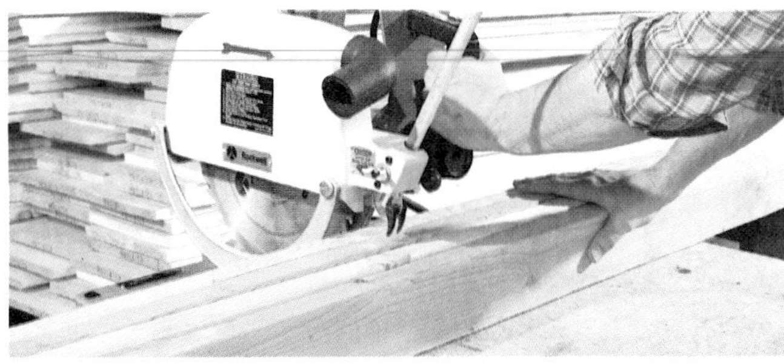

Figure 4-51. Jointing an edge. Top–"Step" the hands along the stock so they will not bear down on the stock while it passes over the cutter head. Bottom–It is best, however, to keep hands at least 6" away from cutter head. Here worker uses a push stick to hold down work. (Photo © Des Moines Public Schools)

- **Do not plane end grain unless the board is at least 12" wide.**
- **The jointer knives must be sharp. Dull knives will vibrate the stock and may cause a kickback.**
- **When work is complete, turn off the machine. Stand by until the cutter head has stopped.**

Special Saws

Special saws have been developed for working with wood construction. If light weight and compact size are important factors, carpenters may use a power miter box or a frame and trim saw for accurate crosscuts and mitering. The motor and blade of the *power miter saw* are supported on a pivot. To operate it, the carpenter sets the angle from a scale marked off in degrees. The cut is made by pulling downward on the handle. A trigger control in the handle turns the motor on and off. See Figure 4-52.

The *frame and trim saw* is supported on a pair of overhead shafts or guides. The support rotates left and right a little more than 45° to make crosscuts and miter cuts. It is capable of all sawing operations except ripping. The saw's

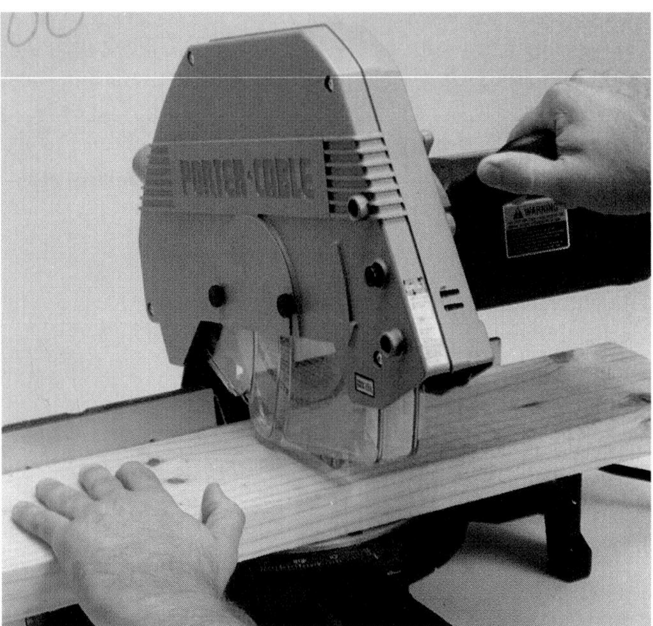

Figure 4-52. The miter saw. Top–Miter saw makes precise mitered cuts to 46 1/2° left and right. (Porter-Cable Corp.) Bottom–Using a power miter to make a 90° cut on a piece of flooring. Angled cuts can also be made either to right or to left as with a regular miter box. (Bullard-Haven Technical School)

capacity is 16" width on crosscuts and 12" on miter cuts of 45°. An extension table allows one worker to cut long stock alone. See Figure 4-53.

Saw Safety

- **Keep guards in place while operating.**
- **Wear safety glasses or a face shield to protect eyes from sawdust and other debris.**
- **Lock the saw securely at the angle of the cut.**
- **Hold stock firmly against the fence.**
- **Keep free hand clear of the cutting area.**
- **Work only with a sharp saw blade.**

Figure 4-53. Sawbuck frame and trim saw being used to make a compound miter cut. It crosscuts, miters, bevels, and makes compound cuts on any stock up to 2 x 12 (Delta International Machinery Corp.)

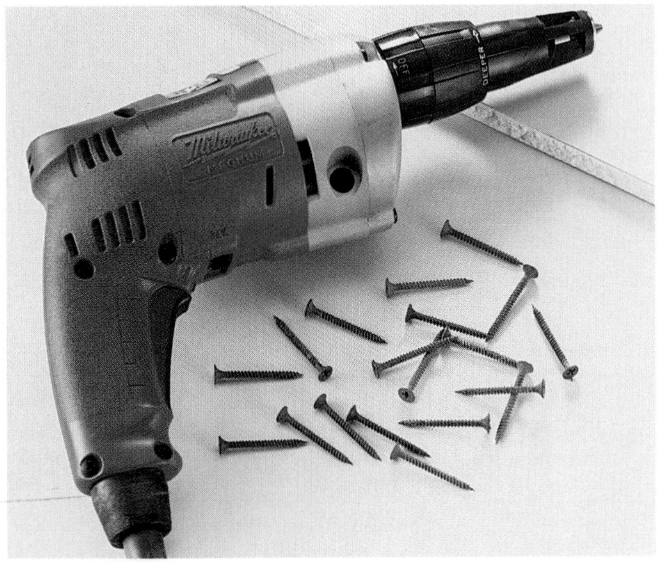

Figure 4-54. Drywall screw gun has variable speeds from 0 to 2500 rpm and trigger control reversing. (Milwaukee Electric Tool Corp.)

Specialty Tools

Drywall screw shooters (also called screw guns) speed up drywall installation and provide better connections, all but eliminating nail "popping." Drywall screws are inserted one at a time into the nose. Locators control depth below the face of the drywall. Usually a belt clip attached to the shooter's housing allows gun to be carried on a tool belt. See Figure 4-54.

A *plate joiner*, often referred to as a biscuit cutter, is a portable tool for cutting slots in the edges of lumber for wood plates or biscuits. The plates strengthen the wood joints used in cabinetmaking. Trim carpenters use plate joiners to assemble casing lumber and joining shelves and facings of cabinets built on site. Figure 4-55 shows the major parts of the plate joiner.

A *panel saw* is sometimes useful for cutting large sheets of paneling or plywood. A typical panel saw is shown in Figure 4-56.

Powder-actuated tools are used to drive various fasteners into concrete and structural steel. The tool may be either direct acting or indirect acting. See Figure 4-57. Both types depend on expanding gases from an exploded cartridge to drive the fastener. Both types have a triggering mechanism similar to that of a gun. To use it, the tool must be pressed firmly against the material being fastened before it will fire. Expanding gases from the exploding powder act on either the fastener or a piston causing the fastener to be driven into the concrete or metal.

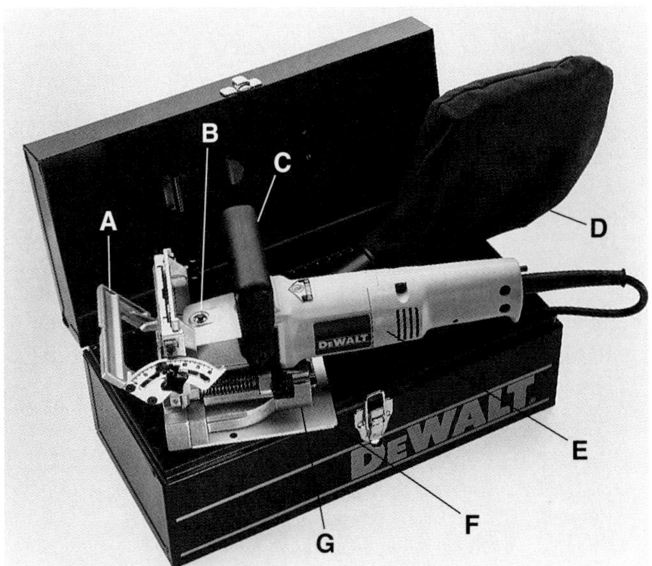

Figure 4-55. A plate joiner and its components. A–Adjustable fence. B–Switch. C–Handle. D–Dustbag. E–Electric motor and housing. F–Base plate. G–Depth adjustment. (DeWALT Industrial Tool Co.)

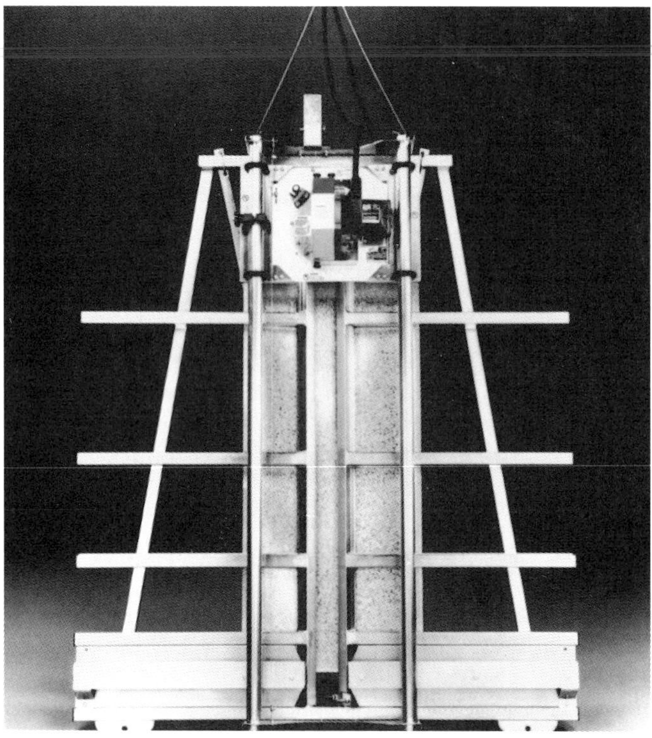

Figure 4-56. A panel saw is used for cutting large sheets of plywood or paneling. Though usually found in a woodworking shop, some carpenters and contractors use them on large residential buildings. (Milwaukee Electric Tool Corp.)

Figure 4-57. A powder-actuated tool with fasteners. The gun ignites blasting powder to drive fasteners into tough material like concrete and steel. (Forslund Bldg. Supply, Inc.)

Power Tool Care and Maintenance

Care of power tools is especially important if they are to function properly while giving long service. Sharp blades and cutters ensure accurate work and make the tool much safer to operate. The good carpenters take pride in their tools' condition and appearance.

Most power tools are equipped with sealed bearings that seldom need attention. Follow the manufacturer's recommendations for lubrication schedules. Gear mechanisms for portable power tools usually require a special lubricant. All equipment will require a few drops of oil on controls and adjustment of bearings from time to time.

Clean and polish bare metal surfaces with 600 wet-or-dry abrasive paper when required. These surfaces can be kept smooth and clean by wiping them occasionally with light oil or furniture polish. Some carpenters apply a coat of paste wax to protect the surface and reduce friction.

Some power tools, especially those with a number of accessories, can be purchased with a case. While making transport easier, such cases keep the accessories organized and protected.

Cutters and blades require periodic sharpening. Most carpenters are too busy and usually do not have the equipment to accurately grind cutters or completely fit saw blades. They usually send these items to a saw shop where an expert job can be performed. Carpenters may, however, lightly hone cutters before grinding operations are required. Also they may prefer to file saw blades several times before sending them in for a complete fitting. A fitting includes jointing, gumming, setting, and filing.

Carbide tipped tools will stay sharp 10 times longer than those with regular steel edges. A special diamond grinding wheel is used to sharpen carbide edges, Figure 4-58.

Figure 4-58. Skilled worker sharpens carbide tipped circular saw blade. (Foley Mfg. Co.)

Important Terms

Belt sander
Chain saw
Continuity monitor
Drywall screw shooter
Electric drill
Finishing sander
Frame and trim saw
Ground fault circuit interrupter (GFCI)
Jointer
Magnetic starters
Nailers
Panel saw
Plate joiner
Portable tools
Powder-actuated tools
Power block plane
Power plane
Power miter saw
Radial arm saw
Reciprocation saw
Rotary hammer drill
Router
Saber saw
Stapler
Stationary tools
Table saw

Test Your Knowledge

1. What is a ground fault interrupter and why should it be used in carpentry?
2. How can you determine the voltage required by a portable power tool?
3. Give the meaning of the term *ground continuity* and explain why it is important.
4. The size of a portable circular saw is determined by the _____________.
5. For general purpose work, a saber saw blade should have about _____________ teeth per inch.
6. When the base of the saber saw rests on a horizontal surface, the blade cuts on the _____________ (*up, down*) stroke.
7. When drilling deep holes, do not withdraw a twist drill until the hole is completed. True or False?
8. The depth of cut of a power plane is adjusted by raising or lowering the _____________.
9. What kind of saw might be useful for demolition work on a remodeling job?
10. A standard router bit is held in a _____________ type chuck.
11. The size of a belt sander is determined by the _____________.
12. To adjust the depth of cut of a radial arm saw, the _____________ is raised or lowered.
13. Name two advantages of a power nailer.
14. It is safe to use an oxygen tank to power a pneumatic stapler. True or False?
15. When crosscutting with the radial arm saw, the blade is _____________ (*pushed away from, pulled toward*) the operator.
16. For regular work, the _____________ of the jointer should be perfectly aligned with the knife edges at their highest point.
17. What cuts can be performed with a frame and trim saw?

Outside Assignments

1. Visit a builder's supply center and study the various portable circular saws on display. Also, obtain descriptive literature concerning the various sizes. After careful consideration, select a brand, model, and size that you believe would be best for rough framing and sheathing work on residential structures. Give your reasons and report to your class. Include specifications and prices of your selection.
2. Visit with a carpenter in your locality and learn what procedures are followed in maintaining and sharpening tools. If he or she uses standard saw blades, learn how they are kept sharp. Ask about the use of hardened tooth and carbide-tipped blades. If he or she has some tools sharpened at a saw shop, find approximate prices. If a tool maintenance center is located in your area, find out what services are available and what they cost. Prepare your notes carefully, then make a report to your class.

Modern high production cutoff saw. Foot pedal controls hydraulic power feed that moves cutterhead through workpiece. (Speed Cut, Inc. Corvallis, OR)

5

Leveling Instruments

Before construction of a foundation or a slab for a building can begin, the carpenter must know where the structure will be located on the property. Most communities have strict requirements—buildings must be set back certain distances from the street and must maintain minimum clearances from other property. The local code must be checked carefully for these requirements before layout begins.

Plot Plan

Many communities require the builder or owner to furnish a *plot plan* before a building permit is issued, Figure 5-1. This plan, as well as a survey of the lot (property), may already have been provided by a surveyor.

If there is a plot plan, it will indicate the location of the structure and indicate distances to property lines. There will be stakes or markers on the property to indicate property lines.

Surveyors should locate the *property lines*, in any event. They should also draw up the plot plan if one is required. The help of an engineer or surveyor will protect the owner and builder against costly errors in measurement.

Establishing Building Lines

Building lines are the lines marking where the walls of the structure will be. They are the lines that must conform to the code requirements on distance of the structure from boundary lines.

Once the property lines are known and marked by the surveyor, the building lines can be found by measuring off the distance with tapes. Be sure to observe proper setbacks and clearances. See Figure 5-2. Check the local code carefully for compliance.

Measuring Tapes

For measurements and layouts involving long distances, *steel tapes*, called *measuring tapes*, may be used. They are housed in winding reels like the one shown in Figure 5-3. Tapes are available in lengths of 50' to 300'. There are various types and the graduations are different from one to the other.

The carpenter will usually select one that is marked off in feet, inches, and eighths. Surveying, on the other hand, requires a tape graduated in feet and decimal parts of a foot.

Using Tapes to Locate Lines

To use a measuring tape for finding building lines, first find the boundary lines. Then measure off the required setback and clearances. Make sure the measurement is taken with the tape perpendicular to the boundary line. There are two ways of checking this:

- Swinging an arc from two dimensions.
- Using the "6–8–10" method.

In the arc method, the tape is extended from one property line to the dimension being measured. Then make a mark on the ground while swinging the tape back and forth. The marking instrument will make a curved line (arc) on the ground. Measure off the other dimension from an intersecting property line and again make an arc at the correct length. The point where the arcs intersect (cross) marks the spot where the intersecting lines are perpendicular.

In the "6–8–10" method, measure off 6' along one property line and 8' on the intersecting property line. Then measure diagonally from one mark to the other. When the corner formed by the intersecting lines is 90° (perpendicular), the diagonal line will be 10' long. Figure 5-4 shows both methods.

Laying Out with Leveling Instruments

In residential construction, it is important that building lines be accurately established in relation to lot lines. It is also important that footings and foundation walls be level, square, and the correct size.

If the building is small, the carpenter's level, framing square, and a rule are accurate enough for laying out and checking the building lines. But, as size increases, special leveling instruments are needed for greater accuracy and efficiency.

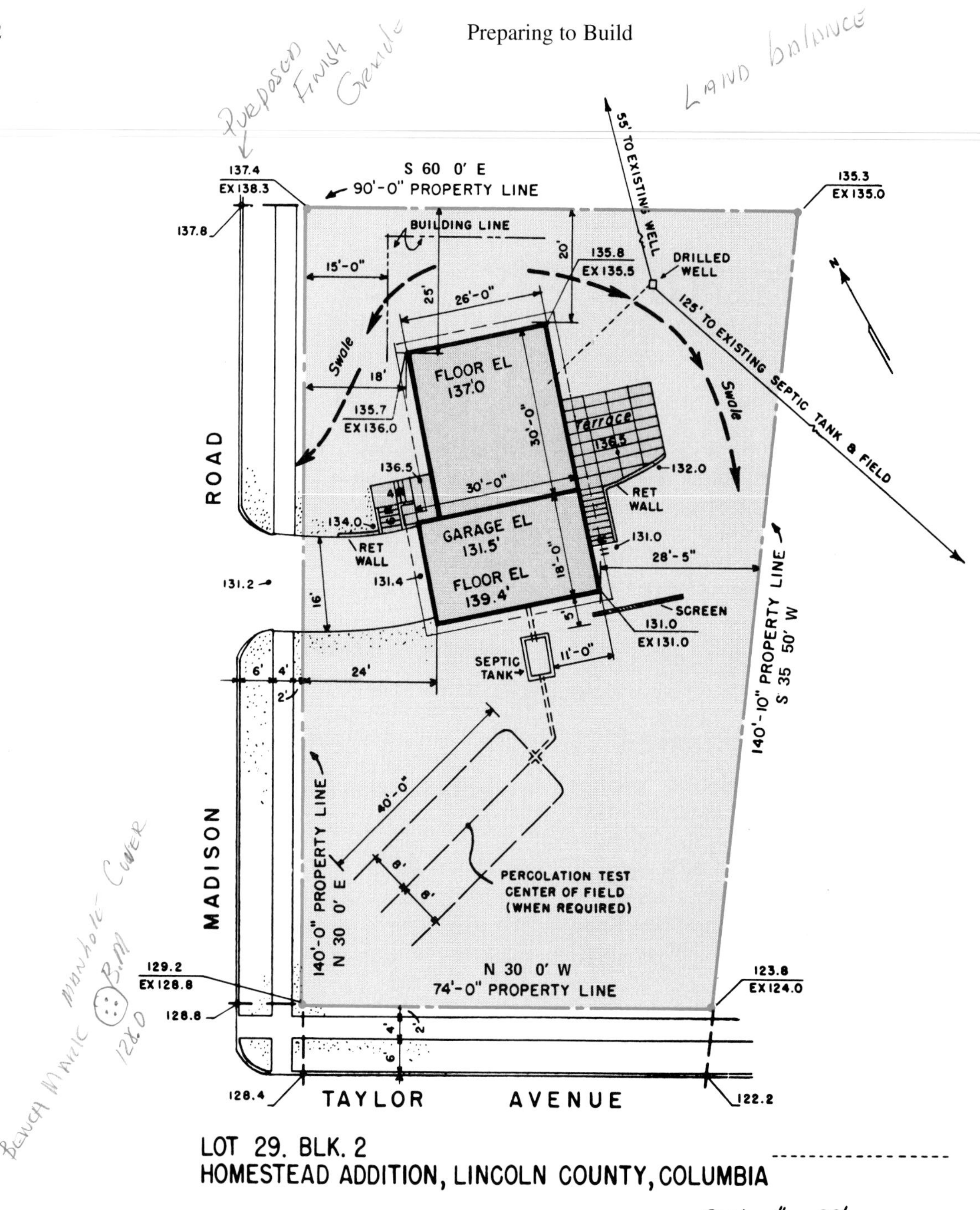

Figure 5-1. Plot plan shows the property boundaries and the building lines. Many communities require a plot plan from the builder or owner before a building permit can be issued.

Leveling Instruments

The level and level-transit are commonly used in laying out and checking construction work. These instruments can also be used for surveying and other space and land-layout jobs. When the job is too large for the chalk line, straightedge, level, and square, leveling instruments should be used.

The instruments include an optical device which operates on the basic principle that a *line of sight* is a straight line that neither dips, sags, nor curves. Any point along a level line of sight will be the same height as any other point. Through the use of these instruments, the line of sight replaces the chalk line and straightedge.

The *builder's level*, also called a *dumpy level* or an *optical level*, is shown in Figure 5-5. It consists of an accurate spirit level and a telescope assembly. These are attached to a circular base. Leveling screws are used to adjust the base after the instrument has been mounted on the tripod. The telescope rotates on the base so that any angle in a horizontal plane can be laid out or measured.

The *level-transit*, Figure 5-6, works like the builder's level. An additional feature permits the telescope to be piv-

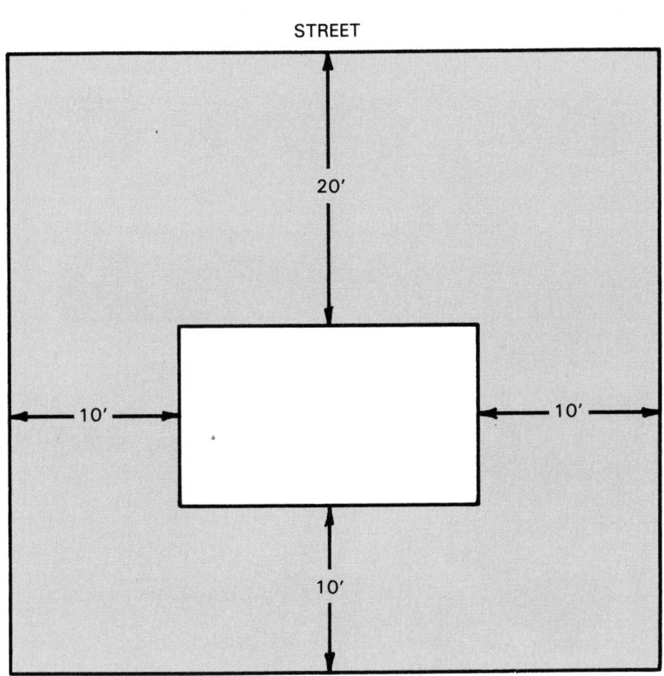

Figure 5-2. Simple rectangular structure laid out by taking measurements using lot lines as reference points.

oted up and down in a vertical plane. Using this instrument, it is possible to accurately measure vertical angles or determine if a wall is perfectly plumb (vertical). Its vertical movement also simplifies the operation of aligning a row of stakes, especially when they vary in height.

In use, both the builder's level and level-transit are mounted on tripods, Figure 5-7. Some models, like the one shown, have legs whose length is adjustable, making the

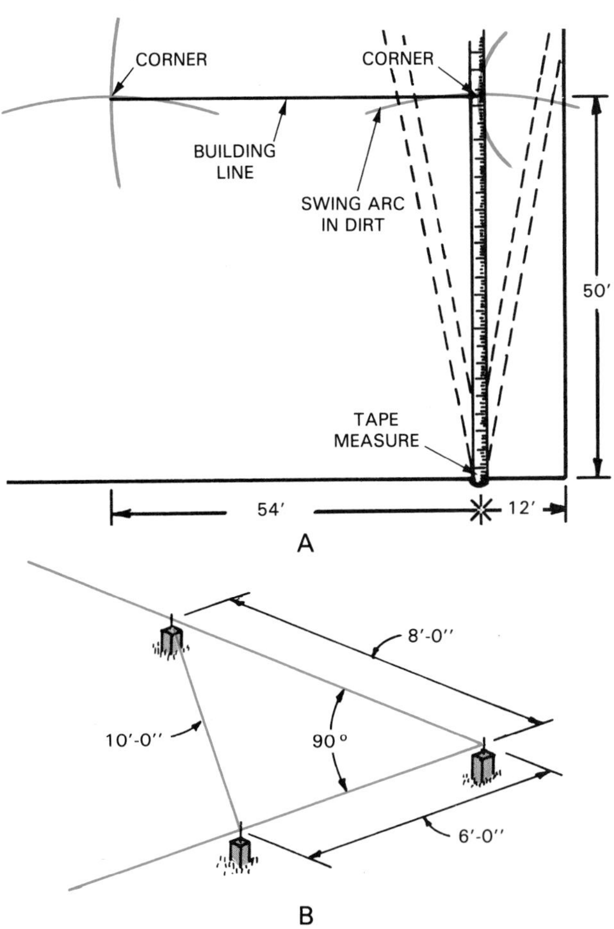

Figure 5-4. Methods for checking that intersecting lines are perpendicular (create an angle of 90° when they meet). A—Swinging an arc. B—Marking off 6' on one line and 8' on the intersecting line will result in a measurement of 10' across the marks if the lines are perpendicular to each other.

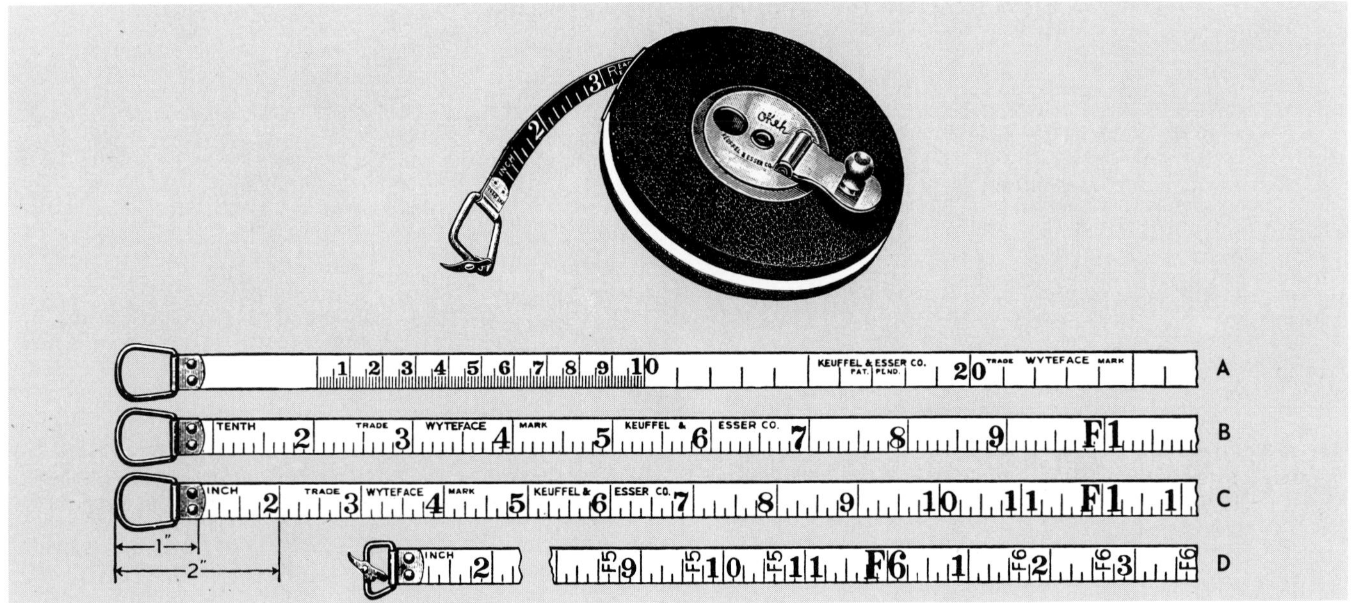

Figure 5-3. Measuring tape. Top—Graduations should be in feet, inches, and eighths. (Keuffel and Esser Co.) Bottom—Measuring tapes are made with different systems and graduations. A—Metric. B—Feet and decimal graduations. C—Feet, inch, and eighth-inch graduations. D—Feet and inches with feet repeated at each inch mark.

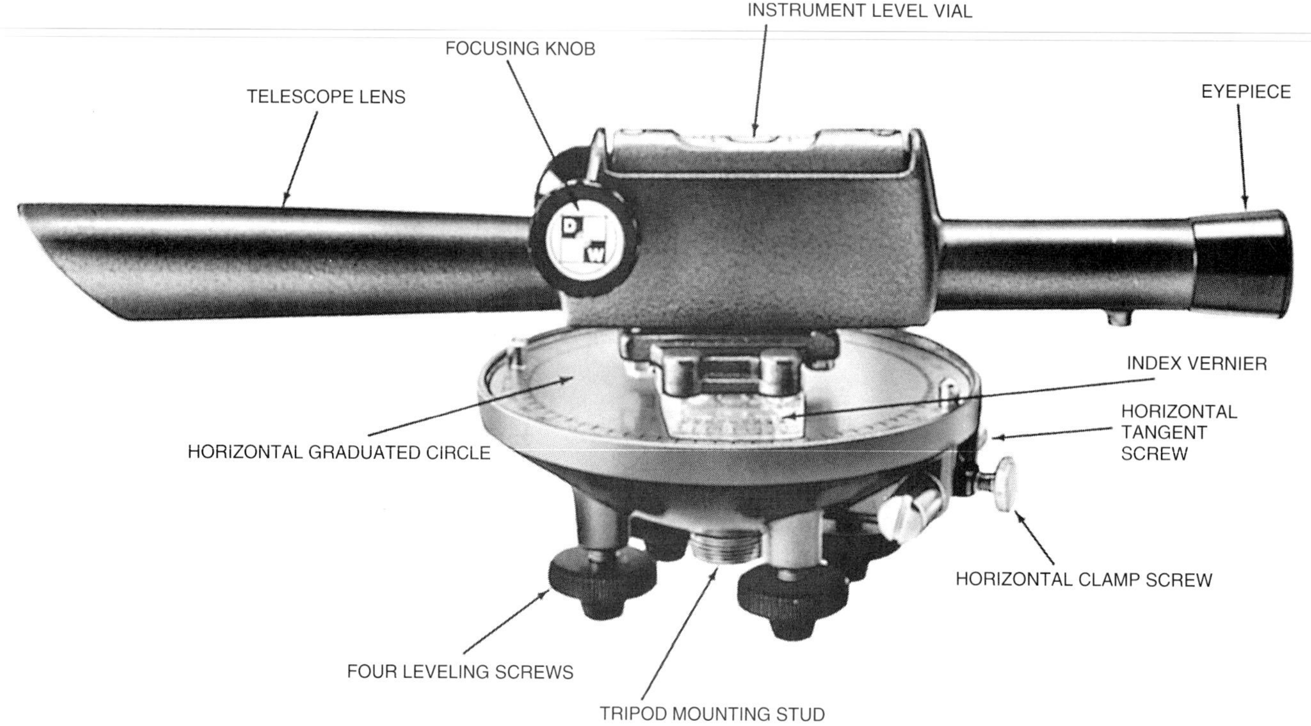

Figure 5-5. Builder's level is used to sight level lines and lay out or measure horizontal lines.

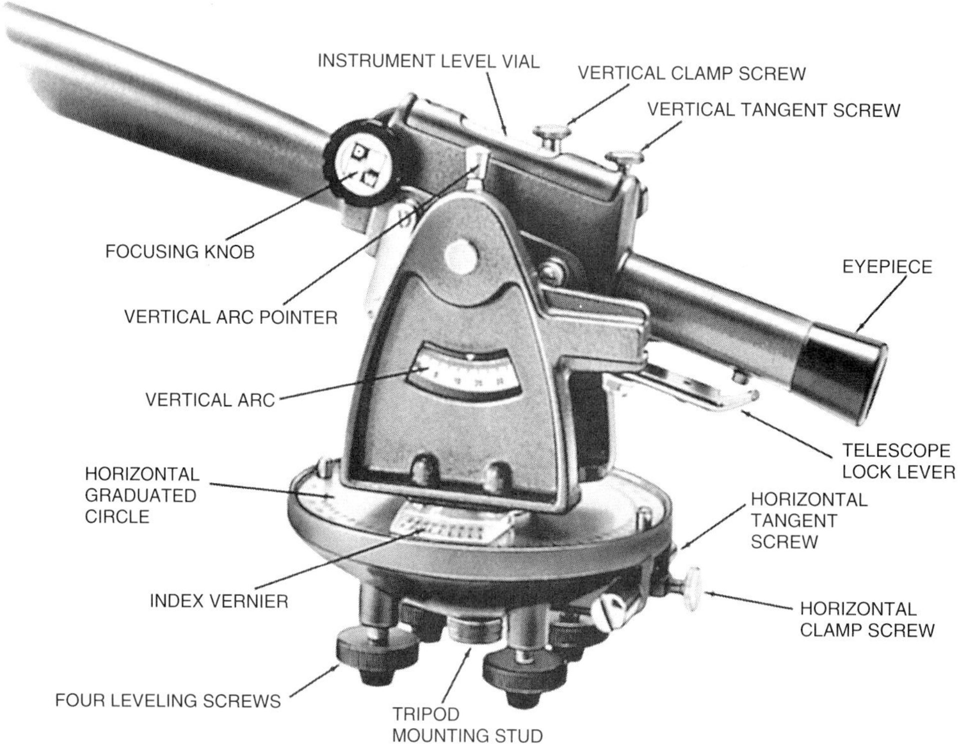

Figure 5-6. Level-transit can be used to lay out or check level and plumb lines. It can also be used to measure angles in either horizontal or vertical planes.

tripod easier to use on sloping ground. This feature also permits the legs to be shortened for handling and storing.

When it is necessary to sight over long distances, a leveling rod is used. See Figure 5-8. It is designed so that differences in the elevation between the position of the level and various positions where the rod is held can be easily read. The rod is especially useful for surveying. Readings can be made by the person operating the level or

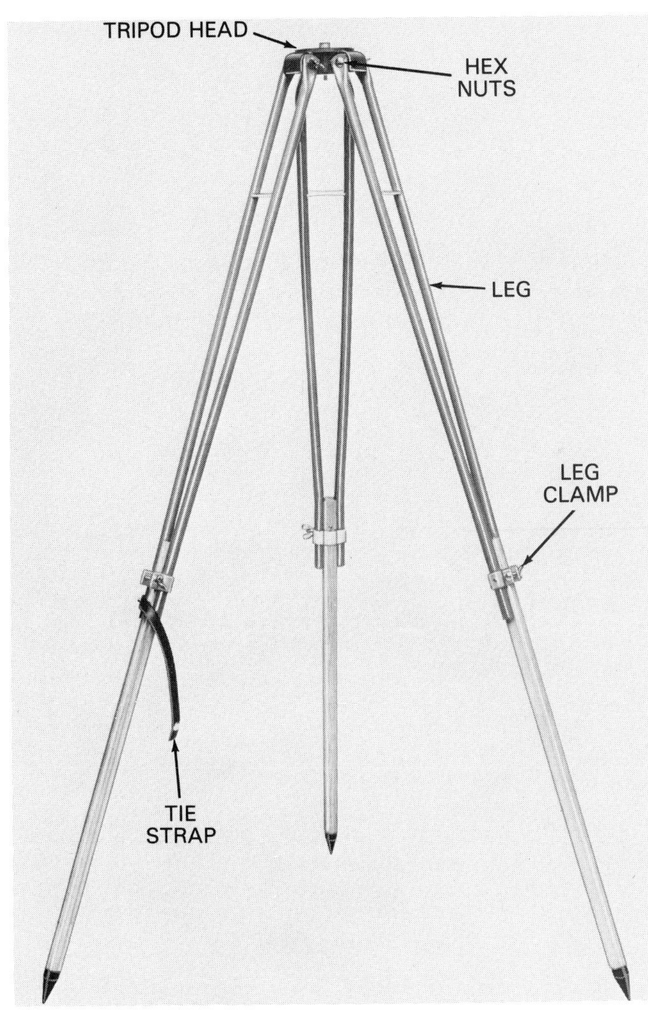

Figure 5-7. Tripod legs hinge at top and are adjustable for use on uneven terrain. (David White Instruments, Div. of Realist, Inc.)

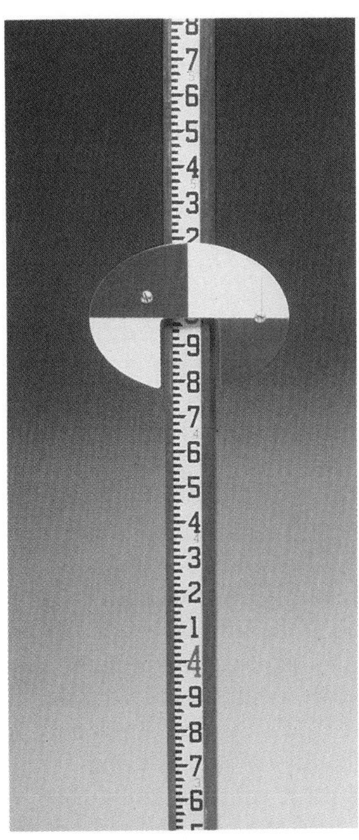

Figure 5-8. Leveling rod with target. The target can be moved up or down to match the line of sight of the level-transit. (David White Instruments, Div. of Realist, Inc.)

the target can be adjusted up and down to the line of sight and then the rod holder (one holding the rod) can make the reading.

The rod shown has graduations in feet and decimal parts of a foot. This is the type used for regular surveying work. Rods are also available with graduations in feet and inches.

When sighting short distances (100' or less), a regular folding rule can be held against a wood strip and read through the instrument. This procedure will be satisfactory for jobs such as setting grade stakes for a footing. Always be sure to hold the strip and rule in a vertical position.

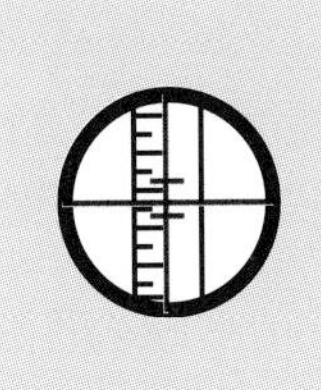

Leveling instruments and equipment will vary somewhat depending on the manufacturer. Always read and study carefully the instructions for a given brand

Care of Leveling Instruments

Leveling instruments are more delicate than most other tools and equipment. Special precautions must be followed in their use so they will continue to provide accurate readings over a long period of time. Some suggestions follow:

- Keep the instrument clean and dry. Store it in its carrying case when it is not in use.
- When the instrument is set up, have a plastic bag or cover handy to use in case of rain. Should it become wet, dry it before storing.
- It is best to grip the instrument by its base when moving it from the case to the tripod.
- Never leave the instrument unattended when it is set up near moving equipment.
- When moving a tripod-mounted instrument, handle it with care. Hold it upright, never carry it in a horizontal position.
- Never over-tighten leveling screws or any of the other adjusting screws or clamps.
- Always set the tripod on firm ground with the legs spread well apart. When set up on floors or pavement, take extra precautions to ensure that the legs will not slip.
- For precision work, permit the instrument to reach air temperature before making readings.

- When the lenses collect dust and dirt, clean them with a camel's hair brush or special lens paper.
- Never use force on any of the adjustments. They should turn easily by hand.
- Have the instrument cleaned, oiled, and checked yearly by a qualified repair station or by the manufacturer.

Setting Up the Instrument

The following procedure is used to set up a tripod-mounted instrument:

1. Set up the tripod so it will be a firm and stable base for the instrument. The base of the legs should be about 3'-6" apart. Make sure the points are well into the ground and the mounted surface is fairly level.
2. Check the wing nuts on the adjustable legs. They should be tight. Tighten the hex nuts holding the legs to the head to the tension desired.
3. Lift the instrument carefully from its case by the base plate. Before mounting the instrument, loosen the clamp screws. On some instruments the leveling screws must be turned up so the tripod cup assembly can be hand tightened to the instrument mounting stud. The telescope lock lever of the level-transit should be in the closed position.
4. Attach the instrument to the tripod. If it is to be located over an exact point, such as a bench mark, attach the plumbing bob and move the instrument over the spot. Do this before the final leveling.

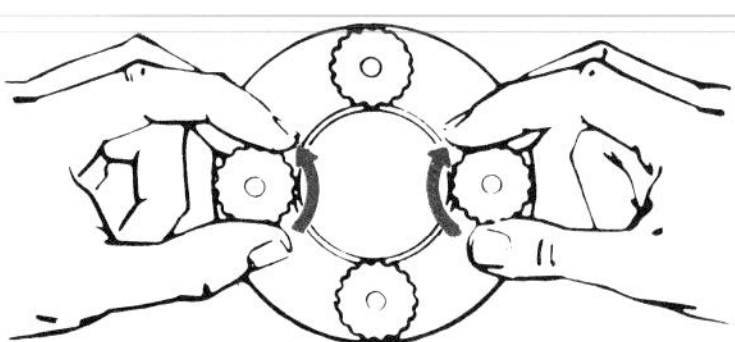

Figure 5-9. Adjust leveling screws to center bubble in level vial.

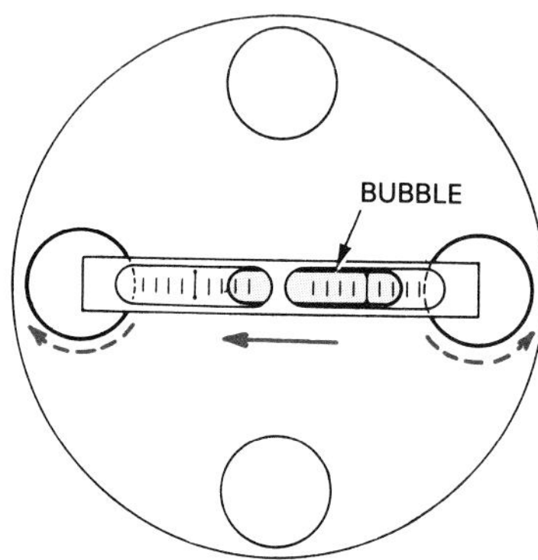

Figure 5-10. The bubble of the level vial will generally move in the same direction as the left thumb.

Leveling the Instrument

Leveling the instrument is a very important operation in preparing it for use. None of the readings taken or levels sighted will be accurate unless the instrument is level throughout the work. To level the instrument:

1. Release the horizontal clamp screw and line up the telescope so it is directly over a pair of the leveling screws.
2. Grasp the two screws between the thumb and forefinger as shown in Figure 5-9. Turn both screws uniformly with your thumbs moving toward each other or away from each other.
3. Keep turning until the bubble of the level vial is centered between the graduations. You will find that on most instruments the bubble will travel in the direction that your left thumb moves. See Figure 5-10. Leveling screws should bear firmly on the base plate. Never tighten the screws so much that they bind.
4. When the bubble is centered, rotate the telescope 90° (so it is over the other pair of leveling screws) and repeat the leveling operation.
5. Recheck the instrument over each pair of screws. When the instrument is level, the telescope can be turned in a complete circle without any change in the bubble.

Sighting

The telescope will magnify the image (object being sighted). Most builder's levels have a telescope with a power of about 20X. This means that the object will appear to be 20 times closer than it actually is. The procedure for sighting is easy to learn:

1. Line up the telescope by sighting along the barrel and then look into the eyepiece, Figure 5-11.
2. Adjust the focusing knob until the image is clear and sharp.
3. When the cross hairs are in approximate position on the object, Figure 5-12, tighten the horizontal-motion clamp.
4. Make the final alignment by turning the tangent screw.

Using the Instruments

Leveling instruments can be used by the carpenter to prepare the building site for excavation and grade leveling. Jobs that can be done with them include:

- Locating the building lines and laying out horizontal angles (square corners).

Figure 5-11. Sighting a level line with a builder's level. Note that both eyes are kept open during sighting. This reduces eyestrain and provides the best view. (David White Instruments, Div. of Realist, Inc.)

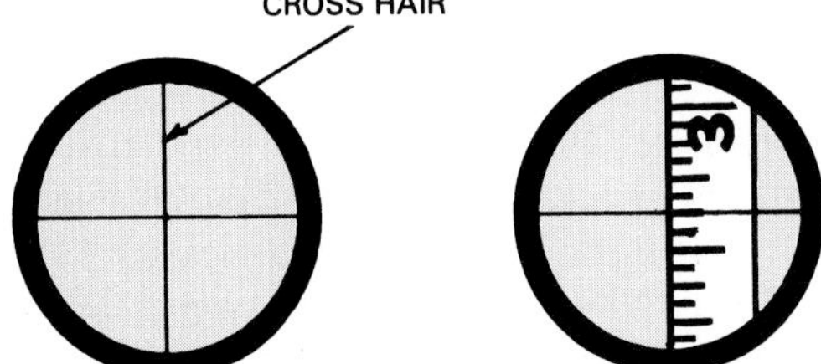

Figure 5-12. View through the telescope. Left—Cross hairs split the image area in half vertically and horizontally. Right—Object in view should be centered on cross hairs.

- Finding grade levels and elevations.
- Determining plumb (vertical) lines.

For layout, the builder's level or level transit must start from a reference point. This can be a stone marker in the ground or a point on a manhole cover or a mark on a permanent structure nearby. The point where the instrument is located is called the *station mark*. It may be the *bench mark* or the corner of the property or a previously marked point that is to be a corner of the building.

The Horizontal Graduated Circle

Laying out corners with the transit requires an understanding of how the horizontal graduated circle is marked. It is divided into spaces of 1°, Figure 5-13. When you

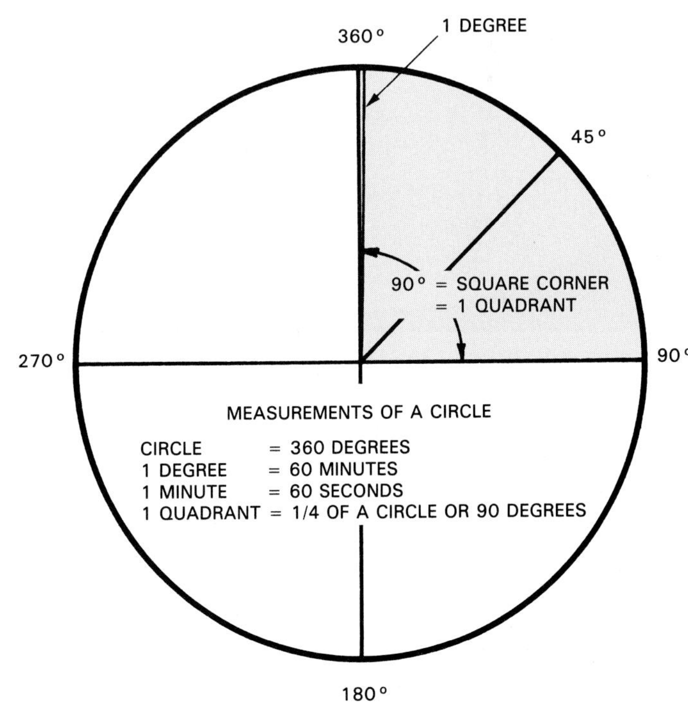

Figure 5-13. Graduated circle of a transit corresponds to the 360° of a full circle. 90° represents a quadrant, which would give you a square corner for a building.

swing the telescope of the builder's level or level-transit, the graduated circle remains stationary, but another scale, called the *vernier scale*, moves. It is marked off in 15-minute intervals. When laying out or measuring angles where there are fractions of degrees involved, you will use this vernier scale.

The upper half of Figure 5-14 shows a section of the graduated circle and the scale. It reads 75°. Notice that the zero mark on the vernier lines up exactly with the 75° mark. Now look at the lower half of Figure 5-14. The zero mark has moved past the mark for 75° but is not on 76°. You need to read along the vernier scale until you find a mark that is closest to being directly over a circle mark. That number is 45. The reading is 75° plus the number on the vernier, 45 minutes.

Vernier scales will not be the same on all instruments. You should study the operator's manual for instructions about the particular model you are using.

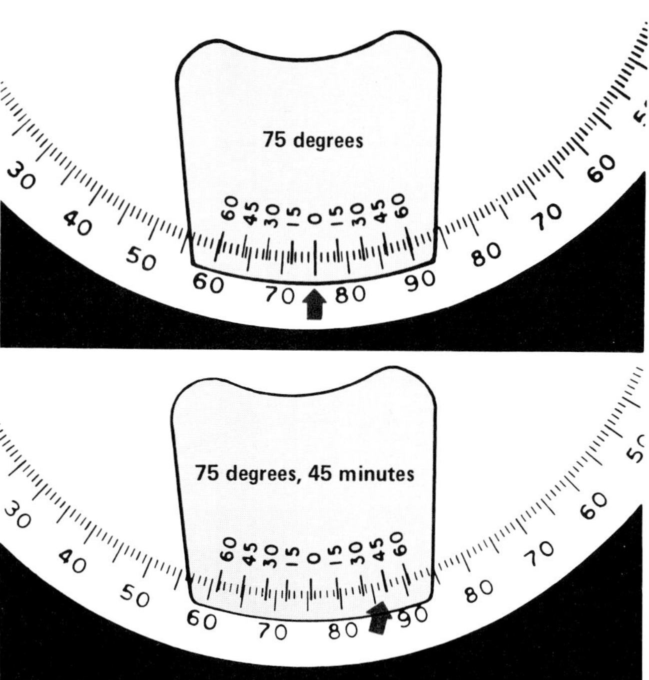

Figure 5-14. Reading the horizontal circle of a transit. Top—When the zero mark of vernier is right on a degree mark, the reading is an even degree. Bottom—When the zero mark is between degrees, read across vernier to find minute mark that aligns with a degree mark. (David White Instruments, Div. of Realist, Inc.)

Laying Out and Staking a Building

Staking out usually begins after one building line has been established. Referring to Figure 5-15, this would be line AB.

Staking a Building

1. Attach a plumb bob to the center screw or hook on the underside of the instrument. Shift the tripod until the point of the plumb bob is directly over the point marking the corner of the building lines (Station A). This is usually marked on a stake, which may have a nail or tack marking the exact corner.
2. Level the instrument before proceeding further. Recheck the plumb bob.
3. Lay out the other corner along the building line using a measuring tape. Place a stake at this point (Station B).
4. From Station A, turn the telescope so the vertical cross hair is directly in line with the edge of a rod held at Station B.
5. Set the horizontal circle on the instrument at zero to align with the vernier zero and swing the instrument 90° (or any required angle).
6. Position the rod along line AC so it aligns with the-cross hairs.
7. Locate the other corner along line AC using a measuring tape.
8. Move the instrument to Station C, sight back to Station A and then turn 90° to locate the line of sight to Station D.
9. Measure the distance to Station D.

If the resulting figure is a rectangle or square, you have completed the layout. However, you may want to move the instrument to Station D to check your work.

In practice, you will find that it is difficult to locate a stake in a single operation. This is especially true when using a builder's level, where the line of sight must be "dropped" with a plumb line to ground level. Usually it is best to set a temporary stake, as in Figure 5-16. Mark it with a line sighted from the instrument. Then, with the measuring tape pulled taut and aligned with the mark, drive the permanent stake and locate the exact point as shown.

All major rectangles and squares of a building line can be laid out using leveling instruments in the manner just described. After batter boards are set and lines attached, the carpenter's level and square can be used to locate stakes for small projections and irregular shapes.

Finding Grade Level

Finding the difference in the grade level between several points or transferring the same level from one point to another is called *grade leveling*. With the instrument level, the line of sight will also be level and the readings, Figure 5-17, can be used to calculate the difference in elevation. When there is a great amount of slope on the building site, the instrument can be set up between the points. The reading is taken with the rod in one position and then the instrument is carefully rotated 180° to get the reading at a second position.

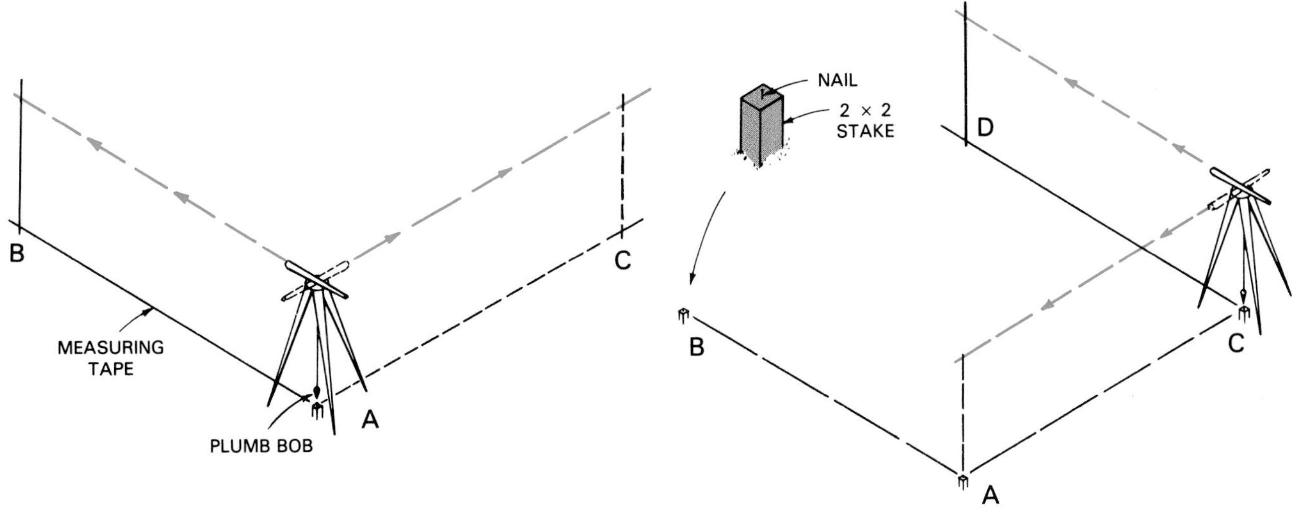

Figure 5-15. Steps for laying out building lines. Left—Locate instrument over stake marking corner and line up "0" on the instrument circle with building line AB. Swing the instrument 90° to establish line AC. Right—Move instrument to point C to establish point D. Rod must be kept vertical with plumb line or carpenter's level.

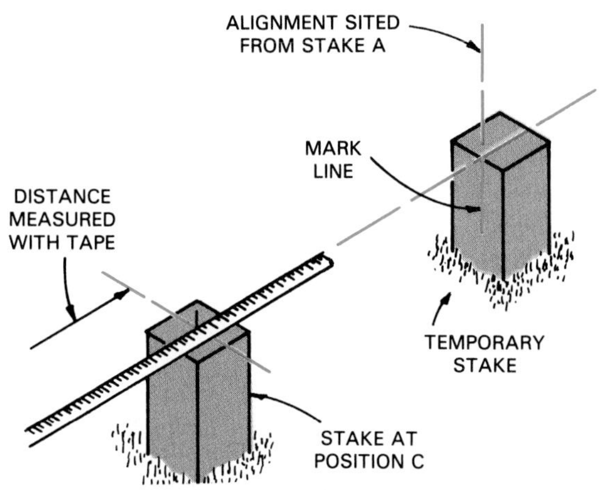

Figure 5-16. Temporary stake may be used to establish an exact point. First set the temporary stake and mark the line of sight on it. Drive second stake and transfer mark from Stake A.

When setting grade stakes for a footing or erecting batter boards, the instrument should be set in a central location, as shown in Figure 5-18. The distances will be about equal and it will reduce the need for changing focus on each corner. An elevation established at one corner can be quickly transferred to other corners or points in between.

Setting Footing Stakes

Grade stakes for footings are usually set to the approximate level by "eye" judgment. They are then carefully checked with the rod and level as they are driven deeper. The top of each stake should be at the required elevation.

Sometimes, reference lines are drawn on construction members or stakes near the work. Then the carpenter transfers them to the forms with a carpenter's level and rule.

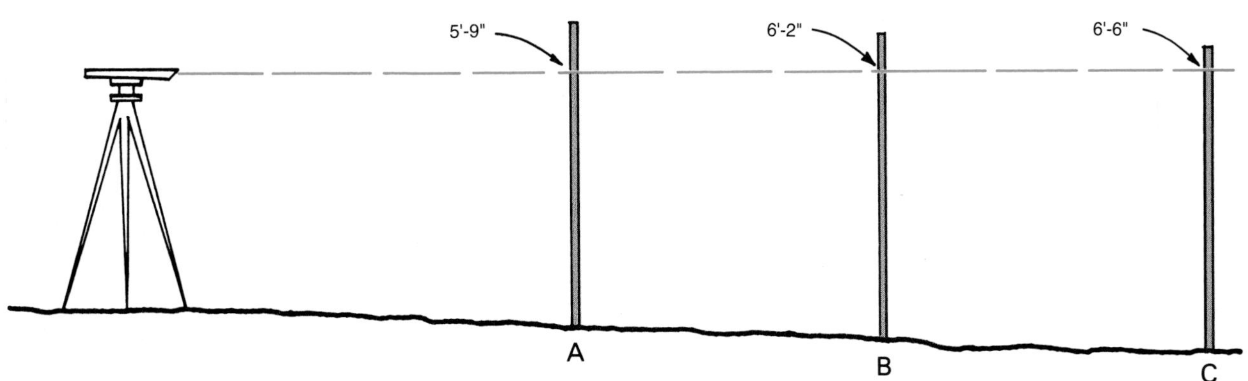

Figure 5-17. Establishing a grade level. Point A is 9" higher than point C.

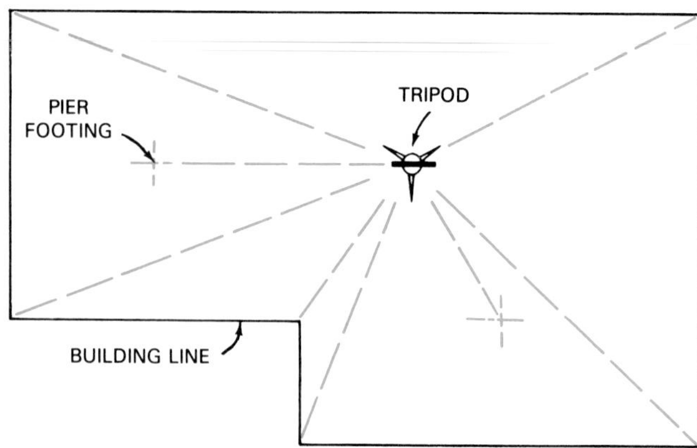

Figure 5-18. A central location for the instrument will make finding and setting grade stakes easier.

There may be situations where the existing grade will not permit the setting of a stake or reference mark at the actual level of the grade. In such cases, a mark is made on the stake with the information on how much fill to add or remove. The letters *C* and *F*, standing for *cut* and *fill*, are generally used. See Figure 5-19 for an example of how stakes are marked.

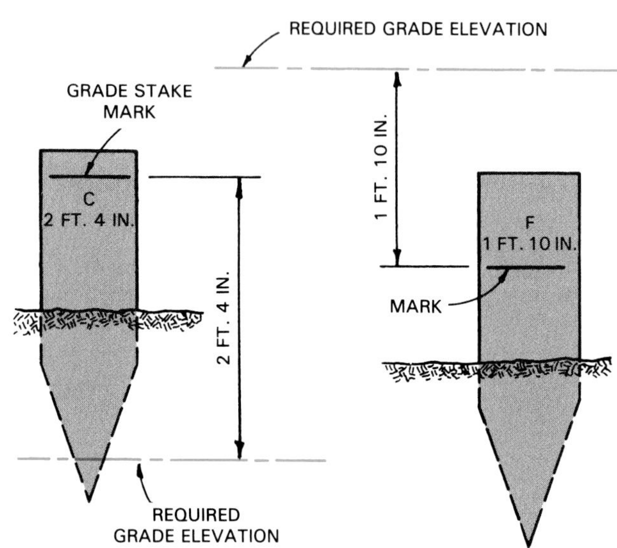

Figure 5-19. Cut and fill stakes are used to tell excavator how much material to remove or add to reach grade level. The letter *C* means cut (remove), the letter *F* means fill (add).

When laying out sloping building plots or carrying a *benchmark* (an officially established elevation) to the building site, it will likely be necessary to set up the instrument in several locations. Figure 5-20 shows how reading from two positions is used to calculate or establish grade levels.

Contour Lines

Contour lines are lines that run through points of equal elevation. Such lines can be laid out on an area of land quickly and easily.

Establishing Contour Lines

1. Set up the instrument at an appropriate point.
2. Directing the rod holder to hold the rod at a beginning point. Sight the rod, and set the target on the horizontal cross hair.
3. The rod holder then moves to the next required location and moves the rod up and down the slope until the target again aligns with the scope.
4. Set a stake at this point and repeat the procedure to locate as many points at the given elevation as needed.

Running Straight Lines with a Transit

Although the builder's level can be used to line up stakes, fence posts, poles, and roadways, more accuracy is gained with the level-transit, especially when different elevations are involved.

Set the instrument directly over the reference point. Level the instrument and then release the lock that holds the telescope in the level position. Swing the instrument to the required direction or until a stake is aligned with the vertical cross hair. Tighten the horizontal circle clamp so the telescope can move only in a vertical plane. Now, by pointing the telescope up or down, any number of points can be located in a perfectly straight line. See Figure 5-21.

Vertical Planes and Lines

The level-transit can be used to:

- Measure vertical angles.
- Lay out and check building walls, flagpoles, or TV antenna masts.

Establishing a Vertical Line

1. Level the instrument.
2. Release the lever that holds the telescope in a horizontal position.
3. Swing the instrument vertically and horizontally until a reference point in the required plane or line is sighted.
4. Lock the horizontal clamp screw. As you tilt the telescope up and down, all of the points sighted will be located in the same vertical plane, Figure 5-22.

Plumb lines can be checked or established by first operating the instrument as shown. Then move it to a second position, usually 90° either to the right or left, and repeat the procedure. To measure or lay out angles in a ver-

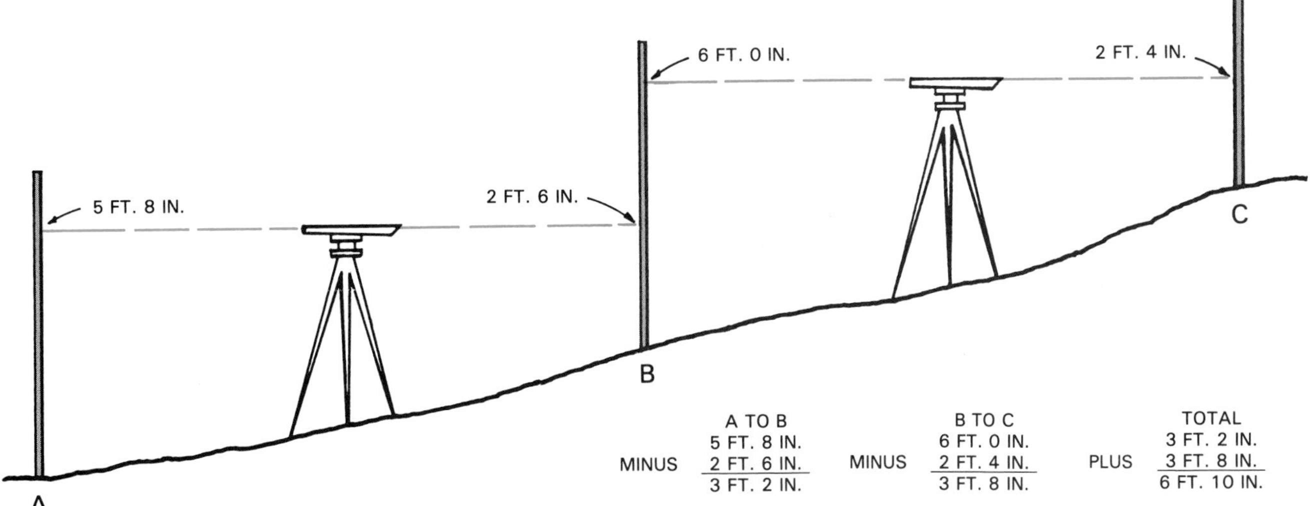

Figure 5-20. When there is a great deal of slope on the property or when long distances are involved, the transit or level-transit will need to be set up in two or more locations.

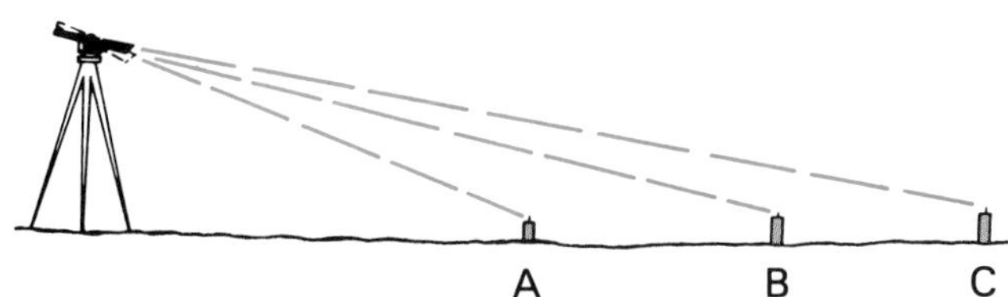

Figure 5-21. How to use the level-transit to align a row of stakes.

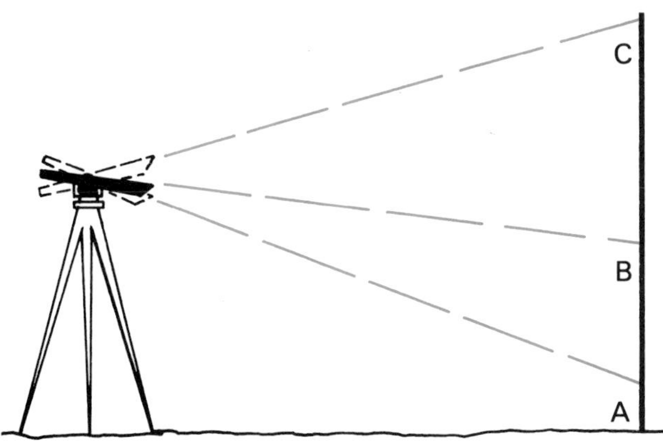

Figure 5-22. A level-transit can be used to lay out or check points in a vertical plane.

tical plane, follow the same general procedure that was used for horizontal angles.

A plumb bob and line may often be the most practical way to check vertical planes and lines. For layouts inside a structure where a regular builder's level or level-transit is impractical, use a plumb line.

Laserplane System

Development of the laser has brought new technology to the building industry. The *laserplane* can be set to emit a laser beam for a full 360° without being tended by an instrument person. Figure 5-23 shows the laser transmitter

Figure 5-23. Laserplane and receiver. Transmitter is at right, receiver at left. (Spectra-Physics Laserplane, Inc.)

(right) and its receiver. During use the transmitter is untended. The receiver can be tended by a rod holder, Figure 5-24, or it can be attached to an excavator or backhoe, Figure 5-25. Operators of excavating equipment can work the entire job site without stopping to check grade levels. The laserplane monitors grade continuously, eliminating overexcavation.

The operator places the bucket or blade cutting edge on the benchmark or the finished elevation. Then the

Figure 5-24. A rod holder moves receiver up or down until it is level with the laserplane. (Spectra-Physics Laserplane, Inc.)

Figure 5-25. Laser receiver can be attached to an excavating machine, speeding up the excavation. (Spectra-Physics Laserplane, Inc.)

receiver is adjusted up or down on the machine and the operator tightens the clamp when the "on-grade" point is reached. The receiver "catches" the laser beam from the rotating transmitter and signals the operator whether measured surface is above, below, or on grade.

Important Terms

Bench mark
Builder's level
Building lines
Contour lines
C, cut
Dumpy level
F, fill
Grade leveling
Laserplane
Level-transit
Line of sight
Measuring tape
Optical level
Plot plan
Property line
Station mark
Steel tape
Vernier scale

Test Your Knowledge

1. What are building lines?
2. Explain how to check perpendicularity of intersecting lines.
3. In the use of leveling instruments, the _____________ replaces the chalk line and straightedge.
4. The builder's level consists of a telescope assembly that is mounted on a _____________ base.
5. For surveying work, a measuring tape with graduations reading in feet and _____________ is usually selected.
6. The most important operation in setting up a builder's level or level-transit is the _____________.
7. When sighting through the telescope, you should adjust the _____________ until the image is sharp and clear.
8. When setting grade stakes for a building footing, the instrument should be set up in a _____________ location.
9. To position a leveling instrument directly over a given point, a _____________ is used.
10. A circle is divided into degrees, _____________, and _____________.

Outside Assignments

1. Study the catalog of a supplier or manufacturer and develop a set of specifications for a builder's level. Be sure it includes a good carrying case. Also select a suitable tripod and measuring tape. Secure prices for all of the items.
2. Through drawings and a written description, tell how you would proceed to lay out a baseball diamond using a level-transit.
3. Make a study of the procedures you would follow and calculations you would make to determine the height of a flagpole, tall building, or mountain, using the level-transit and trigonometric functions. Prepare a report for your class.

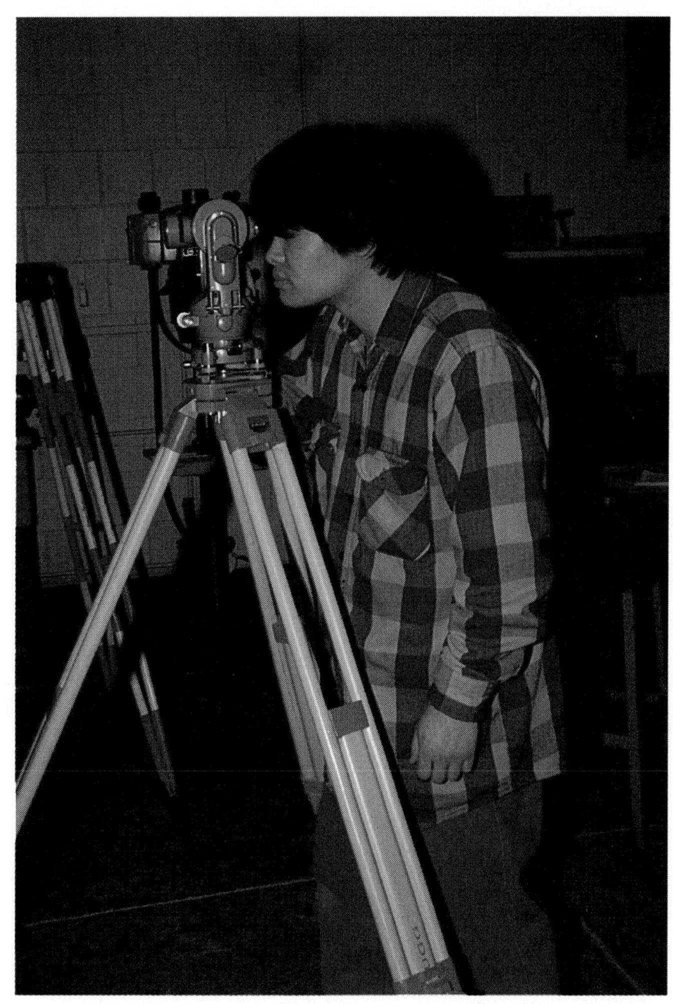

Student practices using laserplane. (St. Paul Technical College)

Surveying team uses a laser instrument to establish boundaries of lots in a subdivision (Cella Barr Associates, Tucson, AZ)

Beautiful renderings in full color help buyers visualize how a new dwelling will look. Plans for this house are shown throughout this unit. (L.F. Garlinghouse Co.)

6

Plans, Specifications, and Codes

In building construction, a good plan and a well-defined contract are important. The old proverb "Early understandings make long friendships" holds true in the construction industry.

Every carpenter must know how to read and understand *architectural drawings* (plans) and correctly interpret the information found in written specifications. Simply put, the plans tell you how to build and the specifications tell you what materials must be used. See Figure 6-1.

Figure 6-1. Carpenters must study plans frequently as they construct a building. (Tony's Construction Inc.)

Copies of the architect's original drawings are usually called *blueprints*, or simply *prints*. At one time, all copies were made on chemically treated paper so that the lines were white on a blue background. The term blueprint is still used today, even though prints may have dark lines on a white background.

Set of Plans

Vast amounts of information are needed to build a modern house or light commercial structure. Since a single drawing could not possibly hold all that information, many sheets of drawings are required. When bound together, these sheets are known as a set of plans.

A contractor will use a set of plans to bid on the construction of a building, Figure 6-2. An estimator will study them in preparing a bid.

Figure 6-2. Contractor (standing) may discuss a building plan with the architect before submitting a bid. (Tony's Construction Inc.)

The carpenter and building contractor will not be the only persons needing a set of plans. The owner receives a set, as well. Tradespeople such as the electrician, plumber, and heating contractor need sets so they can install their systems. Lumber dealers and other suppliers will use them to determine the quantity of materials needed. The carpenter uses only parts of the plan.

Drawings in a Set of Plans

A set of house plans usually includes:

- A plot plan.
- A foundation or basement plan.
- Floor plans.
- Elevation drawings. These show the front, rear, and sides of the building.
- Drawings of the electrical, plumbing, heating, and air conditioning layouts.

Stock Plans

When a set of plans is mass produced to be sold to many clients, it is called a *stock plan*. Today, a wide range of such ready-drawn plans are available for people who want to build. This unit shows a number of drawings taken from a set of stock plans (Plan No. 10372 L.F. Garlinghouse, Topeka, KS). They will be used for reference as we discuss the house plan in more detail.

Scale

Drawings must be reduced so they will fit on the drawing sheet. This must always be done in such a way that the reduced drawing is in exact proportion to the actual size. Such drawings are said to be drawn to *scale*. Residential plan views are generally drawn to 1/4" scale (1/4" = 1'-0"). This means that for each 1/4" on the plan, the building dimension will be 1'.

When certain parts of the structure need to be shown in greater detail, they are drawn to a larger scale such as 1" = 1'-0". Others, such as framing plans, are often drawn to a smaller scale (1/8" = 1'-0"). Figure 6-3 is a partial list of conventional inch-foot scales.

SCALE	WHERE USED	RATIO	RELATIONSHIP TO ACTUAL SIZE
1/32″	SITE PLANS	1:384	1/32″ = 1′0″
1/16″	PLOT PLAN	1:192	1/16″ = 1′0″
1/8″	PLOT PLAN	1:96	1/8″ = 1′0″
1/4″	ELEVATION PLANS	1:48	1/4″ = 1′0″
3/8″	CONSTRUCTION DETAILS	1:32	3/8″ = 1′0″
1/2″	CONSTRUCTION DETAILS	1:24	1/2″ = 1′0″
3/4″	CONSTRUCTION DETAILS	1:16	3/4″ = 1′0″
1″	CONSTRUCTION DETAILS	1:12	1″ = 1′0″

Figure 6-3. Common inch-foot scales used in residential drawings. Construction details may also be shown in 1 1/4" scale.

While carpenters are not responsible for making drawings, many find it helpful if they are able to make a simple sketch for a building or remodeling job. The sketch is often good enough to be used as a guide during the construction. A *pictorial sketch* is a drawing showing three dimensions much like a photograph. It is often needed so the customer can visualize the completed job, Figure 6-4.

Floor and Foundation Plans

Floor plans show the size and outline of the building and its rooms. They also give much additional information that will be useful to the carpenter and other workers in the construction trades.

Floor plans will have many dimension lines to show the location and size of inside partitions, doors, windows, and stairs. Dimension lines will be explained later.

Plumbing fixtures, as well as appliance and utility installations, can also be shown in this view. Figure 6-5 shows the floor plan of the residence shown at the beginning of this unit.

Foundation plans are similar to floor plans and are often combined with basement plans. When shown, the footings are represented as a dashed line. It is assumed that the basement floor is in place and that the grade (ground) covers the footings on the outside. See Figure 6-6.

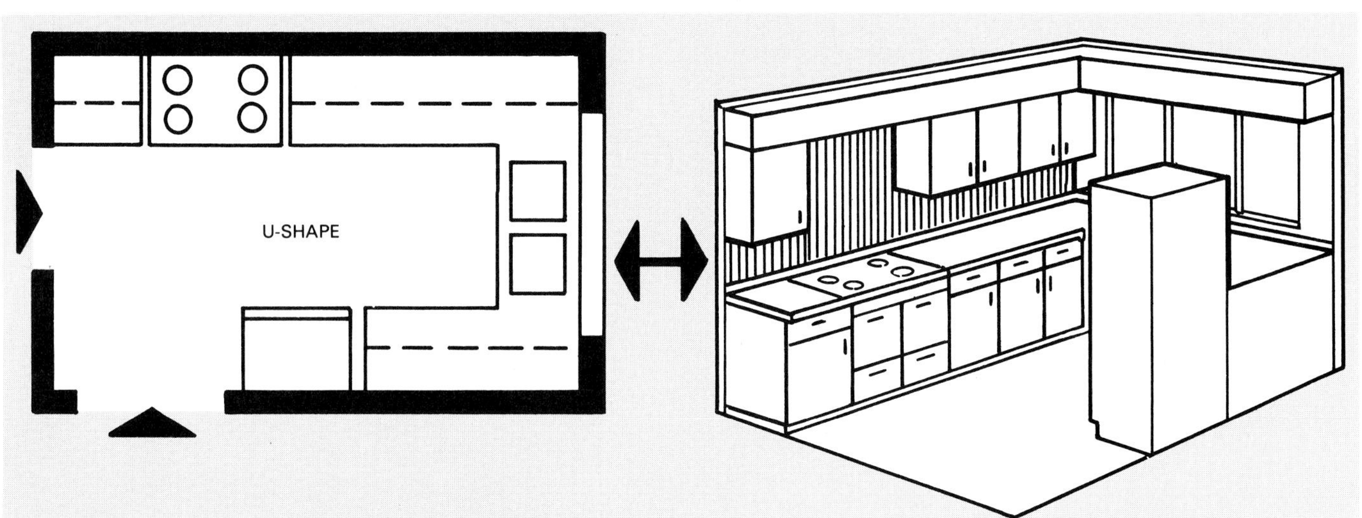

Figure 6-4. It is easier to visualize how the kitchen will look if shown in a picture-like sketch as shown at right. This can easily be done using isometric sketch paper.

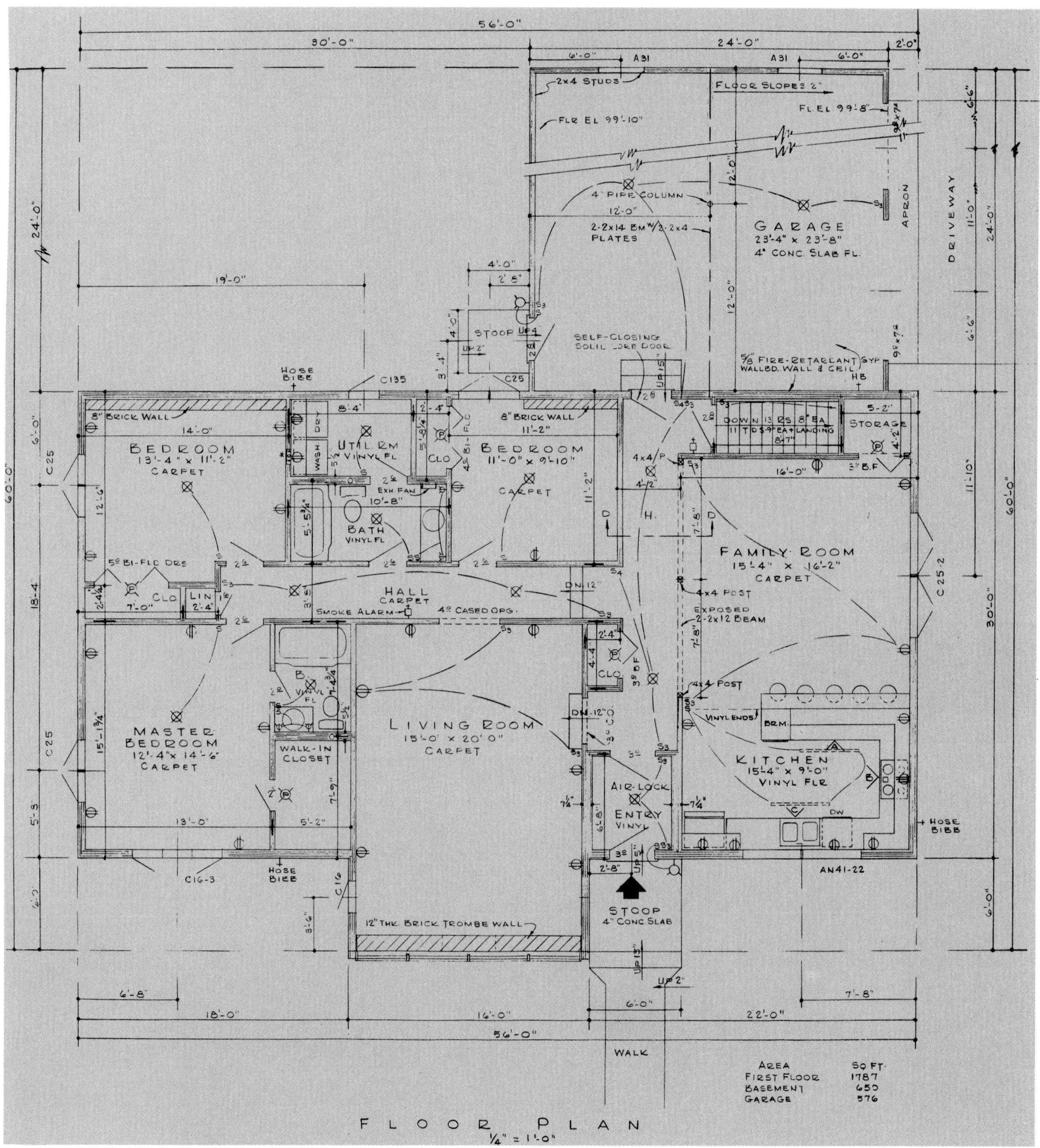

Figure 6-5. Floor plan for residence shown in the rendering at the beginning of this unit. Such plans always include shape and size of rooms, and location of plumbing, wiring, and cabinets. Original was drawn to a scale of 1/4" = 1'-0". This reproduction is much smaller.

Stoop 1 STEP
Porch more 1 STEP

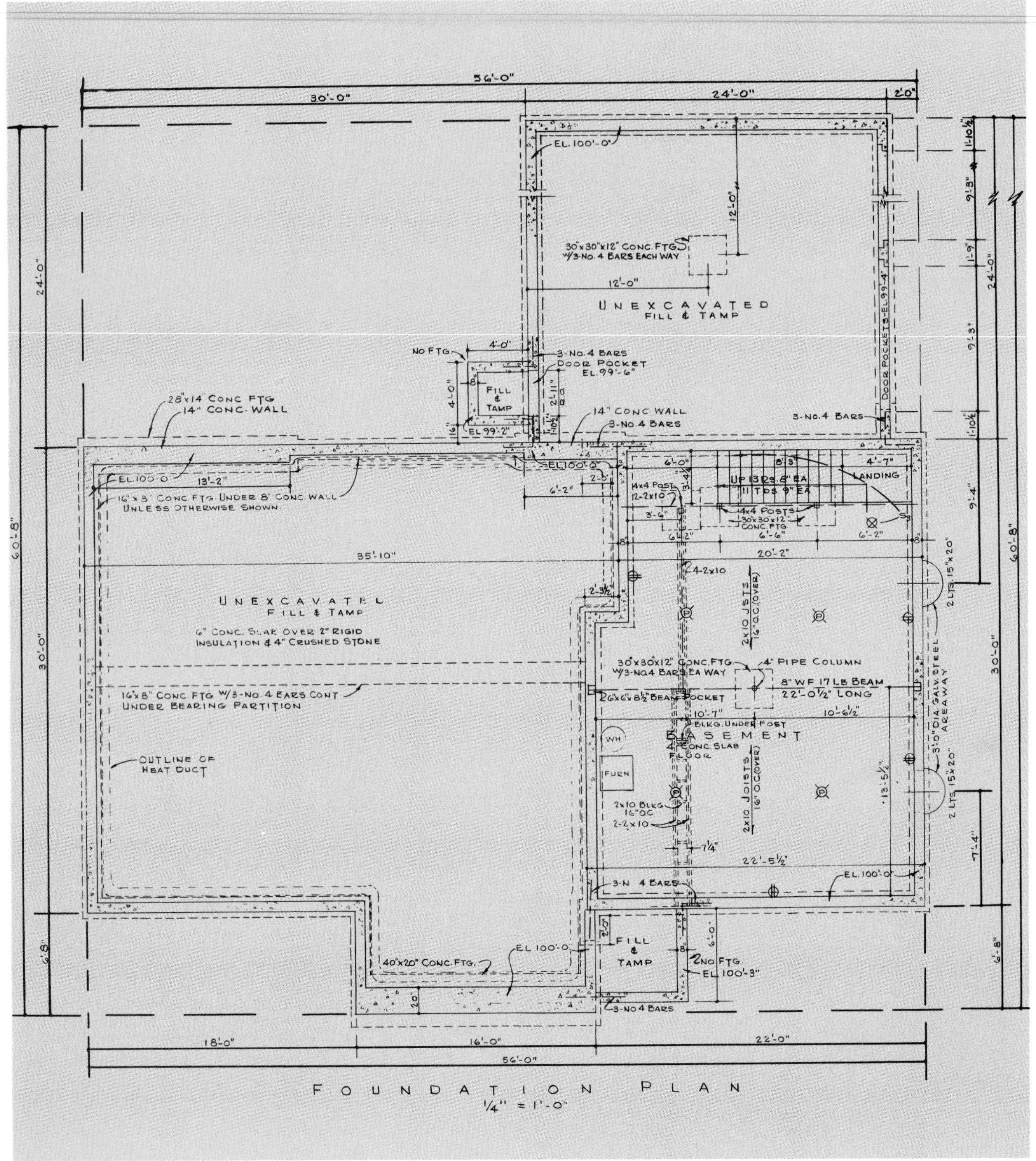

Figure 6-6. Foundation plan shows basement and footings. Dashed lines around the foundation represent the footings.

A complete set of architectural drawings usually includes a plot plan. This is a drawing of the location of the structure on the building site (lot or acreage). It includes the lot lines and the outside lines of the building. See Figure 6-7.

Elevations

Elevations are drawings showing the outside walls of the structure. These drawings are scaled so that all elements will appear in their true relationship. Generally, the various elevations are related to the site by listing them according to the direction they face. However, when plans are not designed for a certain location, the names "front," "rear," "left side," and "right side" are used. Figures 6-8 and 6-9 are typical elevation drawings. By studying the elevation views, the carpenter can determine:

- Floor levels.
- Grade lines.
- Window and door heights.
- Roof slopes.
- Kinds of materials used on wall and roof surfaces.

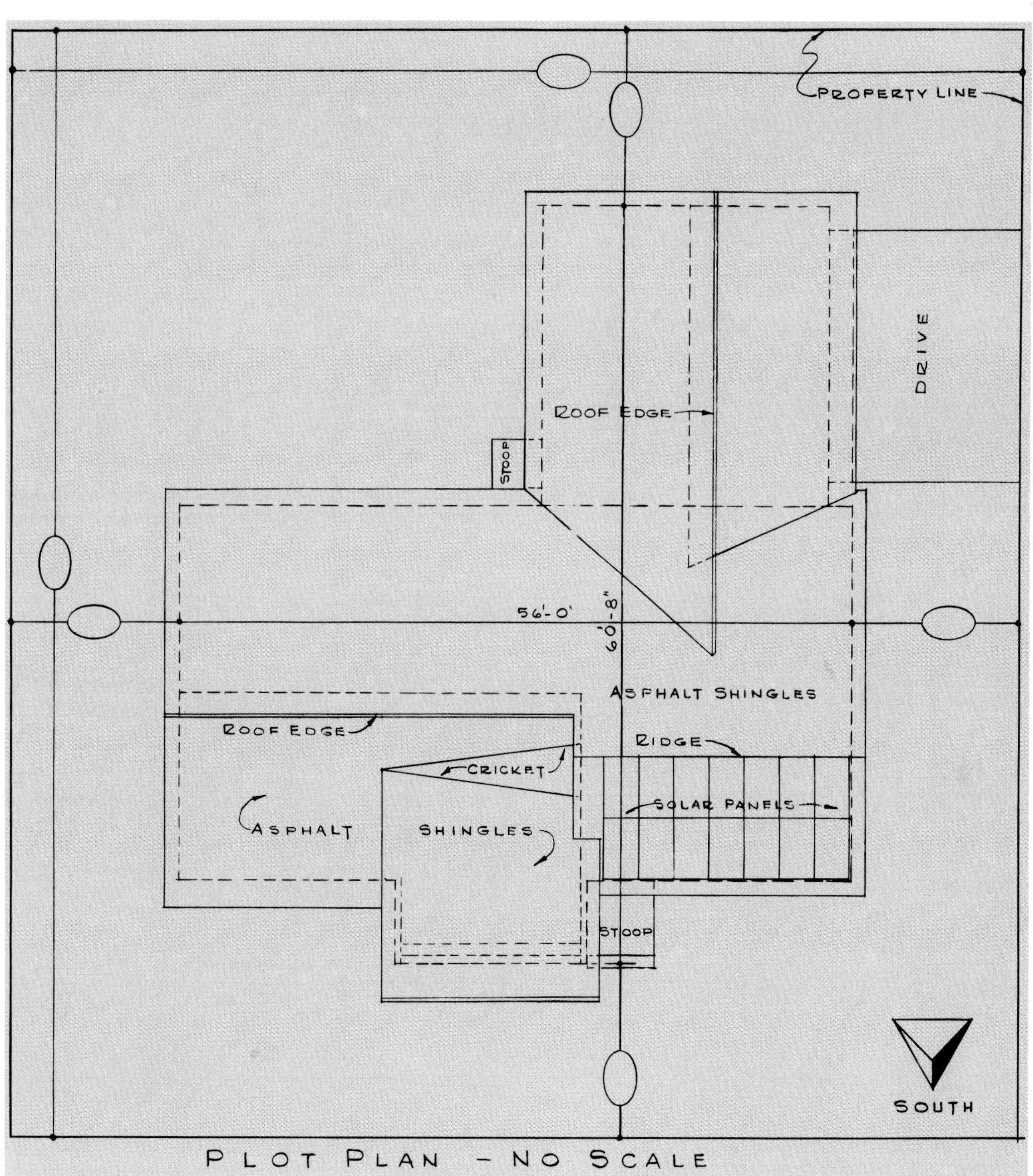

Figure 6-7. Plot plan. Stock plans leave spaces for filling in lot dimensions.

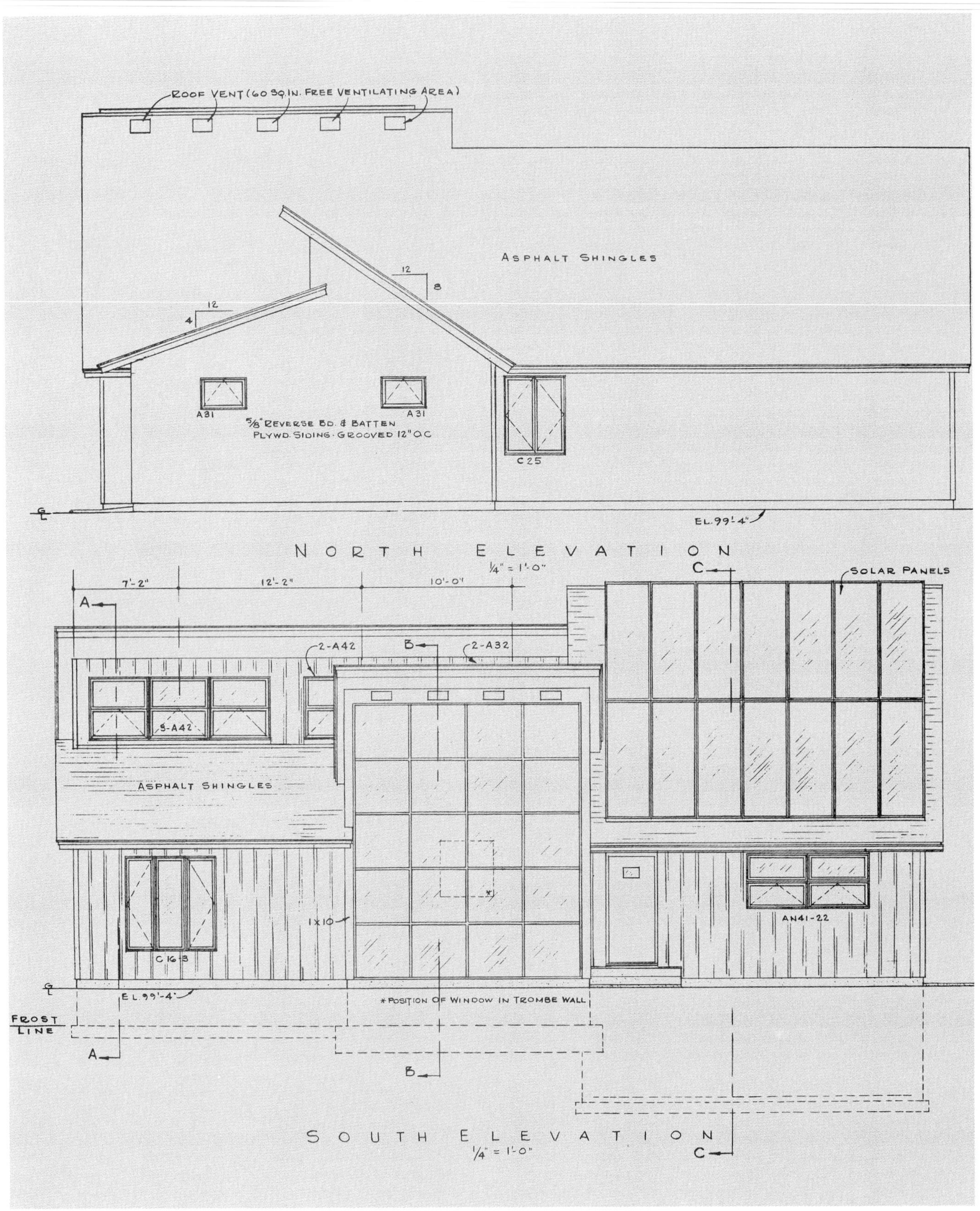

Figure 6-8. In stock plans, elevations are usually marked "front" and "back" because the architect does not know which direction the house will be facing. In this passive solar house, however, the front of the house must face south to catch the sun.

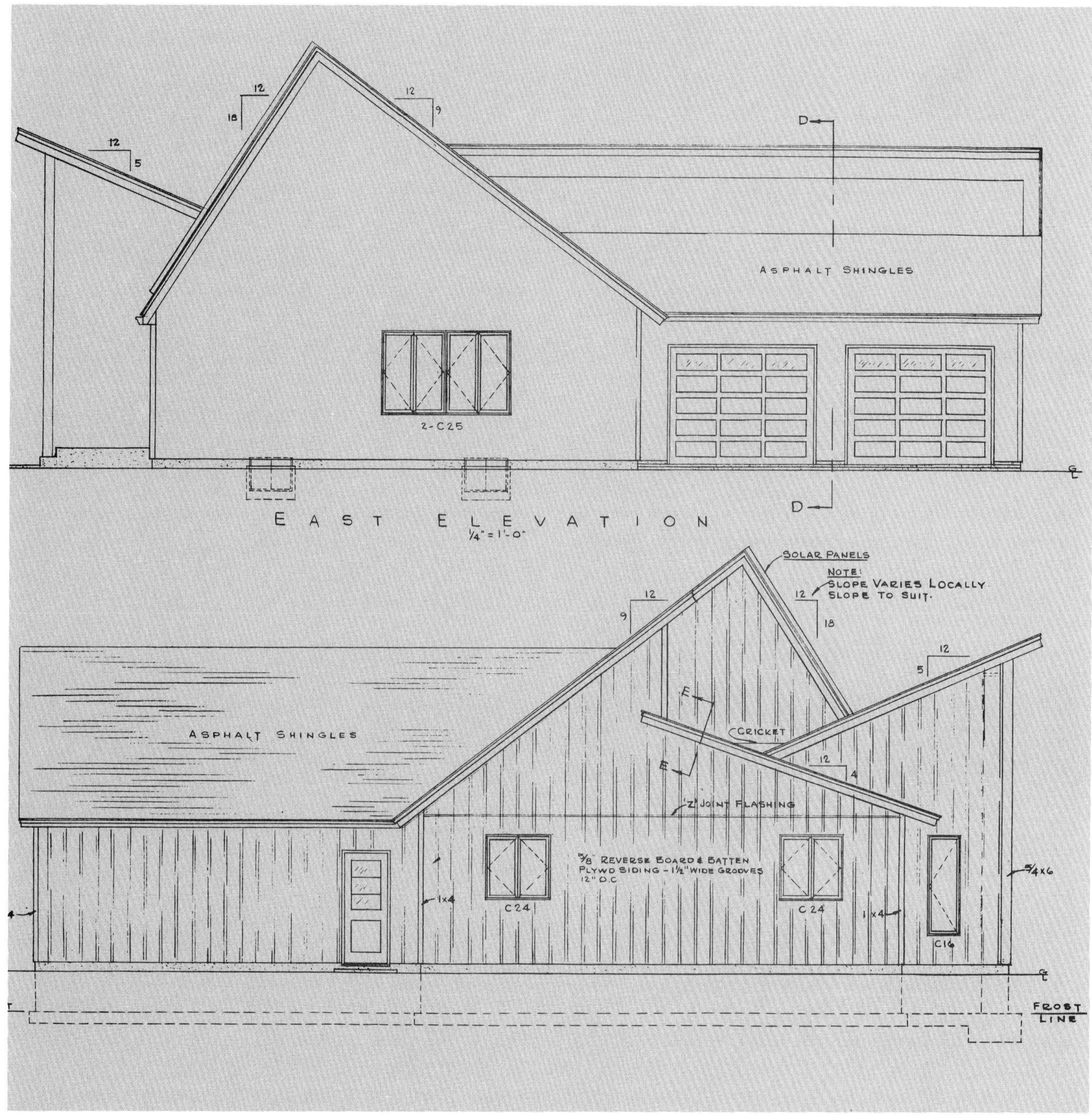

Figure 6-9. Right and left elevation drawings. All parts appear in their true relationship to one another. Note that roof pitches are given.

Framing Plans

Sometimes a house plan will also have drawings showing the size, number, and location of the structural members of a building's frame. These are known as *framing plans.* See Figures 6-10 through 6-13. Separate plans may be drawn for the floors, ceilings, walls, and roof. These plans will specify the sizes and spacing of the framing members. The members are each drawn in as they will be positioned in the building. Openings needed for chimneys, windows, and doors will be shown. Dimension lines are not used but the drawings will be made to scale as with other plan drawings.

Section and Detail Drawings

Looking at a floor plan or an elevation drawing will not show you small parts of the structure or how the parts fit into the total structure. For this, the carpenter needs to consult drawings called sections and details.

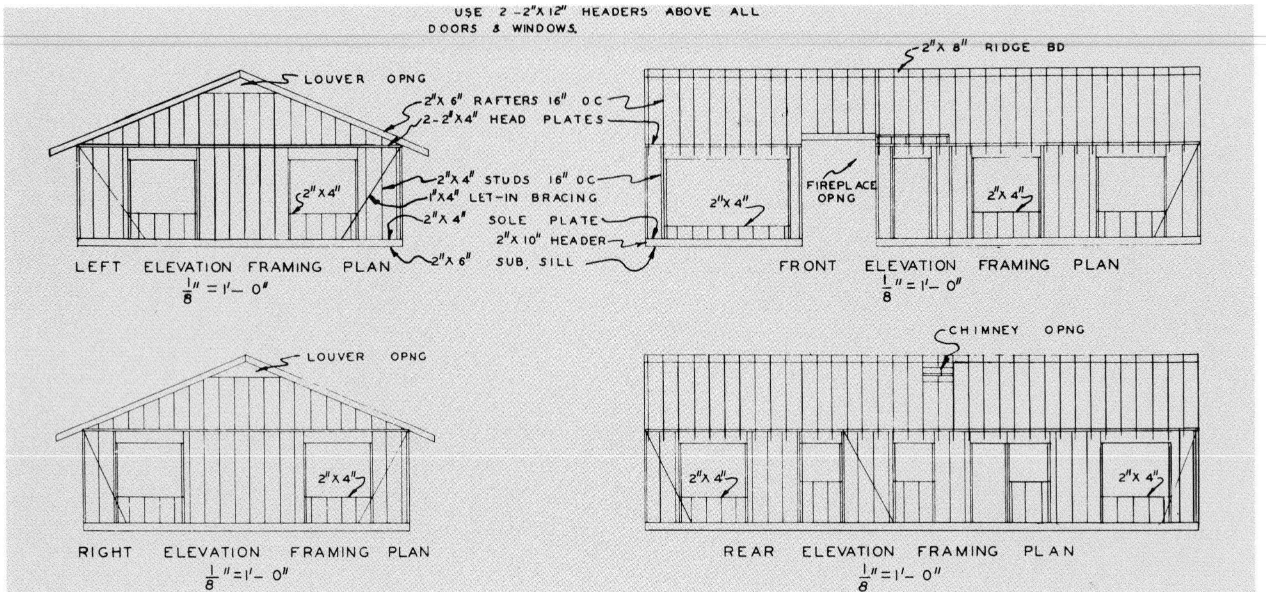

Figure 6-10. Sometimes plans will include drawings for framing. Here elevation drawings help the carpenter determine what dimension lumber is needed for studs, rafters, headers, plates, and sills.

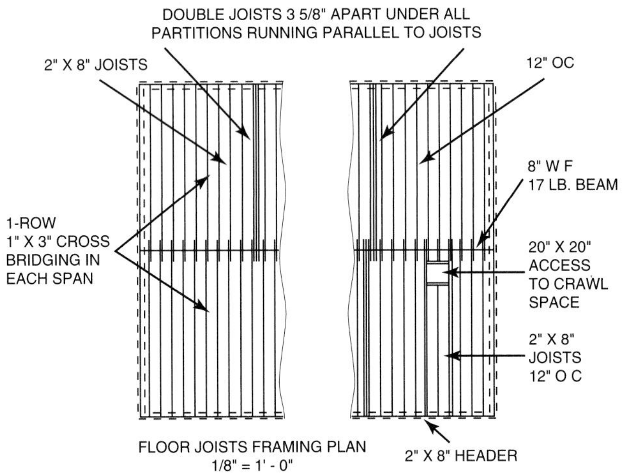

Figure 6-11. This framing plan is for a ranch home.

2"X 6" JOISTS 16" O C
20"X 20" ACCESS TO ATTIC
BEARING PARTITIONS
FIREPLACE OPNG
CEILING JOISTS FRAMING PLAN
1/8" = 1'- 0"

Figure 6-12. Framing plan for ceiling joists in a ranch home.

A *section drawing*, or *section view*, gives important information about size, materials, fastening, and support systems as well as concealed features. Parts of the structure likely to have a section drawing include walls, window and door frames, footing, and foundations.

The section shows how a part of the structure looks when cut by a vertical plane. (Imagine that you are looking at the part after it had been sawed in two and you are facing the cut edge.) Because of the need to show many details, section drawings are made to a large scale.

Figures 6-14 and 6-15 show typical sectional views. Note the attention given to details of size and material. A complicated structure may need many of these sections to show all the details of construction.

Like sections, *details* are used to show the carpenter any important and complicated construction that cannot be included in plan drawings. They are also large scale and show how various parts are to be located and connected. Fireplaces, stairs, and built-in cabinets are examples of items that will be shown in a detail. Refer to Figure 6-16 and Figure 6-17. Some detail drawings are shown full-size, Figure 6-18.

Dimensions

Dimension lines show distances and sizes. In general, all dimensions over one foot are expressed in feet and inches. For example, standard ceiling height is given as 8'-0" rather than 96".

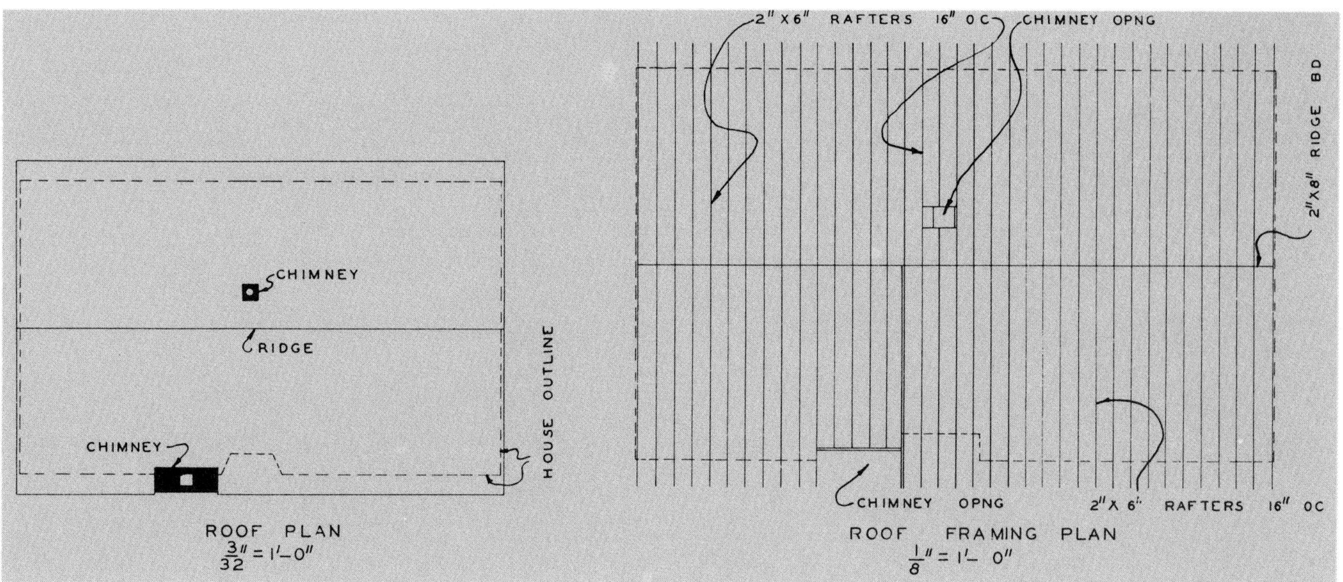

Figure 6-13. This roof and framing plan belong to the ranch home of Figures 6-10 through 6-12. Such a plan helps the carpenter visualize how the rafters must be laid out and where openings will need special framing.

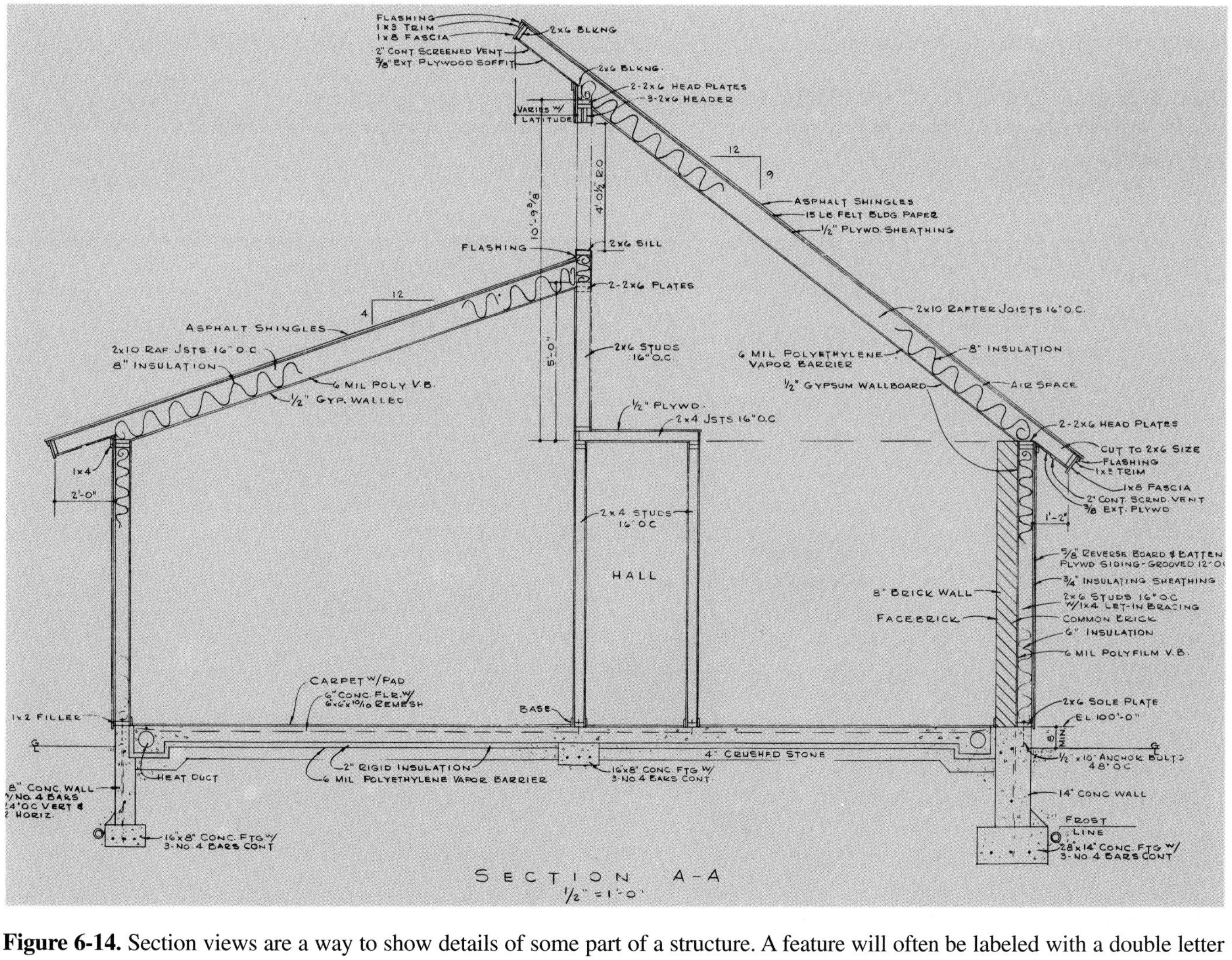

Figure 6-14. Section views are a way to show details of some part of a structure. A feature will often be labeled with a double letter (A-A, for example) which refers to a cutting line in a larger drawing. This one refers to the south elevation drawing in Figure 6-8.

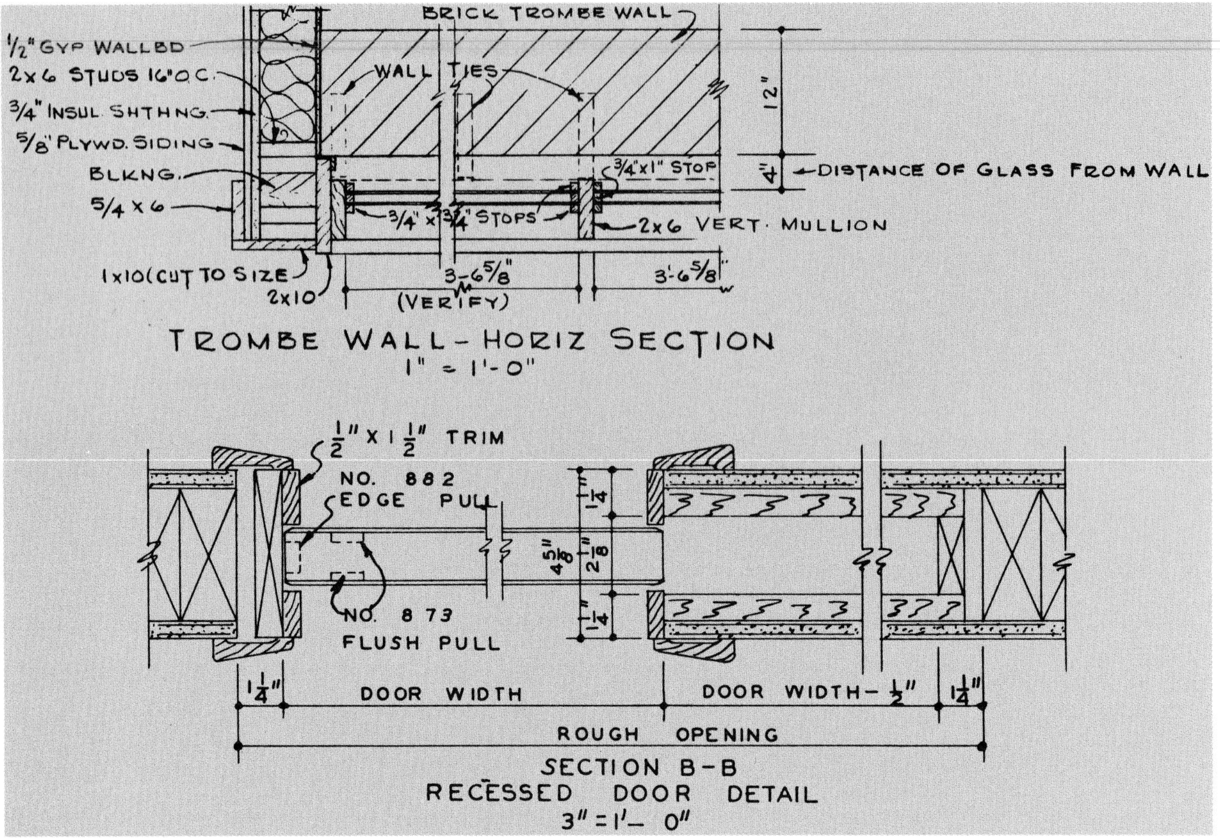

Figure 6-15. Two sample section views. Top–Trombe wall cross section is shown for structural details. (In solar houses, Trombe walls are heavy masonry walls built behind windows to catch and store heat from the sun.) Bottom–Section view of a pocket door.

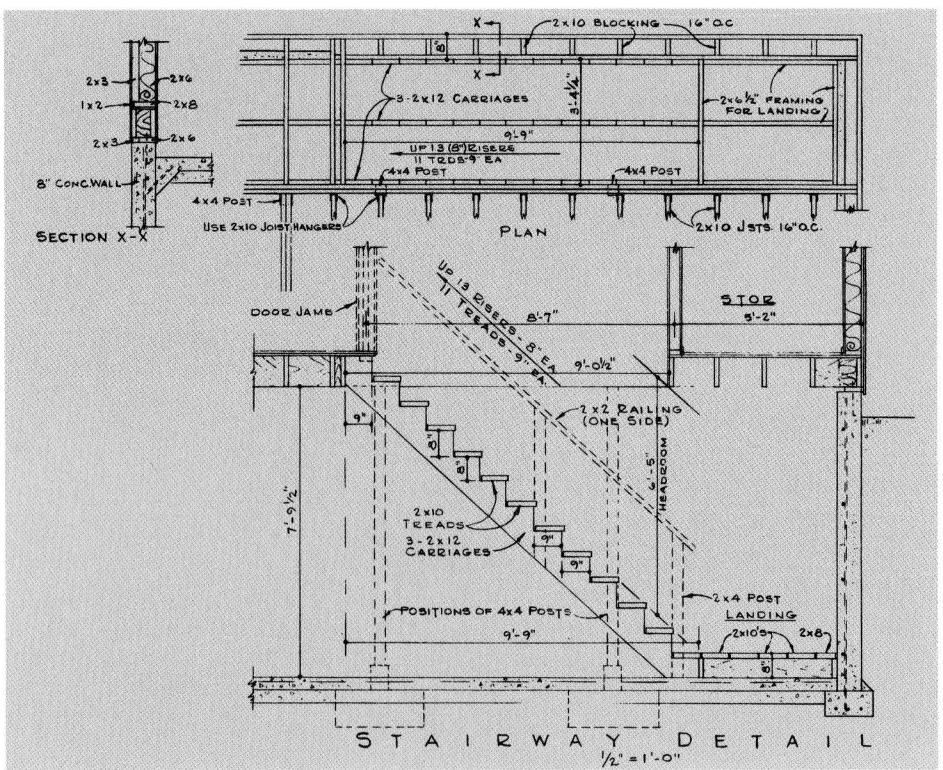

Figure 6-16. More details may need to be given of complicated parts of the construction.

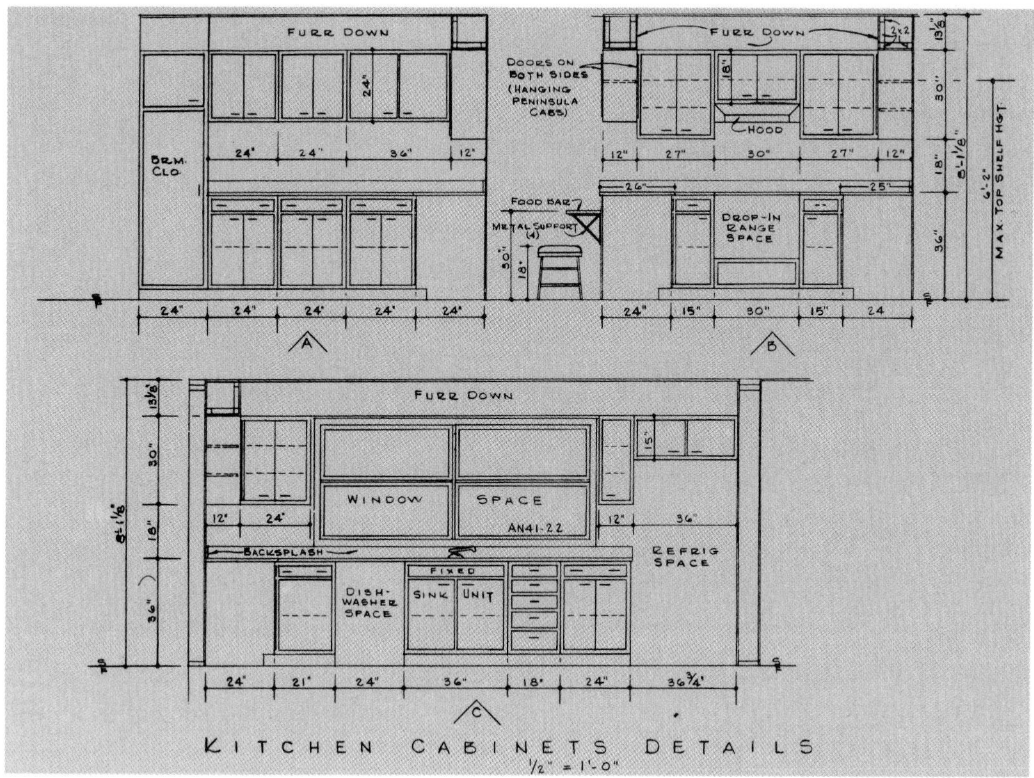

Figure 6-17. Detail drawing of kitchen cabinets. Note the amount of detail given on dimensions.

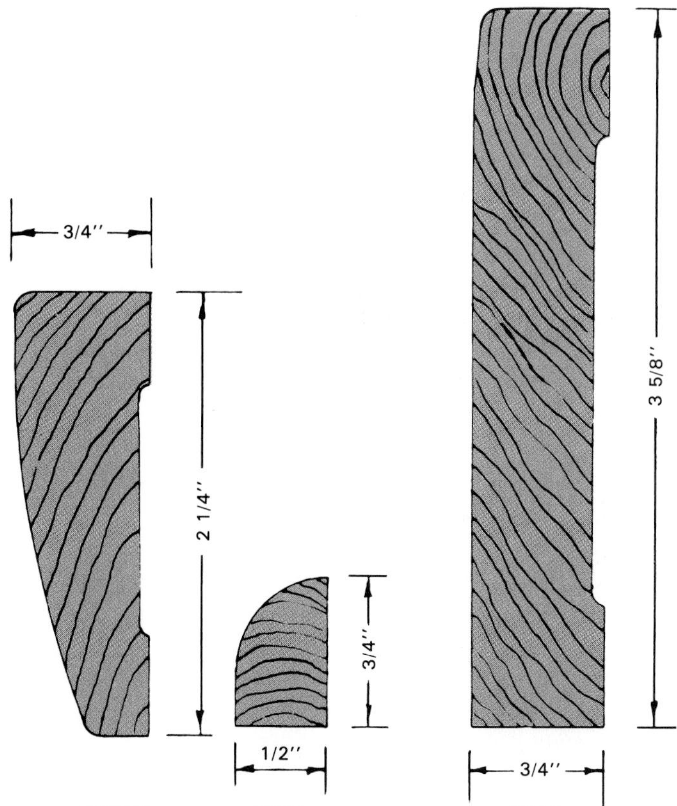

Figure 6-18. Some detail drawings are made full size. The above are cross sections of special millwork needed for finishing a house.

Carpenters using conventional measure prefer to work with feet and inches, since measurement in this form is easier to visualize. When laying out various distances, they often need to add or subtract dimensions. Steps for making calculations are:

Addition
6'-8"
+4'-6"
+2'-4"
+1'-2"
=13'-20" or 14'-8"

Subtraction
8'- 4"
–6'-10"

(since 10 cannot be subtracted from 4, borrow 12" from 8')

Thus:
7'-16"
–6'-10"
= 1'-6"

Lists of Materials

Sets of plans will also include a materials list, Figure 6-19. It is known by other names as well: *bill of materials, lumber list,* or *mill list*. Whatever the list is called, it will include all of the materials and assemblies needed to build the structure.

A materials list usually will include the number of the item, its name, description, size, and the material of which the item is made. Built-in items such as cabinets will be included.

Another part of the materials list is the window and door *schedule*. It gives the quantity needed, sizes of the rough openings, and descriptions. Sometimes the manufacturer is specified. Figure 6-20 shows a door schedule taken from a materials list.

Symbols

Since architectural plans are drawn to a small scale, materials and construction can seldom be shown as they actually appear. Also it would require too much time to produce drawings of this nature. The architect, therefore, uses symbols to represent materials and other items and certain approved short-cuts (called conventional representations). These simplify the illustration of assemblies and other elements of the structure. Generally accepted symbols are illustrated in Figures 6-21 through 6-24.

Abbreviations are commonly used on plans to save space. Refer to the Technical Information section for a listing of common abbreviations.

How to Scale a Drawing

Architectural plans include dimension lines that show many distances and sizes, but carpenters may require a dimension that is not shown. To get this dimension, the carpenter will need to scale the drawing. An architect's scale, Figure 6-25, may be used for this purpose. Each division of an architect's scale represents 1'. The foot division is divided into 12 major parts. Each is equal to 1". Another way to scale a plan is to use a regular folding rule. Can you calculate the distance as shown in Figure 6-26?

Changing Plans

Minor changes in plans, desired by the owner as the job progresses—such as changing the size or location of a window or making a revision in the design of a built-in cabinet—can usually be handled by the carpenter. Sketches or notations should be recorded on each set of plans so there will be no misunderstandings.

Major changes such as the relocation of a loadbearing wall or stairs may generate a chain reaction of problems and should be undertaken only after the necessary plan changes have been made by an architect and approved by the owner.

Specifications

Although the working drawings show many of the requirements for a structure, certain supplementary information is best presented in written form or *specifications* (commonly called *specs*). See Figure 6-27. The carpenter should check and carefully follow the specs.

Headings generally included in specifications for a residential structure include:

- General Requirements, Conditions, and Basic Information.
- Excavating and Grading.
- Masonry and Concrete Work.
- Sheet Metal Work.
- Rough Carpentry and Roofing.
- Finish Carpentry and Millwork.
- Insulation, Caulking, and Glazing.
- Lath and Plaster or Drywall.
- Schedule for Room Finishes.
- Painting and Finishing.
- Tile Work.
- Electrical Work.
- Plumbing.
- Heating and Air Conditioning.
- Landscaping.

Under each heading the content is usually divided into the following sections: scope of work, specifications of materials to be used, application methods and procedures, and guarantee of quality and performance.

Carefully prepared specifications are valuable to the contractor, estimator, tradespeople, and the building supply dealer. They protect the owner and help to ensure quality work. In addition to the items previously described, the specifications may include information and requirements regarding building permits, contract payment provisions, insurance and bonding, and provisions for making changes in the original plans.

Modular Construction

The *modular construction* concept is based on the use of a standard grid divided into 4" squares. See the grid in Figure 6-28. Actually, each individual square (module) should be considered to be the base of a cube so it can be applied to elevations as well as horizontal planes.

All dimensions are based on multiples of 4", including 16", 24", and 48". The last two dimensions are sometimes called the minor and major module. Many building materials and fabricated units are manufactured to coordinate with modular dimensions. This helps to eliminate costly cutting and fitting during construction. A good example of this system is illustrated in standard concrete blocks. They are manufactured in nominal sizes of 8 x 8 x 16". The actual size is 3/8" less in each dimension to allow for bonding (mortar joints). See Unit 7.

Modular dimension standards for components have been developed by the National Lumber Manufacturers Association. The system is called *Unicom* (uniform manufacture of components). The Unicom system helps to make

MATERIALS LIST
PLAN NO. 10372

4900	Face bricks for interior walls
2950	Common bricks for interior walls (bedrooms) & center of trombe wall
7 1/4	Cu. yds. concrete for basement floor slab
20	Cu. yds. concrete for 1st. floor slab (6" thick)
7 1/4	Cu. yds. concrete for garage floor slab and apron
1	Cu. yd. concrete for entrance platforms and steps
14	Cu. yds. concrete for footings
45 3/4	Cu. yds. concrete for foundation walls
224	Lin. ft. 4" plastic drainage tubing
45	Tons 3/4" to 1" crushed stone -- under slabs & around drainage tubing

STRUCTURAL STEEL

1	8WF17 steel beam 22'-01/2" long
1	4" steel pipe columns 7-4 long with plates
1	4" steelpipe columns 8'-3" long with plates (verify)
8	3/8" x 11" x 24" steel plate lintels over trombe wall openings
1	Complete gas vent
3150	Lin. ft. No. 4 reinforcing bars
2500	Sq. ft. 6" x 6" x 10/10 gauge reinforcing mesh
2	Galvanized steel areaways 36" diameter 24" high
64	1/2" anchor bolts 10" long
8	16 x 8 screened foundation vents w/dampers

CARPENTER'S LUMBER

**Note - Recommended framing lumber shall have a mimimum allowable extreme fiber bending stress of 1450 PSI and a mimimum modulus of elasticity of 1,400,000 PSI, except rafter joists over living room have F_b of 1500 and a E of 1,400,000.

Design Live Loads - Floors -- 40 PSF
Ceilings -- 20 PSF
Roof -- 40 PSF

First Floor Joists & Headers

5	2 x 10 x 8
19	2 x 10 x 12
21	2 x 10 x 14
2	2 x 10 x 16

Sub Sills

1	2 x 6 x 10
3	2 x 6 x 12
1	2 x 6 x 14
2	2 x 6 x 16
2	2 x 6 x 20
1	2 x 3 x 16

Figure 6-19. A materials list includes all of the materials and assemblies, such as doors, windows, and cabinets, that will be used in the building. (Only part of the list is shown.)

Doors:	Frames	Openings	Jambs
1	Outside entrance	3-0 x 6-8 x 1 3/4	
1	Outside solid core service	2-8 x 6-8 x 1 3/4	
1	Outside service	2-8 x 6-8 x 1 3/4	
1	Inside	2-8 x 6-8 x 1 3/8	
5	Inside	2-6 x 6-8 x 1 3/8	
2	Inside	2-0 x 6-8 x 1 3/8	
1	Inside	1-6 x 6-8 x 1 3/8	
2	Garage	9-0 x 7-0 x 1 3/8	
1	Combination	2-8 x 6-8 x 1 1/8	
1	Combination	3-0 x 6-8 x 1 1/8	
1	Bi-fold door	5-0 x 6-8 x 1 3/8	
1	Bi-fold door	4-0 x 6-8 x 1 3/8	
2	Bi-fold doors	3-0 x 6-8 x 1 3/8	
2	Sides of door trim	5-0 x 6-8, 1/2 x 2 1/4	
4	Sides of door trim	4-0 x 6-8, 1/2 x 2 1/4	
7	Sides of door trim	3-0 x 6-8, 1/2 x 2 1/4	
5	Sides of door trim	2-8 x 6-8, 1/2 x 2 1/4	
10	Sides of door trim	2-6 x 6-8, 1/2 x 2 1/4	
4	Sides of door trim	2-0 x 6-8, 1/2 x 2 1/4	
2	Sides of door trim	1-6 x 6-8, 1/2 x 2 1/4	

Windows: All *Andersen Perma-Shield casement and awning windows are to be complete with frames, sash, interior trim, exterior trim, screens, storm sash and hardware.

Quan.	No.	
1	C16-3	Perma-shield casement window
5	C25	Perma-shield casement window
1	C16	Perma-shield casement window
1	C135	Perma-shield casement window
1	AN41-22	Perma-shield awning window
5	A42	Perma-shield awning window
2	A32	Perma-shield awning window
2	A31	Perma-shield awning window

*Andersen Corporation
Bayport, MN 55003

2 Steel basement units 2 lites 15 x 20 with screens

Custom Window Material - Trombe Wall

50	Lin. ft. 2 x 10 Surround material
15	Lin. ft. 2 x 8 Surround material
15	Lin. ft. 2 x 4 Blocking
165	Lin. ft. 2 x 6 Mullion Material
588	Lin. ft. 3/4" x 3/4" Stop material
294	Lin. ft. 3/4" x 1" Stop material
36	16 Ga 1 1/2" x 8" Galvanized wall ties
500	Sq. ft. 1/4" Plate glass

Figure 6-20. Part of the door and window schedule from a materials list. Sometimes a manufacturer will be specified.

MATERIAL	PLAN	ELEVATION	SECTION
WOOD	FLOOR AREAS LEFT BLANK	SIDING PANEL	FRAMING FINISH
BRICK	FACE COMMON	FACE OR COMMON	SAME AS PLAN VIEW
STONE	CUT RUBBLE	CUT RUBBLE	CUT RUBBLE
CONCRETE			SAME AS PLAN VIEW
CONCRETE BLOCK			SAME AS PLAN VIEW
EARTH	NONE	NONE	
GLASS			LARGE SCALE SMALL SCALE
INSULATION	SAME AS SECTION	INSULATION	LOOSE FILL OR BATT BOARD
PLASTER	SAME AS SECTION	PLASTER	STUD LATH AND PLASTER
STRUCTURAL STEEL		INDICATE BY NOTE	
SHEET METAL FLASHING	INDICATE BY NOTE		SHOW CONTOUR
TILE	FLOOR	WALL	

Figure 6-21. Symbols are used to represent things that are impractical to draw.

Reinforced

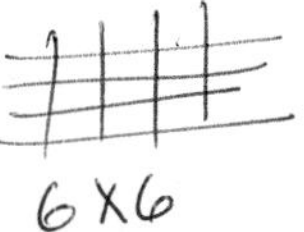

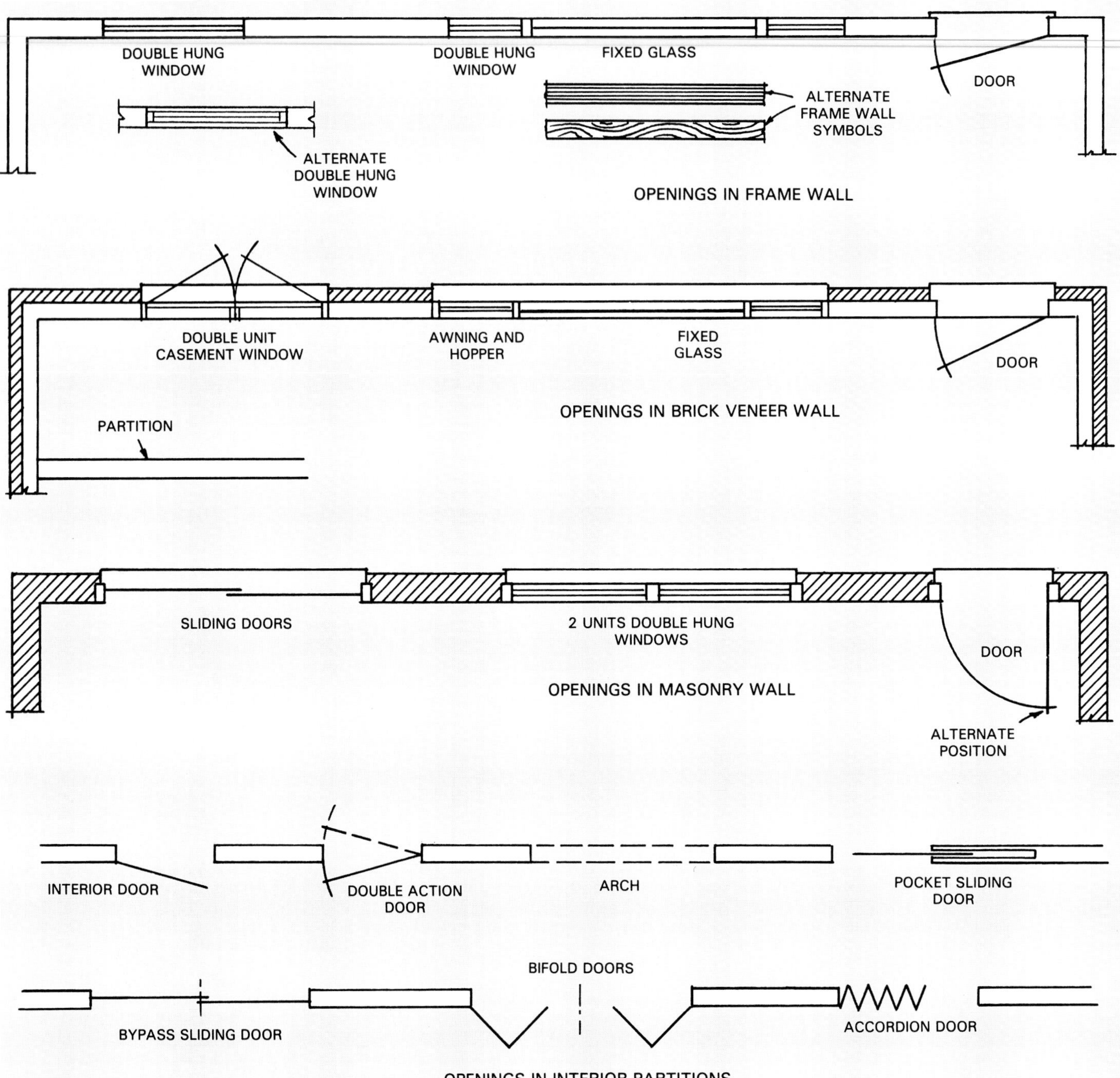

Figure 6-22. How openings in walls are presented in plan views. Handedness of doors (which way they swing), is always shown.

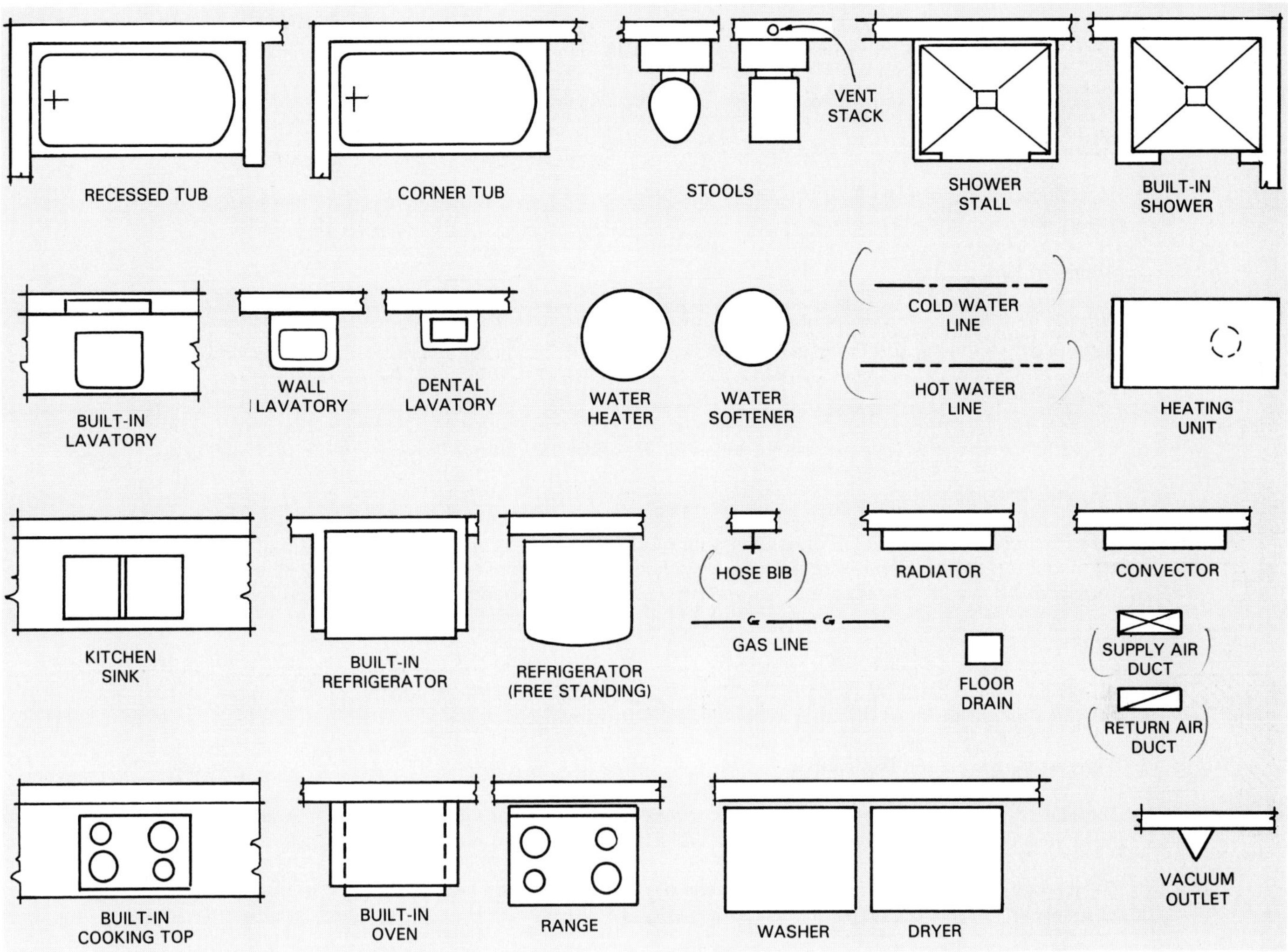

Figure 6-23. These symbols are used to indicate plumbing fixtures, appliances, and mechanical equipment.

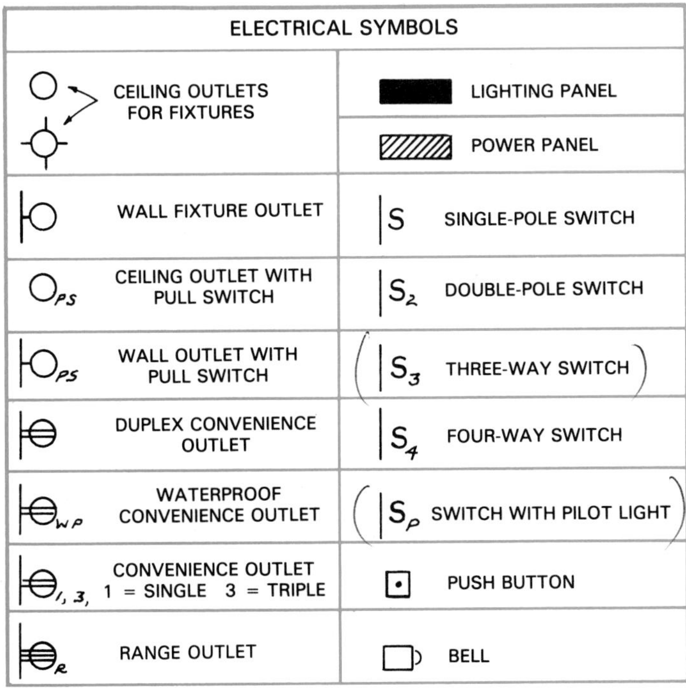

ELECTRICAL SYMBOLS	
CEILING OUTLETS FOR FIXTURES	LIGHTING PANEL
	POWER PANEL
WALL FIXTURE OUTLET	S SINGLE-POLE SWITCH
O_{PS} CEILING OUTLET WITH PULL SWITCH	S_2 DOUBLE-POLE SWITCH
O_{PS} WALL OUTLET WITH PULL SWITCH	S_3 THREE-WAY SWITCH
DUPLEX CONVENIENCE OUTLET	S_4 FOUR-WAY SWITCH
WP WATERPROOF CONVENIENCE OUTLET	S_P SWITCH WITH PILOT LIGHT
1, 3, CONVENIENCE OUTLET 1 = SINGLE 3 = TRIPLE	PUSH BUTTON
R RANGE OUTLET	BELL

Figure 6-24. Some electrical symbols. Carpenters will not be responsible for wiring; they should recognize the symbols.

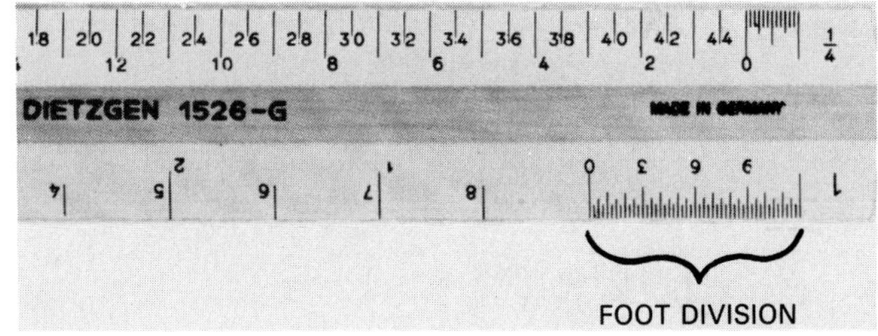

Figure 6-25. Architect's scale may be used to check the dimensions on a scaled drawing. Each division represents 1'. Fractions of a foot are checked against the foot division.

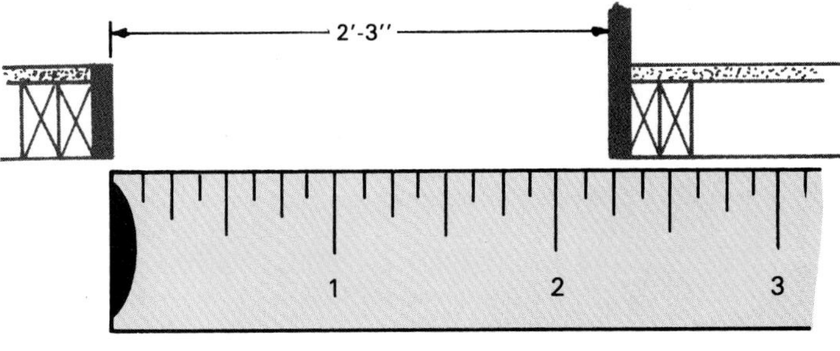

Figure 6-26. How to use a folding rule to scale a plan.

PLAN NO. 12275

DIVISION 5

CARPENTRY, MILLWORK, & HARDWARE

General Conditions:

This contractor shall read the General Conditions and Supplementary General Conditions which are a part of these Specifications.

Scope of Work:

Furnish and install all rough lumber, millwork, rough and finished hardware, including all grounds, furring, and blocking, frames and doors, etc., as shown on the drawings or hereinafter specified.

Rough Lumber:

All lumber used for grounds, furring, blocking, etc., shall be #1 Common Douglas Fir or Pine.

Grounds and Blocking:

Furnish and install suitable grounds wherever required around all openings for nailing metal or wood trim such as casings, wainscoting, cap, shelving, etc.; also building into masonry all blocking as required. All wood grounds must be perfectly true and level and securely fastened in place.

Doors:

Interior doors to be plain sliced red oak for stain finish. Solid core doors to be laminated 1 3/4'' thick, or as called for in the door schedule. See plan for Label doors.

Temporary Doors and Enclosures:

Provide temporary board enclosures and batten doors for all entrance openings. Provide suitable hardware and locks to prevent access to work by unauthorized persons.

At any openings used for ingress and egress of material, provide and maintain protection at jambs and sills as long as openings are so used.

Rough Hardware:

[illegible]rdware such as nails, spikes, anchors, bolts, rods, etc., in connection with rough car-
All finishing [illegible] by this contractor. Use aluminum non-staining nails for all exterior wood.
around openings shall be [illegible]
edges of trim shall be lightly sandpa[illegible]
smoothed and machine sanded at the factory, and [illegible]
carpenters at the building in a first class manner before being set [illegible]
applied.

Finishing Hardware:

An allowance of $1,000.00 shall be included for finishing hardware, which will be selected by the architect and installed by this contractor. Locksets shall match existing design and finish. Adjustment of cost above or below this allowance shall be made according to authentic invoice for hardware for this building from hardware supplier. Hardware supplier shall furnish schedule and templates to frame manufacturer.

Caulking:

Caulk all exterior windows, doors, etc. and openings throughout building with polysulfide based sealant, see specifications under cut stone.

Wood Paneling:

All wood paneling shall be Weyerhauser Forestglo, Orleans Oak 1/4'' prefinished. Apply over 1/2'' drywall on steel studs and furring.

Figure 6-27. One sheet of the specifications for the solar house rendering shown at the beginning of this unit. Other topics included in specifications are plumbing, heating, and wiring.

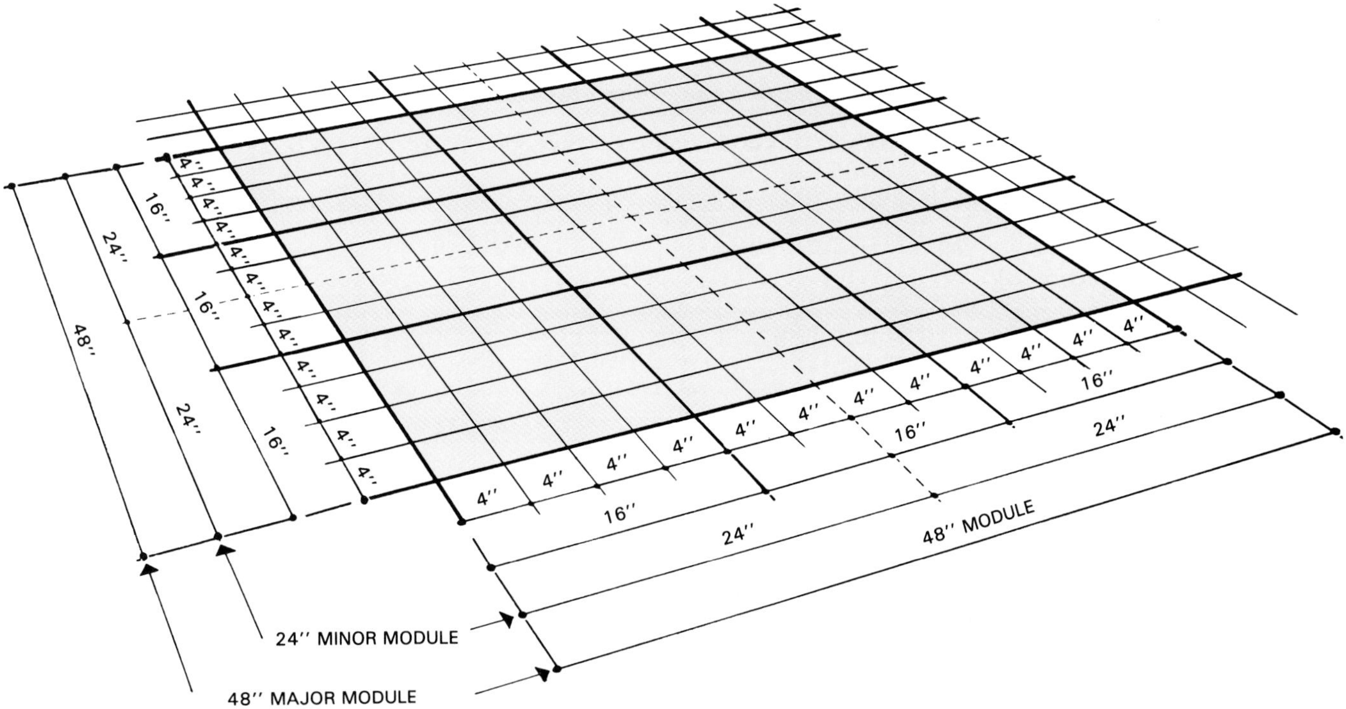

Figure 6-28. Plans for modular structures, to be built with components or by conventional framing methods, are designed to exact grid sizes. (National Forest Products Assoc.)

it possible to apply modern mass-production methods to building construction. For example, a sheet of plywood fits the module exactly at 48" x 96".

A modular system also exists in the SI metric system. It is based on a grid made up of 100 mm squares.

The 100 mm module is about, but not exactly, 4". The ISO standard also recommends that the submultiples of 25, 50, and 75 mm be used as well as the multiples of 300, 400, 600, 800, and 1200. The 600 and 1200 multiples become the minor and major modules. See the reference to metrics in the Technical Information section.

Computers are now frequently used to create a set of plans and even perspective drawings for houses. This is known as *computer-aided drafting and design* (CADD). Working with a variety of architectural software, the drafter creates drawings on the computer screen. Then, with a command to the computer, the drawings are transported to a printer or plotter to be duplicated on paper. Unlike a printer, a plotter has pens of various color to produce a drawing in several colors.

An advantage of CADD is its ability to revise a drawing quickly without the erasures necessary when the drawing is first made on paper. Additional software is available that will "read" the drawings and produce materials lists.

Metric Measurement

While the United States still uses the customary measurement system inherited from England, most nations, including Canada, use the SI metric system of measurement. Since the U.S. is moving toward the metric system, carpenters should be familiar with it. The entire metric system is modular, being based on units of ten. Thus, the millimeter is only 1/1000 of a meter and a centimeter is 1/100 of a meter. A kilometer is equal to 1000 meters. Countries that use metric measure commonly use 300 millimeters as a standard module. This is about 1'.

In architectural drawings, measurements of buildings are given in millimeters, while site measurements are in meters.

Building Codes

A building code is a collection of laws listed in booklet form that apply to a given community. The code covers all important aspects of the erection of a new building and also the alteration, repair, and demolition of existing buildings. The basic purpose is to provide for the health, safety, and general welfare of the occupants of the home being built and other people in the community.

Every carpenter should become thoroughly familiar with local building codes. Work which does not conform must be done over and can add considerably to the expense of construction. See Figure 6-29.

A code sets the minimum standards that are acceptable in a community for design, quality of materials, and quality of construction. It also sets requirements concerning such design factors as:

- Sizes, heights, and bulk of buildings.

13.1.2 Required Room Sizes

No dwelling unit shall be erected or constructed which does not comply with the following minimum room sizes:

(a)	Living Room:	250 sq. feet
(b)	Dining Room:	100 sq. feet
(c)	Kitchen:	90 sq. feet
(d)	Bath Room (Three Fixtures)	40 sq. feet
(e)	Powder Room (Two Fixtures)	24 sq. feet
(f)	First Bedroom:	150 sq. feet
(g)	Each Additional Bedroom:	110 sq. feet
(h)	Den or Library, etc.	100 sq. feet
	(Permitted only when dwelling unit includes two (2) or more bedrooms of the required sizes.)	
(i)	Garage:	253 sq. feet
	(Minimum dimensions shall be 11 feet x 23 feet.)	

13.1.3 Required Rooms

No dwelling unit shall be erected or constructed which does not contain one each of the following rooms: Living Room, Dining Room, Kitchen, Bath Room, Bedroom and Garage. Each room is to be separated from each other room by full height partitioning and doors, except that Living Room, Dining Room and Kitchen need not be fully separated from each other, and the Garage may be a detached building. Required rooms and all additional bedrooms, den and library shall not be located in a basement.

Figure 6-29. Sample of a local building code. Codes will vary from community to community. (Village of Flossmoor, Illinois)

- Room sizes.
- Ceiling heights.
- Lighting and ventilation.

Many local codes contain detailed directions regarding installation of a building's systems. These instructions govern the methods and materials used in installing plumbing, wiring, and heating systems.

A carpenter should be aware that building codes may vary from community to community. What is common practice in one area may not be allowed in another. For example, some communities' codes will require one or more sumps in basement floors and perimeter drainage around the footings. Another community may not mention them. These differences should be carefully noted.

Some items in a code necessarily must be adjusted to local conditions. In northern climates, footings need to be deeper than in southern states; structures in hurricane belts require extra bracing.

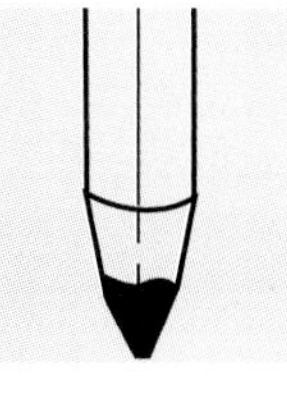

In doing any kind of carpentry work, the importance of closely following all building codes applicable to the job cannot be overemphasized.

Model Codes

Modern research and development have resulted in so many improvements in building construction that it now becomes a tremendous task to prepare and continually update building codes. Because of this, many communities have adopted model codes. Today, four major organizations provide a service of this nature.

The Uniform Building Code, published by the International Conference of Building Officials, is widely accepted. This organization provides annual revisions and the entire code is republished every three years. A short form is available which covers buildings not over two stories in height and containing less than 6000 sq. ft. of ground floor area.

Another organization, the Building Officials and Code Administrators International, Inc., has developed the BOCA–Basic Building Code. An abridged form, designed for residential construction, is revised every three years.

One of the first model codes was introduced by The American Insurance Association (successor to the National Board of Fire Underwriters). This publication is now known as the National Building Code. An abbreviated edition is also available.

A model building code called the Standard Building Code, used in southern states, covers problems in this region. It is prepared under the direction of the Southern Building Code Congress International, Inc.

In addition to building codes adopted by the local community (cities, towns, counties), the carpenter must be informed of certain laws at the state level that govern buildings. Several states have developed building codes for adoption by their local communities. However, for the most part, state codes deal mainly with fire protection and special needs for public buildings.

A carefully prepared and up-to-date code is not sufficient in itself to insure safe and adequate buildings. All codes must be properly administered by officials that are experts in the field. Under these conditions, the owner can be assured of a well-constructed building and the carpenter will be protected against the unfair competition of those who are willing to sacrifice quality for an excessive margin of profit.

Communities have inspectors who enforce the building code. They will make periodic inspections during construction or remodeling. The inspectors are persons who have worked in the construction trades or who are otherwise knowledgeable about construction.

Standards

Building codes are based on standards developed by manufacturers, trade associations, government agencies, professionals, and tradespeople, all of whom are seeking a desirable level of quality through efficient means. A particular material, method, or procedure is technically described through specifications. Specifications become standards when their use is formally adopted by broad groups of manufacturers and builders and/or recognized agencies and associations.

Organizations devoted to the establishment of standards, many of which are directly related to the field of construction, include:

- The American Society for Testing and Materials (ASTM).
- American National Standards Institute (ANSI).
- Underwriters Laboratories, Inc. (UL).

Commercial standards are developed by the Commodity Standards Division of the U.S. Department of Commerce. The chief purpose of the agency is to establish quality requirements and approved methods of testing, rating, and labeling. These standards are designated by the initials CS, followed by a code number and the year of the latest revision.

Building Permits and Inspections

Steps for securing *building permits*, Figure 6-30, will vary from one community to another. Usually, the contractor or building owner will file a formal application with the village or city clerk. The application, Figure 6-31, with one or two sets of plans is given to the clerk. Usually the drawings submitted must include:

- Floor plans.
- Specifications.
- Site plan.
- Elevation drawings.

Sometimes a filing fee and plan review fee are required. These are in addition to the fee required for the building permit itself.

The plans are examined by building officials to determine if they meet the requirements of the local code. Some communities have an architectural committee which will determine if the plans are satisfactory.

Sometimes it will be necessary for the builder or owner to submit supporting data to show how the building design will meet the code. When the plan meets all of the requirements of the building code, a building permit will be issued. Permit fees are based on cost of construction. Usually the range is from $30 to $255 or higher for large structures.

When construction is started, the building permit and an inspection card are posted on the building site. Sometimes the two are combined, as in Figure 6-30.

As work progresses, the building inspector will make inspections and fill out the inspection card for approval of work completed. It is important that the permit and card

VILLAGE OF FLOSSMOOR, ILLINOIS

BUILDING PERMIT

Bldg. Permit No.________________ Elec. Permit No.________________

Plbg. Permit No.________________ Date Issued________________

Street Address________________________________ Lot________ Block________

General Contractor's Name________________________________ Phone________________

Address____

BUILDING	REJECTED	ACCEPTED
1st Inspection; Foundations		
2nd Inspection; Drywall		
3rd Inspection; For Occupancy Permit		
For Building Inspections Call ___-____		
PLUMBING		
Plumbing Contractor:		
1st Inspection; Exterior Rough-In		
2nd Inspection; Drywall		
3rd Inspection; Final		
For Plumbing Inspections Call ___-____		
ELECTRICAL		
Electrical Contractor		
1st Inspection; Drywall		
2nd Inspection for Occupancy Permit		
For Electrical Inspections Call ___-____		
PUBLIC WORKS INSPS. CALL ___-____		
SEWER TAP—SEWER LINES & DRAIN LINES INSPS. CALL ___-____		
WATER TAP—CALL ___-____		
HEATING—For Inspections Call ___-____		
REFRIGERATION (Permit Required) **For Inspections Call ___-____**		

NOTICE!

(1) The approval of the drawings will not sanction nor permit any violation of village zoning or building code.
(2) A complete set of approved drawings along with permit must be kept on the premises during construction.
(3) The permit will become null and void in the event of any deviation from the accepted drawings.
(4) No foundation, structural, electrical, nor plumbing work shall be concealed without approval.
(5) THE BUILDING MAY NOT BE OCCUPIED OR USED FOR STORAGE WITHOUT FIRST OBTAINING AN OCCUPANCY PERMIT.

No work shall be done on any part of the building beyond the point indicated in each successive inspection without acceptance. No structural framework of any part of any building or any underground work shall be covered or concealed without acceptance.

THIS PERMIT MUST BE PROMINENTLY DISPLAYED ON BUILDING AT ALL TIMES

636

Figure 6-30. Construction cannot begin until a building permit is obtained from building officials of the community. Permit and inspection card must be displayed at the building site. This permit is combined with the card. (Village of Flossmoor, Illinois)

VILLAGE OF FLOSSMOOR

APPLICATION FOR BUILDING PERMIT

Type of Work:
Erection ____ Remodel ____ Addition ____ Repair ____ Demolish ____
Construction: Brick ____ Frame ____ Other ____
Color: Roof ____ Brick ____ Trim ____ Siding ____
Proposed use ____
Livable floor area ____

Required Service Entrance Conductors – Full Rated
100 Amp. Service and 100 Amp. Service Switch– .... Wire Size ....
150 Amp. Service and 150 Amp. Service Switch– .... Wire Size ....
200 Amp. Service and 200 Amp. Service Switch– .... Wire Size ....
400 Amp. or Larger Service and 400 Amp. Service Switch– ....
Wire Size ....

Underground Service ☐

MAJOR APPLIANCE CIRCUITS REQUIRED BY CODE

☐ Range .... Amps.	☐ 2 Laundry Circuits–20..... Amps.
☐ Built-in Oven .... Amps.	☐ Heating Plant .... Amps.
☐ Water Heater .... Amps.	☐ Lighting Circuits.......... Amps.
☐ Dish Washer .... Amps.	ELECTRICAL HEATING CIRCUITS
☐ Garbage Disposal .... Amps.	☐ Cable .... Amps.
☐ Sump Pump .... Amps.	☐ Baseboard Units .... Amps.
☐ Clothes Dryer............ Amps.	☐ Electrical Furnace .... Amps.
☐ Bathroom Heater .... Amps.	COMMERCIAL WIRING
☐ Fixed Air Conditioner ..... Amps.	☐ Lighting .... Amps.
☐ Window Air Conditioner ... Amps.	☐ Motors .... Amps.
☐ Electric Door Opener ..... Amps.	☐ Appliances .... Amps.
☐ 2 Kitchen Circuits–20 ..... Amps.	☐ Other .... Amps.

*NOTE – No Lighting or Other Current Consuming Device Shall Be Connected to the 2 Kitchen Circuits, the 2 Laundry Circuits, Heating Plant, Sump Pump or Air Conditioner Circuits.

Street Address: ____
Date ____ Township ____
Real Estate Index No. ____ – – –
Lot: ____ Block ____
Subdivision: ____

Property Owner ____
Present Address ____
Telephone ____
Architect ____
Address ____
Telephone ____
General Contractor: ____
Address ____
Telephone ____ Bond Expir. Date ____
Air Conditioning & Heating Contractor: ____
Address ____
Telephone ____
Electrical Contractor: ____
Address ____
Telephone ____ Licensed ____
Plumbing Contractor: ____
Address ____
Telephone ____ Bond Expir. Date ____
State License No. ____

As owner of the property, for which this permit is issued and/or as the applicant for this permit, I expressly agree to conform to all applicable ordinances, rules and regulations of the Village of Flossmoor.

Contractor or Owners Signature: ____

Estimated cost of building complete, including all materials and labor $ ____

Building Permit No. ____
Electrical Permit No. ____
Plumbing Permit No. ____

Figure 6-31. An application for a building permit must be accompanied by information about the structure that is to be built. In addition to the information given on the form, plans must be submitted. (Village of Flossmoor, Illinois)

always be attached to the building or somewhere on the construction site.

Work on the structure should not proceed beyond the point indicated in each successive inspection. Carpenters on the job must pay close attention to this record. Mechanical work (heating, plumbing, and electrical wiring) may never be enclosed before the installations have been approved by the building inspector. In some communities, a final inspection must be made and an occupancy permit issued. Until then, the building may not be occupied.

Important Terms

Architectural drawings
Bill of materials
Blueprints
Building code
Building permit
Computer-aided drafting and design
Details
Detail drawing
Dimension lines
Drawn to scale
Elevation
Floor plan
Foundation plan
Framing plan
Lumber list
Mill list
Modular construction
Pictorial sketch
Plot plan
Prints
Scale
Schedule
Section drawing
Section view
Specifications (specs)
Stock plan
Unicom system

Test Your Knowledge

1. A set of house plans usually includes what drawings?
2. Residential plan views are usually drawn to a scale of ____________ in. = ____________ ft. ____________ in.
3. Floor plans show the ____________ and outline of the building and its rooms.
4. The plot plan shows the ____________.
5. Elevation drawings show the ____________ walls of the structure.
6. A section view shows how a ____________ of a structure looks when ____________.
7. Dimension lines show ____________ and sizes.
8. Draw symbols which represent these materials and items:
   A. Concrete.
   B. Double hung window.
   C. Interior door.
   D. Refrigerator.
   E. Wall lavatory.
   F. Three-way switch.
   G. Range outlet.
   H. Wall fixture outlet.
9. To obtain a plan dimension not shown, a(n) ____________ scale may be used.
10. Working drawings (plans) provide much information required by the builder. Supplementary information is supplied by written ____________.
11. The modular construction concept is based on the use of a standard grid divided into ____________ in. squares.
12. What does CADD stand for?
13. What modular unit is used in the metric system for building measurements?
14. A building code covers all important aspects of the erection of a building. True or False?

Outside Assignments

1. Obtain a complete set of plans for an average size residence and make a careful study of the views shown. Try to borrow a set from a local builder. Make a list of symbols, notes, and abbreviations that you do not understand. Then go to reference books and architectural standards books to secure the information. Ask your instructor for help if you have difficulty with some of the views.
2. Study the building code in your community. Become familiar with the various sections that are covered and especially note requirements that apply to residential work. Submit a general outline of the material you feel is most important.
3. Make a trip to your city offices and visit with the commissioner or director of building. Be sure to call for an appointment in advance. During your visit obtain information concerning building permits and inspection procedures. Learn the cost and what plans and specifications need to be submitted. Also get information about zoning restrictions and other public ordinances that apply to residential construction. Prepare carefully organized notes and make an oral report to your class.

Architectural drafter prepares a set of plans for a residential structure. Every part of the building from foundation to roof is drawn to exact scale. All of this must be completed before beginning the footings and foundation work which is covered in the next unit. (Grimes-Port-Jones-Schwerdtfeger Architects, Inc.)

Footings, Foundations, and Framing

A large framed structure. Roof is supported by a series of scissor trusses. Diagonal bracing members are needed to provide lateral support. Many of these bracing members will be removed when the walls are finished.

Batter boards are used to ensure that footings are poured in the correct location

7

Footings and Foundations

In the construction of single family dwellings and other structures, carpenters must work with many tradespeople. They also must work closely with the architect and owner in carrying out the total building plan.

On some jobs, the carpenters may be required to lay out the building lines and supervise the excavation. They may also build forms for footings and poured foundation walls. Anyone working in the carpentry trade needs a working knowledge of standards and practices in concrete work. In this unit, some of the material also deals with masonry. It is included because of its close relationship to carpentry.

Figure 7-1. Top—Using a laser level to establish lot lines. It is a good idea to leave this task to an engineer or surveyor. (Cella Barr Associates) Bottom—Site preparation may involve removing stumps and boulders, as well as grading.

Clearing the Site

Preparation of the building site may require grading and/or removal of trees. Grading may be needed before the building lines are laid out. This may require the placement of grade level stakes. The proper establishing of grade is explained in Unit 5. It will usually require the use of a transit or laser level.

If the property is wooded, great care ought to be used in deciding what trees should be removed. Much depends on where the trees are located and their type. In general, evergreens should be used as protection against the cold winter winds. Deciduous (leaf dropping) trees are best used as shade against hot summer sun. Try to place the house to take advantage of the protection offered by trees already on the property.

Mark trees that are to be taken down so the person responsible for their removal takes the right ones. Avoid digging trenches through the root system of trees being retained. It could cause them to die. Likewise, trees usually will not tolerate more than a foot of additional fill around their root systems.

Laying out Building Lines

To protect the owner, lot lines should be located with the help of a registered engineer or licensed surveyor. They can also assist in establishing building lines and grade levels, Figure 7-1. Even if assistance is used, the carpenter should also be familiar with local building code requirements.

It is best to locate building lines with leveling instruments. Follow the procedures described in Unit 5. Lines can, however, be transferred from lot markers. In such cases, it is important that distances be laid out perpendicular to existing lines. Building lines, likewise, must be square. To establish a right angle, the 6–8–10 method can be used. Refer to Unit 5, Figure 5-4.

Following procedures described in Unit 5, locate corners formed by the intersection of the outsides of foundation walls. Mark the positions by driving stakes. Set tacks in the stake tops at the exact spot.

After locating all building lines, check them carefully. Measure the length and, even though they were laid out with a transit, measure the diagonals of squares and rectangles, Figure 7-2. An out-of-square foundation will cause problems throughout construction.

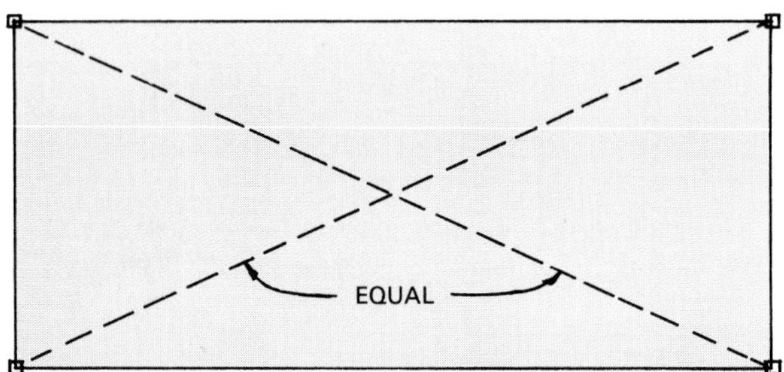

Figure 7-2. Diagonals of a square or rectangle will always be equal. Always measure to diagonal corners to see if building lines are square.

Batter Boards

Batter boards, Figure 7-3, are set up around the building layout stakes. Use 2 x 4s for the stakes and 1 x 6s or wider pieces for the ledgers.

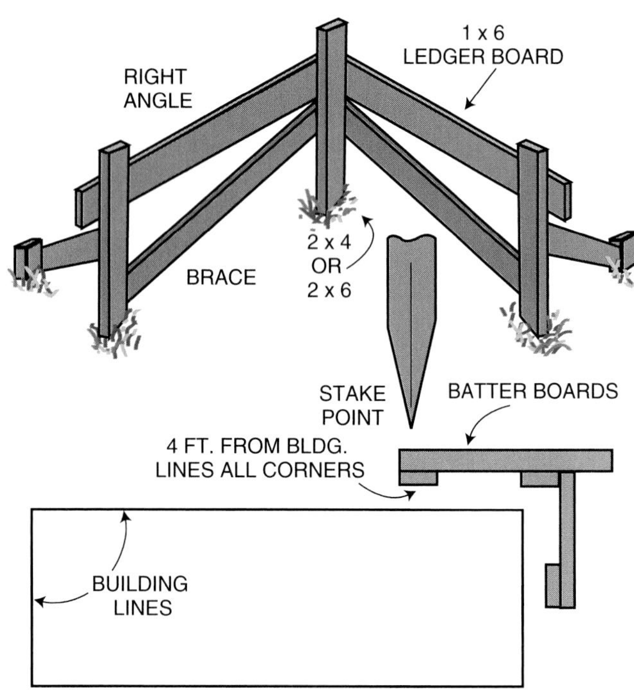

Figure 7-3. Set up batter boards about 4' from building lines on all four corners. In loose soil or when boards are more than 3' off the ground, use braces.

Locate the batter boards 4' or more away from the corners created by the building lines.

Nail the ledger boards to the stakes. They should be level and at a convenient working height, preferably slightly above the top of the foundation. The batter boards should be roughly level with each other. Also, be sure that the ledger boards are long enough to extend well past each corner.

Using lines and a plumb bob or laser level, locate the lines so they intersect over the layout stakes. Mark the tops of the ledger boards where the intersecting lines rest. Make a shallow saw kerf or drive a nail, Figure 7-4. Pull the lines tight and fasten.

If you are using saw kerfs, drive nails into the backs of the ledger boards. You may prefer to wrap them around the ledger and run them through the saw kerf several times.

Sometimes a straight batter board, Figure 7-5, is used when it will not be located at a corner.

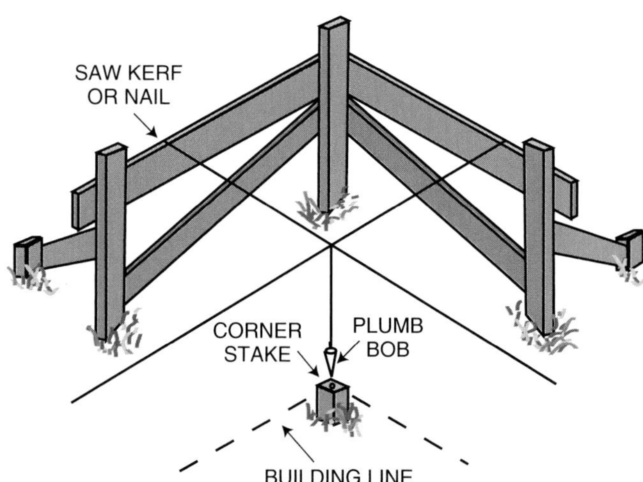

Figure 7-4. Using batter boards and plumb line to establish building lines. Lines must intersect over the tack in the corner stake.

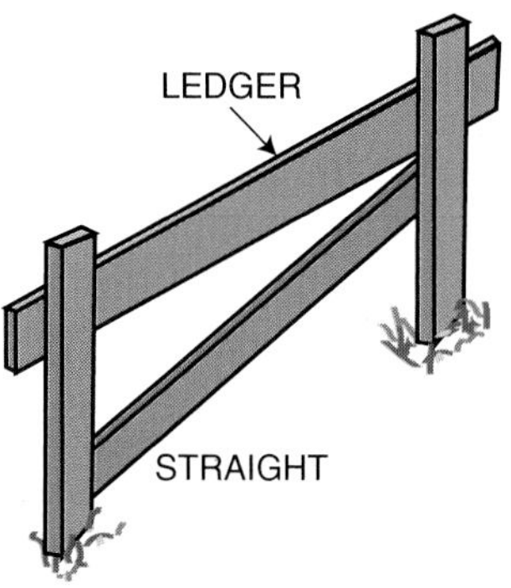

Figure 7-5. Straight batter board is used in some situations as shown in lower half of Figure 7-6.

Excavation

Building sites on steep slopes or rugged terrain should be rough graded before the building is laid out. Top soil should be removed and piled where it will not interfere with construction. This can be used for the finished grade after the building is complete.

Where no grading is needed, the site can be laid out and batter boards erected. See the lower drawing, Figure 7-6.

Stakes marking the outer edge of the rough excavation are set and the lines are removed from the batter boards during the work. For regular basement foundations, the excavation should extend at least 2' beyond the building lines to allow clearance for form work. Foundations for

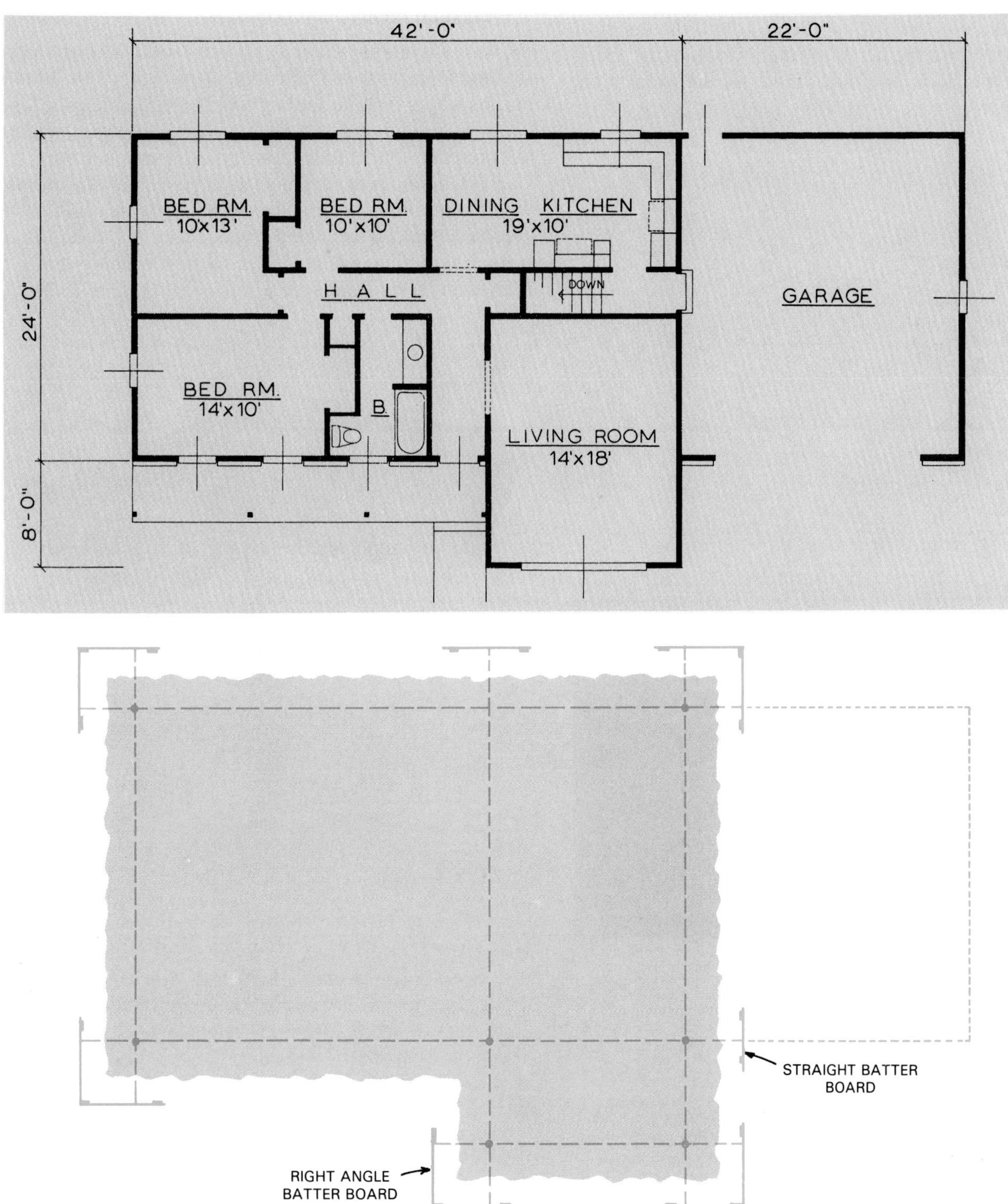

Figure 7-6. Starting the foundation for a home. Top—Plan showing overall dimensions. Bottom—Sketch of building lines set up around area to be excavated. Actual excavation (shown in gray) should be at least 2' outside of building lines.

structures with a slab floor or crawl space will need little excavating beyond the trench for footings and walls.

The depth of the excavation can be calculated from a study of the vertical section views of the architectural plans.

In cold climates, it is important that foundations be located below the frost line. If footings are set too shallow, the moisture in the soil under the footing may freeze. This could force the foundation wall upward, causing cracks and serious damage. Local building codes usually cover these requirements.

It is common practice to establish both the depth of the excavation and the height of the foundation by using the highest elevation on the perimeter of the excavation. This is known as the *control point,* Figure 7-7.

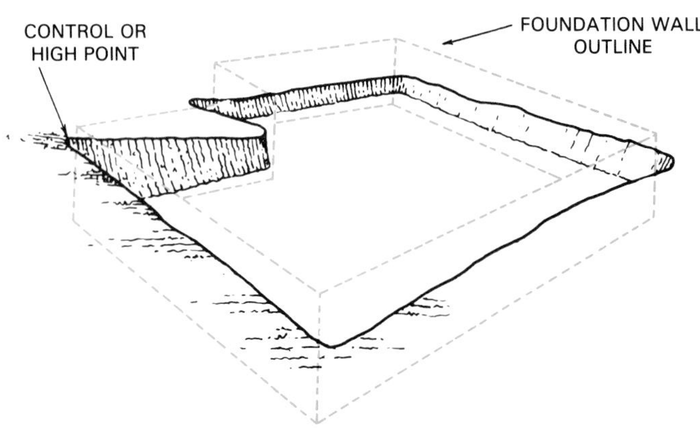

Figure 7-7. The control point. Top—Highest elevation outside the excavation is used to establish depth of excavation. Bottom—Completed excavation.

Foundations should extend about 8" above the finished grade. At this height, wood finish and framing members will be adequately protected from moisture. The finished grade should be sloped away from all sides of the structure so surface water will run away from the foundation.

The depth of the excavation may be affected by the elevation of the site. It may be higher or lower than the street or adjacent property. The level of sewer lines also has an effect. Normally, solving these problems is the responsibility of the architect. Information on grade, foundation, and floor levels is usually included in the working drawings.

Foundation Systems

All structures settle. A properly designed and constructed foundation will distribute the weight to the ground in such a way that the settling will be negligible or at least uniform. Figure 7-8 shows a typical foundation. Figure 7-9 presents several foundation types in simple form.

For light construction, such as residential, the *spread foundation,* illustrated in Figure 7-8, is most common. It transmits the load through the walls, pilasters, columns, or piers. These elements rest on a footing which is nothing more than an enlarged base.

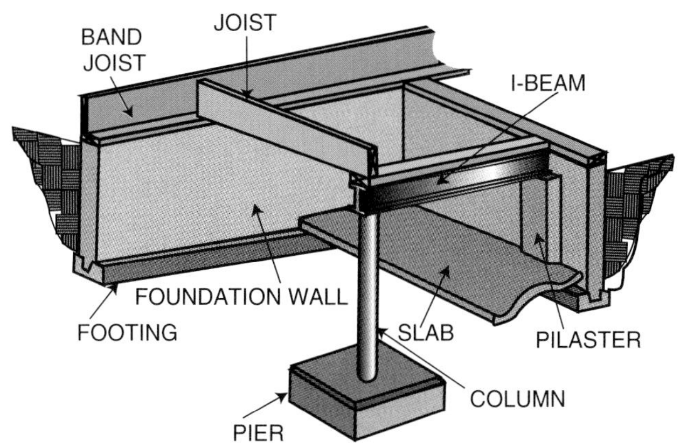

Figure 7-8. Elements of a typical spread foundation are shown. Footings and piers spread the building's weight over a wider area of the soil. This type of foundation is often used when the owner needs a basement.

Footings

Plain footings carry light loads and usually do not need reinforcing. *Reinforced footings* have steel rebar embedded in them for added strength against cracking. They are used when loads must be spread over a larger area or when the load must be bridged over weak spots, such as excavations or sewer lines.

A *stepped footing* is one that changes grade levels at intervals to accommodate a sloping lot. See Figure 7-10. Vertical sections should be at least 6" thick. Horizontal distance between steps should be at least 2'. If masonry units are to be used over the footing, distances should fit standard brick or block modules.

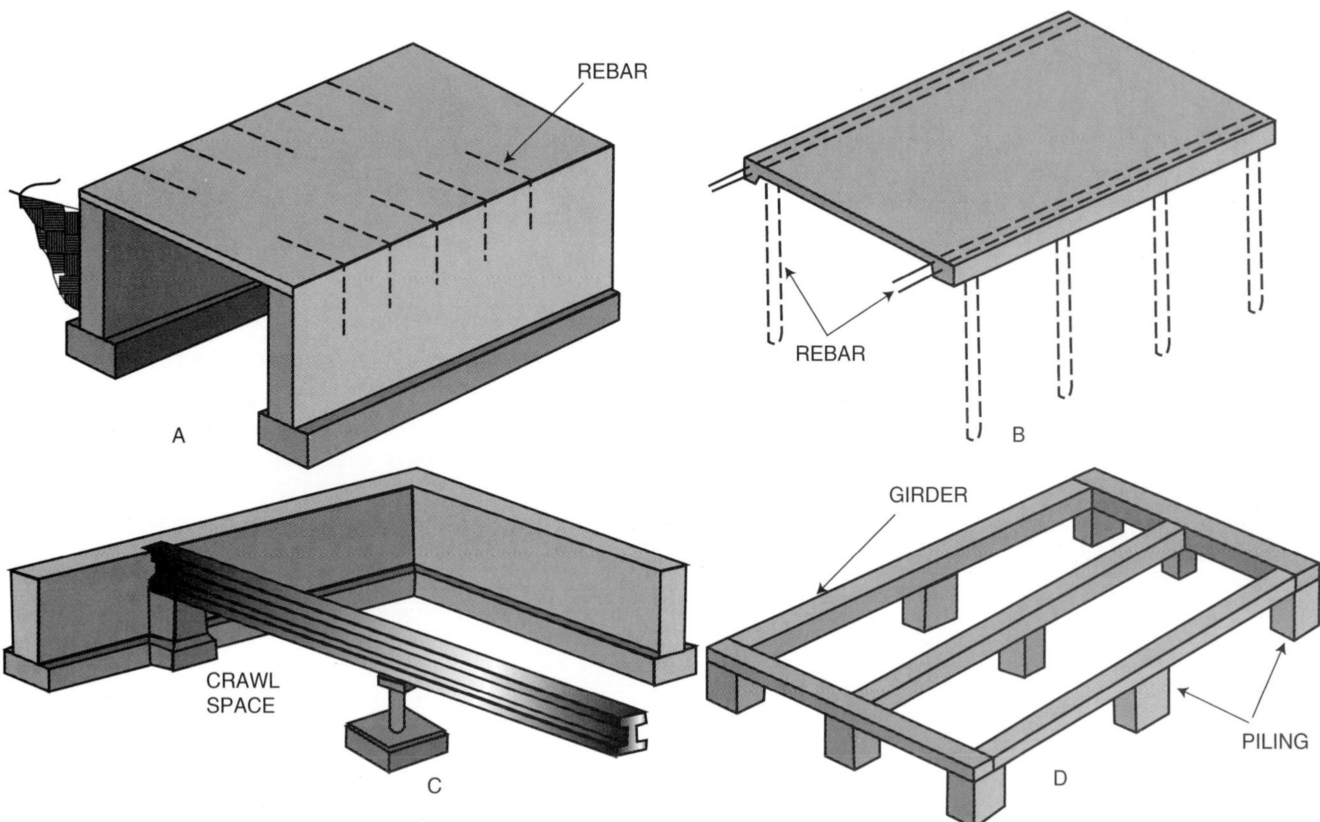

Figure 7-9. Other kinds of foundations. A—Slab-on-foundation is used in cold climates. B—Slab-on-ground foundation is popular in warm climates. Piers (dashed lines) can be used as additional support in unstable soils. C—Crawl-space foundation is similar to basement foundation shown in Figure 7-8. D—Piling-and-girder foundation is used in warm climates where water pipes and drain pipes are not in danger of freezing. In unstable soils, pilings often made of concrete block, may go deep for better support.

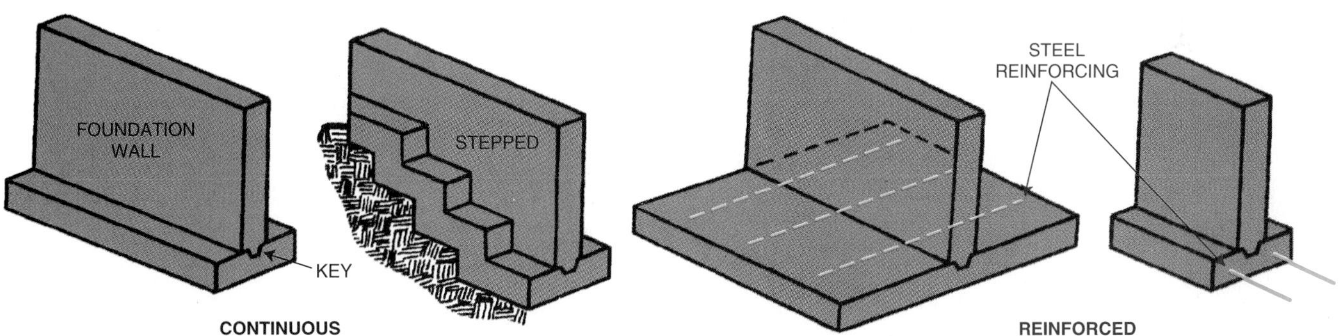

Figure 7-10. Different types of footings are needed for different slope and soil conditions. Vertical runs of stepped footing should not exceed 3/4 of horizontal run between steps.

Slabs

Slab foundations take several forms. The slab can be used with other elements such as walls, piers, and footings, as shown in Figure 7-9(B). This is called a *structurally-supported slab.* A second type is laid directly on top of the ground like those shown in Figure 7-11. These are referred to as *ground-supported slabs.*

Some slabs, particularly in warmer climates, are constructed in one continuous pour. There are no joints or separately poured sections. This is called *monolithic* concrete or a monolithic pour. This type of construction is often used over soils with low bearing capacity, Figure 7-12.

The Council of American Building Officials has modified its model energy code to allow the use of

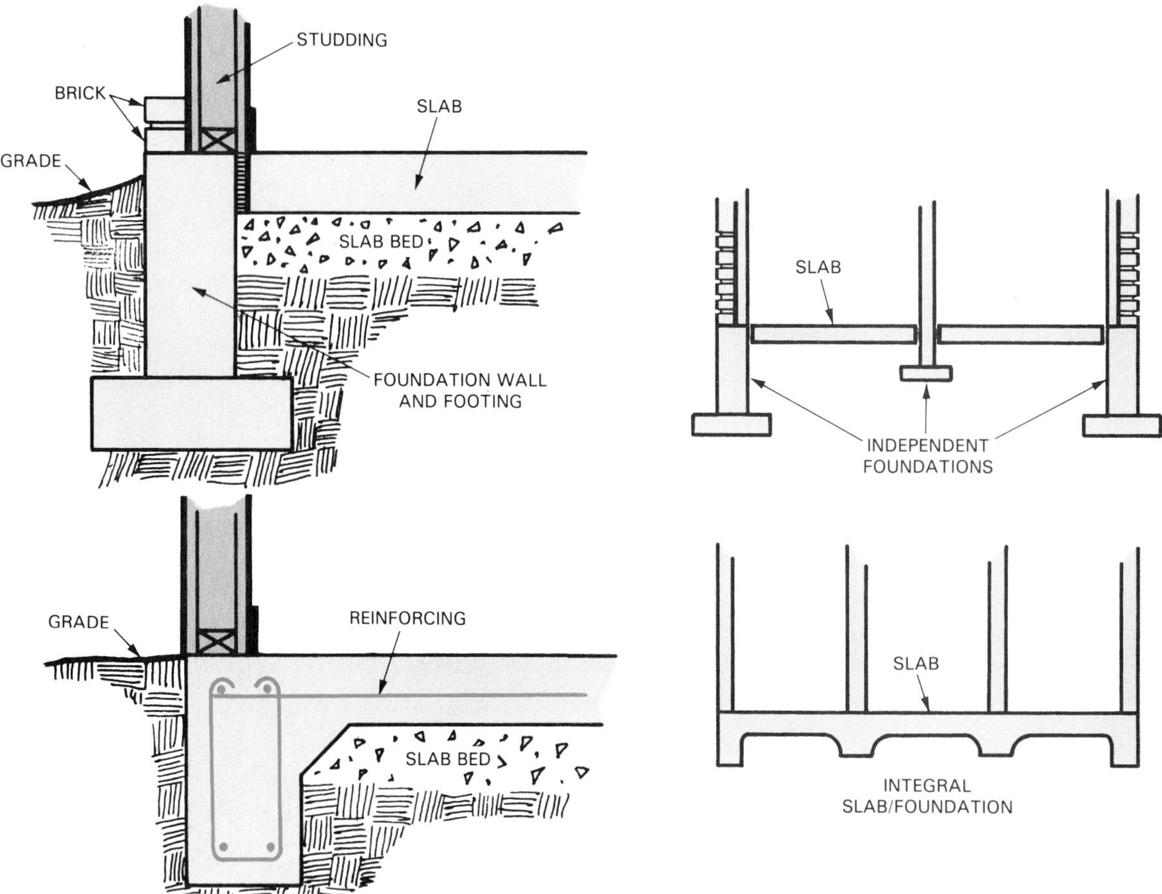

Figure 7-11. Ground-supported slabs rest on the soil under them. Top—In warm climates, ground-supported slab "floats" on the soil. This slab-on-ground has reinforcing around the perimeter. (Pierce Construction Ltd.) Middle—Slab is independent of foundation walls. Bottom—This ground-supported slab has a monolithic structure incorporating a grade beam and is used where soil provides poor support.

shallow foundations if they are frost protected with rigid polystyrene. The method, which has the rigid insulation along the slab edges and foundation corners, allows footings in extremely cold climates, such as North Dakota, to be only 18" below grade.

The change affects CABO's one- and two-family dwelling code. It is said that the method prevents frost heave even where footings are only 4" below grade. The National Association of Home Builders had asked for the code changes which has been used successfully in

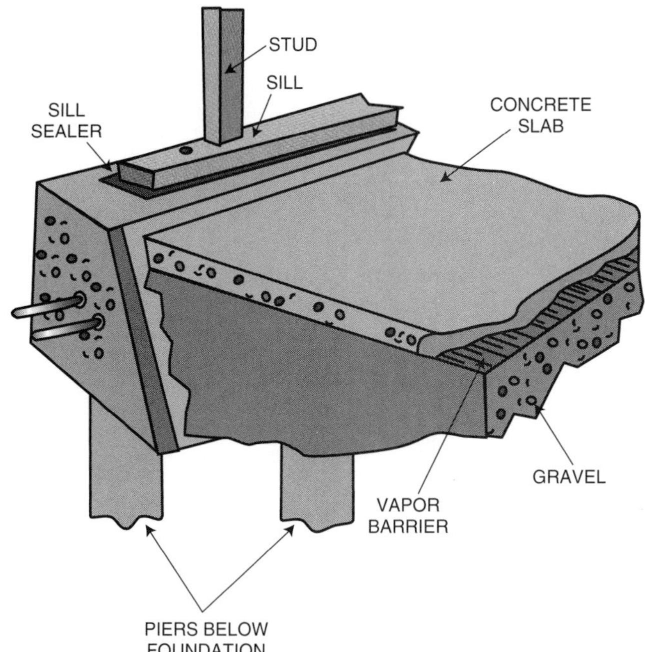

Figure 7-12. A reinforced grade beam foundation rests on concrete piers placed at intervals below the frost line and resting on firm soil. The concrete slab and thickened perimeter are usually separate but may be one continuous slab. The grade beam spans the distance between piers and supports the slab.

Type of soil	Capacity, tons per sq. ft.
Soft clay	1
Wet sand or firm clay	2
Fine, dry sand	3
Hard, dry clay or coarse sand	4
Gravel	6

Figure 7-13. Load carrying capacity of different soil types.

W = WALL THICKNESS
1/2 W
KEY
1/2 W
W
2 W

Figure 7-14. Standard footing design for residential construction. It should be twice the width of the foundation wall that will rest on it. Key designed to anchor wall is often replaced with dowels made of rebar.

Scandinavian countries. Use of the method is expected to spread quickly.

Footing Design

Footings must be wide enough to spread the load over sufficient area. Load-bearing capacities of soils vary considerably. See Figure 7-13. In residential and smaller building construction, the usual practice is to make the footing twice as wide as the foundation wall. See Figure 7-14. The average thickness of a footing is about 8".

Footings under columns and posts carry heavy, concentrated loads and are usually from 2 to 3 ft. square. The thickness should be about 1 1/2 times the distance from the face of the column to the edge of the footing.

Reinforced footings are used:

- In regions subject to earthquakes.
- In situations where the footings must extend over soils containing poor load-bearing material.

Some structural designs may also require the use of reinforcing. The common practice is to use two No. 5 (5/8") rebar for 12" x 24" footings. At least 3" of concrete should cover the reinforcing at all points.

In a single story dwelling where they are independent of other footings, the chimney footings should have a minimum projection of 4" on each side. For a two-story house, chimney footings should have a minimum thickness of 12" and a minimum projection of 6" on each side. Exact dimensions will vary according to the weight of the chimney and the nature of the soil. Where chimneys are a part of outside walls or inside bearing walls, chimney footings should be constructed as part of the wall footing. Concrete for both chimney and wall footings should be placed at the same time.

Footings that must support cast-in-place concrete walls may be formed with a recess forming a keyed joint or steel dowels, as shown in Figure 7-15. A number of typical footing designs for masonry walls are shown in Figure 7-16. These are not working drawings and will need to be adapted to local soil and ground water conditions as well as to local code requirements.

Forms for Footings

After the excavation is completed, the footings are laid out and forms are constructed. Lines are replaced on the batter boards and corner points are dropped with a plumb bob to the bottom of the excavation.

Constructing Forms

1. Drive stakes and establish points at the corners of the foundation walls.
2. Set up a builder's level at a central point of the excavation and drive a number of grade stakes (level with the top of the footing) along the footing line and at approximate points where column footings are

Figure 7-15. Two methods of securing walls on footings. Top—When a concrete foundation wall or a masonry wall is to be erected on a footing, the footing can be keyed with a depression in its center. (Portland Cement Assoc.) Bottom—Many builders use rebar dowels instead of keying the footing. (LiteForm, Inc.)

required. Corner stakes can also be driven to the exact height of the top of the footing.

3. Connect the corner stakes with lines tied to nails in the top of the stakes.
4. Working from these building lines, construct the outside form for the footing. The form boards will be located outside the building lines by a distance equal to the footing extension (usually 4" for an 8" foundation wall). See Figure 7-17. The top edge of the form boards must be level with the grade stakes.
5. Transfer the height from the grade stakes to the form. Use a carpenter's level. With the outside form boards in place, it will be easy to set and level the inside sections using a folding rule or steel tape and a level. Figure 7-18 (top) illustrates this method.

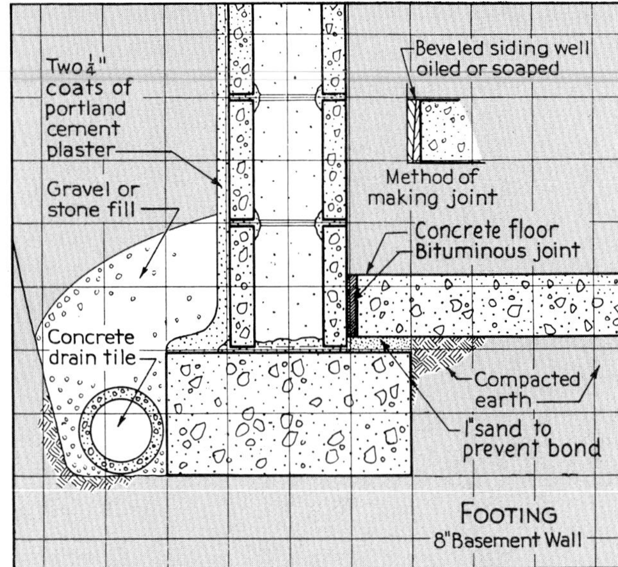

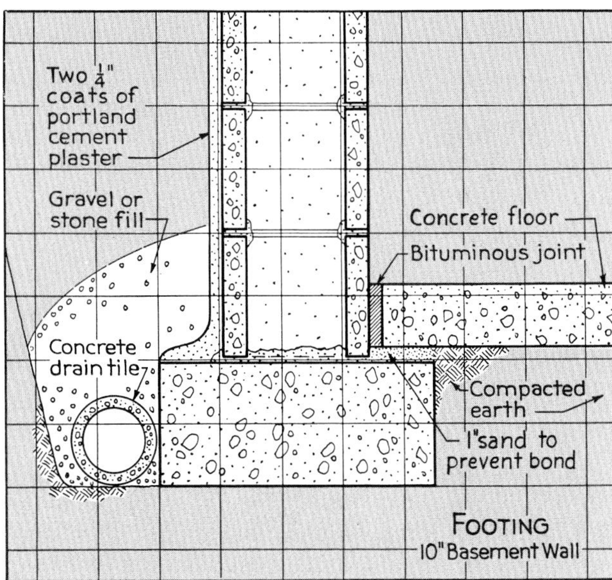

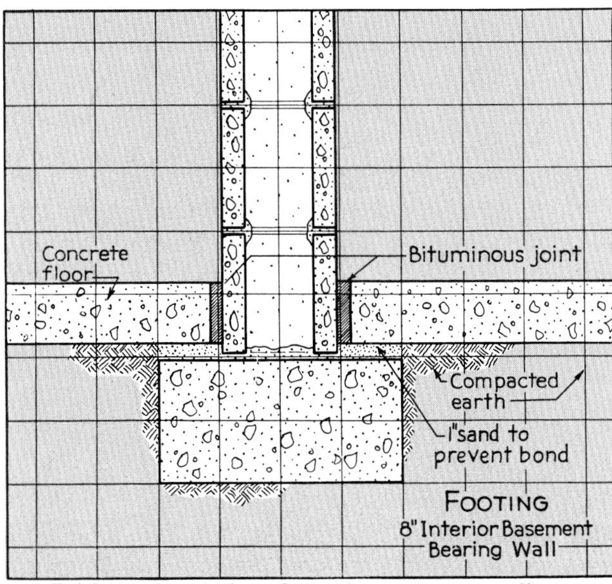

Figure 7-16. Footing designs for various masonry walls.

Another method of building footing forms is shown in Figure 7-18 (middle and bottom). Plastic channel with spreaders can be quickly installed using only a level and an occasional stake to keep the hollow form from shifting as concrete is poured. The form has perforations along its outer edges and it becomes a permanent drain that removes water from around the foundation.

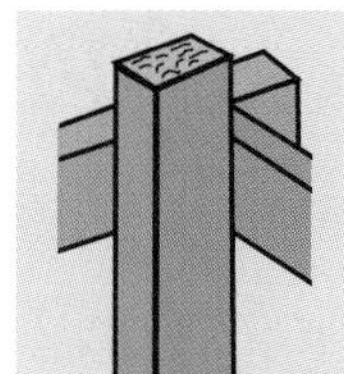

After the excavation is complete, check the batter boards carefully. They may have been disturbed by the excavating equipment. Make necessary adjustments before proceeding with the footing layout.

Forms constructed of 1" boards should be supported with stakes placed 2 to 3' apart. Stakes may be placed farther apart when 2" material is used. Spacers or spreaders used to locate the inside form will save measuring time, Figure 7-19.

Bracing of footing forms may sometimes be desirable. Usually it is necessary only if 1" lumber is used for forming. Attach the brace to the top of a form stake and to the bottom of a brace stake located about 1' away. Refer to Figure 7-19 again.

Stepped footings will require some additional formwork. Vertical blocking must be nailed to the form to contain the concrete until it sets. See Figure 7-20.

Column footings are pads of concrete which will support columns. They carry weight transmitted to the column from the beams, stringers, and joists of a building, Figure 7-21.

Forms for column footings are usually set after the wall footing forms are complete. These are located by direct measurements from the building lines. Forms are leveled to previously-set grade level stakes.

Some hand digging and leveling of the excavation will probably be necessary as forms are set. Loose dirt and debris must always be removed from the ground that will be located under a footing. This is necessary even though the resulting depth will be greater than required.

The top of the footing must be level. The bottom may vary as long as the minimum thickness is maintained.

Form boards are removed once the concrete dries. It is important that they be only temporarily nailed (always from the outside) to stakes and to each other. Duplex (double-headed) nails may be used. If regular nails are used, they should be driven only partially into the wood.

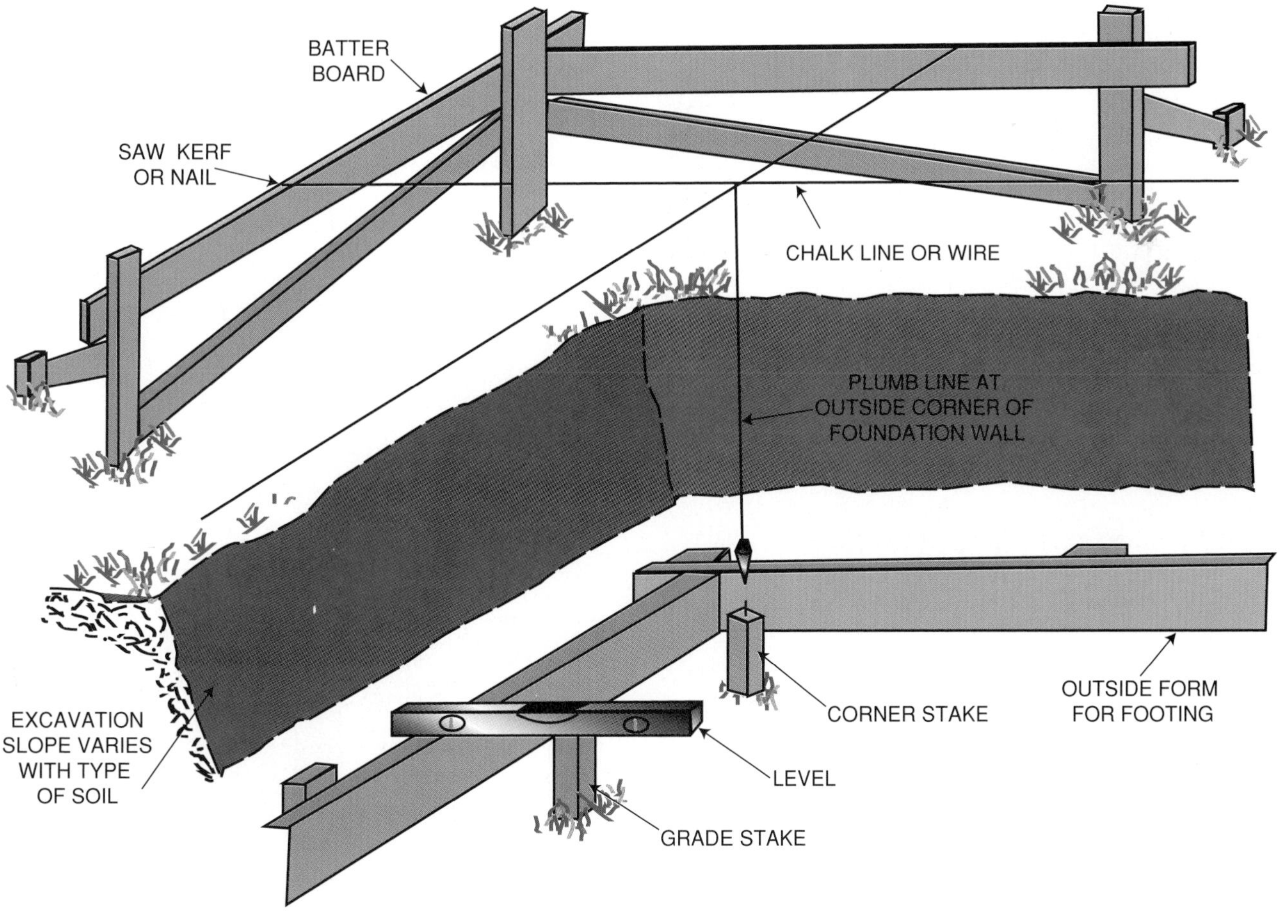

Figure 7-17. Laying out forms for footings. The outside forms are built first. They must be level with grade stakes.

Figure 7-18. Setting forms. Top—Measure distance and set the inside form board. (Portland Cement Assoc.) Middle and bottom—A footing formed up using a patented plastic channel as the form. Form is left in place to also act as a perimeter drain. (Kasten-Weiler Construction)

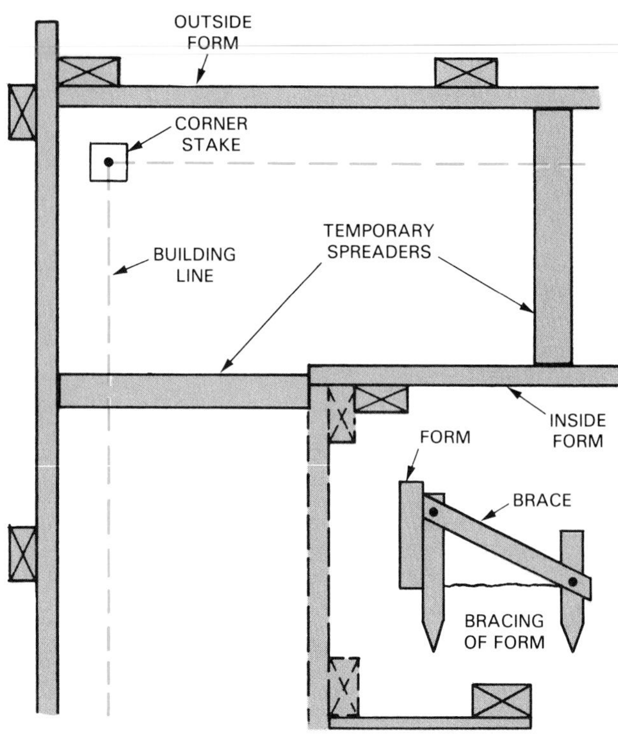

Figure 7-19. Properly located footing form. Note that it extends beyond corner stake and building line. Precut spreaders were used to locate the inside form boards.

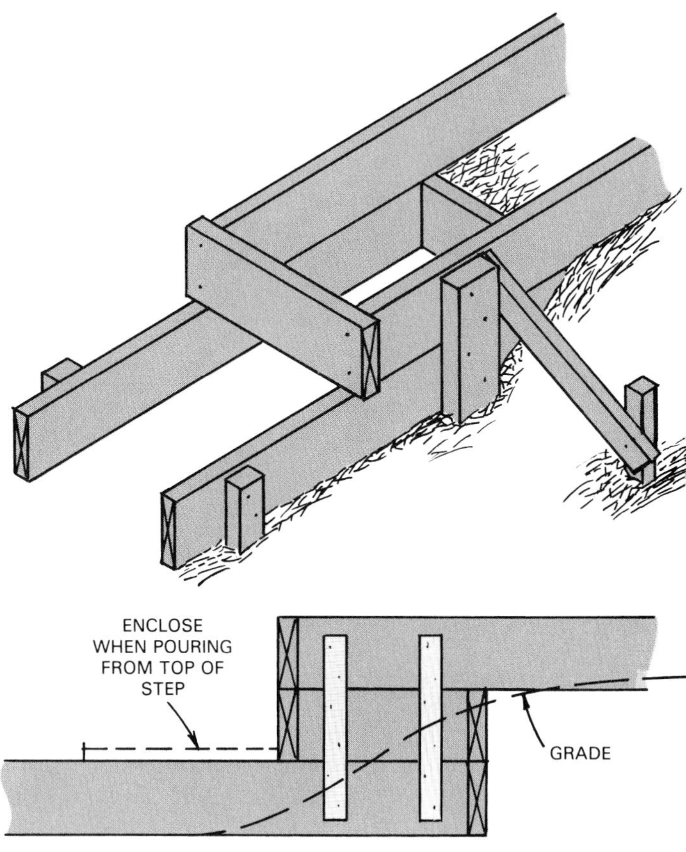

Figure 7-20. Form constructed for a vertical section in a stepped footing. Lower level is poured first and will usually be allowed to set up slightly before step is poured. Top—Basic form. Bottom—Alternate design when more height is needed.

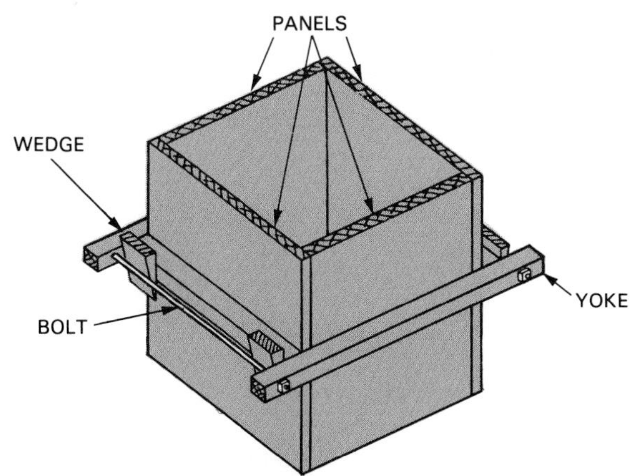

Figure 7-21. Poured column footing. Top—Typical reusable form. Bottom—Actual footing.

Consider problems in form removal. Nail through stakes into the form boards. Do not nail from the inside.

When the forms are completed, check for sturdiness and accuracy. Remove the line, line stakes, and grade stakes. The concrete can now be placed.

Concrete

Concrete is made by mixing the following materials:

- Cement.
- Fine aggregate (sand).
- Coarse aggregate (gravel or crushed stone).
- Water.

To prepare concrete, the aggregate and cement are first mixed together. Then the water is added. The water causes a chemical action (called hydration) to take place and the mixture hardens. The hardening process is not a result of drying. The concrete should be kept moist during the initial hydration process.

The compressive strength of concrete is high, but its tensile strength (stretching, bending, or twisting) is relatively low. For this reason, concrete used for beams, columns, and girders must be reinforced with steel rebar. When it must resist compression forces only, reinforcement is usually not necessary.

Cement

Most *cement* used today is portland cement. It is usually manufactured from limestone mixed with shale, clay, or marl. Each sack of portland cement holds 94 lb. This is equal to one cubic foot in volume. Cement should be a free-flowing powder. If it contains lumps that cannot be pulverized easily between thumb and fingers, it should not be used.

Aggregates

Aggregate may consist of sand, crushed stone, gravel, or lightweight materials such as expanded slag, clay, or shale.

The large, coarse aggregate (seldom over 1 1/2" in diameter) forms the basic structure of the concrete. The voids between these particles are filled with smaller particles; the voids between these smaller particles are filled with still smaller particles. Together they form a dense mass.

Today, practically all concrete is delivered to the building site in ready-mix trucks, Figure 7-22. The mix is purchased by the cubic yard (27 cu. ft.) and is available in a number of different strengths. A minimum order is usually 1 cu. yd.

Erecting Wall Forms

Many different types of wall forming systems are available. There are certain basic considerations that should be understood and applied to all systems.

Figure 7-22. Most concrete is delivered to the construction site already mixed. Here, it is being piped hydraulically to the forms. (Tony's Construction, Inc.)

For quality work, the forms used must be tight, smooth, defect-free, and properly aligned. Joints between form boards or panels should be tight. This prevents the loss of the cement paste, which tends to weaken the concrete and cause honeycombing.

Wall forms must be strong and well braced to resist the side pressure created by the plastic concrete. This pressure increases tremendously as the height of the wall is increased. Regular concrete weighs about 150 lb. per cu. ft. If it were immediately poured into a form 8' high, it would create a pressure of about 1200 lb. per sq. ft., along the bottom edges of the form.

In practice, this pressure is reduced through compaction and hardening of the concrete. It tends to support itself. Thus, the lateral pressure will be related to:

- The amount of concrete placed per hour.
- The outside temperature.
- The amount of mechanical vibration.

Low wall forms, up to about 3' in height, can be assembled from 1" sheathing boards or 3/4" plywood supported by 2 x 4 studs spaced 2' apart. The height can be increased somewhat if the studs are closer together. Refer to Figure 7-23 (top).

For walls over 4' in height, the studs should be backed with wales (2 x 4 horizontal stiffeners) to provide greater strength, Figure 7-23 (bottom).

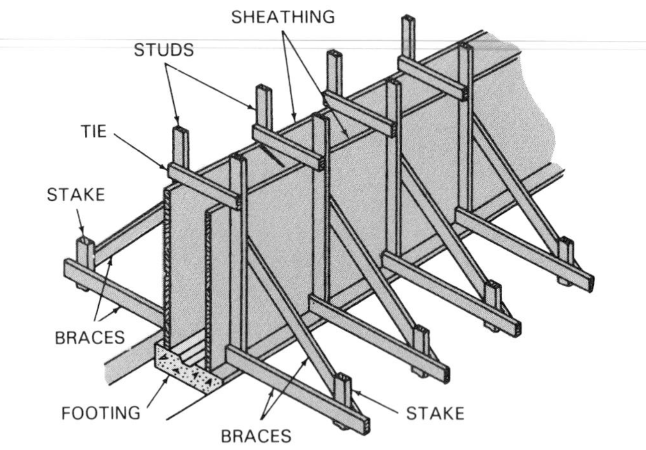

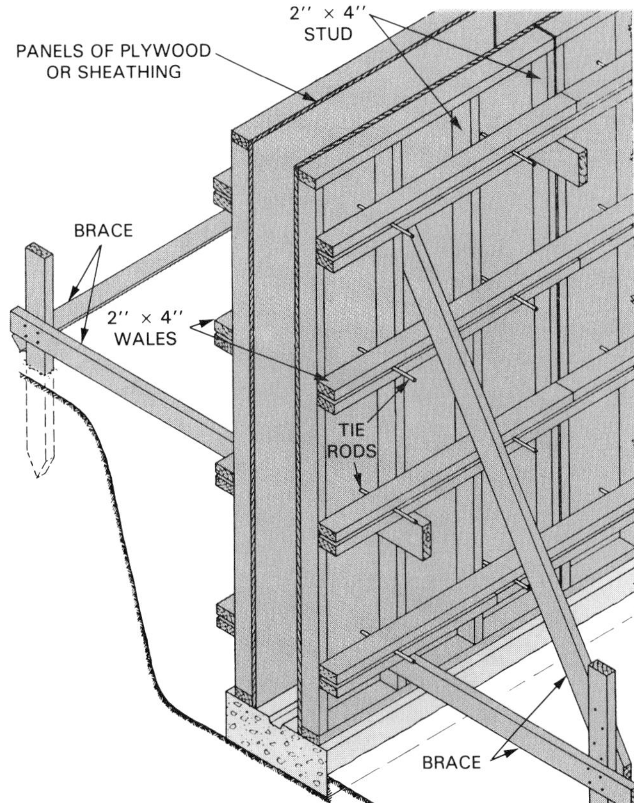

Figure 7-23. Wall form design. Top—Design for wall forms up to 3' high. Use plywood sheathing and space studs 2' apart. Bottom—Prefabricated panels for wall form. For walls over 4' high, wales, the horizontal doubled 2 x 4s, provide greater strength to the form. Note bracing to top and bottom.

Reestablishing the Building Line

Before setting up the outside foundation wall form, you will need to mark the building line on top of the footing. Set up your lines on the batter boards once more. Then drop a plumb line from the intersections (corners) of the building lines to the footing. Mark the corners on the footing. Snap a chalk line from corner to corner on the footing. This line will be the outside face of the foundation wall. As you set up the foundation forms, align the face of the outside form with the chalk line.

Form Hardware

Wire ties and wooden spreaders have been largely replaced with various manufactured devices. Figure 7-24 shows three types of ties. The rods go through small holes in the sheathing and studs. Holes through the wales can be larger for easy assembly or the wales can be doubled as shown.

The snap tie is designed so that a portion of it remains in the wall. Release solution or another type of lubricant is applied to tapered ties prior to pouring the concrete. These solutions allow the ties to be removed after the concrete sets. Bolts in the coil type ties should also be coated with release solution.

Corners of concrete forms can be secured by corner locks. Two types are shown in Figure 7-25.

After the concrete has set, the clamps can be quickly removed and the forms stripped away. A special wrench is used to break off the outer sections of the snap tie rod. The rod breaks at a small indentation located about 1" beneath the concrete surface. The hole in the concrete is easily patched later with grout or mortar.

Other types of patented wall ties are shown in Figure 7-24. The coil type spreader is assembled with a cone of wood, plastic, or metal and a lag screw. The cone provides

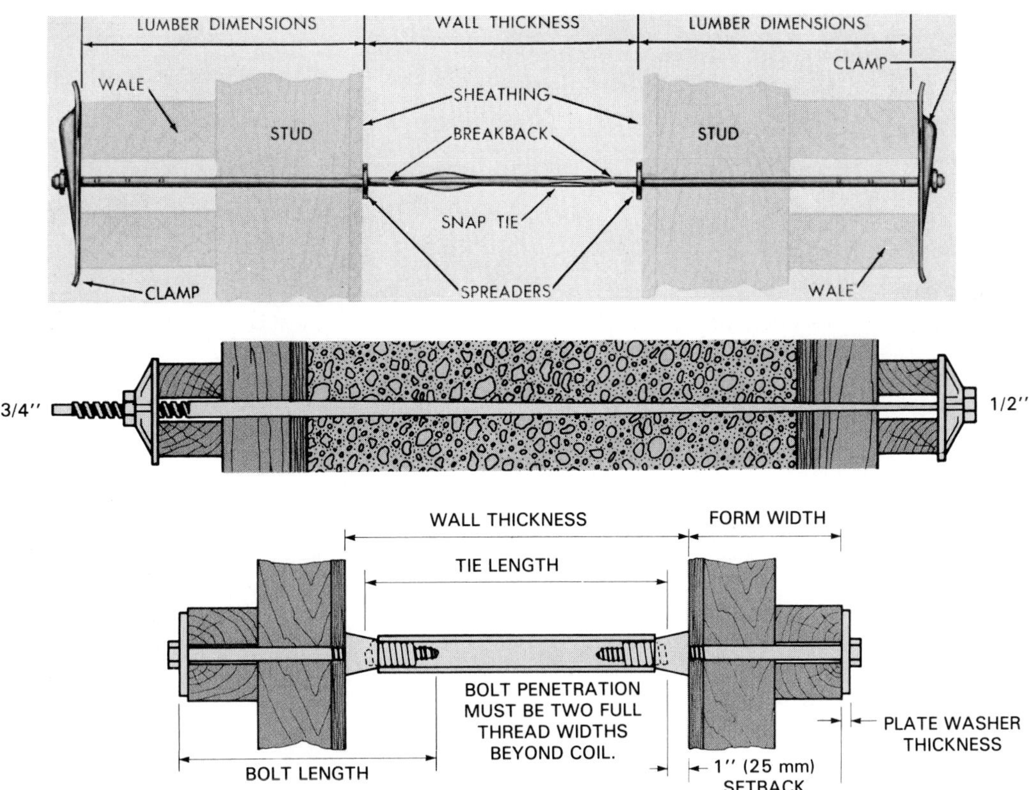

Figure 7-24. Some form ties combine the functions of both ties and spreaders. Top—Snap tie. (Universal Form Clamp Co.) Center—Taper tie is reusable. Bottom—Heavy duty coil tie. (The Burke Co.)

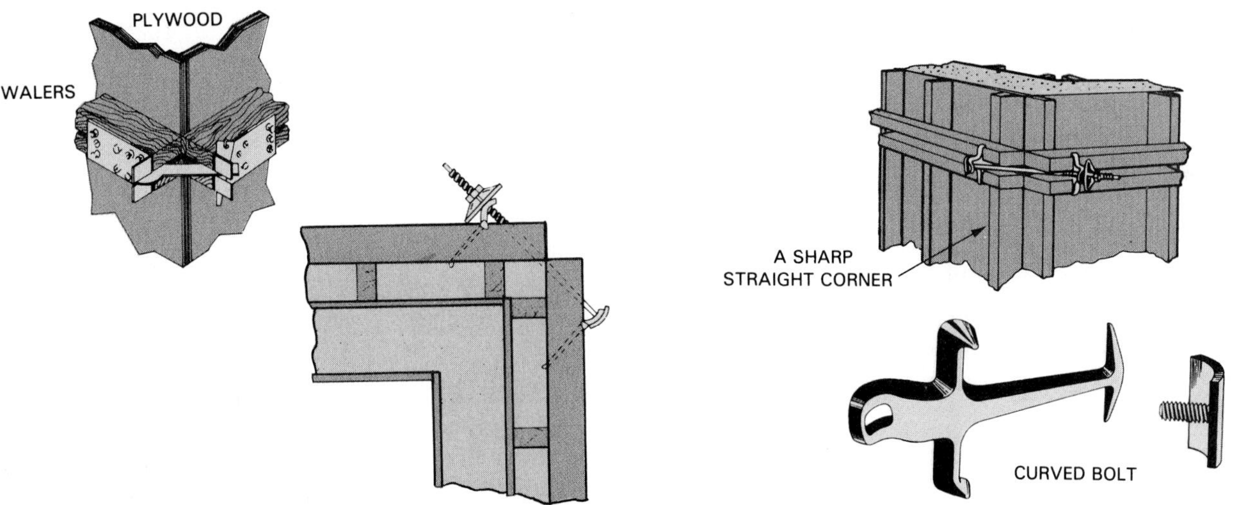

Figure 7-25. Corners of wall forms must be carefully fastened so they will withstand pressure from poured concrete. This can be done by interlocking the wales or using patented locks. Left—A high speed corner lock. Clamps are attached with 6 to 8 penny duplex (double-headed) nails and lag screws. Right—A corner lock requiring no nails or screws. (The Burke Co.)

smooth contact with the form and leaves a recess that is easy to fill.

Taper ties are threaded at both ends for easy removal. The threaded plate is removed from the small end and the rod is pulled free from the other end.

Panel Forms

Today, prefabricated panels are used for most wall forming. The panels, made from a special grade of plywood, are attached to wood or metal frames. See Figure 7-26.

Figure 7-26. Most wall forming is now done with prefabricated forms that are quickly assembled and dismantled. Note hooks supporting wales.

Carpenters can build panels, using 3/4" plywood and 2 x 4 studs to form 2' x 8' or 4' x 8' units. For standard columns and other units, prefabricated forms, as shown in Figure 7-27, will save time.

Tubular fiber forms may be purchased. These are usually found in heavy construction.

Carpenter-built prefabricated forms can be fastened as shown in Figure 7-28. Be sure to add wales and bracing. Select straight lumber for wales.

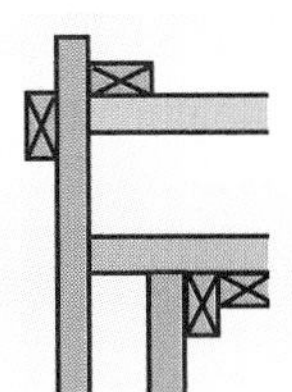

Check formwork carefully before placing the concrete. A form that fails during the pouring will waste material and cause extra work.

Since panel forms are designed to be used many times, they should be treated to prevent the concrete from sticking to the surfaces. Use special form release coatings which are available.

Care should be exercised in removing form components. They should be thoroughly cleaned and then carefully sorted and stacked for movement to the next job or storage.

Manufacturers have developed many forming systems to replace or supplement panel forms built by the carpenter. For residential work, these systems usually consist of steel frames and exterior grade plywood panels. Sometimes the plywood is coated with a special plastic material to create a smooth finish on the concrete and prevent it from sticking to the surface. The panel units are light for easy handling and transporting from one building site to another. Specially designed devices are used to assemble and space the components quickly and accurately. See Figure 7-29.

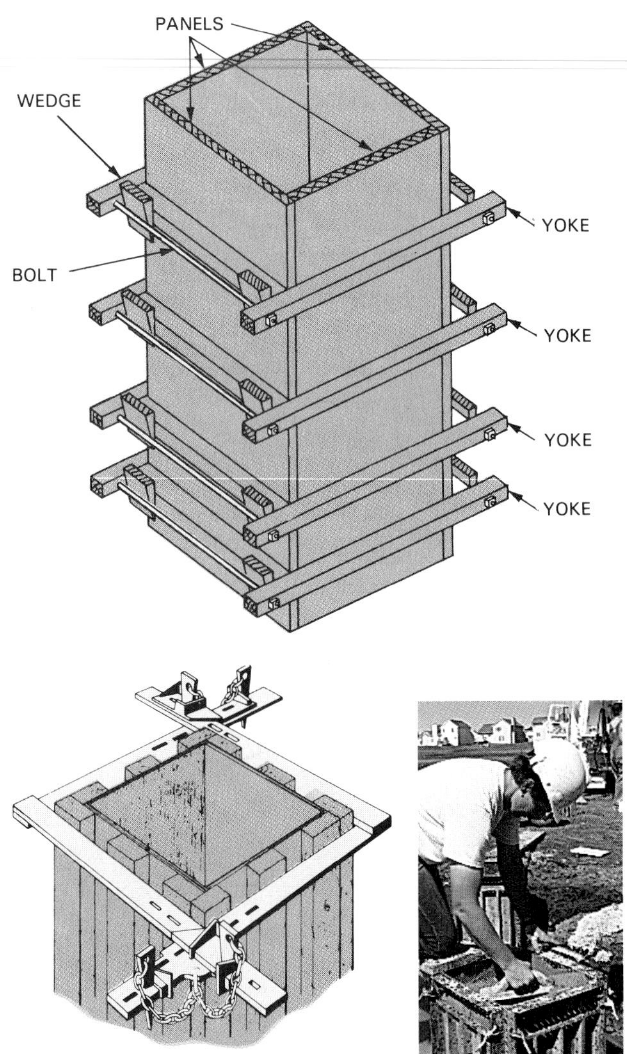

Figure 7-27. Prefabricated column forms save construction time. Top—Yoke and wedge arrangement. Bottom, left—Scissor clamps. (The Burke Co.) Bottom, right—Scissor clamps on a construction site. (Des Moines Public Schools)

Wall Openings

Several procedures are followed in forming openings in foundation walls for doors, windows, and other holes. In poured walls, forms or stops are built into the regular forms. *Nailing strips* may be attached to the form and cast into the concrete. Frames are then secured to these strips after the forms are removed. Figure 7-30 shows several methods of framing openings.

Special framing must also be attached inside the form for pipes or voids for carrying beams. As with windows and doors, formwork for these structures must be attached to the outside wall form before the inside form is erected.

Tubes of fiber, plastic, or metal can be used for small openings. They are held in place by wood blocks or plastic fasteners which are attached to the form. Larger forms made of wood can be attached with duplex nails driven through the form from the outside.

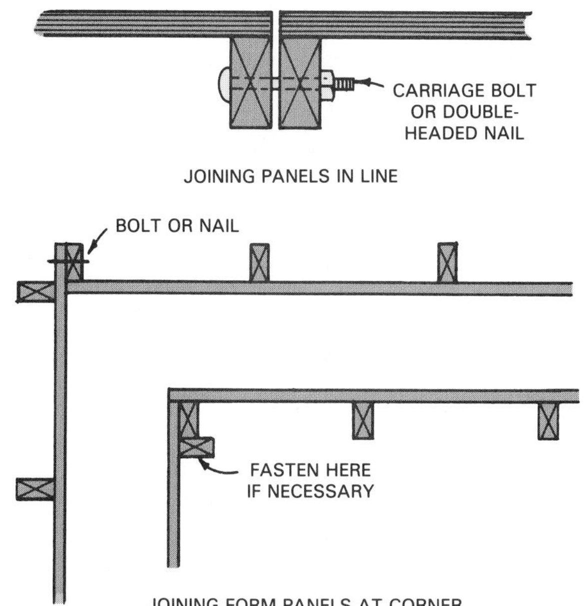

Figure 7-28. Methods of fastening form panels made by the carpenter.

Figure 7-29. Manufactured forms are easier to erect and strip. Top—Form designed for residential foundation. Wedge-bolt connectors are tightened or loosened with a light blow of the hammer. Bottom—Close-up of system having patented corners and form ties. (Universal Form Clamp Co.)

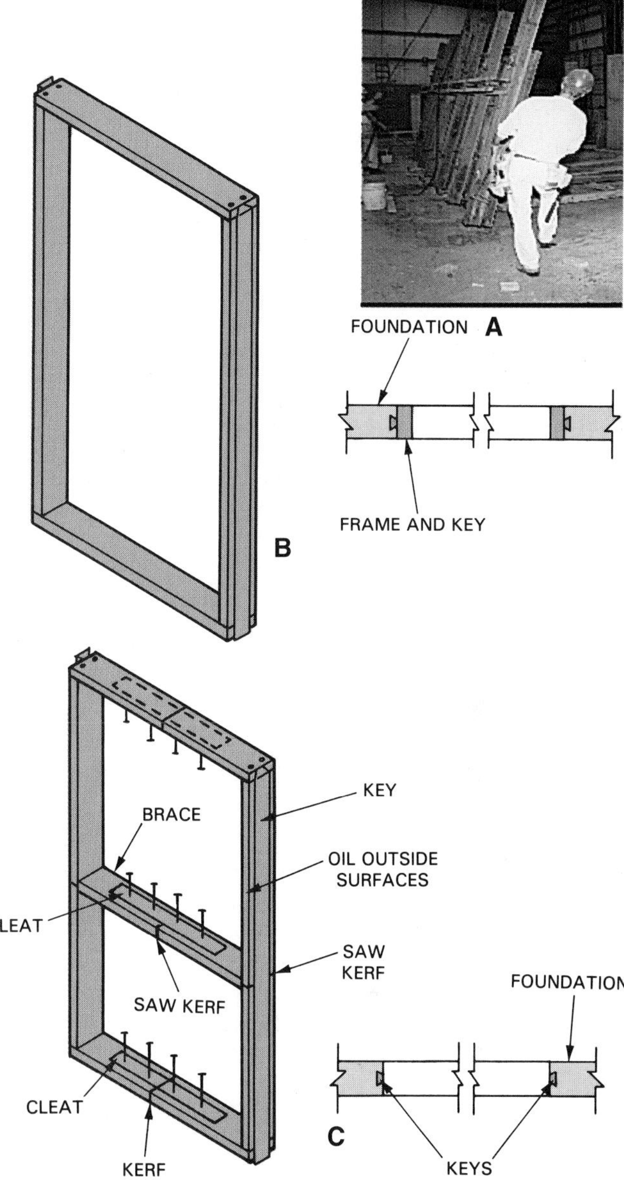

Figure 7-30. Forms used to frame openings in foundation wall can be constructed by the carpenter. A—Trainees have framed in features that will leave voids in a poured wall for later construction. (Construction Training School, St. Louis, Mo.) B—Permanent frame will be left in the wall. C—This frame, called a buck, is designed to be removed. Members are cut partway through for easy removal. Cleats and braces reinforce members at saw cuts. Bucks and frames are nailed to the form with duplex nails driven from the outside.

In concrete block construction, door and window frames are set in place. The masonry units are constructed around them. The outside surface of the frames has grooves into which the mortar flows, forming a key. Basement windows are usually located level with the top of the foundation wall. The sill carries the weight of the structure across the opening, Figure 7-31.

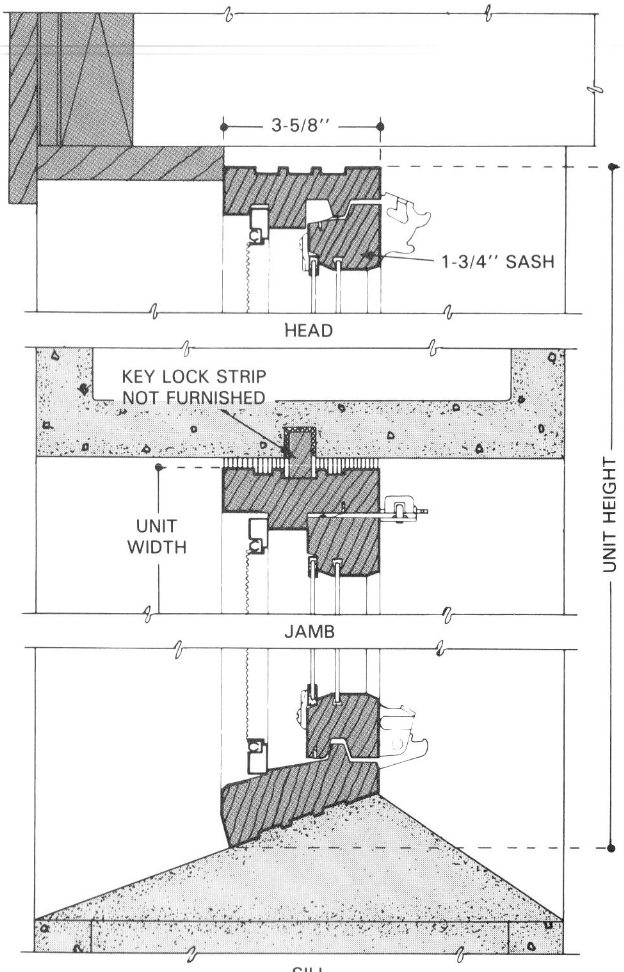

Figure 7-31. Installing windows in basement walls. Left—Detail of basement window unit that can be placed in a concrete or masonry foundation wall. Top right—Same unit installed. (Andersen Corp.) Bottom right—Steel window frame set in a poured wall.

Pilasters

Long walls may have pilasters. A *pilaster* is a thickened section of a concrete or masonry wall that strengthens the wall or provides extra support for beams. Figure 7-32 shows a form set up for pouring a pilaster in a concrete foundation wall.

Placing Concrete

Usually most concrete can be poured directly from the ready-mix truck into the forms. To move the concrete to other areas not accessible to the truck, a wheelbarrow or bucket is generally used, Figure 7-33.

Here are some general guidelines for placing concrete:

- Place concrete near to where it will rest. Never allow it to run or be worked over long distances. To do so could cause segregation. (This is a condition in which large aggregates become separated from cement paste and smaller aggregates.)

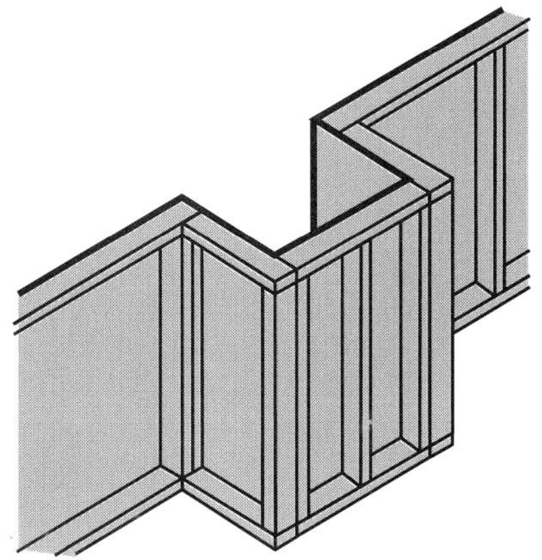

Figure 7-32. Section of foundation form set up for a pilaster. Often the studs at the sides are omitted.

- Place concrete in forms promptly after mixing.
- Concrete for walls should be placed in the forms in

horizontal layers of uniform thickness not exceeding 6" to 12". As the concrete is placed, spade or vibrate it enough to compact it thoroughly. This produces a dense mass.

- Working the concrete next to the form tends to produce a smooth surface. It prevents honeycombing along the form faces. A spade or thin board may be used for this purpose. Large aggregates are forced away from the forms and any air trapped along the form face is released. Mechanical vibrators are effective in consolidating concrete.
- Mechanical vibrators create added pressure on the forms. This factor must be considered in the form design. The vibrator should not be held in one location long enough to draw a pool of cement paste from the surrounding concrete.

Anchors

Wood plates are fastened to the top of foundation walls with 1/2" anchor bolts or straps. They are spaced not more than 4' apart, See Figure 7-34. In concrete walls, they are set in place as soon as the pour is completed and leveled off.

Anchor bolts are set in the cores of a concrete block wall. They should be about 18" long. A piece of metal lath is placed in the second horizontal joint below the top of the wall to hold the grout or mortar. Bolts are installed after the wall is completed. Anchor clips are installed in the same way. See Figure 7-35.

Wall forms help protect concrete from drying too fast. They should not be removed until the concrete is strong enough to carry the loads that will be placed on it. The material should be hard enough so the surface is not damaged by the stripping operation. Hardening of concrete will normally take a day or two.

Figure 7-33. Avoid moving concrete long distances after it is dumped into a form. Use a bucket or wheelbarrow for transport from the ready-mix truck or concrete mixer. Top—Readying a dump from a bucket. Bottom—Wheeling concrete for pouring a basement floor. (Kasten-Weiler Construction)

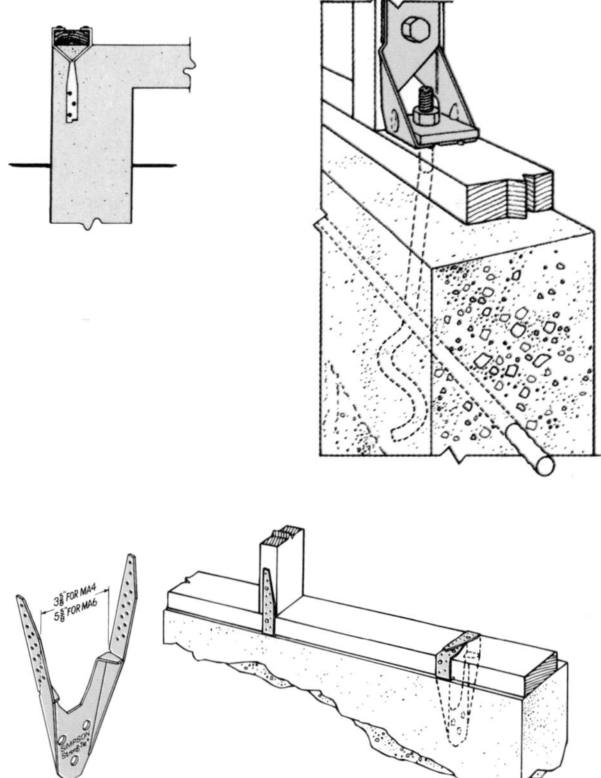

Figure 7-34. Patented anchor clips and bolts are embedded into top of concrete or masonry foundation walls. They secure sill to foundation. (The Panel Clip Co., and Simpson Strong-Tie Co., Inc.)

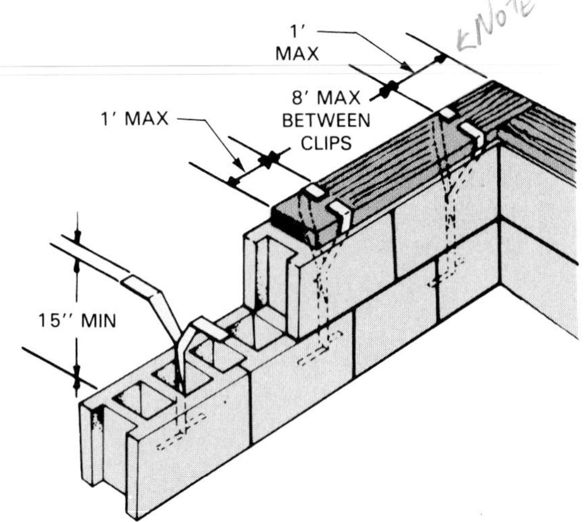

Figure 7-35. Anchor bolts and straps are installed in the same manner. Top—Anchor strap is embedded in the next-to-top block. Bottom—Embedding an anchor bolt in concrete or mortar.

Concrete Block Foundations

In some localities, *concrete blocks* are used for foundation walls and other masonry construction. The standard block is made from Portland cement and aggregates such as sand, fine gravel, or crushed stone. It weighs about 40 or 50 lb.

Lightweight units are made from Portland cement and natural or manufactured aggregates. Among these are volcanic cinders, pumice, and foundry slag. A lightweight unit weighs between 25 and 35 lb. It usually has a much lower U factor. (This is a measurement of the heat flow or heat transmission through materials.)

Blocks should comply with specifications provided by the American Society for Testing Materials. ASTM specifications for a Grade A load-bearing unit requires that the compression strength equal 1000 lb. per sq. inch. Thus, an 8 x 8 x 16 unit must withstand about 128,000 lb., or 64 tons.

Sizes and Shapes

Blocks are classified as solid or hollow. A solid unit is one in which the core (hollow) area is 25 percent or less of the total cross-sectional area. Blocks are usually available in 4", 6", 8", 10", and 12" widths and 4" and 8" heights. Figure 7-36 shows some of the sizes and shapes.

Sizes are actually 3/8" shorter than their nominal (name) dimensions to allow for the mortar joint. For example: the 8 x 8 x 16 block is actually 7 5/8" x 7 5/8" x 15 5/8". With a standard 3/8" mortar joint, the laid-in-the-wall height will be 8" and the length 16".

It is important to limit the moisture content of concrete block. Cover blocks stockpiled at the job site to protect them from the elements, Figure 7-37. They should be secured on pallets or other supports to keep them from contact with the ground. Never wet blocks before or during laying a wall.

Mortar

To build a wall, blocks are bonded one to another with *mortar*. This is a mixture of cement, lime, and masonry sand. These materials give the mortar its strength, ability to retain water, durability, and bonding capabilities.

When water is added, just before use, the mortar retains its plastic properties long enough for the mason or carpenter to lay up the block before the mortar sets. Mortar can be made to resist different pressures, depending on where it is to be used. ASTM has set standards for different uses. Mortar used in foundations must, when set, withstand pressures from 1800 to 2500 psi. Mortar should be used within 1 1/2 to 2 hours after it has been mixed.

Laying Concrete Block

Foundation block should be laid on a good footing. The mason or carpenter first locates the corners on the footing. Then he or she may string out the first course without mortar to check the layout as shown in Figure 7-38. This will avoid unnecessary cutting of block. A chalk line is sometimes used to mark the position of the first course on the footing.

Next, a mortar bed, Figure 7-39, is laid down on the footing for the first course of blocks. Corner block are laid first. Head joints are buttered before being laid. The corners are usually built up first from four to six courses high, Figure 7-40. All corner blocks must be at the right height, in line, level, and plumb. The mason will use the level to check this after laying every three or four blocks. See Figure 7-41.

Adjusting of block must be done before the mortar in the joints stiffens. Each course at the corners is stepped back one-half block from the previous course. Laying the level across the diagonal formed by this stepping will show whether the correct horizontal spacing is being maintained.

With corners built up, laying the wall between corners can begin, Figure 7-42. Note how blocks are "buttered" before being laid. A *mason's line*, Figure 7-43, is used at each course as a reference for proper alignment. As each block is laid, excess mortar should be cut off with the trowel. The mason's level is now used only to check the face of each block to keep it lined up with the face of the wall.

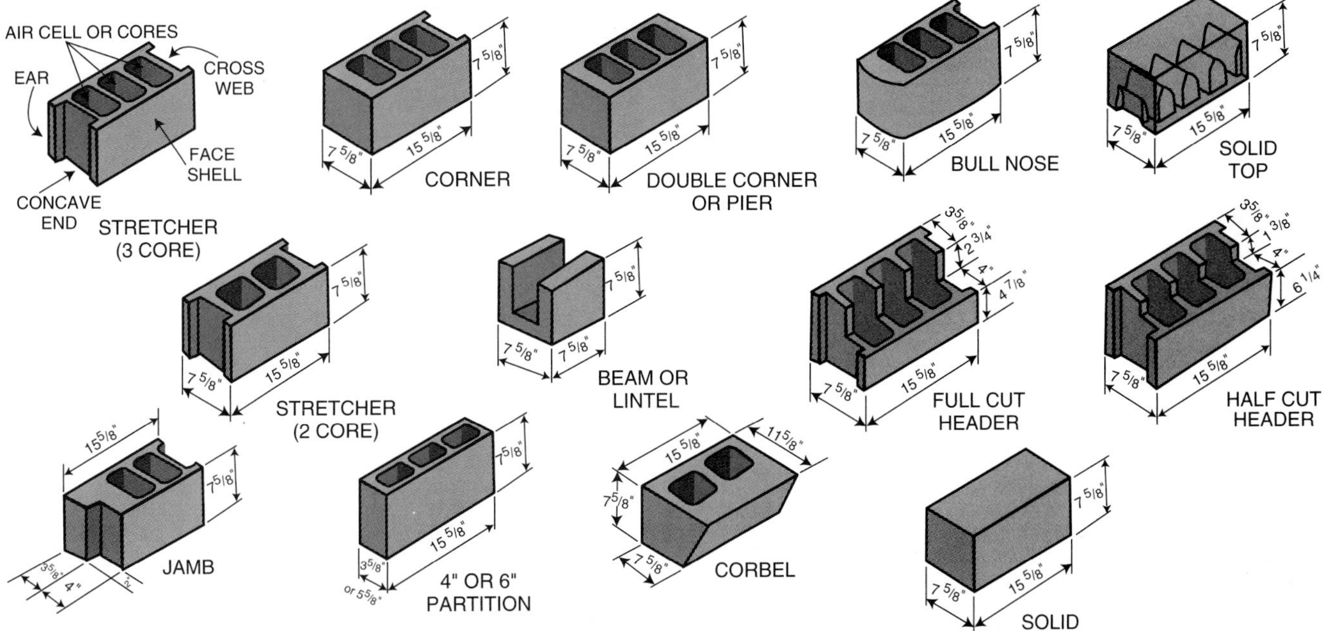

Figure 7-36. Concrete blocks are manufactured for many different purposes. The above shapes are typical. Sizes are specified in this order: width, height, and length. Sizes are nominal, allowing for mortar joints. (Portland Cement Assoc.)

Figure 7-37. Block delivered to a building site need to be covered to protect against rain. (Portland Cement Assoc.)

Figure 7-38. It is a good idea to lay out first course of block dry. It allows adjustments that minimize cutting of block. (Portland Cement Assoc.)

Figure 7-39. A bed of mortar is laid down on the footing for the first course of block. (Portland Cement Assoc.)

The last block to be placed in any course is called a *closure block.* All edges of this block are buttered before the block is carefully placed, Figure 7-44. Sometimes, a *course pole* is used to check the height of each course, Figure 7-45. Horizontal alignment is checked with a board or level, Figure 7-46.

Intersecting bearing walls should be tied into each other every sixth course. A tie bar should be used for this purpose.

Foundation walls should be topped with anchors to secure the sill. (Refer again to Figure 7-35.) Before the top two courses are laid, a strip of metal lath wide enough to cover the core spaces should be laid down. After the top

Figure 7-40. Corners are first laid up five or six courses high. (Portland Cement Assoc.)

Figure 7-41. A mason's level is used to align and level blocks. (Portland Cement Assoc.)

Figure 7-42. If corner blocks are carefully laid, laying the rest of the block wall is simplified. Note buttered blocks and line used as a guide. (Portland Cement Assoc.)

Figure 7-43. Mason's line is used as a guide for laying block between corners. (Portland Cement Assoc.)

course has been laid, fill the cores with mortar and install the anchor bolts. Sometimes solid block is used for the top course. Openings must then be cut for insertion of the anchor bolts.

Figure 7-44. Closure block is buttered at both ends and then carefully placed in the wall. (Portland Cement Assoc.)

Figure 7-45. A course pole can be used to check the height of each course. (Portland Cement Assoc.)

Figure 7-46. The horizontal alignment of bricks and blocks must be carefully checked. If blocks are not aligned, the wall will be weakened.

Corbel blocks are used when brick or stone veneer is to be used as exterior wall finish. The corbeling provides a supporting base for such masonry materials, Figure 7-47.

Lintels

Masonry that is located above the top of openings is supported by a structural unit called a *lintel.* This can be one of the following:

- A precast concrete unit that includes metal reinforcing bars (rebar). See Figure 7-48.
- Steel angle iron, Figure 7-49.

Another method is to lay a course of lintel blocks across the opening (supported by a frame). Then add reinforcing bars and fill the blocks with concrete, Figure 7-50.

Insulating Foundation Walls

In northern climates, when residential plans include a finished basement, insulating the foundation walls may be desirable. There are several ways to reduce the heat flow.

- Lightweight masonry units that have a lower U factor may be used.
- The cores of regular or lightweight blocks can be filled with an insulating material, Figure 7-51.
- Other methods include cavity wall construction or the use of various forms of insulation applied to the interior or exterior surface. Additional information about insulation is included in Unit 14.

Waterproofing

In most localities, the outside of poured concrete or concrete masonry basement walls should be waterproofed below the finished grade, Figure 7-52. Drain tile should also be installed.

Figure 7-47. A standard 8 x 8 x 16 block is being set in place along a course of corbel blocks. The top of the corbel block is a nominal 12" wide and provides a ledge to support a stone or brick veneer facing.

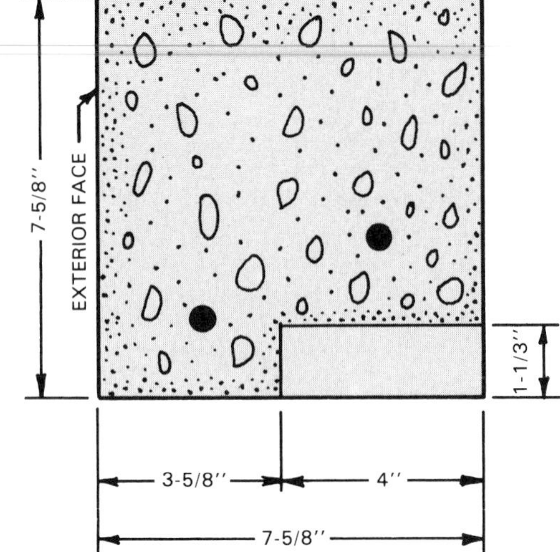

Figure 7-48. This is a cutaway view of a precast solid lintel for a block wall. Note the reinforcing rebar (black dots).

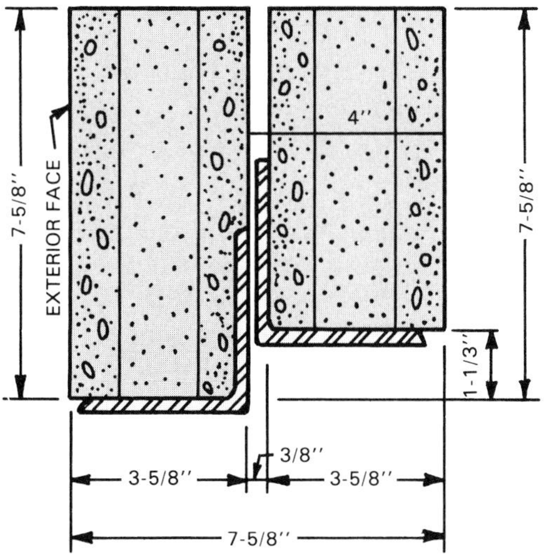

Figure 7-49. Cutaway view of lintel made up of standard block units supported by steel angle iron. Notches are for window or door frames.

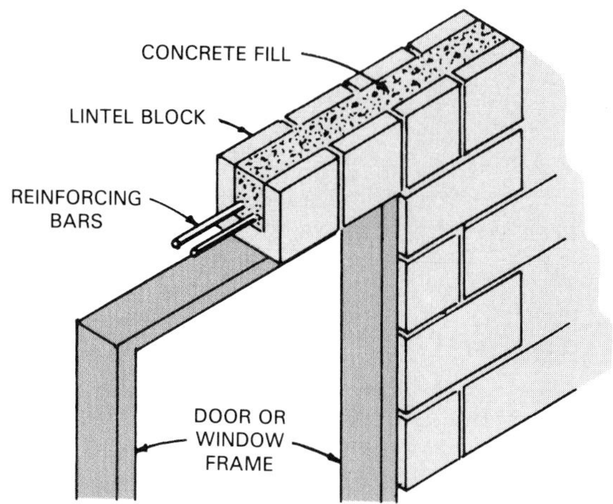

Figure 7-50. Lintel locks are laid over a supporting frame. Then the cavity is filled with reinforced concrete.

Figure 7-51. Loose insulation can be poured into cores of concrete block. (Portland Cement Assoc.)

Masonry (block) walls may be waterproofed by an application of cement plaster followed by several coats of an asphaltic material. The wall surface should be clean and dampened with a water spray just before the first coat of plaster is applied. The plaster can be made of cement and sand (1 to 2 1/2, mix by volume) or mortar may be used. When the first coat has partially hardened, it should be roughened with a scratcher to provide better bond for the second coat.

After the first coat has hardened at least 24 hours, apply a second coat. Again, dampen the surface just before applying the plaster. Both coats should extend from about 6" above the finished grade to the footing.

A cove of plaster should be formed between the footing and wall. This precaution prevents water from collecting and seeping through the joint. The second coat should be kept damp for at least 48 hours.

In poorly drained soils, or when it is important to secure added protection against moisture penetration, the plaster may require coating with asphalt waterproofing or hot bituminous material. See Figure 7-53.

Figure 7-52. Applying cement plaster to a concrete block foundation. Two coats may be required.

To waterproof poured walls, bituminous waterproofing is generally used without cement plaster. Polyethylene sheeting is often used as a waterproofing material. It should cover the wall and footing in one piece.

Figure 7-54 (top) shows a concrete drain tile being laid along the side of a footing. The system should lead away from the foundation to an outlet that always remains open. In some localities, perimeter drains may be connected to a sump pump. The drain is usually installed at a slope of about 1" in 20'. Tiles are spaced about 1/4" apart and the joint covered with strips of roofing felt.

Figure 7-54 (bottom) shows perforated plastic drain piping being installed. Perforated piping is widely used because it requires less labor than standard concrete tile. An alternate system referred to in Figure 7-18, builds drainage into a patented footing form. Called Form-a-drain, it employs a metal channel 4" to 6" high with perforations at intervals. It remains in place as part of the foundation drainage.

As an energy conservation measure, plastic foam board may be added to the outside face of the foundation wall. In such cases, no back plastering or waterproofing should extend any higher than 2" below the final grade. More information on this type of insulation will be found in Unit 14.

Backfilling

After the foundation has cured and waterproofing has been completed, the excavation outside the walls needs to be refilled with earth. This is known as *backfilling*. Since this places considerable back pressure on the wall, the first floor framing and rough flooring are usually installed first. Walls should also be braced, Figure 7-55. This is especially important with concrete block foundations. See Figure 7-56.

If tile lines have been installed, care must be taken to protect the lines from movement. When heavy equipment is used to backfill, the operator must be careful not to damage the walls.

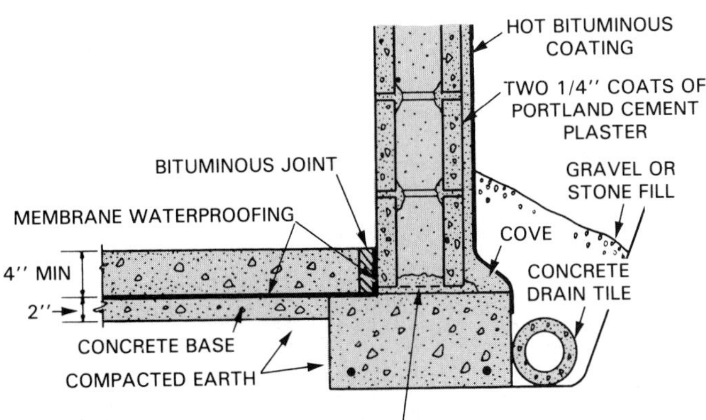

Figure 7-53. Sectional view shows method of waterproofing basements. It is especially effective in very wet soils. Plaster is omitted on outside wall in poured foundations.

Figure 7-54. Installing footing drains. Top—Laying standard cement drain tile. Joints are being covered with roofing felt. Middle—Perforated plastic pipe is delivered in coils. Bottom—Being flexible and already perforated, plastic pipe is easier to install than concrete tile.

Slab-on-Ground Construction

Today, many commercial and residential structures are built without basements. The main floor is formed by placing concrete directly on the ground. Footings and foundations are similar to those for basements. However, they need to extend down only to solid soil and below the frost line. See Figure 7-57.

In slab-on-ground construction, insulation and moisture control are essential. The earth under the floor is called the subgrade and must be firm and completely free of sod, roots, and debris.

A coarse fill, at least 4" thick, is placed over the finished subgrade. The fill should be brought to grade and thoroughly compacted.

This granular fill may be slag, gravel, or crushed stone, preferably ranging from 1/2" to 1" in diameter. The material should be uniform without fines to ensure maximum air space in the fill. Air spaces will add to insulating qualities and reduce capillary attraction of subsoil moisture (action by which moisture passes through fill).

In areas where the subsoil is not well drained, a line of drain tile may be required around the outside edge of the exterior wall footings.

While preparing the subgrade and fill, various mechanical installations should be made. Underfloor ducts, where used, are usually embedded in the granular fill. Water service supply lines, if placed under the floor slab, should be installed in trenches deep enough to avoid freezing. Connections to utilities should be brought above the finished floor level before pouring the concrete.

After the fill has been compacted and brought to grade, a vapor barrier should be placed over the sub-base. Its purpose is to stop the movement of water into the slab.

Among materials widely used as vapor barriers are 55 lb. roll roofing, 4-mil polyethylene film, and asphaltic-impregnated kraft papers. Strips should be lapped 6" to form a complete seal. A vapor barrier is essential under every section of the floor. Instructions supplied by the manufacturer should be carefully followed.

Figure 7-55. Walls should be adequately braced. It is also advisable to install the floor frame before backfilling. This helps the wall resist the pressure of backfilling and grade settling.

Figure 7-56. Before backfilling, pressed steel areaways must be attached around basement windows. Note that drainage is provided.

Perimeter insulation is important. It reduces heat loss from the floor slab to the outside. The insulation material must be rigid and stable while in contact with wet concrete. Figure 7-58 shows a foundation design with perimeter insulation typical of residential frame construction. Note how rigid insulation is applied to the exposed foundation exterior. Thickness of the insulation varies from 1" to 2", depending on outside temperature and type of heating. The insulation can be placed either horizontally or vertically along the foundation wall. Refer to Unit 14 for additional information.

When the insulation, vapor barrier, and all mechanical aspects are complete, reinforcing mesh, if used, is laid, Figure 7-59. Check local building codes for their requirements. Usually a 6 x 6 x 10 gage mesh is sufficient for residential work. This should be carefully located from 1" to 1 1/2" below the surface of the concrete.

Terrazzo, ceramic tile, asphalt tile, wood flooring, linoleum, and wall-to-wall carpeting are coverings appropriate for use on concrete floors. When linoleum, asphalt tile, or similar resilient type flooring materials are to be applied, the concrete surface is usually given a smooth steel-troweled finish. Information on finished wood floors is provided in Unit 16.

Basement Floors

Many of the considerations previously listed for slab-on-ground floors also apply to basement floors. Basement floors are poured later in the building sequence; sometime after the framing and roof are complete and after the

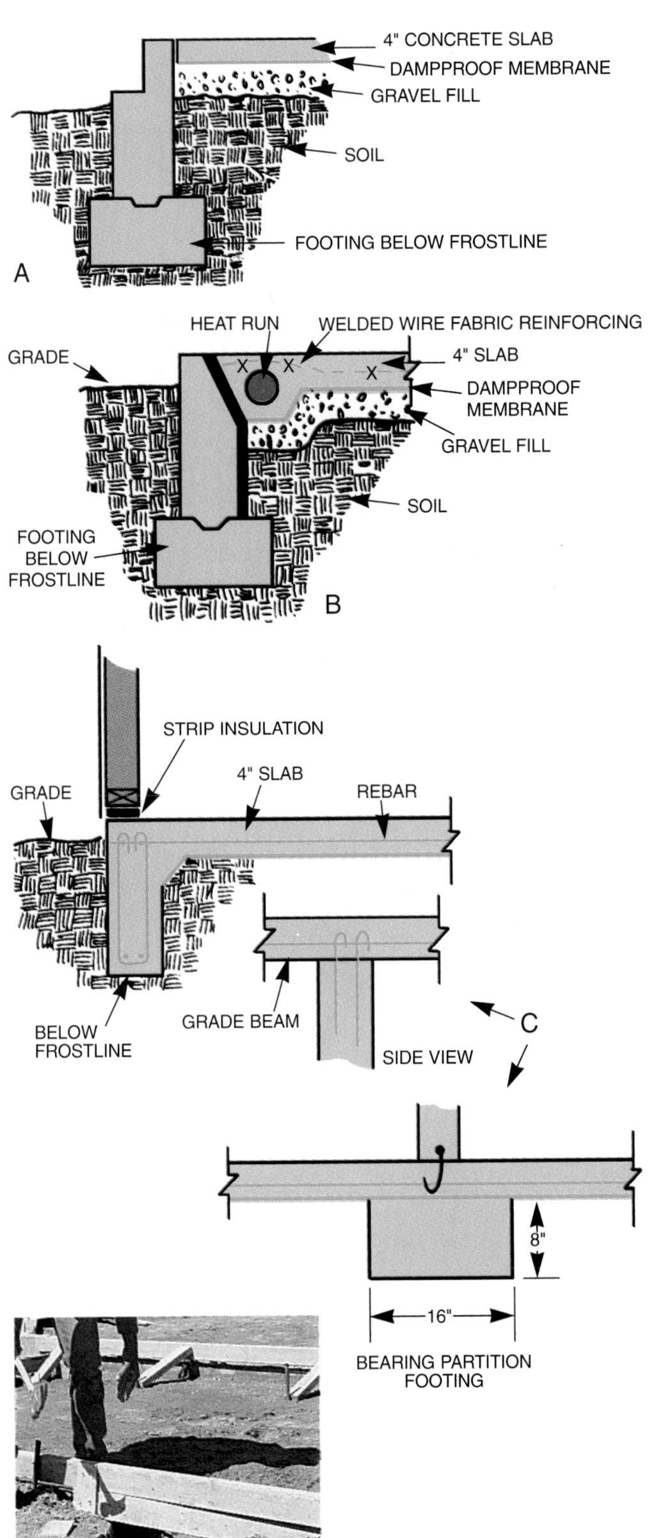

Figure 7-57. Four types of slab-on-ground. A—Unreinforced slab with loads supported by footing and wall is used where soil is coarse and well drained. B—Slab is reinforced with welded wire fabric. Inside of foundation wall is insulated because of perimeter heat duct included in the slab. C—Monolithic slab. This type is used over problem soils. Loads are carried over a large area of the slab. D—In warmer climates, such as southwest United States, frost is not a problem, so slab foundations are used almost exclusively with reinforced perimeter no deeper than 10". (Pierce Construction, Ltd.)

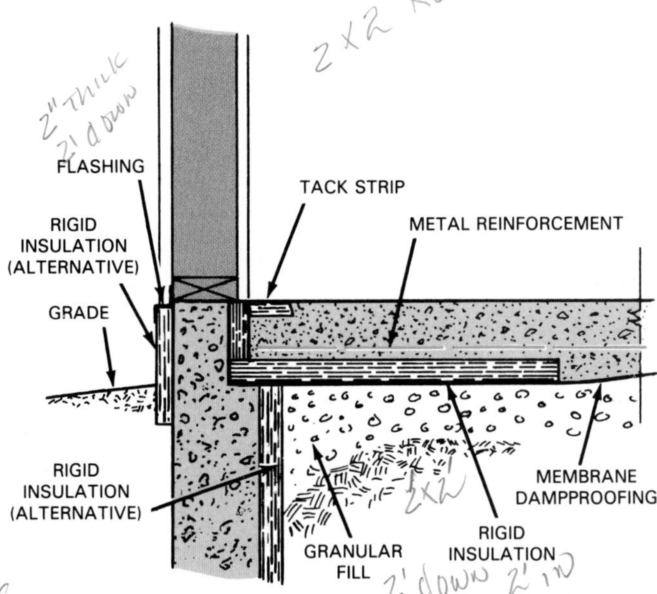

Figure 7-58. Perimeter insulation is important when using slab-on-ground construction in cold climates.

Figure 7-59. Top—In construction of a slab foundation, reinforcing mesh is laid on top of a vapor barrier. As concrete is poured, the mesh is lifted up so that it is 1" to 1 1/2" below surface of finished slab. Bottom—Perimeter of slab is reinforced with rebar. (Witkowski Construction)

plumber has installed waste plumbing and water service lines.

Essentials for a slab foundation:

- Subgrade beneath a slab should be well and uniformly compacted to avoid uneven settling of the slab.
- Finished level of the slab should be at least 8" above grade. Grade should be 6" below siding.
- Never lay a slab without removing top soil.
- Sewer and water lines should be installed and covered with 4" to 6" of crushed rock, well tamped.
- Put down a vapor barrier of 6-mil polyethylene before placing reinforcing and concrete. This not only will keep the slab dry but will keep curing concrete from losing needed moisture to the crushed rock or gravel base.
- Screed concrete immediately after it is poured. Float it when the sheen has disappeared from the surface.
- If a smooth, dense surface is important for application of finish flooring, give the slab a final smoothing with a steel trowel.

Super-Insulated Basement Floors

Most basement floors are poured over a base of gravel, covered with a damp-proof membrane. The typical super-insulated floor also has a 2" layer of rigid foam insulation between the gravel base and the vapor barrier, Figures 7-60 and 7-61.

To prevent cracks, a grid of rebar is installed, as shown in Figure 7-62. This can be held off the surface with patented metal chairs or blocked up with scrap pieces of 2 x 4. If the owner has specified radiant heat in the floor slab, special high resistance conductor is installed in parallel loops from wall to wall, Figure 7-63. This resistance wire is fastened at intervals to the metal reinforcing bars with plastic ties. This work must be completed before the ready-mix concrete arrives.

The concrete is delivered to the basement through a window, stairway, or an open section of the outside wall, Figure 7-64. Sometimes the rough opening for a fireplace located on an outside wall will provide easy access. Concrete should not be moved any great distance once placed in the form. A wheelbarrow is often used to carry it from the ready-mix truck's chute to the form. See Figure 7-65. Working from one side of the form to the other, workers screed the concrete as fast as the form fills, Figure 7-66.

The chief purpose of the *screeding* is to level the surface by striking off (removing) the excess concrete. Striking off commences as soon as the concrete is placed in the form. The screed rides on the edges of the form or previously poured sections of the slab. Low spots found behind the screeded concrete are filled and struck off once

Figure 7-60. Top—Rigid insulation 2" thick is sometimes used under a basement floor. Bottom—Seams in the rigid insulation are sealed with tape. Note that rebar from a section already poured is being held up out of the way. (Kasten-Weiler Construction)

Figure 7-61. A damp-proofing film is rolled out over the rigid insulation. Overlaps will be sealed with tape.

Figure 7-62. Making up grid of rebar. Every intersection is wire tied. High resistance wiring will also be placed at this time. (Kasten-Weiler Construction)

Figure 7-64. Chute from ready-mix truck accesses basement through the rough opening for the fireplace. (Kasten-Weiler Construction)

Figure 7-63. Parallel loop of electrical resistance wiring will heat the floor during the winter season.

Figure 7-65. Concrete is wheeled away from the chute and placed with a wheelbarrow.

more. Two people move the screed across the form. A sawing motion is used as the screed in moved along.

Power equipment may also be used for screeding. See Figure 7-67. Screeding leaves a level but coarse surface. *Floating* provides a smoother finish. It also fills up the hollows and compacts the concrete. The proper tool for floating is the wood float or the long-handled bull float, Figure 7-68. Floating is done when the concrete has begun to set. It must be firm enough so that a person's weight on it produces a footprint no more than 1/4" deep. There should be no bleed water present on the surface. Floating too soon will bring too many fines and more water to the surface. This will produce fine cracks in the surface of cured concrete.

Troweling produces a denser and smoother surface than floating. Trowels are made of metal and have a smooth finish on them. Troweling requires a firmer set than floating. It can proceed when working the surface does not bring up more water.

Figure 7-66. Screeding levels the concrete as the form is filled.

Entrance Platforms and Stairs

Entrance platform foundations should be constructed as a part of the main foundation or firmly attached to the main foundation. Reinforcing bars, placed in the wall when it is constructed, can provide a solid connection. Figure 7-69 shows a method of forming special support brackets for entrance platforms and steps.

Steps may be included and poured as part of the platform, Figure 7-70. When the steps are more than 3' wide, 2" stock should be used to prevent the risers from bowing. Detailed information on stair construction is included in Unit 17. The 2 x 8 riser boards are set at an angle of about 15° to provide a slight overhang (nosing). Also, the bottom edges of the boards are beveled to permit the mason to trowel the entire surface of the tread.

Some concrete steps must be poured against a wall or between two existing walls. A form can be constructed like the one shown in Figure 7-71.

Figure 7-67. Screeding may be done with a power screed. (Portland Cement Assoc.)

Sidewalks and Drives

Usually, sidewalks and drives are laid after the finished grading is completed. If there is extensive fill, wait until the grade has settled.

Main walks leading to front entrances should be at least 4' wide. Those to secondary entrances may be 3' or slightly less.

In most areas, sidewalks are 4" thick and the formwork is constructed with 2 x 4 lumber. Walks and drives are usually laid directly on the soil. If there is a moisture problem and frost action, a coarse granular fill should be put down first.

When joining two levels of sidewalks with steps, it is usually best to pour the top level first. If retaining walls are used, these should be constructed next, along with a segment of the lower sidewalk. Finally the steps are formed and poured as shown in Figure 7-72.

When setting sidewalk forms, Figure 7-73, provide a slope to one side of about 1/4" per foot. Increase thickness to 6" or add reinforcing where there will be heavy vehicular traffic. The concrete should not be permitted to bond against foundation walls or entrance platforms. It should be permitted to "float" on the ground. A 1/2" thick strip of asphalt impregnated composition board is commonly used to form this separation.

Driveways should be 5" or 6" thick with reinforcing mesh included. A single driveway should be at least 10' wide and a double driveway a minimum of 16' wide. Minimum crossway slope should be 1/4" per foot.

Concrete should be placed between the forms so it will be close to its final position. Do not overwork the concrete while it is still plastic. This tends to bring excess

Figure 7-68. Top—Floating the finished slab with a bull float. This tool can be made of wood, steel, or magnesium. (Kasten-Weiler Construction) Bottom—Troweling a smooth finish with a power troweler.

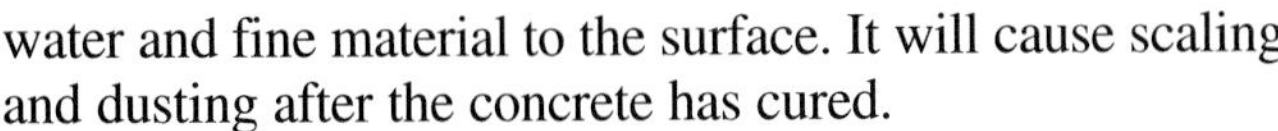

Figure 7-69. Entrance steps can be supported by special brackets cast in the wall when it is poured. Top—Form for the brackets is made by placing sloping 2 x 8s between the sides of the form. Bottom—Brackets are poured at the same time as the foundation. They will support steps and porch slab, preventing settling. (Portland Cement Assoc.)

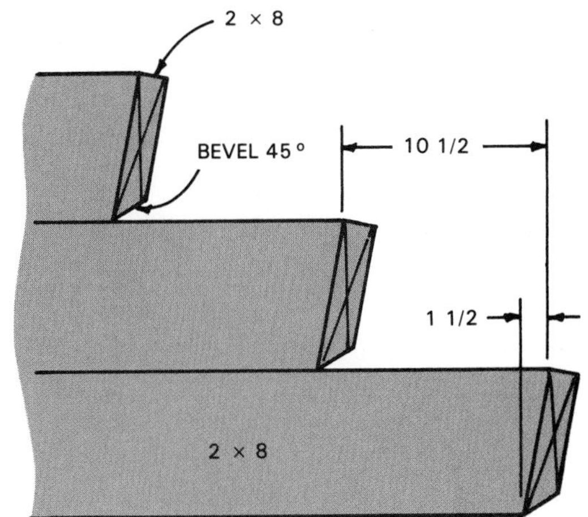

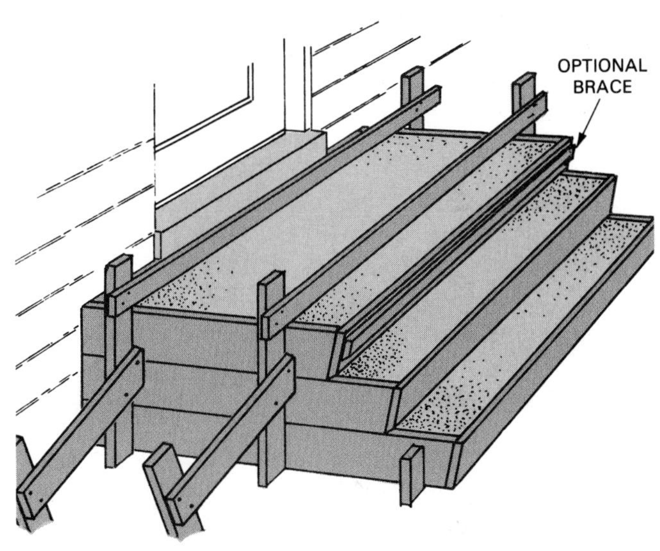

Figure 7-70. Constructing forms for steps. Nosing is formed by tilting form boards inward at the bottom as shown at top.

water and fine material to the surface. It will cause scaling and dusting after the concrete has cured.

Screeding

After the concrete is roughly spread between the forms, screed it immediately as described under laying basement floors. A small amount of concrete should always be kept ahead of the screed, Figure 7-74. After screeding, move a float over the surface. When skillfully performed, this operation removes high spots, fills depressions, and smoothes irregularities.

As the concrete stiffens and the water sheen disappears from the surface, finish edges and cut control joints. These joints should extend to a depth of at least one-fifth of the thickness of the concrete.

For sidewalks and driveways, the distance between the control joints is usually about equal to the width of the slab. Joints can be formed with a groover and straightedge or cut with a power saw. The saw must have a masonry blade. Sawing is done 18 to 24 hours after the concrete is poured.

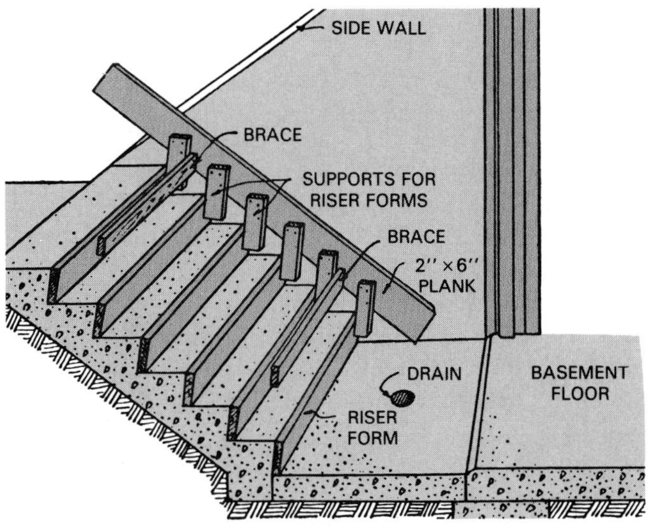

Figure 7-71. This method of building forms can be used when steps are located between two walls already in place.

Figure 7-72. Form is in place for steps between two levels of sidewalk. Sloping 2 x 4s support the risers.

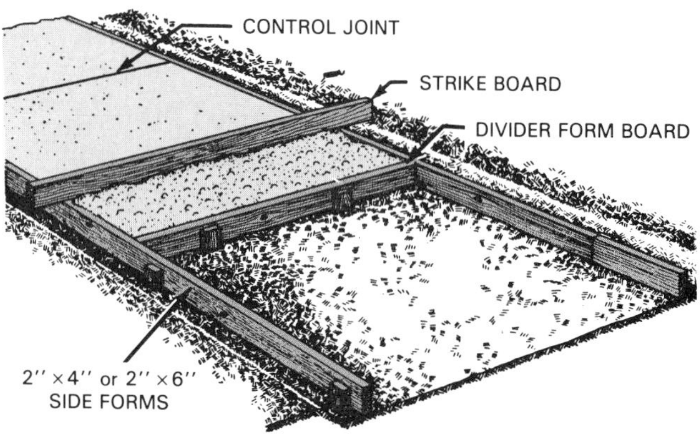

Figure 7-73. Basic setup for pouring a sidewalk. Control joint should be one-fifth the thickness of the concrete.

Edging

Edges of walks and driveways should be rounded by working an edger tool along the forms. The edging tool can also be used to finish the control joints.

Surface finishing operations should be performed after the concrete has hardened enough to become somewhat stiff. For a rough finish that will not become slick during rainy weather, the surface can be stroked with a stiff bristle broom. When a finer texture is desired, the surface should be steel-troweled and then lightly brushed with a soft bristle broom. Several key steps in the pouring and finishing of a concrete drive are shown in Figures 7-75 and 7-76.

For proper curing of concrete, protect it against moisture loss during the early stages of hardening (hydration). Covers of waterproof paper or polyethylene film are commonly used. A convenient alternative is to spray the concrete with a plastic-based curing compound, Figure 7-77. The sprayed material forms a continuous membrane over the surface.

Figure 7-74. Leveling and smoothing the surface of a sidewalk. A small amount of concrete should be kept just ahead of the screed.

Wood Foundations

The Permanent Wood Foundation (PWF) is a special building system that saves time because it can be installed in almost any weather. It provides comfortable living space in basement areas because the stud walls can be fully insulated. All wood parts are pressure treated with a solution of chemicals that make the fibers useless as a food for insects and the fungus that cause decay.

Foundation sections of 2" lumber and exterior plywood can be panelized in fabricating plants or constructed on the building site. Pressure treated wood foundations, Figure 7-78, have been approved by major code groups and accepted by FHA, HUD, and FmHA (Farmers Home Adm.).

For a regular basement, the site is excavated to the required depth (below frostline). Plumbing lines are installed and provisions are made for foundation drainage, following local requirements. Some soils will require a sump (pit for water collection) that is connected to a storm sewer, pump, or other drain.

The subgrade is then covered with a 4" to 6" layer of porous gravel or crushed stone and carefully leveled, Figure 7-79. Footing plates of 2 x 6 or 2 x 8 material are installed directly on this base. The wall sections of the foundation are then erected and securely fastened to the footing plate with noncorrosive fasteners. (All fasteners should be made of stainless steel, silicon bronze, copper, or hot-dipped zinc-coated steel. Special caulking compounds must be used to seal all joints between sections and between footing plates and sole plates.)

Before pouring the basement floor, the porous gravel or crushed stone base is covered with a polyethylene film

Figure 7-75. Spreading concrete for a driveway. Polyethylene film underlayment prevents ground from soaking up water from the concrete. Hook (arrow) is being used to raise reinforcing mesh into the concrete.

(6-mil thick), Figure 7-80. A screed board is used after the basement is poured.

The first floor frame is installed on the double top plate of the foundation wall with special attention given to methods of attachment so that inward forces will be transferred to the floor structure. See Figure 7-81. Where joists run parallel to the wall, blocking should be installed between the outside joist and the first interior joist.

Figure 7-77. A plastic-based material can be sprayed on concrete to form a continuous membrane that aids in proper curing.

Figure 7-78. Installing a wood foundation prefabricated in a plant. All components are pressure treated against rot and insect damage. Outside wall panels located below grade are protected by a 6 mil polyethylene film. (American Plywood Assoc.)

Figure 7-76. Finishing a concrete drive. Left—Using a power screed to compact and level concrete. Engine vibrates the wood frame. Forms (arrow) are patented steel channels supported with steel pins driven into the ground. Right—Large steel trowel is being used to smooth a concrete surface. Angle of trowel's front and rear edges is controlled by twisting the handle.

Figure 7-79. Preparation of subgrade to receive wood foundation. Excavation must extend below frostline. A layer of porous gravel 4 to 6" deep is needed. (American Plywood Assoc.)

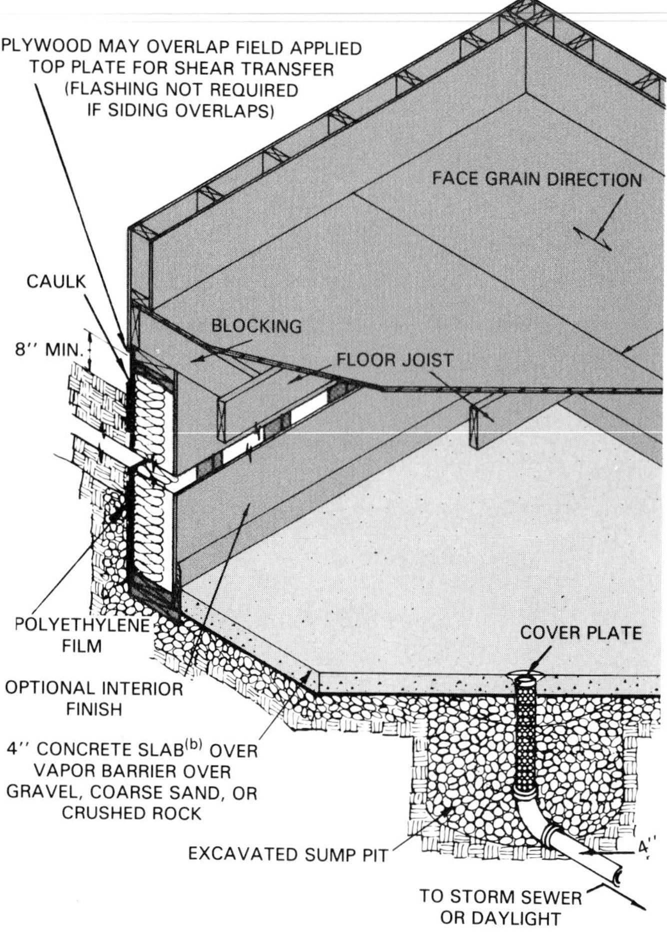

Figure 7-80. Typical wood foundation in cutaway. Note drainage sump which keeps the subsoil dry around the foundation. (American Plywood Assoc.)

The basement floor should be poured and first floor joists installed before backfilling is attempted. This is necessary to avoid shifting and other damage to the foundation. Also before backfilling, place a 6-mil polyethylene moisture barrier to sections of the wall below grade. Bond the top edge of the barrier to the wall at grade level with a special adhesive. Install a treated wood strip over this and caulk it. (Later it will serve as a guide for backfilling.)

Lap vertical joints in the polyethylene film at least 6". Seal joints with the same adhesive.

Do not backfill until the first floor has been installed and the basement floor has cured.

As with any foundation system, satisfactory performance demands full compliance with recommended standards covering design, fabrication, and installation. Standards for the construction of wood foundations are contained in manuals prepared by several associations. Write to the American Plywood Association, P.O. Box 11700, Tacoma, WA 98411 or the National Forest Products Association, 1250 Connecticut Avenue NW, Washington, DC 20036.

Carpenters installing wood foundations should make certain that each piece of treated lumber and plywood carries the mark "AWPB-FDN." This assures them that the materials meet requirements of code organizations and federal regulatory agencies.

Cold Weather Construction

Cold weather may call for some changes in the way concrete and masonry materials are handled and placed. It may be necessary to:

- Heat the materials.
- Cover freshly placed concrete or masonry.
- Erect an enclosure and keep the construction area heated.

When temperatures fall below 40°F (5°C), concrete should have a temperature of 50–70°F (10–21°C) when it is placed. Since most concrete is ready-mix, only the building site will need to be sheltered and heated. However, when blocks are being placed, the materials will also need shelter and heat.

Shelter can be arranged with scaffolding sections, lumber, and tarpaulins. If shelters cannot be built, the bagged materials and masonry units should be wrapped with canvas or polyethylene tarpaulins when the temperature is below 40°F (5°C). Be sure the material is stored so ground moisture cannot reach it. Figure 7-82 lists recommendations for handling of masonry materials in cold weather.

Protecting Concrete

Freshly placed concrete should be protected with a covering. Hydration (chemical reaction of cement and

Figure 7-81. Truss floor joists being installed over a wood foundation. Use of 10d nails is recommended so inward pressure on wall is transferred to the floor system.

WHEN AIR TEMPERATURE REACHES	DO THIS TO MATERIALS	DO THIS TO PROTECT PLACED MASONRY
Below 40 °F	Heat mixing water. Keep mortar temperatures 40 °F to 120 °F	Cover walls and masonry materials to protect from moisture and freezing. Use canvas or plastic.
Below 32 °F	Do all of above but also: Heat sand to thaw frozen clumps. Heat wet masonry units to thaw ice.	Provide windbreak for workers when wind speed is above 15 mph. Cover walls and materials after workday to protect against wetness and freezing. Keep masonry temperature above 32 °F, using heaters or insulated blankets for 16 hours after placing of units.
Below 20 °F	Besides above: Heat dry masonry units to 20 °F.	Enclose structure and heat the enclosure to keep temperature above 32 °F for 24 hours after placing masonry units.

Figure 7-82. Recommendations of Portland Cement Association should be followed for cold-weather construction.

water during hardening) creates heat. The covering will help hold the heat in until the concrete cures.

Avoid pouring concrete on frozen ground. When the ground thaws, uneven settling may crack the concrete. Before pouring, make sure that reinforcing, metals, embedded fixtures, and the insides of forms are free of ice.

Mortar Temperatures

Temperatures of materials used in mortar are important. Water is the easiest to heat. It can also store more heat and will help bring cement and aggregate up to temperature. Generally, it should be cooler than 180°F (82°C). There is a danger that hotter water could cause the mortar to set instantly.

When the air temperature is lower than 32°F (0°C), sand should be heated to thaw frozen lumps. If desired, the sand temperature can be raised as high as 150°F (65°C). A 50 gallon drum, open on one end, or a metal pipe works well for containing fire. Heap the sand over and around the container.

Admixtures

Admixtures are materials other than cement, water, and aggregate which are added to concrete or mortar to change its properties. Cold weather admixtures used for concrete or mortar include:

- Antifreeze to lower the freezing point of the mixture.
- Accelerators to speed up curing. They do not lower the freezing point.
- Air-entraining agents that improve workability and freeze-thaw durability of mortar as it ages.
- Corrosion inhibitors. When reinforcement is placed in winter construction, these materials prevent rust formation.

According to the Portland Cement Association, accelerators and air-entraining agents have proved successful for winter use and are recommended. Other admixtures may help but are not recommended by the Association. Accelerators include calcium chloride, soluble carbonates, silicates, fluosilicates, calcium aluminate, and organic compounds such as triethanolamine.

For mixing of mortar, admixtures must be obtained by the carpenter or contractor. They are added on the job. Ready-mix companies will include them in the mix upon specifications provided by the carpenter, contractor, or architect.

Estimating Materials

Concrete is measured and sold by the cubic yard. A cubic yard is 3' square and 3' high. It contains 27 cu. ft. (3 x 3 x 3 = 27). To determine the number of cubic yards needed for any square or rectangular area, use the following formula. All dimensions should be converted to feet and fractions of feet:

$$\text{cubic yards} = \frac{\text{width x length x thickness}}{27}$$

For example, to find the concrete needed to pour a basement floor that is 30' x 42' x 4":

$$\text{cu. yd.} = \frac{30 \times 42 \times 1/3}{27}$$

$$= \frac{\overset{10}{\cancel{30}} \times \overset{14}{\cancel{42}} \times 1}{\underset{9}{\cancel{27}}\ \underset{1}{\cancel{3}}} = \frac{140}{9}$$

$$= 15.56 \text{ cu. yd.}$$

You should allow extra concrete for waste or slight variations in the cross sections of the form. An additional 5–10% is usually added.

The number of concrete masonry units needed can be estimated by determining the number of units required in each course and then multiplying by the number of courses between the footing and plate.

For example: find the number of 8 x 8 x 16 blocks required to construct a foundation wall with a total perimeter (distance around the outside) of 144' and laid 11 courses high.

$$\text{Total number} = \frac{\text{perimeter}}{\text{unit length}} \times \text{number of courses}$$

$$= \frac{144'}{16"} \times 11$$

$$= \frac{144'}{(4/3)'} \times 11$$

$$= \frac{144 \times 3}{4} \times 11$$

$$= \frac{\overset{36}{\cancel{144}} \times 3}{\underset{1}{\cancel{4}}} \times 11$$

$$= 108 \times 11$$

$$= 1188 \text{ blocks}$$

Another method is to figure the face area of the wall in square feet and divide by 100. This figure is then multiplied by 112.5, which is the number of 8 x 8 x 16 blocks required to construct 100 sq. ft. of wall. This figure can be multiplied by 2.6 to find the cu. ft. of mortar required. See the table in Figure 7-83.

Wall thickness	For 100 sq. ft. of wall		For 100 concrete block
inches	Number of block	Mortar** cu. ft.	Mortar ** cu. ft.
8	112.5	2.6	2.3
12	112.5	3.9	3.5

* Based on block having an exposed face of 7 5/8 x 15 5/8 in. and laid up with 3/8" mortar joints.

* * With face shell mortar bedding —10 percent wastage included.

Figure 7-83. Quantities of concrete block and mortar can be calculated with this chart. (Portland Cement Assoc.)

Important Terms

Admixtures
Aggregate
Anchor bolts
Backfilling
Batter boards
Cement
Chairs
Closure block
Concrete blocks
Control point
Corbel block
Course pole
Floating
Ground supported slab
Hydration
Ledger boards
Lintel
Mason's line
Monolithic concrete
Mortar
Nailing strips
Pilaster
Plain footings
Reinforced footings
Screeding
Spread foundation
Stepped footing
Structurally supported slab
Troweling
Wales

Test Your Knowledge

1. Grading must *(always, sometimes)* be done before building lines are laid out.
2. Carpenters always locate lot lines. True or False?
3. Batter boards should be located ____________ feet or more away from the building lines.
4. Building sites on steep slopes or rugged terrain should be rough-graded before the building is laid out. True or False?
5. In cold climates, foundations should be located below the ____________ line.
6. In residential construction, a safe design is usually obtained by making the width of the footing ____________ as wide as the foundation wall.
7. A ____________ footing is one that changes grade levels at intervals to accommodate a sloping lot.
8. What is a grade beam?
9. Foundation forms constructed of 1" boards should be held in place with stakes placed ____________ feet apart.
10. Loose dirt and debris *(should, should not)* be removed from the ground under a footing.
11. Concrete is made by mixing ____________, ____________, ____________ and water in proper proportions.
12. Concrete hardens by a chemical action called ____________.
13. Each sack of Portland cement holds _______lb.
14. Ready-mix concrete is purchased by the ____________.
15. What is a pilaster?
16. When placing concrete in forms, working the concrete next to the forms tends to produce a ____________ surface along the form faces.
17. Wood sill plates are fastened to the top of a foundation wall with ____________, ____________, or ____________.

18. Explain how to lay up concrete block with a minimum cutting of block.
19. Why are corners laid up several courses high before laying the rest of the wall?
20. List two methods of "topping off" a concrete block foundation.
21. Give the chief purpose of screeding a concrete slab.
22. Explain why a slab is floated after it has been "struck off."
23. Suggest what may occur if a slab is floated while there is still water on its surface.
24. A concrete block specified as an 8 x 8 x 16 block is actually ____________ x ____________ x ____________.
25. A concrete basement wall may be waterproofed by using an application of ____________, ____________, or ____________.
26. List three types of slab-on-ground foundations.
27. In slab-on-ground construction, a ____________ should be laid over the sub-base to stop the movement of ____________ into the concrete slab.
28. In most areas, sidewalks are ____________ inches thick.
29. A wood foundaton is installed on a layer of ____________ that is ____________ to ____________ inches thick.
30. In cold weather it is accepted practice to pour concrete over frozen ground. True or False?
31. How many 8 x 8 x 16 concrete blocks are required to lay 100 square feet of wall surface?

Outside Assignments

1. Visit a ready-mix concrete plant and study the operations. Obtain information about the following: source of aggregates; handling and storing cement and aggregates; equipment used to measure and proportion mixtures; size of truck-mounted mixers; distance trucks can travel without extra charge; cost of a cubic yard of concrete in various psi ratings, and fractional parts of a cubic yard that can be ordered. Prepare a written report.
2. Find a set of house plans that includes a fireplace located on an outside wall. Prepare a scaled (1 1/2"= 1'-0") drawing of the formwork you would use for the footings under the fireplace wall. Show individual form boards and stakes. Include one or more section views to describe the shape of the footing.
3. Study reference materials and booklets prepared by such organizations as the Portland Cement Association or the Perlite Institute. Obtain information about air-entrained concrete, slump tests, lightweight aggregates, ultra-lightweight concrete, thermal conductivity, compression tests, reinforcing, and prestressed concrete units. Prepare an outline and report to the class.
4. Study the building code in your area and learn about such requirements as building setbacks from property lines, design of footings for residential structures, minimum depths for footings and foundations, and basic construction of concrete and masonry foundation walls. Summarize your findings in a written report for your class.

Split-face concrete block provide an attractive alternative to standard block.

Concrete worker in Arizona is using a wood float on a slab-on-ground foundation. (Pierce)

A freshly built free-standing concrete block wall must be braced to prevent collapse from wind pressure or other forces acting against it. (Portland Cement Assoc.)

8

Floor Framing

When the foundation is completed and concrete or mortar has properly set up, floor framing can begin. This task is usually completed before the foundation is backfilled. The floor frame helps the foundation withstand the pressure placed on it by the soil. With the floor frame installed, the site should be brought to rough-grade level. This makes lumber delivery easier and allows better access to the building.

Types of Framing

The building's basic design will dictate the type of framing used. See Figure 8-1. Other factors, such as soil conditions, climate, available materials, the carpenter's experience, and the owner's preference, will influence building methods.

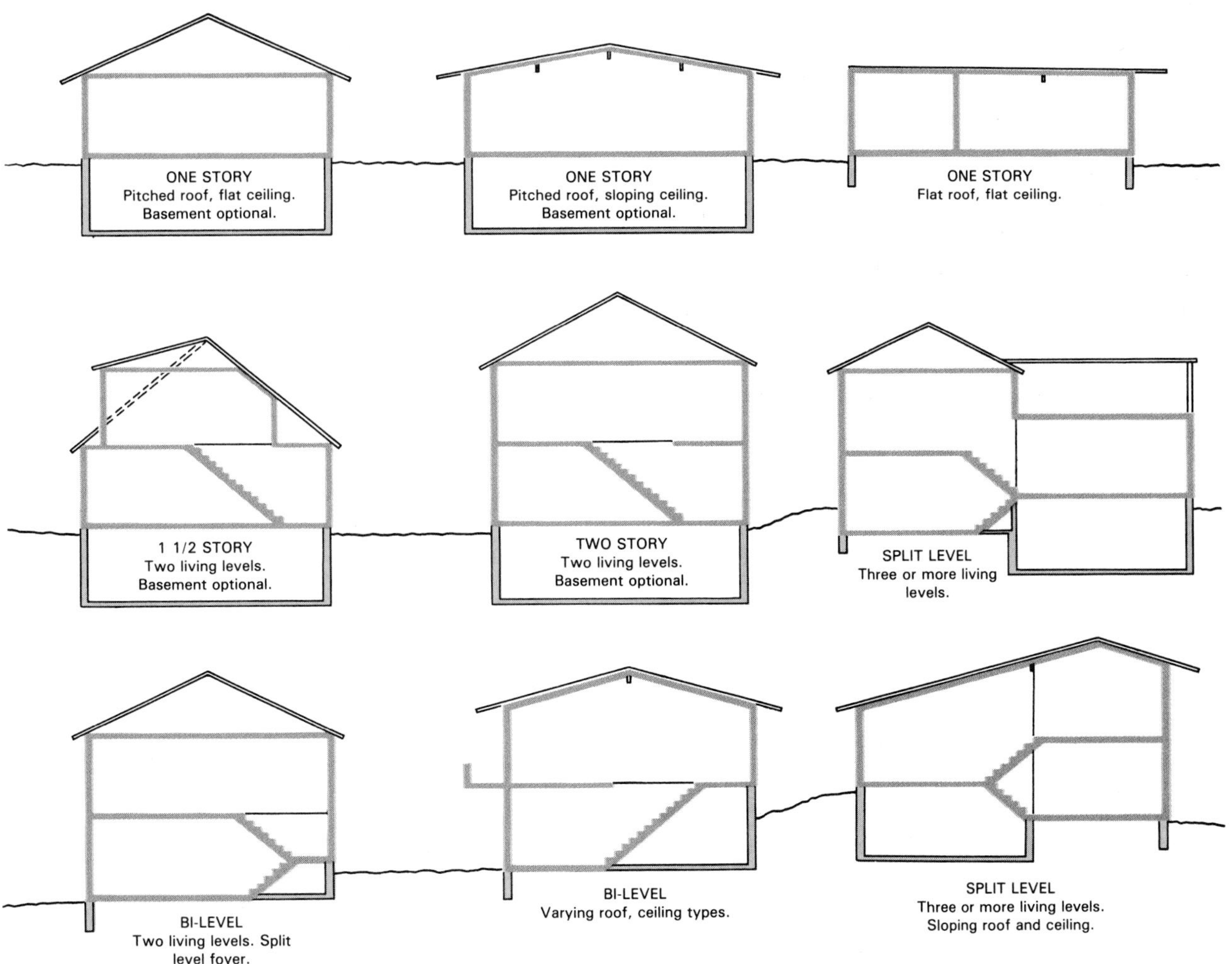

Figure 8-1. How a house will be framed depends on the type of design, how spaces are tied together, and the kinds of materials used. These are basic types.

In some parts of the country, buildings must be constructed with special resistance to wind and rain. In other parts, earthquakes may be the greatest hazard. In cold climates, heavy loads of damp snow may require special roof designs.

All structures should be built to reduce the effects of shrinkage and warping. They must also resist the hazard of fire.

There are two basic types of framing:

- *Platform framing* (also called western framing).
- *Balloon framing* (seldom used any more).

Joists, studs, plates, and rafters are the common structural members in both types of framing. Material with a nominal 2" thickness is used. Post and beam framing, also called plank and beam, is different. Its heavy structural members are 4" or more thick. This kind of construction is covered in Unit 22.

Platform Framing

Most modern residential and light commercial construction uses platform framing. The first floor is built on top of the foundation wall as though it was a platform. It provides a work area for assembling and raising wall sections safely and accurately. Wall sections are one story high. Outside walls and interior partitions support platforms for upper stories. Each floor is framed separately, as shown in Figure 8-2.

Usually, the first-floor platform will rest on a *sill*. This is the wood member that is attached to the foundation by bolts or other metal fasteners that are embedded in the foundation. Sometimes the platform will rest on an *underpinning*. This is a wall made of short studs that extends above a low foundation. It is usually used on a house that is to have a full basement. Sometimes an underpinning is called a *cripple wall*. Refer to Figure 8-2 again.

Figure 8-2. Example of platform framing. Platform (arrow) supports the wall for the next level. Note that the platform is supported by a short wall called an underpinning, cripple wall, or stub wall. (American Plywood Assoc.)

Platform framing is satisfactory for both one-story and multi-story structures. Settlement due to shrinkage occurs in an even and uniform manner throughout the structure.

Typical construction methods used at first and second floor levels are illustrated in Figure 8-3. The only firestopping needed is built into the floor frame at the second floor level. It prevents the spread of fire in a horizontal direction. Here it also serves as solid bridging, holding the joists in a plumb (vertical) position. To save energy, a strip of insulation is usually placed under the sill plate.

The type of framing is usually specified in the architectural plans. There will be sectional views of floors, walls, and ceilings. A typical detail drawing of first floor framing would include not only the type of construction but also the size and spacing of the various members. See Figure 8-4.

Balloon Framing

In balloon framing, now seldom used, the studs are continuous from the sill to the rafter plate. Ends of the second floor joists are supported on a *ribbon*. They are spiked to the stud as well. See Figure 8-5. Firestopping must be added to the space between the studs. This space, which also occurs in load-bearing partitions, permits easy installation of mechanical systems.

In balloon framing, shrinkage is reduced because the amount of cross-sectional lumber is low. Wood shrinks across its width but practically no shrinkage occurs lengthwise. Thus, the high vertical stability of the balloon frame makes it adaptable to two-story structures, especially where masonry veneer or stucco is used on the outside wall.

Girders and Beams

Joists are the supports of the floor frame. They rest on top of the foundation walls. Usually the *span* (distance between the walls) is so great that additional support must be provided. *Girders*, also called beams, resting on the foundation walls and on posts or columns, provide the needed support. Girders may be solid timbers, built-up lumber, or steel beams. Sometimes, a load-bearing partition replaces a girder or beam.

Sizing Girders

1. Find the distance between girder supports.
2. Find the girder load width. A girder must carry the weight of the floors on each side to the mid-point of the joists which rest upon it.

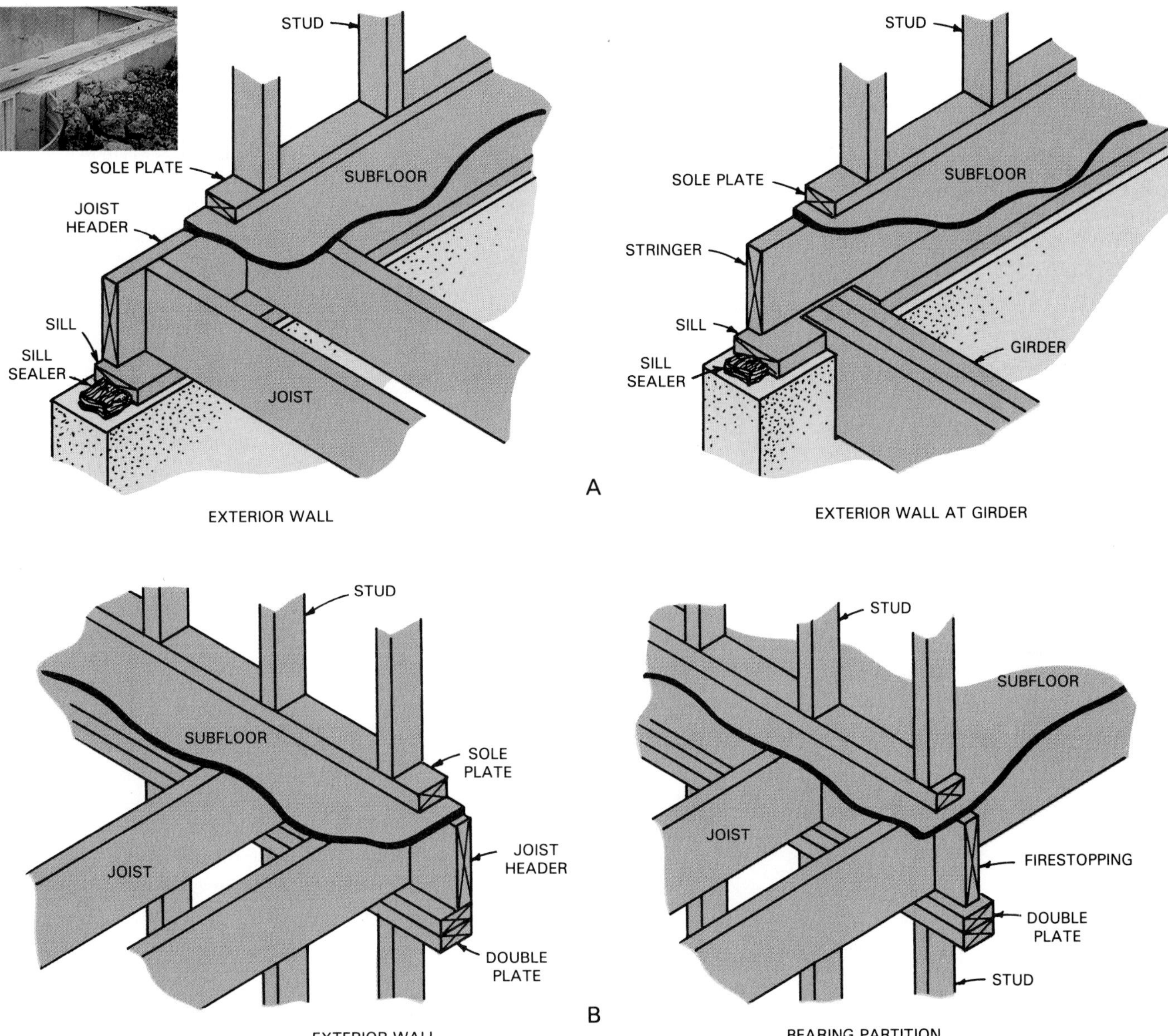

Figure 8-3. Platform framing details. A—First floor. Joist header also acts as a firestop. B—Second floor framing is similar to first. Note solid bridging also acts as firestop.

3. Find the "total floor load" per sq. ft. carried by joists and bearing partitions to girder. This will be the sum of loads per sq. ft. listed in the diagram, Figure 8-6. This does not include roof loads. These are carried on the outside walls unless braces or partitions are placed under the rafters. Then a portion of the roof load is carried to the girder by joists and partitions.
4. Find the total load on the girder. This is the product of girder span x girder load width x total floor load.
5. Select proper size of girder according to the code in your area. The table, in Figure 8-7 is typical. It indicates safe loads on standard size girders for spans from 6' to 10'. Shortening the span is usually the most economical way to increase the load a girder will carry.

Built-up girders can be made of three or four pieces of 2" lumber nailed together with 20d nails. See Figure 8-8. Joints should rest over columns or posts.

Modern engineered lumber, such as glue-laminated beams, may be used for girders. These are factory made by gluing thin strips of wood or veneer together. They can be manufactured to many different sizes and lengths.

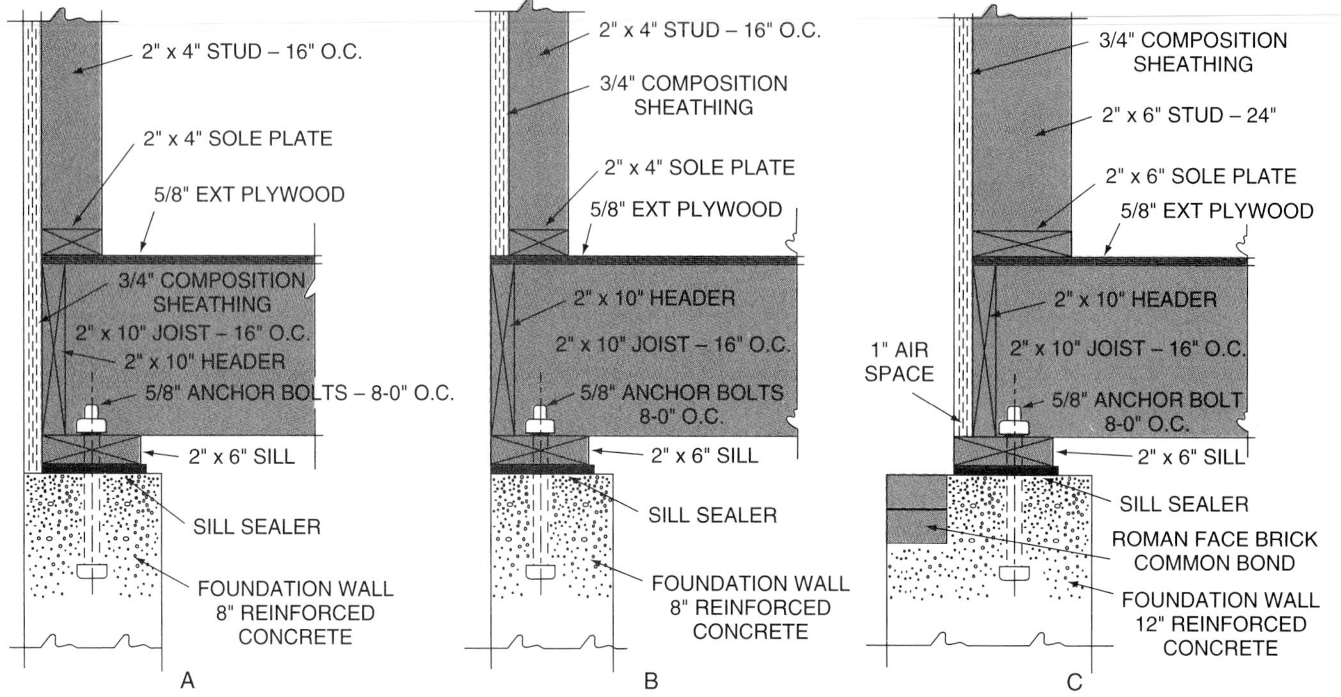

Figure 8-4. Architectural detail drawings show methods of construction as well as materials to use. A—Sheathing brought to foundation. B—Sheathing brought to sole plate. C—Brick veneer construction.

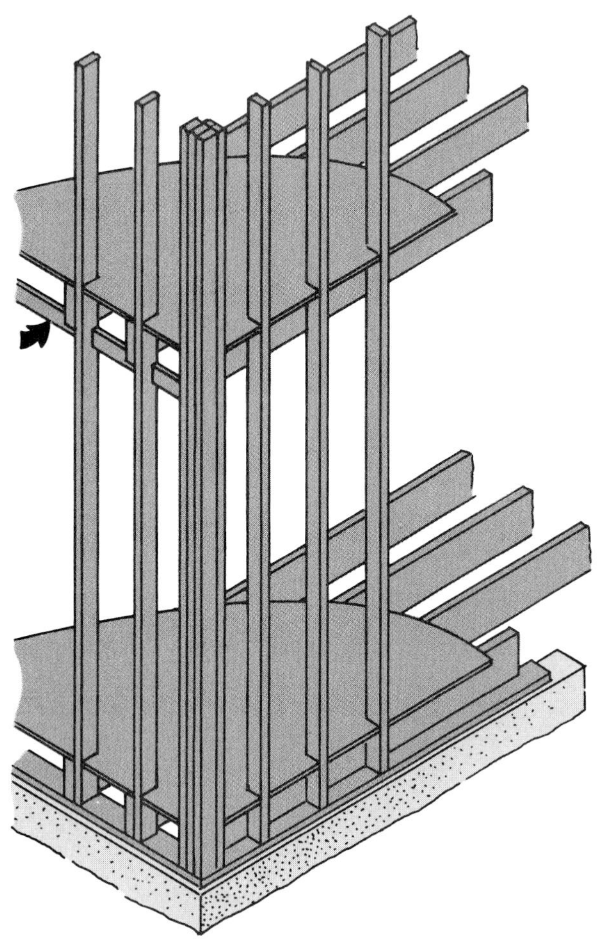

Figure 8-5. Balloon framing. Second floor joists rest on a ribbon (arrow) set into the studs.

Steel Beams

In many localities, steel beams are used instead of wood girders. Sizes depend on the load. The load is calculated in the same way as for wood girders.

Two types of steel beams are illustrated in Figure 8-9. The W (*wide-flange*) is generally used in residential construction. Wood beams vary in depth, width, species, and grade. Steel beams vary in depth, width of flange, and weight.

After the approximate load on a steel beam has been determined, the correct size can be selected from the table, Figure 8-10. This table lists a selected group of steel beams commonly used in residential structures. For example, if the total load on the beam (evenly distributed) is 15,000 lb. and the span between supports is 16'-0", then a W8x18 beam should be used. This specifies an 8" deep beam weighing 18 lb. per lineal foot. The width of the flange is 5 1/4".

Posts and Columns

For ordinary wood posts (shorter than 9' or smaller than 6 x 6), it is safe to assume that a post whose greater dimension is equal to the width of the girder it supports will carry the girder load. For example, a 6 x 6 post would be suitable for a girder 6" wide. For a girder 8" wide, a 6 x 8 or 8 x 8 post should be used.

Adequate footings must be provided for girder posts and columns. Wood posts should be supported on footings which extend above the floor level, as shown in Figure 8-11. To make sure the posts will not slide off their footing,

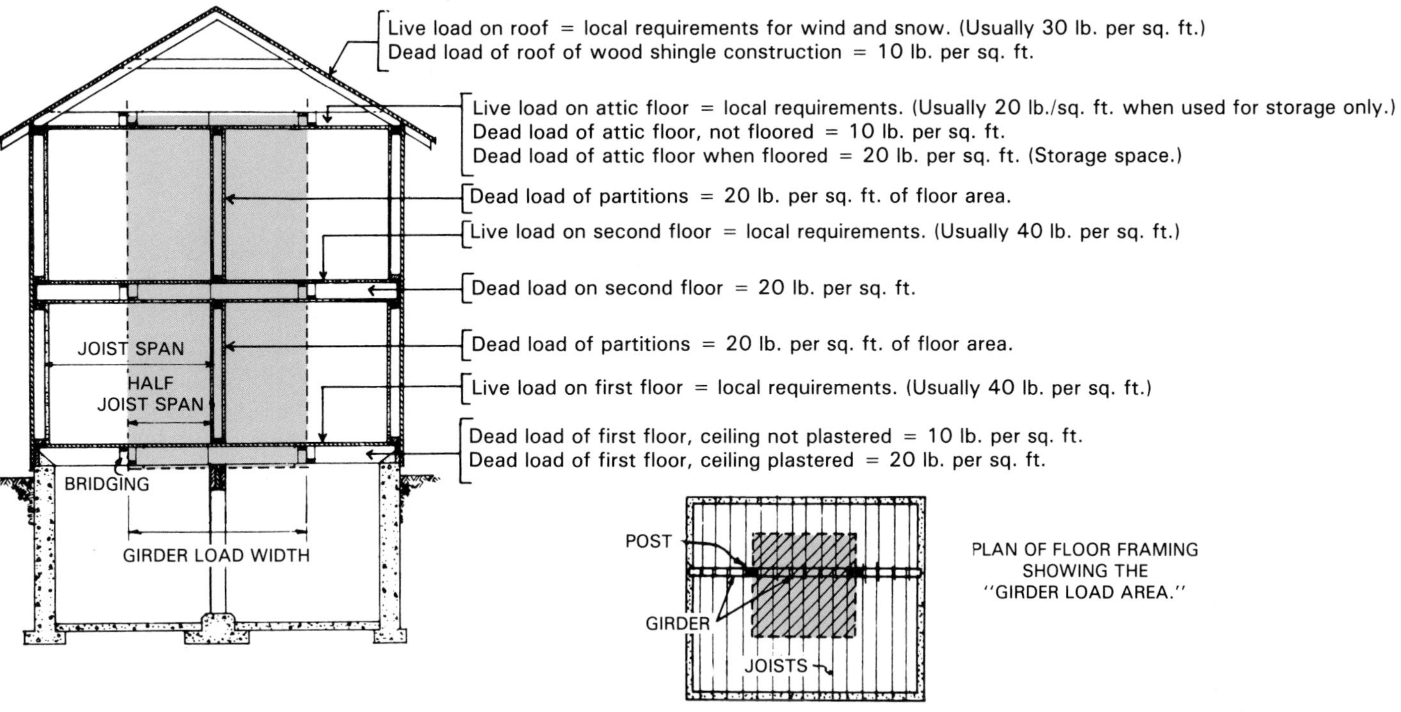

Figure 8-6. Diagram shows method of figuring loads for frame of a two-story home.

GIRDERS	SAFE LOAD IN LB. FOR SPANS FROM 6 TO 10 FEET				
SIZE	6 FT.	7 FT.	8 FT.	9 FT.	10 FT.
6 x 8 SOLID	8,306	7,118	6,220	5,539	4,583
6 x 8 BUILT-UP	7,359	6,306	5,511	4,908	4,062
6 x 10 SOLID	11,357	10,804	9,980	8,887	7,997
6 x 10 BUILT-UP	10,068	9,576	8,844	7,878	7,086
8 x 8 SOLID	11,326	9,706	8,482	7,553	6,250
8 x 8 BUILT-UP	9,812	8,408	7,348	6,544	5,416
8 x 10 SOLID	15,487	14,782	13,608	12,116	10,902
8 x 10 BUILT-UP	13,424	12,768	11,792	10,504	9,448

Figure 8-7. Table indicates typical safe loads for standard size wood girders.

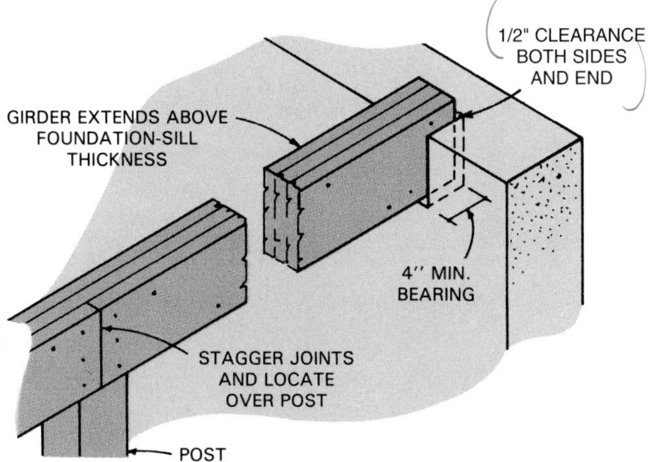

Figure 8-8. Built-up wood girder. Nails should be spaced no farther apart than 32" along top and bottom edges. Metal bearing plate should be placed under girder at foundation wall.

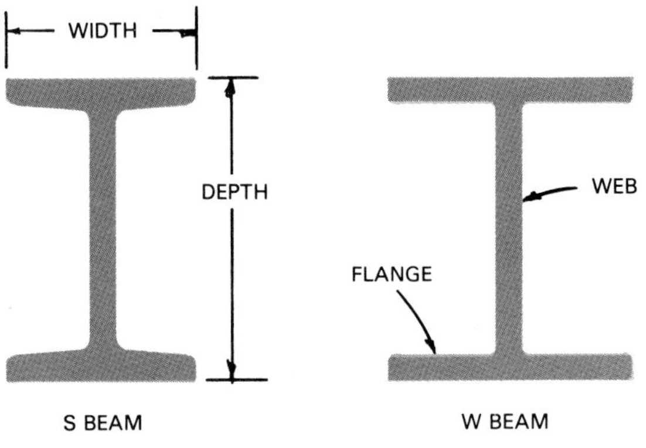

Figure 8-9. Steel beams are commonly used in residential construction. "S" means standard; "W" means wide-flange.

DESIGNATION WT./FT.	NOMINAL SIZE DP. x WD.	SPAN IN FEET 8′	10′	12′	14′	16′	18′	20′	22′	24′	26′
W8x10	8x4	15.6	12.5	10.4	8.9	7.8	6.9	—	—	—	—
W8x13	8x4	19.9	15.9	13.3	11.4	9.9	8.8	—	—	—	—
W8x15	8x4	23.6	18.9	15.8	13.5	11.8	10.5	—	—	—	—
W8x18	8x5 1/4	30.4	24.3	20.3	17.4	15.2	13.5	—	—	—	—
W8x21	8x5 1/4	36.4	29.1	24.3	20.8	18.2	16.2	—	—	—	—
W8x24	8x6 1/2	41.8	33.4	27.8	23.9	20.9	18.6	—	—	—	—
W8x28	8x6 1/2	48.6	38.9	32.4	27.8	24.3	21.6	—	—	—	—
W10x22	10x5 3/4	—	—	30.9	26.5	23.2	20.6	18.6	16.9	—	—
W10x26	10x5 3/4	—	—	37.2	31.9	27.9	24.8	22.3	20.3	—	—
W10x30	10x5 3/4	—	—	43.2	37.0	32.4	28.8	25.9	23.6	—	—
W12x26	12x6 1/2	—	—	—	—	33.4	29.7	26.7	24.3	22.3	20.5
W12x30	12x6 1/2	—	—	—	—	38.6	34.3	30.9	28.1	25.8	23.8
W12x35	12x6 1/2	—	—	—	—	45.6	40.6	36.5	33.2	30.4	28.1

Figure 8-10. Allowable uniform loads for W steel beams. Loads are given in kips (1 kip = 1000 lb.). (Grosse Steel Co.)

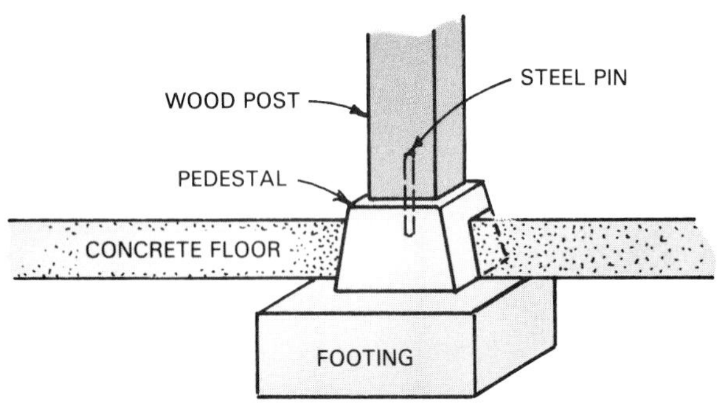

Figure 8-11. Footings for columns must extend above the floor level for moisture protection.

pieces of 1/2" diameter reinforcing rod or iron bolts should be embedded in the footing before the concrete sets. They should project about 3" into holes bored in the bottoms of the posts.

A *post anchor*, Figure 8-12, holds the wood post securely in place. It supports the bottom of the post above the floor, protecting the wood from dampness. The bracket can be adjusted for plumb if the anchor bolt was improperly placed.

When a wood column supports a steel girder, fitting the end of the column with a metal cap is desirable. If wood supports wood, a metal cap should be provided to give an even bearing surface. The metal will also prevent end grain of the post from crushing the horizontal grain of the wood girder.

A built-up wood post may be made by spiking together three 2 x 6s. The pieces should be free from defects and securely nailed together. Otherwise, excessive loading may cause the members to buckle away from each other.

Steel posts are most popular for girder and beam support. The post should be capped with a steel plate to provide a good bearing area. A steel post designed especially for this purpose is shown in Figure 8-13. The post has a threaded hole in the top end. A rod attached to a plate is threaded into this hole. This arrangement allows for adjustment as the wooden beam and other structural members shrink.

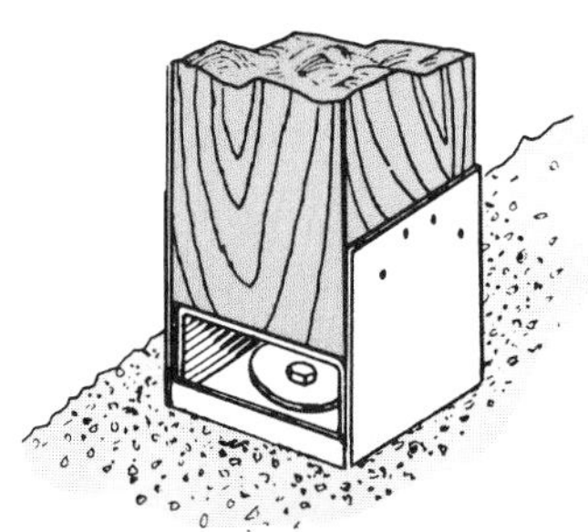

Figure 8-12. Steel post anchor permits lateral adjustment. (Timber Engineering Co.)

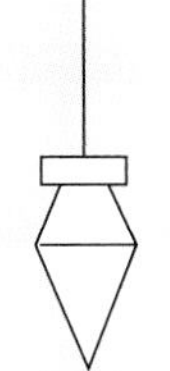

Be sure the tops of posts and columns and also the seats in foundation walls are flat so the girder or beam is well supported with its sides plumb.

Framing Over Girders and Beams

A common method of framing joists over girders and beams is shown in Figure 8-14. The steel beam is placed level with the top of the foundation wall. The 2" wood pad

Figure 8-13. Steel post with threaded top section is easy to install and adjust. Beam supported by the post (arrow) is made of laminated veneer lumber. (LVL)

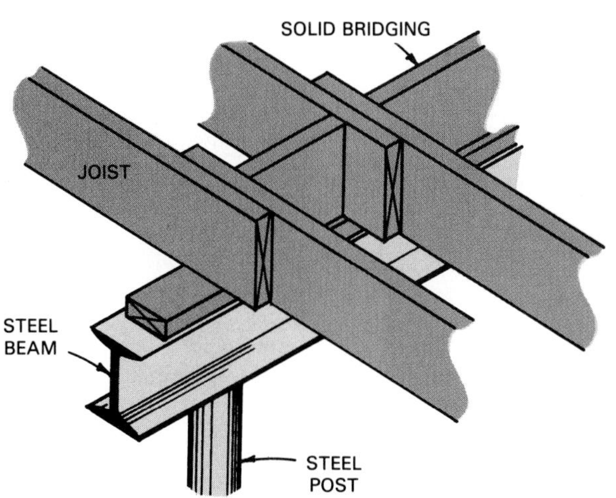

Figure 8-14. Joists supported on top of a steel beam. Top of beam is set flush with top of foundation wall.

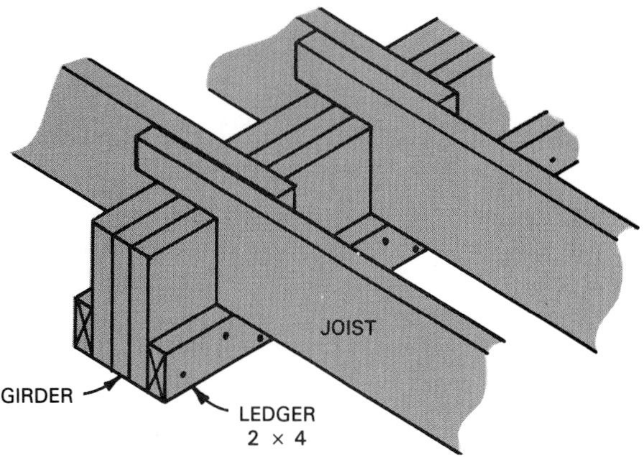

Figure 8-15. This arrangement is used when girder is raised for extra headroom or when ceiling is lowered. Joists should rest on the ledger strip, not on top of the girder.

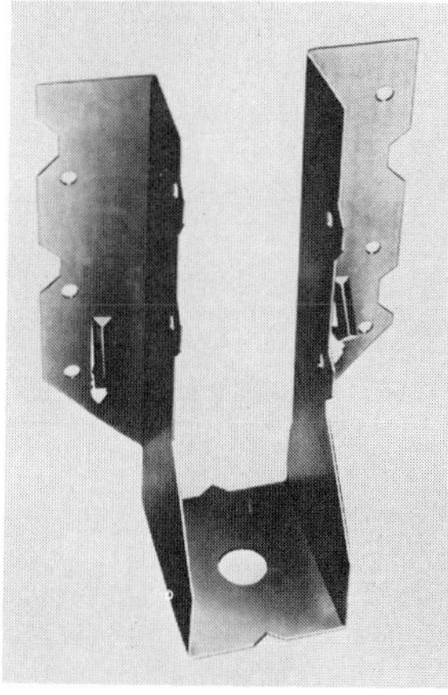

Figure 8-16. Joist and beam hanger is used when bottom of girder must be flush with the bottoms of the joists. (The Panel Clip Co.)

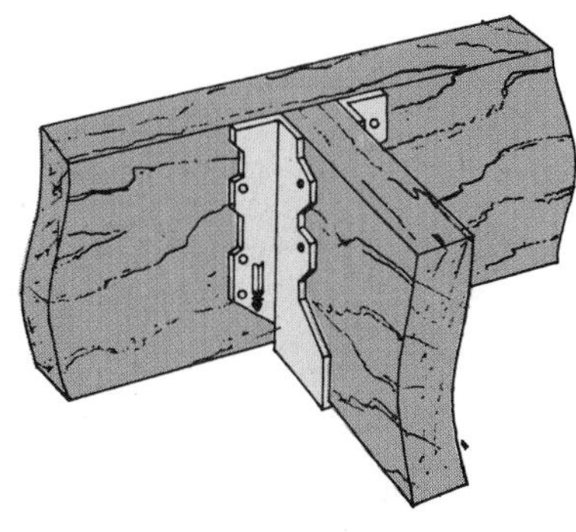

Figure 8-17. Joist attached with a hanger.

then carries the joists level with the sill. When a wooden girder is used, it is usually set so the top is level with the sill.

If ceiling height under joists needs to be moved down, the joists can be notched and carried on a *ledger*, Figure 8-15. When it is necessary for the underside of the girder to be flush with the joists to provide an unbroken ceiling surface, the joists should be supported with *hangers* or *stirrups*. See Figures 8-16 and 8-17.

Framing joists to steel beams at various levels can be accomplished with special hangers in somewhat the same manner as suggested for wood girders. You must make allowance for the fact that joists will likely shrink while the steel beam will remain the same size. For average work with a 2 x 10 joist, an allowance of 3/8" above the top flange of the steel girder or beam is usually sufficient.

A method of attaching joists is shown in Figure 8-18. Notching the joists so they rest on the lower flange of an S-beam is not recommended because the flange surface

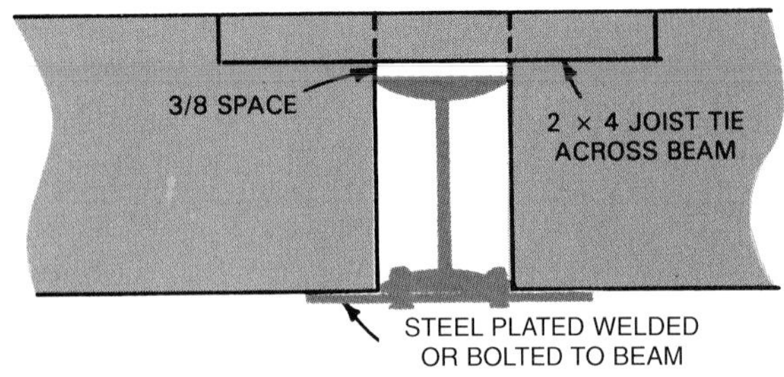

Figure 8-18. Edge view of joists supported on an S-beam. Allow 3/8" space above beam for shrinkage.

does not provide sufficient bearing surface. Wide-flanged beams, however, do provide sufficient support surface for this method of construction. Figure 8-19 shows butt methods of framing over girders.

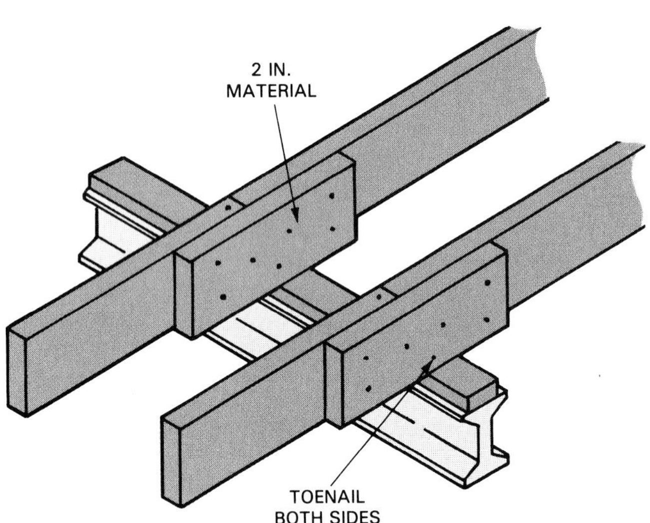

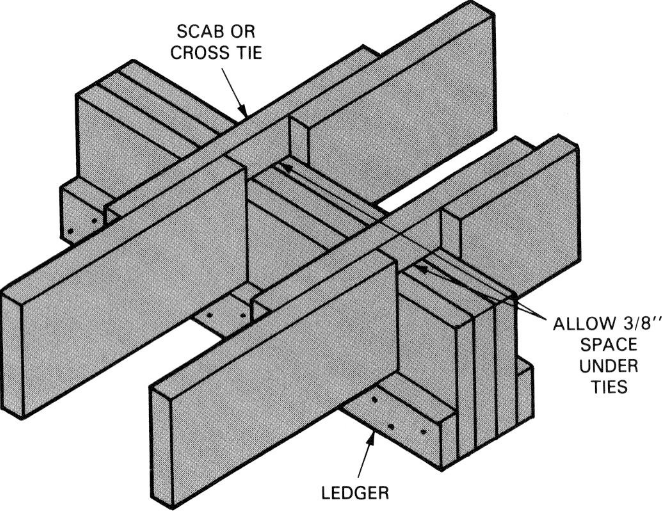

Figure 8-19. Some carpenters like to butt joists over the girder as shown above.

Sill Construction

After girders and beams are set in place, the next step is to attach the sill to the foundation wall. This is the part of the side walls or floor frame that rests horizontally on the foundation. It is also called the sill plate or the mudsill. The latter term originates from the procedure of correcting irregularities in the masonry work by embedding the sill in a layer of fresh mortar or grout.

Sills usually consist of 2 x 6 lumber. However, the width may vary depending on the type of construction. The sills are attached to the foundation wall with anchor bolts or straps. The size and spacing of anchors is specified in local building codes. Figure 8-20 shows a sill plate installed on a foundation wall. A sill sealer has been used to seal the crack between the bottom of the plate and the top of the wall. The seal's purpose is to stop the passage of air.

Figure 8-20. Foundation corner with sill bolted in place. Plastic sill sealer makes the joint between the sill and the foundation draft free.

Termite Shields

If termites are a problem in your locality, special shields should be provided. Termites live underground and come to the surface to feed on wood. They may enter through cracks in masonry or build earthen tubes on the sides of masonry walls to reach the wood.

The wood sill should be at least 8" above the ground. A protective metal shield, not less than 26 gage, should extend out over the foundation wall as shown in Figure 8-21.

In areas where termite damage is great, additional measures should be taken. Sometimes it is necessary to use lumber for lower framing members that has been treated with chemicals. Also, the soil around the foundation and under the structure can be poisoned.

Figure 8-22 is a map of the United States locating various levels of termite infestation. Canada and Alaska are considered to be in region IV. Hawaii and Puerto Rico are in region I.

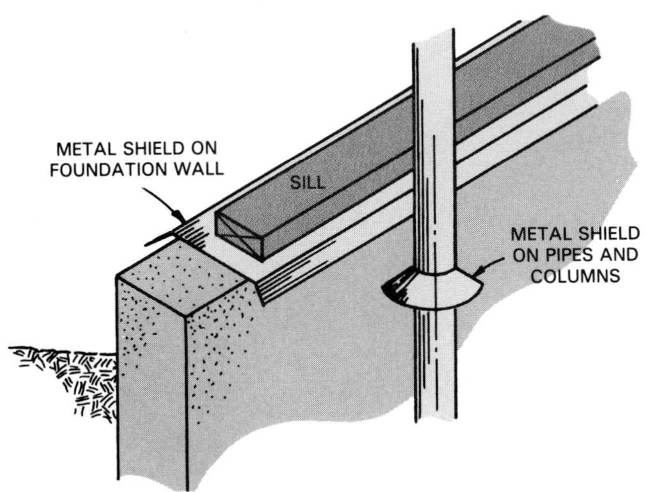

Figure 8-21. Termite shields. Use galvanized sheet iron or other suitable metal.

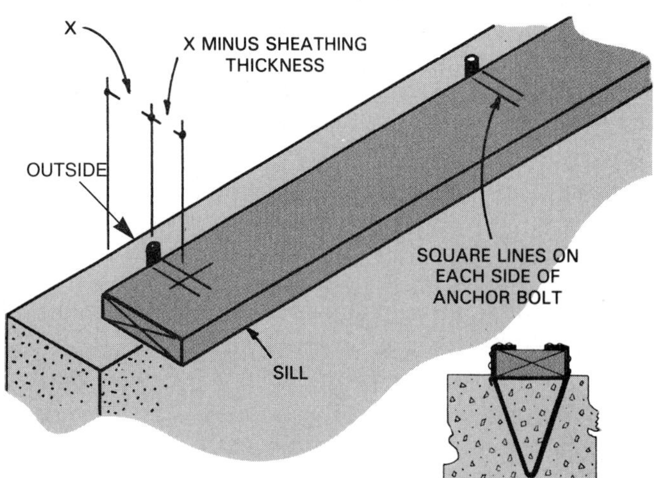

Figure 8-23. Laying out anchor bolt holes. Anchor strap, right, needs no layout. Sill is positioned and strap is nailed to sill. (TECO)

Installing Sills

Figure 8-23 shows two types of sill anchor. With the strap type, position the sill and attach the straps with nails. Some types must be bent over the top of the sill; others are nailed on the sides.

When anchor bolts are used, remove the washers and nuts. Lay the sill along the foundation wall. Remember, the edge of the sill will be set back from the outside of the foundation a distance equal to the thickness of the sheathing.

Draw lines across the sill on each side of the bolts as shown. Measure the distance from the center of the bolt to the outside of the foundation and subtract the thickness of the sheathing. Use this distance to locate the bolt holes. You will probably need to make separate measurements for each anchor bolt.

Since foundation walls are seldom perfectly straight, many carpenters prefer to snap a chalk line along the top where the outside edge of the sill should be located. This will ensure an accurate floor frame, which is basic to all

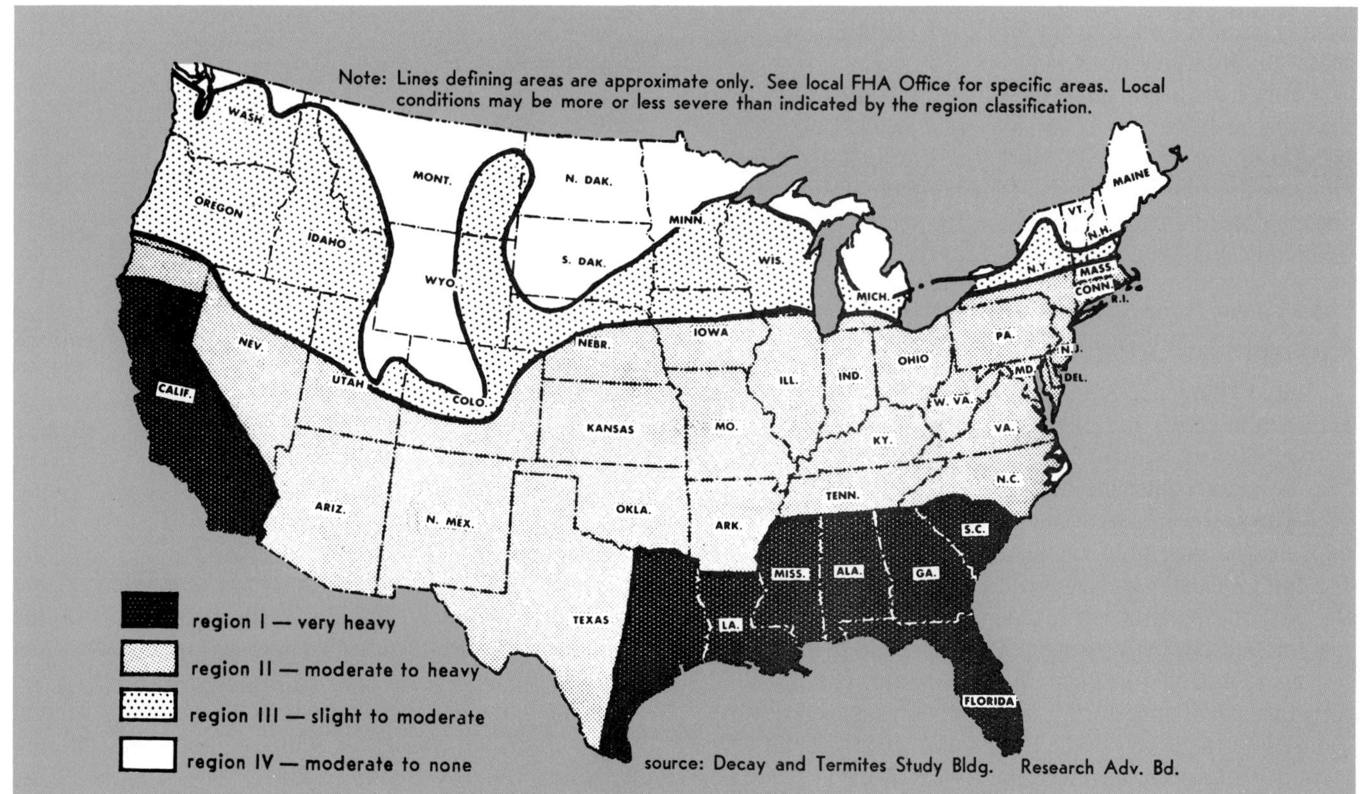

Figure 8-22. Nearly all parts of the United States are infested with termites, which can damage buildings. (Forest Products Lab.)

additional construction. Variations between the outside surface of the sheathing and foundation wall can be shimmed when siding is installed.

After all the holes are located, place the sill on sawhorses and bore the holes. Most carpenters prefer to bore the hole about 1/4" larger than the diameter of the bolts to allow some adjustment for slight inaccuracies in the layout. As each section is laid out and holes bored, position the section over the bolts.

When all sill sections are fitted, remove them from the anchor bolts. Install the sill sealer and then replace them. Install washers and nuts. As nuts are tightened, see to it that the sills are properly aligned. Also, check the distance from the edge of the foundation wall. The sill must be level and straight. Low spots can be shimmed with wooden wedges. However, it is better to use grout or mortar.

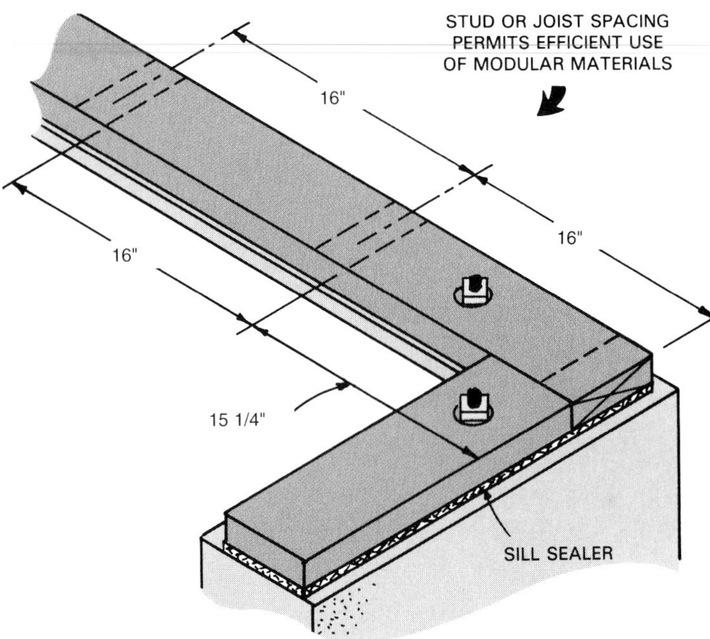

Figure 8-24. After sill is attached to foundation, locations for studs or joists may be marked.

Joists

Floor joists are framing members that carry the weight of the floor between the sills and girders. In residential construction, they are generally nominal 2" lumber placed on edge. In heavier construction, steel bar joists and reinforced concrete joists are used.

The most common spacing of wooden joists is 16" O.C. (on center). However, 12", 20", and 24" O.C. are also used.

The table in the Technical Information section lists safe spans for joists under average loads. For floors, this is usually figured on a basis of 50 lb. per sq. ft. (10 lb. dead load and 40 lb. live load).

Joists must not only be strong enough to carry the load that rests on them; they must also be stiff enough to prevent undue bending or vibration. Building codes usually specify that the deflection (bending downward at the center) must not exceed 1/360th of the span with a normal live load. This would equal 1/2" for a 15'-0" span.

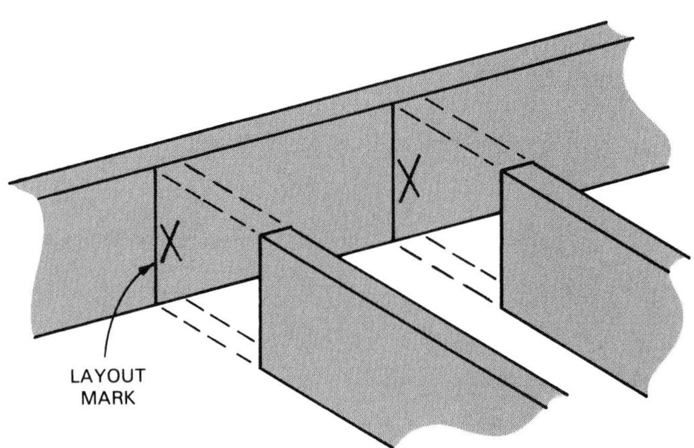

Figure 8-25. How to mark actual location of framing members. Layout marks show where the edge of a joist should be. The "X" indicates which side of the line the joist should be. Crowns (humped edge) of joists should always be turned up.

Laying Out Joists

Study the plans carefully. Note the direction the joists are to run. Also, become familiar with the location of posts, columns, and supporting partitions. The plans may also show the centerlines of girders.

The position of the floor joists can be laid out directly on the sill, Figure 8-24. On platform construction, the joist spacing is usually laid out on the joist header rather than the sill. The position of an intersecting framing member may be laid out by marking a single line and then placing an "X" to indicate the position of the part, Figure 8-25.

Instead of measuring each individual space around the perimeter of the building, it is more accurate and efficient to make a master layout on a strip of wood (called a rod). Use it to transfer the layout to headers or sill. The same rod is then used to make the joist layout on girders and the opposite wall. (When the joists are lapped at the girder, the "X" [location of the joists] is marked on the other side of the layout line for the opposite wall.) In this case the spacing between the stringer and first joist will be different than the regular spacing, Figure 8-26.

Some carpenters set in joists to leave a ledge around all sides of the sill so that sheathing will be flush with the foundation wall. See Figures 8-26(B) and 8-26(C).

Joists are doubled where extra loads must be supported. When a partition runs parallel to the joists, a double

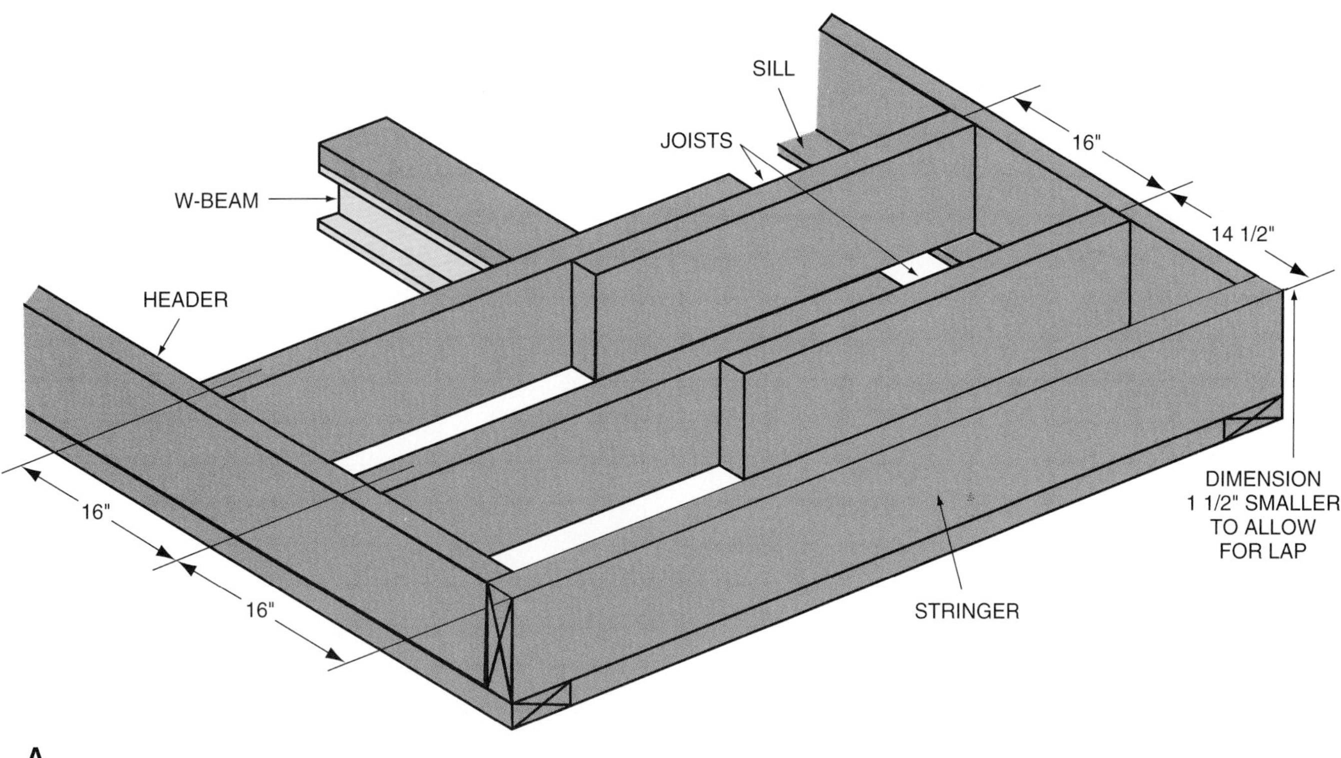

Figure 8-26. A—When joists are lapped over the girder, spacing between first joist and the stringer is different. B—Note that sheathing is flush with sill. C—To achieve this, some carpenters set in the stringers and headers to form a ledge the thickness of the sheathing.

joist is placed underneath. Partitions which are to carry plumbing or heating pipes are usually spaced far enough apart to permit easy access, Figure 8-27.

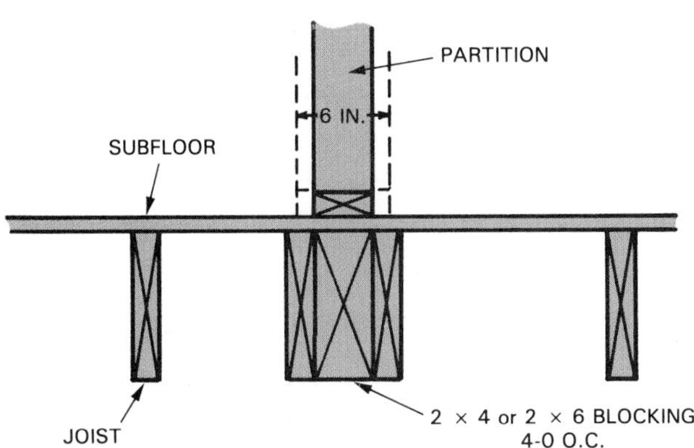

Figure 8-27. Joists under partitions are doubled and spaced to allow access for heating or plumbing runs. If the wall must hold a plumbing stack (vent to roof), the wall will be framed with 2 x 6s.

Joists must also be doubled around openings in the floor frame for stairways, chimneys, and fireplaces. These joists are called *trimmers*. They support the *headers* which carry the *tail* (short) *joists*. See Figure 8-28. The carpenter must become thoroughly familiar with the plans at each floor level so adequate support can be provided.

Select straight lumber for the header joist and lay out the standard spacing along its entire length, Figure 8-29. Add the position for any doubled joists and trimmer joists that will be required along openings. Where regular joists will become tail joists, change the "X" mark to a "T" as shown.

Installing Joists

After the header joists are laid out, toenail them to the sill. Position all full length joists with the *crown* (slight warpage called crook) turned up. Hold the end tightly against the header and along the layout line so the sides will be plumb. Attach the joists to the header using a nailing pattern consisting of three 16d nails, Figure 8-30.

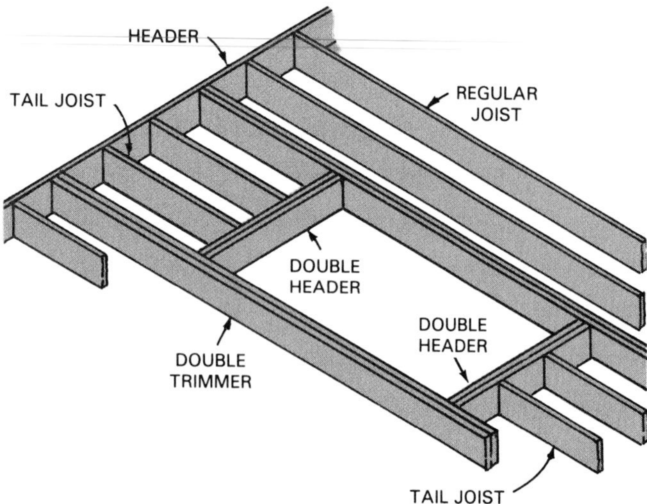

Figure 8-28. Framing members are doubled around floor openings.

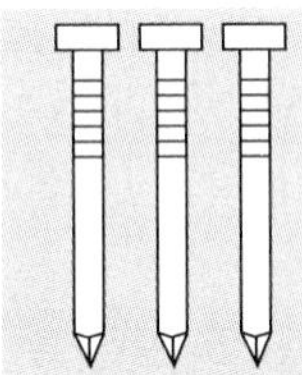

The Uniform Building Code specifies that, in standard framing, nails should not be spaced closer than one-half their length; nor closer to the edge of a framing member than one-fourth their length.

Figure 8-30. Using a pneumatic nailer to attach joists to header. Hold the end of the joist tightly against the header and along the layout line so the sides will be plumb.

Now fasten the joists along the opposite wall. If the joists butt at the girder (join end to end without overlapping), they should be joined with a scarf or metal fastener. If they lap, they can be nailed together using 10d nails. Also use 10d nails to toenail the joists to the girder.

To increase the accuracy of the floor frame, some carpenters first nail the joists to the headers. The headers are then carefully aligned with the sill or a chalk line on the foundation. Then the assembly is toenailed to the sill.

Nail doubled joists together using 12d or 16d nails spaced about 1' along the top and bottom edge. First, drive several nails straight through to pull the two surfaces

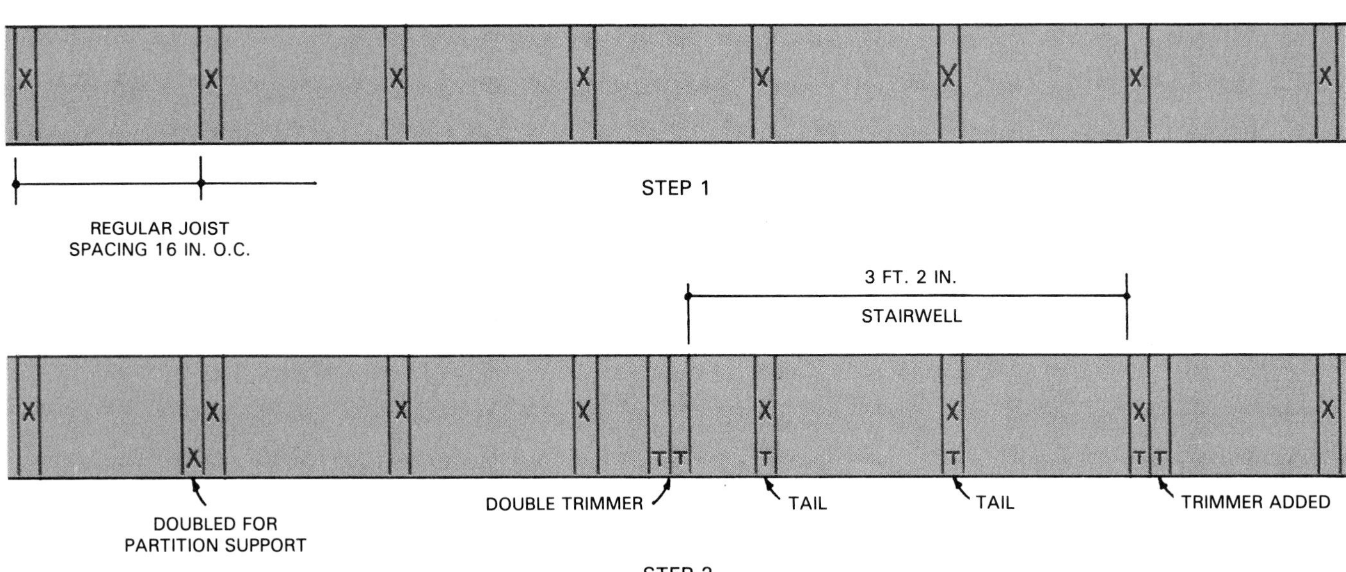

Figure 8-29. Header laid out with joist positions. In step 1, rod was used to mark regular spacing. In step 2, double and trimmer joist positions have been added.

tightly together and clinch the protruding ends. Finish the nailing pattern by driving the nails at a slight angle. Some carpenters lay a bead of caulk along the joint formed by the header and the sill to keep out air, Figure 8-31.

Framing Openings

Place boards or sheets of plywood across the joists to provide a temporary working deck to install header and tail joists. First, set the trimmer joists in place. Sometimes a regular joist will be located where it can serve as the first trimmer. (A *trimmer* is a full-length joist or a stud that reinforces a rough opening.) Figure 8-32 is a plan view of the finished assembly.

Figure 8-31. In cold climates, some carpenters apply a bead of caulking compound (arrow) to seal the joint between the header and sill.

The length of the headers can be determined from the layout on the main header joist. Cut headers and tail joists to length. Make the cuts square and true. Considerable strength will be lost in the finished assembly if the members do not fit tightly together. Lay out the position of the tail joists on the headers by transferring the marks made on the main header in the initial layout.

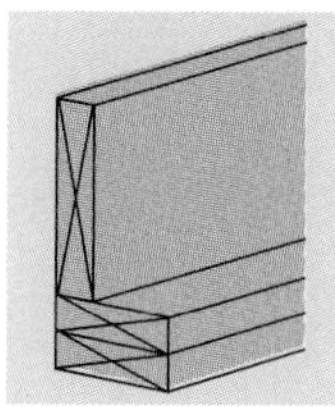

Be accurate in laying out and in cutting floor framing members. The strength of the assembly depends on all of the parts fitting together tightly.

When the assembly of tail joists and first headers is small, they are sometimes nailed together and then set in place. Usually, however, the headers are installed and then the tail joists are attached. One of the tail joists can be temporarily nailed to each trimmer to accurately locate the header and hold it while it is being nailed.

Figure 8-33 illustrates the procedure for fastening tail joists, headers, and trimmers. After the first header and tail joists are in position between the first trimmers, nail the second or double header in place. Be sure to nail through the first trimmer into the second header using three 16d nails at each end. Finally, the second trimmer is nailed to

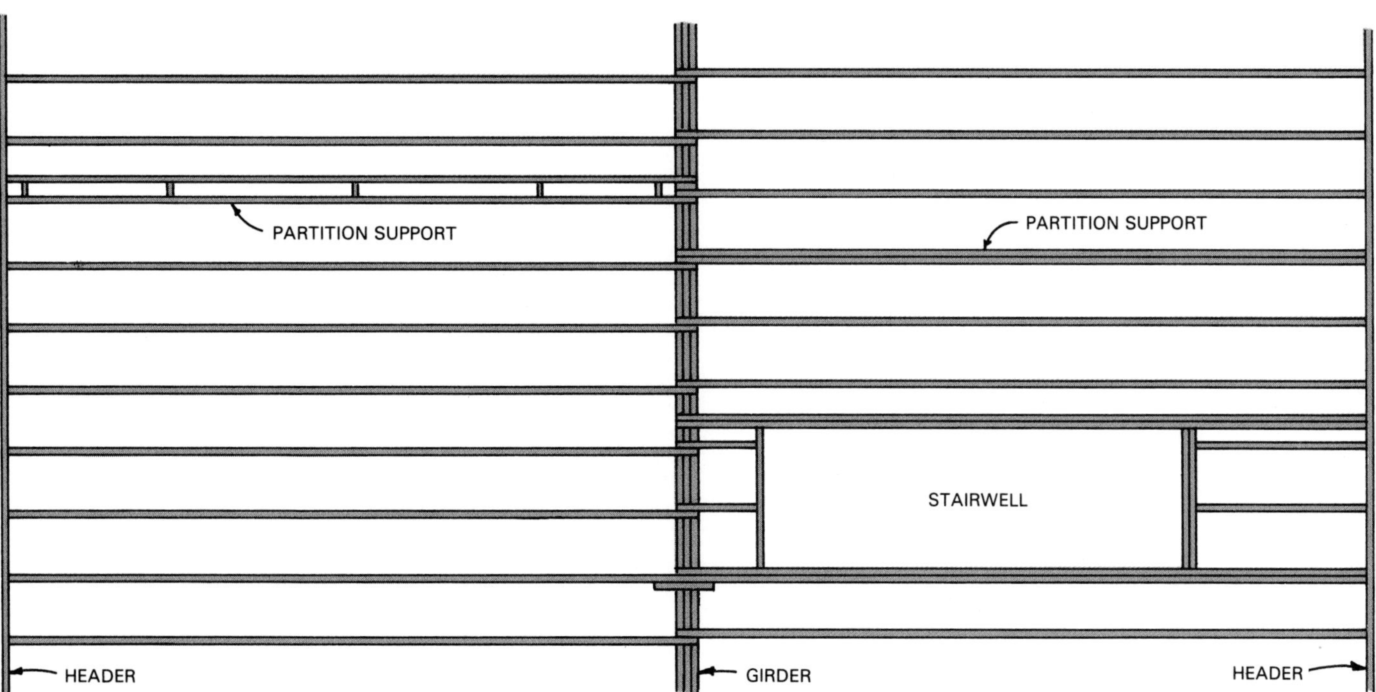

Figure 8-32. Plan view of floor framed for stair opening and partition support. Single header is sometimes used next to girder if the distance between header and girder is short. In some instances, building codes may require a double header.

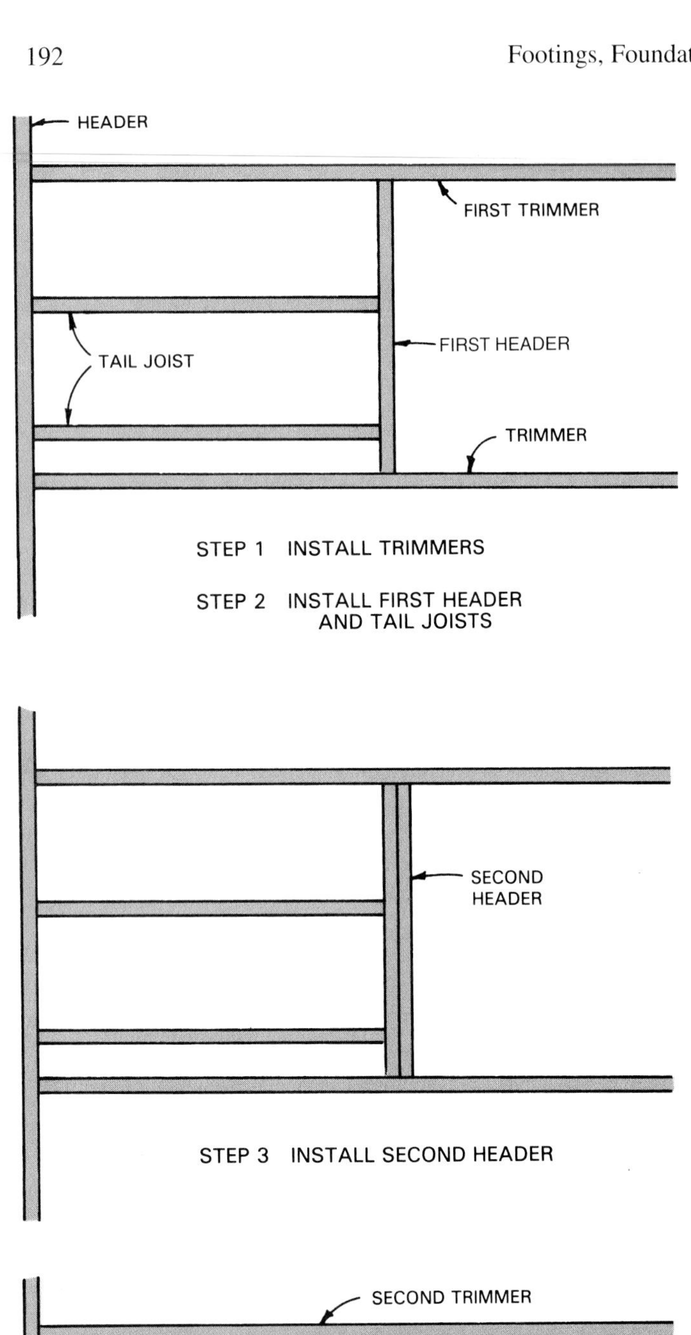

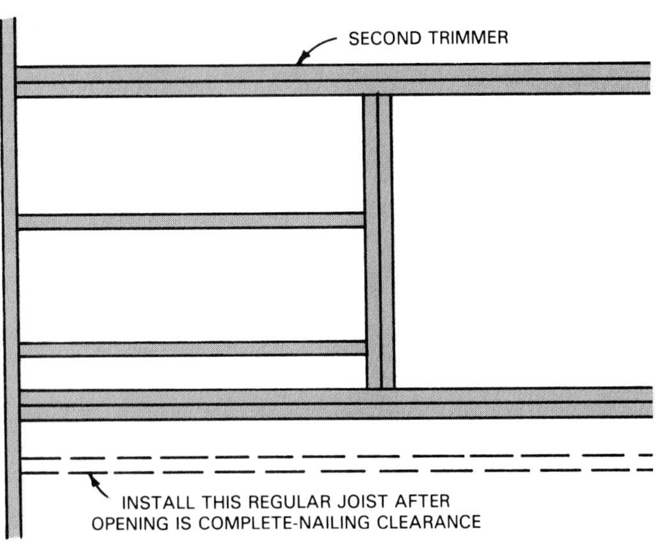

Figure 8-33. These steps can be followed in assembling frames for floor openings.

the first trimmer. A good nailing pattern for the entire assembly is shown in Figure 8-34. To allow more room for nailing, do not install joists adjacent to trimmers until the trimmers are doubled and nailed.

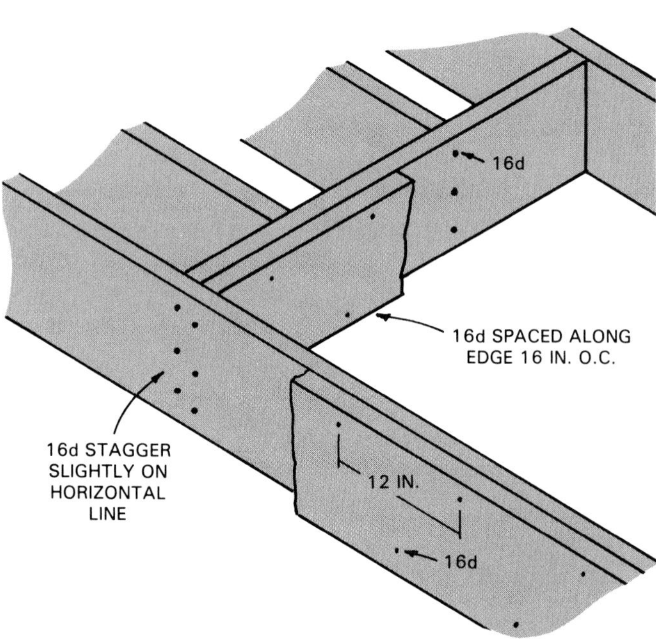

Figure 8-34. Nailing pattern for attaching floor opening members.

This nailing pattern will support a concentrated load of 300 lb. at any point on the floor. It will also hold a uniformly distributed load of 50 lb. per sq. ft. with any spacing and span of tail beams ordinarily used in residential construction, provided the long dimension of the floor opening is parallel to the joist. If the long way of the opening is at right angles to joists, excessive loading may be carried to the junction of headers with trimmers. Anticipated loads should be checked and more nails or additional supports should be provided at these junctions when needed.

Today, metal framing anchors are often used to assemble headers, trimmers, and tail joists, Figure 8-35. They are manufactured from 18 gage zinc-coated sheet steel in a variety of sizes and shapes. Special nails for attaching the anchors are also available. The National Forest Products Association recommends the use of framing anchors or ledger strips to support tail joists that are over 12' long.

Bridging

Some recent studies have shown that *bridging* may be eliminated if the following two conditions are satisfied:

- Joists are properly secured at the ends.
- Subflooring is adequate and carefully nailed.

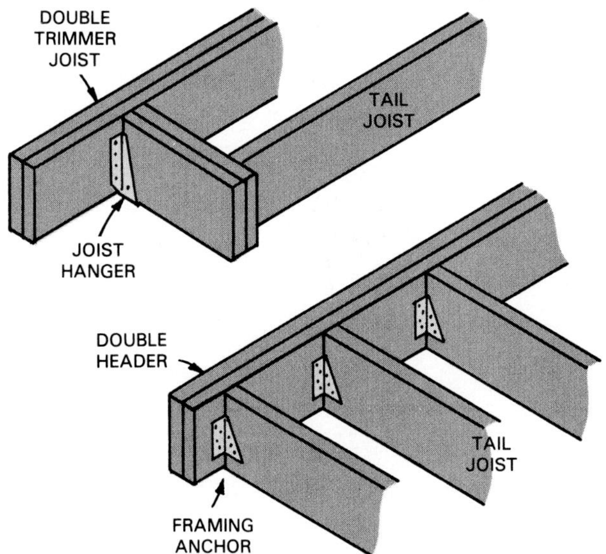

Figure 8-35. Joist hangers and framing anchors are good for assembling floor framing members.

However, many local building codes include bridging requirements and general standards suggest that bridging be installed at intervals of no more than 8'.

Regular bridging, sometimes called herringbone or cross bridging, is composed of pieces of lumber set diagonally between the joists to form an "X." Its purpose is twofold:

- Hold the joists in a vertical position.
- Transfer the load from one joist to the next.

Figure 8-36 shows how the carpenters framing square can be used to lay out a pattern for bridging. Pieces can be cut rapidly on the radial arm saw or a jig can be set up to use a portable electric saw or a hand saw.

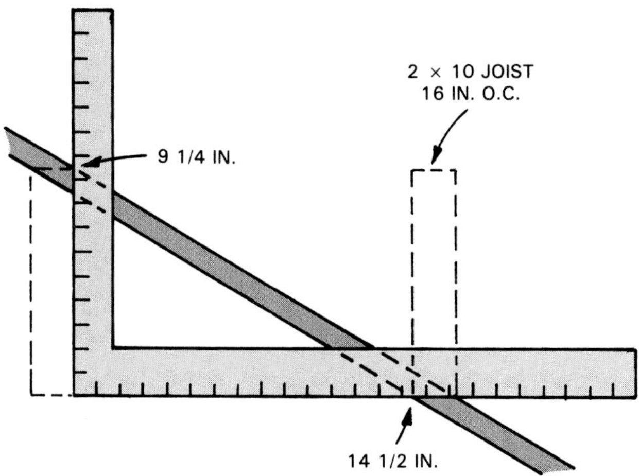

Figure 8-36. Carpenter's square can be used to lay out bridging. Line for lower cut can be secured by shifting the tongue of the square to the 14 1/2" mark on the stock.

Install Wood Bridging

1. Snap a chalk line across the tops of the joists at the center of their span, Figure 8-37.
2. Use two 8d nails to attach the top of the bridging to each side of every joist. Alternate the positioning of the bridging, first on one side of the chalk line and then on the other, Figure 8-38.
3. After the subflooring is complete or before the under surfaces of the floor are enclosed, the lower ends of the bridging is nailed to the joist, Figure 8-39.

Figure 8-37. Mark position of bridging by snapping a chalk line across the tops of the joists at mid-span.

Solid bridging, as the name implies, consists of solid pieces of 2" lumber installed between the joists. This type of bridging is easier to cut and install when there are odd-sized spaces in a run of regular cross bridging. Solid bridging, also called *blocking*, is often installed above a supporting beam, where its chief purpose is to keep the joist vertical. However, it also adds rigidity to the floor.

Several types of prefabricated steel bridging are available. The steel bridging can be installed very quickly. The type shown in Figure 8-40 is manufactured from sheet steel. A V-shaped cross section makes it rigid. No nails are required and it is driven into place with a regular hammer. This design meets FHA Minimum Property Standards, and is approved by the Uniform Building Code.

After the bridging is installed, the floor frame should be checked carefully to see that nailing patterns have been completed in all members. After this is done, the frame is ready to receive the subflooring.

Special Framing Problems

In modern residential construction, the design may include a section of floor that overhangs (sticks out beyond) a lower floor or basement level. When the floor

Figure 8-38. Bridging holds the joists in a vertical position and transfers the load from one joist to the next. In the view above, a section of the subfloor was laid to the point shown and then used as a platform for installing the bridging. Solid bridging (see arrow) was used to connect the system to the header and sill.

Figure 8-39. Under-floor view of regular wood bridging. The lower ends are not nailed until the subfloor is complete or the underside of the floor is enclosed.

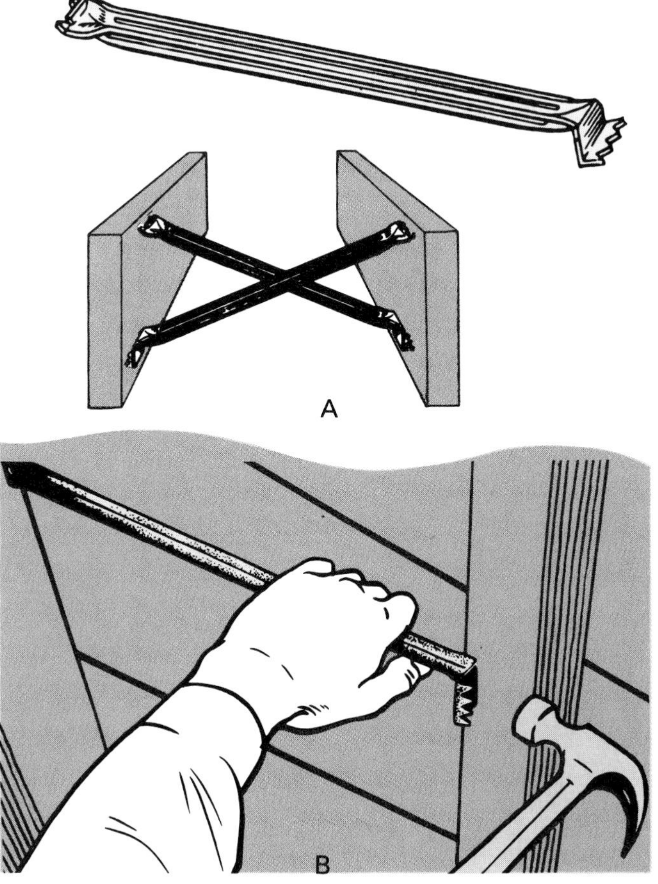

Figure 8-40. Top—Steel bridging. Bottom—Installation method. (Timber Engineering Co.)

joists run perpendicular to the walls, the framing is comparatively easy. It is only necessary to use longer joists. If, however, the floor joists run parallel to the wall, the construction must be framed with *cantilevered* joists as illustrated in Figure 8-41.

The exact spacing and length of the members will depend on the weight of the outside wall. Usually, cantilevered joists should extend inward at least twice as far as they stick out over the supporting wall. Note that since the load at the inside double header is upward, the ledger strip must be positioned at the top.

Entrance halls, bathrooms, and other areas are often finished with tile or stone that is installed on a concrete base. To provide room for this base, the floor frame must be lowered. When the area is not large, this can be done by doubling joists of a smaller dimension, Figure 8-42.

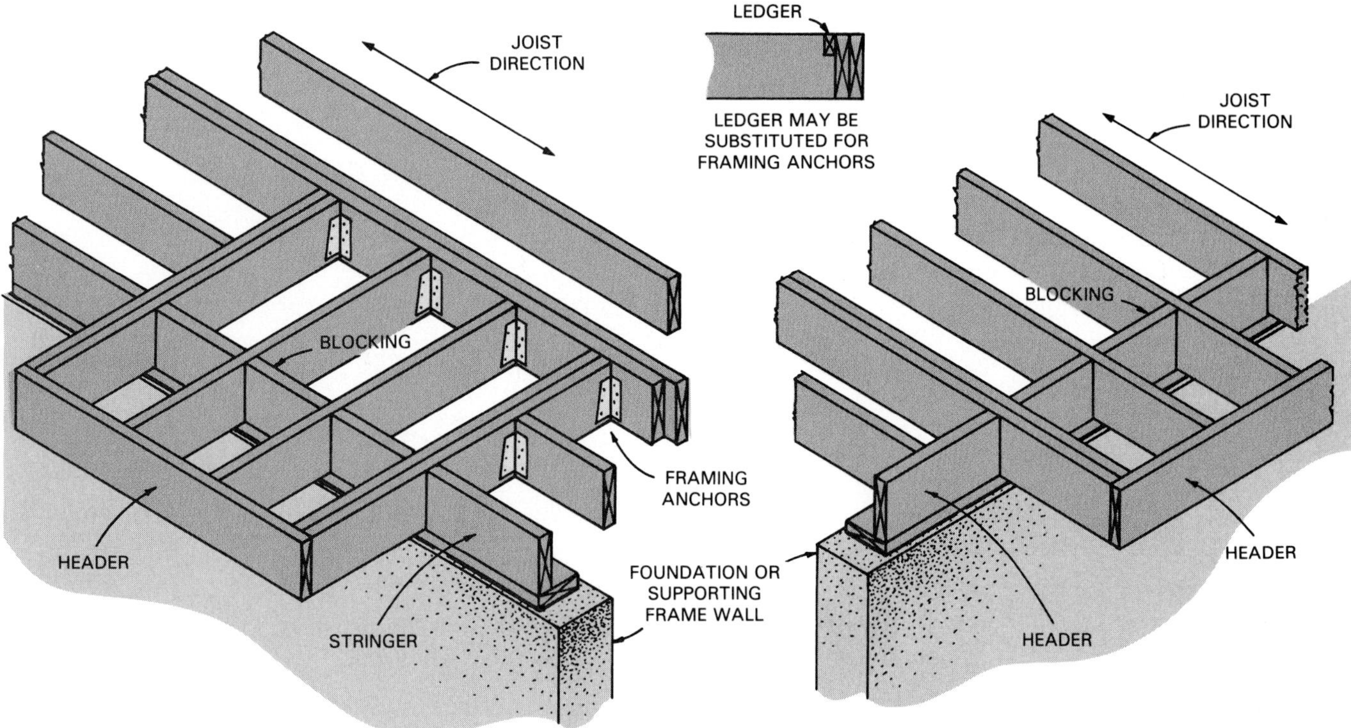

Figure 8-41. Framing methods on overhangs depend on direction the joists run. Blocking holds the joists vertical, adds rigidity, and closes up the space.

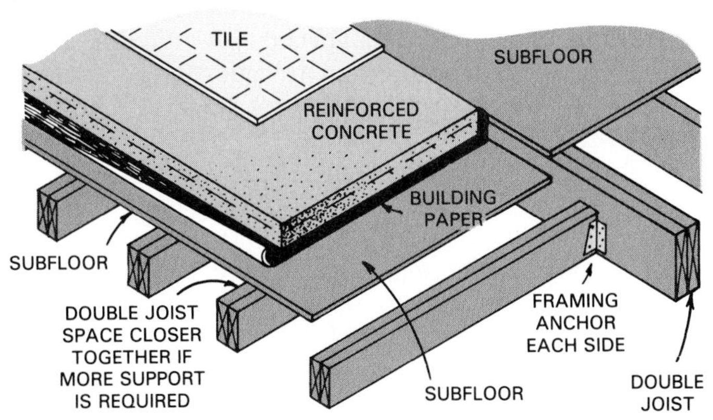

Figure 8-42. Smaller joists are used when a concrete base is needed for tile or stone surfaces.

Additional support can be secured by reducing the spacing. When area is large, steel or wood girders and posts should be added.

Bathrooms must support unusually heavy loads — heavy fixtures and often the additional weight of a tile floor. The fixed dead load imposed by a tile floor will average around 30 lb. per sq. ft. The load from bathroom fixtures adds from 10 to 20 lb. per sq. ft., for a total of 40 to 50 lb. dead load. In addition, it is frequently necessary to cut joists to bring in water service and waste pipes. Special precautions must, therefore, be taken in framing bathroom floors to provide adequate support.

Cutting Floor Joists

Before cutting joists to install plumbing, it is useful to know how stress affects flooring joists. This knowledge will help you determine where to make holes and cut notches.

When the top of a joist is in compression and the bottom in tension, there is a point at which the stresses change from one to the other. At this point, there is neither tension nor compression.

In the usual rectangular joist, this point is assumed to be midway between the top and bottom. Variations in the quality of lumber and other conditions may shift the point slightly; still, this assumption is accurate enough.

If there is neither compression nor tension at the center, it is obvious that a hole—provided it is not larger than one-fourth the total depth of the joist—would have little effect on the strength.

Weight produces the greatest bend if it is at the center of the span. Therefore, a weakness is more likely to reduce the strength of a joist or beam if it is near the center of the span. Considering this, follow these precautions in cutting joists:

- When possible, cut holes at or close to the middle of a joist. If the opening is limited to 1/4 of the total width, the reduction in strength will be insignificant.
- Where it is necessary to cut joists, the cuts should be made from the top. For example: if a 2 x 8 joist is cut to a depth of 4", its strength will be reduced to

that of a 2 x 4. If the cut from the top is 2", it will be equivalent to a 2 x 6. When a joist is cut, the loss in strength must be compensated for by providing headers and trimmers or by adding extra joists. Another way of solving the problem for large plumbing pipes is shown in Figure 8-43.

- If the cut is made elsewhere than at the center of span, the weakening effect will not be as great. Even so, it is advisable to provide as much compensating strength as is lost by the cut.

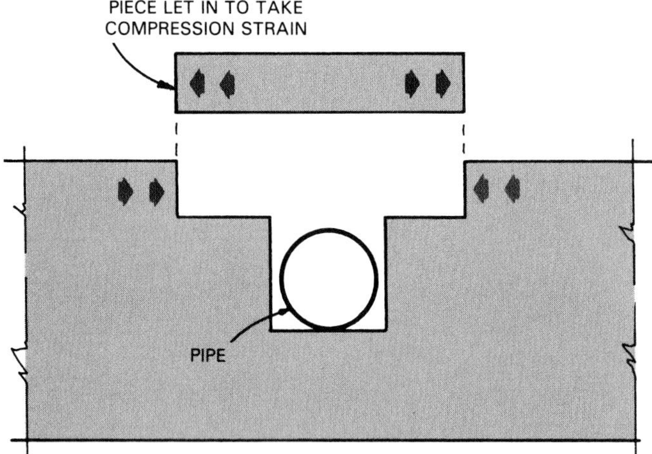

Figure 8-43. Fit a block of wood into a deep notch to restore strength of joist.

Low-Profile Floor Frames

Many home buyers prefer a house with a low silhouette. Standard wood floor construction, whether over a basement or a crawl space, requires adequate distance between the framing members and the ground. This places the first floor level well above the finished grade.

Various framing systems have been devised to bring the floor level closer to the outside grade. These consist of designing the foundation in such a way that the floor frame is surrounded and protected by the wall, Figure 8-44.

In construction of this type, special precautions must be taken to assure that the joists have adequate bearing surface. Allowance should also be made for shrinkage of the wood members.

Figure 8-45 shows a floor framing system over a crawl space. The underfloor space serves as a *plenum* (enclosed space for air under slightly greater pressure than surrounding air, used for heating and cooling a house). The system is easy to build and permits considerable design flexibility. Money is saved by eliminating ductwork and by using smaller joists and beams. With proper insulation, the system can save a great deal of energy.

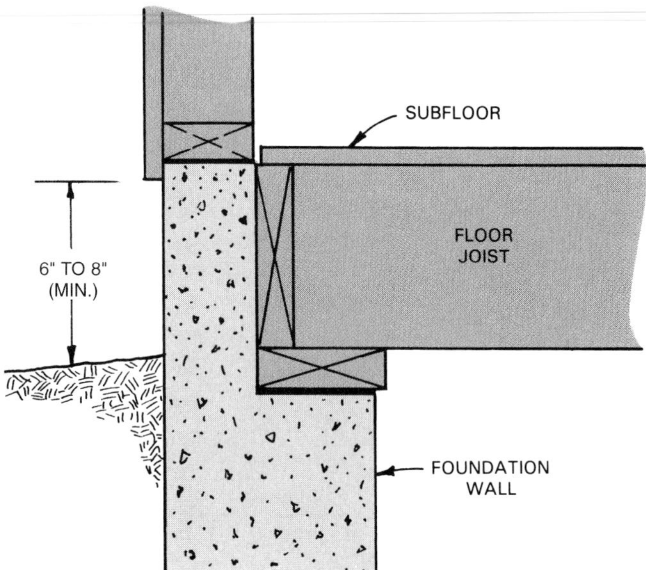

Figure 8-44. Offset in foundation wall reduces distance between the floor level and the finished grade.

Foundations for low-profile floor framing may be constructed of poured concrete, masonry units such as concrete block, or wood. Whichever type of building material is used, proper barriers against moisture and heat or cold must be installed.

Foil-faced insulation should be laid around the entire perimeter with the foil facing upward or inward. It is suggested that the insulation have an insulating factor of about R11. However, local codes and practice should be followed in all cases.

The plastic moisture barrier is put down across the entire floor area of the crawl space. It deflects moisture downward to a layer of gravel or crushed stone. Outside walls should be well insulated as shown in Figure 8-46.

Open Web Trusses

Open web floor trusses are widely used in modern houses and light commercial construction. They provide long clear spans with a minimum of depth. The open webs are lighter and easier to handle. At the same time, they reduce transmission of sound through floor/ceiling assemblies. Open webs make it easy to install plumbing, heating, and electrical systems. See Figure 8-47.

Modern trusses and truss systems are designed with the aid of computers. These designs ensure that loading requirements are met through the use of a minimum amount of material. Engineered jig hardware used in the assembly "builds-in" proper camber in each unit.

Some trusses are fabricated with lumber chords and patented galvanized steel webs. The webs have metal teeth which are pressed into the sides of the chords. They also have a reinforcing rib that withstands both tension and compression forces. See Figure 8-48.

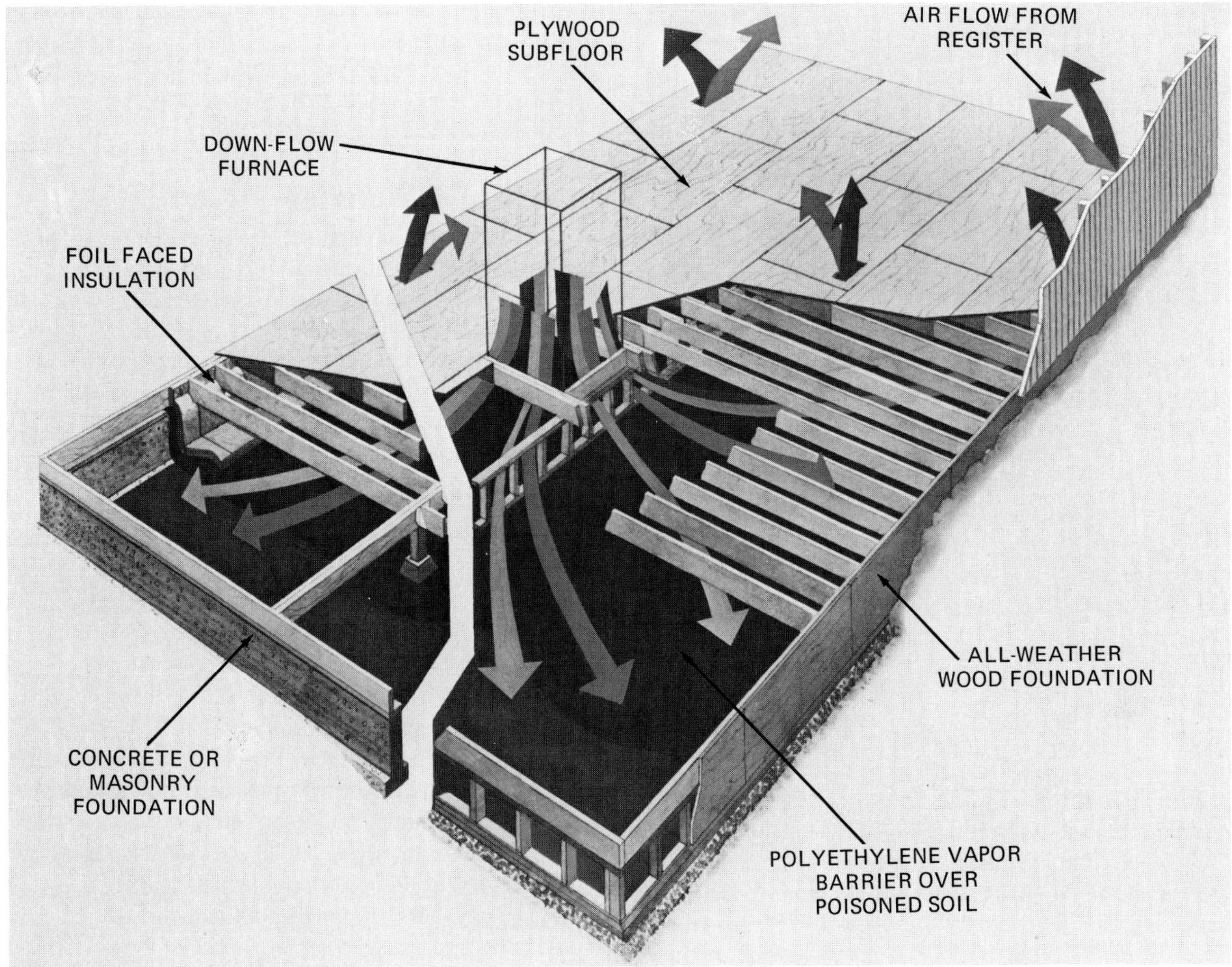

Figure 8-45. Energy-saving construction uses crawl space as a heating/cooling plenum. Arrows indicate flow of heated or cooled air. Either concrete or wood foundation can be used. Design data can be secured from American Plywood Association.

Solid-Web Trusses

Solid-web trusses are generally called *wood I-beams*. The *chords* (top and bottom members of a truss) are made up of either solid lumber or laminated veneer. The web is made of 3/8" plywood or oriented strand board. The web is glued into grooves cut in the chords. No nails are used. Figure 8-49 shows typical wood I-beams. These manufactured joists are not prone to shrinking or warping. This also reduces the occurrence of squeaking floors caused by drying, shrinking lumber. Special techniques must be used to fasten wood I-beams to other frame components. Use of metal connectors, Figure 8-50, may be required.

Subfloors

The laying of the subfloor is the final step in completing the floor frame. Either plywood, shiplap, tongue-and-groove flooring, or common boards can be used. The subfloor serves three purposes:

- It adds rigidity to the structure.
- It provides a base for finish flooring material.
- It provides a surface upon which the carpenter can lay out and construct additional framing.

Subflooring of the board or shiplap type is nailed at each joist with 8d nails. Use two nails in each board when the width is under 6". For widths greater than 6", three nails should be used.

If subfloor is tongued and grooved on ends and edges, end joints need not be made over joists. Subfloor is preferably laid without cracks between boards. If accumulation of water on the subfloor during construction is likely, it may be desirable to leave cracks to permit drainage.

Plywood

In most modern construction, plywood is used for subflooring. It provides a smooth, even base and acts as a horizontal diaphragm that adds strength to the building. Plywood can be installed rapidly and usually ensures a squeak-free floor.

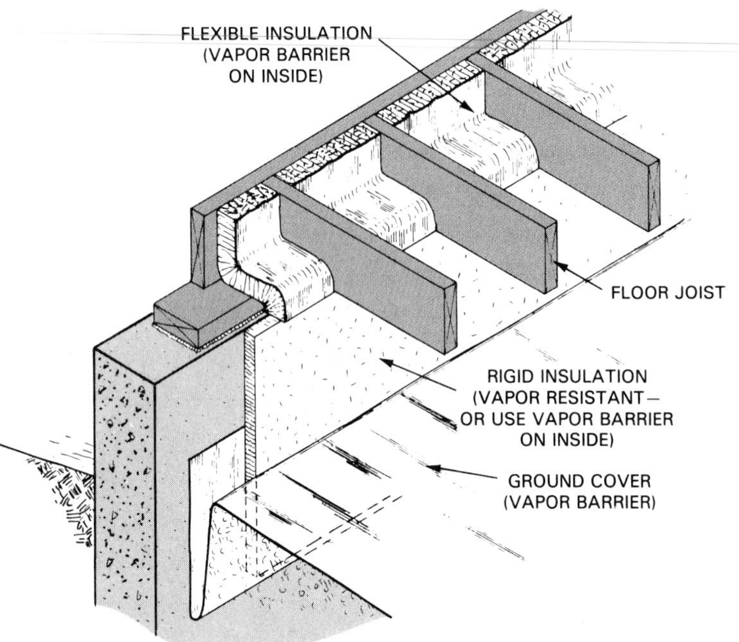

Figure 8-46. Approved method of insulating heated crawl space. When weather permits, the installation should be made before subfloor is laid.

Although 1/2" plywood over joists spaced 16" O.C. meets the minimum FHA requirements, many builders use 5/8" plywood. The long dimension of the sheet should run perpendicular to the joists. Joints should be broken in successive courses, Figure 8-51. For 5/8" or 3/4" plywood, use 8d nails spaced 6" along edges and 12" along intermediate members. See Figure 8-52.

Combined subfloor-underlayment systems utilize a special plywood panel with tongue and groove edges. This single layer provides adequate structural qualities and a satisfactory base for direct application of carpet, tile, and other floor finishes.

Subfloor-underlayment panels are available for joists or beam spacing of 16", 20", 24", or 48". Maximum support spacing is stamped on each panel. A 3/4" thickness is used for 24" O.C. spacing. Be sure to follow instructions supplied by the manufacturer.

Other Sheet Materials

Other sheet materials such as composite board, waferboard (also called waferwood), oriented strand board, and structural particleboard are also approved for use as subflooring. These products have been rated by the American Plywood Association and meet all standards for subflooring. The specifications for application are the same as for plywood. For additional information refer to Unit 1.

Glued Floor System

In a *glued floor system* the subfloor panels are glued and nailed to the joists. Structural tests have shown that stiffness is increased about 25% with 2 x 8 joists and 5/8" plywood. In addition the system ensures squeak-free construction, eliminates nail-popping, and reduces labor costs.

Before each panel is placed, a 1/4" bead of glue is applied to the joists, as shown in Figure 8-53. Spread only enough glue to lay one or two panels. Two beads of glue are applied on joists where panel ends butt together. All nailing must be completed before glue sets.

When laying tongue and groove panels, apply glue along the groove, either continuously or spaced. Use a 1/8" bead so that excessive squeeze-out is avoided. Use a protective strip and a maul to drive sheets into groove of previous subflooring, Figure 8-54. Leave a 1/8" space at all ends and edge joints. Fasten the subfloor immediately (before glue sets) with screws or nails. See Figure 8-55.

Material specifications and application procedures for subflooring are provided in a booklet published by the American Plywood Association.

Figure 8-47. Open truss floor joists are often used for both single family homes, apartment buildings, and light commercial structures. Left—Open trusses made up of 2 x 4s set up for a condominium. Note the 1 x 6 band across the top edge that ties the joists together. Right—Setting up trusses for an apartment complex. (TrusWal Systems Corp.)

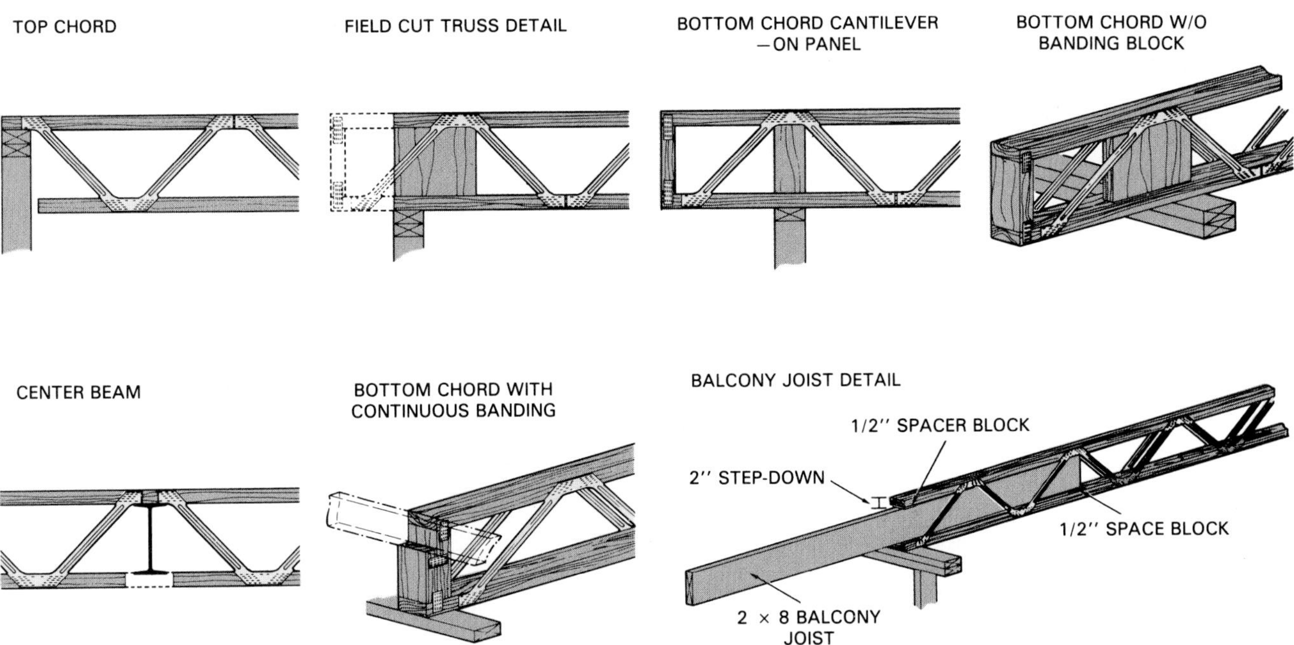

Figure 8-48. Truss construction details. Chords are made of lumber; webbing is a patented galvanized steel design. Trusses provide wide nailing surface because chord is laid flat. (TrusWal Systems Corp.)

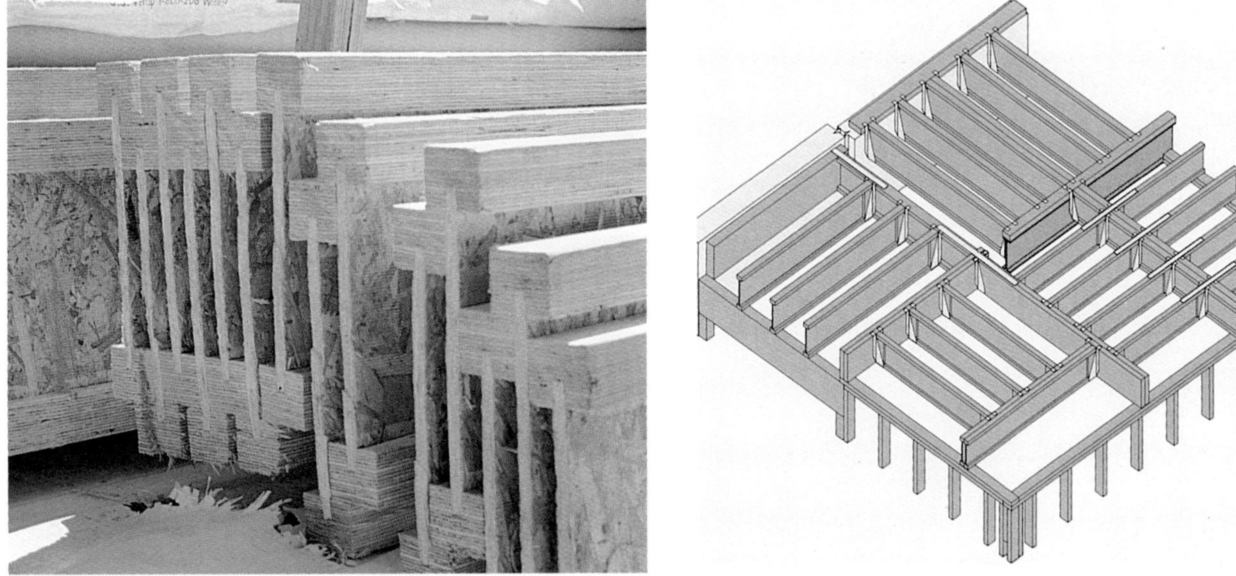

Figure 8-49. Left—Wood I-beams delivered to a construction site. Right—Simulated floor system shows how wood I-beams and other components are connected in any type of floor frame using metal connectors and ties. (Courtesy, Simpson Strong Tie Co., Inc.)

Installing Steel Joists

While steel framing has long been used in commercial construction, its use in residential buildings has become more popular since lumber prices have increased. With some additional equipment, carpenters are able to adapt readily to the different construction methods.

Fastening methods for steel vary. Steel framing members can be used exclusively or combined with wood members. Depending on framing materials, various fastening methods are used. Special nails may be used to attach steel members to wood framing members.

Self-tapping screws are often used to fasten steel to steel or to attach drywall to steel. If steel frames are welded, electric arc welding equipment will be required, Figure 8-56. Pneumatic or electric drills will be used to drive screws. Some builders prefer powder-actuated tools to attach subflooring to steel joists. (Refer to Unit 4 for additional information on power tools.)

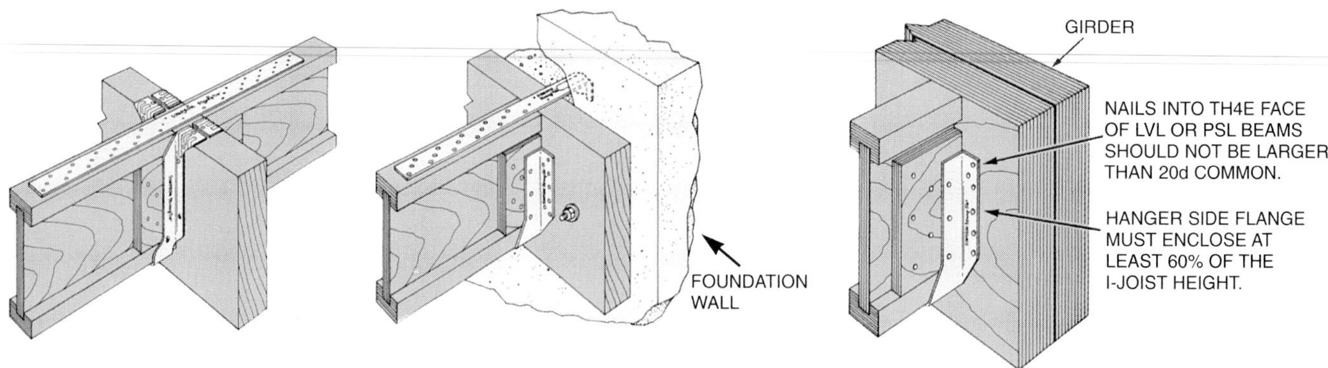

Figure 8-50. Special connectors are used to fasten wood I-beams to other frame parts. Strapping and hangers, rather than toe-nailing, are used to secure them. Note special blocking on the I-beam joists. (Courtesy, Simpson Strong-Tie Co., Inc.)

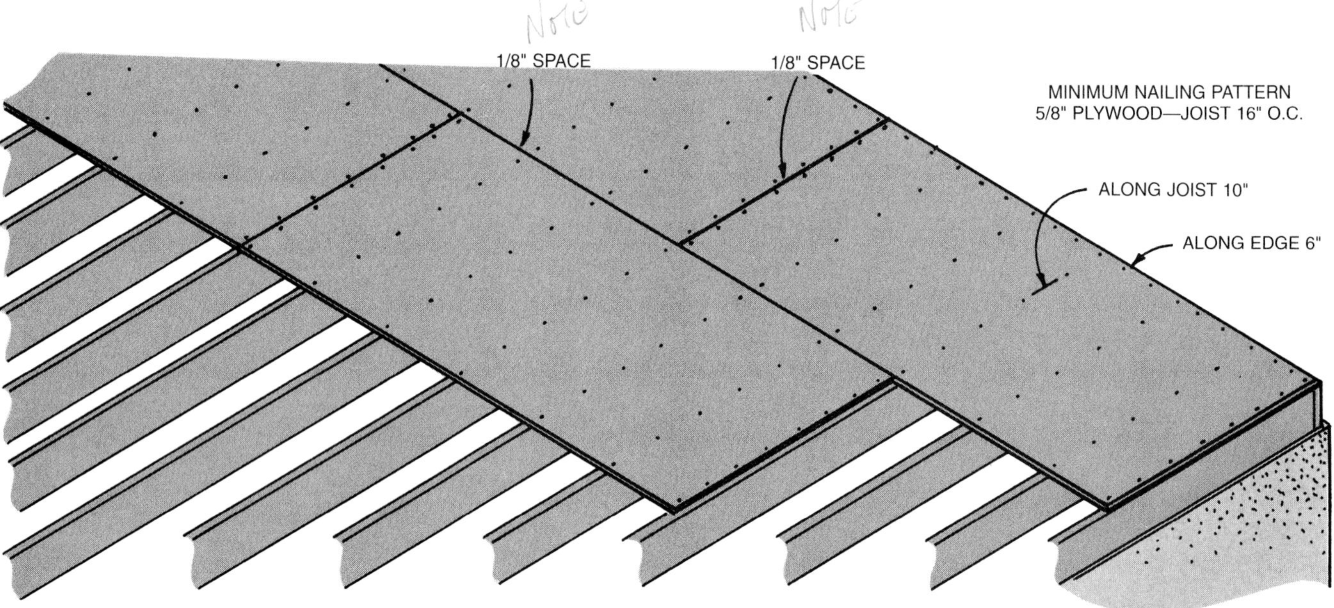

Figure 8-51. Using plywood as a subfloor. Note nailing pattern and how joints are broken for added strength.

Figure 8-52. Using a pneumatic nailer to install 5/8" plywood subflooring. Joists are spaced 16" on center. (ITW Paslode)

Figure 8-53. Applying adhesive for a glued floor system. A 1/4" wide bead of glue to the top of the joist is sufficient. Two beads are applied where panel ends butt.

Figure 8-54. Use a maul and protective strip of wood to close up the tongue and groove joints, leaving a 1/8" gap all around.

Figure 8-55. Fasten the glued subflooring with mechanical fasteners before glue sets. A power screw gun with an extended handle avoids stooping and speeds the operation.

Figure 8-57 (top) shows a floor framed entirely in steel. The headers are fastened to foundations with steel pins. A powder-actuated tool drives the pins through a flange on the headers into the concrete wall. Joists, spaced 16" on center, are first fastened to the headers with screws, then welded. Subflooring, such as waferboard or oriented strand board, can be attached to joists with adhesives and self-tapping screws. Another method involves attaching a metal pan across the joists, Figure 8-57 (bottom), and

Figure 8-56. A portable arc welder powered by a gasoline engine can be pulled to the construction site to weld joints in metal framing.

Figure 8-57. Top—A finished floor frame done in steel. The headers are attached to the foundation wall with steel fasteners driven in by a powder-actuated tool. Bottom—A view from the underside of a metal floor frame. Joists are one piece from wall to wall. Steel I-beam provides support. Floor pan has been attached to joists with screws in preparation for pouring a concrete floor over the joists.

pouring a 2 1/2" to 3" fiber-reinforced concrete floor on top of the joists. Refer to Figure 8-58.

Figure 8-58. A thin concrete base has been poured over the floor. Screeding has been completed and a cement finisher is using a bull float to smooth the surface. The concrete is reinforced with steel fibers to prevent cracks.

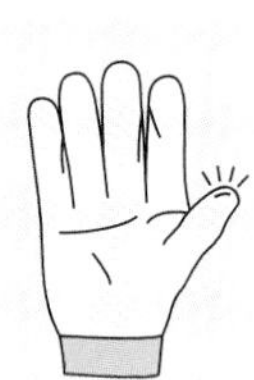

Steel components may have sharp edges. Use care in handling and wear leather gloves. Also, welded joints are hot and care should be exercised. Welded joints should be allowed to cool before handling.

Estimating Materials

If you are required to estimate the number and size of floor joists on a job, first scale the plan and determine the lengths that will be needed. Be sure to allow sufficient length for full bearing on girders and partitions. Average residential structures will require several different lengths.

Multiply the length of the wall that carries the joists by 3/4 for spacing 16" O.C. (3/5 for 20, 1/2 for 24) and add one more. Also add extra pieces for doubled joists under partitions and trimmer joists and headers at openings.

No. of joists = length of wall x 3/4 + 1 + extras

Some carpenters estimate one joist for every foot of wall upon which the joists rest (16" O.C. spacing). The over-run allows for extra pieces needed for doubles, trimmers, and headers.

Header joists are usually figured separately and added to the above figures. Your estimate should include the cross-section size and the number of pieces of each length. For example:

Required floor joists:
40 pcs. — 2 x 10 x 16'-0"
36 pcs. — 2 x 10 x 14'-0"
Required header joists:
4 pcs. — 2 x 10 x 16'-0"
2 pcs. — 2 x 10 x 12'-0"

Procedures for estimating the subflooring will vary, depending on the type of material used. Usually the area is figured by multiplying the overall length and width, and then subtracting major areas that will not be covered. These include breaks in the wall line and openings for stairs, fireplaces, and other items.

This will give the net area and the basic amount of material needed. To this must be added waste and other extras. For example: when 8" shiplap is used, multiply the basic figure by 1.15 and then add another 15% for waste. If the shiplap is laid diagonally to the joist, another 5% should be added. Individual boards are not specified. The amount is simply listed in board feet (equal to the square footage needed) along with a description of the material.

When using sheet materials, there is practically no waste and the net area is divided by 32 (sq. ft. in a 4 x 8 sheet) and rounded out to the next whole number. This will be the number of pieces of plywood required. Be sure to specify the type of sheet material and its span rating.

A new and more complete method of specifying plywood as recommended by the American Plywood Association is included in Unit 1

Important Terms

Balloon framing
Blocking
Bridging
Cantilevered
Chord
Cripple wall
Cross bridging
Crown
Girders
Glued floor system
Hanger
Header
Herringbone
Joists
Ledger
Platform framing
Plenum
Post anchor
Post and beam framing
Ribbon
Sill
Sill sealer
Stirrup
Subfloor
Tail Joist
Trimmer
Underpinning
Western framing
Wood I-beams

Test Your Knowledge

1. The type of framing used in most one-story construction is ____________.
2. When requirements call for the joists to be framed flush with the underside of a wood girder, it is best to use ____________.
3. Standard construction usually requires that the sill be spaced back from the foundation wall a distance equal to the ____________.
4. The studs of a balloon-type frame run continuously from the ____________ to the rafter plate.
5. Name the two types of steel beams used in residential construction.

6. In residential construction, the deflection of first floor joists under normal live loads should not exceed ____________ of the span.
7. A member of the floor frame that runs from the main header to a header for an opening is called a ____________ joist.
8. When framing a floor opening, the double header should be nailed in place before the second ____________ is installed.
9. Cantilevered joists should extend inward at least ____________ times the distance that they overhang the supporting wall.
10. Large holes bored through joists for pipes or wiring should be made at the ____________ (*top, bottom, center*).
11. Shiplap of a nominal width of 8" should be applied with ____________ (*2,3,4*) 8d nails at each joist.
12. When sheet material is used for the subflooring, the short dimension of the panel should run ____________ (*parallel, perpendicular*) to the joist.
13. How might steel headers be fastened to a foundation?
14. Concrete should never be used as a subflooring over steel joists. True or False.

Outside Assignments

1. Obtain a set of architectural plans for a house with a conventional basement. Study the methods of construction specified in the sections and detailed drawings. Then prepare a first-floor framing plan. Start by tracing the foundation walls and supports shown in the basement or foundation plans and then add all joists, headers, and other framing members. Your drawing should be similar to the one in Figure 8-32.
2. Working from a set of architectural plans for a single-story house, develop a list of materials required to frame the floor. Select the type of subflooring, if not specified, and estimate the amount of material needed.
3. From the local building code in your area, find the requirements for floor framing. Prepare a list of the requirements along with sketches that might clarify complicated written descriptions. Make an oral report to your class.

In modern residential construction, walls and ceiling joists form one structural system. The walls support the joists which form the ceiling or the next floor level. Ceiling joists or trusses are supported by the walls. The walls are stiffened and held plumb by the addition of the joists. Top—Carpenter places trusses along outside wall of a two-story house. Note the steel web in the truss joints. (Truswall Systems Corp.) Bottom—A construction foreman inspects the installation of roof joists.

9

Wall and Ceiling Framing

Wall framing includes assembling of vertical and horizontal members that form outside and inside walls of a structure. This frame supports upper floors, ceilings, and the roof. It also serves as a nailing base for inside and outside wall-covering materials. Inside walls are called partitions.

The term *system* commonly means methods and materials of construction. It is used in connection with floors, ceilings, and roofs as well as walls. Included are the design of the framework as well as the surface-covering materials and the methods for applying them. For example, a floor system includes:

- The details of the sill construction.
- Size and spacing of joists.
- The kind of subflooring.
- Application requirements.

Parts of the Wall Frame

The wall-framing members used in conventional construction include sole plates, top plates, studs, headers (also called lintels), and sheathing. Studs and plates are made from 2 x 4 or 2 x 6 lumber while headers usually require heavier material. Bracing made of 1 x 4 stock or steel strips must be built into the wall when the sheathing does not provide enough stiffness.

In one-story structures, studs are sometimes placed 24" O.C. (on center), Figure 9-1. However, 16" spacing is more common. Figure 9-2 shows a typical wall frame with openings for a window and door. Note that extra studs are used at the corner, at the sides of the rough openings for doors and windows, and where an interior wall meets an outside wall.

Height of residential walls varies from one region to another. Walls ten feet high are not uncommon, for example, in warm climates.

Conventional stud spacing of 16" or 24" has evolved from years of established practice. It is based more on accommodating the wall-covering materials than on the actual calculation of imposed loads.

Figure 9-3 illustrates, in more detail, various parts and connections in a wall frame. Full length studs become *cripple studs* when they end because of an opening.

Figure 9-1. Two examples of modern wood construction. Top—Wall stud spacing is sometimes 24" on center. Truss rafters provide framing for ceiling. Plywood sheathing provides enough support so that corner bracing is not needed. (Western Wood Products) Bottom—Framing for a southwestern home. Steel bracing can be seen in the wall. Joists frame ceiling and provide support of second story floor.

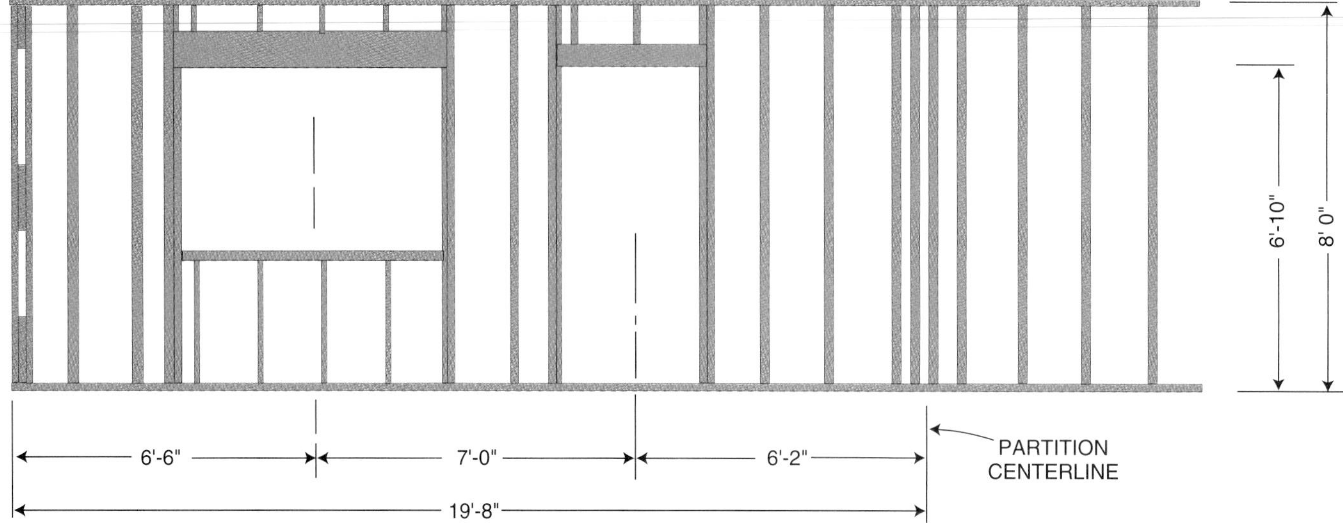

Figure 9-2. Drawing of a typical wall frame section. Eight foot height from bottom of sole plate to top of wall plate is extended by adding a second top plate. The extra 1 1/2" allows for 3/4" of finish flooring and 3/4" for ceiling covering. Typical stud length is 93" . Wall heights are often greater in warmer climates.

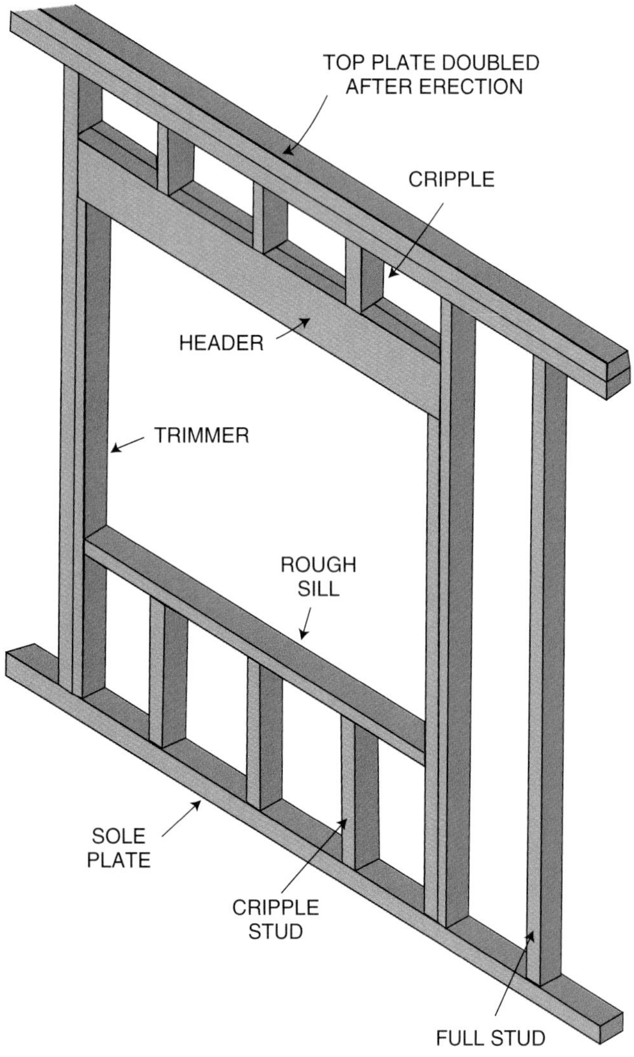

TOP PLATE DOUBLED AFTER ERECTION

CRIPPLE

HEADER

TRIMMER

ROUGH SILL

SOLE PLATE

CRIPPLE STUD

FULL STUD

Figure 9-3. Left—Parts of a typical wall frame. In some modern construction, headers go all the way to the plate. However, rising lumber costs have made this practice less common. Right—Note rough openings for windows with cripples and trimmers in place.

Trimmer studs are shortened studs that stiffen sides of rough openings and bear the direct weight of a header. Normally, carpenters attempt to put headers for doors and windows at the same height so that all of the trimmers will be the same length. All of the trimmers can then be cut at the same time, improving productivity.

Stud intervals remain modular regardless of interruptions by openings. This is so that modular sheet sizes can be installed with minimum cutting or waste.

Wall-framing lumber must be strong and straight with good nail-holding power. Warped lumber will not do the job, especially if the interior finish is drywall. Stud and No. 3 grades are approved and used throughout the country. Species such as Douglas fir, larch, hemlock, yellow pine, and spruce are satisfactory. See Unit 1 for additional information.

Corners

Any of several methods can be used to form the outside corners of the wall frame. In platform construction, the wall frame is usually assembled in sections on the rough floor and then raised. Corners are formed when a sidewall and end wall are joined.

Usually a second stud is included in the sidewall frame. It should be spaced inward the width of a 2 x 4 and stiffened with three or four blocks. When the end wall is erected the complete corner is formed, Figure 9-4(A). An alternate method is to turn the extra stud as shown in Figure 9-4(B).

Only straight studs should be selected for corners. Assemble with 10d nails spaced 12" apart. Stagger them from one edge to the other as shown. Include extra nails to attach the filler blocks.

Some carpenters prefer to build the corners for platform construction separately. They are set in place, carefully plumbed, and braced, before the wall sections are raised. This makes it easier to plumb and straighten the wall sections but does not permit the application of sheathing while the frame is still on the deck.

In climates which require the house to be well-insulated, 2 x 6 studs are commonly used. Figure 9-5 shows typical corner construction.

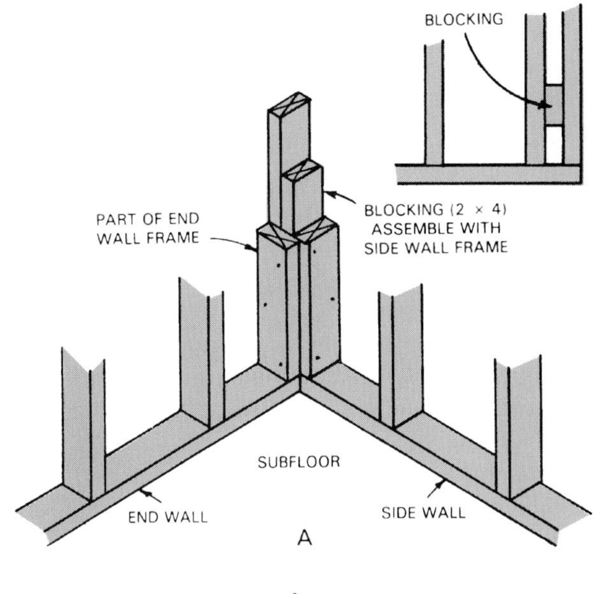

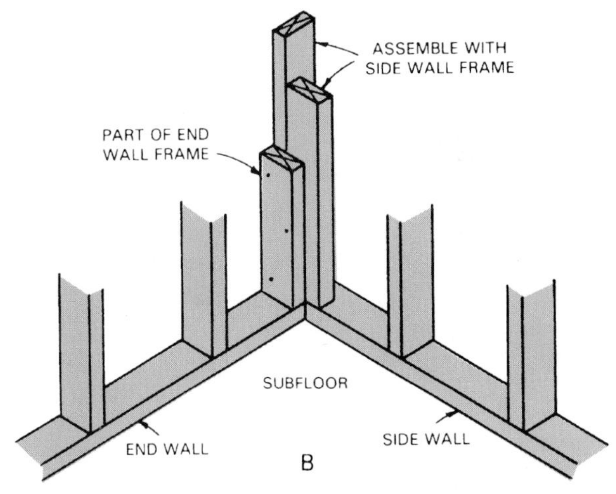

Figure 9-4. Placement of studs to form corners in platform construction. A—Corner built from three full studs and blocking. B—Corner built with three full studs and no blocking.

Figure 9-5. Typical corner construction in 2 x 6 framing. The inside corner is formed with a 2 x 4 (arrow), allowing nailing surface for drywall and stiffening the corner without blocking.

Partition Intersections

Partitions should be solidly fastened to outside walls. This requires extra framing on the outside wall. The framing must not only secure the partition firmly to the outside frame but also provide a nailing surface on inside corners for wall covering such as drywall. Several methods can be used to accomplish these purposes:

- Install extra studs in the outside wall. Attach the partition to them.
- Insert blocking and nailers between the regular studs.
- Use blocking between the regular studs and attach nailers or patented backup clips to support inside

wall coverings at all inside corners. (A nailer is lumber, such as 1 x 6 or 2 x 4, added as a backing at inside corners.) Figure 9-6 shows various methods.

Rough Openings

Study the house plans to learn the size and location of the *rough openings* (often referred to as R.O. on drawings). Plan views with dimension lines. Usually the measurement is taken from corners or intersecting partitions to the centerlines of the openings.

Heights of rough openings are given in elevation and section views. Sizes are shown on the plan view or listed in a table called a door and window schedule. See Unit 6, Figure 6-20.

Headers carry the weight of the building across door and window openings. They are made by nailing together two framing members and placing them in the wall on edge. Plywood spacers inserted between the pieces make the header the same thickness as the wall, Figure 9-7.

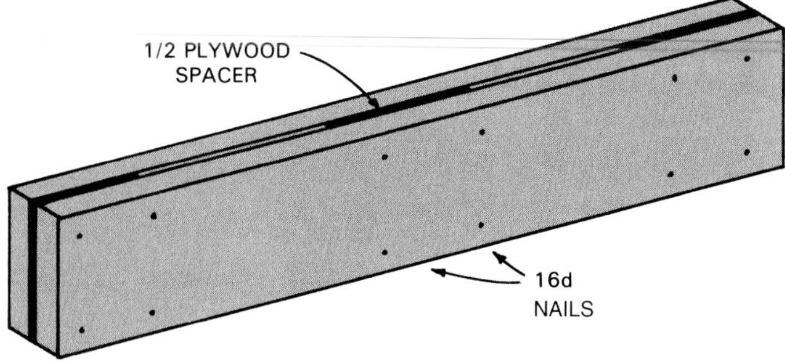

Figure 9-7. Header construction. Plywood spacers are placed 16 to 24" apart on centers.

Header length is equal to the rough opening plus the width of two trimmers (3"). The width of the lumber to be used in the header will depend on the span of the opening. Local building codes may include requirements for head-

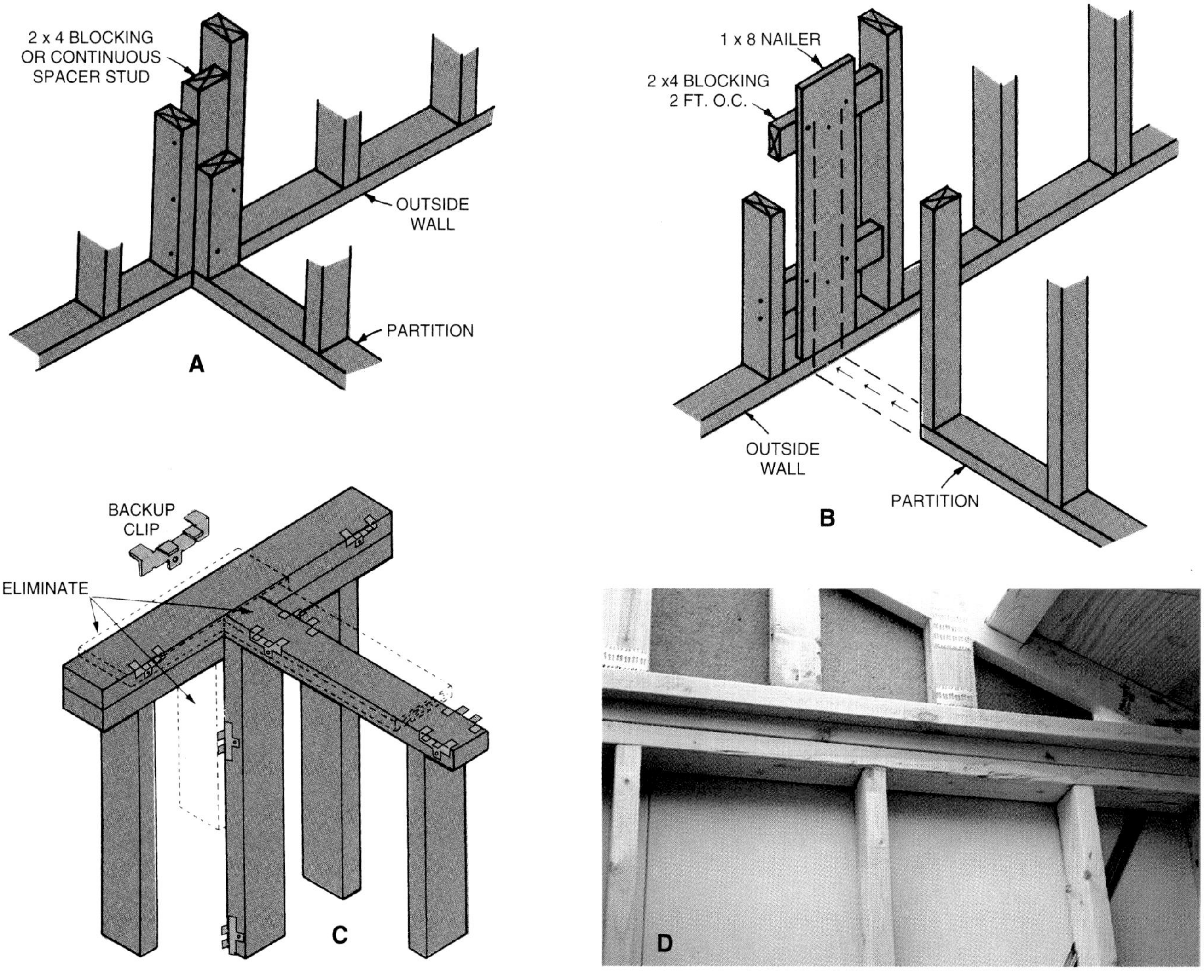

Figure 9-6. Framing details where partitions intersect. A—Using extra studs. Nail studs to blocking with 16d nails in 12" O.C. Use 10d nails to attach partition stud. B—Blocking installed between studs. Use 16d nails. Backing board is attached with 8d nails. C—Backup clips are sometimes used and take place of some framing studs. (TECO) D—2 x 4 used as a nailer.

ers. The table in Figure 9-8 gives the size of headers normally required for various rough opening widths under several load conditions.

Headers are also required across openings in load-bearing partitions. If loads are very heavy or spans are unusually wide, a *flush beam*, or strongback, may be used. In such cases, hangers could be used to attach ceiling for floor joists to the *flush beam*. (These members are discussed later in this unit.) Figure 9-9 shows a truss joist being attached to a doubled laminated-veneer header.

In modern platform construction, extra studs are included around rough openings as shown in the standard assembly, Figure 9-10. The studs and trimmers support the header and provide a nailing surface for window and door casing. Rough sills may be doubled to provide a nailing base for trim.

MATERIAL ON EDGE	SUPPORTING ONE FLOOR, CEILING, ROOF (IN FT. & IN.)	SUPPORTING ONLY CEILING AND ROOF (IN FT. & IN.)
2 × 4	3-0	3-6
2 × 6	5-0	6-0
2 × 8	7-0	8-0
2 × 10	8-0	10-0
2 × 12	9-0	12-0

Figure 9-8. Recommended maximum header spans. Be sure to check local codes.

Figure 9-9. When open-web joists are used for a second story floor frame, solid laminated veneer header may be used to support a cripple joist. (Kasten-Weiler Construction Co.)

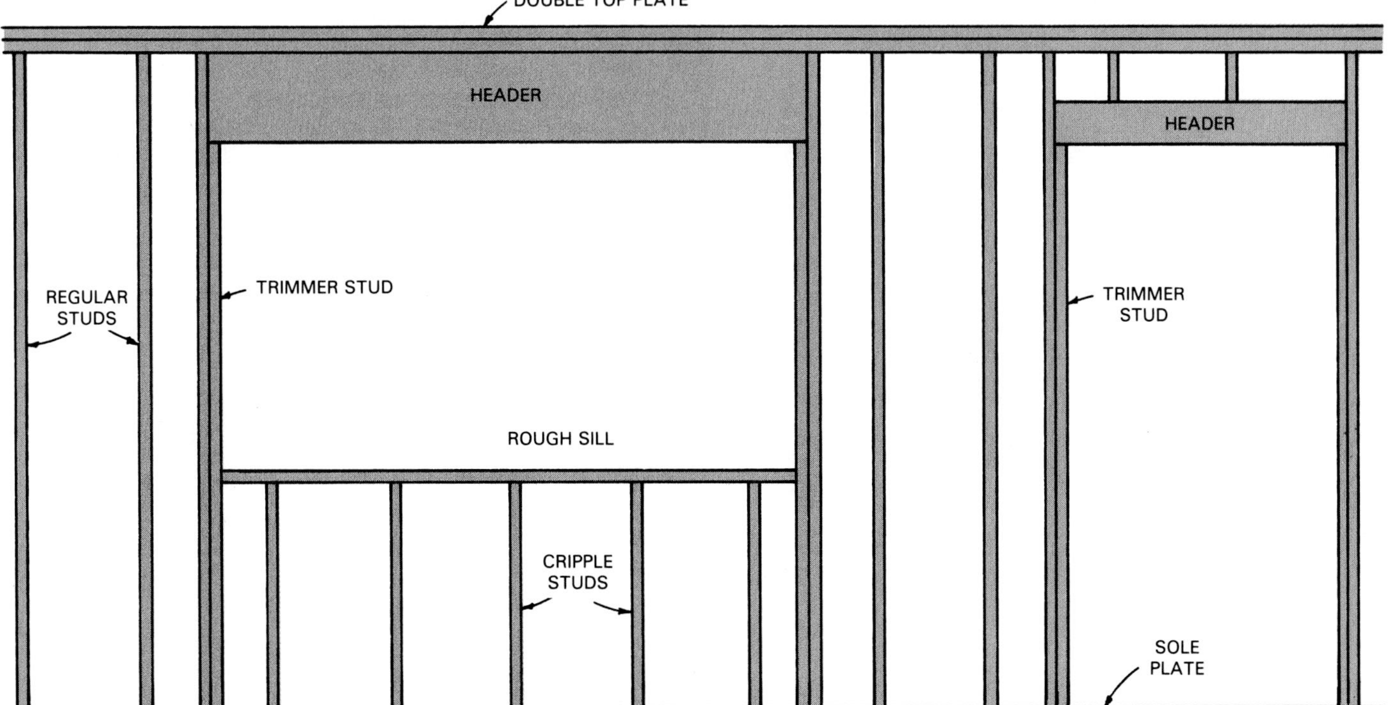

Figure 9-10. Framing door and window openings. Trimmer studs must carry the weight of the header and its load. Though wider than it needs to be, the header for the window saves the labor of cutting and installing cripple studs between the header and plate.

Alternate Header Construction

In large window openings the size of the header will reduce the length of the upper cripple studs to a point where they cannot be easily assembled. They should be replaced with flat blocking.

Another solution is to increase the header size to completely fill the space to the plate. Most builders follow this practice and extend it to include all openings, regardless of the span. They have found that the cost of labor required to cut and fit the cripple studs is usually greater than the cost of the larger headers.

A disadvantage of such construction is the extra shrinkage which, without special precaution in the application of interior wall finish, may cause cracks above doors and windows.

In balloon construction, which was once popular, studs extended from the sole plate to the roof plate. It was common practice to extend headers beyond the rough opening to the next regular stud, Figure 9-11.

Plate Layout

Use only straight 2 x 4 stock for plates. Select two of equal length and lay them side by side along the location of the outside wall. Length should be determined by what can easily be lifted off the floor and into a vertical position after it is assembled. Remember that the weight may include all the framing for rough openings, bracing, and sheathing. If wall jacks or a forklift are available, sections can be made larger. Where they must be lifted by hand, sheathing may be attached after the wall is up.

Laying Out Plates

1. Lay out the plates along the main sidewalls. Align the ends with the floor frame and then mark the regular stud spacing along both plates, Figure 9-12. Some carpenters tack the pieces to the floor with several nails so they will not move while the layout is being made.
2. Study the architectural plans and lay out the centerline for each door and window opening.
3. Measure off one-half the width of the opening on each side of the centerline. Mark the plate for trimmer studs outside of these points. On each side of the trimmer stud include marks for a full length stud. Identify the positions with the letter "T" for trimmer studs and "X" for full length studs.
4. Mark all of the stud spaces located between the trimmers with the letter "C". This designates them as cripple studs.

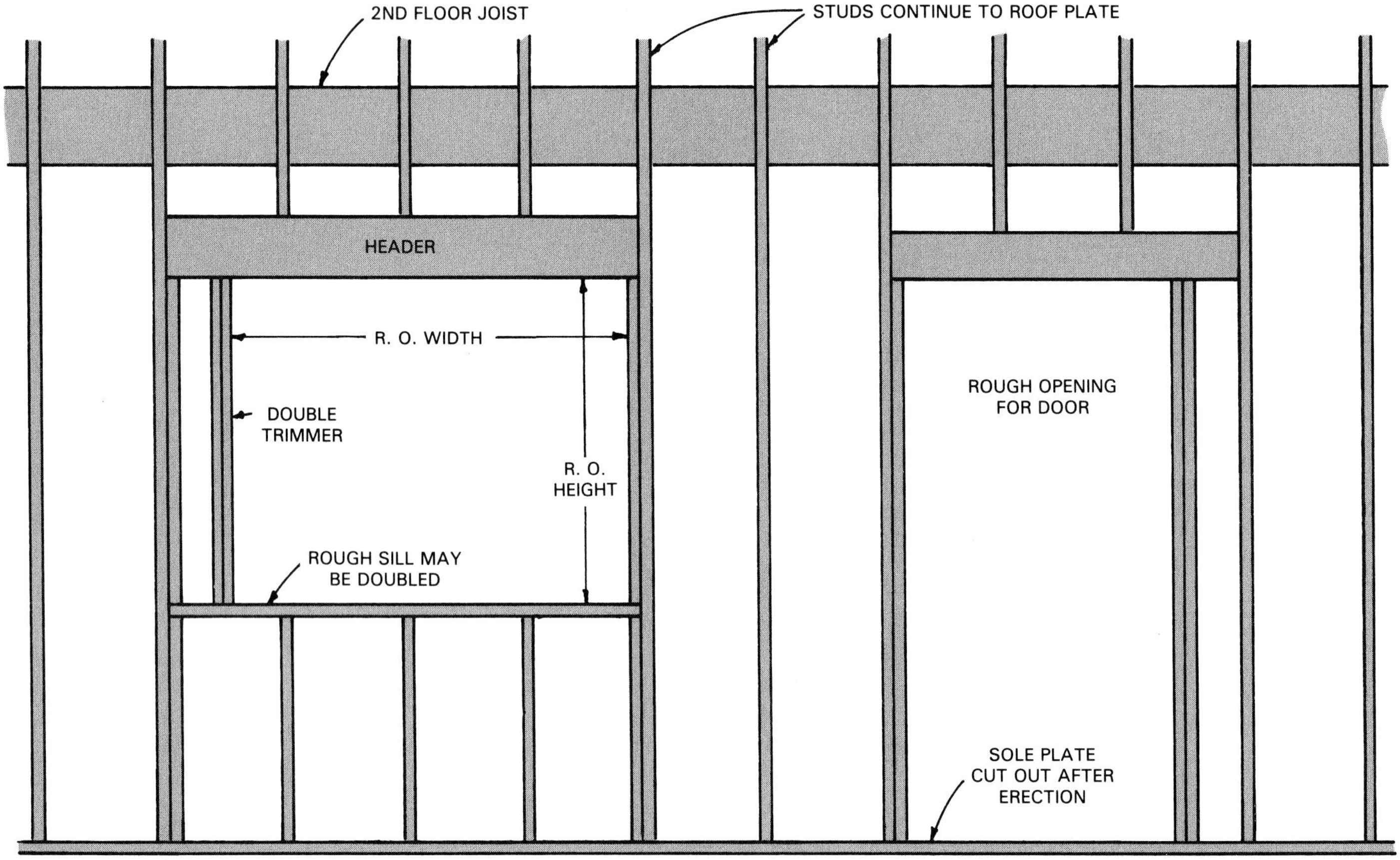

Figure 9-11. Framing of openings for once-popular balloon construction. Note doubled trimmer at right side of door rough opening. It is intended to stiffen the single trimmer.

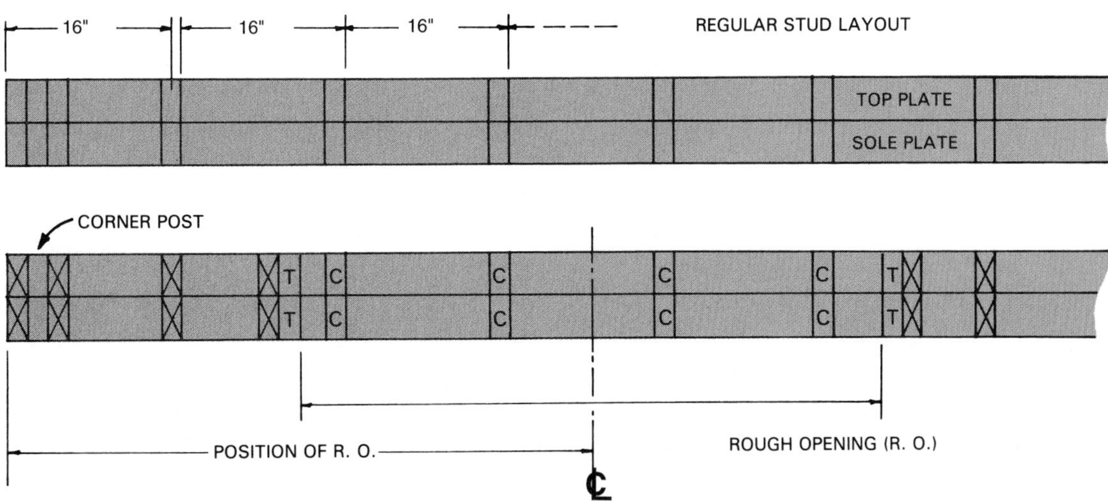

Figure 9-12. Layout of sole and top plates. Top—Regular stud spacing has been marked. Bottom—Layout is converted for a window opening.

5. Lay out the centerlines where intersecting partitions will butt. Add full length studs if required by the method of construction.
6. When blocking between regular studs is used, the centerline will be needed as a guide for positioning the backing strip.
7. Plan the layout of wall corners carefully so they will fit together correctly when the wall sections are erected.

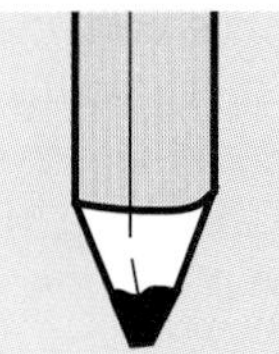

Check over your rough opening layouts carefully. Errors may be difficult to recognize at this point.

A *story pole* is a long measuring stick made up by the carpenter on the job. It represents the actual wall frame with markings made at the proper height for every horizontal member of the wall frame. See Figure 9-13.

When using a story pole, all of the heights for horizontal members are transferred from the drawings to the pole at one time. There is no need to consult the plans several times, as would be needed when no story pole is used. The story pole is particularly useful in split-level construction, Figure 9-14.

A *master stud pattern* is like a story pole but does not include as much information. The layout can be made on either a straight 1 x 4 or 2 x 4. First lay out the distance from the rough floor to the ceiling. This dimension can be taken directly from the story pole or from the plans. Mark off the position of the sole plate and double top plate, Figure 9-15. Now lay out the header. When several header sizes are used, they can be marked on top of each other.

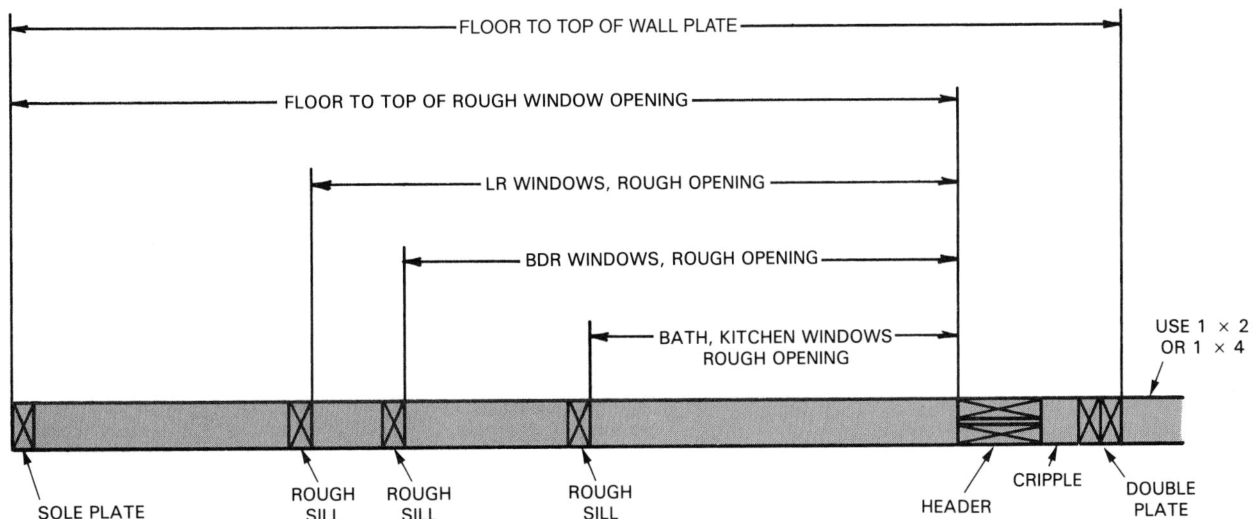

Figure 9-13. Story pole is a handy guide which marks height of every horizontal member and length of every vertical member of the wall frame. It usually extends one story but may include more.

Figure 9-14. Split level homes and homes with different floor levels between the garage and living area will require use of a story pole to check heights.

Lay out the height of the rough openings, measuring down from the bottom side of the header. Then draw in the rough sill.

The length of the various studs (regular, trimmer, and cripple) can now be taken directly from this full-size layout.

When the header height of the doors is different from that of the windows, use the other side of the pattern to keep them separate. In multi-story or split-level structures, a master stud layout will probably be required for each level.

Wall Sections

Constructing a Wall Section

1. Working from the master stud layout, cut the various stud lengths. It is seldom necessary to cut standard full-length studs. These are usually precision end trimmed (P.E.T.) at the mill and delivered to the construction site ready to assemble.
2. Cut and assemble the headers. Their length, and also the length of the rough sill, can be taken directly from the plate layout.
3. Move the top plate away from the sole plate about a stud length. Turn both plates on edge with the layout marks inward. Place a full-length stud, crown up, at each position marked on the top and sole plates. See Figure 9-16.
4. Nail the top plate and the sole plate to the full-length studs using 16d nails, or their equivalent if using a power nailer.
5. Set the trimmer studs in place on the sole plate and nail them to the full-length studs. Now place the header so it is tight against the end of the trimmer and nail through the full-length stud into the header using 16d nails, Figure 9-17.
6. The upper cripples, if used, can be installed after the header is installed. Carpenters commonly run the header all the way to the top plate to save labor. See Figure 9-18.
7. For window openings, transfer marks for cripple studs from the sole plate to the rough sill and assemble the cripples with 16d nails. Some carpenters prefer to erect the wall section before installing the cripples. In this case, the lower ends of the cripple studs are toenailed to the sole plate, Figure 9-19.
8. Add studs or blocking at positions where partitions will intersect.
9. Install any wall bracing that may be required for special installations. Remember, the inside of the wall is face down.

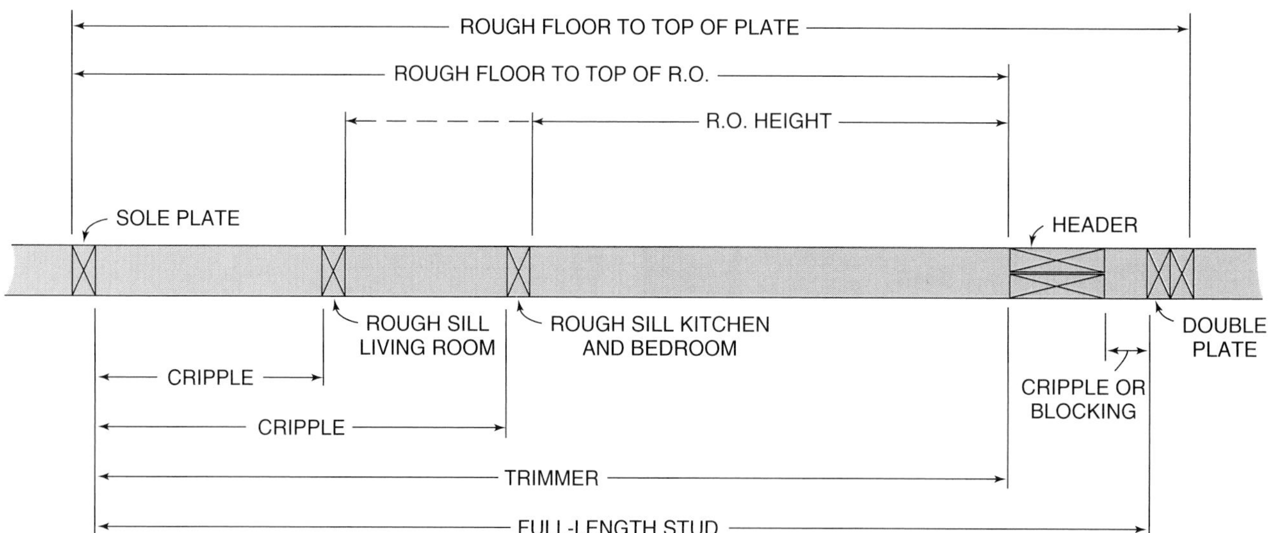

Figure 9-15. Master stud pattern is like a story pole. Stud lengths can be secured from it.

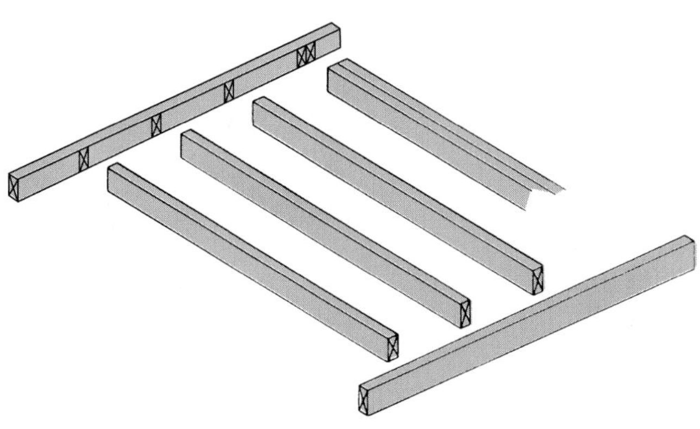

Figure 9-16. Assembling the wall studs and plates. Left—Place a stud at every position marked on the plate. Right—Turn crowns upward and assemble wall frame.

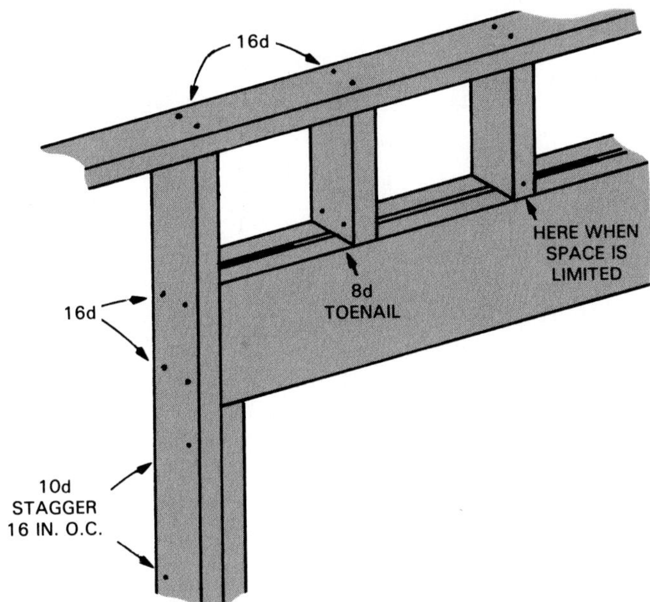

Figure 9-17. Assembling wall section. Older method of installing headers with cripple studs is sometimes used. Note nailing pattern through full-length stud into header.

Figure 9-18. Assembling wall section with header filling space to bottom of top plate. The studs are 2 x 6s spaced 24" O.C.

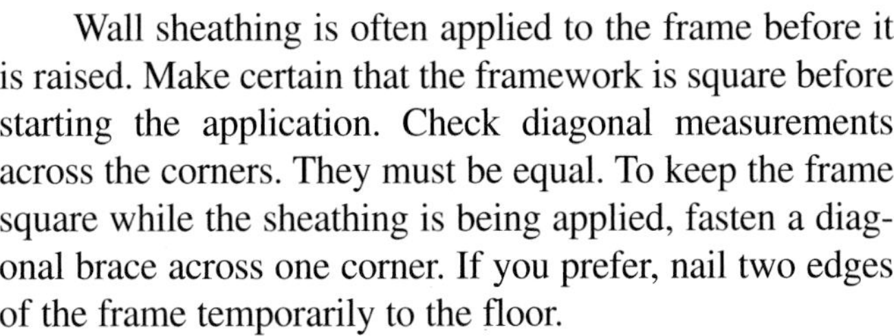

Wall sheathing is often applied to the frame before it is raised. Make certain that the framework is square before starting the application. Check diagonal measurements across the corners. They must be equal. To keep the frame square while the sheathing is being applied, fasten a diagonal brace across one corner. If you prefer, nail two edges of the frame temporarily to the floor.

Erecting Wall Sections

Most one-story wall sections can be raised by hand, Figure 9-20. Larger structures will require the use of a crane or other equipment.

Before raising a section, be sure it is in the correct location. Have bracing at hand and ready to be attached. If the section is large, have extra help available. Make sure each worker knows what to do, Figure 9-21.

When raising sections to which sheathing has not been applied, it is good practice to install temporary diagonal bracing if regular bracing is not included, Figure 9-22. Some carpenters attach temporary blocking to the edge of the floor frame before raising wall sections. It keeps the wall section from sliding off the platform.

Immediately after the wall section is up, secure it with braces attached near the top and running to the subfloor at about a 45° angle. Make final adjustments in the position of the sole plate. Be sure it is straight. Then nail it to the floor frame using 20d nails driven through the subfloor and into the joists, Figure 9-23.

Loosen the braces one at a time and plumb the corners and midpoint along the wall. This can be done with a plumb line, but on one-story construction, a carpenter's

Figure 9-19. Top—Cripple studs can be installed before wall section is erected. Bottom—However, some carpenters prefer to erect wall sections and then install the rough window sill. Cripple studs are toenailed to the sole plate, as shown (arrow).

Figure 9-20. Raising a section of wall. Headers for window and door openings are made from laminated veneer lumber (LVL). (Gang-Nail Systems, Inc.)

Figure 9-21. Top—Raising a section of wall constructed from 2 x 6s. Sole plate is attached and will be nailed to floor frame when in place. Be sure that bracing is at hand and that all workers know what to do. Bottom—Front wall of garage being raised by student carpenters. (North Bennet Street School, Boston)

Figure 9-22. Wall section is in place, nailed to platform, plumbed, squared, and temporarily braced. Note extra studs where planned partition will meet the outside wall (arrow).

Figure 9-23. Waferboard sheathing was added to this wall section before it was raised. Section is being nailed in place through the sole plate. (Blandin Wood Products Co.)

Figure 9-24. Using a carpenter's level to plumb a wall section (arrow). When it is exactly plumb, the second carpenter (at left) will nail the brace to the sill.

Figure 9-25. Temporary bracing intended to square up wall sections can be removed when permanent braces and sheathing are in place.

level is generally used, Figure 9-24. Braces attached temporarily to square a wall section can be removed when permanent braces and sheathing are installed, Figure 9-25.

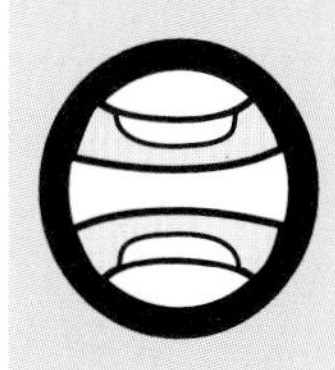

When plumbing a wall with a carpenter's level, hold the level so you can look straight in at the bubble. If the wall framing member or surface is warped, you should hold the level against a long straightedge that has a spacer lug at the top and bottom.

After one section of the wall is in place, proceed to other sections. No particular sequence needs to be followed. Most carpenters prefer to erect main sidewalls first and then tie in end walls and smaller projections. Procedures must be determined on each individual project. Design and construction methods help determine how to proceed.

Partitions

When the outside wall frame is completed, partitions are built and erected. At this stage, it is important to enclose the structure and make the roof watertight. Only bearing partitions (those that support the ceiling and/or roof) are installed at this time. Erection of nonbearing partitions can be put off until later.

Roof or floor trusses are supported by the outsidewalls. When they are used, Figure 9-26, inside partition work is seldom begun before the roof is completed.

Figure 9-26. Open web trusses used here are supported by the outsidewalls alone. Partitions do not need to be installed until the roof is on and the building is closed in. (Kasten-Weiler Construction Co.)

The centerlines of the partitions are established from a study of the plans and then marked on the floor with a chalk line. Plates are laid out; studs and headers are cut; and partitions are assembled and erected in the same way as outsidewalls . Erect long partitions first, then cross partitions. Finally build and install short partitions that form closets, wardrobes, and alcoves.

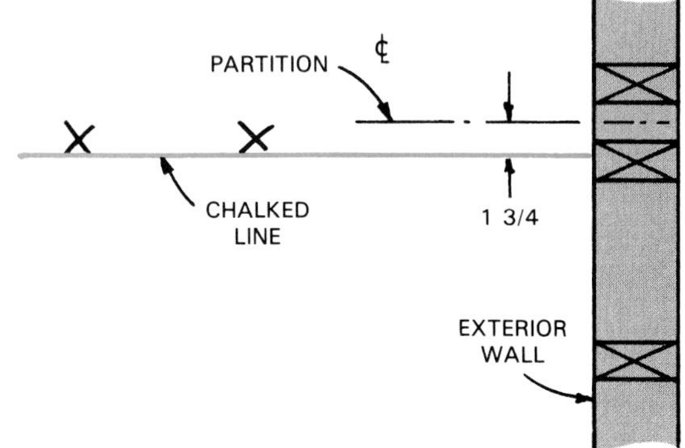

Figure 9-27. Snap a chalk line on floor to mark position of partitions.

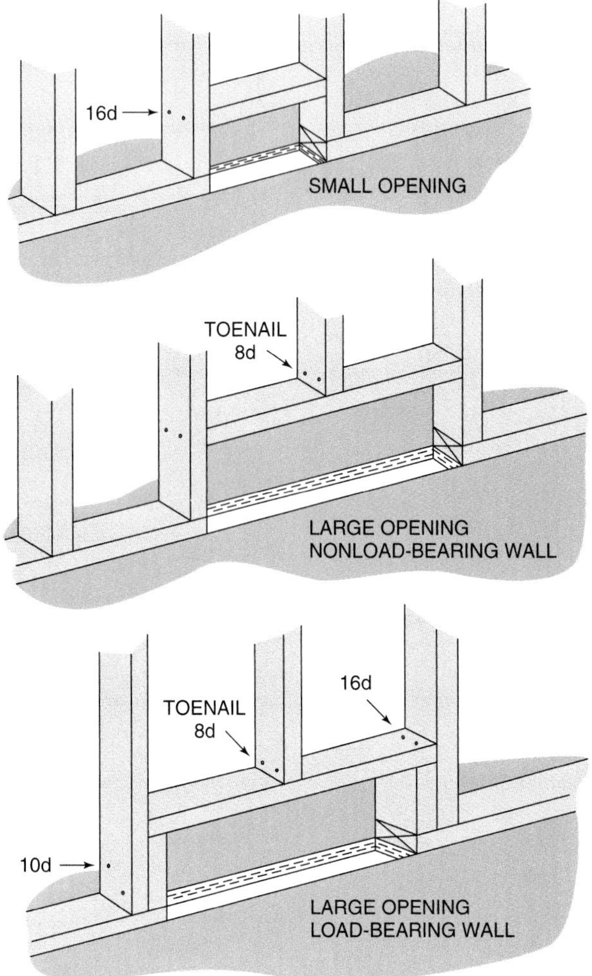

Figure 9-28. How to install special framing for heat ducts. Cripple studs should be added for large openings in load-bearing walls.

The corners and intersections are constructed as described for outsidewalls . The size and amount of blocking, however, can be reduced, especially in nonbearing partitions. The chief concern is to provide nailing surfaces at inside and outside corners for wall-covering material. Refer again to Figure 9-6.

Nonbearing Partitions

Nonbearing partitions do not require headers above doorways and other openings. Many rough openings can be framed with single pieces of 2 x 4 lumber since there is virtually no load on them. Trimmers are usually used for rigidity. They also provide added framework for attaching casing and trim. Door openings in partitions (and outsidewalls) are framed with the sole plate included at the bottom of the opening. After the framework is erected, the sole plate is cut out with a handsaw. Rough door openings are generally made 2 1/2" wider than the finished door size.

The soundproofing of partitions between noisy areas and quiet areas may require a special method of framing. See Unit 14 for information on insulation.

Small alcoves, wardrobes, and partitions in closets are often framed with 2 x 2 material or by turning 2 x 4 stock sideways, thus saving space. This is satisfactory when the thinner constructions are short and intersect regular walls. Snap a chalkline across rough floor to mark position of partitions, Figure 9-27.

During wall and partition framing, various important details can be added. Openings for the installation of heating ducts, Figure 9-28, are easily cut and framed at this time. Bath tubs and wall-mounted stools require extra support, Figure 9-29.

Basic provisions for recessed and surface hung cabinets, tissue-roll holders, and similar items should be added

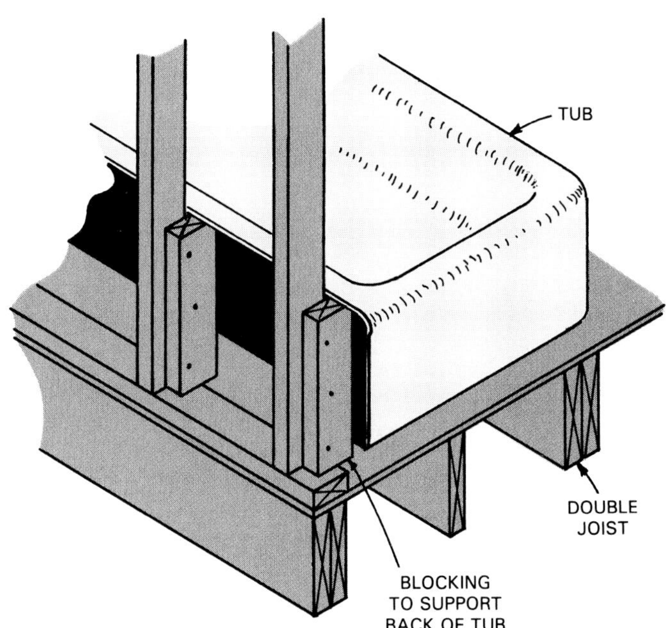

Figure 9-29. Extra joists and blocking are needed to support tubs.

to the framing at this stage. Architectural plans usually provide information concerning their size and location. Wall backing, Figure 9-30, drapery brackets for towel bars, shower curtains, and wall mounted plumbing valves should also be added. Plumbing fixture rough-in drawings, Figure 9-31, will be helpful in locating the backing. For most items, 1" thick backing material will provide adequate support.

Small items that are not critical to the structure can often increase efficiency and quality of work during the finishing stages. For example, corner blocks, Figure 9-32, will make it possible to nail baseboards some distance back from the end. This eliminates the possibility of splitting the wood.

Plumbing in Walls

Where plumbing is run through walls, special construction may be required. Depending on the size of the drain and venting pipes, a partition may have to be made wider. Usually a 6" frame is sufficient. Figure 9-33 shows several methods of construction.

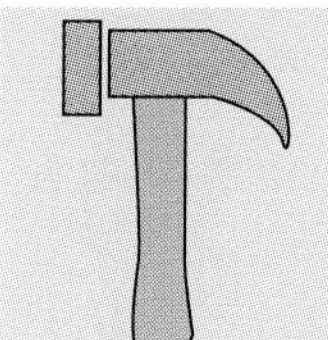

The carpenter must continually study and plan the sequence of the job, so that neither the weather nor work of other tradespeople will cause slowdowns or bottlenecks.

Lateral (horizontal) runs of pipe will require drilling of holes or notching of the studs. A metal strap can be

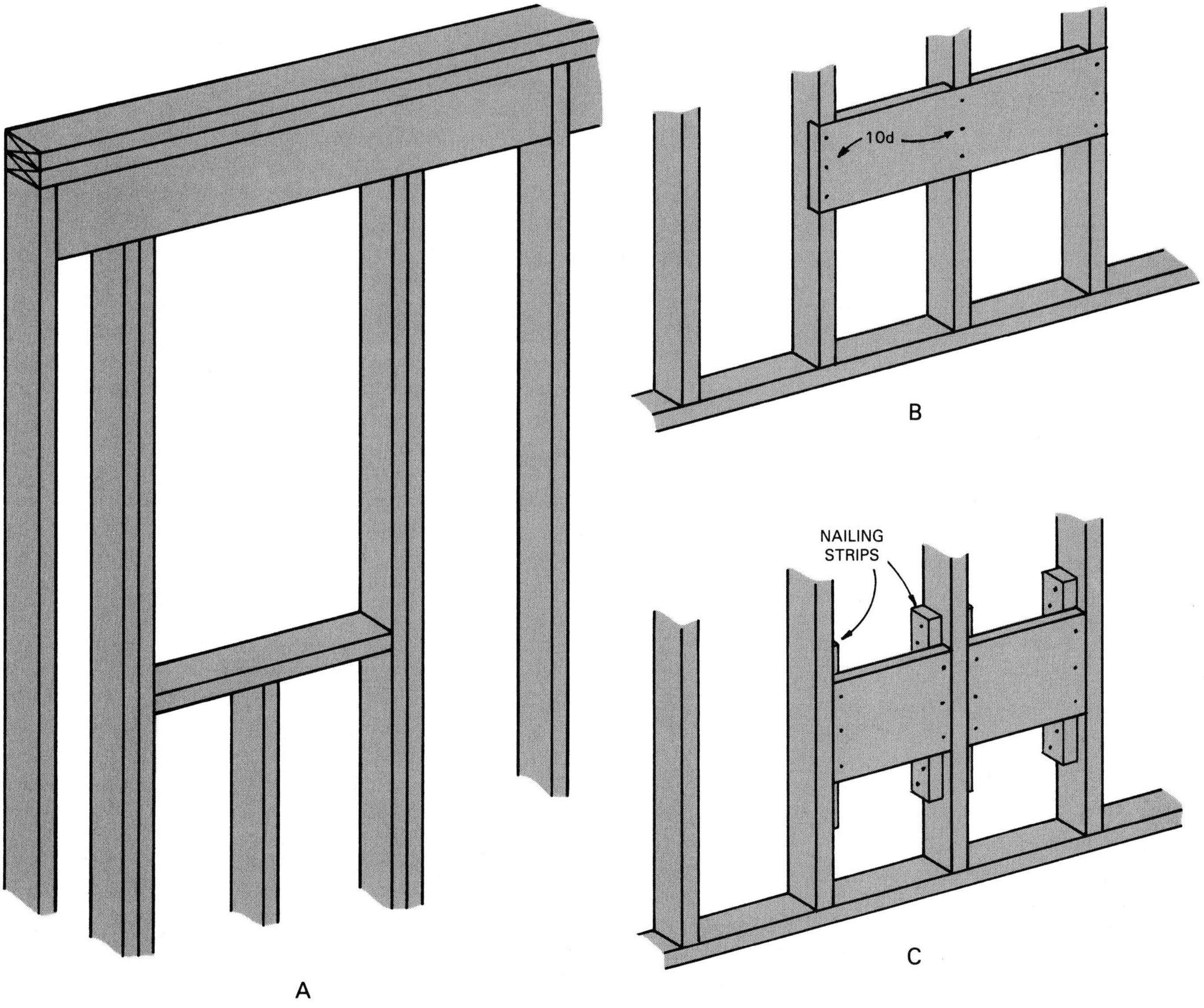

Figure 9-30. Backing for mounting various fixtures and appliances. A—Extending header over windows provides a base for attaching drapery rod brackets. B—Backing let into studs. Never cut back more than 25% of stud width on bearing walls or partitions. C—Backing attached to nailing strips.

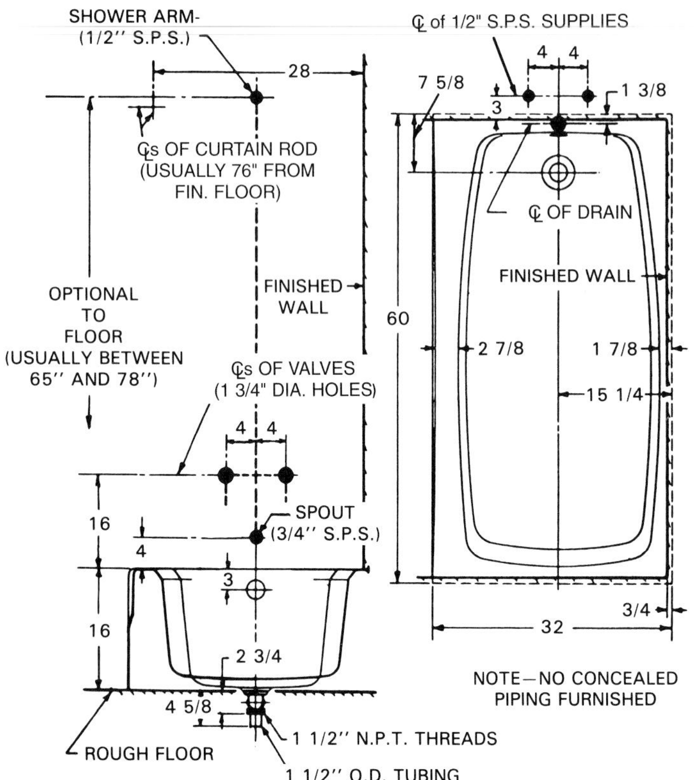

Figure 9-31. Typical rough-in dimensions are helpful reference when locating special backing. (American Standard)

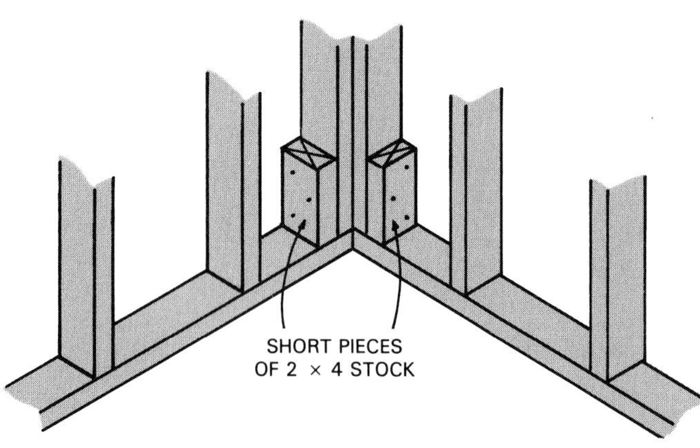

Figure 9-32. Blocking in corners provides better nailing surface for attaching baseboards.

attached to bridge the notch and strengthen the stud, Figure 9-34. The strap also protects the pipe from accidental damage should a nail be driven into the stud at this point. Similar protection should be provided for electrical wiring in walls. Sometimes a wooden block may be used to bridge notch cut for plumbing.

Bracing

Exterior walls usually need some type of bracing to resist lateral (sideway) loads. Some applications of material,

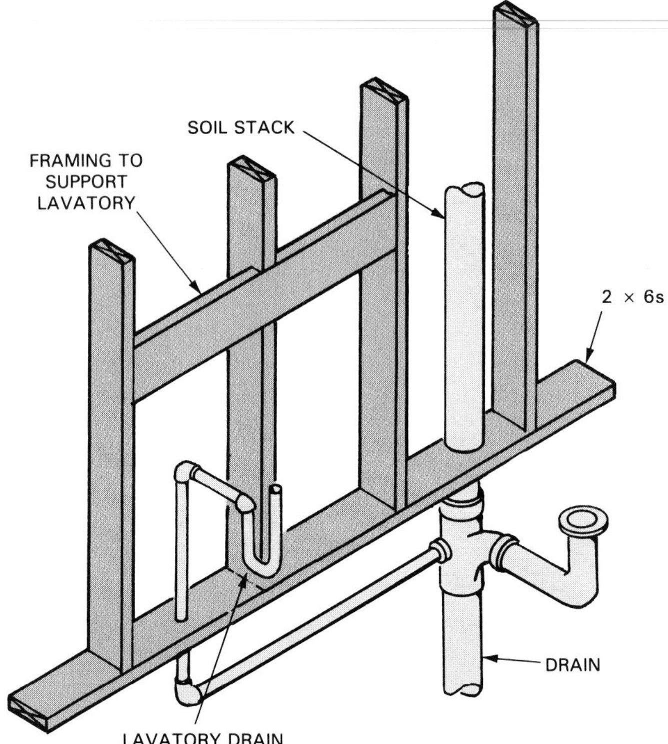

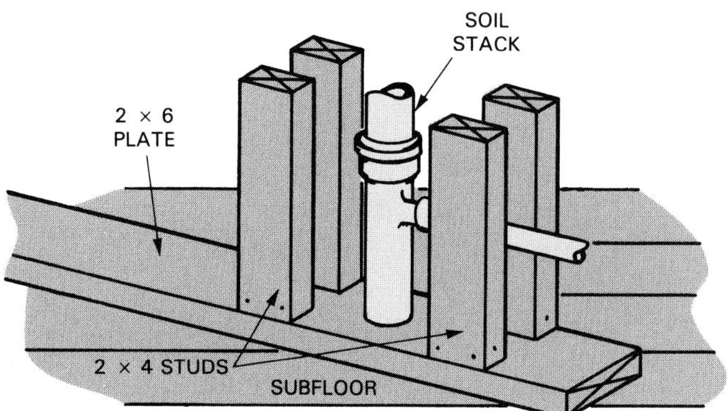

Figure 9-33. A partition containing plumbing pipes may need to be constructed differently to provide room for the plumbing. Top—Wall constructed of 2 x 6 studs and plate. Note special framing for supporting lavatory. Bottom—2 x 4 studs on 2 x 6 plate eliminate need to notch or bore studs for lateral runs.

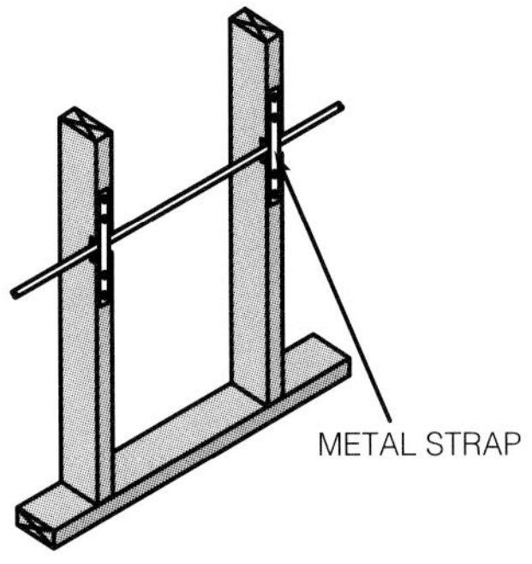

Figure 9-34. Method of reinforcing studs when notched or bored for plumbing. This method is also used to protect electrical wiring from becoming damaged by fasteners.

such as plywood, provide sufficient rigidity and bracing can be eliminated. Always check the exact requirements of the local building code.

A standard installation of let-in corner bracing is shown in Figure 9-35. Note how the bracing is applied when an opening interferes with the diagonal run.

Metal strap bracing is also widely used. It is made of 18 or 20 gage galvanized steel and is 2" wide. One type includes a 3/8" center rib, Figure 9-36.

To install the ribbed strap bracing, snap a chalk line across the erected wall frame. Use a portable saw and make multiple cuts to form the groove for the rib. Drive two 8d nails through the rib and into each framing member. For 2 x 6 studs, use one 16d nail. See Figure 9-37.

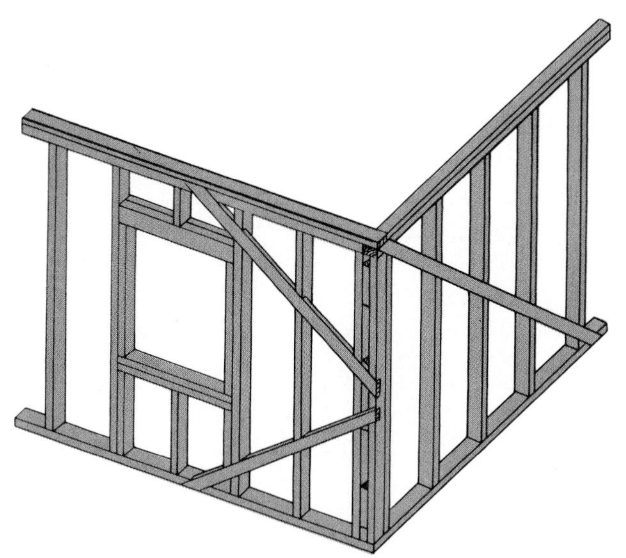

Figure 9-35. Bracing a wall with 1 x 4 lumber. Material is cut away from the studs to receive (let-in) the braces. Use two 8d nails at each connection.

Figure 9-36. Metal strap brace goes from the top plate to the sill. It is easily installed by cutting an angled saw kerf into the studs, top plate, and sole plate. Fasten with 8d nails.

Double Plate

To add support under ceiling joists and rafters, the top plate is doubled. This also serves to further tie the wall frame together. Select long straight lumber. Install the double plate with 10d nails. Place two nails near the ends of each piece. The others are staggered 16" apart.

Joints should be located at least 4' from those in the lower top plate. At corners and intersections, the joints are lapped as shown in Figure 9-38.

A structure that may be shaken by earthquakes or blown by high winds requires a stronger frame. Metal ties are available that strengthen the joints between walls and foundations and rafters and sidewalls. These will help to reduce or eliminate the damage from natural disasters. See Figure 9-39.

Tri-Level and Split-Level Framing

Tri-level and split-level housing presents special challenges in wall framing. Generally, a platform type of construction is used. However, the floor joists for upper levels could be carried on ribbons let into the studs.

The plans will prescribe the type of construction. They should also include careful calculations of distances between floor levels.

When working with split-level designs, the good carpenter will prepare accurate story poles which show full-size layouts of vertical distances and actual sizes of the construction materials.

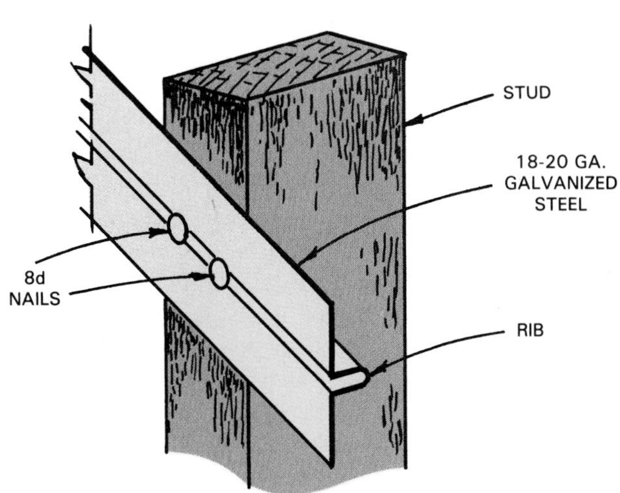

Figure 9-37. Detail of metal strap bracing connection.

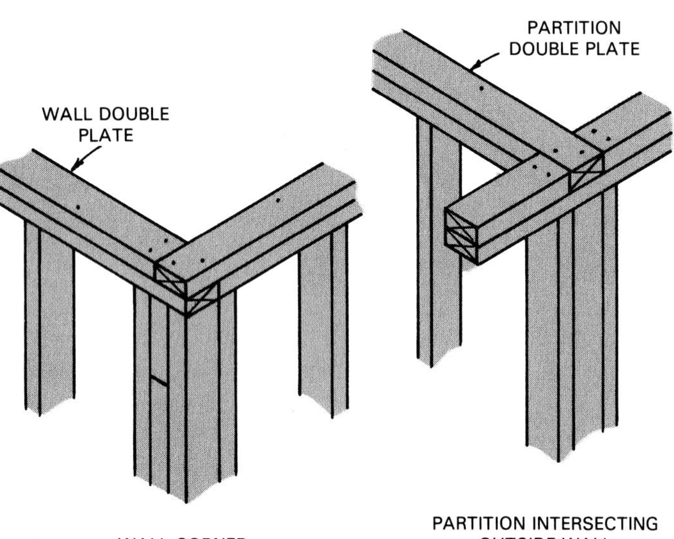

Figure 9-38. Top—Stiffen wall by adding a second 2 x 4 atop the first plate. Bottom—Double plates are lap jointed for strength wherever they intersect.

Figure 9-39. Metal ties can be used to strengthen building frames against damaging winds and earthquakes. Top—Metal fasteners secure studs to foundation. Bottom—Metal fasteners hold studs to wall plate. Fasteners also can be used to secure rafters.

Special Framing

In addition to the floor framing, carpenters will sometimes be asked to build structures with special features. It helps if these features are carefully engineered and detailed by an architect. In this case, construction details are included in the plans. Sometimes they are not, and the carpenter must develop his own plan.

A bay window, though now usually prefabricated by a window manufacturer, might be such a project, Figure 9-40. When asked to build the bay window, the carpenter must visualize the details of construction, lay out and construct the floor frame to carry the project, and build the wall and roof frame.

Framing of the floor extension is best accomplished by extending the floor joists, cantilevering them over the foundation. Joists forming each side of the bay should be doubled to help carry the weight of the structure. A header should be installed over the opening to carry the weight of the structure above, and to provide a nailing surface for the window's ceiling joists.

A cabinet *soffit* is special framing that closes in the space between the ceiling and the tops of cabinets. Soffits are often found in kitchens. See Figure 9-41.

Construction of a soffit begins by placing a chalk line on the studs at the bottom edge of the soffit. This is usually 84" from the finished floor. Next, the carpenter will nail a 2 x 2 along this line, using one 10d nail at each stud. Then he or she will nail another 2 x 2 onto the ceiling along a second chalk line snapped along the ceiling at a distance that is at least the depth of the cabinets.

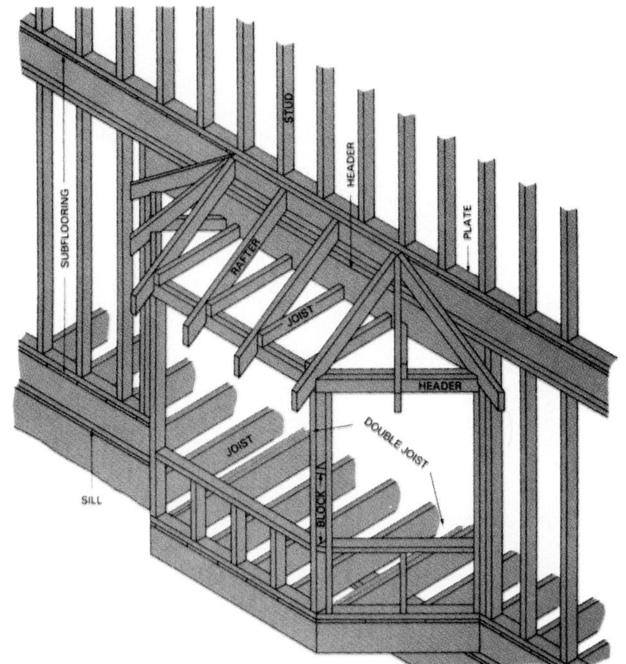

Figure 9-40. Traditional framing for a bay window. Subfloor is not included to show how joists are extended over sill. Joists at either side of the extension should be doubled. (National Forest Products Assoc.)

If the ceiling joists run parallel to the soffit, 2 x 4 blocking will be used to bridge between two joists and provide support for the 2 x 2. The carpenter next cuts a 2 x 4 or 2 x 2 for the lower front edge of the soffit and blocking to frame the front and bottom of the soffit. These pieces are attached with screws or 8d nails.

Wall Sheathing

Wall sections should be covered with *sheathing* before roof framing is started. Sheathing adds rigidity, strength, and some insulating qualities to the wall.

Plywood, waferwood, fiberboard, and rigid insulating panels are widely used for sheathing. These materials are

Figure 9-41. A framed soffit. Note use of 2 x 2 frame against studs and ceiling joists. Special blocking is used at the corner.

available in large sheets that can be attached rapidly. Plywood and waferwood usually provide enough lateral strength that diagonal bracing can be eliminated.

Fiberboard sheathing is made largely from wood fibers with added weather-proofing ingredients. It is commonly available in 4' x 8' sheets. However, sheets as large as 8' x 14' are manufactured. Regular fiberboard sheathing is also available in a 2' x 8' size with tongue and groove or shiplap joints. It is applied horizontally and corner bracing is required. Regular fiberboard sheathing cannot be used as a nailing base for exterior wall finish materials. However, a special nail-base fiberboard is available.

Standard thicknesses of fiberboard sheathing are 1/2" and 25/32". Use 1 1/2" roofing nails for the 1/2" thickness and 1 3/4" for the 25/32" thickness. Provide a 1/8" space between all edge joints when applying fiberboard sheathing.

Diagonal bracing is recommended when using 1/2" fiberboard. Adequate bracing can be provided by installing 1/2" plywood at each corner as shown in Figure 9-42. When gypsum sheathing (usually 1/2" thick) is used, it is necessary to include corner bracing. This type of sheathing is usually installed with 1 3/4" nails spaced 4" around the edge and 8" on intermediate supports.

Either interior plywood or exterior plywood can be used for sheathing, provided it is structural grade. It should be at least 5/16" thick for studs spaced 16" on centers and 3/8" thick for studs spaced 24" on centers. A 1/2" thickness or more is often used so that the exterior finishing material can be nailed directly to the panel.

Sheathing can be applied vertically or horizontally, Figure 9-43. Provide 1/8" space on all edges. Space edge nails 6" apart and field nails 12" apart.

Plain wood boards or shiplap can also be used for wall sheathing. They should be at least 3/4" thick and not over 12" wide. End joints should fall over the centers of the studs. Diagonal bracing is required when the siding is

Figure 9-42. Corner bracing is eliminated when plywood is used to sheath corners of an outside wall. (American Plywood Assoc.)

Figure 9-43. Installing orientated strand board (OSB) sheathing. Space edge nails 6" apart and field nails 12" apart. Strap bracing is not required for this type of sheathing. (Georgia-Pacific Corp)

installed horizontally. When using boards with side and end matching (tongue and groove) for sheathing, the joints need not be over a stud. When applied diagonally, no corner braces are needed. Boards should be slanted in opposite directions on adjoining walls for greatest strength and stiffness. Also, two 8d nails are needed on each stud for 6" or 8" boards. Three nails should be used for wider boards.

Foamed plastic sheathing, Figure 9-44, is made from polystyrene or polyurethane. The use of this type of sheathing has grown dramatically in recent years because of its high insulating properties (reflected in its R-value). Panels

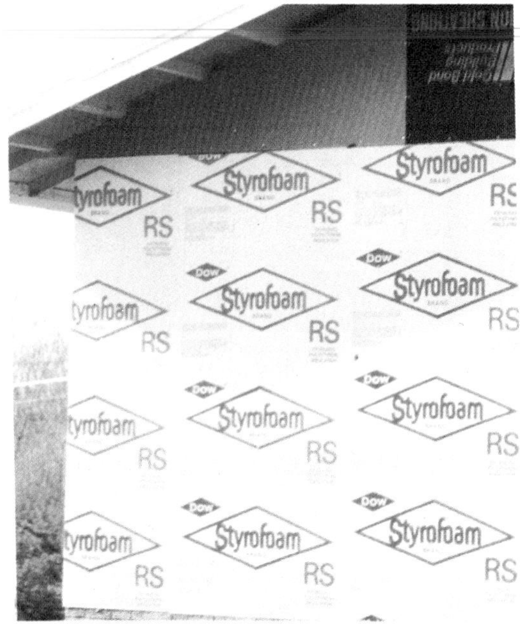

Figure 9-44. Top—Rigid polystyrene foam sheathing. Panels are 1" thick and 8' long. One inch of foam has an R-value of 5.50. Bottom—Fiberglass sheathing being installed. (Owens-Corning)

of this kind of material are not very strong and must be handled with extra care. Corner bracing is required. Always follow the manufacturer's directions for application. Procedures may vary somewhat from one product to another.

Multi-Story Floor Framing

Framing a floor for a second- or multi-story dwellings is similar to framing the first floor. Joists are placed on top of the double plate along with headers—also called band joists. They may be fastened to the plate with steel anchors or toenailed. See Figures 9-45 and 9-46.

Figure 9-45. Upper story framing. Floor frame is fastened to the doubled wall plate of the previous story. A—Framing with lumber. B—Framing with open-web joists. C—Openings in bearing walls must have heavy lintels. D—In platform framing method, the second story of a home is built just like the first story.

Figure 9-46. Installing a header and cripple joist on a second story. Left—Header is in place and joist is being installed. Right—Joist being toenailed to the wall plate. (Kasten-Weiler Construction Co.)

Figure 9-47. Ceiling framing. Top—Lumber ceiling joists for a flat roof. Middle—Second floor joists. Ceiling covering is fastened to the underside. (Kasten-Weiler Construction Co.) Bottom—Underside of truss rafters form a frame for ceiling covering. (Southern Forest Products Assoc.)

Ceiling Framing

A *ceiling frame*, as the name suggests, is the system of support for all components of the ceiling. This frame may be the underside of the floor joists for the next story or an assembly just below the roof. Refer to Figure 9-47.

Basic construction is similar to floor framing. The main difference is that lighter joists are used and headers are not included around the outside.

When trusses are used to form the roof frame, a ceiling frame is not required. The bottom chords of the truss rafters form the ceiling frame. Refer again to Figure 9-47 (bottom).

Main ceiling framing members are called ceiling joists. Size is determined by the span and spacing used. To coordinate with walls and permit the use of a wide range of surface materials, a spacing of 16" O.C. is commonly used. Size and quality requirements must also be based on the type of ceiling finish (plaster or drywall) and what use will be made of the attic space, Figure 9-48. The architectural plans will usually include specifications. These requirements should be checked with local building codes.

Ceiling joists usually run across the narrow dimensions of a structure. However, joists can also be supported by bearing walls. This will make the span even shorter and may result in ceiling joists perpendicular to each other in the same frame, Figure 9-49.

In large living rooms, the midpoint of the joists may need to be supported by a beam. This beam can be located below the joists or installed flush with the joists. In the latter installation, the joists may be carried on a ledger, Figure 9-50 (left). Joist hangers, like those described in Units 1 and 8, can also be used, Figure 9-50 (right). Sometimes a beam is installed above the joists in the attic area. It is tied to the joists with metal straps.

At their outer ends, the upper corners of the joists must be cut at an angle to match the slope of the roof. To lay out the pattern for this cut, use the framing square as

SIZE IN.	SPACING IN.	GROUP A FT. IN.	GROUP B FT. IN.	GROUP C FT. IN.	GROUP D FT. IN.
2 × 4	12 16	9 - 5 8 - 7	9 - 0 8 - 2	8 - 7 7 - 9	4 - 1 3 - 6
2 × 6	12 16	14 - 4 13 - 0	13 - 8 12 - 5	13 - 0 11 - 10	9 - 1 7 - 9
2 × 8	12 16	19 - 6 17 - 9	18 - 8 16 - 11	17 - 9 16 - 1	14 - 3 12 - 4
2 × 10	12 16	24 - 9 22 - 6	23 - 8 21 - 6	22 - 6 20 - 5	19 - 6 16 - 10

Figure 9-48. Spans for ceiling joists are figured for a normal dead load and a live load of 20 psf (pounds per sq. ft.) This permits the attic to be used for storage. Always check local codes. (National Building Code)

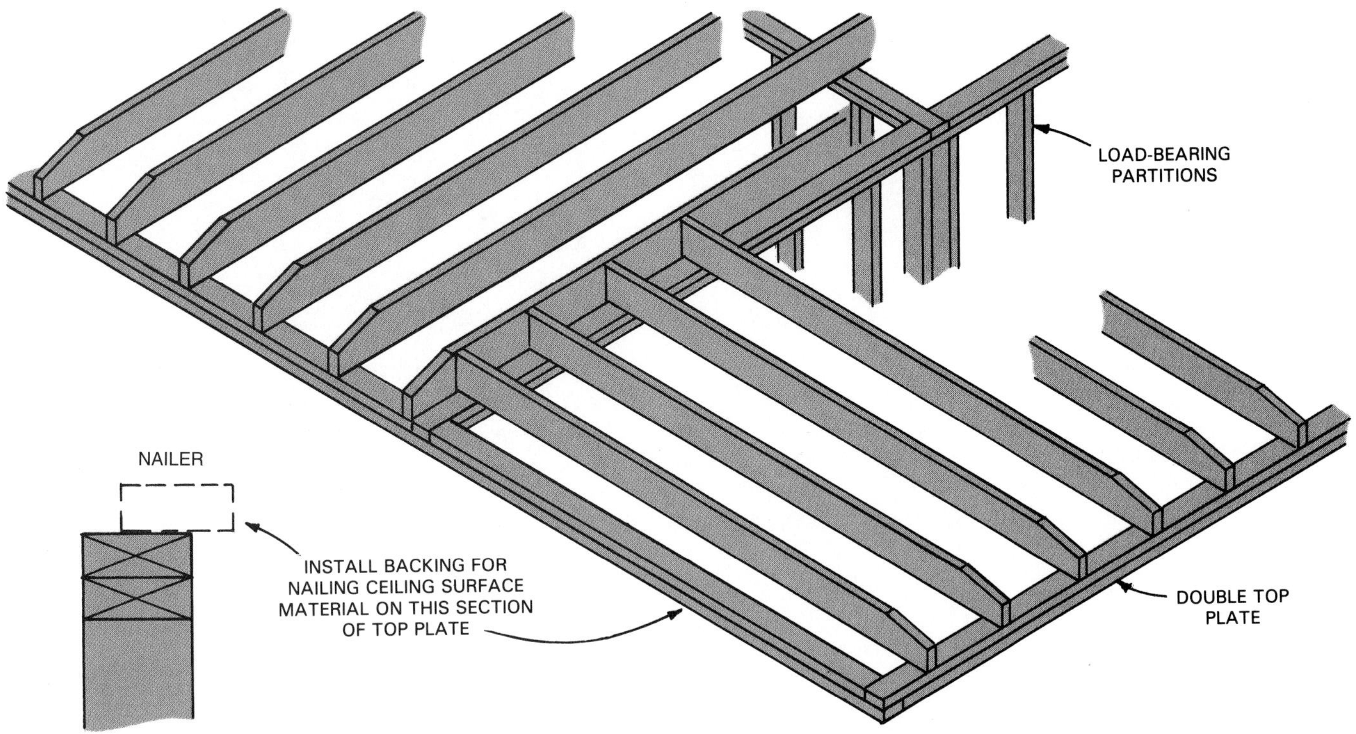

Figure 9-49. Sketch of a ceiling frame. Joists in foreground are turned at right angles to reduce the span.

Figure 9-50. A flush beam is a header in a ceiling that spans a large distance but does not extend below the surface of the ceiling. Left—One method of attaching joists to a flush beam is with a ledger strip. (Western Wood Products Assoc.) Right—Another method is to use hangers attached to the flush beam. Carpenter toenailled the joists to the beam in initial assembly. (Southern Forest Products Assoc.)

illustrated in Figure 9-51. When the amount of stock to be removed is small, the cuts can be made after the joists and rafters are in place. Use a hatchet or saw.

When ceiling joists run parallel to the edge of the roof, the outside member will likely interfere with the roof slope. This often occurs in low-pitched hip roofs. The ceiling frame in this area should be constructed with stub joists running perpendicular to the regular joists, Figure 9-52.

Lay out the position of the ceiling joists along the top plate using a rod. When a double plate is used, the joists do not need to align with the studs in the wall. The layout, however, should put the joists alongside the roof rafters so that the joists can be nailed to them.

Ceiling joists are installed before the rafters. Toenail them to the plate using two 10d nails on each side.

Partitions or walls that run parallel to the joists must be fastened to the ceiling frame. A nailing strip or drywall clip to carry the ceiling material must be installed. Various size materials can be installed in a number of ways. The chief requirement is that they provide adequate support.

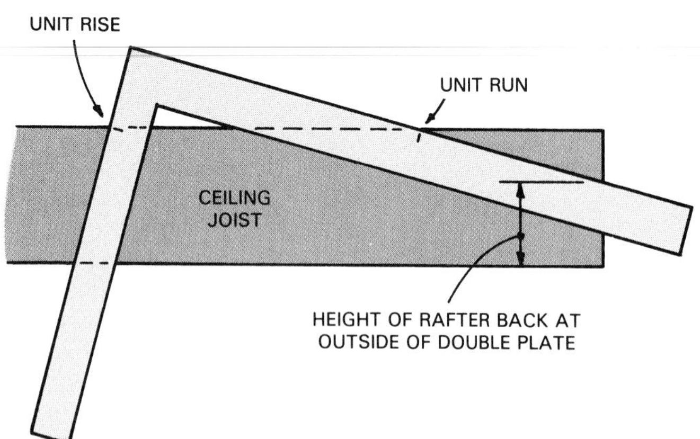

Figure 9-51. A framing square, or "quick" square may be used to lay out the trim cut on the end of a ceiling joist. Using the rise and run of the building, the cut will match the slope of the rafters.

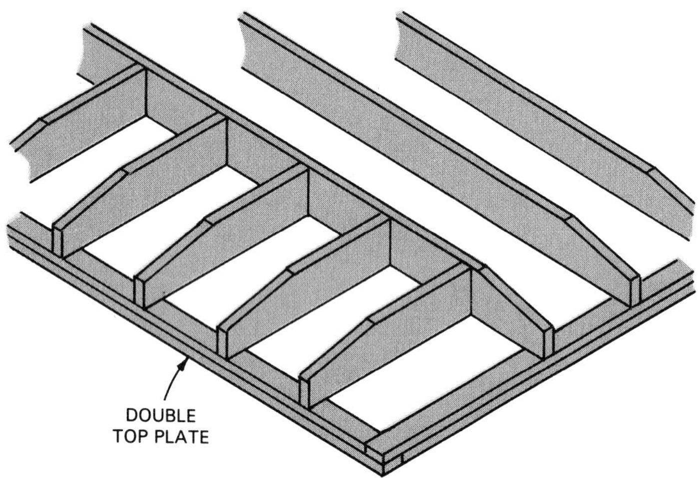

Figure 9-52. Stub ceiling joists butted to a full-length joist. The stub joists will be required for a low-pitched hip roof.

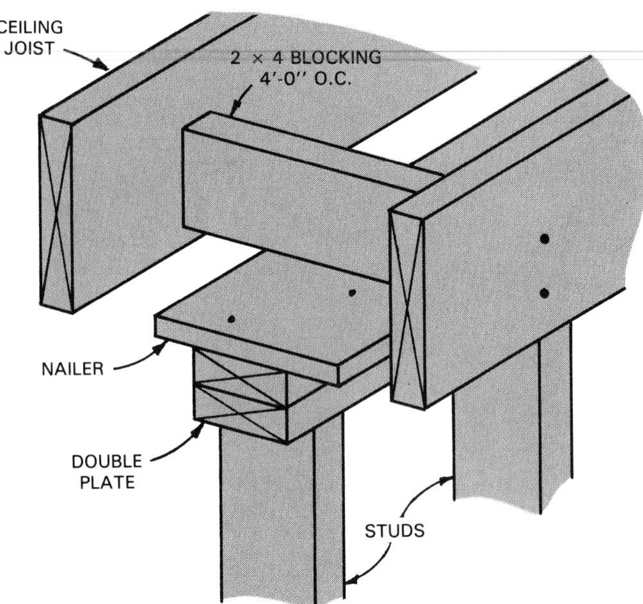

Figure 9-53. Use special blocking to anchor partitions to the ceiling frame when they run parallel to the joists.

Figure 9-53 shows a typical method of making such an installation. Refer once more to Figure 9-6, which shows clip installation for a ceiling.

An access hole (also called a scuttle hole) must be included in the ceiling frame to provide an entrance to the attic area. Fire regulations and building codes usually list minimum size requirements. The building plans generally indicate size and show where it should be located. The opening is framed following the procedure used for openings in the floor. If the size of the opening is small (2 to 3 ft. square), doubling of joists and headers is not required.

Strongbacks

Long spans of ceiling joists may require a *strongback.* This is an L-shaped support that is attached across the tops of joists to strengthen them and maintain the space between them. It also evens up the bottom edges of the joists so the ceiling will not be wavy after the drywall is applied.

To construct a strongback, first mark off the proper spacing (16" or 24" O.C.) on a 2 x 4. Position the 2 x 4 across the tops of the ceiling joists and fasten it with two 16d nails at each joist. Apply pressure against the joists as needed to maintain proper spacing.

Select a straight 2 x 6 or 2 x 8 for the second member. Place it on edge against either side of the 2 x 4 just attached to the joists. Attach one end to the 2 x 4 with a 16d nail. Work across the full-length of the strongback, aligning and nailing. Stepping on either the 2 x 4 or the member on edge will help align each joist. Nail the vertical member to each joist and to the 2 x 4, Figure 9-54.

House Wrap

House wrap comes in 9' wide rolls and is designed to cover cracks at wall joints where air might enter or leave a building. At the same time, it is waterproof but allows water vapor to pass through. (This property is known as vapor permeability.) The typical wrap is made from high-density spun polyethylene fibers that are virtually tear proof. Wrap is stapled to sheathing before exterior doors and windows are installed. At openings, the wrap is trimmed long enough to wrap around rough openings. See Figure 9-55.

Framing with Steel

Steel-framed residential construction is increasing in popularity, Figure 9-56. *Metal studs* can be used with either metal plates or wood plates, as shown in Figure 9-57. Joints are fastened with self-tapping screws or welds,

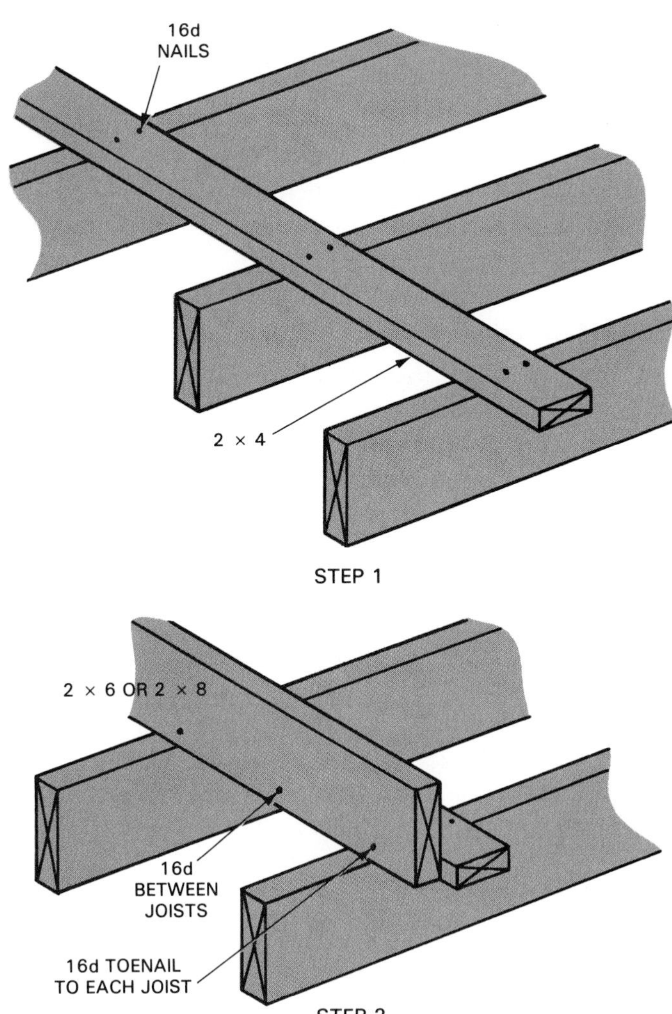

Figure 9-54. Building a strongback: Step 1—Nail a 2 x 4 to the joists. Step 2—Turn a 2 x 6 or 2 x 8 on edge and nail it to the 2 x 4 and to joists.

Figure 9-55. House wrap has been installed on the first story of this house. (Kasten-Weiler Construction Co.)

Figure 9-56. This building is being built entirely with metal framing members. Because steel is light, studs must have bridging to maintain spacing and rigidity. Openings in studs allow for running of plumbing or electrical systems.

Figure 9-57. A wall section combining steel and wood. The wall section is so light, one carpenter can carry it. (H.L. Stud Corp.)

Figure 9-58. Some carpenters first use screws to hold the joint and then secure it with a weld.

Although several types of welding are used on the light gage steel, shielded metal arc welding (SMAW) is the most popular. The equipment is portable and the welds are strong. A 200 A (ampere) "hot box" electric welder or a 200 A gasoline generator is adequate as an electrical power source. Wherever possible, welds are made before walls are erected. See Figure 9-59.

As with wood construction, headers must be installed on load-bearing walls. Figure 9-60 shows header assembly details. Figure 9-61 shows a window roughed out in steel and a first-floor frame completely erected. As with wood wall framing, proper fastening and bracing is important. Wall sections should have horizontal bridging to space studs. Diagonal bracing of sections is required to keep walls square and plumb. Anchoring walls to floors is also important. Refer to Figure 9-62.

Construction of partitions is also similar to wood construction. Diagonal bracing is not required but continuous strap bridging should be installed, Figure 9-63. Rough openings for doors in nonbearing walls can be framed as shown in Figure 9-64. A thermal break or rigid insulation should be applied to the outside face of exterior walls. This will improve the R-value of the outside wall by preventing the transfer of heat through the metal studs and joists. It is the practice of some carpenters to use steel framing for partitions and wood framing for outside walls.

Additional information and technical guidelines are available from the American Iron and Steel Institute, 1101 17 St., NW, Washington DC, 20036-4700. Ask for Residential Construction Guidelines, Publication No. RG-934a, June, 1993.

Estimating Materials

To estimate wall and ceiling framing materials, first determine the total lineal feet by adding together the length of each wall and partition. The plans will include the dimensions of outside walls. These can be added together. Partitions, especially those that are short, may not be dimensioned and you will need to scale the drawing. It is a good idea to place a colored pencil check mark on each wall and partition as its length is added to the list.

For plates, multiply the total figure by three (one sole plate plus two top plates). Add about 10% for waste. Order this number of linear feet of lumber in random lengths or convert to the number of pieces of a specific length. Three pieces of 2 x 4 one foot long equals 2 bd. ft. Thus, the total lineal feet of walls and partitions can be quickly converted to board measure if required. For example:

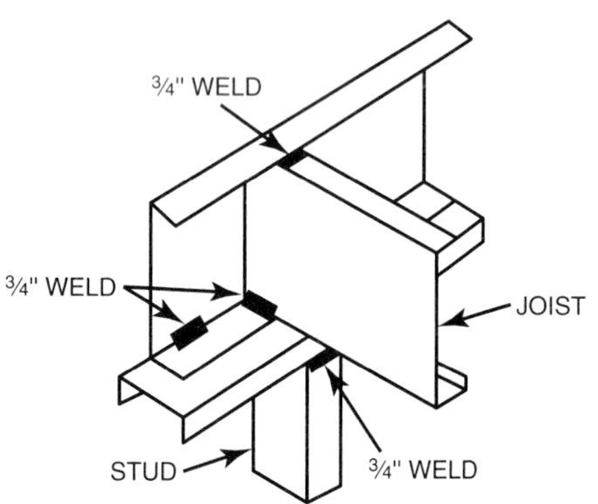

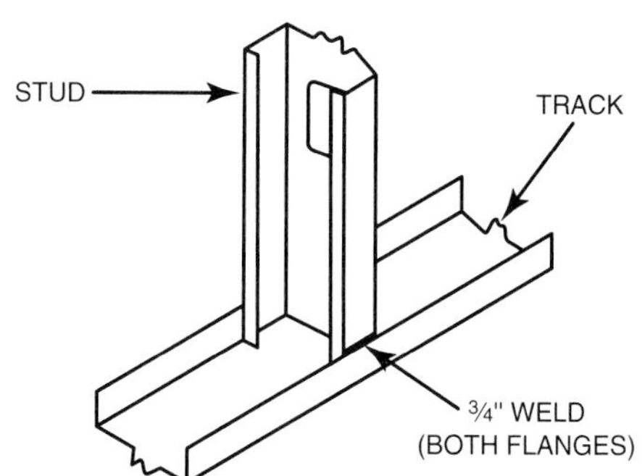

Figure 9-58. Joint assembly details. Left—Wall plate with ceiling joist and header joist assembled and welded. Right—Sole plate and stud assembly. (CEMCO)

Figure 9-59. Carpenter welding cripple studs in a rough opening for a window. Note safety gloves and welding helmet.

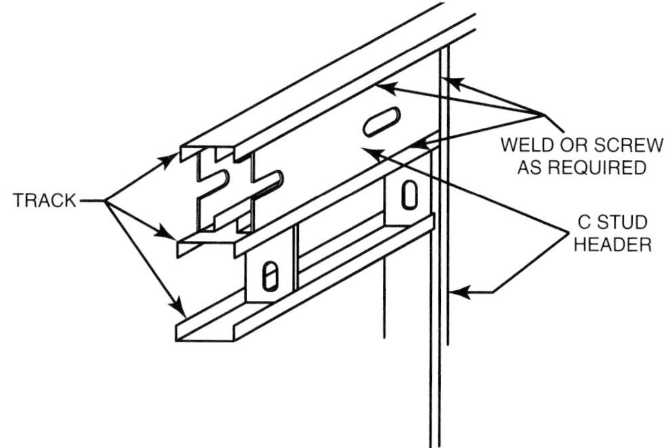

Figure 9-60. Detail is shown for fabricating a header in a load-bearing wall. (CEMCO)

Figure 9-61. Top—Squaring up wall sections and preparing to secure a corner. Note temporary angle brace. Bottom—First-floor wall is erected and preparations are under way for placing ceiling joists.

Wall and partition length	=	240
Total plate material	=	length x 3 + 10%
		240 x 3 + 10%
		792 lineal ft.
		or 57 pcs., 2" x 4" x 14'-0"

Estimating Studs

The total length of all walls and partitions is also used to estimate the number of studs required. When studs are spaced 16" O.C., multiply the total length (in feet) by 3/4 and then add two more studs for each corner, intersection, and opening.

Using the preceding dimensions for the wall length, assume there are 12 corners, 10 intersections, and 20 openings. Find the number of studs needed including outside walls and partitions:

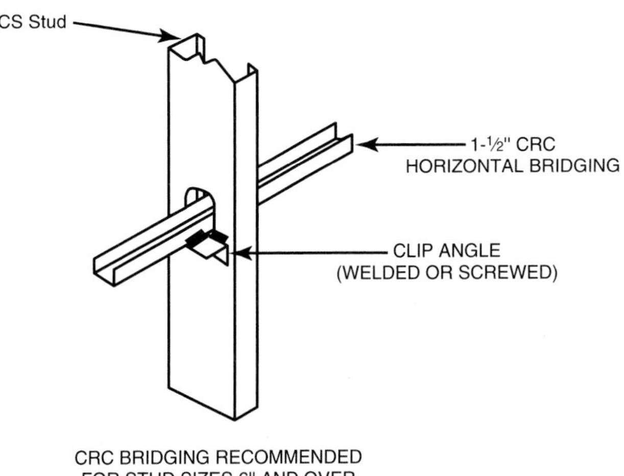

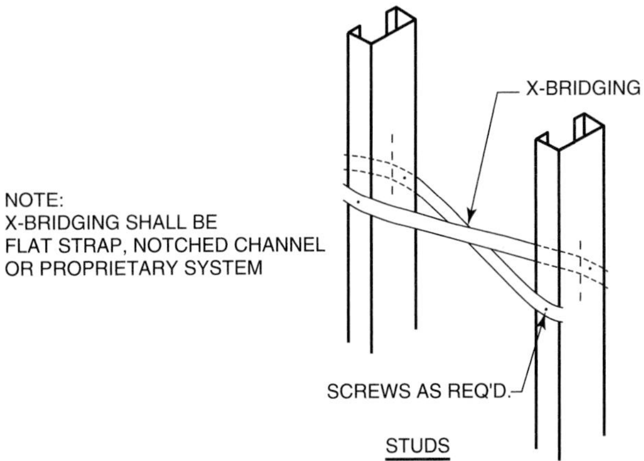

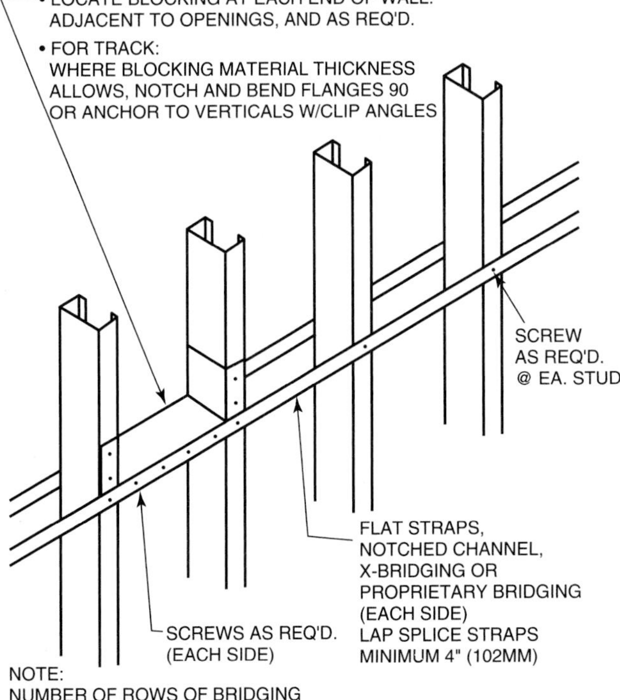

Figure 9-62. Details for spacing and bracing steel wall frames. (CEMCO and Dale Industries Inc.)

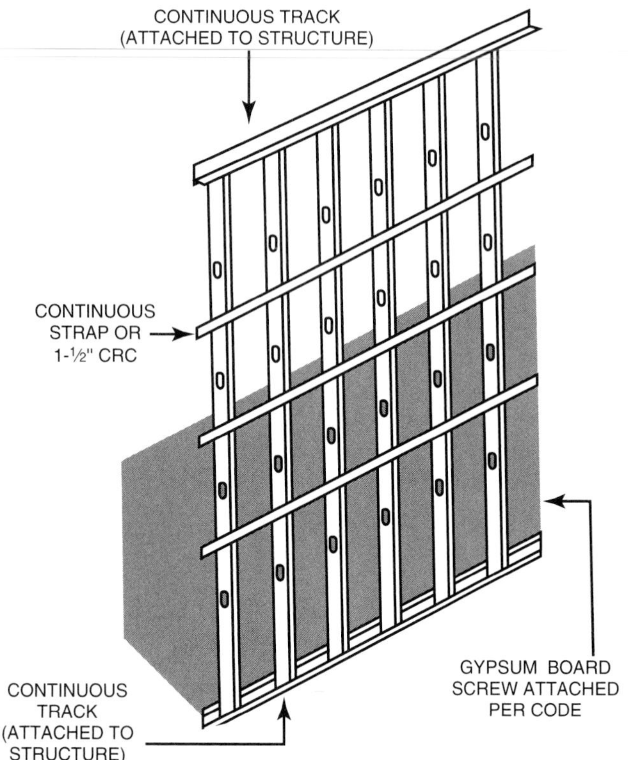

Figure 9-63. Details for nonload-bearing partitions. Note use of continuous bridging. (CEMCO)

$$\text{Total studs} = \text{total length} \times 3/4 + 2\,(\text{corners} + \text{intersections} + \text{openings})$$

$$= \frac{240 \times 3}{4} + 2\,(12 + 10 + 20)$$

$$= \frac{\overset{60}{\cancel{240}} \times 3}{\cancel{4}} + 2\,(42)$$

$$= 60 \times 3 + 84 = 264$$

Many carpenters estimate the number of studs by simply counting one stud (spaced 16" O.C.) for each lineal foot of wall space. About 10% is added for waste. The overrun on spacing provides the extras needed for corners and openings. This method is rapid and fairly accurate. However, it will not estimate enough material for a small house divided into many rooms. On the other hand, too many studs will likely be estimated for a large house with wide windows and open interiors.

Lumber for headers must be calculated by analyzing the requirements for each opening. Use the rough opening width plus the thickness of the trimmers.

Ceiling joists are estimated by the same method used for floor joists. Since ceilings will be relatively free of openings, no extras or waste needs to be included. Because of this, the short method should not be applied to ceiling

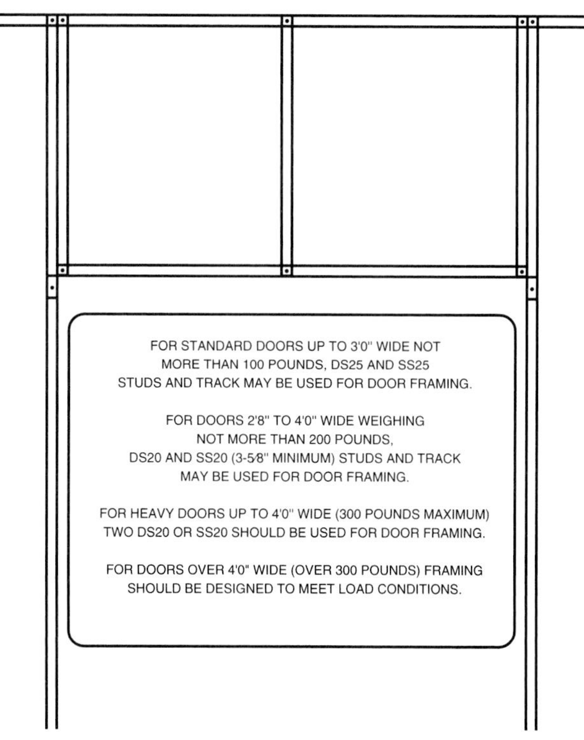

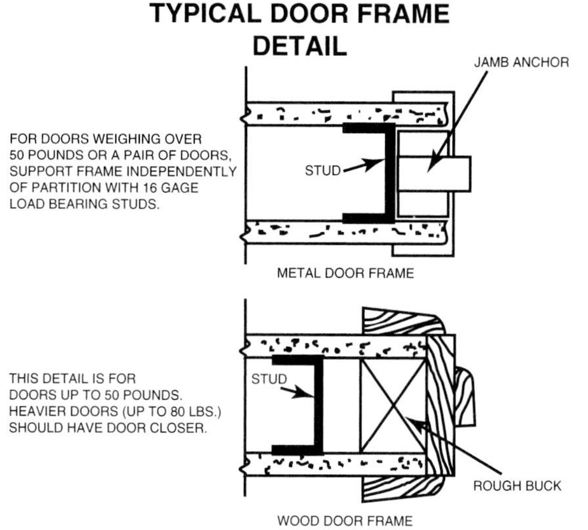

Figure 9-64. Details of rough opening for doors in metal partitions. Door framing may be either metal or wood. (CEMCO)

joists. Use the following formula and include the size of the joists required:

Number of ceiling joists = wall length x 3/4 + 1

Estimating Wall Sheathing

To estimate the amount of wall sheathing, first find the total perimeter of the structure. Multiply this figure by the wall height measured from the top of the foundation when the sheathing extends over the sill construction. The product will be the gross square footage of the wall surface. Now calculate the area of each major opening (windows and doors) and subtract this total from the original figure. Round the opening sizes downward to the nearest foot.

Net area = perimeter x height - wall openings

To the net area to be sheathed add allowances for waste and other extras when common boards or shiplap are used. See Unit 8 for more information. When using sheet materials for sheathing, there is only slight waste. Divide the net area by the square footage per sheet to secure the number of pieces required. For example, if the net area to be sheathed is 1060 sq. ft. and the fiberboard sheets selected are 4' x 9', calculate quantity needed as follows:

Net area = 1060

Fiberboard sheet size = 4 x 9 = 36

$$\text{Number of sheets} = \frac{1060}{36} = 29 + \text{ or } 30$$

L = X x Z
L = 24 x Z
= 24 x 3
= 72

Most problems in estimating, although based on simple formulas, usually contain so many variables that good judgment must be applied to their solution. This judgment is acquired through experience.

Important Terms

Ceiling frame	Nailer
Ceiling joists	Partitions
Cripple studs	Rough opening
Flush beam	Sheathing
Headers	Soffits
House wrap	Sole plate
Metal strap bracing	Story pole
Metal studs	Strongback
	Trimmer studs

Test Your Knowledge

1. What is a sole plate?
2. How many studs are required for a plain wall panel 8' -0" long if they are spaced 16" O.C.?
3. Trimmer studs stiffen the sides of an opening and carry the weight of the ____________.
4. What is a story pole?
5. The first layout to be marked on the plates is the ____________ spacing.
6. A master stud pattern is laid out somewhat like a ____________ ____________.
7. The layout of the cripple studs on the rough sill can be marked directly from the ____________.
8. Most carpenters prefer to erect the ____________ (side, end) walls first.
9. What is the difference between let-in bracing and metal strap bracing?
10. Joints formed along the doubled top plate should be at least ____________ apart.
11. Regular fiberboard sheathing ____________ (can, cannot) be used as a nailing base for exterior wall finish materials.
12. The position of the ceiling joists along the double plate should be coordinated with the____________.
13. The first step in estimating the number of studs required is to figure the total length of all ____________ and ____________.
14. A strongback is needed for ____________.
    A. strengthening a long span of ceiling joists
    B. maintaining proper spacing between ceiling joists
    C. keeping joists even along their lower edges
    D. None of the above.
    E. All of the above.
15. Where and why is house wrap used?
16. What two methods are used to fasten joints in the construction of steel framing?

Outside Assignments

1. Obtain a set of architectural plans for a one-story house. Study the details of construction, especially typical wall sections. Prepare a scale drawing of the framing required for the front walls. Be sure the rough openings are the correct size and in the proper location. Your drawing should look somewhat like the one in Figure 9-2.
2. Working from the same set of plans, develop an estimated list of materials for the wall frame and sheathing. Include the number and length of studs; the number and size of lumber for headers; the material for plates, and type and amount of sheathing. Obtain prices from your local supplier and figure the total cost of the materials.
3. Obtain literature about fiberboard, foamed plastic, and gypsum sheathing. Obtain this material from local lumber dealers or write directly to manufacturers. Also, study books and other reference materials. Prepare a report for the class based on the information you obtain. Include grades, manufacturing processes, characteristics, and application requirements. Discuss current prices and purchasing information. Be prepared to discuss specifications . Relate these "specs" to your local code.

In some situations, such as above, a 4 x 4 or 4 x 6 might be more economical than a built-up corner. Top—Cutting the corner to length. Bottom—Fitting the corner and fastening it after the structure is partially finished.
(Johnson-Manley Lumber Co.)

Carpentry students in Boston raise front wall in place for owner's new garage. Unit is partially sheathed.
(North Bennett Street School)

Using a pneumatic nailer speeds up construction.
(Pierce Construction, Ltd.)

10

Roof Framing

Roof framing provides a base to which the roofing materials will be attached. The frame must be strong and rigid. Besides this, the roof, if carefully designed and proportioned, can contribute a distinctive and decorative feature to the structure.

Roof Types

Roof styles vary widely. Most of them can be grouped into the following types. See Figure 10-1:

- *Gable roof*—Two surfaces slope from the centerline (*ridge*) of the structure. This forms two triangular shaped ends called gables. Because of their simple design and low cost, gabled roofs are often used for homes.
- *Hip roof*—All four sides slope from a central point or ridge. The angles created where two sides meet are called hips. An advantage of this type is the protective overhang formed over end walls as well as over side walls.
- *Gambrel roof*—In this variation of the gable roof, each slope is broken, usually near the center or ridge. This style is used on two-story construction. It permits more efficient use of the second floor level. Dormers are usually included. The gambrel roof is a traditional style typical of colonial America and the period immediately following.
- *Flat roof*—This roof is supported on joists that also carry ceiling material on the underside. It may have a slight pitch (slope) to provide drainage.
- *Shed roof*—This simplest of pitched roofs is sometimes called a "lean-to" roof. The name comes from its frequent use on additions to a larger structure. It is often used in contemporary designs where the ceiling is attached directly to the roof frame.
- *Mansard roof*—Like the hip roof, the mansard has four sloping sides. However, each of the four sides has a double slope. The lower, outside slope is nearly vertical. The upper slope is slightly pitched. Like the gambrel roof, the main advantage is the additional space gained in the rooms on the upper level. The name comes from its originator, architect Francois Mansart (1598-1666). One of the renderings in the opening page of this unit shows a modern variation of the mansard.

There are two basic systems of framing the roof of platform structures: conventional "stick-built" rafters and the truss rafter. Conventional construction is built on-site by the carpenter. The ceiling joists and rafters are laid out, cut, and installed one at a time. *Rafters* are sloped framing

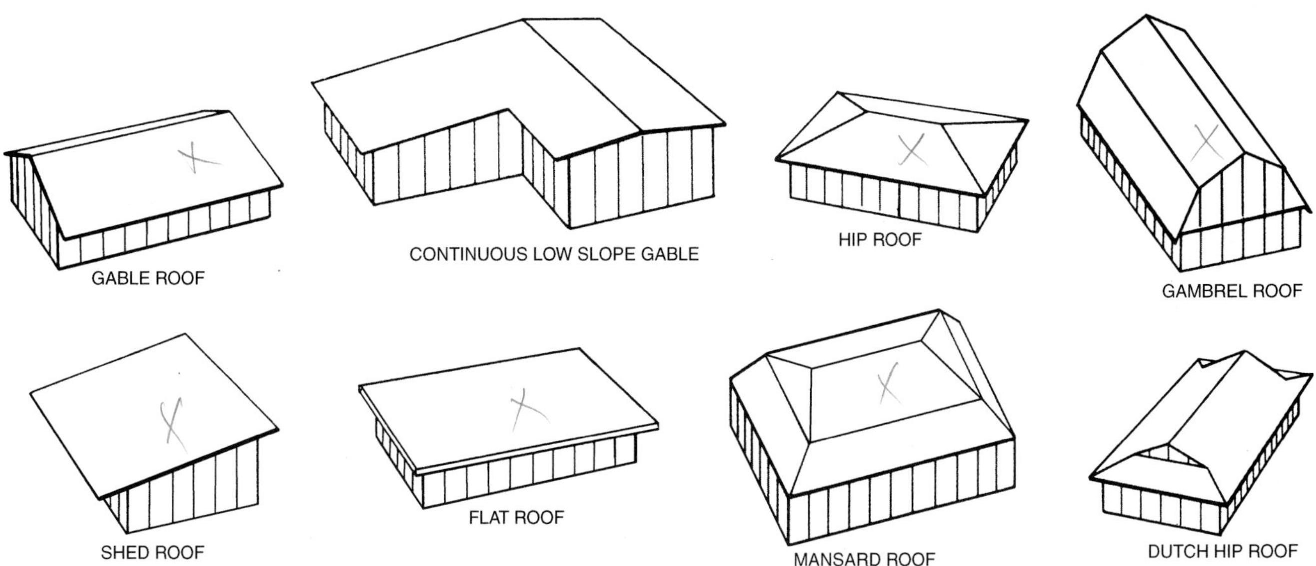

Figure 10-1. These common types of roofs are to be found in residential construction.

members that run from the peak of the roof to the plates of the outside walls. They are the supports for the roof load. As you learned in Unit 9, the ceiling joists tie the outside walls together and support the ceiling materials for the rooms below. They also tie the bottom ends of the rafters, preventing them from pushing outward when loaded.

Roof trusses are engineered and prefabricated assemblies that are usually factory-built and delivered to the building site. They are installed as complete units. The lower chords also serve as ceiling rafters and support the ceiling coverings. Figure 10-2 shows examples of truss construction.

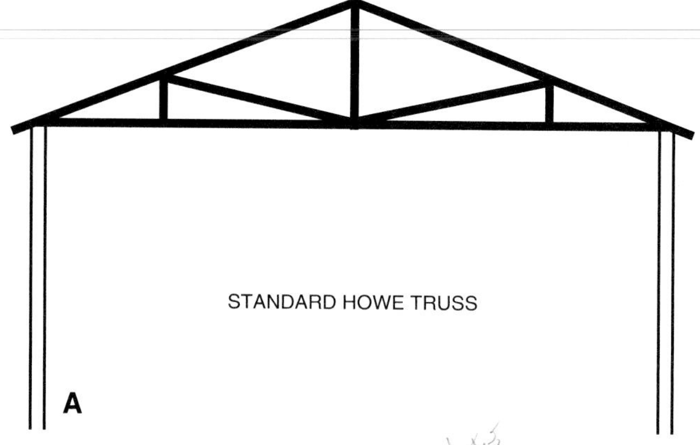

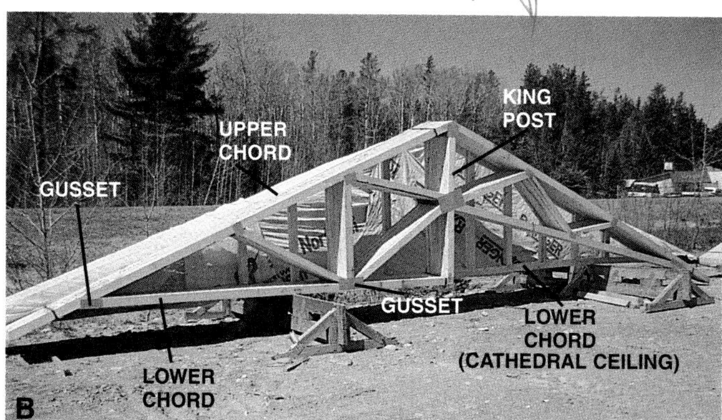

Figure 10-2. A—The standard Howe truss supports weight efficiently. B—Truss rafters come in many styles and sizes. While they can be built on site, it is much faster having them built in a factory. One side of these trusses is designed for a cathedral ceiling. Note names of parts. C—Lower chords of trusses support ceiling materials.

Roof Supports

Roofs, depending on the type of rafter design, are supported by one or all of the following systems:

- Outside walls.
- Ceiling joists (beams which hold the ceiling materials).
- Interior bearing walls.

Parts of Roof Frame

The plan view of the roof shown in Figure 10-3 combines several roof types. The kinds of rafters are identified.

The *common rafters* are those that run at a right angle (plan view) from the wall plate to the ridge. A gable roof has only rafters of this kind.

Hip rafters also run from the plate to the ridge, but only at a 45° angle. They form the support where two slopes of a hip roof meet.

Valley rafters extend diagonally from the plate to the ridge in the hollow formed by the intersection of two roof sections. These sections are usually at right angles to each other.

There are three kinds of jack rafters:

- *Hip jack*—This is the same as the lower part of a common rafter but intersects a hip rafter instead of the ridge.
- *Valley jack*—This is the same as the upper end of a common rafter but intersects a valley rafter instead of the plate.
- *Cripple jack*—Also called a cripple rafter, it intersects neither the plate nor the ridge and is terminated at each end by hip and valley rafters. The cripple jack rafter is also called a *hip-valley cripple jack* or a *valley cripple jack*.

Parts of a Rafter

Rafters are formed by laying out and making various cuts. Figure 10-4 shows the cuts for a common rafter and the sections formed. The ridge cut allows the upper end to fit tightly against the ridge. The *bird's mouth* is formed by a seat cut and plumb (vertical) cut when the rafter extends beyond the plate. This extension is called the overhang or tail. When there is no overhang, the bottom of the rafter is ended with a seat cut and a plumb cut.

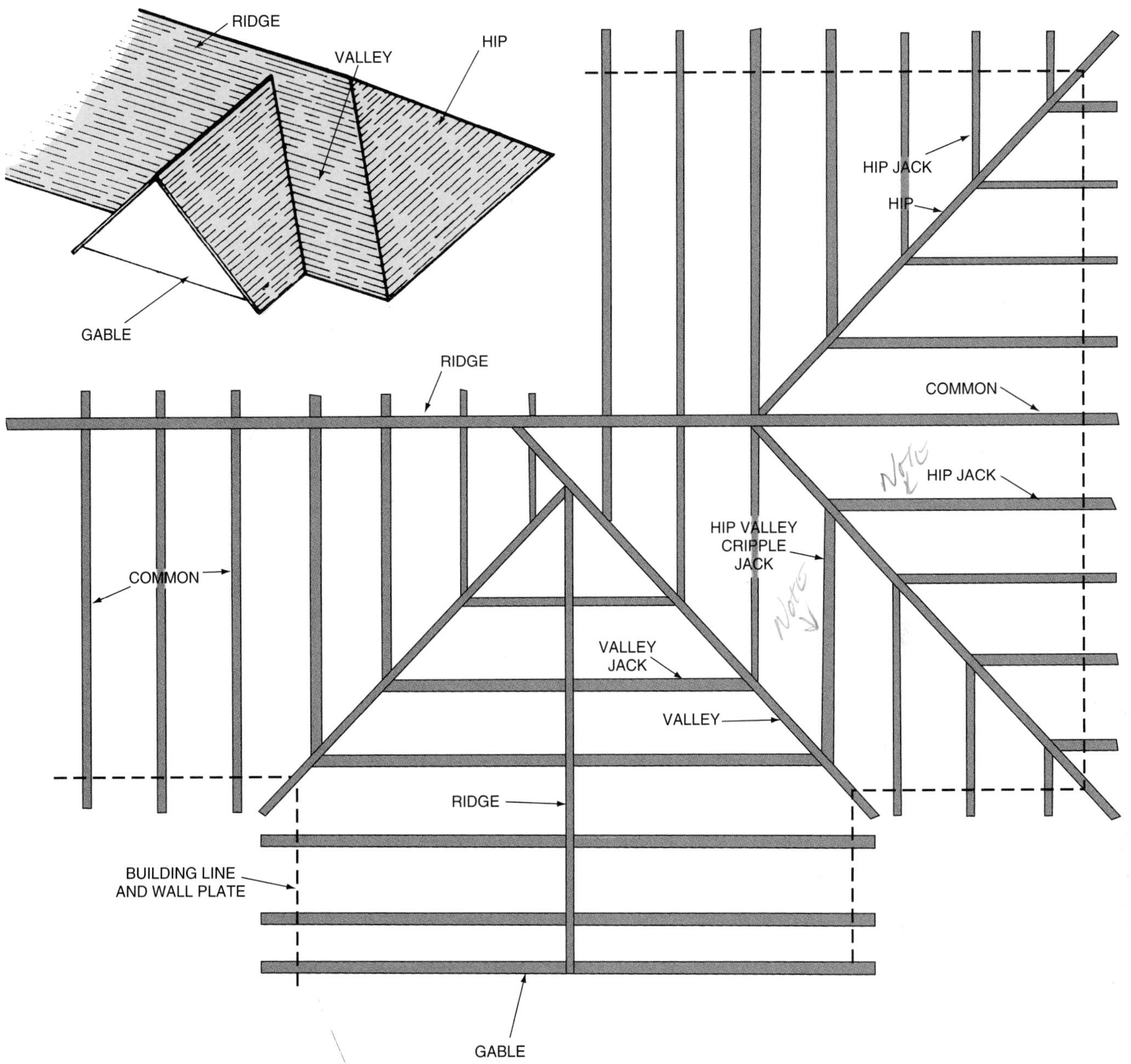

Figure 10-3. A roof frame plan view (means looking down from directly above). Note types of rafters and their names. Inset shows view of a finished roof and names of features.

Layout Terms and Principles

Roof framing is a practical application of geometry. This is an area of mathematics that deals with the relationships of points, lines, and surfaces. It is based largely on the properties of the right triangle:

- The horizontal distance is the *base*.
- The vertical distance is the *altitude*.
- The length of the rafter is the *hypotenuse*.

If any two sides of a right angle triangle, Figure 10-5, are known, the third side can be found mathematically. The formula used is $H^2 = A^2 + B^2$. H is the line forming the third side of the triangle; A is the altitude or height; B is the base. To find the unknown length, one must extract the square root. This is time consuming; the answer could be found using trigonometric functions.

Carpenters use either the tables on the framing square or a layout called the "step-off" method. Either way is fast, simple, and practical. Another option is the "quick" or "super" square. This will be discussed later.

Small hand-held calculators are sometimes used on the job. By entering the rise and the run of a roof, a carpenter will get a readout of the rafter length in feet, inches, and fractions. The same calculator will also calculate lengths of hip and valley rafters.

In rafter layout, the base of the right triangle is called the *run*. It is the distance from the outside of the plate to a point directly below the center of the ridge. The altitude, or

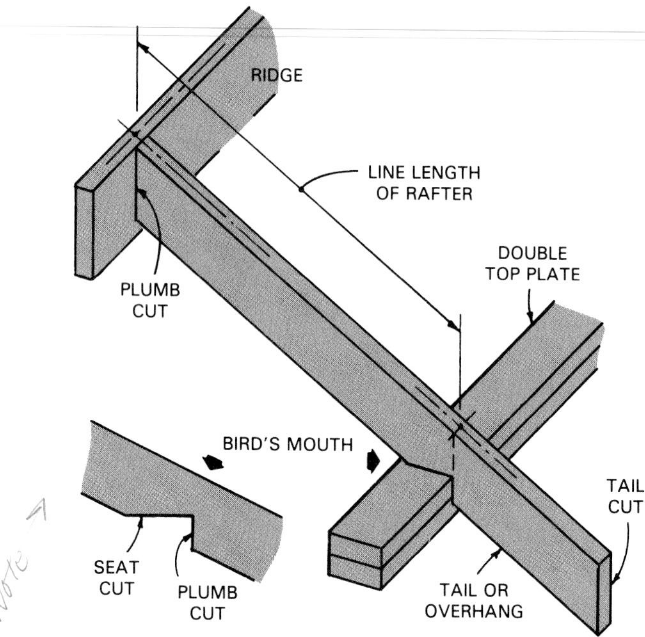

Figure 10-4. Rafter parts. Various cuts and surfaces are important because the rafter must have a snug fit at all joints.

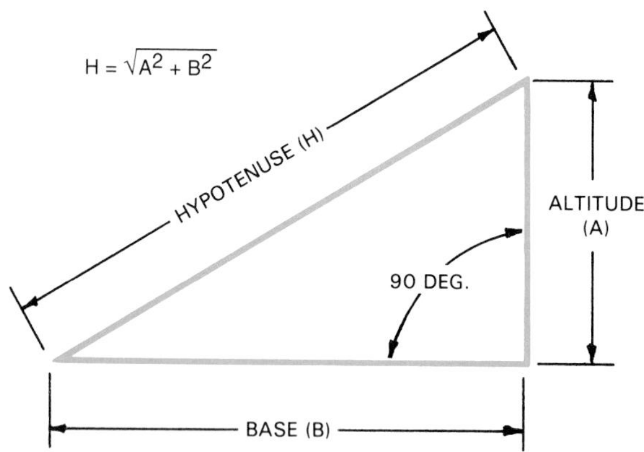

Figure 10-5. Math can be used to find rafter length. The rafter is like the hypotenuse above. If you square the rise and the run of a roof, then add the answers, you have the square of the rafter's length.

rise, is the distance the rafter extends upward above the wall plate. Other layout terms and the relationship of the parts of the roof frame are illustrated in Figure 10-6.

Slope and Pitch

Slope refers to the incline of a roof. It is expressed as the relationship of the vertical rise to the horizontal run. Thus, slope is given as X inches in 12.

For example, a roof that rises at the rate of 4" for each 12" of run is said to have a "4 in 12" slope. A triangular symbol above the roof line in the architectural plans gives this information. The slope of a roof is sometimes called the "cut of the roof."

Pitch is the incline of the roof as a ratio of the rise to the span (which is twice the run). It is given as a fraction. For example, if the total roof rise is 4', and the span is 24', then the pitch is 1/6 (4/24 = 1/6).

Unit Measurements

The *framing square,* also called a *steel square* or *carpenter's square,* is the basic layout tool in roof framing. The side with the manufacturer's name is called the face and the opposite side the *back*. The longer, 24" arm is called the *body*, or *blade*, while the shorter arm (16") is called the *tongue*.

The framing square is not large enough to make the rafter layout with one setting. It is necessary, therefore, to use smaller divisions, called *units*. The foot (12") is the standard unit for horizontal run. The unit for rise is always based on how many inches the roof rises in every foot of run. The unit run and rise is used to lay out plumb and level cuts (Figure 10-7) required at the ridge, bird's mouth, and tail of the rafter.

Framing Plans

When working with simple designs, the carpenter can easily visualize the roof framing. Since the wall framing is already erected, the only additional information needed will be the following:

- The slope of the roof.
- The amount of overhang required.

These items are included in the house plans. When the structure has a complicated roof design, the architect often includes a roof framing plan. If a framing plan is not included, the carpenter should prepare one. This can be done by making a scaled drawing on tracing paper laid directly over the floor plans. Include ridges, overhang, and every rafter. The drawing may be made similar to the one in Figure 10-3. However, you can use a single line to represent each framing member. In your drawing, maintain accurate spacing between rafters. Draw hips and valleys at a 45° angle and make jack rafters parallel to the common rafters.

Rafter Sizes

As in floor and ceiling framing, the strength of the rafter (size) is determined by the spacing and span or length. Local building codes must be consulted for their specifications.

Figure 10-8 shows a table of rafter sizes for various loads and spacings. Stock for hips, valleys, and ridges is usually larger than for other roof framing members.

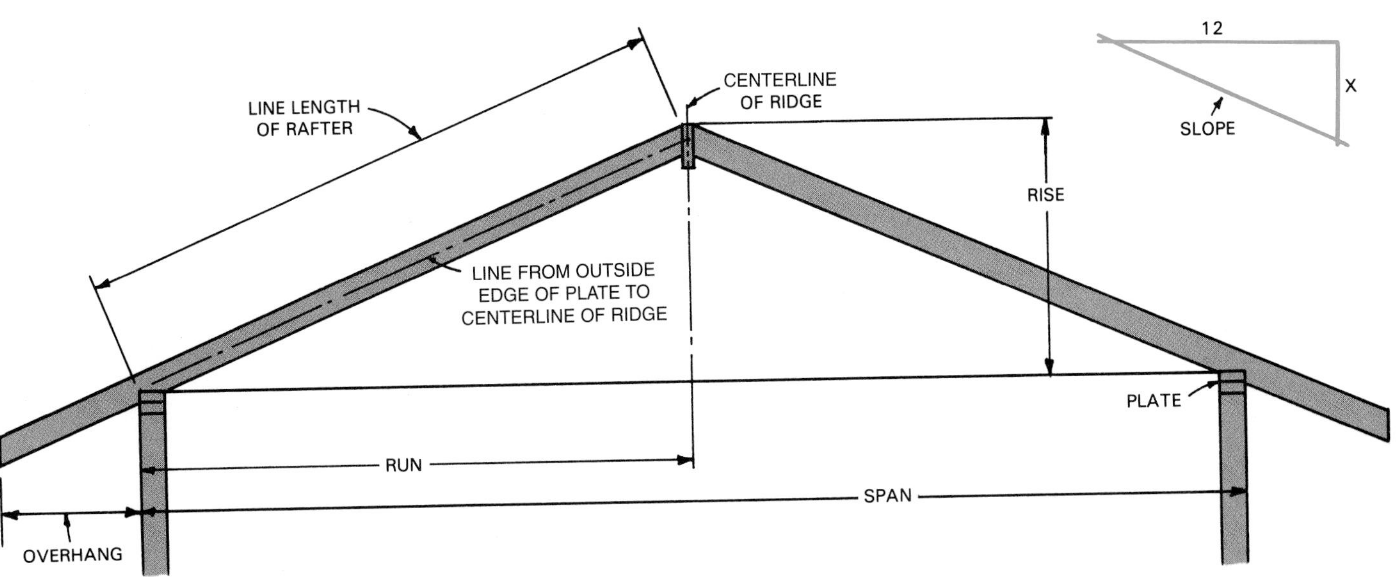

Figure 10-6. These terms are basic in rafter layout. Do you see how the rise, run, and rafter length form a right angle triangle?

For purposes of estimating and ordering material, the rafter lengths can be determined with fair accuracy by making a scaled-down layout with the framing square. Use the back of the square where the outside edge of the blade and tongue are divided into inches and twelfths. Assume the inches to be feet and each twelfth a full inch.

Figure 10-7. How to position a square for marking one unit of measure on a rafter. Run is measured along the blade. Rise (in inches per foot) is measured off the tongue. Same position is used to lay out and mark bird's mouth and plumb cuts.

First draw a triangle, using the unit run and unit rise specified by the slope of the roof. If the total run is more than 12', lengthen the base and hypotenuse. Now lay the blade of the square along the base line until the point of total run on the square falls over the acute angle. Mark the point where the tongue crosses the sloping line, Figure 10-9. Use either the blade or tongue of the square to measure the hypotenuse. This will give the approximate length of the rafter. Add for the overhang. Allow extra material for making cuts at the ridge and tail.

SIZE OF RAFTER (Inches)	SPACING OF RAFTER (Inches)	MAXIMUM ALLOWABLE SPAN (Feet and Inches Measured Along the Horizontal Projection)			
		Group I	Group II	Group III	Group IV
2 × 4	12	10-0	9-0	7-0	4-0
	16	9-0	7-6	6-0	3-6
	24	7-6	6-6	5-0	3-0
	32	6-6	5-6	4-6	2-6
2 × 6	12	17-6	15-0	12-6	9-0
	16	15-6	13-0	11-0	8-0
	24	12-6	11-0	9-0	6-6
	32	11-0	9-6	8-0	5-6
2 × 8	12	23-0	20-0	17-0	13-0
	16	20-0	18-0	15-0	11-6
	24	17-0	15-0	12-6	9-6
	32	14-6	13-0	11-0	8-6
2 × 10	12	28-6	26-6	22-0	17-6
	16	25-6	23-6	19-6	15-6
	24	21-0	19-6	16-0	12-6
	32	18-6	17-0	14-0	11-0

Figure 10-8. Sample table showing maximum runs allowed for rafters sloped 4 on 12 or greater. Groups refer to species of wood. Obtain this kind of information from local building codes.

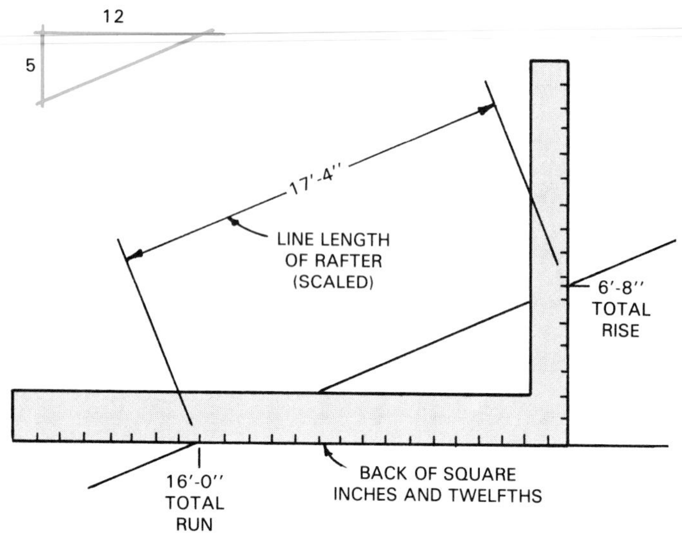

Figure 10-9. Framing square can be used to estimate approximate length of rafters. Use square as a 1" = 1' scale.

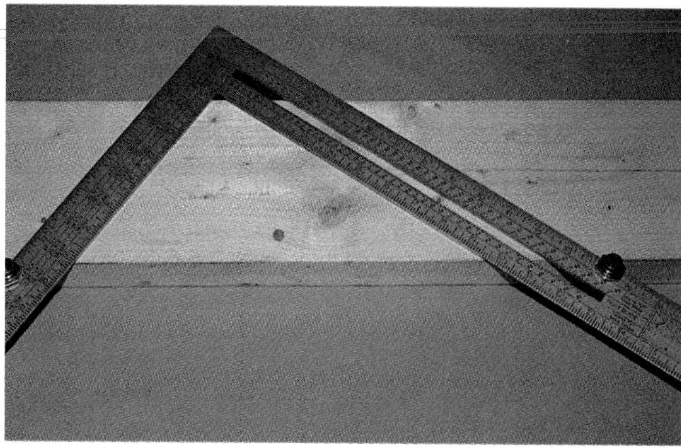

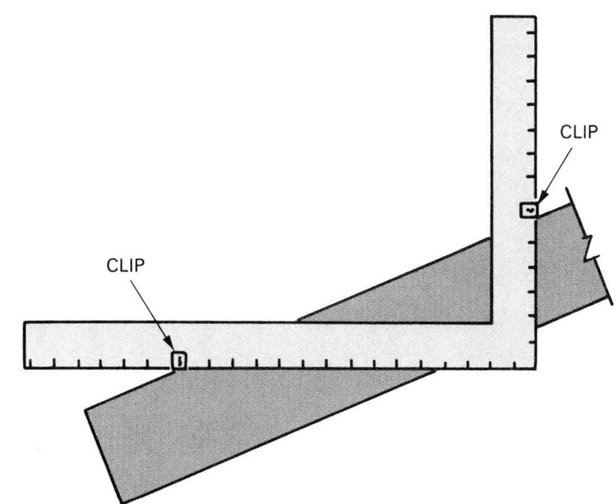

Figure 10-10. Some carpenters attach clips to the square at the rise and run positions. The clips automatically position the square against the rafter edge as each step is laid out.

Laying Out Common Rafters

Rafters can be laid out by the step-off method, the rafter table on the framing square, or with a construction calculator. Any of these methods will give the correct rafter length. The step-off is easy to use and the layout can be double-checked with the rafter tables.

Using any one of the methods, the carpenter will first lay out, check, and cut a pattern rafter. The pattern will be used to mark other rafters of the same size and kind.

For pattern layout, select a piece of lumber that is straight and true. Place it on a pair of sawhorses.

Usually the carpenter stands on the crowned side (top edge) of the rafter. It is easier to handle the framing square from that position.

(In order to describe rafter layouts that can be quickly understood, this position has been reversed in Figures 10-10 and 10-11. The rafter is shown in the position it will have when installed, however.)

Laying Out Rafters (Step-Off Method)

1. Place the framing square on the stock and align the numbers for the unit run (12") and rise with the top edge of the rafter—unit run on the blade and the unit rise on the tongue.
2. Set the rise and run on the square using clips or handscrews. To ensure accuracy, the correct marks on the square must be positioned exactly over the edge of the stock each time a line is marked. Be sure to use a sharp pencil to make the layout lines.
3. Starting at the top of the rafter, position the square and draw the ridge line along the edge of the tongue. Still holding the square, mark the length of the odd unit (8" used in the example).
4. Next, shift the square along the edge of the stock until the tongue setting is even with the 8" mark. Draw a line along the tongue and mark the 12" point on the blade for a full unit, Figure 10-11.
5. Move the square to the 12" point just marked and repeat the marking procedure. Continue until the correct number of full units are laid out. (This number will be the same as the number of feet in the total run [six used in the example]).
6. Form the bird's mouth by drawing a horizontal line (seat cut) to meet the building line so the surface will be about equal to the width of the plate. The size of the bird's mouth may vary depending upon the design of the overhang. In Figure 10-11, note that the square has been turned over to mark these cuts and also to lay out the overhang. This may or may not be necessary, depending on the length of the rafter blank.

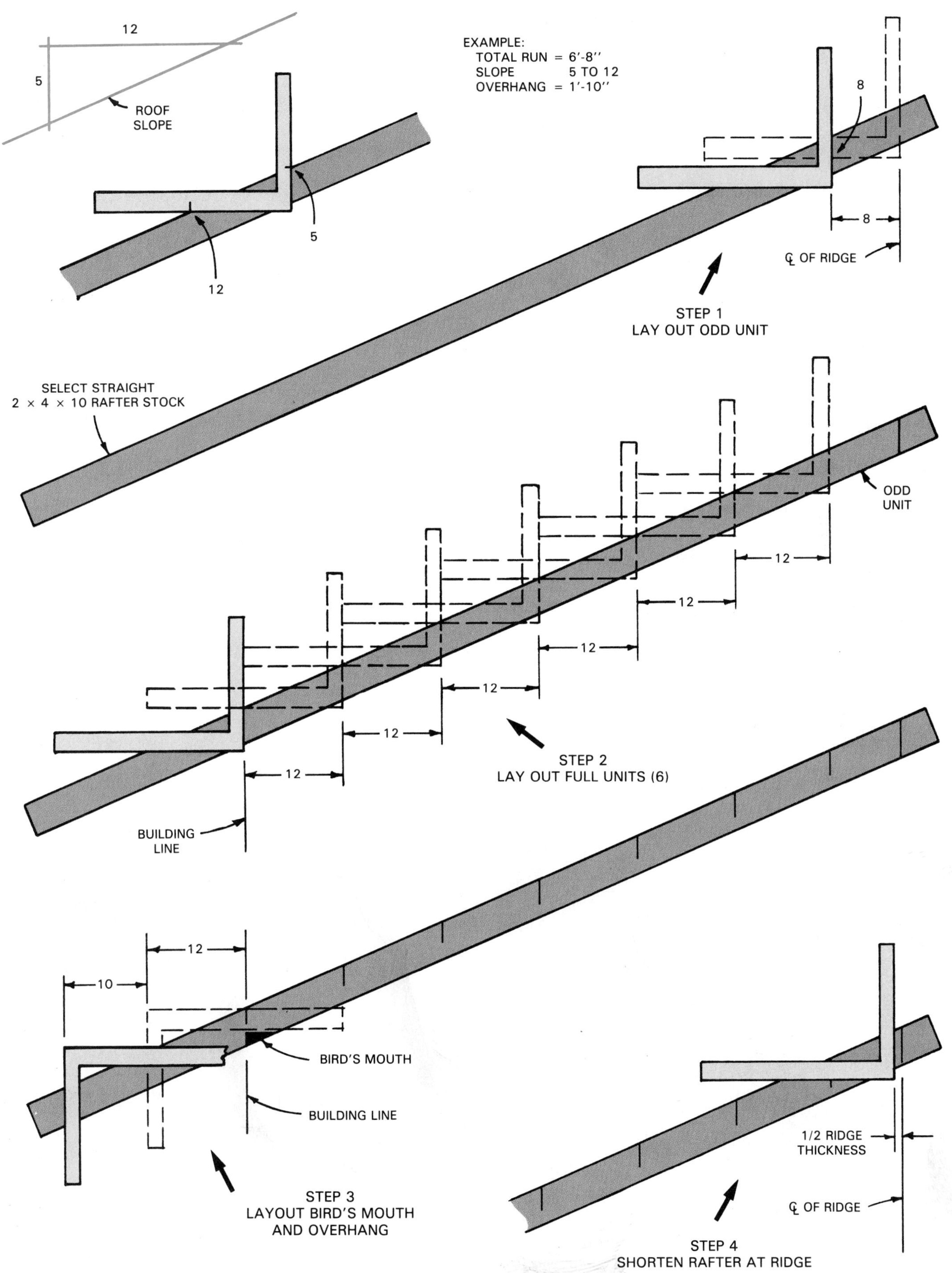

Figure 10-11. Carpenters use this procedure for laying out a common rafter. It is known as the step-off method because each foot of run requires another step of the square.

7. To lay out the overhang, start with the plumb cut of the bird's mouth and mark full units first. Then add any odd unit that remains. The tail cut may be plumb, square, or a combination of plumb and level. Check the cornice details shown in the architectural plans for exact requirements.
8. The final step in the layout consists of shortening the rafter at the ridge. With the square in position, draw a new plumb line back from the ridge line half the thickness of the ridge board, Figure 10-11.
9. Make the ridge, bird's mouth, and tail cuts you have laid out.
10. Label the rafter as a pattern, indicating the roof section to which it belongs.

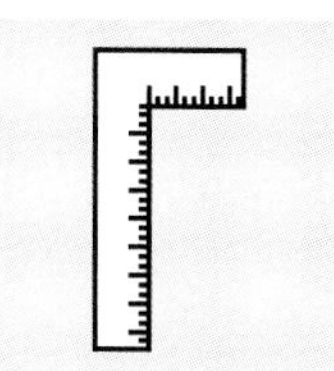

As you lay out rafters, try to visualize how each will appear when it is set in the completed roof frame. Forming the habit of visualizing the rafter in its proper place will help to eliminate errors.

Using the Rafter Table

You can also calculate the length of a common rafter using the table on the framing square. See Figure 10-12. Under the full-scale number that corresponds with the unit rise, secure the number in the first line. This is the line length of the rafter in inches for one foot of run. To find the length of the rafter from the building line to the center of the ridge, multiply the units of run by the figure from the rafter table as in the following examples:

Example No. 1 — Run = 6'-8" Slope = 5 to 12

Run = 6'-8"	= 6 2/3 units
Table No.	= 13
Rafter Length	= 6 2/3 units x 13"
	= 86 2/3"
	= 7'-2 2/3"

Example No. 2 — Run = 10'-4" Slope = 4 to 12

Run = 10'-4"	= 10 1/3 units
Table No.	= 12.65
Rafter Length	= 10 1/3 x 12.65"
	= 130.72"
	= 10' - 10.72"
	= 10'-10 3/4"

These calculations give the line length of the rafter, running from the center of the ridge to the outside of the plate. To make the pattern layout, add the overhang and subtract half the thickness of the ridge board from the length.

Using the "Super Square"

Sometimes a carpenter will use a *super square*, or *quick square*, to find rafter lengths and determine the angle of cuts. This measuring tool, which is shaped like a figure "4," is easier to carry than a rafter square.

To use a super square, you need to know the pitch of the roof. This can be obtained from the plans or it can be determined by some simple math. Rafter charts for every pitch are supplied by the square's manufacturer.

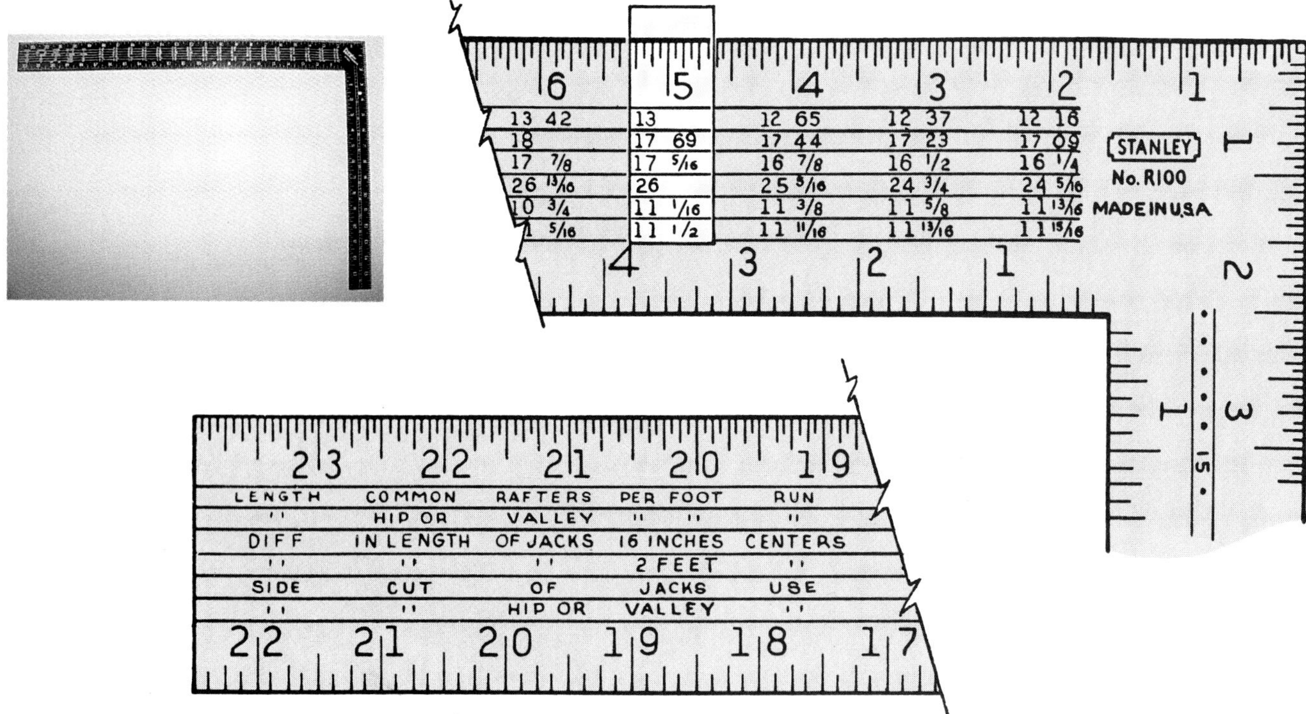

Figure 10-12. The rafter table on a framing square. Left—The blade of the square carries tables for figuring lengths of rafters. Right—For example, if unit rise is 5", you will find that the rafter length for 12" of run is 13"

Laying Out Rafters (Super Square)

1. Once you know the rise per foot of run, refer to the chart for that rise. Read down the chart to the correct run to find the length of the rafter.
2. Mark the length on the rafter.
3. Place the pivot point of the quick square on the length mark while lining up the rise on another scale. (This step will give the angle for marking the plumb cut at the ridge.) See Figures 10-13 and 10-14.
4. Using the rafter pattern, cut the number required.

Some carpenters prefer to stand the completed rafters along the outside wall so they will be easier to reach during the assembly of the roof frame. See Figure 10-15.

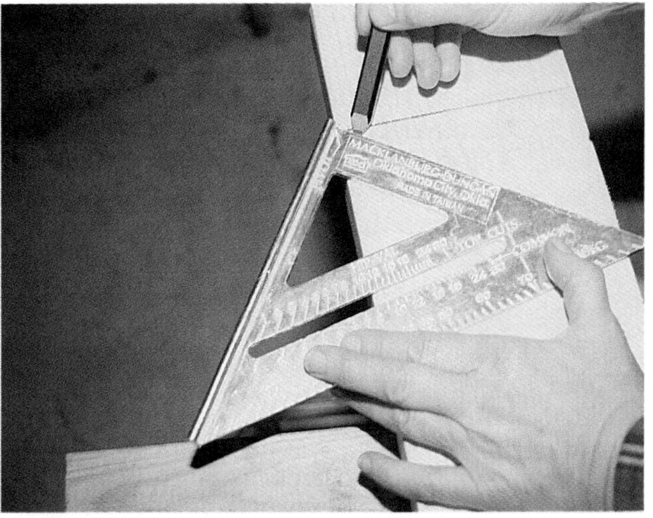

Figure 10-14. Using the quick square to mark plumb cut on a common rafter. Pitch is 8 and plumb angle is 33 3/4° These two marks will be aligned on the top of the rafter.

Erecting a Gable Roof

Lay out the rafter spacing along the wall plate as the ceiling joists are laid out. When rafter spacing is the same as ceiling joist spacing, every rafter will be nailed to a joist. When rafters are spaced 24" O.C. and ceiling joists 16" O.C., the layout is arranged as in Figure 10-16. The plate layout is important; follow the roof framing plan carefully.

To allow more room over the wall plate for additional insulation, some carpenters attach rafters to a 2 x 4 nailed on top of the ceiling joists. See Figure 10-17.

Select straight pieces of ridge stock and lay out the rafter spacing by transferring the marking directly from the plate or a layout rod. Joints in the ridge should occur at the center of a rafter. Cut the pieces that will make up the ridge and lay them across the ceiling joists, close to where they will be assembled with the rafters.

Figure 10-15. Some carpenters like to stand completed rafters against the outside wall where they are easy to reach from above. (Forest Products Laboratory)

8-12 PITCH		
FT. OF RUN	COMMON RAFTER	HIP OR VALLEY RAFTERS
1	1' 2-3/8"	1' 6-3/4"
2	2' 4-7/8"	3' 1-1/2"
3	3' 7-1/4"	4' 8-1/4"
4	4' 9-3/4"	6' 3"
5	6' 0-1/8"	7' 9-3/4"
6	7' 2-1/2"	9' 4-5/8"
7	8' 5"	10' 11-3/8"
8	9' 7-3/8"	12' 6-1/8"
9	10' 9-3/4"	14' 0-7/8"
10	12' 0-1/4"	15' 7-5/8"
11	13' 2-5/8"	17' 2-3/8"
12	14' 5-1/8"	18' 9-1/8"
14	16' 9-7/8"	21' 10-5/8"
16	19' 2-3/4"	25' 0-1/8"

Figure 10-13. Sample of quick square chart for figuring common rafters and valley or hip rafters when pitch of roof is 8 on 12.

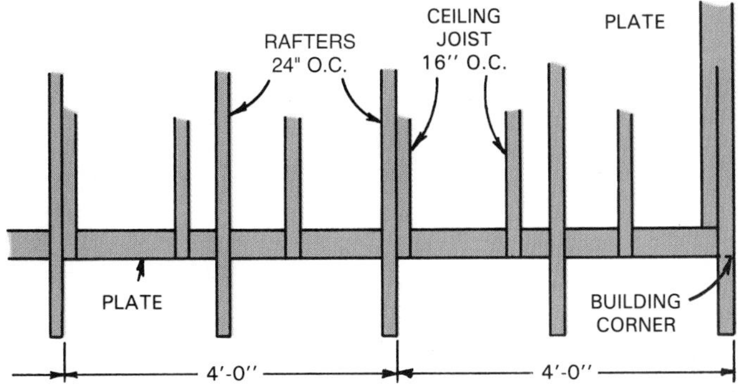

Figure 10-16. Plan view of ceiling joist and rafter layout when rafters are 24" O.C, and joists are 16" O.C. A joist is nailed to every other rafter to act as a tie beam, keeping walls from spreading.

Erecting a Roof

1. Select straight rafters for the gable end and nail one in place at the plate.
2. Install a rafter on the opposite side with a worker at the ridge supporting both rafters.
3. Place the ridge board between the two rafters and nail it temporarily in place.
4. Move about five rafter spaces from the end and install another pair of rafters.
5. Plumb and brace the assembly and make any adjustments necessary in the nailing of the first rafters. Figure 10-18 shows the assembly and nailing pattern at the plate. Special framing anchors, Figure 10-19, are often used.
6. After aligning the ridge, install the remaining rafters. First nail the rafter at the plate and then at the ridge, Figure 10-20. As shown in Figure 10-21, drive 16d nails through the ridge into the rafter. The rafters on the opposite side of the ridge are toenailed. Install only a few rafters on one side before placing matching rafters on the opposite side. This practice will make it easier to keep the ridge straight.
7. Check the ridge periodically to ensure that it remains straight and level.
8. Continue to add sections of ridge and assemble the rafters. Add bracing when required. Always install rafters with the crown (curve or warp) turned upward.

To make the initial assembly, some carpenters prefer to first attach the ridge to several rafters on one side. This assembly is then raised; the rafters are nailed to the plate. Then, several rafters are installed on the opposite side.

Another method of supporting a ridge board may be used. Mark two 2 x 4s for the height of the rise and attach one to either end of the ridge board. The edge of the ridge board should align with the mark. Lift up the assembly and

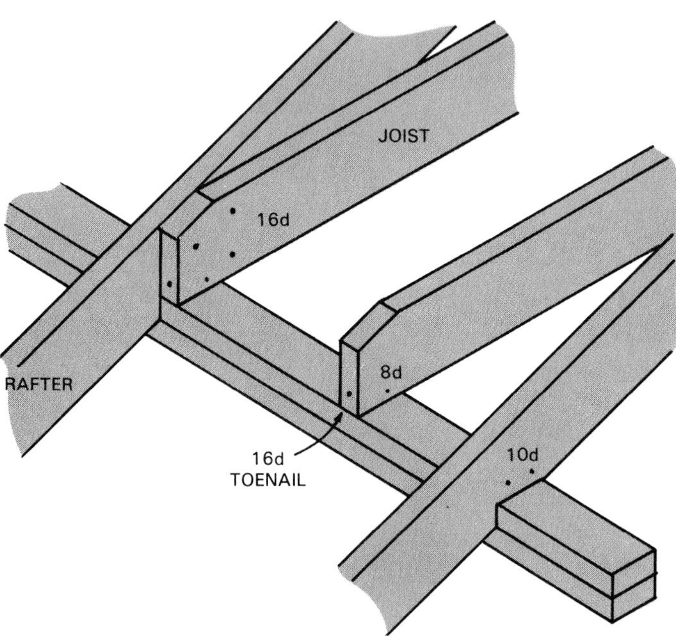

Figure 10-18. Nailing pattern for joists and rafters at wall plate. Opposite sides are toenailed, as well.

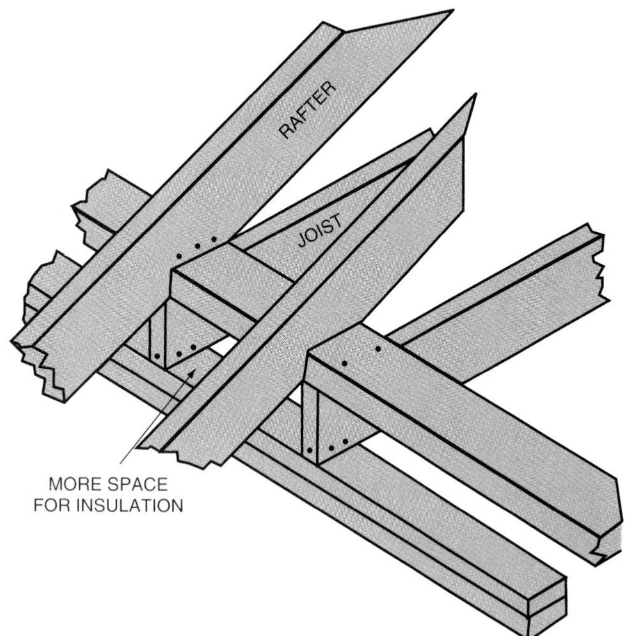

Figure 10-17. Alternate framing method with rafter resting on 2 x 4 plate placed on top of the ceiling rafters. This allows thicker insulation batts over the wall plate without restricting airway to soffit vents. For stronger support of the roof load, it is best to have rafters located over the ceiling joists.

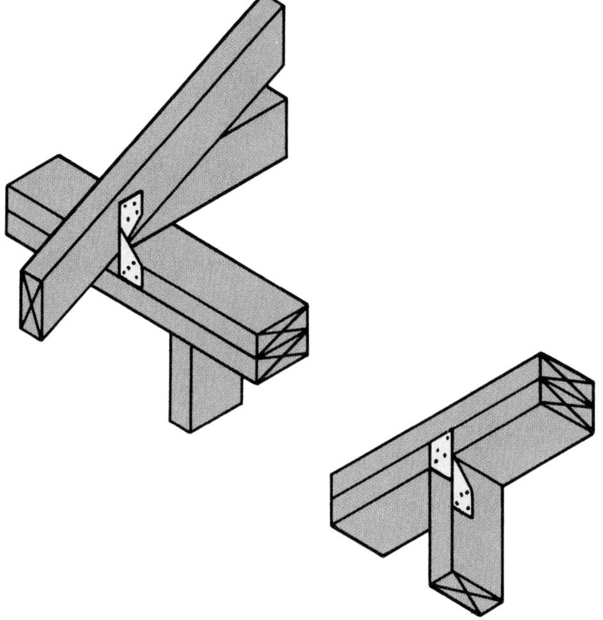

Figure 10-19. Top—Special framing anchors strengthen rafter attachment to walls. Bottom—Same type of clip attaches plate to studding. (Panel Clip Co.)

Figure 10-20. Method of installing common rafters. Ridge end of rafters are supported while bird's mouth is pushed tightly against the plate and nailed. (Southern Forest Products Assoc.)

Figure 10-21. Nailing rafter to ridge board. Align edge of rafter with layout lines. (Southern Forest Products Assoc.)

tack the base of each 2 x 4 to the plate of the bearing wall. Plumb and secure the vertical 2 x 4s with braces. This method requires fewer carpenters.

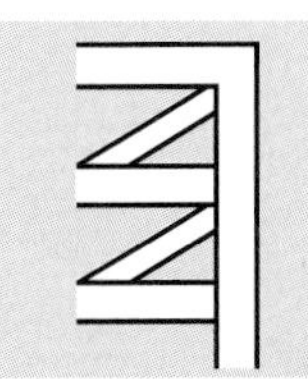

When framing a roof, use extra care to prevent a fall. Erect solid scaffolding wherever it will be helpful. Avoid working directly above another person.

Gable End Frame

The end frames should be assembled after the rafters and ridge have been installed. The end frame consists of vertical studs running from the top plate of the bearing walls to the end rafters. The following procedure should be used to determine the length and location of studs.

Laying Out a Gable End Frame

1. Drop a plumb line from the center of the ridge, above the end wall plate. A stud is placed directly below the ridge, if possible. Often, a ventilator is located in the center of the end frame. If this is the case, mark a distance one-half of the ventilator width to either side of the plumb line. This mark will be the location of the first stud.
2. Position the stud on the top plate and, using a level, be sure it is plumb. Mark the location of the rafter on the side of the stud.
3. Lay out the next stud (16" O.C. is common spacing), plumb it, and mark the rafter location on the stud's side. The difference between the two stud lengths is the common difference (Figure 10-22). This is the difference in length between any two adjacent studs, as long as the spacing remains the same.
4. Using the common difference, continue cutting and placing studs until the edge of the frame is reached. If the rafters on both sides have the same rise and run, the end frame should be *symmetrical* (both sides identical).

The common difference can also be obtained using the framing square (see Figure 10-23).

Finding the Common Difference

1. Set the square on the stud for unit run and rise. Mark a line on the stud along the blade.
2. Slide the square along this line until the number for the stud spacing (16") aligns with the edge.
3. Read the distance along the tongue of the square. This is the common difference.

In modern residential construction, roof designs often include an *extended rake* (gable overhang). Typical framing, as illustrated in Figure 10-24, requires the construction of the gable end frame before the roof frame is completed.

When constructing the gable ends for a brick or stone veneer building, the frame must be moved outward to cover the finished wall. This projection can be formed by using *lookouts* and blocking attached to a ledger. (See Figure 13-3 in the unit on exterior wall finish.)

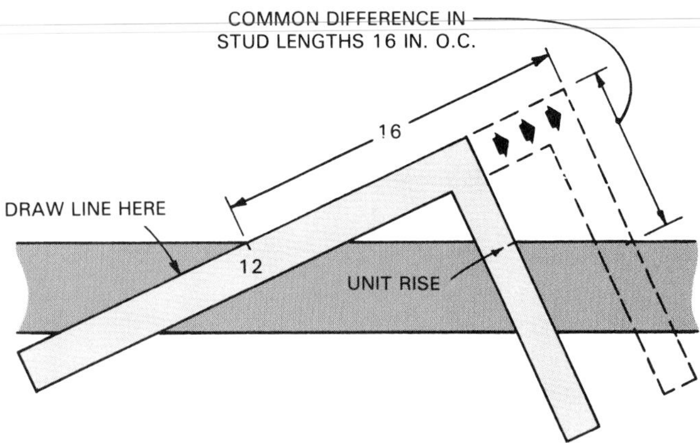

Figure 10-23. Using the framing square to find the length and angle of gable studs. Left—Lining up run and rise on stud. Right—Relationship of the square method to the roof line.

When the top of the veneer is aligned on the sides and ends of the building, the ledger should be attached at the same level as the one used in the *cornice* construction. See Figure 10-25.

Studs are mounted on this projection and are attached to the roof frame in various ways. The architectural plans usually include details covering special construction features of this type.

Hip and Valley Rafters

Hip roofs or intersecting gable roofs will have some or all of the following rafters: common, hip, valley, jack, and cripple.

First, cut and frame the common rafters and ridge boards. The ridge of a hip roof is cut to the length of the building minus twice the run plus the thickness of the rafter stock. It intersects the common rafters, Figure 10-26.

From the corners of the building, lay out along the side walls a distance equal to half the span. These points

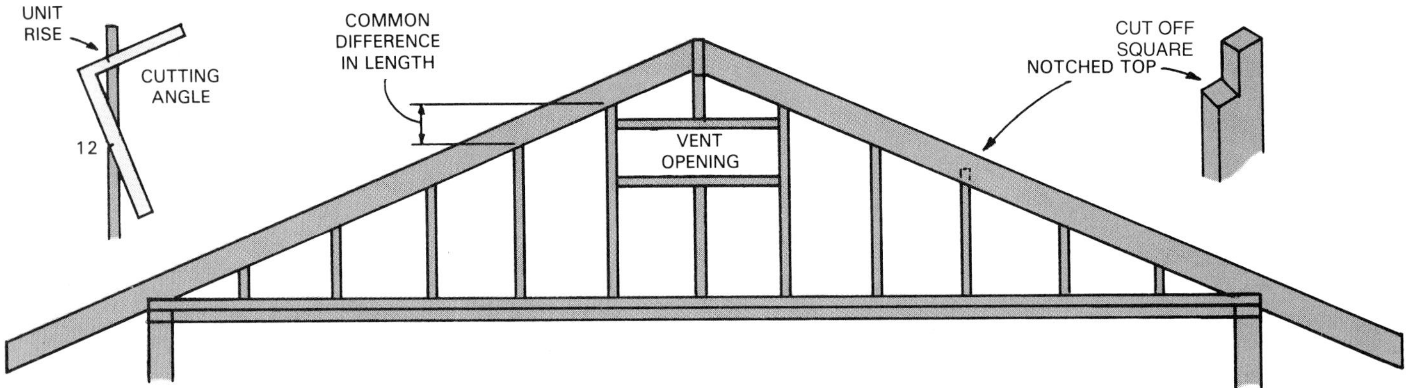

Figure 10-22. Lay out studs for a gable end as shown. Change in length from one stud to another is constant.

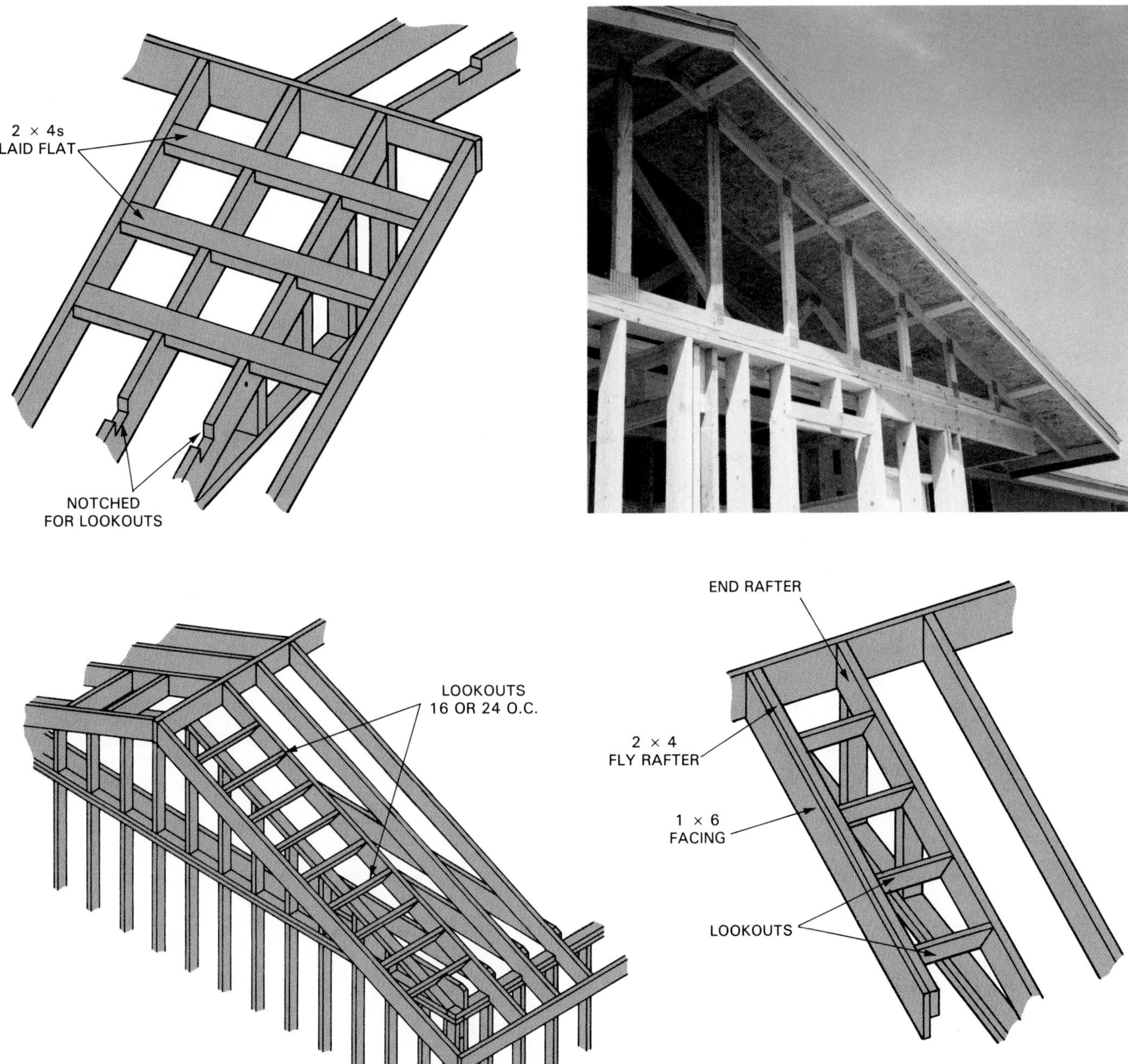

Figure 10-24. Methods of framing gable overhang. Top, left and right—2 x 4 lookouts are laid flat over notched rafters. Bottom, left—Plate atop gable end studs support lookouts laid on edge. Make sure top of plate lines up with bottoms of rafters. Bottom, right—Small overhang with short lookouts supporting 2 x 4 fly rafter and face board.

will be the centerline of the first common rafters. All other rafters are laid out from this position.

Two roof surfaces slanting upward from adjoining walls will meet on a sloping line called a hip. The rafter supporting this intersection is known as a hip rafter.

In a plan view, the hip rafter will be seen as the diagonal of a square, Figure 10-27. The diagonal of this square is the total run of the hip rafter.

Since the unit run of the common rafter is 12", the unit run of the hip rafter will be the diagonal of a 12" square. When calculated, accurately, this is 16.97". For actual application, a rounding to 17" is close enough.

To lay out a hip rafter, follow the same procedure used for a common rafter. Use the 17" mark on the blade of the square instead of the 12" mark, Figure 10-28.

The odd unit must also be adjusted. Its length is found by measuring the diagonal of a square, the sides of which are equal to the length of the odd unit.

Valley rafters will also be the diagonal of a square. The sides are formed by ridges and common rafters. The layout is the same as described for hip rafters with 17" used for the unit run. The length of hip and valley rafters

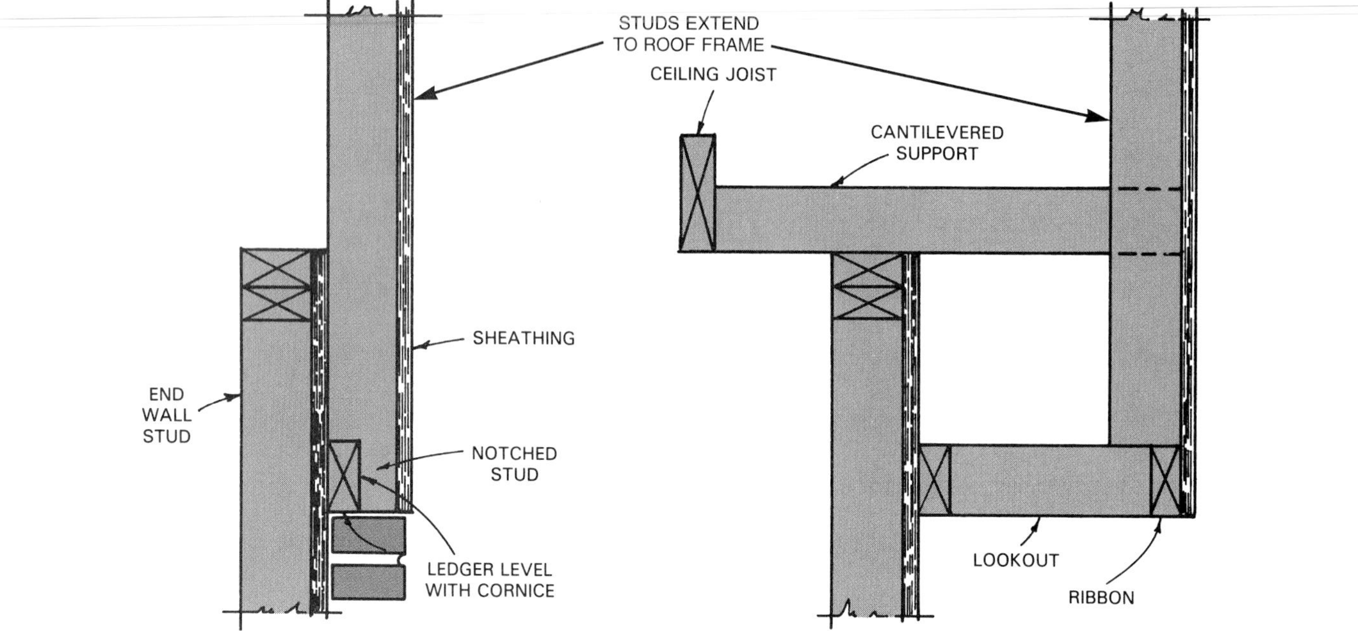

Figure 10-25. Gable-end framing for masonry veneered buildings. Left—Building out framing to cover brick or stone veneer. Right—Alternate framing that extends the gable end farther. Such framing might also be used over doors and other special features.

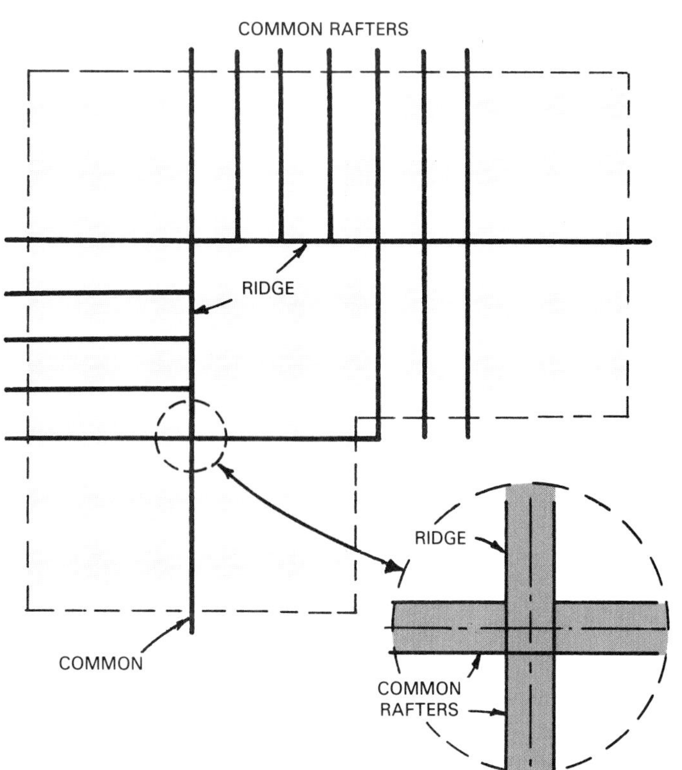

Figure 10-26. The first step in framing a hip or intersecting roof is to install the common rafters and ridges.

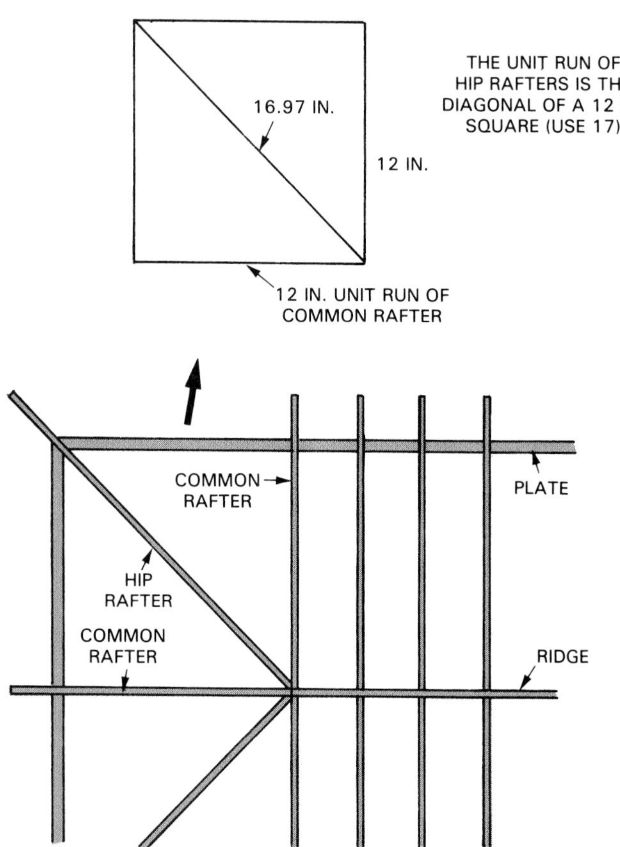

Figure 10-27. A hip rafter is the diagonal of a square formed by the walls and two common rafters.

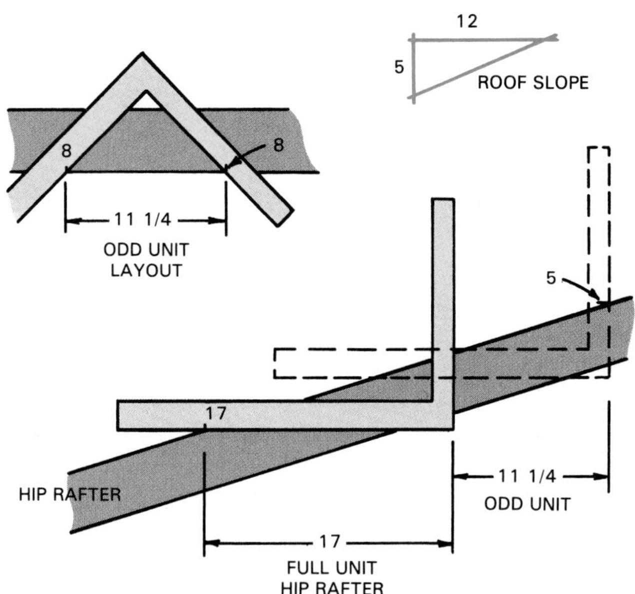

Figure 10-28. Starting the layout of a hip rafter. Slope and odd unit size are the same as those used in common rafter layout.

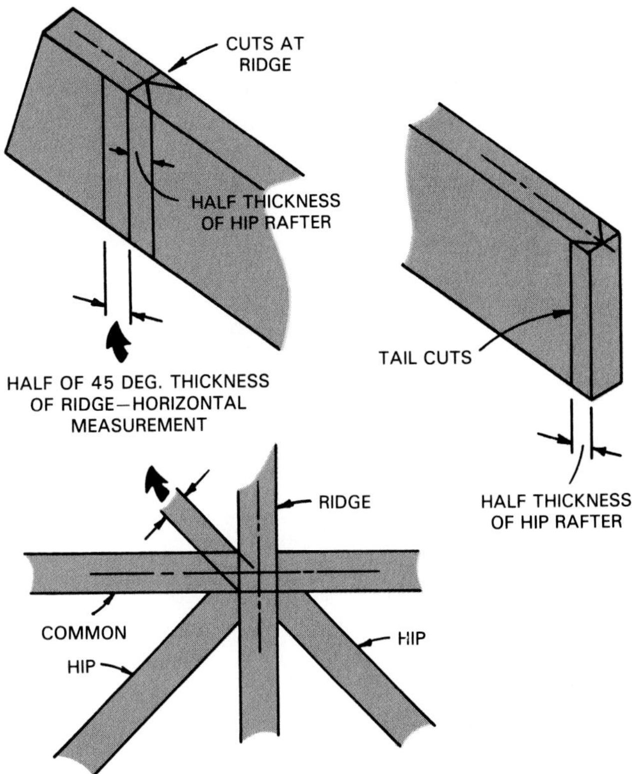

Figure 10-29. Shortening hip rafter and making side cuts. Top, left—Shorten top of hip rafter half the thickness of the ridge and mark; then measure back half the thickness of rafter and mark both sides for angle cuts. Top, right—Note how tail is marked for cutting. Bottom—Drawing explains why angled cut must be half the thickness of the ridge.

can be determined from rafter tables just like common rafters. Using the same figures as were used in a previous example, the calculations are as follows:

Run = 6'-8"	Slope 5 to 12
Run = 6'-8"	= 6 2/3 units
Table No. 2	= 17.69
Hip or Valley Length	= 17.69 x 6 2/3
	= 117.93
	= 9'-9 15/16"
	= 9'-10"

Hip and valley rafters must be shortened at the ridge by a distance equal to half of the 45° horizontal thickness of the ridge, Figure 10-29. The side cuts are then laid out as shown in the illustration.

Another method is to use the numbers from the sixth line of the rafter table—for example, 11 1/2. All of the numbers in the table are based on, or are related to 12, so line up 12 on the blade and 11 1/2 on the tongue along the edge of the rafter as shown in Figure 10-30. Draw the angle for each cut. Now draw plumb lines. Use 17 on the blade and the unit rise on the tongue. Tail cuts at the ends of the rafters are laid out using the same angle.

A centerline along the top edge of a hip rafter is where the roof surfaces actually meet. The corners of the rafter will extend slightly above this line. Some adjustment must be made. The corners could be planed off. However, it is easier to make the seat cut of the bird's mouth slightly deeper, lowering the entire rafter, Figure 10-31.

Using the framing square, align the number 17 and the number for the unit rise on the bottom edge of the rafter. Mark the position with a short line drawn along the body of the square. Measure back (toward the tail) half the thickness of the rafter and mark. Shift the square toward the tail of the rafter. Be sure to maintain the alignment of 17 and the unit rise. When the square aligns with the mark previously made, mark the seat of the bird's mouth. Do not mark the plumb cut until you have read the next paragraph.

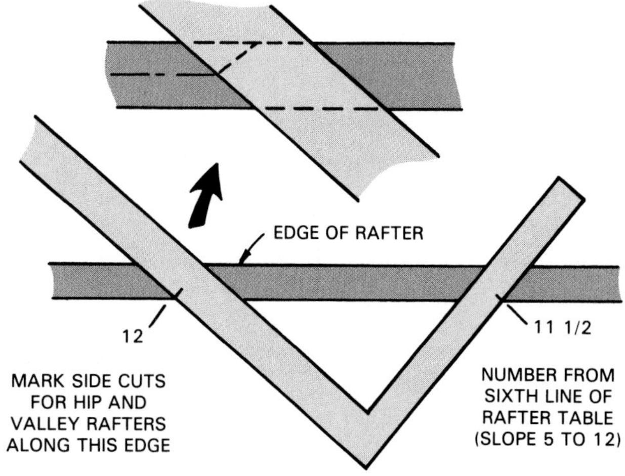

Figure 10-30. Using the framing square to lay out side cuts on hip and valley rafters.

The plumb (vertical) cut of the bird's mouth for valley rafters must be trimmed so it will fit into the corner formed by the intersecting walls. Although side cuts of

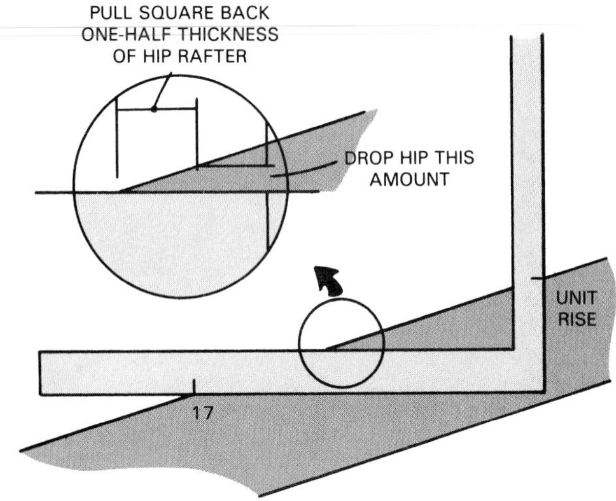

Figure 10-31. Using the square to find and mark distance to drop hip rafter. Bottom of rafter is facing up.

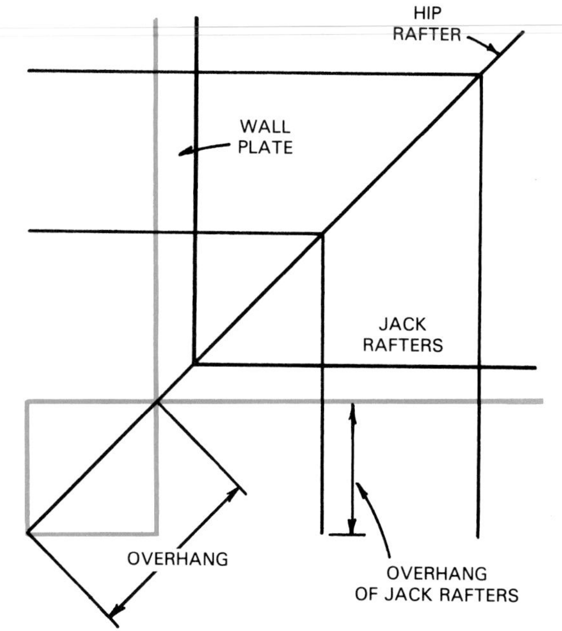

Figure 10-32. Finding the length of hip rafter overhang. You can construct the square full size on paper. Then measure the length of the diagonal. A rafter table on the square will also give this information.

approximately 45° could be made, it is more practical to move the plumb cut toward the tail of the rafter by a horizontal distance equal to the half 45° thickness of the rafter.

The tail of the hip rafter is actually the diagonal of a square formed by extending the line of each of the walls the length of the jack rafters, Figure 10-32. You can find its length by constructing on paper a full-size square. Then measure the diagonal. Mark the plumb cut using the run (17") and the rise (in inches) on the square.

The tail must form a nailing surface for intersecting fascia boards. Refer again to Figure 10-29 for directions on how to make the tail cut.

After hip and valley rafters are laid out and cut, they are installed on the roof. Figure 10-33 is a rafter plan showing their positions.

To use the super square in laying out hip and valley rafters, find the rafter length from the table corresponding to the rise per foot of run. Read down the table to the correct run; then read across to the column for hip and valley rafters. Mark the rafter for the ridge plumb cut, Figure 10-34(A). Next, measure from the top of the rafter down and mark the length obtained from the chart. Use the super square to make the plumb seat mark. Next, rotate the super square until the dashed line is aligned with the seat plumb mark just made. See Figure 10-34(B). Mark the seat cut. Make the adjustments previously described and cut the rafter.

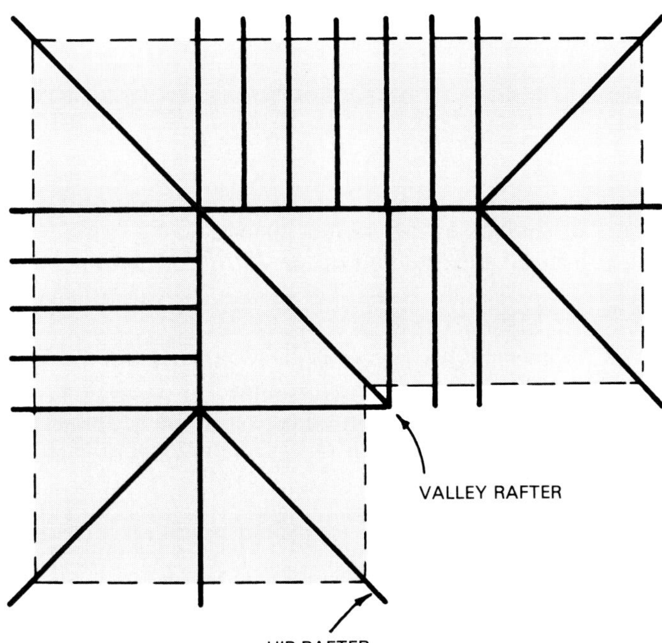

Figure 10-33. Rafter plan shows hip and valley rafters along with common rafters. In actual construction, jack rafters would be cut and installed last.

Jack Rafters

Hip jack rafters are short rafters which run between the wall plate and a hip rafter. They run parallel to common rafters and are the same in every respect except the length from the bird's mouth to the hip rafter. When equally spaced along the plate, the change in length from one to the next is always the same. This consistent change is called the *common difference*, Figure 10-35. You ran into this term earlier in cutting the cripple studs for the gable end frame.

The common difference can be found in the third and fourth lines of the rafter table. For a roof slope of 5 to 12,

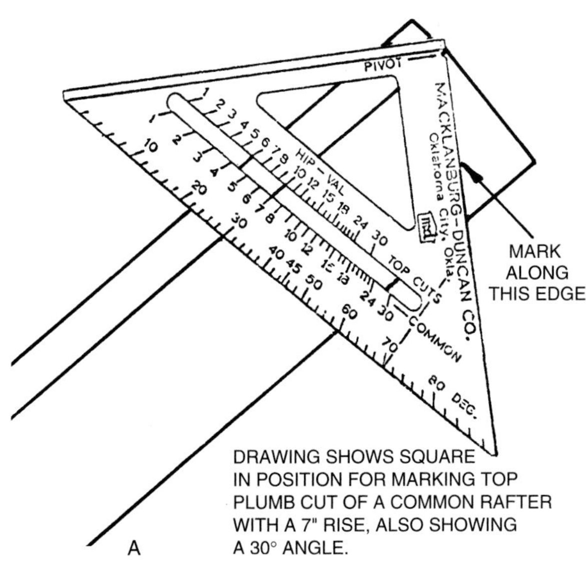

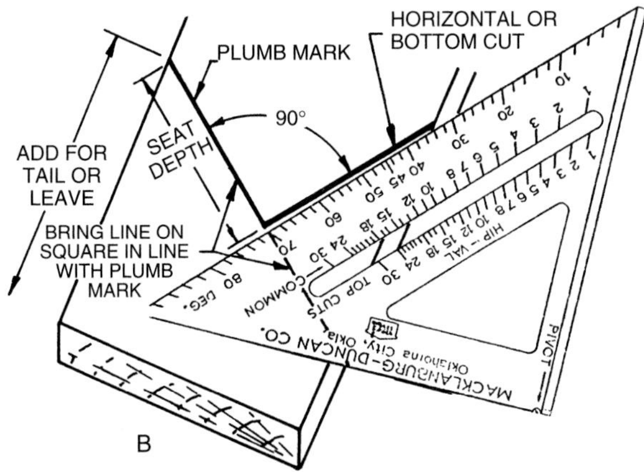

Figure 10-34. Using the super square. A—Plumb cut is being set up for hip or common rafter. Extra allowance must be marked for the miter cut. B—Marking the seat cut for the bird's mouth. (Macklanburg-Duncan)

with rafters specified 24" O.C., the figure from the table on the rafter square is 26". See Figure 10-36.

The common difference can also be found with the layout method illustrated in Figure 10-37. Hold the square along the edge of a smooth piece of lumber according to the unit run and rise of the roof. Draw a line along the blade and then slide the square along this line to a point equal to the rafter spacing. The distance thus laid out along the edge of the lumber will be the required difference in length.

To lay out jack rafters, select a piece of straight lumber. Lay out the bird's mouth and overhang from the common rafter pattern. Next, lay out the line length of a common rafter. This is the distance from the plumb cut of the bird's mouth to the centerline of the ridge. For the first jack rafter down from the ridge, lay out the common difference in length.

Now take off half the 45° thickness of the hip, Figure 10-38. Square this line across the top of the rafter and mark

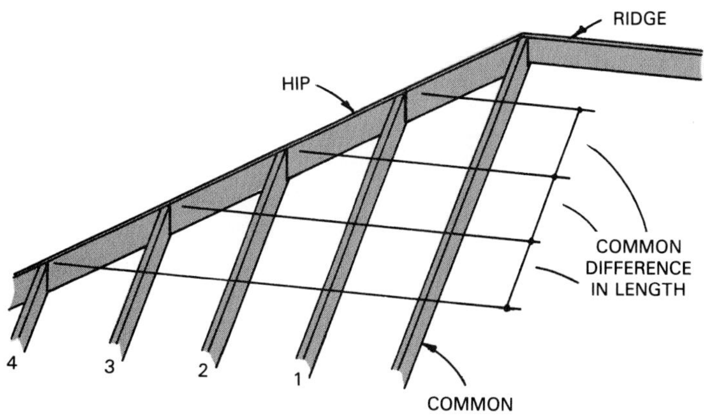

Figure 10-35. When evenly spaced, hip jack rafters have a common (identical) difference in length from one to the next.

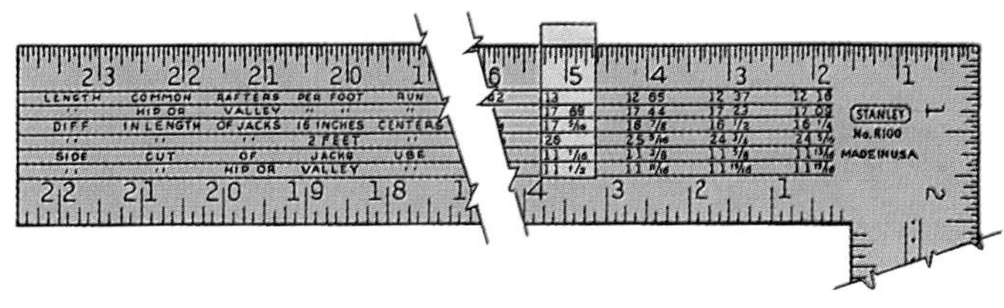

Figure 10-36. Common difference of jack rafters spaced at 24" O.C. can be found on the fourth line of the rafter table.

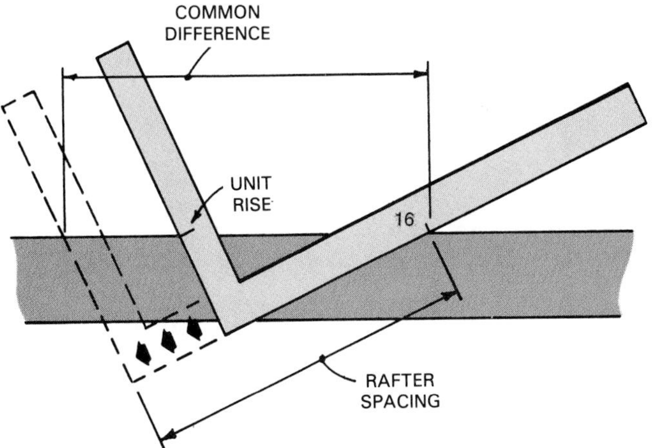

Figure 10-37. Using the framing layout method of determining the common difference of jack rafters. For example, if the stud spacing were 16", one would slide the square in the direction of the arrows until the 16" mark on the blade aligned with the mark made on the rafter.

the center point. Through this point, lay out the side cut as shown in the illustration. Another method is to use a square and the number from the fifth line of the rafter table, Figure 10-39. Mark the plumb lines that will be followed when the cut is made.

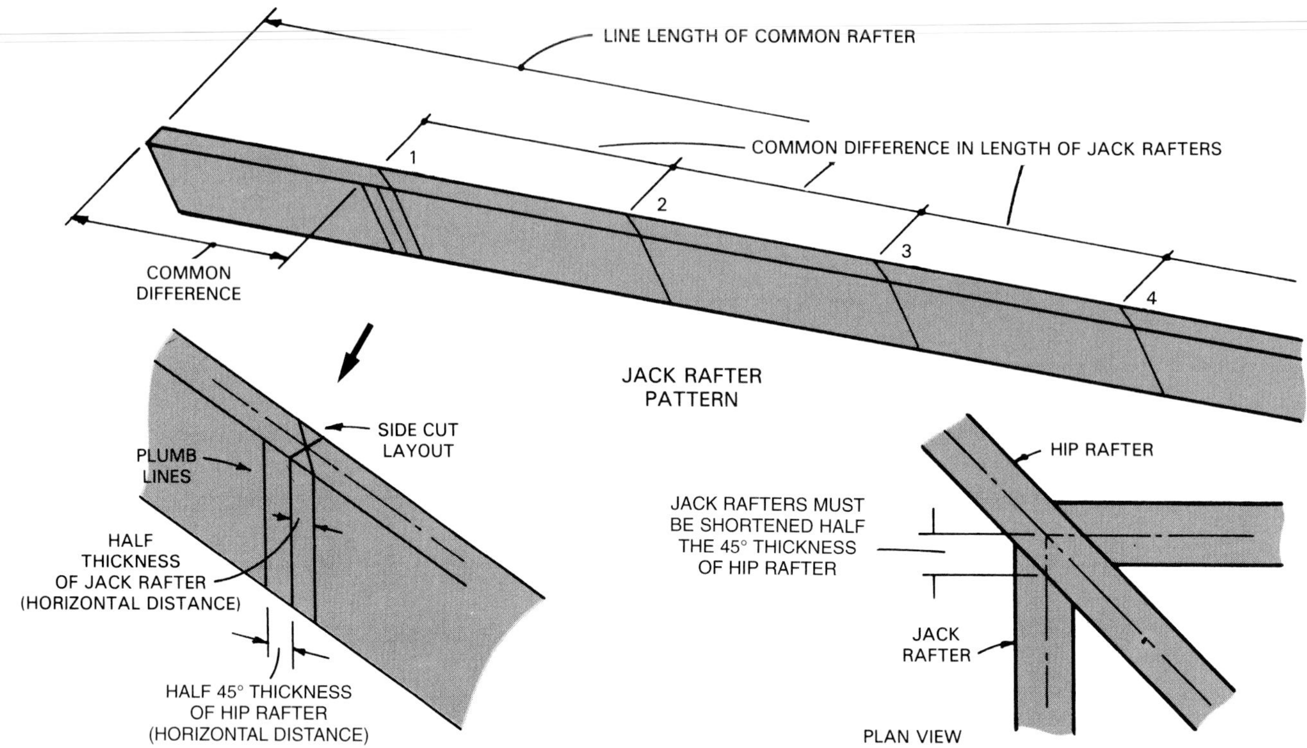

Figure 10-38. This pattern layout is for jack rafters. Note that the layout method is the same as the one used for hip jacks.

For the next hip jack, move down the rafter the common difference and mark the cutting line. A sliding T-bevel is a good tool to use. Continue until jacks are laid out. Then use the T-bevel to mark the jack rafters required.

Each hip in the roof assembly requires one set of jack rafters made up of matching pairs. A pair consists of two rafters of the same length with the side cuts made in opposite directions.

Valley Jacks

Similar procedures are followed in laying out valley jack rafters. For these, however, it is usually best to start the layout at the building line, Figure 10-40, and move toward the ridge. The longest valley jack will be the same as a common rafter except for the side (angle) cut at the bottom.

Use the common rafter pattern and extend the plumb cut of the bird's mouth to the top edge. Lay out the side cut by marking (horizontally) half the thickness of the rafter. If you prefer, use the framing square and apply the numbers located in the fifth line of the rafter table.

The common difference for valley jack rafters is obtained by the same procedure used for hip jacks. Lay out this distance from the longest valley jack to the next. Continue along the pattern until all lengths are marked. Now use this pattern to cut all the valley rafters. They are cut in pairs in the same way as hip jacks.

When all jack rafters are laid out, they should be carefully cut. It requires a great deal of skill to make side cuts with a hand saw. Usually, it is better to use a portable electric saw with an adjustable base and guide. Radial arm saws are designed to do accurate cutting and are well suited for this kind of work, Figure 10-41.

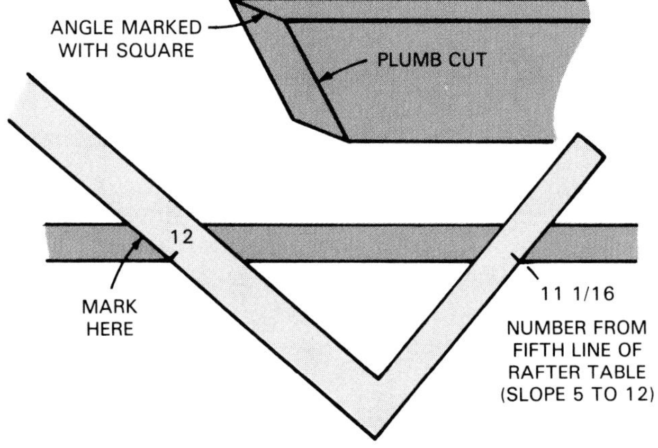

Figure 10-39. This is how to use a framing square to mark the angle cut on the top of the hip jack rafter.

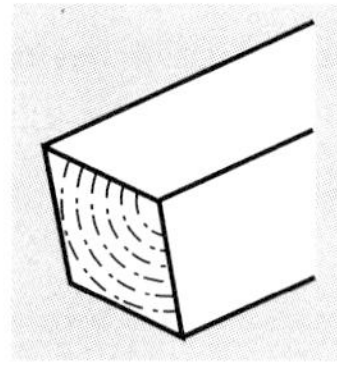

The strength of a roof frame depends a great deal on the quality of the joints. Use special care on the side cuts of jack rafters so the joining surfaces will fit tightly together.

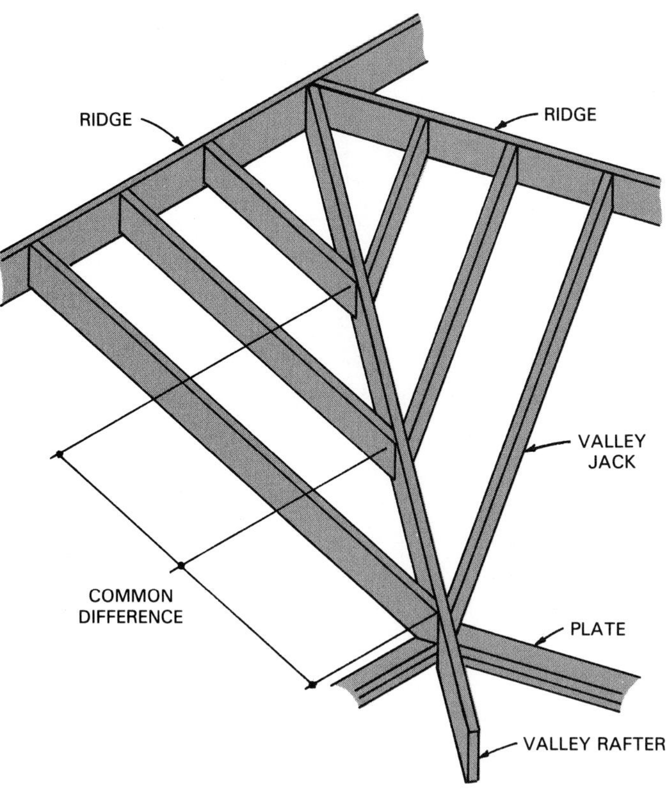

Figure 10-40. Valley jack rafters as assembled. Find the common difference in the same manner as for the hip jacks.

Figure 10-41. Side cuts for jack rafters can be accurately made with a radial arm saw. (Des Moines, Iowa Public Schools)

Erecting Jack Rafters

When all jack rafters are cut, assemble them into the roof frame, Figure 10-42. Nailing patterns will depend on the size of the various members. Use 10d nails. Space the nails so they will be near the heel of the side cut as they go from the jack into the hip or valley rafter.

Jack rafters should be erected in pairs to prevent the hip and valley rafters from being pushed out of line. It is good practice to first place a pair about halfway between the plate and ridge. Carefully sight along the hip or valley to make sure that it is straight and true.

Temporary bracing could also be used for this purpose. Be sure that outside walls running parallel to the ceiling joists are securely tied into the ceiling frame before hip jack rafters are installed. These rafters tend to push outward.

After rafters have been erected and securely nailed, check over the frame carefully. If some rafters are bowed sideways, they can be held straight with a strip of lumber located across the center of the span. Each rafter is sighted and moved as needed. A nail is driven through the strip to hold it in place. When the roof has been sheathed to this point, the spacer strip is removed.

The last material to be attached to the roof frame before the decking is the *fascia*. This is the main trim member that is attached to the plumb-cut ends of the rafters. It conceals the rafter ends, provides a finished appearance, and furnishes a surface to which guttering may be attached.

However, before fascia is attached it will be necessary to stretch a chalk line along the rafter ends to assure that rafter ends are the same length. Any rafters that are too long must be trimmed so the *eave* line is straight.

Fascia may be attached directly to the rafter ends. Some carpenters prefer to install 2 x 4s first to even up the ends of the rafters and provide a solid nailing surface for the 1" fascia board.

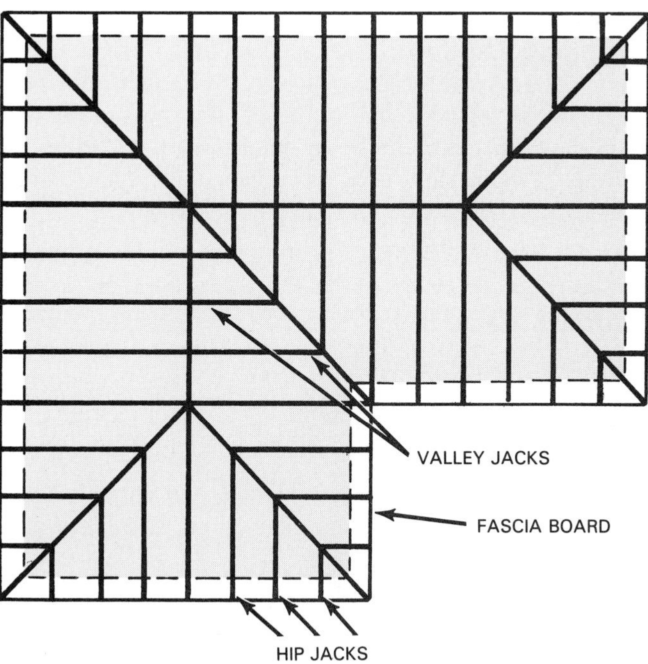

Figure 10-42. Plan view of complete roof frame. All hip and valley jacks are in place and fascia board has been installed.

The upper edge of the fascia board should be cut at an angle to match the slope of the roof. Corners should be mitered and carefully fitted.

Special Problems

When framing intersecting roofs where the spans of the two sections are not equal, the ridges will not meet. To support the ridge of the narrow section, one of the valley rafters is continued to the main ridge. See Figure 10-3 at the beginning of this unit. The length of this extended or supporting valley is found by the same method used in the layout of a hip rafter. It is shortened at the ridge just like the hip but only a single side cut is required. The other valley rafter is fastened to the supporting valley with a square, plumb cut.

A rafter framed between the two valley rafters is called a valley cripple jack. The angle of the side cut at the top is the reverse of the side cut at the lower end. The run of the valley cripple is one side of a square, Figure 10-43. This run is equal to twice the distance from the centerline of the valley cripple jack to the intersection of the centerlines of the two valley rafters. Lay out the length of the cripple by the same method used for a jack rafter. Shorten each end half of the 45° thickness of the valley rafter stock. Make the side cuts in the same way as for regular jack rafters.

Rafters running between hips and valleys are called hip-valley cripple jacks. They require side cuts on each end. Since hip and valley rafters are parallel to each other, all cripple rafters running between them, in a given roof section, will be the same length.

The run of a hip-valley cripple rafter will be equal to the side of a square, Figure 10-44. The size of the square is determined by the length of the plate between the hip and valley rafter. Use this distance and lay out the cripple in the same manner used for a common rafter. Shorten each end by an amount equal to half of the 45° thickness of the hip and valley rafter stock. Now lay out and mark the side cuts. Follow the same steps used for hip and valley jacks. Side cuts, required on each end, form parallel planes.

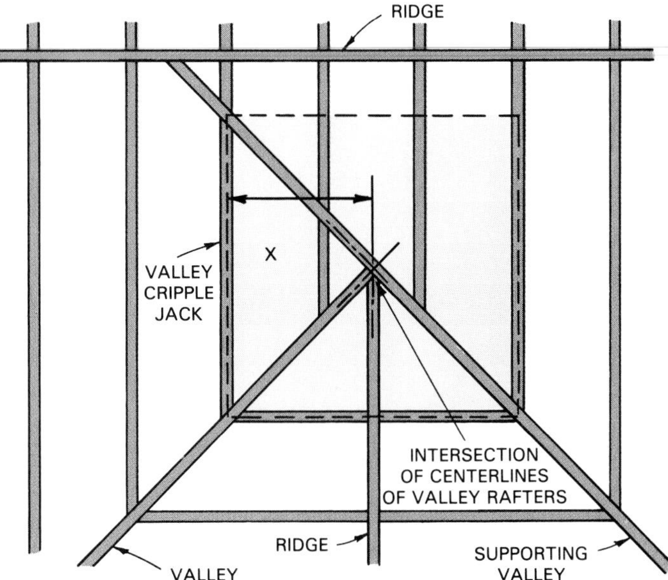

Figure 10-43. The run of a valley cripple jack is twice the distance at X.

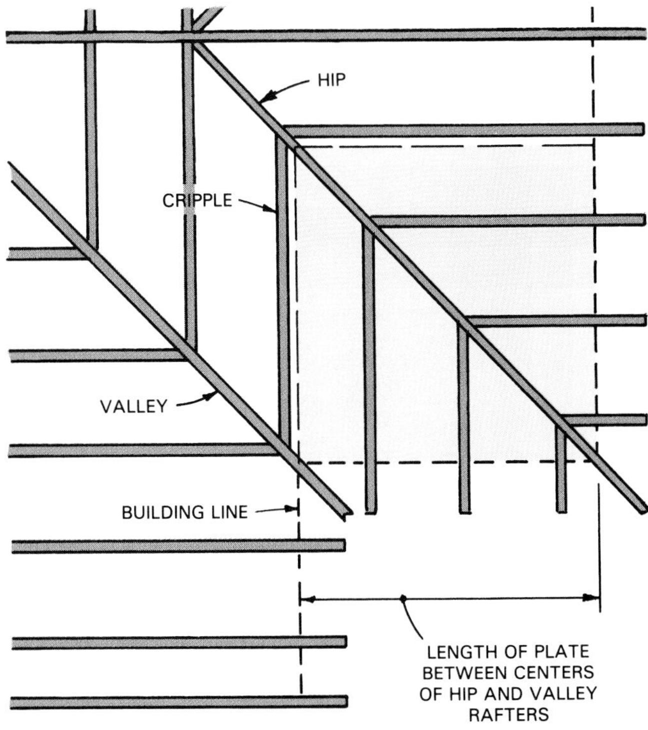

Figure 10-44. The run of a hip-valley cripple jack is equal to the length of the wall plate from the valley rafter to the hip rafter.

Roof Openings

Some openings may be required in the roof for chimneys, skylights, and other structures. To frame large openings, follow about the same procedure used in floor framing.

To construct small-size openings, the entire framework is first completed. Then the opening is laid out and framed.

For a chimney opening, Figure 10-45, use a plumb line to locate the opening on the rafter from openings already formed in the ceiling or floor frame.

Nail a temporary wooden strip across the top of the rafters to be cut. The supporting strip should be long enough to extend across two additional rafters on each side of the opening. This will support the ends of the cut rafters while the opening is being formed.

Now cut the rafters and nail in the headers. If the size of the opening is large, double the headers and add a trimmer rafter to each side.

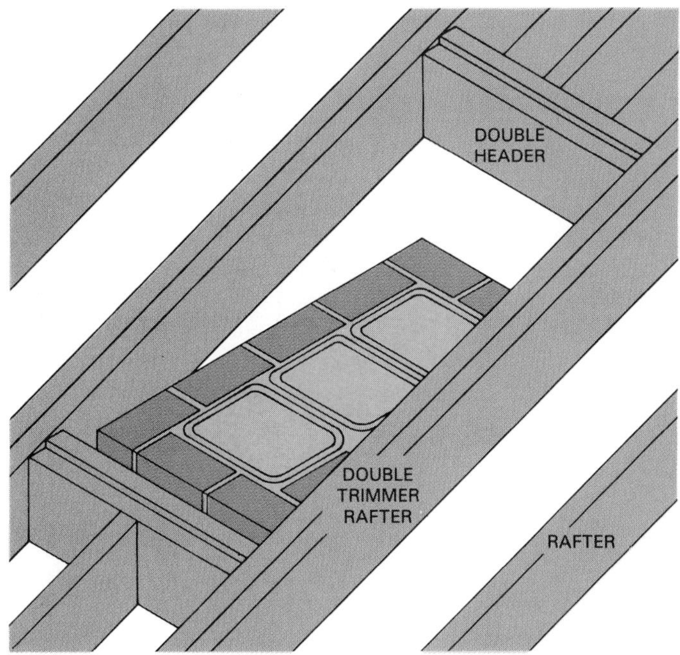

Figure 10-45. Framing an opening in the roof for a chimney. Allow 2" clearance on each side and ends. Note that the headers above and below the chimney are plumbed.

Roof Anchorage

Rafters usually rest only on the outside walls of a structure. They lean against each other at the ridge, thus providing mutual support. This causes an outward thrust along the top plate that must be considered in the framing design.

Sidewalls are normally well secured by the ceiling joists, which are also tied to some of the rafters. End walls, however, will be parallel to the joists. They need extra support, especially when located under a hip roof. Stub ceiling joists and metal straps are one method for reinforcing such walls. Framing anchors can be substituted for the metal straps, especially when subflooring is included in the assembly. See Figure 10-46.

Collar Beams

Collar beams tie together two rafters on opposite sides of a roof, Figure 10-47. They do not support the roof but provide bracing and stiffening to hold the ridge and rafters together. In standard construction, 1 x 6 boards are installed at every third or fourth pair of rafters.

Purlins

Additional support must be provided when the rafter span exceeds the maximum allowed. A *purlin*, usually a 2 x 4, is attached to the underside of the rafters. This member is supported by bracing, also 2 x 4 stock, resting on a plate located over a supporting partition. Bracing under the purlin may be placed at any angle to transfer loads from the mid-point of the rafter to the support below. See Figure 10-48.

Dormers

A *dormer* is a framed structure projecting above a sloping roof surface, and normally contains a vertical window unit. Although its chief purpose is to provide light, ventilation, and additional interior space, dormers also enhance the exterior appearance of the structure.

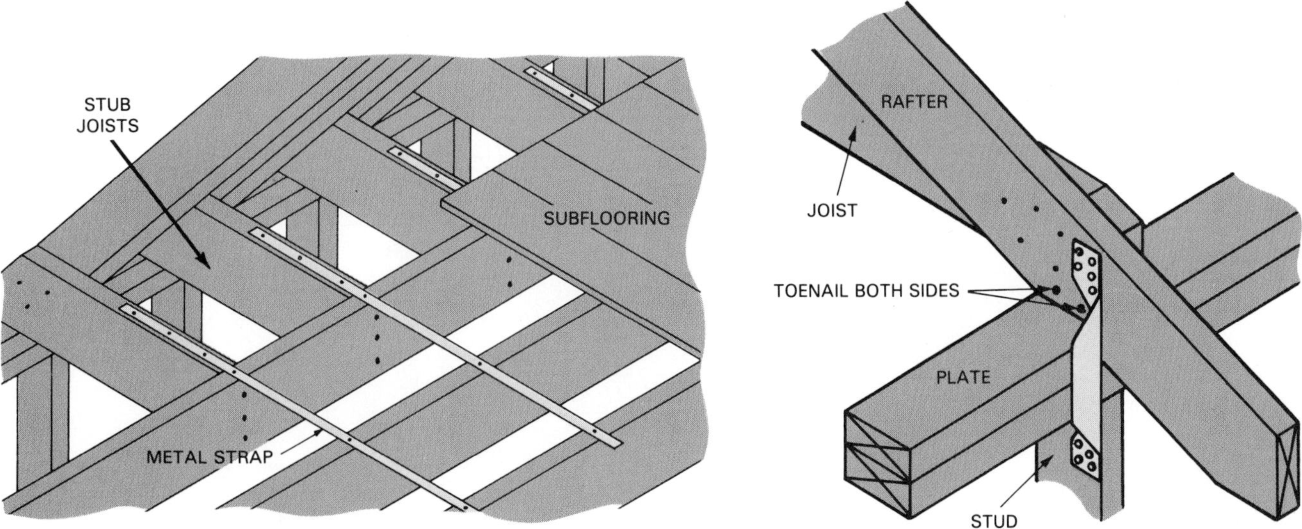

Figure 10-46. Two methods of anchoring rafters—Left. Metal strapping and stub joists can be used to tie down hip roofs to end walls. Right—Tie-down anchors can be used to fasten rafters, plates, and studs together. (TECO)

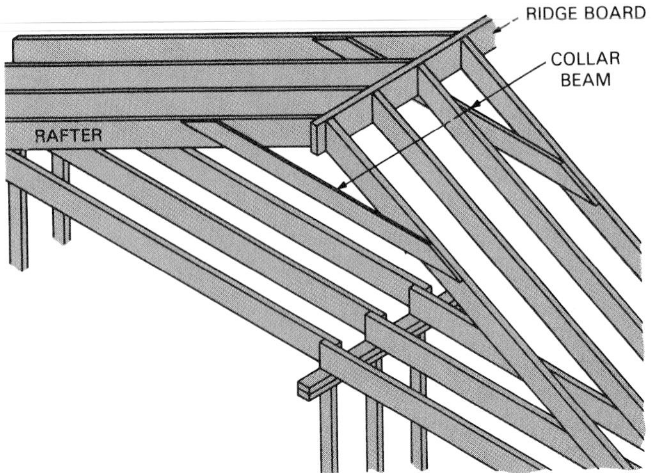

Figure 10-47. Collar beams tie rafters and ridge together, reinforcing the roof frame and securing them against spreading outward.

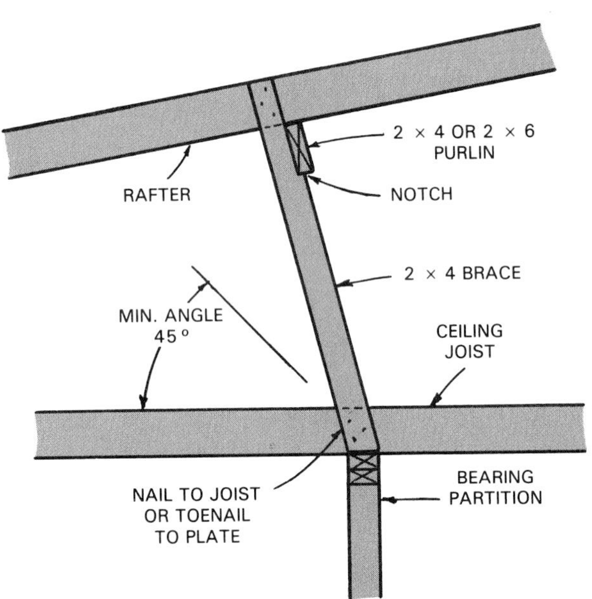

Figure 10-48. Top—A purlin is a plate that supports long runs of rafters at mid-point. Weight is transferred to a bearing partition through 2 x 4 braces. Bottom—Shot of new construction using a purlin. (Hadley-Hobley Construction)

The shed dormer's width is not restricted by its roof design. It is used where a large amount of additional interior space is required, Figure 10-49.

In the simplest construction, the front wall is extended straight up from the main wall plate. Double trimmer rafters carry the side wall.

The rise of the roof is figured from the top of the dormer plate to the main roof ridge. Run will be the same as the main roof. Be sure to provide sufficient slope for the dormer roof.

Gable dormers, Figure 10-50, are designed to provide openings for windows. They can be located at various positions between the plate and ridge of the main roof. See Figure 10-51.

Flat Roofs

Flat roofs provide the long, low appearance often desired in contemporary designs. Improvements in roofing surface materials and methods of application make this type of roof practical.

Methods and procedures used to frame flat roofs are about the same as those followed in constructing a floor. Most designs will require an overhang with the ends of the joists tied together by a header or band. Cantilevered rafters (extend beyond their supports) are tied to doubled roof joists. See Figure 10-52. Corners can be formed as shown or carried on a longer diagonal joist that intersects the double joist.

Builders in the southwestern United States sometimes build traditional homes in the Pueblo or Territorial style. This style of homes includes a flat roof supported by round

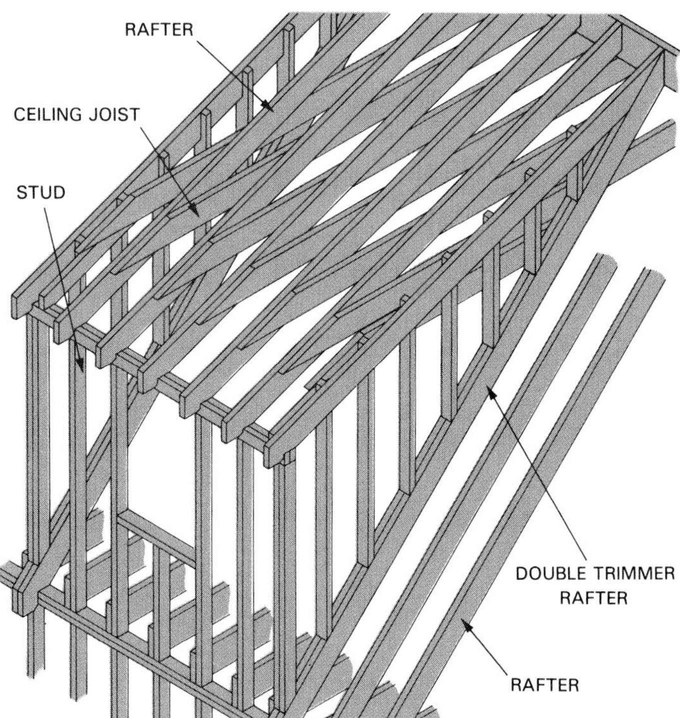

Figure 10-49. Typical framing for a shed dormer. A nailer strip is added along the double trimmer to carry roof sheathing.

Figure 10-50. Framing a gable dormer. A—Rafters have same pitch as the regular roof. Joists will be added to support the ceiling. B—Dormers are sheathed and trimmed ready for shingles and siding. C—Studs are 2 x 6 and rest on a sole plate. Note that they extend through the roof to their own wall plate.

Figure 10-51. View of gable dormers on a Cape Cod. Windows and exterior finishing materials match the rest of the dwelling.

poles called vigas. The vigas are spaced anywhere from 16" to 30", depending on the span. The viga rests on the wall plate and extends through the exterior wall about 2' to 3' A through-bolt secures the vigas to the plate. Usually, the outside wall extends 2' to 3' above the roof as an architectural feature. Figure 10-53 shows a traditional southwestern home under construction.

Mansard Roofs

A mansard roof maximizes the second floor area, Figure 10-54. Second floor joists extend beyond the first floor wall frame and provide support for the mansard rafters.

Gambrel Roof

The gambrel roof is like a gable roof but has two slopes. It is typical in an architectural style known as Dutch

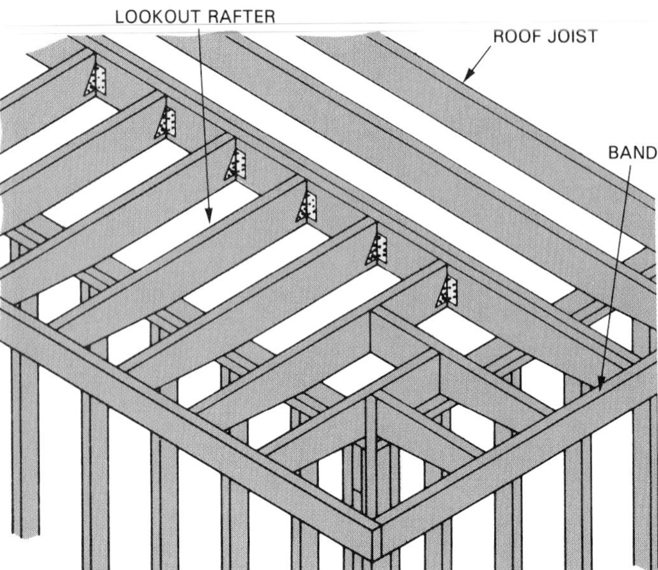

Figure 10-52. Lookout rafters are cantilevered over the outside wall to form an overhang.

Figure 10-53. Flat roofs are a feature of some traditional home styles still built in southwestern U.S. Top—Looking up at the ceiling of a Pueblo style house. Above the viga (pole) is a roof framed in 2 x 12s. Note solid bridging. Bottom—Exterior view of Pueblo style home shows walls extending beyond flat roof. (Pierce Construction Ltd.)

Colonial, Figure 10-55. Often used in two-story construction, it gives added living space with minimal outside wall framing. The upper roof surface usually forms a 20° angle with a horizontal plane while the lower surface forms about a 70° angle. See Figure 10-56.

In residential construction, this type of roof is usually framed with a purlin located where the two roof surfaces meet. Rafters are notched to receive the purlin, which is supported on partitions and/or tied to another purlin on the opposite side of the building with collar beams.

Procedures used to frame a gable roof can be applied to the gambrel roof. The rise and run of each surface is found on the architectural plans. The two sets of rafters are laid out in the same way as previously described for a common rafter.

It may help to make a full-size sectional drawing (if not included in the plans) at the intersection of the two slopes. It is then easier to visualize and proportion the end cuts of the rafters. Figure 10-57 shows basic gambrel roof framing for a small building, such as a garden house or tool shed. Note the use of gussets to join the rafter segments. Also note the simple framing used to form the roof overhang.

Special Framing

Figure 10-58 shows upper floor framing for a 1 1/2-story structure. Generally there is some saving of material since walls and ceilings can be made a part of the roof frame. Knee walls are usually about 5' high. Standard ceiling height is 7'-6".

Low-sloping roofs like the one shown in Figure 10-59 usually require extra points of support. Strength derived from the triangular shapes of regular pitched roofs is greatly reduced. Thus, carefully prepared architectural plans are essential for this type of roof structure.

Roof Truss Construction

A *truss* is a framework that is designed to carry a load between two or more supports. The principle used in its design is based on the rigidity of the triangle. Triangular shapes are built into the frame in such a way that the stresses of the various parts are parallel to the members making up the structure.

Roof trusses are frames that carry the roof and ceiling surfaces. They rest on the exterior walls and span the entire width of the structure. Since no load bearing partitions are required, more freedom in the planning and division of interior space is possible. They permit larger rooms without extra beams and supports. Roof trusses also allow for surface materials to be applied to outside walls, ceilings, and floors before partitions are constructed.

There are many types and shapes of roof trusses. One commonly used in residential construction is the W or Fink truss, which is illustrated in Figure 10-60.

Figure 10-54. Mansard roofs allow second floor ceilings to extend to the outer walls of the lower floor.

Figure 10-55. Gambrel roof provides extra living space on the second story with minimal wall framing.

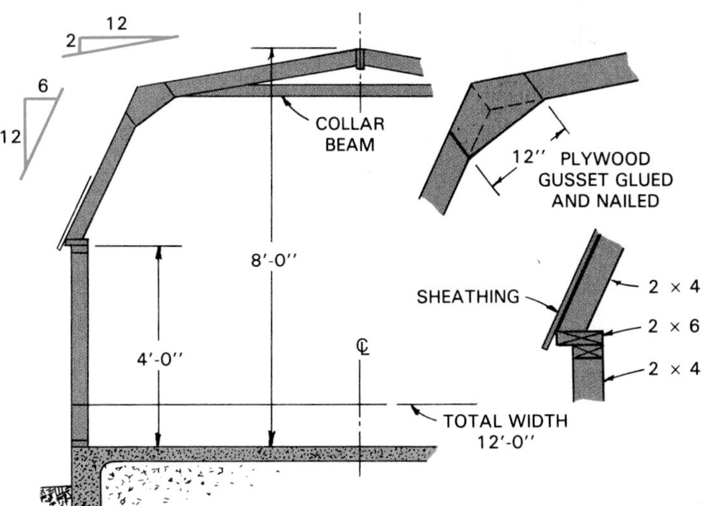

Figure 10-57. Gambrel roof for a small building is easy to design and build.

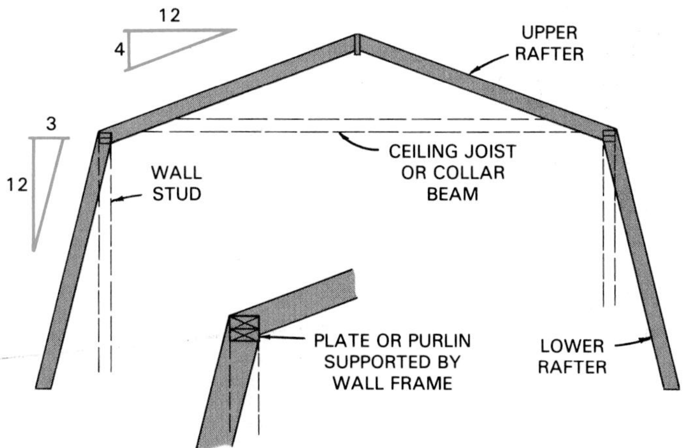

Figure 10-56. Basic framing for a gambrel roof. Rafter patterns can be developed by making a full-size layout on the subfloor.

Most roof trusses are factory built to engineered specifications. See Figure 10-61. If carpenters choose to build their own, the truss must be built to carefully developed and engineered designs. A variety of roof truss designs are shown in the Technical Information section.

Roof trusses must be made of structurally sound lumber and assembled with carefully fitted joints.

Although the carpenter is seldom required to determine the sizes of truss members or the type of joints, she or he should understand their design well enough to appreciate the necessity of first-class work in their construction.

Trusses are pre-cut and assembled at ground level. Spacing of 24" O.C. is common. However, 16" O.C. and other spacing may be required in some designs.

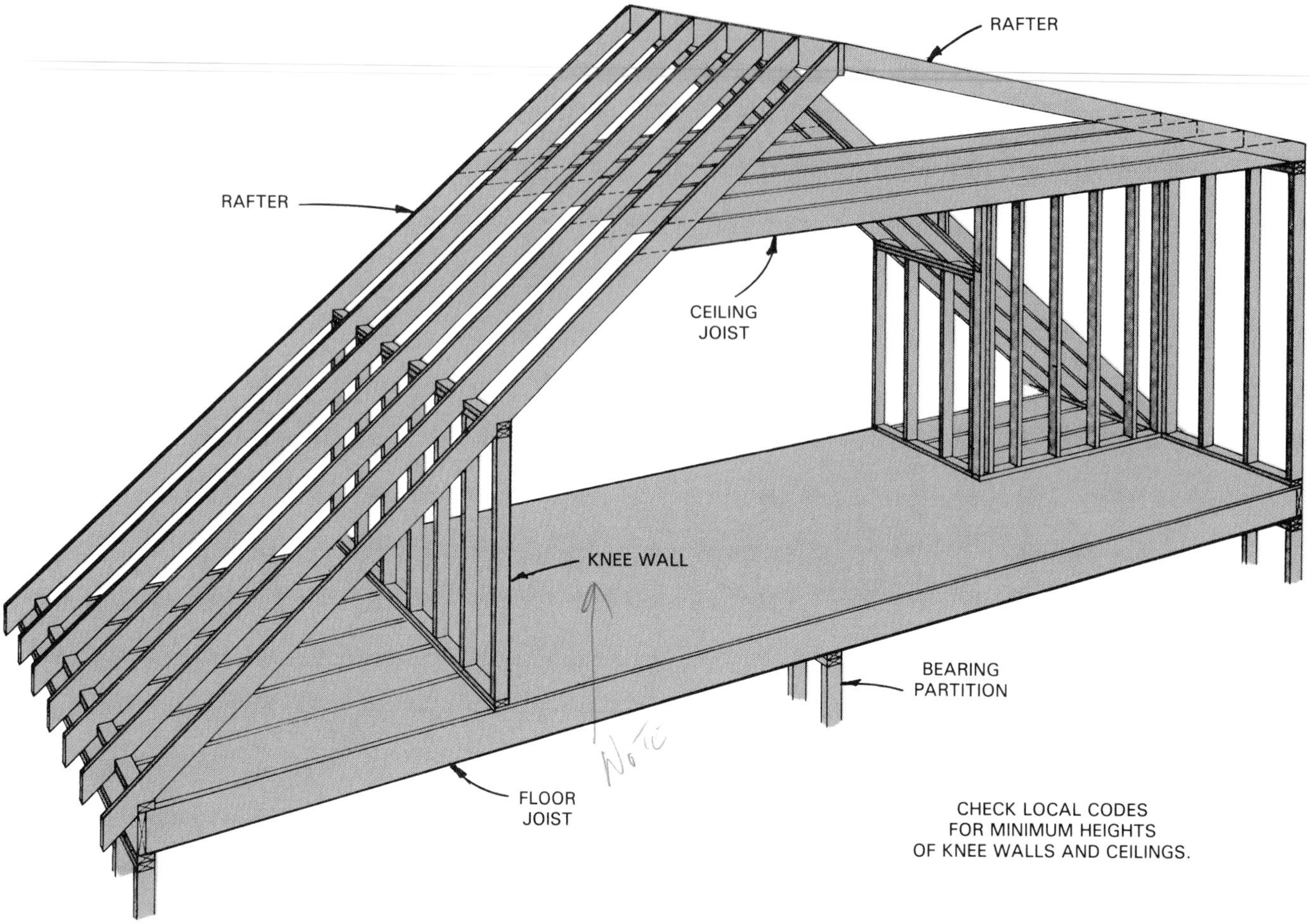

RAFTER

RAFTER

CEILING JOIST

KNEE WALL

BEARING PARTITION

FLOOR JOIST

CHECK LOCAL CODES FOR MINIMUM HEIGHTS OF KNEE WALLS AND CEILINGS.

Figure 10-58. Wall, ceiling, and roof framing for a 1 1/2-story structure. (National Forest Products Assoc.)

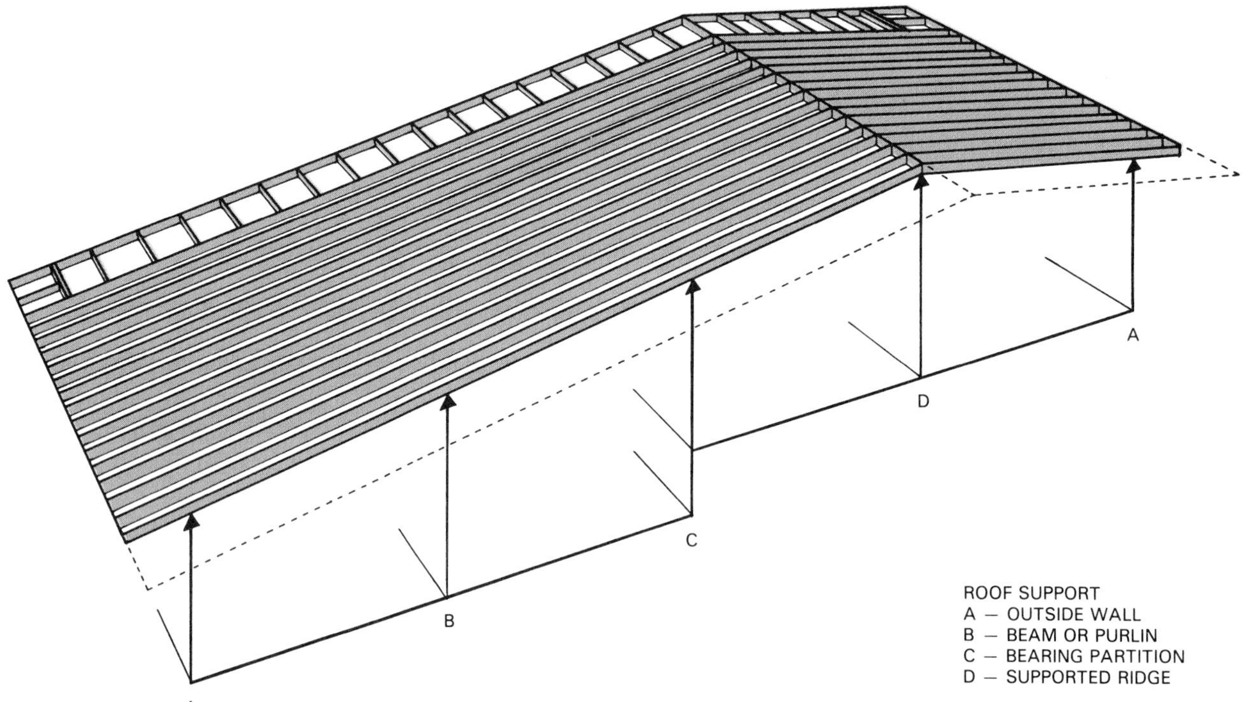

Figure 10-59. Modern roofs with low slope for a split-level design. Arrows indicate support points provided by outside walls, bearing partitions, and purlins or beams.

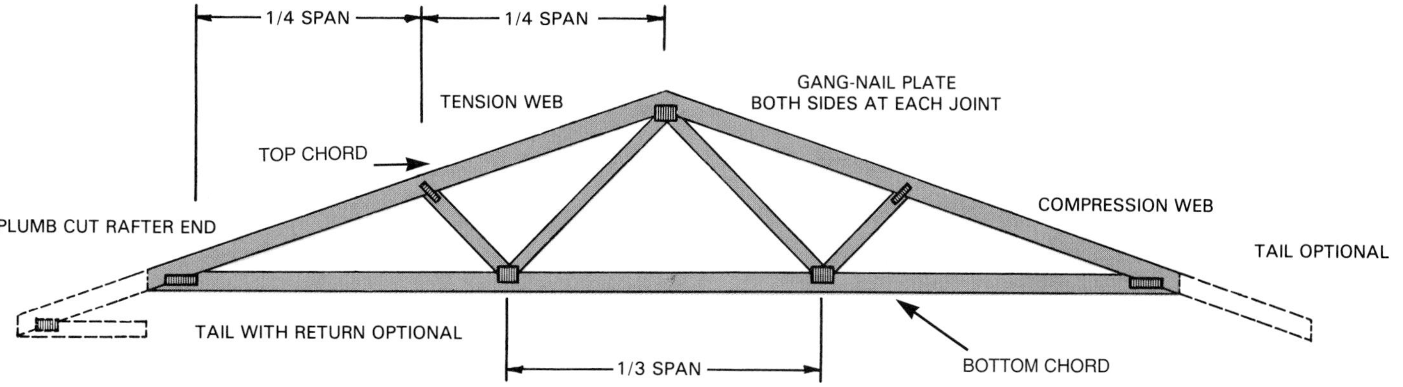

Figure 10-60. A standard W or Fink truss is commonly used in residential construction.

When the truss is in position and loaded, there will be a slight sag. To compensate for this, the lower member (called the bottom chord) is raised slightly during fabrication of the unit. This adjustment is called *camber* and is measured at the midpoint of the span. A standard truss, 24' long, will usually require about 1/2" of camber.

Figure 10-61. Most roof trusses are built in factories or assembly plants. This worker is assembling the chord that forms the ceiling support. (Trussworks, Inc., Hayward, WI)

In truss construction, it is essential that joint slippage be held to a minimum. Regular nailing patterns are usually not satisfactory. Special connectors must be used. Various kinds are available. All of them hold the joint securely and are easy to apply, Figure 10-62. Plywood gussets are applied with glue and nails to both sides of the joint.

When the number of trusses needed is small, they can be laid out and constructed on any clear floor area. First make a full-size layout on the floor, snapping chalk lines for long line lengths and using straightedges to draw shorter lines.

When building many trusses on site, it is worthwhile to use a portable truss assembly unit, Figure 10-63. Such equipment ensures accurate assemblies. Furthermore, it raises the assembly to a more comfortable working height.

Trusses for residential structures can normally be erected without special equipment. Each truss is simply placed upside down on the walls at the point of installation. Then the peak of the rafter is swung upward.

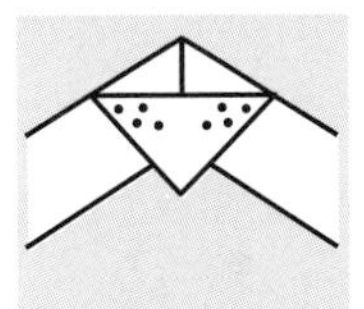

Use extra care when raising roof trusses. The first truss should be held with bracing or guy wires and all succeeding trusses carefully braced to prevent overturning.

Roof trusses up to certain sizes can be lifted and swung into place by hand. On multi-story dwellings or buildings with larger trusses, however, hand installation

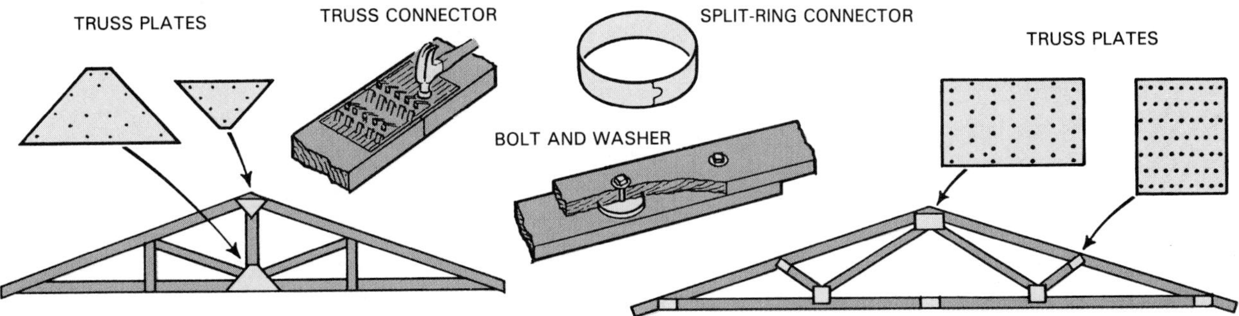

Figure 10-62. Plates and connectors for roof trusses. Truss plates are made in many sizes, shapes, and types. Some are perforated for nails; some require no nails. Split-ring connectors fit into recesses bored in mating joints.

Figure 10-63. Portable truss assembly unit can be easily adjusted to different sizes and designs.

may be impossible or impractical. In such cases a crane should be used. See Figure 10-64.

Roofs framed with trusses need not be limited to gable types. Today, a wide range of configurations can be produced. Designs are based on carefully prepared data covering load and lumber specifications. Computers apply this data to develop specific designs. Further efficiency results from the use of specialized methods, machines, and fasteners. Figure 10-65 shows a variety of roof trusses used on a hip roof.

Before installing roof trusses, always refer to the framing plans. The final erection and bracing of a roof system must be carried out according to plans and specifications if basic design requirements are to be met.

It is the responsibility of the installer to see to the proper storage of truss rafters delivered to the job site. Refer to Figure 10-66. Trusses should not be stored on rough terrain or uneven surfaces that could cause damage to the truss.

Bracing of Truss Rafters

Proper ground bracing and temporary bracing of the truss rafter during erection is vital. Be prepared to ground brace the first rafter erected using either method shown in Figure 10-67.

As additional truss rafters are placed, install lateral and diagonal bracing, as shown in Figure 10-68. This bracing can be removed as roof decking is installed. Additional

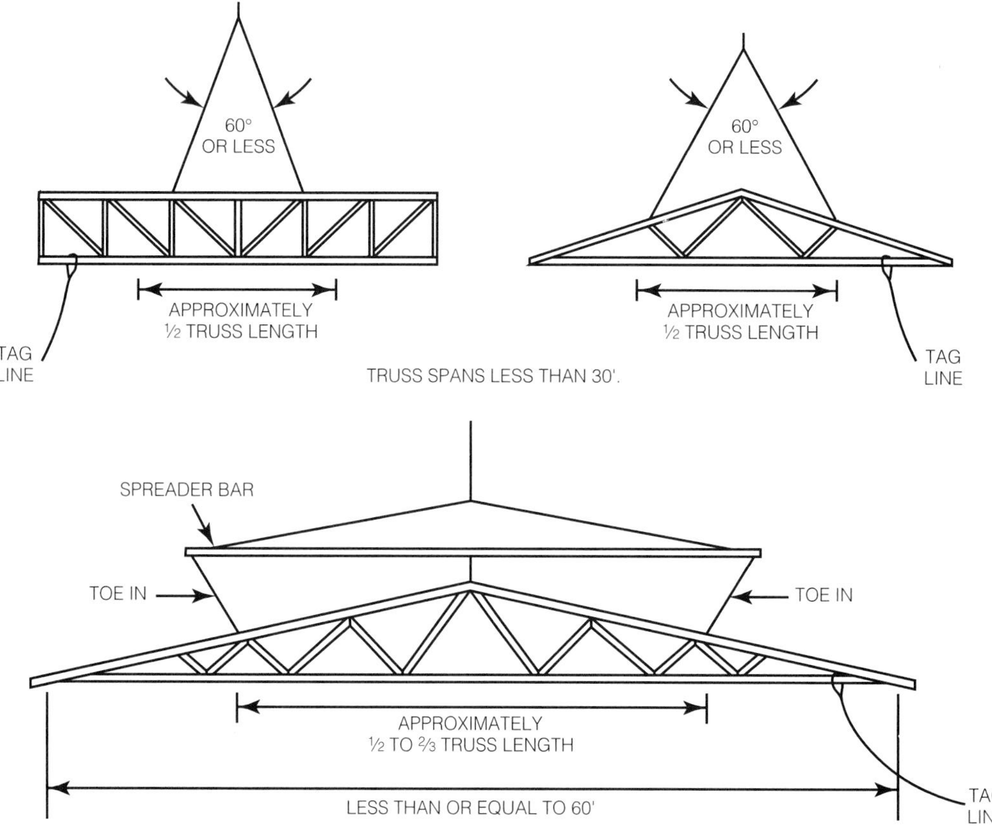

Figure 10-64. Roof trusses are engineered and built to exacting dimensions and specifications so they can be fitted with little or no adjustments. Top—A crane is often needed to lift trusses into position. Bottom—Properly attaching crane cables or hooks to truss rafters will avoid damage to the trusses. Always attach lifting cables at or near a joint. Tag lines should be attached for better control of the truss rafter. (Reproduced from HIB-91, Courtesy of Truss Plate Institute)

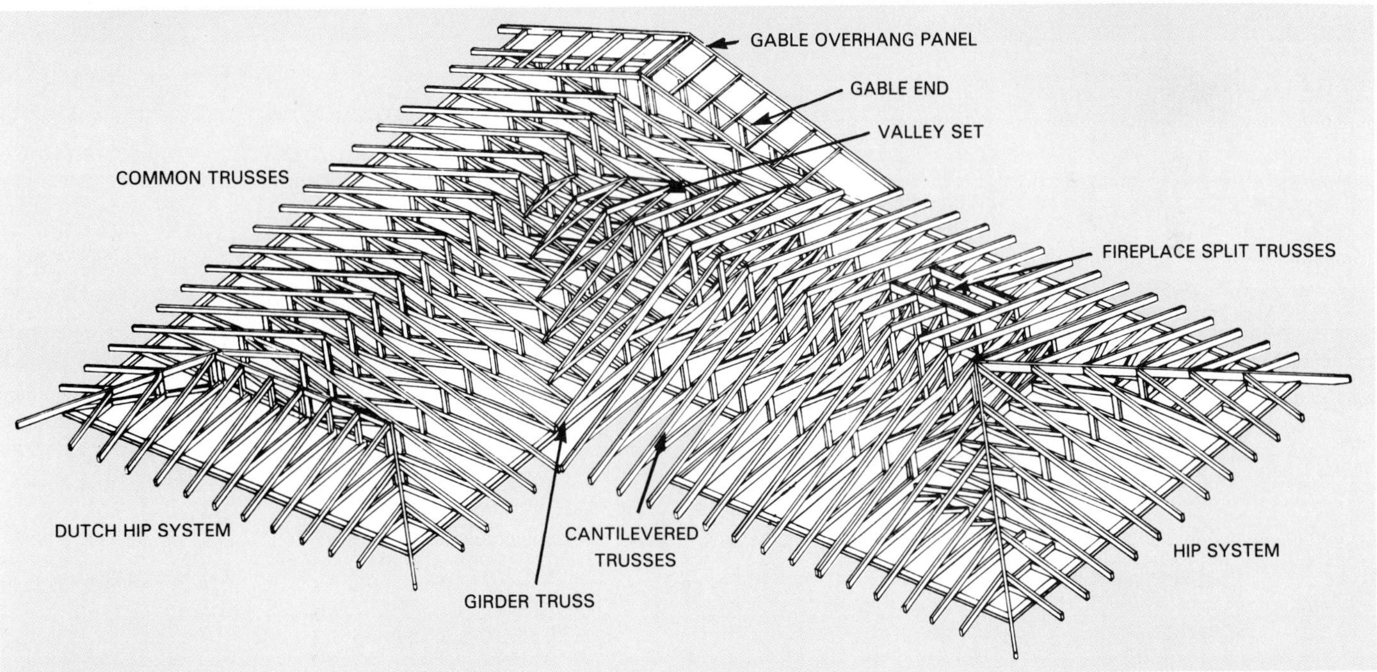

Figure 10-65. Top—Intersecting roof uses many different styles: regular hip, Dutch hip, and gable end. Bottom—Many different prefabricated truss units were used for the roof frame. (TrusWal Systems Corp.)

This safety alert symbol is used to attract your attention! PERSONAL SAFETY IS INVOLVED! When you see this symbol - BECOME ALERT - HEED ITS MESSAGE.

DANGER: A DANGER designates a condition where failure to follow instructions or heed warning will most likely result in serious personal injury or death or damage to structures.

CAUTION: A CAUTION identifies safe operating practices or indicates unsafe conditions that could result in personal injury or damage to structures.

WARNING: A WARNING describes a condition where failure to follow instructions could result in severe personal injury or damage to structures.

HIB-91 Summary Sheet

COMMENTARY and RECOMMENDATIONS for HANDLING, INSTALLING & BRACING METAL PLATE CONNECTED WOOD TRUSSES©

TRUSS PLATE INSTITUTE
583 D'Onofrio Dr., Suite 200
Madison, Wisconsin 53719
(608) 833-5900

It is the responsibility of the installer (builder, building contractor, licensed contractor, erector or erection contractor) to properly receive, unload, store, handle, install and brace metal plate connected wood trusses to protect life and property. The installer must exercise the same high degree of safety awareness as with any other structural material. TPI does not intend these recommendations to be interpreted as superior to the project Architect's or Engineer's design specification for handling, installing and bracing wood trusses for a particular roof or floor. These recommendations are based upon the collective experience of leading technical personnel in the wood truss industry, but must, due to the nature of responsibilities involved, be presented as a guide for the use of a qualified building designer or installer. Thus, the Truss Plate Institute, Inc. expressly disclaims any responsibility for damages arising from the use, application or reliance on the recommendations and information contained herein by building designers, installers, and others.

CAUTION: The builder, building contractor, licensed contractor, erector or erection contractor is advised to obtain and read the entire booklet "Commentary and Recommendations for Handling, Installing & Bracing Metal Plate Connected Wood Trusses, HIB-91" from the Truss Plate Institute.

CAUTION: All temporary bracing should be no less than 2x4 grade marked lumber. All connections should be made with minimum of 2-16d nails. All trusses assumed 2' on-center or less. All multi-ply trusses should be connected together in accordance with design drawings prior to installation.

TRUSS STORAGE

CAUTION: Trusses should not be unloaded on rough terrain or uneven surfaces which could cause damage to the truss.

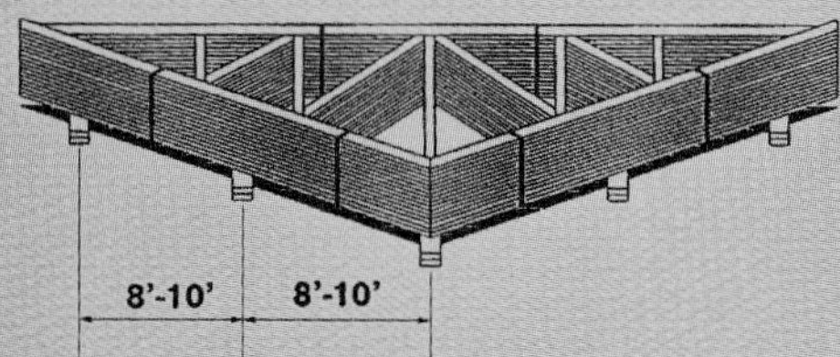

Trusses stored horizontally should be supported on blocking to prevent excessive lateral bending and lessen moisture gain.

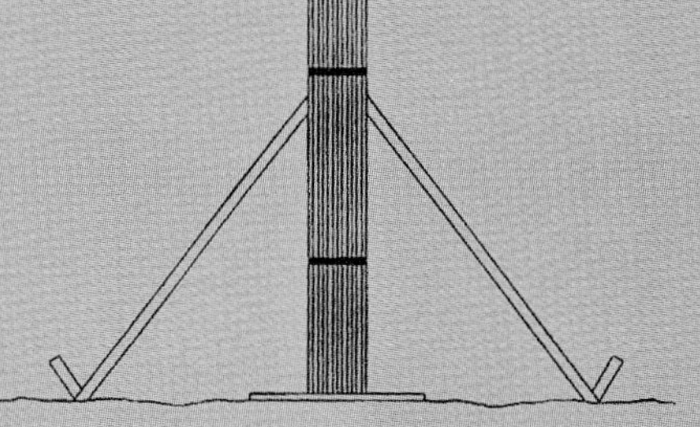

Trusses stored vertically should be braced to prevent toppling or tipping.

WARNING: Do not break banding until installation begins or lift bundled trusses by the bands.

DANGER: Do not store bundles upright unless properly braced.

WARNING: Do not use damaged trusses.

DANGER: Walking on trusses which are lying flat is extremely dangerous and should be strictly prohibited.

Frame 1

Figure 10-66. Follow these cautions for proper storage of trusses on the job site. (Reproduced from HIB-19, Courtesy of Truss Plate Institute)

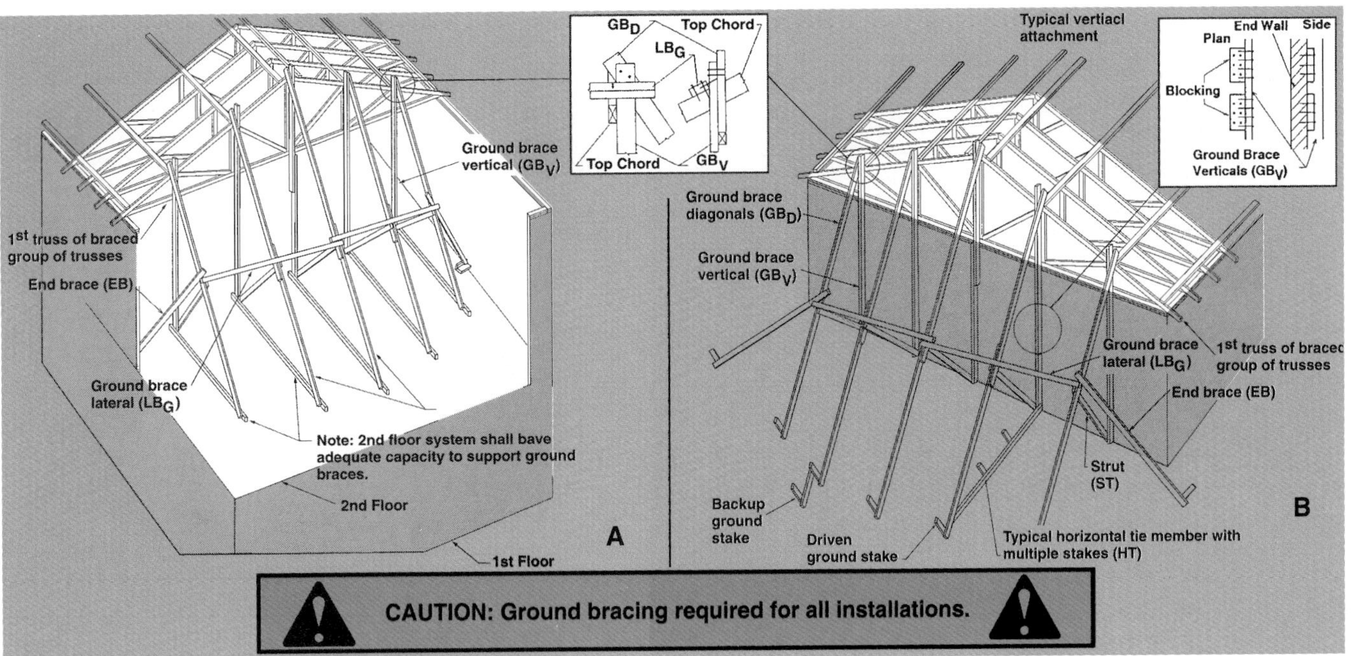

Figure 10-67. Temporarily ground bracing truss rafters. This bracing is adequate for installation of trusses having similar configurations. Consult a registered professional engineer if a different bracing arrangement is wanted. Bracing should be erected before raising the first truss. A—Ground bracing installed in the building's interior. B—Bracing installed outside the building. (Reproduced from HIB-91, Courtesy of Truss Plate Institute) C—When roof trusses are small, a vertical 2 x 4 secured to the wall below provides sufficient bracing.

inside permanent bracing is advisable. Remove the temporary ground bracing before sheathing the end of the roof. Observe the minimum pitch recommendations and bracing suggestions in the chart for Figure 10-68.

The theory of truss bracing is to apply sufficient support at right angles to the plane of the truss to hold each member in its correct position permanently. Figure 10-69 shows a method of lateral and diagonal bracing once the trusses are secured to the wall plate.

When the roof frame is complete, check it carefully to see that all members are secure and that nailing patterns are adequate, Figure 10-70.

Steel Roof Framing

Steel roof trusses are fabricated on-site using the same methods for determining length of rafter, plumb cuts, etc. Members are cut with a portable electric saw, Figure 10-71.

Next, the chords and webs are placed in a jig on the ground, Figure 10-72. The lapped and butt joints are fastened with self-tapping screws, a welding bead or both. Welds are made with arc welding or submerged arc welding equipment, Figure 10-73.

Erection of the steel trusses follows the same procedure as for wood trusses. After they are fastened to the wall plate, permanent bracing is welded to the webs. Figure 10-74 shows prefabricated trusses in place, fastened to the plate, and permanently braced.

Roof Sheathing

Sheathing provides a nailing base for the roof covering and adds strength and rigidity to the frame. Sheathing

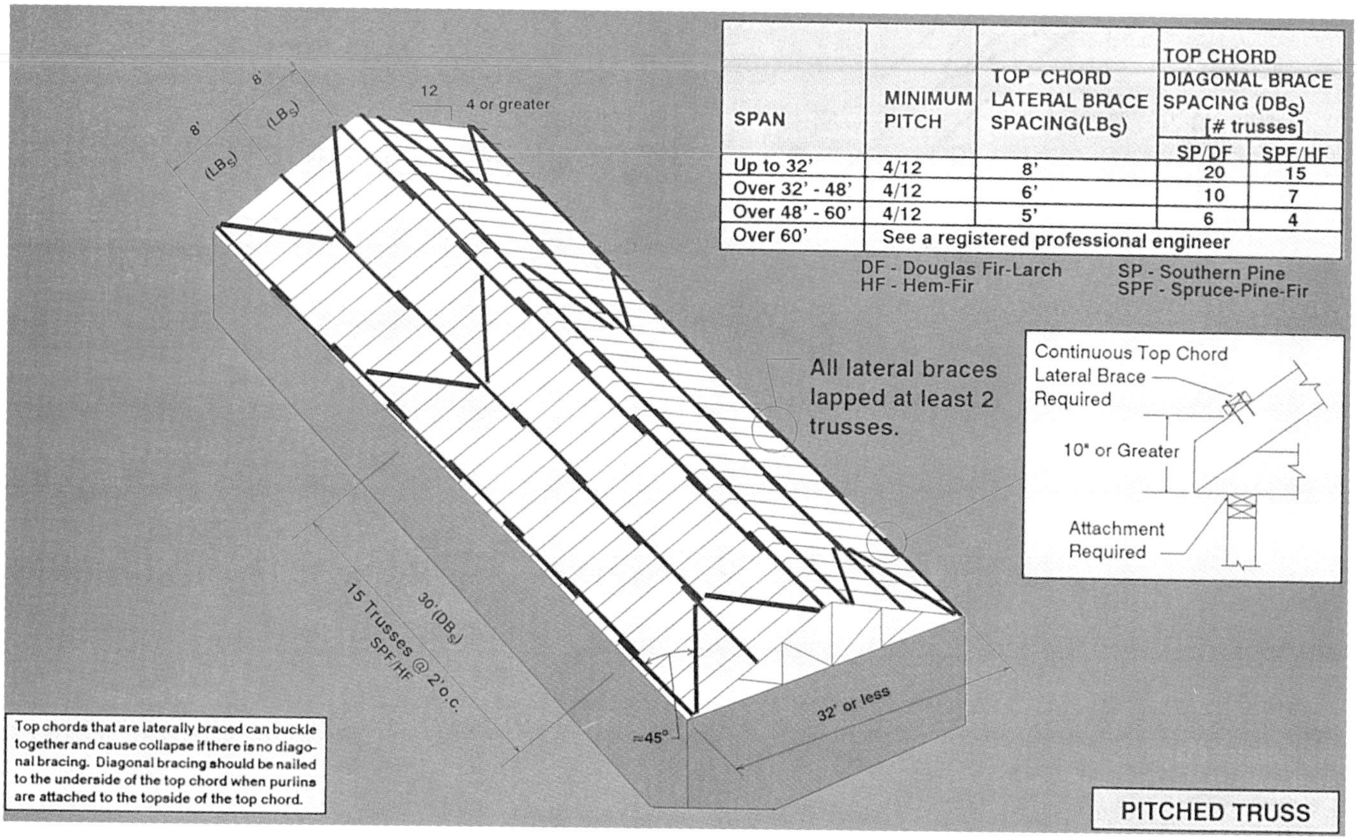
SPAN
MINIMUM PITCH
TOP CHORD LATERAL BRACE SPACING(LBS)
TOP CHORD DIAGONAL BRACE SPACING (DBS) [# trusses]
SP/DF
SPF/HF
Up to 32' 4/12 8' 20 15
Over 32' - 48' 4/12 6' 10 7
Over 48' - 60' 4/12 5' 6 4
Over 60' See a registered professional engineer
DF - Douglas Fir-Larch
HF - Hem-Fir
SP - Southern Pine
SPF - Spruce-Pine-Fir
12
4 or greater
8'
8'
(LBS)
(LBS)
All lateral braces lapped at least 2 trusses.
Continuous Top Chord Lateral Brace Required
10" or Greater
Attachment Required
30'(DBS)
15 Trusses @ 2'o.c. SPF/HF
32' or less
≈45°
Top chords that are laterally braced can buckle together and cause collapse if there is no diagonal bracing. Diagonal bracing should be nailed to the underside of the top chord when purlins are attached to the topside of the top chord.
PITCHED TRUSS

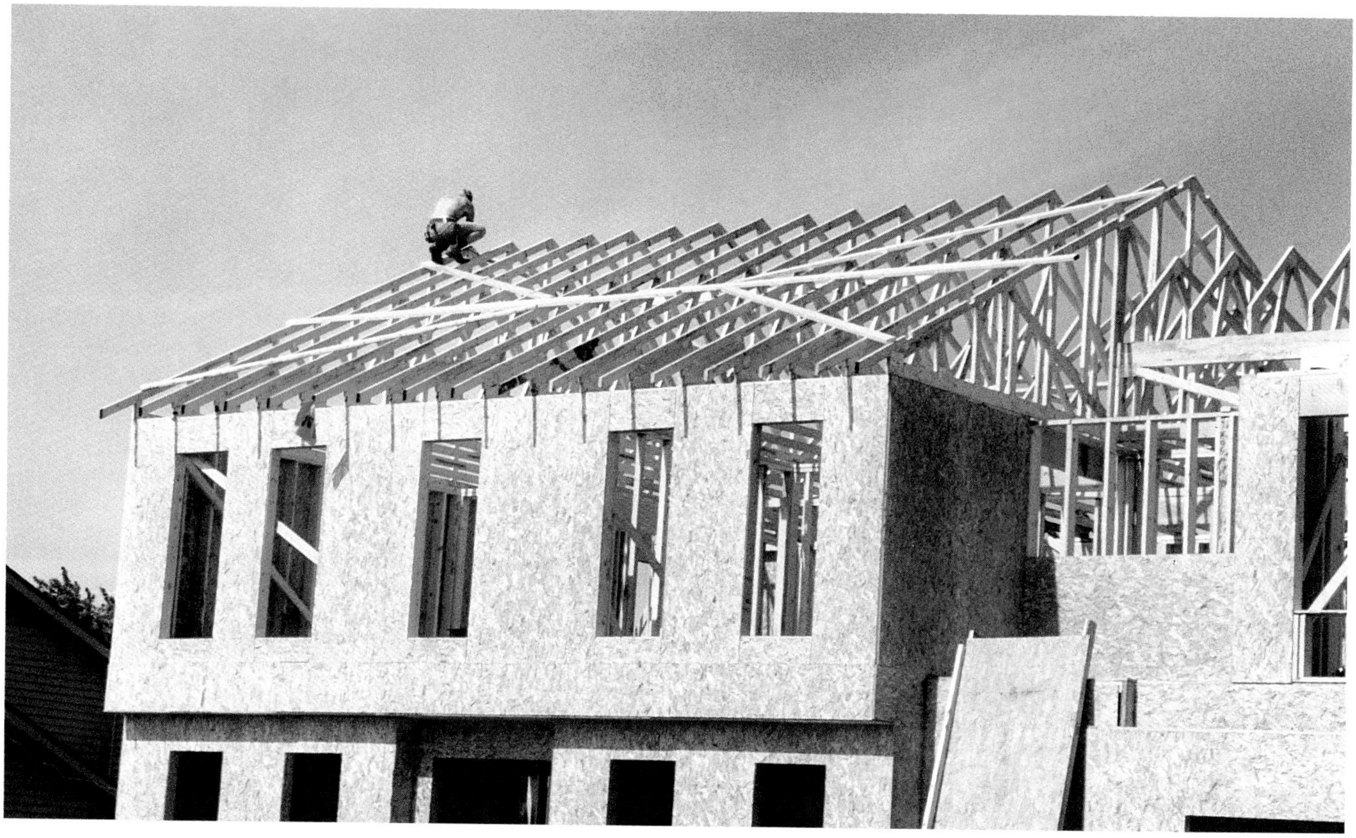

Figure 10-68. Top—Proper temporary bracing for roof trusses. Failure to follow these bracing recommendations can result in building damage as well as severe personal injury. (Reproduced from HIB-91, Courtesy of Truss Plate Institute)
Bottom—A properly braced truss roof.

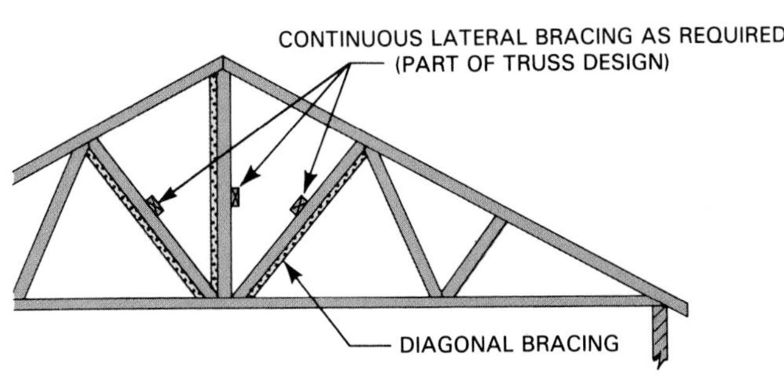

Figure 10-69. Truss rafters may require additional lateral and diagonal permanent bracing.

Figure 10-71. Cutting steel rafters to length using a portable power saw fitted with a carbide blade.

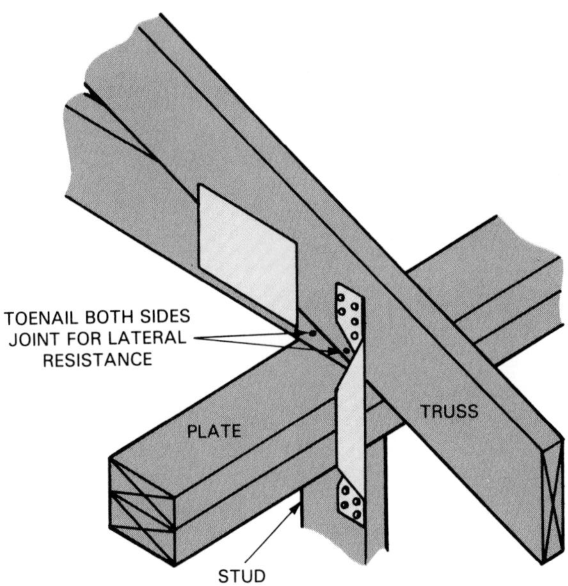

Figure 10-70. Trusses can be attached to plates by toenailing with 10d nails. Metal tie-downs, however, are sometimes used where toenailing would damage gussets or where high winds require stronger fasteners. (TECO)

Figure 10-72. Wooden jig holds steel truss chords and webs as carpenters secure joints with self-tapping screws.

materials include plywood composites, oriented strand board, waferboard, particleboard, shiplap, and common boards, Figure 10-75.

Other special materials are also available. For example, one product consists of panels formed with solid wood boards, bonded together with heavy paper.

Before starting the sheathing, erect the necessary scaffold that will make it easy and safe to install the boards or panels along the lower edge of the roof. This scaffold can also be used later to build the cornice work after the roofing has been completed.

Shiplap and common boards must be applied solid if asphalt shingles or other composition materials are used for the finished roof decking. For wood shingles, metal sheets, or tile, board sheathing may be spaced according to the course arrangement, Figure 10-76. They should be attached with two 8d nails at each rafter. Joints must be located over the center of the rafter. For greatest rigidity, use long boards, particularly at roof ends.

When end-matched boards are used, the joints may be made between rafters. Joints in the next board must not occur in the same rafter space. No boards should be used that are not long enough to be carried on at least two rafters.

Fit sheathing boards carefully at valleys and hips and nail them securely. This will ensure a solid, smooth base for the installation of flashing materials. Around chimney openings, the boards should have a 1/2" clearance from masonry. Framing members must have a 2" clearance. Always nail sheathing securely around roof openings.

Figure 10-73. Welding the metal trusses. This may be done while the truss is in the jig or after it is stacked preparatory to installation.

Figure 10-74. Top—Fastening a gable end stud to the top chord. Middle—Rafter installation is complete when bracing (lighter-colored pieces) has been installed. Bottom—Fascia board must be fastened to the tail of the rafter or truss before sheathing can begin. Be sure to check their length with a chalk line first. Cut off any tails that are too long.

Structural Panels

Structural panels are an ideal material for roof sheathing. They can be installed rapidly, hold nails well, resist swelling and shrinkage, and, because the panels are large, they add considerable rigidity to the roof frame. Plywood is laid with the face grain perpendicular to the rafters. End joints should be directly over the center of the rafter. Small pieces can be used but they should always cover at least two rafter spaces.

For wood or asphalt shingles with a rafter spacing of 16", 5/16" panels are usually recommended. For a 24" span, a 3/8" thickness should be used. Slate and tile shingles require 1/2" thicknesses for 16" rafter spacing and 5/8" for 24" spacing. Panels should be nailed to rafters with 6d nails, spaced 6" apart on edges and 12" elsewhere. If wood shingles are used and the sheathing is less than 1/2" thick, 1 x 2 nailing strips, spaced according to shingle exposure, should be nailed to the sheathing. For a flat deck under built-up roofing, use 1/2" thickness.

When handling large sheets, use extra precaution. They may slide off a roof if they are not properly secured. If sufficient help is available, sheets may be slid up a ladder as shown in Figure 10-77. A special rack, Figure 10-78, will be helpful in moving sheets from the ground to the roof. It will serve for storage until carpenters are ready to install them. On large construction jobs, a power panel elevator may be used to save time.

Installing Sheathing

Before applying sheathing (also called decking) to a roof, check for a level nailing surface. Use a long level or a piece of lumber 6 to 10' long. Shim low trusses or rafters as needed. Blocking can be used to straighten bowed or warped top chords of trusses or rafters. Install sheathing panels with screened surfaces or skid-resistant coatings up. To avoid bows in the sheathing, stand only where framing is supporting the panel as it is fastened.

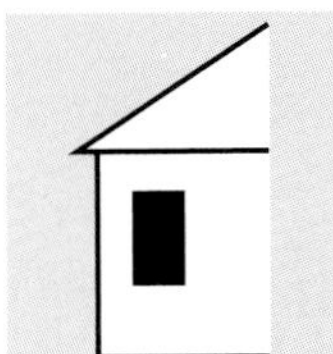

Wear skid-resistant shoes on pitched roofs.

Start sheathing at the eaves and work up toward the ridge. If necessary to square panels on the rafters, drive temporary fasteners at corners. Fasten one edge and then install intermediate fasteners, working inward from the edges. Lay down rows of panels from one edge to the other, maintaining the same fastening sequence. This procedure avoids internal stress in the panels. It is advisable to snap a chalk line to mark the center of rafters or trusses.

When using power nailers or staplers, stand on the panel over the framing to ensure contact with the framing as the fastener is driven. Fastener heads must be driven flush with the panel's surface. Maintain a 1/8" space between edges and ends of panels. A 10d nail can be used to gauge the spacing.

Figure 10-75. Sheathing materials are manufactured for rapid application. Top—Trimming the outer edge of roof sheathing (waferboard) after it has been nailed in place. Bottom—Plywood sheathing is being applied over stick-built rafters. To avoid waste at hip and valley intersections, some cutoffs can be turned over and used on adjoining slopes. (Georgia-Pacific Corp.)

Figure 10-76. Sheathing boards can be spaced when wood shingles, corrugated metal, or tile are the covering materials. (Paslode Co., Div. of Signode Co.)

Figure 10-77. On small jobs, a ladder can sometimes be used as a slide to move roof sheathing to another worker.

It is important to center each panel end on framing. Trim panel ends, if necessary.

When attaching sheathing to steel roof trusses, use self-tapping screws at the same intervals recommended for other fasteners. A screw gun is recommended. See Figure 10-79.

Panel Clips

A patented clip is manufactured to strengthen roof sheathing panels between rafters. See Figure 10-80. The clips are slipped onto the panels midway between the rafter or truss spans. Two clips should be used where supports are 48" O.C.

Clips are manufactured to fit five panel thicknesses: 3/8", 7/16", 1/2", 5/8", and 3/4". An average house requires 250 panel clips.

Estimating Materials

The number of rafters required for a plain gable roof is easy to figure. Simply multiply the length of the building by 3/4 for spacing 16" O.C. (3/5 for 20" and 1/2 for 24") and add one more. Double this figure for the other side of the roof. To determine the length of the rafter, use the 12th scale on the framing square as previously described in this unit. For example, estimate rafters for the following building:

Building Size 28' x 40'. Roof slope 4 to 12.
Overhang 2'-0". Rafter spacing 24" O.C.
Total Rafter Run = 16'-0"
Total Rafter Length = 16'-11"
Nearest Std. Length = 18'-0"
Number of pieces = 2 (Length of wall x 1/2 + 1)
= 2 (40 x 1/2 + 1)
= 2 x 21
= 42
Rafter Estimate: 42 pcs. 2 x 8 x 18'-0"

Figure 10-78. Rack will hold stack of roof sheathing where carpenter on roof can reach it.

Figure 10-79. A screw gun is used to attach sheathing to steel rafters with self-tapping screws.

Use special care in handling sheet materials on a roof, especially if there is a wind. You may be thrown off balance, or the sheet may be blown off the roof and strike someone.

When estimating a hip roof, it is not necessary to figure each jack rafter. The number of jack rafters required for one side of a hip is counted to obtain the number of pieces of common rafter stock. This will normally supply sufficient rafter material for the other side of the hip.

For a short method on a plain hip roof, proceed as if it were a gable roof. Add one extra common rafter for each hip. Also figure and add the hip rafters required.

With complicated roof frames, it is best to work from a complete framing plan. Apply the methods described for plain roofs to the various sections. Make colored check marks on the rafters as they are figured so you will not double up or skip members. In estimating material for the total roof frame, remember to include material for ridges, collar ties, and bracing.

To estimate the roof sheathing, first figure the total surface. Then apply the same procedures as used for subflooring and wall sheathing. Since the total area of the roof surface will also be needed to estimate shingles, building paper, or other roof surface materials, it is worth the extra time to figure the area accurately.

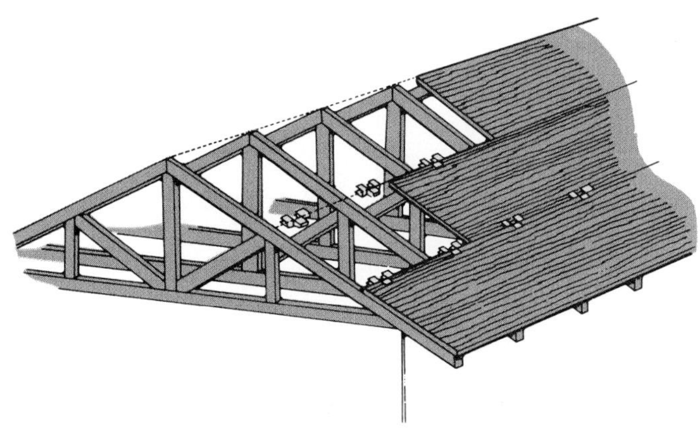

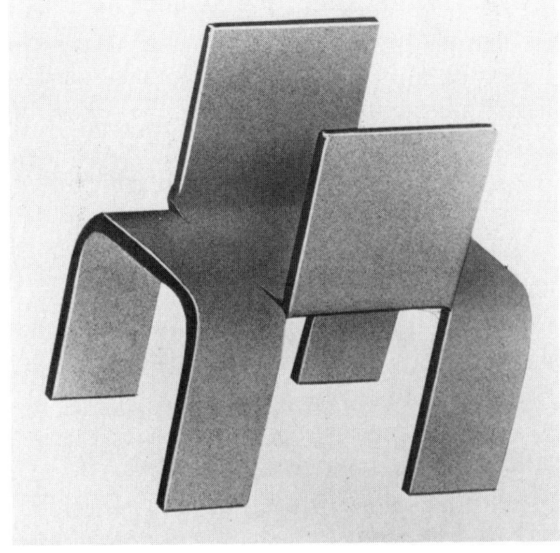

Figure 10-80. Panel clips, sometimes called "H" clips, eliminate blocking on long truss or rafter spans. (Panel Clip Co.)

To figure sheathing for a plain gable roof, multiply the length of the ridge by the length of a common rafter and double the amount. Figure a plain hip roof as though it were a gable roof. However, instead of multiplying the length of a common rafter by the ridge, multiply it by the length of the building plus twice the overhang.

When working with complicated plans and intersecting roof lines, first determine the main roof areas. Multiply common rafter length times the length of the ridge times 2. Now add the triangles that make up the other sections (located over jack rafters). Remember that the area of a triangle equals half the base times the altitude. The altitude of most triangular roof areas will be the length of a common rafter located in or near the perimeter of the triangle. A plan view of the roof lines will be helpful since all horizontal lines (roof edges and ridges) will be seen true length and can be scaled.

Always add an extra percentage for waste when estimating sheathing requirements for roofs that are broken up by an unusually large number of valleys and hips.

Model Construction

Students of carpentry can often get worthwhile experiences through construction of small portable buildings or scaled down models. Working with models requires much time. It is often best to construct only part of a building. See Figure 10-81. Small buildings, such as a play house or storage shed, can be sold to recover costs.

A scale of 1 1/2"=1'-0" will usually make it possible to apply regular framing procedures to the construction of a model that is not too large to handle and store. Cut framing members to their nominal size. For example, a 2 x 4 cut to this scale would actually measure 1/4" x 1/2" while a 2 x 10 would measure 1/4" x 1 1/4".

Make all framing materials from clear white pine or sugar pine. Both have sufficient strength and are easy to work. Use small brads and fast-setting glue for assembly.

Materials other than wood can often be simulated from a wide range of items. For example, foundation work can be built of rigid foamed plastic (Styrofoam) and then brushed with a creamy mixture of Portland cement and water.

Important Terms

Altitude
Back
Base
Bird's mouth
Blade
Body
Camber
Carpenter's square
Collar beams
Common difference
Common rafters
Cornice
Cripple jack rafters
Dormer
Eave
Extended rake
Face
Fascia
Flat roof
Framing square
Gable roof
Gambrel roof
Hip jack rafters
Hip rafters
Hip roof
Hip-valley cripple jacks
Hypotenuse
Lookouts
Mansard roof
Pitch
Purlin
Quick square
Rafters
Ridge
Rise
Roof trusses
Run
Shed roof
Slope
Steel square
Super square
Symmetrical
Tongue
Truss
Units
Valley cripple jack
Valley jack rafters
Valley rafters

Test Your Knowledge

1. A type of sloping roof that simplifies the construction of an overhang for all outside walls is called a ____________ roof.
2. The pitch of a roof is indicated by a fraction formed by placing the rise over the ____________.
3. The tongue of a framing square is ____________ inches long.

Figure 10-81. Models or small buildings can be constructed for carpentry experience. Top—Use a brad pusher to install small pieces. Middle—Model was based on a modular layout. Roof trusses can be constructed in a jig. Bottom—About 30 of these play houses were constructed at a Tucson, Arizona high school by carpentry students. (Santa Rita High School)

4. When laying out a rafter for a run that includes an odd unit, the ____________ (full unit, odd unit) is laid out first.
5. The bird's mouth is formed by a(n) ____________ cut and a plumb cut.
6. The final step in laying out a common rafter is to shorten it at the ____________.
7. When assembling a roof frame, joints in the ridge should occur at the ____________ of a rafter.
8. The part of a gable roof that extends beyond the end wall is called the ____________.
9. Referring to a chart, Figure 10-13, used with a super square, give the length of a common rafter and a hip rafter when the run is 14'.
10. Figures used to make side cuts for hip and valley rafters are found in the ____________ line of the rafter table on the framing square.
11. Hip jack rafters have the same tail and overhang as ____________ rafters.
12. Jack rafters should be erected in ____________ to keep the hip or valley rafters straight.
13. Horizontal ties between rafters on opposite sides of the ridge and usually located in the upper half of the frame are called ____________.
14. The two general types of dormers are ____________ and gable.
15. In residential construction, the gambrel roof is usually framed with a ____________ located where the two surfaces of different slopes are joined together.
16. The adjustment in the lower chord of a roof truss to compensate for sag is called ____________.
17. The sheathing on a roof frame provides a nailing base for shingles and also adds ____________.
18. End joints in the sheathing boards can be made between rafters when ____________ lumber is used.
19. The thickness of plywood required for a sheathing application will vary for different roofing materials and different ____________.
20. To calculate the area of a plain gable roof, multiply the length of the ridge by the length of a(n)________ and then double the product.
21. Explain how to fabricate steel roof trusses on-site.
22. How is sheathing fastened to steel roof trusses?

Outside Assignments

1. Obtain a set of house plans where the design includes a hip roof and/or intersecting sections. It should not have a roof framing plan. Study the elevations and detail sections. Then prepare a roof framing plan. Overlay the floor plan with a sheet of tracing paper. Trace the walls and draw all roof

framing members to accurate scale. Be sure to include openings for chimneys and other items that would be helpful to the carpenter.

2. Prepare an estimate of the framing materials required for the roof used in Activity No. 1. Include the dimensions for all lumber needed. Refer to the detail drawings or specifications to find lumber size requirements. If this information is not included in the plans, find it in the local building code.
3. Working from a set of architectural plans for a residential structure, make a layout for a common rafter in one of the roof sections. Use a good straight piece of stock. If dimension lumber is not available, a piece of 1" material may be used. Make the layout by the step-off method and cover all operations, including the shortening at the ridge. When completed, put on a brief demonstration for the class, showing them the procedure you followed.
4. Study the various types of roof trusses. Learn their names and the basic design patterns. List the advantages and disadvantages of each. Find out where they are most commonly used. Prepare a display board with line drawings of at least eight types and label each.

Closing In

LIFETILE
ARCO

Roofing systems, like many exterior finish item, are not directly required to build a structure. However, they protect the structural elements from the surrounding elements, extending the building's life.

Unlike framing and foundations, which are normally hidden from view, exterior finish items must be visually attractive. This doorway allows entrance into the home and is also pleasing to the eye.

11

Roofing Materials and Methods

Roofing materials protect the structure and its contents from the sun, rain, snow, wind, and dust. In addition to weather protection, a roof should offer some measure of fire resistance and be extremely durable. Due to the large amount of surface that is usually visible, especially on sloping roofs, the materials can contribute to the attractiveness of the building. Roofing materials can add color, texture, and pattern, Figure 11-1.

Figure 11-1. Proper sequence of operations must be followed in installation of wood shingles so that they will shed water. Note that roofing felt, commonly called tar paper, is laid down before each course. (American Plywood Assoc.)

Roof construction consists of a number of operations. Most must follow a definite sequence.

All items that will project through the roof should be built or installed before roofing begins. These structures include chimneys, vent pipes, and special facilities for electrical and communications service. Performing any of this work after the finished roof is applied may damage the roof covering.

Types of Material

Materials used for pitched (sloping) roofs include:

- Asphalt, wood, metal, and mineral fiber shingles.
- Slate and clay or cement tile.
- Sheet materials such as rubberized, single-ply membrane, roll roofing, galvanized iron, aluminum, tin, and copper.

For flat roofs and low-sloped roofs, a membrane system is used. It consists of a continuous watertight surface, usually obtained through built-up roofs or seamed metal sheets.

Built-up roofs are fabricated on the job. Roofing felts are laminated (stuck together) with asphalt or coal tar pitch. Then this surface is coated with crushed stone or gravel.

Metal roofs of this type are assembled from flat sheets. Seams are soldered or sealed with special compounds to ensure watertightness.

When selecting roofing materials it is important to consider such factors as the following:

- Initial cost
- Maintenance costs
- Durability
- Appearance

The pitch of the roof limits the selection. Low-sloped roofs require a more watertight system than steep roofs, Figure 11-2. Materials such as tile and slate require heavier roof frames.

Local building codes may prohibit the use of certain materials because of the fire hazard or because they will not resist the high winds or other elements found in a certain locality.

Roofing Terms

Slope and pitch have already been defined in Unit 10. Several other terms commonly used include the following:

- *Square*—The amount of roofing material needed to provide 100 sq. ft. of finished roof surface. It is the unit of measurement for estimating and purchasing roofing materials.

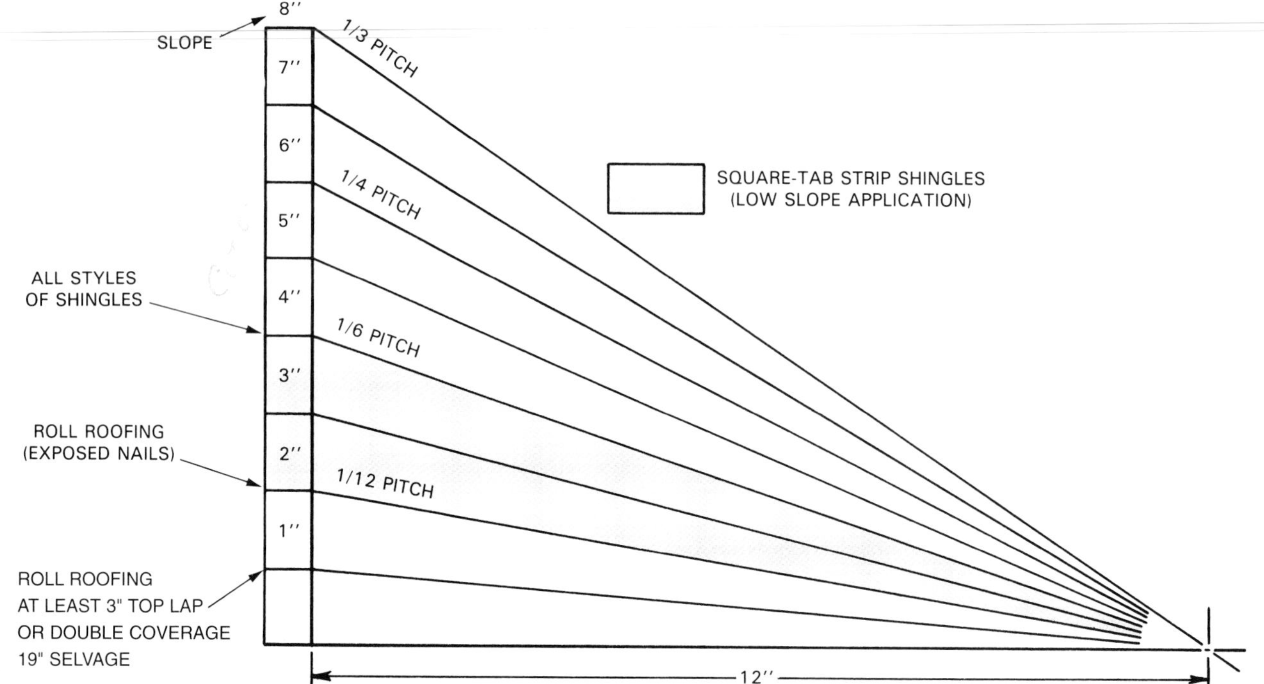

Figure 11-2. Roofing manufacturers list minimum pitch and slope requirements for various asphalt roofing products. (Asphalt Roofing Manufacturers Assoc.)

- *Coverage*—The amount of weather protection provided by the overlapping of shingles. Depending on material and the method of application, shingles may furnish one (single coverage), two (double coverage), or even three (triple coverage) thicknesses of material on the roof.
- *Exposure*—The distance (in inches) between the edge of one course and the edge of the next higher course.
- *Head Lap*—The distance (in inches) from the lower edge of an overlapping shingle or sheet, to the top edge of the shingle or sheet beneath, Figure 11-3.
- *Side Lap*—The overlap length (in inches) for side-by-side elements of roofing. See Figure 11-3.
- *Shingle Butt*—The lower, exposed edge of a shingle.
- *Rake*—The inclined edge of a gable roof.

Preparing the Roof Deck

The roof sheathing should be smooth and securely attached to the frame. It must provide an adequate base to receive and hold the roofing nails and fasteners.

All types of shingles can be applied over solid sheathing. Spaced sheathing (also called skip sheathing) is sometimes used for wood shingles, as it provides ventilation, promoting faster drying after a rain. Solid boards over 6" wide should not be used as sheathing.

Attics should be properly ventilated to remove moisture. Moisture vapor originating in lower stories may sometimes enter the attic. If the vapor becomes chilled below the dew point, it will condense on the underside of the roof deck. This causes sheathing to warp and buckle.

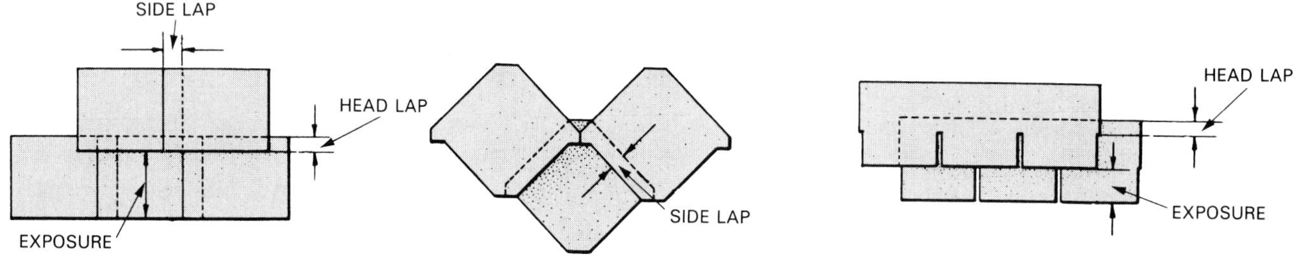

Figure 11-3. Illustrating terms used in application of roofing materials.

To avoid this, louvered openings should be installed in a location that provides adequate ventilation, such as under the eaves in the gable end.

Louvers should provide 1/2 sq. in. of opening per square foot of attic space.

Inspect the roof deck to see that nailing patterns are complete and that there are no nails sticking up. Joints should be smooth and free of sharp edges that might cut through roofing materials. Repair large knot holes over 1" diameter by covering with a piece of sheet metal. Clean the roof surface of chips or other scrap material.

Asphalt Roofing Products

Asphalt roofing products are widely used in modern construction. They include three broad groups: saturated felts, roll roofing, and shingles.

Saturated felts are used under shingles for sheathing paper and for built-up roof laminations. They are made of dry felt soaked with asphalt or coal tar.

Saturated felt is made in different weights, the most common being 15 lb. This number indicates the weight of enough felt to cover 100 sq. ft. of the roof deck with a single layer.

Roll roofing and shingles are outer roof covering. They must be weather-resistant. Their base material is organic felt or fiberglass. This base is saturated and then coated with a special asphalt that resists weathering. A surface of ceramic coated mineral granules is then applied. The mineral granules shield the asphalt coating from the sun's rays, add color, and provide fire resistance.

Data on shingles and other groupings of asphalt products are given in Figure 11-4. Additional products within each group differ in weight and size. For example, the three-tab square butt shingle is available in many qualities and colors and in weights from 215 to 245 lb. per square.

Asphalt shingles are the most common type of roofing material used today. They are manufactured as strip shingles, interlocking shingles, and large individual shingles. Dimensions of a standard three-tab strip shingle are shown in Figure 11-5.

Many of the shingles are available with a strip of factory-applied, self-sealing adhesive. Heat from the sun will soften the adhesive and bond each shingle tab securely to the shingle below. This bond prevents tabs from being raised by heavy winds.

The self-sealing action usually takes place within a few days during warm weather. In winter, the sealing time is considerably longer, depending on the climate.

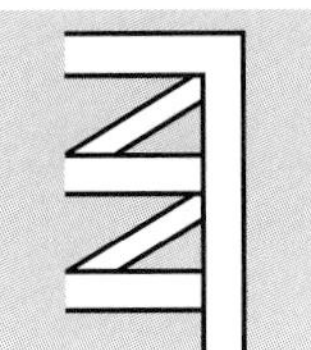

Safety considerations are very important in roofing work. Be sure to erect a secure scaffold that will support the worker at a waist-high level with the eaves. Study Unit 29 for more information and directions.

Underlayment

An *underlayment* is a thin cover of asphalt-saturated felt or other material. It has a low vapor resistance. This underlayment serves the following purposes:

- Protects the sheathing from moisture until the shingles are laid.
- Provides additional weather protection by preventing the entrance of wind-driven rain and snow.
- Prevents direct contact between shingles and resinous areas in the sheathing.

Materials such as coated sheets or heavy felts, which might act as a vapor barrier, should not be used. They allow moisture and frost to gather between the covering and the roof deck. Although 15 lb. roofer's felt is commonly used for this purpose, requirements will vary depending on the kind of shingles and the roof slope.

Do not put down underlayment on a damp roof. Moisture may be trapped and damage the roof.

General application standards for underlayment suggest a 2" top lap at all horizontal joints and a 4" side lap at all end joints. See Figure 11-6. It should be lapped at least 6" on each side of the centerline of hips and valleys.

Drip Edge

The roof edges along the eaves and rake should have a metal *drip edge*. Various shapes, formed from 26 gage galvanized steel, are available. They extend back about 3" from the roof edge and are bent downward over the edge. This causes the water to drip free of underlying cornice construction.

At the eaves, the underlayment should be laid over the drip edge. At the rake, place the underlayment under the drip edge.

Barrier at Eaves

An ice and water barrier is recommended at the eaves in cold climates. It should cover the underlayment and drip edge from the roof's edge to about 24" inside the wall line. Installed on new construction or during reroofing, it prevents leak-through from ice dams or wind blown rain. Available in 36" wide rolls, this material is self-sealing. Some types have adhesive backing and are reinforced with fiberglass. Non-adhering types may require a coat of hot tar on the sheathing to hold it in place. Refer again to Figure 11-6.

This flashing will prevent leaks from water backed up by ice dams on the roof. The lower edge of this strip should be placed even with the drip edge.

Flashing

The installation of roofing materials is complicated by the intersection of other roofs, adjoining walls, and such projections as chimneys and soil stacks. Making these

Table I: Typical Asphalt Shingles

PRODUCT	Configuration	Per Square			Size		Exposure	Underwriters Laboratories Listing
		Approximate Shipping Weight	Shingles	Bundles	Width	Length		
Self-sealing random-tab strip shingle 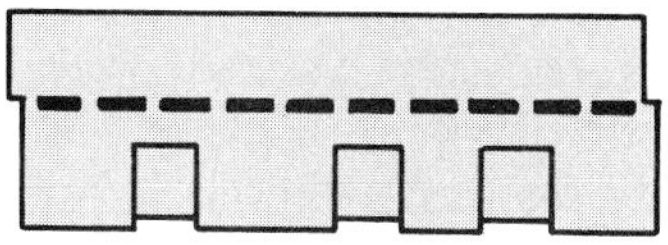 Laminates	Various edge, surface texture and application treatments	285# to 390#	66 to 90	4 or 5	11½" to 14"	36" to 40"	4" to 6"	A or C - Many wind resistant
Self-sealing random-tab strip shingle 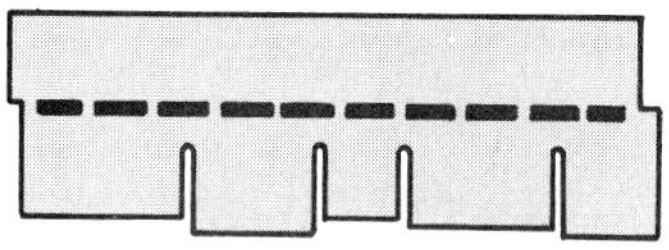 Single-thickness	Various edge, surface texture and application treatments	250# to 300#	66 to 80	3 or 4	12" to 13¼"	36" to 40"	5" to 5⅝"	A or C - Many wind resistant
Self-sealing square-tab strip shingle 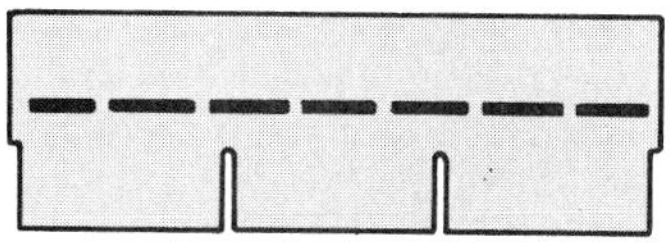 Three-tab	Two-tab or Four-tab	215# to 325#	66 to 80	3 or 4	12" to 13¼"	36" to 40"	5" to 5⅝"	A or C - All wind resistant
	Three-tab	215# to 300#	66 to 80	3 or 4	12" to 13¼"	36" to 40"	5" to 5⅝"	
Self-sealing square-tab strip shingle 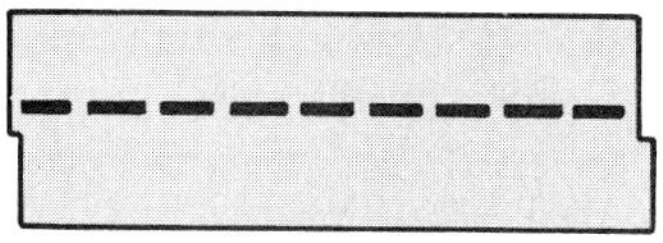 No-cutout	Various edge and surface texture treatments	215# to 290#	66 to 81	3 or 4	12" to 13¼"	36" to 40"	5" to 5⅝"	A or C - All wind resistant
Individual interlocking shingle 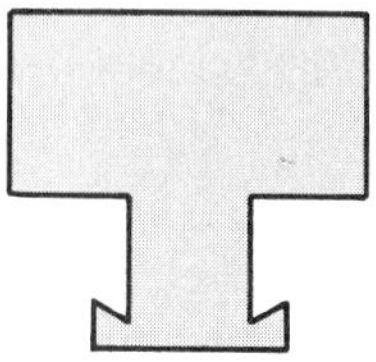 Basic design	Several design variations	180# to 250#	72 to 120	3 or 4	18" to 22¼"	20" to 22½"	—	C - Many wind resistant

Figure 11-4. Follow this chart of specifications for installation data on common asphalt roofing products. (Asphalt Roofing Manufacturers Assoc.)

Table II: Typical Asphalt Rolls

PRODUCT	Approximate Shipping Weight		Squares Per Package	Length	Width	Side or End Lap	Top Lap	Exposure	Underwriters Laboratories Listing*
	Per Roll	Per Square							
Mineral surface roll	75# to 90#	75# to 90#	1	36' to 38'	36"	6"	2" to 4"	32" to 34"	C
		Available in some areas in 9/10 or 3/4 square rolls.							
Mineral surface roll (double coverage)	55# to 70#	110 # to 140 #	½	36'	36"	6"	19"	17"	C
Smooth surface roll	40# to 65#	40# to 65#	1	36'	36"	6"	2"	34"	None
Saturated felt (non-perforated)	60#	15# to 30#	2 to 4	72' to 144'	36"	4" to 6"	2" to 19"	17" to 34"	None

*UL rating at time of publication. Reference should be made to individual manufacturer's product at time of purchase.

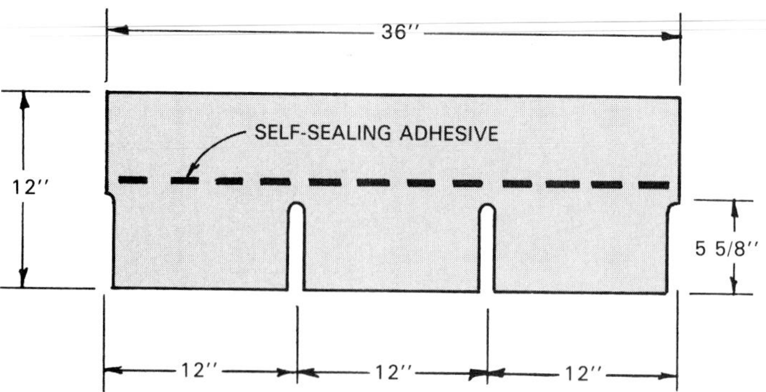

Figure 11-5. Standard three-tab asphalt strip shingle. These are the most common dimensions. Metric sizes measure 336.5 mm by 984 mm (13 1/4" by 38 3/4").

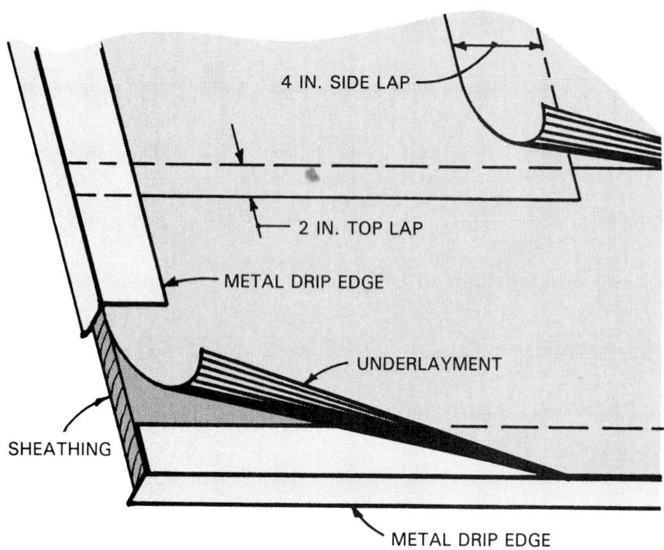

Figure 11-6. Top—Drawing shows proper application of underlayment and metal drip edge in warm climates. Bottom—In cold climates, an ice and water barrier should be laid down and extended 36" up the slope. This material has an adhesive backing. Some have embedments to produce a nonslip surface.

areas watertight requires a special building material called *flashing*. Flashing is water-resistant sheet material designed to keep joints in the roof water tight. Materials used for flashing include: tin-coated metal, galvanized metal, copper, lead, aluminum, asphalt shingles and roll roofing. Some roofers use an ice and water barrier at roof joints.

Installing Open Valley Flashing

A valley is the surface where two sloping roofs meet. Water drainage is heavy at this point. Flashing is one method of sealing valleys against water. For asphalt shingles, 90 lb. mineral-surfaced asphalt roll roofing is recommended. Install it as shown in Figure 11-7.

The first strip, at least 18" wide, is centered in the valley and laid with the mineral surface down. After nailing this strip, a second strip 36" wide is cemented or nailed in place, mineral side up.

Joints are lapped at least 12". If strips are nailed, the lap is sealed with plastic asphalt cement. As each strip is laid, first nail one edge, pressing the material firmly into the valley before nailing the opposite edge.

Before applying the shingles, snap a center chalk line in the valley and one on each side. The outside lines will mark the width of the waterway. This should be 6" wide at the ridge and widen gradually. The lines should move away from the valley at the rate of 1/8" for every foot as they

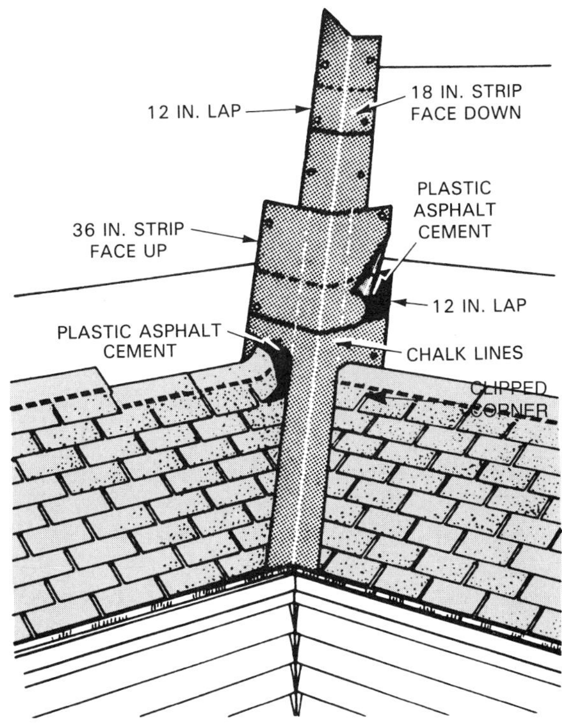

Figure 11-7. Open valley flashing is laid down before shingling begins. Bottom is trimmed to match eave line. Both layers should be 90 lb. mineral surfaced roll roofing. Shingles should be sealed with plastic asphalt cement where they lap the flashing.

approach the eave. (A valley 8' long would be 7" wide at the eave.)

When a course of shingles meets the valley, the chalk line serves as a guide in trimming the last unit. After the shingle is trimmed, cut off the upper corner at about a 45° angle with the valley line. Cement the end of the shingle over the flashing.

Installing Woven Valley Flashing

Some roofers prefer to run shingles across valleys, creating a *woven valley*. It is often used when reroofing. Only asphalt strip shingles may be applied this way.

The strip must be wide enough to straddle the valley with a minimum 12" of material on either side. In order to provide this margin, some of the preceding shingle strips must be cut away. Fasteners must be at least 6" away from the valley centerline. See Figure 11-8.

Installing a Woven Valley

1. Apply a 36" wide strip of 50 lb. (min.) roll roofing across the valley.
2. Lay the first course of shingles along the eave of one roof surface.
3. Extend one strip at least 12" across the valley.
4. Lay the first course on the intersecting roof and extend it across the valley over the previously applied shingle.
5. Succeeding courses are alternated, first along one roof surface and then the other, Figure 11-8.

When laying shingles across the valley, press them firmly into place. Nail at least 6" away from either side of the centerline. Use two nails at the end of each terminal strip.

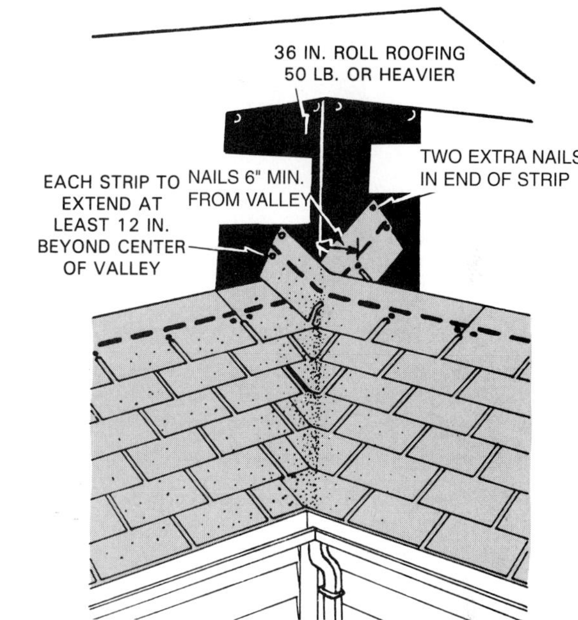

Figure 11-8. Top—Example of woven valley shingling. Bottom—Method of laying woven valley shingles. This method is common during reroofing, but may be used in new construction as well.

Installing Closed-Cut Valley Flashing

Another method of flashing a valley is the *closed-cut valley* method. In this method, the two intersecting roof surfaces are shingled individually.

Installing a Closed-Cut Valley

1. Install the roll roofing and then apply all the shingles on one roof surface. Carry each course across the valley and onto the adjoining roof at least 12".
2. When the first roof surface is complete, apply the first course of shingles along the eaves of the intersecting roof.
3. Where this course meets the valley, trim the shingle along a line 2" back from the centerline of the valley.
4. Trim off the upper corner of the shingle to prevent water from running back along the top edge.
5. Embed the end of the shingle in a 3" wide strip of plastic asphalt cement. Succeeding courses are applied and completed as shown in Figure 11-9.

When reroofing, it is necessary to build up the trough in an open valley to the average level of the roof surface. This can usually be done with strips of beveled wood.

Flashing at a Wall

Where the roof joins a vertical wall, it is best to install metal flashing shingles. They should be 10" long and 2" wider than the exposed face of the regular shingles. The 10" length is bent so that it will extend 5" over the roof and 5" up the wall as shown in Figure 11-10.

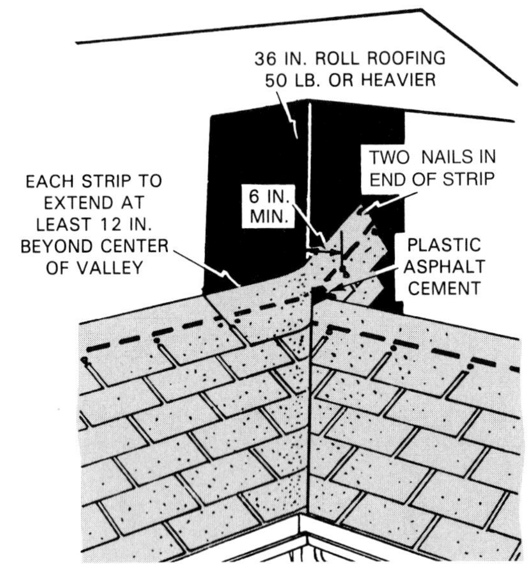

Figure 11-9. Top—View of actual closed-cut valley construction. Bottom—Details of construction. Shingles on the right are cut along a line 2" back from the valley centerline. See arrow on top illustration.

As each course of shingles is laid, a metal flashing shingle is installed and nailed at the top edge as shown. Do not nail flashing to the wall as settling of the roof frame could damage the seal.

Wall siding is installed after the roof is completed and serves as cap flashing, Figure 11-11.

Position the siding just above the roof surface. Allow enough clearance to paint the lower edges.

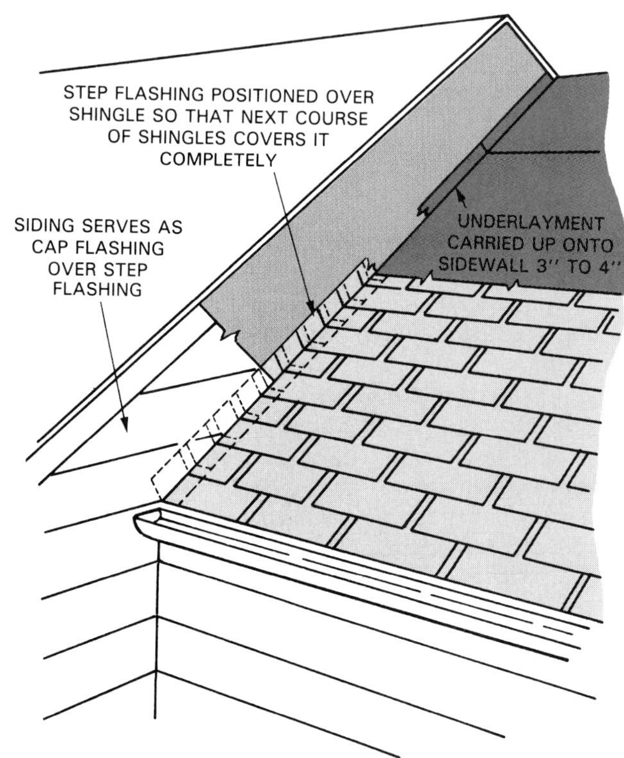

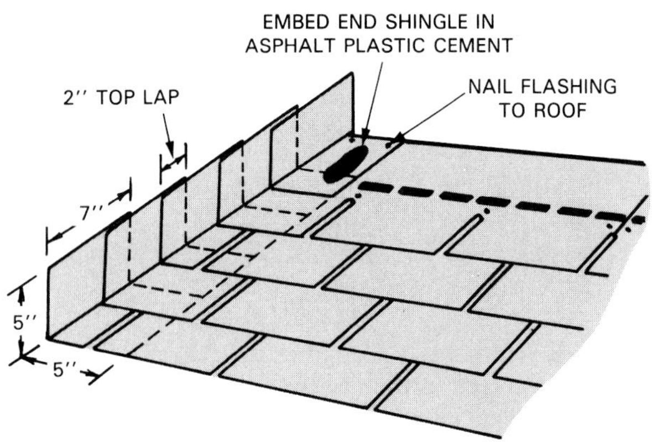

Figure 11-10. Apply metal flashing shingles with each course. Follow these lap and bending directions.

Figure 11-11. Flashing at a wall. Top—Metal flashing shingles have been applied here to waterproof joint between sloping roof and wall. This is generally called step flashing. Bottom—Note use of flashing around the gables and walls.

Chimney Flashing

Flashing around a masonry chimney must allow for some movement caused by settling or shrinkage of the building framework. To provide for this movement, the flashing is divided into two parts:

- The *base flashing,* which is attached to the roof.
- The *cap flashing* (also called *counter flashing*), which is attached to the chimney.

Figure 11-12 (top) shows how mineral-surfaced roofing can be used to form the base flashing. First, cement the front unit into place and then attach the side pieces. The flashing on the high (back) side is installed last. All sections are cemented together as they are applied.

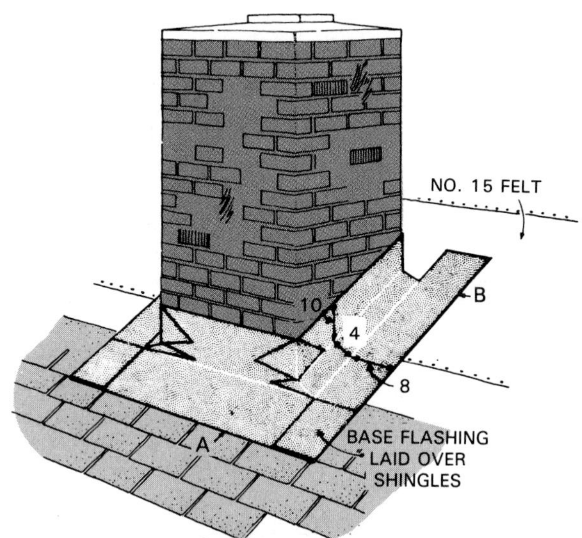

Figure 11-12. Base flashing seals joint between chimney and roof. Top—Mineral-surfaced roofing has been cut and cemented into place. Bottom—Metal flashing is installed around a housing for a prefabricated fireplace chimney. Both installations are ready for cap flashing, which covers the base flashing, sealing out moisture.

Housings for prefabricated chimneys require flashing similar to masonry chimneys, Figure 11-12 (bottom). Some prefabricated chimney units have flashing flanges that simplify their installation. See Unit 21.

Sheet metal is often used for base flashing. It should be applied by the step method previously described in the section on wall flashing.

Cap flashing is sheet metal shaped to cover the top of the base flashing. When used around a masonry chimney, it is set into the mortar joints and bent down over the base flashing. The metal is set into the joints 1 1/2". Cap flashing on the front of the chimney may be one continuous piece; sides must be stepped in sections because of the roof slope, Figure 11-13.

Chimney Saddle

Large chimneys on sloping roofs generally require an auxiliary roof deck on the high side. This structure is called a *saddle* or a *cricket.* It diverts water, preventing ice and snow buildup behind the chimney, which can cause leaks.

Figure 11-14 shows a chimney saddle and suggests a framing design. The frame is nailed to the roof deck and then sheathed. A small saddle could be constructed from triangular pieces of 3/4" exterior plywood.

Saddles are usually covered with corrosion resistant sheet metal. However, mineral-surfaced roll roofing could

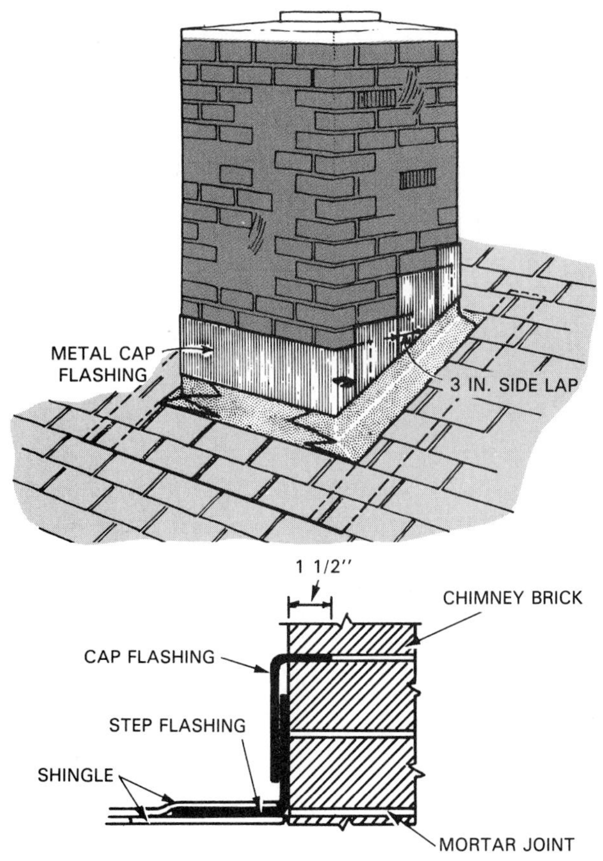

Figure 11-13. Metal cap flashing is set into the mortar joints as the chimney is built. It must go over the top of the base flashing as shown in the cross section.

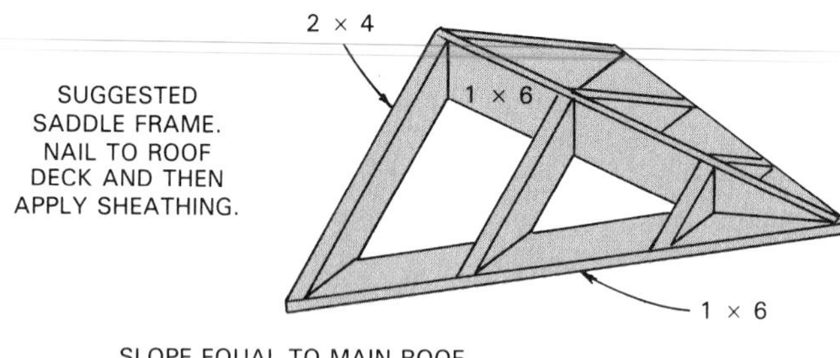

Figure 11-14. Chimney saddle. Small saddles need not be framed. They can be formed from triangular pieces of 3/4" exterior plywood.

be used. Valleys formed by the saddle and main roof should be carefully flashed in the same way as regular roof valleys.

Vent Stack and Skylight Flashing

Pipes and skylights in the roof must also be carefully flashed. Asphalt products can be used successfully for the flashing.

The roofing must be laid up to the stack. Cut and fit the shingles around the stack, Figure 11-15(A). Then carefully cement a flange in place and lay shingles over the top. The flange must be large enough to extend at least 4" below, 8" above, and 6" on each side. Figure 11-15(B) shows the installation of a manufactured flange.

Usually, skylights will have a one-piece flashing that must be fitted under shingles at the top and sides and over shingles at the bottom. A wood curb is sometimes used to raise the skylight above the roof. The flashing procedure is very similar to flashing a chimney.

Strip Shingles

On small roofs, strip shingles may be laid starting at either end. When the roof surface is over 30' long, it is usually best to start at the center and work both ways. Start from a chalk line (running at right angles to ridge) from eaves to ridge.

Asphalt shingles will vary slightly in length (plus or minus 1/4" in a 36" strip). There may be some variations in width. Thus, to achieve the proper placement so shingles will be accurately aligned horizontally and vertically, chalk lines should be used.

When laying shingles from the center of the roof toward the ends, snap a number of chalk lines between the eaves and ridge. They will serve as reference marks for starting each course. Space them according to the type of shingle and laying pattern.

These lines are used in the same way that the rake edge of the roof is used when the application is started at the roof end. The shingles do not need to be cut. Instead, full shingles are aligned with the chalk lines to form the desired pattern.

Chalk lines parallel to the ridge will help maintain straight horizontal lines along the butt edge of the shingle. Usually, every fifth or sixth course should be checked if the shingles are skillfully applied. Inexperienced workers may need to set up chalk lines for every second or third course.

When roofing materials are delivered to the building site, they should be handled with care and protected from damage. Try to avoid handling asphalt shingles in extreme heat or cold.

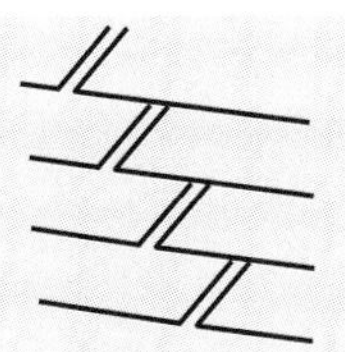

To get the best performance from any roofing material, always study the manufacturer's directions. Make the installation as directed.

Fastening Shingles

Nails used to apply asphalt roofing must have large heads (3/8" to 7/16" diameter) and sharp points. Figure 11-16 shows standard nail designs and suggests lengths for nominal 1" sheathing. Most manufacturers recommend 12 gage galvanized steel nails with barbed shanks. Aluminum nails are also used. The length should be sufficient to penetrate nearly the full thickness of the sheathing or 3/4" through wood boards.

The number of nails and correct placement are both vital factors in proper application of a roofing material. For three-tab square-butt shingles, use a minimum of four nails per strip as shown in the application diagrams. Align each shingle carefully and start the nailing from the end next to the one previously laid. Proceed across the shingle. This will prevent buckling. Drive nails straight so the edge of the head will not cut into the shingle. The nail head should

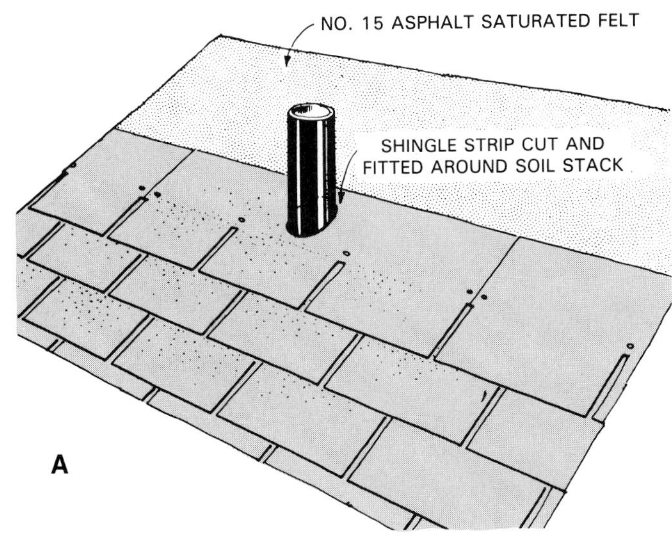

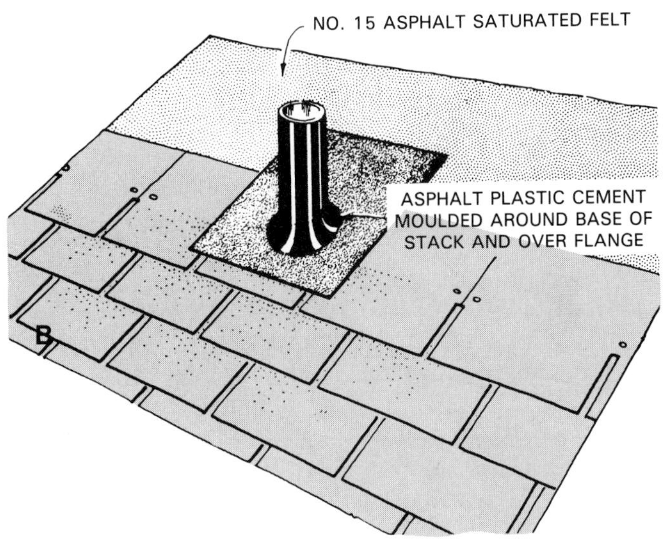

Figure 11-15. Flashing stacks and skylights. A—Lay shingles up to the stack and fit last course around it. B—Install flange and apply shingles over the upper side of the flange. C—This manufactured metal flange has a neoprene gasket. D—Shingling has been completed around this skylight and stack.

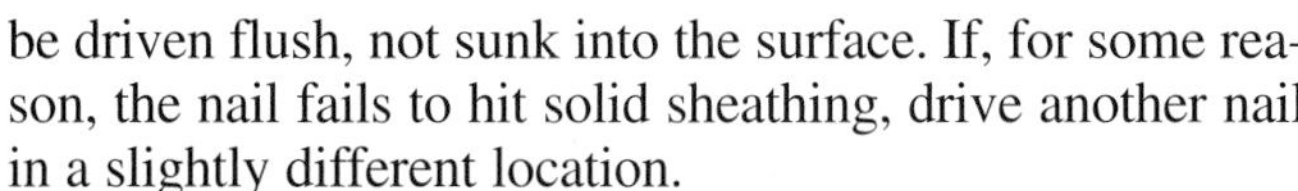
be driven flush, not sunk into the surface. If, for some reason, the nail fails to hit solid sheathing, drive another nail in a slightly different location.

Pneumatic powered staplers are often used to install asphalt shingles, Figure 11-17. Special staples with an extra wide crown should be used.

Always follow manufacturer's recommendations for staples and special power nailing equipment. In general, 16 gage staples with a minimum length of 3/4" should be used to attach asphalt strip shingles to new construction. See Figure 11-18 for cross sections of well set and poorly set staples.

Staple gun pressure can be adjusted for proper staple application. If a staple must be removed from a shingle, repair the hole with asphalt plastic cement according to the manufacturer's directions. Generally speaking, staples should be placed 5/8" below adhesive strips, never on the adhesive itself. Follow staple recommendations in Figure 11-19.

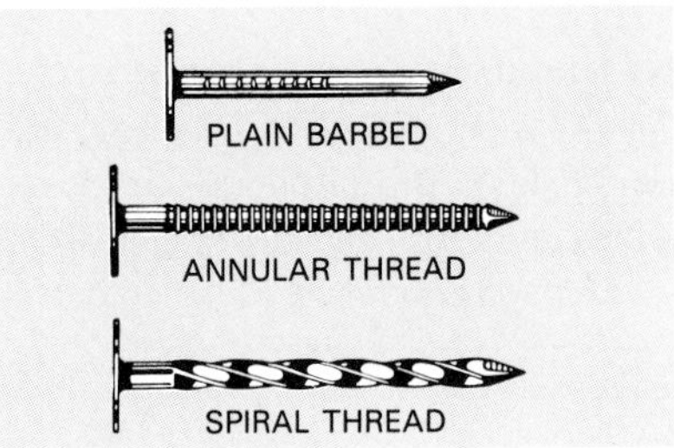

NAILING APPLICATION	1 IN. SHEATHING	3/8 IN. PLYWOOD OR WAFER BOARD
STRIP OR SINGLE (NEW CONTRUCTION)	1 1/4"	7/8"
OVER OLD ASPHALT LAYER	1 1/2"	1"
REROOFING OVER WOOD SHINGLES	1 3/4"	—

Figure 11-16. Nails suited for installing asphalt shingles. They must be long enough to penetrate roofing materials and the decking without going entirely through the decking. Note recommended lengths for different decking and for reroofing.

Figure 11-17. Pneumatic tools are being used to install asphalt shingles. Top—Proper method of handling stapler. (Paslode Co., Div. of Signode Corp.) Bottom—Staple or nail is placed just above the cutout for tabs. In windy areas a fastener is placed above and on either side of the cutouts. (Senco Fastening Systems.)

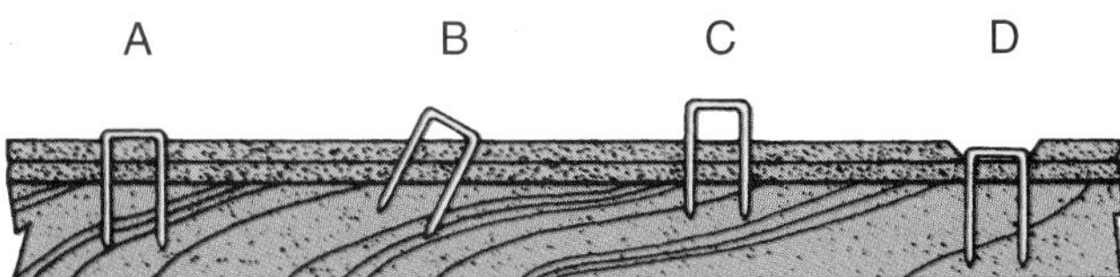

Figure 11-18. Proper stapling of shingles. A—Crown should be parallel to and tight against shingle surface without cutting into it. B, C—Set deeper if you can see daylight under crown. D—If set too deep, staple will cut the shingle. (Senco Products, Inc.)

THICKNESS OF WOOD DECK	MINIMUM STAPLE LEG LENGTH
3/8"	7/8"
1/2"	1"
5/8" and thicker	1 1/4" 1 1/2"

Figure 11-19. Recommendations for staple length when they are applied to new construction. These specifications apply in most parts of the country. However, always check local building codes.

Starter Strip

The purpose of a *starter strip* is to back up the first course of shingles and fill in the space between the tabs. Use a strip of mineral surfaced roofing, 9" or wider, of a weight and color to match the shingles. Apply the strip so it overhangs the drip edge slightly. Secure it with nails spaced 3 to 4" above the edge. Space the nails so they will not be exposed at the cutouts between the tabs of the first course of shingles. Sometimes an inverted (upside down) row of shingles is used instead of the starter strip.

For self-sealing strip shingles, the starter strip is often formed by cutting off the tabs of the shingles being used. These units are then nailed in place, right side up, and provide adhesive under the tabs of the first course.

First and Succeeding Courses

The first course is started with a full shingle. Succeeding courses are then started with either full or cut strips, depending upon the type of shingle and the laying pattern.

Three-tab square-butt shingle strips are commonly laid so the cutouts are centered over the tab in the course directly below, thus the cutouts in every other course will be exactly aligned.

For this pattern, start the second course with a strip from which 6" has been cut. The third course is started with a strip with a full tab removed and the fourth with half a strip. Continue as shown in Figure 11-20.

Reduce the length of shingle in the same sequence for subsequent courses. A pair of tin snips can be used to cut the shingles.

The diagram in Figure 11-21 shows a pattern in which the cutouts break joints on thirds. Called the *four inch method,* this pattern starts the second course with a strip shortened by 4" and the third by 8". The fourth course starts with a full strip.

Using an approved nailing pattern for three-tab shingles is very important in securing the best appearance and full weather protection. Manufacturers recommend that four nails be used as shown in Figure 11-22. When shingles are applied with an exposure of 5", nails should be placed 5/8" above tops of cutouts. Locate one nail above each cutout and one nail in 1" from each end. Nails should not be placed in or above the factory applied adhesive strip.

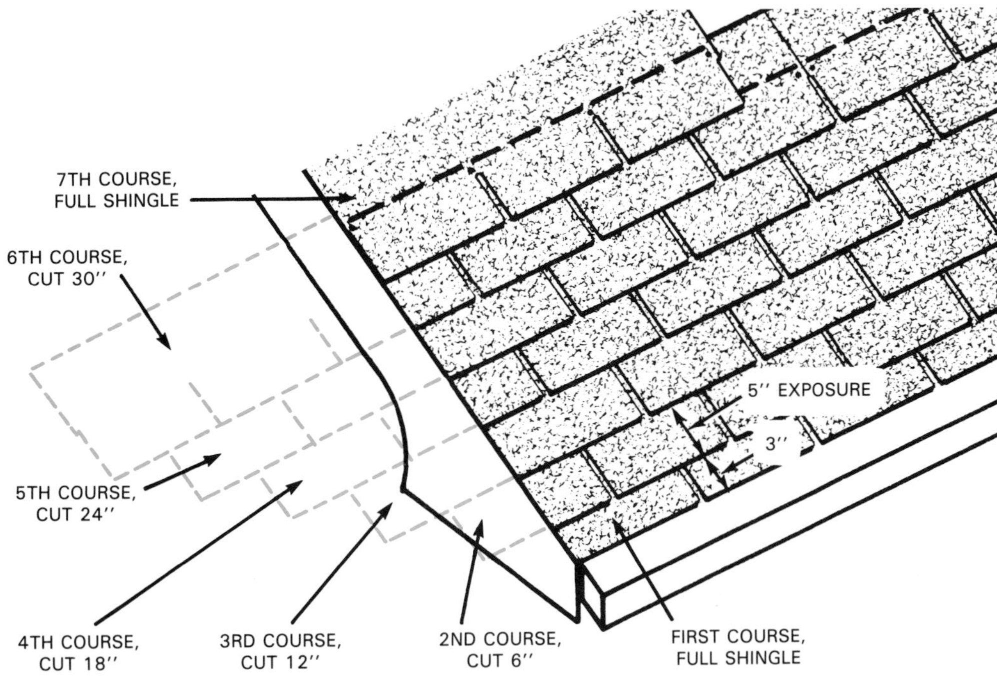

Figure 11-20. These three-tab square-butt shingles are laid so the cutouts are centered over the tabs in the course directly below. This is called the six inch method. (Manville Bldg. Materials Corp.)

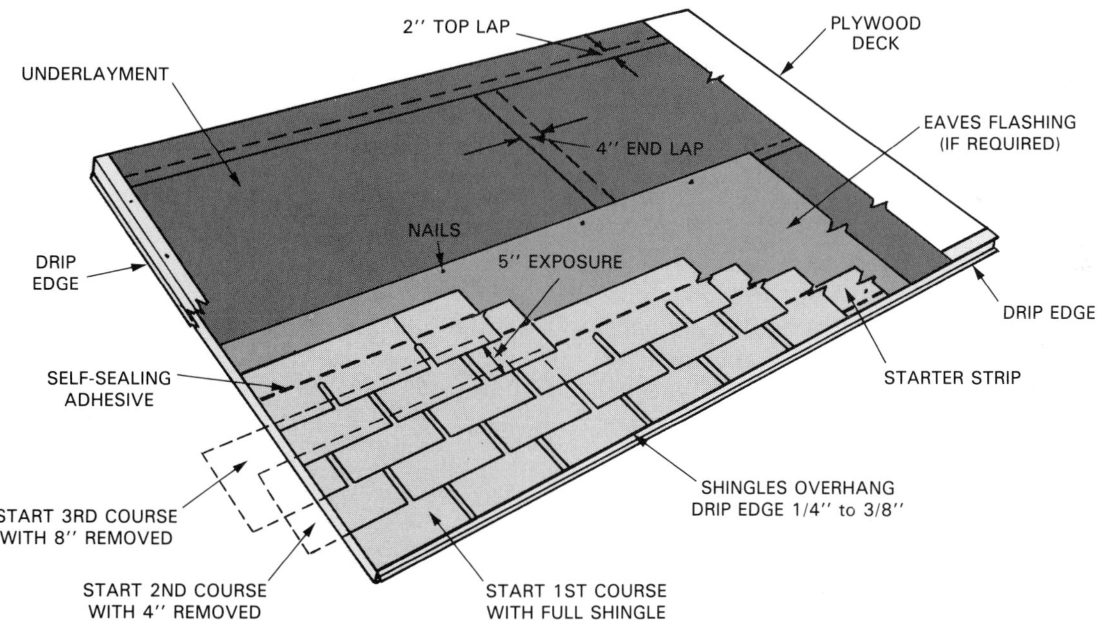

Figure 11-21. In this application, cutouts break joints on thirds. It is also called the four inch method. Eave flashing is usually a self-sealing rubberlike sheet of roll material about 36" wide. Usually there is an adhesive backing. (Bird and Son Inc.)

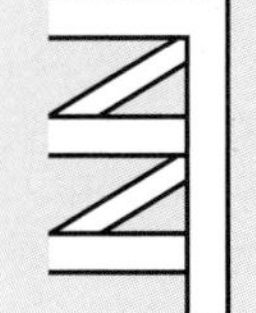

When laying asphalt shingles, it is a good idea to wear soft-soled shoes that will not damage the surface and edge of the shingles. Asphalt products are easy to damage when worked at high temperatures. Try to avoid laying these materials on extremely hot days.

Hips and Ridges

Special hip and ridge shingles are usually available from the manufacturer. The special shingles can be easily made, however. Cut pieces 9" by 12" from either square-butt shingle strips or mineral surfaced roll roofing that matches the color of the shingles.

Figure 11-22. Approved nailing pattern for three-tab square-butt shingles. (Asphalt Roofing Manufacturers Assoc.)

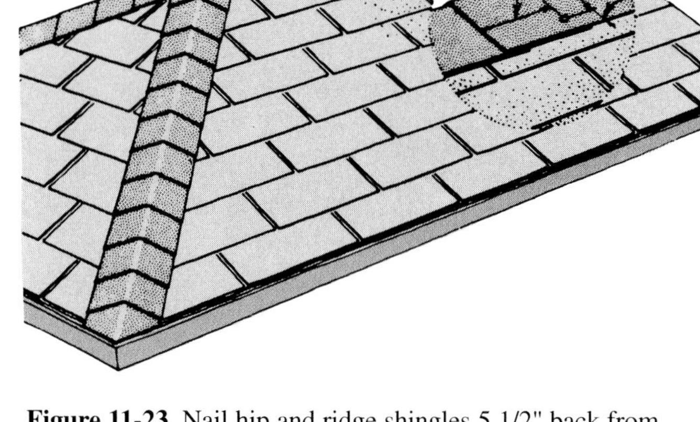

Figure 11-23. Nail hip and ridge shingles 5 1/2" back from edge. Use one nail on each side.

After the shingles are cut, bend them lengthwise in the centerline. In cold weather, the shingle should be warmed before bending to prevent cracks and breaks. Begin at the bottom of the hips or at one end of the ridge. Lap the units to provide a 5" exposure, as illustrated in Figure 11-23. Secure with one nail on each side, 5 1/2" back from the exposed end and 1" from the edge.

Metal ridge roll is not recommended for asphalt shingles. Corrosion may discolor the roof.

Wind Protection

Shingles with factory applied adhesive above tabs are available for windy localities. Only a few warm days are needed to seal tabs to the course underneath. This prevents wind from lifting and damaging tabs. A different nailing pattern is also recommended, Figure 11-24(A). These precautions are more important on roofs with low slopes where it is easiest for wind to get under the shingles.

Self-sealing shingles are satisfactory for roofs with slopes up to about 60°. For very steep slopes, like those used in mansard roofs, Figure 11-24(B), special application steps must be followed. You may have to seal them in place with quick-setting asphalt cement. Follow recommendations provided by the roofing manufacturer.

If regular shingles are used, the tabs can be cemented. Apply a spot of special tab cement about 1" square with a putty knife or caulking gun and then press the tab down. Avoid lifting the tab any more than necessary while applying the cement.

A variety of interlocking shingles are designed to provide resistance against strong winds. They are used for both new construction and reroofing. Details of the interlocking devices and methods of application vary considerably. Always study and follow the manufacturer's directions when installing all types of shingles.

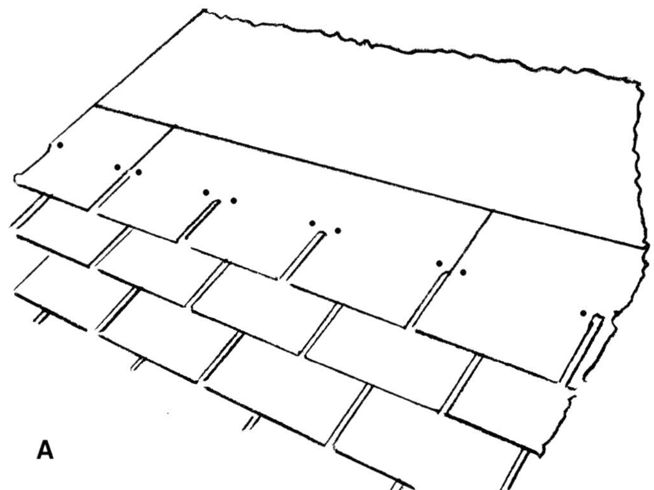

Figure 11-24. Applying shingles in windy localities. A—Nailing pattern for fastening asphalt shingles. B—Mansard roofs may require special application methods if asphalt shingles are used. Wood shingles are often used in windy localities. Check local codes. (Shakertown Corp.)

Individual Asphalt Shingles

Roof surfaces may be laid with an individual asphalt shingle. There are several sizes and designs available. One commonly used is 12" wide and 16" long. Several patterns can be used in its application. See Figure 11-25. Follow the same procedure that was described for strip shingles. Horizontal and vertical chalk lines should be used to insure accurate alignment.

Low-Slope Roofs

When applying asphalt shingles to slopes less than 4 in 12, certain additional procedures should be followed. Slopes as low as 2 in 12 can be made watertight and windtight.

Two layers of felt underlayment should be used on low-slope roofs. Each course should lap the preceding one by 19". In areas where the January daily average temperature is 25°F (-4°C) or colder, cement the two felt layers together from the eaves up the roof to 24" inside the interior wall line of the building.

Rather than cementing the two layers together, an ice and water barrier can be used. Consisting of a thick polymer-modified asphalt membrane, the barrier is available in 36" wide rolls. It is sold under various trade names, such as "WinterGuard", "Water Shield", and "Weather Watch". These materials are self-sealing around nails and deck joints. Many have an adhesive backing, eliminating the need for fasteners. They also make a good underlayment below flashing and could be used over the entire roof surface for low-slope roofs, Figure 11-26.

Shingles provided with factory applied adhesive and manufactured to conform to the Underwriters Laboratories Standard for Class "C" Wind Resistant shingles should then be installed. "Free" tab square-butt strips can be used if you cement all the tabs. See Figure 11-27 for special application methods.

Roll Roofing

Asphalt roll roofing is manufactured in a variety of weights, surfaces, and colors. It is used as a main roof cov-

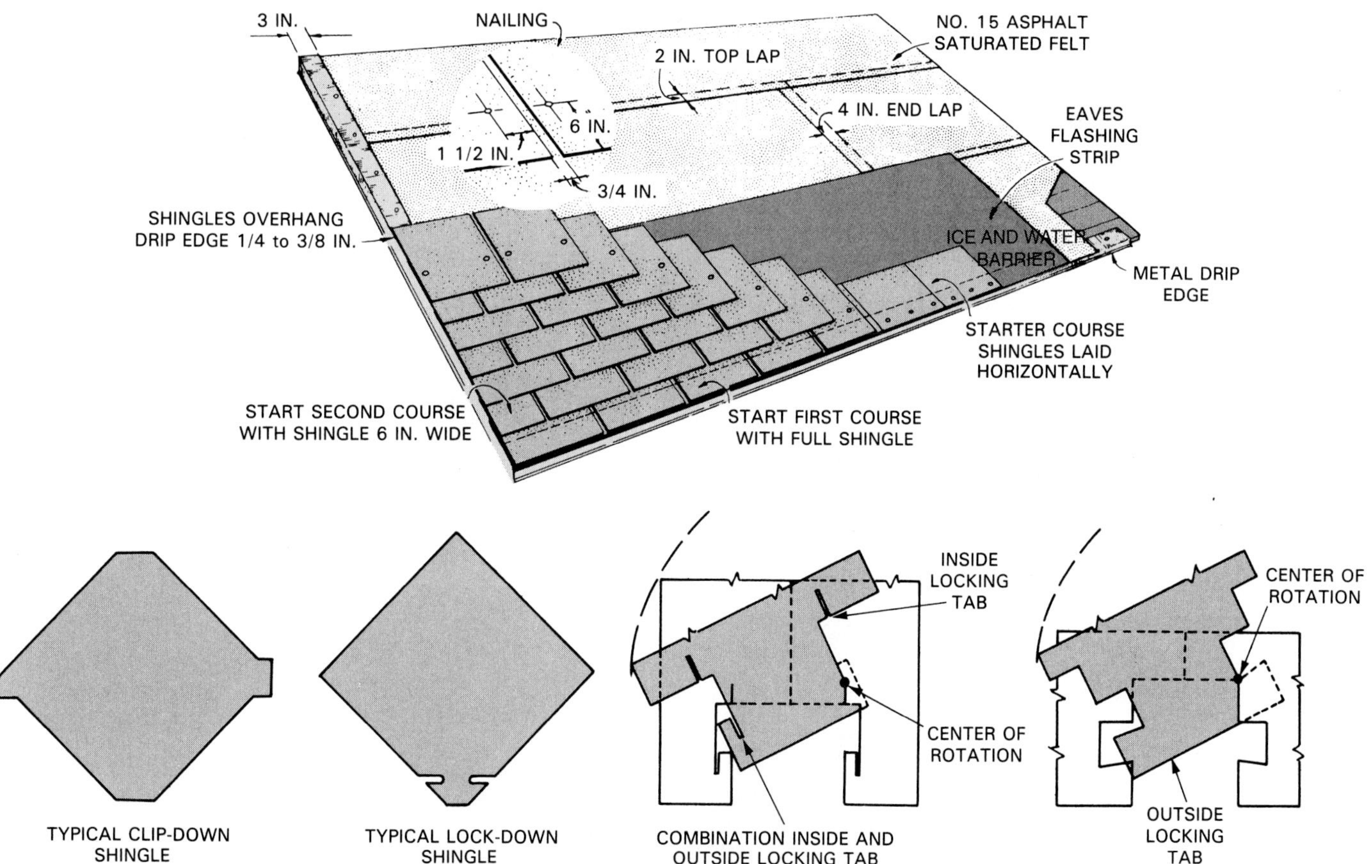

Figure 11-25. Types of individual shingles. Top—Method of installing giant individual shingles. This is called the American method. Bottom left—Two types of hex shingles are intended primarily for application over old roofing. Slope must be 4" per foot or greater. Bottom right—Interlocking devices on individual shingles provide increased wind resistance. (Asphalt Roofing Manufacturers Assoc.)

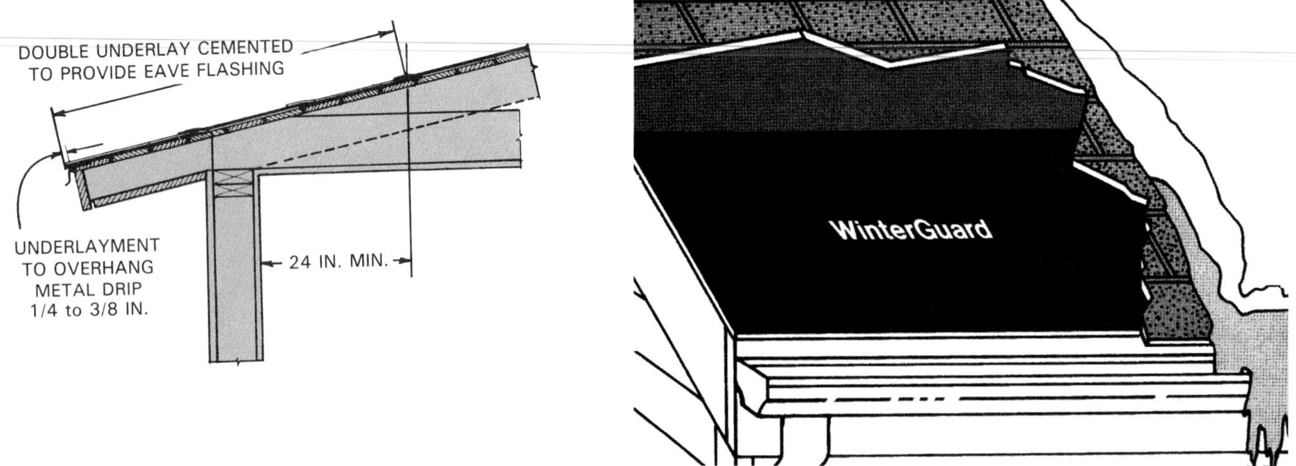

Figure 11-26. Special attention must be paid to underlayments for low-slope roofs. Left—Two plies are cemented together for a watertight eave flashing. Right—Eave flashing consisting of a single layer of a thick polymer-modified asphalt reinforced with a fiberglass mat. This membrane must extend at least 24" upward from the side walls. (CertainTeed Corp.)

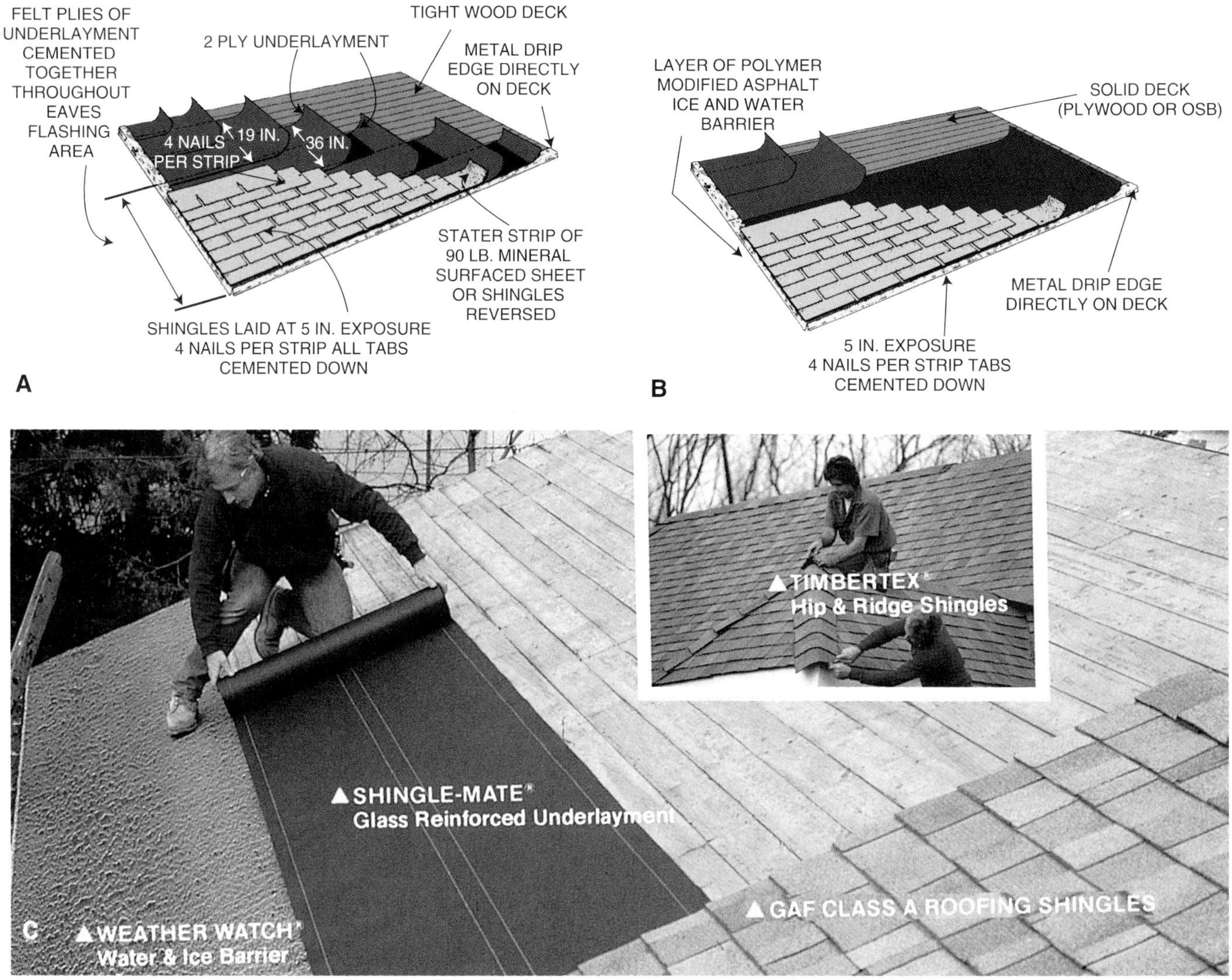

Figure 11-27. Use special application methods for shingling low-slope roofs. A—Double plies of tar paper cemented together. B—Self-sealing layer of polymer-modified asphalt reinforced with fiberglass used as eaves flashing. Rest of underlayment is two-ply tar paper cemented. C—An ice and water barrier has been laid down at eaves and carpenter is rolling out a glass-reinforced underlayment. (GAF Building Materials Corp.)

ering and as a flashing material. For best results, install it at temperatures of 45°F (7°C) or above.

In residential construction, a double-coverage roll roofing provides good protection. The roofing can be used on slopes as low as 1" per foot. The 36" width consists of a granular surfaced area 17" wide and a smooth surface, called a *selvage*, that is 19" wide.

Although double-coverage roll roofing can be applied parallel to the rake, it is usually applied parallel to the eaves, as shown in Figure 11-28. You can make the starter strip by cutting off the granular surfaced portion. Use two rows of nails to install the starter strip—one 4 3/4" below the upper edge and the other 1" above the lower edge.

Cover the entire starter strip with asphalt cement and overlay a full-width sheet. Attach the sheet with a row of nails 4 3/4" from the upper edge and a second row 8 1/2" below the first row. The nail interval should be about 12".

Position each succeeding course so that it overlaps the full 19" selvage area. Nail the sheet in place and then carefully turn the sheet back to apply the cement. Spread the cement to within about 1/4" of the granular surface. Press the overlaying sheet firmly into the cement using a stiff broom or roller. Avoid excessive use of cement. Be sure to follow the manufacturer's recommendations.

Reroofing

When reroofing, a choice must be made between removing the old roofing or leaving it in place. It is usually not necessary to remove old wood shingles, old asphalt shingles, or old roll roofing before putting on a new asphalt roof provided that the following are true:

- The strength of the existing deck and framing is adequate to support the weight of workers and additional new roofing, as well as snow and wind loads.
- The existing deck is sound and will provide good anchorage for the nails used in applying new roofing.
- The additional layer of shingles is allowed by the governing building code.

When putting on new roofing over old wood shingles, all loose or protruding nails should be removed and the shingles renailed in new locations. Renail loose, warped, and split shingles. Replace missing shingles. Cut back shingles at eaves and rakes far enough to allow the application of 4 to 6" nominal 1" thick strips. These strips should be nailed in place with their outside edges projecting beyond the roof deck the same distance as the old wood shingles.

When the old roof consists of square-butt asphalt shingles with a 5" exposure, new self-sealing strip shingles can be applied as shown in Figure 11-29. This application pattern will insure a smooth, even appearance.

The joint between a vertical wall and roof surface should be sealed when reroofing. First, apply a strip of smooth roll roofing about 8" wide. Nail each edge firmly, spacing nails about 4" O.C. As the shingles are applied, asphaltic plastic cement is spread on the strip and the shin-

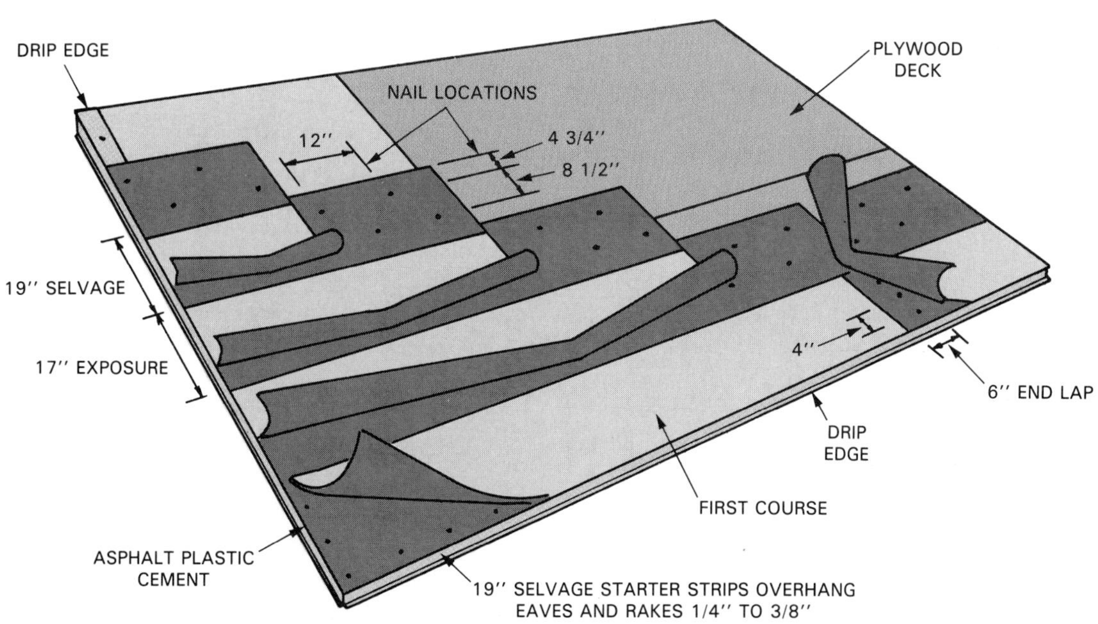

Figure 11-28. One method of applying double-coverage roll roofing parallel to the eaves. (Bird and Son Inc.). The roll can also be laid parallel to the rake, and a membrane of polymer-modified asphalt can be used along the eaves.

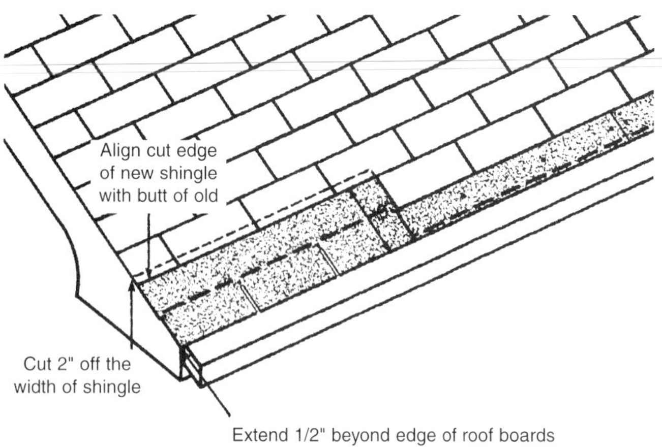

Figure 11-29. Reroofing over old asphalt shingles. Left—Lay down a starter strip the same width as the exposure. Then trim first course to butt against third course of old shingles. (Asphalt Roofing Manufacturers Assoc.) Right—Butt full width strip shingles against next old course. Offset cutouts so they do not fall over the lower old course. (Manville Building Materials Corp.)

gles are thoroughly bedded. To ensure a tight joint, use a caulking gun to apply a final bead of cement between the edges of the shingles and the siding.

When old shingles are to be removed before applying a new roof, it is common to use a flatbladed shovel. Asbestos shingles are hazardous material and must be disposed of properly.

Built-Up Roofing

A flat roof or a roof with very little slope, Figure 11-30, must be covered with a watertight system. Most flat roofs are covered with *built-up roofing,* which is very durable. Companies that manufacture the components (mainly saturated felt and asphalt) provide detailed specifications for making the installation.

On a wood deck, a heavy layer of saturated felt is first nailed down with galvanized nails, Figure 11-31. Nails must have a large head or be driven through tin caps. Each succeeding layer is then mopped in place with hot asphalt. When the felts are all in place, they are coated with hot asphalt and covered with slag, gravel, crushed stone, or marble chips. These materials provide a weathering surface and also improve the appearance. Three to four hundred pounds of the mineral covering is used on a 100 sq. ft. section of roof.

Built-up roofs for residential structures normally have three or four *plies* (layers of asphalt-saturated felts). The asphalt used between each layer and to bed the surface coating are products of the petroleum industry. They begin to flow, very slowly, at a lower temperature. This results in a self-healing property that is essential for flat roofs, where water is likely to stand. A special low temperature asphalt known as "dead-flat" asphalt is used for this type of roof, Figure 11-32.

For sloping roofs, "steep" asphalt (an asphalt with a high melting point) is used. In hot climates, only steep asphalts can be used.

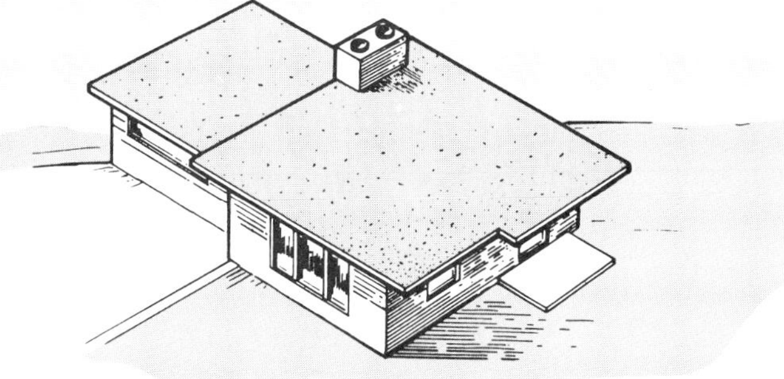

Figure 11-30. Flat and low-pitched roofs must be covered with a watertight membrane.

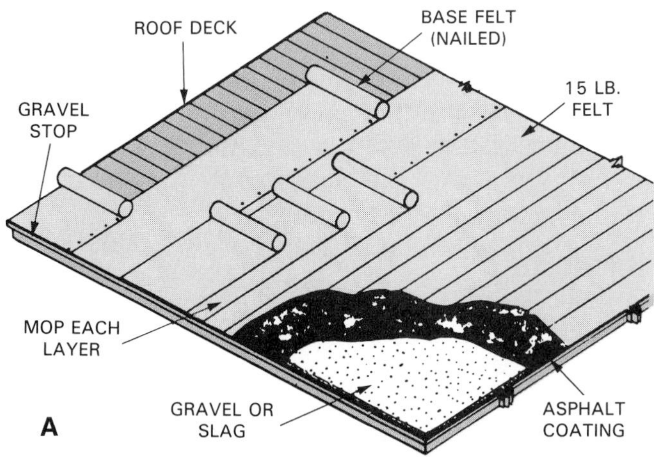

Figure 11-31. Top—The progression shown here is usually followed in constructing a built-up roof. Proper materials are indicated layer by layer. Bottom—Preparation for re-covering a built-up roof.

A gravel stop, usually fabricated from galvanized sheet metal, is attached to the roof deck to serve as a trim member. It also keeps the mineral surface and asphalt in place, Figure 11-33. Gravel stops are installed after the base felt has been laid. Joints between sections of gravel stop are bedded in a special mastic that permits expansion and contraction in the metal.

Flashings around chimneys, vents, or where the roof joins a wall must be constructed with special care. Leaks are most likely to occur at these locations. The best flashing materials include lead jackets, sheet copper, and special flashing cement.

Basic flashing construction is shown in Figure 11-34. The *cant strip* provides support for the felt layers as they curve from horizontal to vertical attitude.

Bare spots on a built-up roof should be repaired. First clean the area. Apply a heavy coating of hot asphalt and then spread more gravel or slag.

Felts which have fallen apart should be cut away and replaced with new felt. The new felt should be mopped in place, allowing at least one additional layer of felt to extend not less than 15" beyond the other layers.

Wood Shingles

Wood shingles are a traditional material used in residential construction. They are available with or without polymer fire-retardant treatment. Building codes often prohibit untreated wood roofing materials. Since wood weathers to a mellow color after exposure, wood shingles provide an appearance that is desired by many homeowners. When properly installed, they also provide a very durable roof.

Figure 11-32. Mopping asphalt between layers of saturated felt. Asphalt is heated to 450°F (230°C). Gravel stop is installed after the base felt is laid.

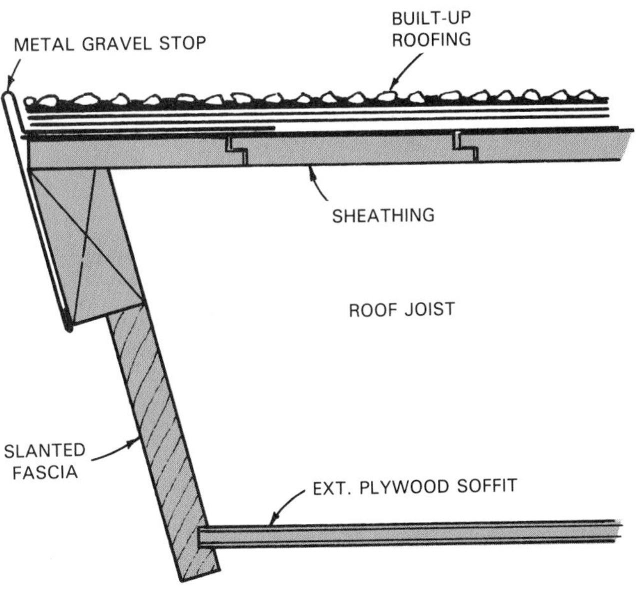

Figure 11-33. A section view through the edge of a flat roof overhang, shows metal gravel stop installation.

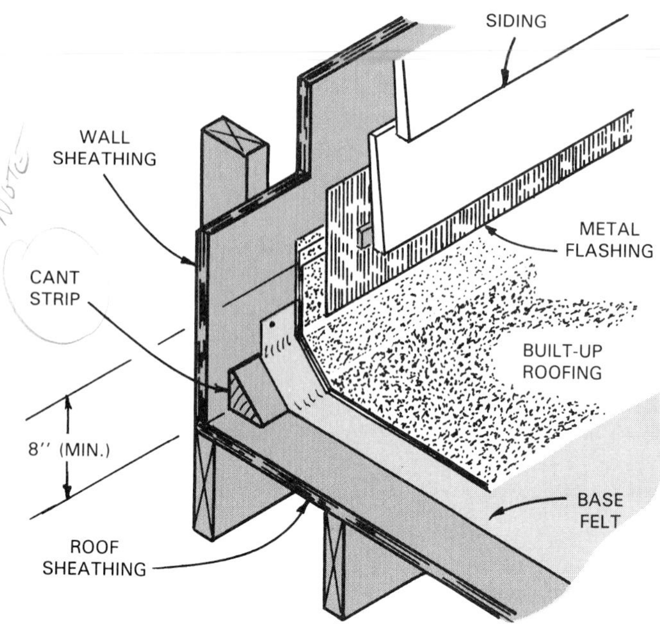

Figure 11-34. Basic construction where a flat roof intersects with a wall. Cant strip is a triangular wooden strip that provides a gently curving base for layers of roofing felt or a rubberized roof membrane.

Wood shingles are made from western red cedar, redwood, and cypress. All are highly decay resistant. They are taper sawed and graded No. 1, No. 2, and No. 3, plus a utility grade. The best grade is cut in such a way that the annular rings are perpendicular to the surface. Butt ends vary in thickness from 1/2 to 3/4", as shown in Figure 11-35. Wood shingles are manufactured in random widths and in lengths of 16, 18, and 24" , Figure 11-36. They are packaged in bundles. Four bundles contain enough shingles to cover 100 square feet of roof using standard application.

The exposure of wood shingles depends on the slope of the roof. When the slope is 5 in 12 or greater, standard exposures of 5, 5 1/2, and 7 1/2" are used for 16, 18, and 24" sizes respectively. On roofs with lower slopes, the exposure should be reduced to 3 3/4, 4 1/4, and 5 3/4". This will provide a minimum of four layers of shingles over the entire roof area. In any type of construction there should be a minimum of three layers at any given point to insure complete protection against heavy wind-driven rain.

Sheathing

Solid sheathing for wood shingles may consist of matched or unmatched 1" boards, shiplap, or plywood. Open, skip, or spaced sheathing, Figure 11-37, is sometimes used because it costs less and permits shingles to dry out quickly. One reason for using solid sheathing is to gain the added insulation and resistance to air infiltration that such a deck offers.

One method of applying skip roof sheathing is to space 1 by 3", 1 by 4", or 1 by 6" boards the same distance apart as the anticipated shingle exposure. Each course of shingles is nailed to a separate board.

Another method uses 1 x 6 lumber as sheathing boards with two courses of shingles nailed to each one.

Underlayment

Normally, an underlayment is not used for wood shingles—except at the eaves—when applied over either spaced or solid sheathing. Refer again to Figure 11-37. If roofing paper is wanted to prevent air infiltration, rosin-sized building paper or "dry" unsaturated felts are suitable. Saturated paper is usually not recommended because it may cause condensation problems.

Underlayment (Fire-Resistant)

Recent tests have shown that flame-spread and burn-through rates for wood shingles and shakes can be reduced. This is achieved by pressure-treating the shingles with fire retardants.

GRADE	Length	Thickness (at Butt)	No. of Courses Per Bundle	Bdls/Cartons Per Square	Description
No. 1 BLUE LABEL	16″ (Fivex) 18″ (Perfections) 24″ (Royals)	.40″ .45″ .50″	20/20 18/18 13/14	4 bdls. 4 bdls. 4 bdls.	The premium grade of shingles for roofs and sidewalls. These top-grade shingles are 100% heartwood. 100% clear and 100% edge-grain.
No. 2 RED LABEL	16″ (Fivex) 18″ (Perfections) 24″ (Royals)	.40″ .45″ .50″	20/20 18/18 13/14	4 bdls. 4 bdls. 4 bdls.	A good grade for many applications. Not less than 10″ clear on 16″ shingles, 11″ clear on 18″ shingles and 16″ clear on 24″ shingles. Flat grain and limited sapwood are permitted in this grade.
No. 3 BLACK LABEL	16″ (Fivex) 18″ (Perfections) 24″ (Royals)	.40″ .45″ .50″	20/20 18/18 13/14	4 bdls. 4 bdls. 4 bdls.	A utility grade for economy applications and secondary buildings. Not less than 6″ clear on 16″ and 18″ shingles, 10″ clear on 24″ shingles.
No. 4 UNDER-COURSING	16″ (Fivex) 18″ (Perfections)	.40″ .45″	14/14 or 20/20 14/14 or 18/18	2 bdls. 2 bdls. 2 bdls. 2 bdls.	A utility grade for undercoursing on double-coursed sidewall applications or for interior accent walls.
No. 1 or No. 2 REBUTTED-REJOINTED	16″ (Fivex) 18″ (Perfections) 24″ (Royals)	.40″ .45″ .50″	33/33 28/28 13/14	1 carton 1 carton 4 bdls.	Same specifications as above for No. 1 and No. 2 grades but machine trimmed for parallel edges with butts sawn at right angles. For sidewall application where tightly fitting joints are desired. Also available with smooth sanded face.

Figure 11-35. Wood shingles are made in several grades and to certain specification for various applications. (Red Cedar Shingle and Handsplit Shake Bureau)

Figure 11-36. Western red cedar shingles provide a durable roof surface. They weather to a soft, mellow gray after exposure and have a pleasing appearance.

Flame penetration time can also be increased by using 1/2". Type X gypsum board under solid or spaced sheathing. For more information see Uniform Building Code Standard No. 32-14.

Flashing

In areas where outside temperatures drop to 0°F (–17°C) or colder and there is a possibility of ice forming along the eaves, an *eaves flashing* strip or ice and water barrier is recommended. The procedure is identical to that used for asphalt shingles. Refer once more to Figure 11-37.

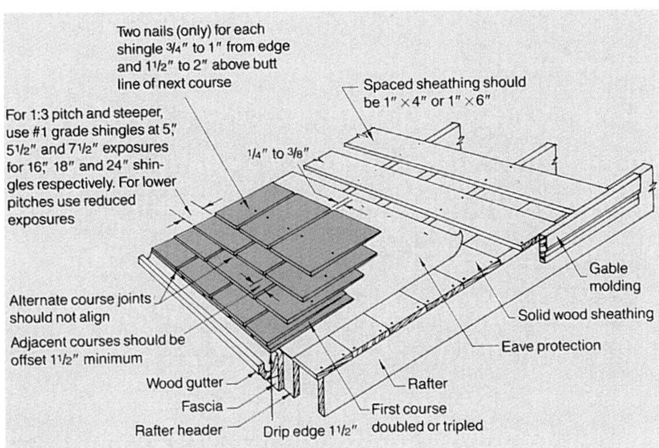

Figure 11-37. General application details for wood shingles. Double or triple layers at eaves and allow butts of first course to project beyond fascia 1 1/2" Note use of both solid and spaced solid wood decking. (Cedar Shake and Shingle Bureau)

It is important to use good materials for valleys and flashings. Materials used for this purpose include tin plate, lead-clad iron, galvanized iron, lead, copper, and aluminum sheets.

If galvanized iron (mild steel coated with a layer of zinc) is selected, 24 or 26 gage metal should be used. Tin or galvanized sheets with less than 2 oz. of zinc per sq. ft. should be painted on both sides with and allowed to dry before being used.

When making bends, care should be taken not to crack the zinc coating. On roofs of 1/2 pitch or steeper, the valley sheets should extend up on both sides of the center of the valley for a distance of at least 7". On roofs of less pitch, wider valley sheets should be used. Minimum extension should be at least 10" on both sides, Figure 11-38. The open portion of the valley is usually about 4" wide and should gradually increase in width toward the lower end.

Tight flashing around chimneys is also essential to a good roofing job. Two methods of installing base and cap flashing around a chimney are shown in Figure 11-39.

In new construction, the cap flashing is laid in the joints when the chimney is built. If the chimney is laid up without flashing, the mortar joints must be chiseled out and the flashing forced in. It may be held in place by nails driven into the mortar. Finally, the joints must be filled or pointed with good mortar.

Nails

Only rust resistant nails should be used with wood shingles. Hot-dipped, zinc-coated nails, which have the strength of steel and the corrosion resistance of zinc, are recommended. Figure 11-40 shows sizes of nails for various jobs.

Most carpenters prefer to use a shingler's or lather's hatchet, Figure 11-41, to lay wood shingles. This has a blade for splitting and trimming. Some have a gauge for spacing the weather exposure.

Applying Shingles

The first course of shingles at the eaves should be doubled or tripled. All shingles, when laid on the roof, should be spaced 1/4"–3/8" apart to provide for expansion when they become rain soaked.

Use only two nails to attach each shingle. The proper placing of these two nails is of considerable importance. They should be near the butt line of the shingles in the next course that is to be applied over the course being nailed, but should never be driven below this line so they will be exposed to the weather. Driving the nails 1 to 1 1/2" above the butt line is good practice. Two inches above is an allowable maximum. Nails should be placed not more than 3/4" from the edge of the shingle at each side. When nailed in this manner, the shingles will lie flat and give good service.

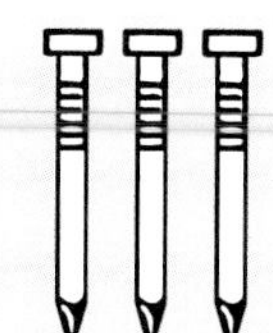

Use care when nailing wood shingles. Drive the nail just flush with the surface. The wood in shingles is soft. It can be easily crushed and damaged under the nail heads.

The second layer of shingles in the first course should be nailed over the first layer so the joints in each course are at least 1 1/2" apart. See Figure 11-42. A good shingler will use care in breaking the joints in successive courses, so they do not match up in three successive courses. Joints in adjacent courses should be at least 1 1/2" apart.

In shingles containing both flat and vertical grain, joints should not be aligned with centerline of the heart grain. Split flat grain shingles in two before nailing. Treat knots and other defects as the edge of the shingle.

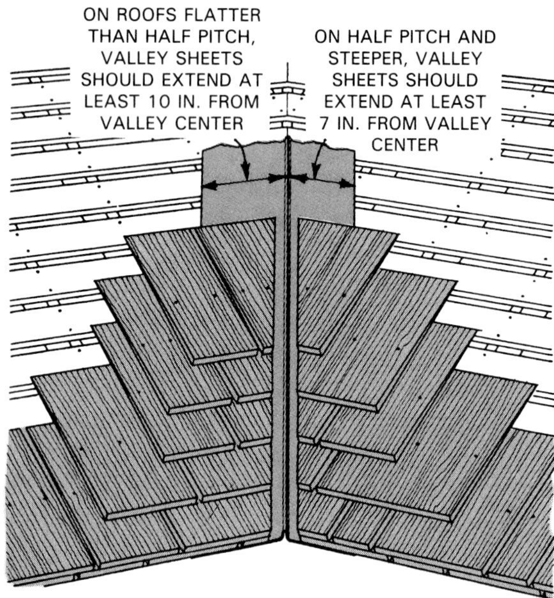

Figure 11-38. Valley flashing for wood shingles is similar to flashing for asphalt shingles. (Red Cedar Shingle and Handsplit Shake Bureau)

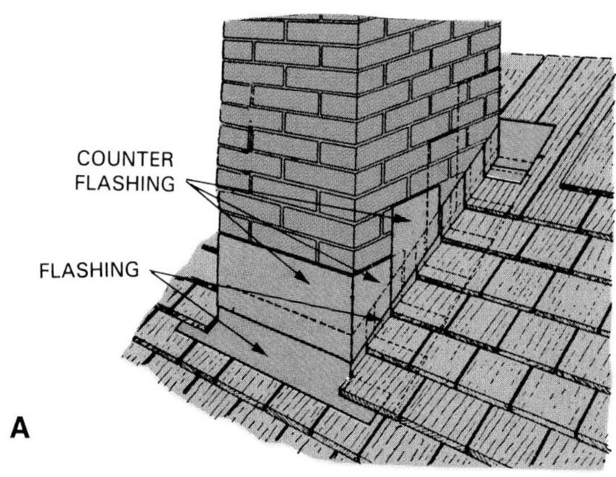

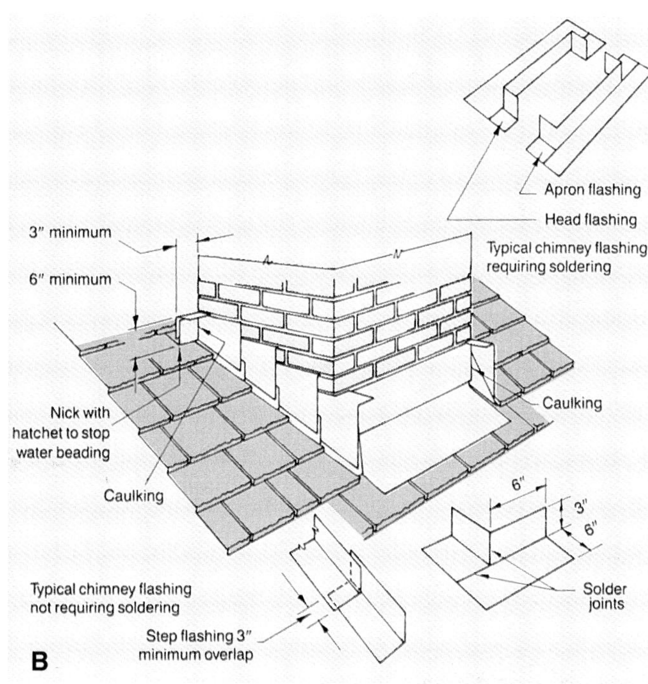

Figure 11-39. Two methods of flashing around a brick chimney. In both methods, flashing should be mortared into joints.

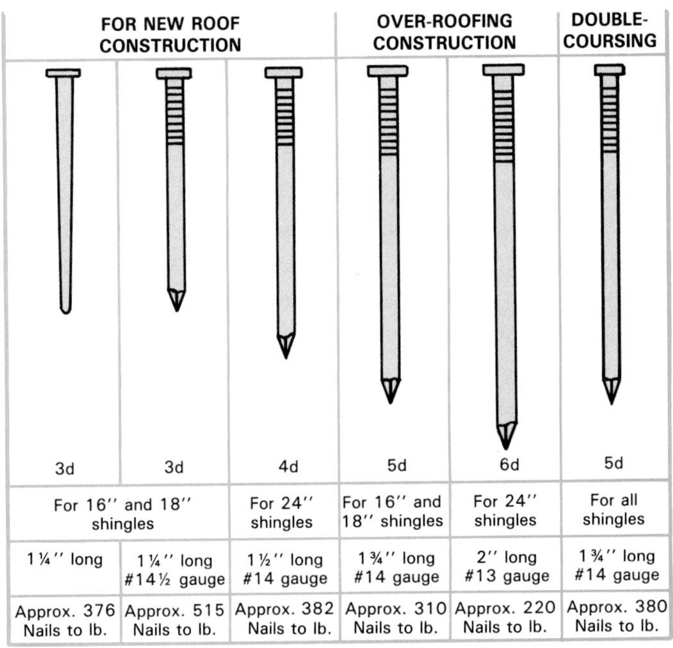

FOR NEW ROOF CONSTRUCTION			OVER-ROOFING CONSTRUCTION		DOUBLE-COURSING
3d	3d	4d	5d	6d	5d
For 16" and 18" shingles		For 24" shingles	For 16" and 18" shingles	For 24" shingles	For all shingles
1¼" long	1¼" long #14½ gauge	1½" long #14 gauge	1¾" long #14 gauge	2" long #13 gauge	1¾" long #14 gauge
Approx. 376 Nails to lb.	Approx. 515 Nails to lb.	Approx. 382 Nails to lb.	Approx. 310 Nails to lb.	Approx. 220 Nails to lb.	Approx. 380 Nails to lb.

Figure 11-40. Use only rust resistant nails of the size recommended for your particular application.

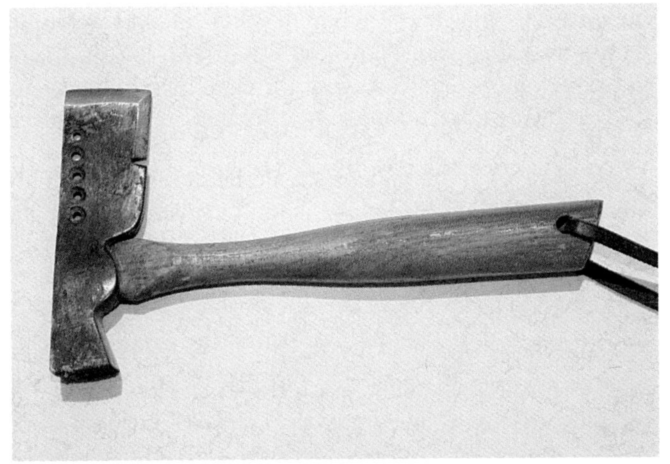

Figure 11-41. Shingler's hatchet is especially suited for laying wood shingles and shakes. (Gene Ahnen, Bayfield, Wisconsin)

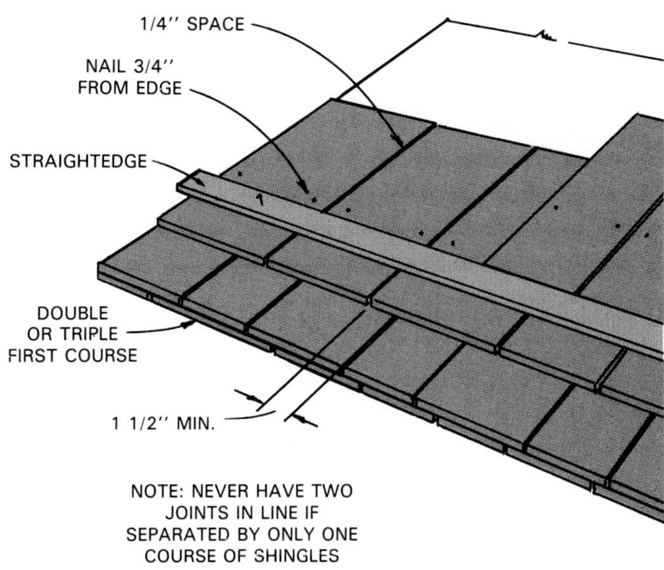

Figure 11-42. A wooden straightedge should be used as a guide in laying wood shingles. Stagger joints and use two nails per shingle.

It is good practice to use a board as a straightedge to line up rows of shingles. Tack the board temporarily in place to hold the shingles until they are nailed. Two shinglers often work together; one distributes and lays the shingles along the straightedge while another nails them in place. As shingling progresses, check the alignment every five or six courses with a chalk line. Measure down from the ridge occasionally to be sure shingle courses are parallel to the ridge.

On a roof section where one end terminates at a valley, shingles for the valley should be carefully cut to the proper angle at the butts. Use wide shingles. Nail the shingles in place along the valley first, Figure 11-43.

Dripping from gables may be prevented by using a piece of 6" bevel siding along the edge and parallel to the end rafter, Figure 11-44.

Shingled Hips and Ridges

Good, tight ridges and hips are required to avoid roof leakage. In the best type of hip construction, Figure 11-45, nails are not exposed to the weather. Shingles of approximately the same width as the roof exposure are sorted out. Two lines are then marked on the shingles on the roof the correct distance back from the centerline of the ridge on each side. On small houses, hip caps may be made narrower.

Factory assembled hip and ridge units are available, Figure 11-46. Weather exposure should be the same as that used for the regular shingles. Be sure to use longer nails that will penetrate well into the sheathing.

Special Effects

By staggering or building up wood shingles, usually in random patterns, shadow lines and texture can be emphasized. This is sometimes a feature used in contemporary as well as traditional architecture. Several applications are shown in Figure 11-47. The ocean wave effect is obtained by placing a pair of shingles butt-to-butt under the regular course and at right angles to the butt line. These cross shingles should be about 6" wide.

Figure 11-43. Mark, cut, and lay wood shingles along a valley first. Then complete each course to the roof's edge.

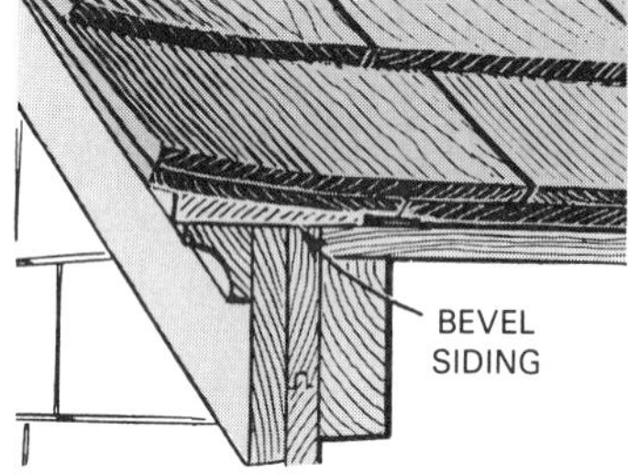

Figure 11-44. Length of beveled siding tilts shingles inward to prevent dripping off the edge.

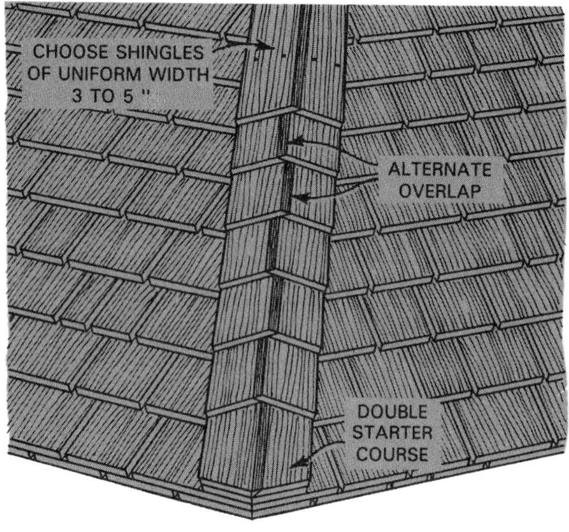

Figure 11-45. How to install wood cap shingles at hips and ridges.

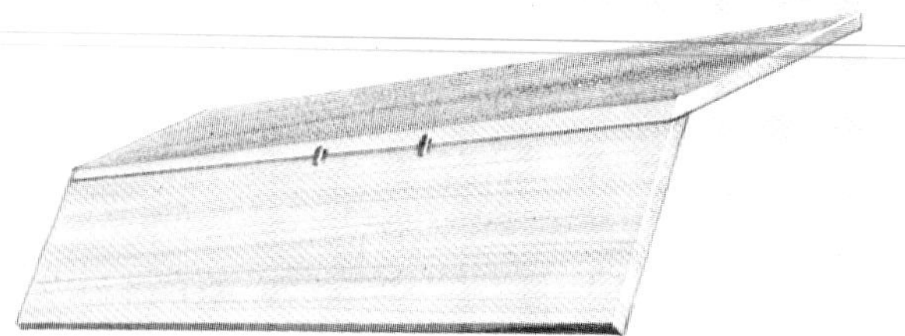

Figure 11-46. Hip and ridge units can be purchased prefabricated. (Red Cedar Shingle and Handsplit Shake Bureau)

The Dutch weave effect is made by doubling or super-imposing extra shingles, completely at random. This effect can be emphasized by using two shingles instead of one. It is generally referred to as a pyramid pattern. Note that joints are always broken by at least 1 1/2".

Reroofing with Wood Shingles

Wood shingles may be applied to old as well as new roofs, Figure 11-48. If the old roofing is in reasonably good shape, it need not be removed.

Before applying new shingles, all warped, split, and decayed shingles should be nailed tightly or replaced. To finish the edges of the roof, the exposed portion of the first two rows of old shingles along the eaves should be cut off with a sharp hatchet. Nail a 1" wood strip in this space. Place the outer edge flush with the eave line. Edges along the gable ends should be treated in a similar manner.

The level of the valleys should be raised by applying wood strips. New flashings should be installed over the strips. Remove old hip and ridge caps to provide a solid base for new shingles. New shingles should be spaced 1/4" apart to allow for expansion in wet weather. Let them project 1/2 to 3/4" beyond the edge of the eaves.

The procedure described for new roofs should be followed. However, longer nails are required. For 16" and 18" shingles, rust-resistant or zinc clad 5d box nails or special over-roofing nails 1 3/4" long, 14 gage, should be used. A 6d, 13 gage, rust-resistant nail is used for 24" shingles.

Usually, no particular attention needs to be given to how the nails penetrate the old roof beneath. It does not matter whether they strike the sheathing strips or not because, with the larger nails that are used, complete penetration is obtained through the old shingles. Enough nails to anchor all the shingles of the new roof will strike sheathing or nailing strips.

New flashings should be placed around chimneys. Do not remove the old flashing. Use high grade non-drying mastics liberally to get a watertight seal between the brick and the metal.

In reroofing houses covered with composition material, whether in the form of roll roofing or imitation shingles, it is usually best to strip off the old material. Otherwise, moisture may condense on the roof deck below. Rapid decay of sheathing could follow.

Figure 11-47. Special patterns for wood shingles are used on contemporary as well as traditional houses. Left—Ocean wave. Center—Dutch weave. Right—Pyramid.

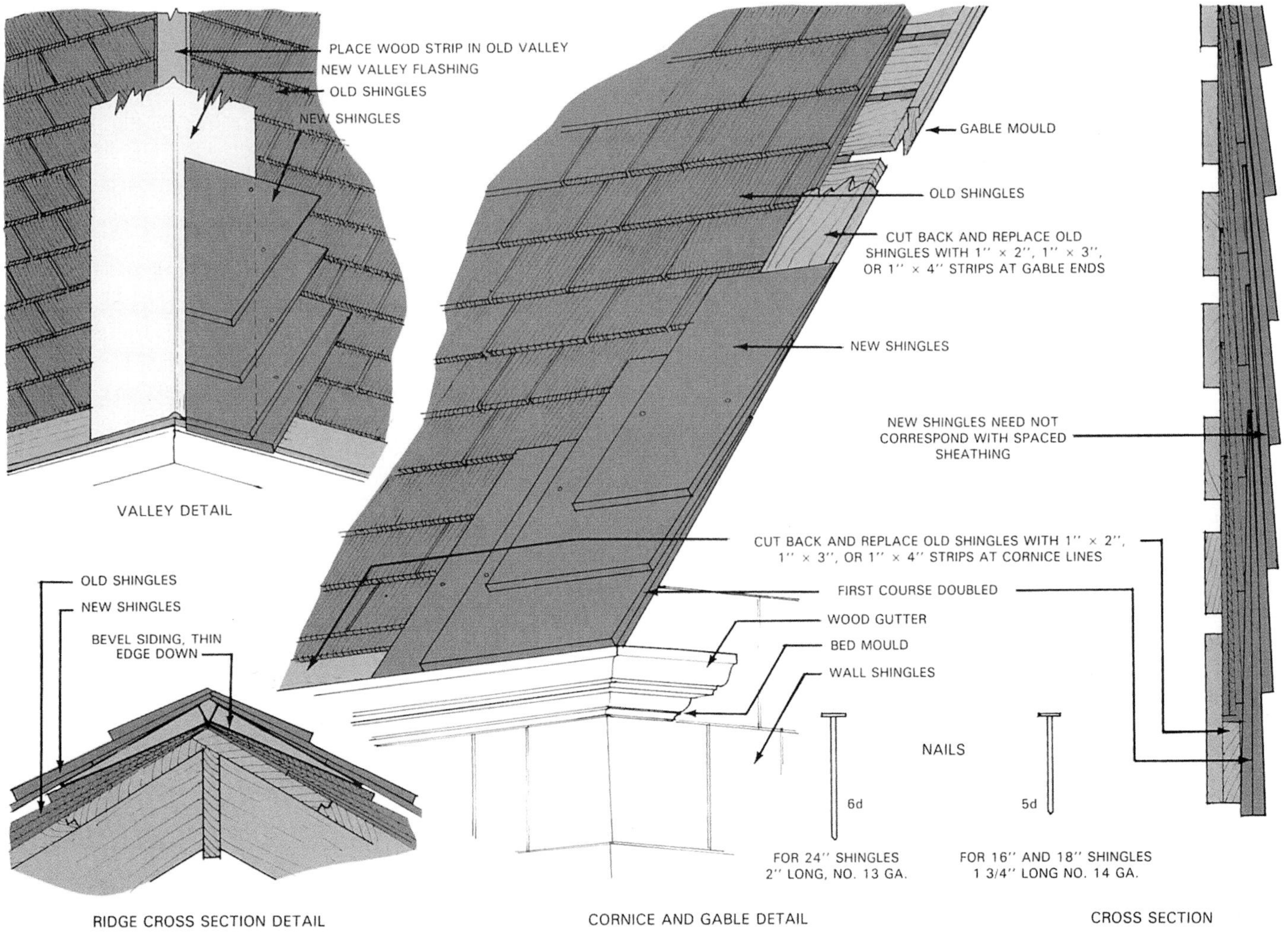

PLACE WOOD STRIP IN OLD VALLEY
NEW VALLEY FLASHING
OLD SHINGLES
NEW SHINGLES
VALLEY DETAIL

OLD SHINGLES
NEW SHINGLES
BEVEL SIDING, THIN EDGE DOWN
RIDGE CROSS SECTION DETAIL

GABLE MOULD
OLD SHINGLES
CUT BACK AND REPLACE OLD SHINGLES WITH 1" × 2", 1" × 3", OR 1" × 4" STRIPS AT GABLE ENDS
NEW SHINGLES
NEW SHINGLES NEED NOT CORRESPOND WITH SPACED SHEATHING
CUT BACK AND REPLACE OLD SHINGLES WITH 1" × 2", 1" × 3", OR 1" × 4" STRIPS AT CORNICE LINES
FIRST COURSE DOUBLED
WOOD GUTTER
BED MOULD
WALL SHINGLES
NAILS
6d
5d
FOR 24" SHINGLES 2" LONG, NO. 13 GA.
FOR 16" AND 18" SHINGLES 1 3/4" LONG NO. 14 GA.
CORNICE AND GABLE DETAIL
CROSS SECTION

Figure 11-48. Detail drawings for reroofing method with wood shingles. Note how valleys and ridges are built up.

Figure 11-49. Wood shakes provide an attractive and durable roof.

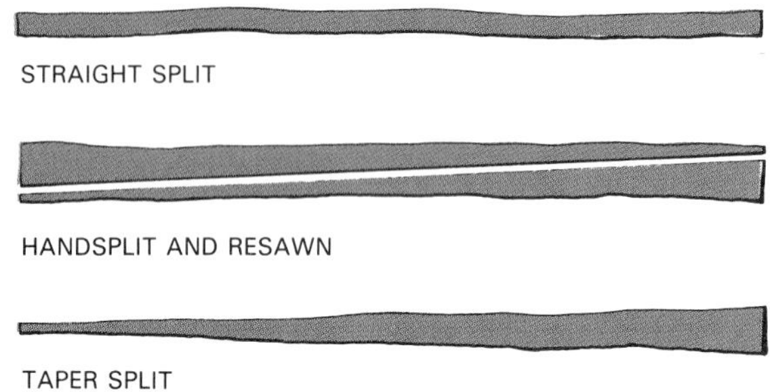

Figure 11-50. There are three basic types of wood shakes.

Wood Shakes

Often called the aristocrat of roofing materials, *wood shakes* provide the most pleasing surface texture, Figure 11-49. They are highly durable and, if properly installed, may outlast the structure itself.

Generally, wood shakes are available as hand-split and resawn, taper split, and straight split, Figure 11-50. Like regular wood shingles, they are available in random widths. Various lengths and thicknesses are standardized as listed in Figure 11-51.

Do not apply shakes to roofs that have too little slope for good drainage. The recommended minimum is 4 in 12. Maximum weather exposure is 13" for 32" shakes, 10" for 24" shakes, and 8 1/2" for 18" shakes.

Start the application by laying down an ice and water barrier or a 36" strip of 30 lb. felt along the eaves. The beginning course is doubled as with wood shingles. A strip of 30 lb. felt must be laid between each course. This must cover the top portion of the shakes and extend onto the sheathing.

For example, if 24" shakes are being laid at a 10" exposure, place the roofing felt 20" above the butts of the shake, Figure 11-52.

Individual shakes should be spaced from 1/4 to 3/8" apart to allow for expansion. These joints should be offset at least 1 1/2" from course to course.

Proper nailing is important. Use rust-resistant nails, preferably the hot-dipped zinc-coated type. The 6d size, which is 2" long, normally is adequate. Longer nails should be used, if necessary, because of unusual shake thickness and/or weather exposure. Nails should be long enough for adequate penetration into the sheathing boards.

Two nails should be used for each shake. Drive them at least one inch from each edge and about one or two inches above the butt line of the following course. Do not drive nailheads into the shakes so that wood fibers are crushed, Figure 11-53.

GRADE	Length and Thickness	18" Pack**		Description
		# Courses Per Bdl.	# Bdls. Per Sq.	
No. 1 HANDSPLIT & RESAWN	15" Starter-Finish 18" x 1/2" Mediums 18" x 3/4" Heavies 24" x 3/8" 24" x 1/2" Mediums 24" x 3/4" Heavies	9/9 9/9 9/9 9/9 9/9 9/9	5 5 5 5 5 5	These shakes have split faces and sawn backs. Cedar logs are first cut into desired lengths. Blanks or boards of proper thickness are split and then run diagonally through a bandsaw to produce two tapered shakes from each blank.
No. 1 TAPERSAWN	24" x 5/8" 18" x 5/8"	9/9 9/9	5 5	These shakes are sawn both sides.
No. 1 TAPERSPLIT	24" x 1/2"	9/9	5	Produced largely by hand, using a sharp-bladed steel froe and a wooden mallet. The natural shingle-like taper is achieved by reversing the block, end-for-end, with each split.
		20" Pack		
No. 1 STRAIGHT-SPLIT	18" x 3/8" True-Edge* 18" x 3/8" 24" x 3/8"	14 Straight 19 Straight 16 Straight	4 5 5	Produced in the same manner as tapersplit shakes except that by splitting from the same end of the block, the shakes acquire the same thickness throughout.

NOTE: * Exclusively sidewall product, with parallel edges.
** Pack used for majority of shakes.

Figure 11-51. Wood shakes are manufactured in four grades with various specifications and sizes. (Red Cedar Shingle and Handsplit Shake Bureau)

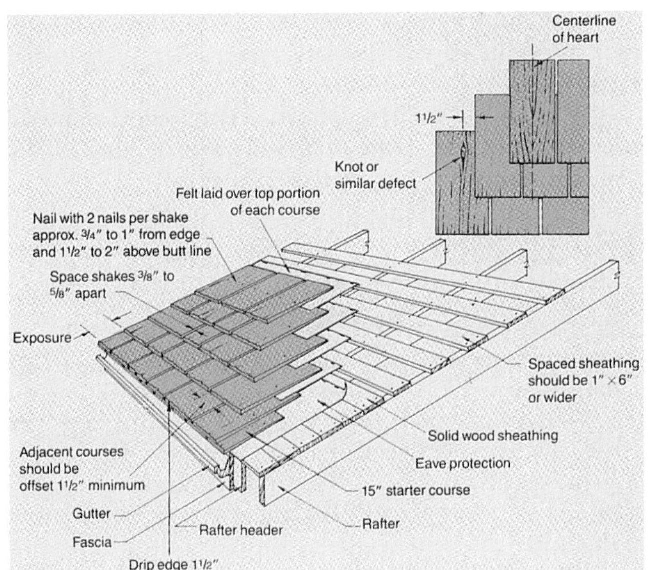

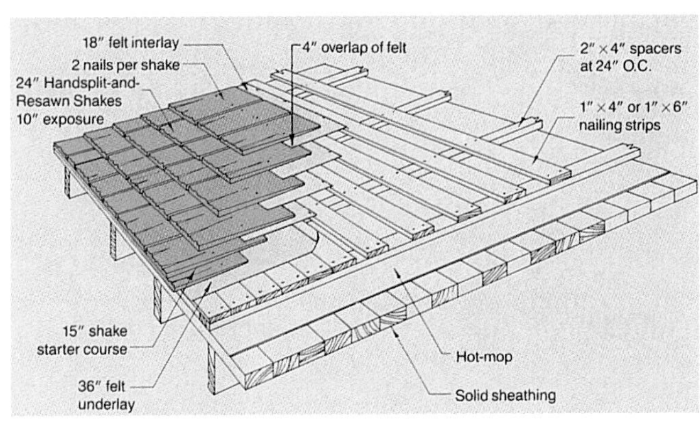

Figure 11-52. Shake application. Left—Protect eaves with ice and water barrier. Lay an 18" wide strip of No. 30 (30 lb.) asphalt saturated felt between each course. Straight-split shakes should be laid with the *froe-end* (end from which the shake has been split) towards the ridge. Right—Recommended method for applying shakes to low-slope roofs. Latticelike framework is embedded in a bituminous surface coating. (Cedar Shake and Shingle Bureau)

Figure 11-53. Putting down wood shakes. Butts can be aligned or laid random as shown here. Be sure to lap joints. (American Plywood Assoc.)

Valleys are laid as recommended for regular wood shingles. Underlay all valleys with 30 Ib. roofing felt. Metal valley flashing must be at least 20" wide or wider.

Chimneys or other structures that project through the roof should be flashed and counterflashed on all edges. Flashing should extend at least 6" under the shakes and should be covered as shown in Figure 11-54.

For the final course at the ridge line, try to select a uniform size of shakes and trim off the ends so they meet evenly. Carefully apply a strip of 30 lb. felt along all ridges and hips. Install shakes that have a uniform width of about 6". Nail them in place following the procedure described for regular wood shingles.

Prefabricated hip-and-ridge units are available. Their use will save time and provide uniformity. See Figure 11-55.

Shakes are also prefabricated in 8' panels. Shingles are bonded to a backing of 5/16" exterior sheathing plywood. Underlayment specifications are similar to those for other shakes. Shake panels can be applied over solid sheathing or furring strips on roofs with a 4 in 12 or steeper slope. See Figure 11-56.

Shakes can be installed on mansard roofs where the pitch is not greater than 20° off the vertical. Underlayment is the same as for shingles.

Flashing should be applied at top and bottom of the roof, at inside and outside corners, and around any openings. Inside and outside corners of wood, vinyl, or aluminum are required. This material must be properly flashed and caulked. See Figure 11-57.

Reroofing over Old Shingles

Nail down old shingles which are badly curled or warped. Generally, underlayment material is not required under asbestos shingles when they are laid over old wood or asphalt shingles. However, missing or badly decayed wood shingles should be replaced and the surface leveled.

If the top edges or corners of the new shingles do not rest on the butts of the old shingles, tilting will result. Wood strips, beveled and of the same thickness as the butts of the old wood shingles, can be used for leveling an old roof deck.

Sometimes at the edge of the roof, the wood shingles and sheathing are in bad condition. In such cases, cut away

Figure 11-54. Base flashing for a vertical projection such as a chimney or wall should extend at least 6" under wood shakes. (Red Cedar Shingle and Handsplit Shake Bureau)

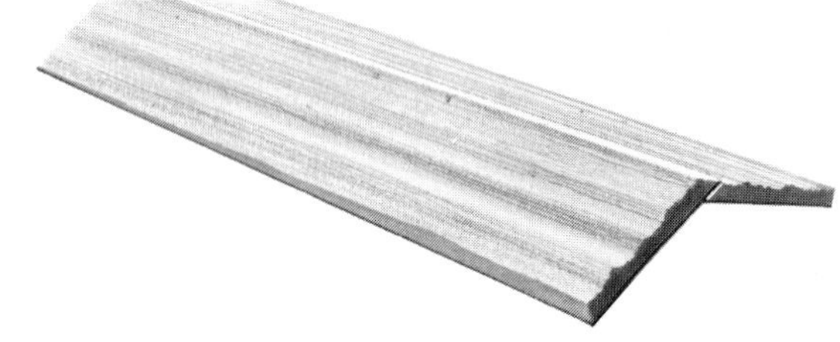

Figure 11-55. Preformed hip-and-ridge units will reduce the labor of attaching the ridge row.

Figure 11-56. Applying shingle panels to a roof. Panels are prefabricated in 8' lengths with 11 1/4" exposure. (Shakertown Corp., Cedar Panel Div.)

the old shingles and lay a new board 4 to 6" wide by 7/8" thick along the edges. This provides a solid base to which the new shingles may be nailed.

Build up old valleys with wood strips to bring the surface flush with the butts of the old wood shingles. Lay waterproof felt over hips, ridges, and at valleys.

Tile Roofing

The most commonly used *roofing tile* are manufactured from either concrete or molded from hard-burned shale, or mixtures of shale and clay. A few are made from metal.

When well made, concrete and clay tile are hard, dense, and very durable. Colors, textures, and shapes come in great variety, Figure 11-58. Most tile are made of concrete and some are glazed. Typical applications are shown in Figure 11-59.

While most applications are on new construction, tile may be applied over old roofs, provided:

- The old covering is in reasonably good condition.
- The roof framing and sheathing is strong enough to support the added weight. (Typically, tile weigh from 5.8 to 10.25 lb. per sq. ft.)

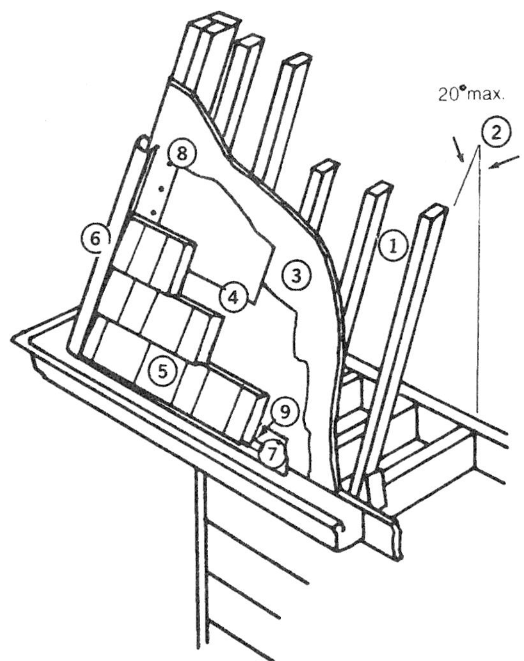

1) 16" o.c. stud spacing.
2) Pitch 20° from vertical maximum
3) 3/8" plywood sheathing
4) 15 lb. asphalt impregnated non-vapor barrier type building paper.
5) 4' Abitibi Shake Shingle Siding
6) Metal Corner Post
7) 6" drip edge flashing
8) Corner flashing
9) 3/8" x 1-1/2" starter strip

Figure 11-57. Applying shakes to a mansard roof requires special flashing. (ABTco, Inc.)

On new construction, roof trusses are engineered to support the weight of the tile. Additional roof framing or bracing may be required on old construction. Sheathing should be at least a nominal 1" thick, bridging no more than 24".

Installing Tile Roofing Units

Requirements for underlayment depend on roof pitch and local climate. Below a 3 in 12 pitch, tile are considered decorative; therefore, a minimum 2 plies of Type 15 felt, hot mopped between layers is recommended. This is topped with vertical battens spaced 24" apart from eave to ridge that are also mopped with asphalt. Then horizontal battens are laid and fastened at intersections with the vertical battens.

Laying Cement and Clay Roofing Tile

Battens are strips of wood installed horizontally on the roof to hold cement or clay tiles in place. Normally, 1 x 2 battens are used over solid sheathing, spaced 24" apart, Figure 11-60. The first row of battens should be placed to allow the tile to overhang eaves by 1 1/2".

Tile roofing is laid from right to left, beginning at the right rake, Figure 11-61. Lugs located on the underside of each tile hook over the battens. Flat tile are usually laid with each course shingle-lapping the previous course. Be careful while moving about on the roof; step only on the lower one-third of the tiles on the overlapped area.

Valley, Hip, and Ridge Installation

Prior to laying tile, valleys must be prepared.

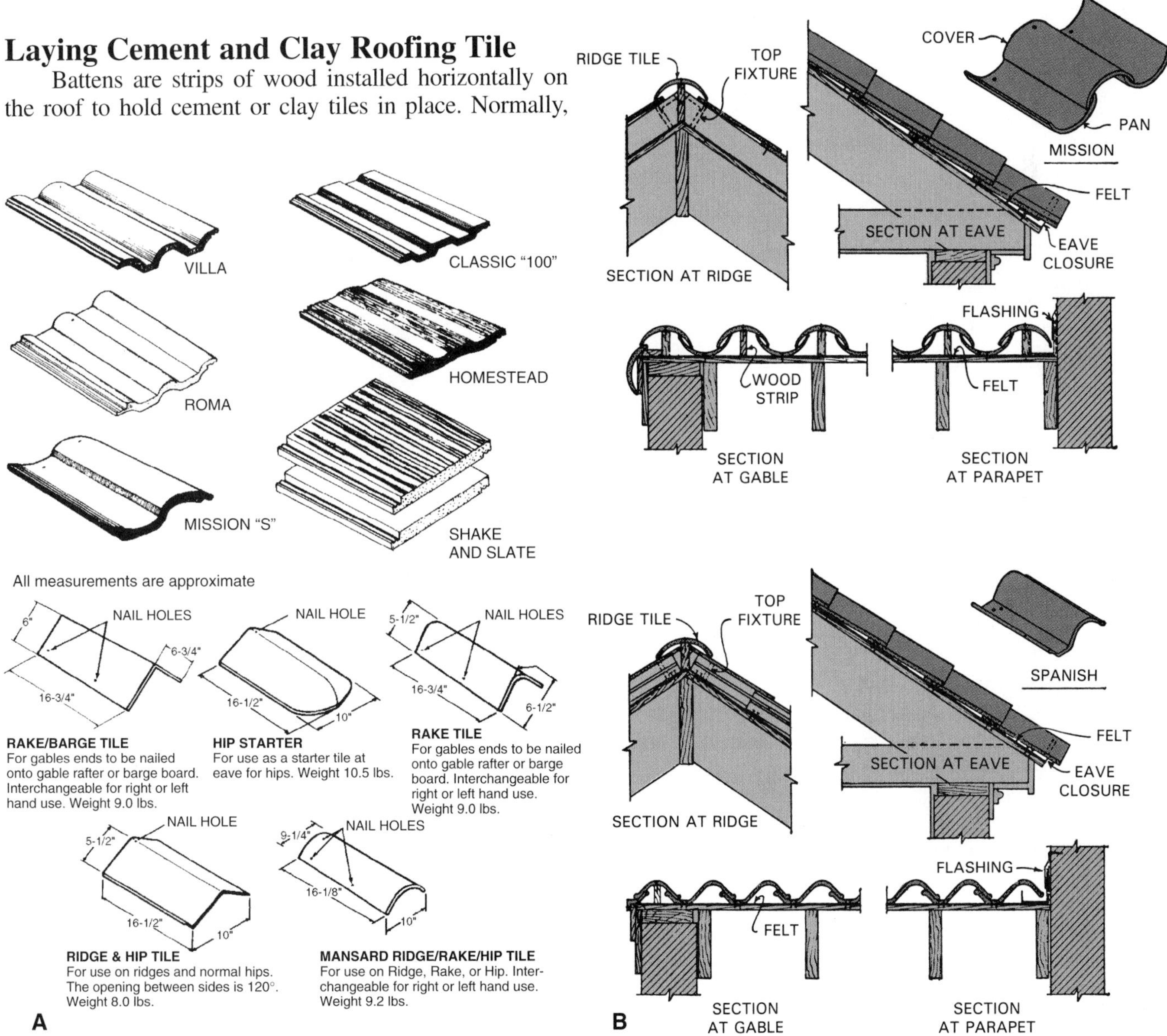

Figure 11-58. Roofing tile. A—Roof tile come in curved "barrel" shapes and flat. Special tile are manufactured for ridges, hips and rakes. (Monier Roof Tile) B—Typical application methods for tile roofs. Top—Two-piece pan and cover commonly known as Mission tile. Bottom—Spanish tile.

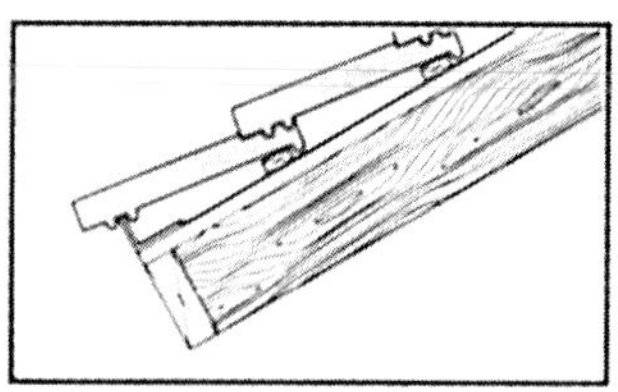

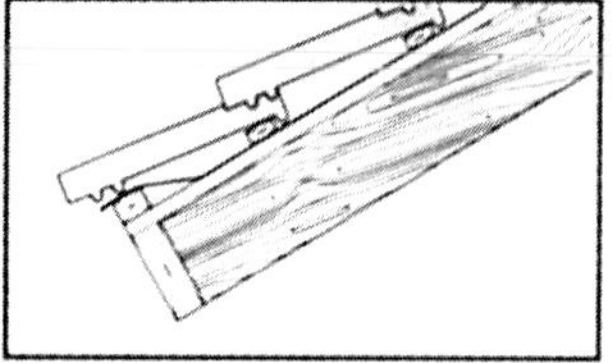

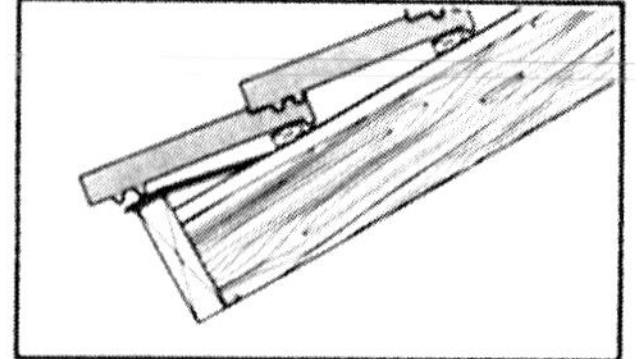

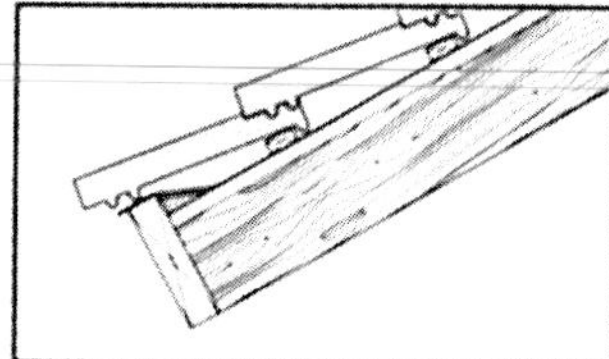

A

Figure 11-59. Tile roofing. A—Typical starter course details. B— Spanish tile, also called barrel tile, are durable and when properly installed, will withstand hurricane-force wind and rain. (Gory Associated Industries, Inc.) C—Flat concrete tile being laid.

Preparing a Valley

1. Lay down a 36" wide sweat sheet, dividing the width across the valley. (A *sweat sheet* is a strip of felt or ice and water barrier.)
2. Install the flashing. Valley flashing should be 28 gage corrosion resistant metal extending at least 11" each way. In the center of the flashing there should be a diverter rib not less than 1" high. Ends should be lapped 6".
3. Cut off the top of the eave riser strip and extend the metal valley flashing slightly over the eave and upward the length of the valley. Edges along the length of the flashing should be turned up 1/2" by 30°. Standard flashing usually comes preformed, ready for installation.
4. Put down underlayment, lapping it over the flashing.

Another method involves laying down the underlayment and weaving it across the valley before putting down the flashing. Figure 11-62 shows proper installation of flashing and underlayment. Figure 11-63 shows proper method of laying tile up to the valley. A circular saw fitted with a masonry cutting wheel is used to make angle cuts on tile. See Figure 11-64.

Figure 11-60. Battens made up of 1 x 2 lumber are installed across the roof and fastened with 8d nails. Battens are required on roof pitches below 3 in 12 and steeper than 7 in 12.

FLAT TILES

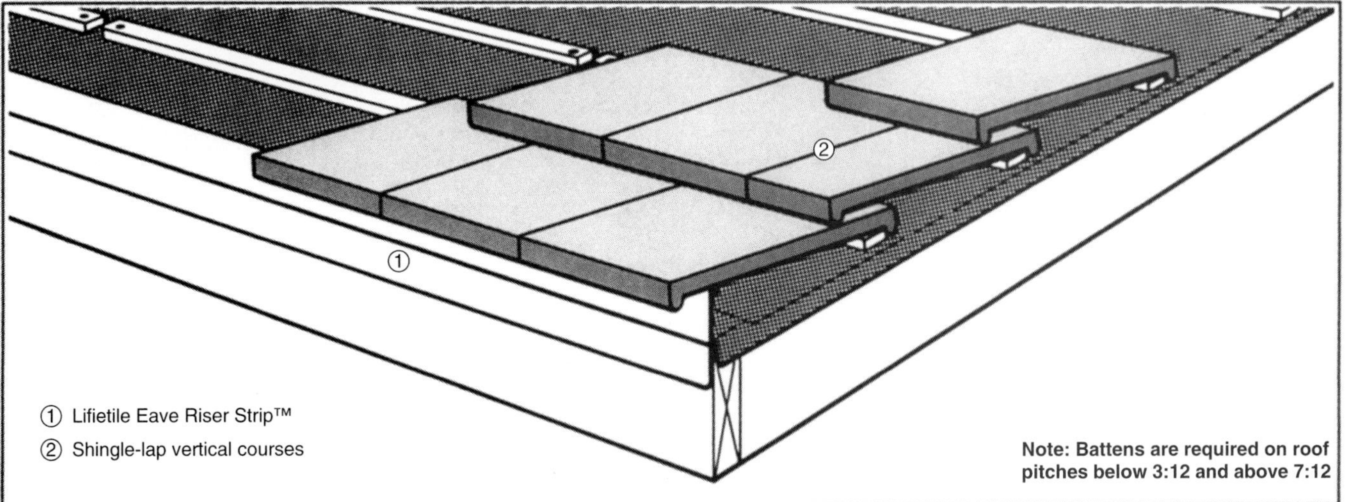

PROFILE TILES

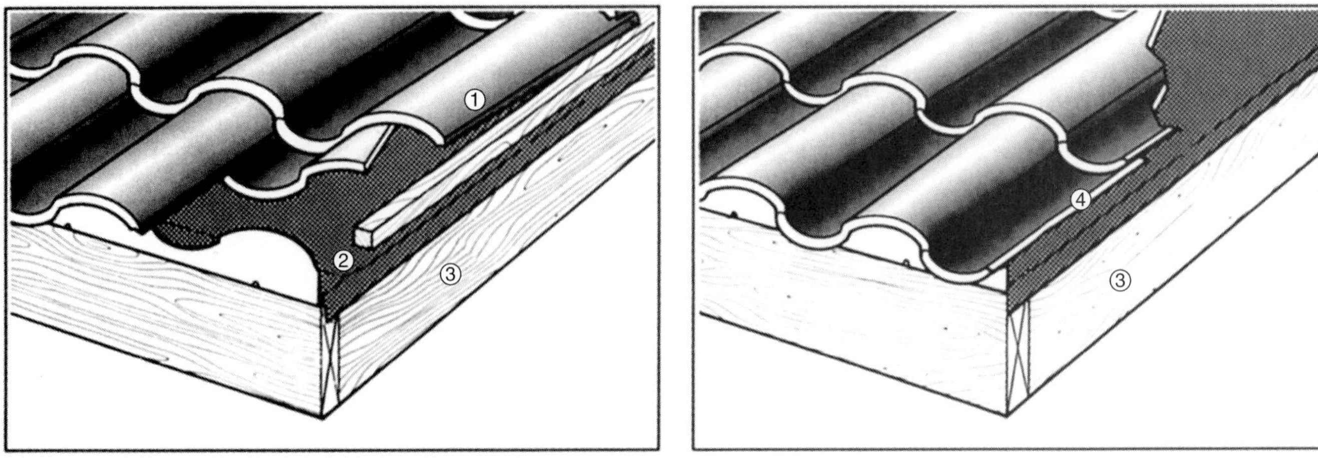

Figure 11-61. Starting to lay tile roofing. Top—Flat tile have joints staggered between courses like shingles. Bottom—Profiled (curved) tile have joints aligned from eaves to ridge.

Hips, Ridges, and Rakes

Nailer boards are wood strips installed on edge at hips and ridges to support trim tiles. Height of the ridge board will vary from 2 to 6". It must be high enough to maintain an even plane of trim tile. Trim tile are attached with one corrosion-resistant 10d nail. See Figure 11-65. Nose ends should be set in a bead of roofer's mastic that also covers the nail head.

Rake tile, Figure 11-66, should be fastened with two nails. The joints between field tile and trim tile should be weatherproofed with a bed of mortar or an approved dry ridge-hip system.

Tiles are extremely heavy and one of two methods should be used to deliver them to the roof. This should be done after the deck has been prepared with flashing, underlayment and battens. Be sure to include flashing around chimneys, pipes, and vents. Figure 11-67 shows roofing tile being delivered to roofs and distributed. Figure 11-68 shows proper installation of vent flashing.

Severe weather conditions and taller structures require extra steps when installing roofing tile. Where wind velocities may exceed 80 mph or where roofs may be more than 40' above the ground, the following guidelines should be followed. Be sure to consult building codes in these cases.

- The heads of every tile must be nailed.
- Noses of eave course must be fastened with special clips.
- Rake tiles must be nailed with two nails.
- Noses of ridge, hip, and rake tiles must be set in a bead of appropriate, approved roofer's mastic.
- Tiles cut too small for nailing should be set with an approved mastic or should be secured to the roof with wire.

General Installation: Typical Valley Details

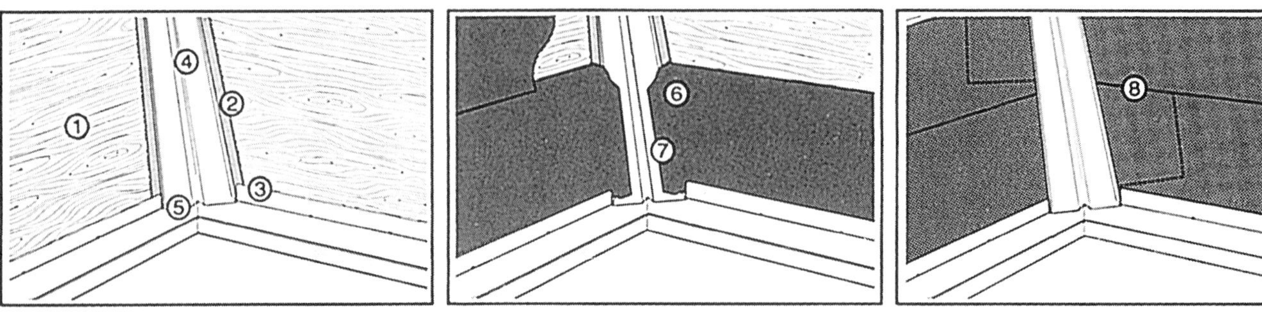

① Decking. ② 36" sweat sheet under valley flashing is recommended. ③ Cut off top of eave riser strip to permit valley drainage. ④ Standard G.I. valley flashing with crimped edges. ⑤ Extend valley flashing beyond eave riser strip. ⑥ Cut top corner of underlayment to insure proper diversion of water into valley flashing. ⑦ Overlap valley flashing with underlayment. ⑧ **Optional weaved underlayment treatment of valley.**

Figure 11-62. Flashing a valley. Standard flashing is available preformed.

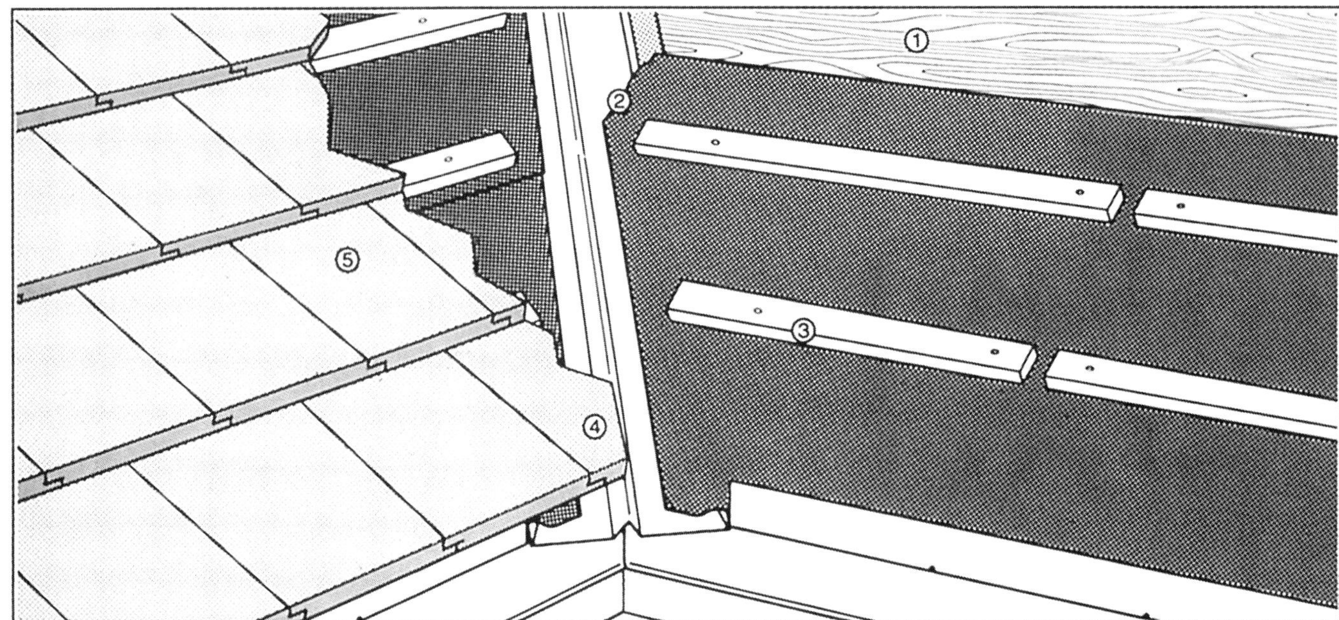

① Decking. ② Cut top corner of underlayment to insure proper diversion of water into valley flashing. ③ Battens are optional on roof pitches between 3:12 and 7:12. ④ Remove lugs under tiles which rest on valley flashing. ⑤ Shingle-lap tile courses on flat tile installation.

Figure 11-63. Proper tile installation at a valley. Lugs must be removed when the top of the tile rests on the flashing.

Figure 11-64. A masonry blade on a circular saw is use to make cuts on clay or concrete tile.

Metal Roofing

Buildings in areas of heavy snow often are given roof coverings of *metal sheeting* because snow tends to slide off before acquiring heavy accumulations. Sheets coated heavily with zinc galvanization (2 oz. per sq. ft.) or aluminum-zinc alloy are approved for permanent structures.

The sheets with lighter coatings of zinc are less durable and are likely to require painting every few years. On temporary buildings, and in cases where the most economical construction is required, lighter metal can be used. It will give satisfactory results if protected by paint.

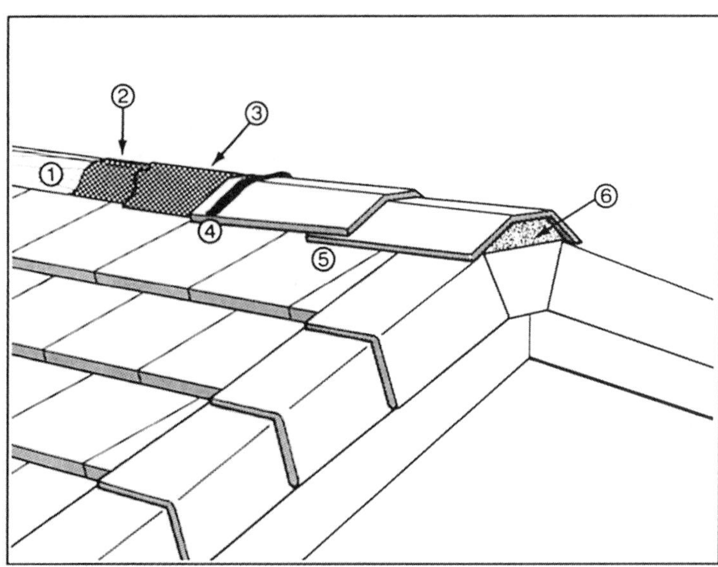

① Ridge nailer to be of sufficient height to maintain even plane of ridge tiles.

② Underlayment carried over or under ridge nailer.

③ Optional second layer of felt over nailer board as weatherblock.

④ Apply continuous bead of approved roofers' mastic at overlapping areas and over nail holes.

⑤ Provide minimum 3" headlap.

⑥ Optional mortar fill end treatment.

Note: Use one 10-penny nail per ridge tile.

General Installation: Hip Trim Details

It should be noted that the hip trim tiles, whether flat or profile, is similar. The options for hip trim tiles lie in the various treatments of end condition detailing. It is good roofing practice to apply hip trim tiles as each field tile course is completed. **Note: Hip nailer boards are to be of sufficient height to maintain an even plane of hip trim tiles. Nailer boards will vary dependent upon the type of tile selected. Suggested nailer sizes are as follows: Espana-2"x8"; Capri-2"x6"; Slate, Shake, Country, Chateau-2"x4". Drive one 10-penny nail per hip tile.**

① Underlayment carried over or under hip nailer.

② Optional second layer of felt over nailer board as weatherblock.

③ Apply contiuous bead of approved roofers' mastic at overlapping areas and over nail holes.

④ Provide minimum 3" headlap.

⑤ Hold back hip nailer 6" from eave edge.

Figure 11-65. Details of ridge and hip trim tile installation. Top—Typical ridge design. Bottom—Typical trim tile installation at a hip.

Figure 11-66. Rake tile are installed with two nails. (Roberto's Roofing)

Slope and Laps

Metal roofing sheets may be laid on slopes as low as 3 in 12 (1/8 pitch). If more than one sheet is required to reach the ridge, ends should lap no less than 8". When the roof pitch is 1/4 or more, 4" of end lap is usually satisfactory.

To make a tight roof, lap sides of corrugated sheets 1 1/2 corrugations, Figure 11-69. Wind-driven rain will likely be forced through single-corrugation lap joints. When using roofing 27 1/2" wide with 2 1/2" corrugations and 1 1/2 corrugation lap, each sheet covers a net width of 24" on the roof.

Sheet metal roofing is available in a form that looks like clay tile. See Figure 11-70. Computer controlled roll-forming machines produce a continual stepped-panel with tile forms about 7" wide and 12" long. Standard sheet size

Figure 11-67. Being extremely heavy, roofing tile should be delivered to the roof by some mechanical means. A—Conveyor system. (Del Webb's Sun City, Tucson, AZ) B—Lift truck. (Headlee Roofing) C—Tile are usually deposited in stacks of 6 at 33" intervals along the roof, skipping every other course. Follow tile manufacturers instructions.

General Installation: Vent Flashing

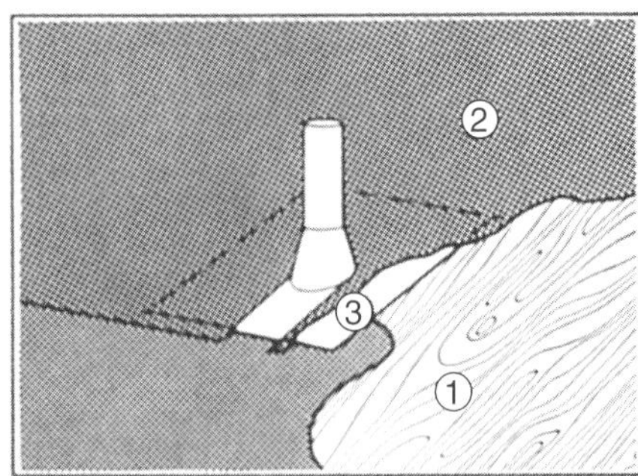

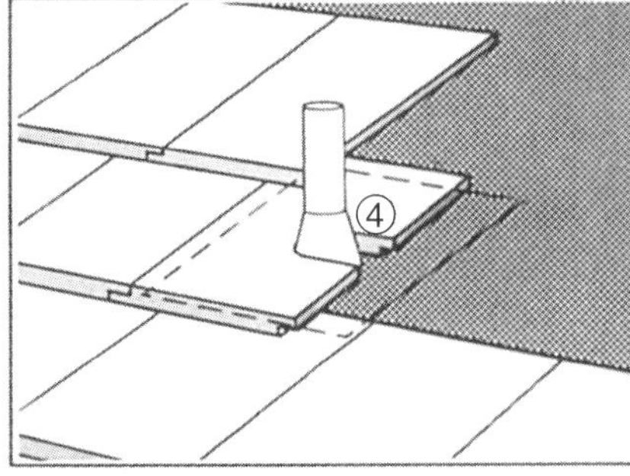

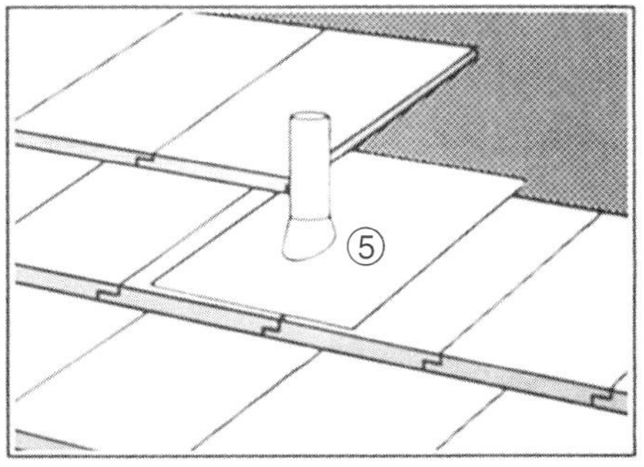

① Decking. ② Underlayment. ③ Standard G.I. base flashing. ④ Notch tile to accept flashing. ⑤ Standard G.I. top flashing. Seal with approved roofers mastic.

Note: For Espana and other higher profile tiles, use lead or other flexible type flashing, minimum 9" skirt around roof projection. Attach to tile with approved roofers' mastic.

Figure 11-68. Vent flashing for tile roofing is similar to the method used for wood or asphalt shingles. With some high-profile tiles, lead or other flexible flashing is used

is 36" wide and from 2 to 20' long. The product is available in a variety of colors and is guaranteed for 20 years.

Sheathing and Nails

If 26 gage sheets are used, supports may be 24" apart. If 28 gage sheets are used, supports should be not more than 12" apart. The heavier gage has no particular advantage except its added strength. A zinc coating for durability is more important than strength for this type of roofing.

For best results, galvanized sheets should be fastened with lead-headed nails or galvanized nails and lead washers. Nails properly located are driven only into tops of the corrugations. To avoid corrosion, use nails specified by the manufacturer.

Aluminum Roofing

Corrugated aluminum roofing, if properly applied, usually makes a long-lasting roof. Seacoast exposure tests reported by the Bureau of Standards indicate this material is capable of resisting corrosion in such localities unless subjected to direct contact with salt-laden spray. Where this is likely to happen, aluminum roofing is not recommended.

Aluminum alloy sheets available for roofing usually have a corrugation spacing of 1 1/4 or 2 1/2". Recommendations for the installation of sheet metal regarding side lap and end lap are applicable to the laying of aluminum sheets.

An important precaution to observe in laying aluminum roofing is to make sure that contact with other kinds of metal is avoided. Where this is not possible, both metals should be given a heavy coating of asphalt paint wherever the surfaces are in contact.

As aluminum is soft and the sheets used for roofing are relatively thin, they should be laid on tight sheathing or on decks with openings no more than 6" wide. Aluminum roofing should be nailed with no less than 90 nails to a square or about one nail for each square foot. It is recom-

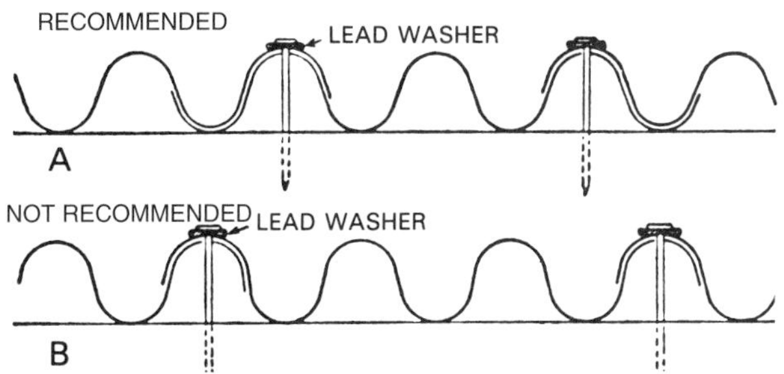

Figure 11-69. Guide to application of corrugated sheet metal roofing. A—Sheets properly laid with one-and-one half corrugation lap. B—Single corrugation lap is not recommended.

Figure 11-70. Sheet metal roofing roll-formed to resemble mission style tile. A variety of colors are available. (Met-Tile Inc.)

mended that aluminum alloy nails be used and that nonmetallic washers be used between nail heads and the roofing.

If desired, the sheathing may be covered with water-resistant building paper or asphalt impregnated felt. Paper that absorbs and holds water should never be used.

To avoid corrosion, aluminum sheets should be stored so that air will have free access to all sides. Otherwise, a white deposit will form. This deposit creates pinholes very quickly.

Terne Metal Roofing

Terne metal roofing is made of copper-bearing steel, heat-treated to provide the best balance between malleability (easy to form) and toughness. It is hot dip-coated with Terne metal, an alloy of 80% lead and 20% tin. The high weather resistance factor (notable in this type of roofing) is due primarily to the lead. Tin is included because the alloy makes a better bond with steel.

Grades are expressed as the total weight of the coating on a given area. This area, by old trade custom, is the total area contained in a box of 112 sheets that are 20" x 28", and amounts to 436 sq. ft. The best grade of Terne coating is 40 lb. It provides a roof surface that will last for many years.

A wide variety of sheet sizes are available, as well as 50' seamless rolls in various widths. This permits its use for many different types of roofs and methods of application. It is used extensively for flashing around both roof and wall openings.

For best appearance and longest wear, Terne metal roofs must be painted. Use a linseed oil-based iron oxide primer for a base coat. Almost any exterior paint and color can be used over this base.

Aluminum Shakes

Aluminum shakes are manufactured of an aluminum-magnesium alloy with a nominal 0.019" thickness. They are available in brown, red, dark gray, white, and natural aluminum. For installation, see manufacturer's instructions.

Zinc-Aluminum Coated Steel Roofing

A typical zinc-aluminum coating is 55% aluminum, 43% zinc, and 2% silicon. This coating is applied by a hot-dip process. Next, a chromate treatment adds corrosion resistance. This is followed by a primer and top coatings of polyesters, silicone, fluorocarbons, or plastisols in a large selection of colors.

However, before coatings are applied, the flat, 29 gage sheets are passed through forming rolls that stiffen and strengthen the sheets by forming ribs in them. See Figure 11-71.

In residential and light commercial construction, zinc-aluminum coated roofing sheets should be applied to solid decking. The recommended substrate (underlayment) is 30 lb. asphalt-saturated roofing felt. Other suitable barrier materials may be substituted. An ice and waterproof membrane should be applied at the eaves in cold climates where ice dams are a problem.

If metal is used for reroofing, it is best to remove old roof coverings. If this is not practical, hot mop a layer of underlayment and then put down 2 by 2 battens at intervals. Mop more asphalt over this construction. If additional roof insulation is needed, sheet insulation can be applied between the battens. Cross battens are laid horizontally across the roof and roofing sheets are fastened to the cross battens. Use only approved self-penetrating, self-tapping screws. Do not overdrive screws; this will cause panel distortion. In some panel designs, screws are concealed by attaching the next panel.

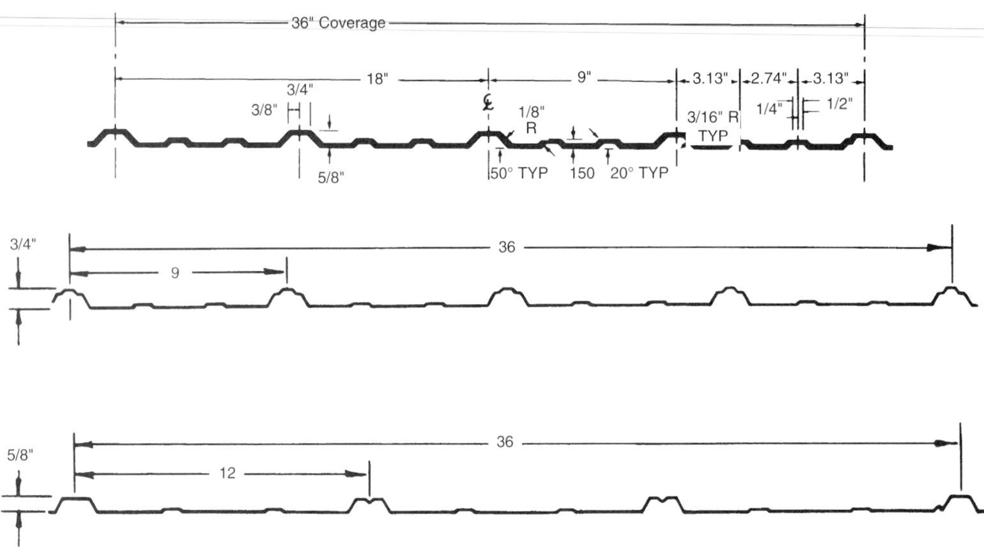

Figure 11-71. Profile of several styles of metal roofing sheets. Ribs help stiffen sheets and provide for leakproof overlapping. (American Building Components)

Valleys, Ridges, and Hips

Valley flashing consists of a layer of ice and water barrier and a preformed metal valley flashing extending at least 9" on either side of the valley. Ends are lapped 12". If other types of flashing are used, they should usually extend up to 20" on either side of the valley. Figure 11-72 shows a typical metal roof on an older house.

Cutting Metal Panels

It will be necessary to cut panels at valleys and hips. Use a circular saw. Cut with the exterior surface turned down. This prevents damage to the finish. Remove steel chips, which can rust and spoil the finish.

Typically, No. 10 by 1" hex washer head screws are installed spaced 18 to 24" O.C. In areas of high winds, closer spacing may be required. Consult local codes. Use a driving tool with a variable speed of 200–2500 rpm and a depth-sensing nose piece.

Various accessories are available for closures and fittings. Their application is shown in Figure 11-73.

Ridge Vents for Asphalt Roofing

Attic space builds up moisture and high temperatures in certain seasons. Venting at the eaves, gable ends, and at or near the ridge helps dissipate heat and moisture. Figure 11-74 shows one system of roof venting. A gap of about 2 or 3" must be left in the sheathing at the ridge to allow air to pass through to the vent. Shingles are secured over the patented vent to keep out weather. Airways, Figure 11-74(D), are often installed on the underside of the sheathing at the eaves to keep insulation from closing off air passage from soffit to ridge.

Gutters

Gutters or eaves troughs collect rainwater from the edge of the roof and carry it to downspouts. Downspouts direct water away from the foundation or into a drainage system.

The term *gutter* refers to a separate unit that is attached to the eave. The term *eaves trough* usually applies to a waterway built into the roof surface over the cornice.

Eaves troughs must be carefully designed and built since any leakage will penetrate the structure. Because of this and the extra cost of construction, they are seldom used in modern residential work.

Wood Gutters

Wood gutters of fir or red cedar are used in many parts of the country. When properly installed and maintained, the life of wood gutters is usually equal to that of the main structure.

Figure 11-72. Metal roofing has been installed during a reroofing. Note valley treatment.

The installation of modern wood gutters is usually made before the shingles are applied. One type is attached to the fascia board after the roof is sheathed.

Some designs are coordinated with the fascia board and are installed with this trim unit before the roof sheathing is complete. In this case, the sheathing overhangs the fascia and gutter, thus reducing the possibility of leakage. When wood gutters are specified, the architectural plans usually include installation details.

Cutting, fitting, and drilling is done on the ground before the various units are set in place. Most wood guttering is primed or prepainted. Always use galvanized or

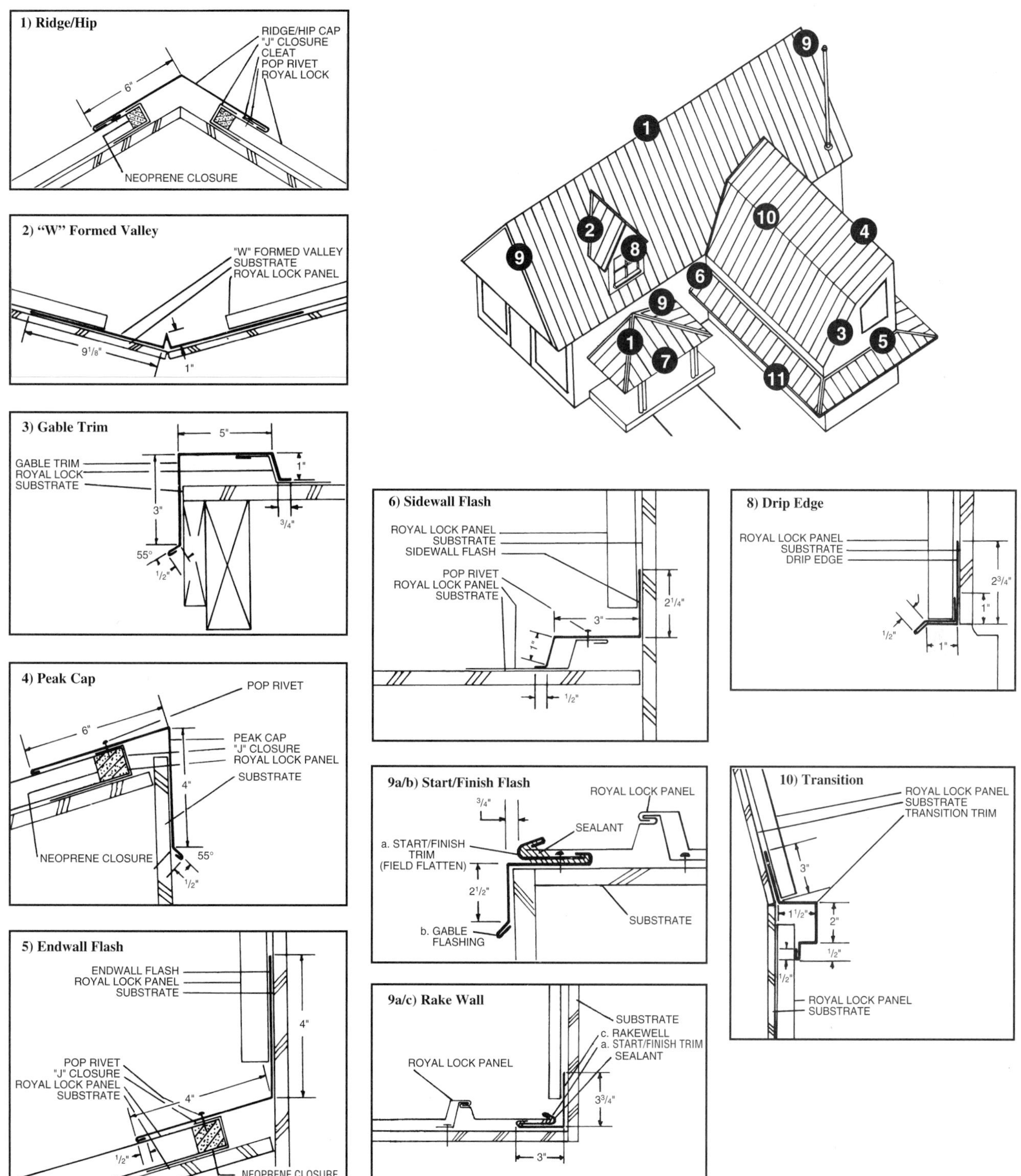

Figure 11-73. Various accessories are available for proper installation of metal roofing. Their application is shown in the individual drawings. (American Building Components)

Figure 11-74. Ridge venting system. A—Openings at ridge allow free passage of air. B—Roll vent is being installed. C—Venting roll is covered with shingles. D—Airway is designed for installation at eaves. It prevents insulation cutting off free passage of air from the soffit vents. (Benjamin Obdyke Inc.)

other types of weatherproof nails and/or brass wood screws.

Gutter ends may be sealed with blocks, returned, and mitered, or butted against an extended rake frieze board. In general, the gutter is treated like cornice molding and should present a smooth, trim appearance.

The roof surface should extend over the inside edge of the gutter with the front top edge at approximately the height of a line extended from the top of the sheathing. For correct appearance, wood gutters are set nearly level. They will drain satisfactorily if kept clean and if adequate downspouts are provided.

Metal and Plastic Gutters

A wide variety of metal gutters are available to control roof drainage. Manufacturers have perfected gutter and downspout systems that include various component parts. Whole systems can be quickly assembled and installed on the building site. Materials consist of galvanized iron and aluminum. Many systems are available in either a primed or prefinished condition to match a wide range of colors. Plastic guttering systems are used extensively on modern residences. The system shown in Figure 11-75 is molded from vinyl.

Gutter systems include inside and outside mitered corners, joint connectors, pipes, brackets, and other items. All are carefully engineered and fabricated. Parts slip together easily and are generally held with soft pop-rivets or sheet metal screws. Figure 11-76 shows standard parts of a typical gutter and downspout system and how they are assembled.

Figure 11-77 shows a gutter system that provides both gutter and fascia in a single unit. This unit also includes a channel to support the outside edge of a prefabricated soffit of either panel or coil stock. The system saves time and material and results in a clean-line appearance. In new construction, it can be mounted directly on the rafter tails, as shown, or it can be attached to an existing fascia in remodeling work.

For best results, gutters and eaves troughs must be sized to suit the roof areas from which they receive water. For roof areas up to 750 sq. ft., a 4" wide trough is suitable. For areas between 750 and 1400 sq. ft., 5" troughs should be used. For larger areas, a 6" trough is recommended. Quality of gutters, like that of flashing, should correspond to the durability of the roof covering. If galvanized steel guttering is used, it should have a heavy zinc coating.

The size of downspouts or conductor pipes required also depends on the roof area. For roofs up to 1000 sq. ft., downspouts of 3" diameter have sufficient capacity if properly spaced. For larger roofs, 4" downspouts should be used.

Gutters having the proper slope stay clean. Metal gutters are usually sloped 1" for every 12 to 16' of length.

Estimating Material

To estimate roofing materials, first calculate the total surface area to be covered. In new construction, the figures used to estimate the sheathing can also be used to estimate the underlayment and finished roofing materials. When these figures are not available, they can be calculated by the same methods used for roof sheathing, described in Unit 10.

Another method used to estimate roof area is to determine the total ground area of the structure. Include all eave and cornice overhang. Convert the ground area to roof area by adding a percentage determined by the roof slope, as follows:

Slope 3 in 12, add 3% of area.
Slope 4 in 12, add 5 1/2% of area.
Slope 5 in 12, add 8 1/2% of area.
Slope 6 in 12, add 12% of area.
Slope 8 in 12, add 20% of area.

Figure 11-75. Modern gutter systems are easy to assemble and install. They are available in either a primed or prefinished condition to match a wide range of colors.

Divide the total square feet of roof surface by 100 to find the number of squares to be covered. For example, if the total ground area, plus overhang, is found to be 1560 and the slope of the roof is 4 in 12, apply the following calculations:

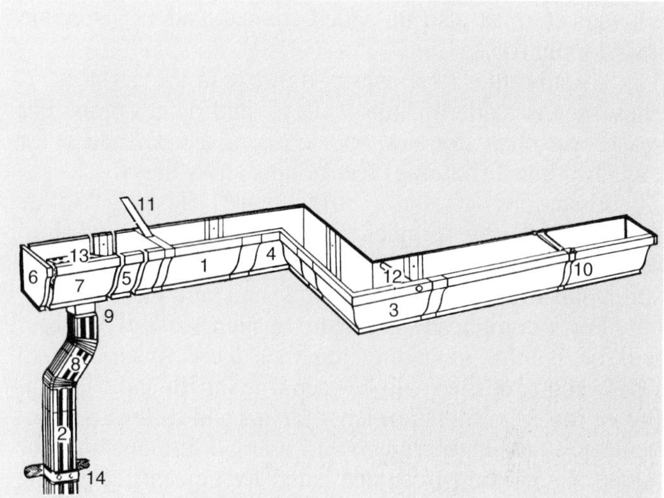

KEY	DESCRIPTION	KEY	DESCRIPTION
1	5" K GUTTER	9	K OUTLET TUBE (With Flange)
2	3" SQUARE CORRUGATED DOWNSPOUT	10	5" K FASCIA HANGER
3	5" K MITER (Outside)	11	5" K STRAP HANGER
4	5" K MITER (Inside)	12	7" SPIKE (Aluminum) 5" FERRULE (Aluminum)
5	5" K SLIP JOINT CONNECTOR	13	5" K STRAINER
6	5" K END CAP LEFT OR RIGHT	14	3" PIPE BAND (Ornamental)
7	5" x 3" K END SECTION WITH OUTLET TUBE	15	TOUCH-UP PAINT SPRAYON TOUCH-UP PAINT (White Only)
8	3" SQUARE CORRUGATED 75° ELBOW OR 60° ELBOW STYLE A AND B (A, B)	16	GUTTER SEAL (Tube or Cartridge)

Figure 11-76. Metal gutter system. Parts can be assembled on the job. (Crown Aluminum Industries)

Roof Area = 1560 + (1560 x 5 1/2%)
= 1560 + (1560 x .055)
= 1560 + 85.80 or 86 = 1646
Number of squares = 16.46 or 16 1/2

After the number of squares is established, additional amounts must be added. For asphalt shingles, it is generally recommended that 10% be added for waste. This, however, may be too much for a plain gable roof and too little for a complicated intersecting roof. Certain allowances must also be added for reduced exposure on low sloping roofs.

Usually the 10% waste figure can be reduced if allowance is made for hips, valleys, and other extras. For wood or asphalt shingles, one square is usually added for each 100 lineal (running) feet of hips and valleys.

Quantities of starter strips, eaves flashing, valley flashing, and ridge shingles must be added to the total shingle requirements. All of these are figured on lineal measurements of the eaves, ridge, hips, and valleys.

For a complicated structure, a plan view of the roof will be helpful in adding together these materials. All ridges and eave lines will be seen true length and can simply be scaled to find their length. Hips and valleys will not be seen as their true length and a small amount must be added. Using the percentage listed for converting ground area to roof area will usually provide sufficient accuracy.

On a reroofing job there is a simple method for determining the roof pitch when it is not known. You can estimate it from the ground with the help of a folding carpenter's rule.

Stand away from the building some distance and fold the rule into a triangle. Hold the folded rule at arm's length and frame the roof inside the triangle. Adjust the triangle until slope of its sides line up with the roof, as in Figure 11-78. Be sure the base of the triangle is level.

Read off the dimensions on the base of the rule that is marked the "reading point" (Figure 11-78). Now, refer to the chart in Figure 11-79 and locate the proper pitch and slope.

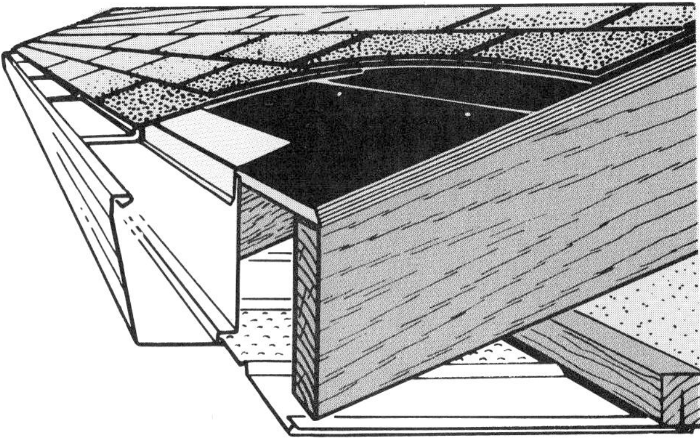

Figure 11-77. This gutter system also acts as fascia and support for soffit. (Omni Products)

Safety

Worker safety is important everywhere on a construction site. Working on a pitched roof can be made safer by observing certain precautions. These include:

- **Wear boots or shoes with rubber or crepe soles. Soles and heels must be in good condition; if worn, discard or have them repaired.**
- **If the roof is steep, workers should wear safety belts tied off to a fall-resistant device. This same precaution applies for shallower pitches if there is danger of serious injury or death from a fall.**
- **Rain, frost, or snow makes a roof slippery; wait until the surface is dry. If work must go on under these conditions, workers should tie off or wear special roof shoes with skid-resistant cleats.**
- **Keep a broom or brush handy to sweep the roof clear of sawdust, loose debris, or dirt.**
- **Install shingle underlayment as soon as possible. Such material reduces danger of slipping.**
- **Install temporary 2 x 4 cleats as toe holds. They can be removed as shingles are installed.**
- **Remove unused tools, cords, and other loose items from the roof; they can be serious hazards. In addition to the foregoing precautions, check local and state OSHA requirements. Be alert to other potential hazards and practice common sense. Taking chances often leads to injuries.**

Important Terms

Asphalt shingles
Backing board
Base flashing
Battens
Built-up roof
Cant strip
Cap flashing
Closed-cut valley
Counter flashing
Coverage
Cricket
Drip edge
Eaves flashing
Eaves trough
Exposure
Feathering strips
Flashing
Head lap
Four inch method
Froe-end
Gravel stop
Gutter
Metal sheeting
Nailer boards
Open-valley flashing
Plies
Rake
Roll roofing
Roofing tile
Saddle
Saturated felt
Selvage
Shingle butt
Side lap
Six inch method
Square
Starter strip
Sweat sheet
Terne metal roofing
Underlayment
Wood shakes
Woven-valley flashing

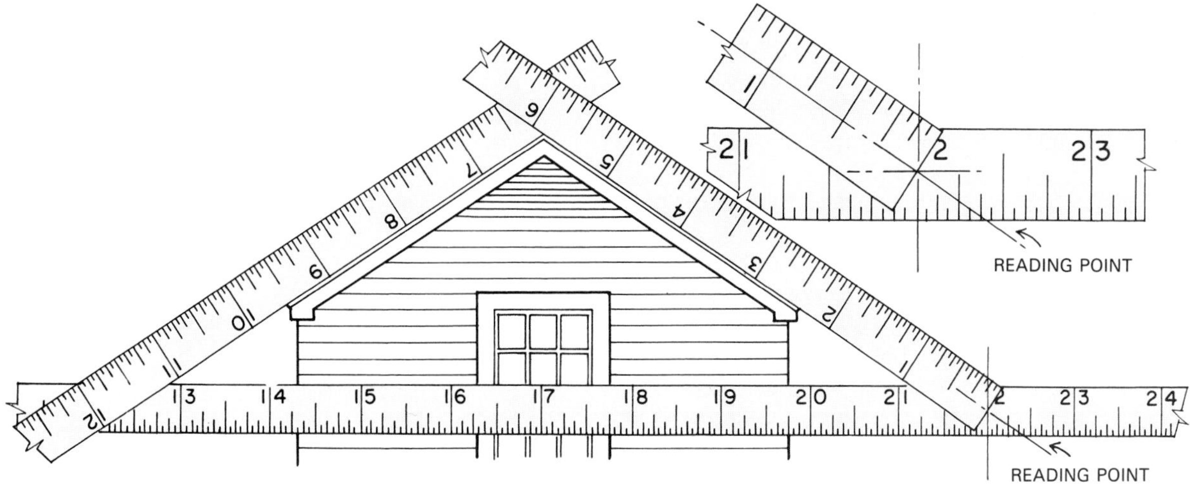

Figure 11-78. Stand a distance from the gable end and frame the roof inside a folded carpenter's rule held at arm's length. Adjust end of rule to get a proper reading at the "reading point." (Asphalt Roofing Manufacturers Assoc.)

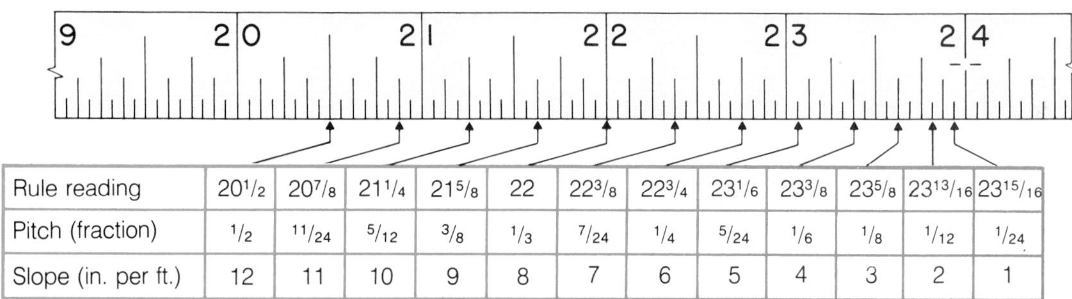

Rule reading	20 1/2	20 7/8	21 1/4	21 5/8	22	22 3/8	22 3/4	23 1/6	23 3/8	23 5/8	23 13/16	23 15/16
Pitch (fraction)	1/2	11/24	5/12	3/8	1/3	7/24	1/4	5/24	1/6	1/8	1/12	1/24
Slope (in. per ft.)	12	11	10	9	8	7	6	5	4	3	2	1

Figure 11-79. Reading point conversions. Locate reading point on the chart and read downward to find pitch and slope. (Asphalt Roofing Manufacturers Assoc.)

Test Your Knowledge

1. The selection of roofing materials is influenced by such factors as cost, durability, appearance, application methods, and _____________ of the roof.
2. In shingle application, the distance between the edges of one course and the next, measured at a right angle to the ridge, is called _____________.
3. Asphalt saturated felt and ice/water barrier are available in rolls that are _____________ wide.
4. Three-tab square-butt shingles are _____________ inches long.
5. The most commonly used underlayment for asphalt shingles is _____________ lb. roofer's felt.
6. The waterway of a valley should diverge (grow wider) as it approaches the eaves at a rate of _____________ inches per foot.
7. The minimum number of nails recommended for the application of each three-tab square butt shingle unit is _____________.
8. Indicate the correct answer(s). When using a power staple gun for attaching asphalt shingles, _____________.
   A. install staples 5/8" below adhesive strips
   B. adjust staple gun pressure for proper staple application
   C. be sure staples are sunk well into shingle surface
   D. if a staple must be removed, repair hole with asphalt cement
   E. gun should have enough pressure to drive the staple flush with the top of shingle
9. What is the size of a metric three-tab asphalt shingle?
10. Chimney flashing consists of two parts: the cap or counter flashing and the _____________ flashing.
11. Asphalt shingles can be used on roofs with a slope as low as _____________ if special application procedures are followed.
12. Where is a chimney saddle used and why?
13. List the mineral materials which may form the top coat of a built-up roof.
14. A formed metal strip, called a _____________ is attached to the edge of built-up roofs.
15. In the best grade of wood shingles, the annual rings run _____________ (*parallel, perpendicular*) to the surface.

16. When laying wood shingles, they should be spaced _____________ inches apart (horizontally) to provide for expansion when they become rain soaked.
17. The vertical joints between wood shingles in adjacent courses should be spaced at least _____________ inches apart.
18. A strip of 30 lb. roofing felt is placed between each course when applying _____________.
19. How does a carpenter determine and provide for the strength of roof trusses that must support a tile roof?
20. Can roofing tile be installed on a roof with less than a 3 in 12 pitch; and, if so, what are the recommendations to prevent failure of the roofing materials?
21. When installing corrugated steel roofing, the joints should be lapped _____________ (*1, 1 1/2, 2*) corrugations.
22. Terne metal roofing consists of sheets of copper-bearing steel coated with an alloy of _____________ and _____________.
23. What is the main advantage of a metal roof in cold climates?
24. The vertical pipes of a gutter system are called _____________ or _____________.

Outside Assignments

1. Study the kinds and qualities of asphalt shingles used in your locality. Obtain manufacturer's literature from a local builders supply center. Prepare a report including information about kinds, grades, and costs. Also include information about materials for underlayment, valley flashing, hip and ridge finish, and fasteners.
2. Report on application procedures used to install a roof on a residence in your community. Visit the building site and observe the methods used. Note the type of sheathing, special preparation of the roof deck, how valley flashing is applied, use of drip edges, type of starter courses, procedures used to align shingle courses, and how ridges are finished. Always be sure to obtain permission from the foreman or head carpenter when visiting a building site.
3. Construct a full-size visual aid showing the application of wood or asphalt shingles. Use a piece of 3/4" plywood about 4' square to represent a lower corner of a roof deck. Apply underlayment (if required), drip edges, starter strips, and then carefully lay the shingles according to an approved pattern. By making only a partial coverage of the various layers, all of the application steps and materials used can be easily studied and observed.

12

Windows and Exterior Doors

Windows and doors are an important part of a structure. The carpenter should:

- Have a basic understanding of the various types, sizes, and standards of construction.
- Be able to recognize good quality in materials, fittings, weather stripping, and finish.
- Appreciate the importance of careful installation of the various units.
- Be an expert in installation.

Placement of the outside doors and the windows starts when the sheathing has been installed, Figure 12-1. Careful installation will assure a close fit so that air infiltration is kept at a minimum.

Manufacture

Today, windows and doors are built in large millwork plants. They arrive at the building site as completed units ready to be installed in the openings of the structure.

Some windows used in residences are made from aluminum and steel. Most of them, however, are made from wood. Since wood does not transmit heat readily, there is less tendency for window frames made of wood to become cold and condense moisture vapor.

Wood will decay under certain conditions and must be treated with preservatives. Exposed surfaces must be painted or clad with metal or plastic. Metal is stronger than wood and thus permits the use of smaller frame members

Figure 12-1. This residential structure has been framed, sheathed, and roofed. It is ready for the installation of windows and exterior doors. Note that the sidewalk and driveway have been poured and finished. This will make it easier to deliver the windows, doors, and other products and materials needed for completion of the structure.

around the glass. Aluminum has the added advantage of a protective film of oxide, which eliminates the need for paint.

Ponderosa pine is commonly used in the fabrication of windows. It is carefully selected and kiln dried to a moisture content of 6 to 12%.

Manufacturing Standards

The control of quality in the manufacture of windows is based on standards established by the National Woodwork Manufacturers Association. These standards cover every aspect of material and fabrication. Included are such details as the projection of the drip cap and the slope of the sill.

For example, NWMA industry standard I.S. 2-74 states that the weather stripping for a double-hung window shall be effective to the point that it will prevent air leakage in excess of 0.50 cfm (cu. ft. per minute) per linear foot of sash crack when tested at a static air pressure of 1.56 psf (lb. per sq. ft.). This is equivalent to the pressure subjected by a 25 mph wind. Standards also cover the methods and procedures used in preservative treatments.

Figure 12-2. Eight light double-hung window is in a traditional style and has an air infiltration rate of less than one-third of the industry standard of 0.5 cfm. (Rolscreen Co.)

Types of Windows

In general, windows can be grouped under one or a combination of three basic headings:

- Sliding.
- Swinging.
- Fixed.

Each of these includes a variety of designs or method of operations. Sliding windows include the double-hung and horizontal sliding. Swinging windows that are hinged on a vertical line are called casement windows while those hinged on a horizontal line can be either awning or hopper windows.

Double-hung

A *double-hung window* consists of two *sash* that slide up and down in the window frame. These are held in any vertical position by a friction fit against the frame or by springs and various balancing devices. Double-hung windows are widely used because of their economy, simplicity of operation, and adaptability to many architectural designs.

Figure 12-2 shows an outside view of a double-hung window unit. Screen and storm sash are installed on the outside of the window.

Horizontal Sliding

Horizontal sliding windows have two or more sash. At least one of them moves horizontally within the window frame. The most common design consists of two sash, both of which are movable. See Figure 12-3. When three sash are used, the center one is usually fixed.

Figure 12-3. Horizontal sliding window is sometimes called "glide-by" unit. (Marvin Windows and Doors)

Casement

A *casement window* has a sash that is hinged on the side and swings outward. Installations usually consist of two or more units, separated by *mullions*. Sash are operated by a cranking mechanism or a push-bar mounted on the frame, Figure 12-4. Latches are used to close and hold the sash tightly against the weather stripping.

The swing sash of a casement window permits full opening of the window. This provides good ventilation. Frequently, fixed units are combined with operating units where a row of windows is desirable.

Crank operators make it easy to open and close windows located above kitchen cabinets or other built-in fixtures. Screen and storm sash are attached to the inside of standard casement windows.

Figure 12-4. Casement windows are hinged on the side and are controlled by a crank or push bar.

Awning

Awning windows, Figure 12-5, have one or more sash that are hinged at the top and swing out at the bottom. They are often combined with fixed units to provide ventilation. Several operating sash can be stacked vertically in such a way that they close on themselves or on rails that separate the units.

Most awning windows have a so-called projected action where sliding friction hinges cause the top rail to move down as the bottom of the sash swings out. Crank and push bar operators are similar to those on casement windows. Screens and storm sash are mounted on the inside.

Awning windows are often installed side by side to form a "ribbon" effect. Such an installation provides privacy for bedroom areas and also permits greater flexibility in furniture arrangements along outside walls.

Consideration of outside clearance must be given to both casement and awning windows. When open, they may interfere with movement on porches, patios, or walkways that are located adjacent to outside walls.

Figure 12-5. Awning windows have one or more sash that are hinged at the top. (Marvin Windows and Doors)

Hopper

The *hopper window* has a sash that is hinged along the bottom and swings inward, Figure 12-6. It is operated by a locking handle located in the top rail of the sash. Hopper windows are easy to wash and maintain. They often interfere with drapes, curtains, and the use of inside space near the window.

Multiple-Use

The *multiple-use window* is a single outswinging sash designed so it can be installed in either a horizontal or vertical position. Figure 12-7 shows two units installed to operate similar to a casement window. These windows are simple in design and do not require complicated hardware.

Jalousies

A *jalousie window* is a series of horizontal glass slats held at each end by a movable metal frame. The metal frames are attached to each other by levers. The slats tilt together in about the same manner as a venetian blind. Jalousie windows provide excellent ventilation. Weathertightness values are low. Their use in northern climates is usually limited to porches and breezeways.

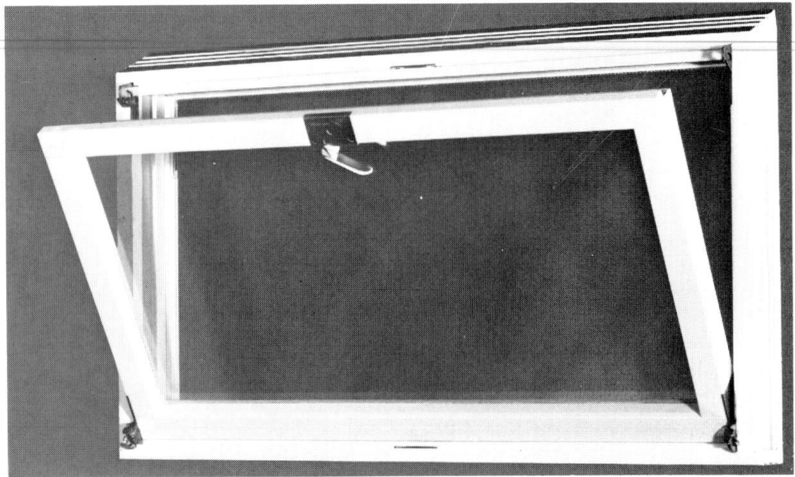

Figure 12-6. Hopper window is hinged at bottom and swings inward. (Andersen Corp.)

Figure 12-7. Multiple-use windows have been installed vertically. (Andersen Corp.)

Fixed

The *fixed window* can be used in combination with any of the movable or ventilating units. Its main purpose is to provide daylight and a view of the outdoors. When used in this type of installation, the glass is set in a fixed sash mounted in a frame that will match the regular ventilating windows. Large sheets of plate glass are often separated from other windows to form "window-walls." They are usually set in a special frame formed in the wall opening.

Window Heights

One of the important functions of a window is to provide a view of the outdoors. The architect will be aware of the various considerations and drawings should reflect the dimensions shown in Figure 12-8. Kitchens to be used by handicapped in wheelchairs will have different window requirements, of course.

In residential construction, the standard height from the bottom side of the window head to the finished floor is 6'-8". When this dimension is used, the heights of window and door openings will be the same. If inside and outside trim must align, 1/2 to 3/4" must be added to this height for thresholds and door clearances. Window manufacturers usually provide exact dimensions for their standard units. See Figure 12-9.

Window Glass

Sheet glass used in regular windows is produced by *floating*. In this process, melted glass flows onto a flat surface of molten tin in a vat more than 150' long. As it flows over the tin, a ribbon of glass is formed that has smooth, parallel surfaces. The glass cools, becoming a rigid sheet and is then carried through an annealing oven on smooth rollers. Finally, the continuous sheet of glass is inspected and cut to usable sizes. Figure 12-10 lists standard thicknesses produced. The letters SS and DS stand for *single-strength* and *double-strength.*

Energy Efficient Windows

Glass areas of a dwelling account for much of the heat loss in winter and heat gain in summer. Glass conducts heat more readily than most other building materials.

The resistance of any material to passage of heat is measured in R-values. When a building material has a low R-value, it means that the material has little resistance to heat passage.

A single pane of glass has an R value of about 0.88. A second pane of glass with a 1/2" air space will increase the R-value to about 2.00. *Storm* sash have long been used to improve the R-value of window space. Normally, it is

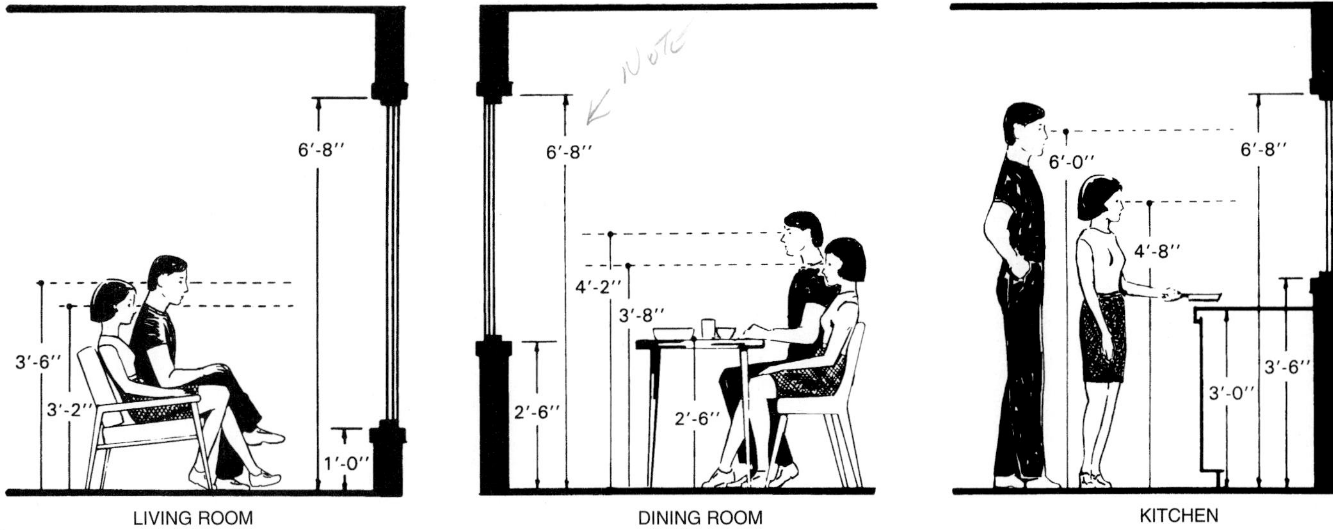

Figure 12-8. Architectural drawings should take into account the standards for window heights. Horizontal framework should be avoided at eye levels shown.

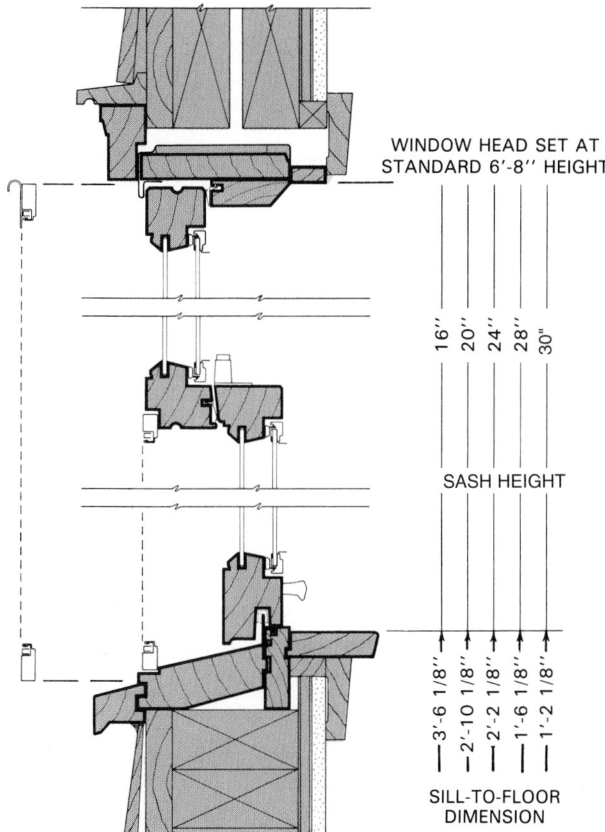

Figure 12-9. Manufacturer's product literature gives window head and sill heights. (Rolscreen Co.)

attached to the outside of the window frame and is removed and stored in the spring. "Triple track" storms did away with the removal and storage problem.

Another variation of the storm sash is the storm panel, Figure 12-11. Panes of glass are mounted in metal frames. The frame can be attached to either the inside or outside of the window sash. It is generally used on horizontal sliding sash and on casement or other hinged sash. When a higher R-value is required, the storm panel can be equipped with sealed double glazing as shown.

Double- and Triple-Sealed Glazing

For movable sash, two or three layers of 1/8" glass are fused together with a 3/16" air space between layers. Double or triple layers of plate glass are used in large fixed units. Special seals are used to trap the air between panes. Air spaces are generally from 1/4 to 1/2". The air is dehydrated (moisture removed) before the space is sealed.

FLOAT GLASS		GLAZING QUALITY	NOMINAL THICKNESS in.	NOMINAL THICKNESS mm
Clear	SS	B	3/32	2.5
	DS	B, Select	1/8	3
			5/32	4
			3/16	5
		Mirror Select	1/4	6
Heavy Duty Clear		Select	5/16	8
			3/8	10
			1/2	12
			5/8	15
			3/4	19
			7/8	22

Figure 12-10. Chart lists standard thicknesses of glass used in residential construction. Metric thicknesses are not exact equivalents of conventional measure. However, they do represent sizes currently produced in other countries. (Libby-Owens-Ford Co., Glass Div.)

Figure 12-11. Cutaway views of energy efficient windows with removable interior panels. A—Triple glazing with 3/4" and 1/4" air spaces (R-3.23). (Rolscreen Co.) B—Sealed double-glazed unit installed in a modern window sash. Flange projecting from the bottom of the window is used to mount and fasten frame in structural opening. Standard insulating glass, as shown, has an R rating of 2.0. High performance insulating glass has an R-3.3 rating. (Andersen Corp.) C—Double-glazed unit has 13/16" airspace for an R-2.43 insulating factor.

Figure 12-11(B) shows sealed double glazing in a standard casement unit.

Double and triple glazing offers the following advantages:

- Lower heat loss in cold weather.
- Downdrafts along window surface are reduced.
- Heat penetration in summer months is reduced.
- Sweating and fogging of windows in cold weather is reduced or eliminated.
- Less outside noise is transmitted through the window.

Low-Emissivity Glazing

Emissivity is the relative ability of a material to absorb or re-radiate heat. Research in the area of glazing technology has resulted in a new method of raising the R-value of double-glazed windows. Commonly referred to as "low-e" or "high performance" windows, these new windows have a clear outer pane, an airspace, and a special coating on the air-gap side of the inner pane.

The special factory-applied coating consists of an extremely thin layer of metal oxide. It reflects infrared (heat wave) radiation, but allows regular light waves to pass through.

During winter months, warm surfaces within a room (wall, floor, furniture) radiate heat waves. When these waves strike the low-emissivity surface of the window, they are reflected back into the room—thus reducing heat loss. During summer months, heat waves from walks, drives, and other outside surfaces are prevented from entering. This lowers air conditioner loads.

Figure 12-12 shows a low-emissivity glass panel being attached to a window sash. The narrow slat blind, which is suspended between the two panes of glass, is also coated with a low-emissivity finish. The total R-value for this kind of window and blind is 4.35.

Argon-Filled Insulating Glass

The R-value of sealed , double-glazed windows can be raised by replacing the air between the glass with argon gas. This gas is heavier than air and has a lower heat conductance factor. Through the use of "low-e" glass surfaces and argon gas, R-values of 4.50 or more are now attainable.

Figure 12-12. Removable interior insulating panel is easy to install. Inside surface of the panel has a low-emissivity coating, which reflects radiant heat. (Rolscreen Co.)

Screens

Ventilating windows require screens to keep out insects. The mesh, usually made from fiberglass, should have a minimum of 252 openings per sq. in. Manufacturers have perfected many unique methods for mounting and storing screen panels. Most modern screens are made with a light metal frame, Figure 12-13.

Muntins

Years ago, window glass was available only in small sheets. By using rabbeted strips called *muntins*, small panes of glass could be used to fill large openings.

Today, even though large sheets of glass are available, muntins are still used for special effects in traditional architecture. They are, however, usually applied as an overlay and do not actually separate or support small panes of glass. Made of wood or plastic and in various patterns, they snap in and out of the sash for easier painting and cleaning, Figure 12-14.

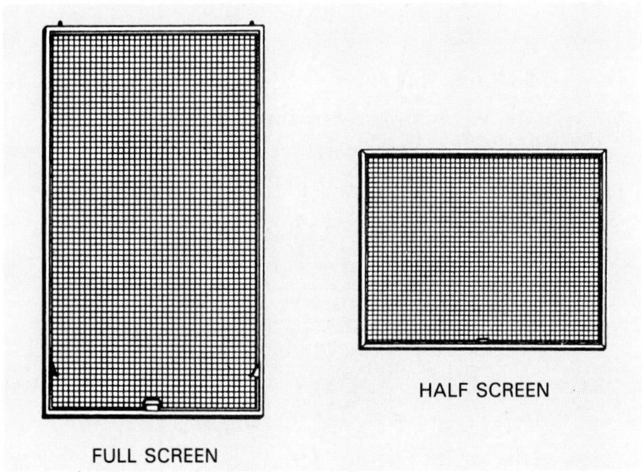

Figure 12-13. Window screens. Half screens are used on double-hung windows and fit under the top sash and between the jambs.

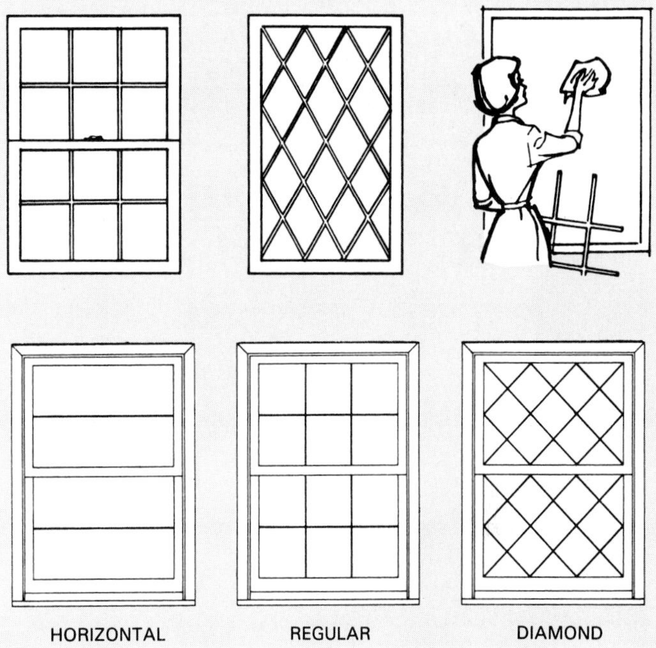

Figure 12-14. Top—Muntin assemblies detach easily for cleaning and painting. Below—Standard muntin patterns.

Parts of Windows

Because much of the actual construction of a window is hidden, sectional views are used to show the parts and how they fit together. It is standard practice to use sections through the top, side, and bottom of a window. See Figure 12-15. These drawings also include wall framing members and surface materials. See Figure 12-16.

A section view of a typical mullion shows how window units fit together. A mullion is formed by the window jambs when two units are joined together as shown in Figure 12-17.

A drip cap, shown in the head section of Figure 12-16 is designed to carry rainwater out over the window casing. When the window is protected by a wide cornice, this element is seldom included.

Windows in Plans and Elevations

The carpenter studies the plans and elevations of the working drawings to become informed about the types of windows and their location. It is common practice to locate the horizontal position of windows and exterior doors by

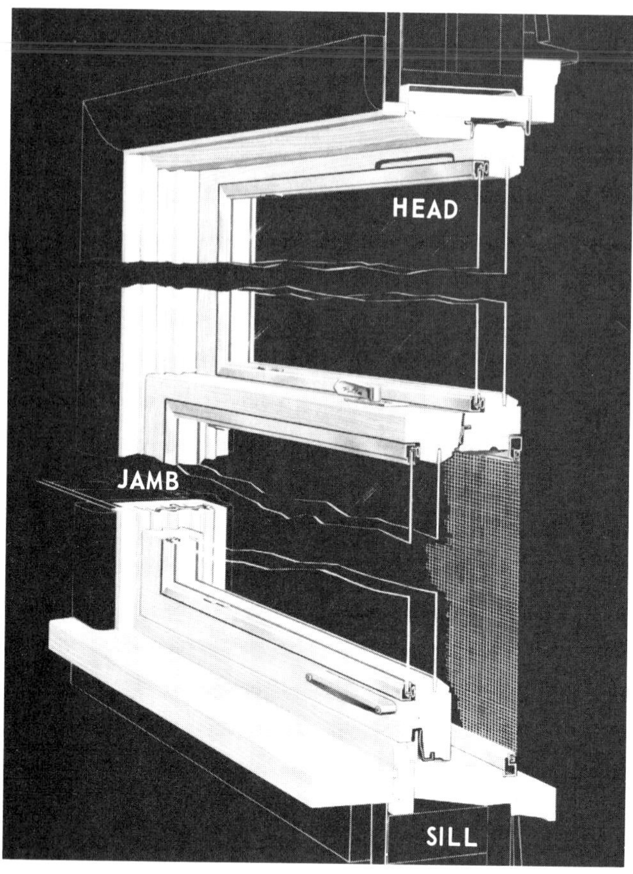

Figure 12-15. Standard sections used to show details of all types of windows. (Rolscreen Co.)

including a dimension line to the center of the opening as shown in Figure 12-18.

In masonry construction the dimension is given to the edge of the opening. The latter method is sometimes used in frame construction.

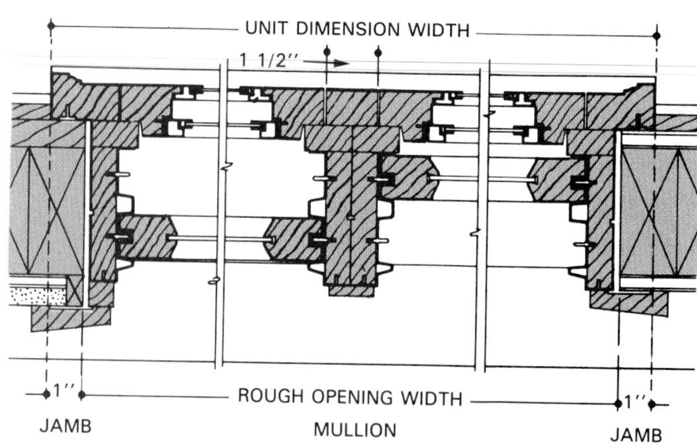

Figure 12-17. Section drawing of a mullion. This is where two window jambs are joined. (Rock Island Millwork)

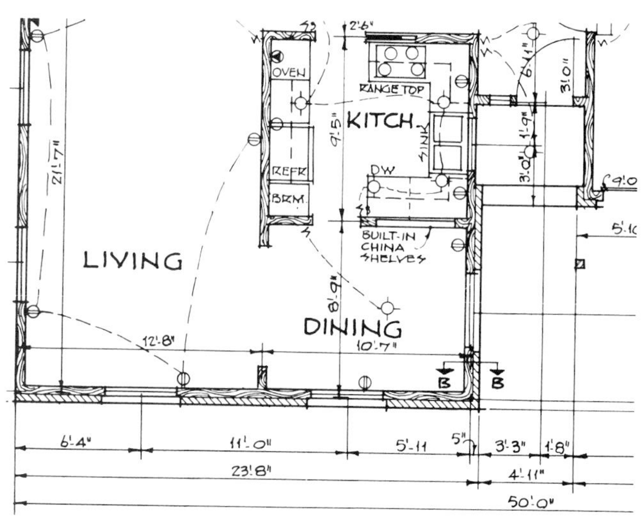

Figure 12-18. Floor plans show location of windows and doors.

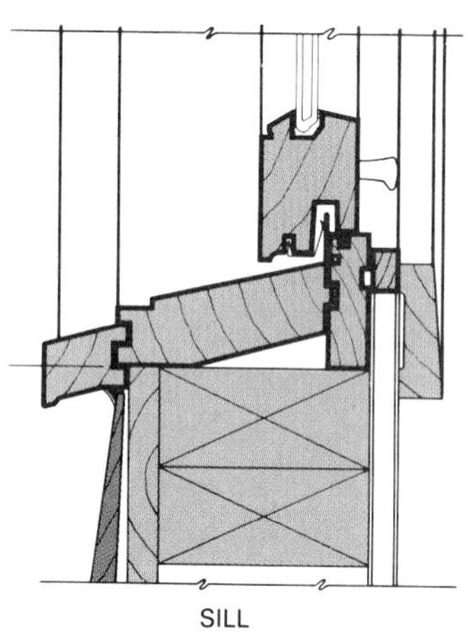

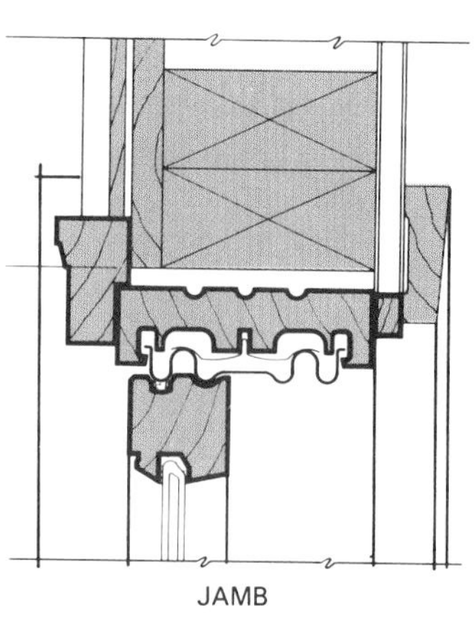

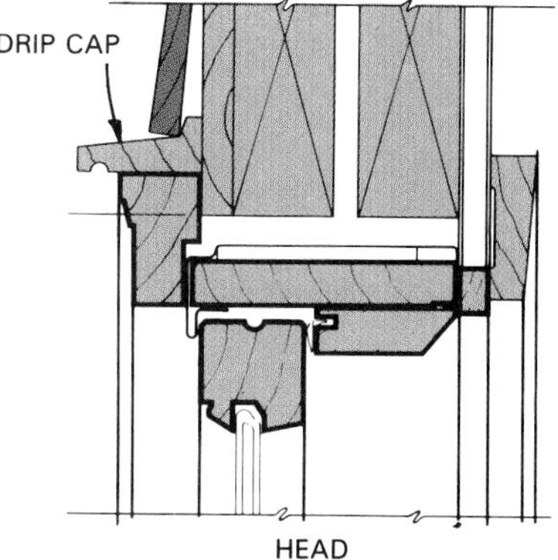

Figure 12-16. Typical detail drawing shows a section through a window. Such details are included with the house plans. They help carpenter to see how windows should be installed.

Elevations will show the type of windows, Figure 12-19, and may include glass size and heights. The position of the hinge line (point of dotted line) will indicate the type of swinging window. Sliding windows require a note to indicate they are not fixed units. Supporting mullions will be included in the plans and also in elevations. See Figure 12-20.

Window Sizes

Besides the type and position of the window, the carpenter should know the size of each unit or combination of units. Window sizes may include several or all of the following:

- Glass size.
- Sash size.
- Rough frame opening.
- Masonry or unit opening.

Figure 12-21 shows the position of these measurements and approximately how they are figured from the glass size. They will vary slightly from one manufacturer to another.

A complete set of architectural drawings should provide detailed information about window sizes. This information is usually listed in a table called a window and door schedule. Refer to Fig. 6-20. It includes, among other things, the manufacturer's numbers and rough opening sizes for each unit or combination. An identifying letter is located at each opening on the plan and a corresponding letter is then used in the schedule to specify the required window unit and the necessary information.

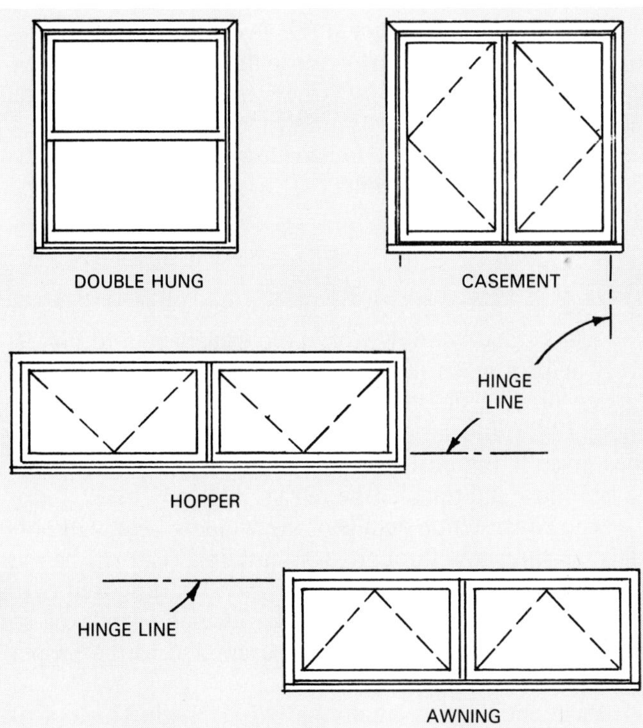

Figure 12-19. How window types are shown in elevation views. Horizontal sliding windows are noted to define them from fixed units.

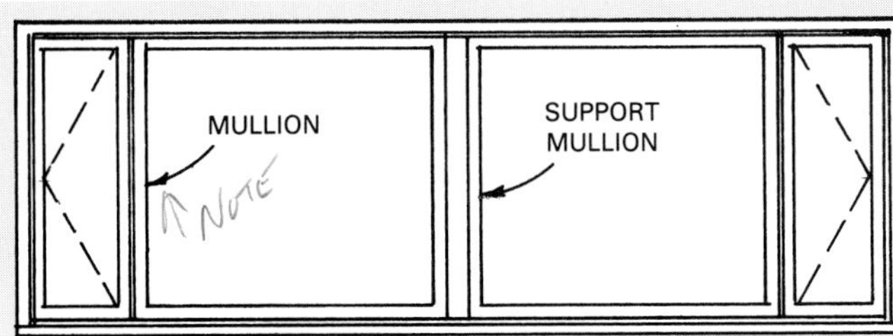

Figure 12-20. Elevation of windows will show supporting and nonsupporting mullions.

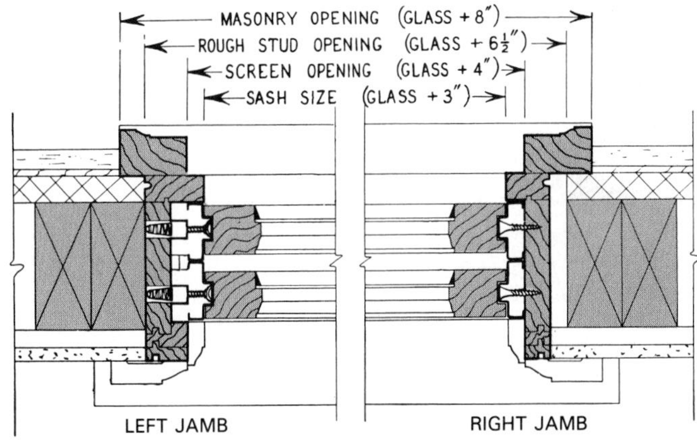

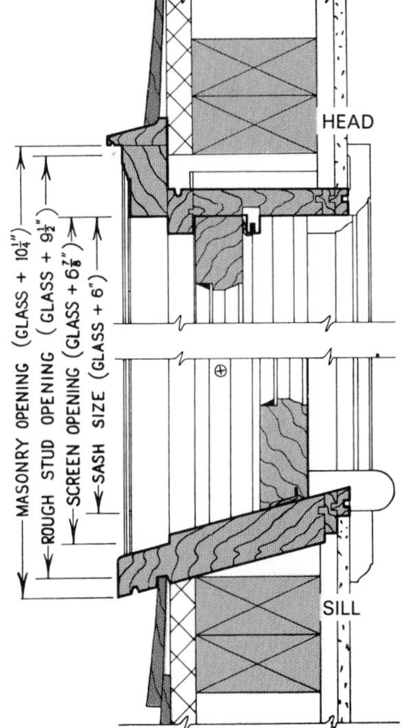

Figure 12-21. Window sizes and location of measurements. The rough opening is larger than the overall size of the frame to permit alignment and leveling when unit is installed.

When this information is not included in the architectural plans, the carpenter will need to study manufacturer's catalogs and other descriptive literature. Sizes for basic units are often given in diagrams as shown in Figure 12-22. The size of the rough opening is of major importance to the carpenter. However, she or he will also need other dimensions—for example, the height of the rough opening above the floor.

Rough openings and other sizes are readily secured from manufacturer's data. An example of sizes for casement combinations is shown in Figure 12-23. For other special combinations, apply the amounts specified in the details included at the bottom of Figure 12-23. Note that the extra allowance for a support mullion is 2".

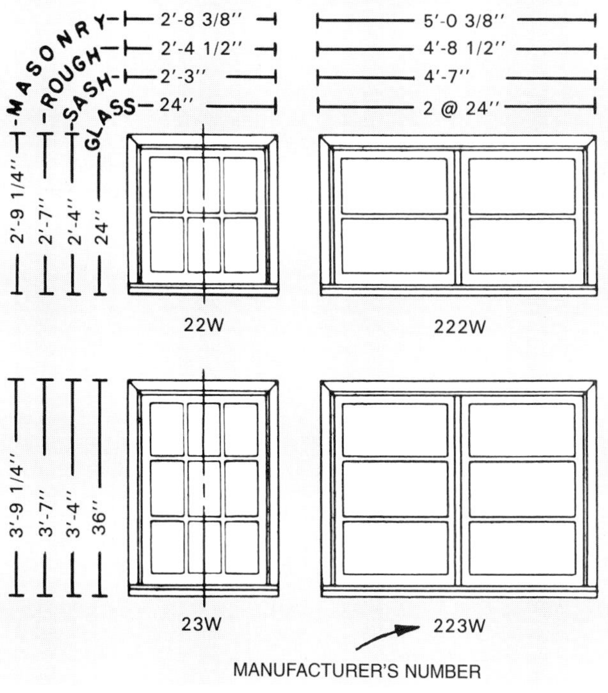

Figure 12-22. This is typical of illustrations used by window manufacturers to give sizes of their standard units. (Rolscreen Co.)

When window and door sizes are given, whether they consist of rough openings, sash size, or other items, the horizontal dimension is listed first and the height second. For example, the unit No. 222W shown in Figure 12-22 requires an R.O. of 4'-8 1/2" x 2'-7".

Detailed Drawings

Sectional drawings are helpful to the carpenter since they show each part of the window and how the unit is placed in a wall structure. Figure 12-24 shows detail drawings for a typical double-hung window. Similar drawings are available for other types of windows from other manufacturers. The architect will often include selected detail views. Whether included in the architectural drawings or made available through manufacturer's catalogs, detailed drawings such as these will be essential in building the rough frame of the wall structure and in installing the window units. See Figure 12-25.

Jamb Extensions

The thickness of window units may be adjusted to various walls. For example, a frame wall with 3/4" sheathing and a standard interior surface of lath and plaster may be 5 1/8" compared with a single layer drywall construction of about 4 3/4". Adjustments in the window frame may be made by applying a *jamb extension*, Figure 12-26. Some manufacturers build their frames to a basic size, such as 4 5/8". Then they equip the unit with an extension as specified by the builder or architect.

Story Pole

A door and window story pole (refer again to Fig. 9-13) is helpful when installing doors and windows. It will help ensure the alignment of doors and window heads, Figure 12-27. It can be marked with the additional dimensions given in Figure 12-28. To use the pole, hold it against the trimmers and transfer the marks.

The construction details of the window head will normally be the same throughout a building. The sill height may vary. Additional positions can be superimposed over other layouts if each one is carefully labeled or indexed so the correct distance will always be applied to the proper opening.

Residential doors are normally 6'-8" high. The tops of windows are usually held at this same height. See Figure 12-29. An extra 1/2" is added to allow for thresholds under entrance doors and clearance under interior doors.

Installing Windows

If rough openings are plumb, level, and the correct size, it is easy to install windows.

Manufacturers furnish directions that apply specifically to their various products. The carpenter should follow them carefully.

When windows are received on the job they should be stored in a clean, dry area. If they are not fully packaged, some type of cover should be used to prevent damage from dust and dirt. Allow wood windows to adjust to the humidity of the locality before they are installed.

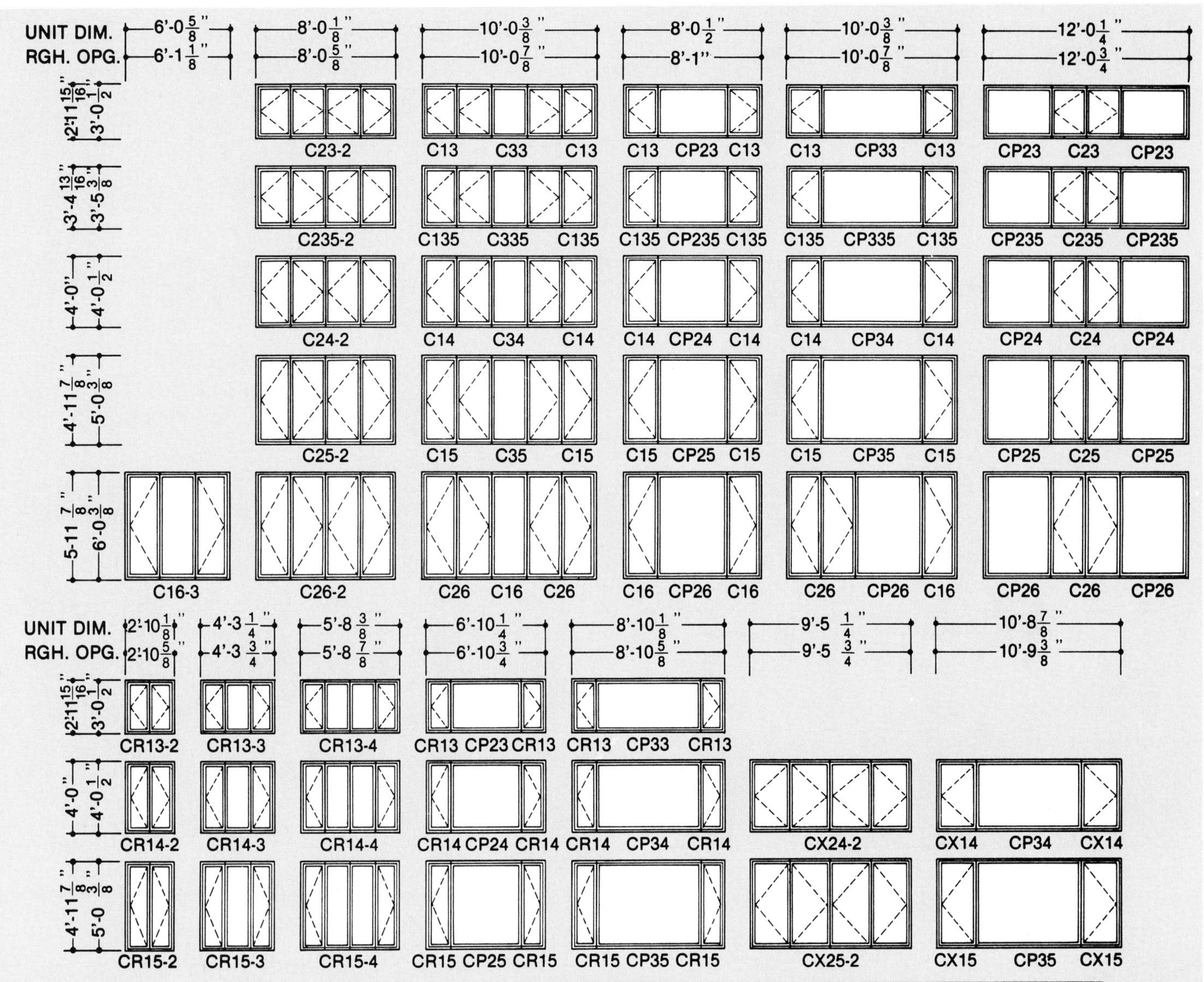

UNIT DIM. / RGH. OPG.	6'-0 5/8" / 6'-1 1/8"	8'-0 1/8" / 8'-0 5/8"	10'-0 3/8" / 10'-0 7/8"	8'-0 1/2" / 8'-1"	10'-0 3/8" / 10'-0 7/8"	12'-0 1/4" / 12'-0 3/4"
2'-11 15/16" / 3'-0 1/2"		C23-2	C13 C33 C13	C13 CP23 C13	C13 CP33 C13	CP23 C23 CP23
3'-4 13/16" / 3'-5 3/8"		C235-2	C135 C335 C135	C135 CP235 C135	C135 CP335 C135	CP235 C235 CP235
4'-0" / 4'-0 1/2"		C24-2	C14 C34 C14	C14 CP24 C14	C14 CP34 C14	CP24 C24 CP24
4'-11 7/8" / 5'-0 3/8"		C25-2	C15 C35 C15	C15 CP25 C15	C15 CP35 C15	CP25 C25 CP25
5-11 7/8" / 6'-0 3/8"	C16-3	C26-2	C26 C16 C26	C16 CP26 C16	C26 CP26 C16	CP26 C26 CP26

UNIT DIM. / RGH. OPG.	2'-10 1/8" / 2'-10 5/8"	4'-3 1/4" / 4'-3 3/4"	5'-8 3/8" / 5'-8 7/8"	6'-10 1/4" / 6'-10 3/4"	8'-10 1/8" / 8'-10 5/8"	9'-5 1/4" / 9'-5 3/4"	10'-8 7/8" / 10'-9 3/8"
2'-11 15/16" / 3'-0 1/2"	CR13-2	CR13-3	CR13-4	CR13 CP23 CR13	CR13 CP33 CR13		
4'-0" / 4'-0 1/2"	CR14-2	CR14-3	CR14-4	CR14 CP24 CR14	CR14 CP34 CR14	CX24-2	CX14 CP34 CX14
4'-11 7/8" / 5'-0 3/8"	CR15-2	CR15-3	CR15-4	CR15 CP25 CR15	CR15 CP35 CR15	CX25-2	CX15 CP35 CX15

MULTIPLE OPENINGS

A number of suggested combinations are shown above using the narrow (no support) mullion.

Additional combinations using support mullion and transom joining can be arrived at by using the dimensions in the formulas listed below.

CASEMENT NARROW MULLION

Joining basic casement units to form multiple units or picture window combinations without vertical support between units.

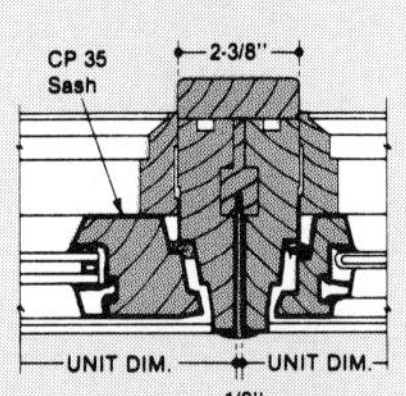

scale: 1½" = 1'0"

Overall Unit Dimension Width — The sum of individual unit dimensions, plus 1/8" for each unit joining.

Overall Rough Opening Width — Add ½" to Overall Unit Dimension Width.

CASEMENT SUPPORT MULLION

Joining basic casement units using a 2 x 4 vertical support between units.

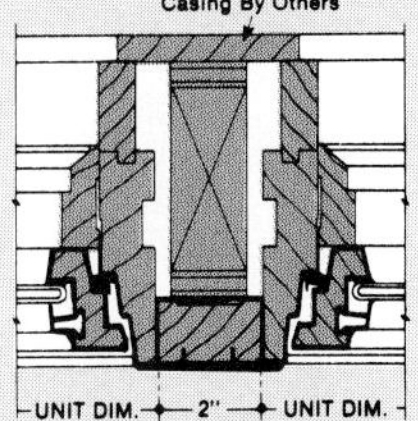

Overall Unit Dimension Width — The sum of individual unit dimensions, plus 2" for each unit joining.

Overall Rough Opening Width — Add ½" to Overall Unit Dimension Width.

CASEMENT TRANSOM

Joining basic casement units by stacking units to form combinations.

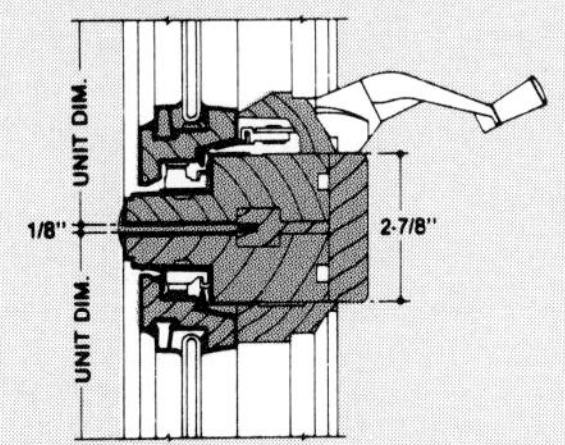

Overall Unit Dimension Height — The sum of individual unit dimension heights, plus 1/8" for each unit joining.

Overall Rough Opening Height — Add ½" to Overall Unit Dimension Height.

Figure 12-23. Manufacturer's catalogs and brochures picture the variety of window units they supply. This page shows sizes of casement and picture window units available. (Andersen Corp.)

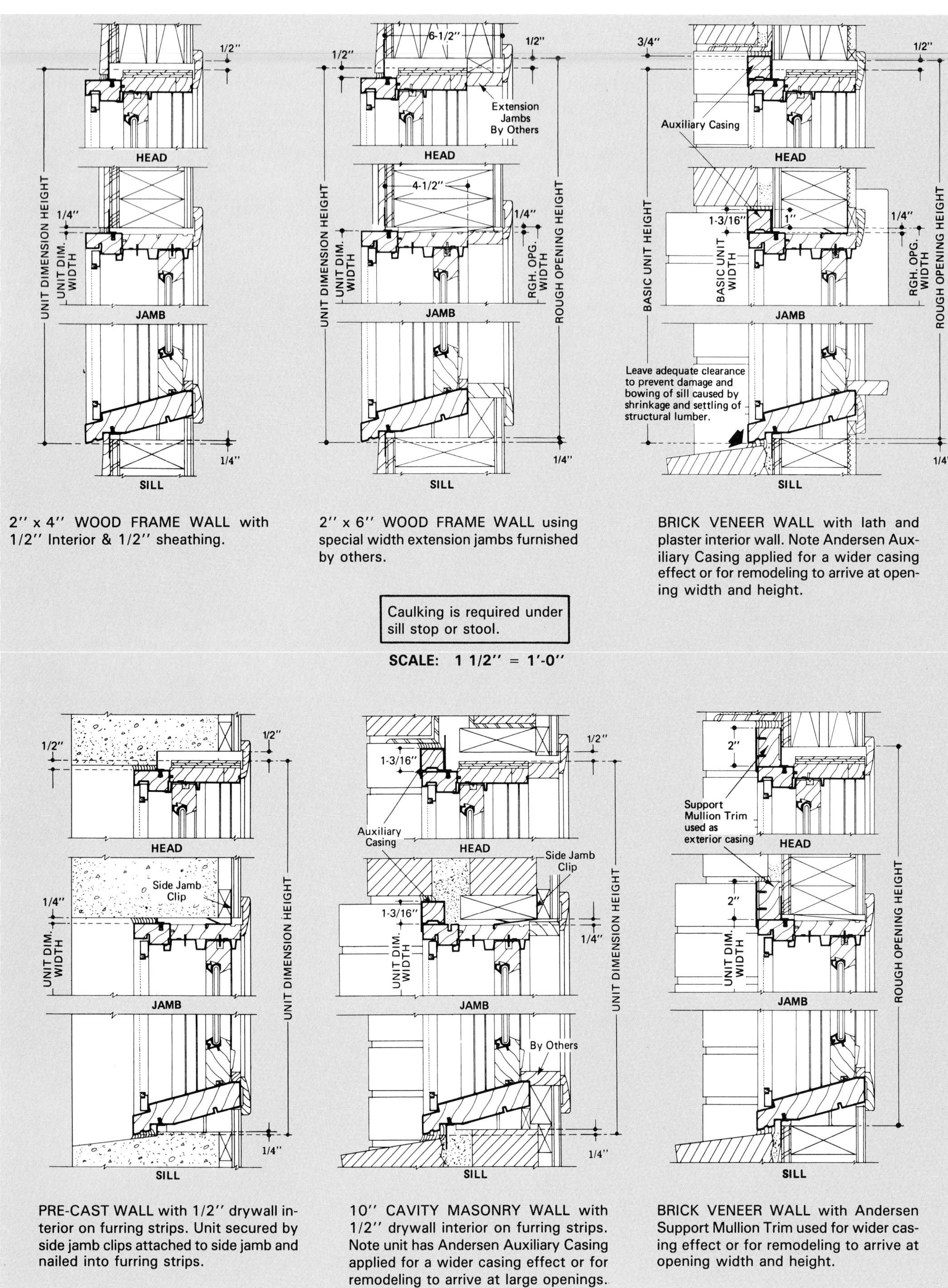

Figure 12-24. Detail drawings show a standard double-hung window installed in different wall structures. (Andersen Corp.)

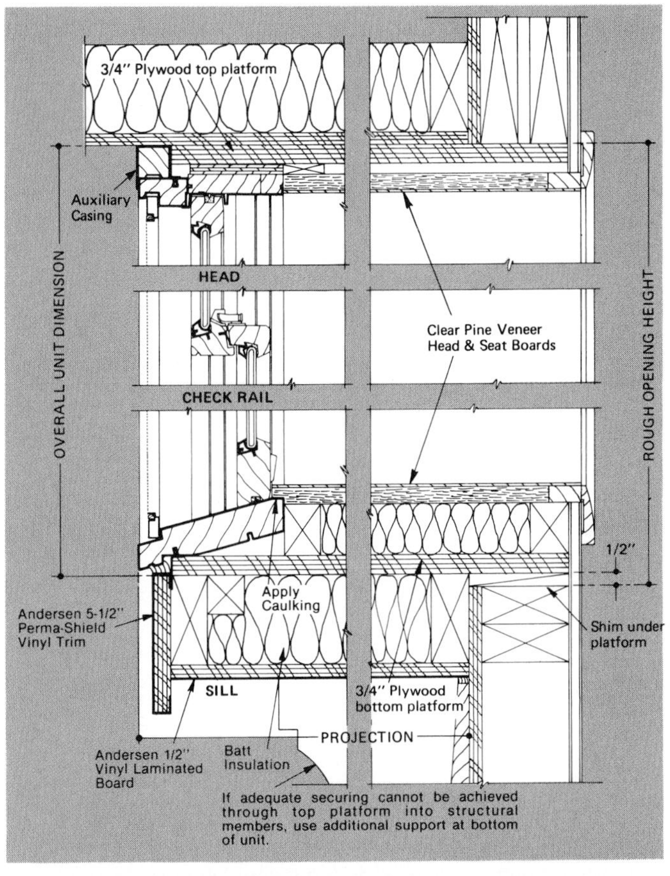

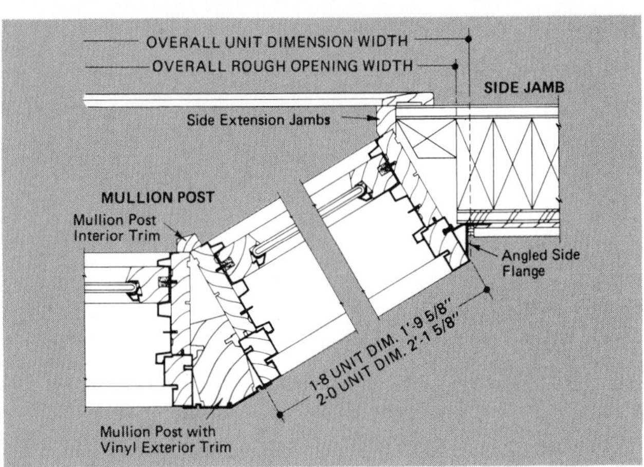

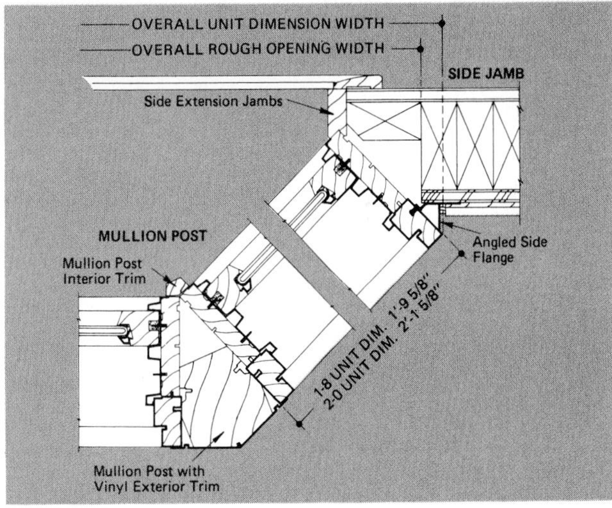

Figure 12-25. Bay window construction is more complicated than standard windows. Details for installation are extremely important for the carpenter. (Andersen Corp.)

Check carton labels and move window units (still packaged) into the various rooms and areas where they will be installed.

Unpack the window and check for shipping damage. Do not remove any diagonal braces or spacer strips until after the installation is complete.

Check the rough opening to make certain it is the correct size. Most windows require at least 1/2" clearance on each side and 3/4" above the head for plumbing and leveling.

If the window has not been primed at the factory, this may be done before installation. Weather stripping and special channels should not be painted. Follow the manufacturer's recommendations.

Most window units and multiple unit combinations are installed from the outside. The unit, when stored on the inside, can be turned at an angle and moved to the outside through the rough opening. Be sure to have plenty of help when handling the window entirely from the inside. See Figure 12-30.

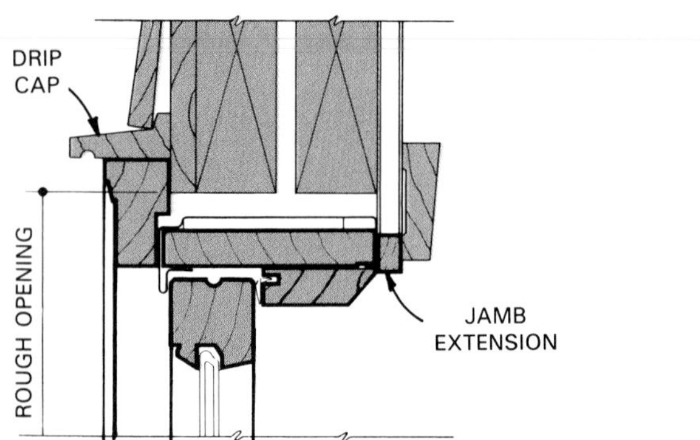

Figure 12-26. Jamb extensions are intended to be applied to a standard window frame to adjust for various wall thicknesses.

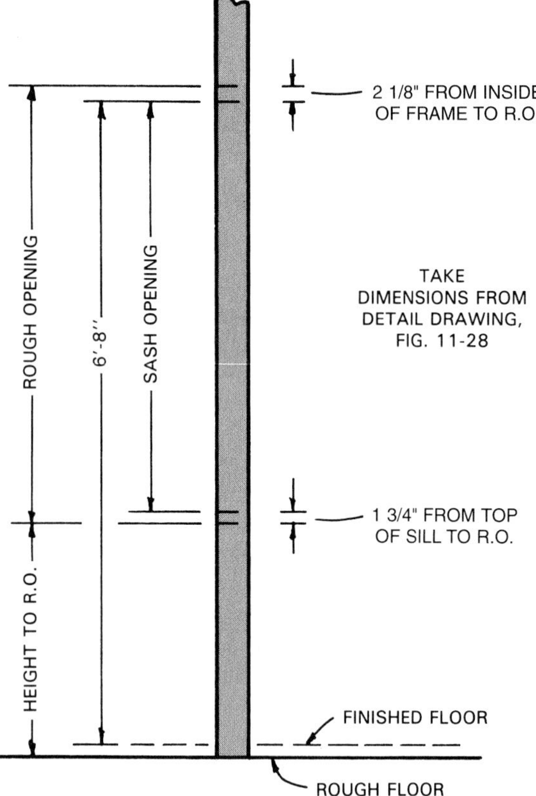

Figure 12-27. Use the story pole once more to check height of certain parts of the window frame.

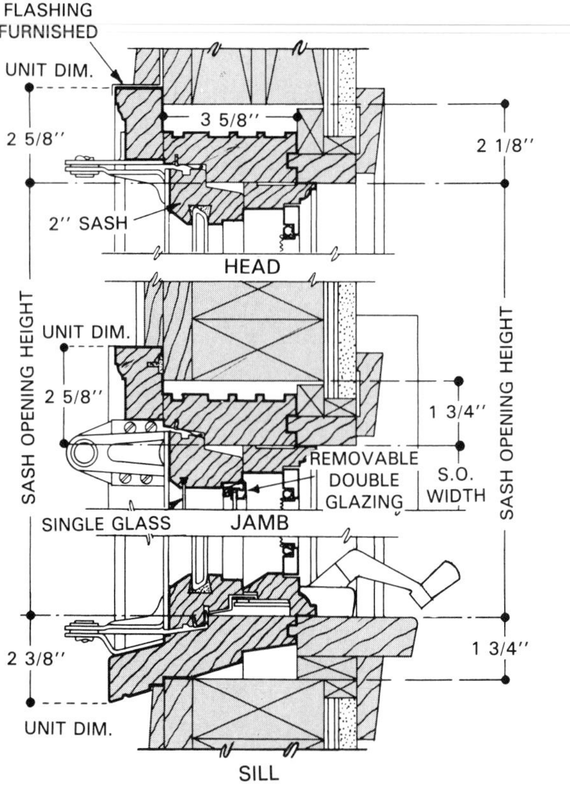

Figure 12-28. Details of a window may also carry information on installation heights of certain window features.

Figure 12-29. Usually, tops of doors and window frames are at same level. (Andersen Corp.)

Installing a Window

1. Place the window sill on the rough opening and then swing the top into place. See Figure 12-31.
2. Use wedge blocks under the sill and raise the frame to the correct height as marked on the story pole. Also adjust the wedges so the frame is perfectly level. There may be a tendency for the sill to sag on multiple units. Use additional wedge blocks to correct this and make the sill perfectly straight.
3. Check to see that the unit is horizontally centered in the rough opening.
4. Secure this position by driving several roofing nails (1 1/2" long) through the lower flange and into the rough sill.
5. Plumb the side jambs with a level, Figure 12-32, and check the corners with a framing square. The sash should be closed and locked in place.
6. Check for front-to-back plumb. Place the level on the outside face of the frame and make sure the window is not tilted outward or inward.

Figure 12-30. Window units can be moved from the inside to the outside for installation. Be sure to have plenty of help.

7. Drive several nails temporarily into the top of the side casings as shown in Figure 12-33.
8. Check over the entire window to see that it is square and level. Use additional wedge blocks (shims) if necessary. Check sash for easy operation and make sure there is an even space between sash and frame.
9. Nail the window permanently in place with 1 3/4" or longer galvanized roofing nails. Space the nails from 12 to 16" O.C. or as specified by the window label. See Figure 12-34.

Figure 12-31. Place the window sill on the bottom of the rough opening and then swing the top into place.

Figure 12-32. Using a carpenter's level to plumb the side jambs. Also check the front-to-back plumb.

Figure 12-33. Partially secure upper corners with 1 3/4" roofing nails and then check over the entire installation.

At this point, many builders prefer to cover the inside of the window with a sheet of polyethylene film to protect it during the application of inside wall surface materials.

Installing Fixed Units

The need for ventilating windows will likely be reduced as more buildings are air-conditioned. Fixed units are less expensive than those that can be opened.

When the fixed glass panel is of medium size, it is usually mounted in a sash and frame, Figure 12-35. Often it is combined with matching ventilating units. The installation of such a unit is the same as for regular windows. They are, of course, larger and heavier and extra precautions should be observed in handling and making the installation.

Figure 12-34. Modern double-hung window held in place by nails that go through the window flange, sheathing, and into the structural frame.

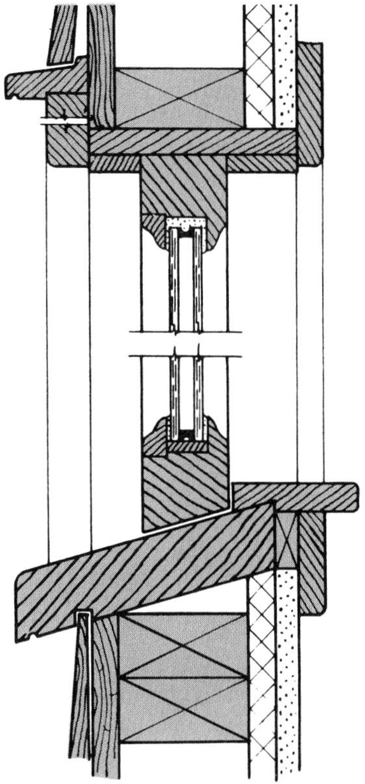

Figure 12-35. Detail of 1" double-glazed window in a conventional sash and frame.

Large insulating units are made of 1/4" glass, which results in considerable weight, Figure 12-36. They are seldom installed in window frames at the factory, although they are sometimes mounted in a sash. Usually, large glass units are glazed (set in opening with glazing sealant) as a separate operation after the frame and/or sash have been installed.

Openings must be square, free of twists, and rugged enough to bear the weight of the glass unit. Use only high-grade wood materials that are dry and free from warp. Special setting blocks and clips, Figure 12-37, may be used to hold the glass in position with clearance on all edges. For wood frames or sash, use two neoprene setting blocks (at least 4" long) located at quarter points along the lower edge. The width of the blocks should be equal to, or greater than, the thickness of the glass unit.

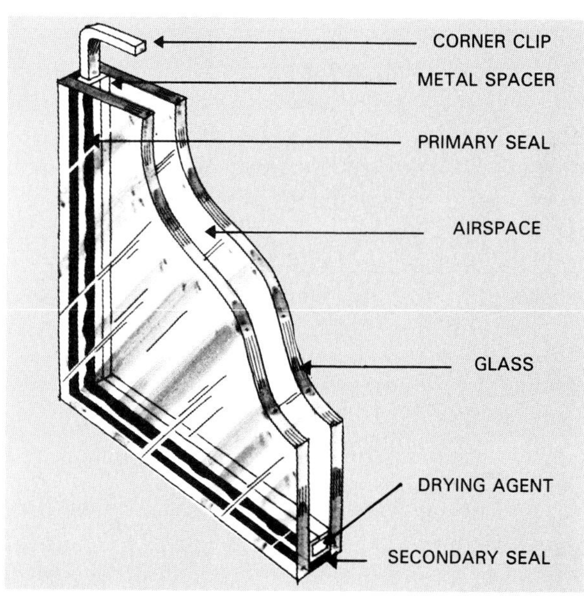

Figure 12-36. Cutaway view of modern double-glazed window unit. Two layers of 1/4" glass are separated by a dry airspace that is hermetically sealed.
(Libby-Owens-Ford Co., Glass Div.)

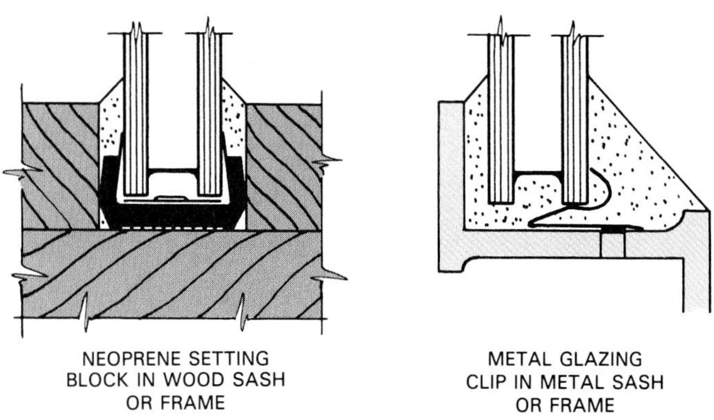

Figure 12-37. Setting blocks and clips. The use of metal clips is limited to welded glass units.

The edge should be completely surrounded with a high-quality, nonhardening glazing sealant. There must be no direct contact between the glass unit and frame. Figure 12-38 shows clearances recommended by one manufacturer.

All glazing systems should be designed and installed to ensure that the seal between the layers of glass (organic type) are not exposed to water for long periods of time. The inclusion of weep holes as recommended by SIGMA (Sealed Insulating Glass Manufacturers Association) should be included.

Insulating glass units cannot be altered in any way on the building site. It is essential, then, that sash and frames be carefully designed and that specified dimensions are followed. Wooden window frames should be treated with a wood preservative.

Figure 12-39 gives step-by-step procedures for building a simple wood frame and installing insulating glass. Some standard thicknesses of insulating glass for fixed window units are listed in Figure 12-40.

Details for the construction of a window wall are shown in Figure 12-41. The photo, Figure 12-42, is an example of how window walls can create pleasing effects by bringing the outdoors inside.

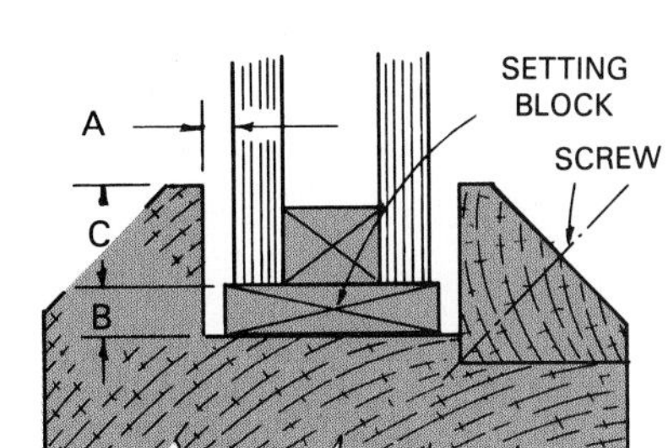

GLASS THICKNESS		DIMENSIONAL TOLERANCE UP TO 48" 1220mm		OVER 48" 1220mm		MINIMUM CLEARANCE FACE (A)		EDGE (B)		BITE (C)	
in.	mm	in.	mm	in.	mm	in.	mm	in.	mm	in.	mm
1/2	12	±1/16	±1.6	+1/8−1/16	+3.2−1.6	1/8	3.2	1/8	3.2	1/2	12.7
5/8	15	+1/8 −1/16	+3.2 −1.6	+3/16 −1/16	+4.8 −1.6						
23/32	18										
3/4[1]	19										
3/4[2]	19	±1/16	±1.6	+1/8−1/16	+3.2−1.6	3/16	4.8	1/4	6.4	1/2	12.7
7/8	22	+1/8 −1/16	+3.2 −1.6	+3/16 −1/16	+4.8 −1.6						
31/32	24										
1	25										

[1] 1/4" (6 mm) Air Space
[2] 1/2" (12 mm) Air Space

Figure 12-38. Recommended clearances for sealed insulating glass. The "C" dimension is commonly referred to as *bite.*

Glass Blocks

Glass blocks, Figure 12-43, have good insulating properties. Used in outside walls, they provide light, help prevent drafts, dampen disturbing noises, cut off unpleasant views, and insure privacy where it is desired. Inside the home, partitions and screens of glass blocks add a pleasant touch to rooms they divide.

Glass blocks are made of two formed pieces of glass fused together to leave an insulating airspace between. They come in several different patterns and are usually available in three nominal sizes: 6 x 6, 8 x 8, and 12 x 12. See Figure 12-44. All are 3 7/8" thick. Special shapes are available for turning corners and for building curved panels. The blocks come in both light-diffusing and light-directing types.

Installing glass blocks is not difficult. Use regular masonry tools. Even though the carpenter will seldom make the actual installation, she or he will be required to build the framework. Therefore, knowledge of the design requirements is helpful.

Installing Small Glass Block Panels

Details for installing glass block panels of 25 sq. ft., and less are given in Figure 12-45. In such panels, the height should not exceed 7' nor the width 5'.

Panels may be supported by a *mortar key* at jambs in masonry, or by wood members in frame construction. No wall anchors and no wall ties are required in the joints. Expansion space is required at the head only.

Installing Large Glass Block Panels

When panel areas exceed 25 sq. ft., additional requirements must be met. See Figure 12-45. Expansion strips (strips of resilient material) are used to partially fill expansion spaces at jambs and heads of larger panel openings.

Panels should never be larger than 10' wide or 10' high. Provide support at jambs with wall anchors or wood members. A portion of each anchor is embedded in masonry and in the glass block mortar joint. Anchors should be crimped within expansion spaces and spaced on 24" centers to rest in the same joints as wall ties. Anchors, which are corrosion resistant, are 2' long and 1 3/4" wide.

Wall ties should be installed on 24" centers in horizontal mortar joints of larger panels and lap not less than 6" whenever it is necessary to use more than one length of tie. Do not bridge expansion spaces. Ties are also corrosion resistant. They are 8' long and 2" wide.

Openings for Glass Blocks

To determine heights or widths of openings required for panels, multiply the number of units by the nominal

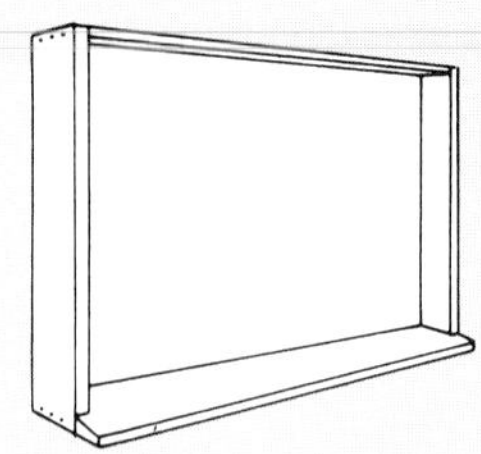

1—Select a good grade of softwood lumber like ponderosa pine. For a medium sized frame use a 1 1/2 in. thickness. Make the frame slightly larger than the insulating glass. See recommended clearances listed in Fig. 11-38. After assembly, apply a coating of water-repellent preservative.

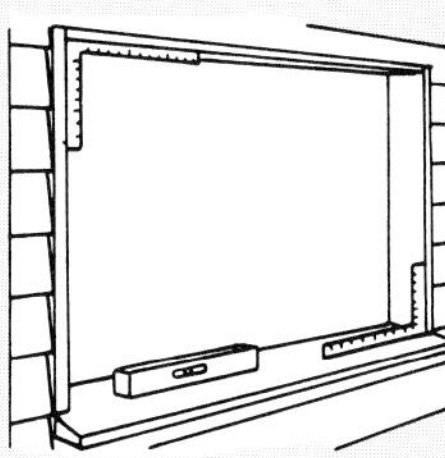

2—Install frame in opening. Place wedge blocks under sill and at several points around perimeter of frame. When sill is level and frame perfectly square, nail the frame securely to structural members of the building.

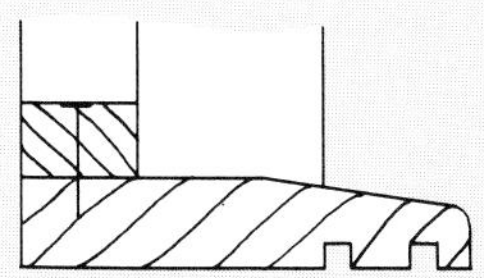

3—Prepare inside stop members according to recommended sizes (Fig. 11-38). Use miter joints in corners and nail down stops. Also prepare outside stops.

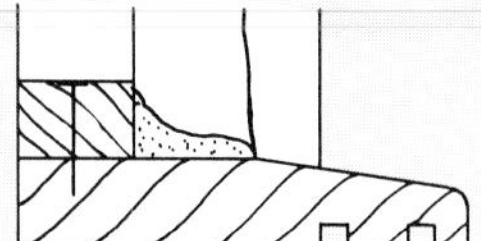

4—Using recommended grade of glazing sealant, apply thick bed to stops, and top, sides, and bottom of frame. Apply enough material to fill space between frame and glass unit when installation is made.

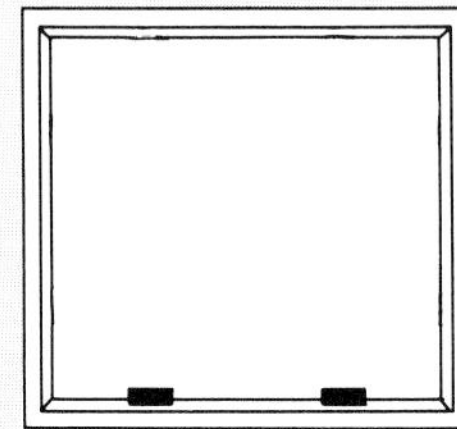

5—Install neoprene setting blocks on bottom of frame or edge of glass unit as shown. Blocks (use only two) should be moved in from the corners a distance equal to one-quarter of frame width.

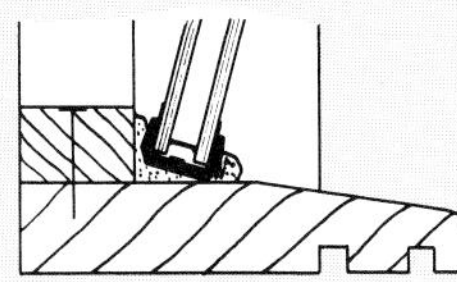

6—Place bottom edge of insulating glass unit on setting blocks as shown. Then carefully press unit into position against stops until proper thickness of glazing sealant is secured around entire perimeter. Small gauge blocks may be helpful.

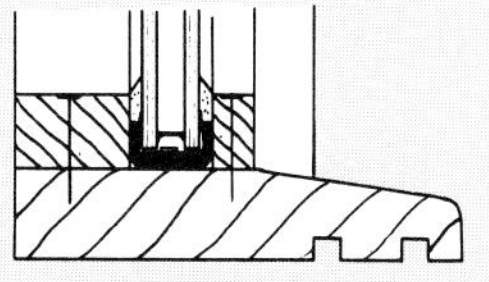

7—Nail on outside stop. Additional sealant may be required to fill joint. Remove excess sealant from both sides of unit and clean surfaces with an approved solvent.

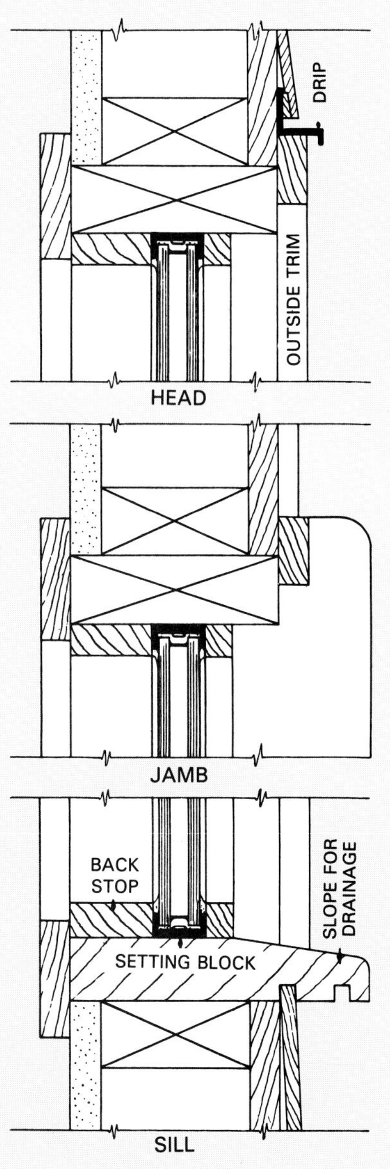

Figure 12-39. Details and method for constructing a frame and installing sealed insulating glass.

UNIT CONSTRUCTION							
SINGLE GLASS THICKNESS		OVERALL UNIT THICKNESS		AIR SPACE		APPROXIMATE WEIGHT	
in.	mm	in.	mm	in.	mm	lb/ft²	kg/m²
1/8	3	1/2	12	1/4	6	3.27	16
		3/4	19	1/2	12		
3/16	5	5/8	15	1/4	6	4.90	24
		7/8	22	1/2	12		
		1	25	5/8	15		
1/4	6	3/4	19	1/4	6	6.54	32
		1	25	1/2	12		

Figure 12-40. Glass manufacturers publish charts of specifications for the glass they produce.
(Libby-Owens-Ford Co., Glass Div.)

block size and then add 3/8". For example, a panel consisting of 8" blocks that was four units wide by five units high would require an opening 32 3/8" wide and 40 3/8" high.

Replacing Windows

Older style windows waste tremendous amounts of heating and cooling energy. Today, many of these windows are being replaced with modern units that have excellent weather stripping and double or triple glazing. Manufacturers provide a wide range of sizes and also a variety of special trim members that are helpful in making the replacement. They also provide detailed instructions for installing their products.

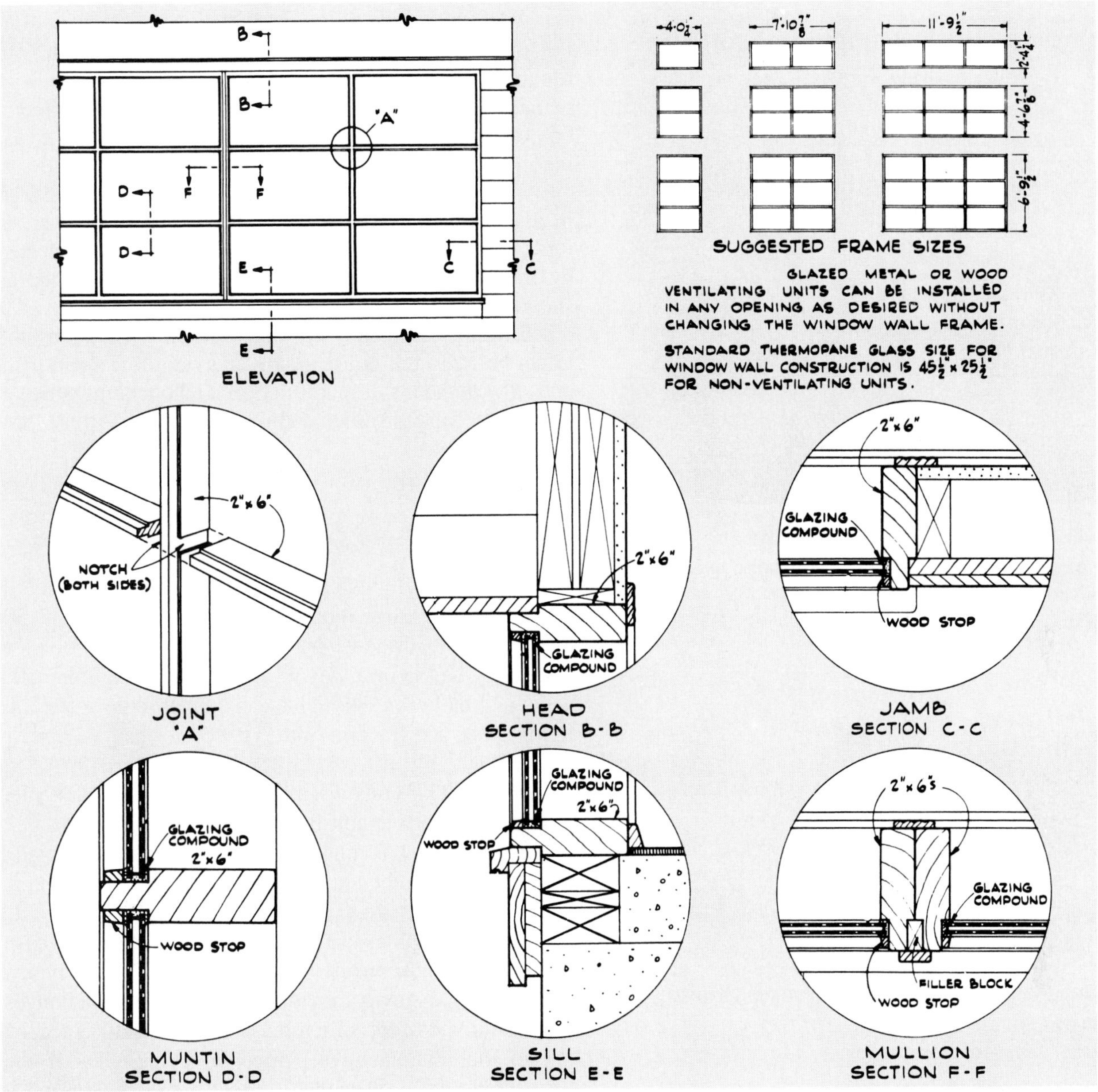

ELEVATION

SUGGESTED FRAME SIZES

GLAZED METAL OR WOOD VENTILATING UNITS CAN BE INSTALLED IN ANY OPENING AS DESIRED WITHOUT CHANGING THE WINDOW WALL FRAME.

STANDARD THERMOPANE GLASS SIZE FOR WINDOW WALL CONSTRUCTION IS 45½" x 25½" FOR NON-VENTILATING UNITS.

JOINT "A" — 2"x6", NOTCH (BOTH SIDES)

HEAD SECTION B-B — 2"x6", GLAZING COMPOUND

JAMB SECTION C-C — 2"x6", GLAZING COMPOUND, WOOD STOP

MUNTIN SECTION D-D — GLAZING COMPOUND, 2"x6", WOOD STOP

SILL SECTION E-E — GLAZING COMPOUND, 2"x6", WOOD STOP

MULLION SECTION F-F — 2"x6's, GLAZING COMPOUND, FILLER BLOCK, WOOD STOP

Figure 12-41. Framing details for an insulated glass window wall.

Figure 12-42. View from a window wall tends to create the feeling of bringing the outdoors inside.

First considerations in window replacement include the selection of type and size of the new units. To determine the size, it is best to remove the inside trim and measure the rough opening. If the trim members are to be reused, pry them off carefully. Remove nails by pulling them through the trim from the back side.

Removing Old Windows

1. Begin by removing the inside stops and lift out the lower sash. Weights or counterbalances may need to be disconnected.
2. Remove the parting stops and lift out the upper sash.
3. With the window sash removed, pry off the outside

Figure 12-43. This glass block window provides a high level of security in a bedroom area. It also reduces noise transmission so that acoustical privacy is maintained. (Pittsburg Corning Corp.)

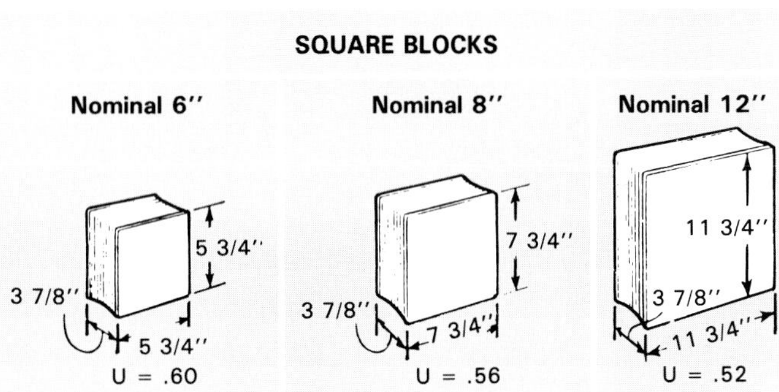

Figure 12-44. Nominal and actual sizes of commonly used glass block.

casing. Make saw cuts through the sill and frame, collapse the members and remove them from the opening as shown in Figure 12-46.

4. Clean the rough opening to remove dirt, plaster, putty, and nails.
5. Carefully measure the opening to determine the thickness of material needed to bring the rough opening to required size.
6. Install the strips as shown in Figure 12-47. Constantly check with a square and level. Use shims where necessary.
7. Check the rough opening for plumb and level.

Insert the new unit, Figure 12-48. With the window resting on the rough sill, make adjustments with shims. The window unit can be attached to the rough opening by nailing through special flanges, metal clips, or exterior casing. Check the manufacturer's directions for their recommendations on specific units.

With the window unit carefully secured in the opening, install outside and inside trim, Figure 12-49. Manufacturers furnish a variety of trim pieces especially made for this purpose. Since the new window unit may not be exactly the same size as the one removed, standard casing may require various widths of filler strips.

All of these exterior trim members should be bedded in a high quality construction mastic to seal against air infiltration. If the top of the window is not well protected by roof overhang, metal flashing should be installed as shown in Figure 13-18.

Finally, completely seal the window by caulking the joints between the house siding and side trim, head trim, and sill. On the inside, carefully pack all openings between the wall and frame with insulation. Apply the trim members.

Skylights

Modern skylights can make a dramatic difference in the appearance and feeling of a room, Figure 12-50. Skylights in bathrooms provide good light and privacy. They may be the best way to secure natural light in stairwells and hallways. Skylights can convert attic space into living space at a fraction of the cost of dormers with regular windows.

Figure 12-51 shows a skylight with sloped double glazing. The glazing consists of tempered glass with a 1/2" airspace. The sash is hinged as shown and can be operated from below with the aid of an extension pole. Fixed units are also available.

Rough openings are framed in about the same way as described for chimneys. A detail drawing of a typical installation is shown in Figure 12-52. Flashing flanges, included in the frame structure, should be installed according to the manufacturer's directions. Skylights of this type should not be installed in roofs with less than a 3 in 12 slope.

In standard frame construction, a shaft is required to connect the ceiling opening with the roof. Figure 12-53 shows several types of shafts. It is very important that the shaft be constructed as airtight as possible. Insulation should be attached to the sides of the shaft to the same thickness as the ceiling insulation.

Exterior Door Frames

Outside door frames are installed at the same time as the windows. Follow similar procedures. Secondary and service entrances usually have frames and trim members to match the windows. Main entrances, however, often contain additional elements that add an important decorative architectural feature. See Figures 12-54 and 12-55.

Exterior doors in residential construction are nearly always 6'-8" high, although 7'-0" sizes are available. Main

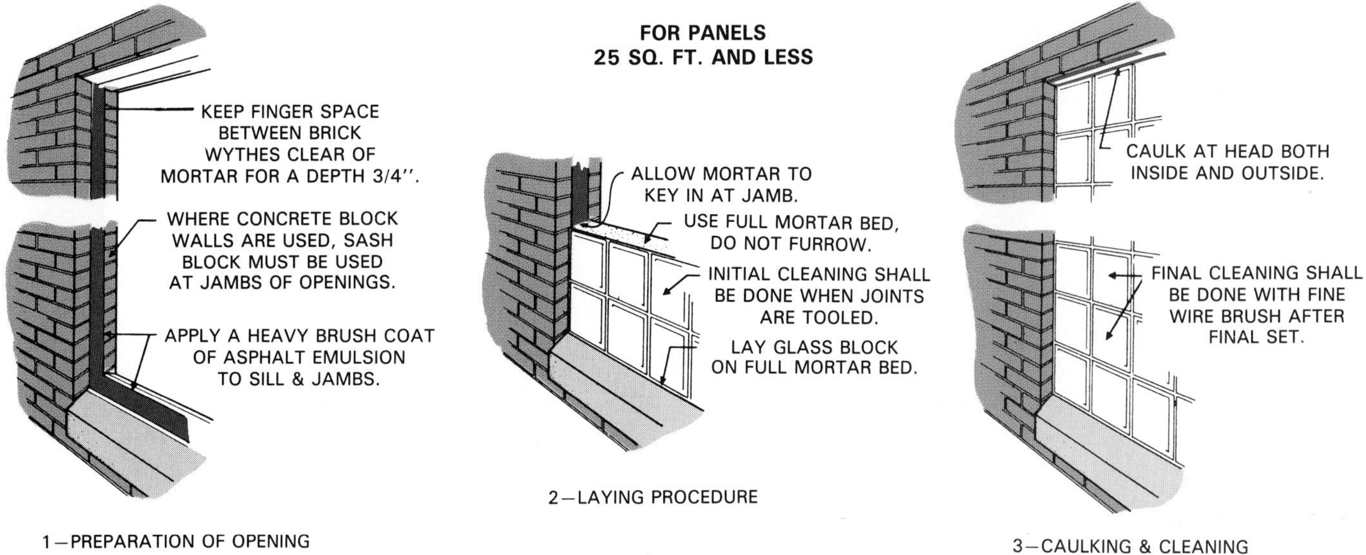

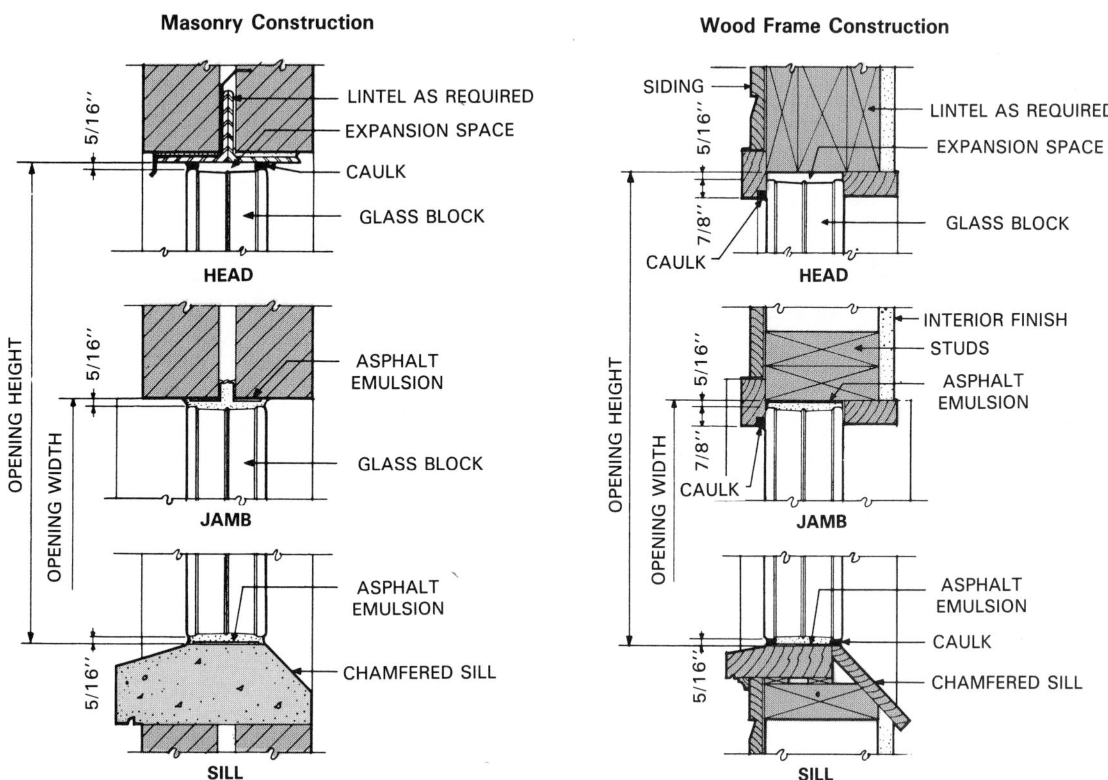

Figure 12-45. Construction details for installing glass block.

entrances usually are equipped with a single door that is 3'-0" wide. Narrower (2'-8" and 2'-6") sizes are used for rear and service doors. FHA Minimum Property Standards specify a minimum exterior door width of 2'-6".

Outside door frames, like windows, have heads, jambs, and sills. The head and jambs are made of 5/4" stock since they must carry not only the main door but also screen and storm doors. Figure 12-56 shows an elevation view of door frames. The doors are in place.

Door frames are manufactured at a millwork plant and arrive at the building site either assembled and ready to install or disassembled (*knocked down*, or K.D.). Sometimes K.D. units are assembled by the dealer or distributor. It is relatively easy to assemble door frames "on the job" when the joints are accurately machined and the parts are carefully packaged and marked.

While details of a door frame may vary, the general construction is the same, Figure 12-57. The head and jambs are rabbeted, usually 1/2" deep, to receive the door.

In residential construction, outside doors swing inward and the rabbet must be located on the inside. Stock door frames are designed for standard wall framing.

Figure 12-46. After removing the sash and trim members, cut through sill and frame and remove them. (Andersen Corp.)

Figure 12-48. Placing the new window unit in the rough opening. Get help if the unit is large. (Andersen Corp.)

Figure 12-47. Installing wood furring to bring the rough opening to the correct size.

Figure 12-49. Installing outside trim furnished by manufacturer.

However, they can also be adapted to stone or brick veneer construction, as shown in Figure 12-58.

Stock frames can be fitted with extension strips, Figure 12-59, that convert the frames to fit greater wall thickness. Extension strips are also used on window frames.

Doorsill design varies considerably. However, the top is always level with the finished floor. Sills may be made of wood, metal, stone, or concrete. The outside stoop at the entrance is placed just below the doorsill or may be lowered by a standard rise height (7 1/2"), Figure 12-60.

Positioning a doorsill so it will be level with the finished floor requires cutting away a section of the rough floor. Part of the top edge of the floor joist must also be cut away. This is done at the time the frame is installed.

The framing of the rough opening (R.O.), however, comes earlier—before the door frame is delivered to the job. The carpenter must check the working drawings carefully. The size of the rough opening will usually be included in the door and window schedule. The height of the opening is shown in detail sections. For standard construction, the R.O. can be calculated by adding about 2 1/2" to the door width and height.

When this information is not included in the working drawings, you should consult the manufacturer's literature. This will contain not only R.O. requirements but also detail drawings. The latter are especially important for front entrances that consist of more complicated structures and for fixed sidelight window units, Figure 12-61.

Figure 12-50. A skylight provides natural overhead lighting for modern kitchens. (Andersen Corp.)

Figure 12-51. Ventilating skylight. This unit is hinged at the top and opens up to 8" at the bottom. (Rolscreen Co.)

Installing a Door Frame

1. Check the size of the rough opening to make sure that proper clearances have been provided. Cut out sill area, if necessary, so the top of the sill will be the correct distance above the rough floor. In some structures it may be necessary to install flashing over the bottom of the opening.
2. Place the frame in the opening, center it horizontally, and secure it with a temporary brace.
3. Using blocking and wedges, level the sill and bring it to the correct height. Be sure the sill is well supported. For masonry walls and slab floors, the sill is usually placed on a bed of mortar.
4. With the sill level, drive a nail through the casing into the wall frame at the bottom of each side.
5. Insert blocking or wedges between the studs and the top of the jambs. Adjust wedges until frame is plumb. Use a level and straightedge as shown in Figure 12-62.
6. Place additional wedges between the jambs and stud frame in the approximate location of the lock strike plate and hinges. Adjust the wedges until the side jambs are well supported and straight. Then, secure the wedges by driving a nail through the jamb, wedge, and into the stud.
7. Finally, nail the casing in place with nails spaced 16" O.C. Follow the same precautions suggested for window frame installation.

After the installation is complete, a piece of 1/4 or 3/8" plywood should be lightly tacked over the sill to protect it during further construction work. At this time, many builders prefer to hang a temporary combination door in order that the interior of the structure can be secured, thus providing a place to store tools and materials.

Setting the threshold and hanging the door is a part of the interior finishing operation and will be described in Unit 18. Exterior door types, designs, and sizes are described in that unit.

When a prehung door unit is installed, the door should be removed from its hinges and carefully stored.

Installing Prehung Door Units

A variety of prehung exterior door units are available. They include single doors, double doors, and doors with sidelights. Millwork plants provide detailed instructions for installing their products.

Installing a Prehung Door

1. First check the rough opening. Make sure the size is correct and that it is plumb, square, and level.
2. Apply a double bead of caulking compound to the bottom of the opening.
3. Set the unit in place, Figure 12-63. Spacer shims located between the frame and door should not be removed until the frame is firmly attached to the rough opening.
4. Insert shims between the side jambs and rough opening. They should be located at the top, bottom, and midpoint.
5. Drive 16d finish nails through the jambs, shims, and into the structural frame members.
6. Manufacturers usually recommend that at least two of the screws in the top hinge be replaced with 2 1/4" screws.
7. Finally, adjust the threshold so it makes smooth contact with the bottom edge of the door.

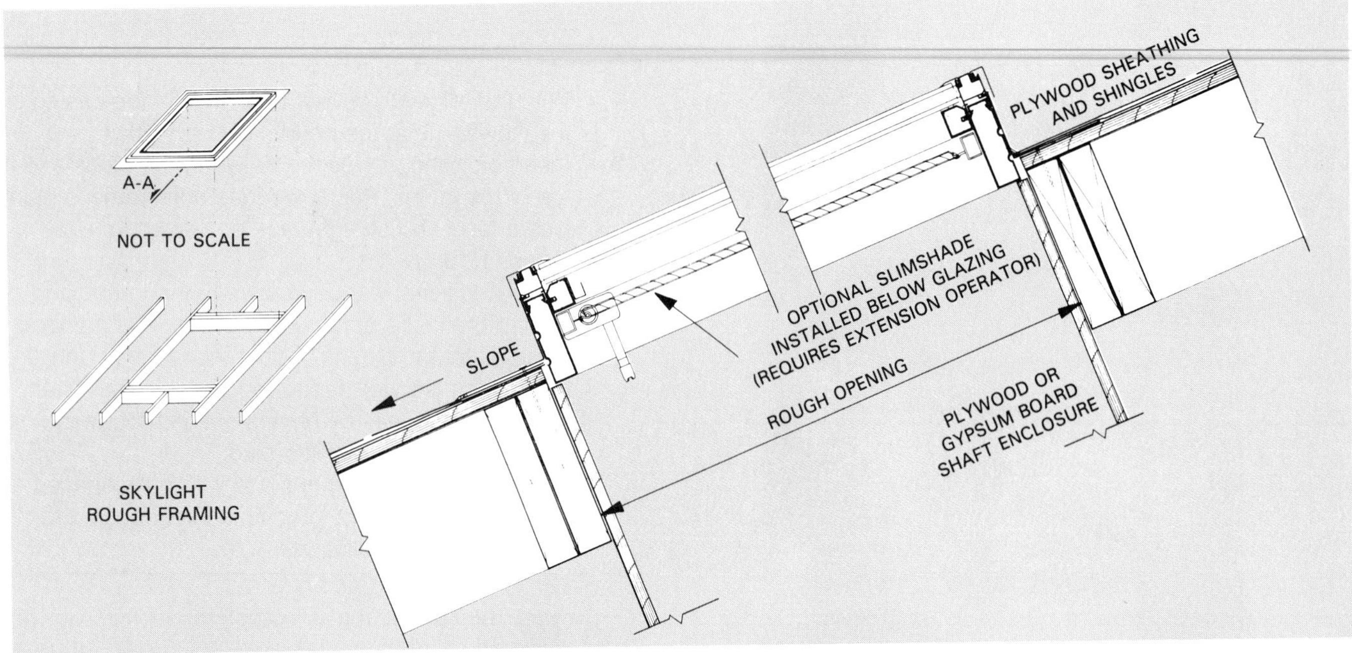

Figure 12-52. Cross section detail for installation of a skylight. (Rolscreen Co.)

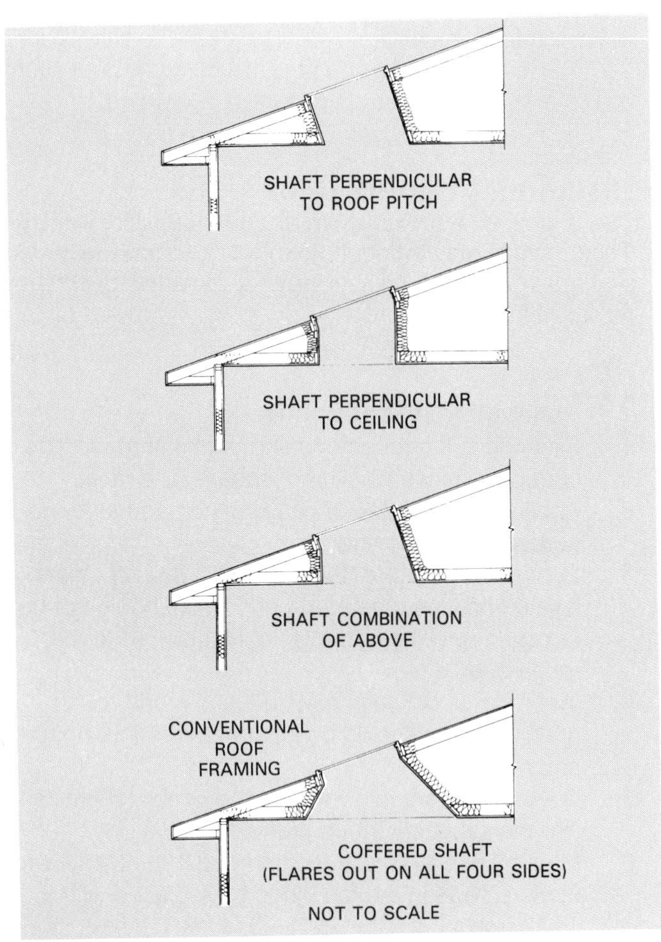

Figure 12-53. Skylight shaft may take any one of several shapes.

Figure 12-54. Wood paneled door with insert and sidelights of leaded glass. Panels are full 1 5/32" thick for strength and heat loss reduction. (C-E Morgan)

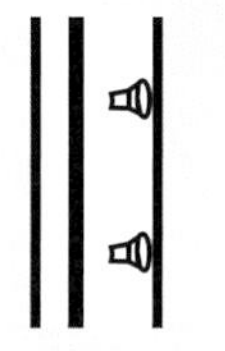

When setting door and window frames, never drive any of the nails completely into the wood until all nails are in place and a final check has been made to make sure that no adjustments are necessary.

Figure 12-55. Main entry door. Traditional style is combined with 0.024" steel face with a thermal break. Cavity is insulated with expanded polystyrene foam. (Pease Industries, Inc.)

Figure 12-56. Frames and doors for main entrances as they appear on elevation drawings.

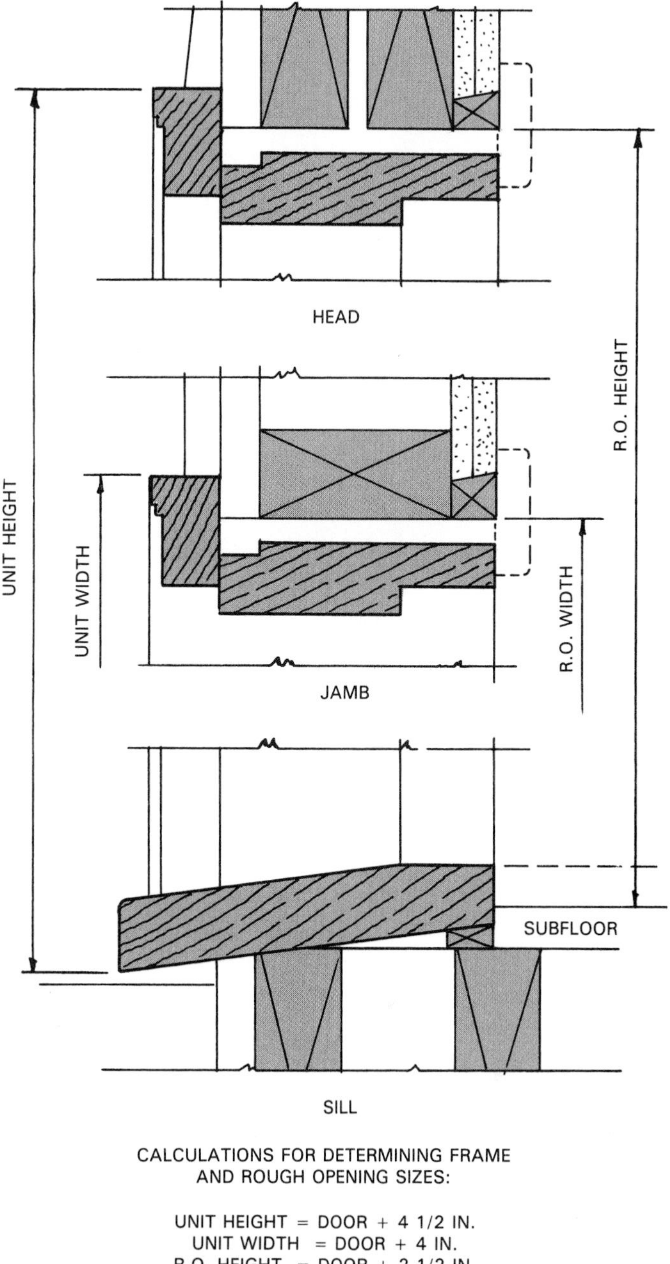

Figure 12-57. Always check drawings for exterior door frame details and sizes. Make sure rough openings are correct size.

Sliding Glass Doors

To accommodate outdoor living, a terrace or patio door is often included in residential designs. The French or casement type door, once used for this purpose, has been largely replaced with the modern sliding glass door.

This type of door, riding on nylon or stainless steel rollers, is easy to operate. When equipped with quality weather stripping and insulating glass, it restricts heat loss and condensation to a level satisfactory even in cold climates. It is available as a factory assembled frame and door unit. Parts and installation details are similar to those for sliding windows.

Sliding glass door units contain at least one fixed and one operating panel. Some may have three or four panels. The type of unit is commonly designated by the number and arrangement of the panels as viewed from the outside. The letters *L* or *R* mean an operating panel. This indicates

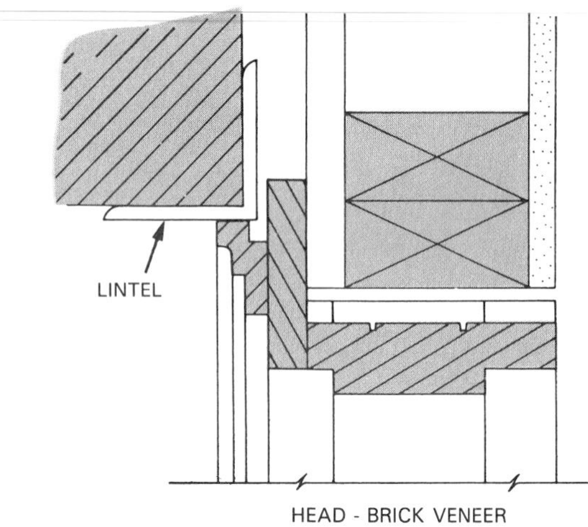

Figure 12-58. Detail on how to install head of door frame in brick veneer construction.

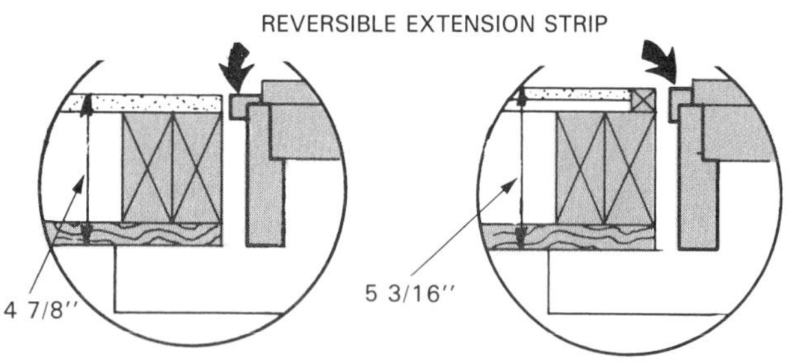

Figure 12-59. Stock frames can be made wider. Reversible extension strips above convert 4 5/8" jambs to either 4 7/8" or 5 3/16". (C-E Morgan)

which way the panel opens. See Figure 12-64. Some manufacturers use the letter *X* to indicate operating panels and *O* for fixed panels. Sliding glass door units must be glazed with tempered (safety) glass.

Construction details will vary from one manufacturer to another, as will unit sizes and rough opening requirements. See Figure 12-65. If detailed drawings or R.O. sizes are not included in the architectural plans, then the carpenter should secure the information from the manufacturer's literature. Figure 12-66 identifies typical parts of a sliding door installation at a sill section.

The installation of sliding door frames is similar to the procedure described for regular outside doors. Before setting the frame in place a bead of sealing compound should be laid across the opening to ensure a weathertight joint. If heavy glass doors are to slide properly, the sill must be level and straight.

Plumb side jambs and install wedges is the same way as you would install regular door frames. After careful checking, complete the installation of the frame by driving galvanized nails through the side and head casings into structural frame members, Figure 12-67.

At this point, many carpenters prefer to check the fit between the metal sill cover and doors rather than installing them. These items are carefully stored away. The opening is enclosed temporarily with plywood or polyethylene film attached to a frame. Then, during the finishing stages of construction, after inside and outside wall surfaces are completed, the sill cover, threshold, doors, and hardware are installed, Figure 12-68.

Manufacturers of sliding glass doors provide detailed instructions for installing their particular product. This material should be read and studied carefully before proceeding with the work.

Figure 12-60. Height of door sill may vary. Left—Raised slightly above porch, stoop, or sidewalk. Right—Elevated a standard riser height above outside surface. (Pease Industries)

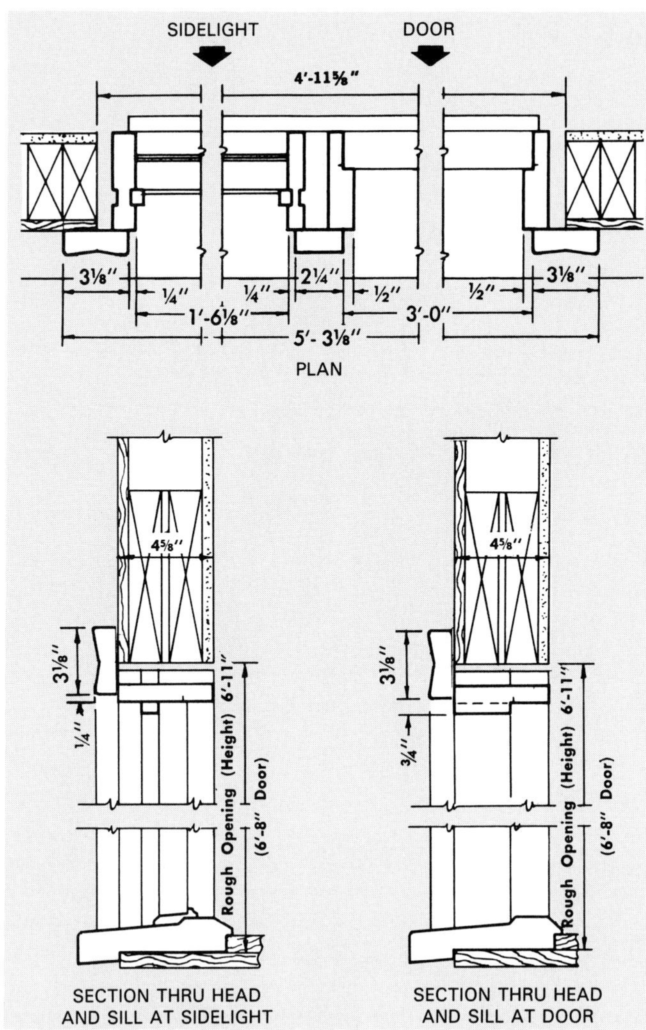

Figure 12-61. Detail drawings of a front entrance door frame with sidelight. (C-E Morgan)

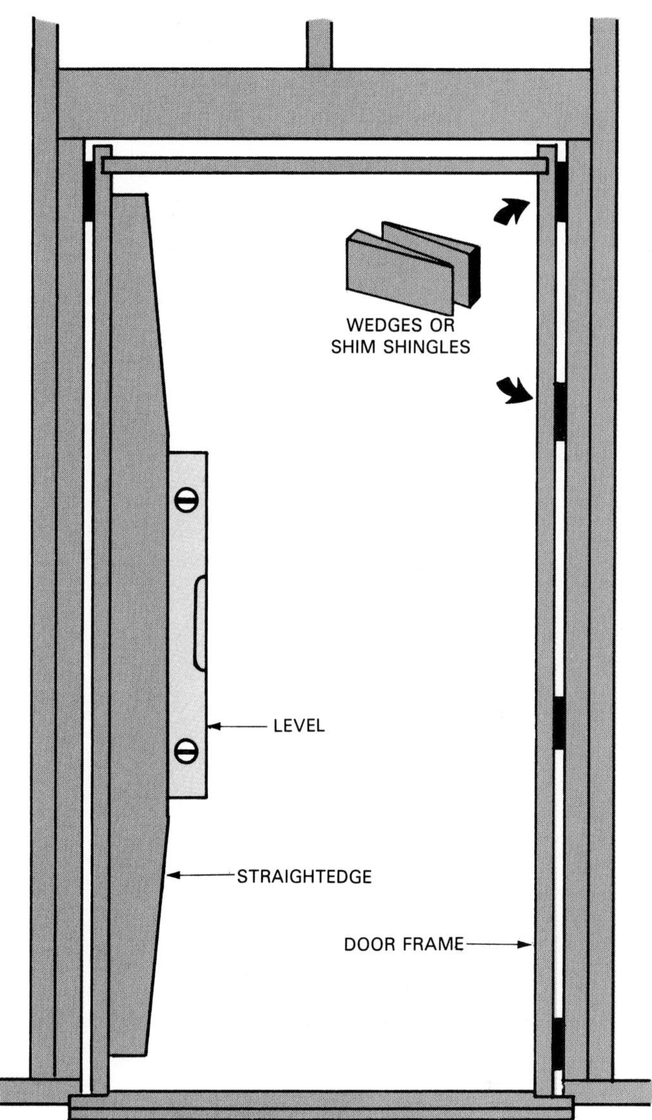

Figure 12-62. How to plumb the door jambs.

After installing large glass units in buildings under construction, it is considered good practice to place a large "X" on the glass. Use masking tape or washable paint. This will alert workers so they will not walk into it or damage it with tools and materials.

Garage Doors

Basically, there are three types of garage doors; hinged or swinging, swing-up, and roll-up. As a result of refinement and perfection in the design of hardware and counterbalancing equipment, the roll-up door has all but replaced hinged and swing-up doors. See Figure 12-69.

Wooden garage doors are constructed much like exterior passage doors. Flush type doors with foamed plastic cores are available for installations that require high levels of sound and thermal insulation. Garage doors are also

Figure 12-63. Prehung door is installed in a rough frame opening. (C-E Morgan)

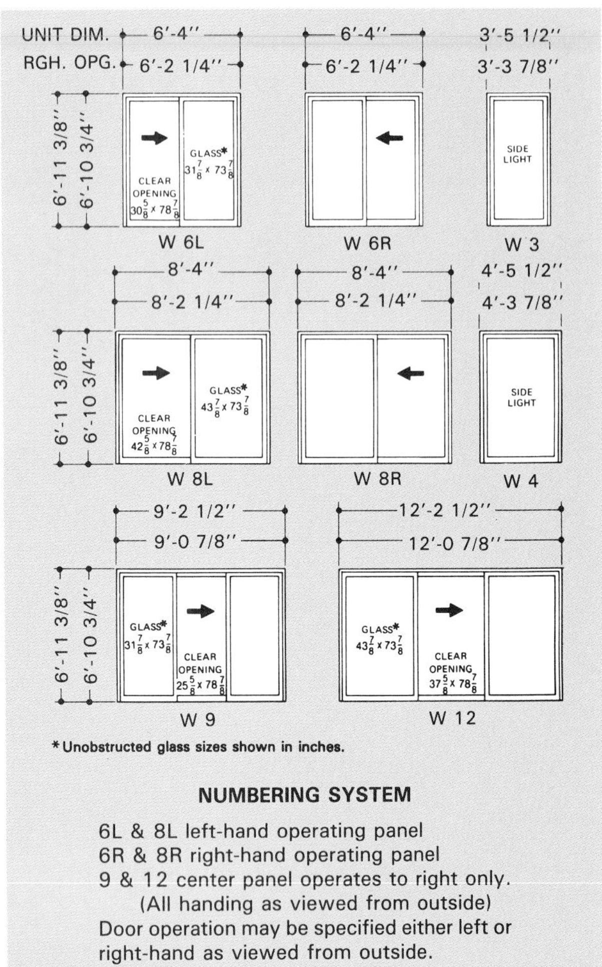

Figure 12-64. Sliding glass door units come in standard sizes and shapes. (Andersen Corp.)

manufactured from steel, aluminum, and fiberglass. Many designs are available to match contemporary or traditional architecture. Stock sizes are listed in Figure 12-70.

Garage Door Frames

Frames for garage doors include side jambs and a head similar to exterior passage doors. No rabbet is required. The frame is usually included in the millwork order along with windows and doors so the outside trim will match. The size of the frame opening is usually the same size as the door. However, the manufacturer's specifications and details should be checked before placing the order.

Figure 12-71 shows a typical jamb section for wood frame or masonry construction. Note the thickness (2" nominal) of the heavy inside frame to which the track and hardware will be mounted. The width of this member should be at least 4" wide with no projecting bolt or lag screw heads.

The rough opening width for frame construction will normally be about 3" greater than the door size. The height of the rough opening should be the door height plus about 1 1/2" as measured from the finished floor.

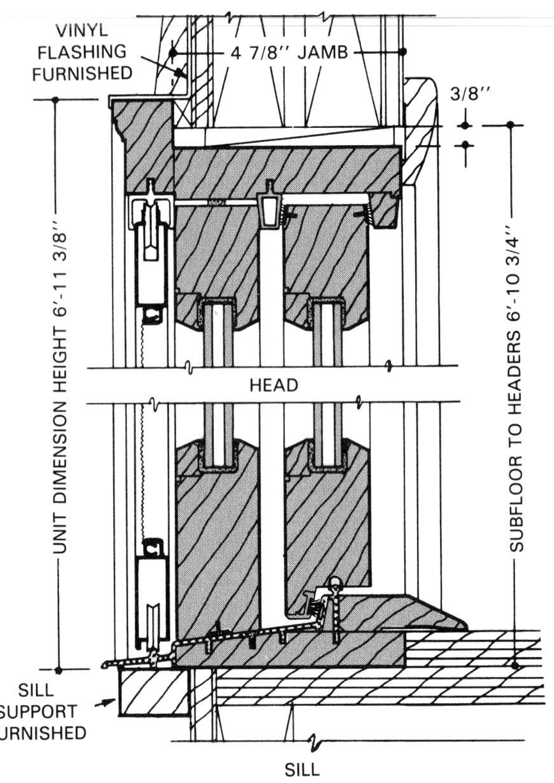

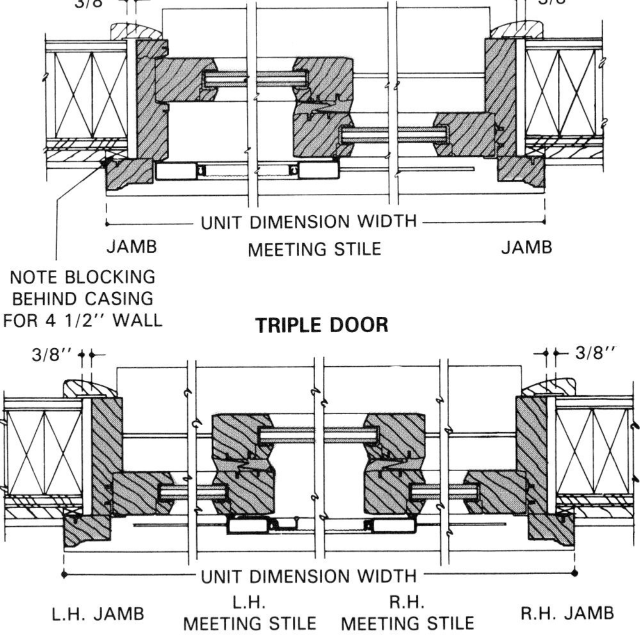

Figure 12-65. Construction details like the above will be different from one manufacturer to the next. Be sure to study this information before installation.

Give careful consideration to the inside height. A minimum clearance between the top of the door and the ceiling must be provided for hardware, counterbalancing

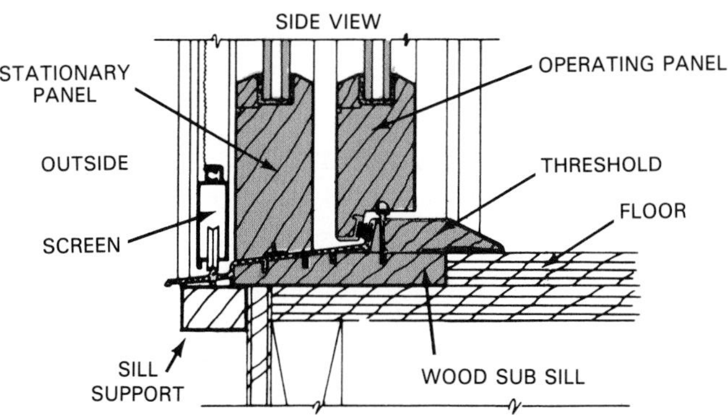

Figure 12-66. Manufacturer's literature often labels main parts of sliding glass doors at sill section.

Figure 12-68. Sliding glass doors are installed during the finishing stages of construction. (Andersen Corp.)

Figure 12-67. After careful alignment of the door frame, drive galvanized nails through the side and head casing into structural frame members.

Figure 12-69. Roll-up garage doors are hinged. Note how sections move upward and tilt to horizontal position. Rollers, attached to each section, run in a special steel track to support door. (Stanley Door Systems)

mechanisms, and the door itself when open. On some special low-headroom designs, this distance may be as little as 6". Always be sure to check the manufacturer's requirements for each door.

When installing garage door frames, follow the general procedure used for regular door frames. Be certain the jambs are plumb and the head is level.

WIDTH	HEIGHT
8′0″	7′0″ or 6′6″
9′0″	7′0″ or 6′6″
10′0″	7′0″ or 6′6″
16′0″	7′0″ or 6′6″
18′0″	7′0″ or 6′6″

Figure 12-70. Stock sizes of garage doors.

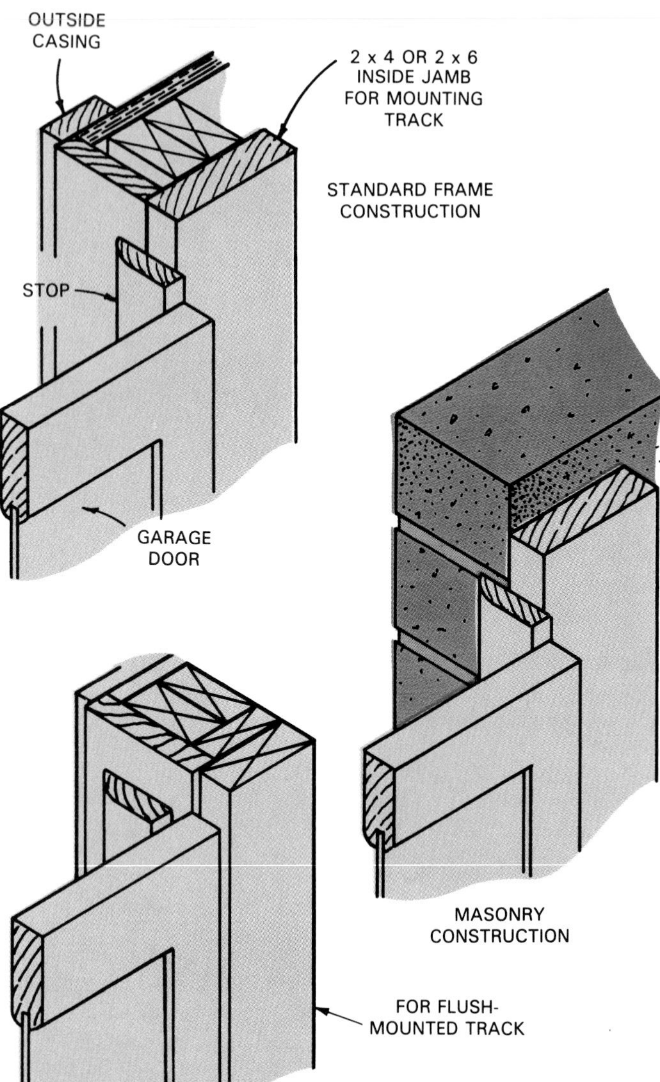

Figure 12-71 Typical jamb construction for garage door frames when roll-up door is used.

Installing a Garage Door

1. Tack stops temporarily in place.
2. Assemble the door sections in the opening and attach hinges.
3. Attach rollers to door.
4. Place track on rollers and attach track to jamb.
5. Mount horizontal track sections.
6. Raise and prop door in open position.
7. Attach counterbalancing mechanism.
8. Open and close door and make necessary adjustments.
9. Reset stops for smooth, tight fit.

Manufacturers always supply detailed directions and procedures for the installation of their products. The carpenter should follow these printed materials carefully. They are usually well-illustrated and easy to understand. See Figure 12-72.

Hardware and Counterbalances

Garage door hardware must be well designed so the door will operate easily. Track, hinges, and bolts should be made of galvanized steel. Gage weight must be heavy enough to last the life of the door.

To offset the weight of the door, various counterbalancing devices are used. Two of the most common types are shown in Figure 12-73. The extension spring is commonly used on residential doors, either single or double. The torsion spring and its mechanism is more expensive but provides a smoother and more consistent action. It is especially recommended for wide doors and those that are heavier in construction.

Today, many residential plans include a two-car garage with a 16' or wider door. The operation of the large door has led to the wide acceptance of electric powered door openers, Figure 12-74.

Installing Bow and Box Bay Windows

Bow and box bay windows, Figure 12-75, are often installed in modern buildings. Being larger as well as projecting a considerable distance outward from the building wall, they require extra support. This support is important, as it prevents sagging of the window and reduces stress on the wall framing.

Braces are one way of providing support. These are placed from the outer edge of the window's base to studs in the wall frame. The bracing is eventually sheathed and wall covering installed.

Another support method is to use cables running from the base of the window to rafter tails or some other support above the window. Kits are provided by the manufacturer, Figure 12-76. Figure 12-77 shows cables attached to rafter tails in a soffit installation. Cleats should be positioned above the "T" nuts; it may be necessary to secure bracing between two rafter tails. In a gable installation, the cleats may be anchored to a lintel, studs, second floor band joist, or sill plate. See Figure 12-78.

Extra support along the base is necessary for larger units. If box or angle bay units have a center sash of 5' x 5' or larger, use a steel angle, channel, or wood member across the bottom, spanning the center sash. Drill holes at either end of the support member for the steel cables. See Figure 12-79.

When preparations are completed, thread cable through the top and bottom platforms, allowing an inch of threads below the bottom platform. Support the unit with jacks, lever arms or some other method. Attach washers and nuts to each cable.

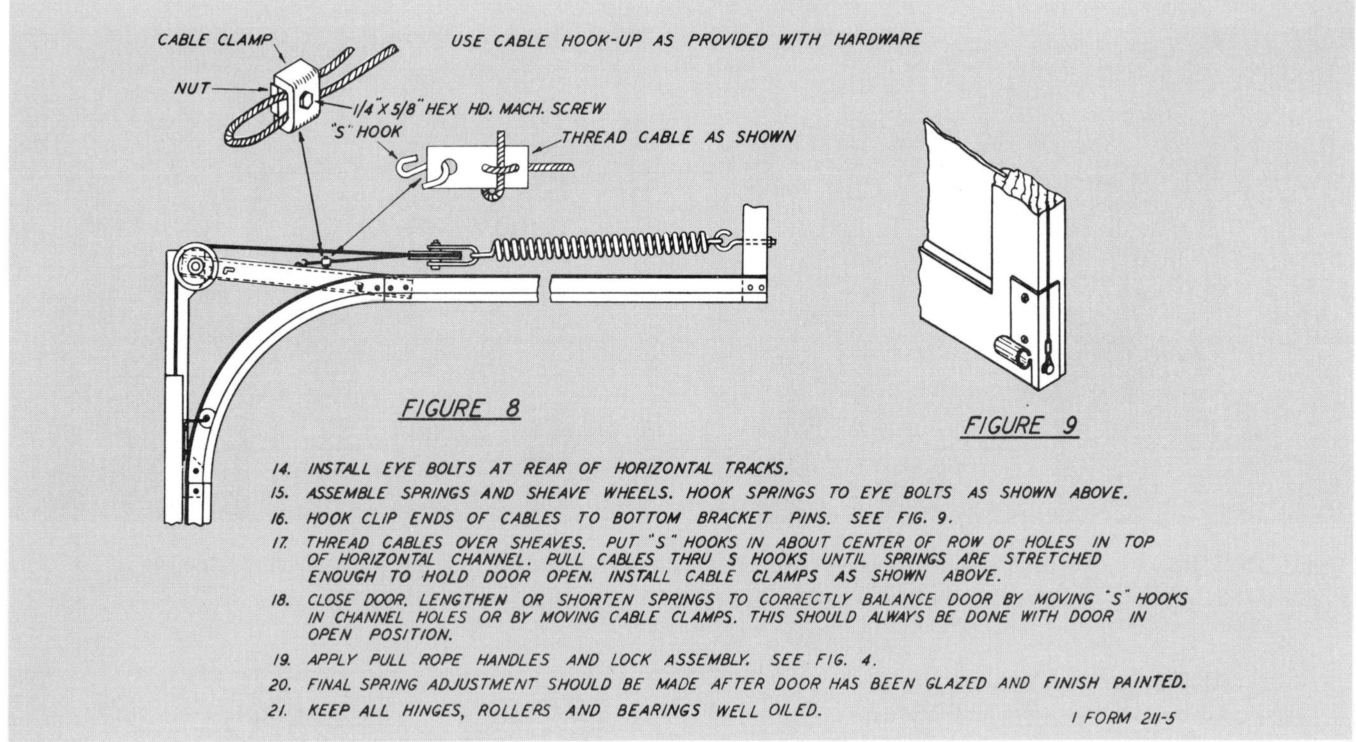

Figure 12-72. Sample of manufacturer's instructions. These are generally well-illustrated and easy to understand.

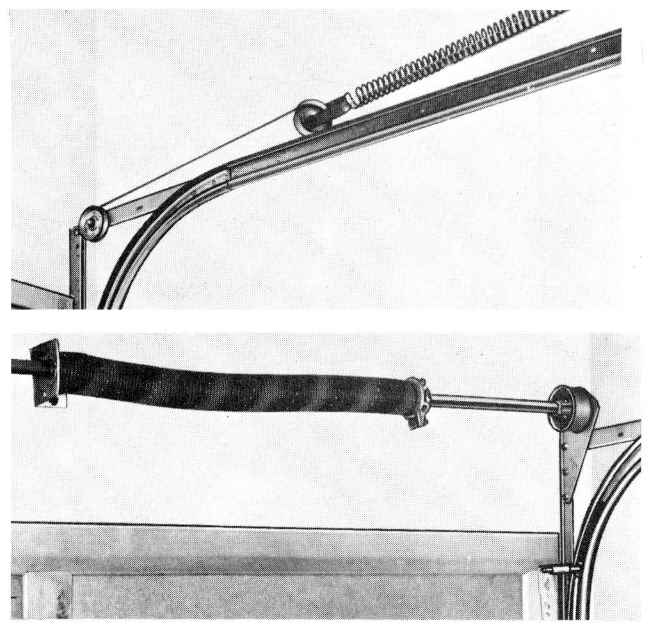

Figure 12-73. Spring counterbalances for garage doors. Top—Extension spring. Bottom—Torsion spring.

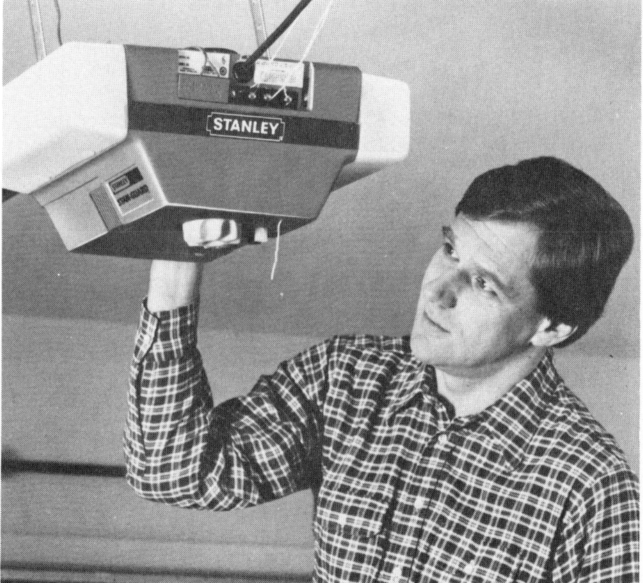

Figure 12-74. Installing an automatic garage door opener. In addition to radio control, it features a smoke and heat alarm, and an intrusion (burglar) warning system.
(Stanley Door System)

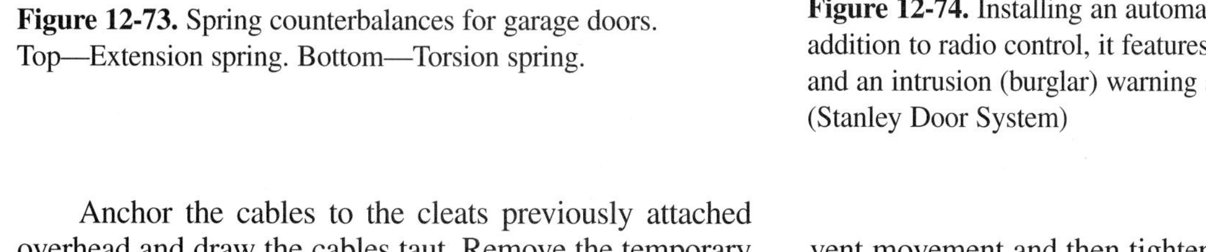

Anchor the cables to the cleats previously attached overhead and draw the cables taut. Remove the temporary support and check unit for plumb, level, and sash reveal. Readjust as necessary. Finally, tighten upper nuts to prevent movement and then tighten the lock nut. Notch head and seat boards and slide them into place. Always follow specific instructions provided by the manufacturer.

Figure 12-75. Bow window being installed in new construction. Left—Two workers slide the unit into the rough opening. Right—When shimmed and plumbed, the unit is secured by driving nails through the exterior trim.

Cable Support System Parts
(Two kits are required for C7 Bow)

A. 2 - Support Cables (9' - 3 1/4")
B. 2 - Cable Cleats
C. 4 - 2" Wood Screws
D. 2 - "T" Nuts
E. 2 - Washers
F. 4 - 1/4" X 28 Hex Nuts

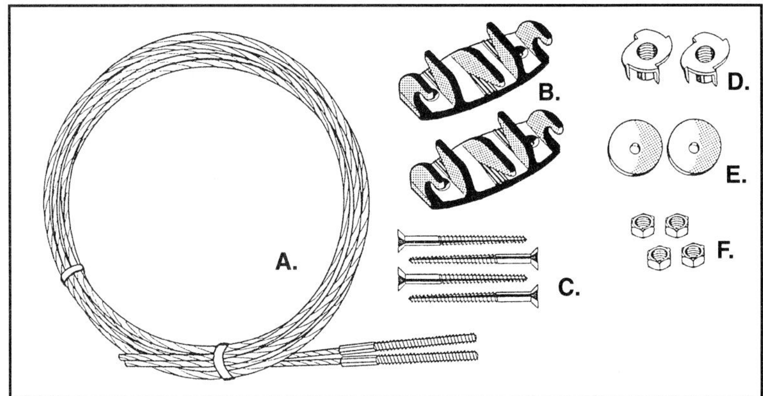

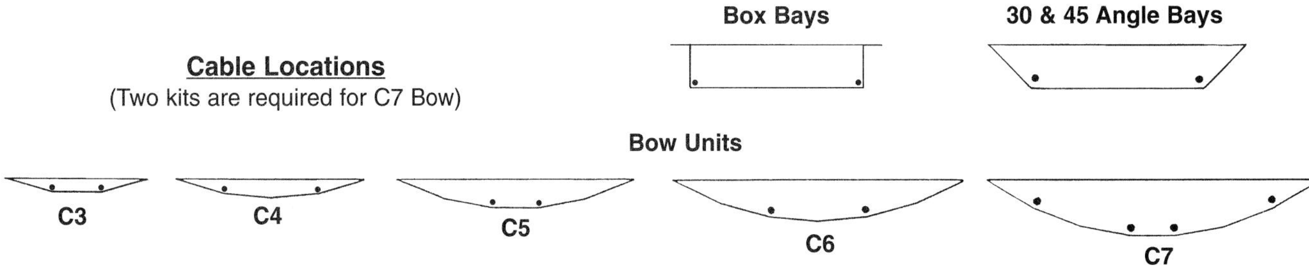

Figure 12-76. Heavier bow and bay windows may require extra supports using cables. (Andersen Corp.)

Important Terms

Awning windows
Bite
Casement windows
Double-hung windows
Emissivity
Fixed window
Floating
Glass block
Hopper windows
Horizontal sliding window
Jalousie window
Jamb extension
Knocked down
Mortar key
Mullions
Multiple-use windows
Muntin
R value
Sash
Storm sash

Test Your Knowledge

1. Today, windows and doors are built in large ____________ plants.
2. The kind of wood most often used to manufacture windows is ____________.
3. A type of window that is hinged on the side and swings outward is called a ____________.
4. The height of a standard residential door is ____________.

Figure 12-77. This illustrates a soffit installation for a bay window. Cables are attached either to rafter tails or to 2" lumber spanning the distance between two rafters. Joist hangers are used to fasten the blocking. (Andersen Corp.)

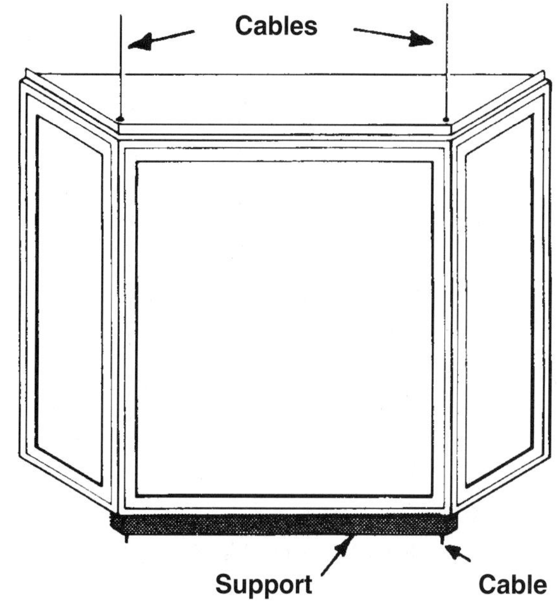

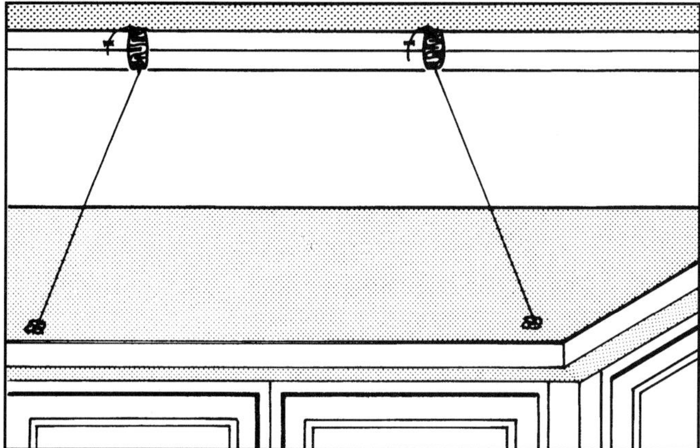

Figure 12-78. In a gable installation, cable can be anchored to a lintel, studs, wall plate, or a second floor joist. (Andersen Corp.)

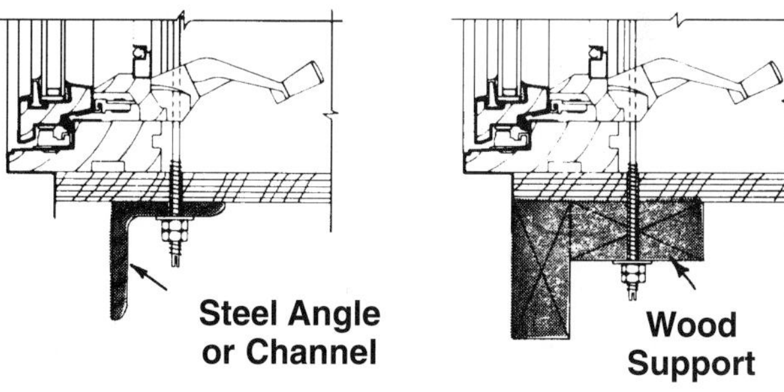

Figure 12-79. Heavier box or angle bay units must have extra support at their bases. Channel or angle iron or wood are recommended. (Andersen Corp.)

5. A single pane of glass in a window has an R-value of about _____________. Adding another pane of glass with 1/2" of airspace increases the R-value to _____________.
6. The side section of a window is called the _____________.
7. When there is little or no roof (cornice) overhang to protect the window, a _____________ should be installed above the head casing.
8. If the R.O. for a window is listed as 3'-6" x 3'-5", the height of the rough opening would be _____________.
9. To adjust for various wall thicknesses, _____________ are applied to standard window frames.
10. After a window unit has been temporarily set in the rough opening, the next step is to _____________ the _____________ at the correct height.
11. Large insulating window units are made from polished glass that is _____________ inches thick.
12. The thickness of standard glass block is _____________ inches.
13. When figuring the size of the opening for glass block panels, multiply the number of units by the _____________ (*actual, nominal*) block size.
14. In brick veneer construction, the masonry over door and window heads is carried by a metal support called a _____________.
15. Outside wood casing for windows and doors is attached with nails spaced ___________ inches O.C.

16. The most popular type of garage doors is the _____________ type.
17. A two-car garage door is usually _____________ ft. wide.
18. The two common types of spring counterbalances for garage doors are extension and _____________.
19. List two methods of supporting bow and bay windows.

Hurricane-proof garage door. Top—Exterior view. Bottom—Inside view shows galvanized steel posts and heavy duty locking hardware. (Stanley Door Systems)

Outside Assignments

1. Prepare a written report on the manufacture of glass. Include such headings as historical development, early production methods, modern processes, float method, drawing method, and grinding and polishing plate glass. Study encyclopedias, reference books, and booklets from glass manufacturers.
2. The rising cost of energy has resulted in special emphasis being placed on window design and construction. Some manufacturers are now producing triple-glazed window units for homes located in northern climates. Gather information about these units, including R-values, prices, special installation directions, and predicted fuel savings.

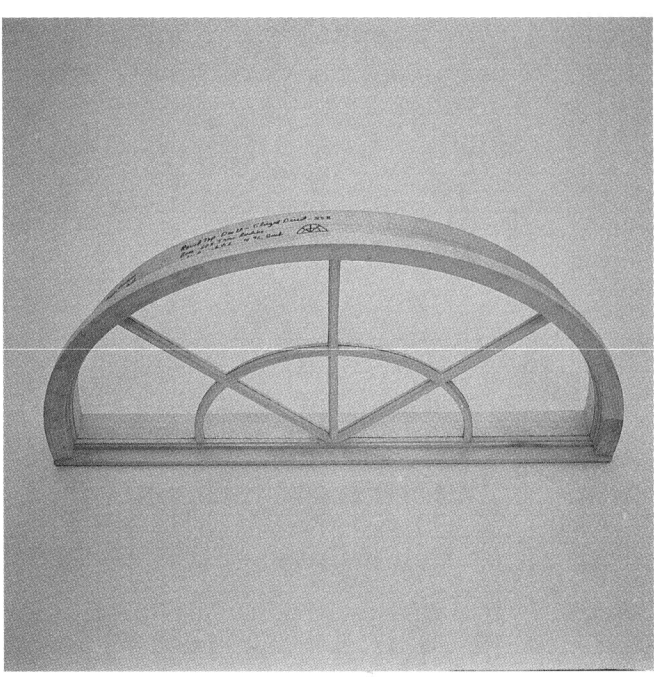

Half round windows are used in combination with other windows to produce arch windows. Such units are also known as circle top windows. Special rough framing is required.

13 Exterior Wall Finish

The term *exterior finish* includes all exterior materials of a structure. It generally refers to the roofing materials, cornice trim boards, wall coverings, and trim members around doors and windows. The installation of special architectural woodwork at entrances, or the application of a ceiling to a porch or breezeway area would also be included under this broad heading.

Previous units have described the application of the finished roof and the installation of the trim around windows and outside doors. This unit will cover the construction and finish of cornice work, and the materials and methods used to provide a suitable outside wall covering.

Cornice Designs

The *cornice* is the overhang of a roof at the eave. It usually includes the fascia board, a soffit for a closed cornice, and any moldings. It provides a finished connection between wall and roof. Style of the house will determine its design. Figure 13-1 shows details of different cornice construction.

Diagrams of several closed or *boxed cornice* designs are illustrated in Figure 13-2. An open cornice is sometimes used, exposing the rafters and underside of the roof sheathing. Wide overhangs are used often in modern buildings. These provide shade for large window areas, protect the walls, and add to attractiveness of the structure.

The *rake* is that part of a roof that overhangs a gable. It is usually enclosed with carefully fitted trim members.

Parts of Cornice and Rake Section

Figure 13-3 shows structural and trim parts of a boxed cornice. The *fascia board* is the main trim member along the edge of the roof. A *ledger strip* (or *frieze board*) is nailed horizontally along the wall to support the lookouts. *Lookouts* are 2 x 4 pieces that are attached between each rafter and the wall as support for the soffit materials.

A nailing strip is sometimes attached to the back of the fascia, between each rafter. Nailing strips, along with the lookouts and ledger, provide a frame to which the *plancier* or *soffit* material can be applied. Soffits can be plywood, hardboard, solid lumber, vinyl, or metal.

The trim used for a boxed rake section, Figure 13-4, is supported by the projecting roof boards. In addition, lookouts or nailers are fastened to the side wall and the roof sheathing. As in cornice construction, these serve as a nailing base for the soffit and fascia.

When the rake projects a considerable distance, the sheathing does not provide adequate support. In such cases, the roof framing should be extended.

Parts of the cornice and rake structure exposed to view are generally called *exterior trim*. In a typical construction, these parts are cut on the job. The properties needed in material used for exterior trim include good painting and weathering characteristics, easy working qualities, and maximum freedom from warp. Decay resistance is also desirable where materials may absorb moisture.

The cedars, cypress, and redwood have high decay resistance. Less durable species may be treated to make them decay resistant. Coat end joints or miters of members subjected to heavy moisture. Special caulking compounds are commonly used for this purpose.

Cornice and Rake Construction

In most construction, the fascia boards are installed on the rafter ends at the time the roof is sheathed. It is important that they be straight, true, and level, with well-fitted joints.

Before attaching fascia, some builders prefer to first nail on a 2 x 4 ribbon to align the tails of the rafters. The fascia is then attached to this ribbon. Corners of fascia boards should be mitered. End joints should meet at a 45° angle, as illustrated in Figure 13-5.

Framing a Cornice

1. Install a ledger strip along the wall if lookouts are to be used. (With light soffit material, lookouts are not needed for support. A metal channel at the wall provides enough support.)
2. With a level, locate and mark points on the wall that are level with the bottom edges of the rafters. Snap a chalk line between these points.

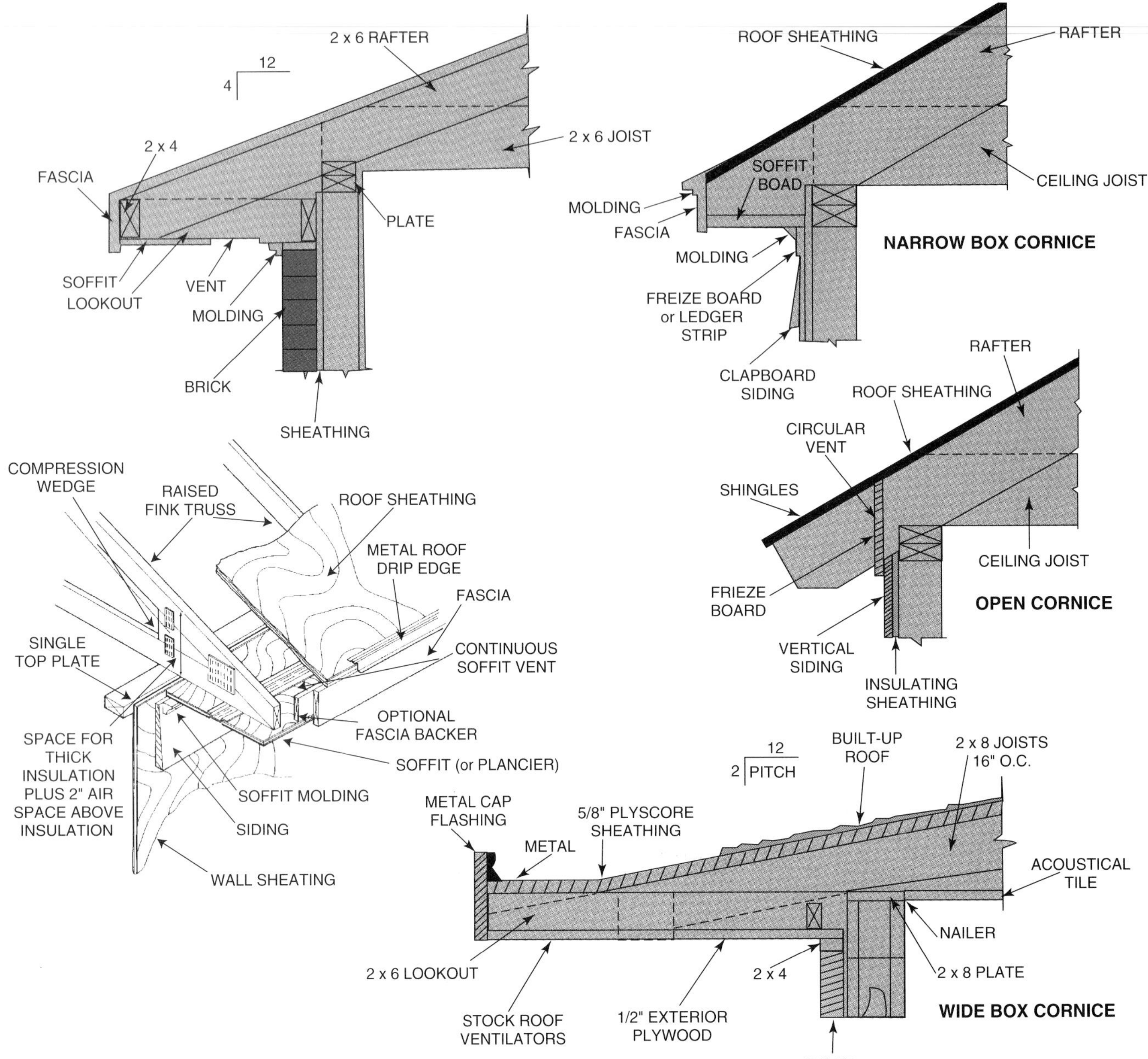

Figure 13-1. Typical cornice details for different architectural styles. The raised Fink truss rafter addresses the need for room to install thick insulation over the outside walls to conserve energy. (Wood Frame House Construction, U.S. Dept. of Agriculture)

3. Nail on the ledger strip (or a metal channel, depending on soffit material).
4. Cut the lookouts. Lookouts are usually made from 2 x 4 stock. Locate them at each rafter or every other rafter, depending on the kind of soffit material used.
5. Toenail one end of the lookout to the ledger and nail the other end to the overhang of the rafter.
6. When using thin material for the soffit, attach a nailing strip along the inside of the fascia to provide a nailing surface. Sometimes the back of the fascia is grooved to receive the soffit material. If the fascia is to be clad with vinyl or aluminum, an "F" channel is installed that covers the fascia and provides a slot for supporting the soffit.
7. Apply the soffit. First cut the material to size and then secure it with rust resistant nails or screws. When regular casing or finish nails are used, they must be countersunk and the holes filled with putty. Do this after the prime coat of paint.

In modern construction a rough fascia is usually nailed to the rafters and then covered with a prefinished vinyl or metal strip. See Figure 13-6.

The rake should be constructed to match the cornice using the same general procedures. After the main trim members are installed, moldings are often set in corners to cover irregularities. In modern construction, however, their

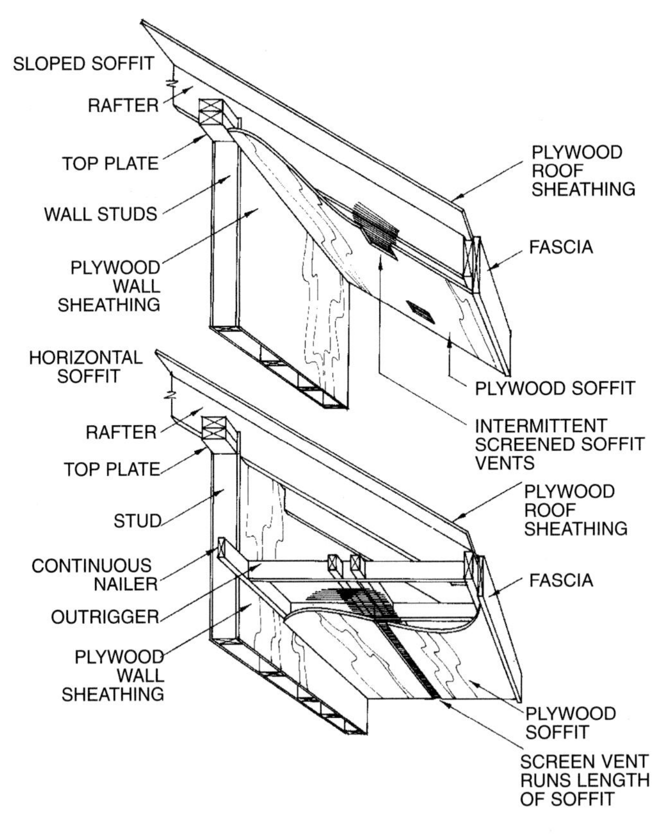

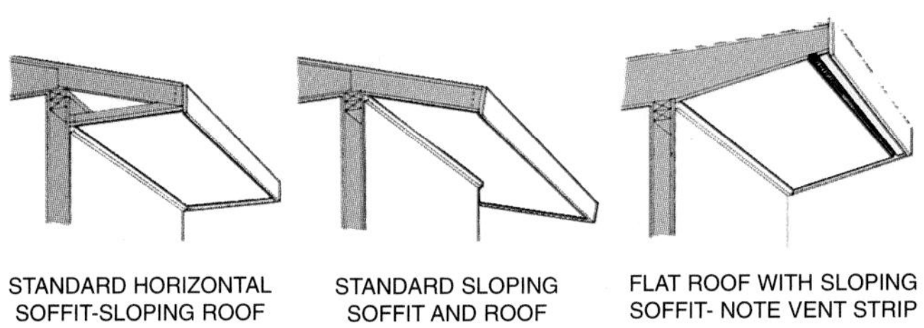

Figure 13-2. Three different cornice designs. Most cornices are boxed and vented. (Council of Forest Industries of British Columbia)

use is minimal in order to maintain a smooth, trim appearance, Figure 13-7.

Most construction makes use of truss rafters. The raised Fink truss is designed to allow thick ceiling insulation over the exterior wall without blocking the airway between the soffit and roof venting. It also allows steeply sloped roofs with wide overhangs that do not interfere with windows and doors. Soffits remain at the same height as interior ceilings regardless of roof slope or projection of the cornice. Soffit material is attached to the bottom chord of the truss which extends to the very end of the rafter. A compression wedge carries the weight of the roof where the top and bottom chords extend over the outside wall. A raised Fink truss is shown in Figure 13-1.

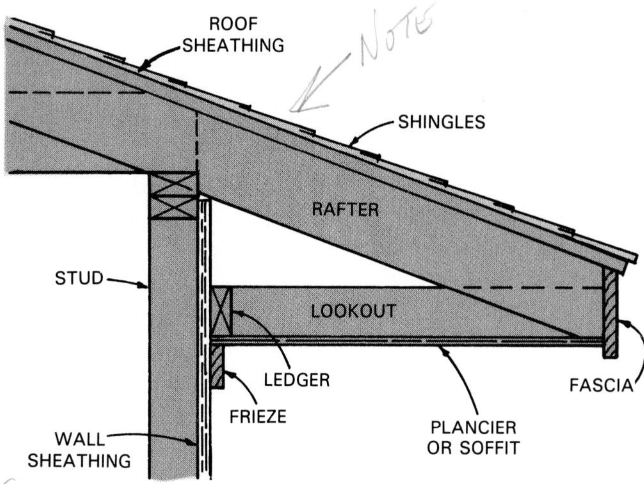

Figure 13-3. Parts of a typical boxed cornice. Soffit materials are often prefabricated and usually are hardboard, plywood, or metal.

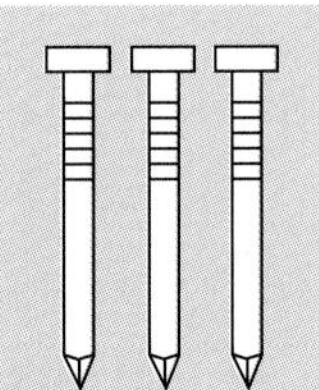

Always use rust resistant nails for outside finish work. They may be made of aluminum or galvanized (or cadmium-plated) steel.

Prefabricated Cornice Materials

Cornice construction is time consuming. Therefore, many builders prefer to purchase prefabricated materials. Various systems are available that provide a neat, trim appearance. One consists of 3/8" laminated wood-fiber panels. These are factory primed and available in a variety of standard widths (12 to 48") and in lengths up to 12'. Panels can be equipped with factory-applied screened vents. See Figures 13-8 and 13-9.

When installing large sections of wood panel, allow some clearance at edges for expansion of the wood. Fasten with 4d rust-resistant nails spaced about 6" along edges and intermediate supports. Start nailing at the edge butted against a previously placed panel. Nail to main supports first and then along the edges. Drive nails flush with the panel surface. See Figure 13-10.

Lookouts may be left out of a soffit system where special supports are attached to the upper surface of the panels. These supports in one system are made of 20 gage steel channels with prongs that make it easy to attach them to the back of the panels. The supports provide rigidity, so the panels only need to be attached at the front and back edges.

Porch and carport ceilings can be covered with factory-primed panels similar to those used for cornice soffits. Standard 4' x 8' and larger units are designed for either 16" or 24" O.C. framing. Always leave a 1/8" space along all edges for expansion.

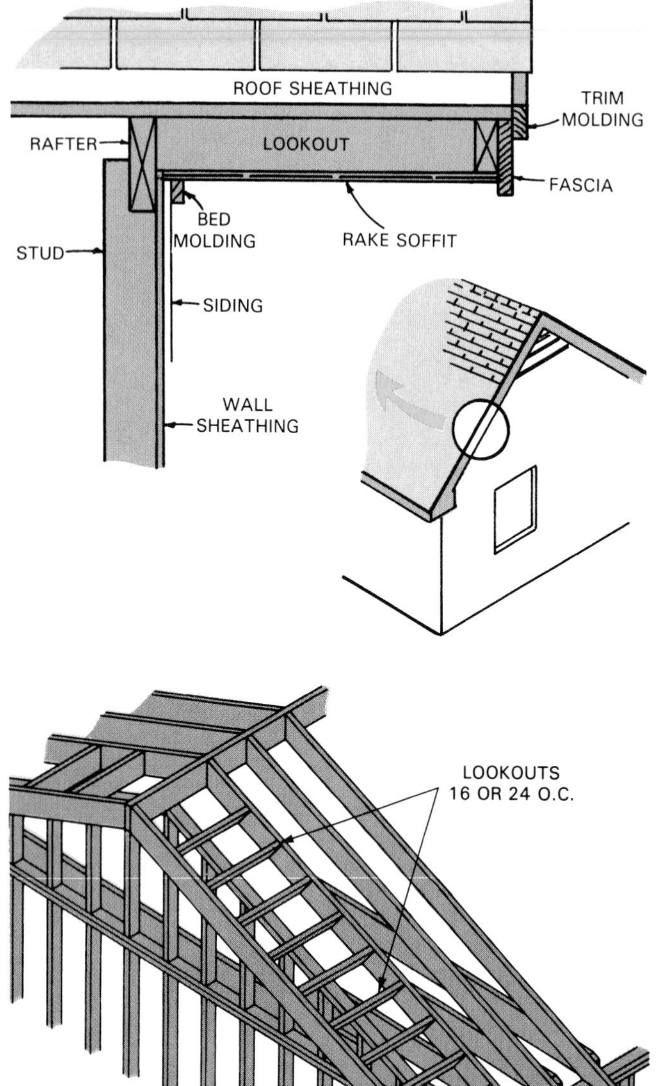

Figure 13-4. Top—Typical boxed rake section. Lookouts provide a nailing surface and support for the soffit. Bottom—Sometimes, the lookouts are cantilevered (ladder framing) to support a wide overhang on the gable.

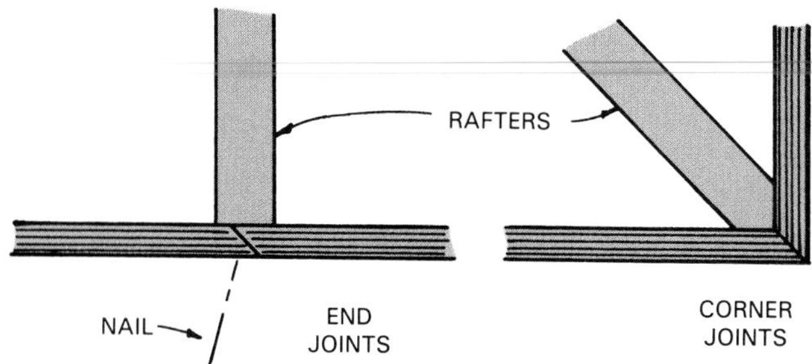

Figure 13-5. Plan view of joints for fascia boards. Miter corners and make matching angled cuts on ends. Note how end joints are nailed.

Figure 13-6. Installing a prefinished metal fascia cover. Follow the manufacturer's recommendations.

Start nailing in the center of the panels and move toward the outside. Space nails about 6" apart. Edges of panels can be secured and joints covered with special H-strips.

Metal Soffit Material

Metal soffit material is provided in rolls and sheets in several widths. Aluminum systems require little maintenance, incorporate venting, are self-supporting, and will never rust. Widths vary from 12 to 48". Edges are held in

Figure 13-7. Two views of a completed cornice and rake. Note continuous vent system in soffit. (Dickinson Homes, Inc.)

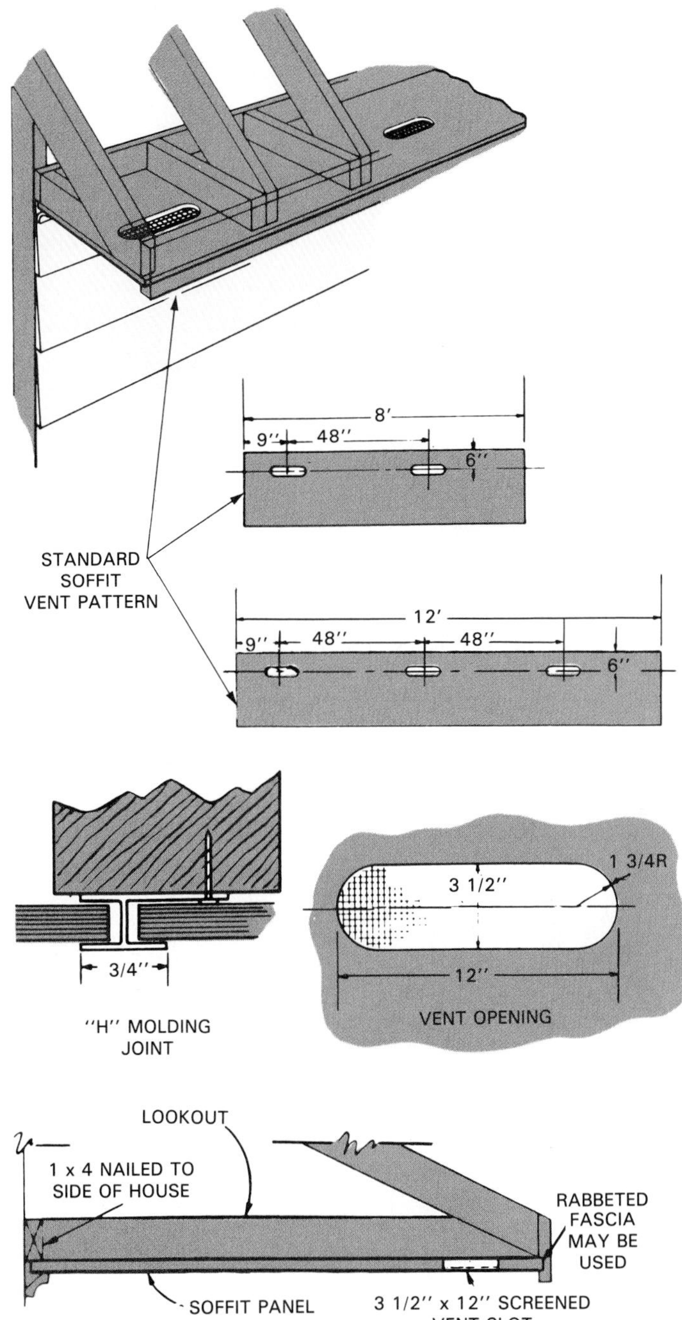

Figure 13-8. Details of prefabricated soffit system. Soffits are manufactured material—usually hardboard or metal.

Figure 13-9. This soffit arrangement has ventilation holes near the outer edge.

Figure 13-10. Soffit can be installed more efficiently with the use of power tools.

U-shaped channels (called *runner guides*) that are attached to fascia and walls.

Soffit systems consist of three basic units:

- Wall hanger strips (frieze strips).
- Soffit panels.
- Fascia covers.

See Figure 13-11. Panels may be vented or unvented. They are available in various widths and lengths.

Hanging Metal Soffit

1. Snap a chalk line along the side wall, level with the bottom edge of the fascia board.
2. Attach the metal U-shaped wall hanger strip along this chalk line.
3. Either attach a hanger strip flush with the bottom of the fascia or nail the metal soffit panels to the bottom of the fascia board as the panels are installed.

4. Insert the panels, one at a time into the strip or strips. See Figure 13-12.
5. After all soffit panels are installed, cut the metal or vinyl fascia cover to fit and install it as shown in Figure 13-13. The bottom edge of the cover is hooked over the end of the soffit panels. It is then nailed into place through prepunched slots located along the top edge. Always study and follow the manufacturer's directions when making an installation of this type.

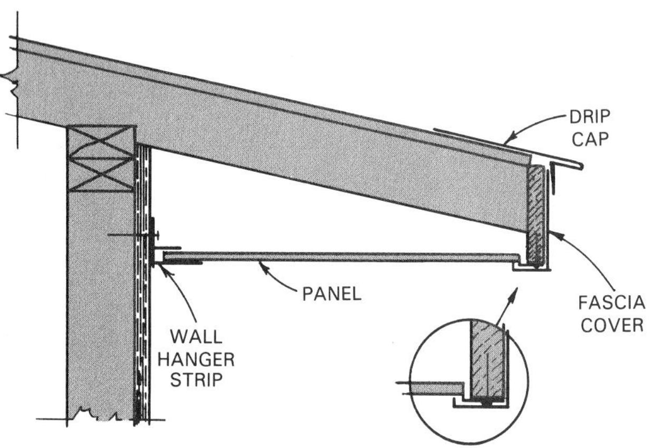

Figure 13-11. Detail of a prefinished metal soffit system. Top—Drawing of system used on wood or aluminum-sided wall. Bottom—Metal soffit system used with brick veneered wall. Note 2 x 6 nailer attached to the wall (arrow).

A continuous soffit ventilation system is shown in Figure 13-14. Designed to be used along with a continuous ridge vent, it comes in 8' lengths that are attached like a cove molding at the junction of the wall and the soffit. It can be used with either aluminum or vinyl soffit cover material. A channel in its upper edge receives the soffit material. The channel will accommodate soffit thicknesses of 1/4 to 1/2".

Some aluminum soffit coverings are sold in 50' coils with widths from 12" to 4'. To install, first attach runners at the wall and at the fascia board. Feed the coil into the runners. On a hip roof with soffits on all sides, leave one end open so that the last run can be fed into the runners. Figure 13-15 shows material and accessories used for soffits.

Figure 13-12. To install metal soffit, slip the inner end of the panel into the wall hanger strip. Either nail the outer end to the bottom edge of the fascia board or use second hanger strip. Panel widths will vary according to amount of roof overhang.

Figure 13-13. If fascia board is to be aluminum or vinyl clad, install fascia cover after all soffit panels are attached. Place the cladding under the drip edge.

Wall Finish

After the cornice and rake section of the roof are covered you can apply siding to the walls. When the structure includes a gable roof, the wall surface material is usually applied to the gable end before the lower section is covered. This permits scaffolding to be attached directly to the wall while siding the gable end.

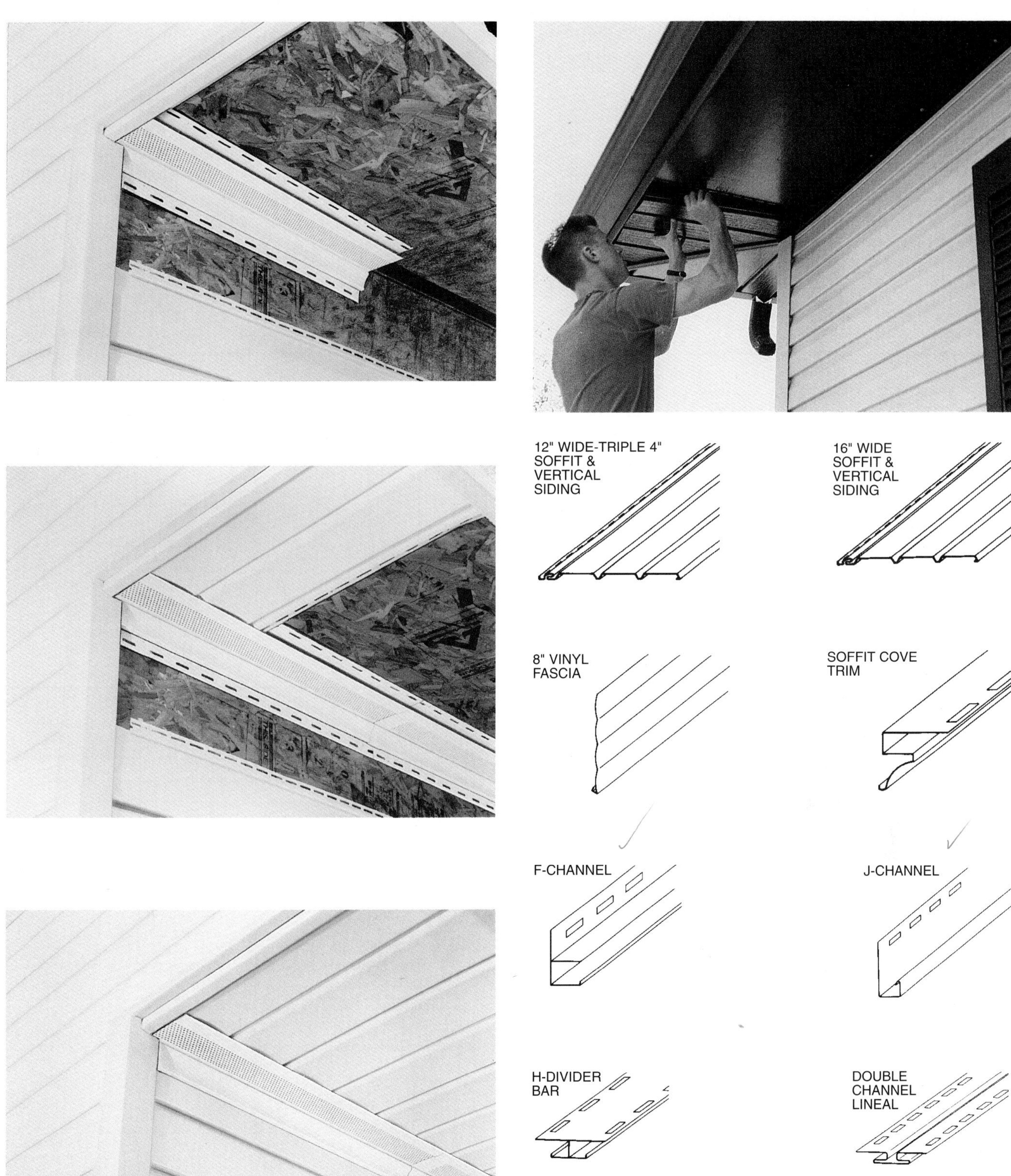

Figure 13-14. This patented covelike molding provides ventilation and has a built-in channel to receive various types of soffit covering. Top—If applied over a covering, cut a 2" slot close to the wall and nail on the continuous vent. Middle—Install soffit and siding. Bottom—Finished soffit. (Tapco)

Figure 13-15. Top—Soffit material is being installed to a wide overhang. Trim pieces are in place. Note "H" channel where corner is mitered. Bottom—Various types of panels and trim pieces are available for cladding fascia and soffits. Panels usually are 10 or 12" exposure in lengths of 12 or 12 1/2' packaged 12, 16, or 20 to a bundle. Trim pieces are generally 12 1/2' long, packaged in bundles of 10, 20, 24, 36, or 40. (CertainTeed Corp.)

All exterior trim members, if not factory-primed, should be given a primer coat of paint as soon as possible after installation.

Horizontal Wood Siding

One of the most common materials used for the exterior finish of American homes is wood siding. The shadow cast on the wall by the butt edge, especially evident in bevel siding, emphasizes the horizontal lines preferred by many homeowners.

Siding is usually applied over a base consisting of sheathing and a house wrap. The sheathing may be oriented strand board, plywood, boards, or rigid insulation board. However, in mild climates or on buildings such as summer cottages, the siding may be applied directly to the studs. Where sheathing is omitted, or where the type of sheathing does not provide sufficient strength to resist a

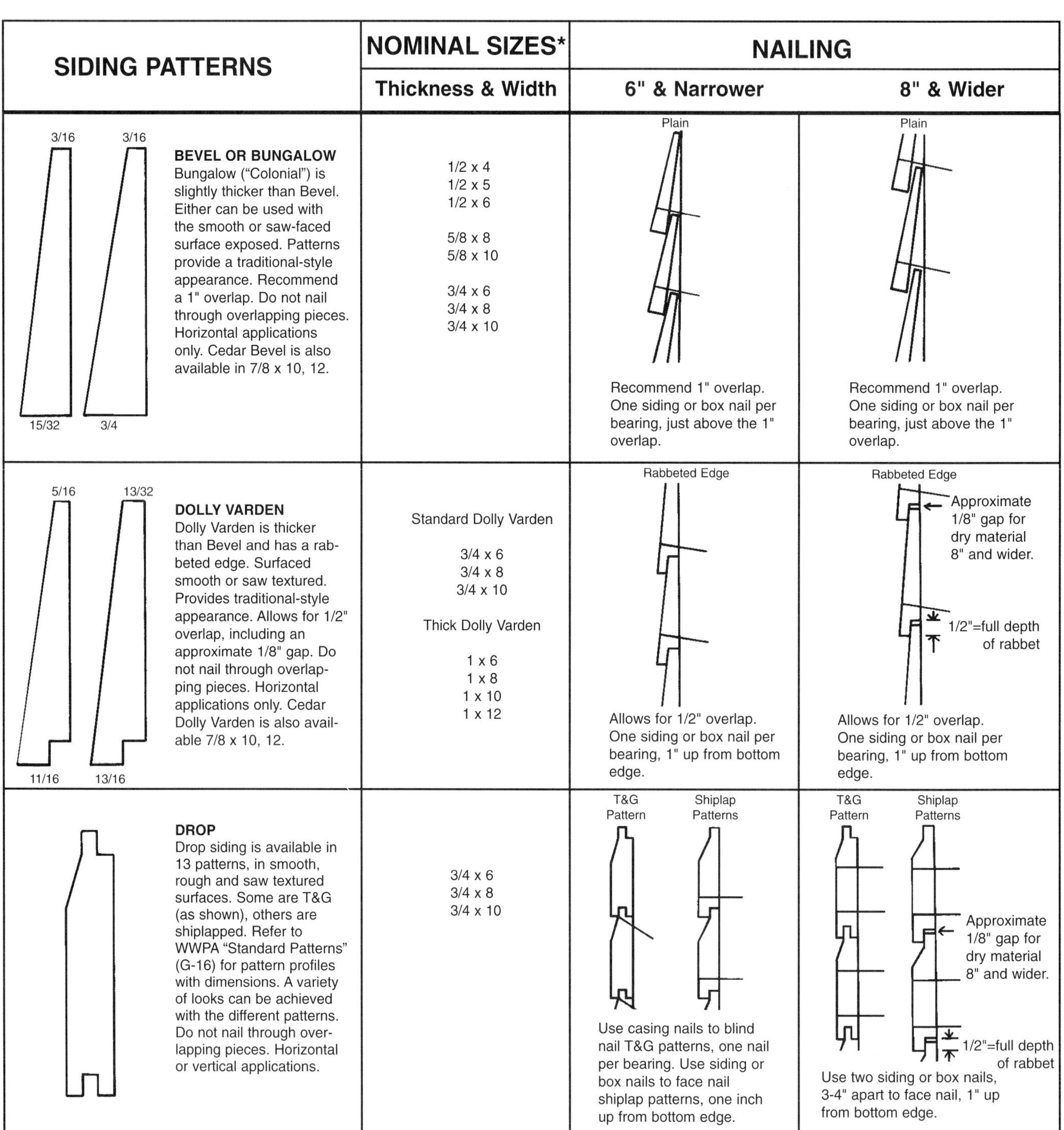

Figure 13-16. Edge views of six different types of horizontal siding. Some types can be installed vertically as well as horizontally. Nominal sizes are used in figuring footage of siding. Shrinkage and machining reduces actual sizes. Note nailing suggestions. (Western Wood Products Assoc.)

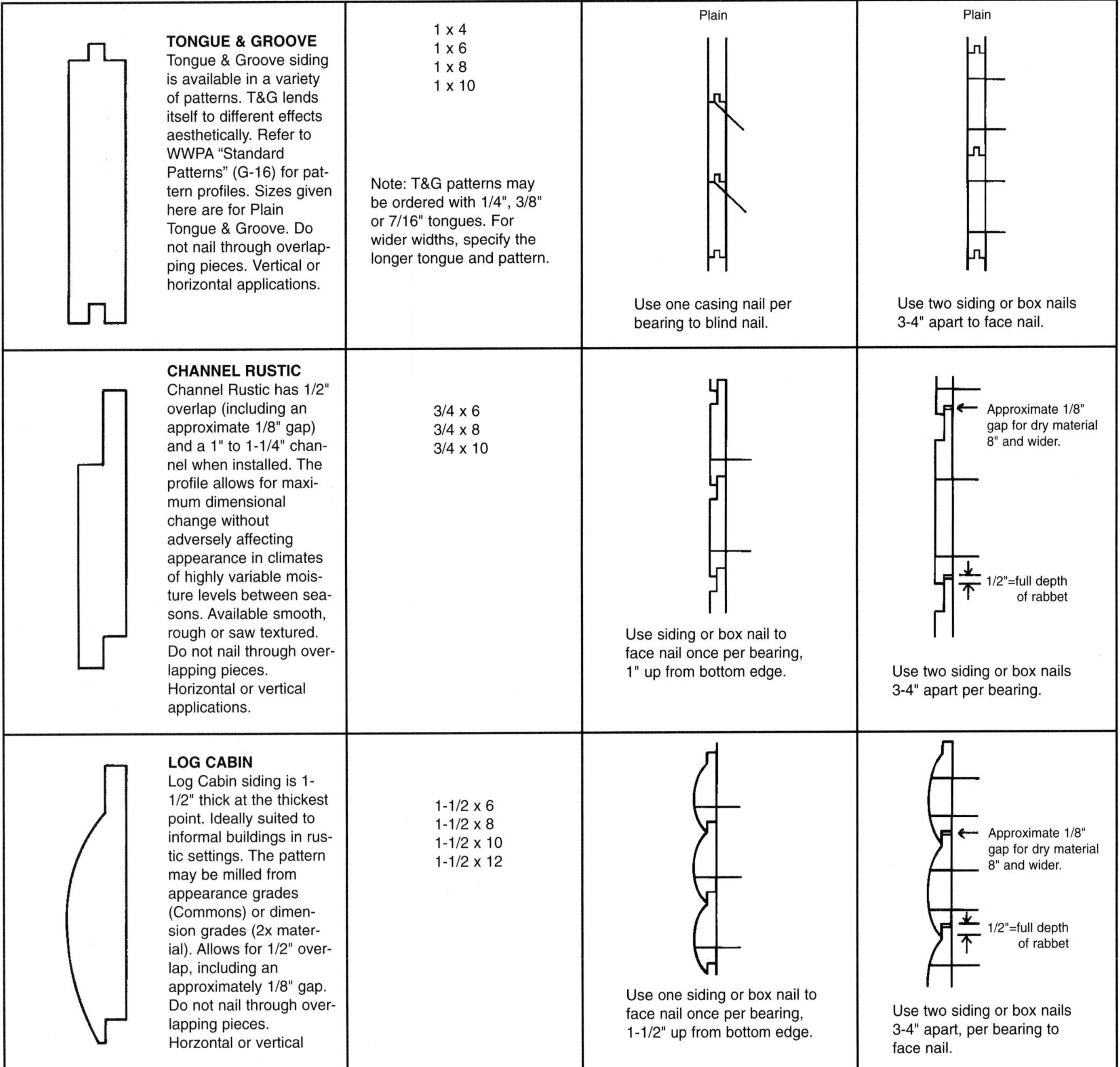

TONGUE & GROOVE Tongue & Groove siding is available in a variety of patterns. T&G lends itself to different effects aesthetically. Refer to WWPA "Standard Patterns" (G-16) for pattern profiles. Sizes given here are for Plain Tongue & Groove. Do not nail through overlapping pieces. Vertical or horizontal applications.	1 x 4 1 x 6 1 x 8 1 x 10 Note: T&G patterns may be ordered with 1/4", 3/8" or 7/16" tongues. For wider widths, specify the longer tongue and pattern.	Plain Use one casing nail per bearing to blind nail.	Plain Use two siding or box nails 3-4" apart to face nail.
CHANNEL RUSTIC Channel Rustic has 1/2" overlap (including an approximate 1/8" gap) and a 1" to 1-1/4" channel when installed. The profile allows for maximum dimensional change without adversely affecting appearance in climates of highly variable moisture levels between seasons. Available smooth, rough or saw textured. Do not nail through overlapping pieces. Horizontal or vertical applications.	3/4 x 6 3/4 x 8 3/4 x 10	Use siding or box nail to face nail once per bearing, 1" up from bottom edge.	Approximate 1/8" gap for dry material 8" and wider. 1/2"=full depth of rabbet Use two siding or box nails 3-4" apart per bearing.
LOG CABIN Log Cabin siding is 1-1/2" thick at the thickest point. Ideally suited to informal buildings in rustic settings. The pattern may be milled from appearance grades (Commons) or dimension grades (2x material). Allows for 1/2" overlap, including an approximately 1/8" gap. Do not nail through overlapping pieces. Horzontal or vertical	1-1/2 x 6 1-1/2 x 8 1-1/2 x 10 1-1/2 x 12	Use one siding or box nail to face nail once per bearing, 1-1/2" up from bottom edge.	Approximate 1/8" gap for dry material 8" and wider. 1/2"=full depth of rabbet Use two siding or box nails 3-4" apart, per bearing to face nail.

Figure 13-16 *continued.*

racking load, the wall framing should be braced as described in Unit 9.

End views of a number of types of horizontal siding are shown in Figure 13-16. *Bevel siding*, most commonly used, is available in various widths. It is made by sawing plain surfaced boards at a diagonal to produce two wedge-shaped pieces. The siding is about 3/16" thick at the thin edge and 1/2 to 3/4" thick on the other edge, depending on the width of the piece.

Wide bevel siding often has shiplapped or rabbeted joints. The siding lies flat against the studding instead of touching it only near the joints as ordinary bevel siding does. This reduces the apparent thickness of the siding by 1/4" but permits the use of extra nails in wide siding and reduces the chance of warping. It is also economical, since the rabbeted joint requires less lumber than the lap joint used with plain bevel siding.

The rabbet, however, must be deep enough so that, when the siding is applied, the width of the boards can be adjusted upward or downward to meet window sill, head casing, and eave lines. The table in Figure 13-16 lists the more common nominal sizes of horizontal siding.

Channel rustic and *drop siding* are usually 3/4" thick and 6, 8, or 10" wide. Drop siding usually has tongue-and-groove joints. Channel rustic or rustic has shiplap type joints.

Drop siding is heavier, has more structural strength, and tighter joints than bevel siding. Because of this, it is often used on garages and other buildings not sheathed.

Wood used for exterior siding should be a select grade free of knots, pitch pockets, and other defects. Edge grain is less likely to warp than a flat grain. The moisture content at the time of application should be what it will reach in service. This is about 12%, except for the southwestern states, where the moisture content should average about 9%.

Siding should be handled carefully when it is delivered to the building site. The wood from which it is made is usually quite soft. The surface can be easily damaged. Try to store siding inside the structure or keep it covered with a weatherproof material until it is applied.

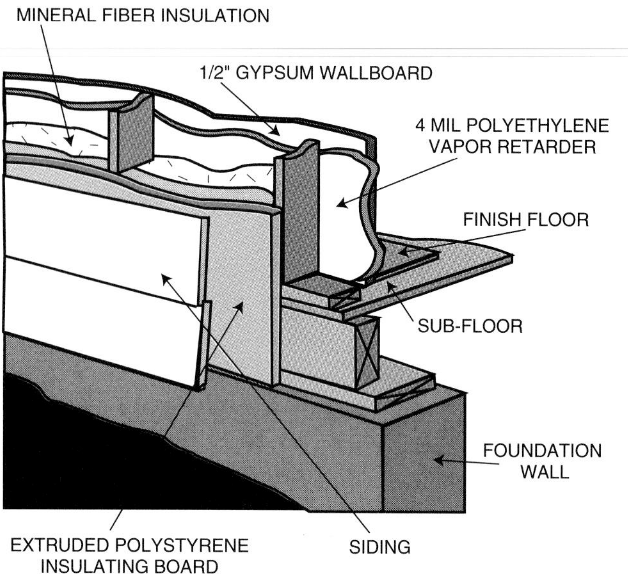

Figure 13-17. Using rigid polystyrene insulation board on an outside wall. Fasten with galvanized roofing nails with a 3/8" head or 16 gage staples with no less than a 3/4" crown. (Amoco Foam Products Co.)

Wall Sheathing and Flashing

Siding can be applied over various sheathing materials. When the sheathing consists of solid wood, plywood, or nail-base fiberboard, the siding is nailed directly to the material at about 24" intervals. End joints in the siding may occur between framing members. Gypsum board and regular fiberboard sheathing cannot be used as a nailing base and the siding should be attached by nailing through the sheathing and into the frame.

Certain special application methods may require wood strips to form a nailing base. Additional information about wall sheathing can be found in Unit 9.

At one time, sheathing paper was applied directly to the studs when sheathing was not used. It was also used to cover common board and shiplap sheathing. It was not considered necessary over plywood sheathing, fiberboard, or treated gypsum sheathing.

In modern construction, several new products may take the place of sheathing and sheathing paper. An insulation board of rigid polystyrene adds substantial R-value to walls, and housewrap seals the walls against air infiltration. The foamed sheets have shiplap edges for tight fit. Sheets are usually available in thickness of 1/2", 3/4", and 1". Figure 13-17 shows a typical application.

Before the application of siding, flashing should be installed where it is required around openings. Metal flashing is usually included over the drip caps of doors and windows, Figure 13-18. In areas not subjected to wind-driven rain, head flashing may be omitted when the vertical height between the top of the finished trim of the opening and the soffit of the cornice is equal to or less than one-fourth of the width of the overhang.

For structures with unsheathed walls, jambs of doors and windows should be flashed with a 6" wide strip consisting of either metal, 3 oz. copper-coated paper, or a 6 mil polyethylene film.

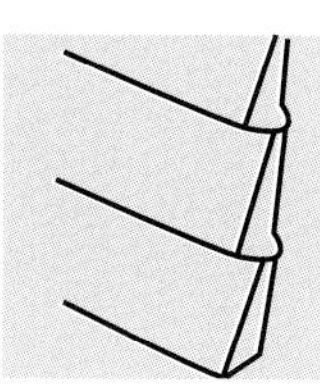

Be sure the structural frame and sheathing are dry before applying the siding material. If excessive moisture is present during the application, later drying and shrinkage will likely cause the siding to buckle.

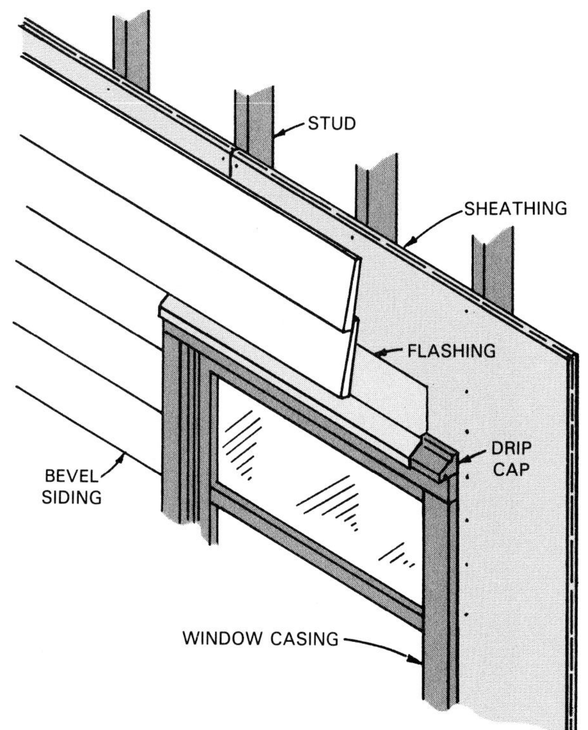

Figure 13-18. Apply metal flashing over the drip caps above windows and doors that are not protected by roof overhang.

Installation Procedures

Wood siding is precision-manufactured to standard sizes. It is easily cut and fitted. Plain beveled siding is lapped so it will shed water and provide a windproof and dustproof covering. A minimum lap of 1" is used for 6" widths, while 8 and 10" siding should lap 1 1/2".

To install horizontal siding, first prepare a story pole.

Preparing a Story Pole

1. Lay out the distance from the soffit to about 1" below the top of the foundation, Figure 13-19.
2. Divide this distance into spaces equal to the width of the siding minus the lap.
3. Adjust the lap allowance (maintain minimum requirements) so the spaces are equal.
4. When possible, adjust the spacing so single pieces of siding will run continuously above and below windows or other wall openings without notching, Figure 13-20.
5. When the layout is complete, mark the position of the top of each siding board on the story pole.

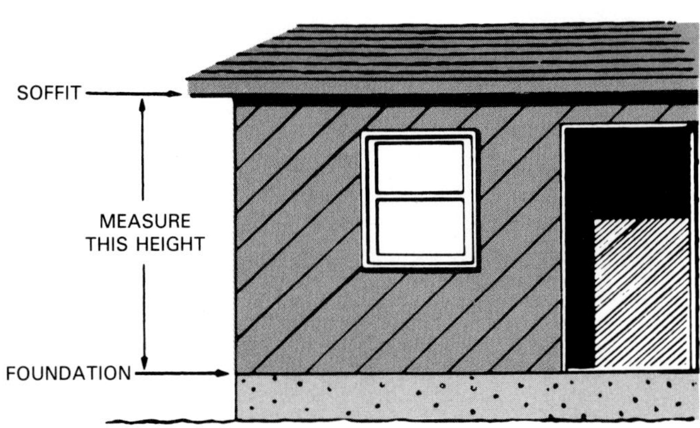

Figure 13-19. Begin layout of story pole by taking the total measurement from beneath the soffit to about 1" below the top of the foundation.

Figure 13-20. Try to adjust courses of siding to come out even, if possible, with tops and bottoms of windows. Avoid notching siding, if possible.

Now hold the story pole in position at each inside and outside corner of the structure and transfer the layout to the wall, Figure 13-21. Also mark the layout along window and door casings.

Figure 13-21. Transfer story pole layout to all inside and outside corners as well as to door and window casings.

Some carpenters prefer to set nails at these layout points, since lines can be quickly attached to them and used to align the siding stock. They can also be used to hold the chalk line if guidelines are laid out by this method.

When snapping a chalk line stretched over a long distance, it is a good idea to hold it against the surface at the midpoint; then snap it on each side.

After the layout has been made, carefully check it. Start the application of bevel siding by first nailing a strip along the foundation line equal to the thin edge of the siding, Figure 13-22. This will provide the proper tilt for the first course. Now apply the first piece. Allow the butt edge to extend below the strip to form a drip edge.

Inside corners can be formed with a square length of wood, or metal corners can be used as shown in Figure 13-23. Although outside corners could be lapped or mitered, in modern construction, metal corners are used almost universally. They can be installed quickly and provide a neat, trim appearance.

Solid lumber can be used for inside corner boards, Figure 13-24. Outside corners are formed with two pieces of lumber. Thickness depends on siding. Attach the assembly to the structure. Corner boards may be plain or molded, depending on the architectural treatment required. After the corner boards are in place, fit the siding tightly against them.

Cut and fit horizontal wood siding tightly against window and door casings, corner boards, and adjoining boards. For quality work, the carpenter first makes the cuts with a fine-tooth saw and then smoothes the ends with a few strokes of a block plane. Square butt joints are used

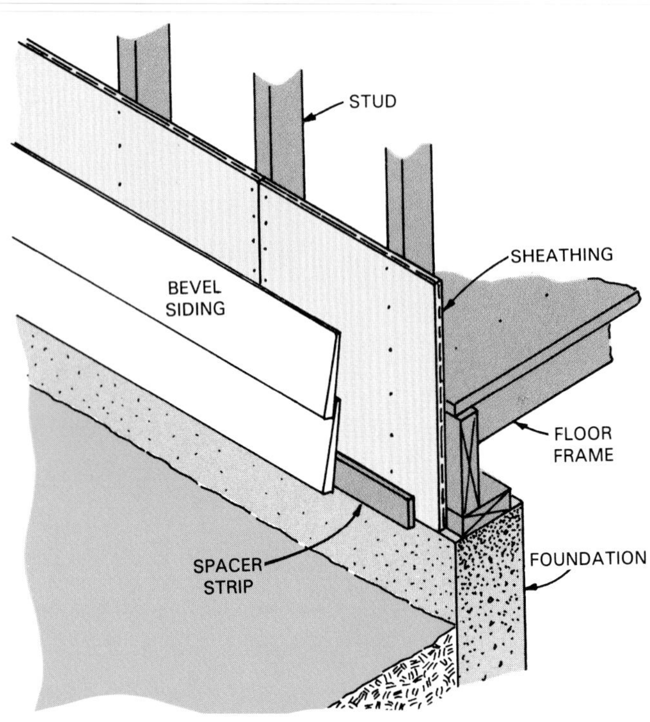

Figure 13-22. Install a spacer strip under the first course of bevel siding to give it the proper tilt to match succeeding courses.

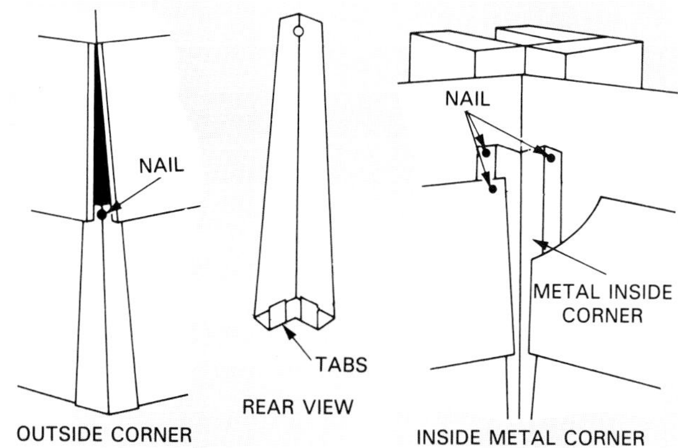

Figure 13-23. Metal corners are almost always used for horizontal siding. Top—Metal corner on finished home. Bottom—Attaching details.

between adjacent pieces of siding and should be staggered as widely as possible from one course to the next.

Wood siding can be given a coat of water-repellent preservative before it is installed, or the water repellent can be brushed on after the installation. Preservatives are sold by lumber dealers and paint stores. They contain waxes, resins, and oils. In addition to this treatment, joints in siding may be bedded in a special caulking compound to make them watertight.

Nailing

To fasten siding, stainless steel, high tensile strength aluminum, hot-dipped galvanized, or other noncorrosive nails are recommended. Avoid fasteners that might produce unsightly rust stains. Even small-headed plain steel nails, countersunk and puttied, are likely to rust.

Horizontal siding should be face-nailed to each stud, Figure 13-25. For 1/2" siding over wood or plywood sheathing, use 6d nails. Over fiberboard or gypsum sheathing, use 8d nails. For 3/4" siding over wood or plywood sheathing, use 7d nails; and over fiberboard or gypsum sheathing, use 9d nails.

For narrow siding, the nail is generally placed about 1/2" above the butt edge. In this location the fastener passes through the upper edge of the lower course.

When applying wide bevel siding, the nail should be driven through the butt edge just above the lap so that it misses the thin edge of the piece of siding underneath. This permits expansion and contraction of the siding boards with seasonal changes in moisture content. It eliminates the tendency for the siding to cup or split when both edges are nailed. Since the amount of swelling and shrinking is proportional to the width of the material, move the nail slightly above the lap when boards are extra wide.

If there is a possibility the material will split when nailing end joints, holes should be drilled for the nails. Figure 13-16 and Figure 13-26 show wood siding nailing patterns for different types of siding.

Although the usual procedure for the installation of horizontal siding is to proceed upward along the wall, some carpenters prefer to start at the top and work down,

Figure 13-24. Inside corners can be formed from solid lumber. When necessary, use two thicknesses to secure the required size.

Figure 13-25. Horizontal siding should be face-nailed over studs. A pneumatic nailer loaded with special, noncorrosive nails will make the job go faster. (Senco Products, Inc.)

Figure 13-27. This is useful for multistory or split level structures where scaffolds are attached to wall. Also, the siding is less likely to be damaged after it is applied. When following such procedure, chalk lines are set at the butt edge of the siding instead of the top edge.

Painting and Maintenance

Wood siding is subject to decay and weathering. Neither will occur if simple precautions are taken. Decay is the disintegration of wood caused by the growth of fungi. These fungi grow in wood when the moisture content is too high.

If the structure is built on a foundation which has been carried well above the ground and the construction is such that water runs off instead of into the walls, decay should never be a problem.

A wide range of finishing materials are on the market. Most fall into one of four general categories: clear water repellents, bleaching oils, stains, and paints. Certain factors need to be considered in selecting a finish. These may include: appearance desired, preparation and maintenance requirements of the finish, location of the structure, climate, and current condition of the siding. More will be said about each of these finishes later.

If the siding is to be painted, a priming coat of paint should be put on as soon as possible. If an unexpected rain should wet unprimed wood siding, the first coat of paint should not be applied until the wood has dried.

Estimating Siding

To determine how much siding is needed, it is necessary to increase the footage to make up for the difference between nominal and finished sizes. More must also be added for the cutting of joints and the overlap in beveled siding. The table in Figure 13-28 provides a factor. The net square footage of the wall surface to be covered should be multiplied by this factor. The following example shows the steps:

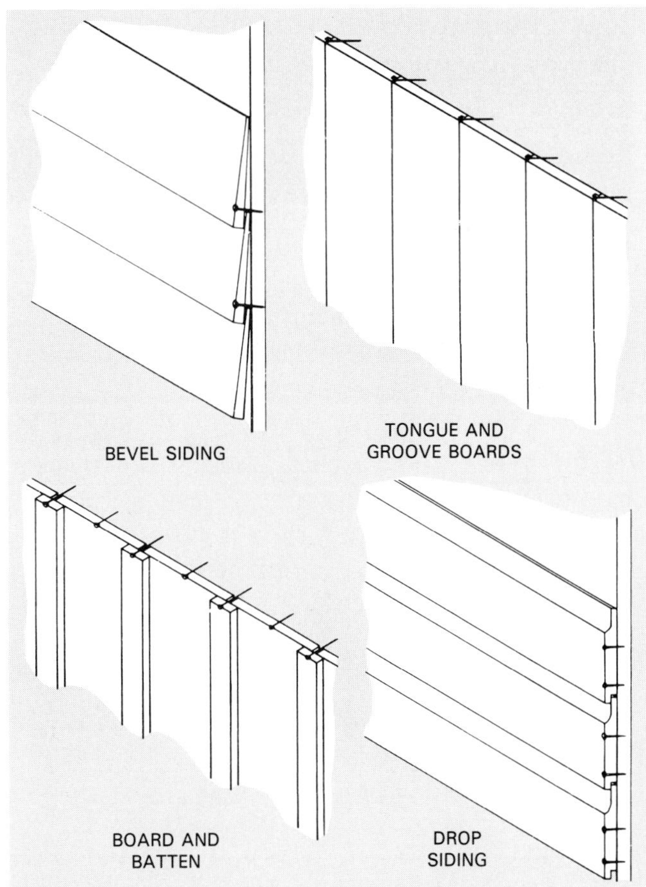

Figure 13-26. Different types of wood siding require different nailing patterns.

Figure 13-27. Scaffolding arrangements. Top—Some carpenters prefer to install horizontal siding from the top down so they can attach scaffolding directly to the wall. Bottom—Others use scaffolding supported on ladder jacks.

TYPE	SIZE (INCHES)	LAP (INCHES)	MULTIPLY NET WALL SURFACE BY
Bevel Siding	1 x 4	3/4	1.45
	*1 x 5	7/8	1.38
	1 x 6	1	1.33
	1 x 8	1 1/4	1.33
	1 x 10	1 1/2	1.29
	1 x 12	1 1/2	1.23
Rustic and Drop Siding (Shiplapped)	1 x 4		1.28
	*1 x 5	———	1.21
	1 x 6		1.19
	1 x 8		1.16
Rustic and Drop Siding (Dressed and Matched)	1 x 4		1.23
	*1 x 5	———	1.18
	1 x 6		1.16
	1 x 8		1.14

*Unusual Sizes.

Figure 13-28. When estimating horizontal wood siding, multiply the net wall surface to be covered by the factor in the last column.

1 x 10 Bevel siding with 1 1/2" lap

Wall height	= 8'
Wall perimeter	= 160'
Door and window area	= 240 sq. ft.
Total area to be covered	= (8 x 160) - 240
	= 1280 - 240
	= 1040 sq. ft.
Siding needed	= 1040 x 1.29
	= 1342 sq. or bd. ft.

Area for gable ends can be calculated by multiplying the height above the eaves by the width and dividing by two. Considerable waste occurs in covering triangular areas and at least 10% should be added to this calculation. When the structure includes many corners due to projections and recesses in the wall line, an additional .05 should be added to the factors shown in Figure 13-28.

Vertical Siding

Vertical siding is commonly used to set off entrances or gable ends. It is also often used for the main wall areas, Figure 13-29. Vertical siding may be plain-surfaced matched boards, pattern matched boards, or square-edge boards covered at the joint with a batten strip.

Matched vertical siding, made from solid lumber, should be no more than 8" wide. It should be installed with two 8d nails not more than 4' apart. Backing blocks should be placed horizontally between studs to provide a good nailing base. The bottom of the boards are usually undercut to form a drip edge.

Board and batten applications are designed around wide square-edged boards, spaced about 1/2" apart. They are fastened at each *bearing* (blocking) with one or two 8d nails. Use 10d nails to attach the battens, Figure 13-30. Locate nails in center of batten so the shanks will pass between the boards and into the bearing.

Figure 13-29. Vertical wood boards make a durable and beautiful siding material. (American Plywood Assoc.)

Board and batten effects are possible with large vertical sheets of plywood or composition material. Simply attach vertical strips over the joints and at several positions between the joints. Figure 13-31 shows an application of this type with solid wood strips being applied to the surface of exterior plywood sheets.

A variation of the board and batten siding is the board on board siding. Refer again to Figure 13-30 to see how this type siding is applied.

Wood Shingles

Wood shingles, Figure 13-32, are sometimes used for wall covering, and a large selection of types is available. Some are especially designed for sidewall application with a grooved surface and factory applied paint or stain.

Shingles are very durable and can be applied in various ways to provide a variety of architectural effects. Handsplit shingles are occasionally used; however, they are expensive and difficult to install.

Most shingles are made in random widths. No. 1 grade shingles vary from 3 to 14" wide. Only a small number of the narrow width are permitted. Shingles of a uniform width, known as dimension shingles, are also available.

For side wall application, follow these recommendations for maximum exposure:

- For 16" shingles — 7 1/2" exposure
- For 18" shingles — 8 1/2" exposure
- For 24" shingles — 11 1/2" exposure

Shingles on side walls are frequently laid in what is called *double coursing*. This is done by using a lower grade shingle under the shingle exposed to the weather. The exposed shingle butt extends about 1/2" below the butt of the under course.

When butt nailing is used, a greater weather exposure is possible. Frequently as much as 12" for 16" shingles, 14" for 18" shingles, and 16" for 24" shingles is satisfactory. Figure 13-33 lists the sizes of a standard side wall shingle with grooved surface. Approximate coverage for various lengths and exposures is included.

In mild climates, sheathing boards can be spaced the distance of the shingle exposure. A high grade shingle provides a satisfactory wall. Housewrap or roofing felt should be used with such construction. Place it either between the shingles and sheathing or between the sheathing and the studding. Spaced sheathing is also satisfactory on implement sheds, garages, and other structures where protection from the elements is the principal consideration.

To obtain the best effect and to avoid unnecessary cutting of shingles, butt-lines should be even with the upper lines of window openings. Likewise, they should line up with the lower lines of such openings.

It is better to tack a temporary strip to the wall to use as a guide for placing the butts of the shingles squarely, rather than to attempt to shingle to a chalk line.

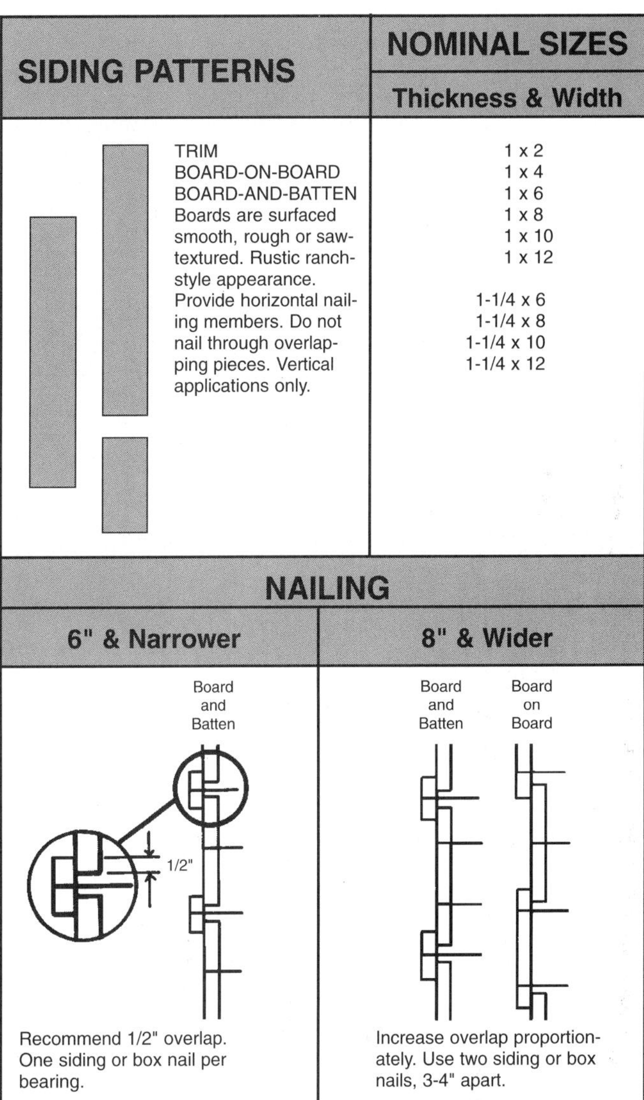

Figure 13-30. Board on board and board on batten siding give a building a rustic appearance. They also provide for expansion and contraction. Top—Board on board application of vertical siding. Bottom—Board and batten being applied. (Western Wood Products Assoc.)

Figure 13-31. Board and batten effect is created by nailing battens over exterior plywood sheets applied vertically.

Single Coursing of Side Walls

The single-coursing method for side wall application is similar to roof application. The major difference is in the exposures employed. In roof construction, maximum permissible exposures are slightly less than one-third of the shingle length. This produces a three-ply covering.

Vertical surfaces of side walls present less weather-resistance problems than do roofs. Accordingly, a two-ply covering of shingles is usually adequate.

In single-coursed side walls, weather exposure of shingles should never be greater than half the length of the shingle, minus 1/2". Thus, two layers of wood will be found at every point in the wall.

For example, when 16" shingles are used, the maximum exposure should be 1/2" less than 8", or 7 1/2".

Single-course side walls, Figure 13-34, should have concealed nailing. This means that the nails must be driven about 1" above the butt line of the succeeding course, so the shingles of this course will adequately cover them. Two nails should be driven in each shingle up to 8" wide, and each nail placed about 3/4" from the edge of the shingle. On shingles wider than 8", a third nail should be driven in the center of the shingle, at the same distance above the butt line as the other nails. Use rust-resistant 3d nails, 1 1/2" long.

Figure 13-32. Wood shingles or shakes are an attractive and durable siding material. (Shakertown Corp.)

Estimating Quantities

In estimating the quantity of shingles required for side walls, areas to be shingled should be calculated in square feet. Deduct window and door areas. Consult the table, Figure 13-33. The coverage of one square (4 bundles) at the exposure to be used should be divided into the wall area to be covered. The figure arrived at will be the number of squares needed. Add 5% to allow for waste in cutting and fitting around openings, and for the double starter course.

Double Coursing of Side Walls

In double coursing, a low-cost shingle is generally used for the bottom layer. This is covered with a No. 1 grade shingle. Many types of shingles are available for the outer course. Prestained shingles, which are available in attractive colors, are particularly suitable. Although wide exposures usually require the use of long shingles, this effect is obtained in double coursing by the application of doubled layers of regular 16" or 18" shingles. The maximum exposure to the weather of 16" shingles double coursed is 12". For 18" shingles it is 14".

The application of shingles on a double-course side wall is illustrated in Figure 13-35. Most procedures used for regular siding can be followed. When the application is made over composition or spaced sheathing, mark the position of the nailing strips on the story pole when it is laid out.

Grade	Length	Thickness (at Butt)	No. of Courses Per Bdl/Carton	Bdls/Cartons Per Square	Shipping Weight	Description
No. 1	16" (Fivex) 18" (Perfections) 24" (Royals)	.40" .45" .50"	33/33 28/28 13/14	1 carton 1 carton 4 bdls.	60* lbs. 60* lbs. 192 lbs.	Same specifications as rebutted-rejointed shingles, except that shingle face has been given grain-like grooves. Natural color, or variety of factory-applied colors. Also in 4-ft. and 8-ft. panels.

NOTE: * 70 lbs. when factory finished.

LENGTH AND THICKNESS	Approximate coverage of one square (4 bundles) of shingles based on following weather exposures												
	3½"	4"	4½"	5"	5½"	6"	6½"	7"	7½"	8"	8½"	9"	9½"
16" x 5/2"	70	80	90	100*	110	120	130	140	150⧗	160	170	180	190
18" x 5/2¼"	....	72½	81½	90½	100*	109	118	127	136	145½	154½⧗	163½	172½
24" x 4/2"	....	....	....	....	....	80	86½	93	100*	106½	113	120	126½

LENGTH AND THICKNESS	10"	10½"	11"	11½"	12"	12½"	13"	13½"	14"	14½"	15"	15½"	16"
16" x 5/2"	200	210	220	230	240†	....	....	....	....	....	....	....	....
18" x 5/2¼"	181½	191	200	209	218	227	236	245½	254½†	....	....	....	....
24" x 4/2"	133	140	146½	153⧗	160	166½	173	180	186½	193	200	206½	213†

NOTES: * Maximum exposure recommended for roofs. ⧗ Maximum exposure recommended for single-coursing on sidewalls. † Maximum exposure recommended for double-coursing on sidewalls.

Figure 13-33. Chart of sizes and coverage for side wall shingles. Shingle thickness is based on number of butts required to equal a given measurement.

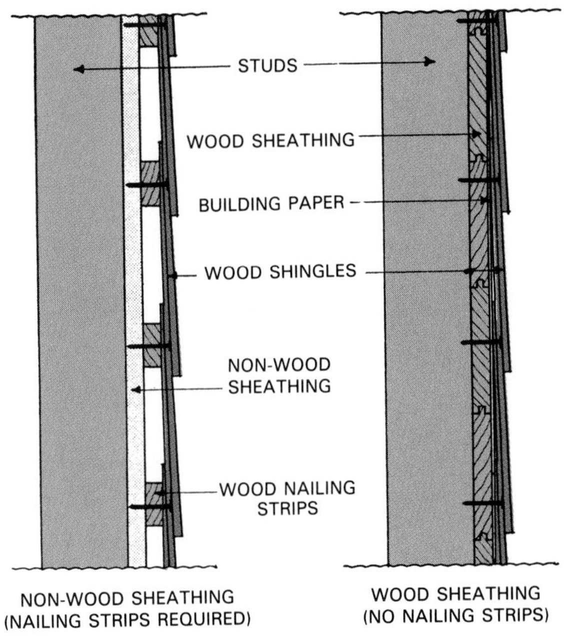

Figure 13-34. Single-course method of applying shingles to sidewalls. Solid backing and nailing base is provided by wood sheathing or nailing strips over nonwood sheathing.

Shingle and Shake Panels

Shingles and shakes for side wall application are available in panel form. The panels consist of individual shingles (usually western red cedar) permanently bonded to a backing. Standard size panels are 8', Figure 13-36. These are available in various textures, either unstained or factory-finished in a variety of colors. Special metal or mitered wood corners are also manufactured.

Shingle panels are applied by following the same basic precautions and procedures described for regular shingles, Figure 13-37. Installation time however, is greatly reduced. Additional on-site labor is also saved when factory-primed or factory-finished units are used. When applying the latter, the installation should be made with nails of matching color, supplied by the manufacturer.

Figure 13-38 shows a completed residential structure where shingle panels were used for the outside wall covering.

When you are making an application of a specialized or prefabricated product such as shingle panels, be sure to follow the recommendations furnished by the manufacturer.

Using Wood Shingles over Old Siding

Shingles or shakes can be applied over old siding or other wall coverings that are sound and will hold nailing strips. First, apply building paper over the old wall. Next, attach nailing strips as previously described. Usually it is necessary to add new molding strips around the edge of window and door casings to trim the edge of the shingles. Figure 13-39 illustrates how nailing strips are applied over an old stucco surface. Figure 13-40 shows shingle panels being applied over stucco. No furring strips are necessary.

Plywood Siding

The use of plywood as an exterior wall covering permits a wide range of application methods and decorative treatments, Figure 13-41. Plywood can also be used alongside other building materials. A few of the ways it can be used inlcude:

- Provide a vertical treatment to gable ends.
- Provide emphasis as fill-in panels above and below windows.
- Establish a continuous decorative band at various levels along an entire wall.

All plywood siding must be made from exterior type plywood. Douglas fir is the most commonly used species. However, cedar and redwood are also available. Panels

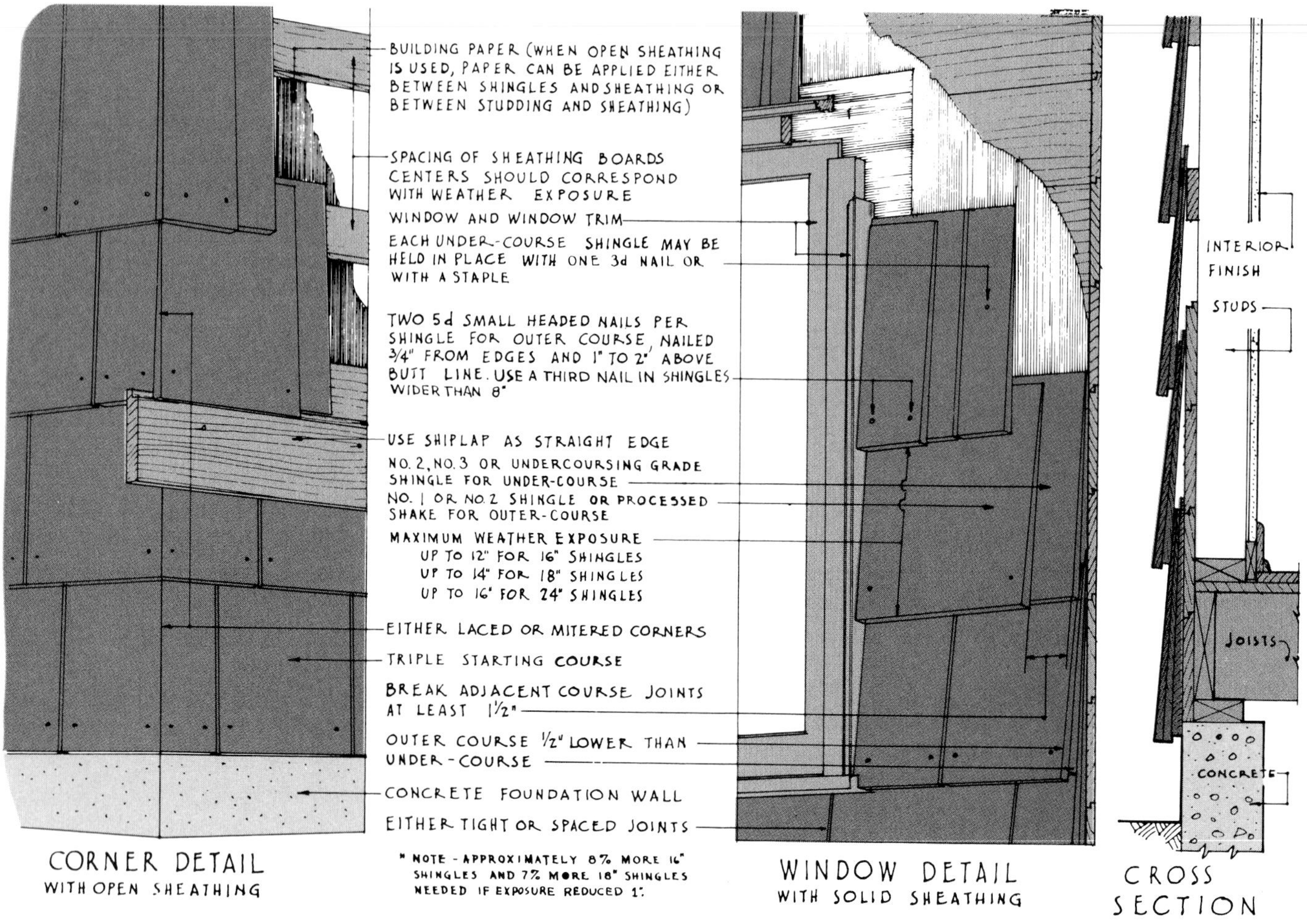

BUILDING PAPER (WHEN OPEN SHEATHING IS USED, PAPER CAN BE APPLIED EITHER BETWEEN SHINGLES AND SHEATHING OR BETWEEN STUDDING AND SHEATHING)

SPACING OF SHEATHING BOARDS CENTERS SHOULD CORRESPOND WITH WEATHER EXPOSURE

WINDOW AND WINDOW TRIM

EACH UNDER-COURSE SHINGLE MAY BE HELD IN PLACE WITH ONE 3d NAIL OR WITH A STAPLE

TWO 5d SMALL HEADED NAILS PER SHINGLE FOR OUTER COURSE, NAILED 3/4" FROM EDGES AND 1" TO 2" ABOVE BUTT LINE. USE A THIRD NAIL IN SHINGLES WIDER THAN 8"

USE SHIPLAP AS STRAIGHT EDGE

NO. 2, NO. 3 OR UNDERCOURSING GRADE SHINGLE FOR UNDER-COURSE

NO. 1 OR NO. 2 SHINGLE OR PROCESSED SHAKE FOR OUTER-COURSE

MAXIMUM WEATHER EXPOSURE
UP TO 12" FOR 16" SHINGLES
UP TO 14" FOR 18" SHINGLES
UP TO 16" FOR 24" SHINGLES

EITHER LACED OR MITERED CORNERS

TRIPLE STARTING COURSE

BREAK ADJACENT COURSE JOINTS AT LEAST 1½"

OUTER COURSE ½" LOWER THAN UNDER-COURSE

CONCRETE FOUNDATION WALL

EITHER TIGHT OR SPACED JOINTS

INTERIOR FINISH

STUDS

JOISTS

CONCRETE

CORNER DETAIL WITH OPEN SHEATHING

* NOTE - APPROXIMATELY 8% MORE 16" SHINGLES AND 7% MORE 18" SHINGLES NEEDED IF EXPOSURE REDUCED 1".

WINDOW DETAIL WITH SOLID SHEATHING

CROSS SECTION

Figure 13-35. Double-coursed shingle siding. Sheathing may be solid or spaced on-center to the nailing line.

Figure 13-36. Applying panelized shakes. Panels are 8' long and consist of two 7" courses bonded to a plywood or veneer core base. (Shakertown Corp.)

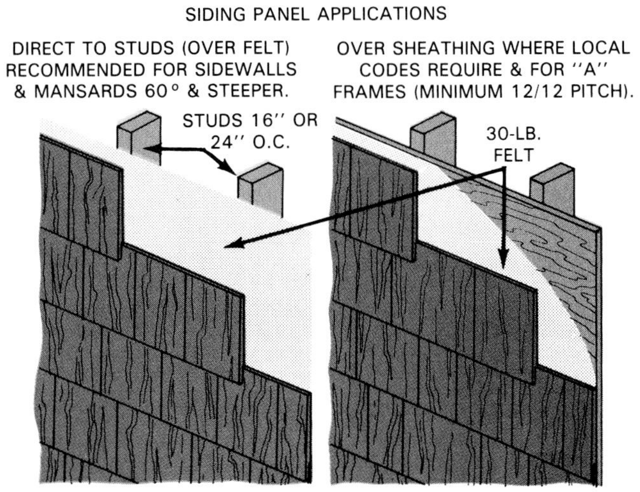

SIDING PANEL APPLICATIONS

DIRECT TO STUDS (OVER FELT) RECOMMENDED FOR SIDEWALLS & MANSARDS 60° & STEEPER.

OVER SHEATHING WHERE LOCAL CODES REQUIRE & FOR "A" FRAMES (MINIMUM 12/12 PITCH).

STUDS 16" OR 24" O.C.

30-LB. FELT

Figure 13-37. Proper application of panelized siding. Panels are self-aligning. (Shakertown Corp.)

Figure 13-38. View of shingle-paneled house. Surfaces may be factory finished, stained, painted, or left natural. (Shakertown Corp.)

Figure 13-39. Re-siding a stucco wall with double-coursed wood shingles. Nailers are spaced to correspond to the nailing lines.

Figure 13-40. Using shingle panels for a re-siding over stucco. No nailing strips are used. (Shakertown Corp.)

Figure 13-41. Examples of two modern plywood siding styles. Top—Rough-sawn surface. Grooves are on 4" or 8" centers. Bottom—Reverse board and batten. Surface may be brushed, rough-sawn, coarse sanded, or otherwise textured. Both styles are manufactured in fir, redwood, southern pine, and other wood species. Edges are shiplapped. (American Plywood Assoc.)

come in either a sanded condition or with factory applied sealer or stain. Information on grading standards may be obtained from Unit 1.

Panel sizes are 48" wide by 8', 9', and 10' long. A 3/8" thickness is normally used for direct-to-stud applications. A 5/16" thickness may be used over an approved sheathing. Thicker panels are required when the texture treatment consists of deep cuts.

Application of large sheets is generally made with the panels in a vertical position. This eliminates the need for horizontal joints.

Figure 13-42 shows a home sided with grooved vertical panels having shiplap edges. For unsheathed walls, plywood thickness should not be less than 3/8" on 16" stud spacing, 1/2" for 20" stud spacing, and 5/8" for 24" stud spacing. See Figure 13-43. Vertical joints must occur over studs; horizontal joints over solid blocking. Standard application requirements are given in Figure 13-44. Battens are an option, but are never used with textured plywood siding.

Plywood lapped sidings may look the same as regular beveled siding. Heavy shadow lines are secured by using spacer strips at the lapped edges.

Application requirements are given in Figure 13-45. A bevel of at least 30° is recommended. The lap should be at least 1 1/2".

Vertical joints should be butted over a shingle and centered over a stud unless wood sheathing at least 3/4" thick is used. Nail siding to each of the studs along the bottom edge and not more than 4" O.C. at vertical joints. Nails should penetrate studs or wood sheathing at least 1".

If plywood lap siding is wider than 12" a wood taper strip should be used at all studs with nailing at alternate studs. Outside corners should butt against corner molding or they should be covered.

Figure 13-46 shows several ways to handle joints between plywood panels. All edges of plywood siding — whether butted, V-shaped, lapped, covered, or exposed — should be sealed with a heavy application of high grade exterior primer, aluminum paint, or oil paint. Special caulking compounds are also recommended.

Because large sheets of plywood and hardboard siding provide tight, draft-free wall construction, it is important to have an effective vapor barrier. This should be between the insulation and the warm surface of the wall.

Figure 13-42. Grooved plywood is applied over a sheathed wall in most instances. Type shown has a rough-sawn surface. The long edges are shiplapped to match the grooves machined in the surface. This siding material is available in thicknesses from 11/32" to 5/8". Depth of grooves varies with panel thickness. (American Plywood Assoc.)

Figure 13-43. Panel installation can be made directly over studs. In such installations, the paneling should be thicker, in this case, 5/8".

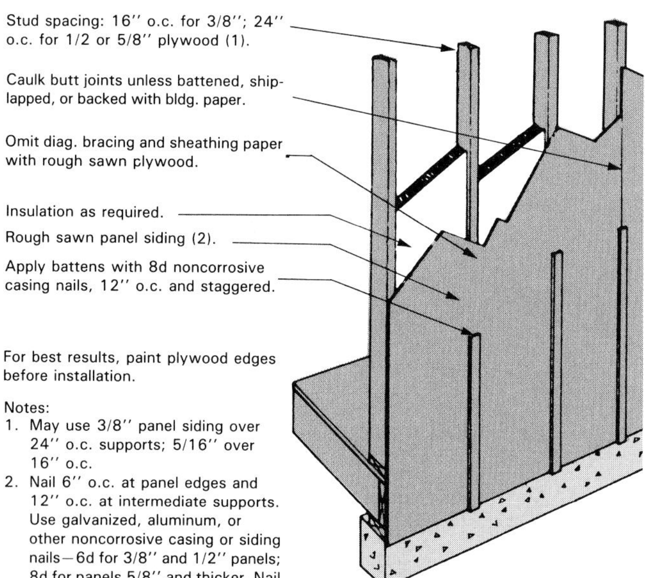

Figure 13-44. Standard requirements have been set for application of vertical siding. (American Plywood Assoc.)

Hardboard Siding

Improved techniques in manufacture have produced *hardboard siding* materials that are durable, easy to apply, and adaptable to various architectural effects. Installation methods are similar to those described for plywood sidings. Hardboard sidings may expand more than plywood. Special precautions should be observed in the application.

Manufacturers usually recommend leaving a 1/8" space where hardboard siding butts against adjacent pieces or trim members.

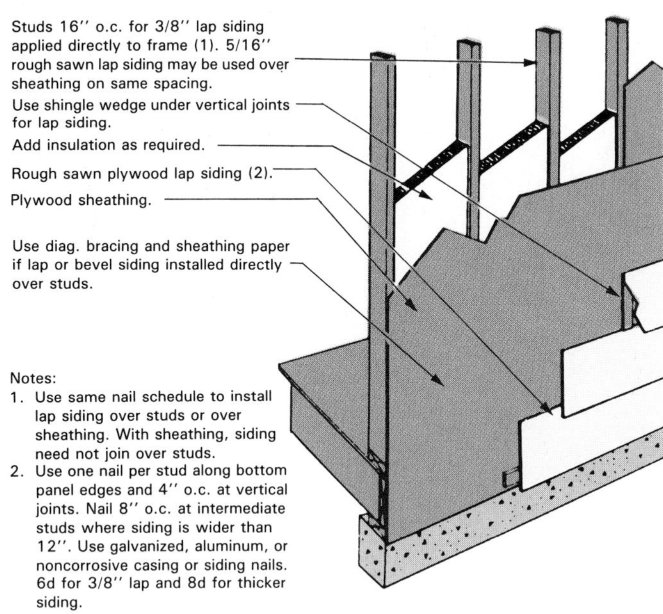

Figure 13-45. Application requirements for lapped plywood siding. Check local codes for variance.

Hardboard siding panels are available in standard widths of 4' and standard lengths of 8', 9', and 10'. Lap siding units are usually 12" wide by 16' long. However, narrower widths can be purchased. The most common thickness is 7/16". Like plywood, hardboard sidings are furnished in a wide range of textures and surface treatments. See Figure 13-47. Most panels, if not prefinished, are given a primer coat at the factory. Prefinished units with matching batten strips and trim members are also available. Figures 13-48 and 13-49 show installation details of prefinished panels.

Prefinished units with matching batten strips and trim members are available.

When applying hardboard siding, follow standard installation procedures described for regular siding materials. Studs should not be spaced greater than 16" O.C. A firm and adequate nailing base is essential. Use a fine-tooth handsaw or power saw equipped with a combination blade to cut the panels. Nails must be galvanized or otherwise weatherproofed. Wood trim and corner boards should be at least 1 1/8" thick. Space nails at least 1/2" in from edges and ends. Use an approved caulking compound at joints.

Siding Systems

Many manufacturers produce siding products that include special designs or devices which simplify the methods of application.

Figure 13-50 shows the installation of a vinyl siding system that has the appearance of bevel siding. Each double strip hooks into the course below and is secured in place by nailing along the slotted top edge. A special corner board unit, Figure 13-51 (top), covers the ends and allows for expansion and contraction resulting from temperature changes. Vinyl siding components can be easily

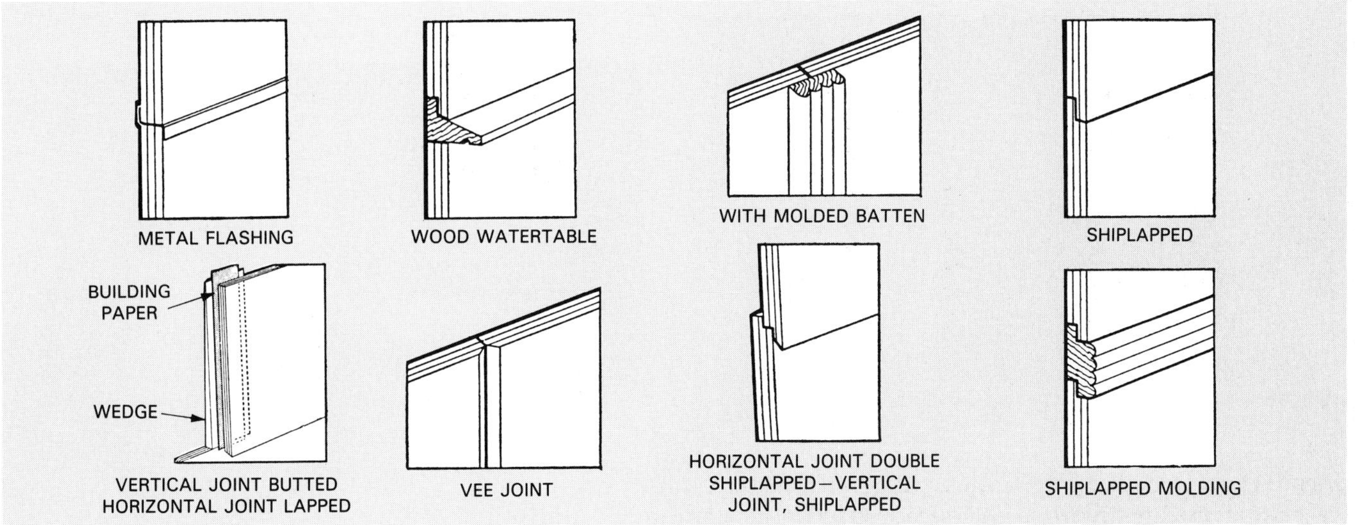

Figure 13-46. Joint details for plywood siding. All edges should be sealed with paint or special caulking compound. (American Plywood Assoc.)

cut with a portable circular saw. See Figure 13-51 (bottom). Use a fine-toothed blade and mount it in reverse. When installing any siding system, always follow the manufacturer's recommendations.

Another system uses an aluminum fastener that not only eliminates face nailing but provides venting to minimize moisture traps behind the siding. A metal starter strip is attached to the bottom of the sheathing or sill plate. The siding unit is then set in place and nailed along the top edge. Special clips are attached to the lower edge of the next course. These clips hook onto the top edge of the panel below and the top edge nailed into place. Study the drawings in Figure 13-52.

Figure 13-53 shows a completed installation of hardboard panelized shakes.

Figure 13-47. This home is sided with shingle panels manufactured from hardboard. Panels are made in 8' and 16' lengths. (Masonite Corp.)

Figure 13-48. Home being sided with a prefinished hardboard. Note use of metal scaffolding to reach second story. Upper panels lap over the first-story panels. (Use of stepladder as shown is not a safe practice.) Also note flanges attaching windows to the wall.

Figure 13-49. Top—Gap of about 1/8" should be allowed around window and door frames for expansion. Use an approved caulk at joints. Drip cap is not needed over windows because of the flange seen in Figure 13-48. Bottom—Drip cap is required over wooden door frame since it has no flange like the window units.

Figure 13-50. Installing a vinyl siding system that has the appearance of bevel siding. Follow the manufacturer's recommendations.

Figure 13-51. Top—Special corner boards cover the ends of the horizontal units and provide for expansion. Bottom—Cutting a siding unit to length using a simple sawing guide.

Aluminum Siding

Aluminum siding offers low maintenance costs. It is factory finished with baked-on enamel, and provides an appearance that closely resembles painted wood siding. It is designed for use on new or existing construction. Aluminum can be applied over wood, stucco, concrete block, and other surfaces that are structurally sound. Basic specifications (alloy and gage) for aluminum siding are established by FHA and the Aluminum Siding Association.

A variety of horizontal and vertical panel styles in both smooth and textured designs are produced with varying shadow lines and size of face exposed to the weather.

Figure 13-53. A completed hardboard shake application. Units are deeply embossed to have the look of real wood. Panel size is 7/16" by 12" by 48". Each bundle contains 10 pieces along with complete installation instructions. (ABTco Corp.)

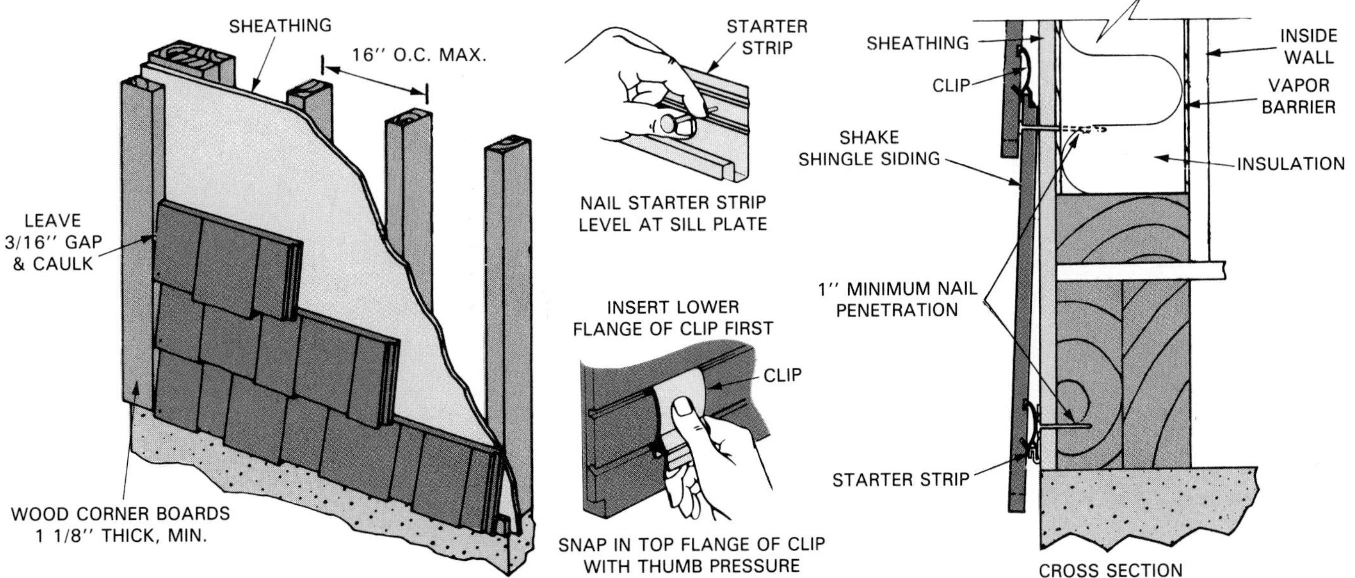

Figure 13-52. Fastening system used for hardboard siding system eliminates exterior fasteners. (ABTco Corp.)

An insulated panel is also produced. It has an impregnated fiberboard material laminated to the back surface.

Manufacturers supply directions for the installation of their products. These should be carefully followed. Figure 13-54 shows a horizontal siding unit being fastened in place. Panels are fabricated with prepunched nail and vent holes and special interlocking design. Standard strips, Figure 13-55, are easily attached around windows and doors to provide a weathertight seal. Special corners and trim members are often formed on the job site, as shown in Figure 13-56.

As a precaution against faulty electrical wiring or appliances that might energize aluminum siding and create an electrical hazard, grounding should be included. The Aluminum Siding Association recommends that a No. 8, or larger, wire be connected to any convenient point on the siding and to the cold water service or the electrical service ground. Connectors should be UL (Underwriters Laboratories) approved.

Figure 13-54. Installing an aluminum siding panel. Lower edge interlocks with previously applied course. (Alcoa Building Products)

Figure 13-55. Special strips are available to cover top nailing surface of aluminum siding. Here it is being applied under windows.

Figure 13-56. Skilled carpenters form aluminum trim pieces for windows and fascia using portable bending brakes on the job site.

Vinyl Siding

Recent developments in the chemical industries have resulted in economical production of a rigid polyvinyl chloride compound that is tough and durable. This material, commonly called vinyl, is extruded into siding units, either horizontal or vertical, and accessories, Figure 13-57. Panel thickness of about 1/20" is available in various widths up to 8".

Like aluminum, vinyl siding is usually installed with a backing board or insulation board behind each sheet, Figure 13-58. This backing board provides a flat surface over old siding, while adding rigidity, strength, and insulation. Panels have interlocking joints that are waterproof. Nail holes are slotted to permit movement during expansion and contraction.

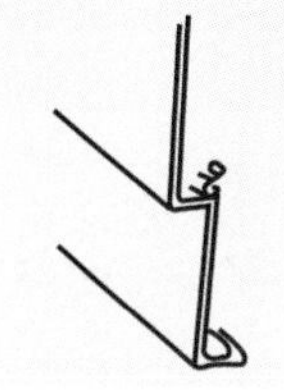

When installing vinyl siding, be sure to read and follow the directions furnished by the manufacturer.

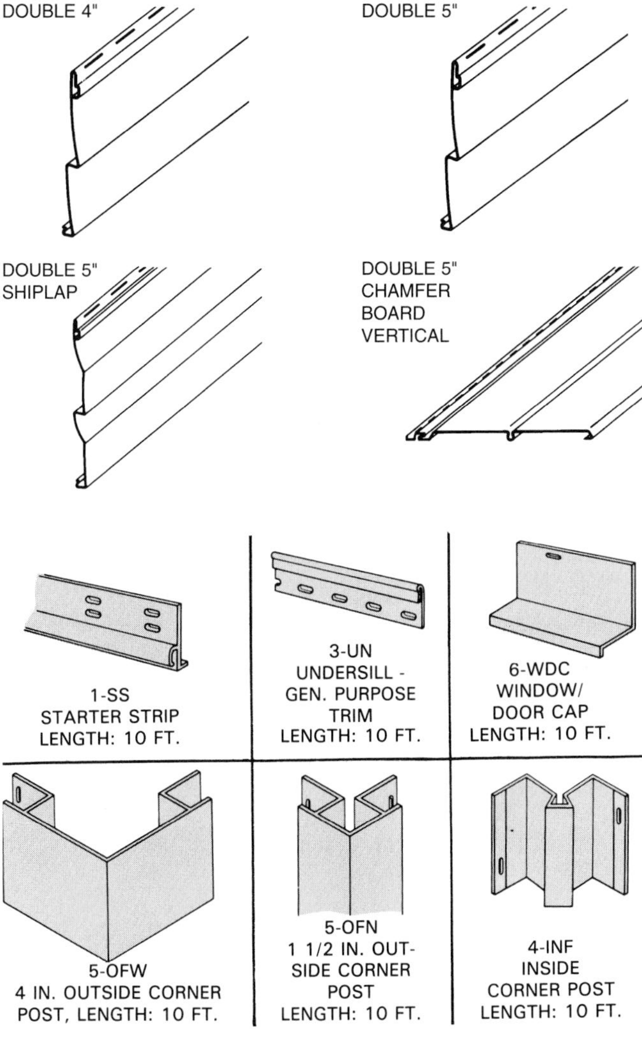

Figure 13-57. Various accessories are made for use with vinyl siding. (Bird and Son, Inc.)

Figure 13-58. Insulation board is a suitable backing for installation of vinyl siding. Top—Extruded polystyrene insulation board has film laminated to both sides to increase its strength, and minimize damage on the job site. It is available in 1/2", 3/4", and 1" thicknesses with interlocking shiplap edges. (Amoco Foam Products Co.) Bottom—Using a pneumatic construction stapler to install insulated wall sheathing. (ITW Paslode)

Installing Vinyl Siding and Soffits

Vinyl siding can be installed with essentially the same tools already in the carpenter's tool box. Certain materials and equipment will also be needed, Figure 13-59.

No special preparation of the exterior walls is needed on new construction. However, when installing new siding over existing siding, preparation requires several steps.

Preparing Old Siding

1. Nail loose boards and trim. Replace rotted boards.
2. Remove old caulk from windows and doors; it interferes with placement of trim.
3. Remove downspouts, lighting fixtures, and moldings where they will interfere with siding installation.
4. Tie back shrubbery and trees that are close enough to be damaged or interfere with work.

Often, carpenters will install rigid sheets of insulation board, Figure 13-60. This provides added insulation and levels walls for smooth installation of siding. Drop-in backing boards are not recommended since they tend to interfere with expansion and contraction of the siding. Backing boards are being developed that fit the siding profile exactly; however, their use should always be checked first with the manufacturer of the siding.

Another method of preparing old siding to receive new vinyl siding is to nail on furring strips. For horizontal siding, vertical strips should be installed at intervals of 16", for vertical siding, install horizontal furring at 16" or 24" intervals.

Care should be taken to cut and attach vinyl siding properly. Vinyl panels can be cut with sheet metal shears, aviation snips, hack saw, utility knife, or power circular saw. If the fine-bladed circular saw is used, reverse the blade for a smoother cut. Always allow a 1/4" expansion gap where siding meets accessories. In cold weather, allow additional room for expansion.

TOOLS

STEEL TAPE
FOLDING RULE
LEVEL (2 FT. MIN.)
STEEL SQUARE
HAND SAW (CROSSCUT)
HACKSAW (FINE-TOOTH, METAL CUTTING)
CHALK LINE
CLAW HAMMER (OR POWER HAMMER/STAPLER)
PORTABLE POWER SAW (FINETOOTH BLADE)
SCREW DRIVER
PLIERS
AVIATION SNIPS
SHEARS
SNAP-LOCK PUNCH
STEEL AWL
LINE LEVEL
UTILITY KNIFE
NAIL SLOT PUNCH

MATERIALS

ALUMINUM TRIM SHEET
ALUMINUM OR GALVANIZED NAILS

1 1/2", general use
2" for re-siding
2 1/2" through siding w/backer board
1-2" trim nails (color matched)

EQUIPMENT

LADDERS/SCAFFOLDS
CUTTING TABLE
PORTABLE BRAKE

Figure 13-59. Certain tools, equipment, and materials are used to install vinyl siding. The brake bender is needed to shape custom trim around window casings, door casings, and window sills.

Figure 13-60. This re-siding project is being installed over insulation board which has shiplap edges. Seams are taped to seal against air infiltration. Siding material is vinyl and trim is metal. Note how ladder and plank are used as scaffolding.

Attaching vinyl siding calls for special care. Do not drive nails tight against the siding; leave a gap of about 1/8" between the nail head and the vinyl. This allows the vinyl to expand and contract. Just as important, it prevents dimpling that could show up as unsightly waves in the finished job. Drive nails straight; never at an angle. Place nails in the center of the nail slot and space every 16". Never pull siding panels taut before nailing; this will deform the panel and cause a bad lap with the panel beneath.

Installing Starter Strips and Trim

Be sure that the first panels are installed level. Find the lowest corner of the building or the lowest corner of the siding if siding over existing siding. Drive a nail partway in and 2 1/4" above the low side. Stretch a chalk line from this nail to another at the next corner, keeping the chalk line level. Snap a line and repeat the same procedure around the entire building. Next, install *starter strips*, as shown in Figure 13-61. Place shims behind the starter strips to bring low spots level. This will avoid a wavy appearance in the finished siding.

Install inside and outside *corner posts* next, Figure 13-62. If splicing is necessary, cut away 1" from all but the outer face of the top of the lower corner post. Lap 3/4" of the upper post over the lower. This allows 1/4" for expansion.

In some re-siding jobs, window sills, heads, and casings are clad with aluminum to reduce maintenance. Cut aluminum from prefinished aluminum sheet and form it on a brake bender. This trim should be installed before installing the vinyl "J" channel. See Figure 13-63.

Installing Siding Panels

The first siding panel snaps into the bottom of the starter strip. Nail according to earlier instructions. Carpenters usually start at the back of the building. Work toward the front, finishing each side before starting the next. Always cover largest areas first; smaller panels can be used up on the smaller surfaces, such as dormers.

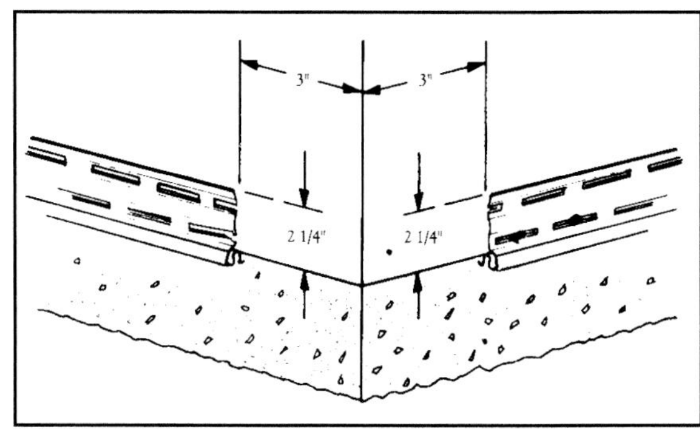

Figure 13-61. Installing the starter strips. A 3" setback at corners allows room for installing trim at inside and outside corners.

Overlap each panel 1", facing the butt edge away from the main traffic areas.

The top course under the eave will require special trim and preparation of the panel. Every manufacturer provides a system for concealing and securing the top siding panel. Generally speaking, this requires a trim piece designed to grip the trimmed top edge of the panel. The following procedure is typical:

Install trim with a double "J" channel against the overhang. To determine how much of the panel to cut off, measure the distance between the top inside slot of the dual undersill trim and the top lock last siding panel and take off 1/4" for expansion. Cut the top panel to this dimension. Punch the top panel with a snap-lock punch to 1/4" below the cut edge. This creates raised points that snap into the double "J" channel.

Gable ends are covered in the same manner as the walls. Use two scrap pieces of panel to make a pattern for cutting the proper angle. Lock one piece into the panel below. Hold the other piece against the gable. Mark a line across the bottom piece and cut. Follow the same procedure for the other side of the gable. Figure 13-64 shows "before" and "after" photos of a re-siding project using vinyl siding. Various elements of a completed house with vinyl siding are shown in Figure 13-65.

Figure 13-62. Corner post installation. Top—Corner trim is installed after starter strips. Normally, starter strips are kept 3" away from the corner to allow for the corner trim and expansion. Start the corner post 3/4" below the bottom of the starter strip. Bottom—The same view as top photo after siding is installed.

Figure 13-63. Top—"J" channel is installed around windows and doors to provide a finished appearance to siding and to seal joints against water. Miter corners as shown. (CertainTeed Corp.) Bottom—Appearance of finished job around window.

Figure 13-64. Done properly, a re-siding enhances the appearance of a building. Top—Old siding appears to be in reasonable shape but lacks "sparkle." Bottom—A vinyl recladding "updates" an older home.

Stucco

When properly applied, stucco makes a satisfactory exterior wall finish. The finish coat may be tinted by adding coloring or the surface may be painted with a suitable material. Where stucco is used on houses more than one story high, the use of balloon framing for the outside walls is desirable.

If platform framing is used, shrinkage of joists may cause distortion or cracks in the stucco.

The base for stucco consists of wood sheathing, sheathing paper, and metal lath, Figure 13-66. The metal lath should be heavily galvanized and spaced at least 1/4" away from the sheathing so the base coat (called scratch coat) can be easily forced through, thoroughly embedding the lath. Metal or wood molding with a groove that "keys" the stucco is applied at edges and around openings. Galvanized furring nails, metal furring strips, and self-furring wire mesh, are available. Nails should penetrate the sheathing at least 3/4". When fiberboard or gypsum sheathing is used, nailing with adequate penetration should occur over studs.

A second coat, applied when the first has dried, is usually given a smoother "pebbled" finish. Figure 13-67 shows scratch and finished coats. Typically, the finished stucco wall is 7/8" thick.

Exterior Insulation Finish Systems

Exterior insulation finish systems (EIFS) are also called "synthetic stucco" and are similar in appearance to stucco. Though relatively new to the construction industry their use has expanded rapidly. Several manufacturers produce these systems which offer excellent insulation quality and a variety of colors.

EIFS are available either as polymer-based (PB) or polymer-modified (PM). The PB, or *soft-coat systems*, are typically thin (1/8"), flexible, and attached with adhesives. PM, or *hard-coat systems*, are thicker (about 1/4") and are mechanically attached. Four or more textures are offered. Figure 13-68 shows a typical EIFS application in cutaway.

Installing an EIFS

An exterior insulation finish system can be installed over wood, concrete, concrete block, and other substrates. If the surface of the substrate is not flat, level it before installing the insulation board. Joints in the insulation board should be offset between sheathing and window openings. Refer to Figure 13-69.

The *base coat* (adhesive) should be at least 3/16" thick, applied in two layers. If the substrate is cement board, coat it with primer before applying the base coat. See Figure 13-70.

Mesh is installed with laps offset from edges of openings, joints, grooves, and corners. Place it diagonally to reinforce corners at all openings. Refer again to Figure 13-69.

If a base coat is required before applying the finish coat, it should be applied in two thin layers of approximately 1/16" each. Final coat, Figure 13-71, can be applied by troweling or spraying.

Brick or Stone Veneer

A *veneer wall* is usually not referred to as a masonry wall, although that is what it is. It is a wall framed in wood or metal to which stone, brick, or even concrete block—are attached rather than siding. However, the weight of the veneer is supported directly by the foundation, Figure 13-72. The foundation must be wide enough to provide a base for the masonry units.

A base flashing of noncorroding metal should extend from the outside face of the wall, over the top of the ledge, and at least 6" up behind the sheathing, as shown. When sheathing is plywood, and an air space of 1" is included, sheathing paper is not required.

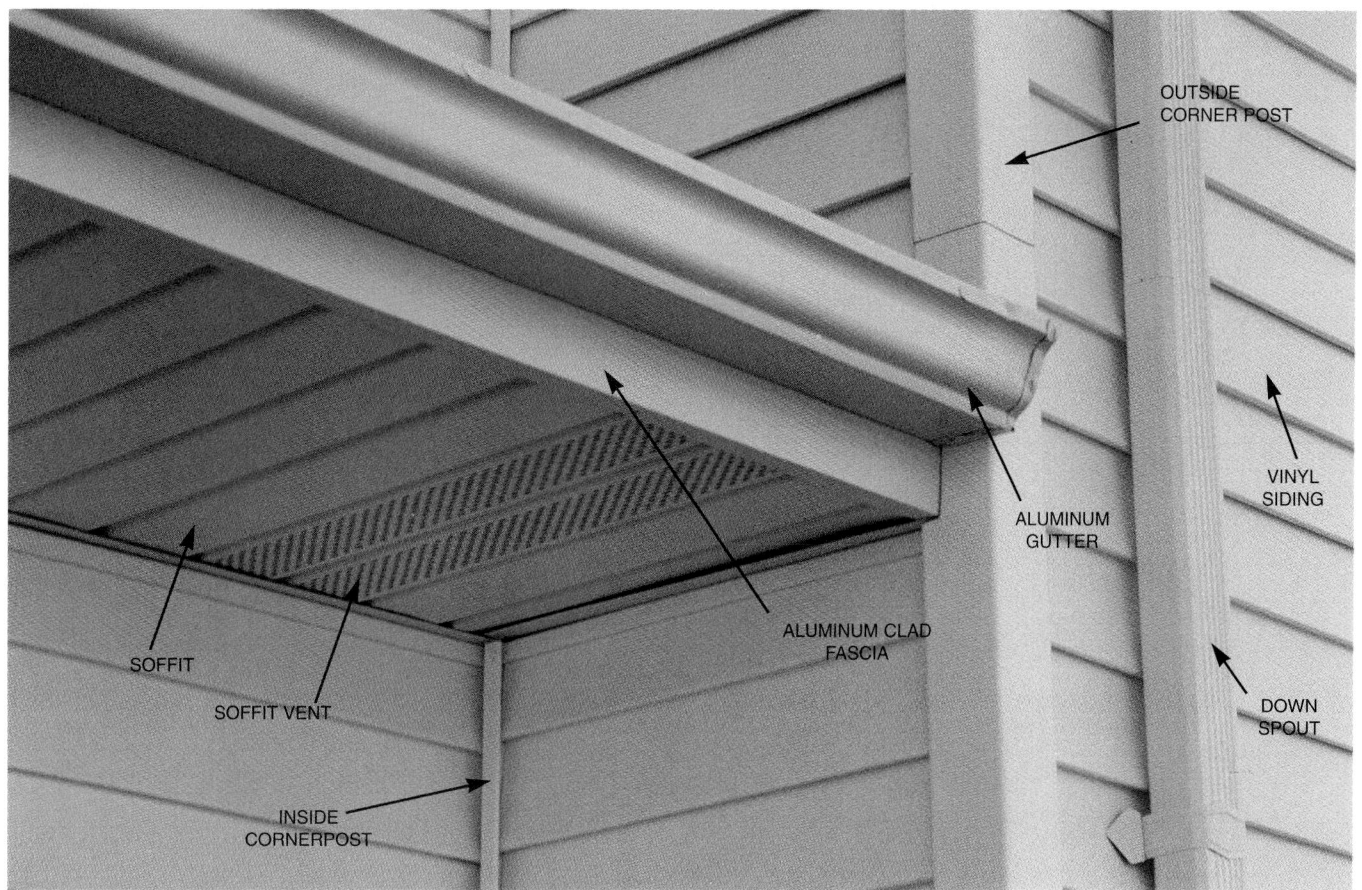

Figure 13-65. A maintenance-free home with different elements of its exterior identified.

Tools

A trowel is the most-used tool in the mason's tool kit. Another important tool is the mason's level. Both are shown in Figure 13-73. The pointed end of the trowel is the *toe* or *point*; the wide end is the *heel*. A mason's level should have both vertical and horizontal vials that can be read from either side. Some are aluminum; some are wood with metal edges designed to withstand the rough environment of the construction site.

The mason usually has two kinds of rules. One is a 6' folding rule that sometimes has a 6" sliding scale on the first section for taking inside measurements. The other is a 10' steel tape. The folding rule usually has markings on its back side giving course heights for various masonry unit sizes and joint thicknesses.

Jointers, or jointing tools, are used to compress, smooth, and shape the surface of the mortar joints. Several jointers are commonly used. Their shape determines the shape and style of the mortar joint. See Figure 13-74. Masons also sometimes use joint rakers to remove a portion of the mortar when a raked mortar joint is desired.

A brick hammer is used to drive nails, strike chisels, and break or chip masonry units. Chisels are used for cutting brick and block. Different sizes and shapes are available. A mason's line is strung level across a wall to keep courses level and walls straight. The line is secured at each end by line holders.

For more accurate and faster cutting of brick or block, masons may use a power-driven masonry blade. See Figure 13-75.

Masonry Materials

Bricks are structural units made to several sizes from clay or shale. This material is mixed with water and then dried in large kilns.

Brick comes in various sizes, some of them modular and some nonmodular. (Modular sizes are based on 4"; this includes dimensions of half of 4 as well as two or three times 4.) See Figure 13-76.

Sizes include allowance for the thickness of the mortar joint. Thus the nominal size is smaller than the actual space the brick will occupy in a wall.

Bricks are engineered for various uses and have the following classifications:

Building or Common Brick—a strong general purpose brick intended for use where strength is more important than appearance. There are three grades:

- SW grade resists freezing and is used often for foundation courses or retaining walls.
- MW grade is used where there may be exposure to below freezing temperatures but in dry locations.
- NW grade is used to back up interior masonry.

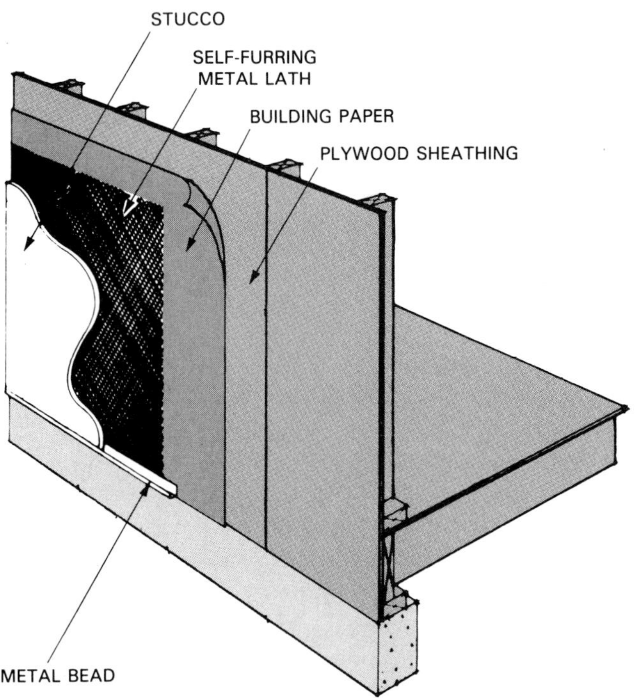

Figure 13-66. Top—Construction worker applies metal lath over an insulating board in preparation for a stucco finish. Middle—Note reinforcing at window corners. (Pierce Construction, Ltd.) Bottom—Construction details for stucco finish. Sheathing paper or housewrap is recommended over both interior and exterior plywood sheathing. (American Plywood Assoc.)

Figure 13-67. Applying stucco. Top—Base coat is going on over metal lath. Sufficient pressure is applied to push some of the stucco through the lath. Bottom—Top coat is being applied. When dry, it will be painted.

Facing Brick—used where appearance is important. In addition, the brick must be strong and durable. There are three types:

- FBX is for general use in exposed interior or exterior walls or partitions. Color and size are uniform.
- FBS is for general use in exposed exterior and interior walls and partitions where wider color variations and sizes are permitted.
- FBA is used to produce architectural effects produced by lack of uniformity in size, color, and texture.

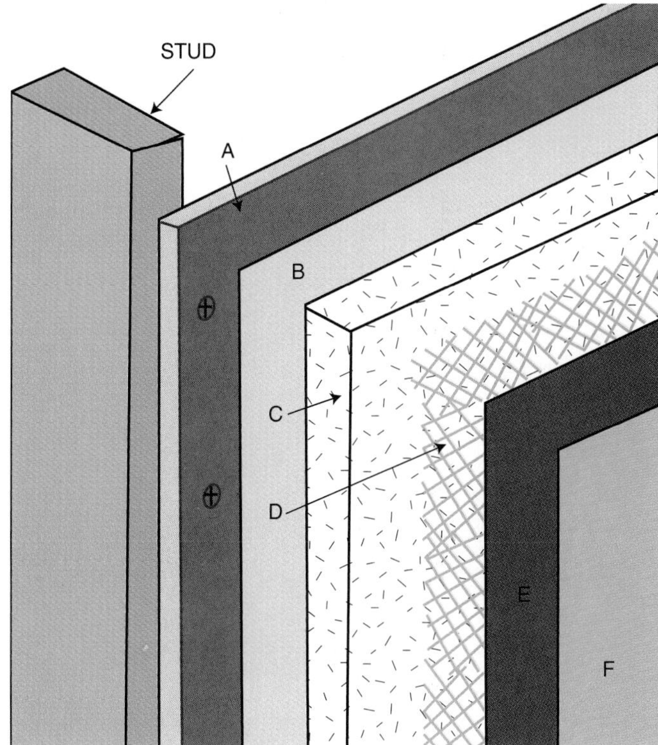

Figure 13-68. Components of an EIFS-sided structure. These materials may be applied over any of several different substrates: plywood, cement board, oriented strand board, glass-faced, etc. A—Cement board. B—Acrylic copolymer adhesive or polymer modified cementitious adhesive. C—Expanded polystyrene insulation board. D—Open weave glass fiber reinforcing mesh. E—Exterior base coat of Portland cement mortar containing dry latex polymers. F—Surface coating based on an acrylic polymer emulsion. (U. S. Gypsum Co.)

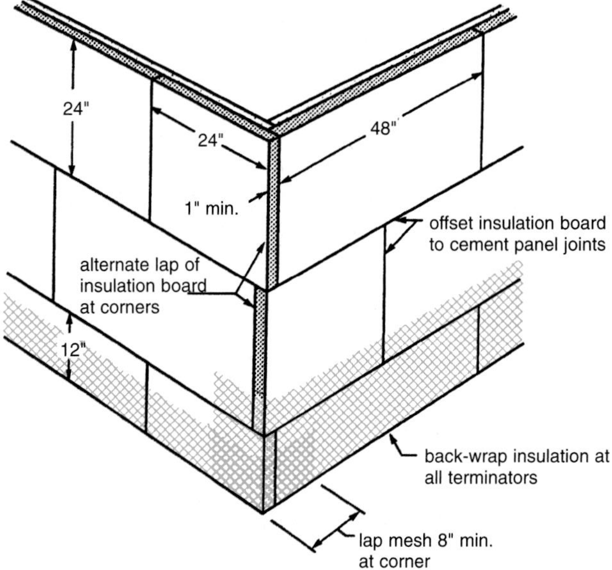

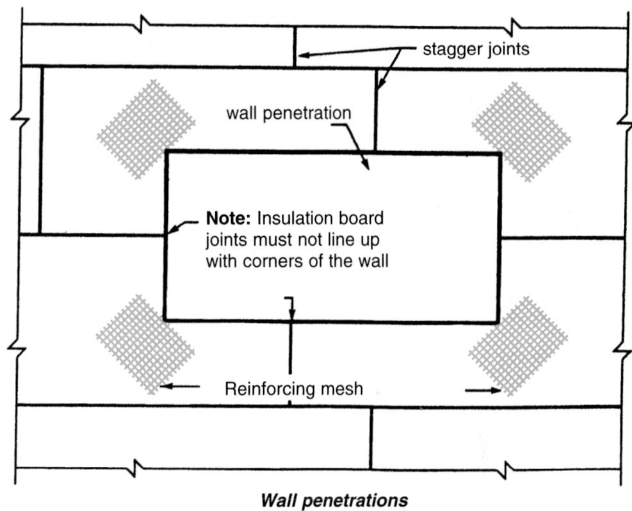

Figure 13-69. Top—Alternate the lap of insulation board at outside corners and offset the vertical joints. Bottom—Avoid lining up joints, either horizontal or vertical, with window or door openings. (U. S. Gypsum Co.)

Mortar is mostly portland cement with the addition of hydrated lime and sand. It is designed to bond bricks and block into a strong, waterproof wall. Various types have been developed to meet the need of various situations. See Figure 13-77.

Corrosion resistant metal ties are used to secure the veneer to the framework. These are usually spaced 32" apart horizontally and 15" vertically. Where other than wood sheathing is used, the ties should be secured to the studs. Weep holes (small openings in the bottom course) permit the escape of any water or moisture that may penetrate the wall. They are spaced about 4' apart.

Select a type of brick suitable for exposure to the weather. Such brick will be hard and low in water absorption. Sandstone and limestone are most commonly used for stone veneer. These materials vary widely in quality. Be sure to select materials known locally to be durable. Procedure for laying brick and stone are essentially the same as laying concrete block. Figure 13-78 shows typical patterns for both brick and stone veneer.

Blinds and Shutters

Some architectural designs may require the installation of blinds and shutters at the sides of window units. These consist of frame assemblies with solid panels or *louvers*, Figure 13-79. In early days of our country, they served an important function since they could be closed over the window. This protected the glass. Closing and locking the shutters also provided some security to the inhabitants.

Today, except where hurricanes are likely to occur, shutters and blinds are decorative. They tend to extend the width of windows, stressing horizontal lines of the structure. Hinges are seldom used and they are usually attached

Figure 13-70. Top—Applying base coat to the insulation board using a trowel. Bottom—Insulation board being applied over an oriented strand board sheathing.

Figure 13-71. Applying final finish to an EIFS. Top—Worker mixes colored stucco finishing materials with an electric tool. Bottom—Other construction workers apply final finish. Window trim is being finished in a darker shade.

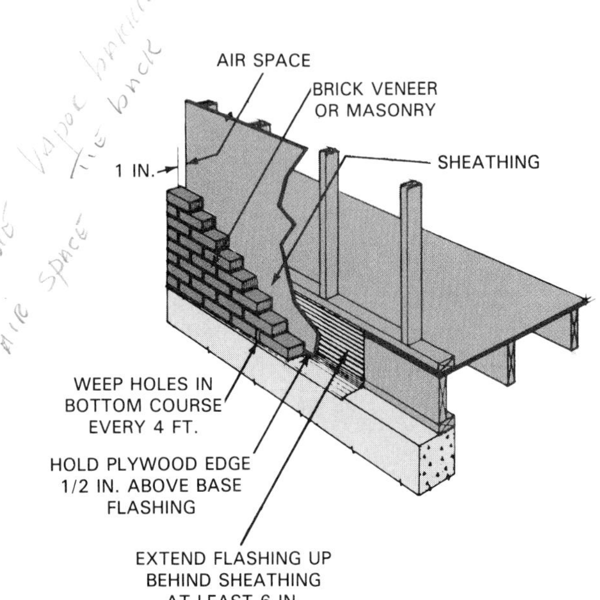

Figure 13-72. Left—General construction details for brick veneered siding with plywood sheathing. One tie should be anchored to a stud for every 2 sq. ft. of brick area. (U.S. Plywood Assoc.) Right—A modern brick veneer wall under construction.

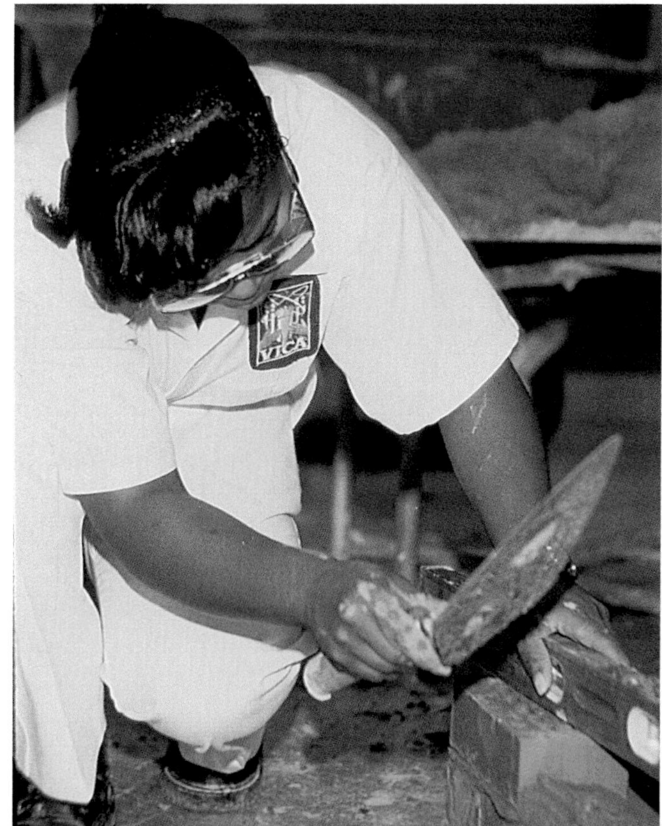

Figure 13-73. Masonry tools. Top—A mason's trowel is used for handling mortar and striking off mortar joints. Bottom—A mason's level is used to keep masonry walls plumb and level. (Vocational Industrial Clubs of America)

to the exterior wall with screws or other fasteners so they can be easily removed for painting and maintenance. Stock sizes include various heights to fit standard window units. Widths range from 14" to 20" with 2" increments (steps).

Figure 13-74. A jointing tool shapes mortar joints before the mortar sets up. Note the steel tape which all masons carry. (VICA)

Figure 13-75. Power-driven saws are sometimes used for rapid, accurate cutting of masonry units. (Custom Precast Company, Cascade, IA)

Important Terms

Aluminum siding
Base coat
Bearing
Bevel siding
Boxed cornice
Channel rustic siding
Corner posts
Cornice
Double coursing
Drop siding
Exterior insulation finish systems (EIFS)
Exterior finish
Exterior trim
Fascia board
Frieze board
Hardboard siding
Hard-coat system
Heel
Ledger strip
Lookouts
Louvers
Mortar
Plancier

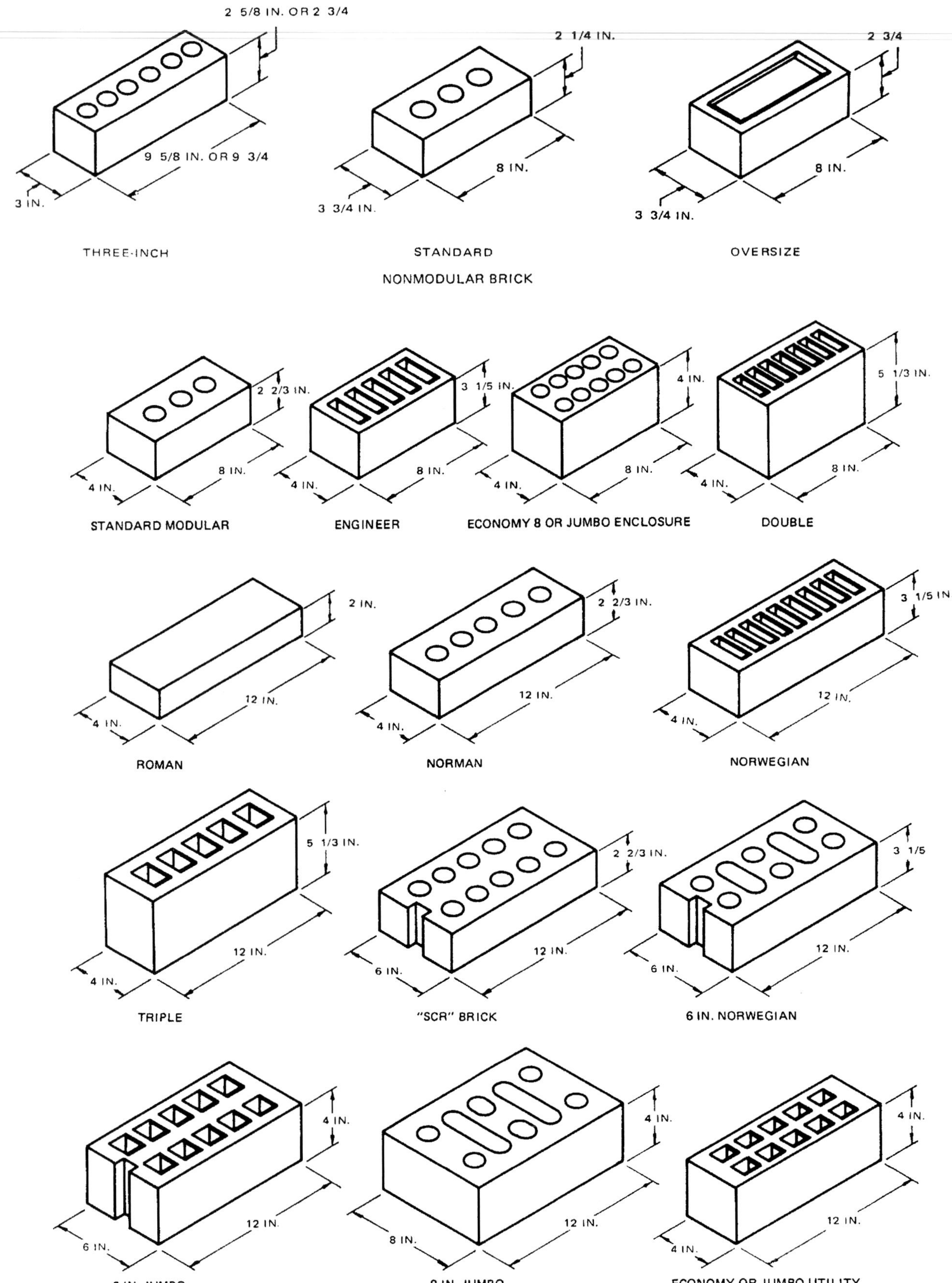

Figure 13-76. Sizes of modular and nonmodular brick are shown in their nominal dimensions.

MORTAR TYPE	PARTS BY VOLUME OF PORTLAND CEMENT* OR PORTLAND BLAST FURNACE SLAG CEMENT**	PARTS BY VOLUME OF MASONRY CEMENT	PARTS BY VOLUME OF HYDRATED LIME OR LIME PUTTY	AGGREGATE, MEASURED IN A DAMP, LOOSE CONDITION
M	1 1	1 (TYPE II) –	– 1/4	NOT LESS THAN 2 1/2 AND NOT MORE THAN 3 TIMES THE SUM OF THE VOLUMES OF THE CEMENTS AND LIME USED.
S	1/2 1	1 (TYPE II) –	– OVER 1/4 to 1/2	
N	– 1	1 (TYPE II) –	– OVER 1/2 to 1 1/4	
O	– 1	1 (TYPE 1 OR II) –	– OVER 1 1/4 to 1 1/2	
K	1	–	OVER 2 1/2 to 4	

ASTM MORTAR TYPE DESIGNATION	CONSTRUCTION SUITABILITY
M	MASONRY SUBJECTED TO HIGH COMPRESSIVE LOADS, SERVER FROST ACTION, OR HIGH LATERAL LOADS FROM EARTH PRESSURES, HURRICANE WINDS, OR EARTHQUAKES. STRUCTURES BELOW GRADE, MANHOLES, AND CATCH BASINS.
S	STRUCTURES REQUIRING HIGH FLEXURAL BOND STRENGTH, BUT SUBJECT ONLY TO NORMAL COMPRESSIVE LOADS.
N	GENERAL USE IN ABOVE GRADE MASONRY. RESIDENTIAL BASEMENT CONSTUCTION, INTERIOR WALLS AND PARTITIONS. CONCRETE MASONRY VENEERS APPLIED TO FRAME CONSTRUCTION.
O	NON-LOAD-BEARING WALLS AND PARTITIONS. SOLID LOAD BEARING MASONRY OF ALLOWABLE COMPRESSIVE STRENGTH NOT EXCEEDING 100 PSI.
K	INTERIOR NON-LOAD-BEARING PARTITIONS WHERE LOW COMPRESSIVE AND BOND STRENGTHS ARE PERMITTED BY BUILDING CODES.

Figure 13-77. Top—Chart shows materials and their proportions in different types of mortar. Bottom—Uses of different types of mortar.

Point
Rake
Runner guides
Soft-coat system
Soffit
Starter strips
Toe
Veneer wall
Wood shingles

Test Your Knowledge

1. In cornice construction, the strip nailed to the wall to support the lookouts and soffit is called a ____________.
2. Some prefabricated soffit systems utilize steel channels to provide rigidity to the soffit material, thus eliminating the need for ____________.
3. Of the various types of horizontal wood siding, ____________ siding usually has the most strength and tightest joints.
4. The best grade of solid wood siding is made from ____________ (edge-grain, flat-grain) material.
5. Head flashing can usually be omitted over an opening if the vertical distance to the soffit is less than ____________ the width of the overhang.
6. Plain beveled siding in a 10" width should be lapped about ____________ inches.
7. Corner boards are installed ____________ (*before, after*) the horizontal siding units are fastened in place.
8. When plain square-edged boards are used for vertical siding, the joint between bonds is usually covered with a ____________ strip.
9. Standard wood shingle lengths for side wall coverage are 16", 18", and ____________ inches.
10. In single-coursed side walls, the weather exposure of wood shingles should not be greater than ____________ their length, minus 1/2".

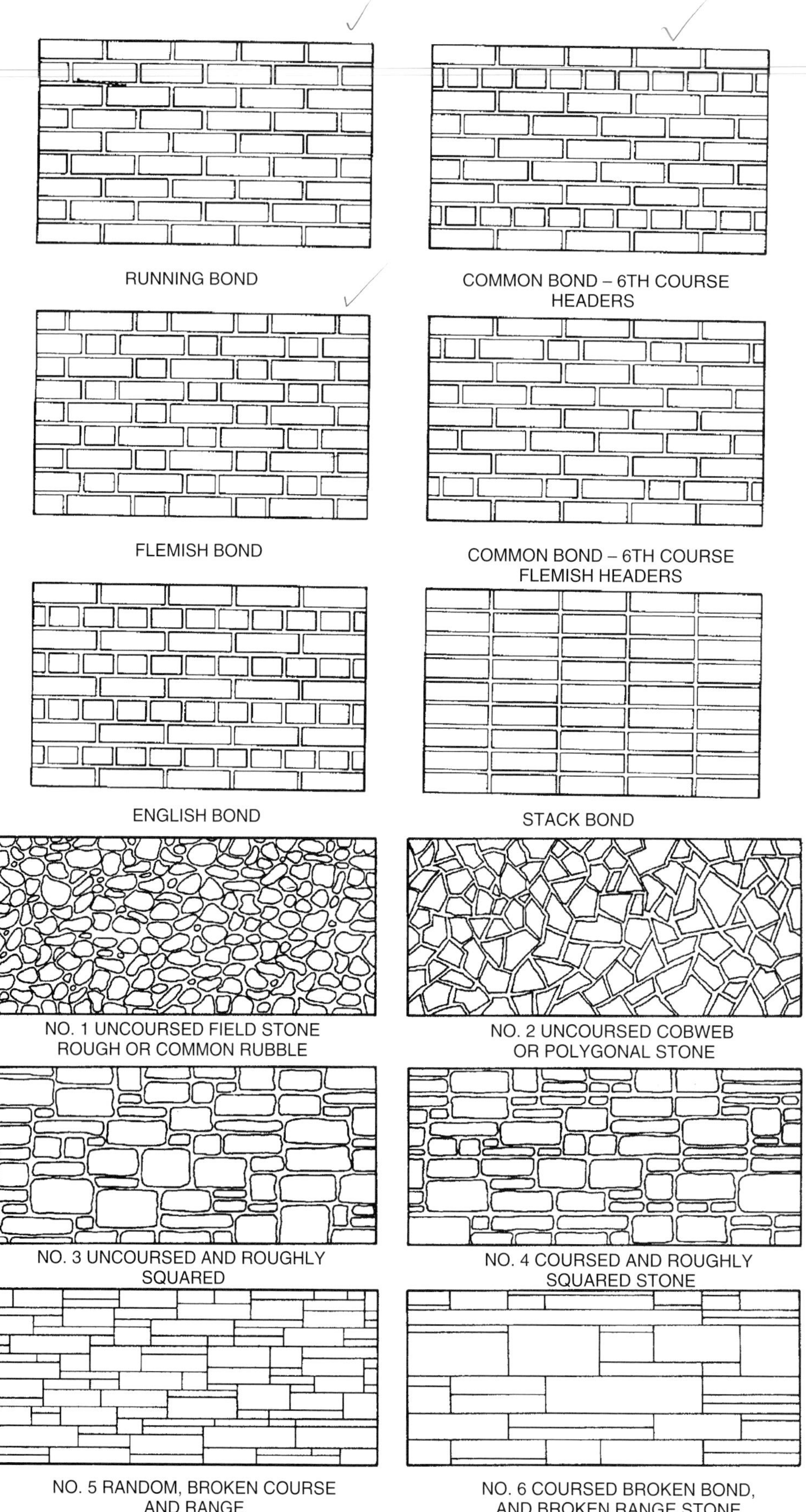

Figure 13-78. These are common patterns for brick and stone veneer siding. Top—The basic structure bonds for brick. Bottom—Common stone patterns.

Figure 13-79. Shutters add a pleasing decorative effect and improve the appearance of window units. (Andersen Corp.)

11. What use can a carpenter make of the fact that grooves in vertical plywood siding are on 4" and 8" centers?
12. When using plywood siding over an unsheathed wall, the thickness should not be less than _____________ inches when studs are spaced 16" O.C.
13. Large sheets of plywood siding usually result in tight construction, therefore it is highly important that a vapor barrier be included on the (*warm, cold*) side of the wall.
14. The most common thickness for hardboard siding material is _____________ inches.
15. As a protection against faulty electrical wiring or appliances, aluminum siding should be _____________.
16. Name four advantages of metal or vinyl soffit materials.
17. Why is rigid polystyrene insulation board sometimes applied to exterior walls?
18. In preparation for re-siding with aluminum or vinyl, what steps would be most essential for a good job?
19. When using aluminum or vinyl siding, install siding before trim. True or false?
20. List the different materials applied to exterior walls and application method for an EIFS installation.
21. List the tools you think necessary to build a brick or stone veneer wall.
22. Small openings located in the bottom course of brick or stone veneer construction that permits moisture to escape are called _____________.

Outside Assignments

1. Study the cornice work on your home or some other residence in your neighborhood. Prepare a detail drawing (use a scale of 1" = 1'-0") showing a typical cross section. Since most of the construction will likely be hidden, you will need to develop your own structural design and select sizes for some of the parts. Study the details shown in various architectural plans and those in Figure 13-1 of this unit. Make your drawing similar to these and be sure to include the sizes of all materials.
2. Visit a builder's supply store or lumber yard. Study the various types of siding carried in stock. Obtain descriptive folders about various prefinished products and special siding systems. Also obtain approximate costs. Write a report that includes a general description of the materials and summarizes the installation procedures. List some advantages and disadvantages of the various types.
3. Visit a building site in your neighborhood where the exterior wall finish is being applied. Be sure to get permission from the builder or head carpenter. Observe the methods and materials being used. Make notes about: type of framing and stud spacing; type of sheathing; type of flashing; type, size, and quality of siding material; layout methods; how units are cut and fitted; nails and nailing patterns; and special joint treatment. Carefully organize your notes and make an oral report to your class.
4. Using the pricing information received from a lumber yard or home improvement center, estimate the cost of materials to cover a 40' x 20' house, single story, with a 4 in 12 roof.

An early adobe home. Walls are built of thick adobe brick.

Modern vinyl siding system being applied to side walls and prefabricated chimney chase. Manufacturers provide instructions for the application of their siding products. These instructions should be followed carefully when making the application.

A wide range of prefabricated materials are available for exterior finish. The view above shows a soffit and fascia system, vinyl siding, and a gutter and downspout. The products are preformed, finished, and ready to install. An exception might include the gutter which is often made up on the site with a special roll-forming machine.

A modern adobe-style home built in the southwest. Though similar in appearance to the adobe brick dwellings of an earlier time, it has a wood frame. (Pierce Construction, Ltd.)

Finishing

The attractiveness of a home is often determined by the quality of its finishes. Care has been taken in the selection of materials used in this staircase.

14

Thermal and Sound Insulation

Insulation, in construction, involves materials that do not readily transmit energy in the form of heat, electricity, or sound. This unit will deal primarily with its use to prevent the transfer of heat either into or out of a building. To a lesser extent, it will discuss use of insulation to prevent sound transmission.

Insulation requirements of homes and other buildings have changed radically since 1973 and the first energy crisis. As energy costs have increased so has the amount of insulation being placed in walls, floors, and ceilings. At the same time, the insulating qualities of doors and windows have been upgraded to hold down energy costs.

Likewise, duct work is likely to be insulated if it is not placed in the heated space itself, Figure 14-1.

Insulation is necessary in buildings where temperature of the buildings must be controlled. In cold climates, the major concern is to retain heat in the building. In warmer regions, insulation is needed to keep heat from entering the building.

A wide range of insulation materials is available to fill the requirements of modern construction. The materials are engineered for efficient installation and come in convenient packages that are easy to handle and store.

Figure 14-1. Interior view of modern residence ready to receive insulation. To make room for 18" (R-60) of insulation in ceiling, wiring has been raised well above ceiling joists. Insulated ducts are already installed in the attic. A strip of 6 mil polyethylene was attached to interior walls before raising and will provide integrated ceiling vapor barrier. No pipes were installed in exterior walls. (Owens-Corning Fiberglas Corp.)

Acoustical treatments and sound control employ many of the same materials used for thermal insulation and are, therefore, described in this unit.

Building Sequence

While carpenters are completing the exterior finish of the structure, other tradespeople are installing mechanical systems on the inside. These systems must be completed before interior walls are insulated and closed in with drywall or plaster. Duct work is installed in the floors, walls, and ceilings for heating and air conditioning. The electrician installs wiring for electrical circuits, attaching conduit or cable to the building's framework. She or he also locates and attaches the boxes that enclose connections for convenience outlets, switches, and lighting fixtures, Figure 14-2.

The plumber installs the water supply piping, drain piping, and venting. These pipes supply fresh water and drain away waste water. Built-in plumbing fixtures (bathtubs, hydro-massage units, and shower stalls) are also installed at this time. They must be carefully covered with building paper or polyethylene film to protect them during the inside finish operations. See Figure 14-3.

Figure 14-2. Before insulation materials are installed (thermal and sound) other construction trades must make their installations.

Figure 14-3. Built-in plumbing fixtures must be installed before wall framing is complete. The tub-shower unit shown above has been covered with a 6 mil polyethylene film to protect it during additional phases of construction.

During the heating and plumbing "rough-in," the tradespersons often need to cut through structural framework. The carpenter should check this work. Framing members may have been seriously weakened and should be reinforced.

How Heat Is Transmitted

Although carpenters are seldom required to design buildings or figure heat losses, they should have some knowledge and understanding of the theory and factors that are involved. Thus, they will be able more fully to appreciate how important it is to carefully select and install thermal insulation materials.

Heat seeks a balance with surrounding areas. When the inside temperature is controlled within a given comfort range, there will be some flow of heat. It will move from the inside to the outside in winter and from the outside to the inside during hot summer weather.

Heat is transferred through walls, floors, ceilings, windows, and doors at a rate directly related to:

- The difference in temperature.
- Resistance to heat flow provided by intervening materials.

Heat transfers by one or more of three methods: conduction, convection, and radiation. See Figure 14-4.

Conduction

Conduction is transmission of heat from one molecule to another within a material or from one material to another when they are held in direct contact. Dense materials, such as metal or stone, conduct heat more rapidly than porous materials, such as wood and fiber products. Any material will conduct some heat when a temperature difference exists between its surfaces.

Convection

Convection is the transfer of heat by another agent, such as air or water. In large spaces, molecules of air can carry heat from warm surfaces to cold surfaces. When air is heated it becomes lighter and rises. Thus, a flow of air (called *convection currents*) is created within the space. Air is a good insulator when confined to a small space or cavity where convection flow is limited or absent. In walls and ceilings, trapped air restricts convection currents and will reduce the flow of heat.

Radiation

Heat can be transmitted by wave motion in about the same manner as light. This process is called *radiation* because it represents radiant energy.

Heat received from the sun is radiant heat. The waves do not heat the space through which they move, but when they come in contact with a colder surface, a part of the energy is absorbed while some may be reflected.

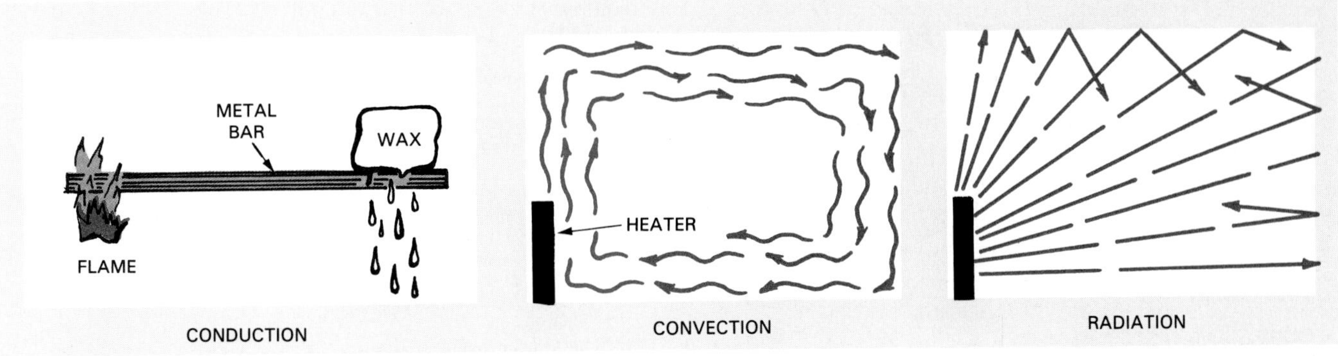

Figure 14-4. Heat is transferred by conduction, convection, and radiation.

Effective resistance to radiation comes about through reflection. Shiny surfaces, such as aluminum foil, are often used to provide this type of insulation.

Actually, heat transmission through walls, ceilings, and floors will be a result of all three of the methods. In addition to this, some heat is lost by convection through cracks around doors, windows, and other openings in the structure.

Thermal Insulation

All building materials resist the flow of heat, mainly conduction, to some degree, depending on their porosity or density. As previously stated, air is an excellent insulator when confined to the tiny spaces or cells inside a porous material. Dense material such as masonry or glass contain few, if any, air spaces and are poor insulators.

Fibrous materials are generally good insulators, not only because of the porosity in the fibers themselves, but also because of the thin film of air that surrounds each individual fiber.

Commercial insulation materials are made of glass fibers, glass foam, mineral fibers, organic fibers, and foamed plastic. A good insulation material should be fireproof, vermin proof, moisture proof, and resistant to any physical change that would reduce its effectiveness against heat flow.

Selection is based on initial cost, effectiveness, durability, and the adaptation of its form to that of the construction and installation methods.

Heat Loss Coefficients

A coefficient is a number that serves as a measure of a property—in this case, the ability to transfer heat. The thermal properties of common building materials and insulation materials are known or can be accurately measured. Heat transmission (the amount of heat flow) through any combination of these materials can be calculated. First, it is necessary to know and understand certain terms.

- *Btu* — The abbreviation for *British thermal unit.* It is the amount of heat needed to raise the temperature of 1 lb. of water 1°F.
- *k* — The amount of heat, in Btu's, transferred in one hour through 1 sq. ft. of a given material that is 1" thick and has a temperature difference between its surfaces of 1°F. It is also called the *coefficient of thermal conductivity.*
- *C* — The *conductance* of a material, regardless of its thickness. It is the amount of heat (Btu's) that will flow through the material in one hour per sq. ft. of surface with 1°F of temperature difference. For example, the C-value for an average hollow concrete block is 0.53.
- *R* — Represents *resistance,* which is the reciprocal (opposite) of conductivity or conductance. A good insulation material will have a high R-value.

$$R = \frac{1}{k} \text{ or } \frac{1}{C}$$

- *U* — Represents the *total heat transmission* in Btu per sq. ft. per hour with 1°F temperature difference for a structure (wall, ceiling, floor) which may consist of several materials or spaces. A standard frame wall with composition sheathing, gypsum lath, and plaster, with a 1" blanket insulation will have a U-value of about 0.11. To calculate the U-value where the R-values are known, apply the following formula:

$$U = \frac{1}{R_1 + R_2 + R_3 + \ldots R_n}$$

Figure 14-5 shows how insulation reduces the U-value for a conventional frame wall. Note that a 5 1/2" thick blanket reduces the U-value from 0.29 to 0.053. This is about an 82% reduction. Actually the U-value for the total wall structure will be slightly higher because the wood studs have a lower R-value than the blanket insulation.

R-values of wall structures with various types and amounts of insulation are shown in Figure 14-6. The R-value provides a convenient measure to compare heat loss in materials and structural designs. However, to determine the total heat loss (or gain) through a wall, ceiling, or floor,

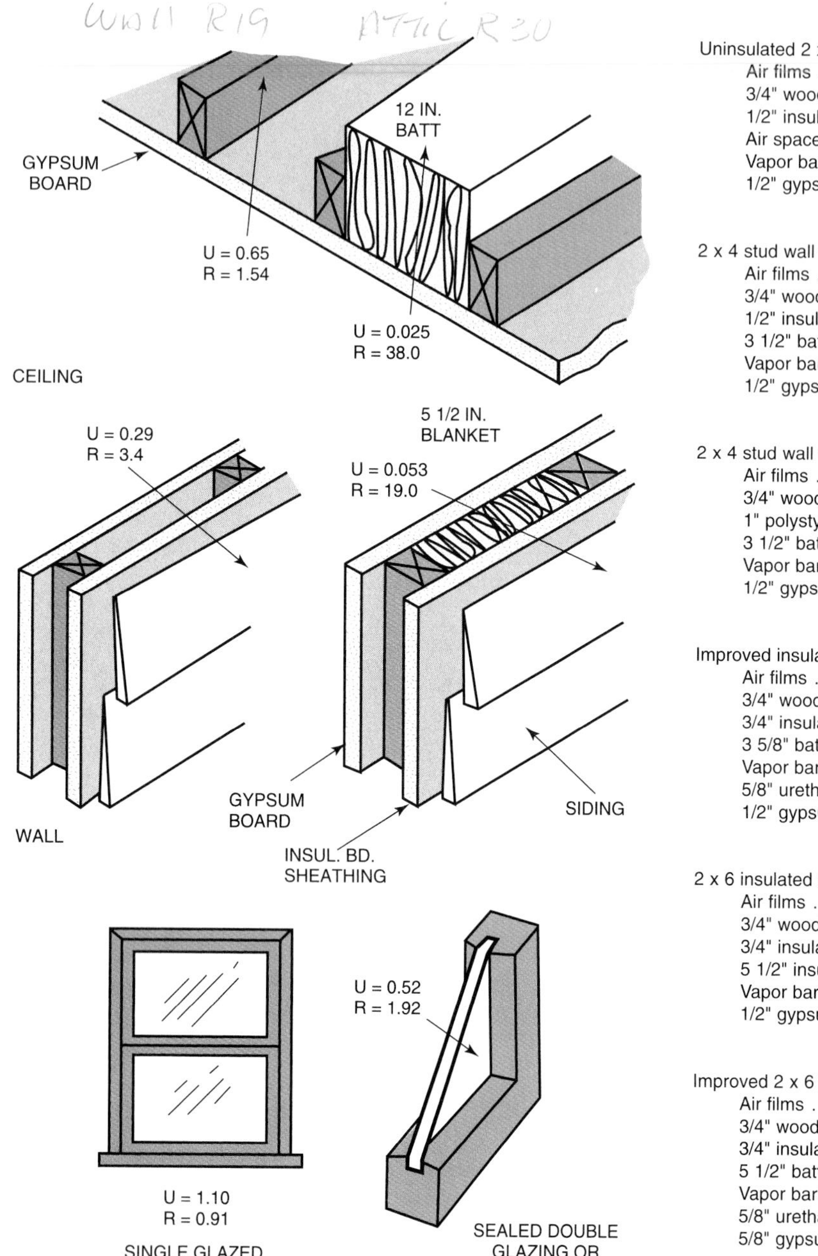

Figure 14-5. Insulation reduces flow of heat. Trapped air, such as is provided by double-glazed windows, is a good insulator.

Uninsulated 2 x 4 stud wall	R =
Air films	0.9
3/4" wood exterior siding	1.0
1/2" insulation board	1.2
Air space	1.2
Vapor barrier	0
1/2" gypsum board	0.5
	4.8 total R

2 x 4 stud wall with batt insulation	R =
Air films	0.9
3/4" wood exterior siding	1.0
1/2" insulation board	1.2
3 1/2" batt or blanket insulation	11.0
Vapor barrier	0
1/2" gypsum board	0.5
	14.6 total R

2 x 4 stud wall with rigid board	R =
Air films	0.9
3/4" wood exterior siding	1.0
1" polystyrene rigid board	5.0
3 1/2" batt or blanket insulation	11.0
Vapor barrier	0
1/2" gypsum board	0.5
	18.4 total R

Improved insulated 2 x 4 stud wall	R =
Air films	0.9
3/4" wood exterior siding	1.0
3/4" insulation board	2.0
3 5/8" batt insulation	13.0
Vapor barrier	0
5/8" urethane insulation board	5.0
1/2" gypsum board	0.5
	22.4 total R

2 x 6 insulated stud wall	R =
Air films	0.9
3/4" wood exterior siding	1.0
3/4" insulation board	2.0
5 1/2" insulating blanket	19.0
Vapor barrier	0
1/2" gypsum board	0.5
	23.4 total R

Improved 2 x 6 insulated stud wall	R =
Air films	0.9
3/4" wood exterior siding	1.0
3/4" insulation board	2.0
5 1/2" batt or blanket insulation	19.0
Vapor barrier	0
5/8" urethane insulation	5.0
5/8" gypsum board	0.6
	28.5 total R

Figure 14-6. Types of wall construction and their R-values. Materials are listed in order from the outside in. Air films refer to the inside and outside film of stagnant air which forms on any surface. It makes a small contribution to the R-value. (Iowa Energy Policy Council)

R-values must be converted to U-values. The total of these U-values (Btu's per sq. ft. per hour with 1°F temperature difference) will be needed to calculate the size of heating and cooling equipment. Directions for calculating U-values are presented in the **Technical Information Section** at the back of the book.

R-values for commonly used insulation and building materials are listed in Figure 14-7. The original source of most data on this subject is the American Society of Heating, Refrigerating, and Air Conditioning Engineers (ASHRAE). R-values can be converted to U-values by calculating the reciprocal (dividing the value into 1).

How Much Insulation?

Insulation is required in any building where a temperature above or below outdoor temperature must be maintained. The amount of insulation recommended has changed in recent years.

Comfort, health, and economy are the three considerations for thermal insulation. Comfortable and healthy indoor temperatures depend not only on the temperature of the air but also on the temperature of the surfaces of walls, ceilings, and floors. It is possible to heat the air in a room

MATERIAL	KIND	INSULATION VALUE
Masonry	Concrete, sand, and gravel, 1 in.	R-0.08
	Concrete blocks (three core)	
	Sand and gravel aggregate, 4 in.	R-0.71
	Sand and gravel aggregate, 8 in.	R-1.11
	Lightweight aggregate, 4 in.	R-1.50
	Lightweight aggregate, 8 in.	R-2.00
	Brick	
	Face, 4 in.	R-0.44
	Common, 4 in.	R-0.80
	Stone, lime, sand, 1 in.	R-0.08
	Stucco, 1 in.	R-0.20
Wood	Fir, pine, other softwoods, 3/4 in.	R-0.94
	Fir, pine, other softwoods, 1 1/2 in.	R-1.89
	Fir, pine, other softwoods, 3 1/2 in.	R-4.35
	Maple, oak, other hardwoods, 1 in.	R-0.91
Manufactured Wood Products	Plywood, softwood, 1/4 in.	R-0.31
	Plywood, softwood, 1/2 in.	R-0.62
	Plywood, softwood, 5/8 in.	R-0.78
	Plywood, softwood, 3/4 in.	R-0.93
	Hardboard, tempered, 1/4 in.	R-0.25
	Hardboard, underlayment, 1/4 in.	R-0.31
	Particleboard, underlayment, 5/8 in.	R-0.82
	Mineral fiber, 1/4 in.	R-0.21
	Gypsum board, 1/2 in.	R-0.45
	Gypsum board, 5/8 in.	R-0.56
	Insulation board sheathing, 1/2 in.	R-1.32
	Insulation board sheathing, 25/32 in.	R-2.06
Siding and Roofing	Building paper, permeable felt, 15 lb.	R-0.06
	Wood bevel siding, 1/2 in.	R-0.81
	Wood bevel siding, 3/4 in.	R-1.05
	Aluminum, hollow-back siding	R-0.61
	Wood siding shingles, 7 1/2 in. exp.	R-0.87
	Wood roofing shingles, standard	R-0.94
	Asphalt roofing shingles	R-0.44
Insulation	Cellular or foam glass, 1 in.	R-2.50
	Glass fiber, batt, 1 in.	R-3.13
	Expanded perlite, 1 in.	R-2.78
	Expanded polystyrene bead board, 1 in.	R-3.85
	Expanded polystyrene extruded smooth, 1 in.	R-5.00
	Expanded polyurethane, 1 in.	R-7.00
	Mineral fiber with binder, 1 in.	R-3.45
Inside Finish	Cement plaster, sand aggregate, 1 in.	R-0.20
	Gypsum plaster, light wt. aggregate, 1/2 in.	R-0.32
	Hardwood finished floor, 3/4 in.	R-0.68
	Vinyl floor, 1/8 in.	R-0.05
	Carpet and fibrous pad	R-2.08
Glass	(see Unit 12 Windows and Exterior Doors)	

Figure 14-7. R-values for commonly used construction and insulation materials. Additional resistance values can be obtained from the ASHRAE Handbook of Fundamentals. It is published by the American Society of Heating, Refrigerating, and Air Conditioning Engineers.

to the correct comfort level and still have cold room surfaces. The human body will lose heat by radiation (or conduction, if there is direct contact) to these colder surfaces.

Rising costs of energy, coupled with the prospect of fuel shortages, prompted the raising of recommended levels of insulation. Today, the amount of insulation for a given structure must be based not only on comfort standards, but also on factors such as insulation costs (material and labor), probable fuel costs in the future, and local climatic conditions. Figure 14-8 shows a map of the continental United States and a chart of recommended insulation for each area according to winter/summer temperatures. In warmer climates, smaller amounts of insulation are needed — not for protection against cold but to keep out heat during summer.

The harshness of climate is measured in degree days. The higher the number of degree days the colder the climate and the more insulation needed.

Normal Number of Degree Days Per Year

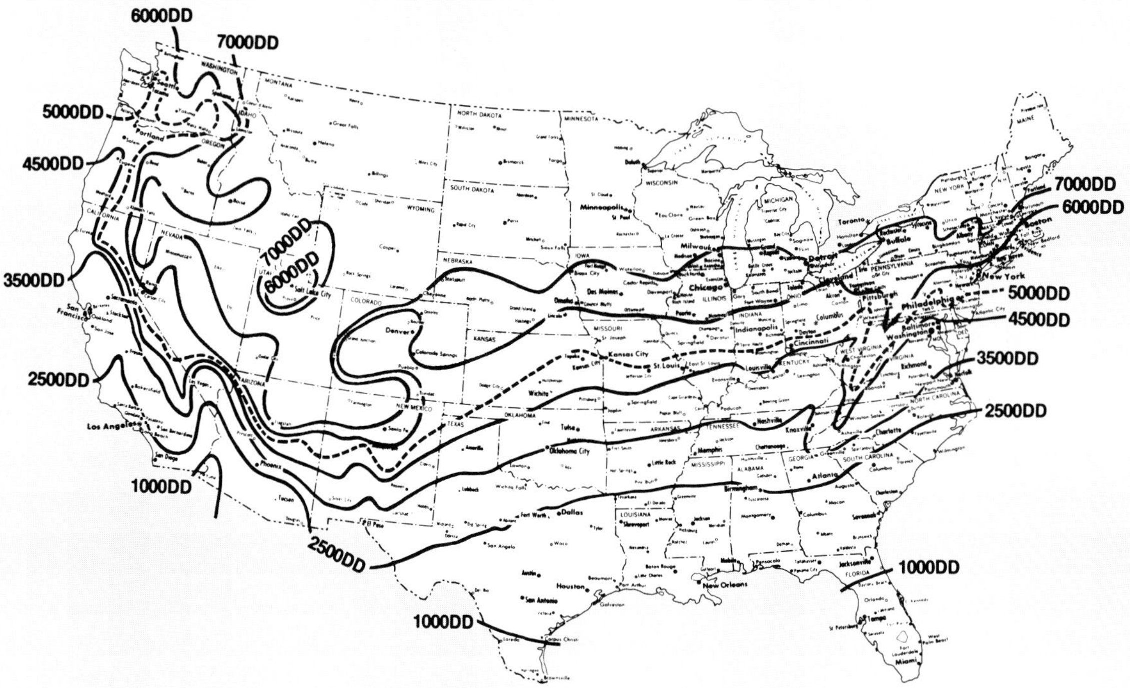

This map is reasonably accurate for most parts of the United States but is necessarily highly generalized, and consequently not too accurate in mountainous regions, particularly in the Rockies.

RECOMMENDED INSULATION R-VALUES

Degree Days	Ceilings	Walls	Floors
Above 7000	38	17	19
6001-7000	38 (30F)	17 (12F)	19 (11F)
4501-6000	30	17 (12F)	19 (11F)
3501-4500	30	12 (17E)	11 (19E)
2501-3000	22 (30E)	11 (17E)	0 (11E)
1001-2500	19 (22E)	11 (12E)	0
1000 and under	19	11	0

FOSSIL FUEL (GAS & OIL) = F; ELECTRIC RESISTANCE HEAT = E; HEAT PUMPS WITH ELECTRIC HEAT = H
R-Values apply to all unless otherwise shown.

Figure 14-8. Map outlines insulation zones for different parts of the United States. R-values listed below map are recommended by the U.S. Department of Energy (DOE). (Owens-Corning Fiberglas Corp.)

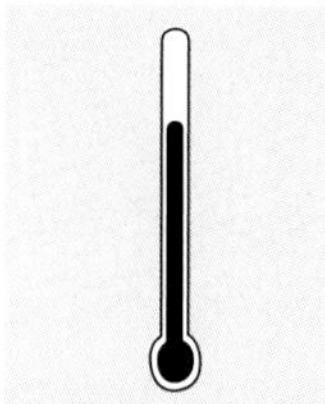

A degree day is the product of one day and the number of °F the mean temperature is below 65°F. Figures are usually quoted for a full year and are used by the heating engineer to determine the design and size of the heating system.

For example: High 60°F, Low 30°F

$$\text{Degree Day} = 65 - \frac{60+30}{2}$$

$$= 65 - \frac{90}{2}$$

$$= 65 - 45$$

$$\text{Degree Day} = 20$$

R-18 plus values in residential wall construction can be obtained by:

- Using 2 x 6 studs, which provide a thicker wall cavity for insulation, Figure 14-9.
- Using 2 x 4 studs sheathed with thick, rigid insulation panels made from foamed polyethylene plastic, Figure 14-10.
- Using a double 2 x 4 frame.

It is important to understand that heat transmission decreases as insulation thicknesses are increased, but not in a direct relationship. This can be noted through a study of U- and R-values for various materials and structures. For example, in a frame wall with a U-value of 0.24, the addition of 1" of insulation will reduce the heat loss by 46%, to U-0.13. A second inch of insulation will reduce the loss by

Figure 14-9. In modern construction, one of the easiest ways to secure R-18 plus values in outside walls is to use 2 x 6 studs.

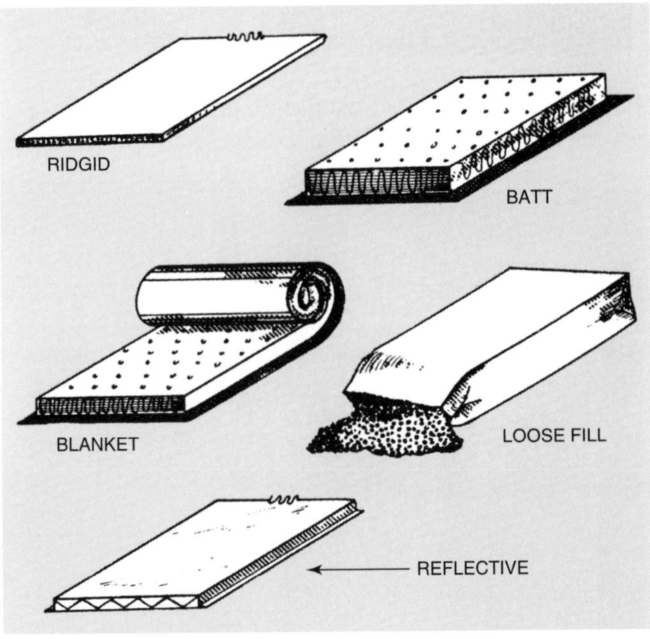

Figure 14-10. Insulation is made in these four basic forms.

16% to U-0.09. And a third inch will reduce the loss by 10% to U-0.065. Additional thicknesses will continue to lower the U-value but at a still lower percentage. At some point it is useless to add more insulation.

Windows provide another example of how U-values decrease. The heat loss through a sash with a single pane of glass will be reduced from U-1.10 to about U-0.52 when a second pane is added. This provides a reduction of about 50%. A third pane (triple-glazing) will reduce the heat loss to U-0.35 and a fourth pane (quad-glazing) will result in a U-value of 0.27. Windows with three and four layers of glass are very expensive. In northern climates, this extra expense is usually justified by lower heating costs and added comfort.

Types of Insulation

Insulation is made in many forms; refer once more to Figure 14-10. It may be grouped into four broad classifications:

- Flexible.
- Loose fill.
- Rigid.
- Reflective.

Flexible insulation is manufactured in two types: blanket or quilt, and batt. Blanket insulation is generally furnished in rolls or strips of convenient length and in various widths suited to standard stud and joist spacing. It comes in thicknesses of 3/4 to 12", Figure 14-11.

The body of the blanket is made of loosely felted mats of mineral or vegetable fibers, such as rock, slag, glass fiber, wood fiber, and cotton. Organic fiber mats are usually treated chemically to make them resistant to fire, decay, insects, and vermin. Blanket insulation is often enclosed in paper with tabs on the side for attachment, Figure 14-12. The covering sheet on one side may be treated to serve as a vapor barrier. In some cases the covering sheet is surfaced with aluminum foil or other reflective insulation. Figure 14-13 shows an unfaced blanket that is easy to install between wall studs. It is held in place by friction. After all the insulation is in place, a vapor barrier is applied over the entire surface of the wall frame.

Batt insulation is made of the same fibrous material as blankets. Thickness can be greater in this form and may range from 3 1/2 to 12", Figure 14-14. They are generally available in widths of 15" and 23" and in 24" and 48" lengths. Batts are available with a single flanged cover or with both sides uncovered.

Loose fill insulation is composed of various materials used in bulk form and supplied in bags or bales, Figure 14-15. It may be poured or blown. Loose insulation is commonly used to fill spaces between studs or to build up any desired thickness on a flat surface.

Loose fill insulation is made from such materials as rock, glass, slag wool, wood fibers, shredded redwood bark, granulated cork, ground or macerated wood pulp products, vermiculite, perlite, powdered gypsum, sawdust, and wood shavings. One of the chief advantages of this type is that when insulating an older structure, only a few boards need to be removed in order to blow the material into the walls. Figure 14-16 shows typical coverage of loose fiber glass insulation blown into walls and attics.

Rigid insulation, as ordinarily used in residential construction, is often made by reducing wood, cane, or other fiber to a pulp and then assembling the pulp into lightweight or low-density boards that combine strength with heat and acoustical insulating properties.

It is available in a wide range of sizes, from tile 8" square, to sheets 4' wide and 10' or more long. Insulating boards are usually 1/2" to 1" in thickness. Boards of greater thickness are made by laminating together boards of standard thickness.

Figure 14-11. Blanket insulation may be kraft faced, foil faced, or unfaced. Top—This fiberglass blanket is 3 1/2" thick (R-11). Middle—This R-25 unfaced blanket insulation is wrapped in film. It is suitable for use in attics or crawl spaces. Bottom—Installation of unfaced blanket insulation in an attic. (Owens-Corning Fiberglas Corp.)

Figure 14-12. Paper faced and flanged insulation. Flange is used to attach blanket to joists and studs.

Figure 14-13. Unfaced blanket fits snugly between studs and is held there by friction. (Owens-Corning Fiberglas Corp.)

Insulating boards are used for many purposes including:

- Roof and wall sheathing.
- Subflooring.
- Interior surface of walls and ceilings.
- Base for plaster and synthetic stucco exterior wall finish.
- Insulation strips for foundation walls and slab floors.

Although expensive, *foamed glass* or *corkboard* makes an excellent rigid form of insulation. A development widely used today is foamed plastic (polystyrene and

Figure 14-14. Batts are sold in various thicknesses. From top to bottom beginning at left: R-38C and R-30C; R-38, R-25, and R-19; R-21, R-15, R-13, and R-11. Thicker batts are designed to give high insulation values for ceilings and attics. (Owens-Corning Fiberglas Corp.)

Figure 14-15. Loose insulation is often used over ceilings in attic areas. Coverage and R-values are listed on the bag.

polyurethane). See Figure 14-17. Because it resists water, it is especially adaptable to masonry work.

Insulating boards should not be confused with ordinary wallboard. The wallboard is more tightly compressed and has less insulating value. Insulating sheathing board ordinarily comes in two thicknesses, 1/2" and 25/32", but is available in 3/4" and 1" thicknesses. It is made in 2' x 8' sheets for horizontal application, and in 4' x 8' sheets and longer for vertical application.

Reflective Insulation

Reflective insulation is usually a metal foil or foil-surfaced material. It differs from other insulating materials in that the number of reflecting surfaces, not the thickness of the material, determines its insulating value. In order to be effective, the metal foil must be exposed to an air space, preferably 3/4" or more in depth.

Aluminum foil is available in sheets or corrugations supported on paper. It is often mounted on the back of gypsum lath. One effective form of reflective insulation has multiple spaced sheets.

Other Types of Insulation

On the market today are insulations that do not fit the classifications covered. Some examples: a confetti-like material mixed with adhesive and sprayed on the surface to be insulated; foams that expand; multiple layers of corrugated paper; and lightweight aggregates like vermiculite and perlite used in plaster to reduce heat transmission.

Lightweight aggregates made from blast furnace slag, burned clay products, and cinders are commonly used in concrete and concrete blocks. They improve the insulation qualities of these materials.

Where to Insulate

Heated areas, especially in cold climates, should be surrounded with insulation by placing it in the walls, ceiling, and floors, Figure 14-18. Refer also to feature illustrations at the beginning of this unit.

It is best to have the insulation as close to the heated space as possible. For example, if an attic is unused, the insulation should be placed in the attic floor rather than in the roof structure.

If attic space or certain portions of the attic must be heated, walls and ceilings should be insulated. If the insulation is placed between the rafters, be sure to allow space between the insulation and the sheathing for free air circulation. The floors of rooms above unheated garages or porches require insulation for maximum comfort.

When a basement is to be used as a living or recreation area, it will be necessary to insulate the walls, Figure 14-19. This is highly recommended as an energy conservation measure, too. Not only does it save heat and provide comfort, it also gives the basement better acoustical qualities.

Whether applied to walls or not, insulation should be installed over the band joists and headers. (These are the floor framing members along the outside of the wall.) See Figure 14-20. This area has little protection against heat loss.

Basementless Structures

Floors over unheated space directly above the ground require the same degree of insulation as walls in the same

SIDEWALL COVERAGE INFORMATION				
NOMINAL R-VALUE	THICKNESS (FRAMING TIMBER)	DENSITY	MINIMUM WEIGHT PER SQ. FT.	MAXIMUM COVERAGE PER BAG**
To obtain a thermal resistance (R) of:	Installed insulation should not be less than: (inches)	Pounds per cubic feet:	Weight per sq. ft. of installed insulation should not be less than: (lbs.)	Contents of bag should not cover more than: (sq. ft.)
R-14	3.5" (2X4)	1.8	0.525	55
R-22	5.5" (2X6)		0.825	35
R-29	7.25" (2X8)		1.088	26
R-37	9.25" (2X10)		1.387	21
R-15	3.5" (2X4)	2.3	0.670	43
R-23	5.5" (2X6)		1.054	27
R-31	7.25" (2X8)		1.389	21
R-39	9.25" (2X10)		1.773	16

ATTIC/OPEN BLOW COVERAGE INFORMATION				
R-VALUE	BAGS PER 1000 SQ. FT.*	MAXIMUM SQ. FT. PER BAG*	MINIMUM WEIGHT PER SQ. FT.	MINIMUM THICKNESS
To obtain a thermal resistance (R) of:	Bags per 1000 sq. ft. of net area:	Contents of bag should not cover more than: (sq. ft.)	Weight per sq. ft. of installed insulation should not be less than: (lbs.)	Installed insulation should not be less than: (inches)
R-60	43.5	23.0	1.307	20.50"
R-50	35.2	28.4	1.057	17.50"
R-44	31.0	32.3	0.928	15.75"
R-38	26.7	37.4	0.803	14.00"
R-30	20.2	49.5	0.606	11.00"
R-26	18.2	54.8	0.547	10.00"
R-22	15.0	66.5	0.451	8.50"
R-19	13.1	76.1	0.394	7.50"
R-13	9.3	107.5	0.279	5.50"
R-11	7.9	126.6	0.237	4.75"

** Coverages per bag do not include framing members, which will increase coverage, depending on 16" or 24" o.c.*
*** **For Sidewall Applications...** To compensate for wall framing, the net coverage per bag should be increased. To calculate for coventional wood stud framing, when framing is 16" o.c. multiply coverage value shown by 1.14. When framing is 24" o.c. multiply coverage value shown by 1.11.*

Figure 14-16. These R-values and coverage charts are for a blown-in type of insulation. (Ark-Seal, Inc.)

Figure 14-17. Left—A sample of extruded polystyrene insulation board. This product is often used as sheathing or is installed over sheathing for its insulating value. It is offered in three thicknesses: 1/2, 3/4, and 1". (Amoco Foam Products Co.) Right—Masonry walls consisting of concrete blocks with a stone or brick facing, can be insulated by inserting polystyrene panels in the cavity as shown. The wall's R-value can be further increased by filling the cores of the concrete block with granular insulation.

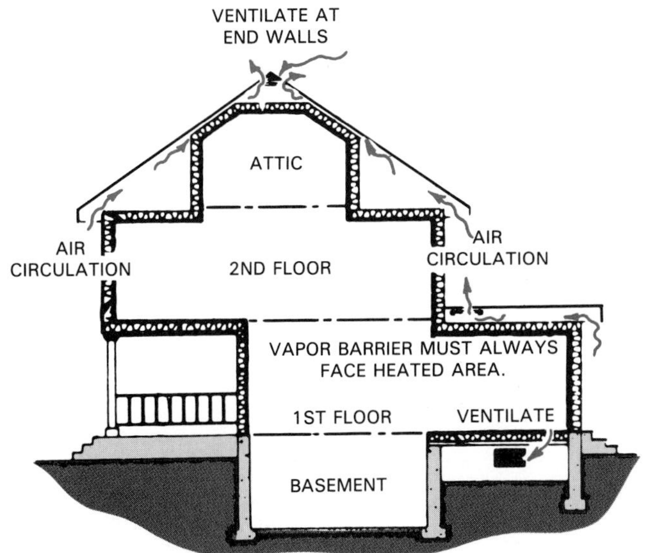

Figure 14-18. Where to insulate residential structures. In many cases, basement walls and crawl space walls will also be insulated.

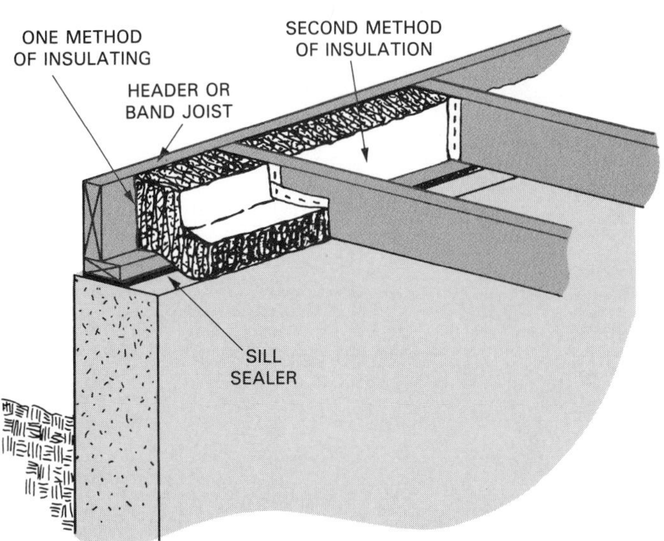

Figure 14-20. Insulate headers, or band joists atop basement walls. As you can see, there is little protection from loss of heat otherwise. (If the interior of the concrete wall is not furred out and insulated, batts may be extended down the wall.)

Figure 14-19. Insulating a basement wall with R-11 unfaced fiberglass. A framework of 2 x 4 studs has been attached to the masonry wall and will provide support for drywall or paneling. (Owens-Corning Fiberglas Corp.)

climate zone. This space, called a *crawl space*, if enclosed by foundation walls, is ventilated and will, therefore, approach the outside temperature.

Figure 14-21 shows crawl space insulation. A vapor barrier should be placed either on top of the insulation, as shown, or between the rough and finish floor.

Moisture coming up through the ground can be controlled by covering it with 6 mil polyethylene plastic film or roll roofing weighing at least 55 lb. per square. The material should be laid over the surface of the soil with edges overlapping at least 4". If the ground is rough, it is advisable to put down a layer of sand or fine gravel before putting down the vapor barrier. Covers of this kind greatly restrict the evaporation of water and less ventilation is needed.

The soil surface beneath the building should be above the outside grade if there is a chance that water might get inside the foundation wall. The soil cover is especially valuable where the water table is continually near the surface, or the soil has high capillarity (absorbs water easily). Be sure the covering is carried well up along the foundation wall.

A crawl space may be unvented and heated along with other parts of the structure. Insulation is installed along the inside of the foundation wall and floor frame. Figure 14-22 shows the general installation of fiber blankets. Extruded polystyrene can provide high R-values for this type of installation. Sections should be carefully cut and fitted. Use a compatible mastic adhesive to hold insulation in place.

Closed crawl spaces are often used as a *plenum* (large duct) to distribute warm (or cooled) air throughout the structure.

Many homes, as well as other structures, are built on concrete slab floors. Such floors should contain insulation and a vapor barrier to prevent heat loss along the perimeter. Very little heat is lost into the ground under the central

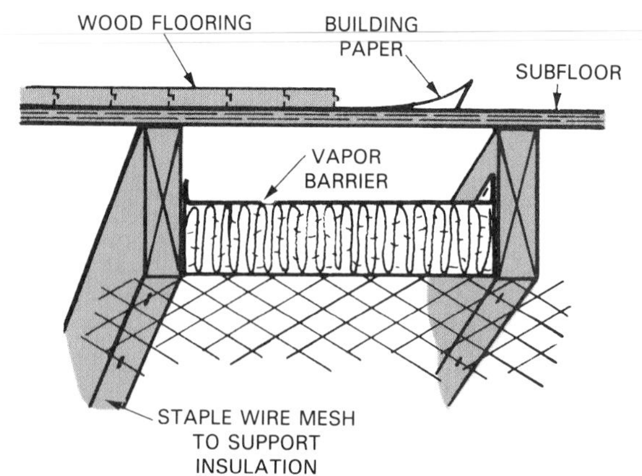

Figure 14-21. Top—Basic construction and insulation requirements over a crawl space. R-19 is required by the FHA Minimum Property Standards. Insulation can be held in place with bowed wire, chicken wire, or even fishing line. Bottom—Wear gloves and protective clothing when working with insulation materials. (Manville Building Products, Inc.)

part of the floor. Only the perimeter needs to be insulated. The vapor barrier, however, must be continuous under the entire floor. The insulation can be installed horizontally (about 2') under the floor or vertically along the foundation walls, as shown in Figure 14-23.

Insulating Existing Foundations

Insulating walls and ceilings of existing structures is fairly simple. However, insulating floors or foundations of basementless structures can be a difficult task. Crawl spaces may not be easy to enter or may not provide room enough to work. Attaching a rigid insulation board to the outside surface of the foundation is usually the best solution.

Figure 14-24 shows extruded polystyrene applied to the exterior wall of a crawl space. Dig a trench along the wall so that the insulation can be installed to a depth of at least 2' below the finished grade. Clean the wall surface and repair cracks.

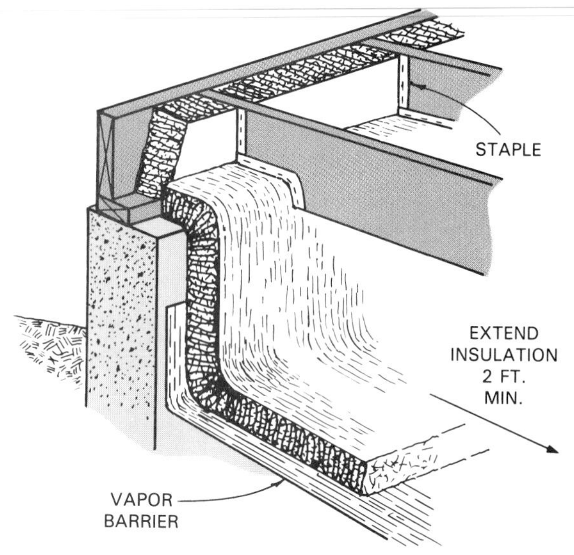

Figure 14-22. Proper insulation of crawl space walls and ground. Staple fiber blankets to wood members. Beads of mastic adhesive can be used to secure blankets to masonry wall.

The insulation board and protective cover will extend outward beyond the wall siding. To waterproof this joint, install a metal flashing as shown in the drawing. Extend the flashing at least 1" upward behind the siding.

Cut the polystyrene panels to size and then attach them to the wall with beads (3/8" diameter) of mastic adhesive. Follow the manufacturer's directions. Only a few minutes are required for the mastic to make an initial set.

Polystyrene panels must be protected against ultraviolet light, wear, and impact forces. An approved method is to cover the surface with special plaster made from cement, lime, glass fibers, and a water resistant agent. This mixture is troweled onto the surface 1/8" to 1/4" thick. It should be extended downward to a level of about 4" below finished grade. The buried portion of the polystyrene panel does not require protection.

Cementitious panels, exterior plywood, or other weatherproof panels can also be used to protect the insulation. These panels can usually be attached with an approved type of mastic adhesive. Their use is practical when only a small area of the insulation board projects above the finished grade.

Polystyrene insulation is available in panels that have a weatherproof coating of fiber-reinforced cement on one side. They are especially designed to insulate the outside surface of foundations. These panels must be cut with a circular saw equipped with a masonry blade.

Existing concrete slab floors that require additional insulation can be handled in about the same manner. Insulation panels should extend downward at least 2' or to the top of the footing.

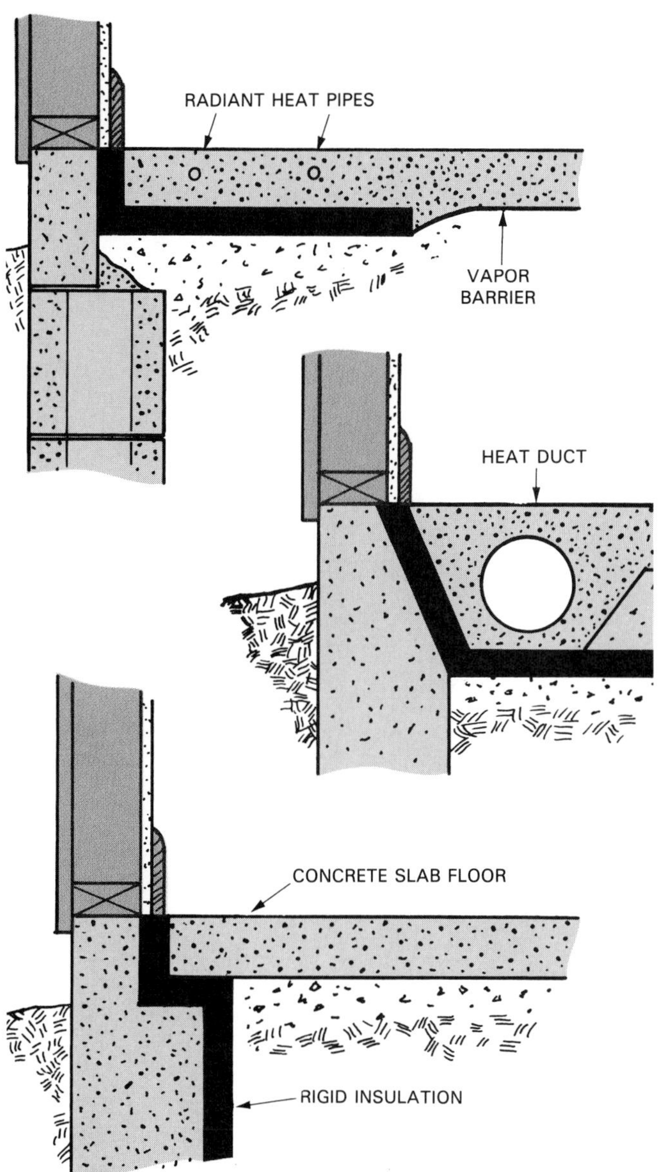

Figure 14-23. Slab floors need insulation along their perimeters. Vapor barriers should be continuous under the entire floor.

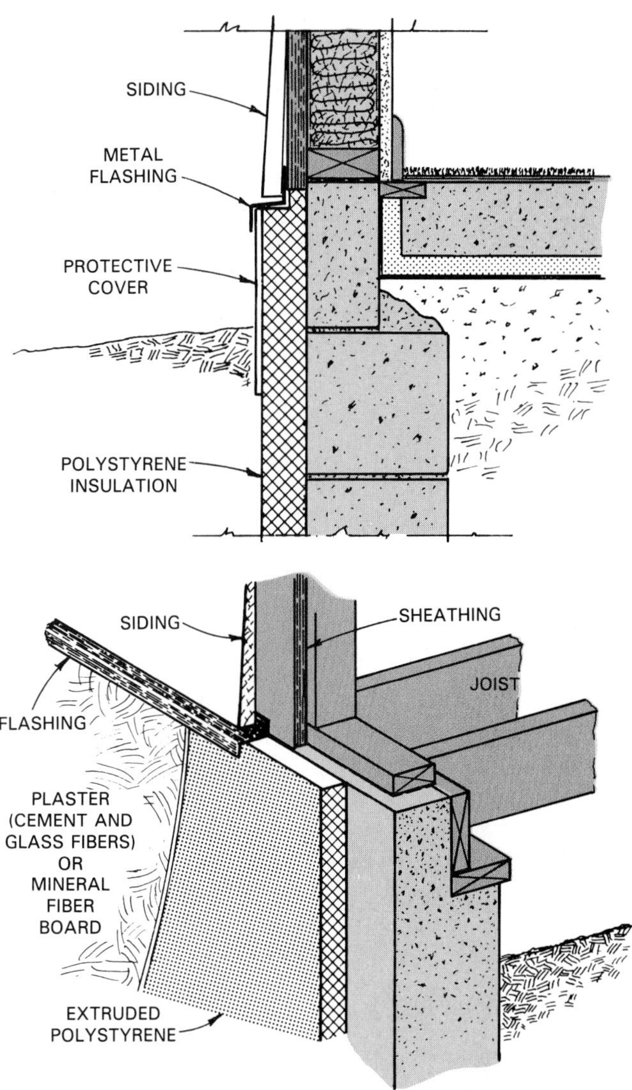

Figure 14-24. Extruded polystyrene can be installed to the outside surface to provide insulation for all types of existing foundations. Top—Slab foundation. Bottom—Crawl space or basement foundation.

Condensation

Water vapor is always present in the air. It acts like a gas and penetrates wood, stone, concrete, and most other building materials. Warm, moisture-laden air within a heated building forms a vapor pressure which constantly seeks to escape and mix with the colder, drier outside air.

Water vapor comes from many sources within a living space. It is generated by cooking, bathing, clothes washing and drying, and by humidifiers.

When warm air is cooled, some of its moisture will be released as condensation. The temperature at which this occurs is called the *dew point.* If you live in a cold climate—any region where the January temperature is 25°F (-4°C) or colder—the dew point can occur within the wall structure or even within the insulation itself. The resulting condensation will reduce the efficiency of the insulation and may eventually damage structural members.

Moisture that collects within a wall during the winter months usually finds its way to the exterior finish in the spring and summer. It causes deterioration of siding and/or paint peeling. The siding is usually a porous material and will allow a considerable amount of moisture to pass. The paint is nonporous and the moisture gathers under the paint film, causing blisters and separation from the wood surface. Recent developments in paint manufacturing have provided products that are somewhat porous. They permit moisture to pass through.

During warm weather, condensation may occur in basement areas or on concrete slab floors in contact with the ground. When warm, humid air comes in contact with cool masonry walls and floors, some of the moisture will condense, causing wet surfaces. Covering these surfaces with insulation will reduce or stop condensation. Operating a dehumidifier in areas surrounded by cool surfaces will also help.

The "dew point" is the temperature at which the air is completely saturated with moisture. Any lowering of the air temperature will cause condensation to occur.

Vapor Barriers

A *vapor barrier* is a membrane through which water vapor cannot readily pass. When properly installed, it will protect ceilings, walls, and floors from moisture originating within a heated space. See Figure 14-25. If you could check the temperature inside an insulated wall, you would find that it is warm on the room side and cool on the outside. The vapor barrier must be located on the warm side to prevent moisture from moving through the insulation to the cool side where it could condense.

Many insulation materials have a vapor barrier already applied to the inside surface. Also, many interior wall surface materials are backed with vapor barriers. When these materials are properly applied, they usually provide satisfactory resistance to moisture penetration.

If the insulating materials do not include a satisfactory vapor barrier, then one should be installed. Vapor barriers in wide continuous rolls include:

- Asphalt-coated paper.
- Aluminum foil, Figure 14-26.
- Polyethylene films.

To prevent accidental puncturing, vapor barriers should be installed after heat ducts, plumbing, and electrical wiring are in place. Cut and carefully fit the barrier around openings such as outlet boxes.

Some architects specify the use of 4 or 6 mil polyethylene film, Figure 14-27. It is applied just before the plaster base or drywall and forms a continuous cover over walls, ceilings, and windows. The covering protects the window unit during plastering operations and permits light to enter. It can easily be trimmed out of the opening just before the finished wood trim is installed.

Ventilation

Proper placement of vapor barriers alone will not protect the structure against moisture. Steps must be taken to provide good attic ventilation. The cold side (outside) of walls should be weathertight but still permit the wall to "breathe." Building paper or housewrap can be used over wood sheathing to reduce infiltration while still permitting moisture in the wall to escape.

It is especially important to ventilate an unheated attic or space directly under a low pitched or flat roof. Figure

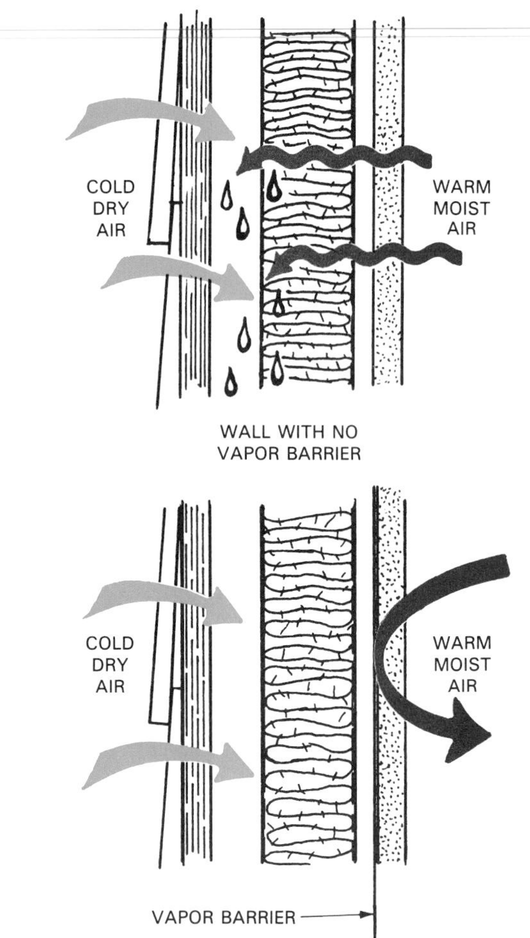

Figure 14-25. Vapor barrier installation. Top—A vapor barrier should always be installed on the warm side of an insulated wall. Bottom—Installing a special infiltration barrier on the outside surface before siding is applied. Its high-tech design prevents air infiltration while still allowing moisture to escape. (Simplex Product Division)

Figure 14-26. Aluminum foil is a good vapor barrier when joints are carefully lapped.

Figure 14-27. Installing a 6 mil polyethylene film over unfaced insulation. Staple it securely to top and bottom plates and around window and door openings.

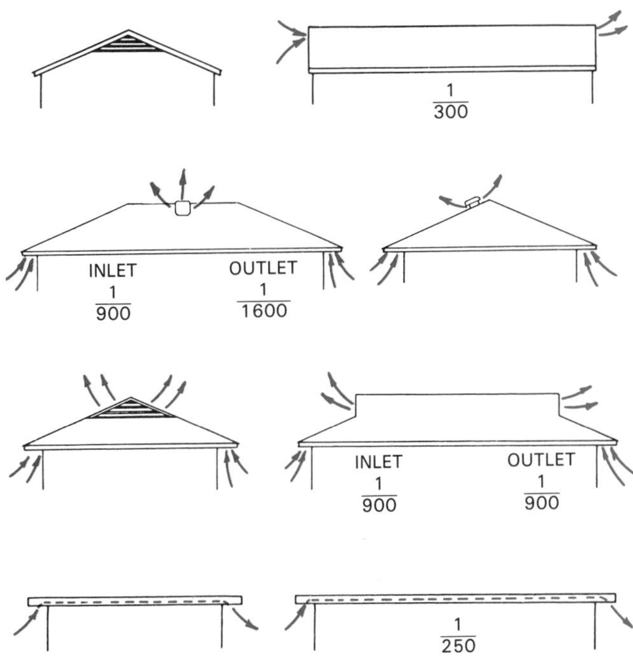

Figure 14-28. Each type of roof has ventilation requirements. Numbers indicate ratio of vent area to total ceiling area.

14-28 shows the most common systems for ventilation and the recommended size of the vent area.

Insufficient insulation or ventilation directly under low pitched roofs used in modern construction may cause a special problem, Figure 14-29. In winter, heat escaping from rooms below may cause snow on the roof to melt and water to run down the roof. At the overhang, the water may freeze again, causing a ledge or dam of ice to build up. Water may back under the shingles and leak into the building.

In using thicker insulation near the cornice, use care not to block the airway from the soffit vent into the attic. Figure 14-30 shows a method of assuring adequate ventilation.

Gable roofs usually have ventilators in the gable ends. Figure 14-31 shows two models that are manufactured in many sizes.

Ventilators located on the roof may leak if not properly installed. Whenever possible, these ventilators should be installed on a section of the roof that slopes to the rear. Sometimes, it may be possible to utilize a false flue or a section of a chimney for attic ventilation, Figure 14-32.

Walls of existing frame buildings that have inadequate "cold side" ventilation can be corrected by installing ventilators as shown in Figure 14-33. Most units are simply a metal tube with a cover that has tiny louvers. To install, simply press them into a hole bored through the siding and sheathing. For maximum ventilation, install one at the bottom and top of every stud space.

Safety with Insulation

The Occupational Safety and Health Act (OSHA) does not have specific recommendations for working with fiberglass. Installing insulation is not particularly hazardous. However, it has proposed a *permissible exposure limit (PEL)*. The American Conference of

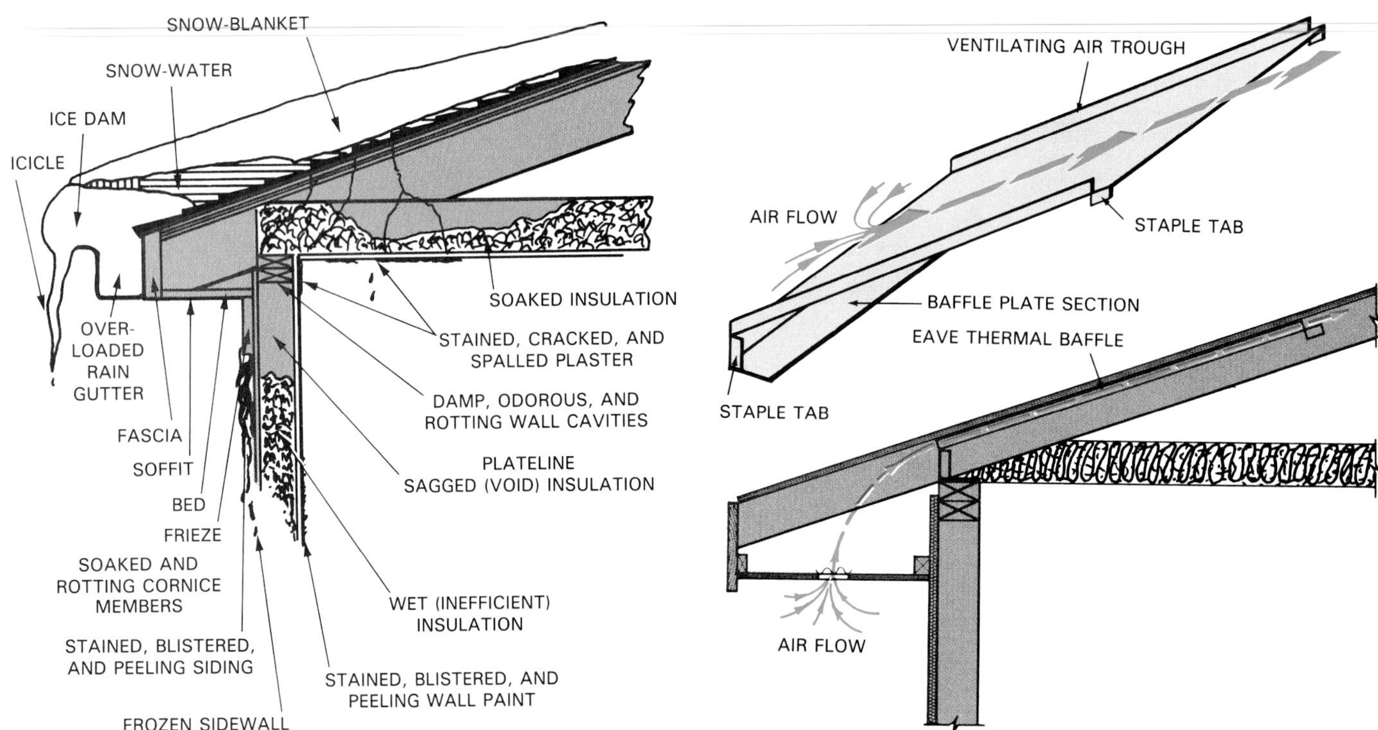

Figure 14-30. Maintaining airway under eaves with special baffle which is attached to rafters. It prevents loose or blanket insulation from shutting off airflow. (Pease Co., Builders Div.)

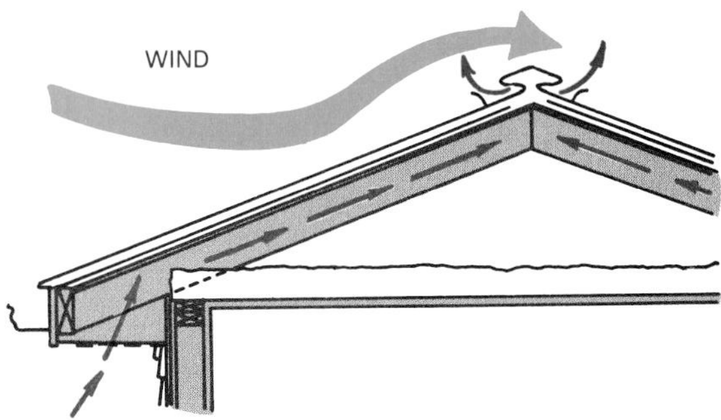

Figure 14-29. How ice dams form on roof with too little insulation and not enough attic ventilation. Top—Escaping heat melts snow, causing runoff. Water freezes in overloaded gutter, damming up water. Bottom—Insulation over the outside wall and good ventilation should solve the problem. (Agricultural Extension Service, University of Minnesota)

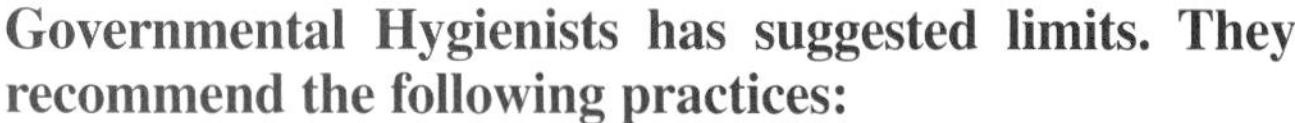

Governmental Hygienists has suggested limits. They recommend the following practices:

- **Wear loose clothing. Long-sleeved shirts or blouses loose at the neck and wrists, caps, and long trousers will prevent most fibers from coming into contact with the skin. Loose clothing will prevent chafing where fibers do contact skin. Gloves may be recommended in some circumstances.**
- **Protect eyes. Use goggles or safety glasses with side shields when applying fiberglass materials overhead or where loose particles or fibers may get into the eyes.**

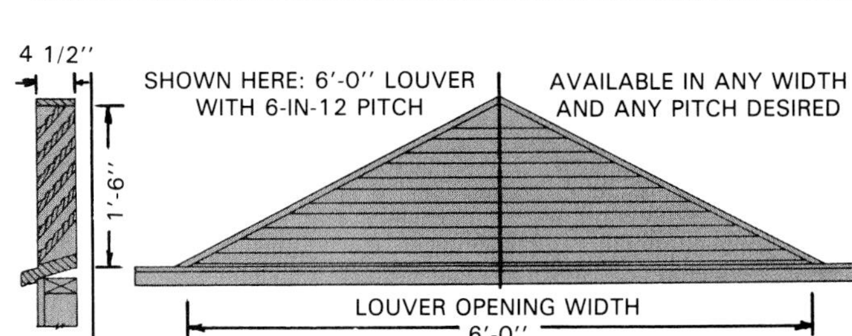

Figure 14-31. Gable-end ventilation. Top—Prefabricated metal gable-end vent. Bottom—Vents can also be constructed from wood, and are available in a variety of shapes. (Ideal Co.)

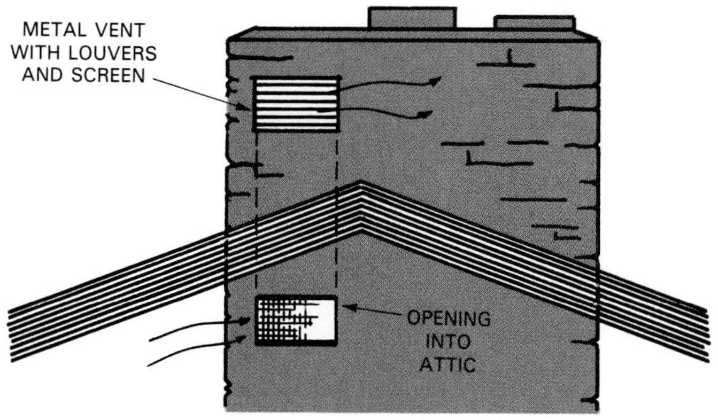

Figure 14-32. A section of a chimney can sometimes be used for attic ventilation.

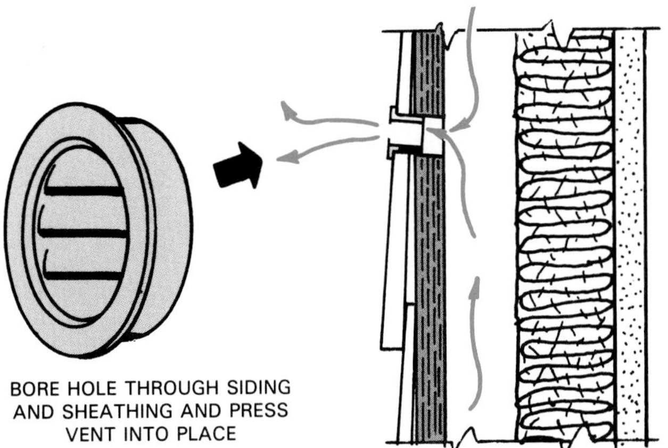

Figure 14-33. Small vents can be installed in siding of existing structures. Sizes range from 1 to 4". A 1" size is usually big enough.

- **Wear a mask covering the nose and mouth. This will prevent or reduce the inhaling of airborne fibers.**
- **Don't rub or scratch the skin. Instead, wash the skin thoroughly but gently with warm water and soap. Barrier creme, applied before working with fiberglass, will minimize the effect of skin contact.**
- **Wash work clothes separately. This practice will remove all possibility of fiber being transferred to other clothing. Rinse the washing machine thoroughly before reuse.**
- **Dispose of scrap materials. Fiberglass scraps allowed to accumulate remain troublesome. Use a vacuum or wet sweeping to pick up dust.**

Installing Batts and Blankets

Insulation materials must be properly installed to perform efficiently. Even the best insulation will not provide its rated resistance to heat flow if the manufacturer's instructions are not followed or if materials are damaged.

Blankets or batts can be cut with a shears or a large knife, Figure 14-34. Measure the space and then cut the insulation 2 to 3" longer. On kraft-faced batts, remove a portion of the insulation from each end so that you will have a flange of the backing or vapor barrier to staple to the framing.

When working with blanket insulation, it is usually best to mark the required length on the floor, unroll the blanket, align it with the marks, and cut the pieces. For wall installation, first staple the top end to the plate and then staple down along the studs, aligning the blanket carefully. Finally secure the bottom edge to the sole plate, Figure 14-35.

Figure 14-34. Fiberglass insulation is being cut with a utility knife. To save time, lay out and cut several at a time.

To install batts in a wall section, place the unit at the bottom of the stud space and press it into place. Start the second batt at the top with it tight against the plate. Sections can be joined at the midpoint by butting them together. The vapor barrier should be overlapped at least 1" unless a separate one is installed. Some batts are designed without covers or flanges and are held in place by friction.

Flanges, common to most blankets or batts, are stapled to the face or side of the framing members. Pull the flange smooth and space the staples no more than 12" apart. The interior finish, when applied, will serve to further seal the flange in place when it is fastened to the stud face. Some blankets have special folded flanges which enable them to be fastened to the face of the framing and also form an air space as shown in Figure 14-36(C).

In drywall construction, specifications may require that the faces of the studs be left uncovered. The flanges of the insulation should fit smoothly along the sides of the framing and the staples should be spaced 6" or closer. Be sure there are no gaps or *fish mouths* (wrinkles). To secure

Figure 14-35. Installing blanket insulation. Top—Using a framing square and utility knife to cut the required length. Bottom—Staple to studs, working from the top to bottom. (Owens-Corning Fiberglas Corp.)

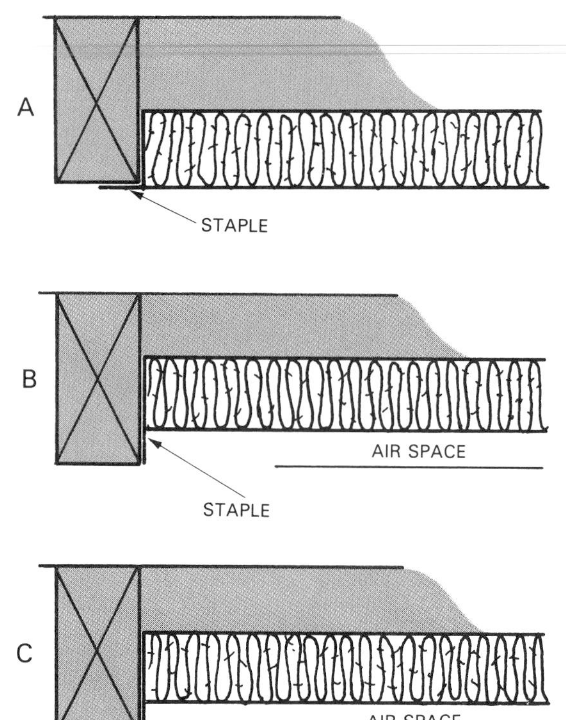

Figure 14-36. Three methods of installing blankets and batts. A—Flush with inside surface of stud. B—With flange stapled along side of stud to form air space. C—With special flange providing air space.

maximum vapor protection, a separate vapor barrier should be applied over the entire wall or ceiling area. Avoid any perforations. Lap joints fully.

New construction will usually require that plumbing be installed in interior walls. On older dwellings you must thread the insulation carefully behind pipes that might be located in the wall, Figure 14-37. If water supply pipes (in cold climates they should never be located in outside walls) are present, it is well to add a separate vapor barrier between the pipe and the interior surface to prevent condensation on the cold pipe.

Ceiling insulation can be installed from below or from above if attic space is accessible. When batts are used they are usually installed from below as shown in Figure 14-38, following the same general procedure recommended for walls. Butt pieces snugly together at their ends and carry insulation right over the outside wall, as shown in Figure 14-39. In cold climates extra thicknesses are recommended, up to an R-49. When constructing "stick-built" rafters, it may be necessary to place a 2 x 4 on top of

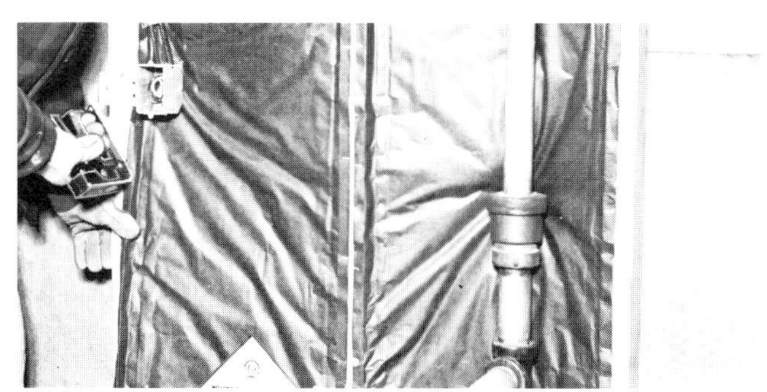

Figure 14-37. Top—On older construction, fit insulation carefully around plumbing located in outside walls. (Acoustical and Board Products Assoc.) Bottom—On new or old construction, fit insulation around electrical boxes carefully to eliminate insulation "voids."

Figure 14-38. Installing ceiling insulation. Top—Hammer tacker is used to install R-38 batt insulation. Vapor barrier has been applied to wall. Mask is recommended to avoid breathing in fibers. Bottom—Vapor barrier is being added to ceiling to prevent condensation in insulation. (Owens-Corning Fiberglas Corp.)

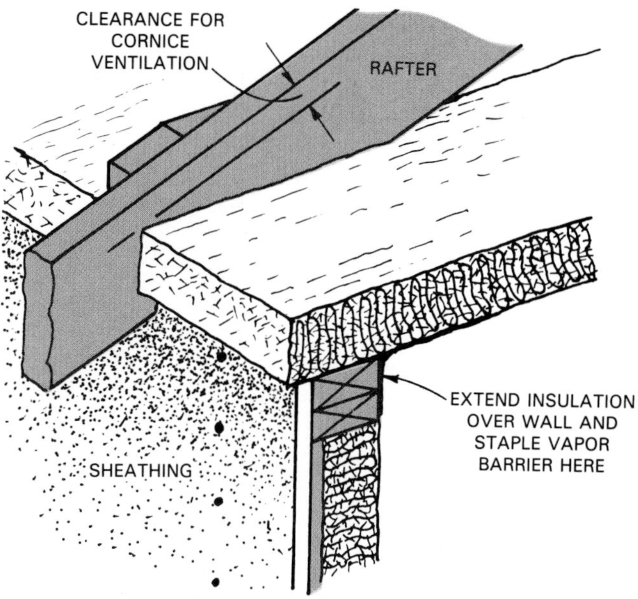

Figure 14-39. Ceiling insulation should extend over top of the wall plate to avoid ice dams on the roof. Be sure to leave an airway between cornice and attic.

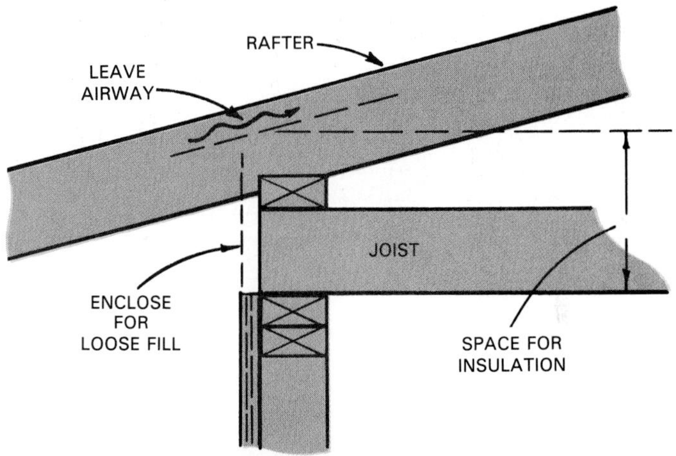

Figure 14-40. Top—One method of framing low-pitched roof to get extra space needed for thicker ceiling insulation. Rafter rests on 2 x 4 added over top of ceiling joists. Bottom—Alternate truss rafter design for insulation/airway clearance.

the ceiling joists; fasten the rafters to it, rather than to the wall plate. See Figure 14-40. A raised roof truss design also provides extra space for insulation. This design is shown in the **Technical Information Section.** Illustrations of both designs can be found in Unit 10, Roof Framing.

In multi-story construction, the floor frame should be insulated as shown in Figure 14-41. Insulation should also be installed in the perimeter of the first floor even though the basement will be heated. Cut and fit pieces so they will fit snugly between the joists and against the header.

Complete large wall and ceiling areas first. Then insulate the odd-sized spaces and areas above and below windows. Small cuttings remaining from the main areas can be used. Take the time to carefully apply the insulation and

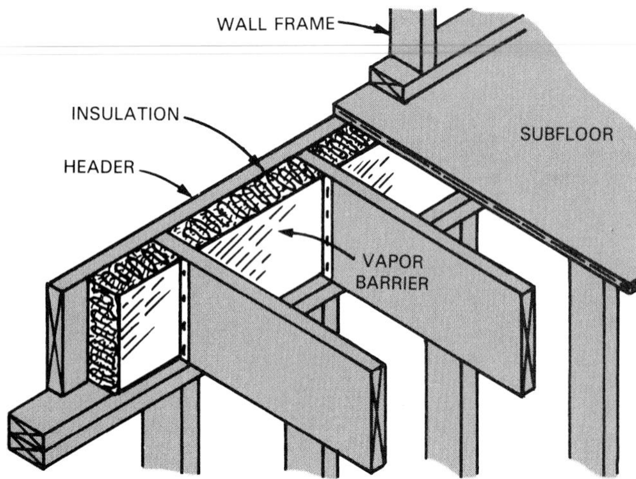

Figure 14-41. Always insulate the perimeter of the floor frame as shown for every floor.

vapor barrier around electrical outlets and other wall openings. Be careful that you do not cover outlet boxes or they may be missed when the wall surface is applied.

For a thorough insulation job, all spaces must be filled. Be sure to include the spaces between window and door frames, and the rough framing, as shown in Figure 14-42. Use cuttings left over from larger spaces. Push the insulation into the small space with a stick or screwdriver. Cover the area with a vapor barrier, Figure 14-43.

Insulate floor projections that carry a chimney chase, bay window unit, or which extend a room over an outside wall. See Figure 14-44. To seal against air infiltration, the sheathing must be tight and the insulation flange and/or vapor barrier must be carefully stapled to the sides of the joist as shown. If weather conditions permit, this segment of insulation could be installed before the subfloor is laid. This would simplify installation since the work could be done from above.

Figure 14-42. Insulating around window and door frames. Carefully fill the cavities to stop heat loss. Use a stick or screwdriver to help place the insulation but use care not to compress the material so that it loses some of its insulating qualities. (Bullard-Haven Technical School)

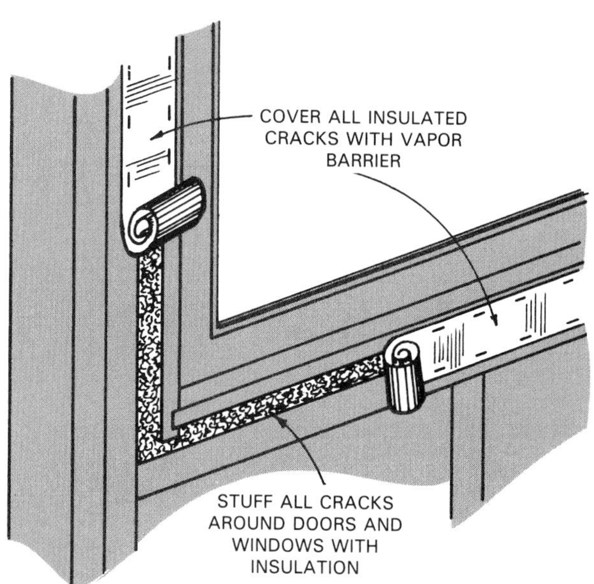

Figure 14-43. Cover insulated cracks with vapor barrier to seal against air infiltration.

Installing Loose Fill Insulation

Loose insulation is placed by pouring or blowing. It is especially adaptable to existing structures because it avoids removing or cutting into wall surfaces.

Ceilings are easily insulated with loose insulation. It can be poured directly from bags into the joist spaces. Leveling is made simple with a straightedge, as shown in Figure 14-45. A vapor barrier should be installed on the underside of the joists before the ceiling finish is applied. It will control the flow of moisture and also prevent fine particles (present in some forms of fill insulation) from sifting through cracks that might develop in the ceiling.

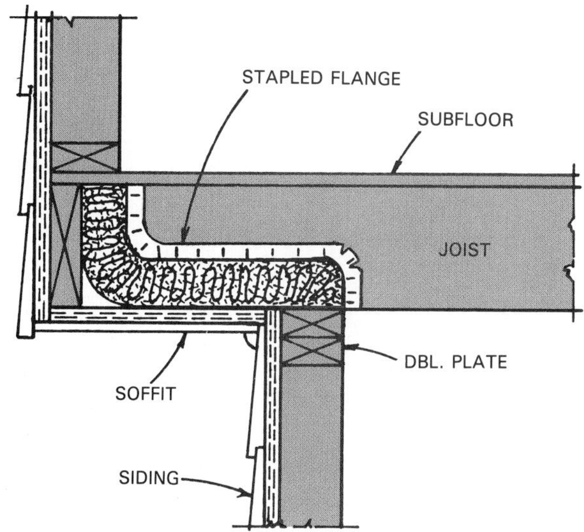

Figure 14-44. Insulate to R-19 floors which project over an outside wall.

Figure 14-46. Blowing loose fill insulation into an attic area. Wear a cap, face mask, goggles, and gloves. A 14" depth will provide an R-30 factor. Colder climates may need an R-49 factor (Owens-Corning Fiberglas Corp.)

Figure 14-45. It is easy to level off poured insulation to a uniform depth with this setup. Strikeoff boards are installed permanently. (Vermiculite Institute)

Ceiling or wall insulation can be blown into place in either new or remodeled structures. Mineral fibers made from rock, slag, or glass are widely used, Figure 14-46. Blown-in fill insulation is also made from cellulose fibers.

One blown-in insulation system mixes a thin coating of binder adhesive into fiberglass fibers, and then sprays the fiberglass into cavities behind a fiber netting. The adhesive eliminates the problem of settling that accounts for voids in the insulating blanket.

The blown-in-blanket material is mixed on the site. It uses a fine-mesh nylon netting, Figure 14-47, that is glued or stapled to the building studs. The netting restrains the bonded fibers injected into each wall cavity, Figure 14-48. A properly filled cavity will have a slight bulge. Using a roller, bring it flush with the studs. Small voids around windows and doors can be filled by hand or sealed with a foamed insulation.

Two methods are suggested for filling the cavities:

- *Two-three hole method* — Insert the tip of the nozzle through the netting 2-3' from the bottom. Fill the cavity to within a foot above the point of insertion. Then, pointing the nozzle tip upward, continue filling until the cavity is about half full. Now, reinsert the nozzle tip 2' from the top. With the tip aimed downward, fill the lower portion; then tip upward to complete the fill. If necessary, insert the tip at remaining voids and underfilled areas to fill the entire cavity.
- *One hole method* — (Use only for cavities 16" O.C., 3 1/2" thick by 8' high.) Attach a 5-8' length of flexible hose to the end of the nozzle. Insert the hose into the center of the cavity and push it until it is about 2' from the bottom. As the cavity fills, pull the hose out. When the cavity is filled half way, reinsert the hose upward to within about 2' of the top, slowly removing the hose as the remaining cavity fills.

Blown-in-blanket can be used to fill cavities in concrete block walls. A 5' length of 2 1/2" PVC pipe is attached to flexible hose. The pipe is dropped into each cavity and slowly removed as the cavity fills. For retrofit of existing walls, 1 1/2" holes are drilled at intervals for access with the nozzle tip.

Figure 14-47. Fine-mesh nylon netting has been attached to studs prior to insulating with adhesive-coated loose insulation. Drywall will be installed to the other side before coated insulation is blown in.

Figure 14-48. A blown-in-blanket insulation operation. Red hose delivers thinned adhesive to insulating fibers. Netting (not visible) holds insulation in cavity. Slight bulge will be flattened flush with studs, using a roller. (Arka-Seal, Inc. International)

The system can also be used to retrofit old buildings, working from the outside. Holes must be drilled in sheathing after sections of siding are removed. Care must be used not to damage the siding, since it must be re-installed later.

Foamed-in-place insulation materials are also on the market. These materials can be installed in open or closed cavities. They expand 100 times after application.

When using loose insulation in an attic, contain the fill around the perimeter of a ceiling area with insulation batts. Install 2' lengths of thick batts next to outside walls. Be sure to provide for air circulation from cornice area.

The National Electric Code requires a 3" space around heat-producing devices. This includes recessed light fixtures and exhaust fans.

The installed R-value for fill insulation will vary depending on the method used (pouring or blowing). Manufacturers include these figures on bag or bale labels. R-values for a 20 lb. bag of mineral wool (pouring) are listed in Figure 14-49.

Fill insulation is often poured into the core of block walls or in the cavity of masonry cavity walls. See Figure 14-50. Thermal resistance is greatly increased. For example, the R-value of a standard concrete block (1.9) is increased to 2.8 when the cores are filled with insulation. A lightweight 8" block will be increased from R-3.0 to R-5.9.

R-VALUE	MIN. THICKNESS	MAX. NET COVERAGE AREA (sq. ft.)	MIN. WT. /SQ. FT. (lbs.)
R-38	11 1/4"	8.5	2.344
R-33	9 3/4"	9.8	2.031
R-30	8 7/8"	10.8	1.849
R-26	7 3/4"	12.4	1.615
R-22	6 1/2"	14.8	1.354
R-19	5 5/8"	17.1	1.172
R-11	3 1/4"	29.5	0.677

Figure 14-49. R-values for mineral wool pouring insulation.

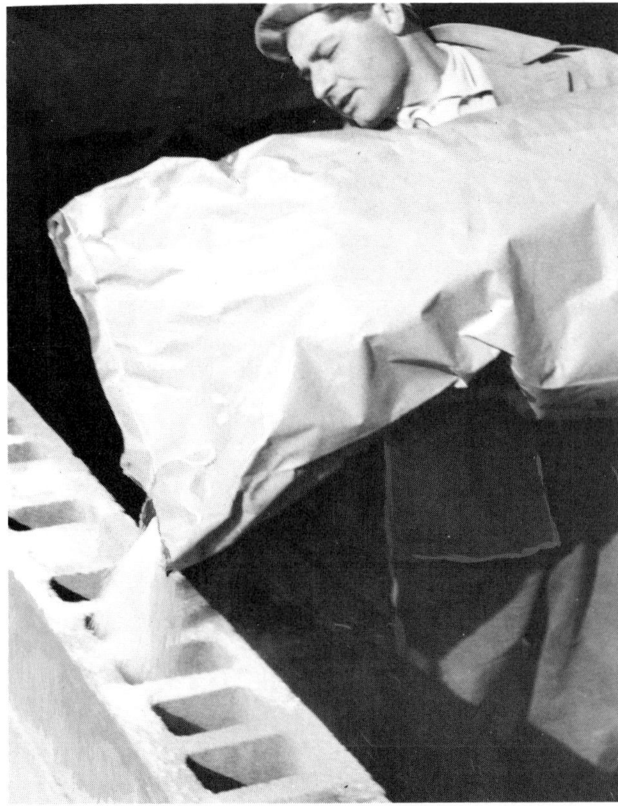

Figure 14-50. Filling cores of concrete block will raise the R-value of the block wall. (Perlite Institute, Inc.)

Installing Rigid Insulation

The installation of slab or block insulation varies with the type of product. Always study the manufacturer's specifications.

Insulating board is widely used for exterior walls. Its application is covered in Unit 9 and Unit 13. Sometimes it is used:

- As the sheathing material for roofs.
- As an insulating material installed over the roof deck.
- As exterior wall insulation applied over sheathing.
- As a base for application of "synthetic stucco" exterior covering.

A number of products are especially designed to insulate concrete slab floors. See Figure 14-51. Waterproof materials, such as glass fibers or foamed plastic, provide desirable characteristics.

Figure 14-52 shows a plastic foam insulation board applied to the interior of a masonry wall. It is bonded to the wall surface with a special mastic. It provides a permanent insulation and vapor barrier.

After the boards are installed, conventional plaster coats can be applied to the surface. Plastic foam (polystyrene) insulation is widely accepted as a rigid insulating material and has been successfully applied to a wide variety of constructions.

Figure 14-51. Perimeter of slab construction should be insulated with rigid insulation. (Owens-Corning Fiberglas Corp.)

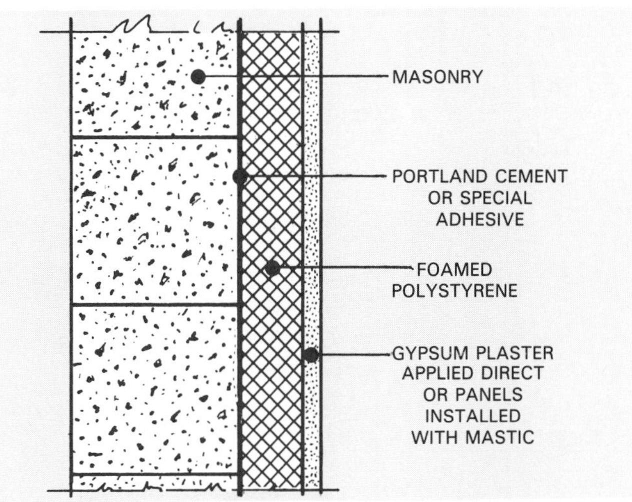

Figure 14-52. Cross section of masonry wall insulated inside with rigid foamed polystyrene.

Insulating Basement Walls

When basements will be used as living space, exterior walls should be insulated. The outside surface of concrete or masonry walls should be waterproofed below grade and should include a footing perimeter drain (See Unit 7).

The inside surface could be finished as shown in Figure 14-52. Studs or furring strips could be used to form a cavity for the insulation and provide a nailing base for surface materials.

In cold climates, a framework of 2 x 4 studs, spaced 16" O.C. is best. See Figure 14-53. Use concrete nails, screws, or mastic to secure the sole plate and fasten the top plate to the joists or sill. Unfaced insulation will require a separate vapor barrier. Figure 14-54 shows one method of insulating the floor perimeter in a basement.

Figure 14-53. In cold climates, basement walls may be given a framework of studs that provides a cavity for insulating batts. (Owens-Corning Fiberglas Corp.)

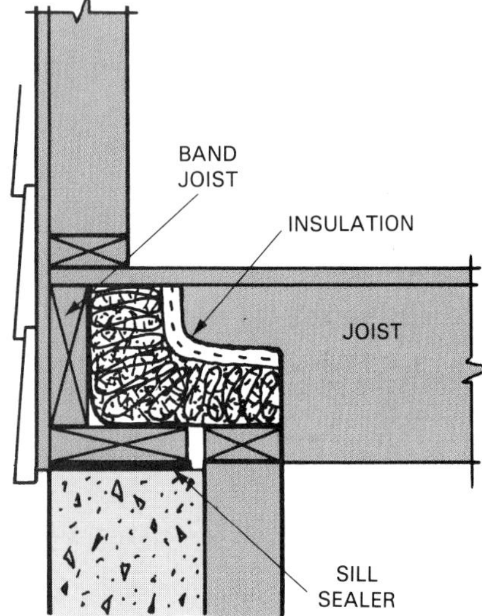

Figure 14-54. Proper method for insulating space between basement wall and first floor.

Insulating Existing Structures

Use special care when insulating an existing structure where no vapor barrier can be installed. Inside humidity should be controlled. "Cold side" ventilation is essential. A satisfactory vapor barrier can be secured by applying to room surfaces one of the following:

- Two coats of oil base paint, rubber emulsion paint, or aluminum paint.
- A vapor barrier wall paper. The application should be carefully made on all surfaces of outside walls and ceiling.

Stopping Air Infiltration

Infiltration refers to air that leaks into buildings through cracks. It occurs around windows and doors, and through other small openings in the structure. Air also leaks out of the building through these cracks. During construction, infiltration can be reduced by assembling materials properly and by sealing joints.

Caulking and sealing is usually required at the following locations:

- Joints between sill and foundation.
- Joints around door and window frames.
- Intersections of sheathing with chimney and other masonry work.
- Cracks between drip caps and siding.
- Openings between masonry work and siding.

Be sure to caulk around the electrical service entrance and around hose bibs. The preferred types of caulking compound include polysulfide, polyurethane, or silicone materials.

Inside the structure, give special attention to recessed light fixtures and any built-in units located in outside walls. Also, seal electrical conduit and plumbing that runs from the attic into walls and partitions located in the living space. Seal conduit where it enters electrical boxes and seal the boxes to the inside wall surface.

Modern windows are built in a factory. Appropriate weather stripping is applied during fabrication. Outside doors, however, are often fitted to the door frame on the job and the weather stripping is installed by the carpenter. Each type will require a different method of installation. Follow the manufacturer's instructions. Figure 14-55 shows a standard type of metal weather stripping.

Sometimes you may think that a window is leaking air when you are near it and feel a slight draft. This is usually due to the air in contact with the cold glass becoming colder and heavier than the rest of the room air. It moves downward to the floor and across the room. Heating registers and convectors are usually positioned to offset or minimize these "down-drafts."

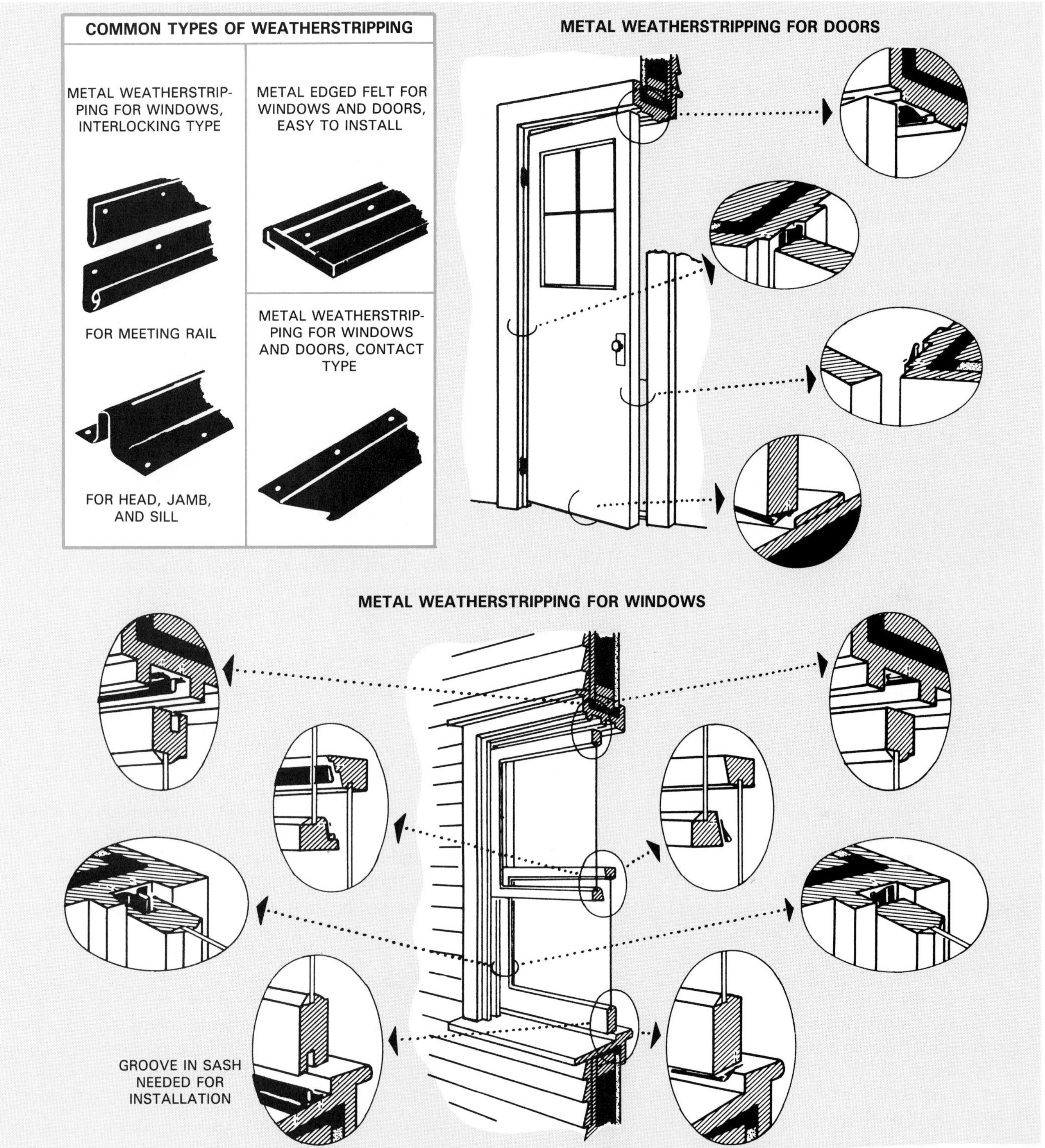

Figure 14-55. Follow manufacturer's instructions for installing door and window stripping.

Estimating Materials

The amounts of insulating materials are figured on the basis of area (square feet). The thickness is then specified as separate data. The size of packages varies considerably, depending on the type and thickness. For example: one manufacturer packages 1 1/2" blankets in rolls of 140 sq. ft. A 3" blanket in the same width comes in rolls of 70 sq. ft.

To determine the amount of insulation for exterior walls, first add up the total perimeter of the structure. Then multiply by the ceiling height. Deduct from the total the area of doors and windows. Many carpenters will deduct only large windows or window-walls and disregard doors and smaller openings. This extra allowance will:

- Make up for loss in cutting and fitting.
- Provide for additional material needed around

plumbing pipes, recessed lighting boxes, and other items.

Here is an example:

Perimeter = 30 + 40 + 36 + 20 + 6 + 20
= 152
Area = 152 x 8 = 1216
Window Wall = 12 x 8 = 96
Net Area = 1216 - 96 = 1120 sq. ft.

Batts are furnished in packages (sometimes called tubes) that contain as much as 100 sq. ft. A 6" batt usually contains 50 sq. ft.

When estimating the amount for floors and ceilings, use the same figures that were listed in calculating the subfloor area. Stairwells and openings for large fireplaces could be deducted. However, here again, these amounts may provide the extra needed for waste and special packing around fixtures.

The same procedure can be used to estimate reflective insulation. Make a greater allowance for cutting and waste, especially when using accordion type. (This type cannot be spliced effectively.) Rolls of reflective insulation hold from 250 to 500 sq. ft.

Rigid insulation is also estimated on the basis of area. It can be calculated from dimensions shown on the working drawings.

For a perimeter insulation strip in a concrete slab floor, multiply the perimeter of the building by the width of the strip. There is little waste on such an installation since even small pieces can be utilized.

Fill insulation comes in bags that usually contain 3 or 4 cu. ft. The cubic feet required can be calculated as follows:

Area = 1200 sq. ft.
Thickness = 4"
= 1/3'
Cu. ft. required = 1200 x 1/3 = 400
*Less 10% = 400 - 40
Net Amount = 360
Number of bags
(4 cu. ft.) = 90
*Allowance for joists 16" O.C.

Manufacturer's directions and specifications will usually list tables that provide a direct reading of the number of bags required for a certain thickness of application. These are especially helpful in estimating amounts needed for such items as filling the core of concrete blocks. See Figure 14-56.

APPROXIMATE COVERAGE							
NUMBER OF BAGS (4 CU. FT.) REQUIRED							
WALL AREA (SQ. FT.)	CORE FILL BLOCK SIZE			CAVITY FILL CAVITY WIDTH			
	6 IN.	8 IN.	12 IN.	1 IN.	2 IN.	2 1/2 IN.	3 IN.
100	5	7	12	2	4	5	6
500	23	33	58	10	21	26	31
1000	46	65	118	21	42	52	62
2000	91	130	236	42	84	104	125
3000	137	195	354	62	124	155	187

Figure 14-56. Table provides estimates of fill insulation needed for masonry walls. (Perlite Institute, Inc.)

Acoustics and Sound Control

Noise is unwanted sound. A by-product of our modern world, it has reached a magnitude that demands sound control in every home and building.

Noise is unpleasant. It reduces human efficiency, and can cause undue fatigue.

Houses, especially those built in cold climates, will have wall and roof structures heavy enough to repel average outside noises. Therefore, noise or sound control will apply mainly to interior partitions, floors, and surface finishes.

Sounds in an average home are generated by conversation, television, radios, stereos, computers, and musical instruments. Vacuum cleaners, washing machines, food mixers, and garbage disposals are examples of mechanical equipment that create considerable noise. Plumbing, heating, and air conditioning systems may be a source of excessive noise if poorly designed and installed. The activities in play rooms and workshops may create sounds that will be undesirable if transmitted to relaxing and sleeping areas.

The solution to problems of sound and noise control can be divided into three parts:

- Reducing the source.
- Controlling sound within a given area or room.
- Controlling sound transmission to other rooms.

The carpenter should have some understanding of how the latter two can be accomplished, and familiar with sound conditioning material and constructions. As in the case of thermal insulations, the carpenter must appreciate the importance of careful work and proper installation methods.

Acoustical Terms

- *Sound* — A vibration or wave motion that can be heard. It usually reaches the ear through air. The air itself does not move but vibrates back and forth in tiny molecular motions of high and low pressure.
- *Decibel* — The unit of measurement used to indicate the loudness or intensity of sound; comparable to the "degree" as a measurement of heat or cold. See Figure 14-57.
- *Reverberation sounds* — These are airborne sounds which continue after the actual source has ceased. They are caused by reflections from floors, walls, and ceilings.
- *Frequency* — Rate at which sound-energized air molecules vibrate; the higher the rate, the more cycles per second (cps). Examples are the low frequency of a bass drum and the high frequency of a flute.

COMMON DESCRIP.	DECI-BELS	THRESHOLD OF FEELING
Threshold of Pain	130	Space Shuttle (180+) (Lift-off) Concorde Boeing 747
	120	
Deafening	110	Thunder, Artillery Nearby Riveter Elevated Train Boiler Factory
	100	
Very Loud	90	Loud Street Noise Noisy Factory Truck Unmuffled Police Whistle
	80	
Loud	70	Noisy Office Average Street Noise Average Radio Average Factory
	60	
Moderate	50	Noisy Home Average Office Average Conversation Quiet Radio
	40	
Faint	30	Quiet Home or Private Office Average Auditorium Quiet Conversation
	20	
Very Faint	10	Rustle of Leaves Whisper Sound Proof Room Threshold of Audibility
	0	

Figure 14-57. Levels of sounds and noises. Those above 90 decibels tend to be disagreeable.

- *Impact sounds* — These are the sounds that are carried through a building by the vibrations of the structural materials themselves. Footsteps heard through the floors of a structure are an example of impact sounds.
- *Masking sounds* — These are the normal sounds within habitable rooms which tend to "mask" some of the external sounds entering the room.
- *Decibels reduction* — An expression used to indicate the sound insulating properties of a wall or floor panel.
- *Sound transmission loss (STL)* — Sound insulating efficiency of wall or floor construction is measured in decibels. Transmission loss is the number of decibels which sound loses when transmitted through a wall or floor. Transmission loss of any wall or floor depends on the materials and techniques used in construction.
- *Sound absorption* — The capacity of a material or object to reduce sound waves by absorbing them is referred to as sound absorption. Acoustical materials, such as acoustical ceiling tile, are designed to absorb sounds within a given area. These are sounds that otherwise would be reflected and cause excessive reverberation and build-up of intensity within that area.
- *Noise reduction coefficient (NRC)* — The sound absorption of acoustical materials is expressed as the average percentage absorption at the four frequencies which are representative of most household noises. These frequencies are 250, 500, 1000, and 2000 cycles per second.
- *Sound transmission class (STC)* — The STC is a single number which represents the minimum performance of a wall or floor at all frequencies. The higher the STC number, the more efficient the wall or floor will be in reducing sound transmission.

Sound Intensity

The number of decibels indicates the loudness or intensity of the sound. Roughly speaking, the decibel unit is about the smallest change in sound that is audible to the human ear. Actually, the decibel has the same relationship to a scale of loudness as the degree has to a thermometer. Reference to Figure 14-57 will show that the rustle of leaves or a low whisper is on the *threshold of audibility.* That is, the sound is barely heard by the human ear. At the top of the scale are painfully loud sounds of over 130 decibels, often referred to as the *threshold of pain.*

One sound level is ten decibels greater than another if its intensity is ten times greater than the other. There is a logarithmic relation, on the decibel scale, to the amount of sound energy involved. If the sounds differ by 20 decibels, the ratio of their intensities is 10^2 or 100 times greater; if by 30 decibels, the ratio is 10^3 or 1000 times greater, and so on up the scale.

Sound Transmission

When sound is generated within a room, the waves strike the walls, floor, and ceiling. Much of this sound energy is reflected back into the room. The rest is absorbed by the surfaces. If there are cracks or holes through the wall (no matter how minute), a part of these sound waves will travel through as airborne sounds. The sound waves striking the wall will cause it to vibrate as a diaphragm, reproducing these waves on the other side of the wall.

Sound transmission through theoretically airtight partitions is the result of such diaphragm action. The sound insulation values of such substances is therefore almost entirely a matter of their relative weight, thickness, and area. In partitions of normal dimensions, this value depends mostly upon weight.

As the sound moves through any type of wall or barrier its intensity will be reduced. This reduction is called Sound Transmission Loss (STL) and is expressed in decibels. A wall with a STL of 30 dB. will reduce the loudness level of sound passing through it from 70 dB. to 40 dB. See Figure 14-58. The transmission loss of any floor or wall will be determined by the materials, design, and quality of constructional techniques.

Although transmission loss rated in decibels is still used, another system of rating sound-blocking efficiency is

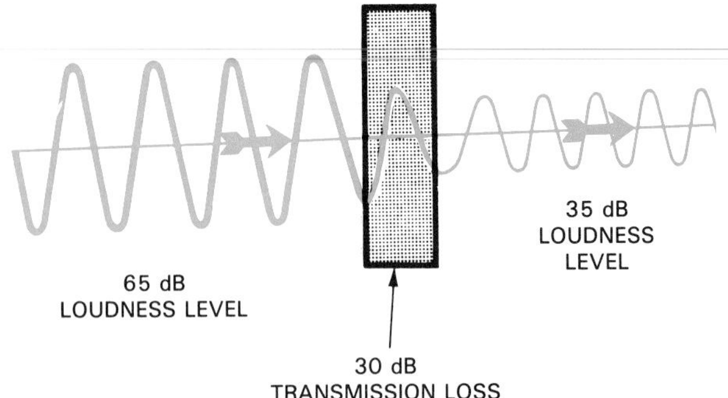

Figure 14-58. Sound transmission loss through a wall. Values will vary somewhat, depending upon the frequency of the sound waves.

widely accepted. It is called the Sound Transmission Class (STC) system. Standards have been established through extensive research by such associations as the Insulation Board Institute and the National Bureau of Standards.

STC numbers have been adopted by acoustical engineers as a measure of the resistance to sound transmission of a building element. Like the resistance in thermal insulation (R), the higher the number, the better the sound barrier. Figure 14-59 shows how a composite STC rating is applied to a given wall construction which has various STL values in decibels through a specified range of frequencies. Figure 14-60 further describes these STC ratings by making a simple application to a wall separating two apartments.

Actually there is another factor present which should be considered when designing any sound insulating panel.

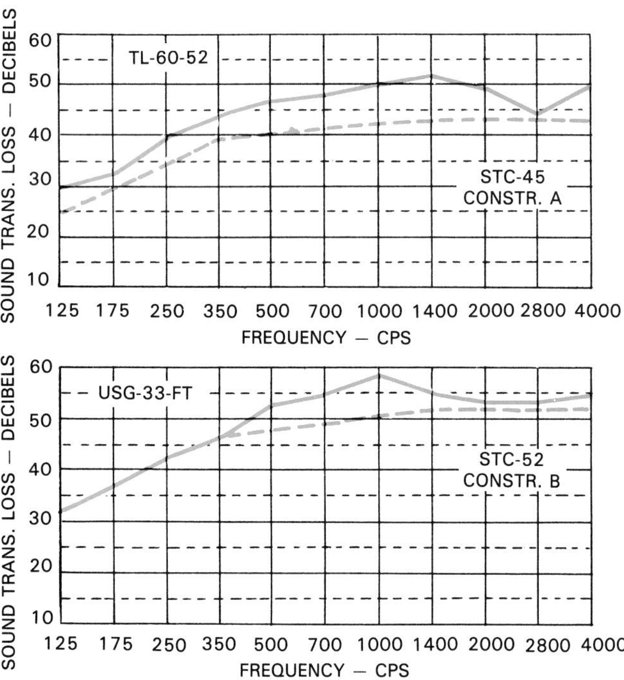

Figure 14-59. Graphs show the TL values in decibels at various frequencies for two STC rated constructions.

30 STC
LOUD SPEECH CAN BE UNDERSTOOD
FAIRLY WELL

35 STC
LOUD SPEECH AUDIBLE BUT NOT
INTELLIGIBLE

42 STC
LOUD SPEECH AUDIBLE
AS A MURMUR

50 STC
LOUD SPEECH NOT AUDIBLE

Figure 14-60. How different STC ratings apply to a partition between two apartments.

These are usually referred to as masking sounds. In theory, an inaudible sound of zero (0) on the decibel scale is for a perfectly quiet room. Since there are noises in every habitable room which tend to mask the sound entering, it is only necessary to reduce sound to the level of the more or less maintained sound level within the space to be insulated. Assume that there is a radio playing soft music in the listening room (about 30 dB.). Thus, sound of less than 30 dB. entering the room would be completely masked.

Wall Construction

How high must an STC rating be for a given wall? This will depend largely on the types of areas it separates.

For example, partitions between bedrooms in an average home usually will not require special soundproofing while those between bedrooms and activity or living rooms should have a high STC rating. Partitions surrounding a bathroom should also have a high STC number. Extra attention should be given to the placement of insulation around pipes.

Figure 14-61 shows a number of practical constructions for partitions and their STC ratings. Some of them include sound deadening board. This is a structural insulation board product designed especially for use in sound control systems. It is made principally from wood and cane fibers in a nominal 1/2" thickness. Standard sizes of sound

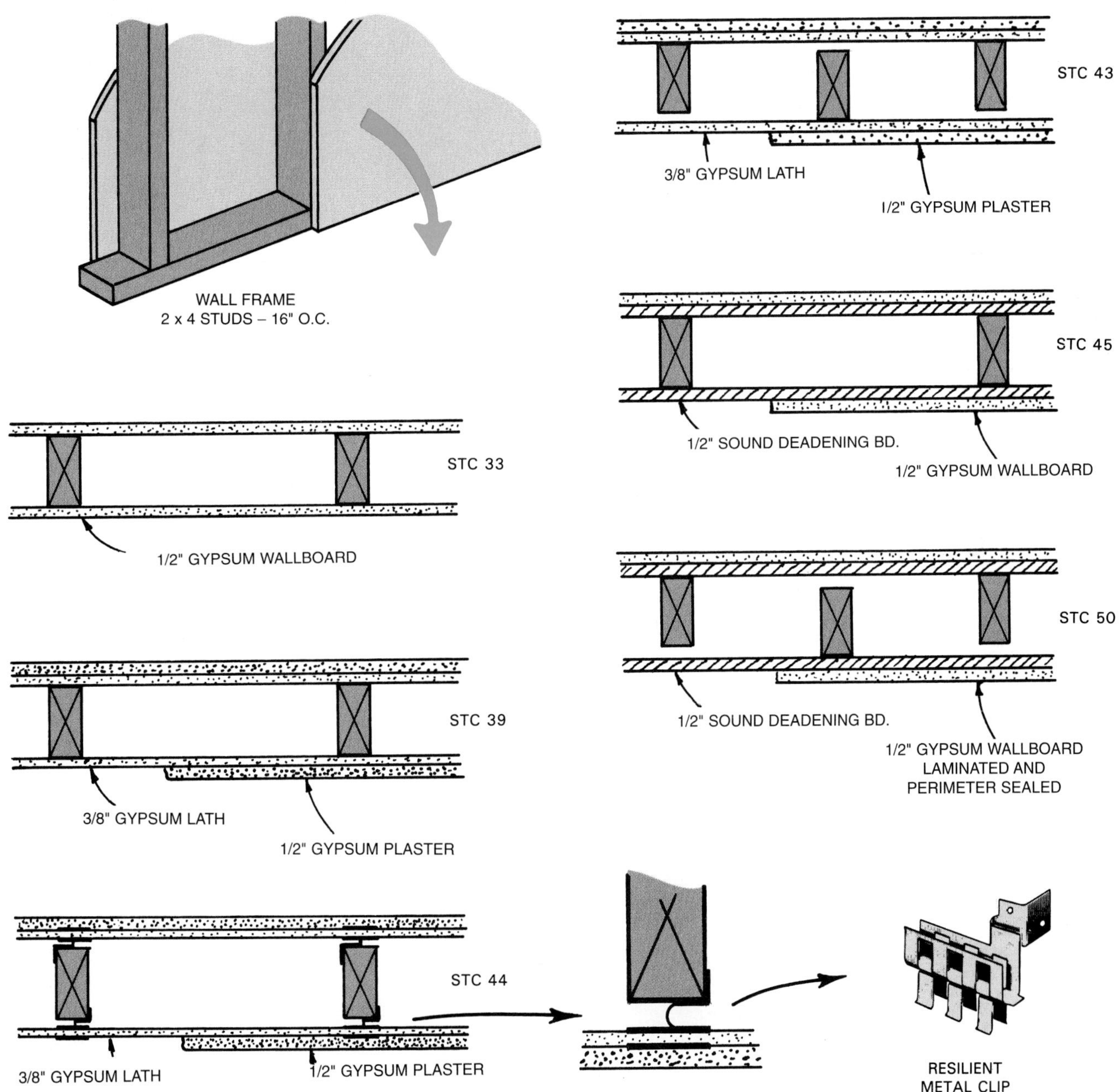

Figure 14-61. Six different methods of constructing interior walls. Sound transmission class (STC) is given for each. An insulation blanket will usually increase the STC rating.

deadening board are the same as regular insulation board materials. It is usually identified with the words: IBI—RATED SOUND DEADENING BOARD on each sheet or package to distinguish it from other insulation board products.

In order to secure the high STC ratings in a wall structure when using a sound deadening board, application details supplied by the manufacturer should be carefully followed. Where nails are used, the size, type, and application patterns are critical. In drywall construction, the joints should be taped, finished, and the entire perimeter sealed. Openings in the wall for convenience outlets and medicine cabinets require special consideration. For example, electrical outlets on opposite faces of the partition should not be located in the same stud space.

Double Walls

Partitions between apartments are often constructed to form two separate walls as shown in Figure 14-62. Standard blanket insulation is installed in about the same manner as for thermal insulation purposes. It should be stapled to only one row of the framing members.

For economical and space-saving construction, strips of special resilient channel are nailed to standard stud frames as shown in Figure 14-63. The base layer of gypsum board is attached with screws. The surface layer is bonded with an adhesive. Since laminated systems like this minimize the use of metal fasteners, they result in a finer appearance along with better sound and fire resistance.

Figure 14-62. A double wall provides a high STC rating for a partition between apartments. Insulation increases the STC. (Owens-Corning Fiberglas Corp.)

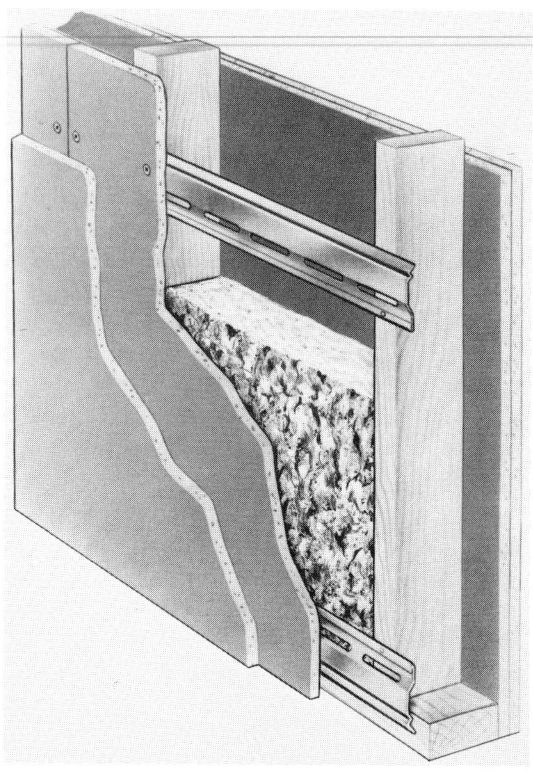

Figure 14-63. Resilient channel can be added to wood studs on one side. Stud spacing is 24" O.C. Double drywall layer with 3" insulation batt gives an STC rating of 50. (United States Gypsum Co.)

Floors and Ceilings

In general, the considerations for soundproofing that were applied to walls can also be applied to floors and ceilings. Floors are subjected to impact sounds. These are the noises you get from activities such as walking, moving furniture, operating vacuum cleaners and other vibrating equipment. Sound control is somewhat more difficult. Often an impact sound may cause more annoyance in the room below than it does in the room where it is generated. The addition of carpeting or similar material to a regular hardwood floor will be effective in minimizing impact sounds.

Sound deadening board, properly installed, will increase the STC rating. See Figure 14-64. The use of patented metal clips to attach the ceiling material is a practical solution. Various suspended ceiling systems also provide high levels of sound control.

Figure 14-65 shows the installation of a floor-ceiling system with an STC rating over 52. It consists of 2 x 10 joists placed 16" O.C. with standard wood subfloor and finished floor. The floor is covered with carpet and pad. Resilient metal channels are attached to joists with 1 1/4" screws. Nails must not be used. Gypsum panels are attached to the channels with screws. The system includes a 3" insulation blanket.

An existing floor can be soundproofed through a change shown in Figure 14-66. *Sleepers* of 2 x 3 wood are

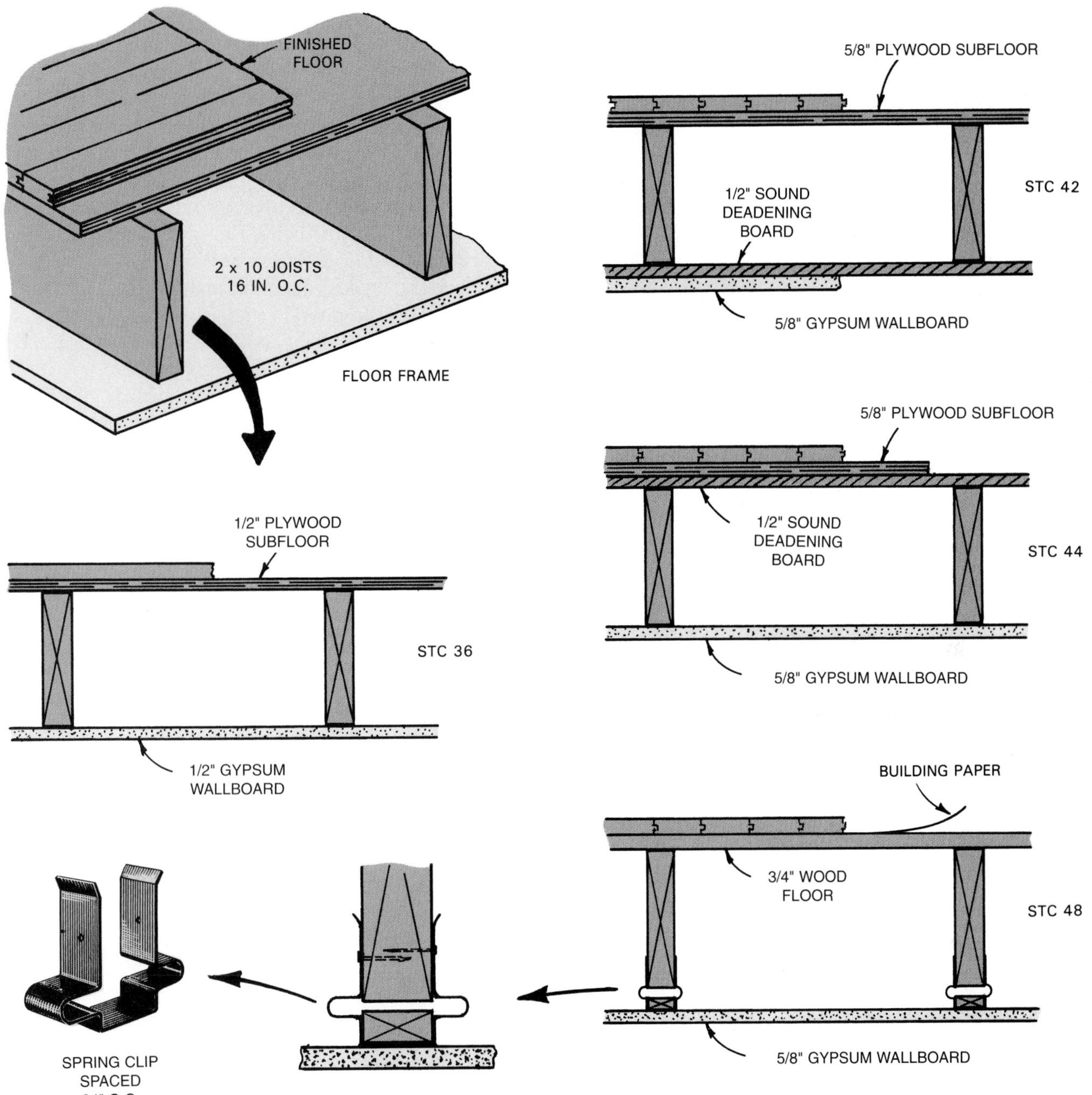

Figure 14-64. Floor and ceiling constructions. Different methods are used to get certain STC ratings.

laid over a glass wool blanket but are not nailed to the old floor. When the new floor is laid, be sure the nails do not go all the way through the sleepers. The only contact between the new and old floor is the glass wool blanket. It will compress to about 1/4" under the sleepers. The system makes the floor resilient in addition to reducing sound transmission.

Doors and Windows

Sound has a tendency to spread out after passing through an opening. Thus, cracks and holes should be avoided in every type of construction where sound insulation is important. Doors between rooms are probably the greatest transmitters of sound. A 1/4" crack around a wood door (1 3/4" thick) would admit four times as much sound of medium intensity as the door itself. Felt, rubber, or metal strips around the jambs and head are desirable. Conditions can be further improved by some form of *draft excluder* at the sill, such as a threshold or felt weatherstop.

Hollow-core interior doors that are well fitted will have a sound reduction value from 20 to 25 dB. Similar double doors, hung with at least 6" air space between, will have a sound reduction factor as high as 40 dB. This factor

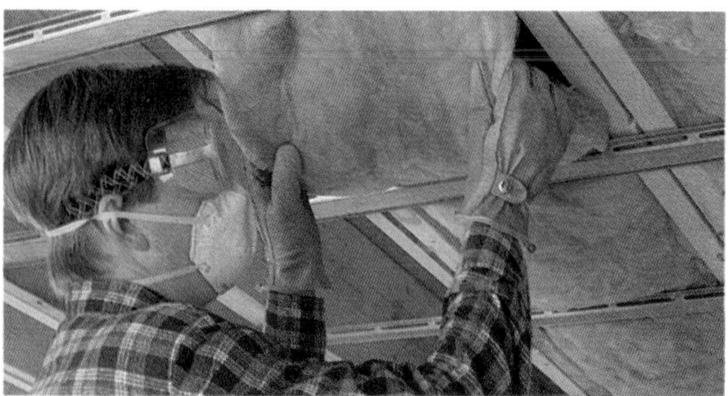

Figure 14-65. Soundproofing the ceiling above a basement room. The resilient channels will carry the ceiling panels. When the floor above is covered with a carpet and pad, an STC of 52 can be attained. (Owens-Corning Fiberglas Corp.)

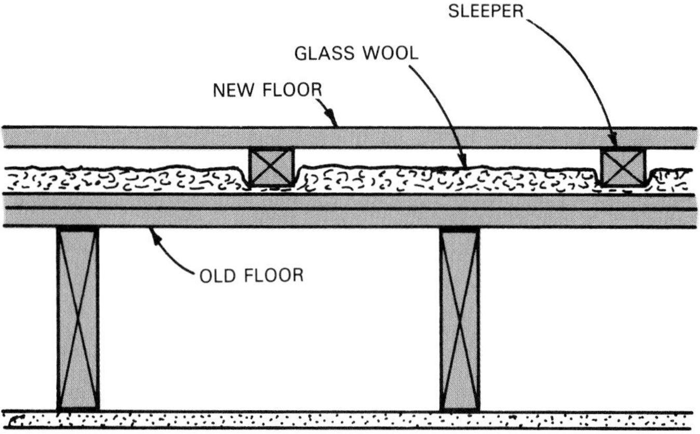

Figure 14-66. Sleepers laid over insulation will soundproof an existing floor.

can be increased by using felt or rubber strips around the stops.

For special installations, "soundproof" doors are available and can be built to suit almost any condition. Special hardware is used on this type of door to prevent sound transmission through the doorknobs.

Similar precautions should be observed with glazed openings. Cracks around these openings may cancel out other efforts to cut down sound transmissions. Windows or glazed openings should be as airtight as is practical. Double or triple glazing will greatly reduce the amount of sound passing through the opening. Glass block have a sound reduction factor of about 40 dB. They are effective where transparent glazing is not needed.

Noise Reduction within a Space

While it is important to design walls and floors that will reduce sound transmission between spaces, it is also advisable to treat the enclosure so that sound will be trapped or reduced at its source. Reducing the noise level within the room will not only cut sound transmission to other rooms, but will improve living conditions within the room. Areas in homes where noise reduction is most important include kitchens, utility rooms, family rooms, and hallways.

There are a number of different types of acoustical material available to the builder. These come in a wide range of sizes from 12" x 12" tiles to 4' x 16' boards. Those with the best acoustical properties will absorb up to 70% of the sound that strikes them. The most common types are perforated or porous fiberboard units, perforated metal pan units, cork acoustical material, and acoustical plaster. All of these materials provide high absorption qualities and, except for the acoustical plaster, have a factory-applied finish. Since the sound absorbing properties of any of these materials depends on its sponge-like quality, they are relatively light and do not require any building reinforcement or structural changes.

How Acoustical Materials Work

The sound absorbing value of most materials depends to a greater or lesser degree on a porous surface. Sound waves entering these pores, or holes, get "lost" and are said to be "dissipated" (scattered) as heat energy. See Figure 14-67.

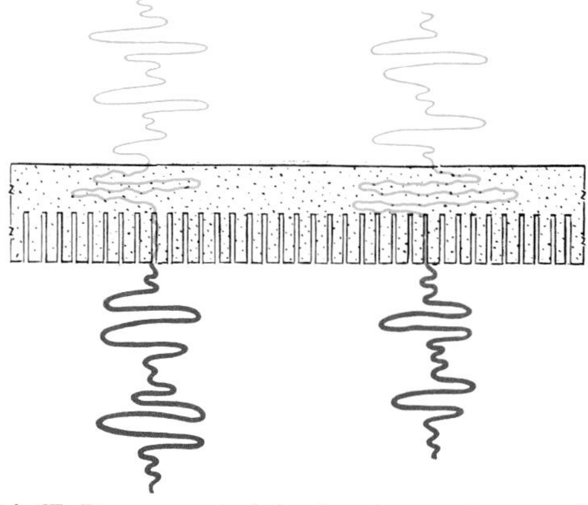

Figure 14-67. Porous materials absorb sound waves. The sound absorption coefficient of a material is the fractional part of the energy of a sound wave that is absorbed at each reflection.

Other materials depend on a similar absorption action to reduce sound. The material used has a vibration point which approaches zero. Heavy draperies or hangings, hair felt, and other soft flexible materials function in this manner.

The efficiency of an acoustical unit or product is measured by its ability to absorb sound waves. Since noise is a mixture of confused sounds of many frequencies, this efficiency is measured by the noise-reduction coefficient (NRC) of a material for the average middle range of sounds. For most installations, this figure can be used to compare the values of one material over another. However, some materials are designed to do a better job for high or low frequency sounds. For special cases such as music studios, auditoriums, and theaters, an acoustical engineer should be consulted. Most manufacturers of acoustical materials furnish this service.

Perforated fiberboard acoustical materials are made in tile shapes of a low density fibrous composition. Holes of various sizes are drilled almost through the tile. The sounds which strike these units are trapped in the holes. The walls of the holes, being relatively soft and fibrous in nature, form tiny pockets which absorb the sound. See Figure 14-68.

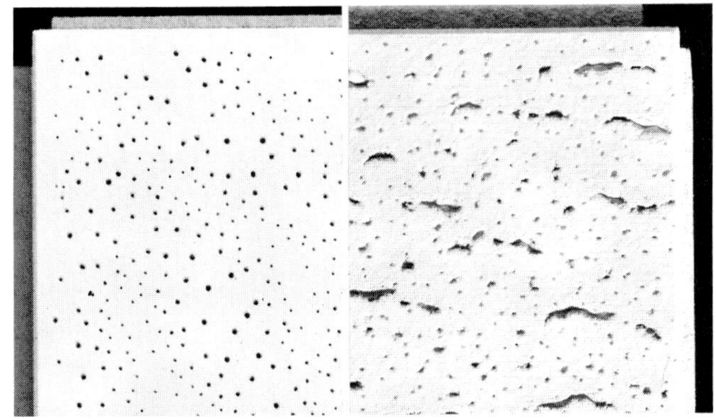

Figure 14-68. Ceiling tile are designed to absorb sound. Left—Drilled holes. Right—Fractured surface. (Wood Conversion Co.)

There are also some fibrous products which are not drilled, but depend on the porosity of the surface and its low vibration point to absorb the sound. Most have a relatively smooth surface, providing a high degree of light reflection without glare.

They are usually installed on the ceiling or upper wall surface by gluing or stapling. Where the surface is in poor condition, furring strips can be used and the material nailed or glued to them. Each manufacturer has specifications on installation.

The perforated metal-pan type performs in somewhat the same manner. The sound enters through the holes in the surface of the pans and is trapped by the backing material which is of a soft resilient nature. It is commonly used in institutional and commercial buildings where ease of maintenance and fire resistance are important factors.

Suspended Ceilings

These factory finished units are installed on suspended metal runners that form a grid. They allow the large panels to simply be dropped into place, Figure 14-69. These systems are used where suspended ceilings are advisable to conceal pipes, electrical wiring, and structural beams. The entire ceiling, or any part of it, may be removed and relocated without damage to the material.

Figure 14-69. 2' x 4' panels install rapidly in a suspended ceiling. (Armstrong World Industries Inc.)

Panels are made from various materials. Some are made from ground cork or glass fibers. The material is pressed into acoustical panels of various sizes and thicknesses. Because of the resistance of both these materials to moisture, they are ideal for use in indoor swimming pools, commercial kitchens, or any place where humidity is a problem. Additional information on suspended ceilings is included in Unit 15.

Acoustical Plaster

A number of lightweight or fibrous materials are used in the preparation of acoustical plaster. For best appearance and highest sound-absorption qualities, the plaster should be sprayed on the surface, Figure 14-70. Acoustical plasters are generally used where an unlined or plain surface wall or ceiling is desired or where curved or intricate planes make the use of sheet materials impractical.

To install acoustical plaster omit the finished plaster coat and instead apply two coats of the acoustical plaster.

Figure 14-70. Acoustical plaster is being sprayed on a ceiling. (Vermiculite Institute)

Installation of Materials

Manufacturer's recommendations on applying all acoustical materials should be carefully followed. If they are not, the sound-deadening materials may not do the job. In many cases, the amount of air space in back of the material is a factor in its sound absorption qualities.

Because most acoustical materials are soft, it is the usual practice to install them on the ceiling or upper portion of sidewalls. For sounds originating in the average room, the ceiling offers a sufficient area for sound absorption materials. Directions concerning the methods and procedures for installing ceiling tile are included in Unit 15.

Maintenance

Painting of perforated boards usually does not lower the efficiency, if properly done. The dirt clogging the pores of the material should be removed first. This may often be done with a vacuum cleaner or by brushing with a soft hair brush. Some acoustical material can be cleaned by washing.

Spray painting is usually preferable to brush painting. A thinner mixture has less tendency to clog the pores of the material. Improperly applied paint will soon fill the pores of the material and destroy its efficiency.

Manufacturers of acoustical tiles and other products have prepared detailed instructions for installation and maintenance. Be sure to follow them carefully.

Important Terms

Acoustical tile
Batt insulation
British thermal unit (Btu)
Coefficient of thermal conductivity (k)
Conductance (C)
Conduction
Convection
Convection currents
Corkboard
Crawl space
Decibel
Decibels reduction
Dew point
Draft excluder
Flexible insulation
Foamed glass
Frequency
Impact sounds
Infiltration
Insulation
Insulating board
Loose fill insulation
Masking sounds
Noise
Noise reduction coefficient (NRC)
Permissible exposure limit (PEL)
Plenum
Radiation
Reflective insulation
Resistance (R)
Reverberation sounds
Rigid insulation
Sleepers
Sound
Sound absorption
Sound transmission class (STC)
Sound transmission loss (STL)
Total heat transmission (U)
Vapor barrier
Water vapor

Test Your Knowledge

1. When heat moves from one molecule to another within a given material, the method of heat transmission is called _____________.
2. Heat can be transmitted by wave motion. This method is referred to as _____________.
3. The selection of an insulation material should be based on its cost, effectiveness, _____________, and the adaptation to the construction.
4. The resistance of a material to heat transmission is represented by the letter _____________.
5. To get an R-value of 23.4, a 2 x 6 insulated stud wall would need a combination of materials. Can you list the materials and give their thicknesses?
6. A standard insulation batt is usually _____________ or _____________ inches long.
7. The temperature at which condensation occurs for a given sample of air is called the _____________.
8. The recommended R-value for insulation in a ceiling below a ventilated attic in northern Minnesota is R-_____________.
9. The vapor barrier in a wall structure should be located on the _____________ (*cold, warm*) side of the insulation.
10. When stapling blanket insulation between studs, start at the _____________ (*top, bottom*).
11. Fill insulation can be poured or _____________ into place.
12. _____________ insulation can be installed either on the outside or inside of exterior walls.
13. Which has a higher insulating value, expanded polystyrene or expanded polyurethane?
14. To obtain R-38 in a ceiling using blown-in-blanket insulation, one would need to produce a thickness of _____________ inches.
15. Heat loss through a double-glazed window may be as much as _____________ times the loss through a well-insulated wall.
16. Air leakage around windows and doors is called _____________.
17. The unit of measure used to indicate the loudness or intensity of a sound is called the _____________.
18. A wall with a STL of 40 dB. will reduce a loudness leveling through it from 70 dB. to _____________ dB.
19. A method of rating the sound-blocking efficiency of a wall, floor, or ceiling structure is called the _____________ system.
20. Sound-deadening board is made largely from wood and _____________ fibers.
21. To ensure a high STC rating, electrical outlets on opposite faces of a partition should not be located in the same _____________.
22. A resilient channel added to one side of a wood stud

wall and double drywall on each side needs ____________ thickness of insulation for an STC rating of 50.

23. The efficiency of an acoustical ceiling unit is expressed in an NRC rating, which is an abbreviation for ____________.

Outside Assignments

1. Obtain samples of various thermal insulating materials, such as glass, mineral, organic fibers, and foamed plastic materials. Include loose fill insulations made from such material as vermiculite. Enclose fibrous and granular materials in small envelopes made of polyethylene plastic film. Mount samples on a display board with descriptive titles including k and R factors.
2. Prepare a brief study of the most common types of heating and cooling systems used in your region. Make several drawings showing how they operate and how they are controlled. Study local building codes and outline the basic requirements that must be followed in the installation. Visit with a heating contractor and obtain basic information concerning the heating and cooling load calculations for residential structures. Get information about manufacturers, approximate costs, and installation procedures. Give a report in your class.
3. From a study of reference books and trade magazines, prepare a report on radiant heating systems. Include information about panels that are heated with electricity as well as those heated with hot water. Place special emphasis on the methods of installation since structural design may need to be modified when these systems are employed. Also, collect information concerning types and amounts of insulation materials. Discuss special application procedures that may be required.
4. Using a tape recorder, carefully record the various sounds produced by equipment and devices found in a modern home. Include laundry equipment, dishwashers, food mixers, garbage disposals, plumbing fixtures, and vacuum cleaners. Also include such noises as walking on hard surfaced floors and closing of passage or cabinet doors. Play the recording for your class. Then lead a discussion on how to control each sound through proper design and the use of special materials and construction.
5. Modern residential construction emphasizes the use of vapor barriers in walls, ceilings, floors, and weather-stripped windows and doors. This nearly airtight quality can result in high levels of humidity and air pollution. These conditions are potential hazards to health and damage to interior surfaces and fixtures. Modern air-to-air heat exchangers can provide desirable standards of ventilation. They transfer heat from outgoing air to fresh air entering the structure. Study trade magazines and visit local heating/cooling contractors to learn about this kind of equipment. Prepare a written report.

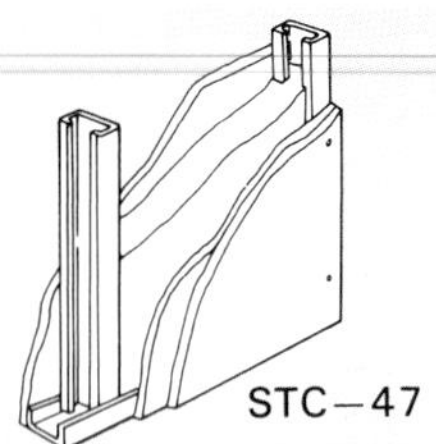

Unbalanced wall, 2 1/2'' metal studs 24'' o.c.; double layer 1/2'' Type X gypsum board one side, single layer 1/2'' Type X gypsum board other side; one thickness R-8 Fiberglas insulation.

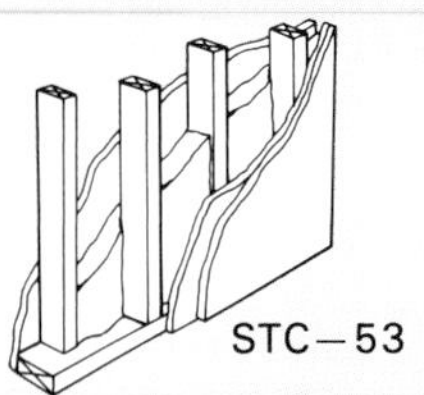

Unbalanced wall, staggered wood studs 24'' o.c.; double layer 1/2'' gypsum board one side, single layer other side; one thickness R-11 Fiberglas insulation.

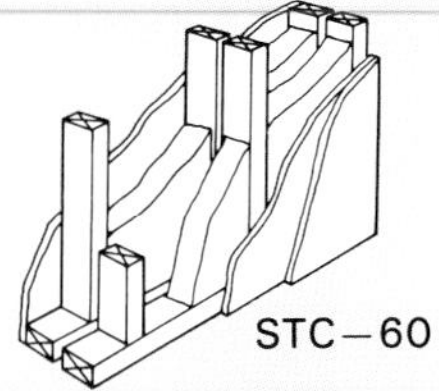

Unbalanced wall, double wood studs 16'' o.c.; double layer 1/2'' gypsum board one side, single layer other side; two thicknesses R-11 Fiberglas insulation.

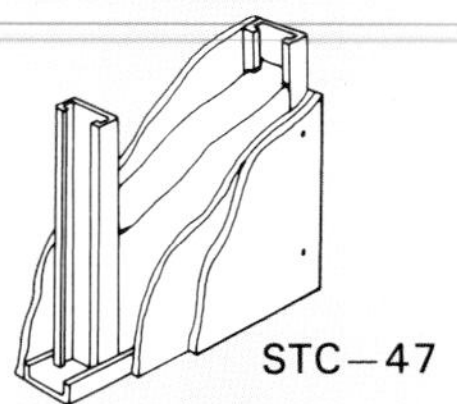

3 5/8'' metal studs 24'' o.c.; double layer 1/2'' Type X gypsum board one side, single layer other side; one thickness R-11 Fiberglas insulation.

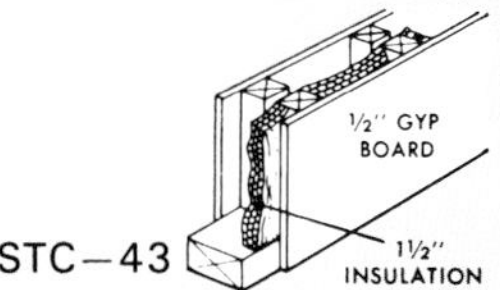

Slit-stud wall with 1 1/2'' blanket insulation hung from top plate. Center cuts in wide face of stud should extend to about 3'' from each end.

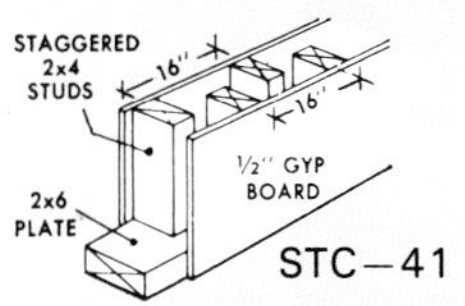

Staggered studs provide complete separation between wall faces. Staggering 2 × 3s on 2 × 4 plate is almost as effective.

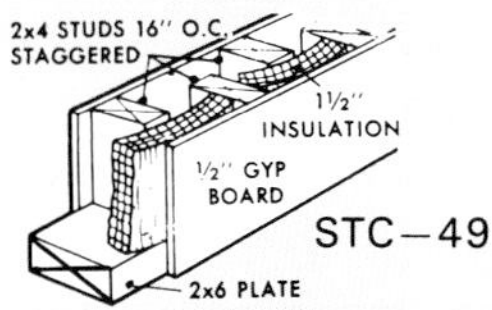

Blanket insulation between staggered studs performs as well as weaving continuous insulation behind studs.

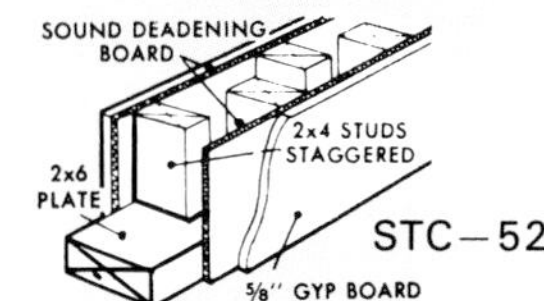

Sound deadening board absorbs noise, dampens wall vibration, adds surprisingly to effectiveness of staggered-stud wall.

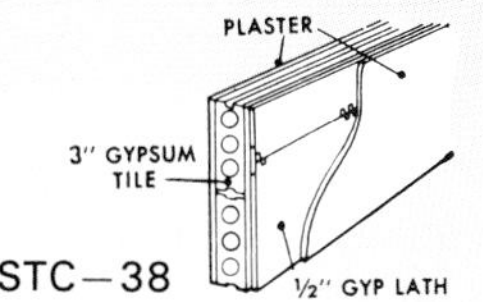

Fireproof wall of 3'' gypsum tile has been considered a good acoustical wall, yet its rating is surprisingly low.

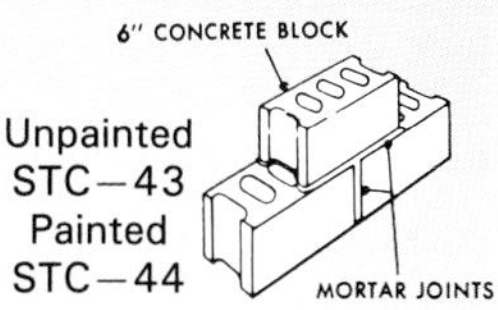

Concrete block wall becomes more effective when pores are sealed with two coats of paint.

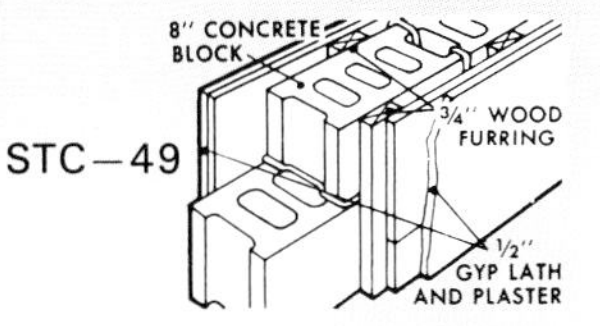

Concrete block with furred gypsum lath and plaster both sides has excellent properties in mid and high frequencies.

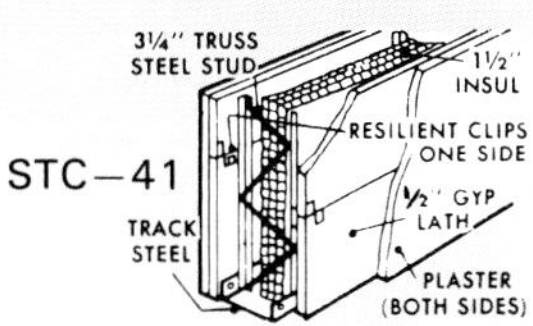

Steel truss studs, 3 1/4'' wide with gypsum lath and plaster both sides, require insulation to be effective.

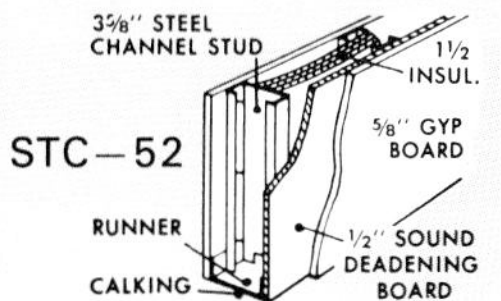

Steel channel studs with insulation plus sound deadening board on one side rate high. Caulking seals gaps.

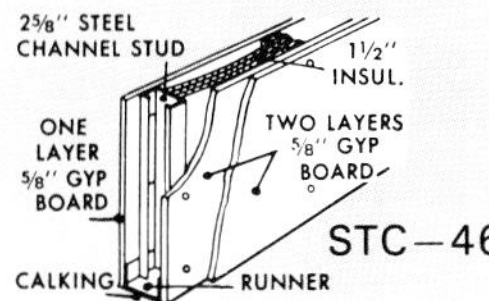

Extra layer of 5/8'' gypsum board on one side of 2 5/8'' channel studs gives fair performance with insulation.

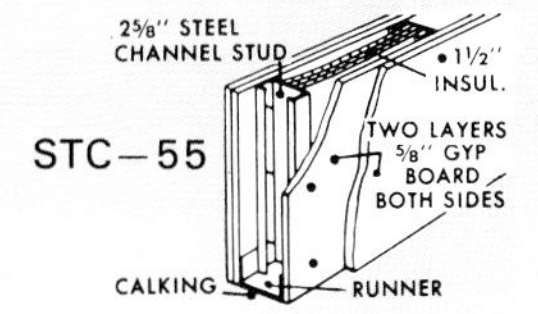

Laminated 5/8'' gypsum board on both sides of 2 5/8'' steel channel studs plus insulation gets very high rating.

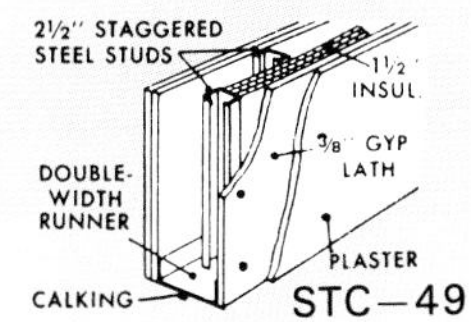

Staggered steel studs with 3/8'' gypsum lath and plaster compares favorably with staggered wood studs and insulation.

Structural designs of wall and their sound transmission factors (STC). Other designs using wood or steel framing are shown in the Technical Information Section.

15 Interior Wall and Ceiling Finish

Interior finishing is the installation of cover materials to walls and ceilings. This stage of construction can start after utilities, heating, and insulation are installed. Exterior doors must be hung and windows installed. They will protect the finishing materials from the weather.

Interior walls can be covered with any one of a number of materials:

- *Gypsum wallboard* — Commonly called *drywall*, gypsum wallboard is a laminated material with a gypsum core and paper covering on either side. It usually comes in 4' x 8' sheets. However, it is also available in 7', 9', 10', 12', and 14' lengths. It is sold in the following thicknesses: 1/4", 5/16", 3/8", 1/2", and 5/8". Gypsum wallboard is used on both walls and ceilings.
- *Gypsum wallboard for plaster veneering* — This is a base of gypsum board, usually 1/2" thick. It is applied as a backing for a thin coat of plaster.
- *Predecorated gypsum paneling* — This is the same as gypsum wallboard. However, decorative vinyl finishes have been applied and edges have received special treatment so that no other finishing work need be done after the panels have been installed. The finishes are tough and easily cleaned.
- *Plywood and particleboard* — Figure 15-1. Plywood is fabricated in 4' widths. Lengths include 7', 8', 9', and 10'. Usually, the sheets are prefinished in a variety of colors and patterns. Surface material may be either a hardwood or a softwood. Panels are manufactured in thicknesses of 1/4", 3/8", 7/16", 1/2", 5/8", and 3/4". Surfaces can be embossed, stained, or color toned. Some have veneers of paper with a wood grain printed on them. Flakeboard or waferwood is sometimes used for rustic finishes.
- *Hardboard and fiberboard* — These are produced from wood fibers in sizes and thicknesses similar to plywood. The face finish is simulated to look like wood. Other decorative patterns are also applied. Sheets may be embossed and grooved to simulate random planking, leather, or wallpaper. Surfaces may also be coated with plastic. Variations of fiberboard are used as ceiling coverings.
- *Solid wood paneling* — These are boards or pieces of solid wood. Widths of boards will vary from 2 to 12" and thicknesses are either 1" or 2". Faces may be rough-sawed, plain, or molded in a variety of patterns. Lengths vary from 4 to 10'. Shingles, usually considered a siding or roofing material, are occasionally used on interior walls.

Figure 15-1. Plywood and other wood materials are used as wall coverings. A—Plywood fabricated to look like diagonal planking. (American Plywood Assoc.) B—Plywood panel cut to look like planking. Surface is prefinished. C—Waferwood paneling used as wall covering in a den. (Louisiana-Pacific)

- *Plaster* — For many years the most popular wall covering, plaster is made of powdered gypsum to which other materials are added to improve drying time. A plastered wall system includes a base support, such as metal or gypsum lath, over which is applied coats of wet plaster.
- *Cement board* — Available under several different brand names, cement board is a versatile fiber-reinforced cement panel material that is used as a base (underlayment) for finishing materials used on walls (exterior and interior), floors, and countertops. It is fireproof, not damaged by water, and resists impact. Some of these products are lightweight with fiberglass reinforced matting and silicone treated cores and surface coatings to make them moisture resistant. Figure 15-2 shows a sample of a cement board product.
- *Special finishes* — These include a variety of products and materials: brick, stone, glazed tile, plastic tile, and plastic laminates. They are used either as an accent material or to provide a wear resistant surface. They are often found in kitchens and bathrooms.

Figure 15-2. A small sample of a fiber-reinforced cement underlayment. It can be used as backing for ceramic tile, marble, and plastic laminates. Dot at edge is a guide for nailing. (James Hardie Building Products)

Ceilings can be covered with many of the same materials used for walls. Composition tiles are especially suitable because they are easy to install.

Before beginning the wall and ceiling application, check over the framework. Be certain that sufficient backing has been installed for fixtures and appliances. Refer to Figure 9-30. Nailers must be included at all vertical corners. Nailers must also be included at all intersections between walls and ceilings. See Figure 9-53. Special equipment, like the prefabricated fireplace shown in Figure 15-3, must be installed before the interior wall surface is applied.

Drywall Construction

Drywall materials, such as gypsum wallboard, Figure 15-4, are the most common coverings used in modern construction. Most builders prefer to use drywall because it

Figure 15-3. Left—Wall framing around a prefabricated fireplace. Multiple steel wall construction and special firebox linings permit "zero-clearance." This means that fireplace's metal housings can touch wood framing members. Right—Framing around heating and air-conditioning ducts in a basement area. When cabinets will be hung from the overhead framework, use screws or threaded nails.

Figure 15-4. Transporting drywall into a new residence.

TYPE	THICKNESS (IN.)	EDGES
Regular (ASTM C36, FS SSL30d)	1/4 5/16 3/8 1/2 5/8	Tapered Square Square Tapered Bevel
Fire Resistant Type "X" Wallboard	1/2 5/8	Square Tapered Bevel
Insulating Wallboard (Aluminum Foil on Back Surface)	3/8 1/2 5/8	Square T & G Tapered Round
Regular Backing Board (ASTM C442, FS SSL30d)	1/4 3/8 1/2	Square Square T & G
Foil-Backed Backing Board	3/8	Square
Fire Resistant Type "X" Backing Board	1/2 5/8	T & G
Coreboard (Homogeneous or Laminated)	3/4 1	Square T & G Ship Lap
Predecorated	3/8 1/2 5/8	Bevel Round Square Bevel Square

Figure 15-5. Main types of gypsum wallboard. Water resistant types are still being used in shower areas in bathrooms; however, cement board is now preferred.

saves time. Regular plaster requires considerable drying time.

Either type of finish presents advantages and disadvantages. Drywall construction, for example, requires that studs and ceiling joists be perfectly straight and true, otherwise the wall surface will be uneven. This can be corrected in ceiling joists by installing a strongback. Where steel or engineered wood joists and studs are used in framing, this is not a problem. The wood framing material must also have a moisture content very near to that which it will eventually attain in service (around 15%). This will help prevent "nail pops" and joint cracks. Use of screws for drywall will also solve this problem, to a great extent.

Gypsum wallboard has a fireproof core. Figure 15-5 lists a variety of thicknesses, edge joint designs, and types. *Backing board* is a gypsum board with a gray liner paper on both sides. It is used as the base sheet on multilayer applications. It is not suited for finishing and decorating.

Figure 15-6 shows several standard edge designs for gypsum wallboard. Its tapered edges form a shallow depression between adjacent sheets. This depression is brought level with drywall tape and filler. The result is a smooth, uninterrupted surface.

Drywalling requires some special tools for marking, cutting, installing, and finishing. See Figure 15-7.

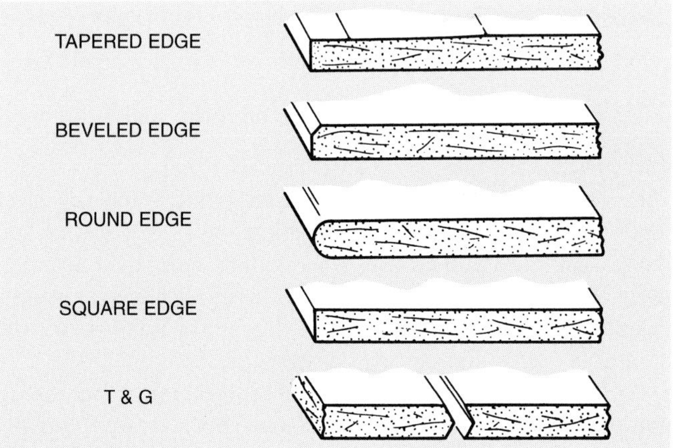

Figure 15-6. Gypsum wallboard is manufactured with several different edge styles.

Single Layer Construction

In single layer construction, use 1/2" or 5/8" gypsum wallboard. The 5/8" sheet is preferred for high quality construction. Cover ceilings first, then the walls.

There are two methods of arranging the drywall sheets:

- Parallel — Long edges of panels run in the same direction as studs and joists.
- Perpendicular — Long edges of panels are at right angles to studs and joists.

The second method is generally preferred for several reasons:

- There is less footage of joints to treat.
- Panels bridge more studs and joists, making the building's frame stronger.
- The strongest dimension of the panel runs across the frame.
- There are fewer problems with irregularities in alignment and spacing of the frame.
- Horizontal joints are easier to treat because they are lower on the walls.

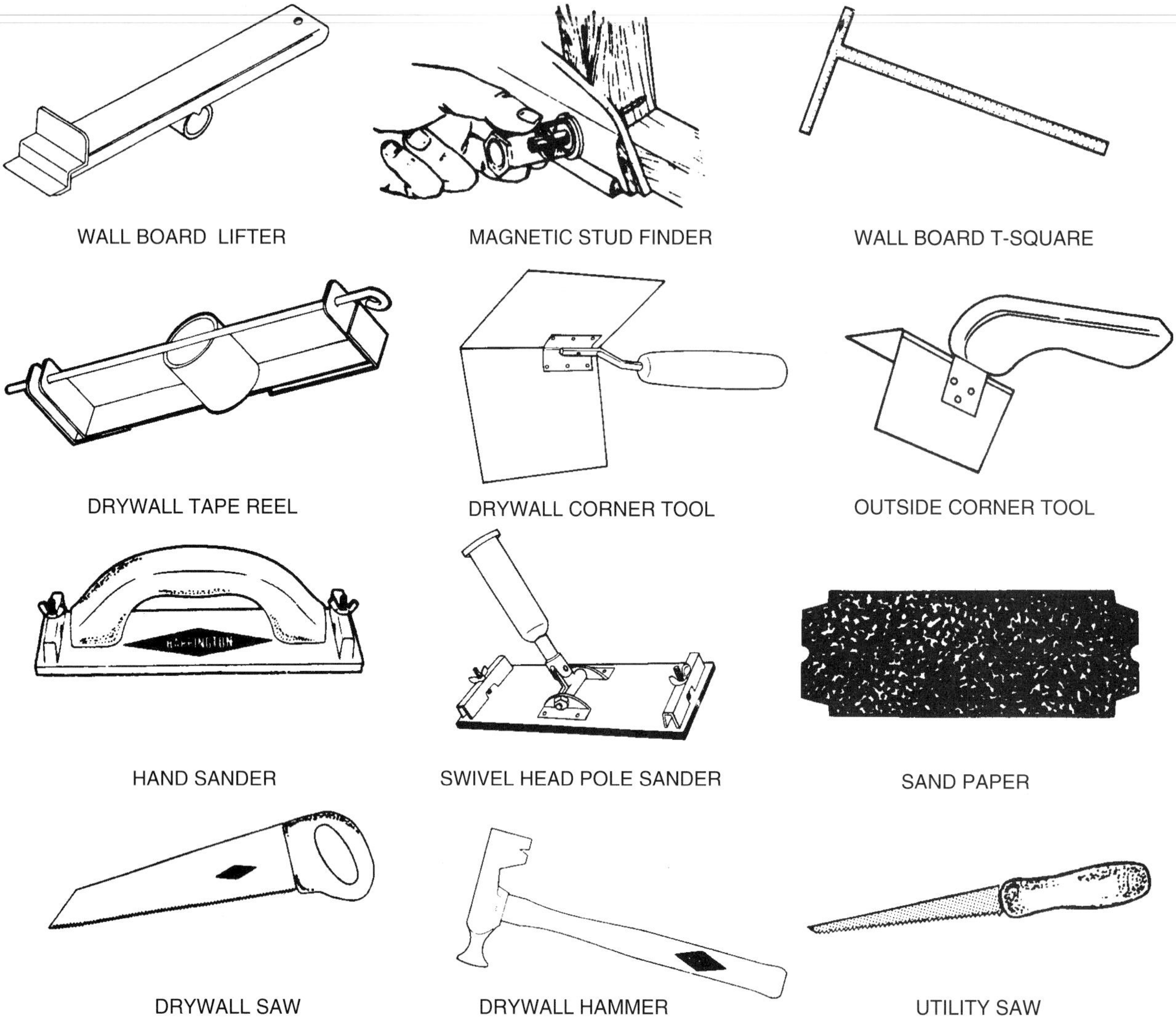

WALL BOARD LIFTER MAGNETIC STUD FINDER WALL BOARD T-SQUARE

DRYWALL TAPE REEL DRYWALL CORNER TOOL OUTSIDE CORNER TOOL

HAND SANDER SWIVEL HEAD POLE SANDER SAND PAPER

DRYWALL SAW DRYWALL HAMMER UTILITY SAW

Figure 15-7. These tools are used for preparing and installing drywall. Note names. (Harrington)

In either method, vertical wall joints must fall over and center on studs. Figure 15-8 shows both parallel and perpendicular applications. As a general rule, the carpenter will use whichever method results in the fewest joints. Stagger end joints and locate them as far away from the center of walls and ceilings as possible.

Butt wall panels loosely against the ceiling panels. In parallel application, use a wallboard lifter to raise the panel to the ceiling. A lifter is a lever device operated by the foot, refer again to Figure 15-7. Stepping on it while one end is under the drywall will raise the panel enough to press it against the ceiling.

In horizontal applications the top wall panels are installed first. This is done so that any gaps or cut edges come at the floor where trim will cover them.

Measuring and Cutting

All measurements should be carefully taken from the spot where the wallboard will be installed. Usually it is best to make two readings, one for each side of the panel. Following this procedure will eliminate errors. It also allows for openings and framing that are not plumb or square. Use a steel tape to take measurements.

Straight cuts across the width or length of a board are made by first scoring the face with a sharp knife pulled along a straightedge used as a guide. The scoring cut should penetrate the paper and enter the gypsum core.

Support the main section of the sheet close to the scored line. With one hand holding the sheet firmly on the table or bench, snap the core by pressing downward sharply on the overhang. Support the cutoff with the other hand. Score the backing paper, as shown in Figure 15-9, and snap the cutoff upward. When necessary, the cut can be smoothed with a file or with coarse sandpaper mounted on a block of wood.

When scoring wallboard, always use a sharp knife that will make a “clean” cut through the paper face and penetrate slightly into the core.

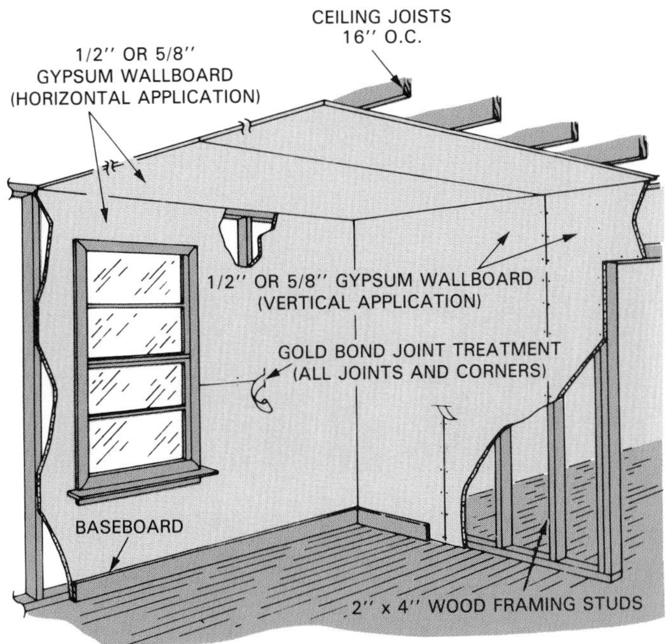

Figure 15-8. Single layer drywall application. Left-hand wall shows horizontal application; right-hand is a vertical application.

Figure 15-9. Marking and cutting drywall. Top—A drywall square can be used to mark a square cut line on the panel. (Construction Training School, St. Louis) Bottom—Cut the paper on the back side of the drywall after breaking the gypsum core.

Irregular shapes and curves can be cut with a coping saw, compass saw, or electric saber saw. Figure 15-10 shows an opening for a convenience outlet being made with a portable power saw.

Nails and Screws

Nail spacing will vary depending on the materials being used. For single layer construction, they are spaced no farther apart than 7" on ceilings and 8" on walls. Keep nails at least 3/8" from ends and edges. Annular ring nails with a 1/4" head and 1 1/4" long are generally recommended. See Figure 15-11 for drywall fasteners.

All wallboard must be drawn tightly against the framing so there can be no movement of the board on the nail shank. During nailing, press the board tightly against the stud or joist to avoid breaking through the wallboard face. Figure 15-12 shows proper countersinking of nails. Figure 15-13 shows how large panels can be held with braces while nailing them to a ceiling.

Start nailing at the abutting edge. Nail the *field* (area between edges) first. If perimeter is nailed first, a sag in the ceiling panels will never draw up tight to joists.

Drive nails straight and true. Use extra care during the final strokes so the nailhead rests in a slight dimple formed by the crowned head of the hammer or stapler. Be careful not to break the paper face of the board.

When applying drywall to ceilings it will be necessary to stand on a platform or use stilts to reach your work. See Figure 15-14.

The double nailing method of attachment ensures firm contact with framing. Panels are applied as required for conventional nailing, except that nails in the field of the board should be spaced 12" on center. After the panel is secured, another nail is driven approximately 2" from the first. If necessary, the first nail should receive another blow to assure snug contact.

Gypsum wallboard screws were shown in Figure 15-11. They provide firm, tight attachment to wood or metal framing, Figure 15-15. Special self-tapping screws are used for metal-framed wallboard systems. Fasteners of this type must be driven so the screw head rests in a slight dimple formed by the driving tool. The paper face of the wallboard should not be cut. Neither should the gypsum core be fractured.

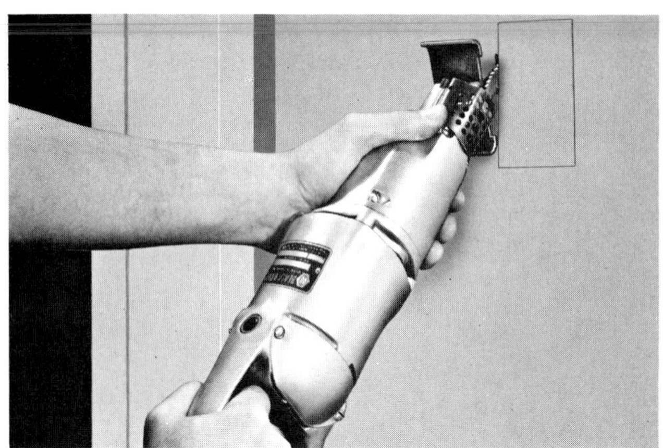

Figure 15-10. Inside cuts, curves, and irregular shapes in drywall can be cut with a portable saber or reciprocating saw.

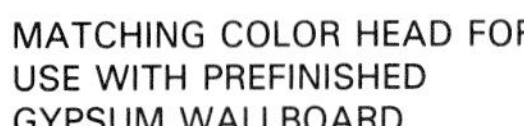

Figure 15-11. Gypsum wallboard and gypsum lath fasteners. Others are available.

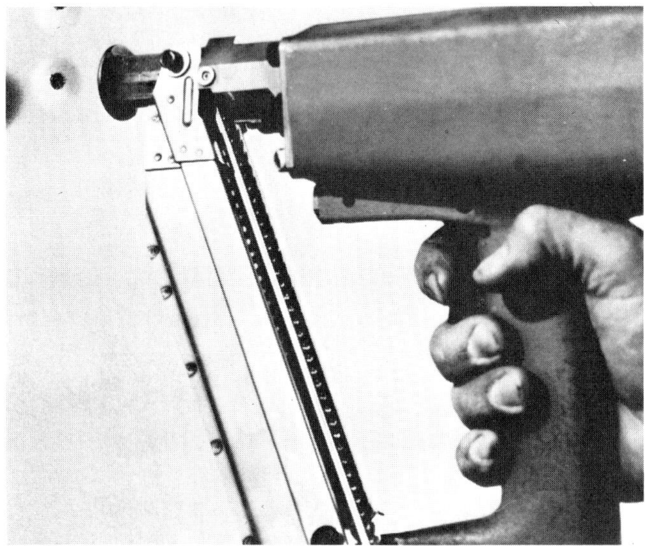

Figure 15-12. Drywall nailer forms proper dimple and drives nail below surrounding surface without breaking paper facing. Recommended air pressure is 80 psi. (ITW Paslode)

Figure 15-13. Top—Braces, also called shores, will hold ceiling panel in place while it is being nailed. (The Flintkote Co.) Bottom—After ceilings are drywalled, apply drywall to walls. Try to get a close fit in corners.

Since screws hold the wallboard more securely than nails, ceiling spacing can be extended to 12" and side walls to 16".

Adhesive Fastening

When wallboard is fastened with a special adhesive, there are no depressions from mechanical fasteners to be filled later. Adhesives also produce a sturdier wall that is more resistant to impact sounds. Caution: Some adhesives have flammable solvents; use care where open flames are present.

Follow manufacturer's directions for use. Apply adhesives only when temperatures are between 50 and 100°F (10–38°C). Keep adhesive containers closed; evaporation of solvent can affect the adhesive's performance and ability to bond.

Apply a continuous bead of adhesive to the center of all studs, joists, or furring as shown in Figure 15-16. Where two pieces of wallboard join on a framing member, use a zigzag bead pattern. The bead should be 1/4 to 3/8" wide. Then, when the board is in place it will be held by a band at least 1" wide and 1/16" thick.

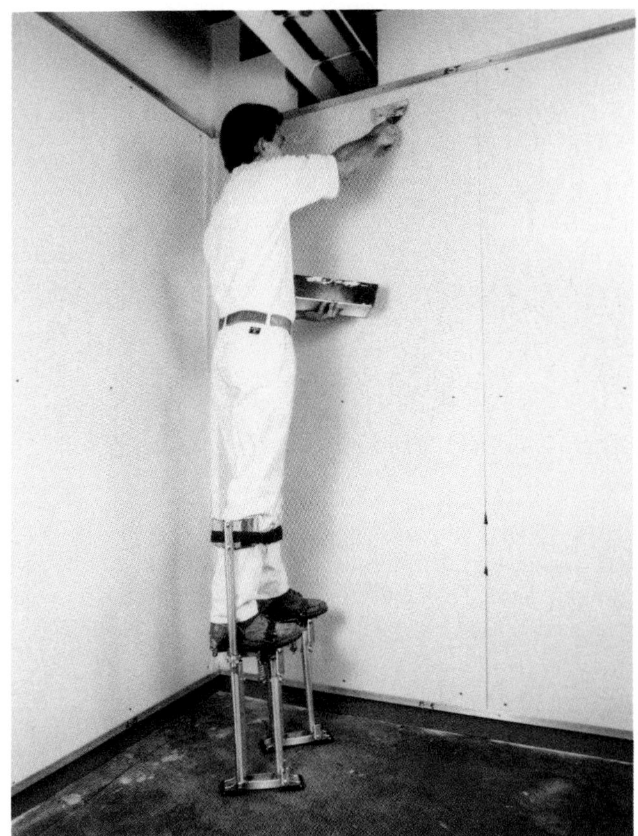

Figure 15-14. Stilts are very helpful when working on ceilings and upper walls. The pair shown is adjustable in height and provides a "walking action" that reduces fatigue. (Goldblatt Tool Co.)

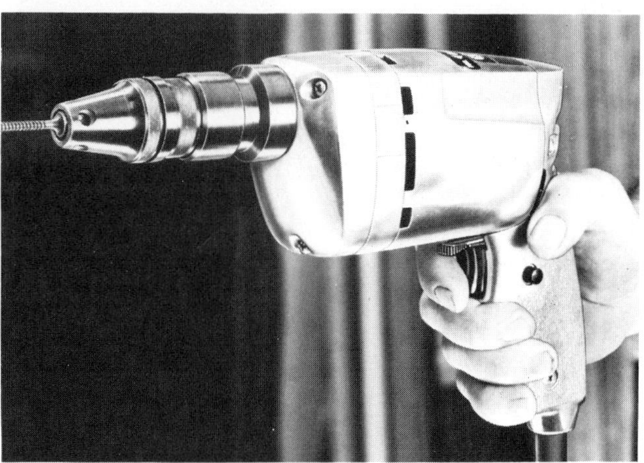

Figure 15-15. Top—An electric drill with a special depth-adjusting clutch is used to drive drywall screws into steel studs. Clutch disengages when nose strikes panel surface. (St. Paul Technical School) Bottom—Close-up of drill with depth-adjusting clutch.

Use temporary nailing or bracing to ensure full contact of the wallboard. This allows the adhesive to develop proper bonding strength.

Joint and Fastener Concealment

To conceal joints, first apply a bedding coat of joint compound into the depression formed by the tapered edges of the board over all butt joints. Use a 5" or 6" joint knife. Center the reinforcing tape over the joint and smooth it out to avoid wrinkling or buckling. Press tape into compound by drawing the knife along the joint with enough pressure to remove excess compound. Apply a skim coat over the tape. See Figure 15-17.

After the embedding coat is completely dry, apply a second coat over the tape. Feather the edges approximately 1/2 to 3/4" beyond the edges of the first coating. When this coat is completely dry, a third coat is applied with the edges feathered out about 2" beyond the second coat. After the last coat is dry, sand lightly, if necessary. Fasteners are also concealed with compound, each coat being applied at the same time the joints are covered.

Figure 15-16. Apply a bead of adhesive 1/4" wide to framing members.

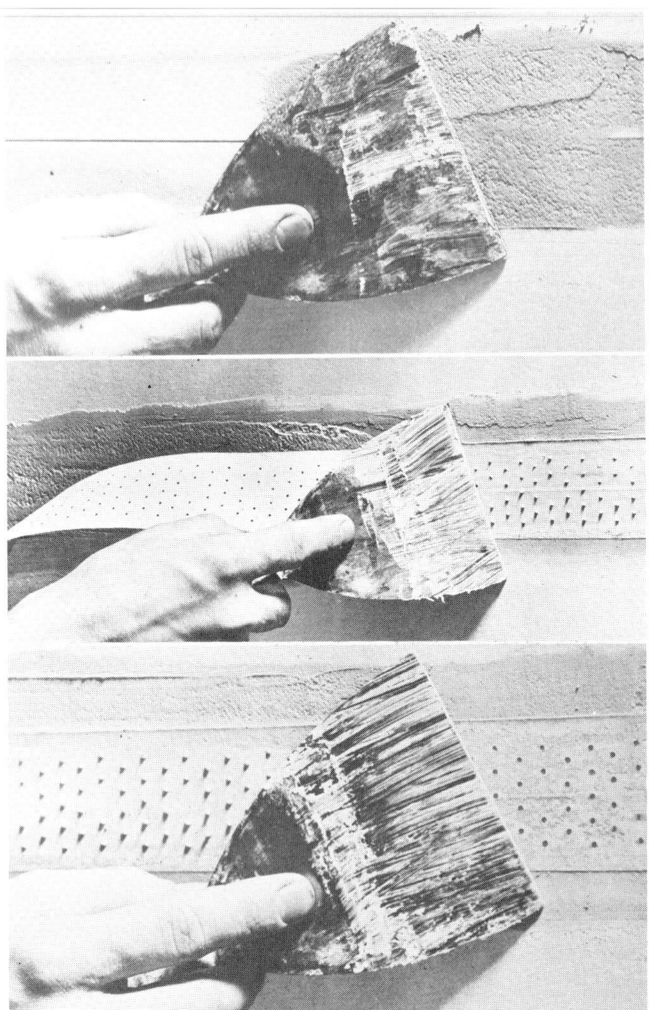

Figure 15-17. Taping wallboard joint with broad taping knives. Top–First apply compound to channel at joint. Middle—Embed tape. Be sure it is centered over the joint.
Bottom—Immediately apply a skim coat over tape and smooth edges. Use broader knife. (Gold Bond Building Products)

Figure 15-18. Using a mechanical taping tool speeds up drywall construction. The tape unwinds from the roll (arrow) as the tool is moved along the joint. Compound is forced out of the hollow tube as pressure is applied to the bottom end.

Figure 15-19. Using a broad taping knife to smooth a taped joint. The excess compound is removed from the knife by pulling it over the edge of the mud pan the taper is holding. Compound is commonly referred to as "mud."

A mechanical taping tool is used to apply the compound and tape, Figure 15-18. The joint is then smoothed with a broad knife, as shown in Figure 15-19.

Pressure-sensitive glass-fiber tape reduces the time required to conceal and reinforce joints and interior angles. It has an open weave (100 meshes per sq. in.) which provides excellent reinforcing and keying of plaster or compound coats. Simply use hand pressure to attach it to the wall. Bond it with a finishing knife or trowel. The tape is also easily applied to inside corners.

Apply two coats of fast-setting joint compound over the tape and sand smooth. When the surface is to be covered with a texture paint, you can finish the joints with a single coat of compound. Figure 15-20 shows a texture finish being applied with a spray gun.

Rollers and brushes can also be used to apply the finish. A wide variety of finish coatings are available for drywall construction. Some are in a powder form ready to be mixed with water. Others are ready to use from the containers. Always read and follow the manufacturer's recommendations.

Corners

Outside corners are reinforced with a metal corner bead. The bead is installed after horizontal and vertical joints have been taped. See Figure 15-21 (top). Fasten the

Figure 15-20. Using special spray equipment to apply a textured finish to a drywall surface. Various patterns can be applied by using different sizes of orifices and air nozzles.

Figure 15-21. Top—Metal corner bead installed. Nails should penetrate the wall frame. Bottom—Finished corner. Outlet box must be covered before spraying a surface finish.

bead by nailing through the wallboard and into the framing. After installation, the bead is concealed with joint compound in about the same manner as regular joints. Figure 15-21 (bottom) shows the finished wall surface. Figure 15-22 shows various styles of corner bead.

At internal corners, both horizontal and vertical reinforcing tape is used. First, apply a bedding coat of joint compound to both sides of the corner. Then fold the tape along the centerline and smooth it into place. Remove excess compound and finish surfaces along with the other joints.

Metal channel trim is available to finish and reinforce edges around doors, windows, and other openings. Figure 15-23 shows several shapes and sizes.

Steel Frame Application

Single layer application to steel members is similar to wood frame application. This type of construction is being

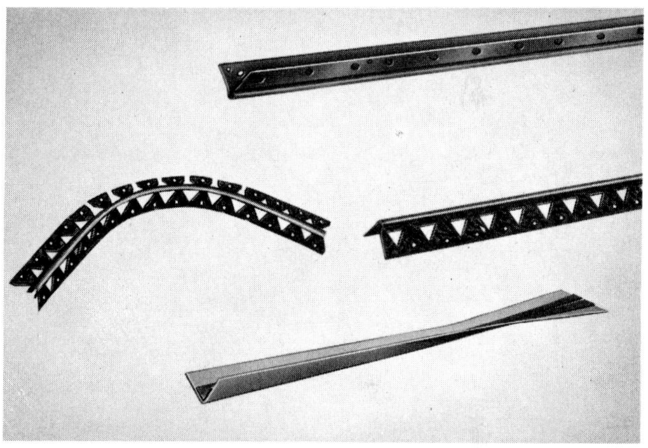

Figure 15-22. Metal corner beads are made in many different styles. Bead at bottom of photograph has paper flanges.

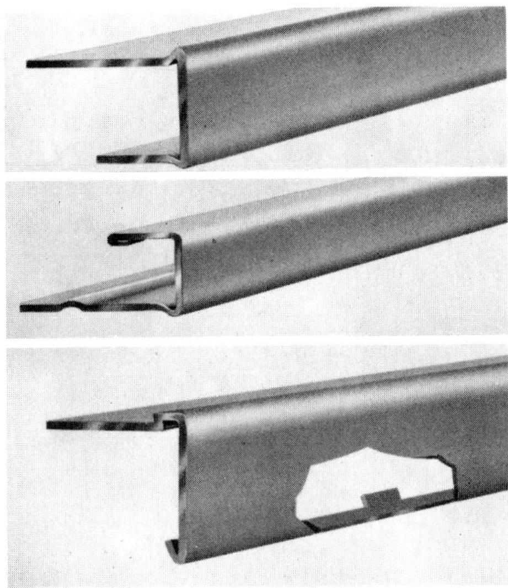

Figure 15-23. Metal channel trim is installed to protect drywall edges around openings such as doors and windows.

widely used because it is inexpensive, lightweight, and highly resistant to combustion, if not fireproof. Arrange panels either parallel or perpendicular to framing and attach with 1" Type S screws. Panels' leading ends or edges must first be attached to the open edge of the stud or joist flange.

Double Layer Construction

Double layer (also called two-ply) wallboard applications over wood framing ensure a strong wall surface. Fire protection and sound insulation qualities are also improved. This method is adaptable to either the use of predecorated panels or standard beveled wallboard with treated joints.

The base layer, Figure 15-24, may be a regular gypsum wallboard or a specially designed base called backing board. It is made up the same as regular gypsum board except that the coverings are a gray liner paper. It is not suitable for decorating and should not be used as a top surface.

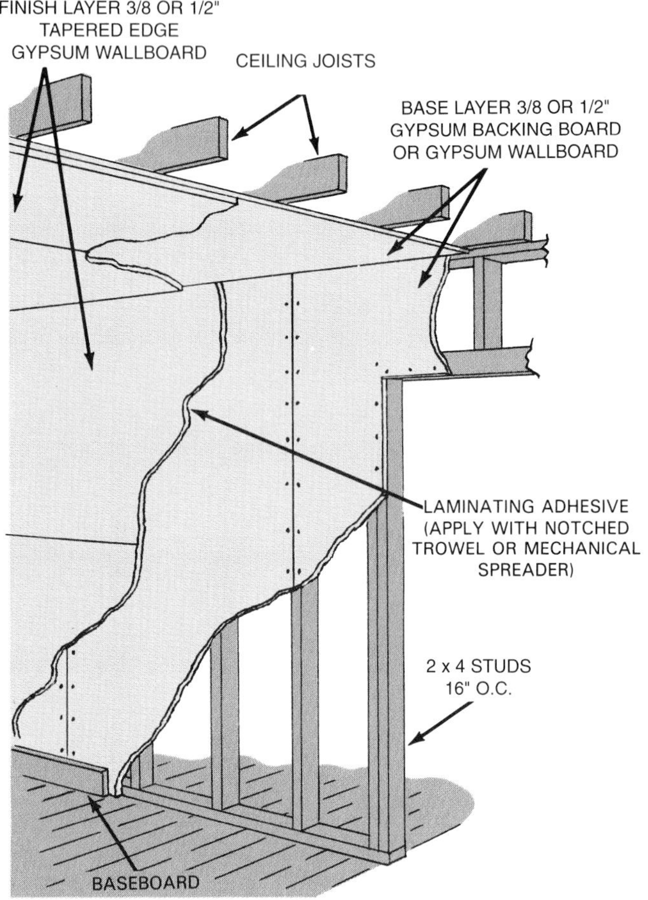

Figure 15-24. Cutaway of a double layer gypsum wallboard construction. Finish layer may be applied at right angles to the base (as shown) or running parallel.

For areas where there is likely to be moisture coming into direct contact with the wall, there are highly water resistant backing boards and some impervious to water. Their use is recommended in shower areas as a base for tile and other protective coverings.

Sound-deadening backing board is sometimes used for the base of double layered walls. It is specified where high STC (sound transmission control) ratings are needed.

Attaching the Layers

Base layers are applied to framing with staples, nails, or power-driven screws. The finish layer is laminated to the base layer with an adhesive or joint compound. Joints of the finish layer should be offset at least 10" from the joints of the base layer. Finish layers can be applied parallel to the base layer or at right angles.

Adhesive is usually applied to the entire surface. However, strip lamination is used in some applications. (This method has ribbons of adhesive spaced at regular intervals.)

Many methods of applying adhesive are acceptable. Trowels and powered devices are available. Whatever method is used, the spacing and size of the bead of adhesive must provide the required spread when panels are pressed into position. Figure 15-25 shows a notched spreader being used to apply adhesive to the entire back surface of a finish layer panel. Strip lamination is often used for sidewall panels, Figure 15-26. The application can be made either on the base surface or on the face panel. Temporary bracing may be used to hold panels in position until bonding has taken place. Fasteners are recommended as follows:

- On ceiling applications when using a laminating adhesive — Space fasteners 16" O.C. along ends and edges. At mid-width, use one fastener for every framing member.

Figure 15-25. Drywaller is using notched spreader to apply adhesive for double layer construction. Spreader forms 1/4" beads spaced 2" apart. (U.S. Gypsum)

- When laminating with compound — Provide permanent supplementary or temporary (usually overnight) fastening until the compound has dried.

When nails are used, they should provide a minimum penetration of 3/4" into the wood framing members. Consult a nail chart.

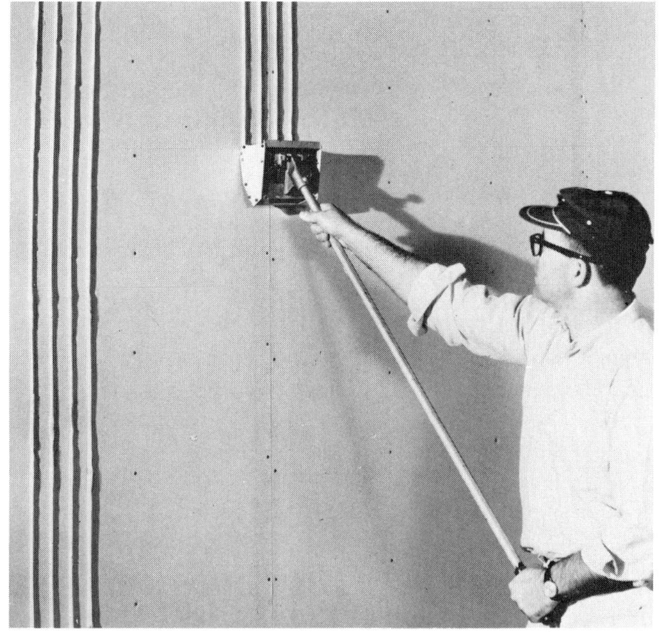

Figure 15-26. Strip laminating with a mechanical spreader. Adhesive may be applied to the base, as shown, or to the back of the finish layer.

Finishing Double Layer Wallboard

If the wallboard is to receive other covering material, joints should be taped, nails concealed, and corners finished in the same way as single layer construction. If a veneer plaster is to be applied, use reinforcing tape and a single bedding coat of compound over joints. Tape inside corners and apply a special bead to outside corners as shown in Figure 15-27. Always use a single length that extends from floor to ceiling.

Backing Board

Special backing is available as a base for tile in areas where walls are frequently wet, for example, bathrooms and showers. One type, known as cement board, is manufactured from a slurry of portland cement reinforced with polymer-coated fiberglass mesh embedded in both sides. Some backing board products are rigid while others are somewhat flexible. They are manufactured under such trade names as "Hardibacker," "Durock" and "DensShield."

These materials, depending on the brand, can be used in many different applications, on both interior and exterior surfaces. Applications recommended include: floor underlayment for tile, resilient coverings, carpeting, or thin brick; floor or wall heat shields for stoves; backing for tiles on walls in shower stalls or tub surrounds; base for countertops; and base for exterior finishes such as ceramic tile, thin brick, or synthetic stucco.

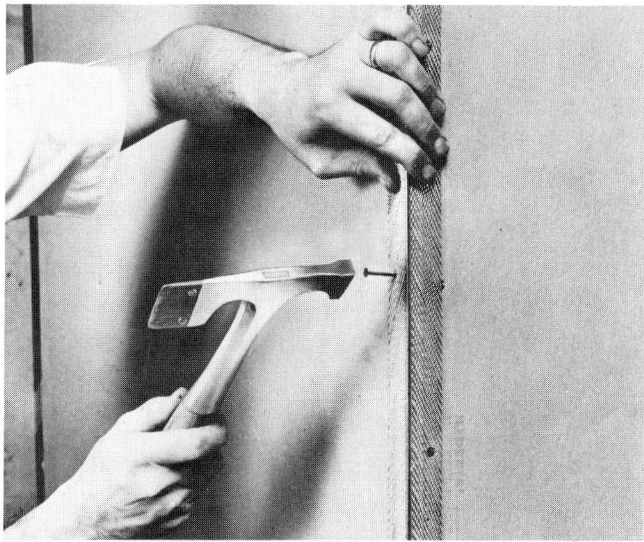

Figure 15-27. Installing special corner bead for veneer plaster coating. Bead provides 1/16" grounds for one-coat system. For two-coat veneer, use bead with 3/32" grounds. (U.S. Gypsum)

Backing board can be worked with ordinary carpenter's tools. Panels can be fastened with nails, screws, or staples. Usual thicknesses are 1/4, 7/16, and 1/2". Panel dimensions are 3' x 5' and 4' x 8'. A sample of backing board was shown in Figure 15-2.

Cement Board

To cut *cement board*, score it several times with a tungsten-tipped knife. Use a straight edge or drywall square as a guide, Figure 15-28.

Use the straight edge to distribute topside pressure at the score and snap upward. Rough edges can be smoothed with a rasp or coarse sandpaper. A power circular saw or a

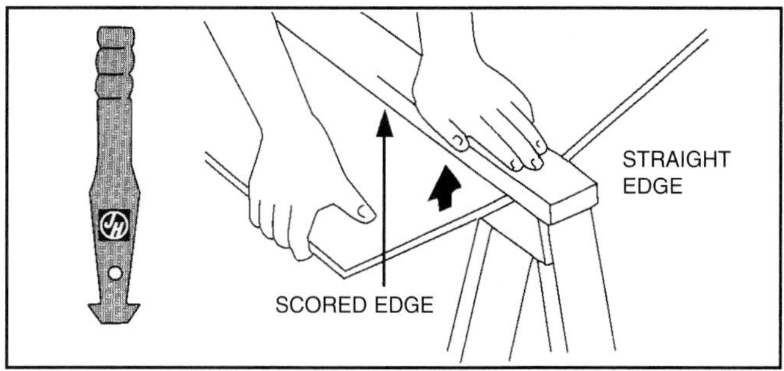

Figure 15-28. Cutting cement-based backing board. It is used for a variety of applications where a highly water resistant base is required in areas where a surface is frequently wet. This material can be fastened with screws, nails, or staples. (James Hardie Building Products, Inc.)

handsaw will also make satisfactory cuts, Figure 15-29. Small holes should be outlined with a series of drilled holes. Use a tungsten carbide-tipped masonry bit. On larger holes, score all sides and then make a diagonal score across the opening. Break out both types of holes from the face side with a hammer. See Figure 15-30.

Cement board is installed in about the same way as moisture resistant (MR) board, Figure 15-31 shows board being installed for a shower and a tub surround. A third illustration shows ceramic tile being installed over a 1/2" thick portland cement board with a polymer-coated glass-fiber mesh bonded to the front and back surface.

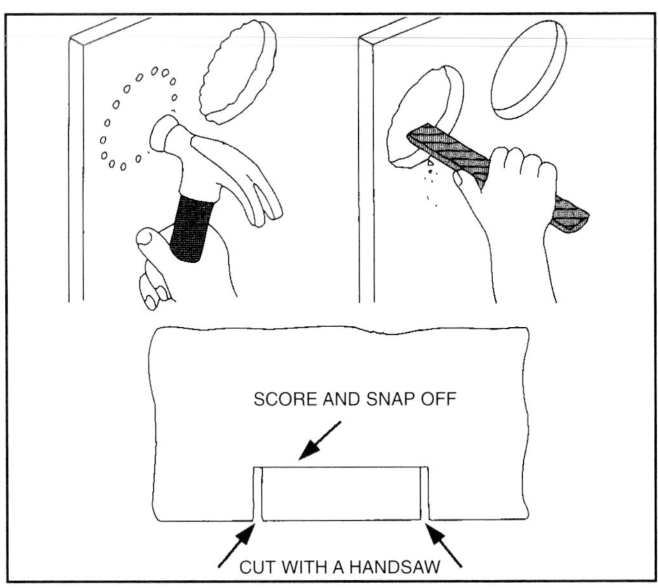

Figure 15-30. Cutting holes. On small holes, make a series of small holes and break out waste with a hammer. Score larger openings and saw where possible. (James Hardie Building Products, Inc.)

Figure 15-29. Working cement board is similar to working drywall. Top—After double-scoring the board, snap upward in the direction of the score for a clean break. (Georgia-Pacific) Bottom—Sawing with a hand or power saw is an alternate method. (USG Corp.)

Moisture Resistant (MR) Wallboard

This type of gypsum wallboard was processed to withstand the effects of moisture and high humidity. Its facing paper is light green so it can be easily identified. Though still in use, it is not as water resistant as cement board. Standard thicknesses include 1/2" and 5/8". Standard width is 4' and lengths range from 8' through 12'.

Moisture resistant wallboard can be used as a base under ceramic tile or other nonabsorbent finishing materials in showers and tub alcoves. Figure 15-32 shows a detail of the construction at the edge of a tub. Stud spacing should not be greater than 16" O.C. Note the *furring strip*. It ensures alignment between the tub lip and the wallboard. Also note the 1/4" space that should be maintained along the tub edge.

Veneer Plaster

Veneer plaster is a high-strength material applied as a coat less than 1/8" thick. Because of the composition and thinness of the coat, it dries very rapidly. Trim and decoration work may proceed after a minimum drying time of 24 hours.

A special gypsum board is used for the base. Its face surface consists of several layers of paper. The outer layer absorbs moisture rapidly and makes it easier to apply the plaster coat. The inner layer keeps the gypsum core dry and rigid. To identify the face surface, note that the outer layer is rolled over the long edge. Other than this, the

Figure 15-31. Top—Using screws, carpenters prepare to install cement board backing in a shower stall. (Georgia-Pacific) Middle—Tub surround installation. Bottom—Setting ceramic tile in a tub alcove. The wall surface has been covered with 1/2" cement board. (USG Corp.)

materials and methods are nearly the same as those for regular drywall construction.

Veneer plaster can be applied as a one- or two-coat system. Either system can be given a smooth or textured surface. Corner bead, trim, and grounds must be carefully set for a 1/16" thickness in one-coat applications and 3/32" for two-coat applications. Figure 15-33 shows a two-coat application of veneer plaster.

For best performance and workability of veneer plaster, the manufacturer's directions should be carefully followed. Proper mixing is especially important and is usually accomplished with a cage-type paddle mounted in an electric drill.

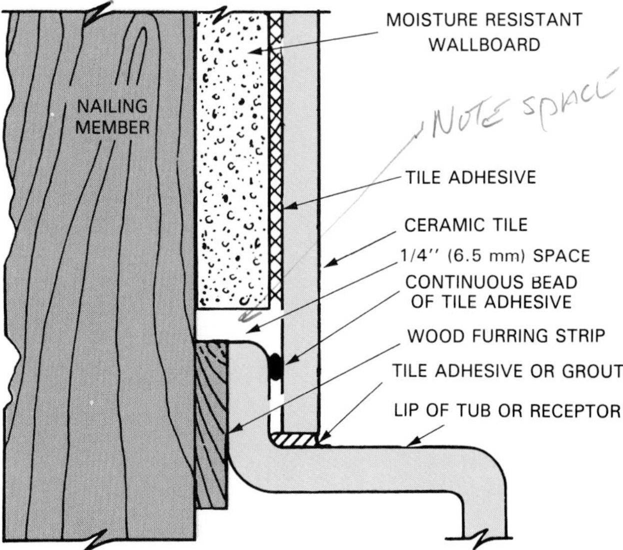

Figure 15-32. Moisture-resistant (MR) gypsum wallboard is designed for tub and shower areas. Note detail around tub edge. (Gold Bond Bldg. Products)

Predecorated Wallboard

A variety of predecorated gypsum wallboard is available. This is usually applied vertically because of the difficulty involved in successfully matching and finishing butt joints.

Wall surfaces must be dry before installation can begin. Panels should be unpacked and stood on their long edges, exposing both sides to room air for 24 to 48 hours before being attached. Prefinished panels can be attached to furring strips, studs, or solid surfaces. On remodeling jobs, remove wallpaper and repair loose paint and damaged plaster.

The use of an adhesive to bond the panels to a base layer is common practice. However, matching colored nails are available. For best results, manufacturers recommend using both gluing and nailing for 1/4" panels. Be sure to drive colored nails with a plastic-headed hammer, rawhide mallet, or a special cover placed over the face of a regular hammer. Nails should be spaced 8" apart and should never be closer than 3/8" from the ends or edges of

Figure 15-33. Application of a two-coat veneer plaster system. Top—Spreading the first coat with a standard trowel. The pallet held in the right hand of the plasterer is called a hawk. Bottom—Applying the finish coat.

the wallboard. Adhesive contacting the decorated surface must be removed immediately with a soft cloth and mineral spirits.

When adhesives alone are used for fastening, check each panel after about 15 to 30 minutes to ensure that the adhesive is set. Press firmly along edges and framing members. Cover a block of wood with a soft cloth or a rubber mallet and tap along all areas where adhesive was applied.

Avoid tight fit between floors and ceilings. A space of 1/16" should be allowed between panels for expansion and to avoid buckling.

To trim edges and joints of predecorated panels, you can use aluminum moldings made to match the finished surface of the wallboard, Figure 15-34. Cut them with a hacksaw and attach with flat-head wire nails spaced 8 to 10" apart.

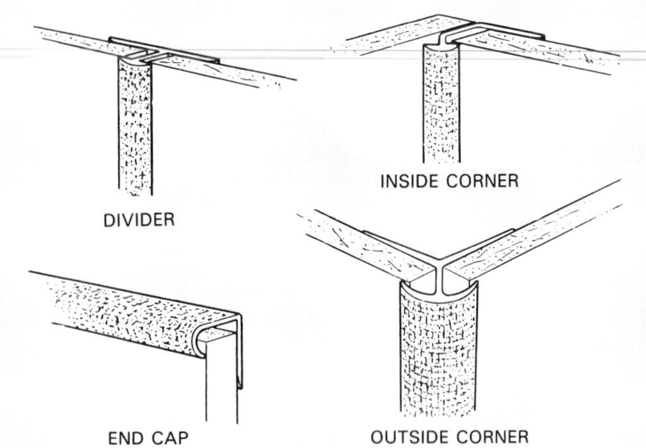

Figure 15-34. Moldings can be used to cover raw edges of predecorated gypsum wallboard.

When attaching divider strips, first place the molding on one panel that is carefully aligned, nail the exposed flange in place, then insert the next panel. Figure 15-35 shows an installation that has been completed.

Figure 15-35. This bathroom has walls of prefinished paneling done in geometric patterns. (ABTco, Inc.)

Wallboard on Masonry Walls

Gypsum wallboard can be installed over metal or wood furring strips attached to a masonry wall. Where the structure consists of an interior wall that is straight and true, the panels can be laminated directly to the masonry surface with a special adhesive.

Exterior walls must be thoroughly waterproofed and insulation should be included if the structure is located in a cold climate. Figure 15-36 shows a masonry wall application made with furring strips. The insulation may be rigid plastic foam or batts. When wood furring strips are used, they should be a nominal 2" wide and 1/32" thicker than the insulation. The plastic foam slabs are usually bonded to the masonry surface with adhesive. Wallboard joints and nail holes are concealed, following the finishing steps previously described.

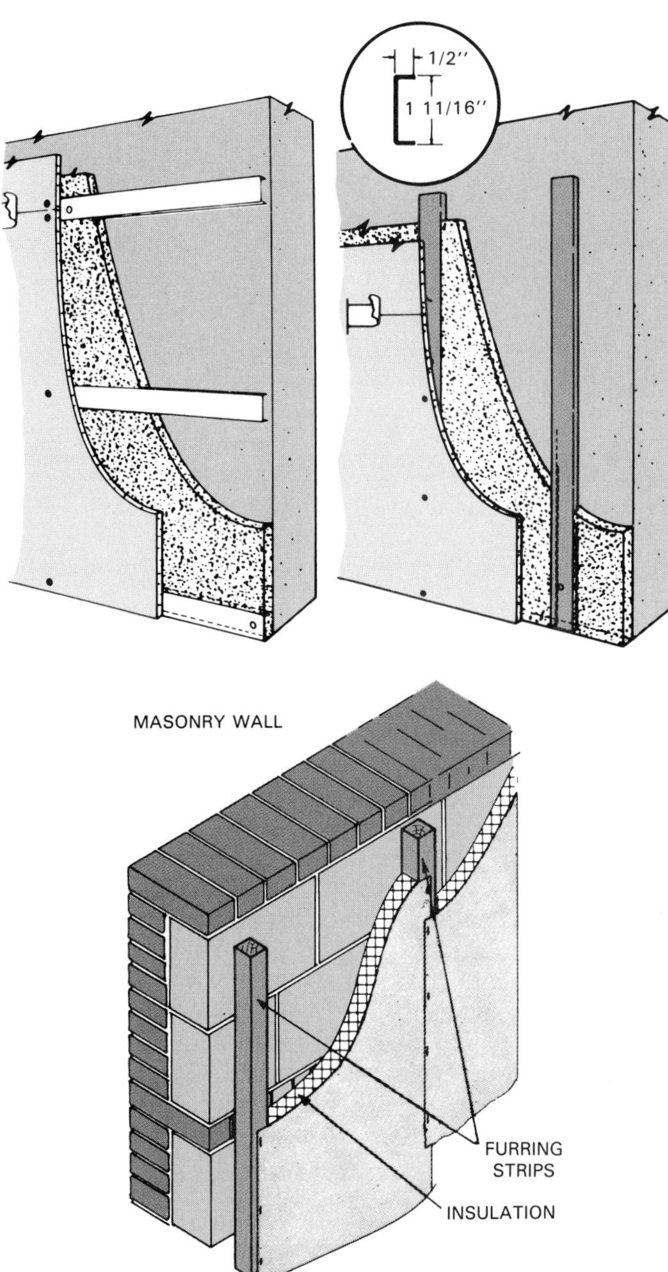

Figure 15-36. Two methods of preparing masonry walls for interior finish. Top—Wallboard can be attached to metal furring channels. Rigid insulation is used. Bottom—Wood furring strips and blanket insulation.

Installations of wallboard on exterior masonry walls, especially those below grade, must be carefully done. Be sure to follow recommendations furnished by manufacturers.

Installing Plywood

Most of the plywood used for interior walls has a factory-applied finish that is tough and durable, Figure 15-37. Manufacturers can furnish matching trim and molding that is also prefinished and easy to apply. Color-coordinated putty sticks are used to conceal nail holes.

Joints between plywood sheets can be treated in a number of ways. Some panels are fabricated with machine shaped edges that permit almost perfect joint concealment. Usually, it is easier to accentuate the joints with grooves or use battens and strips, Figure 15-38.

Figure 15-37. Plywood for interior walls is prefinished in a variety of finishes. (ABTco, Inc.)

Before installation, the panels should become adjusted (conditioned) to the temperature and humidity of the room. Prefinished plywood should be removed from cartons and carefully stacked horizontally. Place 1" spacer strips between each pair of face-to-face panels. Do this at least 48 hours before application.

Plan the layout to reduce the amount of cutting and number of joints. Figure 15-39 shows two application designs. It is important to align panels with openings whenever possible. If finished panels will have a grain, stand the panels around the walls and shift them until you

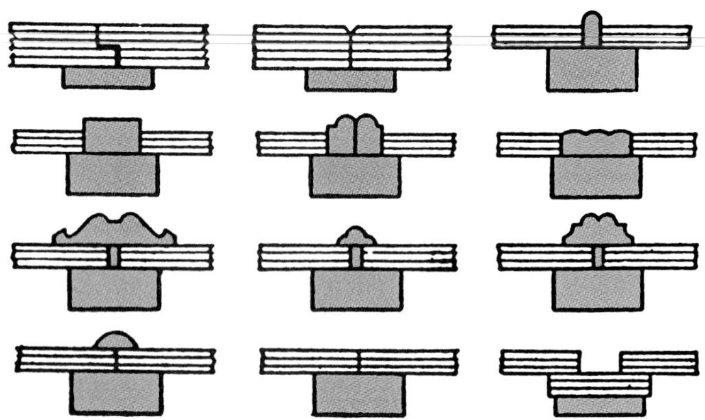

Figure 15-38. Different styles of battens can be used to conceal joints in plywood paneling.

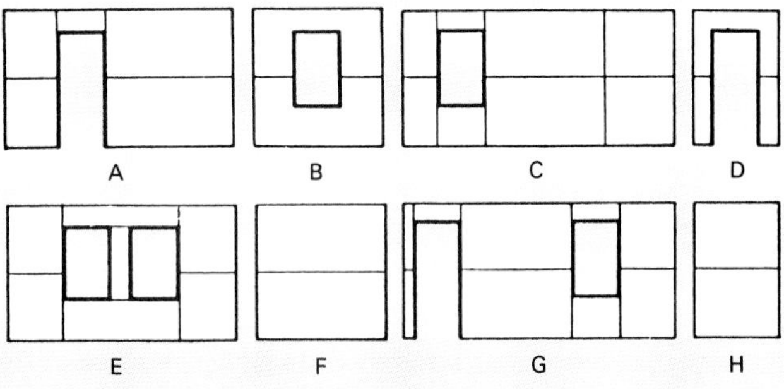

DESIGN 1 is a simple two-panel horizontal arrangement. The single continuous horizontal joint is placed midway between door and ceiling. Vertical joints at openings (elevations A, C, E, and G), again, key panel design. However, they may be omitted where panel length exceeds wall element width as in Elevations B and D.

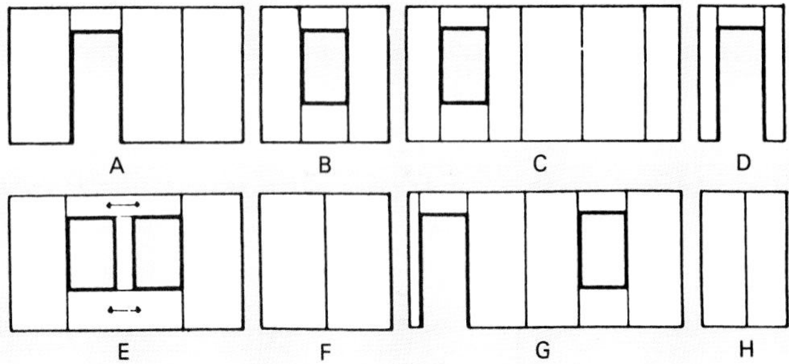

DESIGN 2, a vertical panel arrangement is another illustration of the basic principle of initiating panel design by lining up vertical joints with wall openings. The plain wall space is then divided vertically in widths proportionate to that of openings. In a vertical panel arrangement where width of a door or window opening exceeds panel width, panels may be placed horizontally as shown by arrows in Elevation E. Such combinations of vertical and horizontal arrangements may be used in the same room with pleasing effect.

Figure 15-39. Two plywood wall paneling arrangements commonly used. The basic rule is to "work from the openings." First, line up vertical joints above doors and above and below windows. Divide the remaining plain wall space into an orderly pattern as stud location allows.

have the most pleasing effect in color and grain patterns. To avoid confusion, number the panels in sequence after their position has been established.

When cutting plywood panels with a portable saw, mark the layout on the back side. Support the panel carefully and check for clearance below. Make the cut as shown in Figure 15-40. The cutting action of the saw blade will be upward against the panel face. Splintering will be minimal. This is even more important when working with prefinished panels.

Figure 15-40. A portable power saw can be used to cut prefinished plywood panels. Work should be performed with the back of the sheet facing up to avoid damage to the finish.

Plywood can be attached directly to the wall studs with nails or special adhesives. Use 3/8" plywood for this type of installation. When studs are poorly aligned or when the installation is made over an existing surface in poor condition, it is usually advisable to use furring. Nail 1 x 3 or 1 x 4 furring strips horizontally across the studs. Start at the floor line and continue up the wall. Spacing depends on the panel thickness. Thin panels need more support. Install vertical strips every 4' to support panel edges. Level uneven areas by shimming behind the furring strips. Figure 15-41 shows furring strips being attached to a concrete block wall. Prefinished plywood panels are being installed. A special panel adhesive can be used, Figure 15-42. The panels are simply pressed into place with no sustained pressure being required.

Begin installing panels at a corner. Scribe and trim the edge of the first panel so it is plumb. Fasten it in place before fitting the next panel. Allow about 1/4" clearance at the top and bottom. After all panels are in place, molding is used to cover the space along the ceiling. Baseboards will conceal the space at the floor line.

On some jobs, 1/4" plywood is installed over a base of 1/2" gypsum wallboard. This backing is recommended for several reasons:

- It tends to bring the studs into alignment.
- It provides a rigid finished surface.
- It improves the fire resistant qualities of the wall.

Figure 15-41. Paneling can be attached directly to masonry walls, but furring strips produce a better looking wall. Strips where panels join should be spray painted a color close to the tones in the paneling. This avoids unsightly off-color cracks at joints.

Figure 15-42. Applying adhesive to furring strips. Use a 1/4" bead. Follow adhesive manufacturer's directions. (Borden Inc.)

The plywood is bonded to the gypsum board with adhesive. In general, you should follow the same procedure as was described in double-layer drywall construction. See Figure 15-43.

Hardboard

Through special processing, *hardboard*, also called *fiberboard*, can be fabricated with a very low moisture absorption rate. This type is often scored to form a tile pattern and is used in bathrooms and kitchens. Panels for wall application are usually 1/4" thick.

Since hardboard is made from wood fibers, the panels will expand and contract slightly with changes in humidity. They should be installed when they are at their maximum size. There will be a tendency for them to buckle between the studs or attachment points if installed when moisture content is low. Manufacturers of prefinished hardboard panels recommend that they be unwrapped and then placed separately around the room for at least 48 hours before application.

Figure 15-43. Plywood paneling produces an attractive wall covering and is available in many different patterns. This style simulates board paneling. (American Plywood Assoc.)

Procedures and attachment methods are similar to those previously described for plywood. Special adhesives are available, as are metal or plastic molding in matching colors. Drill nail holes for the harder types. Figure 15-44 shows an attractive application of hardboard.

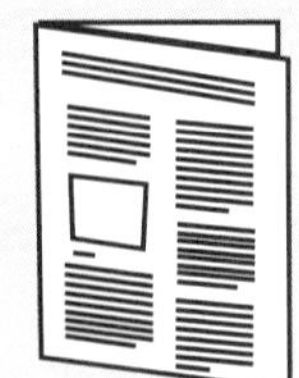

When applying any of the various factory finished wallboard, plywood, or hardboard materials, always follow the recommendations furnished by the manufacturer.

Figure 15-44. Prefinished interior hardboard wall paneling is made to look like pine. Note the wainscotting in the dining area. (Masonite Corp.)

Plastic Laminates

Plastic laminates are sheets of synthetic material that are hard, smooth, and highly resistant to scratching and wear. Although basically designed for table and countertops, they are also used for wainscotting and wall paneling in homes and commercial buildings.

Since the material is thin (1/32 to 1/16"), it must be bonded to other supporting panels. Contact bond cement is commonly used. Recently, manufacturers have developed prefabricated panels with the plastic laminate already bonded to a base or backing material. Figure 15-45 shows the installation of a prefabricated panel consisting of a 1/32" plastic laminate mounted on 3/8" particleboard. Edges are tongue and grooved so that units can be blind-nailed into place. Matching corner and trim moldings are available. Figure 15-46 shows a completed installation.

Figure 15-45. These wall panels are manufactured from plastic laminate with a particleboard base. (Formica Corp.)

Figure 15-46. An office wall. Plastic laminated paneling is used for its durability.

Solid Lumber Paneling

Solid wood paneling makes a durable and attractive interior wall surface and may be appropriately used in nearly any type of room. A number of different species of hardwood and softwood are available. Sometimes, grades that contain numerous knots are used to secure a special appearance. Defects, such as the deep fissures in pecky cypress, can provide a dramatic effect.

Softwood species most commonly used include pine, spruce, hemlock, and western red cedar. Boards range in widths from 4" to 12" (nominal size) and are dressed to 3/4". Board and batten or shiplap joints are used, but tongue and groove joints combined with shaped edges and surfaces are more popular.

Paneling patterns are often reversible, offering two choices in a single panel. Some patterns are smooth on one side and saw-textured on the other. Figure 15-47 shows several patterns offered. Dozens of variations are possible, depending on species of wood, texture, finish, and how the paneling is applied to a wall.

When solid wood paneling is applied horizontally, furring strips are not required and the boards are nailed directly to the studs. Inside corners are formed by butting the paneling units flush with the other walls.

If random widths are used, boards on adjacent walls must match and be accurately aligned. Vertical installations require furring strips at the top and bottom of the wall and at various intermediate spaces, Figure 15-48. Sometimes 2 x 4" pieces are installed between the studs to serve as a nailing base. Even when heavy tongue and groove boards are used, these nailing members should not be spaced more than 48" apart.

Narrow widths (4"–6") of tongue and groove (T&G) paneling are *blind nailed* (nailed so heads of nails do not appear on finished surface). This eliminates the need for countersinking and filling nail holes and provides a smooth blemish-free surface, which is especially important when clear finishes are used. Use a 6d finish nail and drive it at a 45° angle into the base of the tongue and on into the bearing point.

Exterior wall constructions, where the interior surface consists of solid wood paneling, should include a tight application of building paper or housewrap. This will prevent the infiltration of wind and dust through the joints. In cold climates, insulation and vapor barriers are important.

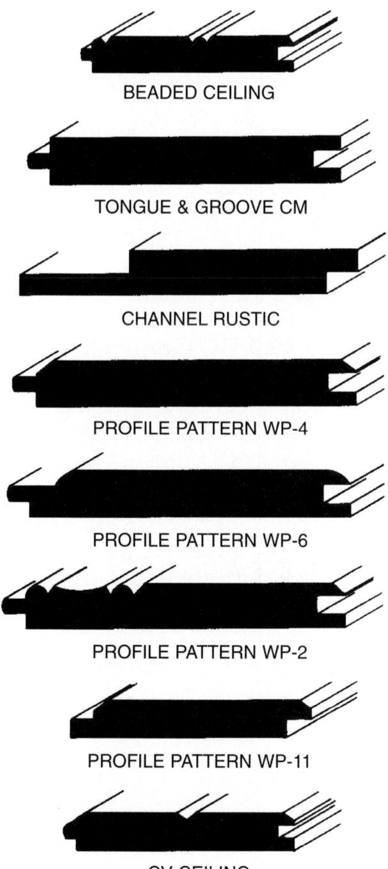

Figure 15-47. Smooth-surfaced solid wood paneling is available in several pattern profiles including those shown here. (Western Wood Products Assoc.)

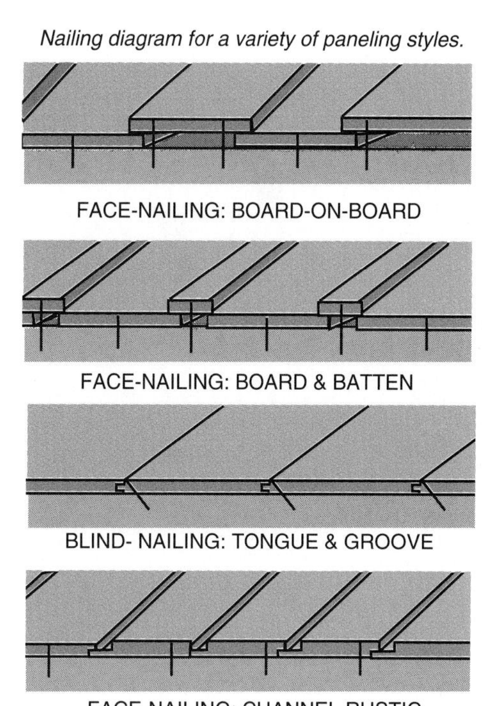

Figure 15-48. Top—Nailing techniques for a variety of paneling styles. (Western Wood Products Assoc.) Bottom—Installing solid wood paneling. Boards are edge and end matched. (Forest Products Laboratory)

Installing Solid Paneling at an Angle

Though more difficult and time consuming to install, angled paneling is handsome and goes well with today's informal designs. Diagonal paneling angles in only one direction. Angles of 22 1/2°, 30°, and 45° are most popular. Chevron paneling is installed in a "V" or inverted "V" pattern. Herringbone paneling alternates the direction of the angle at regular intervals. All three types are shown in Figure 15-49 along with charts on coverage.

To establish the proper angle for diagonal paneling, first establish a vertical line in the middle of the wall (use a plumb line). Draw an intersecting horizontal line about a foot off the floor. Mark off 3' from the intersection in each direction; draw a diagonal line through the two points. For other angles, use a protractor to establish the diagonal line.

For chevron and herringbone panel application, it is important to have plumb nailing bases wherever the diagonals change direction. For chevron paneling this nailing

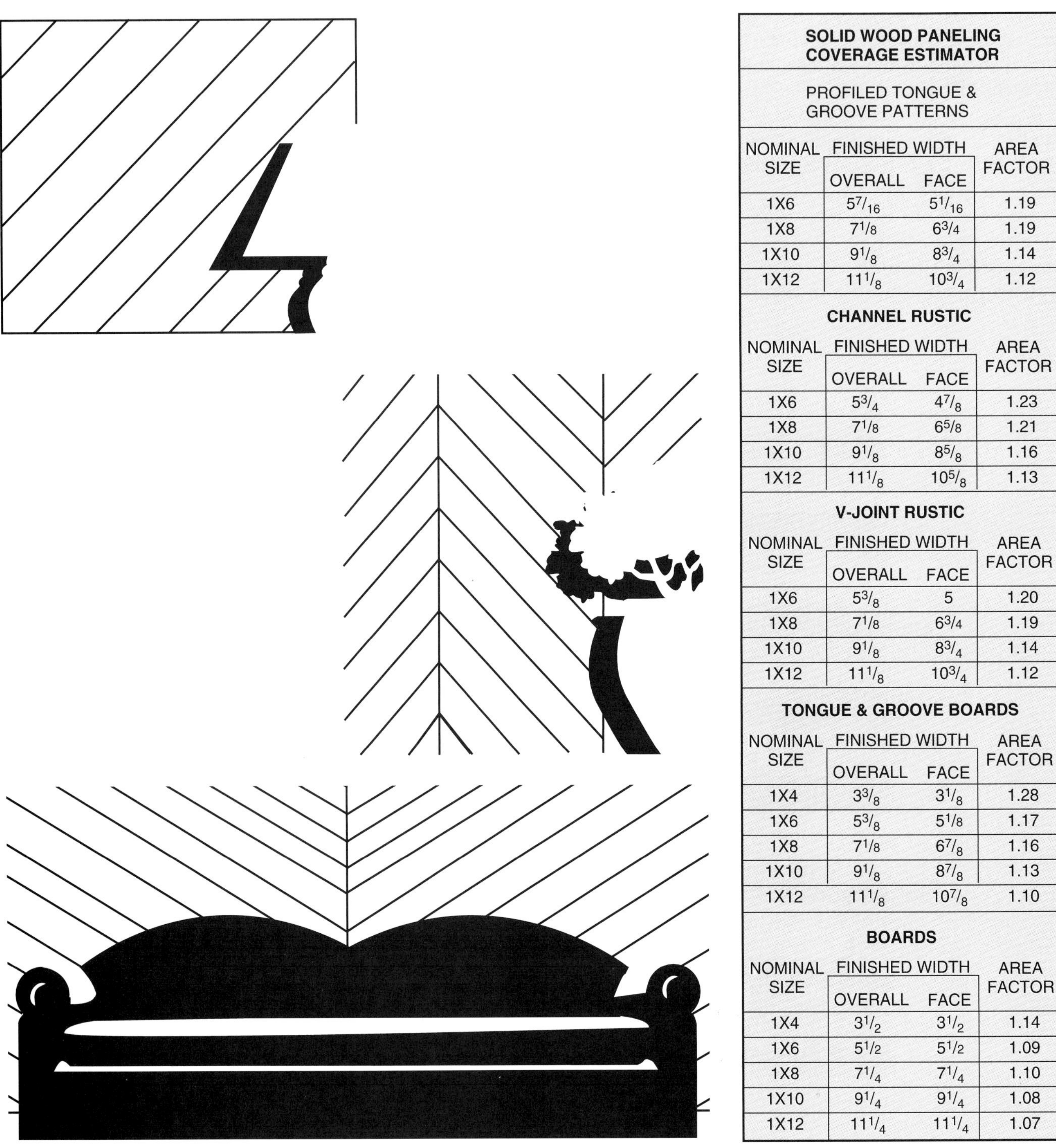

SOLID WOOD PANELING COVERAGE ESTIMATOR

PROFILED TONGUE & GROOVE PATTERNS

NOMINAL SIZE	FINISHED WIDTH OVERALL	FINISHED WIDTH FACE	AREA FACTOR
1X6	5 7/16	5 1/16	1.19
1X8	7 1/8	6 3/4	1.19
1X10	9 1/8	8 3/4	1.14
1X12	11 1/8	10 3/4	1.12

CHANNEL RUSTIC

NOMINAL SIZE	FINISHED WIDTH OVERALL	FINISHED WIDTH FACE	AREA FACTOR
1X6	5 3/4	4 7/8	1.23
1X8	7 1/8	6 5/8	1.21
1X10	9 1/8	8 5/8	1.16
1X12	11 1/8	10 5/8	1.13

V-JOINT RUSTIC

NOMINAL SIZE	FINISHED WIDTH OVERALL	FINISHED WIDTH FACE	AREA FACTOR
1X6	5 3/8	5	1.20
1X8	7 1/8	6 3/4	1.19
1X10	9 1/8	8 3/4	1.14
1X12	11 1/8	10 3/4	1.12

TONGUE & GROOVE BOARDS

NOMINAL SIZE	FINISHED WIDTH OVERALL	FINISHED WIDTH FACE	AREA FACTOR
1X4	3 3/8	3 1/8	1.28
1X6	5 3/8	5 1/8	1.17
1X8	7 1/8	6 7/8	1.16
1X10	9 1/8	8 7/8	1.13
1X12	11 1/8	10 7/8	1.10

BOARDS

NOMINAL SIZE	FINISHED WIDTH OVERALL	FINISHED WIDTH FACE	AREA FACTOR
1X4	3 1/2	3 1/2	1.14
1X6	5 1/2	5 1/2	1.09
1X8	7 1/4	7 1/4	1.10
1X10	9 1/4	9 1/4	1.08
1X12	11 1/4	11 1/4	1.07

Figure 15-49. Left—Three patterns for angled solid paneling. Right—Tables are available for estimating coverage of various patterns of solid paneling. (Western Wood Products Assoc.)

base (stud or furring strip) is located midway on the wall. For herringbone paneling, Figure 15-50, locate a plumbed vertical nailer every 36". Center an additional nailing surface at 18".

To start chevron and herringbone patterns, install triangles of paneling at centerlines. These should be glued or blind nailed. For tight, well-matched joints, mark the cutting angle for each board as it is installed. Use a level or straight edge as a guide. Manufacturers, or their trade associations provide detailed instructions for such installations.

%

Interior solid wood paneling may cause problems resulting from expansion and shrinkage. Be sure the material used has a moisture content about equal to that which it will attain in service. This should be about 8 to 10% for most parts of the United States. Prior to installation, store the paneling in the room and allow room air to reach both sides of the panels.

Plaster

Through the years, gypsum *plaster,* Figure 15-51, has provided qualities that are desired in a wall and ceiling finish: beauty, durability, economy, fire protection, structural rigidity, and resistance to sound transmission. It is also highly adaptable because it can be readily applied to curved or irregular surfaces.

Plaster is made from gypsum, one of the common minerals found in the earth. Fire protection engineers, recognizing the fire resistance qualities of plaster, have developed accurate ratings. These are listed by the National Bureau of Standards and the National Bureau of Fire Underwriters. These ratings are used as a basis for establishing requirements in various building codes.

When plaster is used for the interior wall and ceiling surface, the carpenter usually applies the plaster base and installs the *grounds* that serve as guides for the plasterer. (Grounds are strips of wood or metal placed along floors and around wall openings as a thickness guide for the plasterer.) Carpenters, in addition to a thorough knowledge of the requirements and methods of making this installation, should also have a general knowledge of how the plaster coats are applied.

Plaster Base

A plaster finish requires some type of base. For many years, *wood lath* was used for this purpose. However, in modern construction, sheet materials and metal lath have replaced it. Plaster is sometimes applied directly to masonry surfaces or special gypsum block units.

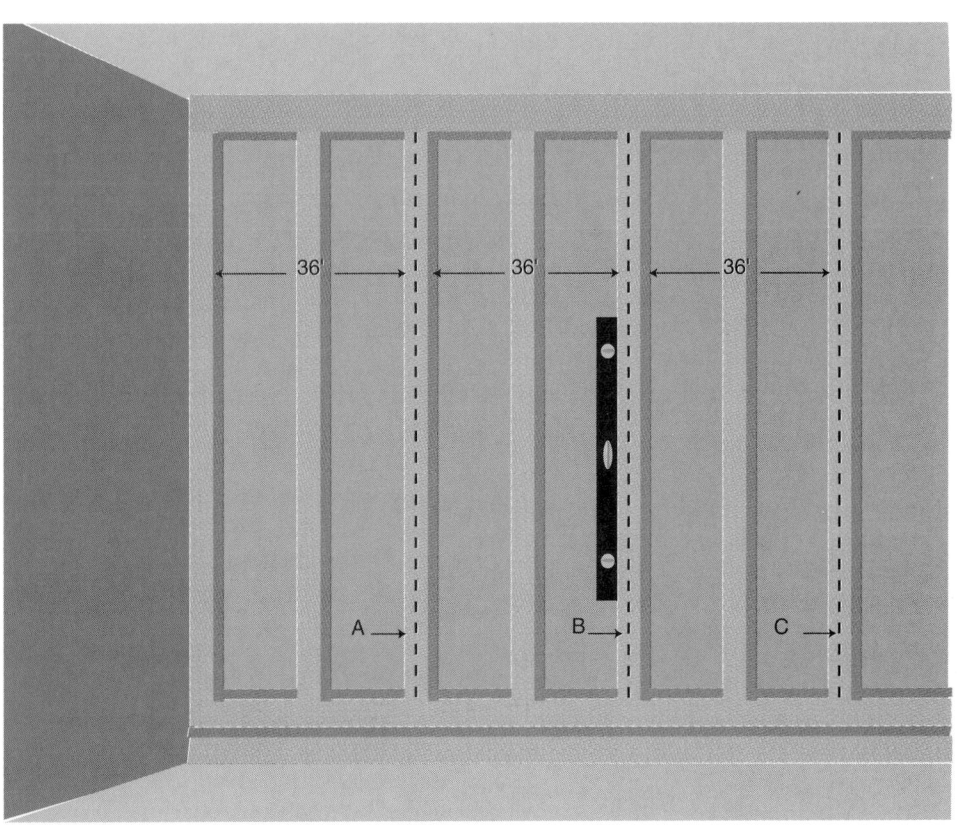

Figure 15-50. Herringbone paneling requires furring strips every 18" O.C. Plumb lines should be drawn every 36" O.C. (Western Wood Products Assoc.)

Figure 15-51. Plaster is always applied over a supporting base, in this case, perforated gypsum lath. Old style wood lathe has been replaced by several new products. (Cemco)

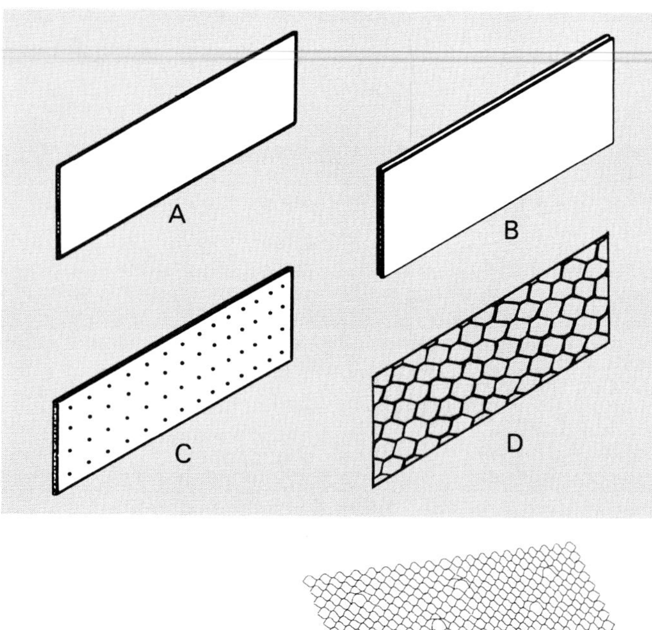

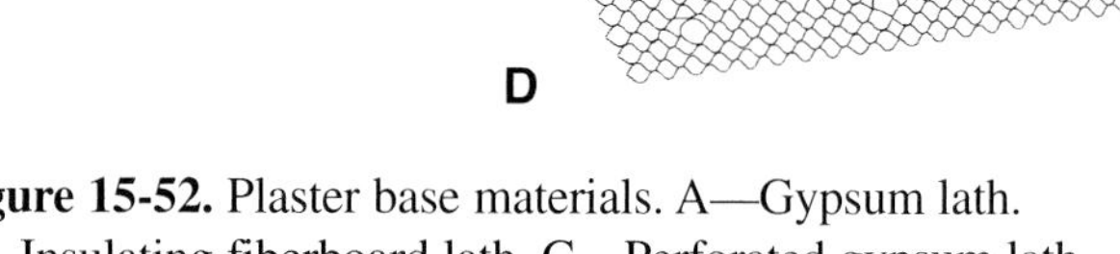

Figure 15-52. Plaster base materials. A—Gypsum lath. B—Insulating fiberboard lath. C—Perforated gypsum lath. D—Expanded metal lath.

Today, commonly used plaster bases include *gypsum lath* and expanded metal, Figure 15-52. A standard gypsum panel measures 16" x 48" and is applied horizontally to the framing members of the structure. It consists of a rigid gypsum filler with a special paper cover.

For a stud or ceiling joist spacing of 16" O.C., 3/8" lath is used. For a 24" spacing, 1/2" is required.

Gypsum lath is also made with a backing of aluminum foil vapor barrier. Lath with perforations improves the plaster bond and extends the time the wall surface will remain intact when exposed to fire. Some building codes specify this type.

Insulating *fiberboard lath* is also used as a plaster base. It comes in a 3/8" and 1/2" thickness with a shiplap edge. Fiberboard lath has considerable insulation value and is often used on ceilings or walls adjoining exterior or unheated areas. Figure 15-53 shows sizes available.

Expanded metal lath consists of a copper alloy sheet steel, slit and expanded to form openings for keying the plaster. The two most common types are diamond mesh and flat rib. Standard size pieces are 27" wide by 96" long. Metal lath is usually dipped in black asphaltum paint, or is galvanized, to help it resist rust.

TYPE	THICKNESS (Inches)	WIDTH (Inches)	LENGTH (Inches)
Plain	3/8 1/2	16 16	48 or 96 48
Perforated	3/8 1/2	16 16	48 or 96 48
Insulating	3/8 3/8 or 1/2	16 24	48 or 96 as requested to 12 ft.
Long Length	1/2	24	as requested to 12 ft.

Figure 15-53. Gypsum and insulating lath are available in these sizes.

Installing Lath

Before lath is applied, the framing should be inspected for proper spacing and alignment. Check to see that corners and openings have nailing members to support the ends of the lath.

Units should be applied with the long dimension at right angles to the framing. Stagger the end joints between adjacent courses, Figure 15-54. Turn the folded or lapped paper edges of gypsum lath toward the framing. Edges and ends of lath should be in moderate contact.

Gypsum lath 3/8" thick is usually applied with 13 gage gypsum lathing nails 1 1/8" long. Nail sizes for other installations are given in Figure 15-55. The lath must be nailed at each stud or joist crossing. Insulating lath is installed in the same manner as gypsum lath, except that 13 gage 1 1/4" blued nails should be used. Nails can be driven with pneumatic staplers as shown in Figure 15-56. In

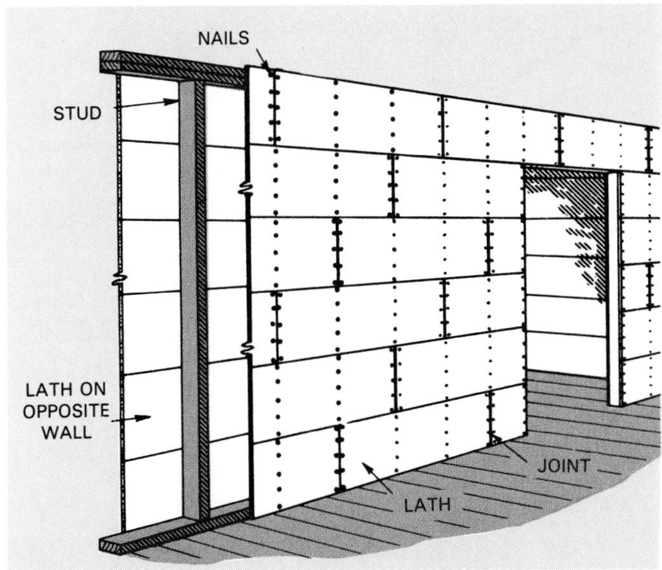

Figure 15-54. Application of gypsum lath to wood studs. Joints must be staggered. Avoid joints at corners of window and door openings.

TYPE FRAMING	BASE THICKNESS		FASTENER	MAX. FRAME SPACING		MAX. FASTENER SPACING	
	in	mm		in	mm	in	mm
Wood	3/8	9.5	Nails—13 ga., 1 1/8" long, 19/64" flat head, blued	16	406	5	127
			Staples—16-ga. galv. flattened wire flat crown 7/16" wide, 7/8" divergent legs				
	1/2	12.7	Nails—13 ga., 1 1/4" long, 19/64" flat head, blued	24	610	4	102
			Staples—16-ga. galv. flattened wire, flat crown 7/16" wide, 1" divergent legs				
USG Steel Stud	3/8	9.5	1" Type S Screws	16 24	406 610	12	305
TRUSSTEEL Stud	3/8	9.5	Clips	16	406	16	406

Figure 15-55. Follow these nailing requirements for gypsum lath.

modern construction, staples are often used to attach the lath.

Gypsum lath is easily cut to size by scoring one or both sides with a pointed or edge tool. Break the lath along the line. Be sure to make neatly fitted cutouts for plumbing pipes and electrical outlets.

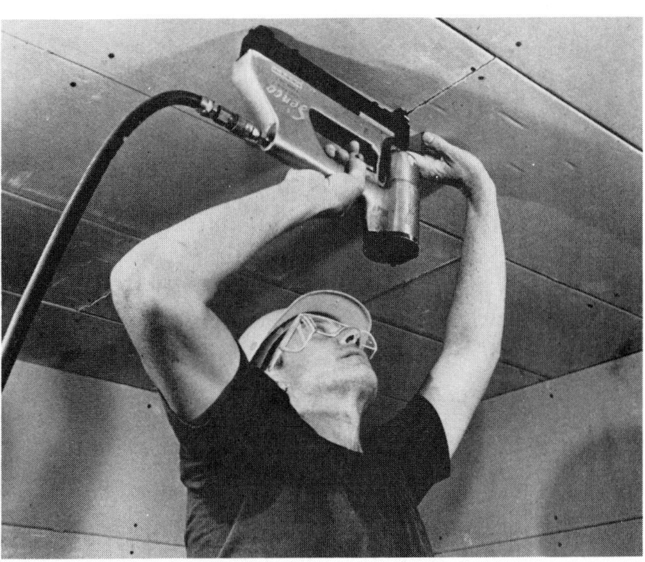

Figure 15-56. Pneumatic powered nailer is being used to attach gypsum lath.

When applying gypsum lath, drive the nails so the head will be just below the paper surface. Avoid additional hammer blows that will crush the gypsum core.

Metal Lath

In commercial construction, *metal lath* is common. In residential work it may be used only in shower stalls or tub alcoves. The metal lath provides a rigid wall surface when the construction is properly designed.

Metal lath must be of the proper type and weight for the support spacing. Sides and ends are lapped and corners

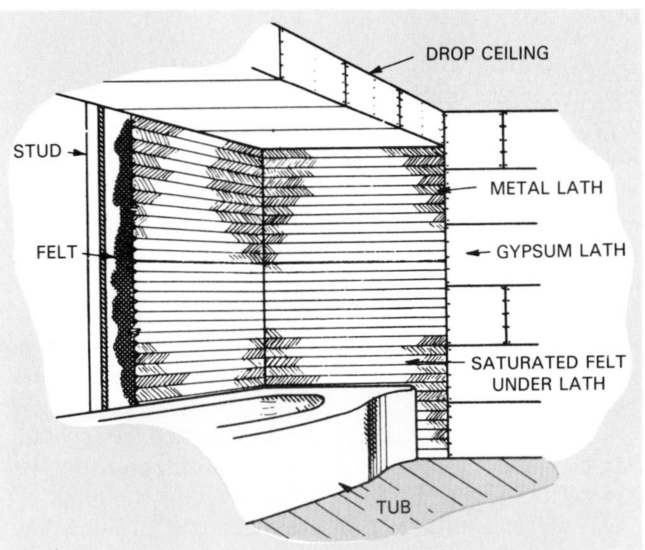

Figure 15-57. Metal lath is installed over layer of waterproof felt paper in areas subject to moisture. This can be eliminated if cement board is used under the lath.

are returned (overlapped). Studs are usually covered first with a 15 lb. asphalt-saturated felt, Figure 15-57. Portland cement plaster is often used as the first coat when the surface will be finished with ceramic tile. Gypsum plaster is used for top coats.

Reinforcing

Since some drying will nearly always occur in wood frame structures, shrinkage can be expected. This will likely cause plaster cracks to develop around openings, in corners, or wherever there is a concentration of cross-grain wood.

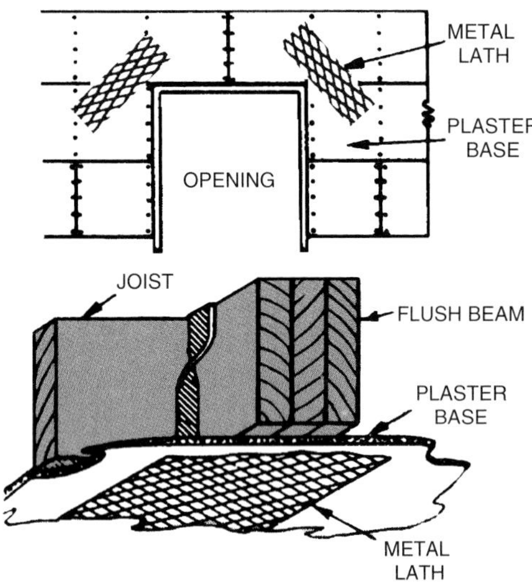

Figure 15-58. Reinforcing the plaster base. Top—Place metal lath around the jamb area of large openings, especially where large headers are used. Bottom—Providing reinforcing under flush beams.

To minimize cracking, expanded metal lath is often used in key positions over the plaster base. Strips 8" wide are applied at an angle over the corners of doors and windows as shown in Figure 15-58. Tack or staple these into place lightly so they become a part of the plaster base only. If nailed securely to the framing, warping, shrinking, and twisting of the frame will be transmitted into the plaster and cause cracks.

Metal lath should be used under and around wood beams that will be covered with plaster. Be sure to extend the edges of the reinforcing well beyond the structural element being covered.

Inside corners may be reinforced with a specially formed metal lath or wire fabric sometimes called "Cornerite," Figure 15-59. Minimum widths should be 5", 2 1/2" on each surface of the internal angle. In some plaster base systems that employ a special attachment clip, Cornerite is not recommended.

Outside corners where no wood trim will be applied are reinforced with metal corner beads. They must be carefully applied and plumbed or leveled since they serve not only as a reinforcement for the plaster but as a ground (guide) for its application.

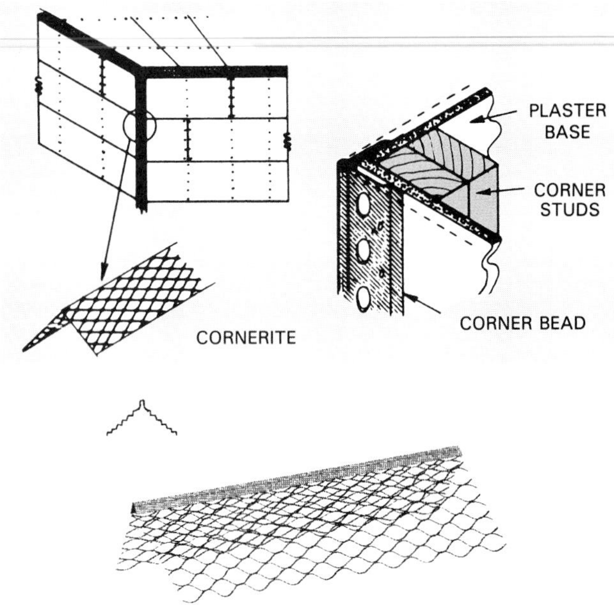

Figure 15-59. Cornerite, or corner beads. Top—Prefabricated reinforcing is made for both inside and outside corners. Bottom—This corner bead is to be shaped over uneven corners. (Cemco)

When applying corner bead, use a straightedge and level or plumb line. Also use a spacer block to check the distance between the corner of the bead and the surface of the plaster base. This distance must be equal to the thickness of the plaster coats.

Plaster Grounds

Plaster grounds are usually wood strips as thick as the plaster base and plaster. For average residential construction this would be 3/8" plus 1/2". Grounds are installed before the plaster is applied. The plasterer uses them as a

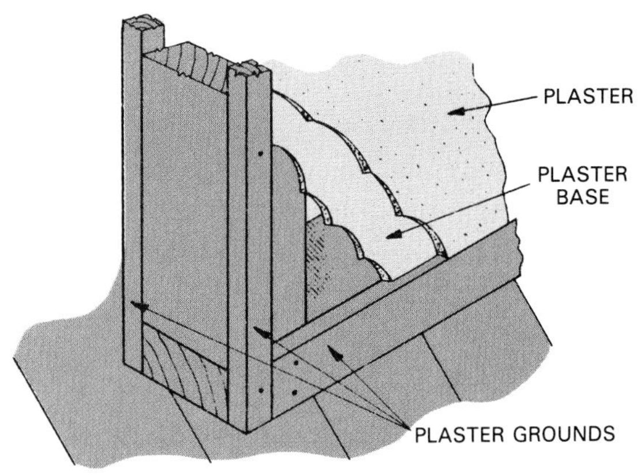

Figure 15-60. Plaster grounds are installed around openings and sometimes along the floor.

gauge for thickness of the plaster. They also help keep the plaster surface level and even. Later, the grounds may become a nailing base for attaching trim members. They are used around doors, windows, and other openings. Sometimes they are included at the bottom of walls along the floor line, Figure 15-60. In some wall systems, especially large commercial and institutional buildings, metal edges and strips serve as grounds.

Windows are usually equipped with jamb extensions that are adjusted for the various thicknesses of materials used in the wall structure. These jamb extensions serve as plaster grounds and the carpenter seldom needs to make any changes.

Grounds around some openings are removed after the plastering is complete. Those used at door openings must be carefully set (plumbed) and conform to the width of the door jamb to ensure a good fit of the casing. The width of standard interior door jambs is usually 5 1/4".

Carpenters often construct a jig or frame that is temporarily attached to the door opening. Grounds can then be quickly nailed in place along the straight edges of the jig. Instead of using two strips, some carpenters prefer to use a single piece of 3/8" exterior plywood ripped to the same width as the door jamb. Such grounds, if carefully removed, can be reused on future jobs, Figure 15-61.

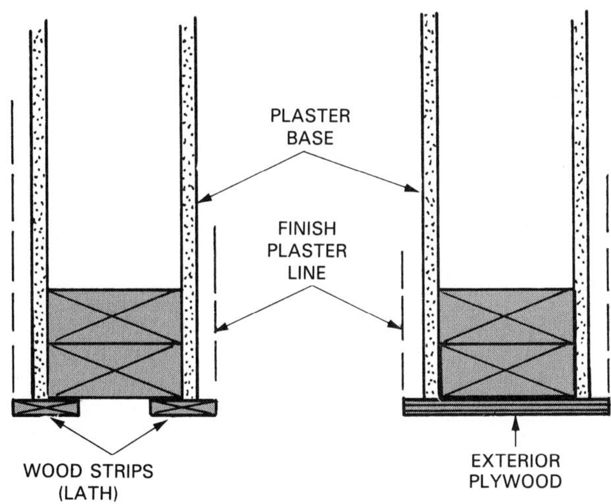

Figure 15-61. Note method of installing removable grounds at door openings.

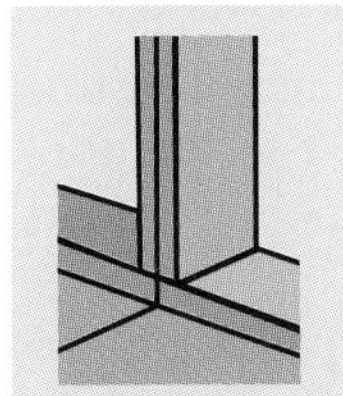

When the plaster base is complete, mark lines on the subfloor at the centerline of each stud. After the plaster has been applied and is dry, these marks are transferred to the wall to help when installing baseboards, cabinets, and fixtures.

Plaster Base on Masonry Walls

Gypsum plaster can be applied directly to most masonry surfaces. However, to prevent excessive heat loss through outside walls, furring strips are usually installed.

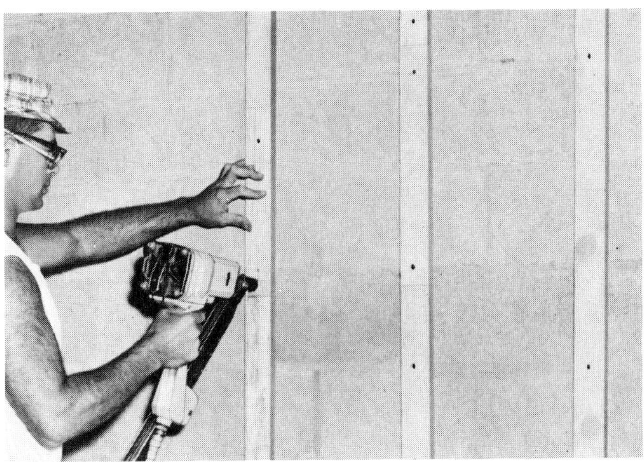

Figure 15-62. Usually, furring strips are attached to concrete or masonry walls to form a base for lath. Pneumatic nailer with 6d round headed nails is being used. (Duo-Fast Fastener Corp.)

These carry the plaster base materials and provide an air space. Strips are attached in the following ways:

- Case hardened nails driven into the mortar joints.
- Metal or wood nailing plugs placed in the wall during construction.
- Newly developed adhesive materials and fasteners.

Figure 15-62 shows strips being attached with a pneumatic nailer. When heat loss needs to be further reduced, blanket insulation can be installed. Some carpenters use plastic foam panels that are attached directly to the wall with adhesives. If furring strips are not used, the plastic foam can serve as a plaster base. Follow manufacturer's recommendations when making an installation of this kind.

Plastering Materials and Methods

Plaster is applied in two or three coats. The base coats are prepared by mixing gypsum with an aggregate, either at the gypsum plant or on the job. These aggregates may consist of wood fibers, sand, perlite, or vermiculite. Sand is the most commonly used material.

Discard plaster that has started to set. Wash out the mixer with clean water after each batch has been prepared. Keep tools and equipment clean.

In three-coat work, the first application, called the *scratch coat*, is applied directly to the plaster base, Figure 15-63. It is cross-raked, or scratched, after having "taken up" (stiffened). This coat is then allowed to set and partially dry.

The second, or *brown coat*, is then applied and leveled with the grounds and screeds. A long flat tool called a *darby* and a rod (straightedge) are used. When the brown coat has set and is somewhat dry, it is time to apply the third or finish coat.

In two-coat work, the scratch coat and brown coat are applied almost at the same time. The cross raking of the scratch coat is omitted. The brown coat of plaster is usually

Figure 15-63. Base coat of plaster being applied with a trowel. Care must be used around electrical outlets.

applied (doubled-back) within a few minutes. This application method is the one most frequently used over gypsum or insulating lath plaster bases commonly used in residential construction.

Minimum plaster thickness for all coats should not be less than 1/2" when applied to regular gypsum or insulating lath bases. A 5/8" thickness is usually required over brick, tile, or masonry. When plaster is applied to metal lath, it should measure 3/4" in thickness from the backside of the lath.

Figure 15-64. Plastering machine is being used to apply a base coat. (Gypsum Assoc.)

A plaster job should be inspected constantly for basecoat thickness. Unless grounds and screeds are used on ceilings or large wall areas, it is extremely difficult to keep the thickness uniform. Should the thickness be reduced, the possibility of checks and cracks is much greater. A 1/2" thickness of plaster possesses almost twice the resistance to bending and breaking as a 3/8" thickness.

Recent developments in the plastering trade center around the plastering machine, Figure 15-64. It not only saves a great deal of labor but also improves the quality of the plaster application. The lapsed time between mixing and application is shortened. The machine also makes possible the control of the plaster coat's density.

The final or finish coat (about 1/16" thick) consists of two general types: the sand-float or textured surface and the putty or smooth finish. In the sand-float finish, special sand is mixed with gypsum or lime and cement. After the plaster is applied to the surface it is smoothed with a float to produce various effects, depending on the floating method and the coarseness of the sand.

A smooth finish is produced by applying a puttylike material consisting of lime and gypsum or cement, Figure 15-65. It is troweled perfectly smooth, like concrete.

Figure 15-65. A puttylike mixture of lime and gypsum or cement is used for the final or finish coat of plaster.

Ceiling Tile

Ceiling tiles are used for both old and new construction. They can be installed over engineered metal strips, wood furring strips, solid plaster, drywall, or any smooth, continuous surface.

There is a considerable range in the types of material used to make ceiling tile. Some types are fiberboard tile, mineral tile, perforated metal tile, and glass-fiber tile.

When selecting a product, consider its appearance, light reflection, fire resistance, sound absorption, maintenance, cost, and ease of installation.

A standard size tile is 12" x 12". However, the tiles are available in larger sizes; for example, 24" x 24" and

16" x 32". A wide range of surface patterns and textures are manufactured. Figure 15-66 shows a typical fiberboard tile design that would be appropriate for a residential installation.

Figure 15-67 illustrates an overlapping, tongue and groove edge that provides a wide flange to receive staples. This type of joint also permits efficient installation when an adhesive is used to hold the tile in place.

Figure 15-66. Typical fiberboard ceiling tile is a foot square with a face that has acoustical properties.

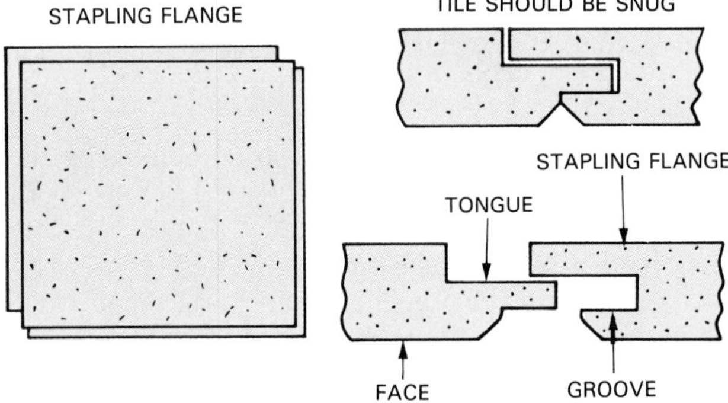

Figure 15-67. Tongue and groove joint used on standard ceiling tile. Flanges will receive staples while groove holds one edge of the next tile.

Layout Procedure

First, measure the two short walls and locate the midpoint of each one. Snap a chalk line, Figure 15-68, to establish the centerline. In the middle of this line, establish a chalk line that is at right angles to the long centerline. All tiles are installed with their edges parallel to these lines.

For an even appearance, the border courses along opposite walls should be the same width. For example, in a room 10'-8" wide, use nine full tiles and a border tile trimmed to 10" on each side.

After determining the width of the border tile, snap chalk lines parallel to the centerlines that will provide a guide for the installation of these border tile or the furring strips that will support them.

Figure 15-68. Use a chalk line to mark the center of the ceiling tile layout.

Installing Furring

Furring should be nailed to the ceiling joists on new construction. If applied over an old ceiling surface, be sure to locate and mark the joists before attaching the strips. Place the first furring strip flush against the wall at right angles to the joists and nail it with two 8d nails at each joist, Figure 15-69. Nail the second strip in place so that it will be centered over the edge of the border tile. Use the line previously laid out. All other strips are then installed on center (for 12" x 12" tile, locate strips every 12"). Figure 15-70 shows completed furring.

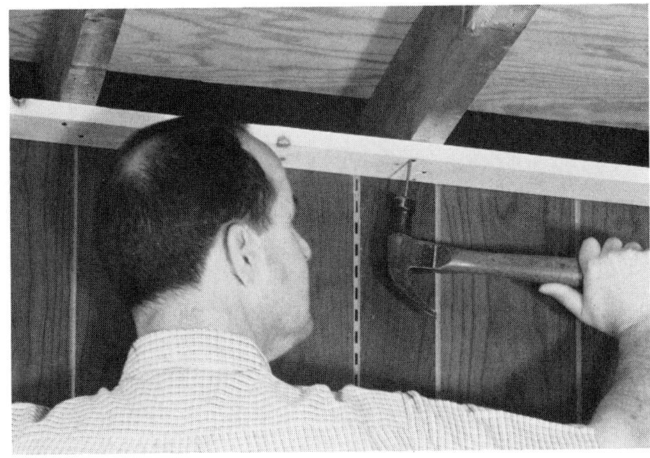

Figure 15-69. Install the first furring strip. Use two 8d nails at each joist.

Figure 15-70. Completed installation. Note that two strips along wall are closer together. This is to take care of the narrower width of border tile.

Figure 15-72. Furring strips can be leveled by driving tapered shims between the joist and the strip.

Some carpenters prefer to start from the center and work each side toward the walls. If the furring strips are uniform in width, a spacer jig, Figure 15-71, may speed up the job. Double-check the position of the furring strips from time to time. Make sure they will be centered over the tile joints.

It is essential that the faces of the furring strips be level with each other. Check alignment with a carpenter's level, straightedge, or by stretching a line across the strips. To align strips, drive tapered shims between the strips and the joist as shown in Figure 15-72. If only one or two joists extend below the plane of the others, it may be best to notch these joists before installing the furring strips.

If pipes or electrical conduit are located below the ceiling joists, it may be necessary to double the furring strips, Figure 15-73. The first course is spaced 24 to 32" on center and attached directly to the joists. The second course is then applied at right angles to the first. Space according to the width of the tiles as previously described. Large pipes or ducts that project below the ceiling joists should be boxed in with furring strips before the tile is installed. Wood or metal trim can be used to finish corners and edges.

Figure 15-71. A spacer jig can be used to locate furring strips.

Figure 15-73. Doubling the furring at right angles provides space for conduit located below ceiling joists.

There is an alternate furring method. It consists of installing gypsum wallboard to the joists or an existing ceiling surface. The tile is then attached to the wallboard either with adhesive or special staples.

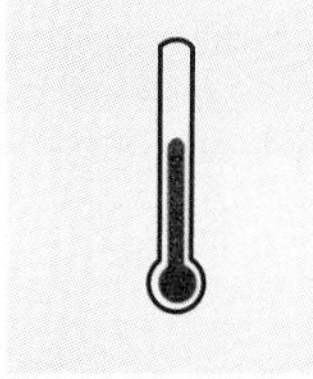

Most manufacturers recommend that fiberboard tile be unpacked in the area where they will be installed, at least 24 hours before application. This will allow the tile to adjust to room temperature and humidity.

Installing Tile

Check over the furring carefully and then snap chalk lines on the strips to provide a guide for setting the border

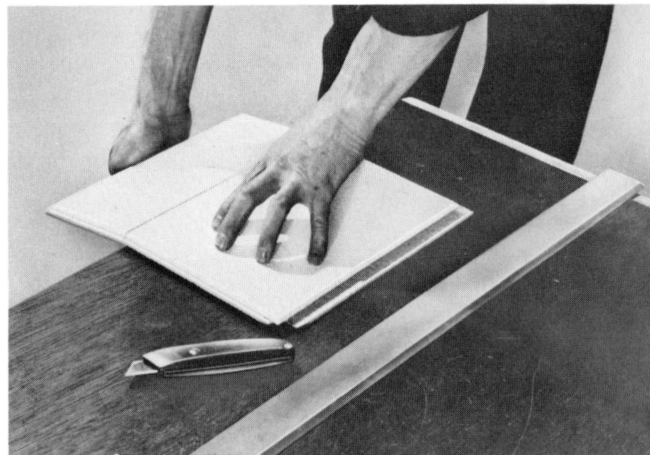

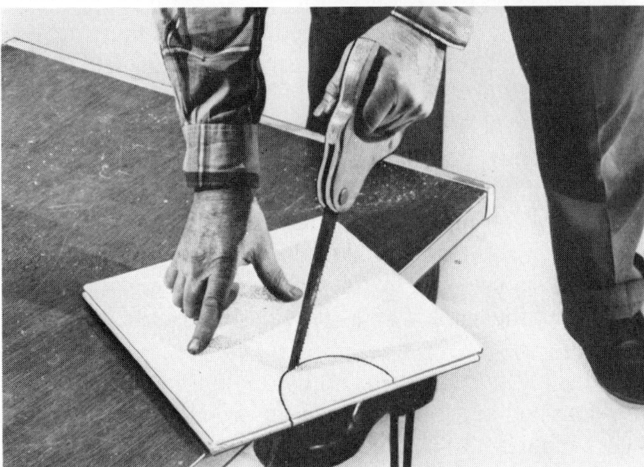

Figure 15-74. Top—Snapping a border tile after it has been scored. Remove the tongue edge. Leave the wide flange for stapling. Bottom—Using a compass saw for cutting curves. (Flintkote Co.)

Figure 15-75. Staples can be used to attach tile to furring strips. (Duo-Fast Corp.)

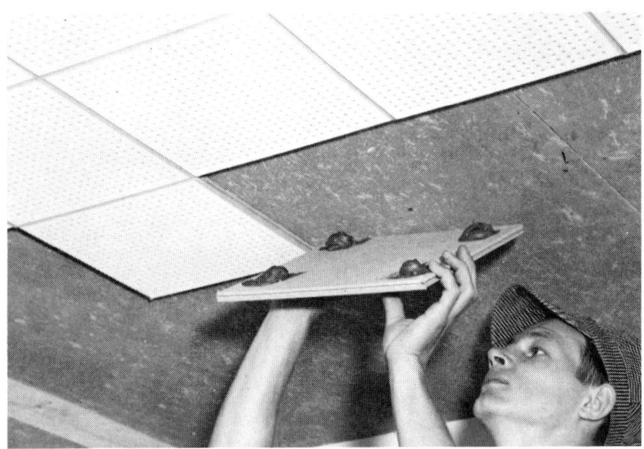

Figure 15-76. Installing tile directly to a finished ceiling with putty-type adhesive.

tile. Do this along each wall and be certain the lines are correct and in accord with the initial layout previously described. Double-check to see that the chalk lines form 90° angles at the corners.

To cut the border tile, first score the face deeply with a knife drawn along a straightedge. Break it along this line by placing it over a sharp edge. See Figure 15-74. Irregular cuts around light fixtures or other projections may be made with a coping or compass saw. Power tools can also be used. Some tiles are made of mineral fiber which rapidly dulls regular cutting edges.

Start the installation with a corner tile and then set border tile out in each direction. Fill in full-size tile. When you reach the opposite wall, trim the border tile to size. If border tile are full, remove the stapling flanges and face-nail the tile. Locate these nails close to the wall so they will be covered by the trim molding.

Figure 15-75 shows tile being attached with a stapler. Be sure the tile is correctly aligned. Hold it firmly while setting the staples.

For 12" x 12" tile, use three staples along each flanged edge. Use four staples for 16" x 16" tile. Staples should be at least 9/16" long.

There are several types of adhesive designed especially for installing ceiling tile. The thick putty type is applied in daubs about the size of a walnut, Figure 15-76. Apply adhesive to each corner of 12" x 12" tile and about 1 1/2" away from each edge. Now position the tile. Slide it back against the other tile as shown in Figure 15-77. This motion, along with firm pressure, will spread the adhesive so the daubs will be about 1/8" thick.

Some adhesives are thinner and are applied with a brush. In remodeling work, be sure the old ceiling is clean and that any paint or wallpaper is adhering well. Always follow the recommendations and directions provided by the manufacturer of the products being used.

Be sure to keep your hands clean while handling ceiling tile.

Figure 15-77. Carefully align new tile with previously set tile.

Metal Track System

Another system uses 4' metal tracks to replace the wood furring strip. The track is nailed to the old ceiling or to joists at 12" O.C. intervals, Figure 15-78. Tongue and grooved panels are slipped into place. A clip snapped into the track slides over the tile lip. No other fasteners are used, Figures 15-79 and 15-80.

Figure 15-78. Attach predrilled track to ceiling at 12" intervals. (Armstrong World Industries, Inc.)

Suspended Ceilings

When heating ducts and plumbing lines interfere with the application of a finished surface, a *suspended ceiling* is practical. In other instances it provides a simple way of lowering high ceilings. Modern installations consist of a metal framework especially designed to support tile or panels, Figure 15-81.

The height of the suspended ceiling must first be determined. Then a molding must be attached to the perimeter of the room, Figure 15-82. Use a level chalk line or a laser level as a guide.

Carefully calculate room dimensions, lay out positions of main runners, and then install screw eyes (4' O.C.) in existing ceiling structure. Panels next to the wall (border panels) may need to be reduced in width to provide a symmetrical (same on both sides of room) arrangement. Plan the layout the same way as previously described for regular ceiling tile.

Install the main runners by resting them on the wall molding and attaching them to the wires tied to the screw eyes, Figure 15-83. Use chalk lines or string stretched between the wall molding to ensure a level assembly. Sections of runners are easily spliced. Odd lengths can be cut with a fine-tooth hacksaw or aviation snips.

After all main runners are in place, recheck the level and adjust the wires as necessary. Install cross tees next. See Figure 15-84. Check the required spacing for the tiles

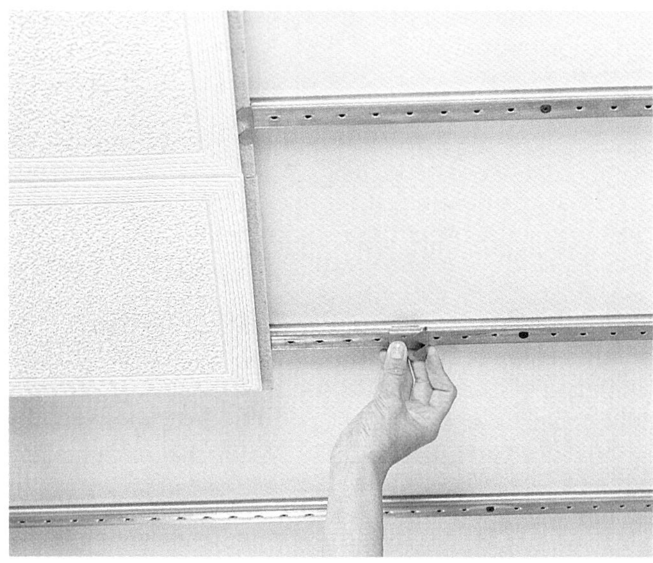

Figure 15-79. Track system uses clips to hold ceiling panels. Use one clip for each panel. Border panels must have two clips. Top—Snap clip into track. Bottom—Slide it snugly against tile. (Armstrong World Industries, Inc.)

Figure 15-80. As each tile is installed, slide the tile firmly against adjacent tile. Lip will hold the tile while patented clip is installed. (Armstrong World Industries, Inc.)

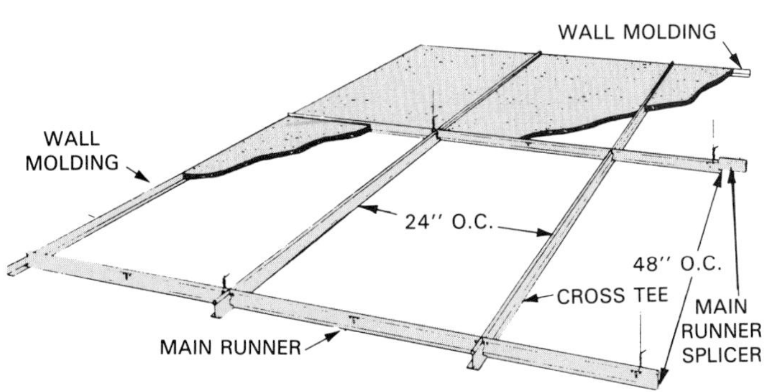

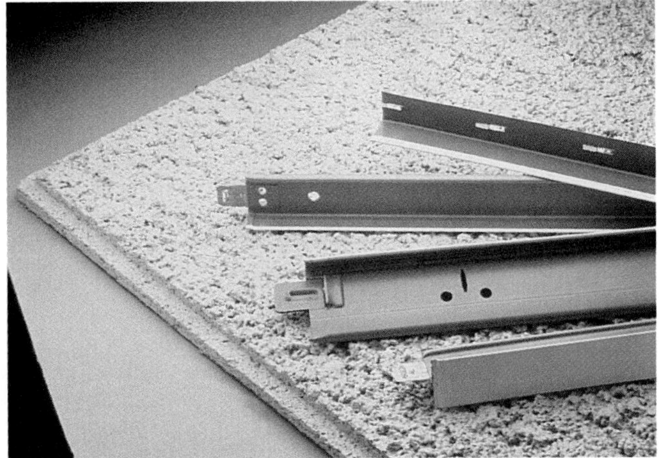

Figure 15-81. Suspended ceilings. Top—This sketch is typical of the framework used for suspended ceilings. Bottom—Runners and tile for a suspended ceiling. (USG Interiors, Inc.)

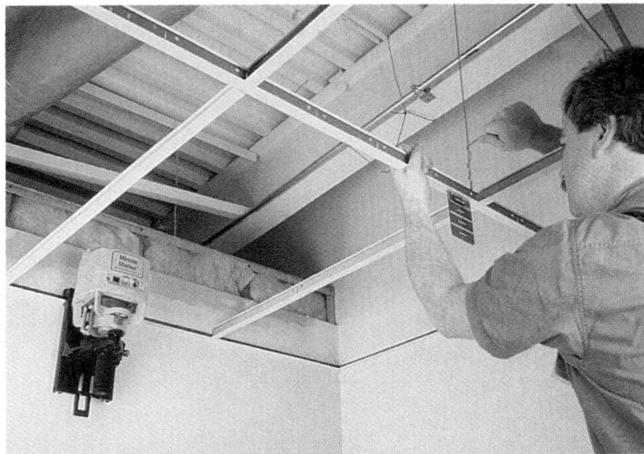

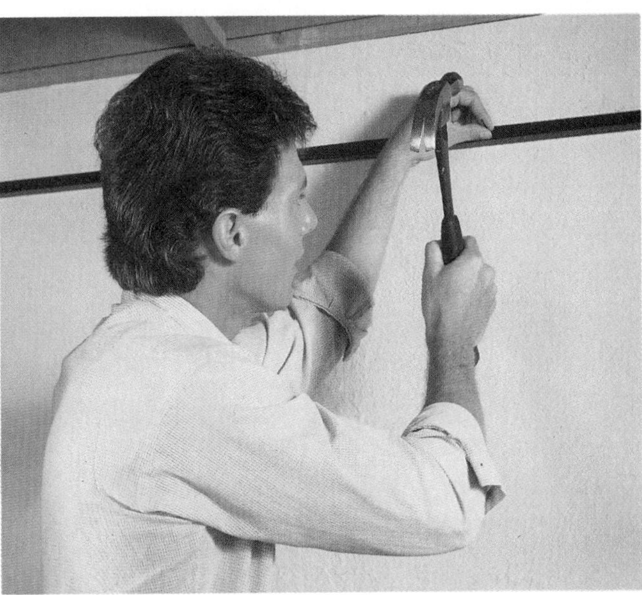

Figure 15-82. Top—Working alone, one installer with a laser level can level a suspended ceiling quickly. (Spectra-Physics Laserplane, Inc.) Bottom—After locating and marking the ceiling's level on all walls, attach metal molding to the walls. (Armstrong World Industries, Inc.)

or panels. Simply insert the end tab of the cross tee into the runner slot.

When the suspension framework is complete, panels can be installed. Each panel is tilted upward and turned slightly on edge so it will "thread" through the opening, as illustrated in Figure 15-85. After the entire panel is above the framework, turn it flat and lower it onto the grid flanges.

A concealed suspension system is shown in Figure 15-86. The tongue and groove joint is similar to the joint on regular ceiling tile. Joints are interlocked with the flange of the special runner when the installation is made, Figure 15-87.

Figure 15-83. Attach eye screws to joists and hang metal runner from joists. Use a chalk line or laser level to check level.

Figure 15-85. Ceiling panels are tilted, slipped through the framework, and lowered into place. (USG Corp.)

Figure 15-84. Cross tees are installed between main runners. (Armstrong World Industries, Inc.)

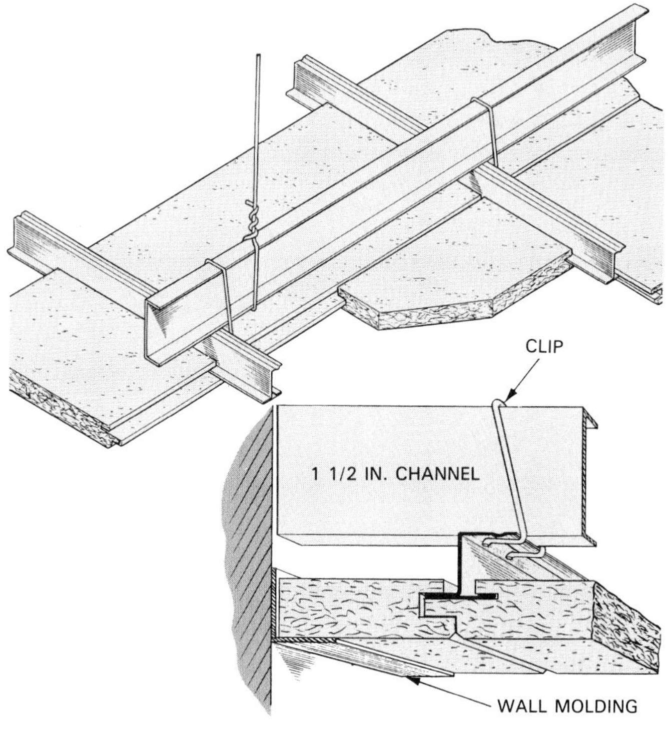

Figure 15-86. Top—Channels of suspension system are concealed by the ceiling panels. Bottom—Tongue and groove joint detail.

Estimating Materials

The amount of wall or ceiling covering materials required is estimated by first calculating the total number of square feet of wall and ceiling area. Regular size door and window openings are disregarded. These provide allowances for waste. However, larger window walls and picture windows may be subtracted.

Determining Area of Rooms

For materials that do not come in sheets, you need to know the area you are covering. Ceiling area is usually the same as the floor area. It is much easier to take floor dimensions and multiply the length times the width.

To find area of walls, add all the wall lengths together and multiply by the wall height.

Sheet Materials

When ordering sheet materials such as wallboard or paneling, be sure to specify the length. Always plan to use

Figure 15-87. Ceiling panel joints interlock with supporting flange. (Wood Conversion Co.)

the longest practical sheet. This holds butt joints to a minimum or eliminates them. Divide the total length of the walls by the width of the sheets to find the number of sheets needed.

Estimate each room separately. Take dimensions directly from the walls or carefully scale the plan. Consult tables and charts of suppliers for estimates of joint compound, adhesives, and nails.

When working with expensive hardwood panels, it is usually advisable to make a scaled layout. Look at the diagrams in Figure 15-39 for horizontal and vertical arrangements.

Estimating Solid Paneling

Estimates for solid paneling are based on its nominal and unfinished size. Seasoning and planing bring it to its dressed size. Forming of joints may further reduce the actual face widths.

For example, a 1 x 6 tongue and groove board has a face width of 5 1/8". First calculate the square footage of the wall to be covered. Then multiply by various factors taken from lumber tables:

- For 1 x 6 tongue and groove boards, use 1.17.
- For 1 x 8 tongue and groove boards, use 1.16.

On standard vertical applications, add 5% for waste when required lengths can be selected or when the lumber is end matched.

Estimating Gypsum Lath

Since gypsum lath is produced in smaller sections than full sheets, you will need to use different methods of figuring quantity needed. Figure the area of the ceiling and add to this the area of the walls (length of walls x height).

For example:

Ceiling area (same as floor area)	= 1250 sq. ft.
Total length of walls	= 14 + 12 + 14 + 12 + 10 = 62
Wall area	= 62 x 8 = 496 sq. ft.
Total area	= 496 + 1250 = 1746 sq. ft.
Standard lath bundle	= 64 sq. ft.
Bundle estimate	= 28

Plasterers usually base their prices and estimates on the number of square yards. Convert square footage to square yards by dividing by 9 (1 sq. yd. = 9 sq. ft.).

Quantities of ceiling tile are estimated by figuring the area (square footage) to be covered. Round out any fractional parts of a foot in width and length to the next larger full unit when making this calculation.

Add extra units when it is necessary to balance the installation pattern with border tile along each wall. When using 12" x 12" tile, the number required will equal the square footage plus the extra allowance described. Standard 12" x 12" ceiling tile are packaged 64 sq. ft. to a carton.

Important Terms

Backing board	Gypsum board
Brown coat	Gypsum lath
Cement board	Hardboard
Chevron paneling	Herringbone paneling
Darby	Interior finishing
Diagonal paneling	Metal lath
Double layer construction	Moisture resistant board
Drywall	Plaster
Fiberboard	Scratch coat
Fiberboard lath	Single layer construction
Field	Suspended ceiling
Furring strips	Wood lath
Grounds	

Test Your Knowledge

1. List nine different materials that are used as interior wall and ceiling coverings.
2. In single layer dry wall construction, the gypsum wallboard thickness should be ____________ or ____________ inches.
3. Being ____________, cement board is well suited as a base for tile in a shower stall or above a bathtub.
4. When nailing wallboard in place, the nails are spaced closer together on the ____________ (*walls, ceilings*).
5. In addition to the skim coat applied over the tape, wallboard joints usually require ____________ (*1, 2, 3*) coats of joint compound.
6. When making a double layer wallboard application, joints running parallel to each other should be offset at least ____________ inches.
7. Cement board, which is used as a backing board in various applications where surfaces are often wet, is available in three sizes: ____________, ____________, and ____________.

8. Solid wood wall paneling should have a moisture content of about ____________ to ____________ percent for most areas of the United States.
9. Prefinished plywood wall paneling should be "room conditioned" for at least ____________ hours before application.
10. Using a base layer of gypsum wallboard under plywood paneling improves alignment and rigidity, and also makes the construction more ____________.
11. A standard size panel of gypsum lath measures 3/8" x 16" x ____________.
12. To reinforce the plaster base at the corners of window and door openings, a strip of ____________ is used.
13. Strips of wood or metal installed at the edge of openings to provide a guide for the plasterer are called ____________.
14. The most common aggregate used in plaster mixes is ____________.
15. In standard three-coat plaster applications, the brown coat is applied ____________ (*first, second, last*).
16. When applied to regular gypsum lath, the total thickness of the plaster coats should be not less than ____________ inches.
17. Some solid lumber paneling can be reversed for a different surface pattern. True or False?
18. Name the three patterns used for installing solid wood paneling at an angle.
19. Standard ceiling tile have a wide stapling or nailing flange on ____________ (*1, 2, 3, 4*) edges.
20. For a symmetrical ceiling tile pattern in a room 11'-6" wide, use 10 full (12" x 12") tile and a ____________ inch wide border tile.
21. The metal framework of a modern suspended ceiling is supported mainly by ____________ tied to the building structure directly above.
22. It is general practice for plasterers to base their cost estimates on the number of ____________ of ceiling and wall areas.

Outside Assignments

1. Obtain and study literature from companies that manufacture gypsum products. Learn about the history and development of plastering and modern processes used in making plaster, gypsum lath, and gypsum wallboard. Get samples of the products commonly used in your locality, along with approximate prices and costs of installation. Prepare a written report or outline the information carefully and report to your class.
2. Visit a local drywall contractor and learn about some of the special problems and remedies typical to this type of wall finish. Obtain information on the following topics: care and handling of materials; repairing and adjusting warped studs and framework; repairing damaged boards and surfaces; cause and remedies of tape blisters; and definition and prevention of nail pops. Organize the information under appropriate headings and make a presentation to your class.
3. From various reference books, including manuals on architectural standards, learn about methods and construction details for ceiling systems that include radiant heating. Cover both hot water and electrical systems. Prepare scaled drawings (larger than actual size) of sections through various ceiling constructions, showing the heating pipes and elements. Include notes concerning material specifications and critical temperatures for the various constructions.
4. Suspended ceilings in institutional and commercial buildings often include a ventilation system that provides both heating and cooling. The space above the ceiling serves as a plenum. Air enters the room through small slots or holes either in the tile or at special joints between the tile. Study this method of air distribution and prepare a written report with drawings and other illustrations.
5. Take the dimensions of a room and calculate the number of sheets needed to cover it with drywall. Be sure to include the ceiling.
6. From your calculations determine the cost of the drywall in Assignment 5. Consult a local lumber yard for the cost per sheet or discuss per-sheet cost with your instructor.

16

Finish Flooring

Finish flooring is any material used as the final surface of a floor. A wide selection of materials is manufactured for this purpose.

Hardwoods and softwoods are available as strip flooring in a variety of widths and thicknesses, and as random width planks or unit blocks. Many new and improved materials are being used in the production of composition (resilient) flooring. Notable among these are the vinyl plastics which have largely replaced asphalt tile, rubber tile, and linoleum.

Flagstone, slate, brick, and ceramic tile are frequently selected for special areas such as entrances, bathrooms, or multipurpose rooms. Floor structures usually need to be designed to carry this type of finish surface.

When the finish flooring is wood, it is usually laid after wall and ceiling surfaces are completed, and before interior door frames and other trim are added. The floor surface should be covered to protect it during other inside finish work. Sanding and finishing of the floor surface becomes the last major operation as the interior is completed.

Where prefinished wood flooring is laid, it must be covered and protected. As with wood, the installation of resilient flooring and prefinished wood flooring must be among the last steps in interior finishing. Select and install finish flooring carefully.

Wood Flooring

Wood is popular as flooring in residential structures. Wood flooring, especially hardwood, has the strength and durability to withstand wear while providing an attractive appearance, Figure 16-1.

Oak is a widely used species. However, maple, birch, beech, and other hardwoods also have desirable qualities. Softwood flooring includes such species as pine and fir. These are fairly durable when produced with an edge-grain surface.

Types of Wood Flooring

Three general types of wood flooring are used in residential structures: strip, plank, and block. As the name implies, *strip flooring* consists of pieces cut into narrow strips. It is laid in a random pattern of end joints.

Most strip flooring is tongue and groove both on the sides and ends. This design is also referred to as *side-and-end matched.* Another feature of modern strip flooring is the undercut. See Figure 16-2. This is a wide groove on the bottom of each piece that enables it to lie flat and stable even when the subfloor surface is slightly uneven.

Plank flooring, Figure 16-3, provides an informal atmosphere. It is particularly appropriate for colonial and ranch style homes.

Plank floors are usually laid in random widths. The pieces are bored and plugged to simulate the wooden pegs originally used to fasten them in place.

Today, this type of floor has tongue and groove edges. It is laid in about the same manner as regular strip floors.

Block flooring looks like conventional parquetry. This is an elaborate design formed by small wood blocks. Modern unit blocks consist of short lengths of flooring, held together with glue, metal splines, or other fasteners. Square and rectangular units are produced. Generally, each block is laid with its grain at right angles to the surrounding units, Figure 16-3. Blocks, called laminated units, are produced by gluing together several layers of wood.

Sizes and Grades

Today, hardwood strip flooring is generally available in widths ranging from 1 1/2 to 3 1/4". Standard thicknesses include 3/8", 1/2", and 3/4".

Solid planks are usually 3/4" thick. However, greater thicknesses are available. Widths range from 3" to 9" in multiples of 1". Unit blocks are also commonly produced in a 3/4" thickness. Dimensions (width and length) are in multiples of the widths of the strips from which they are made. For example, squares assembled from 2 1/4" strips will be 6 3/4" x 6 3/4", 9" x 9", or 11 1/4" x 11 1/4".

Uniform grading rules are established by manufacturers working with the U.S. Bureau of Standards and such organizations as the National Oak Flooring Manufacturers' Association or the Maple Flooring Manufacturers Association. Grading is based largely on appearance. Consideration is given to knots, streaks, color, pinworm holes, and sapwood. The percentage of long and short pieces is also a consideration.

Oak, for example, is separated into two grades of quarter-sawed stock and five grades of plain-sawed stock. In descending order, plain-sawed grades are: clear, select and better, select, No. 1 common, and No. 2 common.

Figure 16-1. Oak strip flooring is very durable and has a natural beauty that is brought out by clear finishes. The floor above is made of strips that are 3/4" thick by 2 1/4" wide. Today, a wide variety of hardwood flooring is available with a factory applied finish. (Robbins/Sykes)

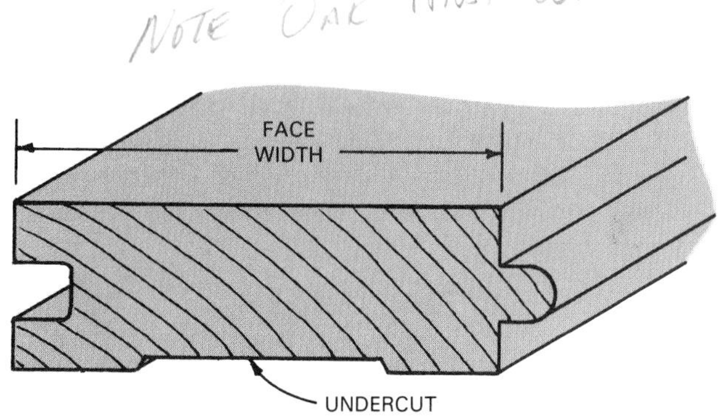

Figure 16-2. Typical section of strip flooring. Ends (not shown) are also tongue and grooved.

White and red oak species are ordinarily separated in the highest grades. A chart on grades for hardwood flooring is included in the reference section.

Manufacturers recommend that wood flooring be delivered four or five days before installation. Avoid transporting it in rain or snow. In damp or foggy conditions, protect the lumber with a tarp. Pile the flooring loosely throughout the structure. Do not store where floors are less than 18" above the ground and where there is poor air circulation under the floor. A minimum inside temperature of 70°F should be maintained. Avoid damp buildings; make sure concrete work and plaster are dried thoroughly before bringing in flooring. This period of "conditioning" permits the wood to match its moisture content with that in the building.

Subfloors

In conventional joist constructions, most building codes specify a sound subfloor. It adds considerable strength to the structure and serves as a base for attaching the finish flooring.

For regular strip or plank flooring, the subflooring should consist of good quality plywood or boards 1" thick and not more than 6" wide. Space the boards about 1/4" apart and face nail them solidly at every bearing point. Inadequate or improper nailing of subfloors usually results in squeaky floors.

In modern construction, plywood is commonly used for the subfloor. It must be installed according to recognized standards. Refer to Unit 8 for sizes and installation methods.

Installing Wood Strip Flooring

Normally, laying of flooring should be the last finishing operation. Before installing wood strip floors, check over the subfloor to make certain it is clean and that nailing patterns are complete. Put down a good quality building paper. It should extend from wall to wall, with a 4" lap as shown in Figure 16-4. The location of the joists should be chalk lined on the paper. This is especially important when plywood is used as the subfloor.

Over the heating plant or hot air ducts, it is advisable to use a double-weight building paper. Insulation may be attached directly to the underside of the subfloor. This extra precaution will prevent excessive heat from reaching

Figure 16-3. Two types of wood floors. Left—Random planks have walnut plugs. Pieces are 3/4" thick and are 2 1/4" and 3 1/4" wide. Lengths are random. (Robbins Inc.) Right—Unit block parquet flooring. Blocks are 6" x 6" and accurate with tongue and groove edges. Available with a factory applied finish and a special backing for do-it-yourself installation. (Tibbias Flooring Inc.)

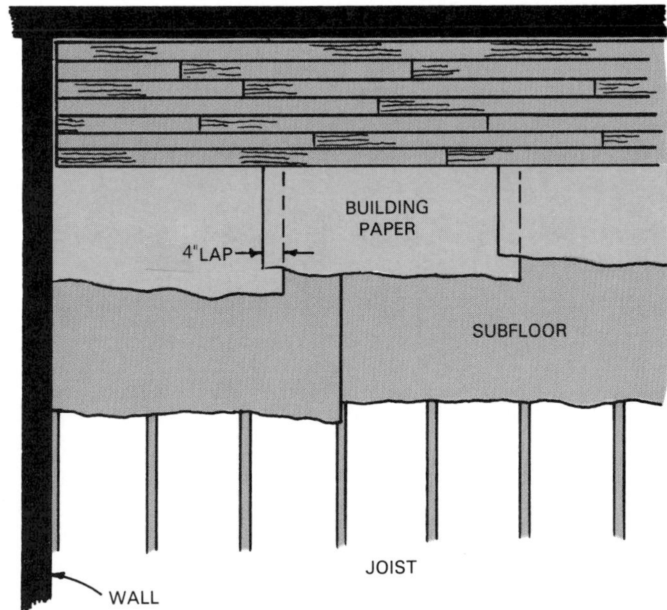

Figure 16-4. Cutaway view showing various layers of material under strip flooring.

the finish flooring causing cracks and open joints. Lay strip flooring at a right angle to the floor joist. The flooring will look best when this direction aligns with the longest dimension of a rectangular room. Since floor joists will normally span the shortest dimension of the living room, this will establish the direction the flooring runs in other rooms.

Stagger joints so that several are not grouped in one small area. Shorter pieces should be used up in closets and in areas where there is least traffic. Long strips should be used at entrances, or starting, or for finishing off a room.

%

The moisture content usually recommended for flooring at the time of installation is 6% for the dry southwestern states, 10% for the more humid southern states, and about 7% or 8% for the remainder of the country.

Nailing

Floor squeaks or creaks are caused by the movement of one board against the other. The problem may be in either the subfloor or finished floor. Using enough nails of the proper size will reduce these undesirable noises. When possible, the nails used for the finish floor should go through the subfloor and into the joist. When plywood is used for the subfloor, place a nail at each joist and one in between. Nail sizes are specified for different types of floors in Figure 16-5.

Flooring Nominal Size, Inches	Size of Fasteners	Spacing of Fasteners
3/4 × 1 1/2 3/4 × 2 1/4 3/4 × 3 1/4 3/4 × 3" to 8" plank	2" machine driven fasteners; 7d or 8d screw or cut nail.	10"-12"* apart 8" apart into and between joists.
Following flooring must be laid on a subfloor.		
1/2 × 1 1/2 1/2 × 2	1 1/2" machine driven fastener; 5d screw, cut steel or wire casing nail.	10" apart
3/8 × 1 1/2 3/8 × 2	1 1/4" machine driven fastener, or 4d bright wire casing nail.	8" apart
Square-edge flooring as follows, face-nailed—through top face		
5/16 × 1 1/2 5/16 × 2	1", 15-gauge fully barbed flooring brad. 2 nails every 7 inches.	
5/16 × 1 1/3	1", 15-gauge fully barbed flooring brad. 1 nail every 5 inches on alternate sides of strip.	

Figure 16-5. Nail chart suggests fastener size and spacing for application of strip flooring.

Start the installation in one of the rooms. Lay the first strip along either sidewall. Select long pieces. If the wall is not perfectly straight and true, set the first course along a chalk line. Place the groove edge next to the wall and leave at least 1/2" for expansion. This space will be covered later by the baseboard and base shoe.

Make sure the first strip is perfectly aligned. Then face nail it as illustrated in Figure 16-6. In face nailing, the nail head must be set (sunk below face) and the hole filled.

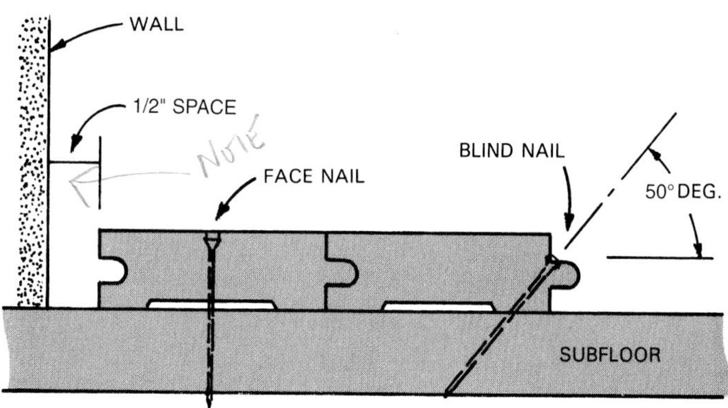

Figure 16-6. Cutaway drawing shows how to nail the starter strip. Note, also, how to nail succeeding strips.

Succeeding strips are blind nailed with the nail penetrating the flooring where the tongue joins the shoulder. The nail is driven at an angle of about 50°. Use a nail set to finish the driving so that the edge of the strips will not be damaged.

Each strip should fit tightly against the preceding strip. When it is necessary to drive strips into position, use a piece of scrap flooring as a driving block.

Cut and fit a number of pieces and lay them ahead of the installed strips as shown in Figure 16-7. Use different lengths. Match them so they will extend from wall to wall with about 1/2" clearance at each end.

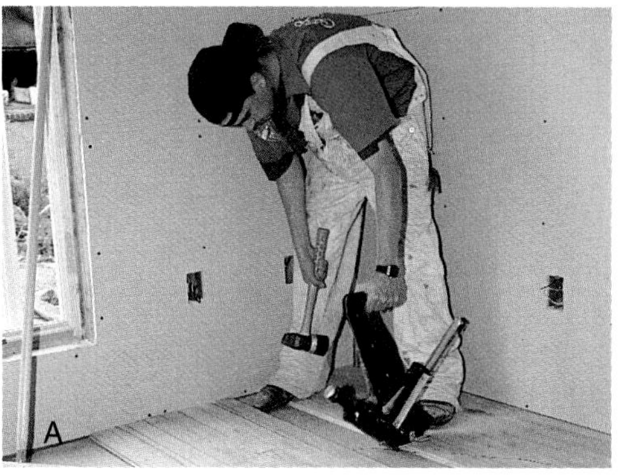

Figure 16-7. Strip floor installation. Top—A portable nailer is useful for installing strip flooring. (Bullard-Haven Technical School) Bottom—A finished wood strip flooring. (Chickasaw Custom Oak Strip Floor)

Joints in successive courses should be 6" or more from each other. Try to arrange the pieces so the joints are well distributed. Blend color and grain for a pleasing pattern. Pieces cut from the end of a course should be carried back to the opposite wall to start the next course. This is the only place it can be used since its leading end has no tongue or groove.

Flooring strips should run uninterrupted through doorways and into adjoining rooms. When there is a projection into the room, such as a wall or partition, follow the procedure shown in Figure 16-8:

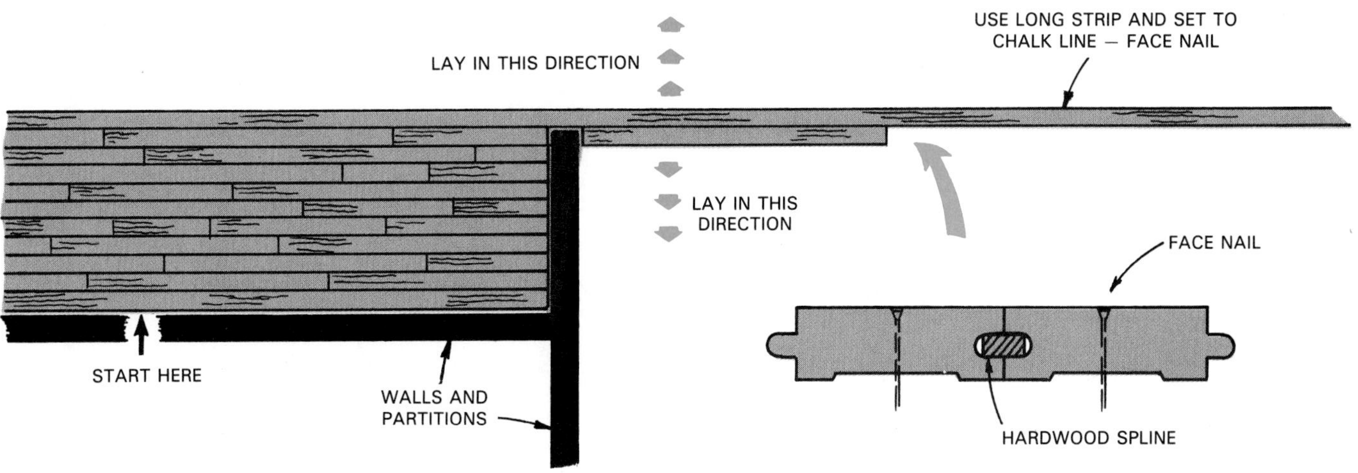

Figure 16-8. Follow this procedure when laying strip floors around a wall or partition.

Laying around Projections

1. Lay the main area to a point even with the projection.
2. Extend the next course all the way across the room. Set the extended strip to a chalk line and face nail it in place.
3. Form a tongue on the grooved edge by inserting a hardwood spline.
4. Install the flooring in the second room in both directions from this strip.

If a large area is to be laid with strip or plank flooring, it sometimes helps to set up a starter strip at or near the center. Use a spline in the groove of the starter strip as previously described. Be sure to measure and accurately align the starter strip with the walls on each side of the room.

When the floor has been brought to within 2 or 3 feet of the far wall, the room should be checked again to find out if the strips are parallel to the wall. If not, dress off the grooved edges slightly at one end until the strips have been adjusted to run parallel. This is necessary in order that the last piece may be aligned with the baseboard.

Multiroom Layout

When flooring installation is carried throughout the major part of a building, study the floor plan to determine the most efficient procedure to follow.

Figure 16-9 shows a typical residential floor plan with a method for the installation of strip flooring. A setup line (chalk line) is first laid out two flooring widths (plus 1/2") from the partition as shown. The starter courses are aligned with the setup line before face nailing. Installation is continued across the living room, across the hall, into and through bedrooms No. 1 and No. 2.

A spline is set in the groove of the starter strip and the floor is laid from this point into the dining room and bedroom No. 3. In bedroom No. 2, a splined groove is also used to lay the floor back into the closet.

In small closet areas, such as shown in bedroom No. 1, it is usually impractical to reverse the direction of laying with a splined groove and the pieces are simply face nailed in place.

The last strip laid along the wall of a room will usually need to be ripped so it will have the required clearance (1/2" minimum). The last several strips must be face nailed. Do not use ripped strips where they might detract from the appearance. It is recommended that full length strips always be used around entrances and across doorways.

In general, plank flooring is installed by following the same procedures used for strip flooring. In addition to the regular blind nailing, screws are set and concealed in the face of wide boards.

Estimating Strip Flooring

To determine the number of board feet of strip flooring needed to cover a given area, first calculate the area in square feet. Then add the percentage listed for the particular size being used. See Figure 16-10. The figures listed are based on laying flooring straight across the room. They provide an allowance for side-matching, plus 5% for end matching. Where there are many breaks and projections in the wall line, add additional amounts. For example:

Total area	= 900 sq. ft.
Flooring size	= 3/4" x 2 1/4"
Bd. ft. of flooring	= 900 + 38 1/3% of 900
	= 900 + 345
	= 1245
Number of bundles	= 52 (24 bd. ft. to bundle)

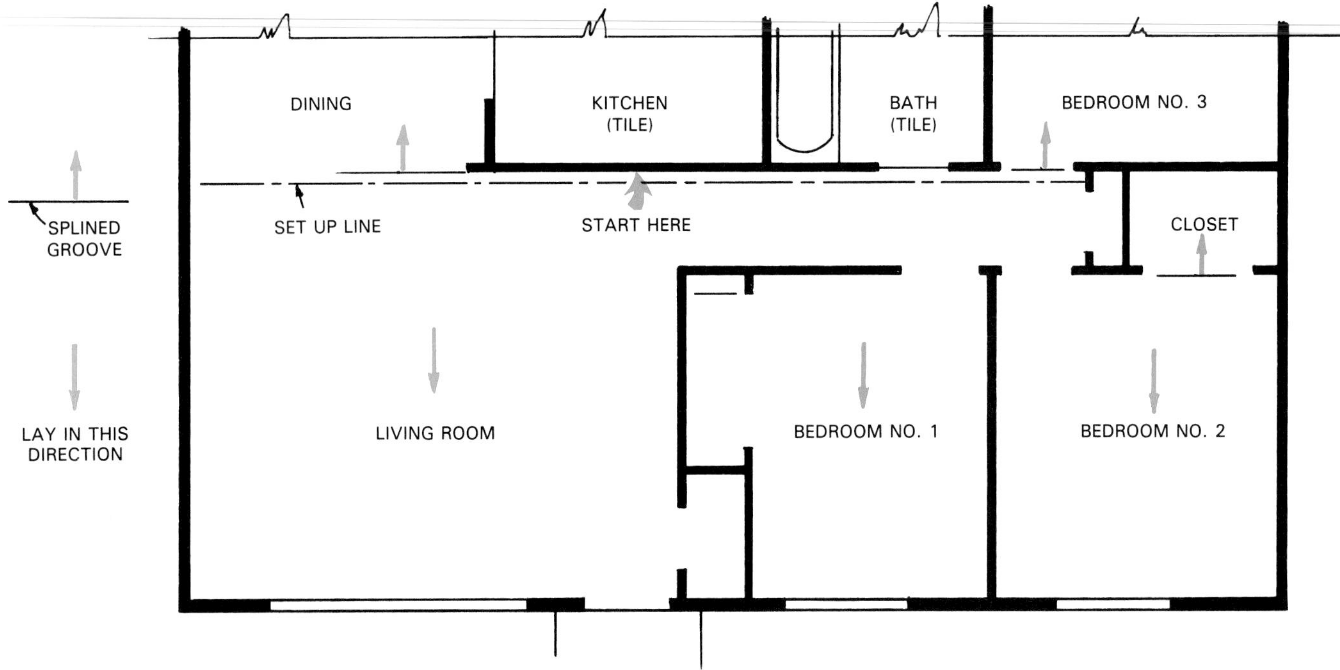

Figure 16-9. Floor plan shows correct procedure for laying strip flooring throughout a number of rooms.

FLOORING BOARD SIZE	ADDITIONAL PERCENTAGE
3/4" x 1 1/2"	55%
3/4" x 2"	42 1/2%
3/4" x 2 1/4"	38 1/3%
3/4" x 3 1/4"	29%
3/8" x 1 1/2"	38 1/3%
3/8" x 2"	30%
1/2" x 1 1/2"	38 1/3%
1/2" x 2"	30%

Figure 16-10. Add above percentages to total area to be covered with strip flooring.

Wood Flooring over Concrete

Finished wood flooring systems can be successfully installed over a concrete slab. When the concrete floor is suspended, with an air space below, a moisture barrier is usually not required. For slab-on-grade installation, put down sleepers (also called screeds). These are wood strips attached to or embedded in the concrete surface. The sleepers serve as a nailing base for the flooring material.

Beware of damp floors. Moisture from a concrete floor not completely dry or cured will damage wood flooring.

To test for moisture presence, place a flat, noncorrugated rubber mat on the slab. Weight it down so the moisture cannot escape. Allow the mat to remain overnight. Moisture in the concrete will show as water marks when the mat is removed.

If the concrete is placed directly on grade either at or below ground level, an approved membrane moisture barrier must be installed between the flooring and the concrete. Figure 16-11 illustrates, in a general way, a system of waterproofing and sleeper installation that will provide a base for 3/4" strip flooring. A coat of asphalt mastic (thinned with an appropriate solvent) is first evenly applied over the concrete surface. This is covered with polyethylene film. Another coat of a special asphalt mastic is then applied as shown. Then wood sleepers are set in place. The sleepers are 2 x 4 lumber about 30" long, laid flat and running at a right angle to the flooring. Lap them at least 3" for 2 1/4" flooring and 4" for 3 1/4" flooring. Spacing between centers is usually 12 to 16".

A newer system of laying strip floors over a concrete slab-on-grade offers savings over older methods. It consists of a double layer of 1 x 2 wood sleepers nailed together, with a moisture barrier of 4-mil polyethylene film placed between them, Figure 16-12. This is accepted by FHA.

The sequence of photographs in Figure 16-13 illustrates the basic steps for installation.

Figure 16-11. Preparing a concrete floor for installation of wood flooring. Top—Moisture proofing a concrete slab-on-grade with asphalt mastic and polyethylene film. Bottom—2 x 4 sleepers set in asphalt mastic provide nailing base for wood flooring. (Bruce Hardwood Floors)

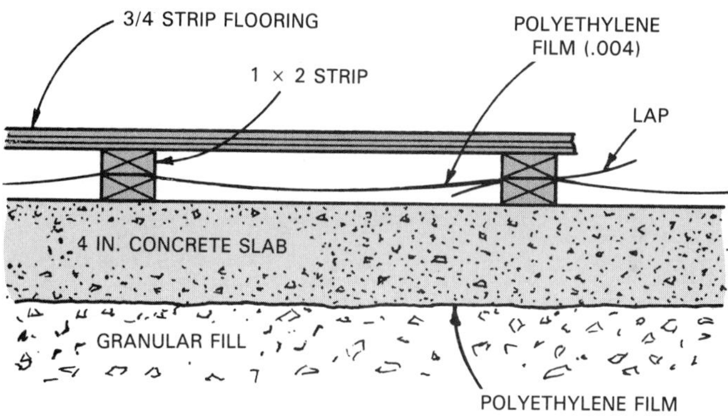

Figure 16-12. Strip flooring system over concrete slab. Sleepers are doubled 1 x 2 strips.

Laying over Concrete

1. Clean and prime the floor.
2. Then snap chalk lines 16" apart.
3. Put down bands (rivers) of adhesive along the layout lines.
4. Embed treated wood sleepers in the adhesive and secured with 1 1/2" concrete nails or special concrete screws, about 24" apart. (If screws are used, drill holes down through the sleepers into the concrete. Use a carbide-tipped masonry drill and a hole shooter.) Strips could also be attached with powder-actuated equipment that will drive the nails through the wood into the concrete.
5. Place a layer of 4 or 6 mil polyethylene film over the strips. Join sheets by forming a lap over a sleeper.
6. After the vapor barrier is placed, a second layer of untreated strips are nailed to the bottom sleepers with 4d nails about 16" apart. Install strip or plank flooring as previously described. No two adjoining flooring strips should break joints between the same two sleepers.

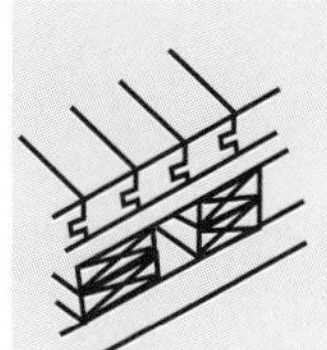

Installation of wood flooring over concrete must be done carefully. Secure detailed specifications from manufacturers of the products to be used. Organizations such as the National Oak Flooring Manufacturers' Association can also provide instructional materials and advice.

Wood Block (Parquet) Flooring

Block or parquet flooring is flooring made up of squares of wood in various patterns. It requires different application methods than strip flooring. Blocks are usually made up on squares of 6", 8", or 12". Block flooring is produced in different ways:

- Unit blocks are glued up from several short lengths of flooring.
- Laminated blocks are made by bonding three plies of hardwood with moisture resistant glue.

Parquet flooring may use intricate patterns of small pieces in producing the squares. Each block is tongue and grooved, Figure 16-14. Installation of all block type flooring is basically the same. Allow some clearance on unit blocks for expansion. Rubber strips are sometimes used for this purpose. Generally, it is recommended that unit blocks be installed with a 1" space at the wall.

Block flooring is nearly always laid in a mastic. The mastic is spread evenly over the subfloor to a thickness of 3/32". See Figure 16-15. Some installations may require a layer of 30 lb. asphalt-saturated felt. The felt is laid in a mastic coat and then the top surface is coated with mastic to receive the flooring. If blocks are to be nailed, nail through the tongued edges as with strip or plank floors. Lay blocks in this coat, as in Figure 16-16.

Installation can be made either on a square or a diagonal pattern. Chalk lines are first snapped equal distances from the sidewalls if laying is done from a center point. Layout procedures are similar to those described for resilient flooring materials included later in this unit. See Figure 16-17.

Since block flooring and adhesive materials vary from one manufacturer to another, detailed information

APPLY RIVERS OF MASTIC ALONG CHALKED LAYOUT LINES.

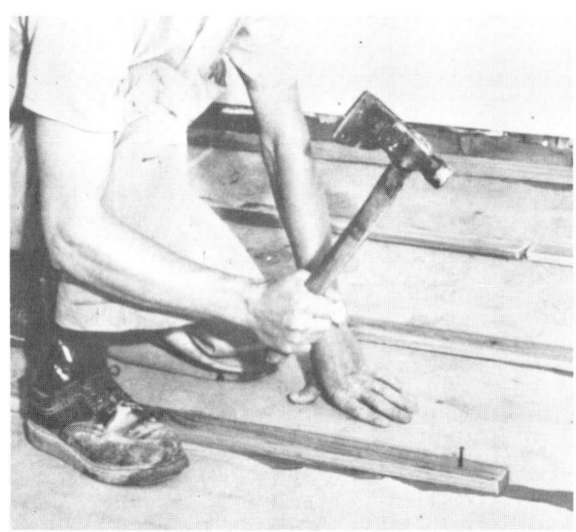

NAIL TREATED SLEEPERS OVER MASTIC.

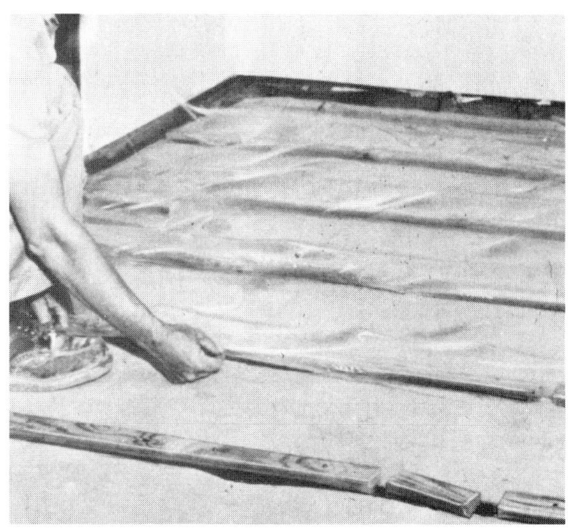

SPREAD POLYETHYLENE FILM OVER BOTTOM SLEEPERS. LAP FILM ONLY OVER A SLEEPER.

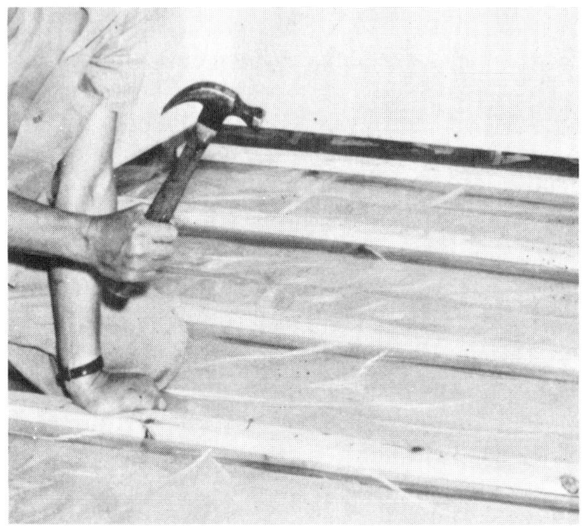

NAIL SECOND LAYER OF SLEEPERS ON TOP OF FIRST SLEEPERS.

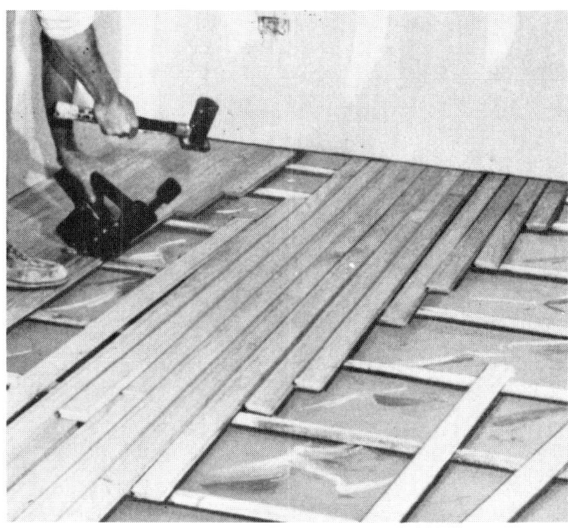

NAIL STRIP FLOORING TO SLEEPERS.

Figure 16-13. Follow these steps for installing strip hardwood flooring over concrete slab. (National Oak Flooring Manufacturers Assoc.)

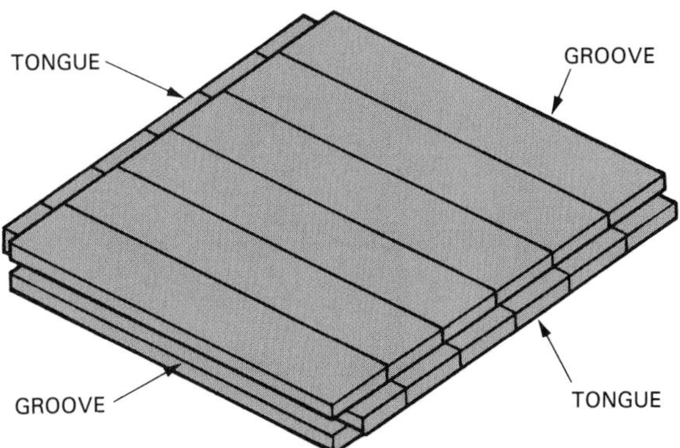

Figure 16-14. Typical block or parquet flooring. Edges are always tongue and grooved.

Figure 16-15. Spreading mastic over plywood subfloor. Floor should be smooth, clean, and dry.

concerning installation procedures should always be secured from the manufacturer of the product.

Prefinished Wood Flooring

Most of the flooring materials described are available with a factory applied finish, Figure 16-18. They are generally considered to be superior to that which can normally be applied on-the-job. Another advantage of prefinished flooring is that it speeds construction. Floors are ready for service immediately.

Disadvantages arise from the fact that the installation must be made with care. Although holes resulting from face nailing can be covered with special filler materials, face nailing must be held to a minimum when working with prefinished materials. Hammer marks caused by careless blind nailing are extremely difficult to repair.

Figure 16-16. Installing a factory-finished parquet block floor. The individual blocks are laid in a mastic. Always follow the manufacturer's directions. (Bruce Hardwood Floors)

A prefinished floor must be the last step in the interior finish sequence. Other trim that abuts the finished floor must be set beforehand with proper allowances. For example, interior door jambs and casing require spacer blocks equal to the floor thickness. Place these under the units while they are being installed.

Unless prefinished baseboards are installed after the floor is laid, they must also be placed to allow necessary clearance. Other special provisions will need to be made around built-in cabinets and at stairways and entrances.

Underlayment

Flooring materials such as asphalt, vinyl, linoleum, and rubber will usually reveal rough or irregular surfaces in the flooring structure upon which they are laid. Conventional subflooring does not provide a satisfactory surface. An *underlayment* of plywood, hardboard, cement board or other manufactured product is required. On concrete floors, a special mastic material is sometimes used when the surface condition does not meet the requirements of the finish flooring.

An underlayment also prevents the finish flooring materials from checking or cracking when slight movements take place in a wood subfloor. When used for carpeting and resilient materials, the underlayment is usually installed as soon as wall and ceiling surfaces are complete.

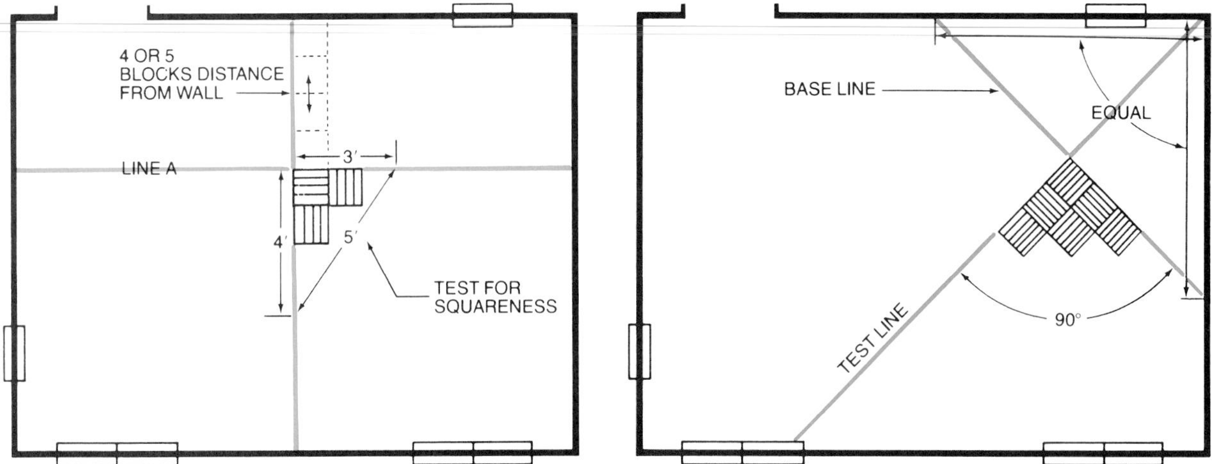

Figure 16-17. Laying out chalk lines for parquet flooring patterns. Left—Square pattern. Right—Diagonal pattern. (Oak Flooring Institute)

Figure 16-18. Prefinished flooring. Three piece units are 3/4" x 4 1/2" x 9". Edges have a slight bevel. Units can be installed in mastic over concrete or wood subfloors. (Memphis Hardwood Flooring Co.)

Cementitious Underlayment

Cementitious underlayment has the advantage of being dimensionally stable and is particularly suited for use under ceramic tile floors. It is not necessary to leave expansion room between panels or at walls. Further, it is impervious to water. *Cement board*, as it is often called, comes in 4' x 4' panels 5/16" thick. See Figure 16-19.

Figure 16-19. Installing cement board underlayment to bathroom floor using screw fasteners. No clearance is needed between panels as the cementitious product is not subject to expansion and contraction due to moisture absorption. (USG Corp.)

Hardboard and Particleboard

Both of these products meet the requirements of an underlayment board. The standard thickness for hardboard is 1/4". Particleboard thicknesses range from 1/4" to 3/4".

This type of underlayment material will bridge small cups, gaps, and cracks. Larger irregularities should be repaired before application. High spots should be sanded down and low areas filled. Panels should be unwrapped and placed separately around the room for at least 24 hours before they are installed.

To apply, start in one corner and fasten each panel securely before laying the next. Some manufacturers print a nailing pattern on the face of the panel. Allow at least a 1/8" to 3/8" space along an edge next to a wall or any other vertical surface.

Stagger the joints of the underlayment panels. See Figure 16-20. The direction of the continuous joints should be at right angles to those in the subfloor. Be especially careful to avoid alignment of any joints in the underlayment with those in the subfloor. Leave a 1/32" space at the joints between hardboard panels. Particleboard panels are butted lightly.

Underlayment panels should be attached to the subfloor with approved fasteners. These fasteners include ring grooved and screw shank nails.

Spacing of nails for particleboard varies for different thicknesses. Be sure to drive nail heads flush with the surface. When fastening underlayment with staples, use a type that is etched or galvanized and at least 7/8" long. Space staples not over 4" apart along panel edges.

Special adhesives will also bond underlayment to subfloors. They eliminate the possibility of nail popping under resilient floors.

Plywood

Install plywood underlayment smooth side up. See Figure 16-21. Since a range of plywood thicknesses are available, vertical alignment of the surfaces of various finish flooring materials is easy to attain, Figure 16-22.

Follow the procedures described for hardboard. Turn the grain of the face ply to run at a right angle to the framing supports. Stagger end joints. When applying 1/4" plywood, use 3d ring grooved or screw shank nails. Edge spacing should not be greater than 3". Field spacing should be 6" each way.

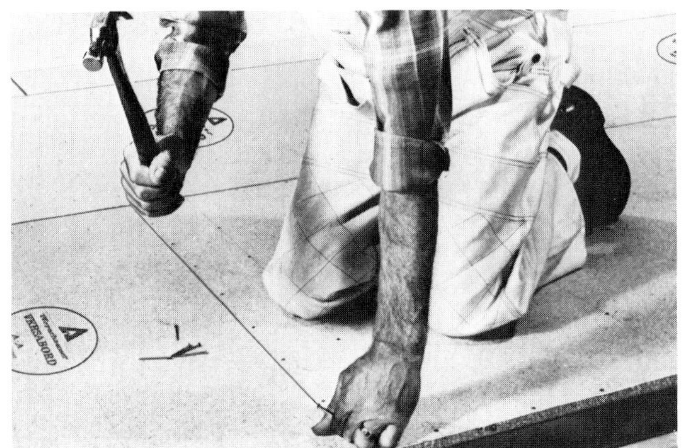

Figure 16-20. Installing particleboard underlayment. Keep nails 1/2" to 3/4" from edge. End joints must be staggered. (Weyerhaeuser Co.)

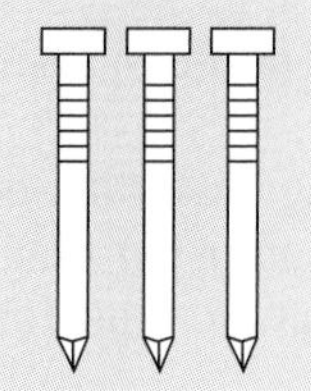

When preparing a base for resilient flooring materials by applying an underlayment over a solid board subfloor, do not drive long nails through the underlayment, subfloor, and on into the joists. Subsequent shrinkage of the subfloor and framing (especially in new construction) will likely cause these nails to rise above the surface and form "blisters" in the floor surface.

Resilient Floor Tile

After the underlayment is securely fastened, sweep and vacuum the surface carefully. Check to see that surfaces are smooth and joints level. Rough edges should be removed with sandpaper or a block plane.

The smoothness of the surface is extremely important, especially under the more pliable materials (vinyl, rubber, linoleum). Over a period of time these materials will "telegraph" (show on the surface) even the slightest irregularities and rough surfaces. Linoleum is especially susceptible

Figure 16-21. Installing 1/4" plywood underlayment. Use 3d ring grooved or screw shank nails. APA recommends 3" spacing along the edge and 6" throughout the field.

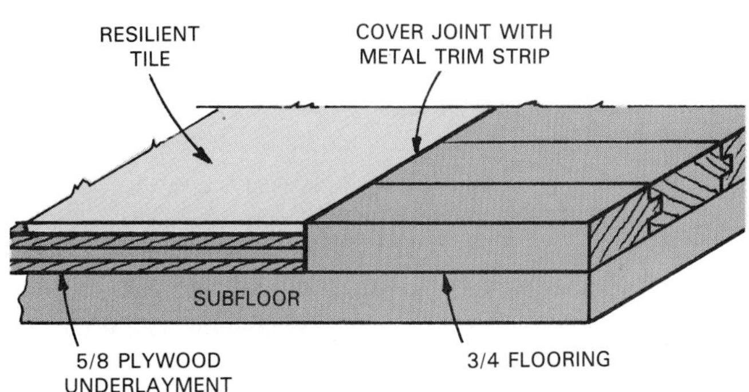

Figure 16-22. Alignment of resilient floor tile with strip flooring is easier when using a plywood underlayment.

and for this reason a base layer of felt is often applied over the underlayment when this material, either in tile or sheet form, is to be installed.

Because of the many resilient flooring materials on the market, it is essential that each application be made according to the recommendations and instructions furnished by the manufacturer of the product.

Figure 16-23. Top—Snapping a chalk line to establish the centerline of a room. This is needed before installing resilient tile. (Armstrong World Industries, Inc.) Bottom—Use a carpenter's square to lay out a centerline at right angles to the first centerline.

Installing Resilient Tile

1. Start a floor tile layout by locating the center of the end walls of the room. Disregard any breaks or irregularities in the contour. Establish a main centerline by snapping a chalk line between the two points. When snapping long lines, remember to hold the line at various intervals and snap only short sections.
2. Lay out another centerline at right angles to the main one. Use a carpenter's square or set up a right triangle (base 4', altitude 3', hypotenuse 5'). A chalk line can be used or you can draw the line along a straightedge. See Figure 16-23.
3. With the centerlines established, make a trial layout of tile along the centerlines as shown in Figure 16-24.
4. Measure the distance between the wall and last tile. If the distance is less than 2" or more than 8", move the centerline closer to the wall by half the tile's dimension. This adjustment will eliminate the need to install border tiles that are too narrow.
5. Check the layout along the other centerline in the same way. Since the original centerline is moved exactly half the tile size, the border tile width will remain uniform on opposite sides of the room.
6. Remove the loose tile.
7. Clean the floor surface and spread the adhesive over one-quarter of the total area, Figure 16-25.
8. Lay the tile.

Make the adhesive spread even with the chalk line but do not cover it. Be sure to use the type of spreader (trowel or brush) recommended by the manufacturer of the adhesive.

The spread of adhesive is very important. If it is too thin, the tile will not adhere properly. If too heavy, the adhesive will creep up between the joints.

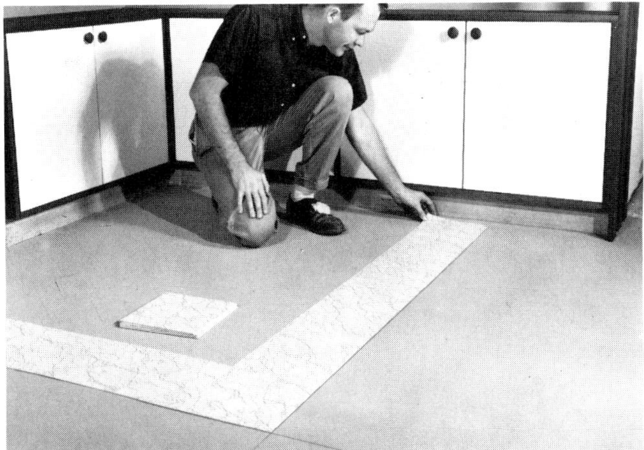

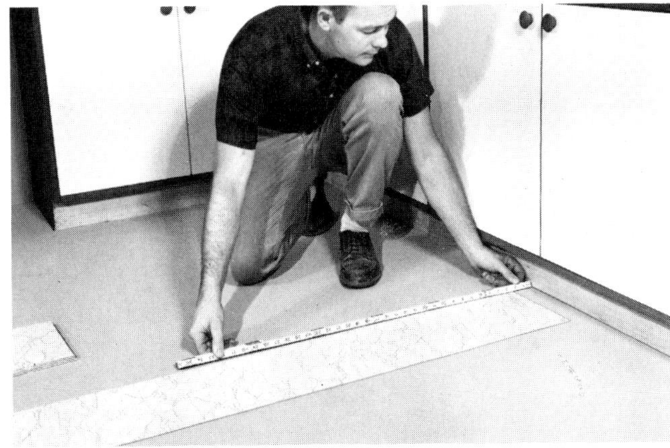

Figure 16-24. Making a trial layout. Top—Lay down tiles along both centerlines to walls. Bottom—Measure width of border tile and adjust centerline, if necessary.

Figure 16-25. Carefully spread adhesive up to the centerlines on one quadrant (1/4) of the floor area. Do not cover up the centerlines.

Allow the adhesive to take an initial set before a single tile is laid. The time required will vary from about 15 minutes to a much longer time, depending on the type of adhesive used. Test the surface with your thumb. It should feel slightly tacky but should not stick to your thumb.

Laying Tile

Start laying the tile at the center of the room. Make sure the edges of the tile align with the chalk line. If the chalk line is partially covered with the adhesive, snap a new one or tack down a thin, straight strip of wood to act as a guide in placing the tile.

Butt each tile squarely to the adjoining tile, with the corners in line, Figure 16-26. Carefully lay each tile in place. Do not slide the tile; it may cause the adhesive to work up between the joints and prevent a tight fit. Take the time to see that each tile is positioned correctly. There is usually no hurry since most adhesives can be "worked" over a period of several hours.

Asphalt and vinyl-asbestos tile do not need to be rolled. Rubber, vinyl, and linoleum are usually rolled after a section of the floor is laid. Be sure to follow the manufacturer's recommendations.

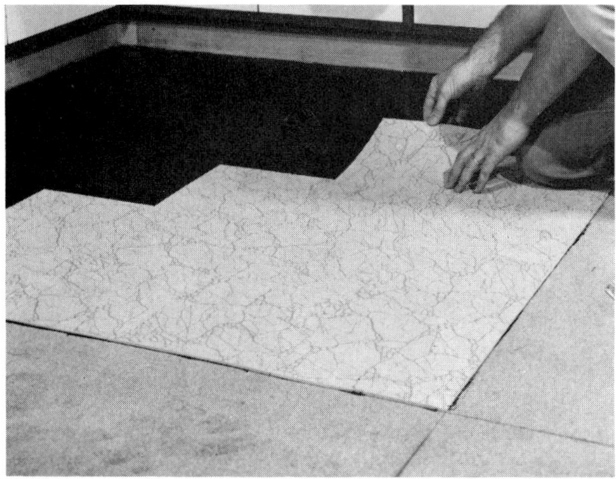

Figure 16-26. Laying tile. Align joining edges first and then lower the rest of the tile.

After the main area is complete, set the border tile as a separate operation. To lay out a border tile, place a loose tile (the one that will be cut and used) over the last tile in the outside row. Now take another tile and place it in position against the wall and mark a sharp pencil line on the first tile, Figure 16-27.

Cut the tile along the marked line, using heavy duty household shears or tin snips. Some types of tile require a special cutter or they may be scribed and snapped. Asphalt tile, if heated, can be readily cut with snips.

Various trim and feature strips are available to customize a tile installation. They are laid by following the general procedures previously described for the regular tile, Figure 16-28.

After all sections of the floor have been completed, cove base can be installed along the wall and around fixtures as shown in Figure 16-29. A special adhesive is available for this operation. Cut the proper lengths and make a trial fit. Apply the adhesive to the cove base and press it into place.

Check over the completed installation carefully. Remove any spots of adhesive. Work carefully using cleaners and procedures approved by the manufacturer.

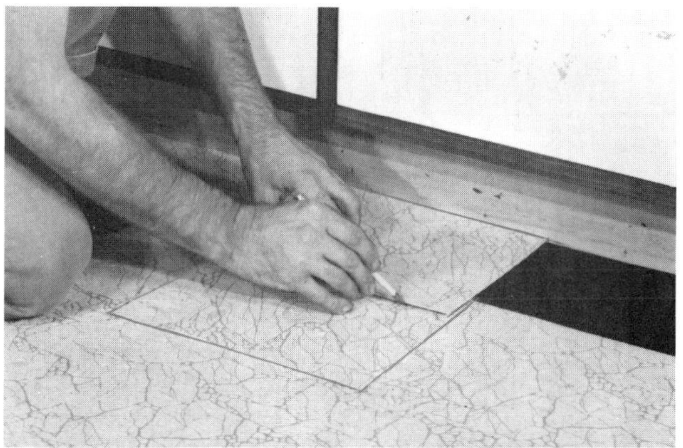

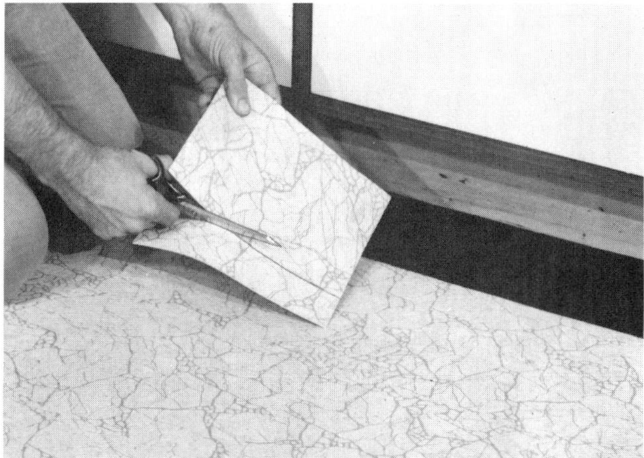

Figure 16-27. Finishing off the borders. Top—Mark the border tile. Bottom—Cut tile with shears.

Installing Resilient Tile over Concrete

Before installing resilient tile over concrete, see to proper preparation of the surface. It must be dry, smooth, and structurally sound. There must be no depressions, pits, scale, or foreign deposits. Remove any paint, varnish, oil or wax. Trisodium phosphate mixed with hot water will remove most paints except those with a chlorinated rubber or resin base. These may be removed with grinding.

Joints and cracks must be filled with a latex underlayment or crack filler. (Follow manufacturer's instructions for on-grade or sub-grade applications.) Dusty or chalked surfaces on suspended concrete floors should be covered with a single coat of primer before spreading adhesive. Chalking or dusty floors on or below grade may be a sign of leaching. Carry out a bond and moisture test before going ahead.

To make a bond and moisture test, install 3' squares of the flooring material at different places. If they are securely bonded after 72 hours the floor is dry enough for application of the tile.

Self-Adhering Tiles

Self-adhering tiles are easy to install. Remove the paper from the back of the tile, position the tile on the floor, and press it down.

Figure 16-28. Setting a feature strip creates a custom effect.

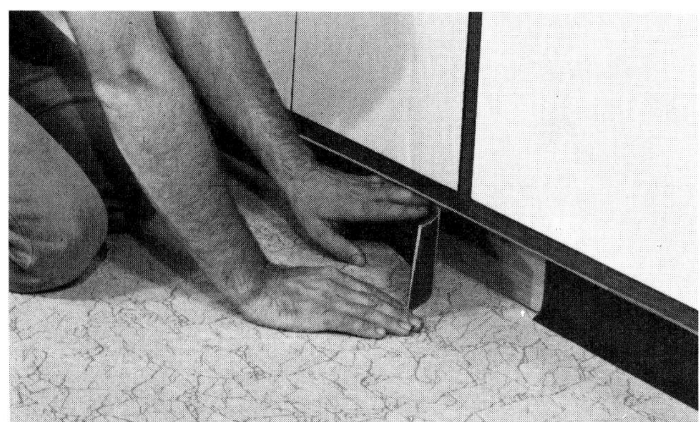

Figure 16-29. Cove base is installed last. Apply special adhesive to the cove.

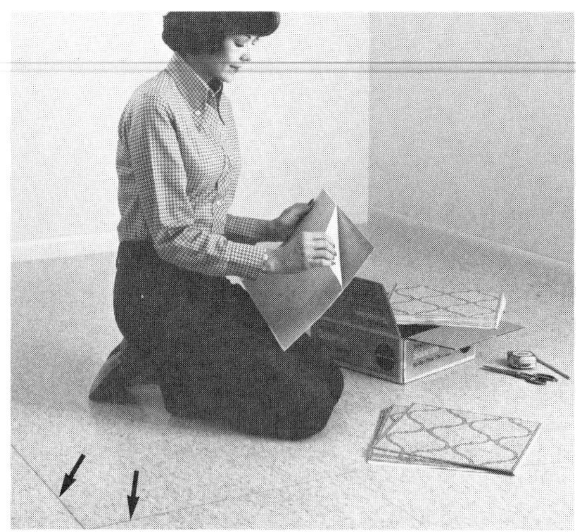

Remove release paper from tile back. Room centerlines (arrow) have been laid out.

Locate position and lay down the tile. Press firmly. If reverse side carries arrows, all should point in same direction.

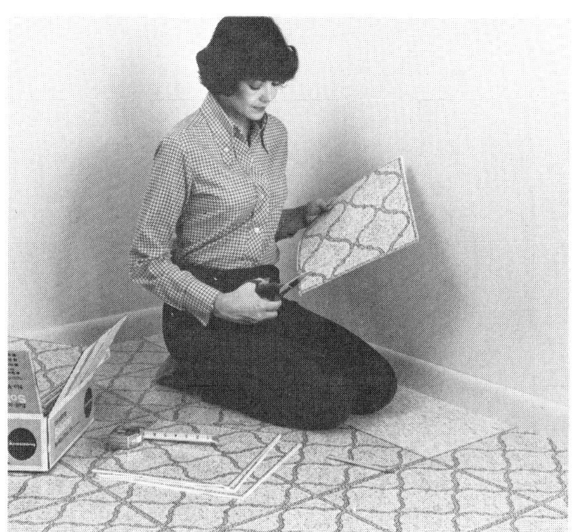

Border tiles can be cut with household shears. Mark cutting line as shown in Figure 16-27.

Figure 16-30. How to install self-adhering vinyl tile. (Armstrong World Industries, Inc.)

It is very important that floors be dry, smooth, and completely free of wax, grease, and dirt. Generally, tiles can be laid over smooth-faced resilient floors. Embossed tile, urethane finish, or cushioned floors should be removed. Figure 16-30 shows basic steps in the installation of a self-adhering tile floor over an existing sheet linoleum floor.

Tiles should be kept in a warm room (at least 65°F (18°C)) for 24 hours before and during installation. The room should be kept warm for one week after installation. This will ensure a firm bond to the subfloor surface. Always study and follow the manufacturer's directions.

Sheet Vinyl Flooring

Recent developments in vinyl flooring have produced a material that is extremely flexible. This property makes installation much easier. Since sheets are available in 12' widths, many installations can be made free of seams.

Flexible vinyl flooring is fastened down only around the edges and at seams. It can be installed over concrete, plywood, or old linoleum.

To install, spread the sheet smoothly over the floor. Let excess material turn up around the edges of the room. When there are seams, carefully match the pattern. Fasten the two sections to the floor with adhesive. Trim edges to size as shown in Figure 16-31.

After all edges are trimmed and fitted, secure them with a staple gun as shown in Figure 16-32, or use a band of double-faced adhesive.

Always study the manufacturer's directions carefully before starting the work. Figure 16-33 shows a completed installation.

Ceramic Floor Tile

Ceramic tiles are made from a mixture of clay and other materials that include ground shale, gypsum, talc, vermiculite, and sand; but sometimes they are made from pure clay. During manufacture, the dry ingredients are mixed with water and other materials and shaped into a bisque. This is the body of the tile. The bisque is formed by extrusion, formed by die pressing, cut from a sheet, or formed by hand. The bisques are allowed to air dry for a short time before being placed in a kiln and fired at temperatures ranging from 1900° to 2200°F (1000° to 1200°C).

The types of tile are grouped according to their permeability (the tendency of the tile to absorb water). The following list is organized from most to least permeable:

- Nonvitreous.
- Semivitreous.
- Vitreous.
- Impervious.

Both permeable and impermeable tile are used on floors, walls, and countertops, Figure 16-34. Floor tiles are made in a variety of colors and shapes.

Tile can also be organized by their characteristics. Those following are the most common:

- *Paver tile*—Glazed or unglazed, they are at least 1/2" thick. The smaller ones are 4" x 6" and larger ones are 12" square.
- *Quarry tile*—Being very dense, this tile is ideal for floors. Generally semivitreous or vitreous, they are always unglazed and range in thickness from 1/2" to 3/4".
- *Mosaic tile*—This type includes any tile that are 2" square or even smaller. They are usually vitreous and are from 3/32" to 1/4" square. See Figure 16-35.

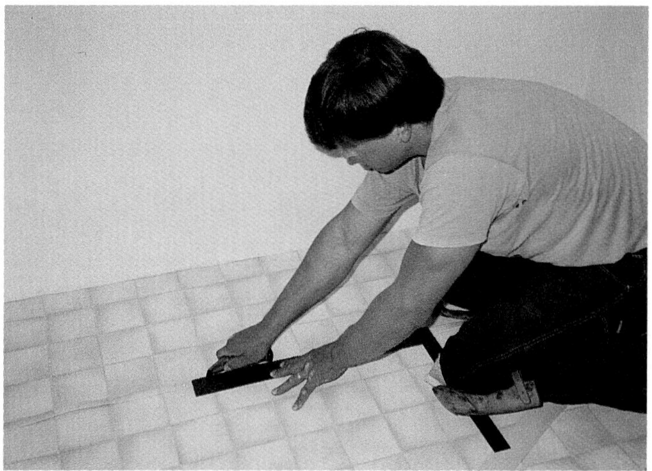

Figure 16-31. Trim flexible vinyl sheet linoleum with a utility knife drawn along a straightedge. Be sure straightedge is parallel to wall.

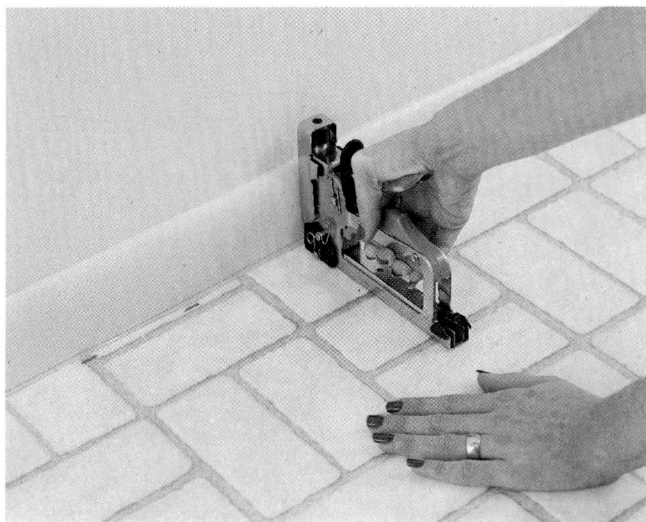

Figure 16-32. Attach trimmed edge to floor with a staple gun. Base shoe will be used to cover the staples. (Armstrong World Industries, Inc.)

Figure 16-33. Completed installation of a sheet vinyl floor in a kitchen-dining area. (Armstrong World Industries, Inc.)

Figure 16-34. Tile comes in a variety of colors and textures. This versatile product provides durable and attractive surfaces not only for floors but for walls and countertops. (American Olean Tile Co.)

- *Lugged tile*—This name applies to any kind of tile with protrusions that ensure exact spacing around each tile. The lugs are eventually concealed by grouting.

Figure 16-36 shows several examples of tile designed and manufactured specifically for floors, walls, and counters. Within each tile group there exists a variety of colors.

Tiles are attached with various types of adhesives. Cement mortar, the preferred adhesive 40 years ago, is still in use. Today, the most used are mastics, dry-set, and latex-portland cement mortars.

Mastics are the least expensive adhesive and come ready to use. Mastics do not have the strength or flexibility of mortar. Further, they should not be used where the tile will be subjected to heat (i.e., around fireplaces). It is also thought that mortars are a better choice of adhesive where tile will be exposed to water.

Mortar mixes, though still basically mixtures of sand and portland cement, now contain additives that improve bonding. Dry-set can be mixed with plain water or a liquid modified with latex, acrylic, or epoxy resins. While expensive, epoxy dry-set mortar has high bonding strength and resistance to impact. Furthermore, it can be applied to nearly any surface — plastic laminate, plywood, and even steel.

Installing Ceramic Tile

Surfaces to be tiled should be tight and even. It may be necessary to check nailing patterns of substrates or to add an underlayment such as cement board or other types of backing board.

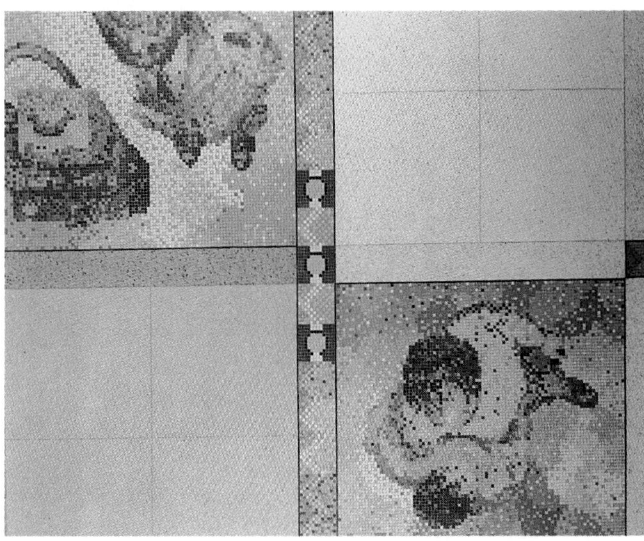

Figure 16-35. Examples of installations using mosaic tile. These tiles are suitable for interior walls, countertops, ceilings and floors in all types of residential or commercial buildings. They are also recommended for exterior walls and floors in any climate. They are also suggested for use as swimming pool linings and decking. (American Olean Tile Co.)

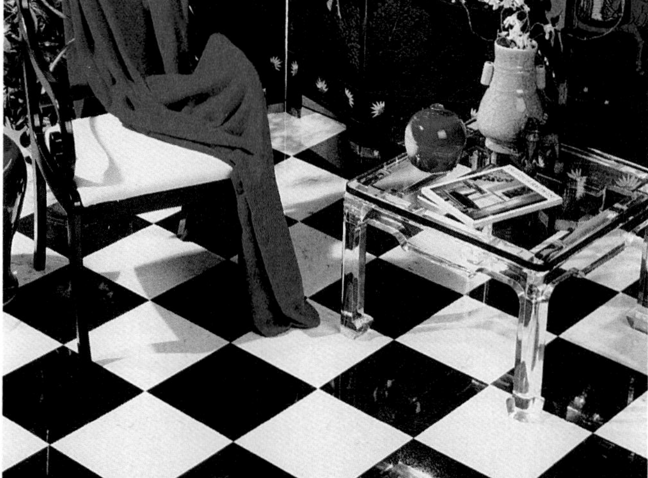

Figure 16-36. Various flooring applications with tile. Top—A vitreous tile with a Mexican saltillo look. Suitable for walls, counters, and floors in residences and light commercial buildings. Middle—Quarry stone in granite and marble. These materials vary in hardness, veining, color, and finish. Bottom—Matte finished, white body glazed tile for use in floors, walls, and counters in residential and light commercial buildings. Not recommended for use where subjected to standing water or accumulations of grease. (American Olean Tile Co.)

Cement backing board sheets should be installed in a "brick" pattern. Do not allow four corners to meet at one point. Keep panels 1/8" away from walls. Wherever possible, place cut edges to the outside.

Lay down a bed of mastic to a minimum thickness of 3/32" to give support to the backing board. Panels should be installed before the mastic films over. Allow edges of sheets to touch but do not force them together tightly. Fasten with 1 1/4" corrosion resistant roofing nails or screws 1 to 1 1/4" long. Space fasteners 6" on center around the perimeter, staying in 3/8" to 3/4" from edges. Drive fastener heads flush with the surface. (Some backing board has printed dots as a guide for fastener placement.)

Expansion joints, Figure 16-37, should be used with ceramic tile in certain situations:

- Over existing structural joints in the floor.
- Where floor dimension is over 15' in one or both directions.
- At doorways where tile is carried through to the next room.
- Where direction changes occur (such as in L-shaped rooms).

It is important for good adhesion that the mortar or mastic completely cover the back of the tile. A notched trowel should be used to apply a full coat to the underlayment. Use the notched edge to spread the material evenly.

Two methods of installation are used: thick-bed and thin-set. In thick-bed, the mortar is 3/4" to 1 3/4" thick. The mortar is allowed to set and cure before tiles are adhered with a dry-set adhesive. This method produces a surface that is strong and unaffected by frequent and prolonged wetting.

In thin-set installations, the adhesive is applied in a thin layer to the substrate. Thickness may vary from 1/32" to 1/8". Thin-setting is least expensive, easier to install, and quicker, Figure 16-38.

To determine quantities of tile needed, first determine the square footage of the area or areas to be covered. Measure length and width and multiply. Add an additional 10% to compensate for breakage, cuts, and extras for future repairs. Manufacturer's supply charts, Figure 16-39, which are available where tile is sold.

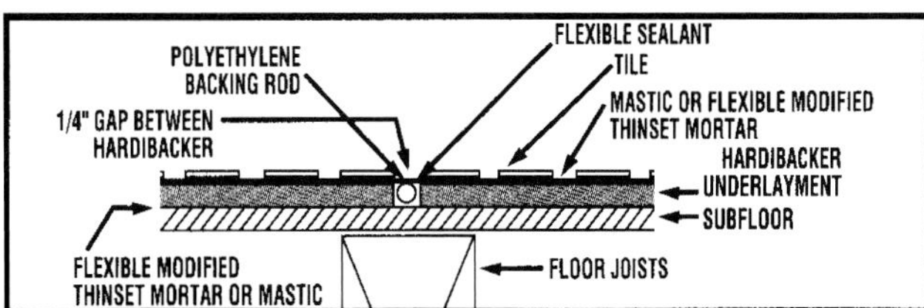

Figure 16-37. Expansion joints are recommended in certain applications of ceramic tile. (James Hardie Building Products, Inc.)

Important Terms

Blind nailed
Block flooring
Cement board
Cove base
Face nail
Finish flooring
Lugged tile
Mosaic tile
Parquet flooring
Parquetry
Paver tile
Plank flooring
Quarry tile
Screed
Self-adhering tile
Setup time
Side-and-end matched
Sleepers
Spline
Strip flooring
Undercut
Underlayment

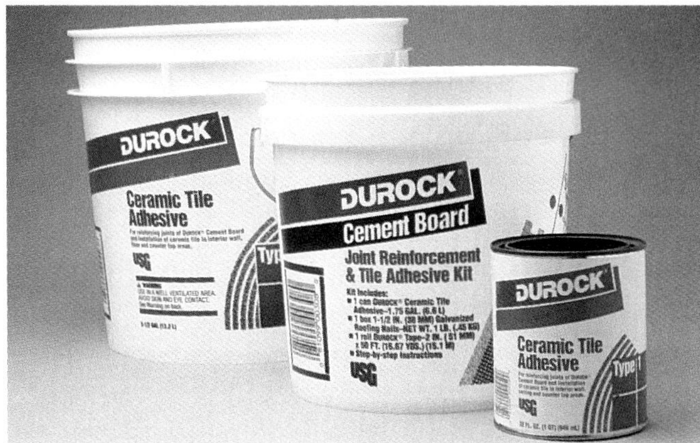

Figure 16-38. An organic adhesive for setting ceramic tile over cement backing board and other structurally sound interior substrates. The adhesive can also be used for reinforcing cement panel joints and bonding cement panels to subfloors. (United States Gypsum Co.)

SQUARE FEET OF COVERAGE		
NUMBER OF CARTONS	TILE SIZE	
	4 1/4" x 4 1/4" 6" x 6"	4" x 4" 8" x 8" 10" x 10" 12" x 12"
	CARTON SIZE	
	12.5 SQ. FT. PER CARTON	11.1 SQ. FT. PER CARTON
1	12	11
2	25	22
3	37	33
4	50	44
5	62	55
6	75	66
7	87	77
8	100	88
9	112	99
10	125	111

Figure 16-39. Coverage chart is an aid to determining tile needed. (American Olean Tile Co.)

Test Your Knowledge

1. The species of wood most commonly used for finish flooring in residential structures is ____________.
2. Standard thicknesses of hardwood flooring include 3/8", ____________, and 3/4".
3. When laying strip floors that start against a wall, lay the first course with the ____________ (*tongue edge, groove edge*) turned toward the wall.
4. When the starter strip is located in the center of an area, install a hardwood ____________ in the grooved side to permit laying in both directions.
5. How many board feet are there in a standard bundle of hardwood flooring?
6. The two types of block flooring most commonly used in residential installations are unit blocks and ____________ blocks.
7. Why does cementitious underlayment require no expansion gap between panels?
8. Hardboard panels used for underlayment are laid with a space of ____________ inches at each joint.
9. Approved metal fasteners for underlayment include ring-groove nails, divergent staples, and ____________ nails.
10. A resilient flooring material that is most likely to show small irregularities in the base is ____________.
11. Before cutting asphalt tile, they should be ____________.
12. Which ceramic tile would you likely place on a bathroom floor?

Outside Assignments

1. Study reference books located in the library or pamphlets obtained from a local building supply firm. Prepare a written report describing the procedures and processes used in the manufacturing of hardwood flooring. Include kiln drying requirements and moisture content standards. Also include grading rules that are applied to the species and qualities commonly used in your geographical area.
2. Prepare an oral report on resilient flooring materials. Obtain samples from a floor covering contractor or home furnishing store. Include information about thicknesses, colors, tile sizes, and approximate costs. Include a list of the advantages and disadvantages of the various types. If time permits, include information about the composition (basic materials used in manufacture) of each type and general requirements for installation.

Flooring materials of wood, composites, and ceramic materials are very popular in residential and light commercial construction. (Memphis Hardwood Flooring Co., Armstrong World Industries, Dal-Tile Corp.)

17

Stair Construction

A stair is a series of steps, each elevated a measured distance, leading from one level of a structure to another. When the series is a continuous section without breaks formed by landings or other constructions, the term "run of stairs" or *"flight of stairs"* is sometimes used. Other terms that can be properly substituted for stairs include "stairway" and "staircase."

In residential buildings, the popularity of the one-story structure has, for many years, minimized requirements for stair construction. Regular carpenters could usually handle the relatively simple task of constructing the service stairs leading from the first floor to the basement level. However, revival of traditional styling along with split-level and multilevel designs have made fine stair construction an important skill.

Because of European influence, main stairs have often been the chief architectural feature in an entrance hallway or other area. However, in modern dwellings, where public rooms are usually on the first floor, there is a trend to move the stairs to a less conspicuous location. Stair construction requires a high degree of skill. See Figure 17-1. The quality of the work should compare with that found in fine cabinetwork.

Today, the parts for main stairways are usually made in millwork plants and then assembled on the job. Even so, the assembly work must be performed by a skillful carpenter who understands the basic principles of stair design and who knows layout and construction procedures.

Main stairways are usually not built or installed until after interior wall surfaces are complete and finish flooring or underlayment has been laid. Basement stairs should not be installed until the concrete floor has been placed.

Carpenters build temporary stairs from framing lumber to provide access until the permanent stairs are installed. These are usually designed as a detachable unit so they can be moved from one project to another.

Sometimes permanent carriages are installed during the rough framing and temporary treads are attached. (*Carriages,* or *stringers,* are the inclined supports that carry the treads and risers.) Later, as the interior is finished, these treads are replaced with finished parts.

Figure 17-1. A main stairs can provide an attractive architectural feature to a residence. Its design and construction has long been considered one of the highest forms of joinery. (C.E. Morgan)

Types of Stairs

Basically, stair types are divided into service stairs and main stairs. Either of these may be closed, open, or a combination of open and closed. See Figure 17-2. The types usually listed are straight run, platform, and winding. The platform type includes *landings* where the direction of the stair run is usually changed. Such descriptive terms as L type (long L and wide L), double L type and U type are commonly used. See Figure 17-3.

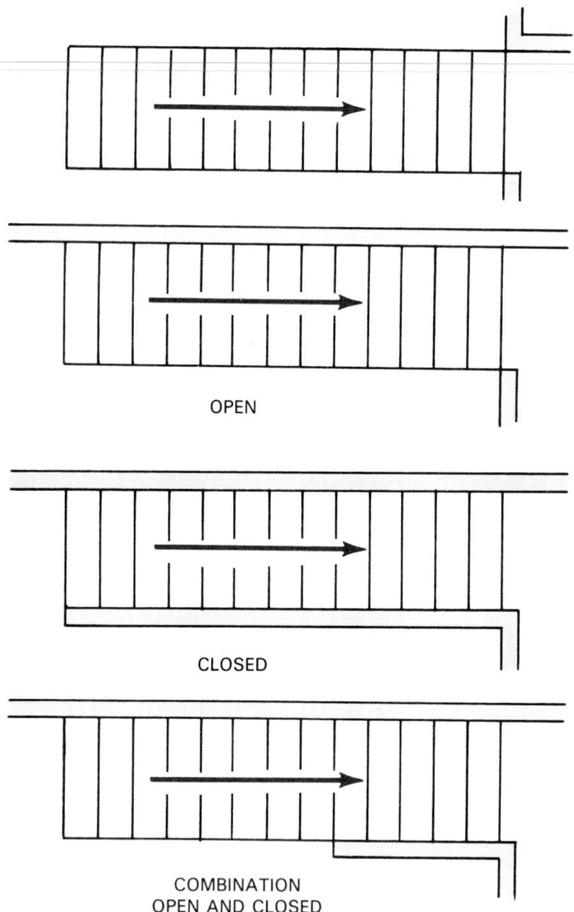

Figure 17-2. Simplified drawings of open and closed stairways. Heavy colored lines represent walls. A stair is called "open" even though one side is enclosed by a wall.

The *straight run* stairway is continuous from one floor level to another without landings or turns. It is the easiest to build. Standard multistory designs require a long stairwell. This often presents a problem in smaller structures. A long run of 12 to 16 steps also has the disadvantage of being tiring. It offers no chance for a rest during ascent.

In modern split-level designs, the runs are short and generally straight. They connect directly to the next floor level. Usually stair runs of this type are located so that headroom is automatically provided by the stair run directly above, Figure 17-4.

Winding stairs, also called "geometrical," are circular or elliptical. They gradually change directions as they ascend from one level to another. These often require curved wall surfaces that are difficult to build. Because of the expense, such stairs are seldom used except in prestigious homes.

Parts and Terms

Refer to Figure 17-5 as your read the definitions:

- *Stairwell*—The rough opening in the floor above to provide headroom for stairs. Thus, the run of the stairs must be known when floors are framed.
- *Stringer*—One of the inclined sides of a stair that supports the risers and treads. A housed stringer, Figure 17-6, has treads and risers let into its sides so that their edges are concealed. A built-up stringer is straight-edged with blocking supporting the risers and treads. A cut-out stringer has notches for the treads and risers cut into its top edge. Also called a carriage or a string.
- *Riser*—The vertical surface between two treads.
- *Tread*—The horizontal face of one step.
- *Nosing*—The part of a tread that extends over the riser beneath it.
- *Newel*—A post that supports the railing, especially at the start of the stairs but also at points where the stairs angle in a different direction and at platforms.
- *Handrail*—An angled piece supported by a wall or a railing intended to be grasped by persons climbing or descending stairs.
- *Balusters*—Vertical members (spindles) supporting the handrail on open stairs. The bottom edges rest on the steps or on a lower rail. Their main purpose is to prevent anyone, children especially, from slipping under the railings and falling to the floor below. Codes usually require spacing of no more than 6", although 4" is required in some localities.
- *Platform*—A horizontal area placed between two flights of a stair.
- *Landing*—The floor at the top or bottom of each story where a flight of stairs begins or ends.
- *Winders*—Wedge-shaped treads installed where stairs turn.
- *Unit run*—The width of a tread minus the nosing.
- *Unit rise*—The height of a riser; vertical distance between two treads.
- *Total run*—The sum of all unit runs.
- *Total rise*—The sum of all unit rises.
- *Headroom*—The vertical distance from a step to a ceiling above.

Stairs are basically sets of risers and treads supported by stringers. The relationship between the riser height and the tread width will determine how easily the stairs may be negotiated. Research has indicated that the ideal riser height is 7" while the ideal tread witdth is 11".

Headroom is measured from a line along the front edges of the treads to the enclosed surface or header above. This distance is usually specified in local building codes. Refer, again, to Figure 17-5. FHA requires a minimum headroom of 6'-8" for main stairs and 6'-4" for basement or service stairs.

Stairwell Framing

Methods of stair building differ from one locality to another. One carpenter may cut and install a carriage

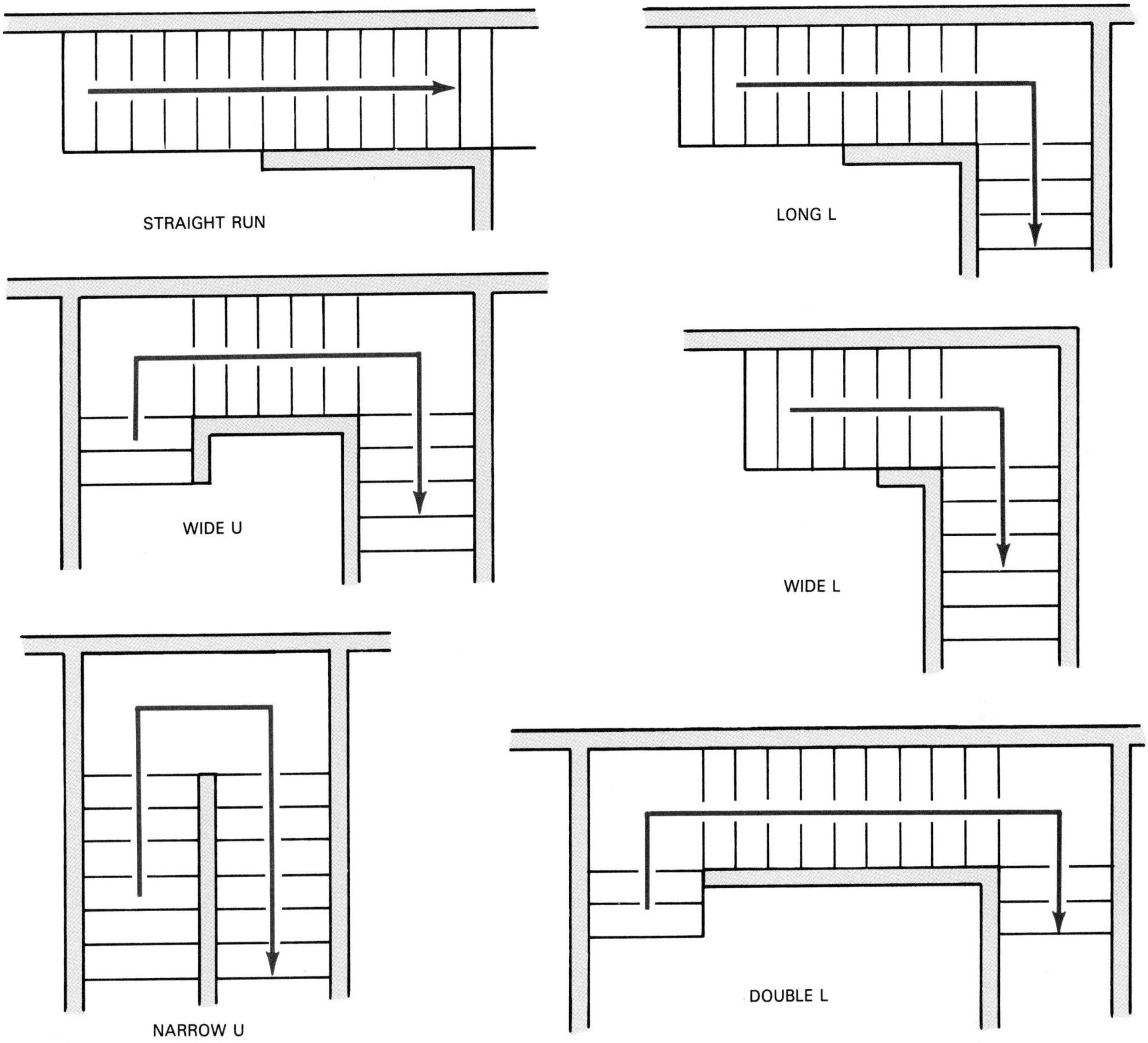

STRAIGHT RUN

LONG L

WIDE U

WIDE L

NARROW U

DOUBLE L

Figure 17-3. Terms used to define different stair types.

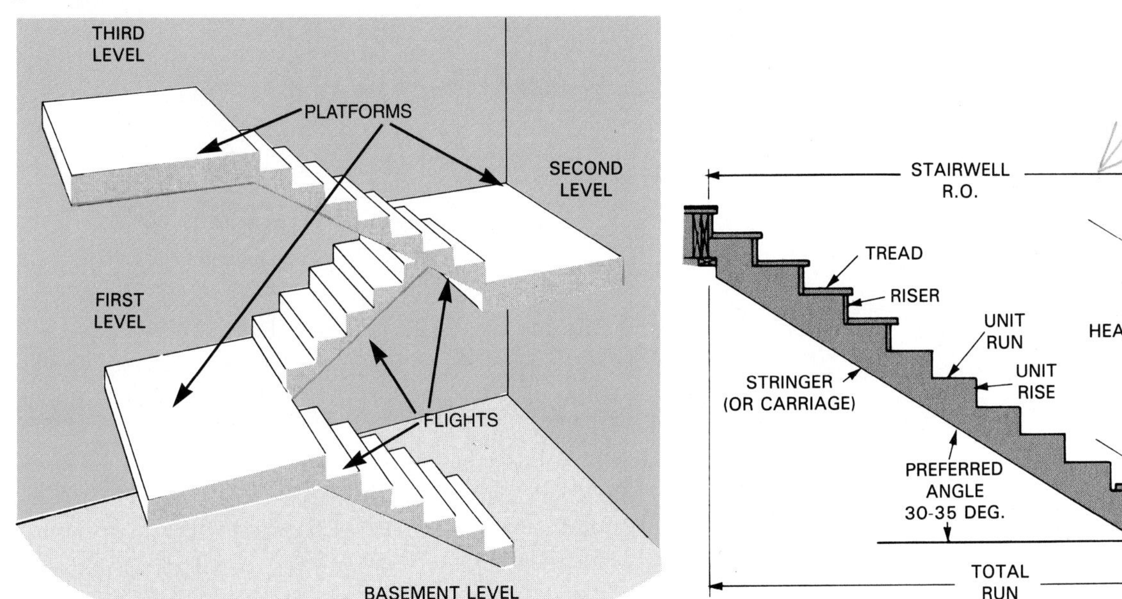

Figure 17-4. Stair runs are often made one above the other to get headroom. This one is designed for a split-level home.

Figure 17-5. Basic stair parts and terms. Total number of risers is always one greater than the total number of treads.

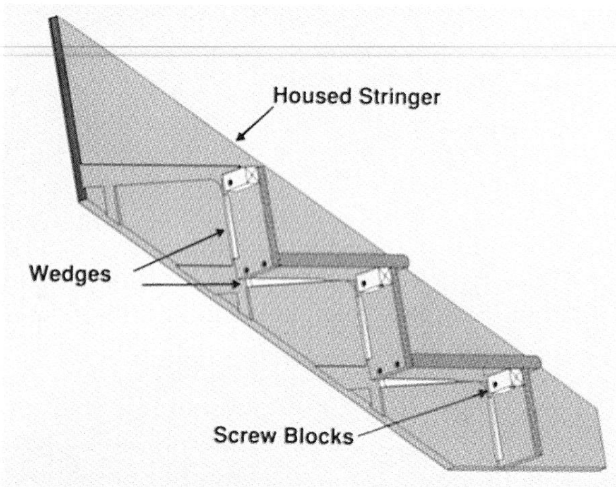

Figure 17-6. A housed stringer has grooves machined into it with a router. The grooves hold the treads and risers. The wedges secure the joints and prevent squeaking. Screws and glue are also used for sturdier construction. (L. J. Smith, Inc.)

2. Stair carriage is a cut-out type. There may be two or three carriages. These were installed during rough framing. The 2 x 4 spacer (arrow) gives clearance for installation of the wall finish.

1. Carpenter is making a plumb cut on a housed (or closed) stringer. Grooves are cut in the stringer to receive the treads and risers.

3. Newel post is being installed by cutting through the subflooring and anchoring the bottom end to a floor joist. Closed stringer (arrow) is in place and exposed stair components have been finished. Flakeboard subflooring has been installed.

Figure 17-7. This series of photos shows various stages in building stairs. Some stair components are cut on site; others are preformed and ready to install.

4. Handrail has been installed and the carpenter is cutting and placing the prefinished balusters.

5. Carpenter is fastening lower rail of a banister to the floor. Since the banister is made of oak, it is necessary to drill nail holes. Glue is also applied to each joint.

6. Stairs are completed except for the installation of carpeting over padding. Often, tackless carpeting strip are fastened to the rear of the tread and to the bottom of the riser.

Figure 17-7 continued.

(stringer) during the wall and floor framing. Another may put off all stairwork until the interior finishing stages. Figure 17-7 shows several stages of stair-building.

Regardless of procedures followed, the rough openings for the stairwell must be carefully laid out and constructed. If the architectural drawings do not include dimensions and details of the stair installation then the carpenter will need to calculate the sizes. She or he must follow recognized standards and local code restrictions.

Trimmers and headers in the rough framing should be doubled, especially when the span is greater than 4'. Headers more than 6' long should be installed with framing anchors unless supported by a beam, post, or partition. Tail joists over 12' long should also be supported by framing anchors or a ledger strip. Refer to Unit 8 for additional information on framing rough openings.

Providing adequate headroom is often a problem, especially in smaller structures. Installing an auxiliary header close to the main header, Figure 17-8, will permit a slight extension in the floor area above a stairway. When a closet is located directly above, the closet floor is sometimes raised for additional headroom.

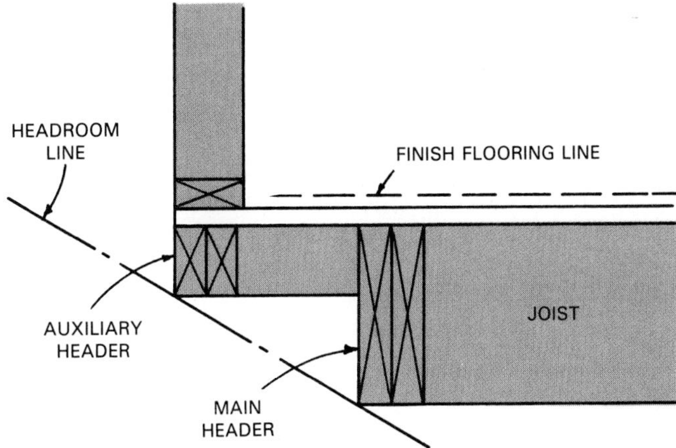

Figure 17-8. Extending upper floor area with a shallow auxiliary header. Partition over the auxiliary must be nonsupporting.

Stair Design

Most important in stair design is the mathematical relationship between the riser and tread. There are three generally accepted rules for calculating the rise-run or riser-tread ratio as follows. It is wise to observe them:

- The sum of two risers and one tread should be 24 to 25".
- The sum of one riser and one tread should equal 17 to 18".
- The height of the riser times the width of the tread should equal between 70" and 75".

A riser 7 1/2" high would, according to Rule 1, require a tread of 10". A 6 1/2" riser would require a 12" tread.

In residential structures, treads (excluding nosing) are seldom less than 9" or more than 12" wide. In a given run of stairs it is extremely important that all of the treads and all risers be the same size. A person tends to measure (subconsciously) the first few risers and will probably trip on subsequent risers that are not the same.

When the rise-run combination is wrong, the stair will be tiring and will cause extra strain on leg muscles. Further, the toe may kick the riser if the tread is too narrow.

A unit rise of 7 to 7 5/8" high with an appropriate tread width will combine both comfort and safety. Main or principal stairs are usually planned to have a rise in this range. Service stairs are often steeper but risers should be no higher than 8". As stair rise is increased the run must be decreased. See Figure 17-9.

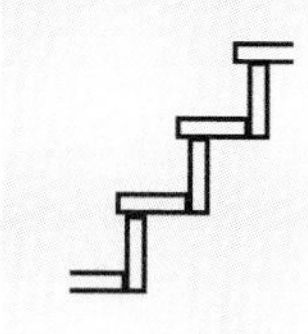

In a given run of stairs, be sure to make all the risers the same height and all the treads the same width. An unequal riser, especially one that is too high, may cause a fall.

A main stair should be wide enough to allow two people to pass without contact. Further, it should provide space so furniture can be moved up or down. A minimum width of 3' is generally recommended, Figure 17-10. FHA permits a minimum width, measured clear of the handrail, of 2'-8". On service stairs, the requirement is reduced to 2'-6". Furniture moving is an important consideration and extra clearance should be provided in closed stairs of the L and U type; especially those that include winders, Figure 17-11.

Stairs should have a continuous rail along the side for safety and convenience. A *handrail* (also called a stair rail) is used on open stairways that are constructed with a low partition or banister.

In closed stairs, the support rail is called a *wall rail*. It is attached to the wall with special metal brackets. Except for very wide stairs, a rail on only one side is sufficient. Figure 17-12 illustrates the correct height for a rail.

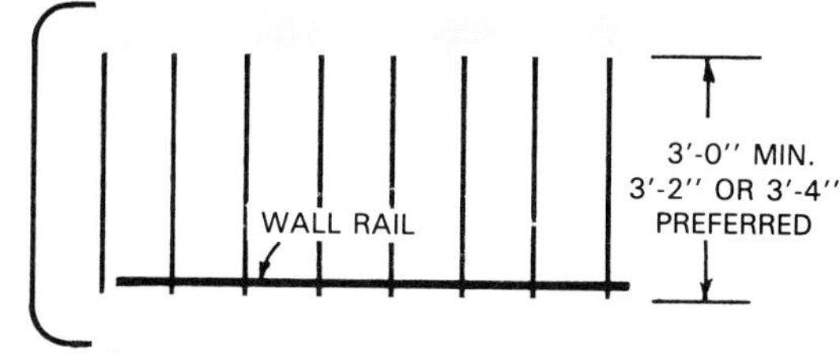

Figure 17-10. A main stair should be at least 3' wide for easy movement of people and furniture.

Figure 17-11. This L-shaped stairs is spacious enough for moving furniture up and down. (C.E. Morgan)

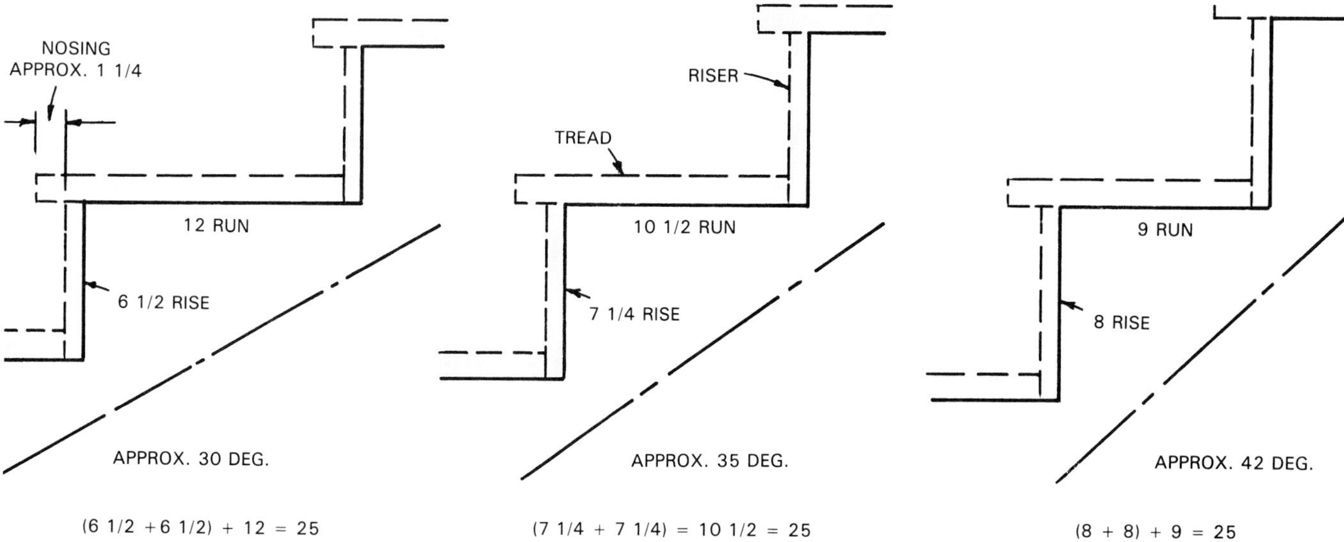

Figure 17-9. Be careful about rise-run relationships in stair design.

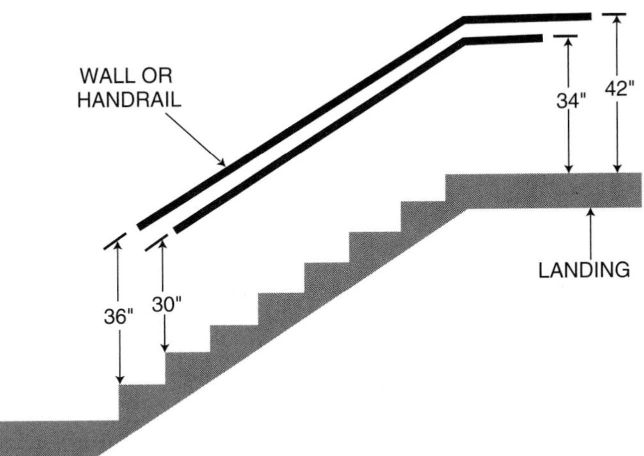

Figure 17-12. A handrail height of 30" at rake (slope) and 34" at landings have been an accepted standard. Recently, building codes in some places have been adopting heights of 36" at the rake and 42" at landings. (C.E. Morgan)

Building Officials and Code Administrators (BOCA) have proposed a change in the National Building Code affecting allowable tread width and height of risers. The present standard for residences is an 8 1/4" rise and a 9" tread. A change to a 7" rise and an 11" tread has been proposed. The BOCA code book, which is updated every three years, has no legal force in law until it is adopted by state or local governments.

Under the current standard, stairways have 13 risers at 8 1/4" each and 12 treads, each 9" wide; the new proposed standard would have 16 risers, each 7" high, and 15 treads, each 11" wide.

A complete set of architectural plans should include detail drawings of main stairs, especially when the design includes any unusual features. For example, the stair layout in Figure 17-13 shows a split-level entrance with open-riser stairs leading to upper and lower floors. An exact description of tread mountings, overlap, nosing requirements, and height of the handrail is not included. These items of construction are the responsibility of the carpenter,

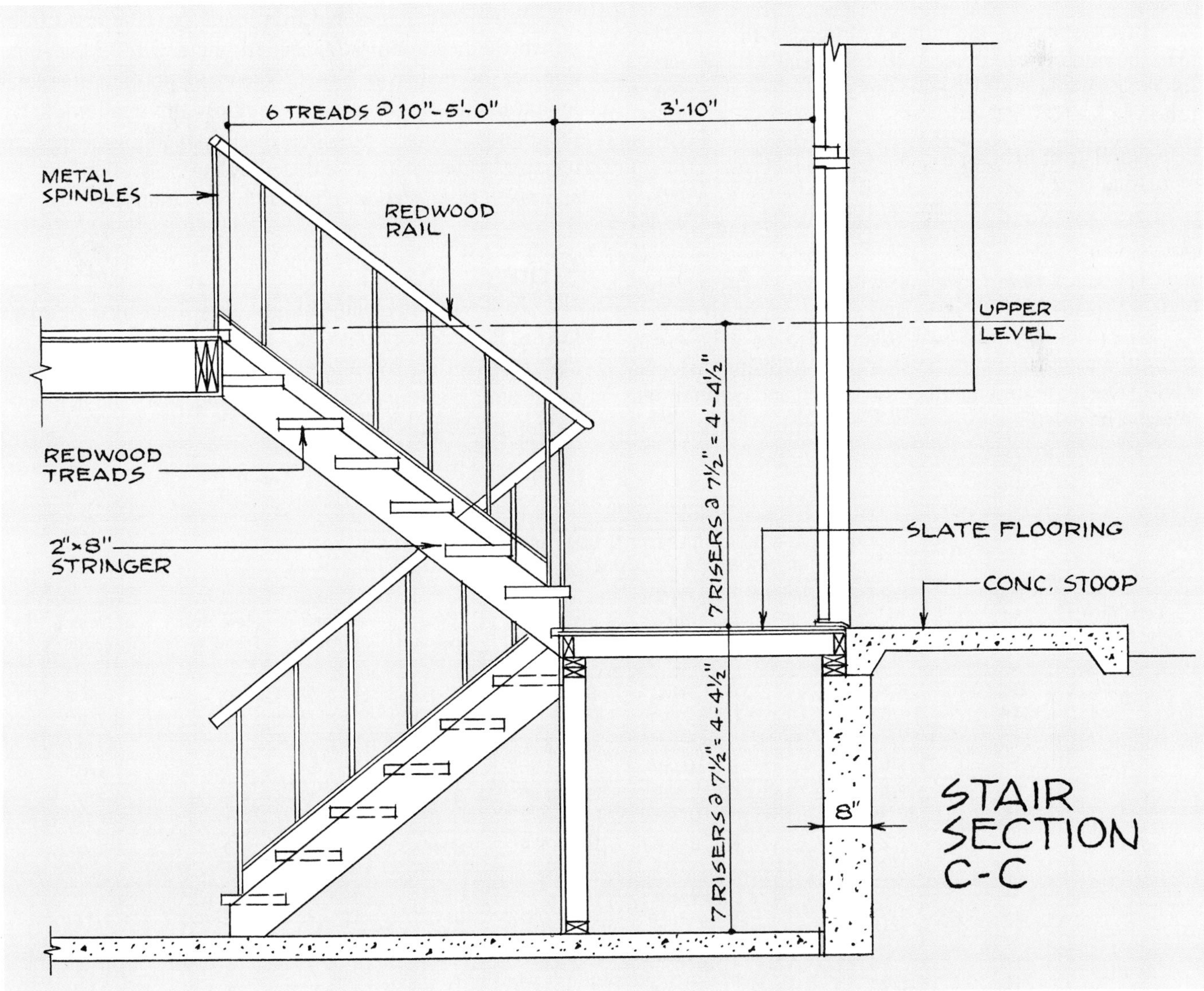

Figure 17-13. Architectural drawings will show stair layouts like this. Note information given for riser-tread ratios.

who must have a thorough understanding of basic stair design and how to lay out and make the installation.

All stairs, whether main or service, will be shown on floor plans. When details of the stair design are not included in the complete set of plans, the architect will usually specify on the plan view the number and width of the treads for each stair run. Sometimes the number of risers and the riser height are also included.

Stair Calculations

To calculate the number and size of risers and treads (less nosing) for a given stair run, first divide the total rise by 7 to determine the number of risers. (Some divide by 8. Either number is accurate enough.) For example: if the total rise for a basement stairway is 7'-10", or 94", dividing by 7 would yield 13.43. Since there must be a whole number of risers, select the one closest to 13.43 and divide it into the total rise to determine the unit rise:

94" ÷ 13	= 7.23 or 7 1/4"
Number of risers	= 13
Riser height	= 7 1/4"

In any stair run, the number of treads will be one less than the number of risers. A 10 1/2" tread will be correct for the example and the total run would be calculated as follows:

Number of treads	= 12
Total run	= 10 1/2" x 12
	= 126"
	= 10'-6"

The stairs in the example will have 13 risers 7 1/4" high, 12 treads 10 1/2" wide, and a total run of 10'-6".

Since the example was assumed to be a basement stairs, the total run could be shortened by using a steeper angle. Decrease the number or risers and shorten the treads. The calculations are:

94" ÷ 12	= 7.83 = 7 5/6"
Number of risers	= 12
Height of risers	= 7 5/6"
Tread width selected	= 9"
Number of treads	= 11
Total run	= 9" x 11"
	= 99"
	= 8'-3"

Some manufacturers supply tables for arriving at rise and run, riser, and tread ratios. See Figure 17-14.

Stairwell Length

The length of the stairwell opening must be known during the rough framing operations. If not included in the architectural drawings, it can be calculated from the size of the risers and treads.

It is necessary to know the headroom required. Add to this the thickness of the floor structure and divide this total vertical distance by the riser height. This will give the number of risers in the opening.

When counting down from the top to the tread from which the headroom is measured, there will be the same number of treads as risers. Therefore to find the total length of the rough opening, multiply the tread width by the number of risers previously determined. Some carpenters prefer to make a scaled drawing (elevation) of the stairs and floor section to check the calculations.

Stringer Layout

To lay out the stair stringer, first determine the riser height. Place a story pole (straight strip of 1 x 4 lumber) in a plumb position from the finished floor below through the rough stair opening above. On the pole, mark the height of the top of the finished floor above.

WELL OPENINGS BASED ON MIN. HEAD HGT. OF 6'-8" DIMENSIONS BASED ON 2" x 10" FLOOR JOIST									
Total Rise Floor to Floor H	Number of Risers	Height of Riser R	Number of Treads	Width of Run T	Total Run L	Well Opening U	Length of Carriage	Use Stock Tread Width	Dimension of Nosing Projection
8'-0"	12	8"	11	9-1/2"	8'-8 1/2"	9'-1"	11'-4 5/8"	10-1/2"	1"
	14	6-7/8"	13	10-5/8"	11'-6-1/8"	10'-10"	13'-8-1/2"	11-1/2"	0-7/8"
8'-4"	13	7-11/16"	12	9-13/16"	9'-9-3/4"	10'-0"	12'-5-1/2"	10-1/2"	11/16"
	14	7-1/8"	13	10-3/8"	11'-2-7/8"	11'-0"	13'-7-5/8"	11-1/2"	1-1/8"
8'-6"	13	7-7/8"	12	9-5/8"	9'-7-1/2"	9'-2"	12'-5-1/4"	10-1/2"	0-7/8"
	14	7-5/16"	13	10-3/16"	11'-0-1/2"	10'-8"	13'-7"	11-1/2"	1-5/16"
8'-9"	14	7-1/2"	13	9-1/4"	10'-0-1/4"	9'-5"	12'-10-3/4"	10-1/2"	1-1/4"
	14	7-1/2"	13	10"	10'-10"	10'-1"	13'-6-1/2"	11-1/2"	1-1/2"
8'-11"	14	7-5/8"	13	9-3/8"	10'-1-7/8"	9'-5"	13'-1-1/4"	10-1/2"	1-1/8"
	14	7-5/8"	13	9-1/16"	9'-9-7/8"	9'-0"	12'-10"	10-1/2"	1-7/16"
	14	7-5/8"	13	10-1/4"	11'-1-1/4"	10'-2"	13'-10-1/4"	11-1/2"	1-1/4"
9'-1"	14	7-13/16"	13	9-11/16"	10'-6"	9'-5"	13'-5-3/4"	10-1/2"	13/16"
	15	7-1/4"	14	10-1/4"	11'-11-1/2"	10'-8"	14'-7-3/4"	11-1/2"	1-1/4"

Figure 17-14. A chart such as this can be used to determine number of risers and treads and their dimensions. (C.E. Morgan)

Set a pair of dividers to the calculated riser height and step off the distances on the story pole. There will likely be a slight error in the first layout so adjust the setting and try again.

Continue adjusting the dividers and stepping off the distance on the story pole until the last space is equal to all the others. Measure the setting of the dividers. This length will be the exact riser height to use in laying out the stringers.

For a simple basement stair, select a straight piece of 2 x 10 or 2 x 12 stock of sufficient length. Place it on sawhorses to make the layout. Begin at the end that will be the top and hold the framing square in the position shown in Figure 17- 15. Let the blade represent the treads and the tongue represent the risers. For example, if the risers are 7 5/6", align that mark with the edge of the stringer; if the treads are 10 1/2", align that mark on the blade with the edge of the stringer.

Draw a line along the outside edge of the blade and tongue. Now move the square to the next position and repeat. The procedure is similar to that described for rafter layout in Unit 10. Accuracy can be assured in this layout by using framing square clips or by clamping a strip of wood to the blade and tongue.

Extreme accuracy is required in laying out the stringer. Be sure to use a sharp pencil or knife and make the lines meet on the edge of the stock.

Continue stepping off with the square until the required number of risers and unit treads have been drawn, Figure 17-16.

The stair begins with a riser at the bottom, so extend the last tread line to the back edge of the stringer as shown. At the top, extend the last tread and riser line to the back edge.

One other adjustment must be made before the stringer is cut. Earlier calculations which gave the height of the riser did not take into account the thickness of the tread. Therefore the total rise of the stringer must be shortened by one tread thickness. Otherwise the top tread will be too high. The bottom of the stringer must be trimmed as shown in Figure 17-17.

Treads and Risers

The thickness of a main stair tread is generally 1 1/6" or 1 1/8". Hardwood or softwoods may be used. FHA requires that stair treads be hardwood, vertical grain softwoods, or flat-grain softwoods covered with a suitable finish flooring material.

Lumber for risers is usually 3/4" thick and should match the tread material. This is especially important when the stairs are not covered. In most construction, the riser drops behind the tread, making it possible to reinforce the joint with nails or screws driven from the back side of the stairs.

Where the top edge of the riser meets the tread, glue blocks are sometimes used. A rabbeted edge of the riser may fit into a groove in the tread. A rabbet and groove joint may also be used where the back edge of the tread meets the riser.

Stair treads must have a nosing. This is the part of the tread that overhangs the riser. Nosings serve the same purpose as toe space along the floor line of kitchen cabinets. They provide toe room.

The width of the tread nosing may vary from about 1 1/8 to 1 1/2". It should seldom be greater than 1 3/4".

In general, as the tread width is increased, the nosing can be decreased. Figure 17-18 illustrates a number of nosing forms. Cove molding may be used to cover the joint between riser and tread as well as conceal nails used to attach the riser to the stringer or carriage.

Basement stairs are often constructed with an open riser (no riser board installed). Sometimes an open riser design is built into a main stair to provide a special effect. Various methods of support or suspension may be used.

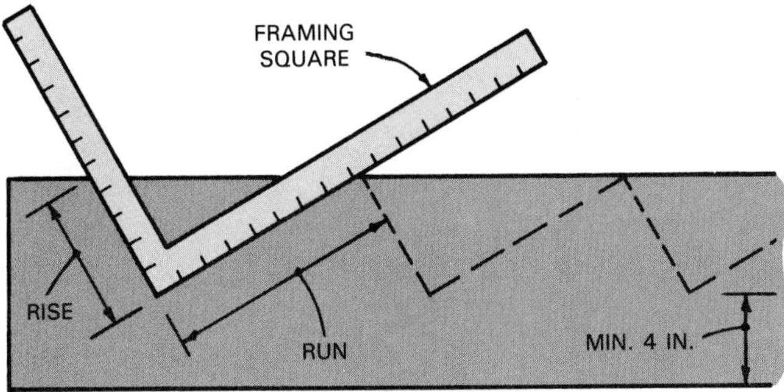

Figure 17-15. Use a framing square to lay out a stringer. Method is almost identical to laying out rafters.

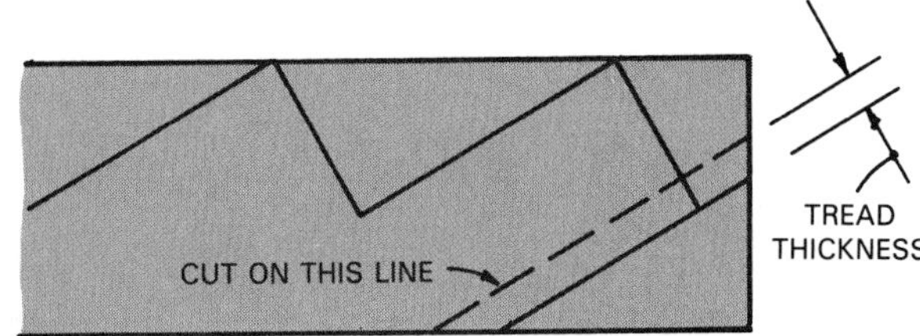

Figure 17-17. Trim the bottom end of the stringer to adjust for tread thickness.

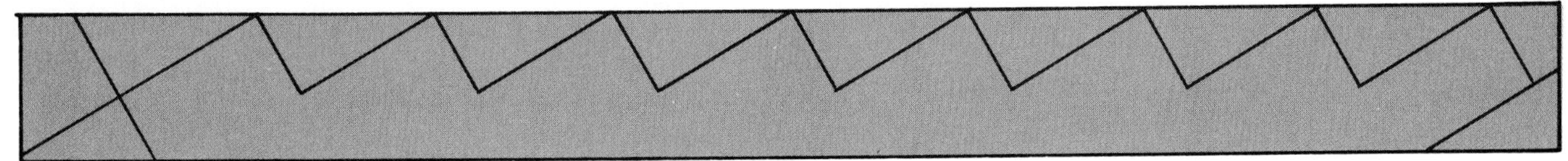

Figure 17-16. Completed stringer layout will look something like this.

Often custom-made metal brackets or other devices are needed. Figure 17-19 shows basic types of riser designs. A slanted riser is sometimes used in concrete steps since it provides an easy way to form a nosing.

Types of Stringers

Treads and risers are supported by stringers or carriages that are solidly fixed to the wall or framework of the building. For wide stairs, a third stringer is installed in the middle to add support.

The simplest type of stringer is formed by attaching cleats on which the tread can rest. Another method consists of cutting dados into which the tread will fit, Figure 17-20. This type is often used for basement stairs where no riser enclosure is called for.

Standard cut-out stringers (type used in layout description) are commonly constructed for either main or service stairs. Prefabricated treads and risers are often used for this type of support. An adaptation of the cut-out stringer, called semihoused construction, is illustrated in Figure 17-21. The cutout stringer and backing stringer may be assembled and then installed as a unit or each part may be installed separately.

A popular type of stair construction has a stringer with tapered grooves into which the treads and risers fit. It is commonly called housed construction. Wedges, with an application of glue, are driven into the grooves under the tread and behind the riser, Figure 17-22. The treads and risers are joined with rabbeted edges and grooves or glue blocks.

This type of construction produces a stair that is strong and dust tight. It will seldom develop squeaks.

Housed stringers can be purchased completely cut and ready to install. They can be cut on the job, using an electric router and template.

To assemble the stairs, the housed stringer is spiked to the wall surface and into the wall frame. The treads and risers are then set into place. Work from the top downward, using wedges and glue. This type of stringer shows above the profiles of the treads and risers and provides a finish strip along the wall. The design should permit a smooth joint where it meets the baseboard of the upper and lower levels.

Winder Stairs

Winder stairs, Figure 17-23, present stair conditions that are frequently regarded as undesirable. In fact, some localities do not allow them. Their use, however, may

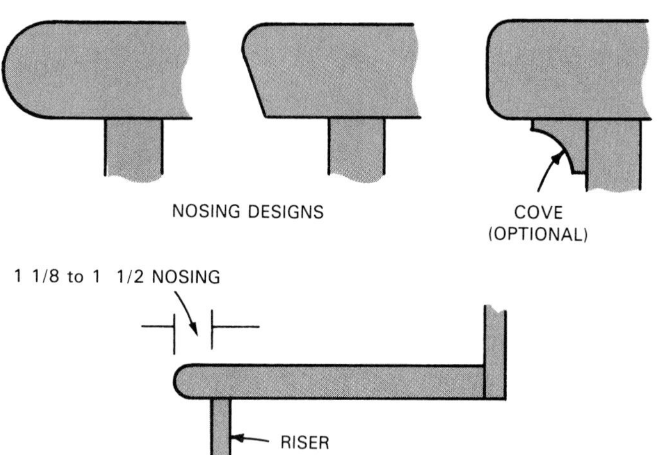

Figure 17-18. Tread nosings commonly take these shapes.

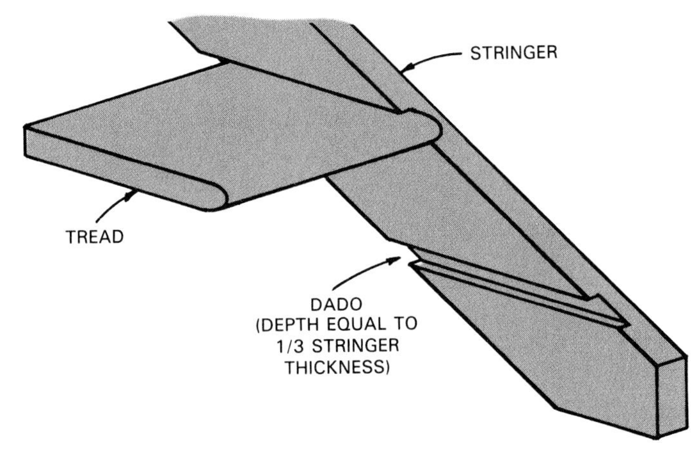

Figure 17-20. Open riser stairs. Treads are set into dados cut in the stringer.

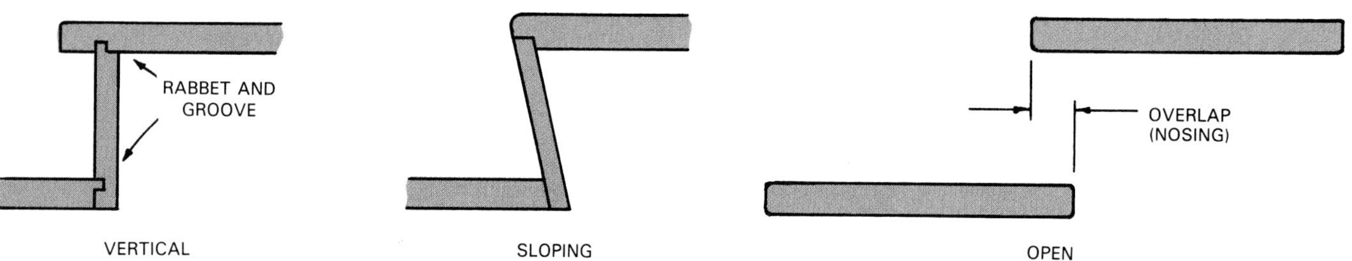

Figure 17-19. Basic stair riser shapes. For the open riser, the tread should overlap the riser at least 2".

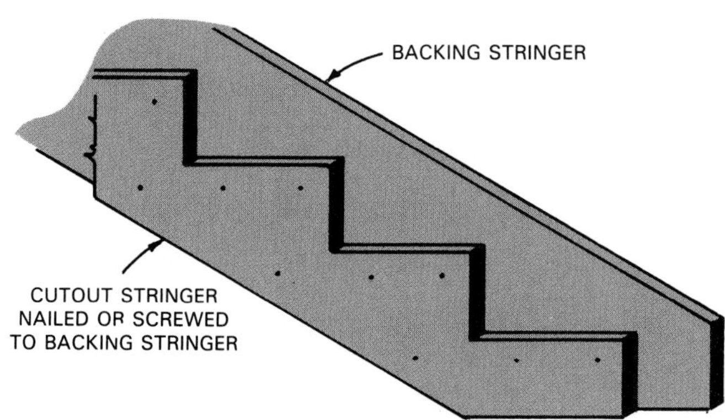

Figure 17-21. This is a semihoused stringer.

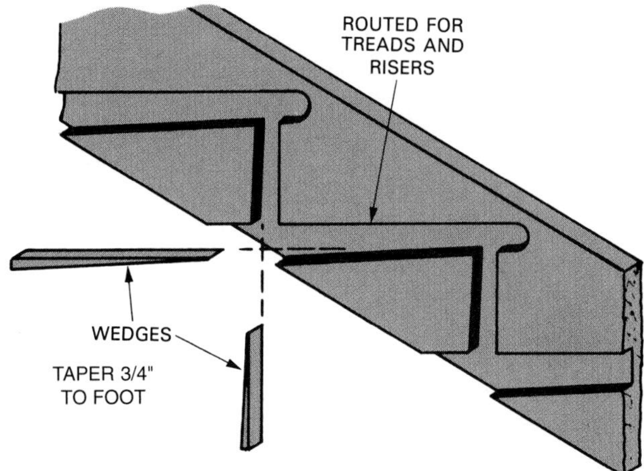

Figure 17-22. In a housed stringer, risers and treads are let into the stringer. This type of housing is difficult to make.

sometimes be necessary where space is limited. Check local building codes.

It is important to maintain a winder-tread width along the line of travel that is equal to the tread width in the straight run.

An adaptation of the standard winder layout is illustrated in Figure 17-24. Here, if you extend the lines of the risers, they meet outside the stairs. This provides some tread width at the inside corner. Before starting the construction of this type of stairs, the carpenter should make a full-size or carefully scaled layout (plan view). The best radius for the line of travel can then be determined.

Open Stairs

Main stairs that are open on one or both sides require some type of decorative enclosure and support for a handrail. Typical designs consist of an assembly of parts called a *balustrade*, Figure 17-25. The principal members of a balustrade are newels, balusters, and rails. They are usually factory made and assembled on the job by the carpenter.

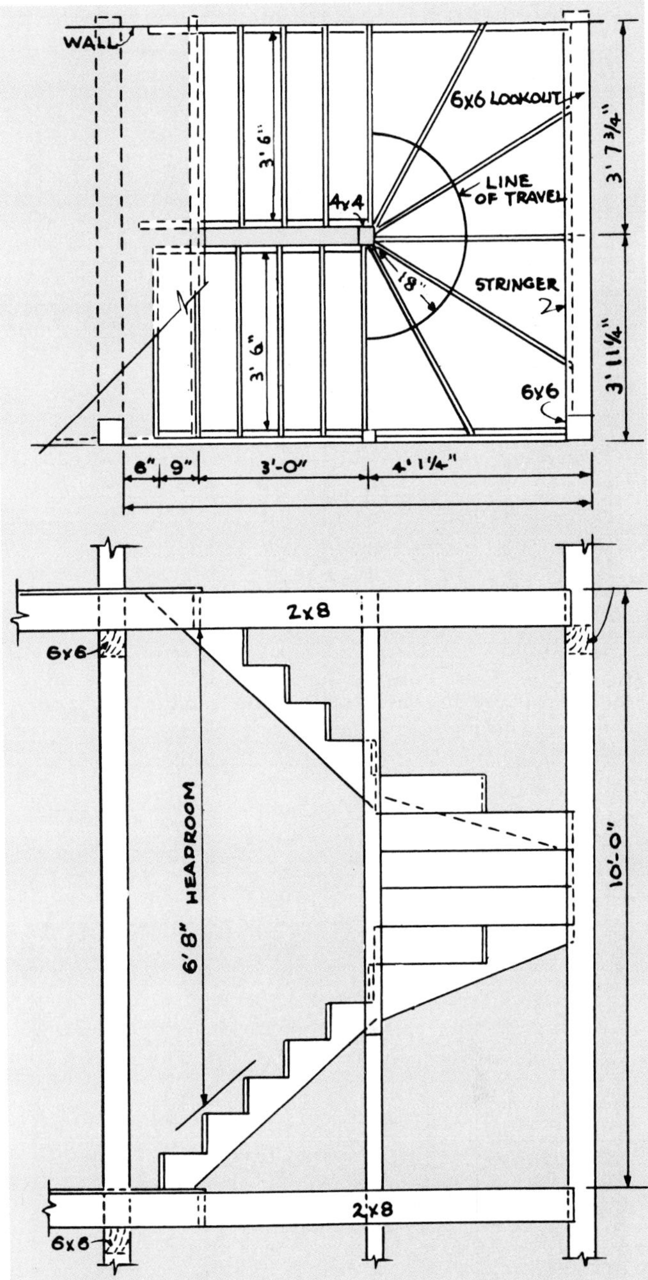

Figure 17-23. Typical drawing of a winder stairs. Tread width on winding section should be the same at line of travel (near middle of stairs) as tread in straight run.

The starting newel must be securely anchored either to the starter step or carried down through the floor and attached to a floor joist. *Balusters* are joined to the stair treads using either a round or square mortise. Two or three may be mounted on each tread.

Using Stock Stair Parts

While many parts of a main staircase could be cut and shaped on the job, the usual practice is to use factory-made parts. These are available in a wide range of stock sizes and

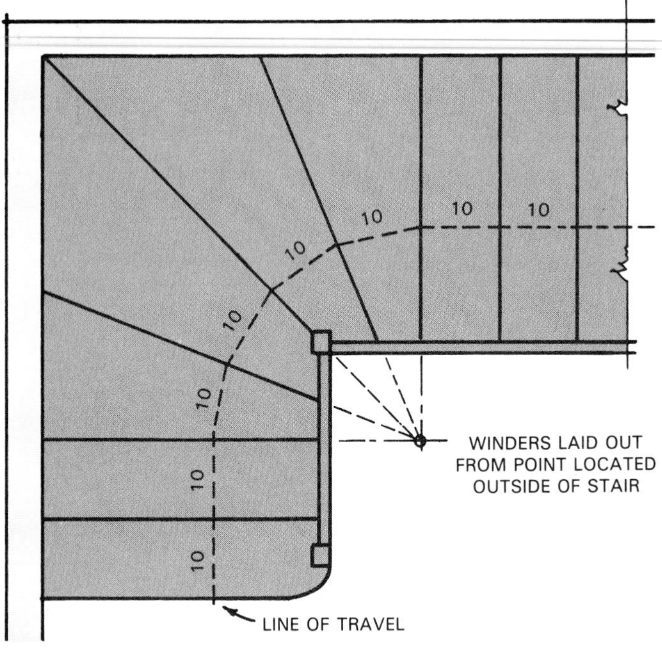

Figure 17-24. Laying out a winder stairs with lines representing the tread nosings converging outside the construction. When winders must be used it is best to place them near the bottom of the run.

can be selected to fill requirements for most standard stair designs. See Figures 17-26 and 17-27. Stair parts are ordered through lumber and millwork dealers and are shipped to the building site in heavy, protective cartons along with directions for fitting and assembly.

A completely prefabricated stairway and a factory assembly are shown in Figure 17-28. Stringers are made in two sections for easier shipping. The system is available in lengths up to 18 steps and widths of 36" and 48".

Figure 17-29 shows some suggested assemblies of balustrades, using stock parts. Hardware, especially designed for stairwork, is illustrated in Figure 17-30.

Spiral Stairways

Metal spiral stairways eliminate framing and save space. Units are available in aluminum or steel in a variety of designs to fit requirements up to 30 steps and heights up to 22'-6". The Uniform Building Code permits use of the spiral stairway for exits in private dwellings or in some other situations when the area served is not more than 400 sq. ft. See Figures 17-31 and 17-32.

Figure 17-25. Parts of an open stair. An assembly including a newel, balusters, and rail is called a balustrade.
A. Half newel. B. Handrail. C. Baluster. D. Newel post. E. Tread. F. Shoerail. G. Riser. H. Stringer. I. Nosing. J. Rosette.

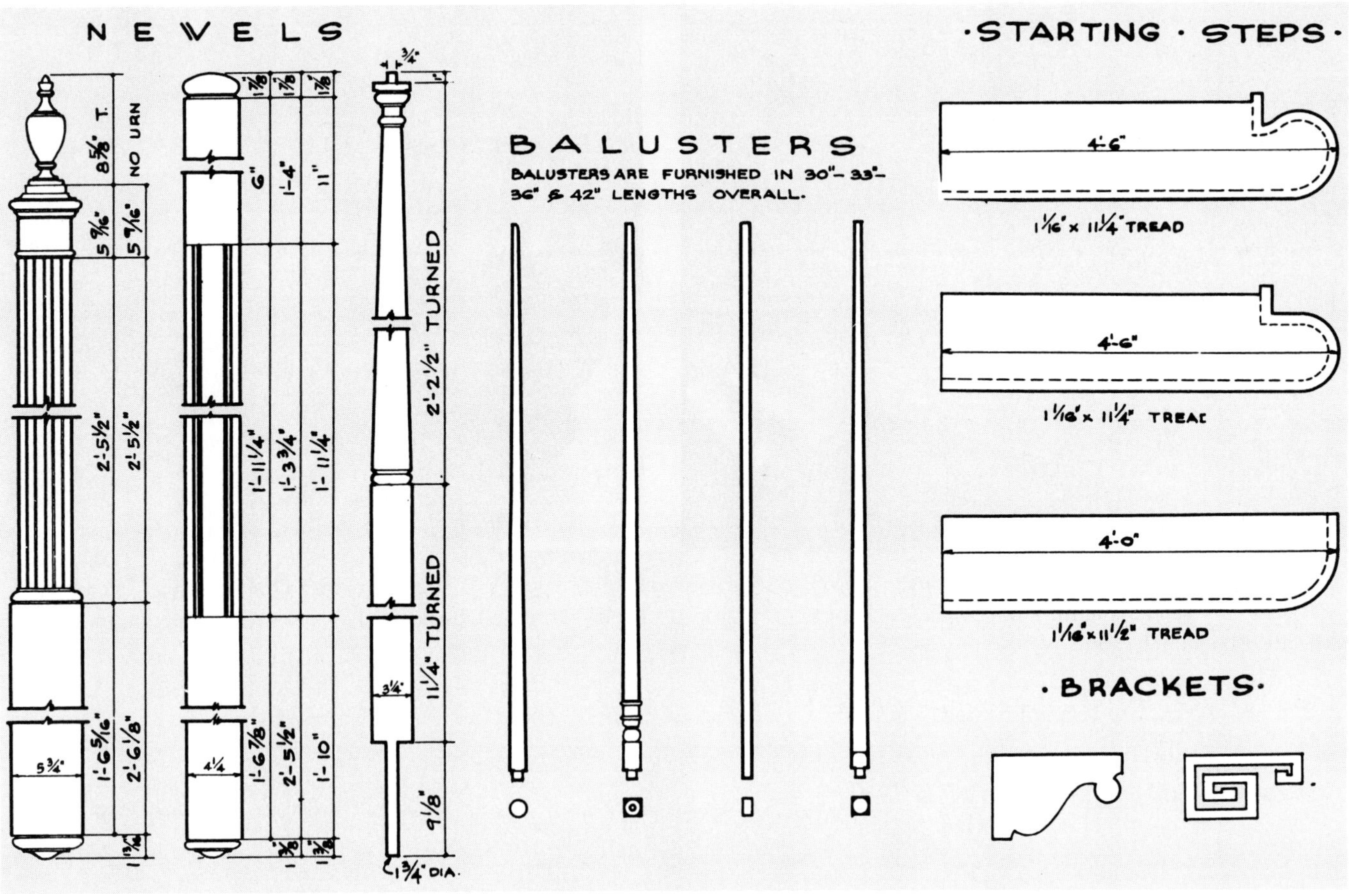

NEWELS

8 5/8" T. NO URN
5 9/16" 5 9/16"
2'-5 1/2" 2'-5 1/2"
5 3/4"
1'-6 5/16" 2'-6 1/8"
7/8" 1 7/8" 1 7/8"
6" 1'-4" 11"
1'-11 1/4" 1'-3 3/4" 1'-11 1/4"
4 1/4
1'-6 7/8" 2'-5 1/2" 1'-10"
3/4"
2'-2 1/2" TURNED
11 1/4" TURNED
3 1/4"
9 1/8"
1 3/4" DIA.

BALUSTERS

BALUSTERS ARE FURNISHED IN 30"- 33"- 36" & 42" LENGTHS OVERALL.

·STARTING · STEPS·

4'-6"
1 1/16" x 11 1/4" TREAD
4'-6"
1 1/16" x 11 1/4" TREAD
4'-0"
1 1/16" x 11 1/2" TREAD

· BRACKETS·

Figure 17-26. Typical stock parts above are commonly available for stair construction.

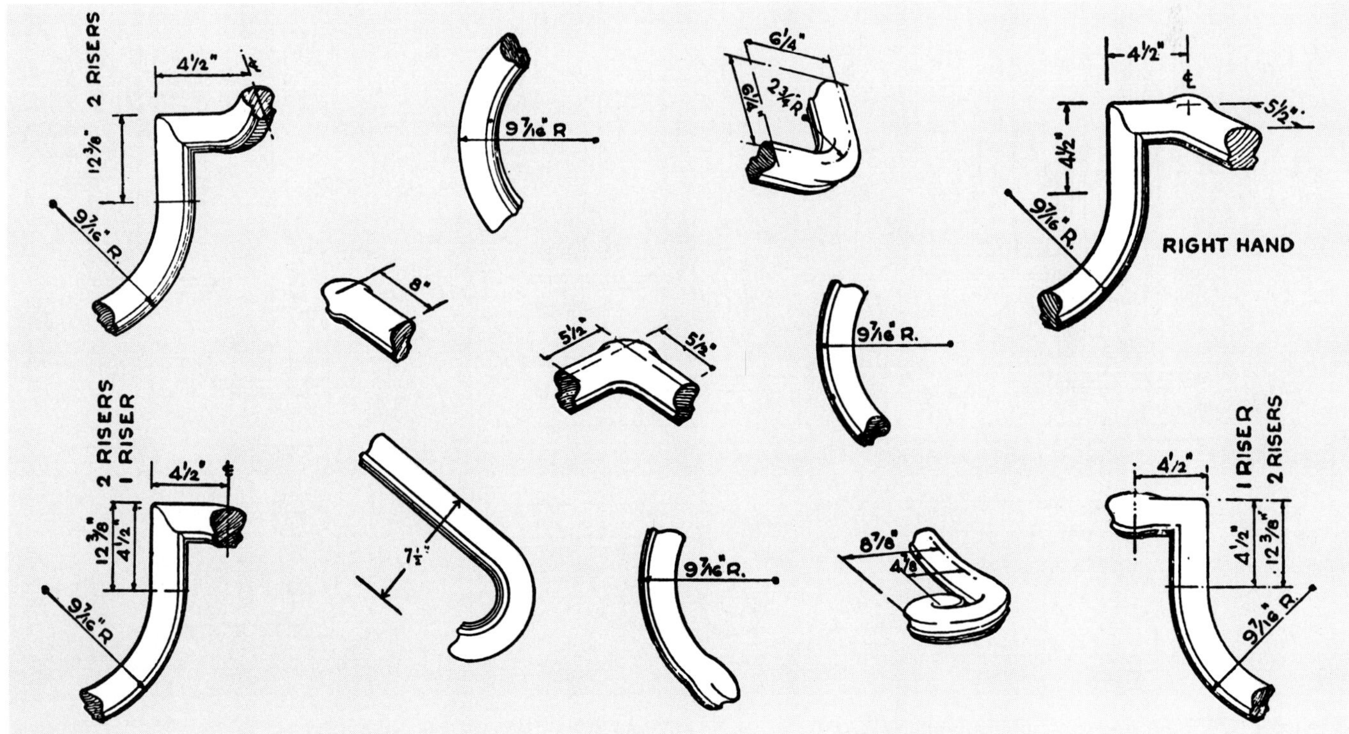

Figure 17-27. Preformed handrails and stock parts for special shapes can be purchased. (C.E. Morgan)

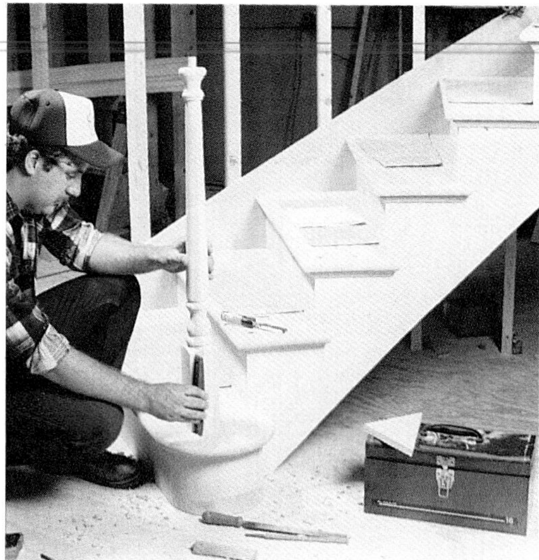

Figure 17-28. Left—Parts for prefabricated stairway system. Sections of stringers lock together with a common tread. Treads and risers fit into dovetails in the stringers and lock together in grooves. Right—Installed mock-up shows assembled section mounted on substringers. (Visador Co.)

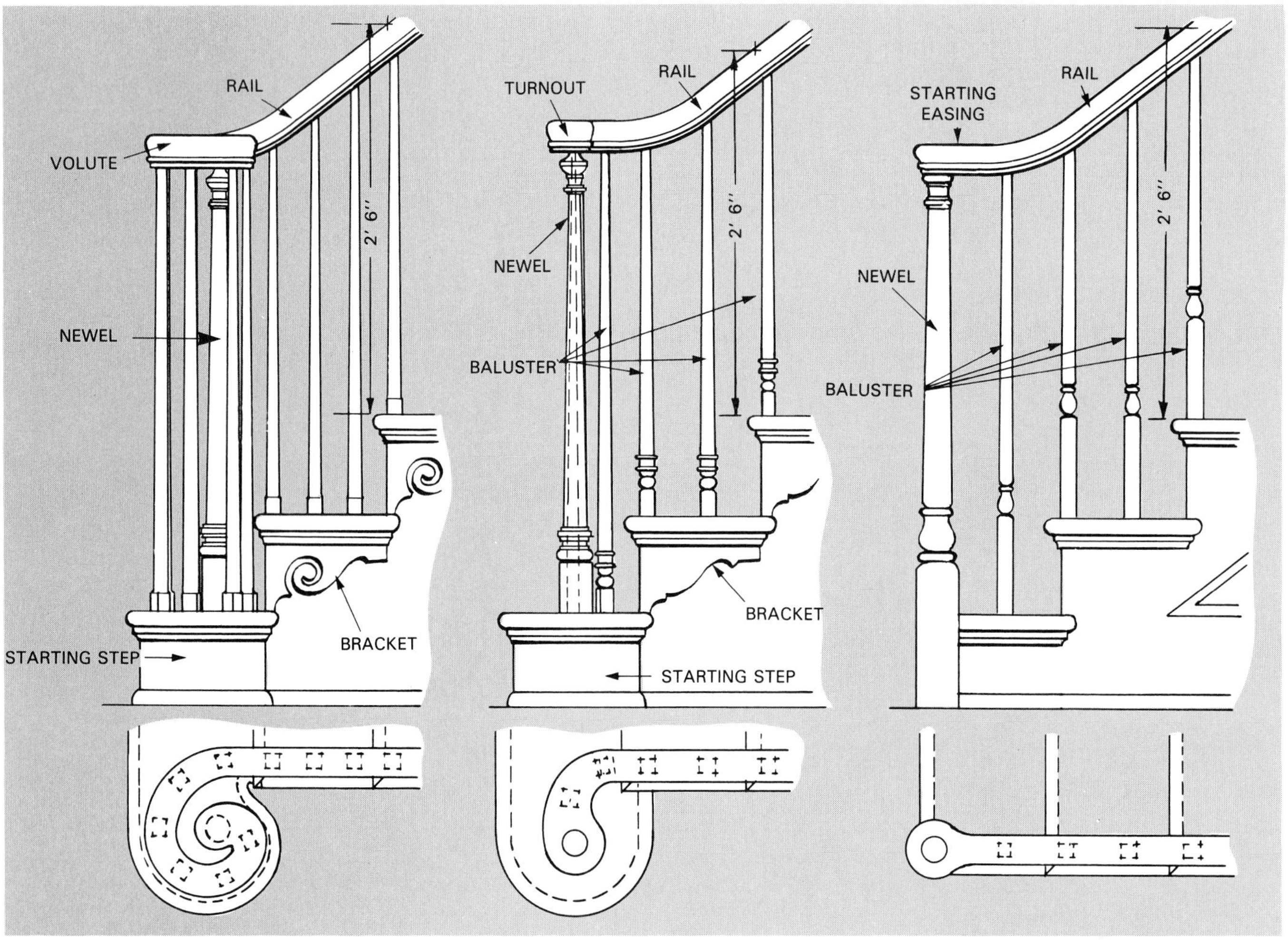

Figure 17-29. Balustrade assemblies produced from stock parts. (Colonial Stair and Woodwork Co.)

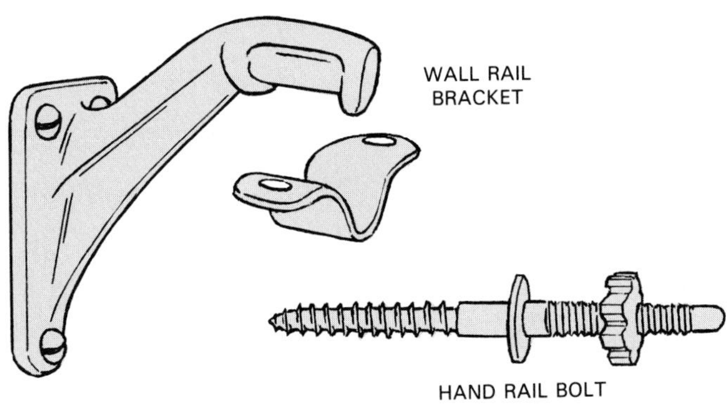

Figure 17-30. Hardware for rails. Rail bolt is concealed in the center of a joint. Nut is engaged and tightened through a hole bored in the underside.

Figure 17-31. Prefabricated metal stairway can be installed quickly in a finished opening. (Columns, Inc.)

Disappearing Stair Units

Where attics are used primarily for storage and where space for a fixed stairway is not available, hinged or disappearing stairs are often used. Such stairways may be purchased ready to install. They operate through an opening in the ceiling and swing up into the attic space when not in use, Figure 17-33. Where such stairs are to be provided, the attic floor should be designed for regular floor loading and the rough opening should be constructed at the time the ceiling is framed.

Important Terms

Baluster
Balustrade
Built-up stringer
Carriage
Cut-out stringer
Flight of stairs
Handrail
Headroom
Housed stringer
Landing
Newel
Nosing
Platform
Rise
Riser
Run
Semihoused stringer
Stairwell
Straight run
Stringer
Total rise
Total run
Tread
Unit rise
Unit run
Wall rail
Winding
Winders

Test Your Knowledge

1. The platform type of stairway includes ____________ where the direction of the stair runs is usually changed.
2. The minimum headroom for a main stairway as specified by FHA is ____________.
3. The riser height of a service stairs should not exceed ____________ inches.
4. One of the rules used to calculate riser-tread relationship states that the sum of two risers and one tread should be ____________.
5. The front edge of the tread that overhangs the riser is called the ____________.
6. A stairs in a split-level home has six risers with a tread width of 11". The total run of the stairs is ____________.
7. Winders are allowed by all building codes. True or false?
8. A semihoused stair stringer is formed by attaching a ____________ stringer to a backing stringer.
9. Wedges used to assemble risers and treads in housed stringers should have a taper of ____________ inches per foot.
10. The three principal members of a balustrade are called newels, rails, and ____________.
11. Study the following instructions for installing a housed stringer stairs. Determine if the instructions are correct for proper assembly and attachment of the stair parts. If not, suggest correct procedure.
    A. Set treads and risers into the stringers. Work from the bottom up using wedges and glue.
    B. Place the assembly into the stair well and spike the stringers to the wall surface and into the wall frame.

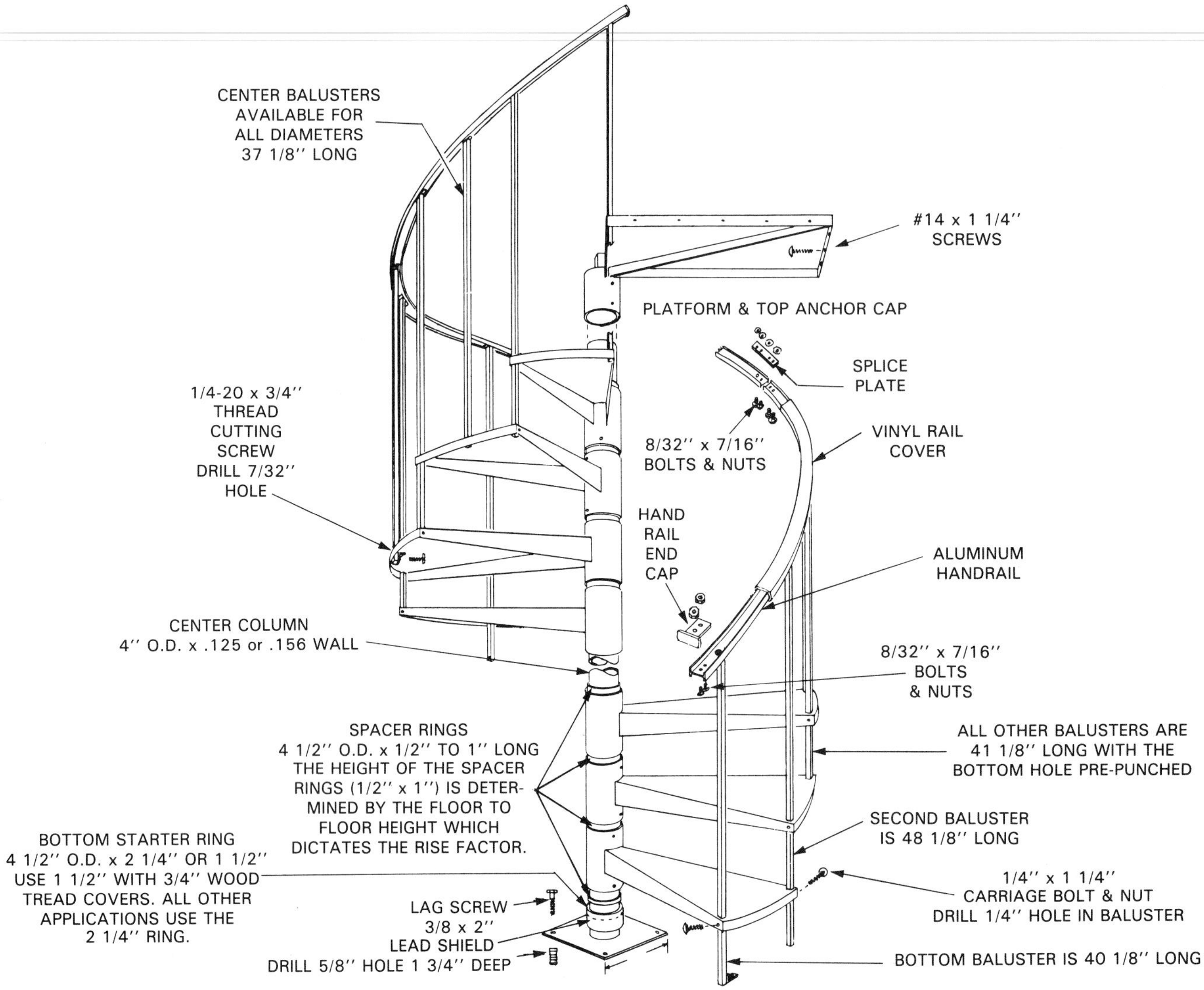

Figure 17-32. Assembly drawing of spiral stairway.

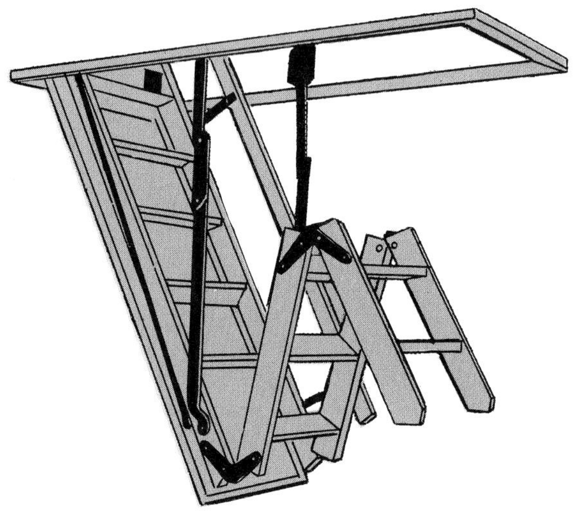

Figure 17-33. Disappearing stair unit is designed to fold into the ceiling. (Rock Island Millwork)

Outside Assignments

1. Obtain a set of architectural plans where the main or service stairway is not drawn in detail. Study the stair requirements carefully and then prepare a detail drawing somewhat like the diagram in Figure 17-5 or the drawing in Figure 17-13. Use a scale of 1/2" equals 1'. Select and calculate the riser-tread ratio carefully and be sure the number and size of risers is correct for the distance between the two levels. Check the headroom requirements against your local building codes and determine the stairwell sizes. Submit the completed drawing and size specifications to your instructor.
2. Study a millwork catalog and become familiar with the stock parts shown for a main stairway. Working from a set of architectural plans or a stair detail that you may have drawn, prepare a list of all the stair parts you would need to construct the stairway.

Include the number of each part needed and also its size, quality, kind of wood, and catalog number. Take your list to a building supply dealer and obtain a cost estimate for the materials. Be prepared to discuss the materials and costs with your instructor and the class.

A curved staircase with a closed string on one side and an open string on the other. Carpenter must frame and cover the curved wall.

Freestanding staircase is factory made entirely of wood. (Arcways, Inc.)

This freestanding circular stair was completely fabricated in a manufacturing plant and then dissassembled and shipped to the building site. (L.J. Smith)

18 Doors and Interior Trim

This unit will deal with the methods and materials of an important part of interior finish. It will include:

- Installing door frames.
- Hanging doors.
- Fitting trim around openings.
- Fitting trim at intersections of walls, floors, and ceilings.

This aspect of carpentry requires great skill and accuracy. Well-fitted trim greatly enhances the appearance and desirability of the home.

Moldings

Moldings are decorative wood or plastic strips. They are designed to provide essential functions as well as provide decoration. For example, window and door casings cover the space between the jamb and the wall covering. They also make the installation more rigid.

A wide range of types, patterns, and sizes of moldings are used in modern homes. See Figure 18-1. Common shapes and where to use them are shown in Figure 18-2 and Figure 18-3. In addition to those shown, a complete list would also include the following: cove molding, brick molding, battens, glass beads, drip caps, picture molding, and screen mold. Information on molding patterns along with a numbering system and grading rules is included in a manual that is available from Western Wood Products Association, Portland, Oregon.

Interior Door Frames

The door frame forms the lining of the door opening. It also covers the rough edges of the partition.

The frame consists of two side jambs and a head jamb, Figure 18-4. Interior frames are simpler than exterior frames. The jambs are not rabbeted and no sill is included. See Unit 12.

Figure 18-1. A few of the many moldings used in a modern dwelling. The various shapes produce shadows and highlights.

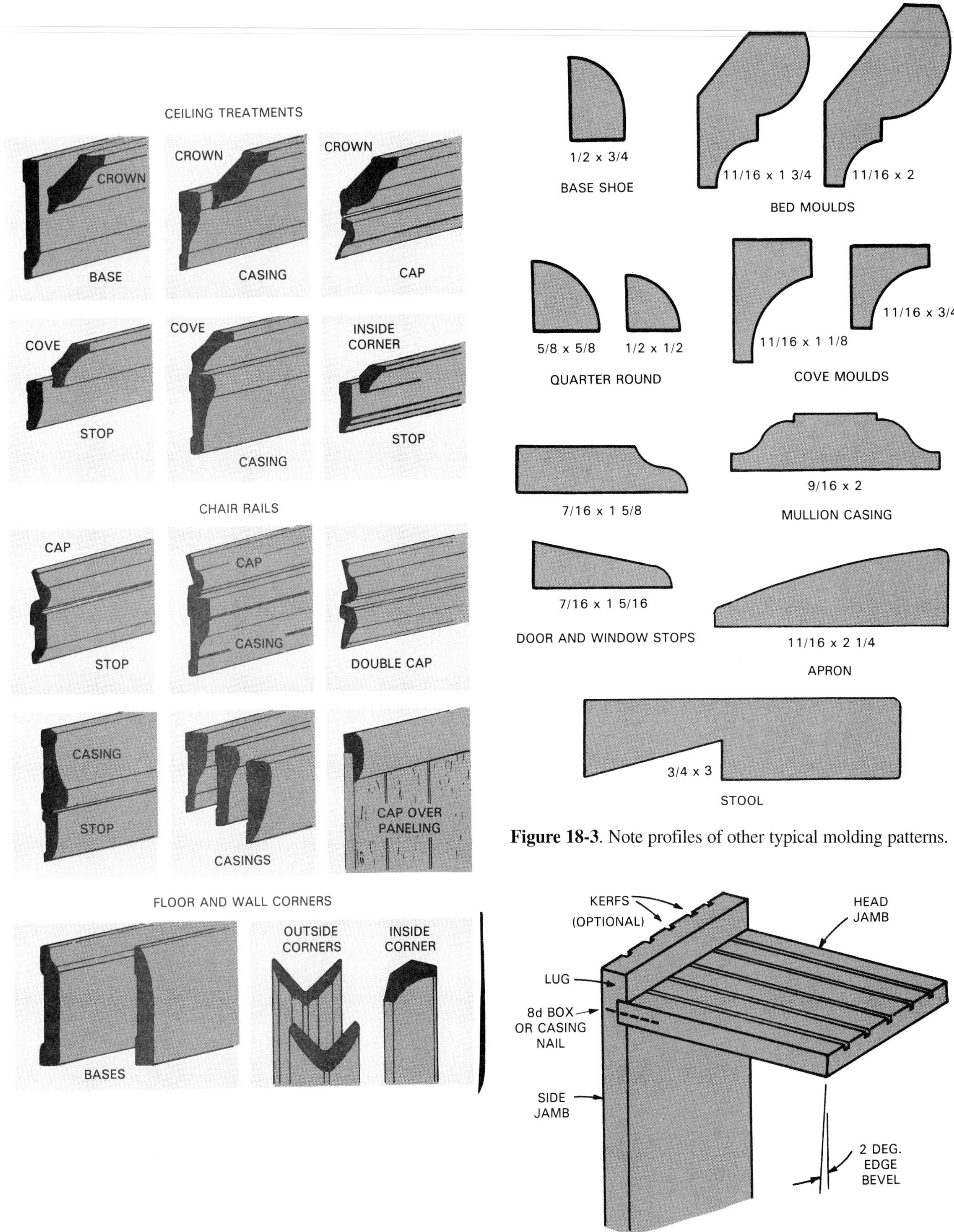

Figure 18-3. Note profiles of other typical molding patterns.

Figure 18-2. Moldings perform various functions depending upon where they are used. (ABTco, Inc.)

Figure 18-4. Section view of interior door frame. Parts are listed. Note that edges of all jambs are beveled slightly so trim will fit snugly to it.

Standard jambs for regular 2 x 4 stud partitions are made from nominal 1" material. For plaster walls, the jambs are 5 1/4" wide. For drywall, the jambs are 4 1/2" wide. The backside is usually kerfed to reduce the tendency toward cupping (warping). The edge of the jamb is beveled slightly so the casing will fit snugly against it with no visible crack.

Side jambs are dadoed to receive the head jamb. The side jambs for residential doorways are made 6'-9" long (measured to the head jamb). This provides clearance at the bottom of the door for flooring materials.

Interior doorjambs are sometimes made to be adjustable. They are designed to fit walls of different thicknesses. One, the three-piece type, depends upon a rabbet joint and a concealing doorstop. A second type is made in two pieces. See Figure 18-5.

Modern door frames are usually cut, sanded, and fitted in millwork plants. This allows quick assembly on the job. Door frames should receive the same care in storage and handling as other finished woodwork.

Before assembling the door frames, check the length of the head and side jamb. Determine if they are correct for the opening.

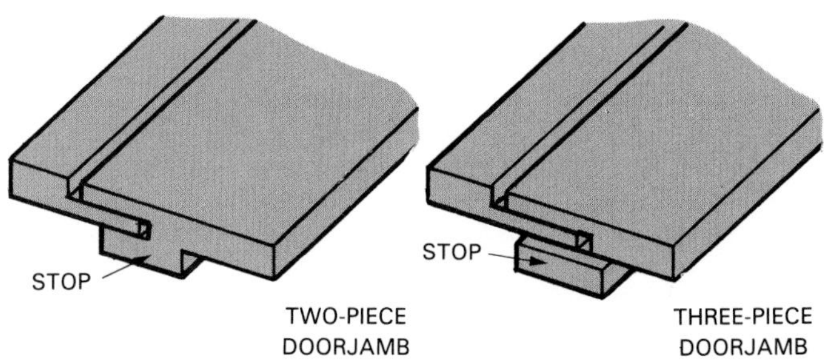

Figure 18-5. Adjustable doorjambs will fit any thickness of wall. Stop will conceal joint.

Installing Door Frames

1. Nail the jambs together using 8d casing or box nails.
2. If there is not enough vertical clearance in the rough opening, trim away part of the lug. (This is the part of the side jamb that extends beyond the head jamb.)
3. Place the frame in the opening. Let the side jambs rest on the finish flooring or on spacer blocks of the right thickness. (This spacer is needed only if the final flooring surface has not been laid.) Level the head jamb. Trim the bottom of the side jamb if it is too high.
4. Place a 1 x 6 spreader between the side jambs at the floor level, Figure 18-6. It should be the same length as the horizontal distance between the side jambs measured at the top.
5. On each side jamb, draw a light pencil line in from the edge of the door side, a distance equal to the door thickness plus 7/8". All nailing of the jambs is done along this line. Later it will be covered by the doorstop, Figure 18-7.
6. Center the frame in the opening. Secure it with double-shingle wedges at the top and bottom on each side. Plumb the jambs with a straightedge and level, or a long carpenter's level. Make adjustments in the double-shingle blocking until each side is correct.
7. Fasten the top and bottom of each side jamb with an 8d casing nail.
8. Complete the blocking by placing more double-shingle wedges in back of each jamb as illustrated in Figure 18-6. On the hinge jamb, locate one block 11" up from the bottom and one 7" down from the top. Set a third block halfway between these two. Continue to check the jamb with a straightedge while adjusting the wedges; then nail through the blocking into the studs. Generally, it is best to use two 8d nails, and stagger them about 1/2" on either side of the nailing line.

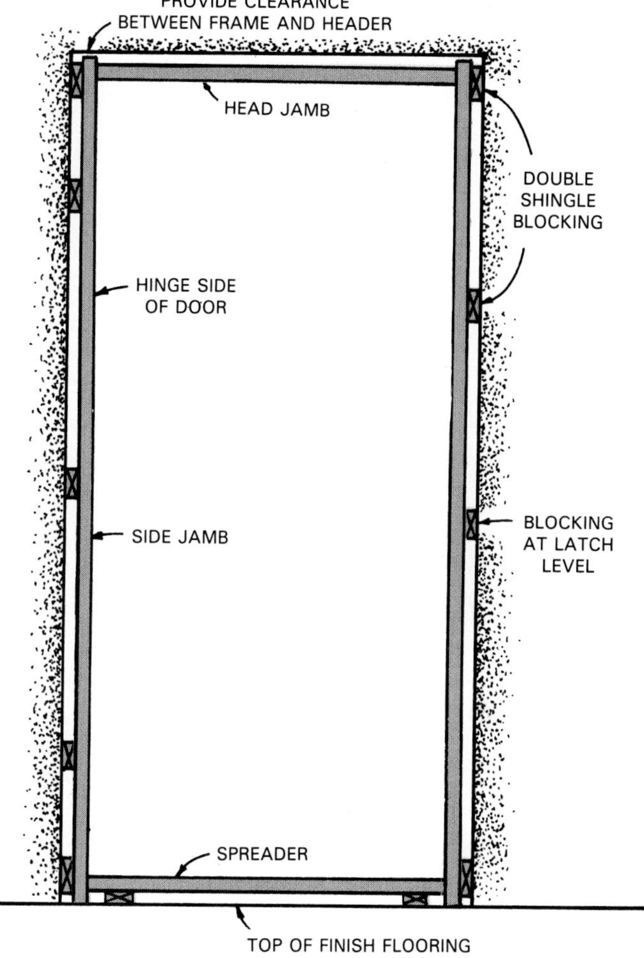

Figure 18-6. Door frames must be set into rough opening using spreader and frequent shingle blocking.

When setting a door frame, do not drive any of the nails "home" until all blocking has been adjusted and the jambs are straight and plumb.

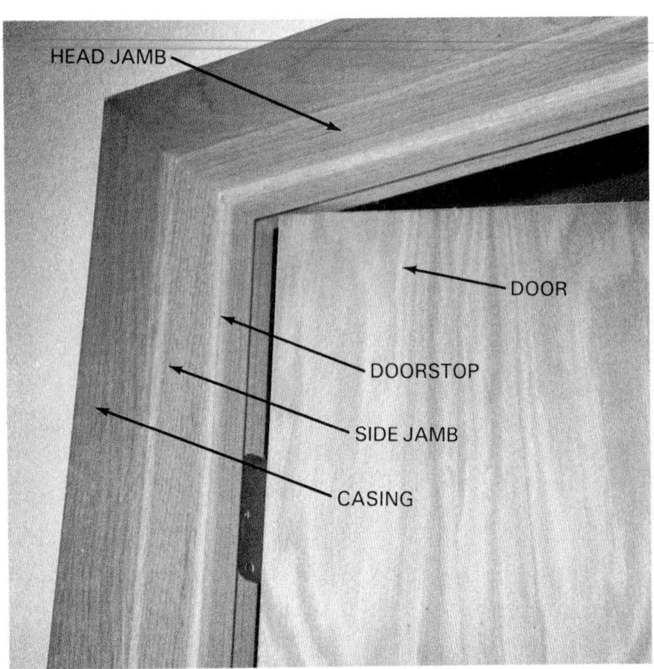

Figure 18-7. Parts of a standard inside passage door frame. The doorstop covers jamb nailing.

Door Casing

Door casing is applied to each side of the door frame to cover the space between the jambs and the wall surface. This secures the frame to the wall structure and stiffens the jambs so they will carry the door. Figure 18-8 shows a section view through a doorjamb. Note that the casing covers the blocking and is attached to the jamb and wall surface.

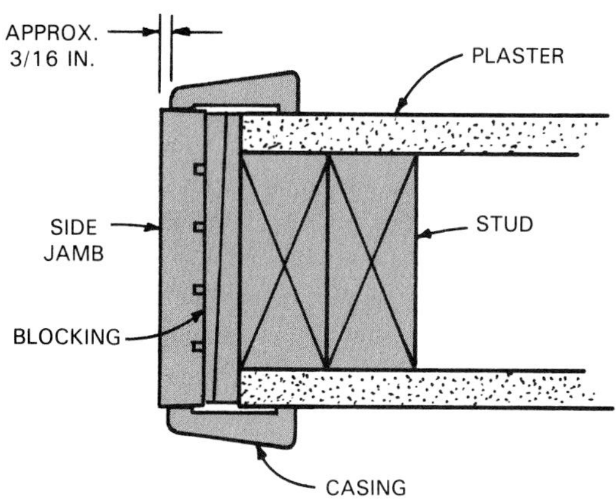

Figure 18-8. Section shows position of casing that is attached to the doorjamb.

Installing Casing

1. Select the casing material and align it around the opening. Some carpenters prefer to draw a light pencil line on the edge of the jamb 1/4" back from the face. Check the bottom end of the side casing to see that it is square and will rest tightly on the finished floor.
2. With the side pieces held in place, mark the position of the miter joint at the top. Use a miter box, power miter saw, or a wood trimmer, Figure 18-9, to make an accurate cut.
3. Nail the side casing temporarily with casing or finish nails. Mark, cut, and fit the head casing. If the miters do not fit properly, trim them with a block plane.
4. Finally, drive nails home and complete the nailing pattern. Use 4d or 6d nails along the jamb edge and drive 8d nails through the outer edge into the studs. Each pair should be spaced about 16" O.C.

Figure 18-10 shows the completed casing at the miter joint. Nails have been set. When using hardwood casing, it is advisable to drill nail holes.

Using Plinth Blocks

A *plinth block* is a decorative corner trim used mainly on door frames and window frames. It allows the finish

Figure 18-9. Several different tools can be used to make miter cuts. Top—Power saw can make compound cuts. Bottom—Wood trimmer makes a smooth, accurate cut for mitered joints. This cut can also be made with the table saw miter or a small miter box.

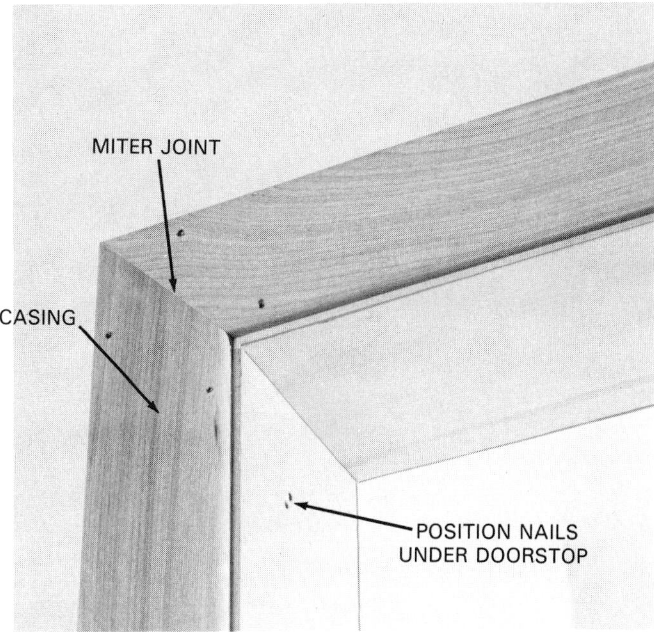

Figure 18-10. Installing casing. Top—Use a nail set to drive casing nails below surface. Holes will be filled before finishing. Bottom—Close-up of casing applied to a door frame. Note how tightly miter joint fits.

carpenter to make butt joints instead of the more difficult mitered corners. Installation of the casing is slightly different from that used with mitered corners. Usually, the head casing and plinths are installed first; then the side casings are fitted, marked and cut. See Figure 18-11.

Panel Doors

There are two general types of doors: panel and flush. The panel door is also referred to as a stile and rail door. This type of construction is used in sash, louver, storm, screen, and combination doors. Sash doors are similar to panel doors in appearance and construction but have one or more glass lights in place of the wood panels.

Figure 18-11. Upscale homes often have plinth blocks installed at corners of casings. Top—Here, head casing and plinth are installed first. Middle—Attaching side casing with a pneumatic nailer. Bottom—Close-up of an installed plinth.

A *panel door*, Figure 18-12, consists of *stiles* and *rails* with panels of plywood, hardboard, or solid stock. The rails and stiles are usually made of solid material; however, some are veneer applied over a lumber core. In some cases, the doors are molded from a wood fiber, Figure 18-13. A variety of designs are formed by changing the number, size, and shape of the panels.

TOP RAIL
MULLION
PANEL
LOCK OR INTER-MEDIATE RAIL
STILE
BOTTOM RAIL

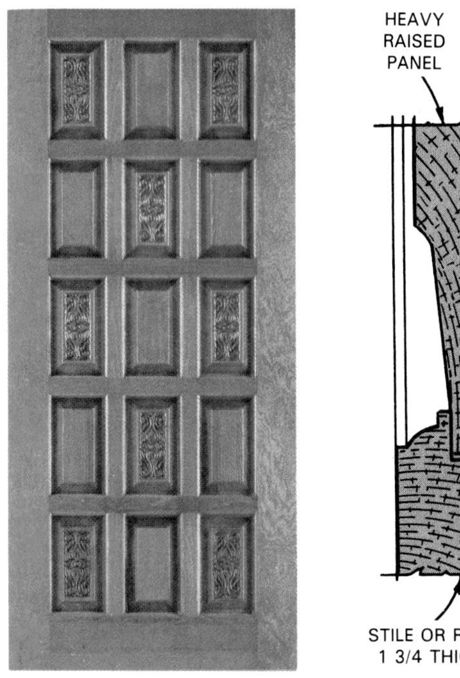

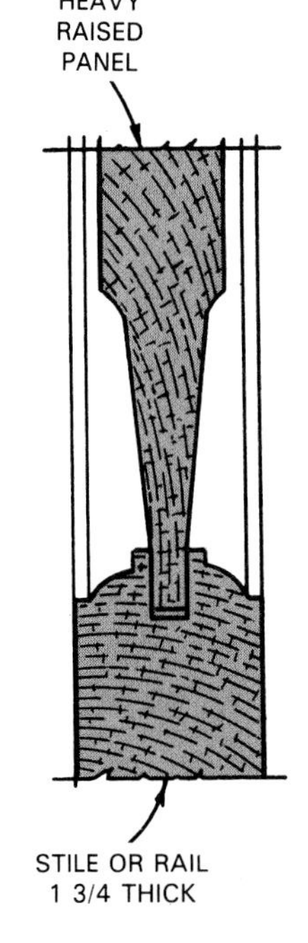

Figure 18-12. Top—Panel doors are constructed of several intricate parts which are carefully fitted and glued. Bottom—Section through stile and panel shows joint detail. (C.E. Morgan)

Special effects are secured by installing raised panels, which add line and texture. This panel is formed of thick material that is reduced around the edges where it fits into the grooves in the stiles and rails.

Flush Doors

A *flush door* consists of a wood frame with thin, flat sheets of material applied to both faces. Improvements in the manufacture of plywood and adhesives have made it

Figure 18-13. Molded doors are made to simulate panel doors. Top—Doors come primed. (A section has been left unprimed to show "hardboard" base.) Bottom—Finished doors cannot be distinguished from wood panel doors.
(Masonite Div., Building Products Group)

possible to produce a flush door that is strong and durable. Today, flush doors account for a high percentage of wood doors.

Face panels, also called skins, are commonly made of 1/8" plywood. However, hardboard, plastic laminates, fiberglass, and metal are also used. Figure 18-14 shows a flush door constructed with metal face panels.

Figure 18-14. Flush type entrance door with wood rails and stiles and a skin of 24 gage steel. Core is foamed-in-place polyurethane. (Stanley Door Systems)

Flush doors are made with solid or hollow *cores.* Cores of wood or various composition materials are used in solid (or slab) construction. See Figure 18-15. The frame is usually made of softwood that matches the color of the face veneers. The most common type of solid core construction uses wood blocks bonded together with the end joints staggered. See Figure 18-16.

Flush doors with hollow cores are widely used for interior doors, and may also be used for exterior doors if the construction is bonded with waterproof adhesives. Hollow flush doors do not ordinarily provide as much thermal (heat) and sound insulation as solid core doors. Usually their fire resistance rating is lower. Some modern exterior doors (flush type) have compression-molded fiberglass face panels. They are attached to a wooden

Figure 18-16. Forming solid core for flush door with wood blocks bonded together in an electronic clamping and curing machine. (Andersen Corp.)

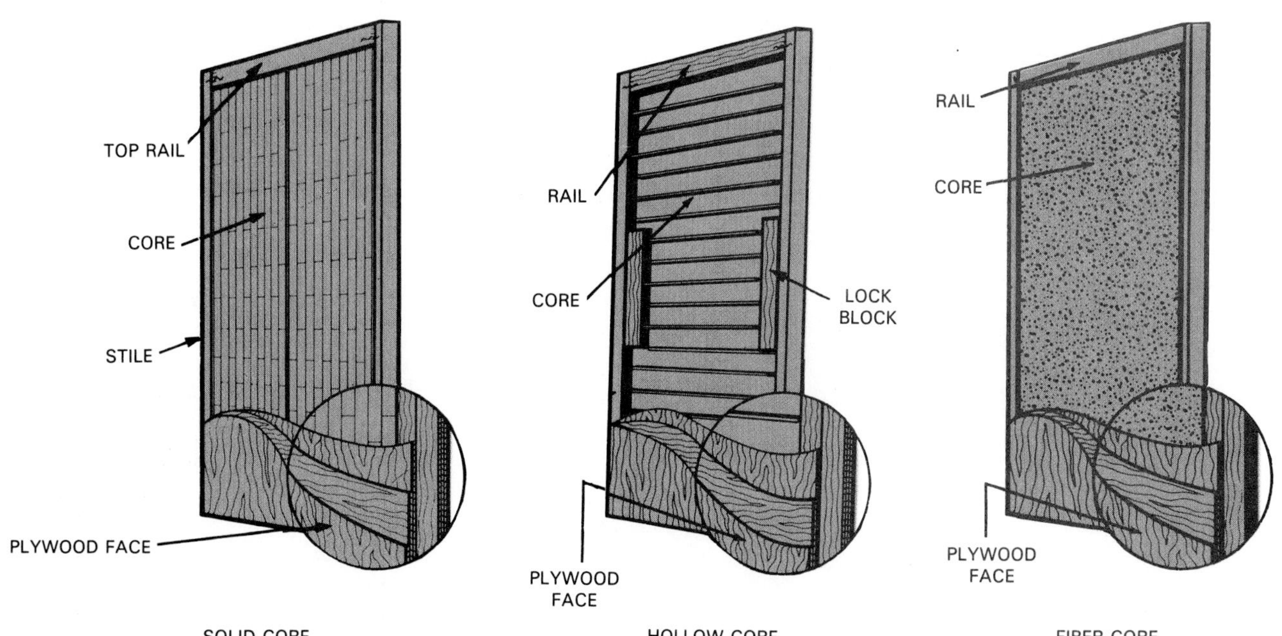

Figure 18-15. Basic types of cores used in modern wood flush door construction. The basic hollow core types include lattice (also called mesh or grid), ladder, and implanted blanks. Frames are usually made of softwood.

frame and can be formed to reproduce various traditional and contemporary designs. The core is a high-density polyurethane foam having a high R-value. Unlike steel-faced doors, the fiberglass door can be trimmed for a precision fit. The surface, textured like wood, can be stained or painted, Figure 18-17.

Figure 18-17. Top—Plane is being used to trim edge of a fiberglass door. Bottom—Oil stain is used to highlight simulated wood grain pattern and texture. (Therma-Tru)

Sizes and Grades

Standard thickness for exterior doors is 1 3/4"; interior passage doors are 1 3/8". Widths of 2'-8" and 3'-0" are most commonly used for exterior doors, Figure 18-18.

Residential doors, both interior and exterior, have a standard height of 6'-8" but 7'-0" doors are sometimes used for entrances or special interior installations. A 7'-0" height is usually considered standard for commercial buildings.

Interior door widths vary with the installation. FHA specifies a minimum size of 2'-6" for bedrooms and a 2' width for bathrooms. Closet doors may also be 2' wide. Interior door sizes and patterns are illustrated in Figure 18-19.

Grades and manufacturing requirements for doors are listed in Industry Standards developed by the National Woodwork Manufacturing Association (NWMA) and the Fir and Hemlock Door Association (FHDA). The purpose of these standards is to establish nationally recognized dimensions, designs, and quality specifications for materials and work.

Door Installation

Check the architectural drawings and door schedule to determine:

- The correct type of door for the opening.
- The direction it will swing.

Figure 18-18. Exterior door designs made with face panels of 24 gage steel. Decorative moldings are plastic bonded to the steel surface. Standard widths include: 2'-6", 2'-8", 3'-0", and 3'-6". Standard heights are 6'-8" and 7'-0". (Stanley Door Systems)

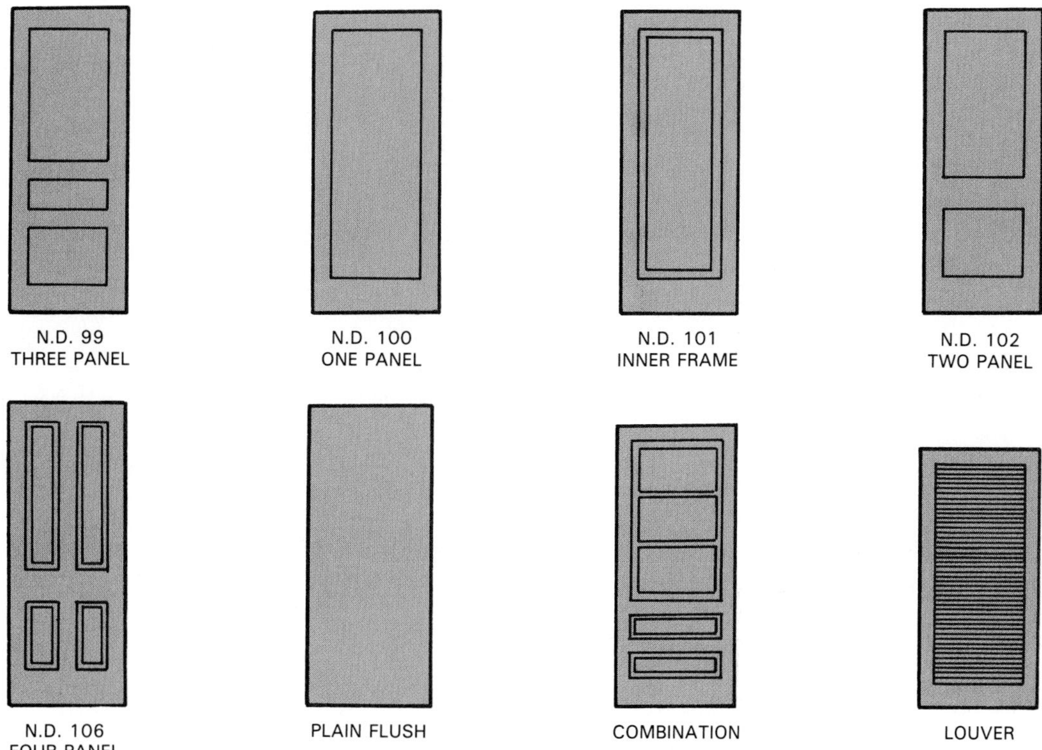

CONSTRUCTION DETAILS

Design No.	Stiles	Top Rail	Cross Rail	Lock Rail	Intermediate Rails	Mullions or Muntins	Bottom Rail	Panels
N.D. 99	4 3/4"	4 3/4"	4 5/8"	4 5/8"	......	......	9 5/8"	Flat
N.D. 100	4 3/4"	4 3/4"	......	......	......	......	9 5/8"	Flat
N.D. 101	4 1/4" Face	4 1/4" Face	......	......	......	......	9 1/4" or 9 1/2" Face	Flat
N.D. 102	4 3/4"	4 3/4"	......	8"	......	......	9 5/8"	Flat
N.D. 106	4 3/4"	4 3/4"	......	8"	......	4 5/8"	9 5/8"	Raised
N.D. 107	4 3/4"	4 3/4"	......	......	4 5/8"	......	9 5/8"	Raised
N.D. 108 (1)(2)(3)	4 3/4"	4 3/4"	......	8"	3 7/8" or 4 5/8"	3 7/8" or 4 5/8"	9 5/8"	Raised
N.D. 111 (1)(2)(3)	4 3/4"	4 3/4"	......	8"	3 7/8" or 4 5/8"	3 7/8" or 4 5/8"	9 5/8"	Raised

Figure 18-19. Sizes and patterns commonly used for interior doors. Table lists construction details of various parts. (National Woodwork Manufacturers Assoc.)

Mark the doorjamb that will receive the hinges. Also mark the edge on which they will be mounted.

Doors may be trimmed to fit but should not be cut to fit smaller openings. If a large amount of the perimeter is removed, the structural balance of the door may be disturbed. Warping may result. Cutouts for glass inserts in flush doors should never be more than 40% of the face area and the opening should not be within 5" of the edge.

Handle doors carefully. Do not soil unfinished doors. If doors must be stored for more than a few days, stack them horizontally on a clean flat surface and keep them covered. Doors should be conditioned for several days so they will reach the average prevailing moisture content before they are hung or finished.

First, trim the door to fit the opening. Most doors are carefully sized at the millwork plant, leaving only a slight amount of on-the-job fitting and adjustment. The amount of planing necessary can be laid out with a rule.

Clearances should be 3/32" on the lock side and 1/16" on the hinge side, Figure 18-20. A clearance of 1/16" at the top and 5/8" at the bottom is generally satisfactory. If the door is to swing across heavy carpeting, increase the bottom clearance. Thresholds are used under exterior doors. Where weather-stripping is used around exterior door openings to reduce infiltration, additional clearance is needed. About 1/8" on each side and on the top edge is enough. The threshold for exterior doors may be installed before or after the door is hung.

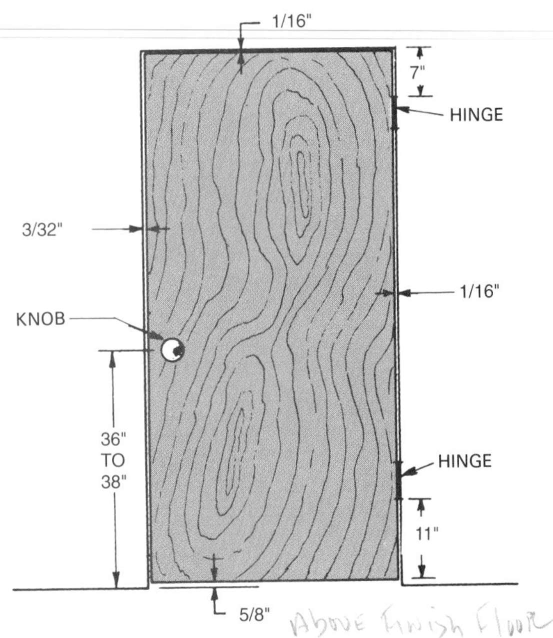

Figure 18-20. Recommended clearances around interior doors. Some carpenters use a quarter to check clearance at top and lock side. For higher quality construction, a third hinge is located midway between top and bottom hinges.

Door trimming can be done with a hand plane, but the modern carpenter generally uses a power plane. While being planed, the door may be held securely on edge either by clamping to sawhorses or by using a special door holder.

After the door is brought to the correct size, plane a bevel on the lock side to provide clearance for the edge when it swings open. This bevel should be about 1/8" in 2" (approximately 3 1/2°). Narrow doors require a greater bevel than wide doors since the arc of swing is smaller. The type of hinge and the position of the pins should be considered in determining the exact bevel required.

After the bevel is cut and the fit of the door is checked, use a block plane to soften (round) corners on all edges of the door. Smooth with sandpaper.

Millwork plants can furnish prefitted doors that are machined to the size specified with the lock edge beveled and corners slightly rounded. Doors can also be furnished with gains cut for hinges, and/or holes bored for lock installation.

Installing Hinges

Gains are the recesses cut into the edge of a door to receive the hinges. Gains are best cut with an electric router. Used with a door-and-jamb template, the router saves time and ensures accuracy.

The template is positioned on the door as shown in Figure 18-21. After the gains are routed, the template is attached to the doorjamb where matching gains are made.

Adjustments can be made for various door thicknesses and heights, as well as for different sizes of butt hinges. See Figure 18-22. The design of most templates makes it nearly impossible to mount them on the wrong side of the door or jamb.

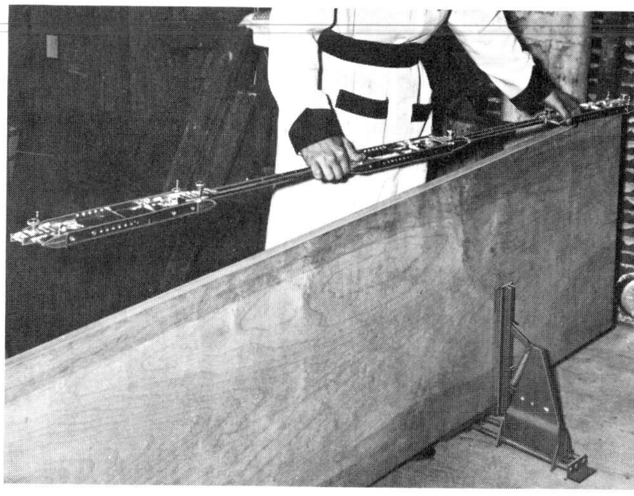

Figure 18-21. Setting routing template over hinge edge of door in preparation for cutting hinge gain.

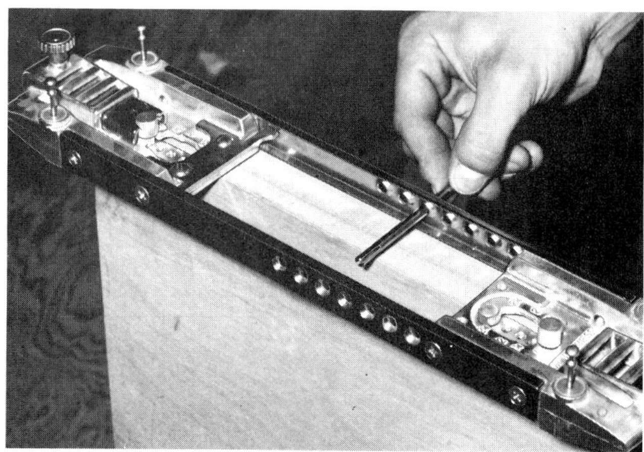

Figure 18-22. Hinge templates can be adjusted for different door sizes as well as for different hinge sizes.

Figure 18-23 shows the cut being made with template and router. For this type of equipment, hinges with rounded corners may be used. It will save the time required to square the corner with a wood chisel.

Once the gains have been cut, the door can be hung.

Hanging a Door

1. Place the hinge in the gain so the head of the removable pin will be up when the door is hung. Drive the first screw in slightly toward the back edge to draw the leaf of the hinge tightly into the gain. To speed up the work, use an electric drill with a special screwdriving attachment, Figure 18-24.
2. Follow the same procedure to attach the free leaf of the hinge to the jamb. Set only one or two screws in each hinge leaf. Check the fit before installing the remaining screws.

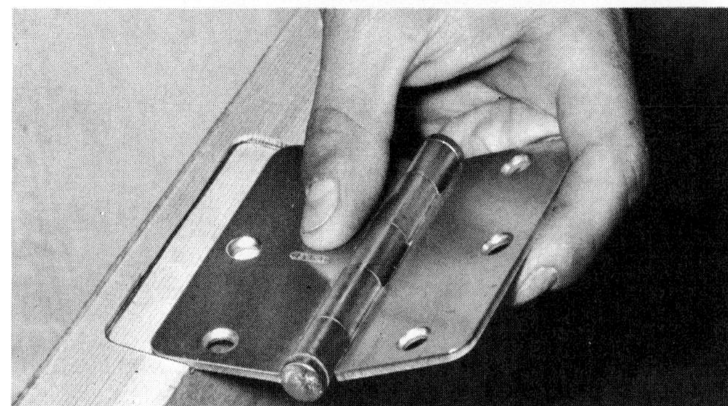

Figure 18-23. Cutting gains. Top—Router and template are being used to cut a gain for a round cornered hinge. Bottom—Checking the completed gain. Depth should equal thickness of hinge leaf.

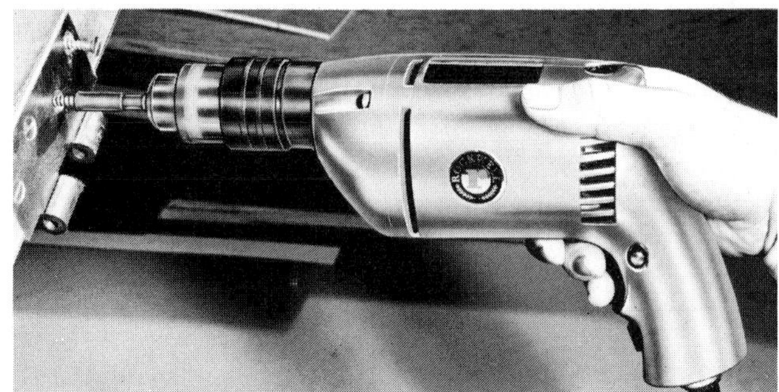

Figure 18-24. Electric drill with special chuck for driving screws. Screws should be started slightly toward back of the hole to draw the leaf tightly into the gain.

3. After all hinges are installed, hang the door and check clearance on all edges. Required corrections should be made by planing the door edges or adjusting the depth of the hinge gains.

Minor adjustments can be made by applying cardboard or metal shims behind the hinge leaf, Figure 18-25. As shown, the center of the hinge can be shifted so more clearance is provided along the lock jamb (View A) or the space can be closed with the shim applied on the outside edge, (View B). When adjusting a door that has been in service, it may be necessary to use longer screws.

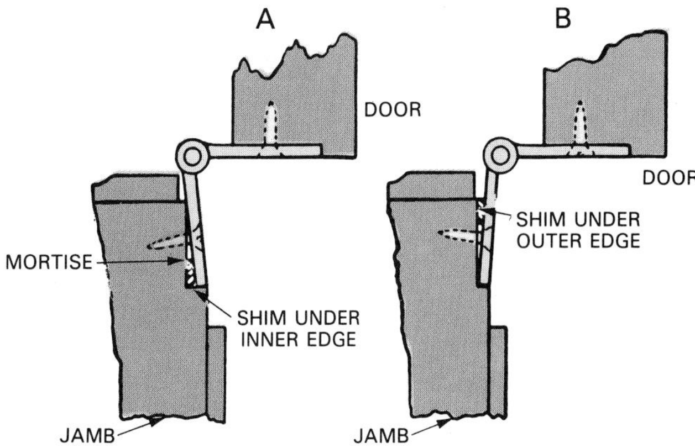

Figure 18-25. Cardboard or metal shims can sometimes be used to make minor adjustments in door clearance. Shims should be placed between leaf and jamb. Metal only is used on light commercial construction.

Doorstops

A *doorstop* is a narrow strip of wood attached to the side and top jambs to stop the door when it closes. The doorstops are usually the last trim members to be installed. Many carpenters cut and tack the stops in place before installing the lock. Permanent nailing comes after the lock installation has been completed.

With the door closed, set the stop on the hinge jamb with a clearance of 1/16". The stop on the lock side is set against the door except in the area around the lock. Here allow a slight clearance for humidity changes and decorating. Set the stop on the head jamb so it aligns with the stops on the side jambs. Use miter joints and attach with 4d nails spaced 16" O.C. Using power operated nailers, Figure 18-26, will save time.

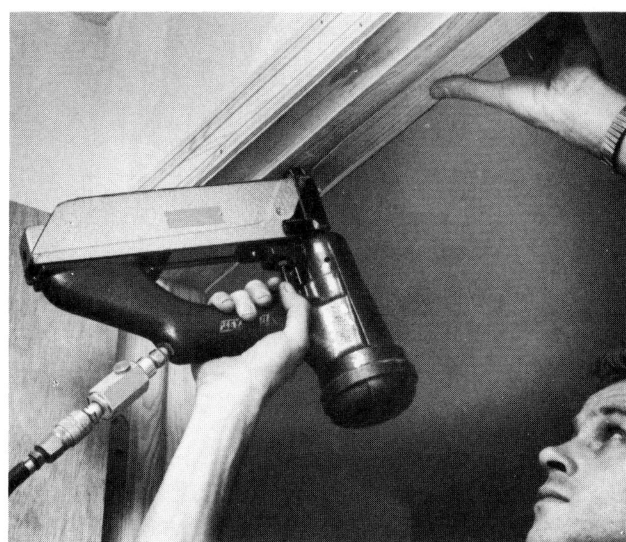

Figure 18-26. Installing doorstop. Side stops are attached first, then the head stop as shown. Use butt or miter joints.

Door Locks

Four types of passage door locks are illustrated in Figure 18-27. Cylindrical and tubular locks are used most often in modern residential work because they can be installed easily and quickly. Unit locks are installed in an open cutout in the edge of the door and need not be disassembled when installed. Such locks are commonly used on entrance doors for apartments and some commercial buildings where locks must be changed from time to time.

Cylindrical locks have a sturdy, heavy-duty mechanism that provides security for exterior doors, Figure 18-28. They require boring a large hole in the door face, a smaller hole in the edge, and a shallow *mortise* for the front plate. Often, the doors come with these holes already bored at the factory. The tubular lock is similar but requires a smaller hole in the door face.

When ordering door locks, it is often necessary to describe the way in which the door swings. This is referred to as the "hand of the door." Hand is determined by facing the outside of the door. The outside is the street side of an entrance door and the corridor side of an interior door.

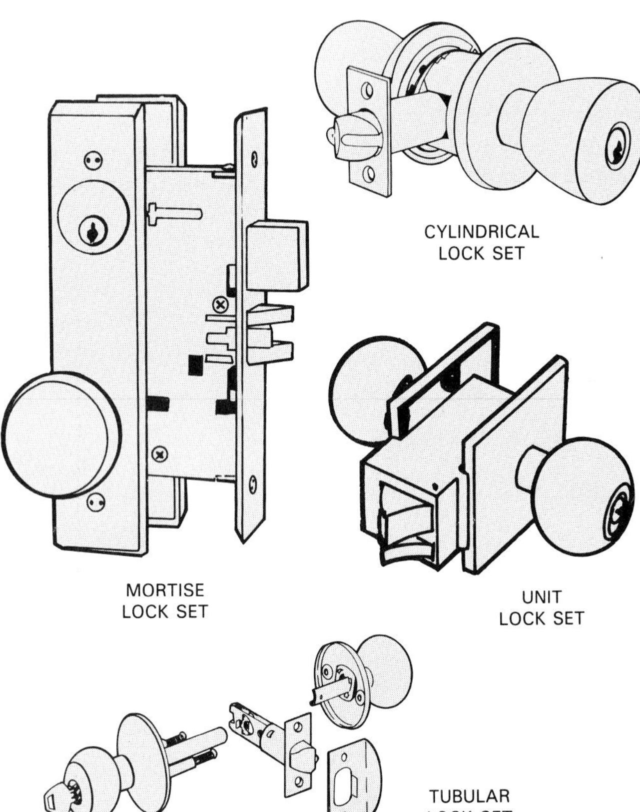

Figure 18-27. Basic types of door lock sets. Mortise lock sets provide high security and are often found in apartment buildings. Cylindrical and tubular locks are most often chosen for residential with the cylindrical being more secure for entrance doors. Unit lock sets are best for apartments where locks are frequently changed.

Figure 18-29 illustrates standardized procedure in determining this specification.

Dead bolts or deadlocks are extra locks which provide additional security against unauthorized entry. Units are made with single cylinder and double cylinder action. (Double cylinder dead bolts require key use on both sides of the door. They offer more security than single cylinders since thieves cannot open the door — even from the inside.) Figure 18-30 shows two types of dead bolts.

Installing a Lock

1. Open the package that contains the lock set and check the contents. Instructions furnished by the manufacturer should be carefully followed.
2. Open the door to a convenient position and block it with wedges placed underneath.
3. Measure up from the floor a distance of 38" (36" is sometimes used) and mark a light horizontal line. This will be the center of the lock.
4. Position the template furnished with the lock set on the face and edge of the door. Lay out the centers of the holes, Figure 18-31. Continue to follow the instructions included with the lock set.
5. Use of boring jigs, Figure 18-32, assures accurate work. A template layout is not required since the jig is designed to make holes in the correct locations. Either hand-operated or power-driven bits can be used to bore the holes.
6. The shallow mortise on the edge of the door can be laid out and cut with standard wood chisels. A faceplate mortise marker, also called a marking chisel, Figure 18-33, is faster and more accurate. After the perimeter is cut with this device, the wood inside can be quickly removed with a standard wood chisel of appropriate width.

Thresholds and Door Bottoms

Exterior doors require a trim unit called a *threshold* to seal the space between the bottom of the door and the doorsill. For many years oak or other hardwoods have been machined to a special shape for this purpose. Modern wooden thresholds are usually equipped with rubber or vinyl sealing strips.

A wide range of threshold designs is made from aluminum extrusions. These are available in a clear anodized finish or gold color. Special vinyl strips are inserted into the threshold, Figure 18-34, or to matching units attached to the door. Thresholds of this type are effective in providing an under-the-door seal. Manufacturers furnish detailed instructions for making the installation.

Interior doors do not require a threshold but may be equipped with various sealing strips such as the automatic door bottom shown in Figure 18-35. This prevents air

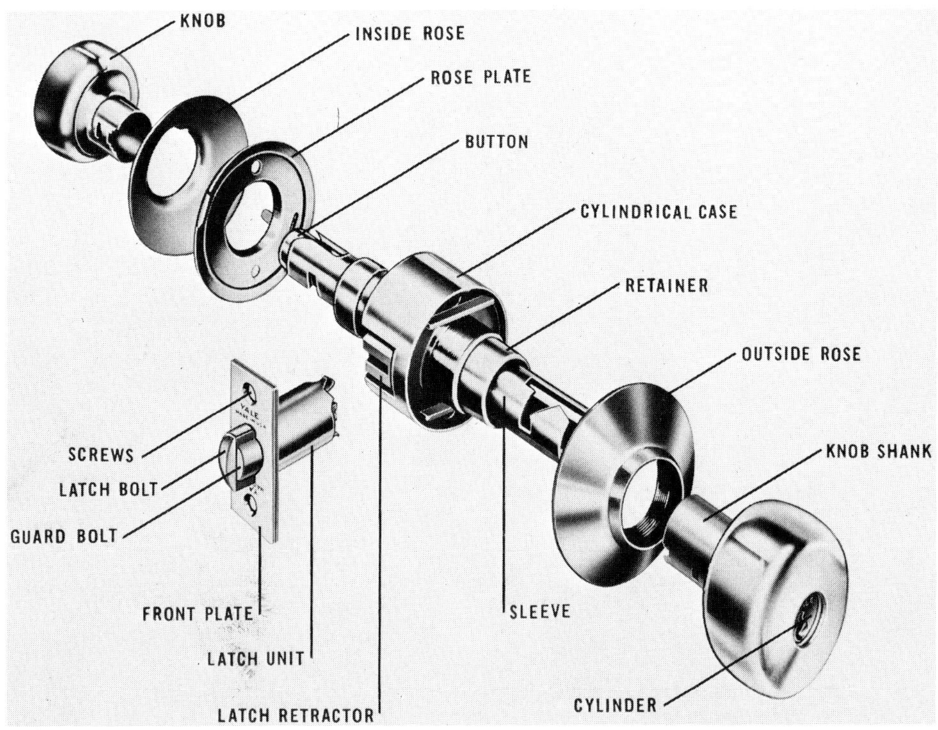

Figure 18-28. Parts of a cylinder lock set. (Eaton Yale & Towne Inc.)

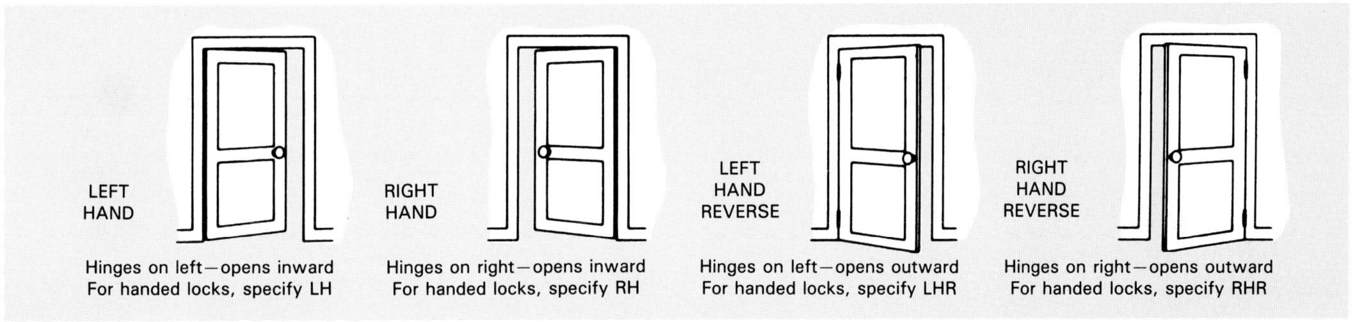

Figure 18-29. Determining the "hand of a door." Use abbreviations suggested.

Figure 18-30. Two styles of dead bolts. Both have bolts with full 1" throw. Left—Rim cylinder safety lock attaches to the inside of door. Right—Tubular dead bolt. (National Lock Hardware and Kwikset)

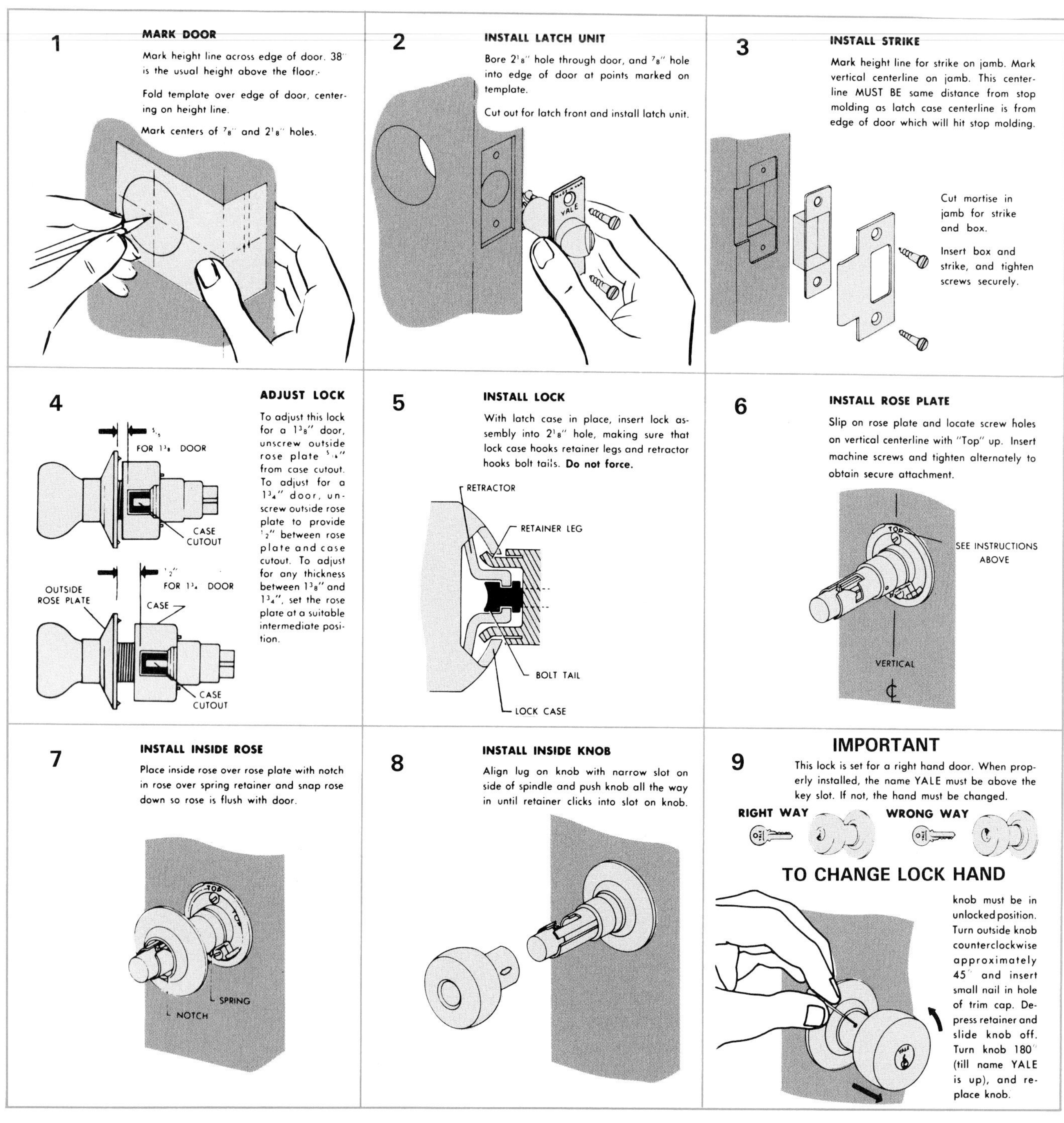

Figure 18-31. Manufacturers furnish detailed instructions for installation of their lock sets.

movement and reduces sound transmission. The unit has a movable strip that drops to the floor when the door is closed and lifts when the door is opened.

Prehung Door Unit

A *prehung door unit* consists of a door frame with a door already installed, Figure 18-36. The frame includes both sides of casing. Lock hardware may or may not be installed although machining for its installation has usually been completed. Quite often the door is prefinished. The unit is carefully packaged and shipped to the job.

Several frame designs are available. Figure 18-37 shows the door side of a standard split jamb unit being set in place. The split jamb permits some variance in wall thickness.

Inside doors normally swing into the room that they serve. Therefore, the prehung door package should be

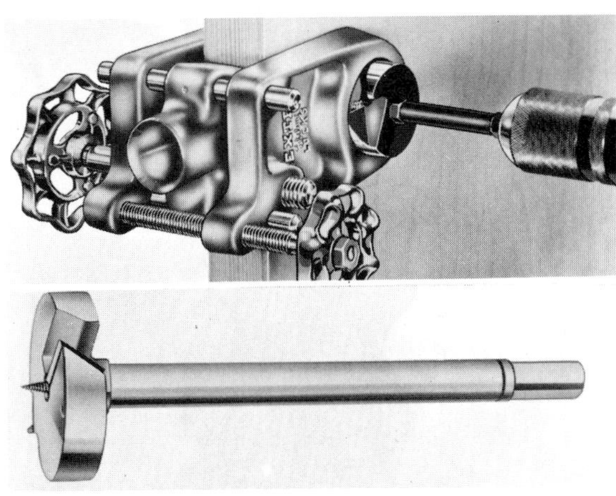

Figure 18-32. Installing locks. Top—Boring jig for door hardware saves time and ensures accuracy. Bottom—Boring bit. (Dexter Industries Inc.)

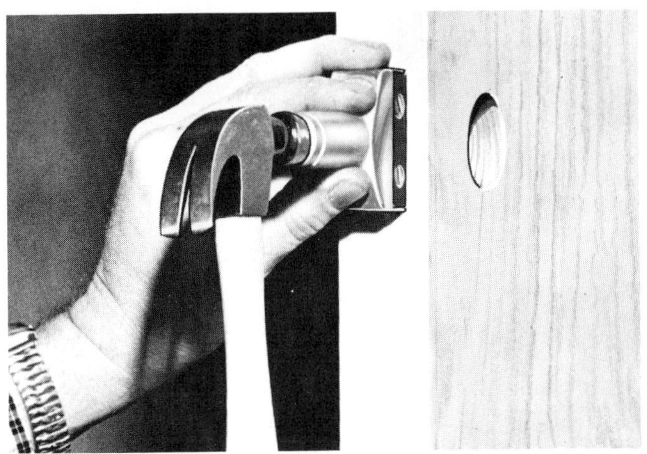

Figure 18-33. Using a faceplate mortise marker is more accurate and faster than use of a template.

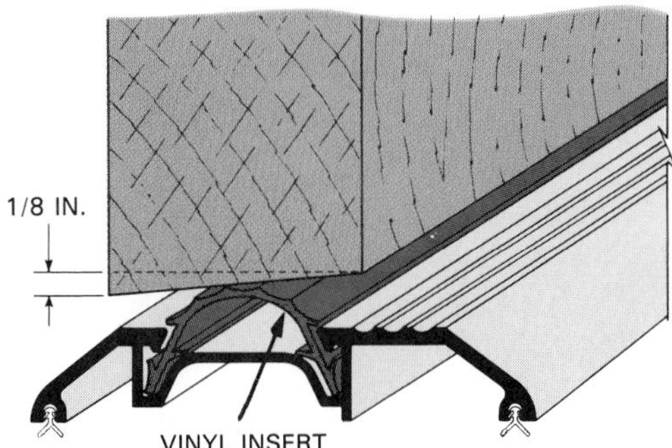

Figure 18-34. Aluminum threshold with vinyl seal strip. These are frequently found in residential construction. Bottom edge of door should be beveled. (Pemko Mfg. Co.)

opened on the room-side of the rough opening. Figure 18-38 shows a general procedure to follow in making the installation. Always study and follow the manufacturer's directions which are included in the package.

Figure 18-35. Automatic door bottom. Sealing strip moves upward when the door is opened.

Figure 18-36. Prehung doors are fabricated in millwork plants. Top—Gains are cut for receiving hinges. Bottom—Doorjamb are being installed. (Stan Greer Millwork)

Sliding Doors (Pocket Type)

A sliding *(pocket-type)* door offers a space-saving feature since it is opened by simply sliding it into an opening in the partition. The door frame consists of a split side

Figure 18-37. Installing a prehung door unit. Side jambs should rest on finished floor or spacer blocks of equal thickness. Left—Setting the door into the rough opening. Right—Snug door against drywall and plaster, plumb and nail in place through the trim.

1. Remove door unit from carton and check for damage. Separate the two sections. Place tongue side (not attached to door) outside of the room.

2. Slide frame section that includes the door into the opening. Side jambs should rest on finished floor or spacer blocks.

3. Carefully plumb door frame and nail casing to wall structure. Be sure all spacer blocks are in place between the jambs and door.

4. Move to other side of wall and install shims between side jambs and rough opening. Shims should be located where spacer blocks make contact with door edges. Nail through jambs and shims.

5. Install remaining half of door frame. Insert the tongue edge into the grooved section already in place. Nail casing to wall structure.

6. Nail through stops into jambs. Remove spacer blocks and check door operation. Make any adjustments required. Drive extra nails where shims are located. Install lock set.

Figure 18-38. General procedure for installing prehung interior door units. (Frank Paxton Lumber Co.)

jamb attached to a framework built into the wall. The rough opening in the structural frame must be large enough to include the finished door opening and the pocket, Figure 18-39. The pocket framework and track is installed during the rough framing stages. Pocket-frame units are available from millwork plants in a number of standard sizes.

Manufacturers that specialize in builder's hardware have developed steel pocket door frames, Figure 18-40. They are easy to install and provide a firm base for wall surface materials.

Figure 18-41 shows a typical track and roller assembly for a pocket door. The hanger (wheel assembly) snaps into the plate attached to the top of the door. It can be easily adjusted up or down to plumb the door in the opening.

Sliding Doors (Bypass Type)

Standard interior door frames can be used for a bypass sliding door installation. When the track is mounted below the head jamb, the height of a standard door must be reduced and a trim strip installed to conceal the hardware, Figure 18-42. Head jamb units are available with a recessed track that permits the doors to ride flush with the underside of the jamb.

Cutaway views of standard bypass track and hangers are shown in Figure 18-43. Each type can be used with either 3/4" or 1 3/8" doors. Note the hanger adjustment that raises or lowers the door for alignment after installation.

Figure 18-40. Top section of steel framework unit for pocket door. Kit includes track, hangers, and guides. (Ekco Building Products Co.)

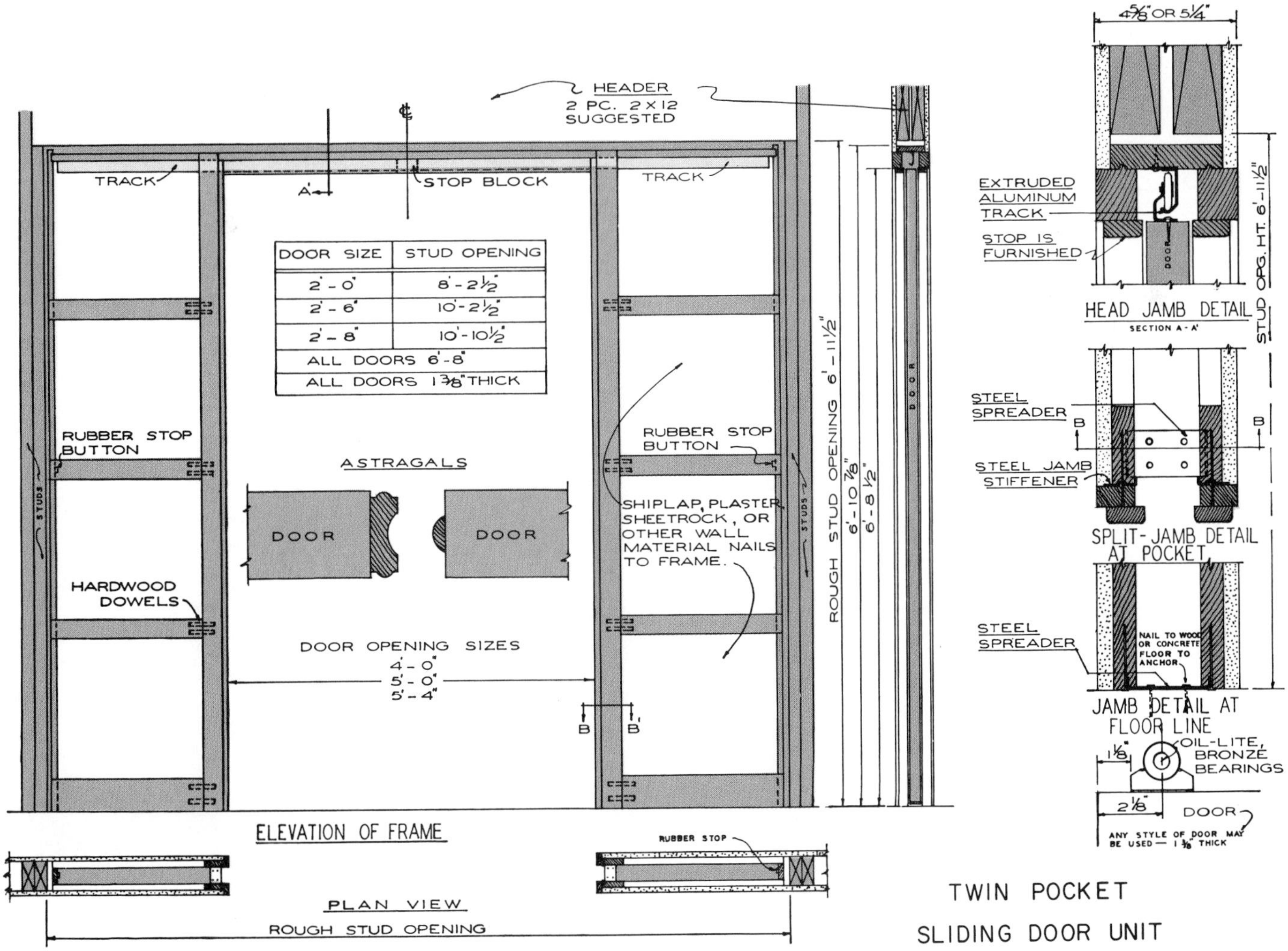

DOOR SIZE	STUD OPENING
2'-0"	8'-2½"
2'-6"	10'-2½"
2'-8"	10'-10½"
ALL DOORS 6'-8"	
ALL DOORS 1⅜" THICK	

Figure 18-39. Structural details of a pocket door. (Ideal Co.)

Figure 18-41. Cutaway shows track and roller assembly for pocket door.

Sliding door hardware for bypass doors is packaged complete with track, hangers (rollers), floor guide, screws, and instructions for making the installation.

A disadvantage of bypass sliding doors is that access to the total opening at one time is not possible. They are, however, easy to install and are practical for wardrobes and many other interior wall openings. Figure 18-44 shows a completed installation of bypass sliding doors.

Folding and Bifolding Doors

These types of folding doors consist of pairs of doors hinged together. Folding action is guided by an overhead track. A complete unit may consist of a single pair of doors or two or more pairs of doors. See Figure 18-45. Folding door units are well suited to wardrobes, closets, pantries, and certain openings between rooms.

The opening for bifold units is trimmed with standard jambs and casing. Figure 18-46 illustrates an installation of hardware. Pivot brackets and center guides have self-lubricating nylon bushings. The weight of the doors is supported by the pivot brackets and hinges between the doors, not by the overhead track and guide. Two-door units generally range from 2' to 3' wide while four-door units are available in widths from 3' to 6'.

When the total opening for a four-door unit is greater than 6' (two-door unit over 3') it is usually necessary to install heavier hardware. Also, a supporting roller-hanger is used instead of a regular center guide. This type of heavy-duty hardware is also used to carry multipanel folding door units that serve as room dividers.

Folding door hardware is supplied in a package that includes hinges, pivots, guides, bumpers, aligners, nails, and screws. Instructions for the installation are also included. Millwork plants can supply matching doors with prefitted hardware.

Multipanel Folding Doors

Multipanel folding doors are built from narrow panels with some type of hinge along the edges. One manufacturer produces a design where the hinge action is provided by steel springs threaded through the panels. The entire door assembly is supported by nylon rollers located in an overhead metal track. The track is wood-trimmed to match the door. Figure 18-47 shows how the track is installed in either a wood framed or plastered opening.

Door panel surfaces are available in a variety of materials and finishes. The best grade of panels are made from

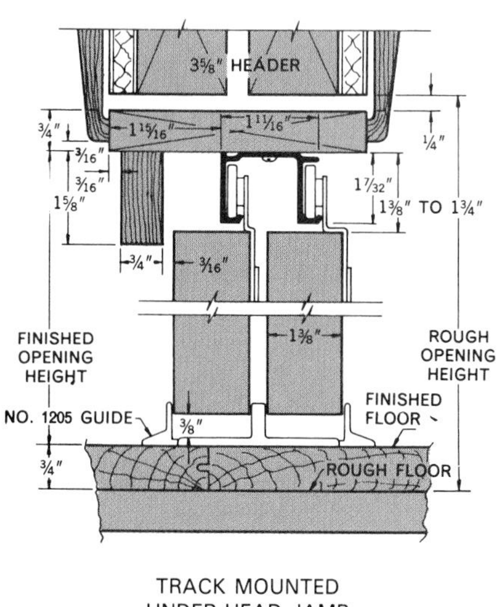

TRACK MOUNTED UNDER HEAD JAMB 1 3/8 IN. DOORS

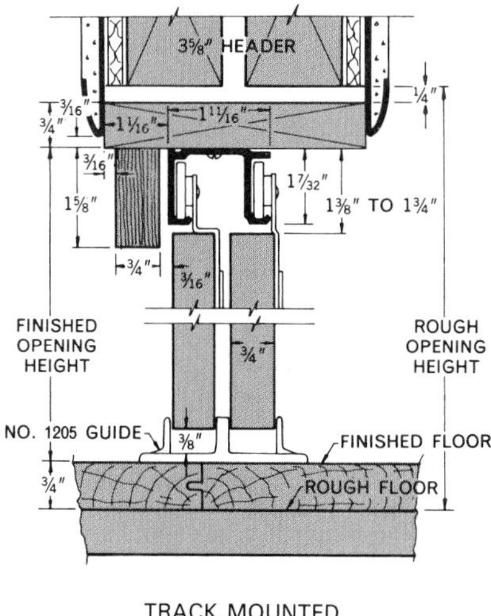

TRACK MOUNTED UNDER HEAD JAMB 3/4 IN. DOORS

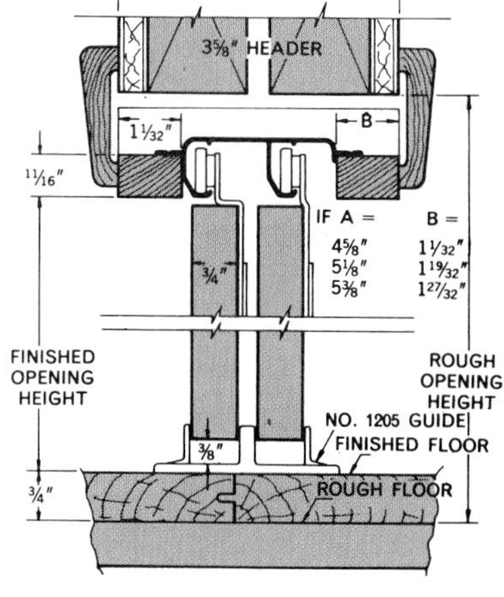

TRACK RECESSED IN HEAD JAMB 3/4 IN. DOORS

Figure 18-42. Drawings of typical bypass sliding door installations.

Figure 18-43. Cutaway views of bypass sliding door installations. Top—Track mounted below head jamb. Note trim member needed to conceal track. Bottom—Track is recessed into head jamb. Hangers include vertical adjustment. (Ekco Building Products Co.)

Figure 18-44. Bypass doors used for closet space. The doors open from either side. (C.E. Morgan)

Figure 18-45. Bifold or multipanel folding doors are often used to close off wardrobe space. Top—Bifold doors. (C.E. Morgan) Bottom—Multipanel folding doors. (LTL Home Products, Inc.)

genuine wood veneers, bonded to wood cores. Panels are also made of stabilized particleboard wrapped with woodgrain embossed vinyl film.

An important advantage of this type of folding door is its space-saving feature. As the door is opened, the panels fold together forming a "stack" that does not require clear room space. Bifold doors and regular passage doors must have clearance in the room as they are opened and closed. Figure 18-48 shows general details of construction with the door in both opened and closed positions.

When open, the door requires only a small amount of space. For example, the stack dimension for an 8' opening is only 11 1/2". When it is desirable to clear the entire opening, the stack can be housed in a special wall cavity.

Folding doors are available in a variety of sizes. Figure 18-49 shows an installation for a wardrobe that includes hanging space, dresser, and drawers.

Folding doors can be used to separate room areas, laundry alcoves, and general storage space. Manufacturers furnish door units in a complete package that includes track, hardware, latches, and instructions for making the installation.

Window Trim

Interior window trim consists of casing, stool, apron, and stops, Figure 18-50. Millwork companies select and

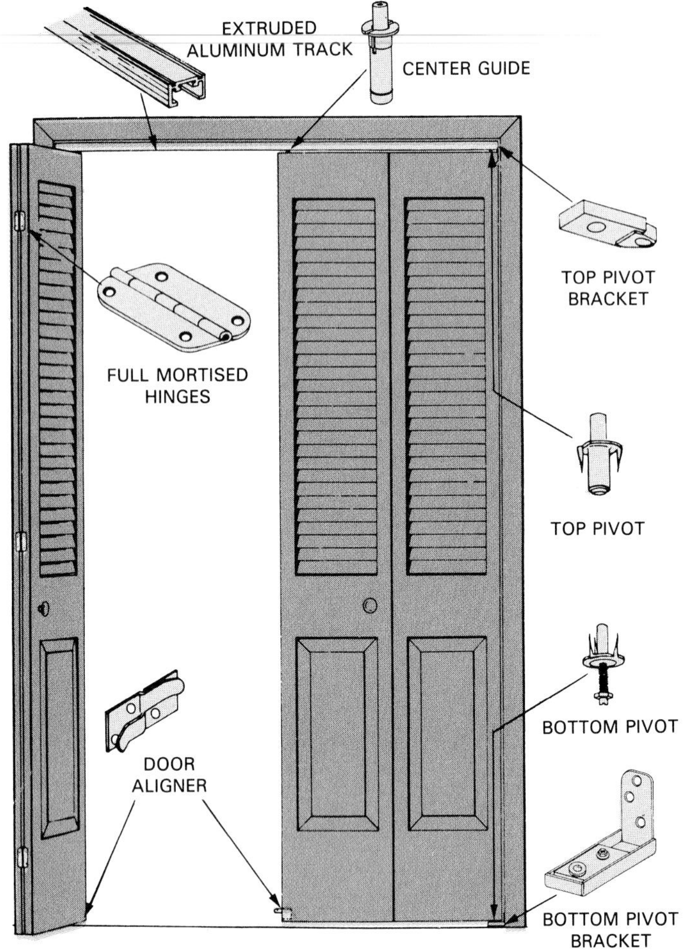

Figure 18-46. Typical hardware used for folding door installation. The hinged units are supported by brackets. No weight is carried by center guide. (Ideal Co.)

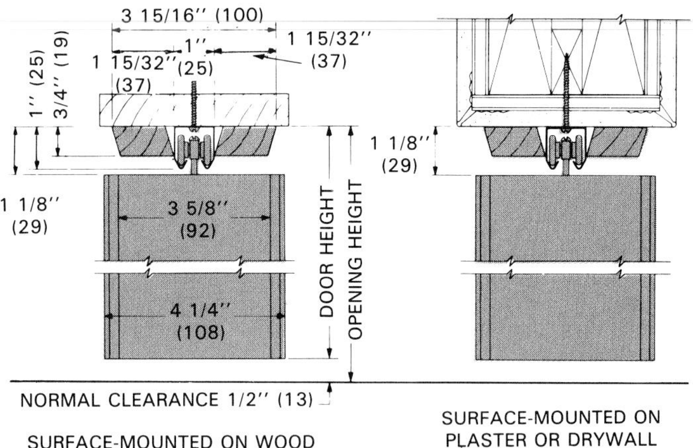

Figure 18-47. Typical head sections show installation of overhead track for folding doors. Metric sizes (millimeters) are shown in parentheses. (Rolscreen Co.)

package the proper length trim members to finish a given unit or combination of units.

Installing Trim

1. Mark the *stool* for cutting — hold it level with the sill and mark the inside edges of the side jambs. Also mark a line on the face where it will fit against the wall surface. For a standard double-hung window this line will usually be directly above the square edge of the stool rabbet (notch where it fits over the top of the sill).
2. Carefully cut out the ends of the stool and check the fit. You will have to open the lower sash slightly to slide the stool into position. Bring down the window carefully on top of the stool and draw the cutoff line so the sash will clear the stool. Allow about 1/16" between the front edge of the stool and the window sash. Position a piece of side casing and measure beyond it about 3/4". Mark the cutoff lines for the ends of the stool.
3. Cut the ends, sand the surface, and nail the stool into place. Some installations require that the stool be bedded in caulking compound or white lead.
4. Set a length of side casing in position on the stool, Figure 18-51, and mark the position of the miter on the inside edge. Some carpenters prefer to set the casing back from the jamb edge about 1/8", following about the same procedure as previously described for door casing.
5. Cut the miter.
6. Nail side casing in place. Drill nail holes if the casing is made of hardwood.
7. Cut and install head casing.
8. Cut apron to length of outer edges of side casing.
9. Sand any visible saw cuts on apron and nail in place.

When the profile (edge) of the apron is curved, the ends should be returned or coped so that the shape of the ends will be the same as the sides. The returned end is commonly used and formed with miter cuts as illustrated in Figure 18-52. Completed trim at the lower right-hand corner of a double-hung window is shown in Figure 18-53. Note the spacing of the various members and the end contour of the apron which has been returned.

In modern construction, the stool and apron are sometimes eliminated. Instead, a piece of beveled sill liner is installed to match the window jamb. Regular casing is then applied around the entire window. See Figure 18-54.

Baseboard and Base Shoe

The *baseboard* covers the joint between the wall surface and the finish flooring. It is among the last of the interior trim members to be installed since it must be fitted to the door casings and cabinetwork. *Base shoe* is used to seal

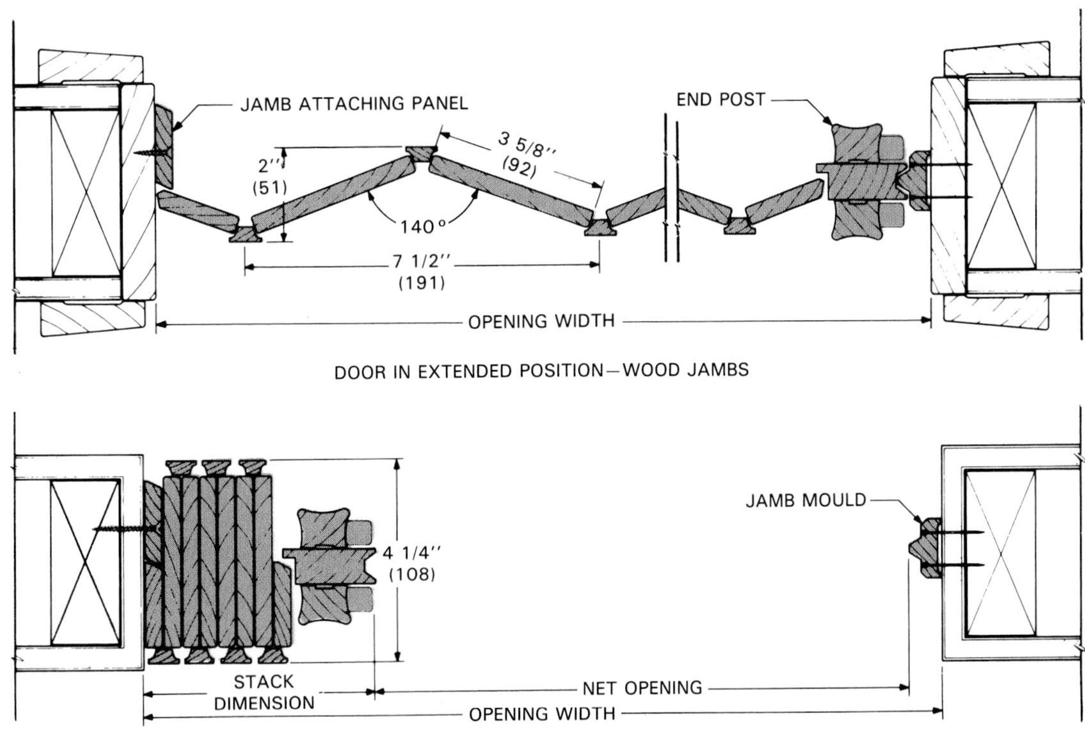

Figure 18-48. General details show operation of folding door. Metric sizes are shown in parentheses. (Rolscreen Co.)

Figure 18-49. Application for use of folding door includes closing off a wardrobe. Doors stack on either side of the wardrobe opening. (Rolscreen Co.)

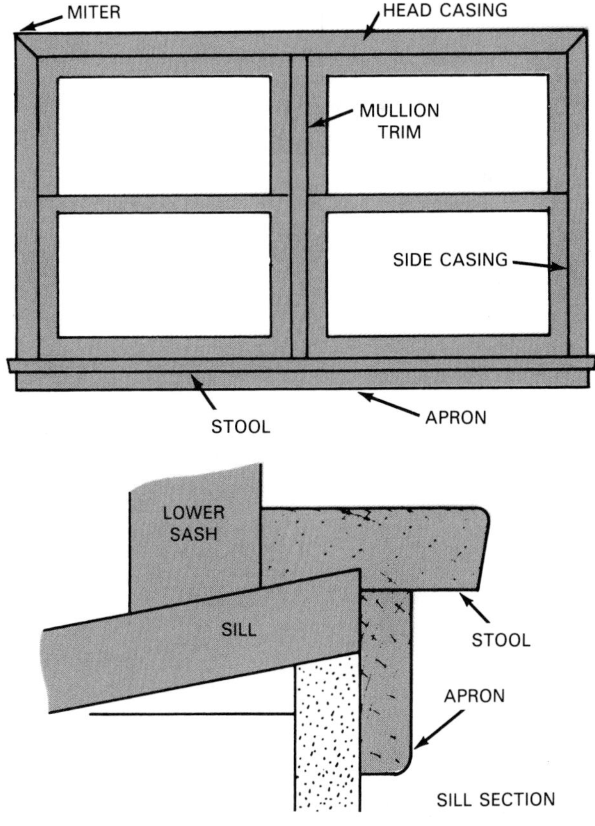

Figure 18-50. Trim members used for the standard double-hung window.

Figure 18-51. Measuring and fitting side casing. Bottom must be squared and should rest firmly on the stool before marking miter cut at top.

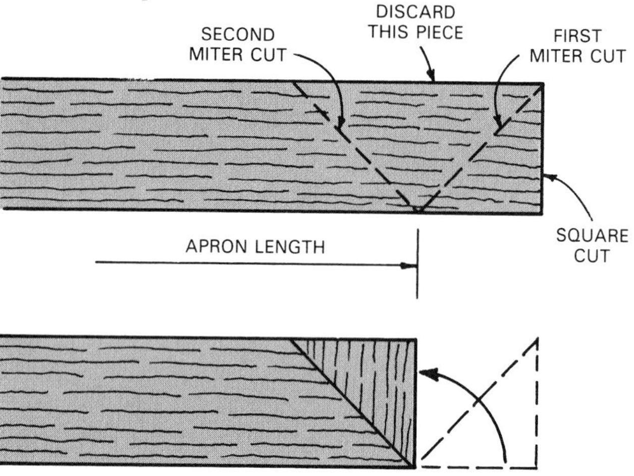

Figure 18-52. How to mark and cut a returned end. Use glue to attach the end piece.

Figure 18-53. View of finished right-hand corner of a double-hung window. Note nailing pattern, especially for stool. Nail holes will be filled during finishing operations. Some carpenters nail the stool to the wall by toenailing it from underneath. This eliminates the nail hole.

the joint between the baseboard and the finished floor, Figure 18-55. It is usually fitted at the time the baseboard is installed but is not nailed in place until after surface finishes (lacquer, varnish, or paint) have been applied. Base shoe is often used to cover the edge of resilient tile or carpet.

Baseboards run continuously around the room between door openings, cabinets, and built-ins. The joints at internal corners should be coped. Those at outside corners are mitered. See Figure 18-56.

Select and place the baseboard material around the sides of the room. Sort the pieces so there will be the least amount of cutting and waste. Where a straight run of baseboard must be joined, use a mitered-lap joint (also called a *scarf joint*). Be sure to locate the joint so it can be nailed over a stud. See Figure 18-57.

Figure 18-54. Modern windows often have no apron or stool. Top—Lower corner of window shows how casing can be used to replace the stool and apron. It is commonly known as picture frame trimming. Bottom—Using a cordless drill to make holes for attaching hardwood window casing.

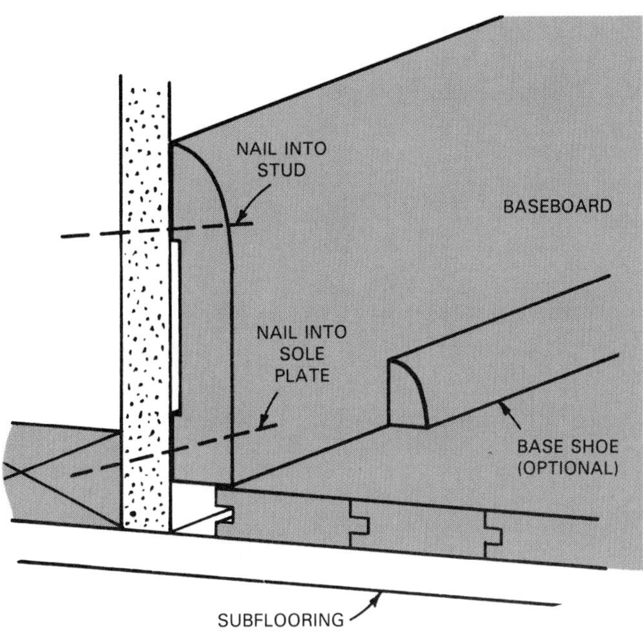

Figure 18-55. Cutaway of baseboard and base shoe. They conceal the gap between flooring and finished wall.

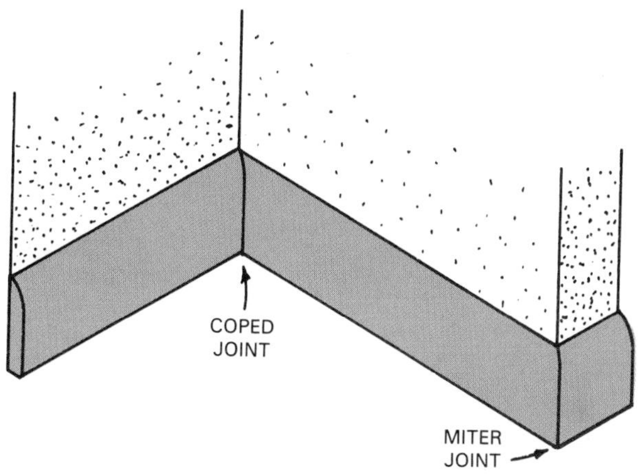

Figure 18-56. Coped joints are suitable for inside corners; however, use mitered joint for outside corners.

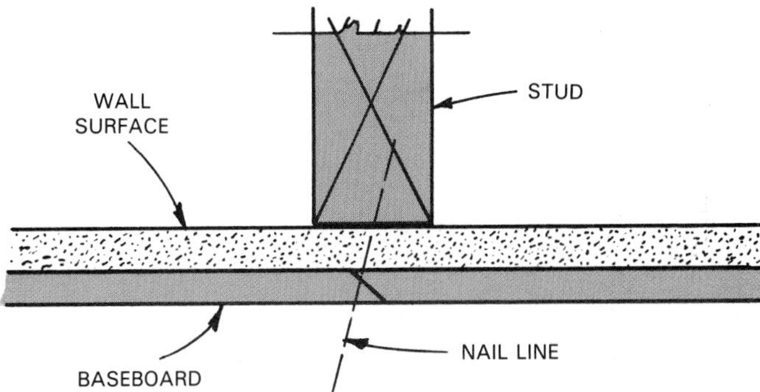

Figure 18-57. Cut a scarf joint when joining a straight run of baseboard. Be sure to drill nail holes when using hardwood material.

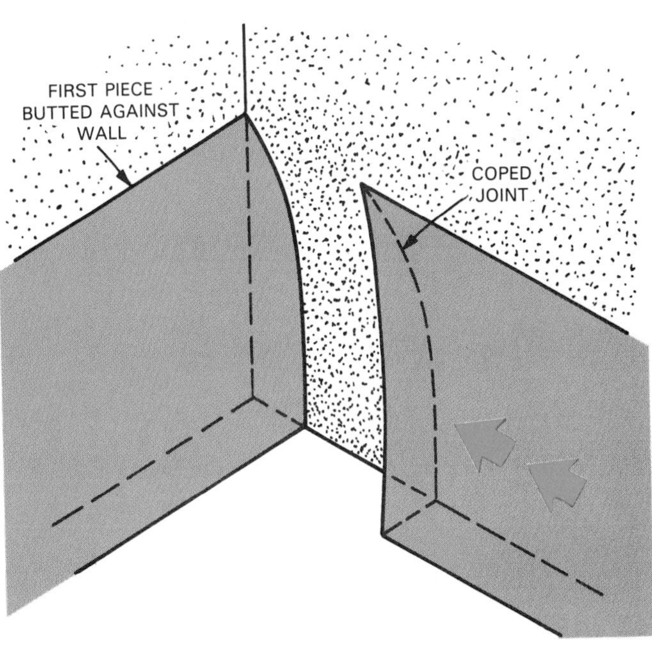

Figure 18-58. Coped joint follows contour (shape) of the piece it will butt against.

Baseboard installation is easy if stud locations have been marked, first on the rough floor ahead of plaster and then on the wall surface before the finish flooring or underlayment is applied.

If the stud positions have not been marked, tap along the wall with a hammer until a solid sound is heard or use a stud finder. Drive nails into the wall to locate the exact position of the stud, then mark the location of others by measuring the stud spacing (usually 16" O.C.).

Cut and fit the first piece of baseboard so it makes a tight butt joint with the intersecting wall surface. The next piece, running from an internal corner, is joined to the first with a coped joint, Figure 18-58.

To form this type of joint, first cut an inside miter on the end of the baseboard. Then, using a coping saw, cut along the line where the sawed surface of the miter joins the curved surface of the baseboard, Figure 18-59. Make the cut perpendicular to the back face. This forms an end profile that will match the face of the baseboard. Some carpenters undercut the coped end slightly to ensure a tight fit at the front edge.

The coped joint takes a little longer to make than a plain miter, but it makes a better joint at an inside corner. It will not open when the baseboard is nailed into place. Neither will a noticeable crack appear if the wood shrinks after installation.

All outside corners are joined with a miter joint. Hold the baseboard in position and mark at the back edge. Make the cut using a miter box or a power miter box, Figure 18-60.

Before installing a section of baseboard, check both ends to make sure the cut and fit is correct. To install, hold the board tightly against the floor and nail it in place with finishing nails long enough to penetrate well into the studs. The lower nail is angled slightly down so it is easier to drive and will enter the sole plate. Set the nail heads as shown in Figure 18-61.

Baseboards are normally butted against the door casing as illustrated in Figure 18-62. The edge of the casing is designed with sufficient surface to accommodate the slightly thinner dimension of the baseboard. Base shoe, if used, is ended at the casing with a miter cut as shown.

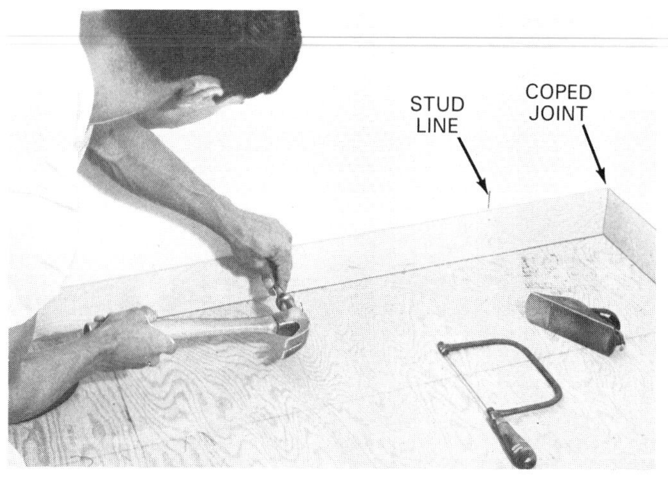

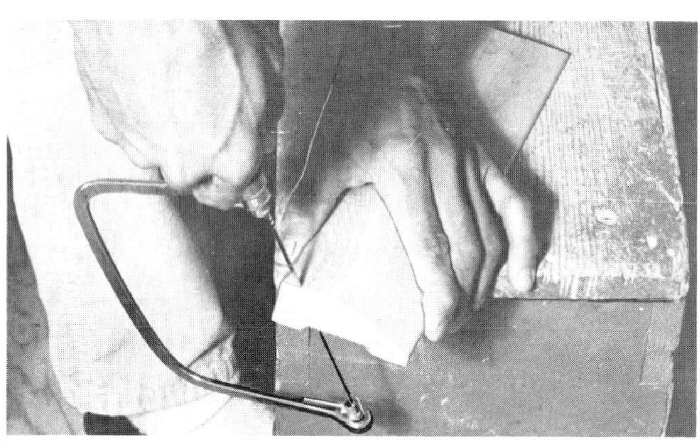

Figure 18-59. Form a coped joint by making a perpendicular cut along the front edge of an inside miter joint.

Figure 18-61. Nailing the baseboard. Top—Set nails after installing a section of baseboard. Right-handed workers usually prefer to install pieces in a counterclockwise direction. Bottom—Using a pneumatic nailer to attach shoe base.

Figure 18-60. Using a power miter box to cut miter joint for outside corners. Hold piece securely against fence for an accurate cut.

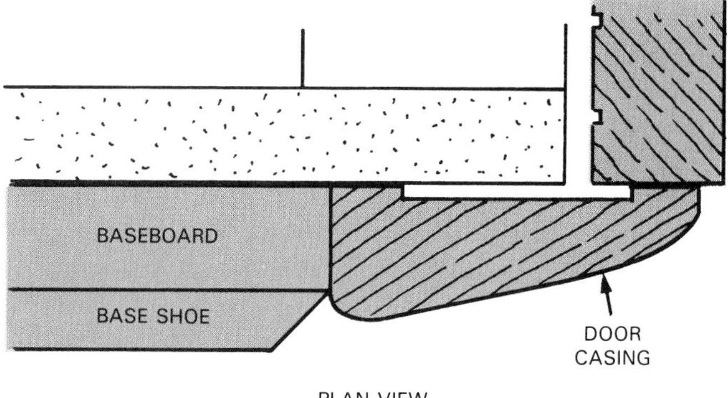

Figure 18-62. Baseboard is butted against door casing.

Important Terms

Baseboard
Base shoe
Casing
Cores
Dead bolt
Door frame
Doorstop
Flush doors
Gains
Head jamb
Lug
Moldings
Mortise
Panel doors
Plinth block
Pocket sliding doors
Prehung door unit
Sash door
Scarf joint
Side jambs
Stile and rail door
Stool
Threshold

Test Your Knowledge

1. Interior doorjambs for a standard framed wall with a plastered finish should be ____________ inches wide.
2. If a door frame is slightly high for the opening, part of the ____________ should be trimmed.
3. When nailing the doorjambs, locate the nails so they will be covered by the ____________.
4. To cover the space between the wall surface and the doorjambs, door ____________ is applied to each side of the frame.
5. What is a plinth block?
6. A panel door consists of rails, panels, and ____________.
7. The face panels of flush doors are usually made of 1/8" plywood and are commonly called door ____________.
8. Three types of hollow core door construction are: lattice, ____________, and implanted blanks.
9. A standard size bedroom door is ____________ wide and ____________ high.
10. When setting the doorstop on the hinge jamb, provide a clearance of ____________ with the door.
11. If you are facing the outside of an entrance door and it swings toward you with the hinges on your left, it is called a ____________ *(LH, RH, LHR, RHR)*.
12. The two most commonly used types of residential door locks are cylindrical and ____________.
13. The type of sliding door that opens into an opening in the wall or partition is called a ____________.
14. After the stool has been set, the next step in trimming a single window unit is to install the ____________.
15. The last trim member applied to a window unit is the ____________.
16. When installing baseboards, a ____________ joint should be used at internal corners.

Outside Assignments

1. Visit a residential building site in your community where the inside finish work is in progress. Note the type of trim being applied and the procedures being followed. Possibly one of the carpenters would discuss with you the advantages and disadvantages of prehung door units. Prepare carefully organized notes of your observations and make an oral report to the class. Obtain permission from the supervisor or head carpenter before making the visit.
2. Develop a door schedule for a preliminary or presentation drawing of a residence. Sizes of doors will usually not be shown and you will need to make the selection. A door schedule should include the location, width, height, thickness, type and design, kind of material, and quality requirements.
3. To extend the assignment in No. 2, obtain cost estimates. Refer to a builders supply catalog or visit a local lumber dealer.
4. Build a full-size sectional mock-up of a partition with a doorway. Include all parts—studs, wall finish, doorjamb, casing, baseboard, and door. Carry the section up about 16" above the floor and extend the partition out from the door only about the same amount. A 3/4" thickness of particleboard could serve as the base and represent the finish flooring. It will be best to glue most of the parts together since the structure may not withstand much nailing.

Special cutter produces miter cut on side jambs. (Stan Greer Millwork)

Prehung doors are stocked at lumberyards and home improvement centers in a variety of styles and prices.

19 Cabinetmaking

Cabinetwork, as used in the interior finish of a residence, refers to kitchen and bathroom built-in storage and, in a general way, to such work as closet shelving, wardrobe fittings, desks, bookcases, and dressing tables. The term *built-in* emphasizes that the cabinet or unit is located within or attached to the structure. The use of built-in cabinets and storage units is an important development in modern architecture and design.

Modern kitchen cabinets present an attractive appearance and help to increase kitchen efficiency, Figure 19-1. Here, and in other areas of the home, storage units should be designed for the items that will be stored.

Space must be carefully allocated. Drawers, shelves, and other elements should be proportioned to satisfy specific needs. Three types of cabinetwork used in homes are:

- That which is built on the job by the carpenter.
- Custom-built units constructed in local cabinet shops or millwork plants.
- Mass-produced cabinets from factories that specialize in this area of manufacturing.

Except in large housing projects, combinations of these types of cabinetwork are found on most jobs. Even when most of the cabinets are factory produced, the carpenter is responsible for the installation. This task requires skill and careful attention to detail.

Figure 19-1. Modern kitchen cabinets are attractive and provide efficient storage space in shelving and drawers. They also provide important working counter surface around sinks, cooking tops, and ovens. Some include a snack bar. (Merillat Industries, Inc.)

Drawings for Cabinetwork

Architectural plans usually include details of built-in cabinetwork. The floor plan shows cabinetwork location. Elevations, usually drawn to a larger scale, provide detailed dimensions. A typical drawing of a base cabinet, desk, and room divider is illustrated in Figure 19-2.

Drawings of this type are scaled and, thus, the need for extensive dimensioning is eliminated. The drawings serve as a construction guide to the carpenter. They are also followed in the selection of factory-built components. When built on the job, the detail of joints and structure becomes the responsibility of the carpenter, who should be skilled in cabinetmaking.

Architectural drawings may include more specific details concerning the cabinetwork. This is justified on large commercial contracts. Here a number of individuals will be associated with the project. It also becomes important when the cabinetwork contains special materials and constructions.

Figure 19-3 shows kitchen cabinets designed for an apartment complex. Written specifications will define the type of joinery and quality of materials.

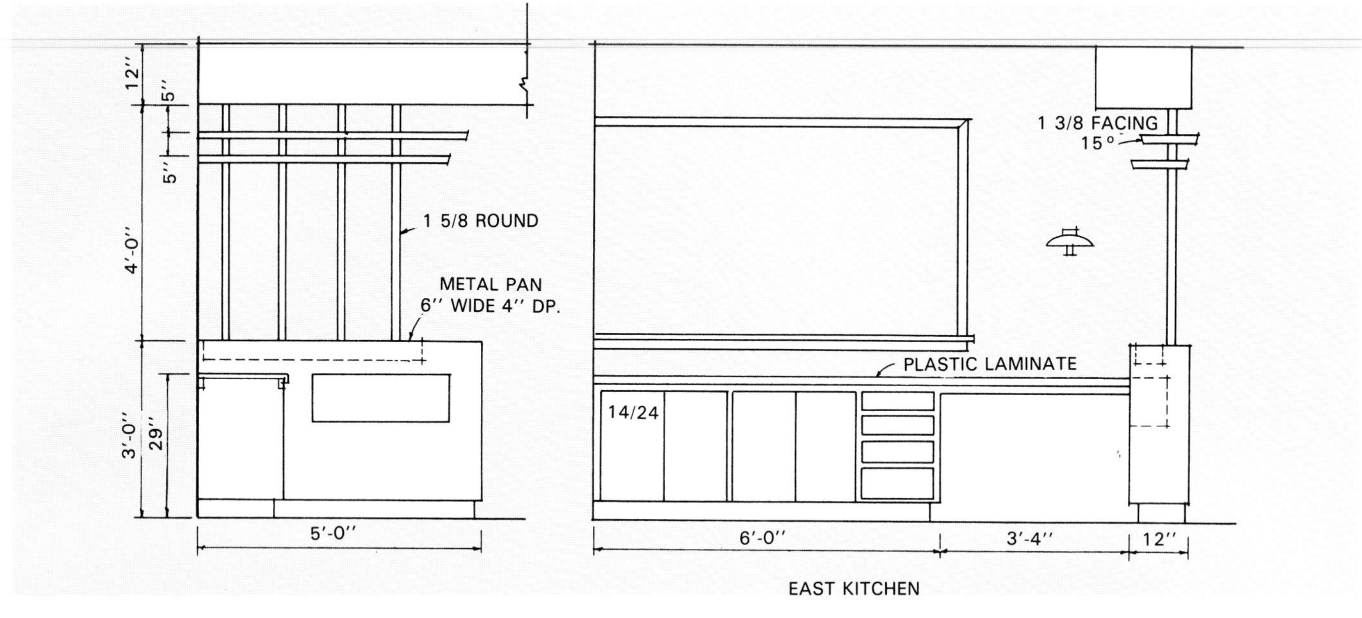

Figure 19-2. Detail drawings. Top—Typical built-in cabinetwork detail. Since the drawings are carefully scaled, many dimensions have been eliminated. Bottom—Completed cabinetwork. Drawers include a recess under the front edge that serves as a finger pull.

Standard Sizes

Overall heights and other dimensions of built-in units are usually included in the architectural plans. However, the carpenter should be familiar with basic design requirements.

Base cabinets for kitchens are usually 36" high and 24" deep. The countertop extends about 1" beyond the base cabinets, Figure 19-4. The vertical distance between the top of the base unit and the bottom of the wall unit may vary from 15 to 18". FHA specifies a minimum height of 24" when the wall cabinet is located over a cooking unit or sink.

Cabinets with built-in lavatories in bathrooms or dressing rooms are normally 31" high, Figure 19-5. Depth varies depending on the type of fixture. The counter surface may be plastic laminate, a synthetic stone product, or tile. Some will provide knee room; others do not.

Standards for closets and wardrobes vary widely. The determining factors are the items to be stored. The minimum clear depth for clothing on hangers is 24". When hooks are mounted on doors or the rear wall, this distance

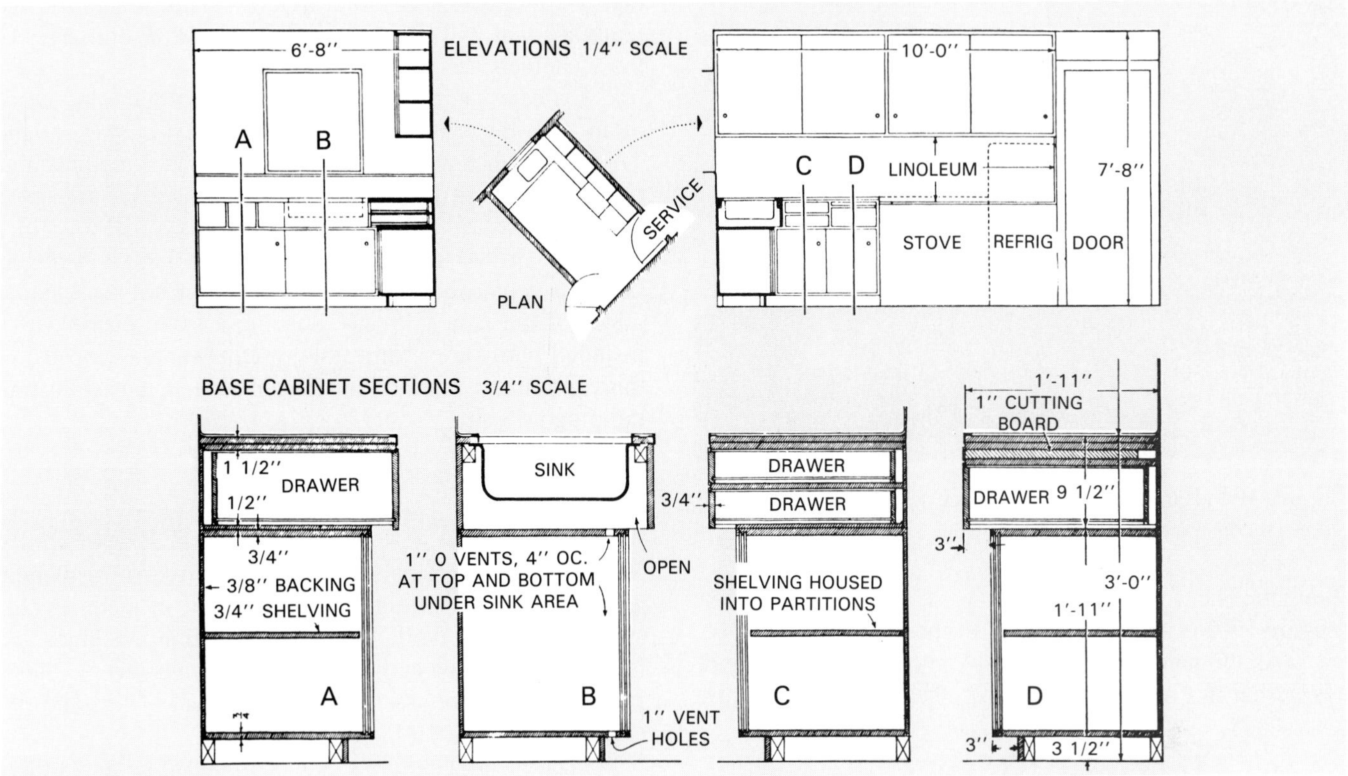

Figure 19-3. These drawings are for cabinetwork in a large apartment building. (Architectural Woodwork Institute)

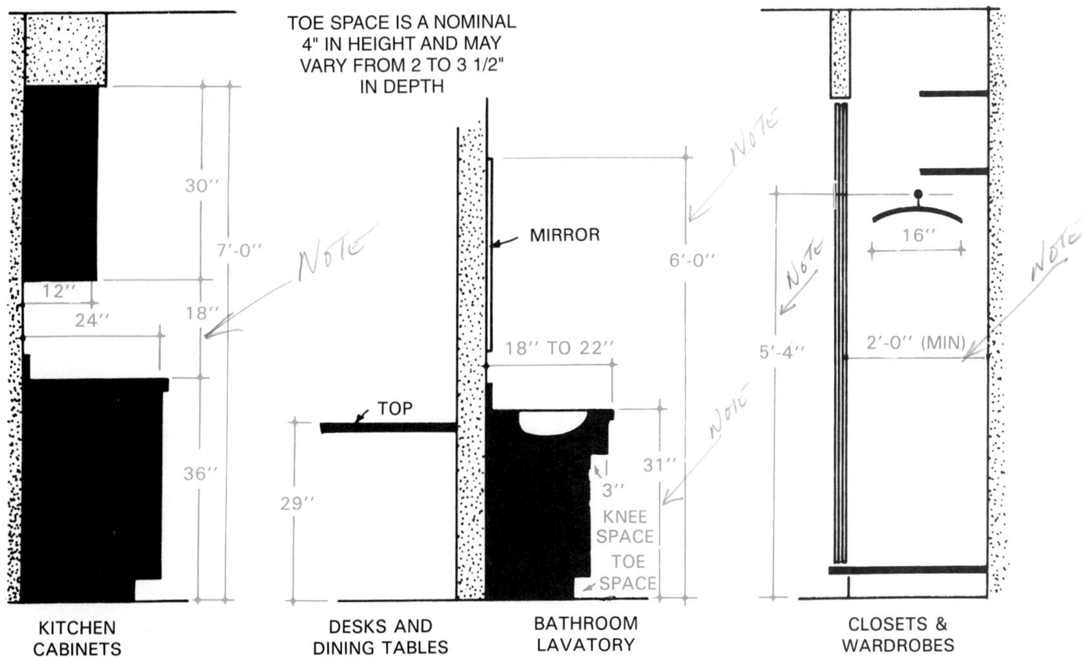

Figure 19-4. Dimensions are standardized for cabinetwork.

must be extended. Refer to architectural standards manuals for additional information.

Types of Construction

Cabinets may have a frame, a frame-with-cover, or may be frameless. In frame construction, a *skeleton* of solid lumber supports shelves or work surface. This type of construction is often used in constructing furniture but may be used for book shelves or other built-ins.

Frame-with-cover construction, also called *face-frame construction,* is similar to frame construction except that panels of plywood or fiberboard cover the frame. Since support is provided by the frame, panels are light-

Figure 19-5. Bathroom vanity designed for a modern home. Type of construction is face frame. (Dal-Tile Corp.)

weight. The front edges are faced with solid lumber. Figure 19-6 is an example of this type.

As the name implies, *frameless construction,* Figure 19-7, has no support except for the panels, which are

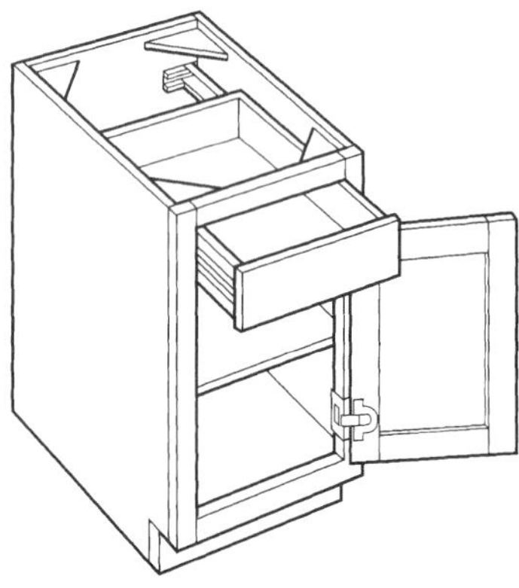

Figure 19-6. Drawing shows face-frame construction. Edges of cabinet front are covered with solid lumber. Often, doors are framed panels. Also, Figure 19-5 is a good example of this type of construction. (Merillat Industries, Inc.)

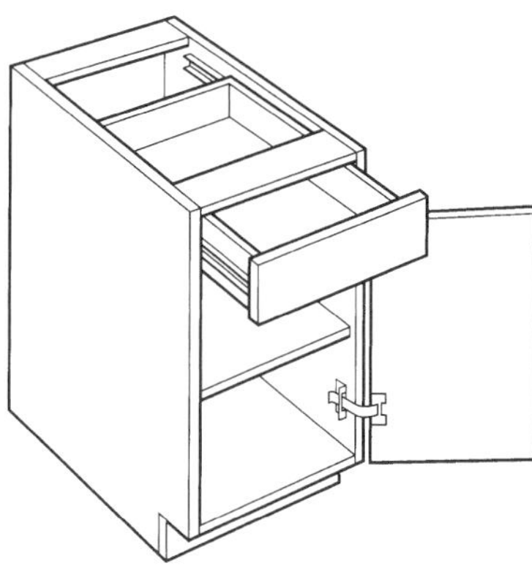

Figure 19-7. Example of frameless construction. Note that front edges of this storage unit are not faced. This style is sometimes called "European" since it originated in Germany. (Merillat Industries, Inc.)

heavy enough to carry the weight of the assembly and materials that will be stored in the cabinet. Front edges of these cabinets are not framed.

Frameless construction, which is most popular today, often uses the standardized *32 mm construction system.* This system has a set of standard hole sizes and spacings on panels that are designed to be used on various case types and sizes. Central to this system are two vertical rows of 5 mm holes drilled in side panels on 32 mm centers. The vertical row centerlines are 37 mm from the front and the rear edges of the panel. Precise spacing assures proper fit of installed hardware. While the system was developed initially for mass production, it can be adapted to custom building as well.

Factory-Built Cabinets

Today, a major part of the cabinetwork for residential and commercial buildings is constructed in factories that specialize in this work. Modern production machines and tools can save time and produce high-quality work. Mass-produced parts are assembled with the aid of jigs and fixtures. See Figures 19-8 and 19-9.

Factory-built cabinets may be obtained in one of three forms:

- Disassembled.
- Assembled but not finished (in-the-white).
- Assembled and finished.

Disassembled or "knocked-down" cabinets consist of parts cut to size and ready to be assembled on the job. The assembled but unfinished cabinet is ready to set in place. Hardware is included but not installed. All surfaces are sanded and ready for finish. After installation, finishing materials and procedures can be coordinated with doors and inside trim—thus ensuring an exact match.

Figure 19-8. Mass production of cabinetwork. View shows final assembly operations as cabinets are carried on a conveyor. (Kitchen Kompact, Inc.)

Figure 19-9. Operator removes workpiece from automatic shaper. Revolving table carries workpieces past a cutter head that forms profile and shapes edge in a single pass. (Conestoga Wood Specialties, Inc.)

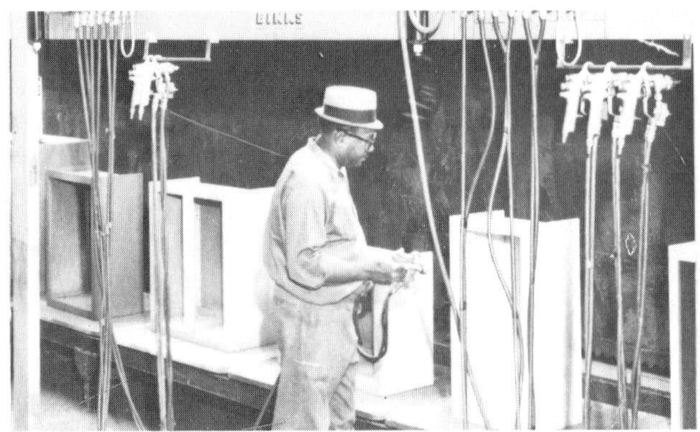

Figure 19-11. Cabinet finishing. Top—Spray finishing kitchen cabinet units as they move through a modern spray booth. (Binks Mfg. Co.) Bottom—Worker applies a glaze coating. Air washer at right helps trap the overspray before it can be exhausted to the atmosphere. (DeVilbiss Co.)

The assembled and finished cabinets are widely used because they save time during finishing stages of construction. Manufacturers offer a variety of shades and colors which are applied by experts, Figure 19-10. Because of the controlled conditions and special equipment, finishing materials with high resistance to moisture, acids, and abrasion can be applied, Figure 19-11. After the finishing process is complete and hardware has been installed, the units are carefully packaged and shipped to the distributor or directly to a construction site.

Manufacturers of cabinetwork offer a variety of standard units, especially in kitchen cabinets. Most kitchen layouts can be made entirely from these units. However, when necessary, special custom built units can be ordered. Sometimes factory-built cabinets are combined with units constructed in custom cabinet shops, Figure 19-12, or with various units built on the job. Figure 19-13 shows line drawings of standard base and wall units offered by one cabinet manufacturer. In addition to these, other standard

Figure 19-10. Highly skilled wood finishers hand rub stain coating on premium quality cabinet doors. (Riviera Kitchens, an Evans Products Co.)

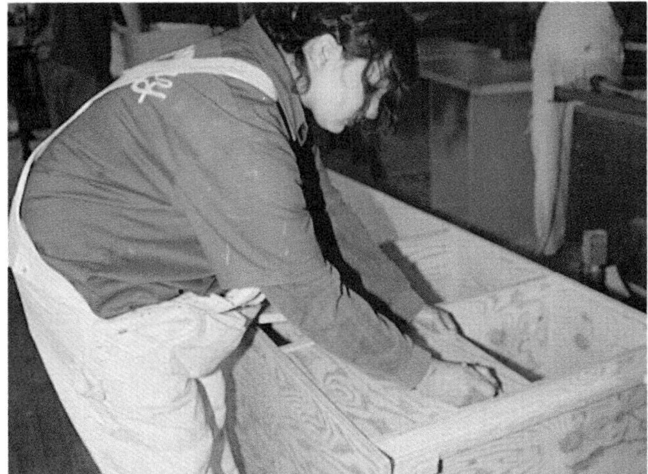

Figure 19-12. Cabinetmaking student learns custom building and is shown taking a measurement for a face frame. (Bullard-Haven Vocational Technical School)

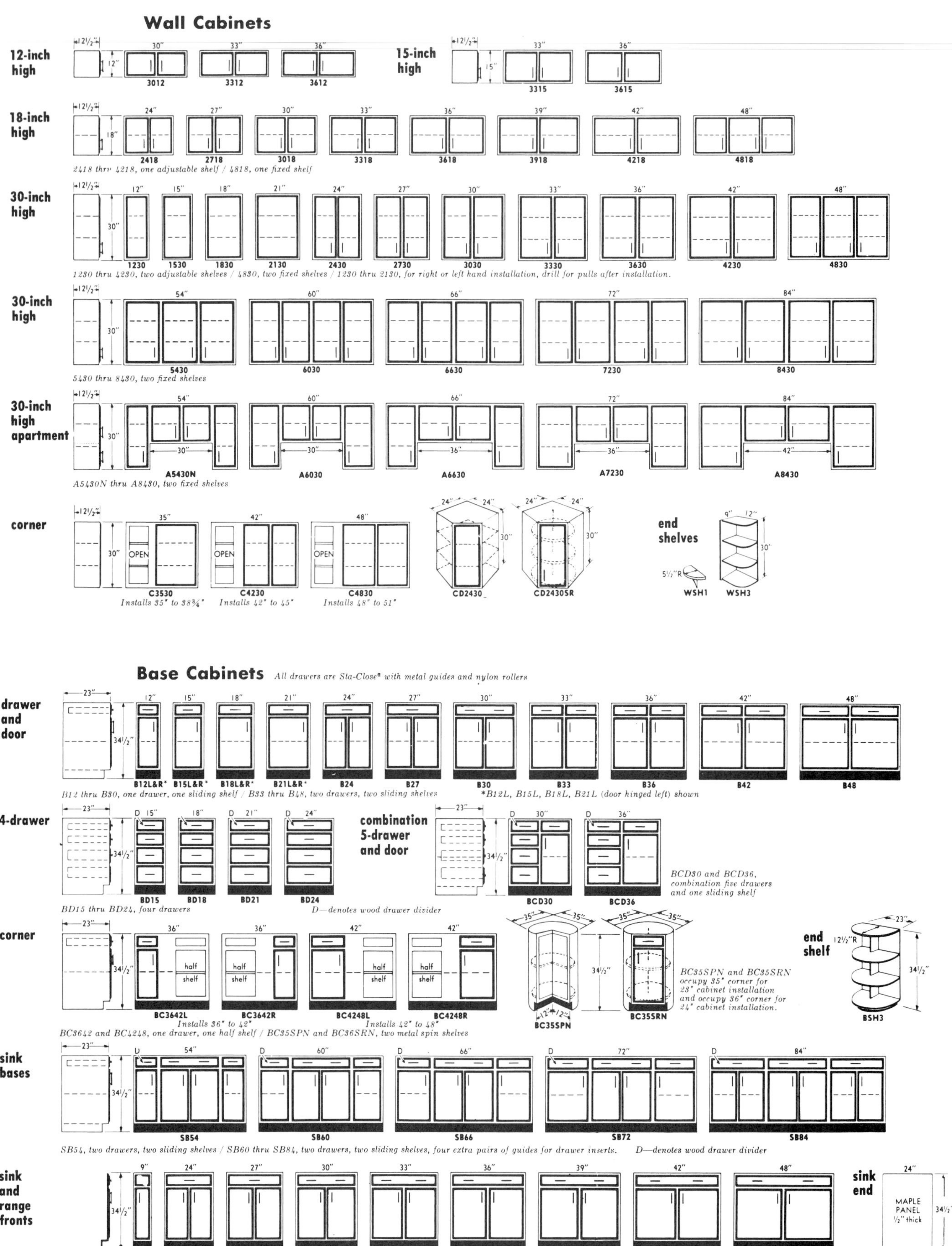

Figure 19-13. Cabinet manufacturers provide information about their standard cabinet units. Dimensions are given as well as code numbers for different units. (I-XL Furniture Co.)

units include oven and utility cabinets, peninsular base units, and bathroom vanity and sink cabinets.

Cabinet Materials

A wide range of materials are used in factory-built cabinets. Low-priced cabinets are usually made from panels of particleboard with a vinyl film applied to exposed surfaces. The vinyl is printed with a wood grain pattern and has the appearance of genuine wood.

High quality cabinets are made from veneers and solid hardwoods including such species as oak, birch, ash, and hickory. Hardboard, particleboard, and waferboard may be used for certain interior panels, drawer bottoms, and as the base for plastic laminate countertops. Frames are assembled with accurately made joints. Dovetail joints are generally used in drawer assemblies. See Figure 19-14. A completed installation along one side of a kitchen is shown in Figure 19-15.

A variety of storage features can be added to standard cabinet units as shown in Figure 19-16. Examples include revolving shelves (lazy Susan), special compartments and dividers in drawers, slide-out breadboards, and slide-out shelves. Canned goods storage units provide extra convenience with swing-out shelving with wood or wire racks. A wire rack provides special lid storage. Other available storage features include: a slide-out wire rack for under-sink storage, swing-out storage trays, file cabinets, wall tambour storage, pull-out ironing boards, and swing-out base multistorage wire shelves.

Cabinet Installation

Before cabinets can be installed, the carpenter must check walls and floors for uneven spots. He or she must locate these spots and then shim or scribe cabinets to make the installation plumb, true, and square.

Floors can be checked with a long straightedge or straight 2 x 4 and a level. The floor should be checked within 22" of the walls where base cabinets will be installed. A level line should be snapped on the wall from the high point around the wall as far as the cabinets will extend. This is called the base level line.

To check the wall, first mark the outlines of all cabinets on it. Use a straightedge to check for low and high points. High spots must be removed by scraping or sanding. Low spots can be shimmed with thin pieces of wood or wood shingles. See Figure 19-17.

There are two basic procedures for installing factory-built cabinets:

- Some manufacturers recommend that the wall cabinets be installed first. When layouts are made and wall studs located with a stud finder or by tapping the wall with a hammer, the wall units are lifted into position. They are held with a padded T-brace (sometimes called a "story stick") that allows the worker to stand close to the wall while making the installation. See Figure 19-18. After the wall cabinets are securely attached and checked, the base cabinets are moved into place, leveled, and secured.
- Following a second procedure, base cabinets are installed first. The tops of the base cabinets can then be used to support braces (story sticks) that hold the wall units in place. This procedure is illustrated in Figures 19-19 and 19-20.

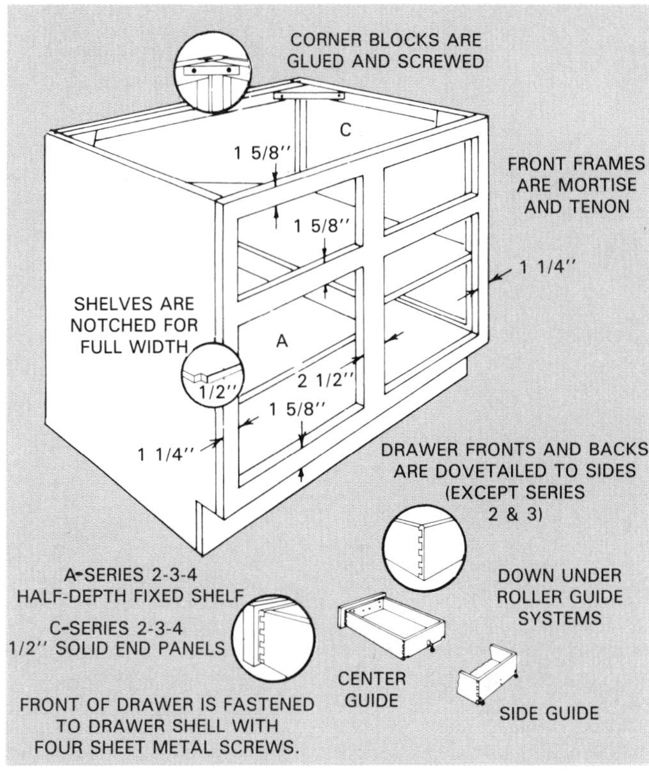

Figure 19-14. Drawings show construction details of factory-built kitchen base cabinet. (Brammer Mfg. Co.)

Figure 19-15. Installation view of modern factory-built cabinets along one wall. Note the built-in cooking top and oven. (Riviera Kitchens, an Evans Products Co.)

Figure 19-16. Special features are available in a variety of styles from most cabinet manufacturers. They add important conveniences for organized storage. (KraftMaid Cabinetry, Inc.; Merillat Industries, Inc.)

Floors and walls are seldom exactly level and plumb. Therefore, as previously indicated, shims and blocking must be used so the cabinets are not racked or twisted. Doors and drawers cannot be expected to operate properly if the basic cabinet is distorted by improper installation.

Screws should go through the hanging strips (if used) and into the stud framing. Never use nails. Toggle bolts are required when studs are inaccessible. Join units by first clamping them together, Figure 19-21, and then, while aligned, install bolts and T-nuts.

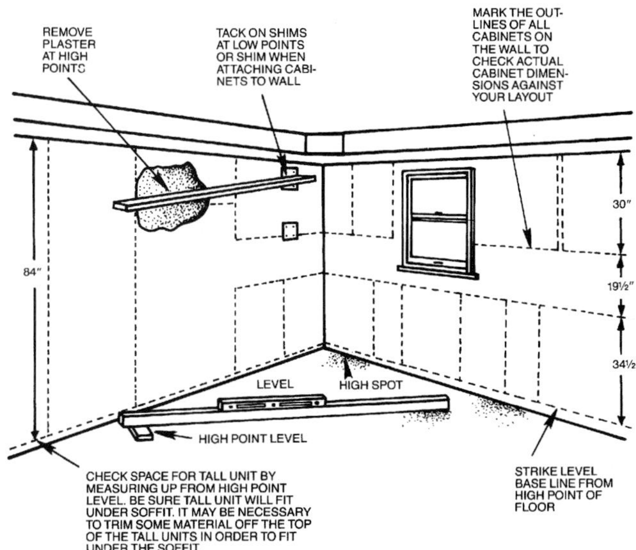

Figure 19-17. Before cabinet installation begins, mark on the wall the outline of every cabinet. Then check for high spots, plumb of walls, and levelness of floor. (KraftMaid Cabinetry, Inc.)

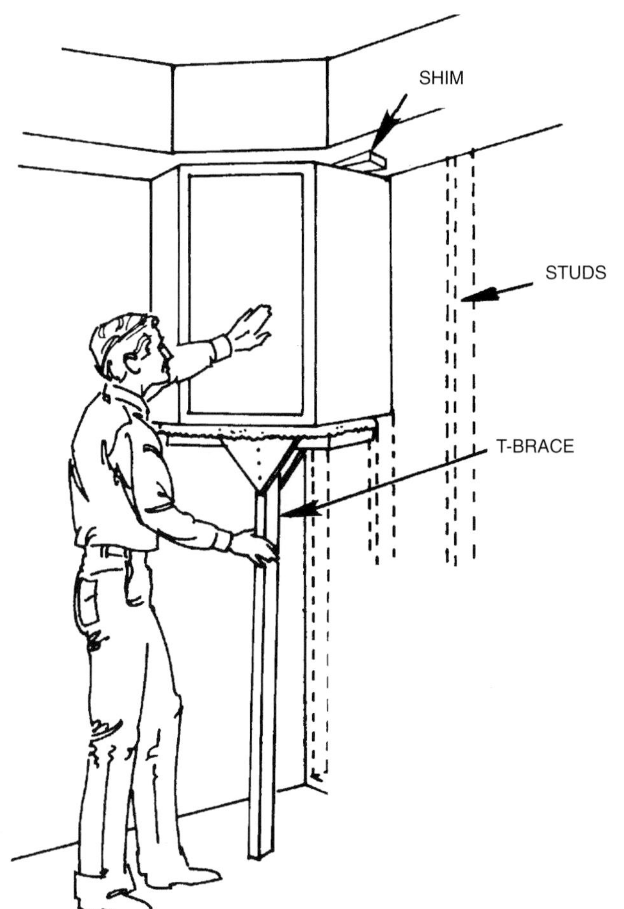

Figure 19-18. Using a "story stick" to support a wall cabinet. Locate studs and mark their location in the inside of the cabinet. (KraftMaid Cabinetry, Inc.)

Figure 19-22 shows the installation of a mitered countertop. Miters must be accurately cut and fitted, a difficult procedure requiring great skill. The joint is secured and drawn tight from the bottom with bolts that are let into the substrate.

Cabinets for Other Rooms

Some manufacturers of kitchen cabinets build storage units for other areas of the home. Built-in units provide drawers and cabinets that save space. They improve efficiency and also offer greater convenience. For example, the cabinets shown in Figure 19-23 include storage for towels, soap, dental and pharmaceutical supplies, cosmetics, and many other items needed in the bath area.

In other rooms, carefully planned built-in cabinets provide efficient storage and display for items used in work, recreation, and hobby activities. Their custom-built appearance usually adds to room decor. They may even eliminate the need for some movable pieces of furniture.

Figure 19-24 shows a free-standing buffet installed in a dining room. The term *modular* refers to units assembled from parts having standard sizes. Thus, a variety of finished assemblies can be made from the same parts.

Building Cabinets

Two different procedures are commonly used when building cabinets on the job. The first consists of cutting the parts and assembling them "in place" a piece at a time. The carpenter or cabinetmaker attaches each piece to the floor, wall, or to other members. After the basic structure is assembled, facing strips, doors, and other fittings are marked, cut to size, and attached.

In the second procedure, the entire unit is first assembled. Then the finished piece is fixed to the floor or attached to the wall.

The structure is formed with end panels, partitions, and backs. Horizontal frames join the parts. See Figure 19-25.

Basically, this is the method used in custom cabinet shops or for mass-production of cabinets in factories. When units of this type are built on the job, they can be much larger (longer sections) than is practical in shops or factories. The latter type must be of a size easily moved to the building site.

Master Layouts

Before cutting out the parts for a cabinet, it will be helpful to prepare a *master layout* on plywood or cardboard, Figure 19-26. This is especially true when following the second procedure described above.

Several layouts may be required. These should be drawn as section views, where the structure is complicated with drawers, pull-out boards, and special shelving.

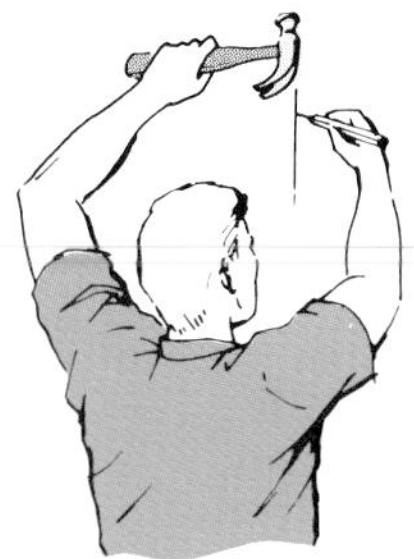

1.

Locate the position of all wall studs where cabinets are to hang by tapping with a hammer. Mark their position where the marks can easily be seen when the cabinets are in position.

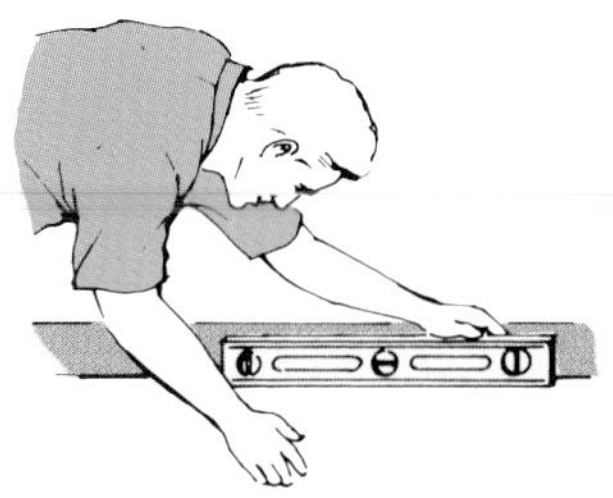

2.

Find the highest point on the floor with a level. This is important for both base and wall cabinet installation later. Remove the baseboard from all walls where cabinets are to be installed. This will allow them to go flush against the walls.

3.

Start the installation with a corner or end unit. Slide it into place then continue to slide the other base cabinets into the proper position.

4.

When all base cabinets are in position, fasten the cabinets together. This is done by drilling a 1/4″ diameter hole through the face frames and using the 3″ screws and T-nuts provided. To get maximum holding power from the screw, one hole should be close to the top of the end stile and one should be close to the bottom.

5.

Check the position of each cabinet with a spirit level, going from the front of the cabinet to the back of the cabinet. Next shim between the cabinet and the wall for a perfect base cabinet installation.

6.

Starting at the high point in the floor, level the leading edges of the cabinets. Continue to shim between the cabinets and the floor until all the base cabinets have been brought to level.

7.

After the cabinets have been leveled, both front to back and across the front, fasten the cabinets to the wall at the stud locations. This is done by drilling a 3/32″ diameter hole 2 1/4″ deep through both the hanging strips for the 2 1/2″ x 8 screws that are provided.

8.

Fit the counter top into position and attach it to the base cabinets by predrilling and screwing through the front corner blocks into the top. Use caution not to drill through the top. Cover the counter top for protection while the wall cabinets are being installed.

9.

Position the bottom of the 30″ wall cabinets 19″ from the top of the base cabinet, unless the cabinets are to be installed against a soffit. A brace can be made to help hold the wall cabinets in place while they are being fastened. Start the wall cabinets installation with a corner or end cabinet. Use care in getting this cabinet installed plumb and level.

10.

Temporarily secure the adjoining wall cabinets so that leveling may be done without removing them. Drill through the end stiles of the cabinets and fasten them together as was done with the base cabinets.

11.

Use a spirit level to check the horizontal surfaces. Shim between the cabinet and the wall until the cabinet is level. This is necessary if doors are to fit properly.

12.

Check the perpendicular surface of each frame at the front. When the cabinets are level, both front to back and across the front, permanently attach the cabinets to the wall. This is done by predrilling a 3/32″ diameter hole 2 1/4″ deep through the hanging strip inside the top and below the bottom of the cabinets at the stud location. Enough Number 8 screws should be used to fasten the cabinets securely to the wall.

Figure 19-19. Basic steps for installing factory-built cabinets. Spacers, sometimes called "story sticks," support the upper cabinets at the proper height until they can be secured to a wall. (See No. 10 above.) Some installers, however, prefer to hang wall cabinets first. (Haas Cabinet Co., Inc.)

1. After locating studs and other framing members, lift the wall cabinet into position. Use a brace or "story stick" to hold the cabinet while checking proper position and installing fasteners. Follow the same procedure as described in Figures 19-19.

2. Start the installation of base cabinets with corner or end units. Check their position and then add additional units. Fasten them together as recommended by the manufacturer. Level the assembly and fasten to the wall.

3. On-site construction of a counter. It consists of two layers of underlayment. First layer can be plywood or particleboard; second layer can be the same or it may be backing board. Above, a 2" solid wood facing strip is being glued and nailed to the edge.

4. Using a portable belt sander, level and smooth the top and facing strips. When a counter extends some distance beyond a supporting frame, a backing sheet should be glued to the underside.

5. Using a portable router, trim the top ends of the plastic laminate facing strip. The edge must be level with the substrate. After the top sheet of laminate is installed, the router or a file can be used to trim the edge.

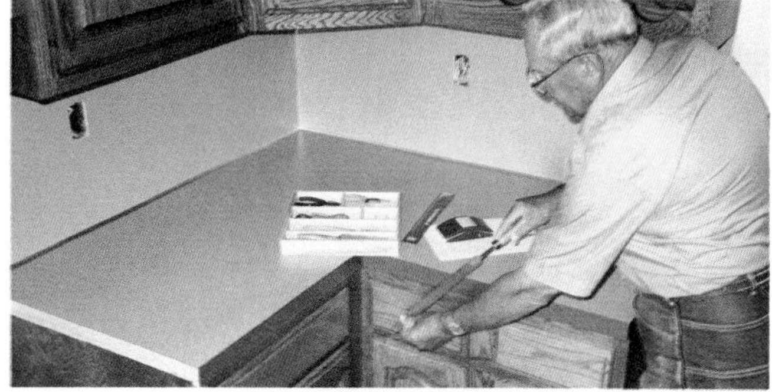

6. The corners and edges of the counter should be beveled. This smooths them and makes them longer lasting. The bevel can be formed with a smooth mill file as shown above. Routers can be equipped with guides and special bits to perform this task.

Figure 19-20. Basic steps for installing factory-built cabinets and constructing a built-in-place countertop.

Follow the overall dimensions provided in the architectural plans and details. Draw each member full size. Show all clearances that may be required.

This master layout will be valuable when cutting side or end panels to size and locating joints. It can also be referred to for the following:

- Exact sizes and locations of drawer parts.
- Other detailed dimensions not included in the regular drawings.

Basic Framing

When following the assembled-in-place procedure, some of the basic layout can be made directly on the floor and/or wall surface. Each end panel and partition is represented with two lines. Be sure these lines are plumb since they can be used to line up the panels when they are installed.

Construct the base first, as illustrated in Figure 19-27. Use straight 2 x 4s and nail them to the floor and to a strip

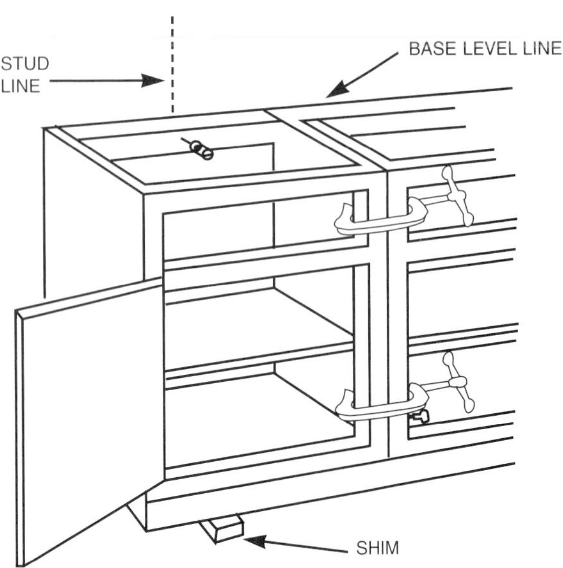

Figure 19-21. Installing base cabinets. Shim between the base and floor to level and bring cabinets up to the base level line. Use clamps to hold base units together until all adjustments are made and units are fastened together and to studs with screws. (KraftMaid Cabinetry, Inc.)

attached to the wall. If the floor is not level, place shims under the various members of the base. Exposed parts can be faced with a finished material or the front edge can be made of a finished piece such as base molding.

After the base is completed, cut and install the end panels, Figure 19-28. Attach a strip along the wall between

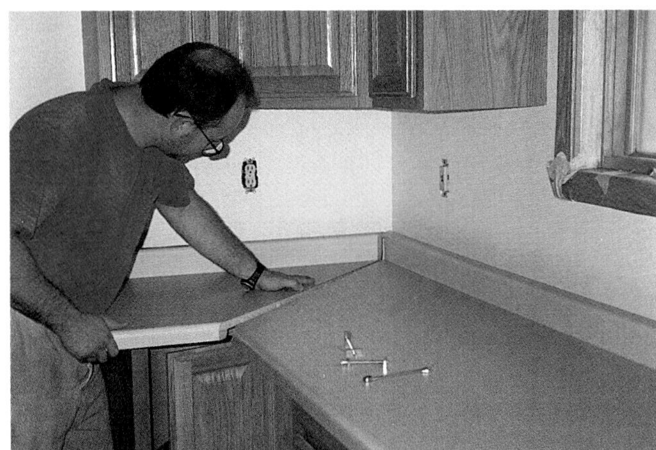

Figure 19-22. Fitting a mitered countertop is a job for a skilled carpenter or woodworker.

Figure 19-23. Built-in bathroom unit includes lavatory countertop, base cabinets, and wall cabinets, allowing ample storage. (Haas Cabinet Co., Inc.)

the end panels and level with the top edge. Be sure the strip is level throughout its length. Nail it to the wall studs.

Next, cut the bottom panels and nail them in place on the base. Follow this with the installation of the partitions which are notched at the back corner of the top edge so they will fit over the wall strip.

Figure 19-24. A free-standing buffet features curved doors. It is mass produced in a cabinet factory. (KraftMaid Cabinetry, Inc.)

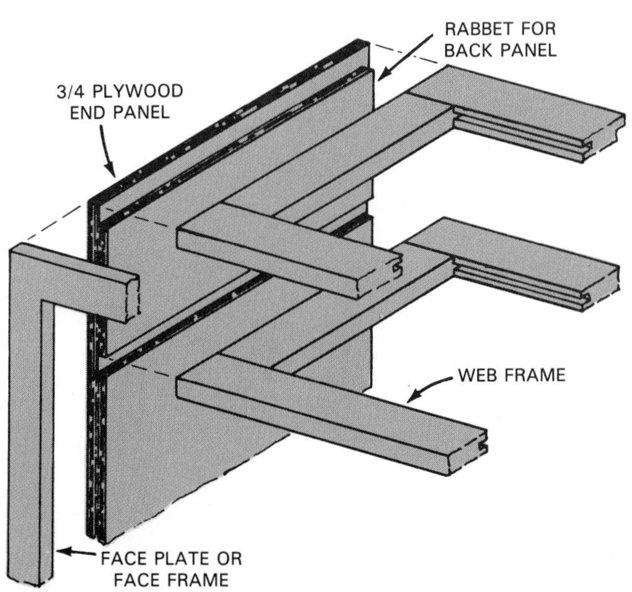

Figure 19-25. Typical frame construction of cabinets built on site as a separate unit. Terminology is recommended by Architectural Woodworking Institute.

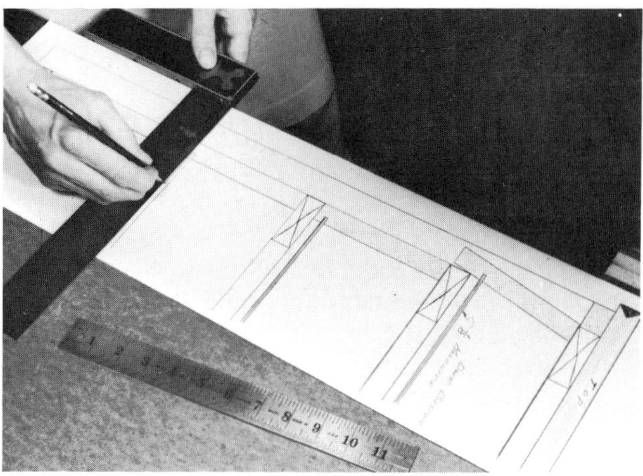

Figure 19-26. Making a full-size master layout of a base cabinet.

Figure 19-27. Constructing the base. Note the layout lines marked on the wall for an end panel and partition.

Figure 19-28. Marking an end panel while it is held in place. This is typical of the method used when constructing cabinets by the assembled-in-place procedure.

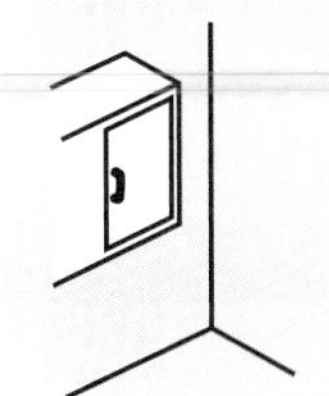

The carpenter may prefer to install the wall cabinets before the base units. This makes it easier to work on the wall cabinets since he or she can work directly below them.

Plumb the front edge of the partitions and end panels. Secure them with temporary strips nailed along the top as shown on the left-hand cabinet in Figure 19-29.

Wall units are constructed about the same way as the base units. Make layout lines directly on the wall and ceiling. Attach mounting strips by nailing through the wall surface into the studs. At inside corners, end panels can be attached directly to the wall, Figure 19-30.

During the basic framing operations, use care in constructing the openings for built-in appliances. They must be the correct size. Secure rough-in drawings and specifications, Figure 19-31. These are furnished by the manufacturer of the appliances. Follow them carefully. Figure 19-32 shows a completed cabinet with a built-in range.

Facing

Finished *facing strips* are applied to the front of the cabinet frame. In factories and cabinet shops, these strips are often assembled into a framework (called a face plate or face frame) before they are attached to the basic cabinet structure. The vertical members are called *stiles* and the horizontal members are called *rails*.

For assembled-in-place cabinets, each piece is cut and installed separately. The size is laid out by positioning the facing stock on the cabinet and marking them as shown in

Figure 19-29. Base cabinet partially completed. "Lazy Susan" (also called carousel shelving) unit must be installed at this stage.

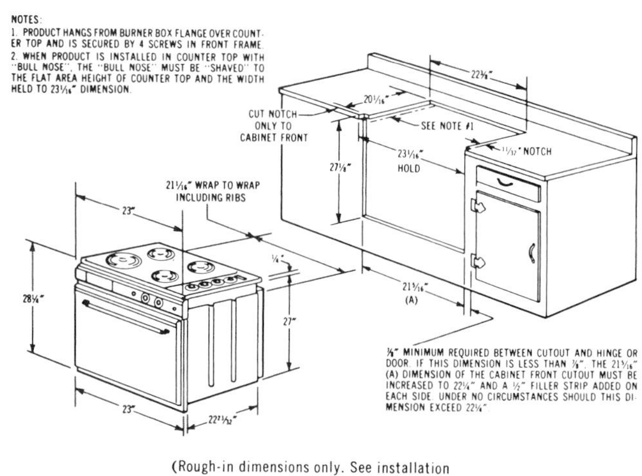

Figure 19-31. Typical drawing shows rough-in dimensions for a built-in range unit. (Whirlpool Corp.)

Figure 19-30. Wall cabinet construction. Left—End panels of a wall cabinet are in place. Right—Completed framing with facing partially applied.

Figure 19-32. Installation of a built-in range has been completed.

Figure 19-33. Then the finished cuts are made. A marked part can be used to lay out duplicate pieces.

Generally, stiles are installed first and then the rails, Figure 19-34. Sometimes an end stile is attached plumb and rails are installed to determine the position of the next stile.

The parts are glued and nailed in place with finishing nails. When nailing hardwoods, drill nail holes where splitting is likely to occur.

Many kinds of joints can be used to join the stiles and rails. A practical design is illustrated in Figure 19-35. The depth of the gain or dado is maintained at about 3/8" so it will be covered by the lip of the door or drawer. Lap joints and dowel joints are also commonly used to make this attachment.

Drawer Guides

Drawer guides are devices that keep drawers in place and allow them to be slid in and out. Three common types of drawer guides are used:

- Corner guides.
- Center guides.
- Side guides.

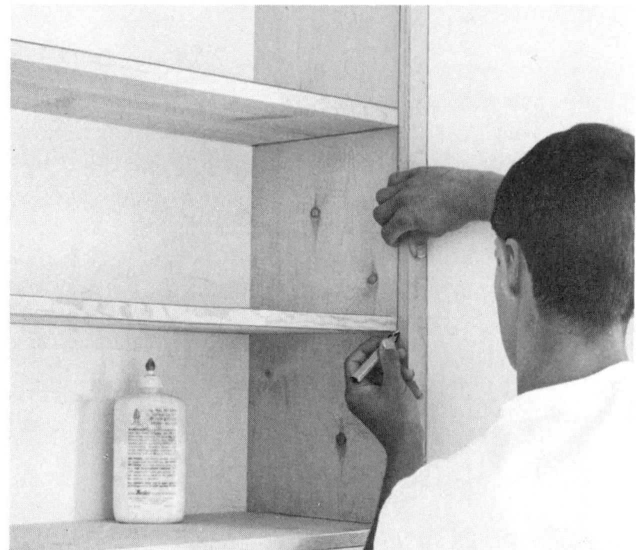

Figure 19-33. Lengths of face frame are marked while held in place against the cabinet frame.

Figure 19-34. Face frame pieces are being applied to a base cabinet. Stiles and rails are glued and nailed in place.

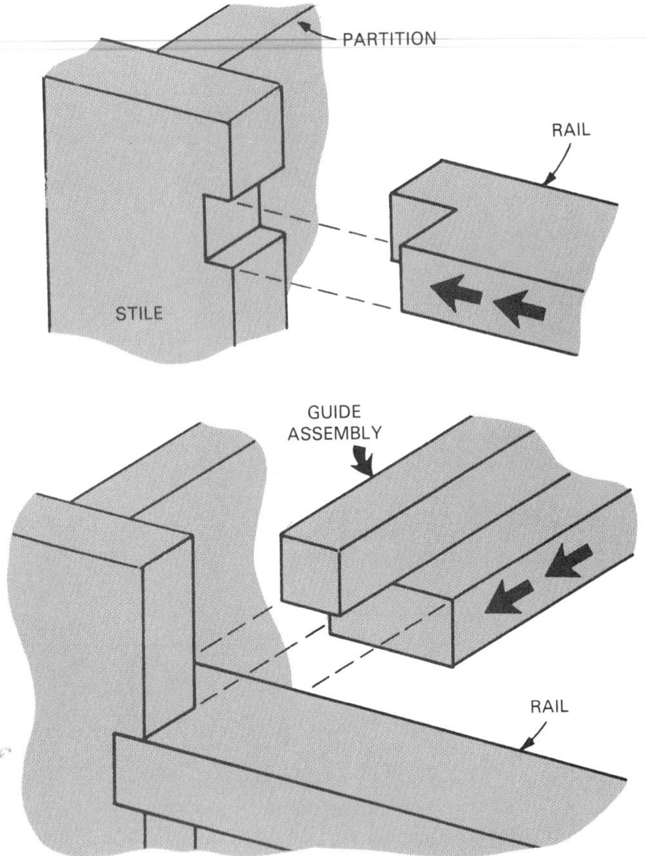

Figure 19-35. Practical joinery for assembled-in-place cabinet facing. Top—Stile and rail joint. Bottom—Drawer guide assembly fits behind facing and rests on rail.

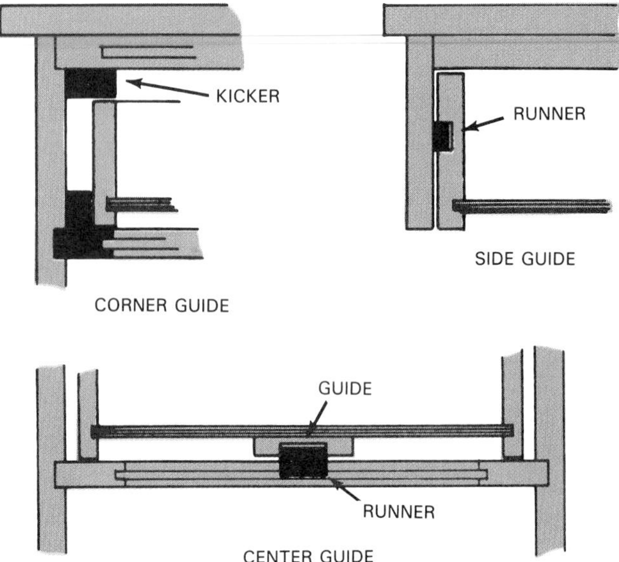

Figure 19-36. There are three basic types of drawer guides.

Figure 19-36 shows all three. The corner guide may be formed in the cabinet by the side panel and frame. It may be necessary to add a spacer strip to hold the drawer in alignment with the front facing.

An adaptation of the corner guide is used in the cabinet frame shown in Figure 19-37. Note that the guide unit located in the center is designed to carry a drawer on each side.

Center guides are often used in cabinetwork and consist of a strip or runner fastened between the front and back rails. A guide which is attached to the underside and back of the drawer rides on this runner. The runner is attached to the frame or rails with screws. It can be adjusted so the clearance on each side is equal and the face of the drawer aligns with the front of the structure.

In drawer openings where there is no lower frame, a side guide may be used. Grooves are cut in the drawer side before it is assembled and matching strips are fastened to the side panels. This type of guide can be used for shallow drawers or trays in wardrobe units where framing between each drawer opening is unnecessary and would waste space.

The drawer carrier arrangement may require a *kicker*. One is shown in Figure 19-37. It keeps the drawer from tilting downward when it is opened. The kicker may be located over the drawer side or centered to hold down the drawer back.

Wooden drawer guides should be carefully fitted. The parts should be given a coat of sealer. When the sealer on the moving parts is sanded lightly and waxed, the drawer will work smoothly.

Figure 19-38 shows how to attach a commercially made back bearing for a center slide. For large drawers, or those that will carry considerable weight, special drawer slides, Figure 19-39, may be used. Some of these support weights up to 50 lb. Units are available in several sizes.

Drawers

There are two general types of drawers:

- Flush.
- Lip.

Flush drawers are drawers that fit into the drawer opening. They must be carefully fitted. *Lip drawers* have a rabbet along the top and sides of the front. This style overlaps the opening and is much easier to construct. The rabbet is used mainly in built-in cabinetwork.

Sizes and designs in drawer construction vary widely. Figure 19-40 shows standard construction with several types of joints commonly used.

The joint between the drawer front and the drawer side receives the greatest strain. It should be carefully designed and fitted. Often, in high quality work, the corners are dovetailed and the bottom grooved into the back, front, and sides. The top edges of the drawer sides are rounded.

Flush drawer fronts may be constructed with the sides set about 1/16 to 1/8" deeper into the fronts so that a slight lug (projection) is formed. After the drawer is assembled, this lug is trimmed to form a good fit.

Figure 19-37. Lavatory cabinet in a bathroom area. Left—General view of framing with facing applied. Right—Close-up view showing drawer guides. The back of the drawer will slide along the kicker. It prevents the drawer from tilting downward when opened.

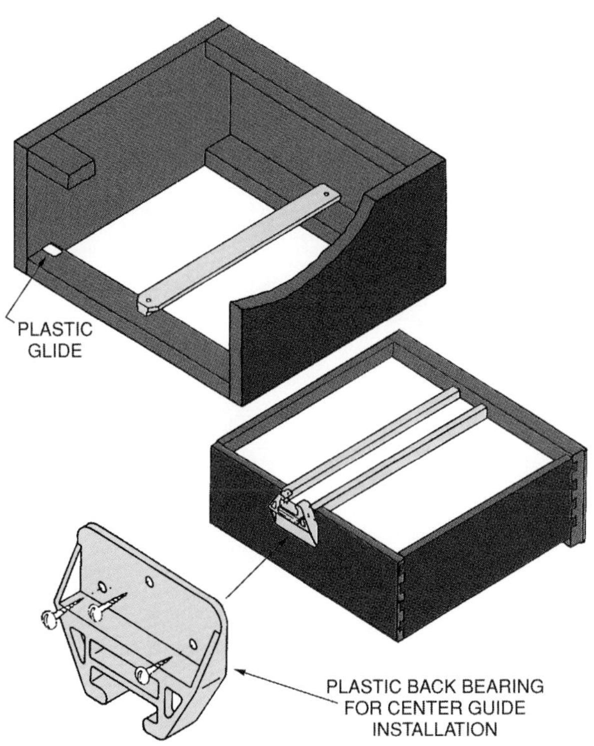

Figure 19-38. This center guide installation uses a patented back bearing. (Ronthor Plastics Div., U.S. Mfg. Corp.)

Constructing a Drawer

1. Select the material for the fronts. Grain patterns should match or blend with each other. Solid stock or plywood may be used. A variety of prefabricated drawer fronts are offered in many standard sizes. See Figure 19-41.
2. Cut the drawer fronts to the size of the opening (if flush type) allowing a 1/16" clearance on each side and on the top. This clearance will vary depending on the depth of the drawer and the kind of material. Deep drawers with solid fronts will require greater vertical clearance. For lip drawers, add the depth of the rabbets to the dimensions.
3. Select and prepare the stock for the sides and back. A less expensive hardwood or a softwood can be used. Plywood is usually not satisfactory for these parts. The surfaces should be sanded either before or after cutting to length.
4. Select material for the bottom. Hardboard or plywood should be used. Trim to final size after the joints for the other drawer members are cut.
5. Cut groove for the bottom in the front and sides.
6. Cut joints in the drawer fronts that will hold the drawer sides. Figure 19-42 shows a sequence for cutting a locked joint. For drawer fronts with a lip, cut the rabbet first, then the joint.
7. Cut the matching joint in the drawer sides. Be sure to cut a left and right side for each drawer.
8. Cut the required joints for the drawer sides and backs.
9. Trim the bottom to the correct size. Make a trial assembly as shown in Figure 19-43. Make any adjustments in the fit that may be needed. The parts should fit together smoothly. If they are too tight they will be difficult to assemble after glue is applied.
10. Disassemble and sand all parts. Rounding of the top edge of the sides can be done at this point. Stop the rounding about 1" from each end.
11. Make the final assembly. Any of a number of procedures can be used. One method is to first glue the bottom onto the front and then glue one of the front corners as shown in Figure 19-44. Be sure the bottom is centered. The side grooves are usually not glued. Turn the drawer on its side and glue the back into the assembled side. Glue on the remaining side

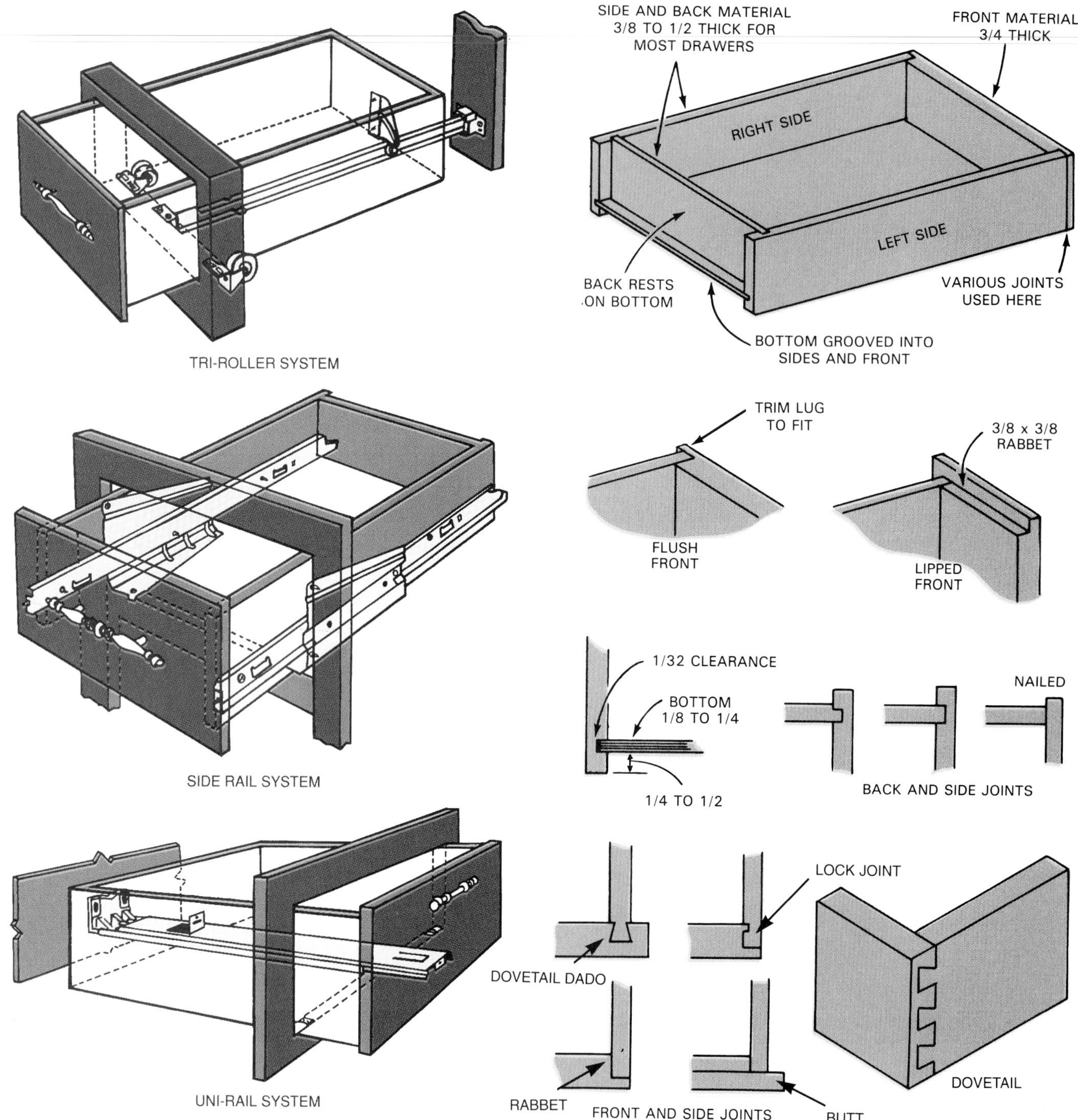

Figure 19-39. Manufactured drawer slide assemblies are available in several designs. (Amerock Corp.)

Figure 19-40. Standard drawer construction. Variety of joints may be used.

as shown. Clamps or a few nails can be used to hold the joints together until the glue has completely hardened.

12. Carefully check the drawer for squareness. Then drive one or two nails through the bottom into the back. If the bottom was carefully squared, you should have little trouble with this operation. Wipe off the excess glue.
13. After the glue has cured, fit each drawer to a particular opening. Trim and adjust drawer guides. Place an identifying number or letter on the underside of the bottom. This "label" will make it easy to return the drawer to its proper opening after sanding and finishing operations. The inside surfaces of quality drawers should always be sealed and waxed.

Shelves

Shelves are widely used in cabinetwork, especially in wall units. In some designs, it may be necessary to fit and glue the shelves into dados cut into the sides of the cabinet.

Figure 19-41. Prefabricated drawer fronts are made in many designs. Top—Raised panel effect. Bottom—Ogee edge. (Frank Paxton Lumber Co.)

Figure 19-42. Cutting lock joint in drawer front. Left—Cut groove first. Right—Trim tongue to length in second cut.

(Refer again to Figure 19-30.) This adds strength to the joint.

If possible, make the shelves adjustable. Then the storage space can be used for various purposes.

Figure 19-45 presents several methods of installing adjustable supports. Lay out the shelf support system carefully and accurately so the shelves will be level. Usually, it is best to do the cutting and drilling before the cabinet is assembled. If shelf standards are the type set in a groove, it is absolutely necessary to cut the groove before assembly. Some patented adjustable shelf supports are designed to mount on the surface.

Standard 3/4" shelving should be supported every 42" or closer. This applies especially to shelves that will carry heavy loads. The front edge of plywood shelving should be overlaid with a strip of wood material that matches the cabinet wood. This may be solid stock or thin strips which may be glued to the plywood edge.

Doors

Either swinging or sliding doors may be used with built-in cabinets. Swinging doors are commonly used for kitchen cabinets. Figure 19-46 shows the six basic styles of cabinet doors.

Doors may be made of plywood, preferably with a lumber core, or they may consist of a frame with a panel insert. Today, fine cabinet doors often have a frame covered on each side with thin plywood. This construction is similar to a hollow core door. Because of its tendency to warp, solid stock is seldom used, except on very small doors.

Many carpenters use prefabricated cabinet doors. These are available in various styles and in a wide range of

Figure 19-43. Top—Make trial assembly of the drawer parts to be certain they fit. Bottom—Parts are laid out and ready to be glued.

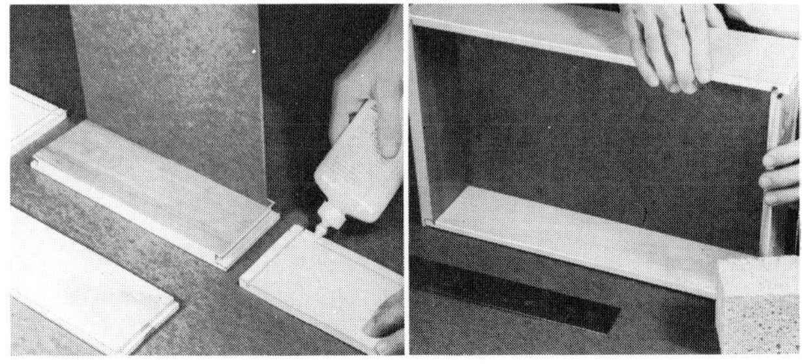

Figure 19-44. Gluing drawer parts together. Left—Bottom in place and glue being applied to a front corner. Right—Placing the second side in position after drawer has been turned on its side.

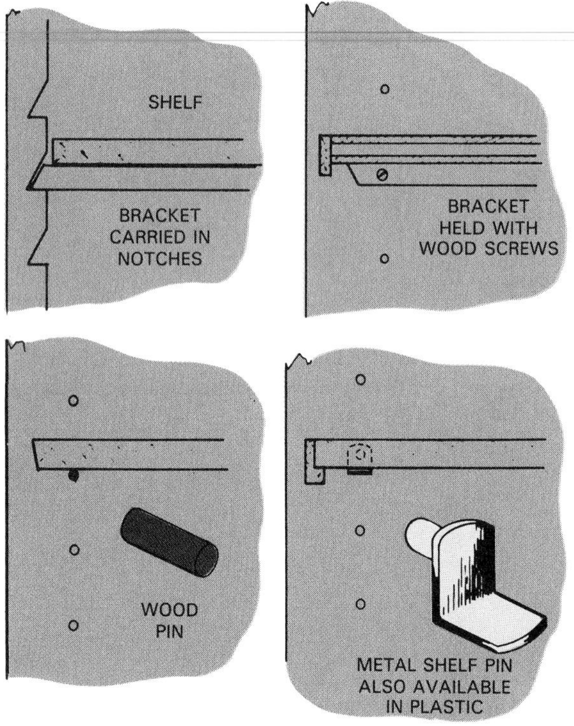

Figure 19-45. Methods for supporting shelves. When plywood is used for shelving materials, it should be faced with solid stock.

stock sizes. Doors are delivered to the job, cut to the exact dimensions with lipped edges (if specified) already machined.

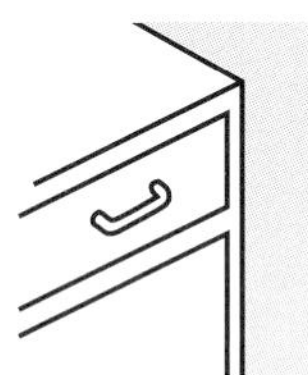

When specifying the size of the opening or the size of a cabinet door, always list width first and height second. This is standard practice among carpenters and is commonly followed by lumber suppliers and millwork plants.

Some manufacturers offer panel doors with special "see-through" inserts, Figure 19-47. These are useful for display of items on cabinet shelves. Simulated lead glass and expanded metal inserts are of this type.

Swinging doors for cabinets are of three basic types depending on how they fit into or over the door opening. Figure 19-48 shows the basic shapes of the three types:

- *Flush door*—The door fits into the opening and does not project outward beyond the frame.
- *Overlay door*—Though its edges are square like the flush door, it is mounted on the outside of the frame wholly or partly concealing it.
- *Lipped door*—The door is rabbeted along all edges so that part of the door is inside the door frame. A lip extends over the frame on all sides concealing the opening.

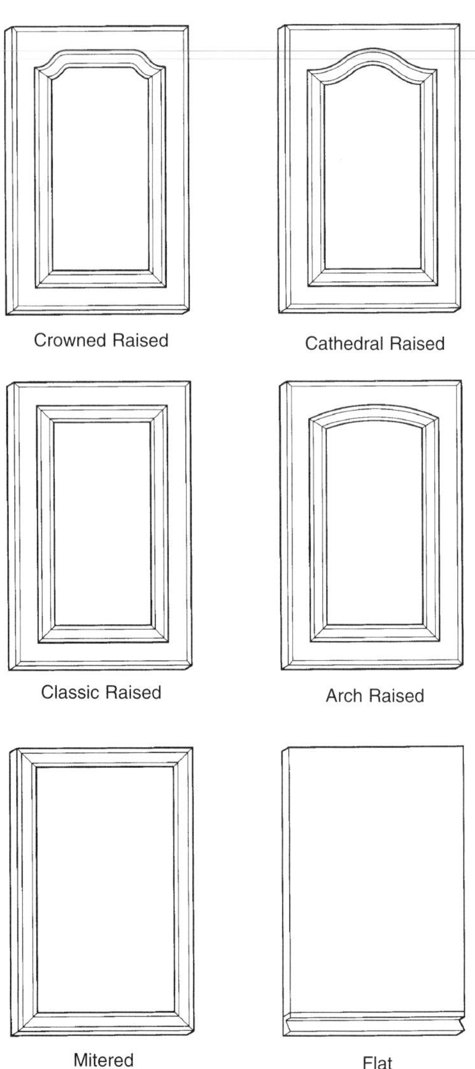

Figure 19-46. These six basic cabinet door styles allow a homeowner to make a personal statement, from traditional to Southwest to neoclassic. (Merillat Industries, Inc.)

Installing Flush Doors

The flush door is usually installed with butt hinges. However, surface hinges, wrap-around hinges, knife hinges, or various semiconcealed hinges can be used. Figure 19-49 shows several styles that are good for cabinetwork.

The size of a hinge is determined by its length and width when open. Select the size to suit the size of the door. Large wardrobe or storage cabinet doors should have three hinges.

Flush doors must be carefully fitted to the opening with about 1/16" clearance on each edge. The total gain (space) required for the hinge can be cut entirely in the door. However, for fine work, it is best to cut equally into door and stile. On large doors, it is practical to use the portable router for this operation.

Figure 19-47. Door inserts of simulated lead glass and expanded metal allow display of items stored on cabinet shelves. (Brammer Mfg. Co.)

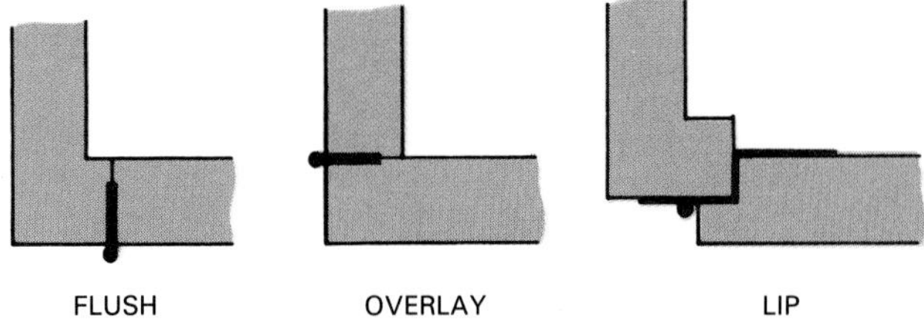

Figure 19-48. Cabinet doors are of three basic types.

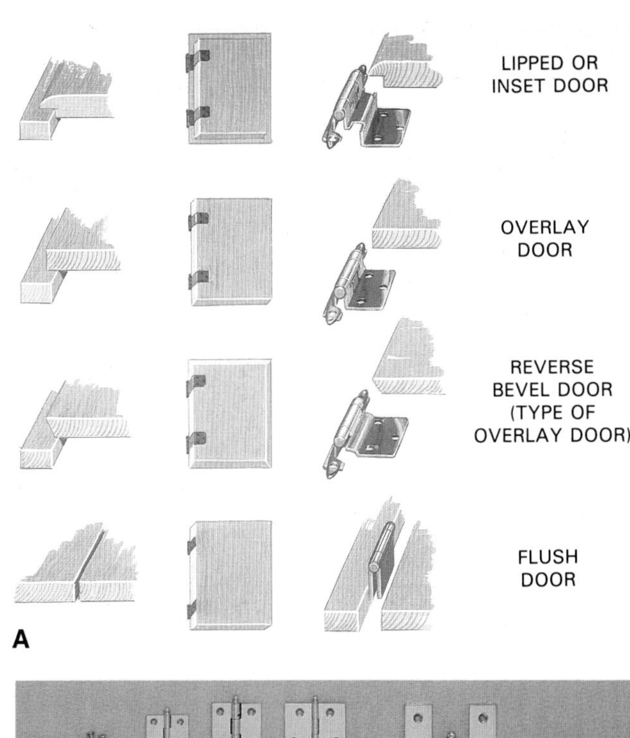

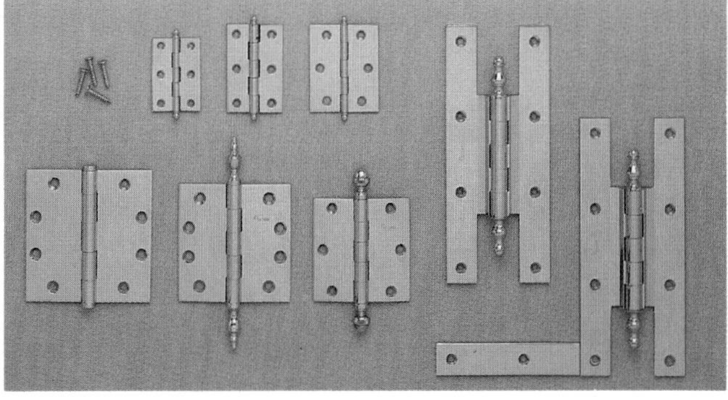

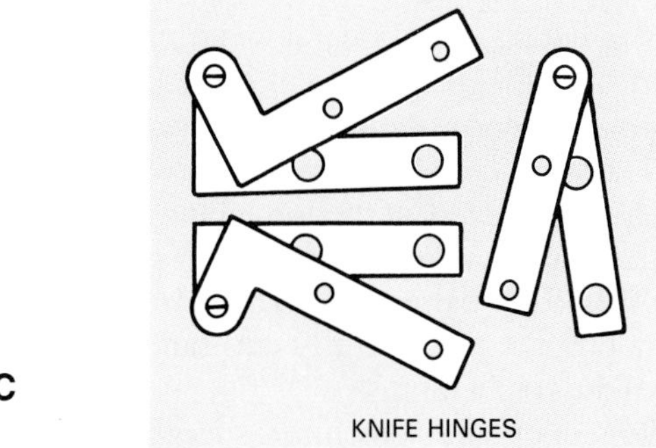

Figure 19-49. Various hinge types. A—Different hinges are used for different doors. B—Brass hinges for flush doors. C—Knife hinges (Baldwin Hardware Corp. and Amerock Corp.)

Install the hinges on the door and then mount the door in the opening. Use only one screw in each hinge leaf. Adjust the fit and then set the remaining screws. It may be necessary to plane a slight bevel on the door edge opposite the hinges.

Stops are set on the door frame so the door will be held flush with the surface of the opening when closed. The stops may be placed all around the opening or only on the lock or catch side.

Overlay doors provide an attractive appearance in some contemporary styles of cabinetwork. Butt hinges can be used; however, it is usually best to make the installation with a type that is designed for the purpose, Figure 19-50.

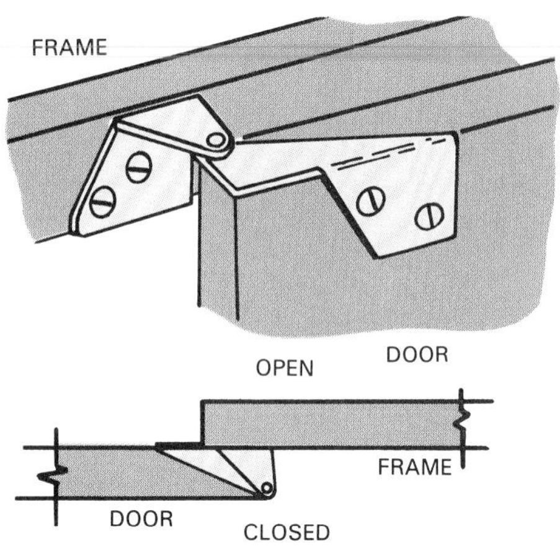

Figure 19-50. This pivot hinge is designed for use on overlay doors.

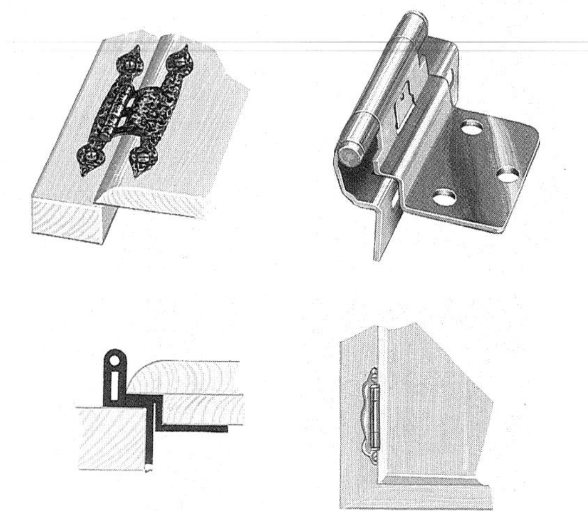

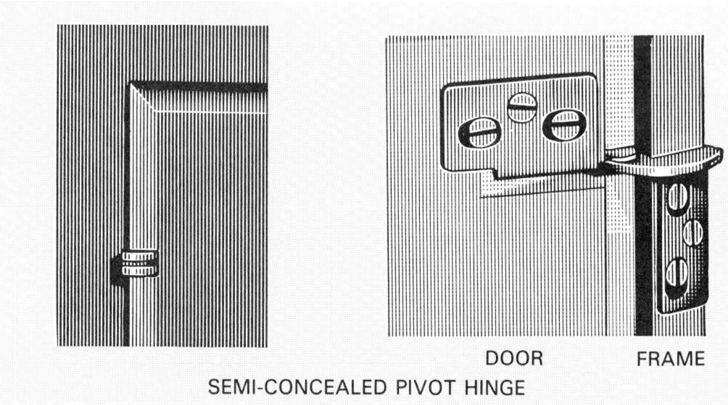

Figure 19-51. Hinges for lipped doors. Note how pivot hinges conceal all but the pin. (Amerock Corp.)

Cutting and Fitting Lipped Doors

Lipped doors are easier to cut and fit than flush doors because the clearance is covered. They must be installed with a special offset hinge. See two styles in Figure 19-51.

Select the door blanks for a given series of openings. Try to match grain and color when a natural finish will be applied. Cut the doors to the size of the openings plus the amount required for the lip all around. Allow clearance of about 1/16" on each edge. Make an additional allowance for the hinges which are not gained into the door or frame.

The lip is formed by cutting a rabbet along the edge. This can be accomplished on the table saw or with a shaper.

Check the fit of each door in its proper opening and mark its position in small letters on the back. Remove all machine marks with abrasive paper.

Soften edges and corners. Hardware may be prefitted and then removed before the finishing operations. Figure 19-52 shows lip doors hung after finish has been applied.

Figure 19-52. Lipped doors are installed on this base cabinet.

Sliding Doors

Sliding doors are often used where the swing of regular doors would be awkward or cause interference. They are adaptable to various styles and structural designs, Figure 19-53.

Construction details of a sliding door arrangement are shown in Figure 19-54. Grooves are cut in the top and bottom of the cabinet before assembly. The doors are sometimes rabbeted so the edge formed will match the groove with about 1/16" clearance. Cut the top rabbet and groove deep enough that the door can be inserted or removed by simply raising it into the extra space. The doors will slide smoothly if the grooves are carefully cut, sanded, sealed, and waxed. Avoid too much finish.

Sliding glass doors are sometimes specified in cabinets. They are heavy and require a special plastic or roller track. Follow the manufacturer's recommendations for installation.

A wide range of sliding door track and rollers are available. Figure 19-55 illustrates a self-lubricating plastic track. It is easy to install and provides smooth operation for furniture and cabinet doors. Overhead track and rollers are used for large wardrobe doors and passage doors.

Sliding glass doors are adaptable to wall units and add variety to an installation. Figure 19-56 shows bypass (1/4" plate glass) sliding doors for a cabinet in a kitchen-dinette area.

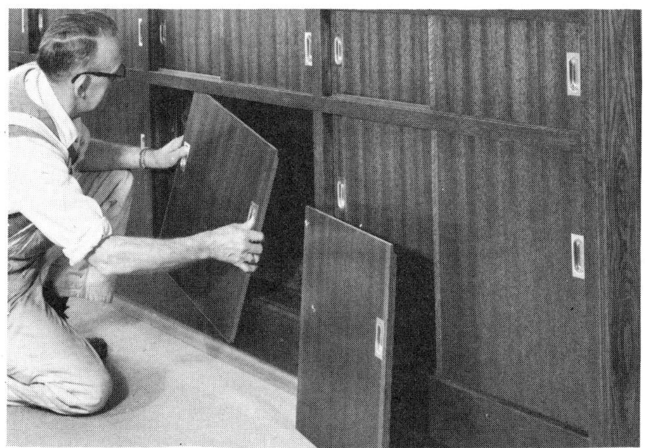

Figure 19-53. Sliding doors in a storage cabinet. They can be adapted to various styles.

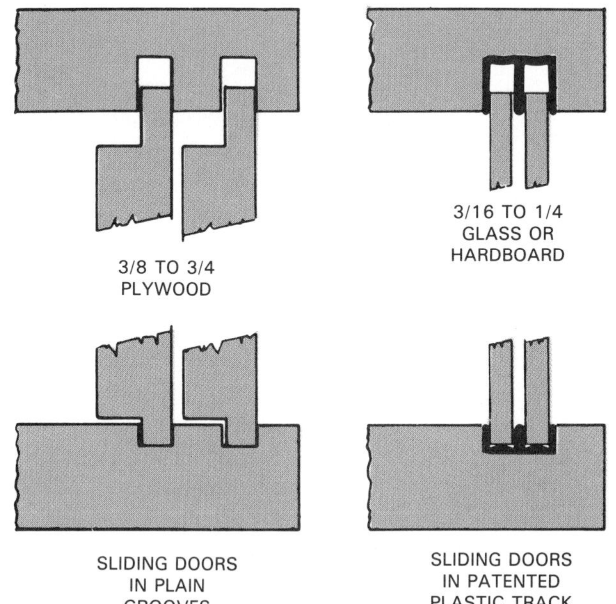

Figure 19-54. Sliding door details. Thicker door stock must be rabbeted to fit the track.

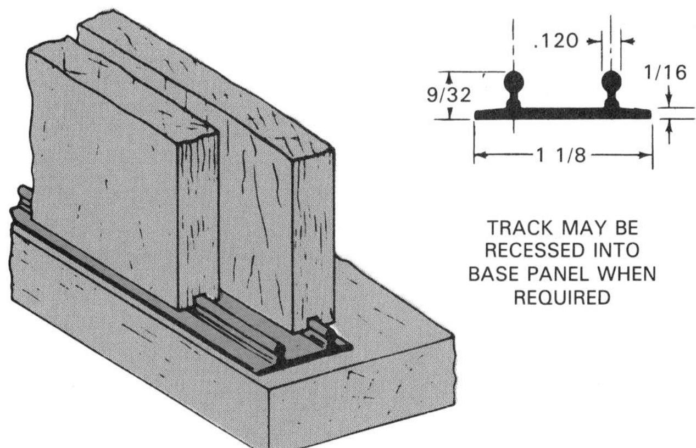

Figure 19-55. Plastic door track is self-lubricating. (Kentron Div. North American Reiss)

Figure 19-56. Sliding plate glass doors are featured in a wall cabinet.

Counters and Tops

In modern cabinetwork, a high-pressure type of *plastic laminate* is commonly used as a surface for counters and tops. This material, usually 1/16" thick, offers high resistance to wear and is unharmed by boiling water, alcohol, oil, grease, and ordinary household chemicals, Figure 19-57.

Although the laminate is very hard, it does not possess great strength and is serviceable only when bonded to plywood, particleboard, waferwood, or hardboard. This base or core material must be smooth and dimensionally stable. Hardwood plywood (usually 3/4" thick) makes a satisfactory base. However, some plywoods, especially fir, have a coarse grain texture which may telegraph (show through). Particleboard, which is less expensive than plywood, provides a smooth surface and adequate strength.

When the core or base is free to move and is not supported by other parts of the structure, the laminated surface may warp. This can be counteracted by bonding a backing sheet of the laminate to the second face. It will minimize moisture gain or loss and provides a balanced unit with

Figure 19-57. Heat resistance factors make plastic laminates an ideal surface material for counters with built-in cooking units. (Formica Corp.)

identical materials on either side of the core. For a premium grade of cabinetwork, Architectural Woodwork Institute standards specify that a *backing sheet* be used on any unsupported area exceeding 4 sq. ft. Backing sheets are like the regular laminate without the decorative finish and are usually thinner. A standard thickness for use opposite a .060" (1/16") face laminate is .020".

Working Laminates

Plastic laminates can be cut to rough size with a handsaw, table saw, portable saw, or portable router, Figure 19-58. Use fine-tooth blades and support the material close to the cut. Laminates 1/32" thick, which are used on vertical surfaces, can be cut with tin snips.

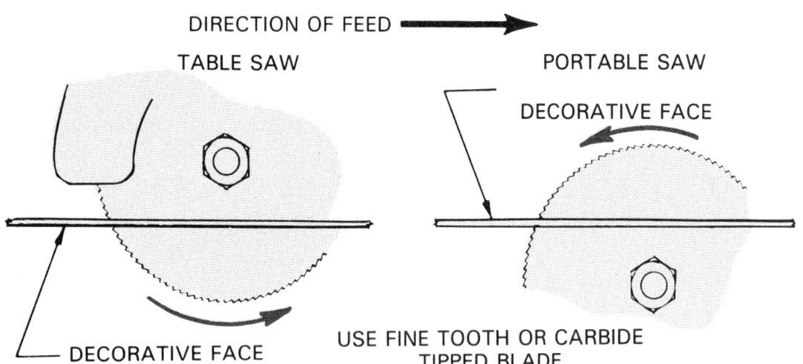

Figure 19-58. Plastic laminate can be cut with circular saws.

It is best to make the roughing cuts 1/8 to 1/4" oversize. Trim the edges after the laminate has been mounted. Handle large sheets carefully because they can be easily cracked or broken. Be careful not to scratch the decorative side.

Contact bond cement is used to apply the plastic laminate because no sustained pressure is required. It is applied with a spreader, roller, or brush to both surfaces being joined. On large horizontal surfaces, it is best to use a spreader, Figure 19-59.

For soft plywoods, particleboard, or other porous surfaces, the spreader is held with the serrated edge perpendicular to the surface. On hard, nonporous surfaces, and plastic laminate, hold the edge at a 45° angle, as shown. A single coat should be sufficient.

An animal hair or fiber brush may be used to apply the adhesive to small surfaces or those in a vertical position. Apply one coat, let it dry thoroughly and then apply a second.

All of the surface should be completely covered with a glossy film. Dull spots, after drying, indicate that the application was too thin and that another coat should be applied.

Stir the adhesive thoroughly before using and follow the manufacturer's recommendations. Usually, brushes and applicators must be cleaned in a special solvent.

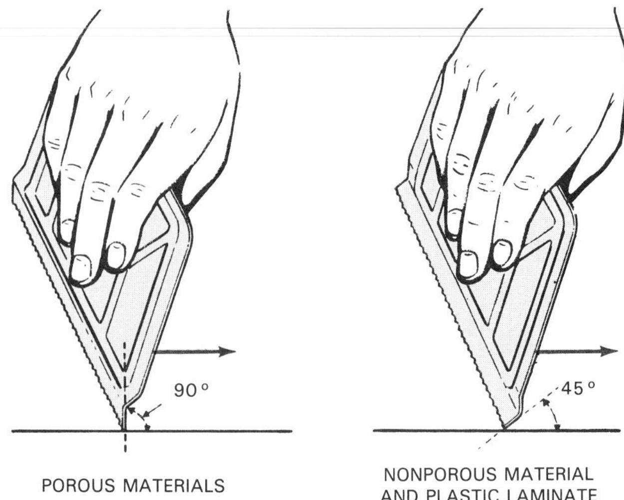

Figure 19-59. Metal spreader may be used to apply contact bond cement.

Some types of contact cement are extremely flammable. Nonflammable types may produce harmful vapors. Be sure the work area is well ventilated. Follow the manufacturer's directions and observe precautions.

After the adhesive has been applied to both surfaces, let it dry (usually at least 15 minutes). You can test the dryness by pressing a piece of paper lightly against the coated surface. If no adhesive sticks to the paper, it is ready to be bonded. This bond can usually be made any time within an hour. Time varies with different manufacturers. If the assembly cannot be made within this time, the adhesive can be reactivated by applying another thin coat.

Adhering Laminates

Bring the two surfaces together in the exact position required because they cannot be shifted once contact is made. When joining large surfaces, place a sheet of heavy wrapping paper, called a *slip-sheet*, over the base surface. Then slide the laminate into position. Withdraw the paper slightly so one edge can be bonded and then remove the entire sheet and apply pressure.

Total bond is secured by the application of momentary pressure. Hand rolling provides satisfactory results if the roller is small (3" or less in length), Figure 19-60. Long rollers apply less pressure per square inch. Work from the center to the outside edges and be certain to roll the entire surface. In corners and areas that are hard to roll, hold a block of soft wood on the surface and tap it with a rubber mallet.

Trimming and smoothing the edges is an important step in the application of a plastic laminate. A plane or file may be used, but many carpenters prefer an electric router, equipped with an adjustable guide, Figure 19-61. When

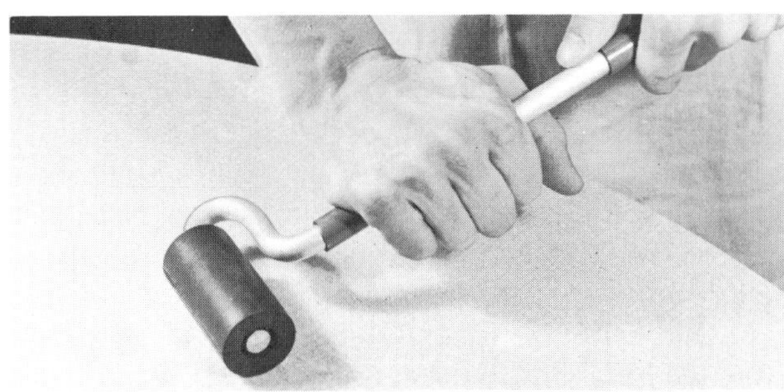

Figure 19-60. Apply pressure to the laminate with a hand roller.

Figure 19-61. Using an electric router to trim an overhanging edge. The guide ensures an accurate cut. (Black & Decker Mfg. Co.)

making cutouts for sinks or other openings, an electric saber saw, as illustrated in Figure 19-62, is practical.

The corner and edges of a plastic laminate application should be beveled, Figure 19-63. This makes it smooth to touch and durable. The bevel angle can be formed with a smooth mill file, Figure 19-64. Stroke the file downward and use care not to damage the surfaces of the laminate.

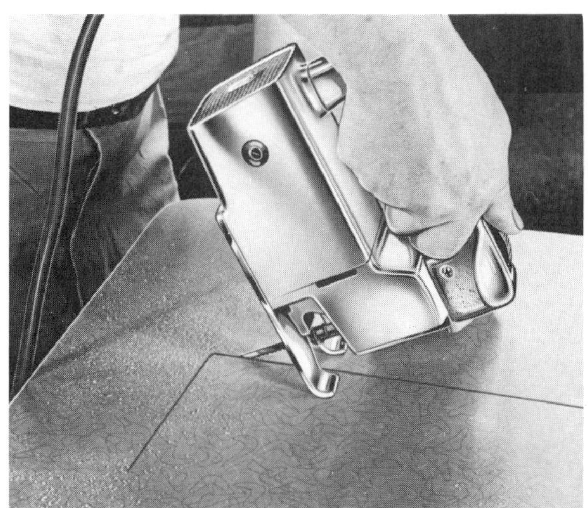

Figure 19-62. Using an electric saber saw to cut an opening for a sink. View shows start of cut.

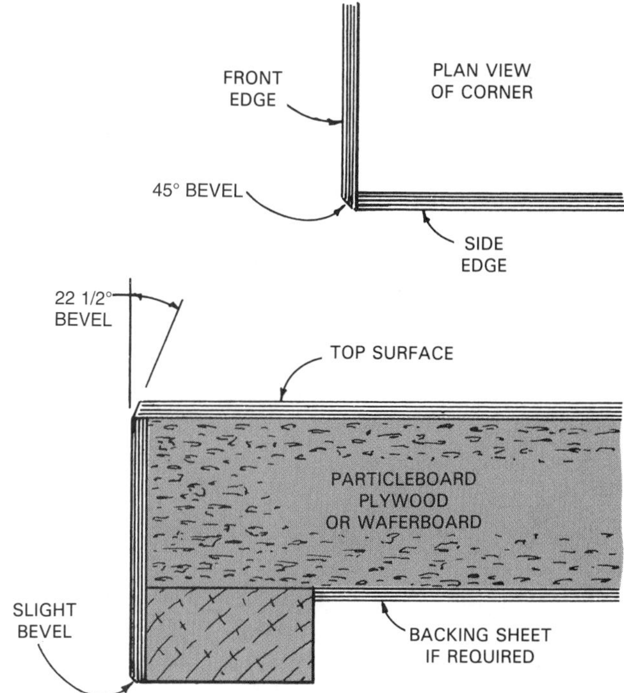

Figure 19-63. Bevels for plastic laminate corners are very important in the production of quality work.

Figure 19-64. A mill file can be used to produce a beveled corner on plastic laminate. (Bullard Haven Vocational-Technical School)

Some routers can be equipped with an adjustable guide and a special bit that will make this cut, Figure 19-65. Final smoothing and a slight rounding of the bevel should be done with a 400 wet-or-dry abrasive paper.

When working with plastic laminates, be especially careful that files, edge tools, or abrasive papers do not

Figure 19-65. Special router bit also makes a beveled corner. (Black & Decker Mfg. Co.)

damage the finished surfaces. Such damage is difficult to repair.

A completed countertop for the bathroom built-in unit pictured previously in construction views is shown in Figure 19-66. A special metal cove and cap strip have been used to trim the "back splash." FHA requires a minimum back splash height of 4" where kitchen counters join the wall surface.

Figure 19-66. Plastic laminate counter surface is completed for a bathroom built-in unit.

Cabinet Hardware

After counters and tops have been covered and surface finishes applied, hardware is installed. This consists of knobs, pulls, and various other metal fittings. Care should be used in the selection of an appropriate size, style, material, and finish. Figure 19-67 shows several designs of quality hardware for drawers and doors.

Some hardware (mainly hinges) should be prefitted before the final surface finish is applied. Pulls and catches can be installed afterward. Use care in the layout. Drill holes with sharp bits that will not splinter the surface. Drilling jigs, like the one shown in Figure 19-68, are fast and accurate.

Drawer pulls usually look best when they are located slightly above the centerline of the drawer front. Normally, they should be centered horizontally and level. *Door pulls* are convenient when located on the bottom third of wall cabinets and the top third of base cabinets.

Swinging doors require some type of catch to keep them closed. Several common types are illustrated in Figure 19-69. An important consideration in their selection is the noise they produce. In general, the catch should be placed as near as practical to the door pull. The package usually contains instructions for installation. Follow these carefully.

Other Built-In Units

Previous descriptions have been largely directed toward kitchen cabinets. The same general procedures can be applied to wardrobes, room dividers, and various built-in units for living rooms and family rooms. See Figure 19-70.

Built-in features are popular for these reasons:

- The demand for storage space.
- The general requirement that all space be organized and used as efficiently as possible.

Figure 19-67. Cabinet drawer and door hardware. (Amerock Corp.)

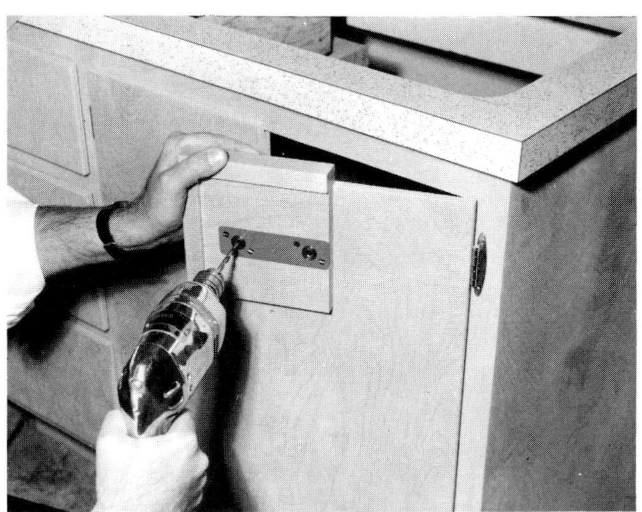

Figure 19-68. A drilling jig can be used to locate holes for door pulls.

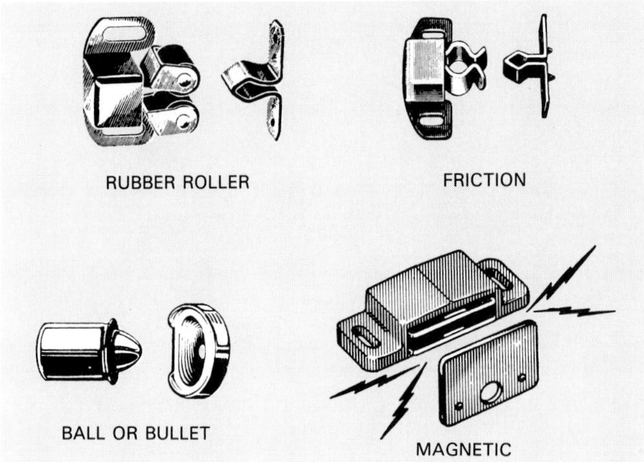

Figure 19-69. Cabinet doors catches use various devices to hold doors shut.

- The attractive customized appearance that can result from well designed and carefully constructed units.

Study the details of the construction provided in the architectural plans or develop a carefully prepared working drawing using the actual dimensions secured from the wall, floor, and ceiling surfaces. When drawings are provided, check the space available to make sure that it agrees with the drawings.

Wall and floor surfaces are seldom perfectly level or plumb. Slight adjustments will be needed as the cabinetwork is constructed. Room corners or the corners of alcoves designed for built-in units will likely not be square. Here, the cabinetwork will need to be trimmed or strips and wedges used to bring the work into proper alignment.

Keep a square and level constantly at hand, especially during the rough-in stages, Figure 19-71. Do not carry inaccuracies from the wall or floor into the cabinet framework. Doing so will make it difficult to keep the cabinet facing plumb and level. Errors will also be a source of annoyance during the hanging of doors and fitting of drawers.

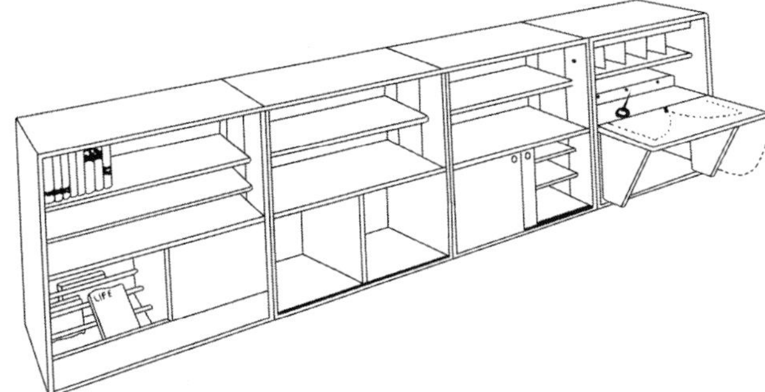

Figure 19-70. Built-in units for living rooms add storage and work areas.

Built-in units vary in so many ways that no particular procedure can be recommended for their construction. For example: a bookcase that also serves as a room divider, Figure 19-72, could be constructed during the regular interior finishing stages, as shown. It could also be built as a detachable unit and installed just before interior painting and decorating. Figure 19-73 shows a partial view of built-in units located in a bedroom area. Here the various components (headboard, bookcase, desk) were constructed and surface finishes applied in a basement area. Then they were moved into the room and attached to the walls after walls were painted and the carpet was laid.

Sequence of Interior Finish

Interior finish, after wall, ceiling, and floor surfaces are complete, involves a wide variety of work as described in this unit, Unit 18, and Unit 20. The sequence in which the work is performed may vary considerably. Much depends upon the type of materials and construction used. The carpenter should organize the work to prevent bottlenecks (delays). Appropriate times should be established for the delivery of materials. When delivered too far ahead, they will interfere with other work and may be damaged. Work schedules need to be carefully planned and followed as closely as possible. The application of trim and the construction of cabinetwork generally proceed at the same time.

Figure 19-74 illustrates the sequence on a job where assembled-in-place cabinetwork was used. The top view shows the basic framework for a built-in desk in a kitchen area. It is installed on the underlayment and before the baseboard is set. In the center view, the facing and top have been added. Note that the baseboard is in position. Drawers are built and fitted and then surface finishes are applied to the desk, baseboard, and other trim. The plastic laminate top surface is installed. Then, the floor covering is laid. Finally, as shown in the bottom view, hardware is fitted and the base shoe is set in place.

Figure 19-71. This built-in unit will serve as a wardrobe and room divider in a bedroom area. (American Plywood Assoc.)

Figure 19-72. Attaching facing strips (stiles) to a bookcase. This unit could serve as a room divider, as well.

Figure 19-73. Built-in bookcase, desk, and storage in a bedroom area. Plastic laminate top matches grain pattern and color of wood. Door features a woven wood pattern.

Figure 19-74. Sequence of interior finish. Top—Basic framing for a built-in desk. Center—Facing and top complete. Bottom—Drawers fitted, plastic laminate applied, surface finish applied, and floor covering laid.

Important Terms

Backing sheet
Base level line
Built-in
Center guides
Door pulls
Drawer guides
Drawer pulls
Face frame
Face-frame construction
Face plate
Facing strips
Flush door
Flush drawers
Frame construction
Frameless construction
Frame with cover construction
Kicker
Lip drawers
Lipped door
Master layout
Modular
Overlay door
Plastic laminate
Rails
Shelves
Skeleton
Sliding doors
Slip-sheet
Stiles
32 mm construction system

Test Your Knowledge

1. List the three types of cabinetwork used in modern homes.
2. Dimensions of cabinets are usually found in the ____________ for the house.
3. In cabinetry using ____________ construction, front edges are not face framed.
4. Central to the 32 mm system are two vertical rows of 6 mm holes drilled in side panels. True or False?
5. To assure that cabinets will not sag or rack, the carpenter installing them will (select most correct answer):
   A. Check the cabinets for squareness.
   B. Check doors and drawers for binding and fit.
   C. Check walls and floors for high spots.
   D. Check floors to see if they are level.
   E. Check walls for plumb.
6. A padded T-brace that supports wall cabinets during installation is known as a ____________.
7. Base cabinets are always installed before wall cabinets. True or False?
8. The base unit of a standard kitchen cabinet is ____________ inches high.
9. When carpenters draw a full-size sectional view of a cabinet, they generally refer to it as a ____________.
10. When building assembled-in-place base cabinets, the partitions are installed ____________ (*before, after*) the bottom is secured to the base.
11. The vertical members used to face a cabinet are called ____________.
12. Generally, stiles are installed ____________ (*before, after*) the rails on assembled-in-place cabinets.
13. Three common types of drawer guides include: side guides, corner guides, and ______________.
14. Stock for kitchen cabinet drawer sides is usually ______________ to ______________ inches thick.
15. Standard shelving that is 3/4" thick should be carried on supports spaced no greater than ______________ inches.
16. Three types of swinging doors that may be used in cabinetwork include: lip, flush, and ______________.
17. When making a plastic laminate installation on a table or counter, it is recommended that a ______________ be used on any unsupported area greater than four square feet.
18. When exerting pressure to bond a plastic laminate by hand, it is best to use a (*long, short*) roller.
19. Drawer pulls are usually located slightly ____________ (*above, below*) the centerline.

Outside Assignments

1. Using a straightedge or straight 2 x 4, check a wall and floor for high spots, plumbness, and level. Report findings to your instructor.
2. Secure a set of architectural plans that include detail drawings of the kitchen cabinets. After a study of the details, prepare a master layout of a typical base cabinet. Since a complete view would be rather large, you may prefer to show only the front 4-6". Use a sharp pencil. Show all the parts and clearances. Present your drawing to the class with explanations of the various structural parts.
3. If you have access to shop equipment, construct a typical cabinet drawer. Select appropriate joints that can be used for on-the-job built cabinets. Cut and fit the parts carefully. You may prefer not to glue the drawer together so that when you make a presentation to your class, it will be easier to show the joints. Another alternative is to glue the drawer, then cut it apart in several places so joint sections could be easily viewed.
4. Study the various methods and devices used to install sinks in countertops. From manufacturer's literature, prepare a sectional drawing about four times actual size that you can use in making a presentation to your class. Check into the possibility of borrowing sample rims and brackets from a building supply store that can be passed around to the group during your explanation.
5. Measure up a kitchen counter for a new one. Have your instructor check your measurements.
6. Study the sketch below of a typical kitchen. Produce a sketch of it on a 1/2"=1' grid, filling in the details on lights and receptacles. Select lengths for each wall and then sketch in a kitchen layout. From a manufacturer's catalog select units that will fit your sketch. Prepare a list of units that would need to be purchased. If prices are available, compute the cost.

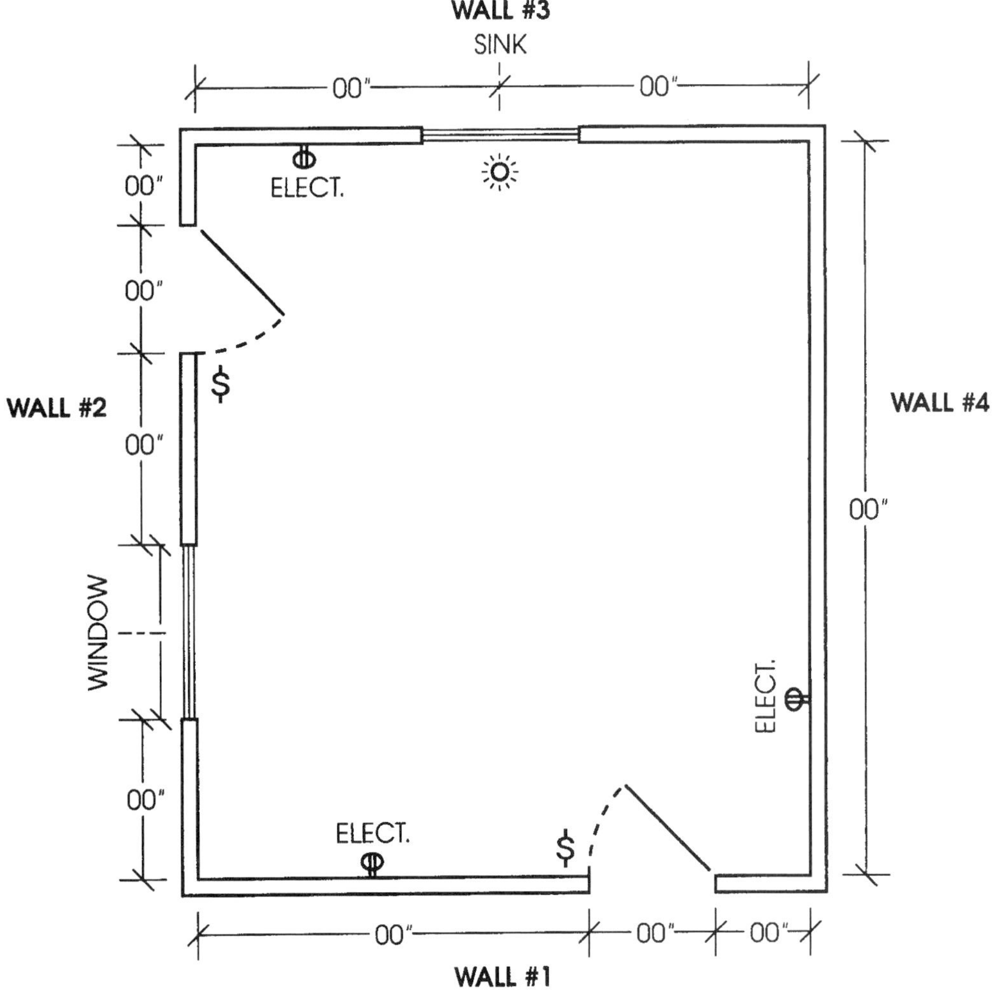

Kitchen floor plan for Problem 6. (KraftMaid)

20 Painting, Finishing, and Decorating

Wood and certain other covering materials used on the inside and outside of the home require protective coatings against soiling, rot, and other types of deterioration caused by the environment. Unfinished woods discolor, shrink, swell, check, and warp if left unprotected. Moreover, these coatings can improve the appearance of the covering materials and create more pleasant surroundings.

The term *coatings* covers all types of finishes, whether designed for wood or other materials, such as metals and drywall. Coatings include paints, stains, varnishes and various synthetic materials both clear and colored. Included would be *anodizing,* a coating process used to color and preserve the surface of aluminum.

Safety

Certain rules of safety apply to working with paints and finishes:

- **Wear safety glasses when applying finishing materials.**
- **Wear rubber gloves, goggles, and a rubber apron when applying bleaches and acids.**
- **When working with or spraying thinners and reducers such as naphtha, lacquer thinner, and enamel reducer, keep the work area well ventilated. The fumes are highly toxic.**
- **Store chemicals and soiled rags in safe containers.**
- **Wear an approved respirator when using toxic chemicals, Figure 20-1.**
- **Never smoke while sanding or applying finishes. Not only will mixtures of sanding dust and smoke create a health hazard but the combination could cause a fire.**
- **Wash the hands well after applying finishes to remove any toxic materials.**
- **Be sure there is a sink, shower, or eye wash station in case of a splash or spill that lands on the skin or in the eyes.**
- **Keep a fire extinguisher handy in the finishing area.**
- **When preparing older structures for repainting, a test for lead in the old paint should be made. Sanding of leaded paints requires special precautions to prevent breathing in the lead particles that would be released. Inadvertent ingestion of lead could cause serious health problems. Removal of lead should not be attempted except by professionals trained for this purpose.**

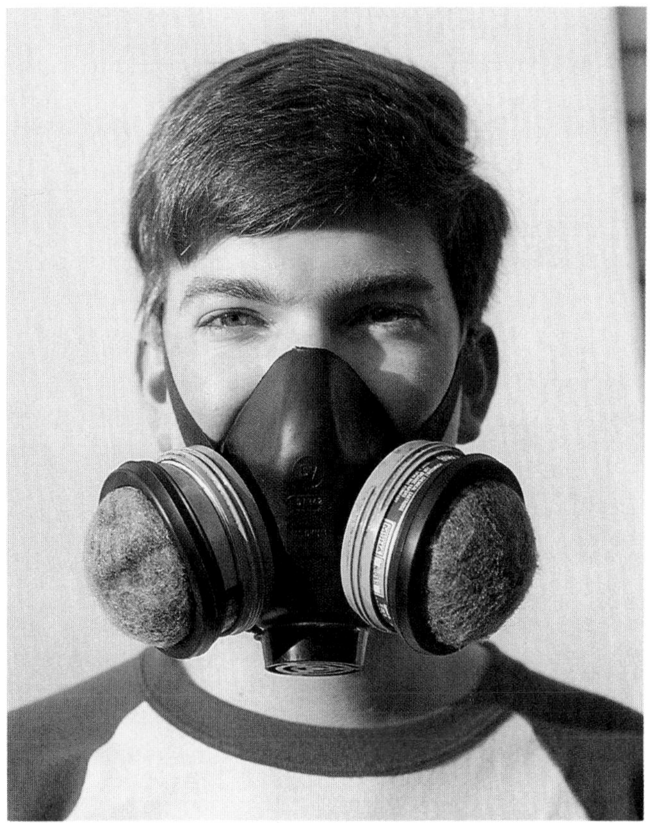

Figure 20-1. This respirator is approved for use with any painting task up to and including spraying of lacquer.

Painting and Finishing Tools

Various tools may be needed to tackle any painting or finishing task. These will include ladders, scaffolding, compressor, paint sprayer, sanders, brushes, rags, paint roller and pan, paint pads, tack rags, putty knives, and paint guards.

Brushes

Brushes are sold in many sizes and grades and for various purposes. The most common brushes are shown in Figure 20-2.

Brushes have several parts, as shown in Figure 20-3. The bristles used in brushes are either synthetic or animal hair. Nylon and polyester bristles, manufactured from petroleum products, are used widely for certain types of brushes. Natural-fiber brushes are made from the hair of hogs, especially Chinese hogs, which grow long hair. Hog bristles are oval in cross section and have naturally tapered and flagged (split) ends. These flagged tips provide paint-holding ability through capillary action.

Nylon and polyester bristles are flagged during manufacture. These bristles will outlast natural bristles and are especially suited for water-based finishes. Horse, oxen, and fitch hair are also used in brushes as is tampico, the fiber of cactus.

Quality brushes have a smooth taper from the ferrule to the tip. These are formed by varying the length of the hairs. Inside the brush, completely surrounded by the hair is a plug which may be wood, metal, fiber, or plastic. It creates a reservoir to hold the paint.

Wire brushes, Figure 20-4 are useful for cleaning plaster off subflooring, abrading rust from metal, and removing old peeling or flaking paint.

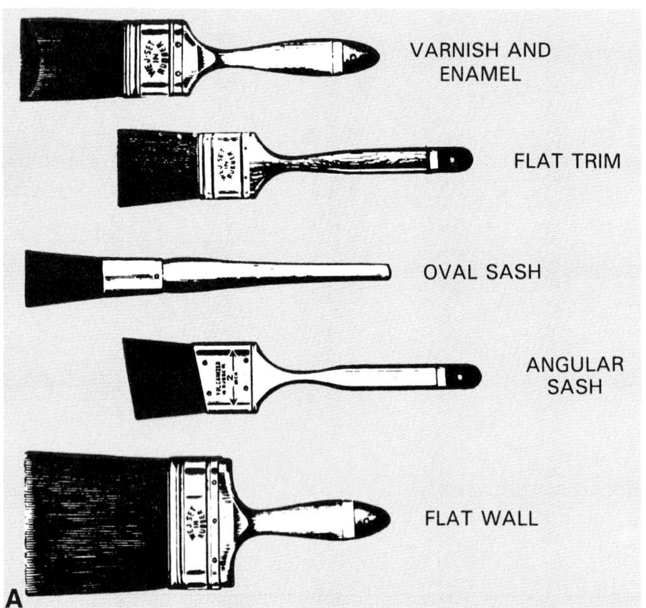

Figure 20-2. Brushes. A—Most painting or finishing tasks can be accomplished with these five kinds of brushes. B—Good painting and decorating stores carry a wide variety of brushes for all purposes. (John Dutcher Glass and Paint)

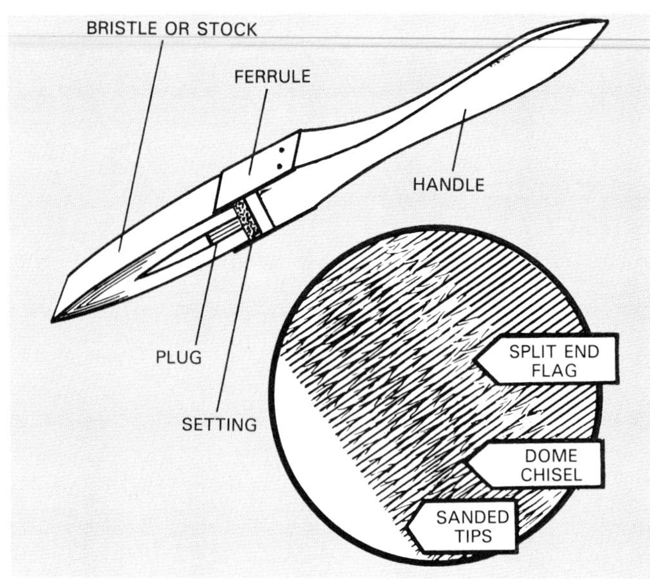

Figure 20-3. Parts of a brush. Inset shows the tip of a quality brush, pointing out its features.

Rollers, Pans, and Pads

Paint rollers are sold in various widths. They are designed to hold tube-shaped pads that are available in many different types, naps, and sizes. Figure 20-5 shows various rollers, roller covers, and pads. Rollers are ideal for painting large surfaces. They can be used to apply many types of finishes (interior and exterior). Applications include oil paint, water base paints, floor and deck paints, masonry paint, and aluminum finishes. For smooth surfaces, roller covers are made with a short nap; the rougher the surface to be painted, the longer the nap should be.

Pans for roller painting, Figure 20-6, have a sloped bottom with the deepest part designed to hold a supply of paint. The pan has a textured sloping surface for rolling off excess paint on the roller.

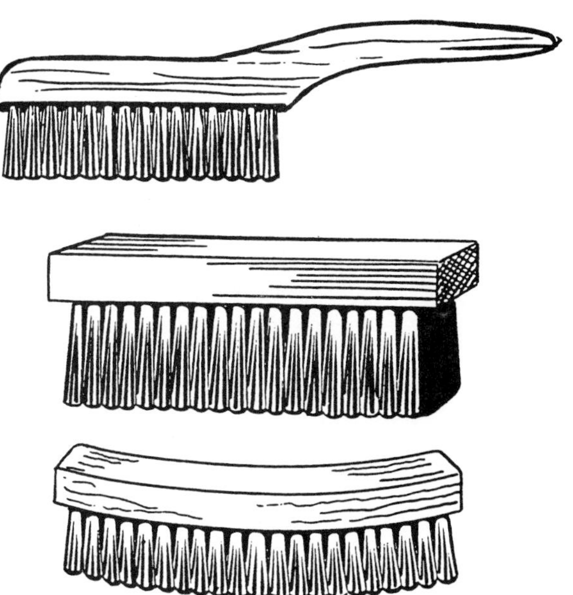

Figure 20-4. Wire brushes are useful for removing rust and other unwanted foreign materials from a surface to be painted.

Mechanical Spraying Equipment

Nearly all types of paint, stain, and varnish can be applied with a compressor and spray gun, Figure 20-7. It is the most practical and the quickest method of painting very large surfaces. See Figure 20-8.

Paint spray guns are either suction or pressure type. The suction type, Figure 20-9, works like an atomizer. It depends on air passing over a tube extending into the paint. This action sucks the paint into the air stream. A suction-feed gun is designed for spraying light-bodied materials such as shellacs, stains, varnishes, lacquers, and synthetic enamels. A pressure-feed gun diverts some air into the paint container. The pressure created forces finishing material into the air stream. Pressure-feed guns are intended for spraying heavy-bodied paints.

Spray guns may have either an internal mix or an external mix spraying head, Figure 20-10. The external mix head is used to apply fast-drying materials and provides greater control of the spray pattern. The internal mix head is used for slow-drying materials and is best used when air pressures are low.

Figure 20-5. Paint rollers. A—Rollers and roller coverings are made in different widths. Covers are designed in different naps for smooth to rough surfaces. B—Pads and mitts are also used for paint application.

Interior
Wall

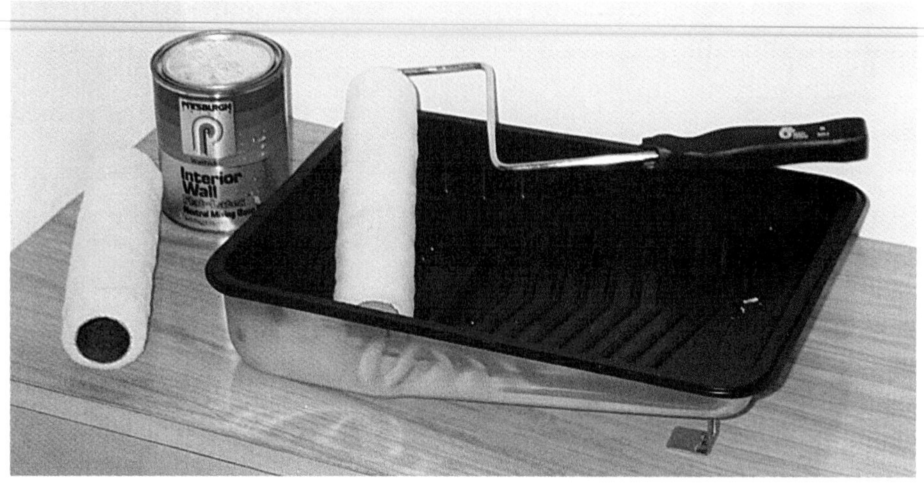

Figure 20-6. A painting pan is designed for even spreading of paint on a roller covering.

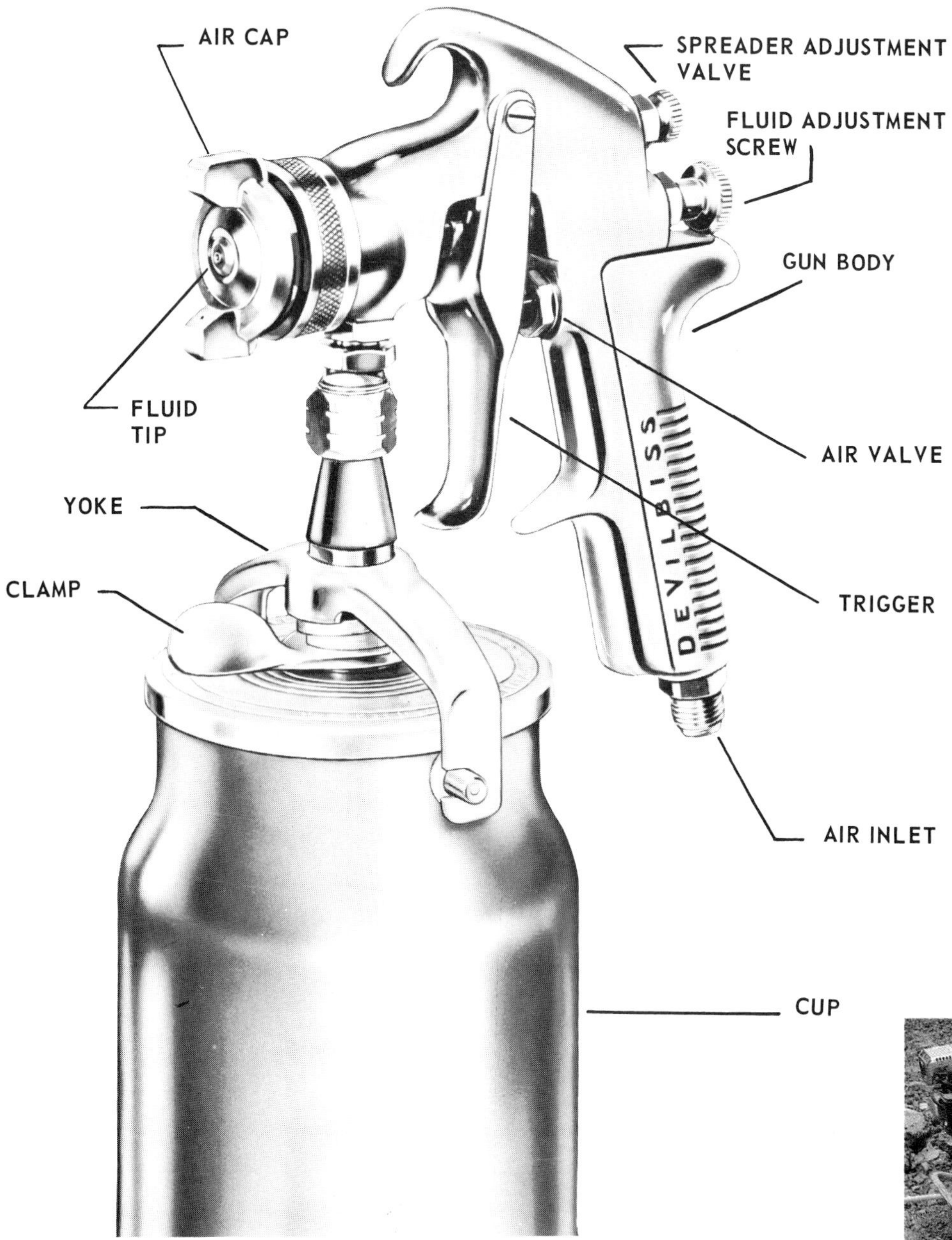

AIR CAP
SPREADER ADJUSTMENT
VALVE
FLUID ADJUSTMENT
SCREW
GUN BODY
FLUID
TIP
AIR VALVE
YOKE
TRIGGER
CLAMP
DEVILBISS
AIR INLET
CUP

Figure 20-7. Equipment for paint spraying.
Left—A spray gun uses compressed air to apply paint to large surfaces.
Bottom—Compressor may be powered by a gasoline engine like this unit or by an electric motor (where electric power is available).

BOSTITCH

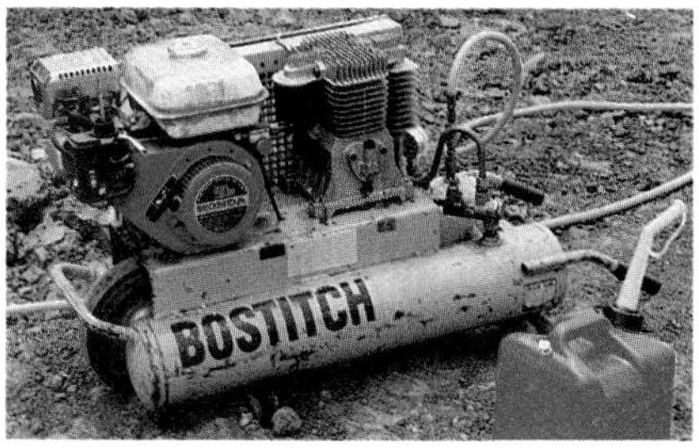

Figure 20-8. Spraying is a practical and efficient method of coating large surfaces with paint. (Greco Painting)

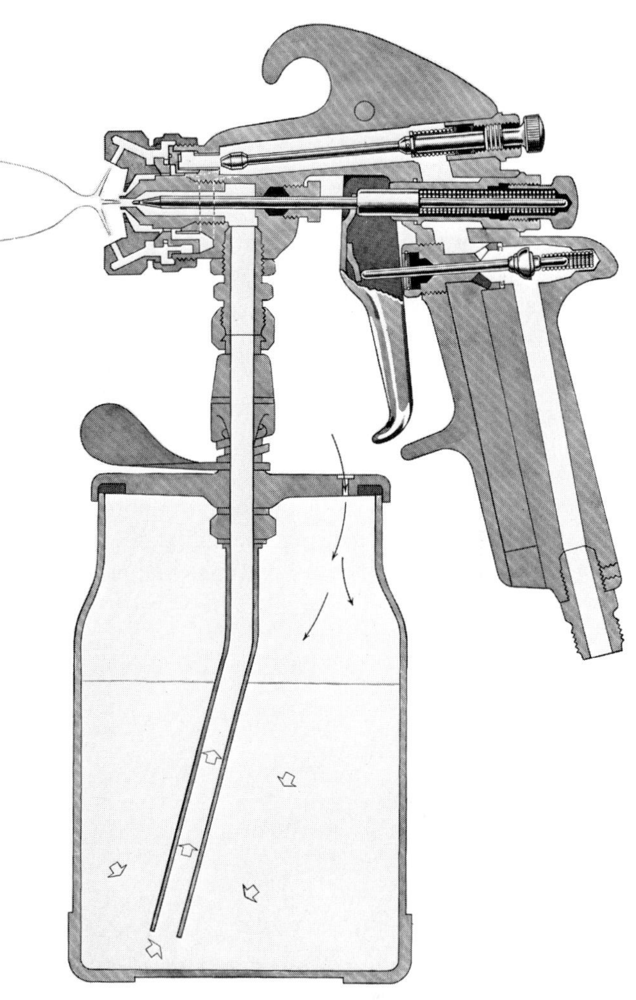

Figure 20-9. A section view of a nonbleeder, external-mix, suction-feed spray gun. (DeVilbiss Co.)

Great care must be taken to clean painting tools after each use. Dried paint may be difficult or even impossible to remove. Of course, such a tool is useless unless the paint can be removed.

Ladders and Scaffolds

Both ladders and scaffolds are widely used for painting and decorating. They provide a safe method of reaching high places with painting tools. New OSHA rules regarding safety in high places may apply to painting. Review information in Unit 29, Scaffolds and Ladders.

Sanders

Electric sanders speed the work of removing old paint, foreign materials, or roughness from a surface, Figure 20-11. Belt sanders are useful for removing material rapidly. Orbital and vibrating sanders are best for fine smoothing,

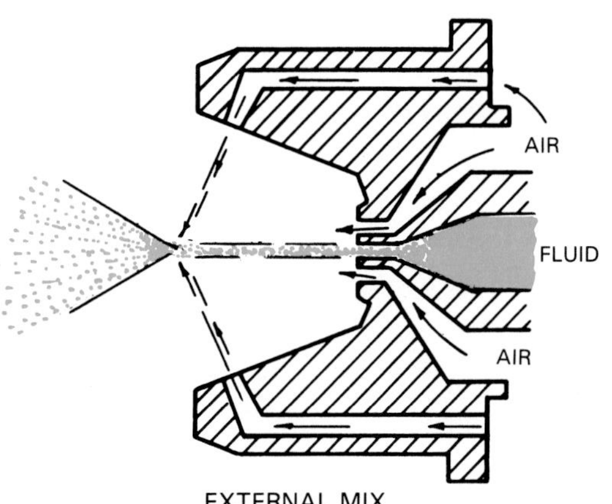

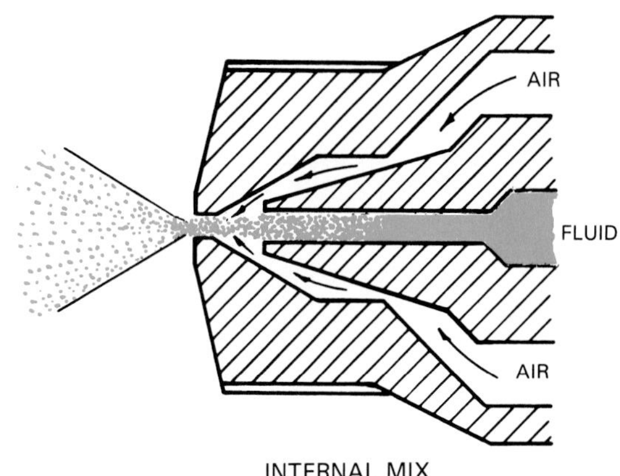

Figure 20-10. There are two basic types of spray heads. The head, or nozzle, consists of an air cap and a fluid tip.

Figure 20-12. New shapes and sizes make most sanding locations accessible. Hand sanders are available, as well.

Painting, Finishing, and Decorating Materials

Materials for painting, varnishing, and paperhanging include paints, vehicles or binders, solvents and thinners, varnishes, stains, fillers, sealers, bleaches, pumice, sandpaper, steel wool, wallpaper and wallpaper paste.

Paints, Varnishes, and Stains

Paints, Figure 20-13, varnishes, and stains, Figure 20-14, are adhesive coatings in a manner of speaking. They are spread onto the surface in a thin, unbroken film that is designed to protect and beautify the surface. *Coatings* is a term that covers all products that protect and beautify a surface. Usually *paint* is reserved for oil-based products. While *coatings* designates all types of protective coverings, the term applies especially to newer resin-based materials.

Paint's principle ingredients are *pigments,* which provide color and opacity, and *vehicles,* which are oils or resins that make the paint a fluid and allow it to be applied to a surface. The vehicle also includes a *binder* which remains behind after the oils or resins are dried. The binder then holds the particles of pigment together.

Varnishes, lacquers, and *sealers* are clear coatings applied over wood. They not only protect the wood surface but enhance the full beauty of the grain and color of the wood. Sealers are also used as an undercoat to keep stains and other clear coatings from being absorbed too deeply into the porous wood. They also "tie down" stains and fillers that have been applied. Sealers, Figure 20-15, also improve adhesion of topcoats and prevent bleeding of certain stains through topcoats.

Figure 20-11. This disk sander is designed to remove paint from flat surfaces. (Porter-Cable)

Figure 20-12. These orbital sanders are designed for fine sanding to produce a very smooth surface.

Figure 20-13. Paint stores are usually well stocked with paints and aids that help buyers in selection of colors. A—Overall view of paint stock. B—Close-up view showing labeling that aids user in his or her selection.

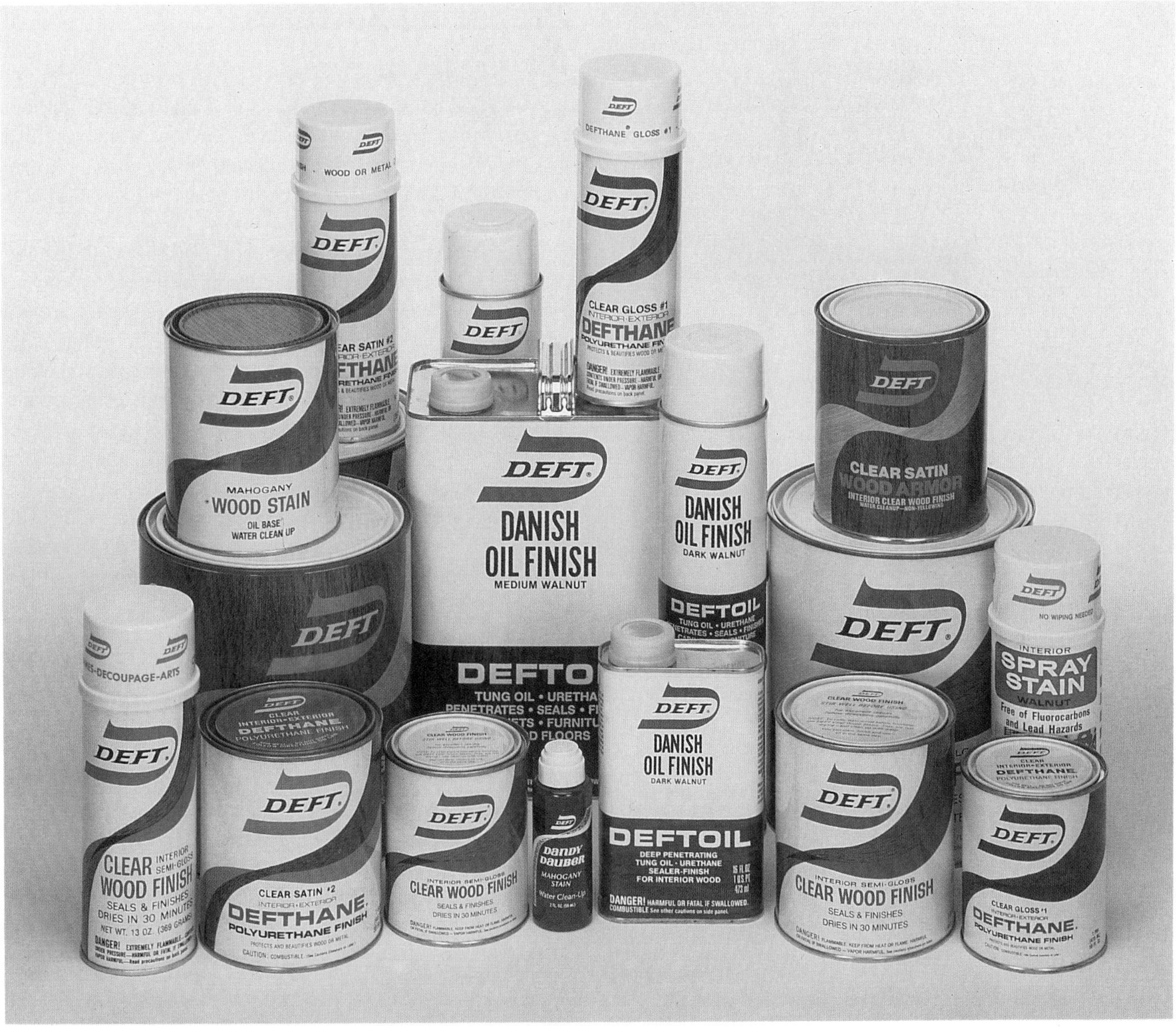

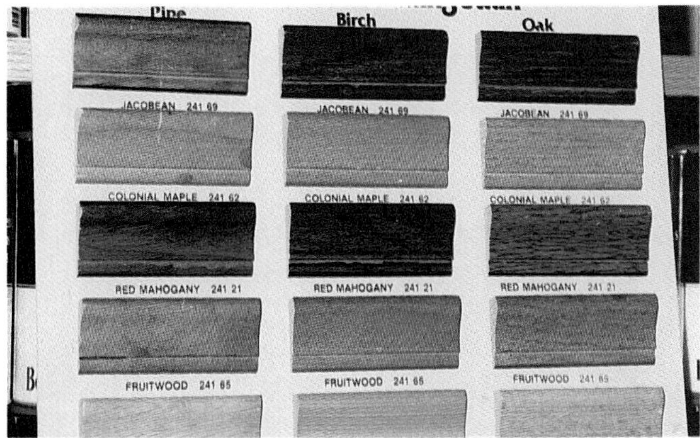

Figure 20-14. Top—These are a few of the finishes available for cabinets, woodwork, and furniture. Most are easy and quick to use while producing fine finishes. Left—Manufacturers use sampler charts showing how stains appear on different types of wood.

Fillers are heavy-bodied liquids that are used to fill open grain before application of stains and topcoats. *Putty* is a plastic, doughlike material used to fill holes and large depressions in wood surfaces. It is also used to fill nail holes before painting.

Stains are grouped as spirit stains, oil stains, and water stains. *Spirit stains* set up and dry rapidly because their base material is either alcohol or acetone. Their use is generally limited to spray applications. *Oil stains* are either penetrating or pigmented. Penetrating stains are brushed on and the excess is removed with a dry cloth. Some finishers prefer to use a cloth pad as an applicator, Figure 20-16. Pigmented stains are designed to be applied with either a brush or a pad. For deeper tones, the stain is brushed on and allowed to dry without wiping. *Water stains* are a mixture of powders and water.

Figure 20-15. Shellac is often used as a surface sealer to prevent stains from striking in too deeply. It may also be used as a surface finish where it is not likely to come into contact with water.

Sandpaper is used to smooth surfaces prior to painting or varnishing. Steel wool and pumice are used to smooth and enhance the surface of clear coatings once they have dried. Sandpaper may be used between coats to provide a "tooth" for better adhesion.

Color Selection

Color selection is important when working with paints. A *color wheel,* Figure 20-17, is useful in selecting colors that go well together. There are a few terms that are important to understand when working with color schemes.

- *Hue*—A particular name of a color, like red or orange.
- *Primary colors*—Red, blue, and yellow.
- *Secondary colors*—Violet, green, and orange.
- *Intermediate colors*—Red-violet, blue-violet, blue-green, yellow-orange, red-orange, and yellow-green.
- *Related colors*—Hues that are side-by-side on the color wheel.
- *Complementary colors*—Hues opposite each other on the color wheel.
- *Value*—The lightness or darkness of a color.
- *Tint*—Color nearer white in value (pure color with white added).
- *Shade*—Color nearer black in value (pure color with black added).

Paints are available in factory-mixed colors or as bases to which color may be added at the paint store. See Figure 20-18. Selections are made from paint chips. Following instructions from the paint manufacturer, the store mixes in coloring materials to match the chip exactly.

Preparing Surfaces for Coating

A good paint or varnishing job is not possible without proper preparation of the surface. On new work, removal of any foreign matter, such as plaster, grease, dirt and pitch, and light sanding to get a smooth surface may be all that is necessary.

Renewal or recoating of old finishes will usually require much more work. All loose paint must be removed by scraping, wire brushing, and sanding. Edges around bare spots must be feathered. Work with the grain while sanding. Sometimes, complete removal of the paint may be necessary. Such is the case with extensive blistering, cracking, and alligatoring of the paint film. A heat gun is useful in such cases, Figure 20-19. Cracks and nail holes must be filled with putty and then sanded smooth.

Cracks that persist or reopen in plaster or drywall should be repaired by spanning the crack with fiberglass tape and a compatible compound. Repaired areas should be primed before being repainted.

Interior Painting

Cover floors and furniture to protect them from drips or spatters. Remove all attachments to walls and trim that are not to be painted. This includes wall hangings, pictures, switch and receptacle plates, and hardware.

Mask off areas such as windows and trim. This prevents spatters that may be difficult to remove.

Surfaces must be clean. Normally a thorough dusting is enough. However, surfaces soiled with grease or glossy

Figure 20-16. Stains can be applied with a cloth, pad, or brush. (Deft, Inc.)

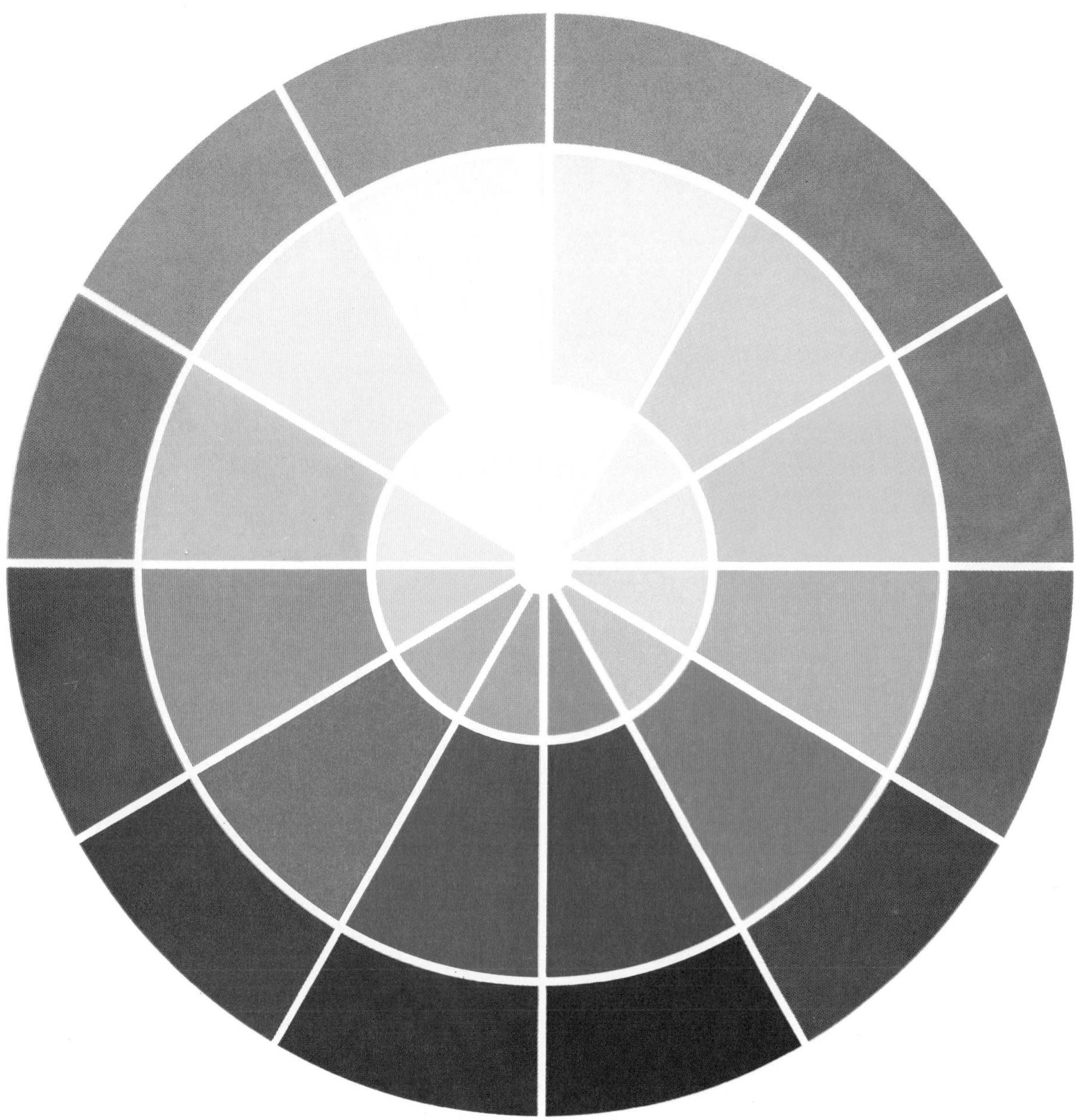

Figure 20-17. A color wheel may be used to assist in picking color combinations that are pleasing and complementary.

surfaces will need to be washed with a strong washing compound such as trisodium phosphate (TSP) to degrease and dull the surface. Rinse with clear, hot water. Pay particular attention to surfaces surrounding sinks and stoves, as well as surfaces frequently touched (around switches, doorknobs, handrails, and the edges of doors).

There are also special chemical deglossers that will prepare glossy painted surfaces for recoating. These are especially effective over latex paint.

When painting an entire room, start with the ceiling, then walls, and finally wood trim and doors. Assemble tools and mix paints thoroughly, even if they have recently been shaken on a paint store vibrator. Stir rapidly to mix settled pigments. Use a stirring paddle or a power mixer.

Mask off edges that are not to be painted. Standard masking tape is available in different widths. Some tape will peel off cleanly after being attached a week or more.

Start with the ceiling (as mentioned) if it is to be painted. Either rollers or brushes may be used, although most painters prefer rollers because they allow the painter to progress more quickly. An extension handle attached to the roller will be useful and easier than climbing up-and-down a stepladder or a platform.

Figure 20-18. More on selecting color. Top—Chips representing 1000 different colors assist the buyer is selection. Information is included on the chip to aid the store owner and his staff in mixing the color correctly. Bottom—Many stores carry color coordinated fabrics to go with paint or other wall coverings. (John Dutcher Glass and Paint)

A small brush will be necessary to paint into the corners formed by ceiling and wall. (This is known as cutting in.) Paint large surfaces in narrow strips so that you are always painting against a wet edge. Dry edges that are overlapped will show up as lap marks or will have a different sheen when the paint is dry.

When painting walls, cut in the edges formed by intersecting surfaces or by baseboards. This provides a narrow painted strip of paint at the ceiling line, around doors and windows, and along baseboards. These edges cannot be reached with the roller without depositing paint where it is not wanted.

If painting with a brush, dip the bristles into the paint one-third of their length. Tap the brush gently against the side or edge of the paint container to release paint that might otherwise drip. Paint stores have available a grid of expanded metal that hangs inside five gallon containers for striking off excess paint from brush or roller. See

Figure 20-19. A heat gun may be employed to remove old paint. An open flame should never be used for this purpose. (Bosch)

Figure 20-20. A stiff wire can be strung across a one gallon can for removing excess paint from a brush.

Loading a roller should be done carefully to avoid dripping paint. If using a tray, pour a small amount of paint into the deep end. Dip the roller into the paint and work in the paint by rolling it back and forth on the slanting end of the tray. Start rolling in one corner of the wall or ceiling, painting a narrow strip. Overlap each stroke, working first in one direction and then in another, until the surface is covered. Work roller gently, avoiding quick, choppy strokes that cause paint spatters. Continue rolling until the newly painted surface is covered and roller needs reloading. Figure 20-21 shows proper rolling and brushing techniques. Unless repainting with the same color, a second coat will look more professional than one coat.

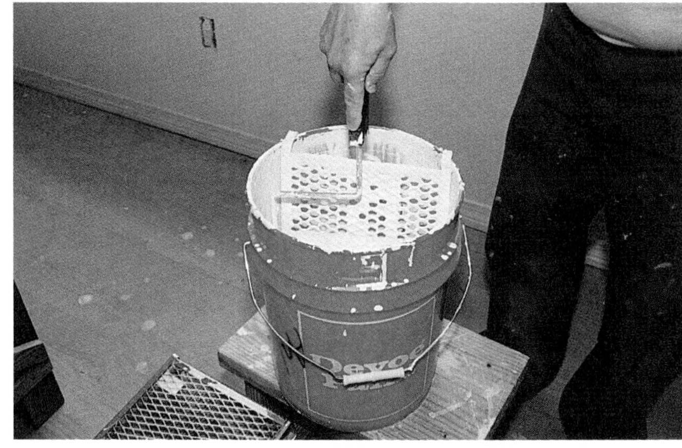

Figure 20-20. A grid of expanded aluminum can be used in a five gallon can to remove excess paint from a roller or brush.

Trim that is to be painted should be done only after walls are thoroughly dry. A 1 1/2" brush is suitable for windows; use a 2" brush for other trim. Be careful not to overload the brush so that drips and runs are a problem.

On multipaned windows, start painting on center mullions first, working outward in either direction. Trim should be painted last.

Figure 20-21. Roller and brush technique. Top—Rolling paint onto a wardrobe shelf. Middle—Cutting in the edge between ceiling and wall. Bottom—Painting with a roller.

Use masking tape to avoid spreading paint onto the window glass. Accidental spatters on the glass, however, can be removed later with a razor blade when the paint is dry.

Doorjambs and casing should be painted before the door. To avoid laps and brush marks when painting paneled doors, paint the molded edges of the panels first, starting with the top panel. Then fill in the panels, stroking with the grain. Next, paint the rails (horizontal cross boards). Finish by painting the stiles (vertical side boards). Refer to Figure 20-22. If the door swings away from the side being painted, paint the hinge edge the same color; if it swings toward the side being painted, paint the lock edge. If both sides of the door are to have the same color, all edges should be painted.

If preferable, multipaneled doors can be painted with a roller. First brush the recessed panels and decorative edges; then roll flat, wide surfaces. Finally, brush out roller marks.

Paint baseboards last. A plastic, metal, or cardboard guard should be used to protect the finished floor or carpeting when painting the lower edge of the baseboard. As an alternative, painter's gummed masking paper can be used to protect hard-surfaced finish flooring.

Exterior Painting

In painting new wood, all knots and pitch-containing areas must first be sealed with a shellac or knot sealer. The first coat of finish should be a pigmented primer or a sealer. This coat should be allowed to dry thoroughly before finish coats are applied.

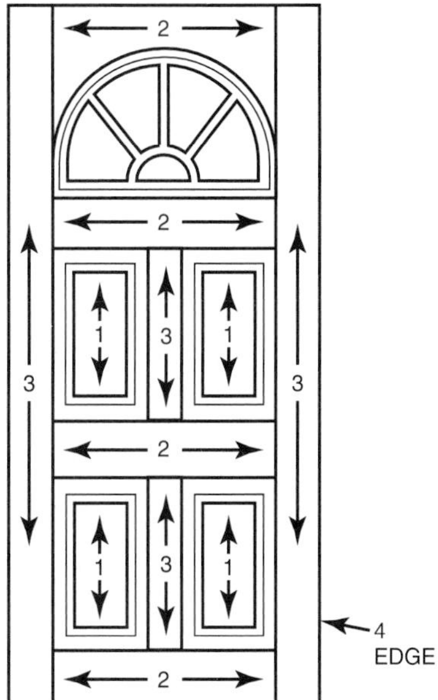

Figure 20-22. Efficient method of painting a paneled door. Paint the panels first, then rails, then stiles. Arrows indicate direction of the brushing.

Previously painted surfaces may need to be scraped or sanded to remove loose paint, Figure 20-23. Any imperfections should be repaired and holes puttied. This is very important on heavily weathered paint if the paint job is to last even one season. All bare spots should be primed with an oil based primer before final coating. On extremely weathered surfaces, sand the surface down to "bright" wood. Many painters then apply a 50/50 mixture of turpentine (or mineral spirits) and linseed oil, allowing it to dry 48 hours before priming.

Painting with a brush is easiest and the paint covers best if the brush is properly held. See Figure 20-24 for proper grasp in different situations.

Top coatings may be with exterior oil paint, oleoresinous finishes, alkyd coatings, or water borne coatings. The thickness and smoothness of a paint coat is controlled to an extent by the amount of paint carried in the brush. Do not dip the brush more than half the length of the bristles.

Oil paints are applied in a slightly different way than other coatings. The paint is daubed on in spots; then it is spread out with full-arm level strokes. Each brush load should be brushed out with sweeping strokes. Use the tip for brushing and gradually lift the brush at the end of each stroke. This will give a thin featheredge at the end of each stroke.

The lower edges of clapboard siding should be painted first. Rough surfaces require additional brushing in all directions. This will assure that pores and crevices are well covered with paint. Poking at the surface with the brush tip will eventually destroy the flags on the bristle ends. Use an old brush if poking is necessary.

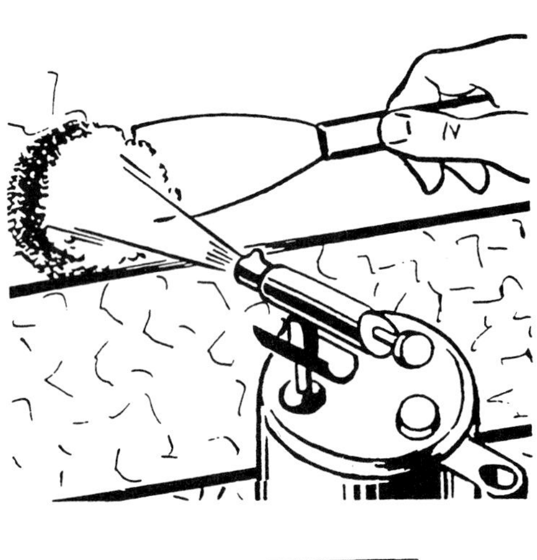

Figure 20-23. All loose paint must be removed before repainting. Weathered wood should be sanded down to "bright" wood.

Alkyds and *waterborne paints* require less brushing and dry rapidly. Thus, there is less time for brushing out than with oil paints. If too rapid drying becomes a problem, inhibitors may be added; "Flowenol" may be used for latex paint, "Penetrol" for oil and alkyd paints.

As with interior painting, it is important to paint to a wet edge. The paint band should be narrow enough so that the next band can be completed before the first has dried. This will eliminate lap marks. Start painting at the highest point and move downward in horizontal bands. Never stop painting partway across a band. Always end up a day's work at a window, door, the edge of a board or a corner.

Roller and Pad Application

Topcoats of waterborne finishes can be applied over oil-based primers using a roller. Work the roller across one board at a time, working from top to bottom.

A brush is handy for cutting in the edges of boards. Several edges can be cut in at once as long as the edges do not dry before the boards are rolled.

Paint pads have a napped fabric over a foam base. The pad is secured in a plastic holder. Special pans use about 1/2 gallon of paint. To fill the pad, the pan is tipped to flood a grid area with paint. The pad is pressed on the grid to load it. Three or four boards can be completed to a stopping point before moving to a new area.

Exterior spray painting is sometimes preferred to painting with brushes, rollers, or pads. It is most economical where there are large surfaces that are to be painted in one color. Proper mixing of paints is important. Some painters "box" the paint by intermixing different containers of paint. This procedure assures that the paint will be uniform in color and texture. See Figure 20-25. Areas that are to receive no paint must be carefully masked as shown in Figure 20-26. Operators should wear protective clothing completely covering the skin, Figure 20-27. A respirator should always be worn.

Spray painting requires certain safety precautions to avoid the hazards of explosions, fire, and personal injury. Spraying should be done in a well ventilated area. Flammable liquids should be kept in safety containers. Fumes and fine spray are toxic and should be avoided. Avoid spray painting skin, it could cause permanent disability. Keep a fire extinguisher handy.

Working with Stains and Clear Finishes

Before a wood surface can be stained and finished, all surface dents must be removed or repaired. Minor defects

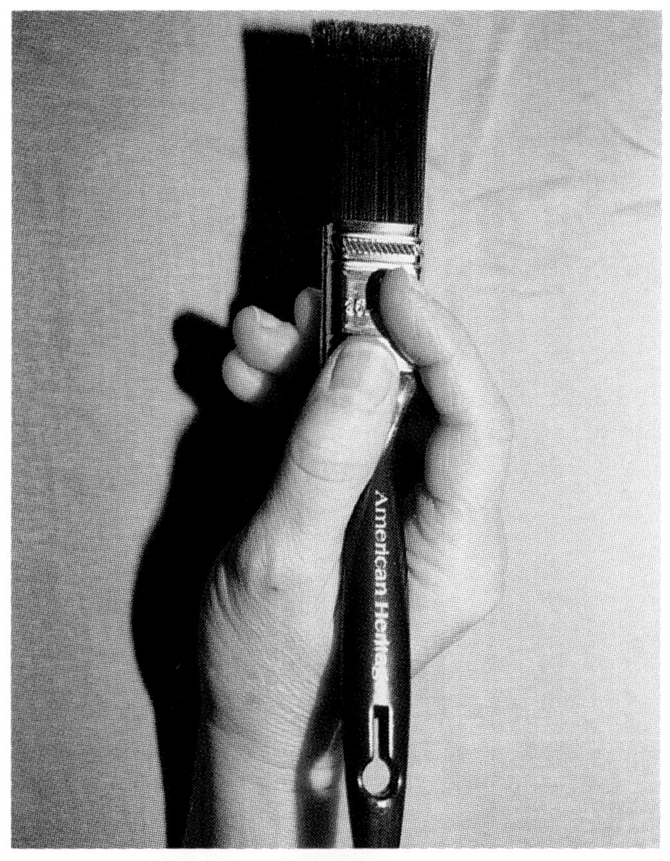

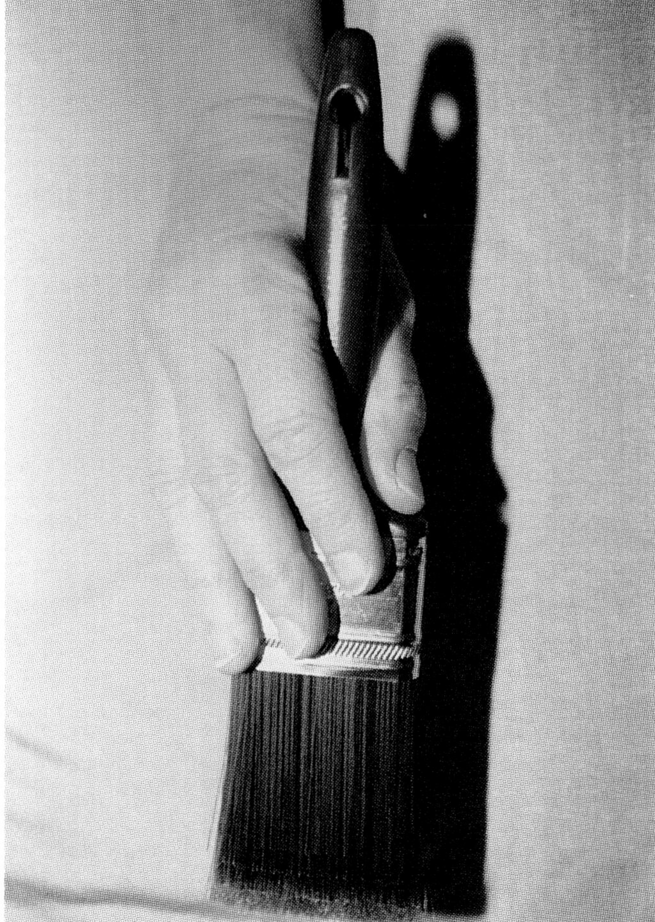

Figure 20-24. Proper method of holding brushes.

such as machine marks, small scratches, and excess glue can be removed with light sanding. More serious conditions, such as dents and holes will require repairs.

Dents can generally be removed, if not too large, by applying a drop of warm water to the spot. Deeper dents can be lifted with a damp cloth and a hot iron, Figure 20-28. Place the damp cloth over the dent and apply the hot iron to the cloth. Move the iron around to avoid burning the wood. Sand the dried surface smooth.

Figure 20-25. When painting large surfaces such as the outside of a house, it is important to box the paint so that color will be uniform throughout. (Greco Painting)

Figure 20-26. When using a sprayer to apply paint, it is important to mask off surfaces not being painted. Top—Masking a window. Bottom—Masking at the foundation.

Figure 20-27. Stuccoed house being spray painted. Painter is wearing protective clothing and headgear, as well as a mask.

Figure 20-28. Small dents may be steamed out using a damp pad and heat from an iron.

Holes include open joints, cracks, splits, and gouges. Fillers are needed to repair such defects. Basic fillers include plugs, plastic wood, wood putty, stick shellac, and wax sticks, Figure 20-29.

Plugs are wooden pieces shaped to fit and forced into the defect. *Plastic wood* is a combination of wood powder and a plastic hardener. *Wood putty* is a mixture of wood and adhesive in powder form. *Stick shellac* is colored shellac. It must be applied with a heated knife. *Stick wax* is a colored wax that closely resembles crayons. Wax is applied by rubbing the stick over the defect and forcing the wax into it.

Items to be finished should be smooth, free of dust, dirt, glue, and grease. Hardware should be removed, Figure 20-30.

Figure 20-29. Various materials are used for filling holes and open defects in wood. Included are a wood plug, plastic wood, wood putty, stick shellac, and wax sticks.

Wood bleaching is the removal of some of the natural color of the wood. Bleaches can be purchased ready-mixed and usually consist of two solutions. Mixing is done in a glass or porcelain container. The solution is applied to the bare wood with a rubber sponge or a cotton rag. Wear gloves and a face shield for this operation. Most bleaches are caustic and, therefore, dangerous.

Surfaces repaired of defects are ready for staining and final finishing. While not necessary for every surface, stains are used to change the color of natural wood and, when used on exterior surfaces, have a protective quality. They can be applied by wiping, brushing, or spraying, Figure 20-31. Excess, if any, is wiped off and the stain should be allowed to dry for 24 hours. Then it is covered with a sealer. The sealer coat "ties down" the stain and filler. Some sealers are ready for final finishes in about 30 minutes. Follow directions on the container.

Clear finishes are brushed or sprayed, Figure 20-32. Brushes should be the finest available. Flow the finish on rather than brushing it out as with paint. Be sure not to leave "holidays" (bare spots). Do not apply in too heavy a coat or sags and drips will mar the finish.

Dried finish coats can be smoothed with fine steel wool and treated to a rubbing with wax. Allow at least 48 hours of drying time before attempting rubbing the final coat.

Estimating Coatings

Estimating materials needed is rather easy. First, find the area that is to be covered by measuring the height and width and then multiplying the two measurements. Check the label on the paint container to see how many square feet it will cover. Divide that number into the total square feet to find the number of containers needed. If two coats are needed, double the number of containers.

Figure 20-30. For a professional job, hardware should be removed from a door before it is stained or varnished. (The Stanley Works)

Problems with Coatings

Painters often experience problems with previously painted exteriors. When repainting previously painted exteriors, it is helpful to recognize certain defects in the old coatings. Figure 20-33 illustrates some of the following defects:

- *Alligatoring and checking*—Alligatoring is a severe case of checking. It is the result of applying relatively hard (containing less oil) finishing coats over primers that are softer (more oily). To avoid this situation, make coats progressively more flexible from primer to finishing coats. Allow more drying time between coats and follow the manufacturer's recommendations for thinning.
- *Cracking and scaling*—These conditions are usually the result of using paint lacking in elasticity. It becomes hard and brittle and unable to contract and expand with the wood. The conditions are aggravated by too-thick application. To avoid this condition, use high quality paint and avoid buildup of thick layers.
- *Excessive chalking*—Oils in the finish coat render a glossy finish. Ultraviolet rays destroy the oils and

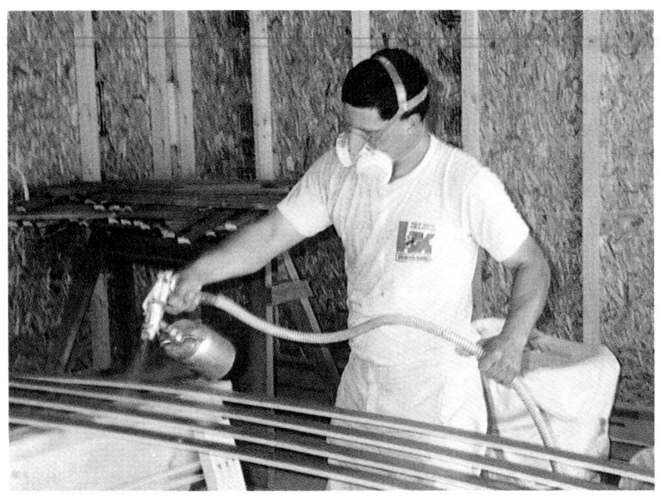

Figure 20-31. Stain, varnish, and clear finishes can be applied to trim before it is installed.

leave loosely bound pigment on the surface. Some chalking is desirable to keep the paint clean. Heavy chalking wears away the paint and leaves the surface unprotected. Poor quality paint is the major cause. Before repainting, remove the chalked paint by scrubbing or wire brushing.

- *Blistering and peeling*—Blistering always precedes peeling. This condition is due to water pressure under the coatings pushing the paint film from the substrate. The only cure is to eliminate the source of the moisture. Moisture barriers such as aluminum paint or mylar film will stop the migration of interior moisture-laden air to the outside. Repair of external leaks may be necessary. Install proper flashing and caulk joints and cracks. Proper grading will eliminate ground moisture coming in contact with siding. Proper ventilation of attics and crawlspaces will also reduce moisture buildup. Finally, allow surfaces to thoroughly dry before painting.

Figure 20-32. With proper masking, stains and finishes can be applied after trim has been installed.

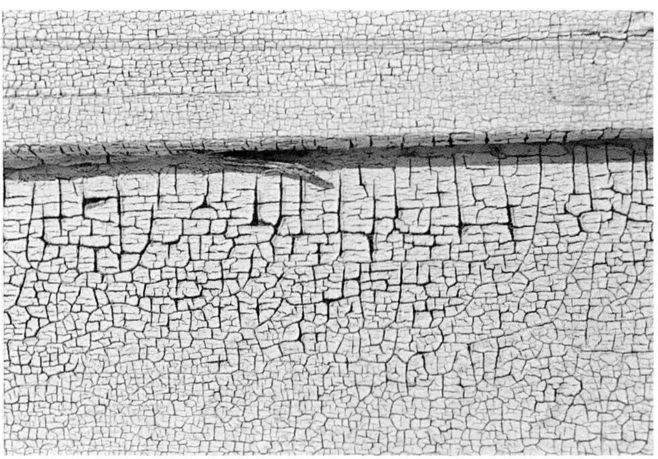

Figure 20-33. Various types of paint failure. A great deal of work will be required before these surfaces can be repainted.

- *Mildew*—This is a form of plant life. It flourishes on soft paints. Eliminate mildew by washing the affected surface with a mild alkali such as washing soda, trisodium phosphate, or sodium metasilicate. These cleaners are sold under such trade names as "Spic and Span" and "Oakite."

Improper application causes certain defects in varnishes and other clear finishes.

- *Crawling*—The finish gathers up into small globules and does not cover the surface. This may be caused by applying the finish over a wet, cold, waxy, or greasy surface. Other causes are mixing of different kinds of varnish, using too much drier, thinning with naphtha instead of turpentines or other approved thinners.
- *Blistering*—Caused by too much moisture in the wood.
- *Blooming (clouding)*—May be caused by high humidity in the room where finishing is done. Another cause is the use of water and pumice instead of oil and pumice for rubbing the finish.
- *Runs, curtains, and sags*—Caused by too-heavy application.
- *Specks in surface*—Results from hardened particles in the finish (needs straining), dirty brushes, improperly prepared surfaces, or airborne dust in the room.

Wall Coverings

Hanging wall coverings, or *wallpapering,* is an important phase of decorating. Doing professional quality work demands a thorough knowledge of materials and procedures.

Most of the wall coverings used today are made from vinyl resins. Vinyls consist of a continuous flexible film that has been applied to a paper or fabric backing. Such materials have the advantage of being long-lasting and washable. Moreover, they can be made highly decorative. More exotic wall hangings used by restorers are the Victorian era embossed wall coverings, Lincrusta and Anaglypta.

Traditional wall coverings are made from cotton, linen, hemp, wood, and wastepaper. More expensive coverings include the grass cloths, flocks, and the foil-based materials. Lincrusta is a rigid product made from a linseed oil mixture with a deep relief. Anaglypta is a similar, lightweight embossed material made from cotton pulp.

Wall coverings are measured by the roll. A single roll contains from 27 to 29 sq. ft. With trimming, about 21 to 23 sq. ft. are available in actual coverage.

Usually, for convenience sake, the coverings come in bolts containing two or three single rolls. Most are 20, 21 1/2, or 22" wide. English-made papers are generally 22" wide and rolls are 36' long. French papers are generally 18" wide and 27' long in single rolls. Other foreign coverings vary in length and width. The amount in a single roll is only an approximation.

Always keep a record of wall covering purchases or retain a label with the collection, pattern number, and dye lot or printing run number. This information is important if you need to reorder to finish a job.

Estimating Wall Covering

Take measurements of the room, rounding off to the next larger foot or half foot. Make a sketch of the room showing doors, windows, fireplace, built-in shelving, etc. Write down the measurements of wall lengths and heights, excluding baseboards and moldings. Subtract areas that will not be covered. Take this information to your dealer or use the chart in Figure 20-34.

Paperhangers require a number of hand tools, a paste table, and an expandable plank. Tools include shears for cutting paper, several sizes of rollers, smoothing and paste brushes, several knives, a razor blade and holder, a broad knife, a 6 or 7' straightedge, plumb bob, chalk line and blue chalk. See Figure 20-35.

Preparing to Hang Wall Covering

Old wallpaper is not a good foundation for new paper and should be removed. Some old coverings are "strippable" and can be easily removed. Otherwise, a steamer is the modern way of removing it.

New walls, whether plastered or drywalled, should be thoroughly dried before covering. Plastered walls should be treated with a solution that neutralizes the lime and alkali in the plaster. *Sizing* is a glue solution that is brushed or rolled onto a newly plastered or drywalled surface to seal the pores and prevents the paste from being absorbed into the wall.

Preparing the Wall Covering

Before wall coverings are cut to length, the paperhanger will check that there is enough to do the job. More will have to be purchased if there is too little to finish the job. Care must be taken that the new covering is from the same run; other runs may vary in color and ruin the job.

Wall coverings are brittle and must be softened before being applied to walls. On prepasted papers, this is done by placing the prepared strip in a water tray for a time specified by the manufacturer before it is placed on the wall. On papers not prepasted, it is done by application of the paste which contains water. The paste may be purchased from paint stores or made from cooked flour or starch and water.

When using coverings with a design, it is necessary to match the strips on the paste table before cutting them. Do not cut the strips too short; allow about 2" excess at both top and bottom. Trimming to exact length is done after strips are on the wall.

Wall, Room & Border Estimating Chart For English/Metric Single Rolls

Single Rolls for Wall Area(s)

DISTANCE OF WALL IN METERS	/ FEET	CEILING HEIGHT 8 FEET or 2.4 METERS	9 FEET or 2.7 METERS	10 FEET or 3.0 METERS	11 FEET or 3.4 METERS	12 FEET or 3.7 METERS
1.8	6	3	3	3	4	5
2.4	8	3	4	4	5	6
3.0	10	4 TO 5	5	5	6	7
3.7	12	5	5	5	7	8
4.3	14	5	5	7	7	8
4.9	16	5 TO 7	7	8	8	9
5.5	18	7	8	8	8	10
6.1	20	8	8	9	10	11
6.7	22	8 TO 9	9	10	11	12
7.3	24	9	9	10	12	13
7.9	26	9 TO 1O	10	12	13	14
8.5	28	10	12	13	14	16
9.1	30	10	12	13	15	16
9.8	32	12	13	14	16	18
10.4	34	13	14	15	16	18
11.0	36	14	14	16	18	20
ROOM SIZE IN METERS	/ FEET					
2.4 X 2.4	8 X 8	12	13	14	16	18
2.4 X 3.0	8 X10	14	14	16	18	20
3.0 X 3.0	10 X 10	14	16	18	20	22
3.0 X 3.7	10 X 12	16	18	20	22	24
3.0 X 4.3	10 X 14	18	20	22	24	26
3.7 X 3.7	12 X 12	18	20	22	24	26
3.7 X 4.3	12 X 14	20	22	24	26	28
3.7 X 4.9	12 X 16	20	22	26	28	30
3.7 X 5.5	12 X 18	22	24	28	30	32
3.7 X 6.1	12 X 20	24	26	30	32	34
4.3 X 4.3	14 X 14	20	22	26	28	30
4.3 X 4.9	14 X 16	22	24	28	30	32
4.3 X 5.5	14 X 18	24	26	30	32	34
4.3 X 6.1	14 X 20	24	28	32	34	38
4.3 X 6.7	14 X 22	26	30	32	36	40
4.9 X 4.9	16 X 16	24	26	30	32	34

After finding the number of rolls needed for the wall(s) or room you are hanging, deduct one half-roll for each standard size door or window. Design repeat of the pattern and length of roll may increase your rollage needs. Your dealer can advise you if the wall covering selected will significantly affect your rollage needs.

Figure 20-34. This chart may be used to determine how much wallpaper is needed for a room. (National Decorating Products Assoc.)

Booking the Strips

Booking is a method of pasting and folding wall coverings. Turn all cut unpasted strips face down on the paste table. Paste one strip at a time. Paste two-thirds of the strip. Then, with clean hands, book the strip. This is an important step as it allows the paste to soak into the wall covering. To book, pick up the corners of the pasted end and fold it over only the pasted portion (pasted side in). Do not go beyond the pasted portion and do not crease the fold. Then pasting and folding is repeated on the other one-third. Edges must be even. If there is salvege on the covering, trim it at this time. Finally, roll the booked strip loosely and let it sit.

Allow the booked strip to "relax" for three to five minutes before hanging. Check manufacturer's recommendation.

There are products that will activate the paste on prepasted papers. These watery products are rolled onto the back of the covering. One gallon will normally cover 10 single rolls.

Hanging Wall Covering

1. Using a plumb bob or a level, make a plumb starting line on the wall as shown in Figure 20-36. This is used as a guide for hanging the first strip.
2. Overlap the top edge of the first strip 2" onto the ceiling and align the side edge with the plumb line.
3. Unfold the top one-third of the strip. Smooth the opposite side against the wall.
4. When the entire top portion is smoothed against the wall, as in Figure 20-37, the second portion is unfolded and aligned with the plumb line.
5. Smooth out the strip from the center to the edges, working from the top down. Use brushing strokes in a sequence like that shown in Figure 20-38. Use care not to stretch the wet covering. When satisfied that the strip is properly aligned, smooth the entire surface again to remove remaining air bubbles.
6. Trim the ceiling and baseboard ends with a razor knife. Use a wall scraper or a broad knife as a guide. Hold the scraper or broad knife tightly against and parallel to the wall as you trim with the razor knife, Figure 20-39. Move the scraper or broad knife along the strip but keep the razor knife in contact with the wall. Replace the razor after each strip.
7. Rinse the strip to remove excess adhesive. Remove adhesive from the ceiling and baseboard at the same time. Change rinse water frequently and use a good quality natural or synthetic sponge.
8. Each succeeding strip is matched to the previous strip. Butt the edge against the previous strip for a tight seam. Brush and smooth in the same way. Be sure to wipe off excess adhesive as before.

Seams are usually butted, but two other seam types are used: the wire edge and the lapped. After hanging every three or four strips, go over the seams with a roller. Keep the roller clean and wipe away adhesive squeezed out of the seams. Do not press too hard; it will cause stretching or shrinking problems later. Seam rollers should never be used on flocked or embossed wall coverings.

It is recommended that the first strip be started at a corner with a portion of the strip wrapping around the corner. The corner lap should be at least 1/2".

Hanging Wall Covering around Openings

Never attempt to precut a strip to fit around a window or door. Apply the paper over the opening and use a razor

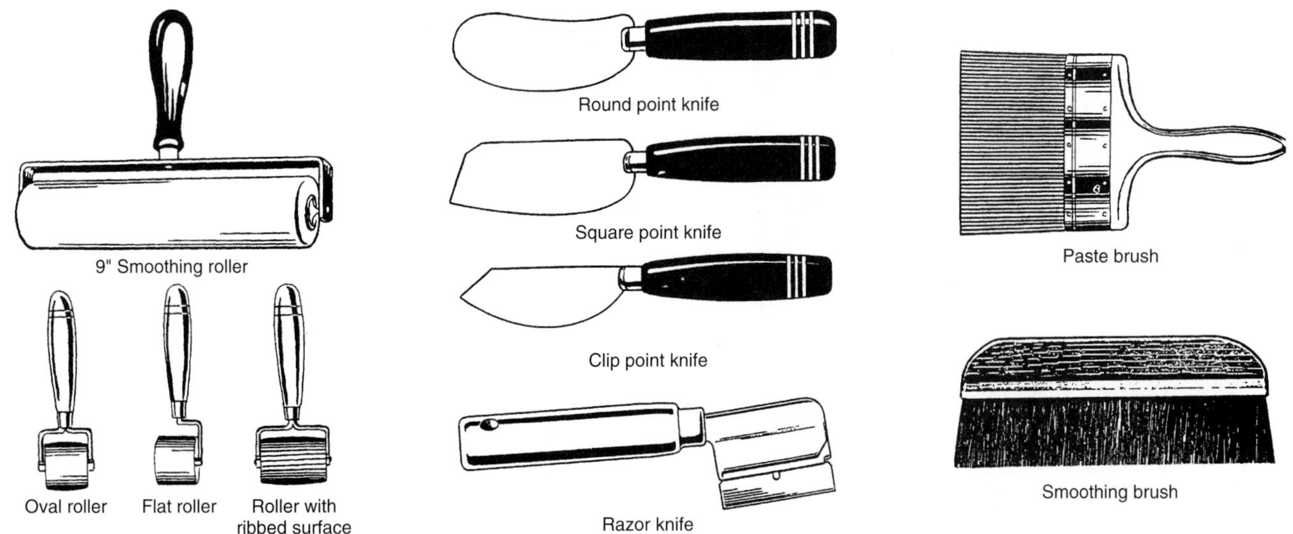

Figure 20-35. These tools are recommended for anyone doing wallpapering.

Figure 20-36. A plumb line is being set up using a level and a straightedge. This is usually made near a corner so the first roll of paper wraps around the corner at least 1/2". A chalk line can also be snapped for the plumb line.
(National Decorating Products Assoc.)

Figure 20-37. Attach the first strip of wallpaper as shown, aligning or centering it on the plumb line.
(National Decorating Products Assoc.)

knife to trim the excess, Figure 20-40. This method is also used when papering around light switches and receptacles.

NOTE: Make sure circuit breaker has been tripped before working around any electrical device.

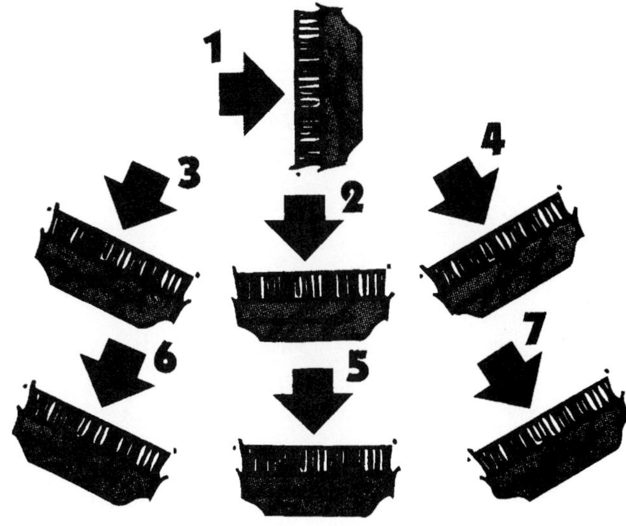

Figure 20-38. Strips should be brushed out in sequence shown to remove air pockets and adhere them firmly to the wall.

Figure 20-39. Trim off each strip at the ceiling and baseboard using a broad knife and a razor knife.

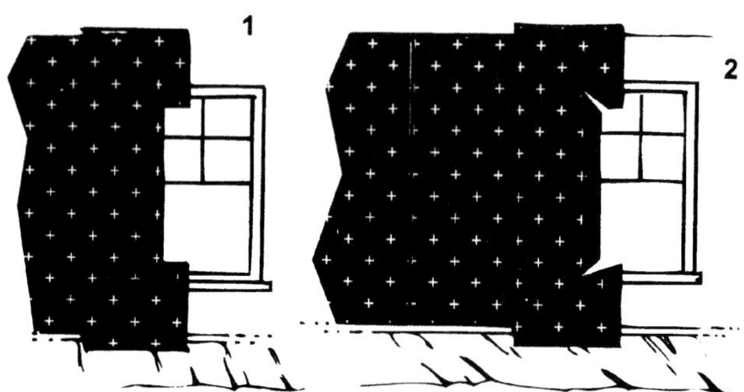

Figure 20-40. Proper procedure for papering around windows and doors. Before making final trim, cut into all corners at a 45-degree angle. (National Decorating Products Assoc.)

Important Terms

Alkyds
Anodizing
Binder
Booking
Coatings
Color wheel
Complimentary colors
Cutting in
Fillers
Hue
Intermediate colors
Lacquers
Oil paint
Oil stain
Pigments
Plastic wood
Plugs
Primary colors
Putty
Related colors
Sealers
Secondary colors
Shade
Spirit stain
Tint
Value
Varnish
Vehicle
Wallpapering
Water stains
Waterborne paint
Wood bleaching
Wood putty

Test Your Knowledge

1. Which of the following are reasons for painting either interiors or exteriors?
   A. Improve appearance.
   B. Protect from weather.
   C. Protect surfaces against soiling.
   D. All of the above.
   E. None of the above.
2. It is considered safe to remove leaded paint if wearing an approved respirator. True or False?
3. Bristles of paint or varnish brushes are made from either synthetic materials or _____________.
4. _____________ brushes are useful for removing rust from metal surfaces or removing peeling or flaking paint.
5. The external mix head of a spray gun is intended for spraying _____________ coating materials.
6. Which spraying head provides greater control of the spray pattern?
7. _____________ are the ingredients that give color and hiding quality to a paint.
8. What product is used to level open-grained wood surfaces?
9. Red, blue, and green are called the _____________ colors.
10. Preparing for painting a new wood surface is _________*(more, less)* work than preparing a deteriorating painted surface.
11. Why is masking tape used during painting?
12. List the sequence used to paint an entire room.
13. Painting inside corners first is called _____________.
14. In preparing to paint new wood, the first step is to:
    A. Apply primer coat.
    B. Seal knots and pitch pockets.
    C. Apply a thin coat of linseed oil.

15. To estimate paint quantity, what two things must be known?
16. Cracking and scaling of paint is usually the result of using paint lacking in ____________.
17. Traditional wallpaper is made of what materials?

Outside Assignments

1. Measure a room and determine the number of gallons of paint required to paint it.
2. Research a wood-finishing system and prepare a report on the proper procedure for finishing an unpainted pine cabinet to a desired wood tone. Indicate both equipment and materials.
3. Select a room, take measurements, and determine the number of rolls of wallpaper needed to wallpaper all of the walls.
4. Take photos of paint defects and bring them to a class discussion on probable cause of the paint failure.

A carpentry student paints a playhouse constructed as a class project. (Santa Rita High School)

Painting around trim requires the use of a brush. (Sherwin-Williams)

Painting wood exteriors not only preserves the structure from weathering but greatly enhances the appearance.

Special Construction

NURCO
NURCO

This steel I-beam is the main support member in this post-and-beam system.

A system-built house factory. The assembly is more efficient than building on-site because the environment is controlled—there is no wind, rain, or snow inside the warehouse. (Wausau Homes Inc.)
Photo on previous page also provided by Wausau Homes Inc.

21 Chimneys and Fireplaces

A *chimney* is a vertical shaft which exhausts the smoke and gases from heating units, fireplaces, and incinerators. When properly designed and built, a chimney may also improve the outside appearance of a home.

The fireplace once was the only source of heat for dwellings. Today the fireplace is popular even though its efficiency as a source of heat is low compared with modern heating systems. The desire for a fireplace results from the cheerful, homelike atmosphere it creates and the value it adds to the home.

The rising cost of energy has led to many improvements in fireplace design. Masonry fireplaces are often equipped with glass doors to save heat. They may also include ductwork to heat room air as it is circulated around the firebox. More important, they may have provisions to draw combustion air from outside. A wide variety of prefabricated fireplaces are available. They are easy to install and are usually less expensive than masonry units.

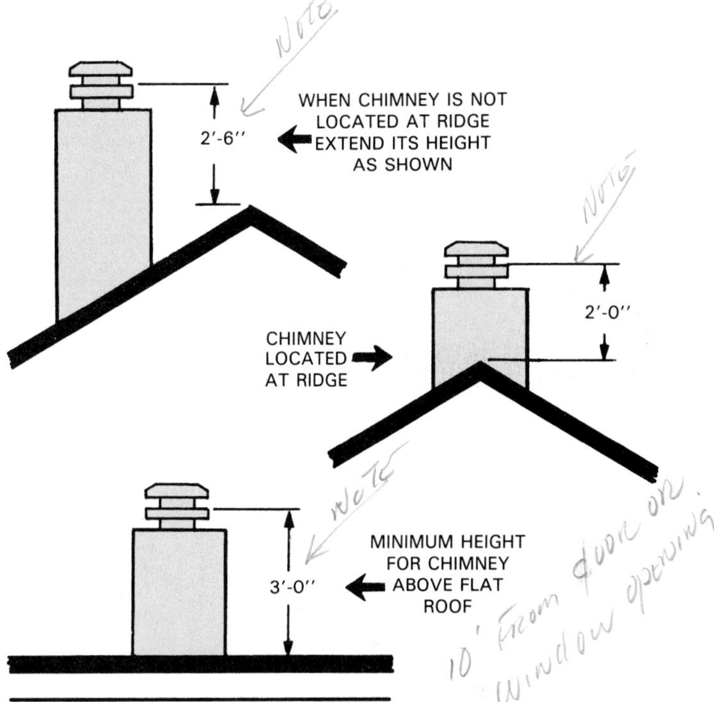

Figure 21-1. Maintain these minimum chimney heights above roof. Check building codes in your area.

Masonry Chimneys

A masonry chimney usually has its own footing and is built in such a way that it provides no support to, nor receives support from, the building frame. Footings should extend below the frost line and project at least 6" beyond the sides of the chimney. Walls of a chimney with clay *flue lining* should be at least 4" thick. Foundations for a chimney or fireplace, especially when located on an outside wall, may be combined with those used for the building structure.

The size of a chimney will depend on the number, arrangement, and size of the *flues* (vertical openings in the chimney). The flue for a heating plant should have enough cross-sectional area and height to create a good draft. This permits the heating equipment to develop its rated output. Always follow the heating equipment manufacturer's recommendations when deciding on flue sizes.

Building codes require that chimneys be constructed high enough to avoid downdrafts caused by the turbulence of wind as it sweeps past nearby obstructions or over sloping roofs. Minimum heights generally required are illustrated in Figure 21-1. The chimney should always extend at least 2' above any roof ridge that is within a 10' horizontal distance.

Combustible materials, such as wood framing members, should be located at least 2" away from the chimney wall, Figure 21-2. The open spaces between the framework and the chimney can be filled with mineral wool or other incombustible material.

Flue Linings

The National Board of Fire Underwriters recommends *fireclay* flue linings for masonry chimneys. Most local building codes also require them.

Linings are available in square, rectangular, and round shapes. Each shape is available in several sizes as listed in Figure 21-3. Note that different methods of measurement are used for the three types:

- Outside measurements for the old standard.
- Nominal outside dimensions for the modular.
- Inside dimensions for the round linings.

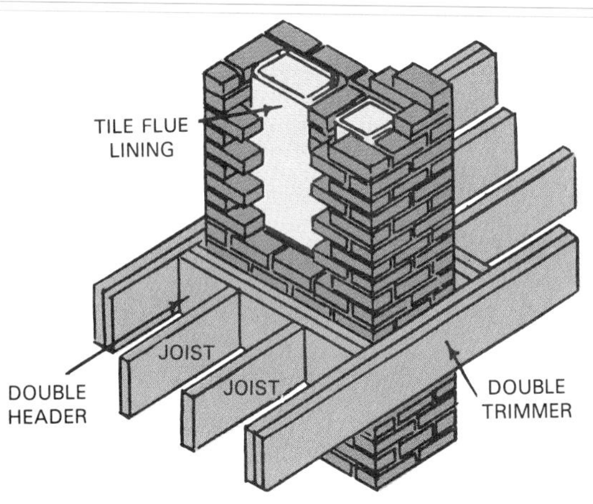

Figure 21-2. Floor framing around chimney. Note tile flue lining. When a flue column is next to another, joints of flue lining should be staggered vertically a distance of at least 7".

STANDARD LINERS		MODULAR LINERS		ROUND LINERS	
Outside Dimensions (In.)	Ares of Passage (Sq. In.)	Nominal Outside Dimensions (In.)	Areas of Passage (Sq. In.)	Inside Diameter (In.)	Areas of Passage (Sq. In.)
8 1/2x8 1/2	52.56	8x12	57	8	50.26
		8x16	74		
8 1/2x13	80.50			10	78.54
		12x12	87		
8 1/2x18	109.69	12x16	120	12	113.00
13x13	126.56	16x16	162	15	176.70
13x18	182.84	16x20	208		
18x18	248.06	20x20	262	18	254.40
		20x24	320	20	314.10
		24x24	385	22	380.13
				24	452.30

Figure 21-3. Flue liner dimensions and clear cross-sectional areas in square inches. The National Building Code requires that the wall of the flue liner be a minimum of 5/8" thick.

Small and medium flue linings are usually available in 2' lengths. Flue rings are placed during the assembly of the chimney to provide a form for the masonry.

Although unlined fireplace chimneys with masonry walls (minimum thickness 8") will operate well, glazed flue lining is usually recommended. The smooth inside surface of a glazed lining is less likely to attract pitch and tar, which may eventually restrict the passage.

As a rule, a single flue should be used for only one heating unit. However, it is often permitted to connect the vent from a gas-fired hot water heater to a furnace flue. Do not combine larger flues.

Construction

Allow the chimney footing to cure for several days to give it proper strength. Use care in starting the masonry work. Make all lines level and plumb.

Each section of flue lining is set in place before each part of the chimney wall is built. The lining is a guide for the brick work. Joints in the flue lining are bedded in mortar or fireclay. Use care in placing the lining units so they are square and plumb. Brick work is carried up along the lining. Then another lining unit is placed.

Where offsets or bends are necessary in the chimney, miter the ends of the abutting sections of lining. This prevents reduction of the flue area. The angle of offset should be limited to 60°. The center of gravity of the upper section must not fall beyond the centerline of the lower wall.

Chimneys are often *corbeled.* This means that successive courses extended outward for several courses. This corbeling is done just before the chimney projects through the roof, Figure 21-4. The larger exterior appearance is often more attractive. Also, breakage due to wind is less likely for thicker masonry. Corbeling should not exceed a 1" projection in each course. The final size should be reached at least 6" below the roof framing.

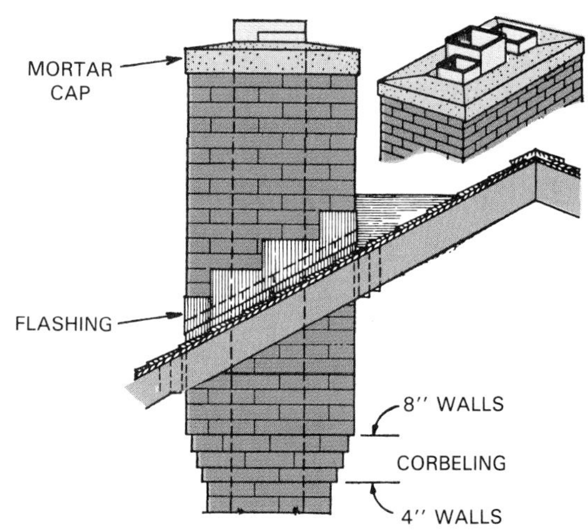

Figure 21-4. Enlarging the top of a chimney helps it resist wind breakage and usually provides a more attractive outline.

Openings in the roof frame should be formed before constructing the chimney. Refer to Unit 10. Water leakage around the base can be prevented by proper flashing. Corrosion resistant metal such as sheet copper is often used for this purpose. The flashing is attached to the roof surface and extends up along the masonry. Cap or counter flashing is bonded into the mortar joints and is then lapped down over the base flashing. Refer to Unit 11 for further details concerning the installation of flashing.

At the top of the chimney, the flue lining should project at least 4" above the top brick course or cap. Surround the lining with cement mortar at least 2" thick. Slope the cap so wind currents are directed upward and water will drain away. When several flues are located in the same construction, extend them to different heights. Space them no closer than 4" apart horizontally.

Masonry Fireplaces

Figure 21-5 (top) shows a cutaway view that reveals the parts of a masonry fireplace. The chimney and fireplace are combined in a single unit which often includes a flue for the regular heating equipment.

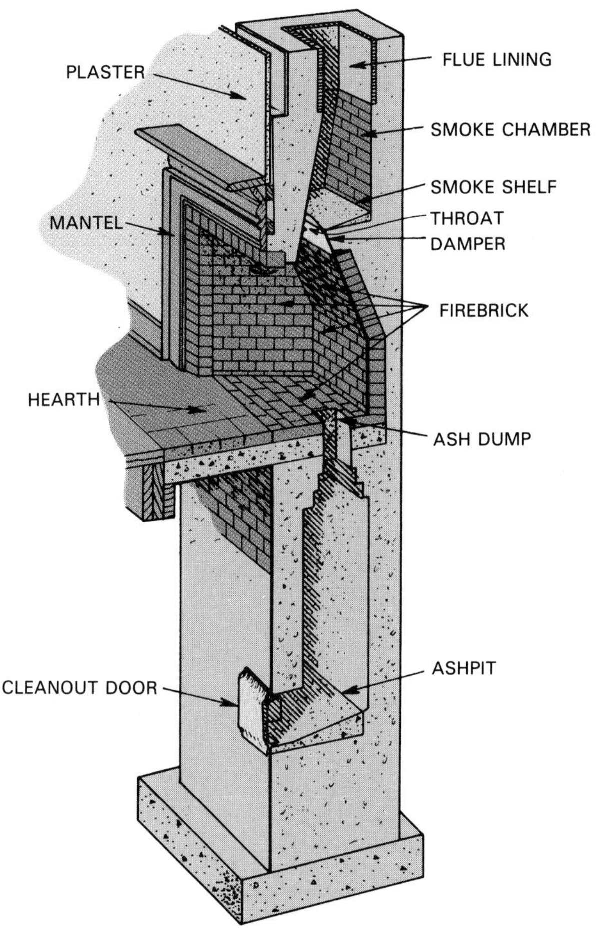

Figure 21-5. Masonry fireplace. Top—Major components of a fireplace. The carpenter who does remodeling should know about many kinds of structures. Bottom—Sometimes a fireplace's footing is tied into the foundation.

The *hearth* consists of two parts: one is located in front (front hearth) and the other is below the fire area (back hearth). The latter, along with the sides and back wall, are lined with firebrick that will withstand direct contact with flame. The side and back walls are sloped to reflect heat into the room.

A *damper* is located above the fire to control combustion. The damper also prevents loss of heat from the room when the fireplace is not being used. The throat, smoke shelf, and smoke chamber are all important parts of the fireplace and must be carefully designed to ensure good operation.

Architectural plans often include details of the fireplace construction. See Figure 21-6. The drawings include overall dimensions and can be scaled (measured) to find other lengths.

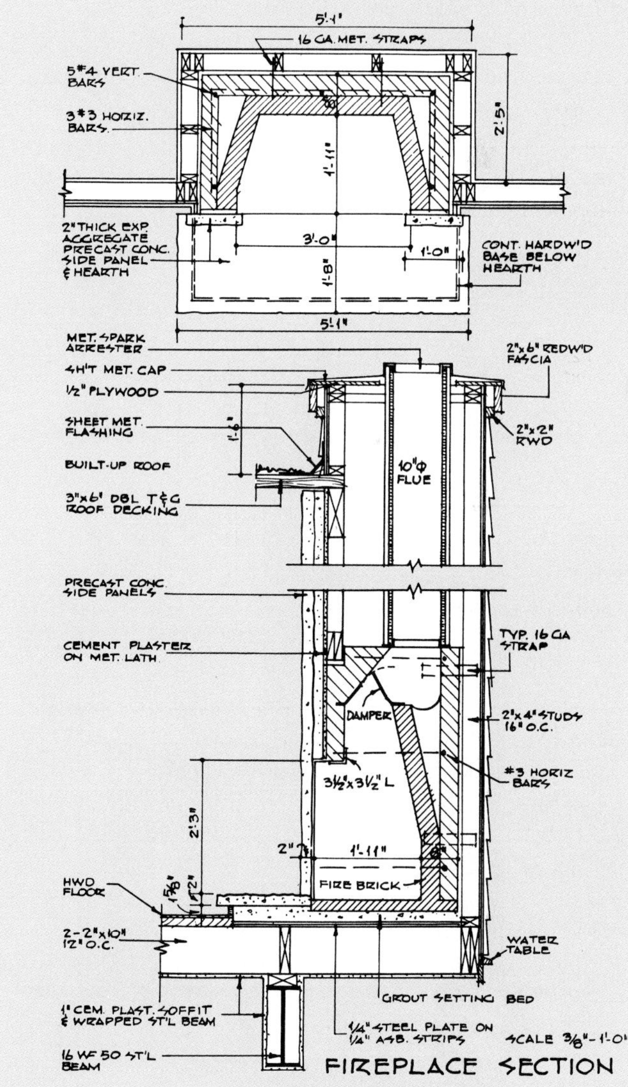

Figure 21-6. Fireplace details are included in the architectural plans. Note the 2 x 4 framework on the outer edge in both views. A chimney frame of wood that juts out from a wall is called a *chase.*

Design Details

The size of the fireplace opening should be based on the size of the room and should be matched with the style of architecture. Some authorities suggest that it accommodate firewood 2' long (standard cordwood cut in half). Figure 21-7 illustrates standard procedure in listing fireplace dimensions.

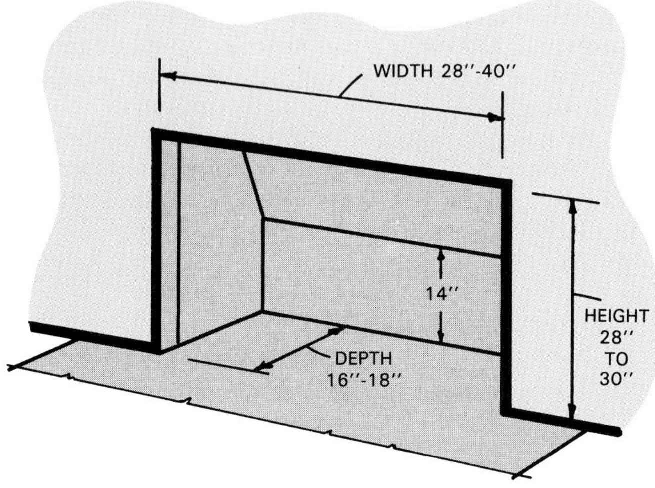

Figure 21-7. Fireplace opening dimensions with range of sizes commonly used.

Masonry structure and wood framing details are shown in the detail drawings of Figure 21-8. This design includes a flue for the furnace. The table in Figure 21-9 lists some recommended dimensions for openings and parts for many fireplace sizes.

Hearth

The hearth, including the front section, must be completely supported by the chimney. The support is constructed by first building shoring (temporary support) and forms. Then concrete (3 1/2" minimum thickness) with reinforcing is poured. In the best construction, a cantilevered design is secured by recessing the back edge of the rough hearth into the rear wall of the chimney.

An *ash dump* may be provided in the rear hearth for clearing ashes, if there is space below for an ash pit. The dump consists of a metal frame with pivoted cover. The basement ash pit should be of tight masonry and include a clean-out door.

When the floor structure consists of a slab-on-grade, an ash pit may be formed by a raised hearth as illustrated in Figure 21-10. This design may be used when the fireplace is located on an outside wall. In some designs, especially when no ash pit is included, the rear hearth is lowered several inches so ashes are contained in this area.

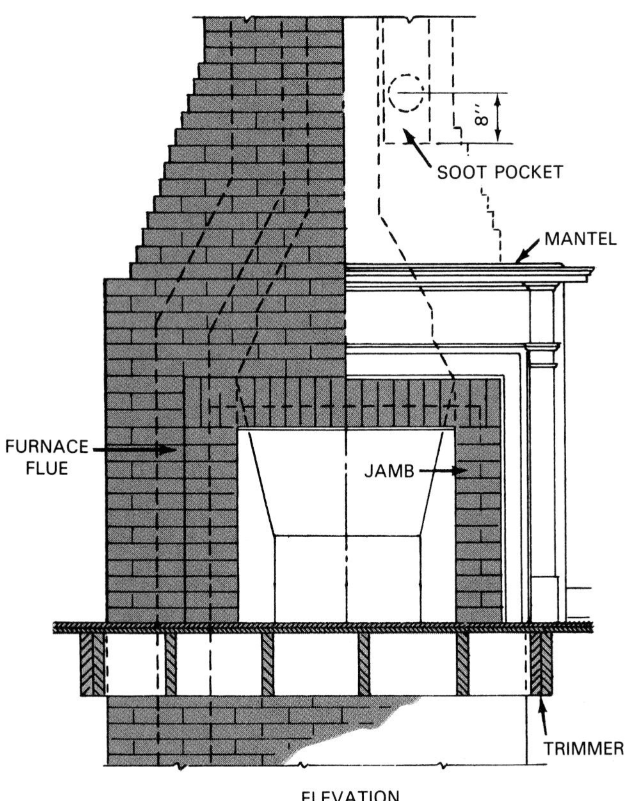

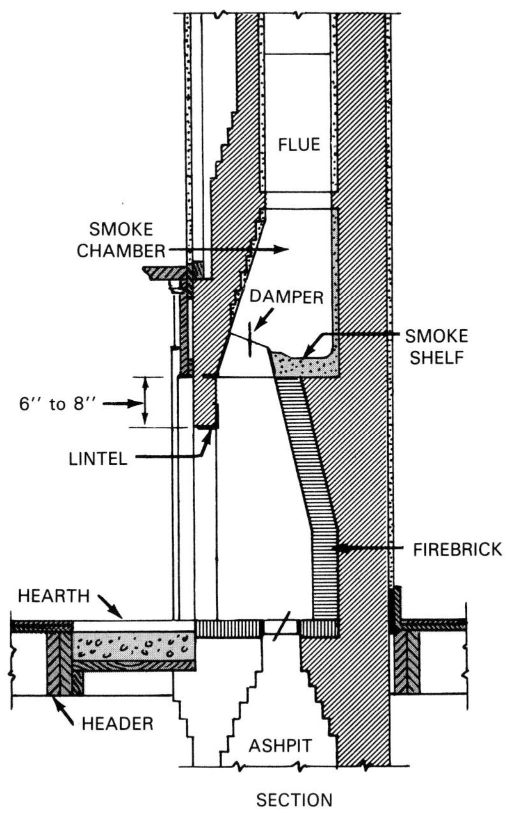

Figure 21-8. Elevation and section views show details of construction for typical masonry fireplace. The carpenter must understand how wood is joined to masonry before doing remodeling or new construction.

Fireplace Opening		Depth	Minimum Back (Horizontal)	Vertical Back Wall	Inclined Back Wall	Outside Dimensions of Standard Rectangular Flue Lining	Inside diameter of Standard Round Flue Lining
Width	Height						
Inches	Inches	Inches	Inches	Inches	Inches	Inches	Inches
24	24	16-18	14	14	16	8 1/2 by 8 1/2	10
28	24	16-18	14	14	16	8 1/2 by 8 1/2	10
24	28	16-18	14	14	20	8 1/2 by 8 1/2	10
30	28	16-18	16	14	20	8 1/2 by 13	10
36	28	16-18	22	14	20	8 1/2 by 13	12
42	28	16-18	28	14	20	8 1/2 by 18	12
36	32	18-20	20	14	24	8 1/2 by 18	12
42	32	18-20	26	14	24	13 by 13	12
48	32	18-20	32	14	24	13 by 13	15
42	36	18-20	26	14	28	13 by 13	15
48	36	18-20	32	14	28	13 by 18	15
54	36	18-20	38	14	28	13 by 18	15
60	36	18-20	44	14	28	13 by 18	15
42	40	20-22	24	17	29	13 by 13	15
48	40	20-22	30	17	29	13 by 18	15
54	40	20-22	36	17	29	13 by 18	15
60	40	20-22	42	17	29	18 by 18	18
66	40	20-22	48	17	29	18 by 18	18
72	40	22-28	51	17	29	18 by 18	18

Figure 21-9. Recommended dimensions for a wide range of fireplace sizes.

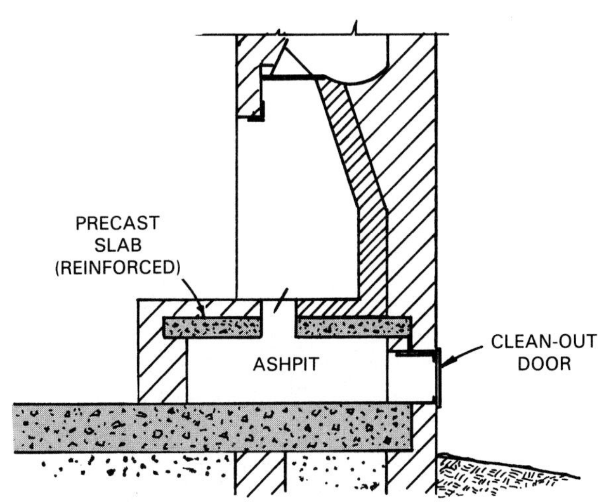

Figure 21-10. A raised hearth provides space for an ash pit in slab-on-grade construction. This type of ash pit is not used in regions where snow reaches door height.

Side and Back Walls

The side and back walls of the combustion chamber continue upward to the level of the damper. The wall must be lined with firebrick at least 2" thick. The firebrick are set in a special clay mortar that will withstand the heat. The total thickness of the walls, including the firebrick, should not be less than 8".

Side walls are angled to reflect heat into the room. This angle (also called *splay*) is usually laid out at 5" per foot. The back wall extends vertically from the base a distance slightly less than one-half of the opening height and then slopes forward. This slope directs the smoke into the throat of the fireplace. The slope also keeps an area clear for the smoke shelf that will be located above.

Damper and Throat

Part of the fireplace throat is formed by the damper, Figure 21-11. A stationary front flange is angled a small amount away from the masonry of the front wall. The back flange is movable.

The damper affects a flow called the downdraft. In Figure 21-11, arrows indicate the upward flow of hot air and smoke from the throat into the front side of the smoke chamber. The rapid upward passage of hot gases creates a downward current *(downdraft)* on the opposite side. One purpose of the damper is to change the direction of the downdraft so smoke is not forced into the room.

The damper is installed so that masonry work above will not interfere with its full operation. Also, the ends have a slight clearance in the masonry to permit expansion. A standard damper design is shown in Figure 21-12. Manufacturers provide data and recommendations concerning model and size for specific installations.

Correct *throat* size controls the efficiency of a fireplace. The highest efficiency occurs when the cross-sectional area of the throat is equal to that of the flue. The length of the throat along the face of the fireplace is the

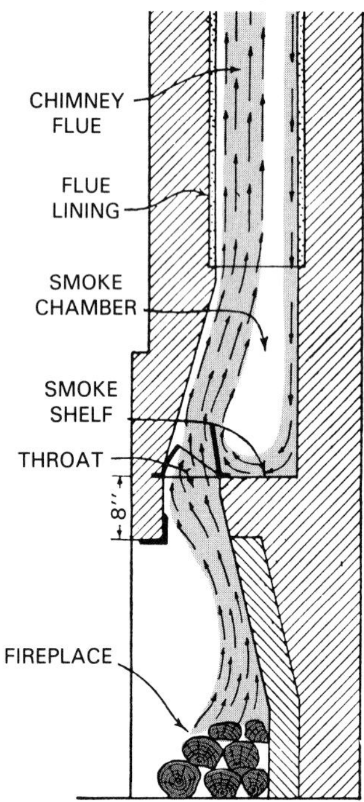

Figure 21-11. Details of operation of fireplace. Arrows indicate air and smoke currents. Note how damper keeps downward current from entering room.

Figure 21-12. Top view of a standard fireplace damper.

same as the opening width. Horizontal depth is much less to get the proper cross section.

Smoke Shelf and Chamber

The *smoke shelf* helps the damper to change the flow direction of the downdraft. The deeper the shelf behind the damper, the better the fireplace works. The depth may vary from 4" to 12", depending on the depth of the fireplace. Some smoke shelves are curved to reduce turbulence in the airflow. The length along the face for all types is equal to the full width of the throat.

The *smoke chamber* is the space extending from the top of the throat up to the bottom of the flue. The area at the bottom of the chamber is quite large, since its width includes that of the throat plus the depth of the smoke shelf. This space holds accumulated smoke temporarily if a gust of wind across the top of the chimney momentarily cuts off the draft. Without this chamber, smoke would likely be forced out into the room. A smoke chamber also lessens the force of the downdraft by increasing the area through which it passes. Side walls are generally drawn inward one foot for each 18" of rise. The surfaces of most smoke chambers are plastered with about 1/2" thick cement mortar.

Flue Size

The cross-sectional area of the flue is based on the area of the fireplace opening. As a general rule, the flue area should be at least 1/10 of the total opening when a lining is used. This applies to chimneys that are at least 20' high. A somewhat larger flue may be required in the lower chimney heights normally used in modern single-story construction. The upward movement of smoke in a low chimney does not reach a high velocity—thus a greater cross-sectional area is required.

One recommended method of calculating the area for a flue is to allow 13 sq. in. of area for the chimney flue to every square foot of fireplace opening. For example: if the fireplace opening equaled 8.25 sq. ft., then the flue area should equal at least 107 sq. in. If the flue is to be built of brick and unlined, it would probably be made 8" x 16", or 128 sq. in., because brickwork can be laid to better advantage if the dimensions of the flue are multiples of 4". If the flue is lined, and lining is strongly recommended, the lining should have an inside area of at least 107 sq. in.

Construction Sequence

Masonry fireplaces are nearly always built in two stages. The first begins during the rough framing of the structure. Masonry work is carried up from the foundation and the main walls of the fireplace are formed. After a steel lintel is set above the opening, the damper is installed and the smoke chamber built. Then the chimney is carried upward through the roof and the exterior masonry is completed. These steps usually occur before the roof deck is laid.

The second and finishing stage of the fireplace takes place after plastering or other type of wall surface is complete and during the application of interior trim. Decorative brick or stone may be set over the exposed front face. The surface of the front hearth can be finished at this time, since the reinforced concrete base was placed during the rough masonry construction. Install wood trim (mantel) when masonry work is complete.

Special Designs

Contemporary fireplace designs often have openings on two or more sides. For these, follow the same principles in planning as previously described for conventional designs. When calculating the flue area, the sum of the area of all faces must be used. Figure 21-13 illustrates a corner fireplace design in which the flue area is based on the total

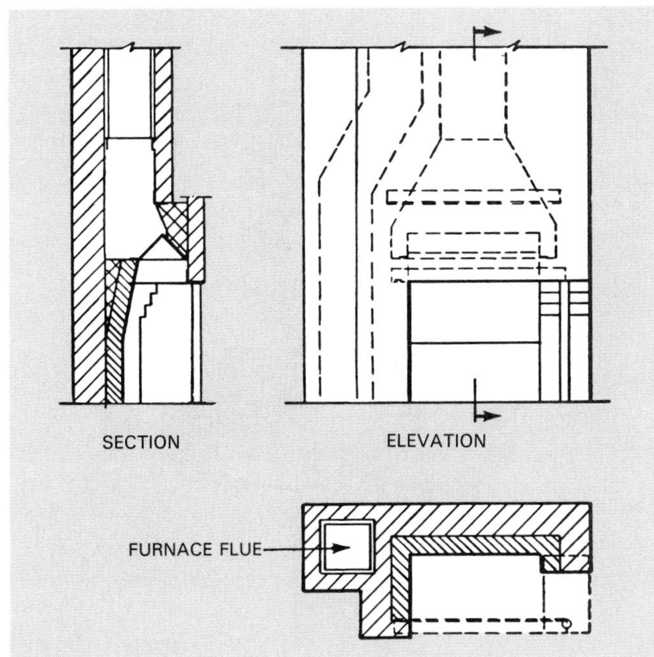

Figure 21-13. General design for a projecting corner fireplace.

of the front face opening, plus the end face opening. In this particular construction, the side walls are not splayed. However, the rear wall is sloped in the usual manner.

Multiface fireplaces must incorporate a throat and damper with requirements similar to standard designs. Special dampers with square ends and sides are available for two-way fireplaces that serve adjoining rooms.

Built-In Circulators

The heating capacity of a fireplace can be increased by using a factory-built metal unit as shown in Figure 21-14. The sides and back are double walled, providing a space where air is heated. Cool air enters this chamber near the floor level. When the air is heated, it rises and returns to the room through registers at a higher level.

Modern built-in circulators (also called modified fireplaces) include not only the firebox and heating chamber, but also the throat, damper, smoke shelf, and smoke chamber. Since all of these parts are carefully engineered, proper flue draw is assured when the installation is made according to the manufacturer's directions and flue size is adequate.

To install a circulator unit, first position it on the hearth and then build the brick and masonry work around the outside, Figure 21-15. Steel lintels are required across the top of the opening, as shown, and may be required in other locations to provide support. The unit itself should not be used for support of any masonry work.

When installing, follow specifications furnished by the manufacturer. Some type of fireproof insulating material is usually placed around the metal form not only to prevent the movement of heat but also to provide some

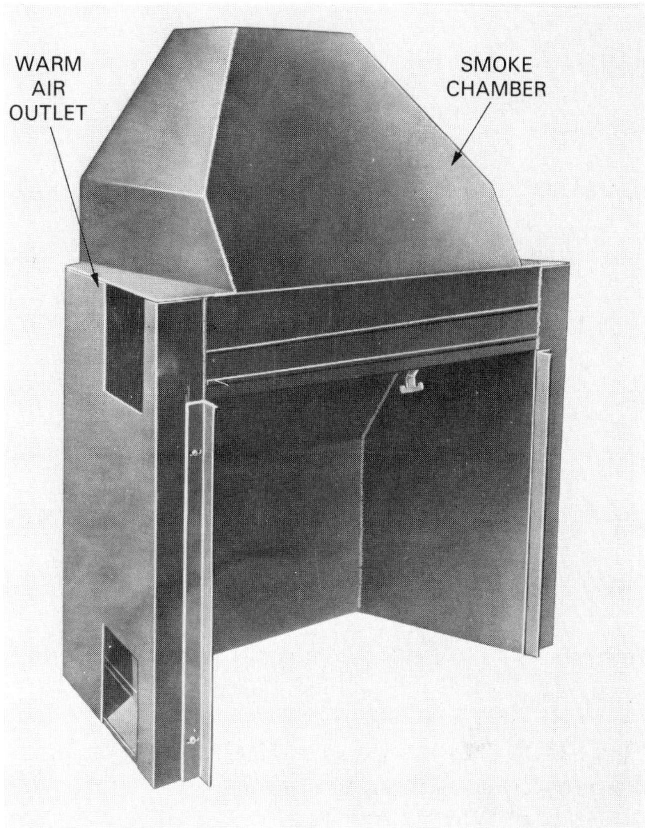

Figure 21 14. Modern fireplace circulator made of heavy gauge metal. The sides and back consist of a double wall. (Majestic Co.)

expansion space between the metal and masonry.

In any type of fireplace, correct operation will depend somewhat on an adequate flow of air into the building to replace the air exhausted through the flue. Usually infiltration around doors and windows is sufficient. However, where weather stripping is tight, some inlet should be provided. Energy efficiency suggests that combustion air be piped to the fireplace from the outside.

Prefabricated Chimneys

Lightweight chimney units that require no masonry work are available. They provide flues for heating equipment or fireplaces and can be installed in one-story or multistory structures, Figure 21-16. Prefabricated chimneys usually consist of double or triple-walled sections of pipe that are assembled to form the flue. Special flanges and fittings are used to fasten the flue to the building frame and provide proper clearance from wood members. Figure 21-17 shows details of a typical installation.

The roof-top section of a prefabricated chimney unit must be sealed to prevent roof leaks. Figure 21-18 shows the basic parts of a simple pipe projection. Bed the flashing unit in mastic with the roofing material overlapping the top and side edges. The storm collar diverts rain water

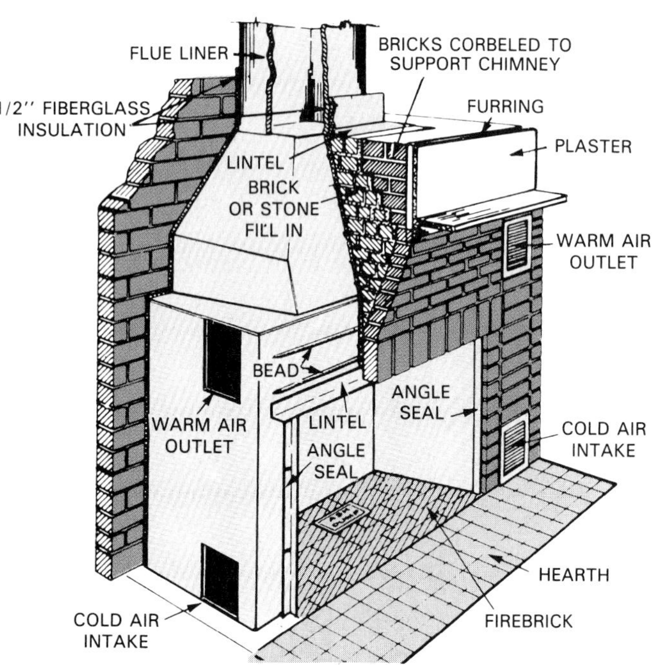

Figure 21-15. Cutaway view shows how masonry is installed around circulator unit. (Majestic Co.)

Figure 21-16. Prefabricated chimneys serve fireplaces and furnaces in a multiple-unit housing complex. The carpenter can often install the chimney sections in addition to building the support framework for the chase. (Preway Inc.)

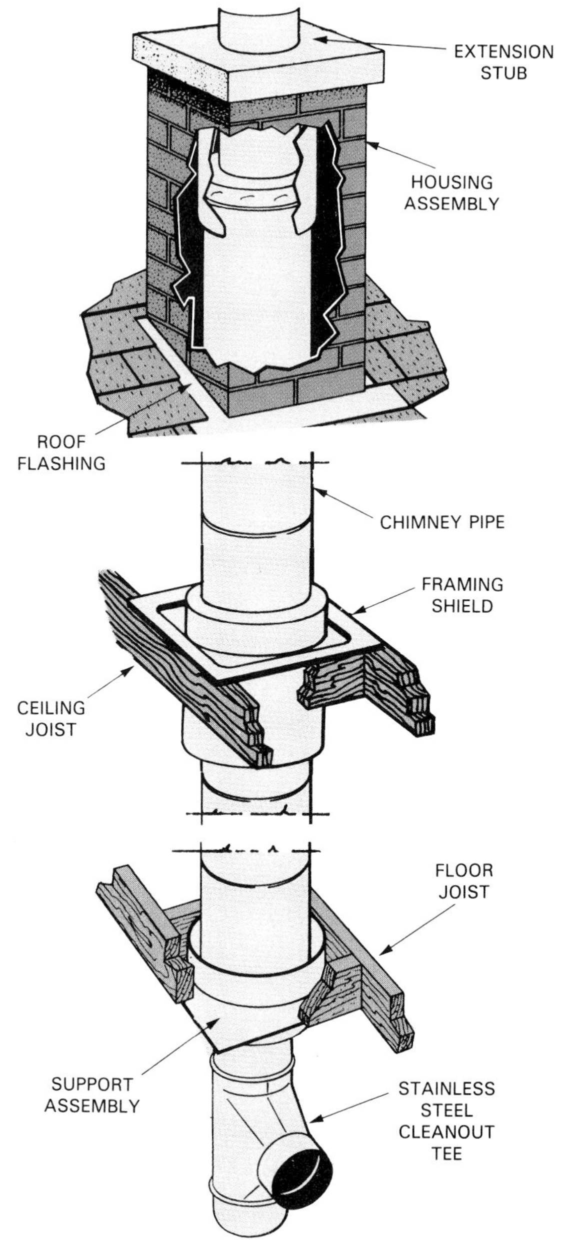

Figure 21-17. Prefabricated chimney. Pipe sections consist of a triple metal wall. Inside diameters range from 6" to 14". Various types of rain caps can be mounted on extension stub. (Wallace Murray Corp.)

from the pipe to a conical section of the flashing. Some type of cap should be installed on the top of the flue to keep out rain or snow.

Many types of prefabricated *termination tops* (part of the chimney that extends above the roof) are available. For the best appearance they should blend with the architectural style of the building. Figure 21-19 shows several typical designs. Be sure to follow the manufacturer's directions for assembly and installation.

When a prefabricated chimney is used for venting smoke and gases that may contain corrosive acids (incinerators or solid fuel boilers), the inside pipe should be made of stainless steel or have a porcelain coated surface.

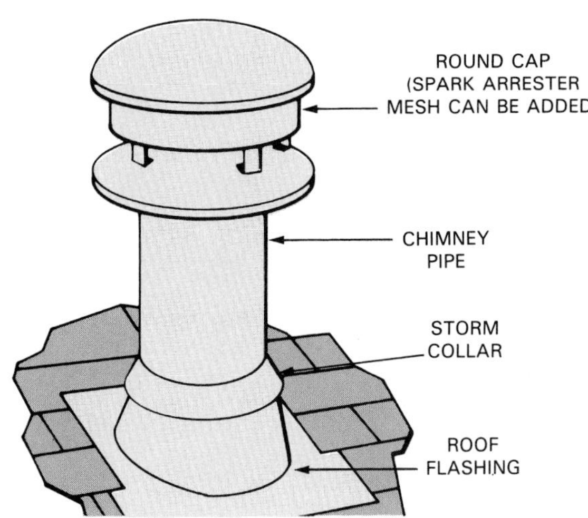

Figure 21-18. Basic parts of a rooftop projection for a standard flue pipe.

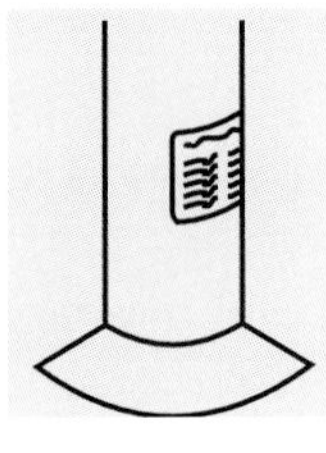

A prefabricated chimney should have a label showing that it has been tested and listed by the Underwriters' Laboratories, Inc. or other nationally recognized testing organizations. It must also conform to local building codes. Be sure to make the installation in strict accordance with the manufacturer's instructions.

Prefabricated Fireplaces

Many of the fireplaces in today's new homes are factory-built units. These units are generally less expensive than masonry units.

Manufacturers of prefabricated fireplaces provide a wide range of designs that are easy to install with ordinary tools, Figure 21-20. In some cases, multiple steel wall construction and special firebox linings permit *zero clearance*. This means that outside housings can rest directly on wood floors and touch wood framing members. Follow the manufacturer's recommendations when installing.

Other precautions should be taken if gas or electric starters are used or if gas logs are installed. Check local fire codes before starting work.

Prefabricated fireplaces operate like the metal circulator built into masonry fireplaces. Room air enters intakes at floor level and flows through chambers around the firebox. As the air is heated, it rises. This motion carries the air through the grillwork at the top of the unit and back into the room. Some units are equipped with circulating fans that are controlled by switches. See Figure 21-21.

Some units can be attached to vertical ducts that carry the heated air to various locations within the room or to an adjoining room. See Figure 21-22. Some prefabricated fireplaces have blowers that increase the flow of air and thus improve heating efficiency.

Figure 21-23 shows a partially completed installation; Figure 21-24 shows framing and masonry built around a prefabricated unit. In new construction, the basic frame is usually built at the same time as exterior walls and partitions. The facing side of the frame should remain open until the fireplace is installed. If support is required above

Figure 21-19. Rooftop housings and termination caps (rain caps) for prefabricated chimneys. Left—Traditional design of simulated brick made from embossed metal. Center—Simulated stucco with wood trim. Right—Wooden framework covered with vinyl siding system to match exterior walls.

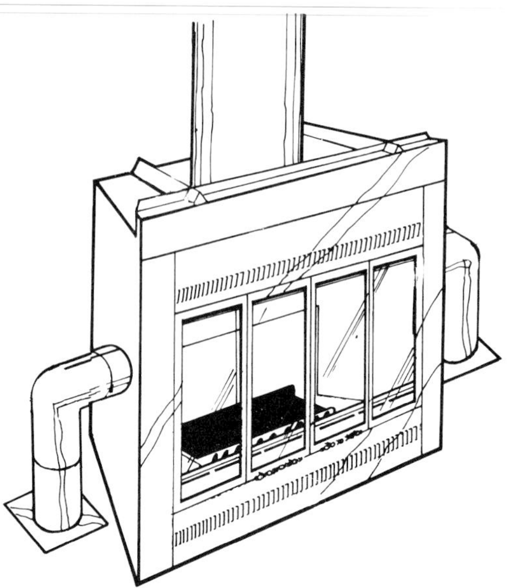

Figure 21-20. A prefabricated fireplace consists of a metal shell, firebrick, ductwork, and a face panel. Two persons can lift most units. Installation may take as little as one-half hour. (Preway Inc.)

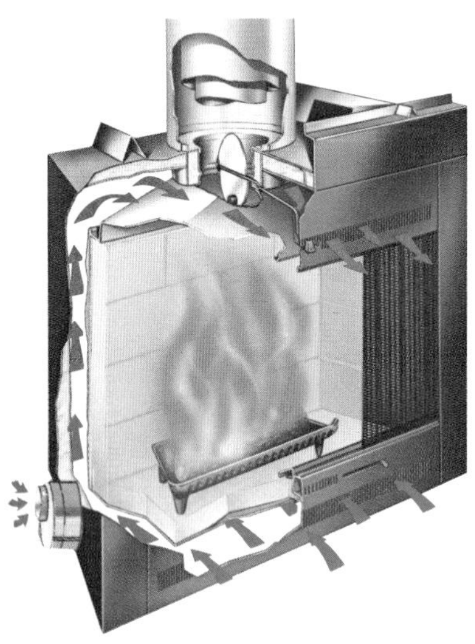

Figure 21-21. Cutaway view with glass removed shows basic operation of prefabricated fireplace. Room air (blue arrows) is drawn from floor level into heating chamber. There it is warmed and returned to the room (red arrows). Air for burning can be piped from outside and connected to inlet at lower left. Note triple-walled flue. (Preway Inc.)

this opening, it should be framed with a header like that for a door. After the fireplace unit is installed, any front framing below the header can be added.

Manufacturers provide detailed drawings and instructions for the installation of their fireplace units. These

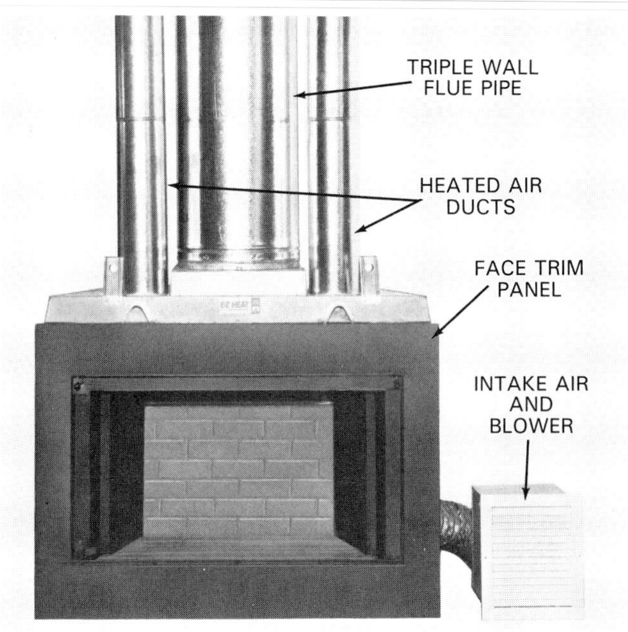

Figure 21-22. Prefabricated fireplace with ducts to return heated air to one or more rooms. Blower is optional. (Superior Fireplace Co.)

should be carefully studied and followed throughout installation.

Many materials, including both wood and masonry, can be used to finish the wall and trim the fireplace opening. According to FHA specifications, wooden parts should not be placed closer than 3 1/2" to the edge of the opening. Greater clearance is required when the parts project more than 1 1/2". For example: a wooden mantel shelf must be 12" above the opening.

Figure 21-23. A partially completed installation. Prefabricated insert is surrounded with concrete block. Note opening above lintel. A grill will be installed for circulation of heated air.

Figure 21-24. Top—Cutaway view shows basic wood framing for fireplace and ductwork. (Superior Fireplace Co.) Middle—Brickwork is being laid for the hearth and mantel. Bottom—Brickwork as it appears when completed.

Free-standing fireplaces are usually fabricated with double walls somewhat like built-in units, Figure 21-25. Proper support exists in many rooms in the home and the units are easy to install. A single-walled pipe runs from the unit to the ceiling where it is fitted to a triple-walled section.

Figure 21-25. This free-standing fireplace is made of heavy gage steel with porcelain finish. It requires a 36" clearance from combustible wall. (Malm Fireplaces Inc.)

Figure 21-26 shows a wall-mounted unit which is partially supported by the floor. Triple-wall construction is used around the firebox so the fireplace unit can be placed against any combustible material. Free-standing fireplaces are available with many of the same features of built-in models. They can be equipped with a blower to force the circulation of air around the firebox. Air for combustion can be drawn from the outside.

Before selecting and installing any type of factory-built fireplace, be sure to check local building codes. Failure to comply with these requirements can result in the installation being disapproved by the local building inspector.

Chimneys for Prefabricated Fireplaces

Chimney (flue pipe) systems for prefabricated fireplaces are designed for specific units and usually are sold as a package along with the fireplace. Figure 21-27 shows the general assembly of standard components that run through the ceiling and roof structure.

Special attention must be given to clearance and support. Openings through the ceiling and roof should be carefully framed to provide the correct size openings for fire-stop spacers and support boxes. Figure 21-28 shows the installation of a fire-stop spacer in an attic area.

The chimney pipe must be supported at the roof. If not, the entire weight of the pipe will rest on the fireplace. By supporting the chimney pipe at the roof, its weight is supported more even and the pipe is more stable.

When installed on an exterior wall, the chimney and fireplace unit can be located in a special projection called a *chase.* This is a box-like structure that is built as a part of the floor, wall, or roof frame, Figure 21-29 and Figure 21-30. The outer walls of the chase are usually finished to match the outer walls of the house as shown in Figure 21-31. Insulate the walls of the chase and seal them against infiltration. Be sure insulation does not touch the flue pipe.

When a chase is not used, the fireplace and chimney are located within the room confines. This reduces the amount of available living space. By using a chase, the room area is not reduced by the presence of the fireplace.

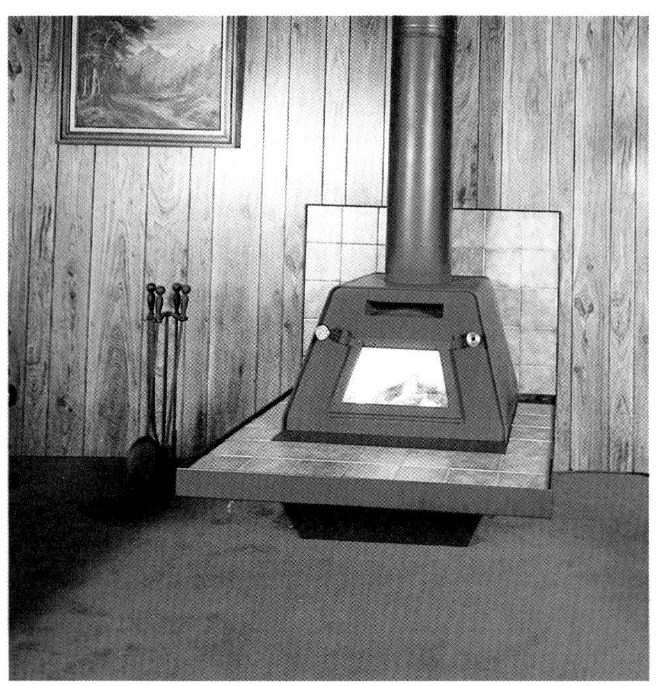

Figure 21-26. Wall-mounted fireplace. Hearth corners are cantilevered. It includes a 300 cfm (cubic feet per minute) blower and ceramic glass door. (Malm Fireplaces Inc.)

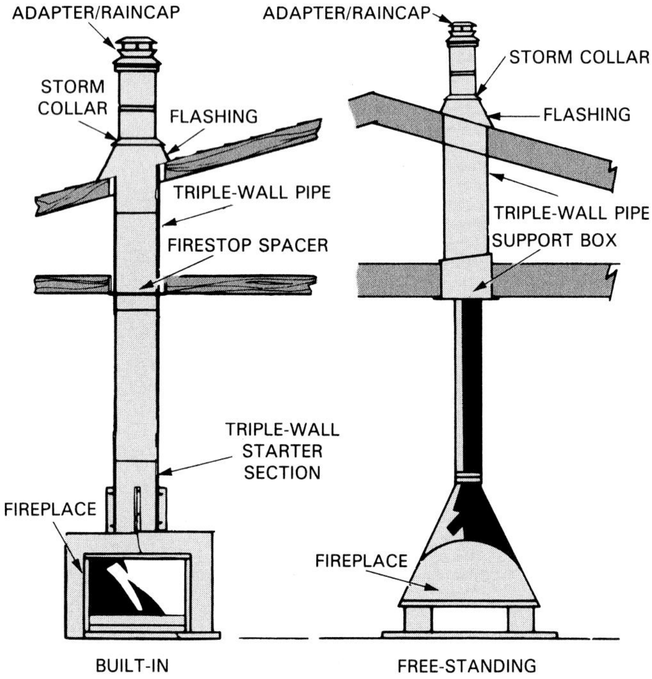

Figure 21-28. Installing fire-stop spacer after cutting and framing ceiling opening. Triple-walled pipe is supported by fireplace in room below. It is a good practice to close off vertical channels of any kind in a building frame. (Majestic Co.)

ADAPTER/RAINCAP
STORM COLLAR
FLASHING
TRIPLE-WALL PIPE
FIRESTOP SPACER
TRIPLE-WALL STARTER SECTION
FIREPLACE
BUILT-IN
ADAPTER/RAINCAP
STORM COLLAR
FLASHING
TRIPLE-WALL PIPE
SUPPORT BOX
FIREPLACE
FREE-STANDING

Figure 21-27. Chimney system for prefabricated fireplaces. (Preway Inc.)

Glass Enclosures

Glass enclosures improve the efficiency of fireplaces and can be installed on either masonry structures or prefabricated units. They reduce the amount of heated room air that escapes up the chimney, even when there is no fire burning. Efficiency is improved because the rate of burning can be controlled by adjusting draft vents and dampers.

Figure 21-29. Typical framing for a chimney chase. The roof framework will hold the chase steady on three sides. (American Plywood Assoc.)

Figure 21-30. Another type of chase construction. To make use of standard length studs, the chase is built in two sections. Uprights are joined at the midpoint with overlapping horizontal strips. (American Plywood Assoc.)

Figure 21-31. View of a completed chase. Very little weight is supported at the base of the framework since the fireplace is a metal shell.

Important Terms

Ash dump	Flue lining
Chase	Hearth
Chimney	Smoke chamber
Corbeled	Smoke shelf
Damper	Splay
Downdraft	Termination top
Fireclay	Throat
Flue	Zero clearance

Test Your Knowledge

1. The overall size of a chimney will depend on the materials from which it is constructed and the number and size of the _____________.
2. Wood framing should be located at least _____________ inches away from a masonry chimney.
3. The offset angle in a chimney should be limited to _____________ degrees.
4. To prevent heat loss from a room when the fireplace is not in operation, the fireplace must be equipped with a _____________.
5. The length of the throat along the face should be equal to the _____________ of the fireplace opening.
6. All surfaces of the smoke chamber should be finished with a 1/2" thickness of _____________.

7. As a general rule, the cross-sectional flue area should equal about _____________ of the total area of the fireplace opening.
8. Prefabricated chimneys consist of pipe sections with double or _____________ walls.
9. When a prefabricated metal chimney is used for an incinerator, the inside pipe should be coated with porcelain or made of _____________.
10. An outside air intake on a prefabricated fireplace provides air for _____________ *(burning, circulating).*
11. A chase is made rigid with decorative sheathing and with support from wall and _____________ framing.

Outside Assignments

1. Write a report on the historical development of the fireplace. Start with the simple designs used by primitive humans and highlight developments through the years. Place special emphasis on the materials used for building fireplaces and chimneys. Is it possible to determine:
   A. When metal parts became popular?
   B. When safe designs with wood structures became common? Ask your school librarian or media specialist to suggest books that would be helpful.
2. Get a copy of the building code that applies to your locality and study the sections that deal with chimneys, fireplaces, and venting systems. Make a list of the specific requirements concerning residential structures. Include information about masonry and prefabricated metal chimneys for heating plants, incinerators, and fireplaces. Also include requirements and restrictions for the use and installation of prefabricated fireplaces. Make a report to your class.
3. Interview a fireman on proper installation and use of fireplaces. Prepare a list of safety rules from the interview.

Modern prefabricated fireplace. Room air is drawn from both sides into sealed heating chamber and flows back into room thorugh top louvers. Air fire is piped in from outside. Bi-fold glass doores limit the loss of heated room air through the chminey. (Majestic Co.)

22

Post-and-Beam Construction

Post-and-beam construction consists of large framing members—posts, beams, and planks. Because of their great strength, these members may be spaced farther apart than conventional framing members. Frames of this type are similar to "mill construction" once used for barns and heavy-timbered buildings. It is often used today for upscale residential building since it permits greater flexibility in contemporary and traditional designs than conventional framing methods. See Figure 22-1 for a comparison of lumber sizes and construction methods.

Post-and-beam construction is also known as plank-and-beam construction. It is often combined with conventional framing, Figure 22-2. For example, the walls might be built conventionally and the roof framed with beams and planks. In such a case, the term *plank-and-beam* could be applied to the roof structure only. Similarly, it would be correct to refer to a heavy-timbered floor structure as a "plank-and-beam" system.

Advantages

Advantages of this construction are the distinctive architectural effect created by the exposed beams in the

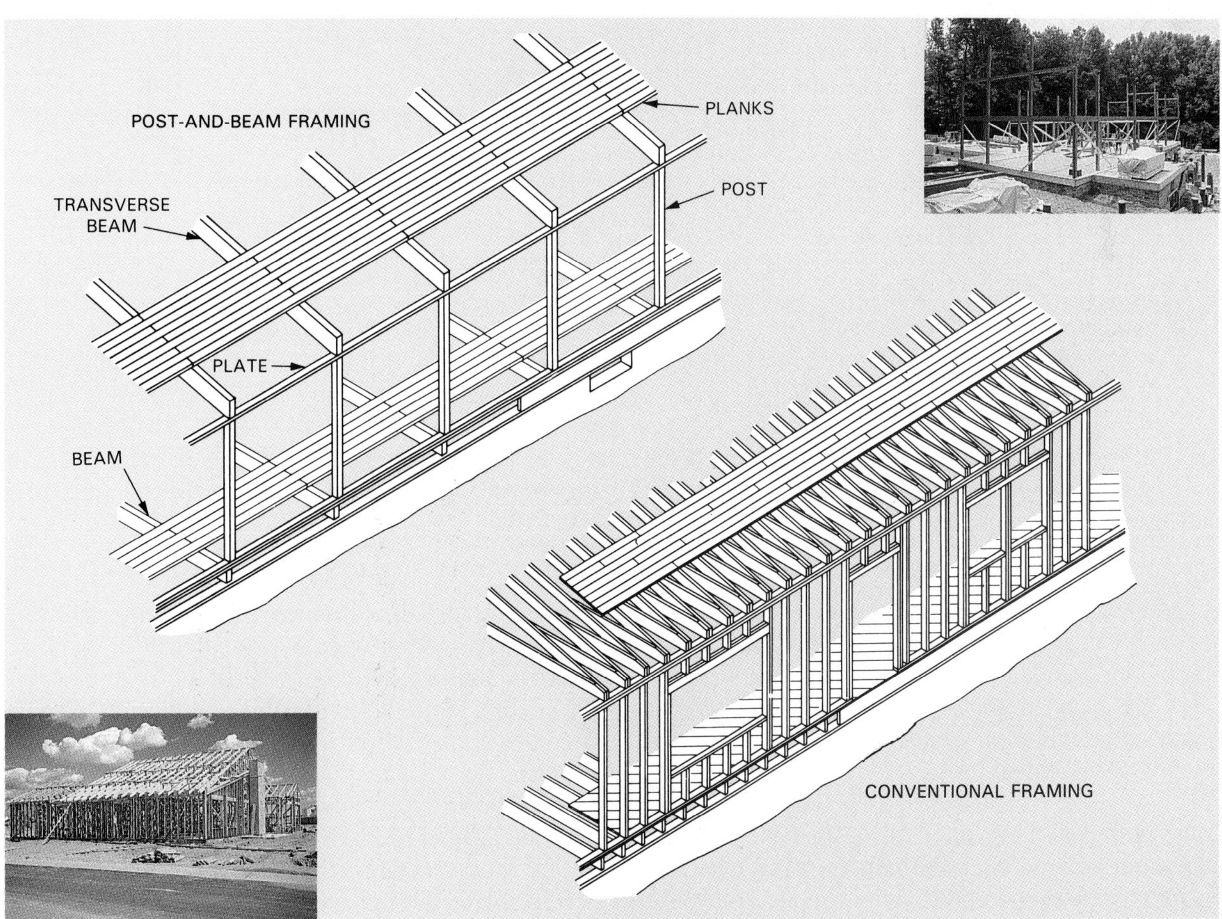

Figure 22-1. Comparison of typical post-and-beam framing and conventional framing. In post-and-beam construction, framing around large windows and doors is simplified since headers can be eliminated. (National Forest Products Assoc.)

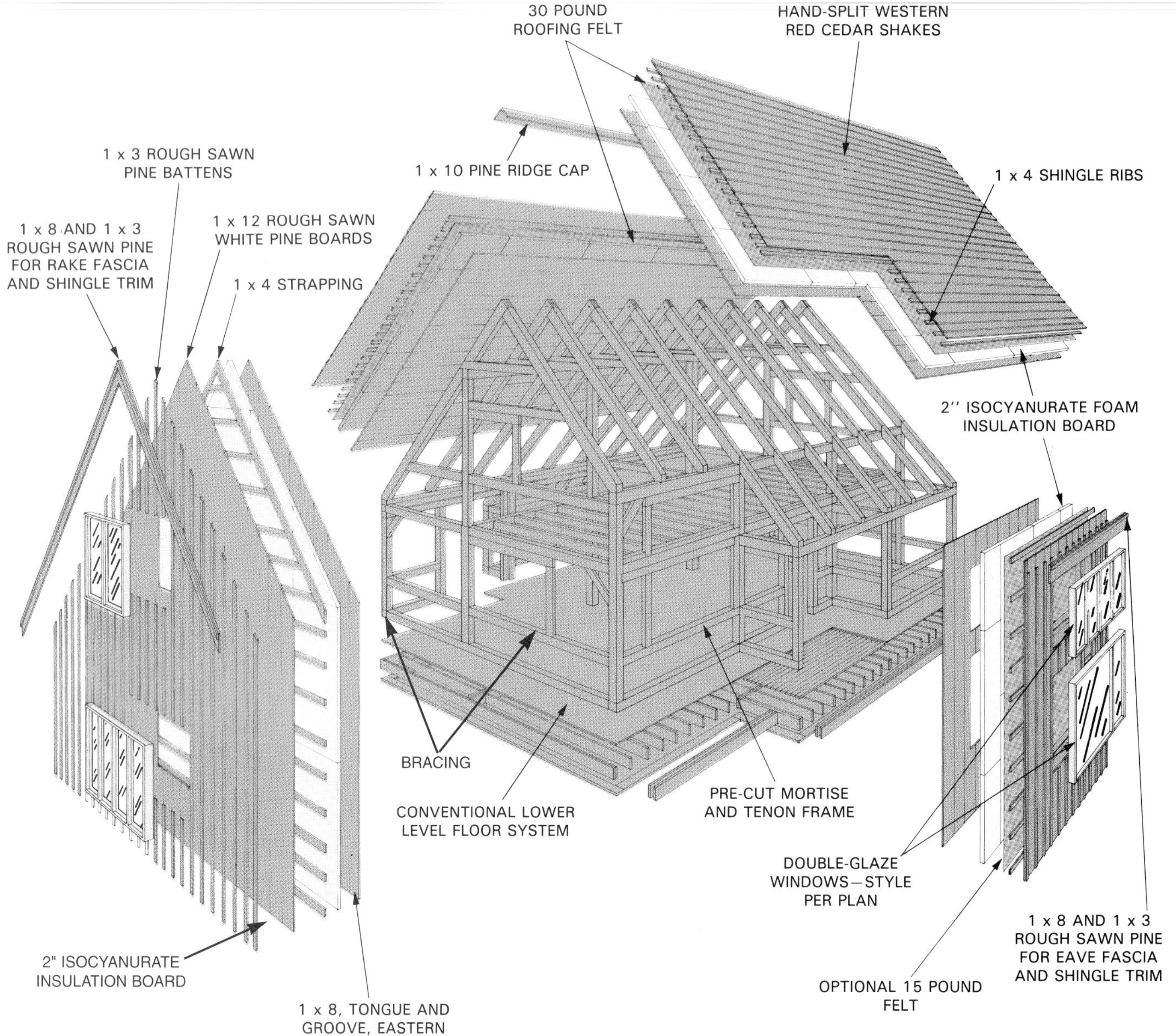

Figure 22-2. This rustic variation of the post-and-beam house follows the traditional design of colonial barn framing. Conventional flooring system is used on the first level; heavy beams are used on the second floor. Note angle braces in the end wall. Pine siding also stiffens walls against racking. (Timberpeg)

ceiling and the added height. See Figure 22-3. The underside of the roof planks may serve as the ceiling surface, thus providing a saving in material.

Post-and-beam framing may also provide some saving in labor. The pieces are larger and fewer in number and can usually be assembled more rapidly than conventional framing.

One of the chief structural advantages is the simplicity of framing around door and window openings. Loads are carried by posts spaced at wide intervals in the walls. Large openings can be framed without the need for headers, Figure 22-4. Window walls, characteristic of contemporary architectural styling, can be formed by merely inserting window frames between the posts. Another advantage is that wide overhangs can be built by simply extending the heavy roof beams.

In addition to its flexibility in design, post-and-beam construction also provides high resistance to fire. Wood beams do not transmit heat and collapse like unprotected metal beams. Exposure of wood beams to flame results in a slow loss of strength.

Most limitations of post-and-beam construction can be resolved through careful planning. The plank floors, for example, are designed to carry moderate uniform loads. Therefore, extra framing must be provided under bearing partitions, bath tubs, refrigerators, and other places where heavy loads are likely. Extra members may be needed to provide lateral stability to the frame and walls. This might be provided with various types of bracing. It is more common, however, to enclose some of the wall area with large panels and use conventional stud constructions as shown in Figure 22-5. Absence of concealed spaces in outside walls and ceilings, makes installation of electrical wiring, plumbing, and heating somewhat more difficult.

Figure 22-3. Exposed ceiling beams provide an attractive architectural feature while adding to ceiling height. (Georgia-Pacific Corp.)

Figure 22-4. Huge expanses of windows are possible with post-and-beam construction, since there is no need for headers. (Yankee Barn Homes, Inc.)

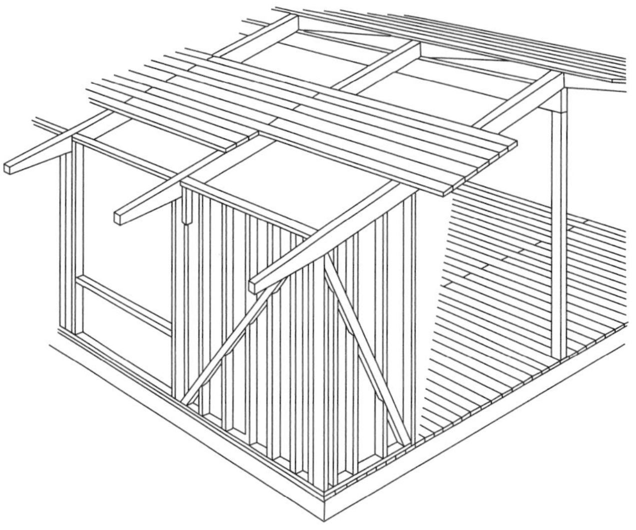

Figure 22-5. Conventional stud wall with let-in bracing can be used to provide lateral stability. Refer to Figure 22-2 for another type of bracing sometimes used.

Foundations and Posts

Foundations for post-and-beam framing may consist of continuous walls or simple piers located under each post. Refer once more to Unit 7 on foundation types. Either type of foundation must rest on footings that meet the requirements of local building codes.

Posts must be strong enough to support the load and also large enough to provide full bearing surfaces for the ends of the beams. In general, posts should not be less than 4 x 4 nominal size. They may be made of solid stock or built up from 2" pieces. Where the ends of beams are joined over a post, the bearing surface should be increased with bearing blocks as shown in Figure 22-6.

When posts extend any great height without lateral bracing, a heavier cross-sectional area is required to prevent buckling. Requirements are usually listed in local building codes through an l/d ratio (also called slenderness ratio). The "l" represents the length in inches and the "d" stands for the smallest cross-sectional dimension (actual size). For example, a 4 x 4 piece 8' long would have a ratio of about 27. This is within the limits usually prescribed.

Distance between posts will be determined by the basic design of the structure. This spacing must be carefully engineered. Usually posts are spaced evenly along the length of the building and within the allowable free span of the floor or roof planks. With today's emphasis on modular dimensions, construction costs will be cheaper if post-and-beam positions occur at standard increments (increases) of 16", 24", and 48".

In single-story construction, a plate is attached to the top of the posts in about the same way as conventional framing. The roof beams are then positioned directly over the posts as shown in Figure 22-7.

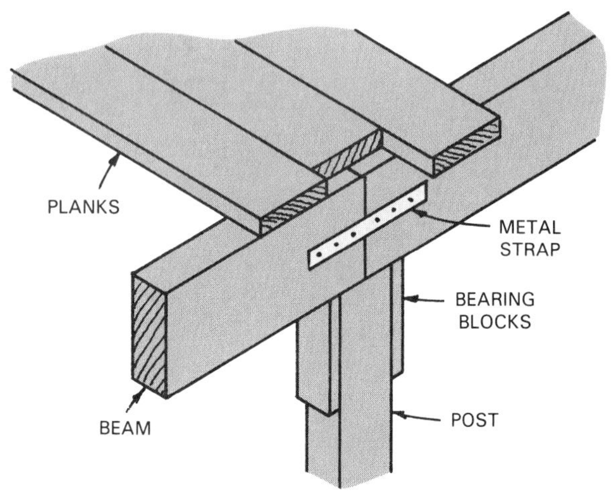

Figure 22-6. Provide adequate bearing surface where beams are joined over a post. Heavy steel plate may replace the bearing blocks.

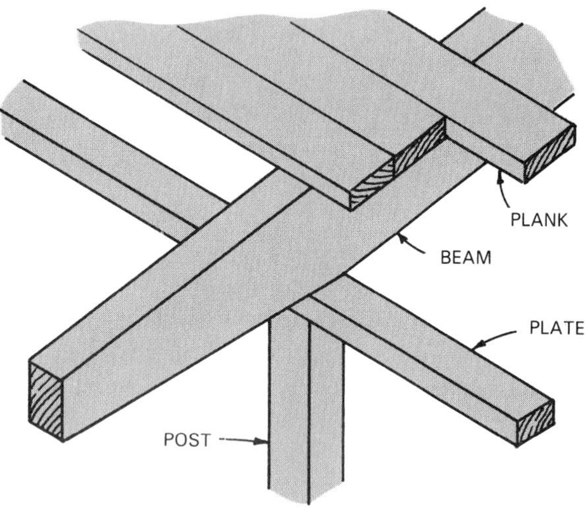

Figure 22-7. Roof beams must rest directly over supporting posts to prevent sagging.

Floor Beams

Beams for floor structures may be solid, glue-laminated, or built-up, Figure 22-8. Sometimes the built-up beams are formed with spacer blocks between the main members. Box beams, described in another section of this Unit, may also be used.

For one-story structures, where under-the-floor appearance is not critical, standard dimension lumber can be nailed together to form any size of beam. Figure 22-9 illustrates how 2 x 8s have been nailed together and then installed 4' on center. The decking being applied is 1 1/8" plywood with a special tongue-and-groove edge.

Design of sills for a plank-and-beam floor system can be similar to regular platform construction, Figure 22-10. When it is desirable to keep the silhouette of the structure low (floor level near grade level) the beams can be supported in pockets in the foundation wall.

Beam Descriptions

In general, it is best to use solid timbers when beam sizes are small and when a rustic architectural appearance is desired. Where high stress factors demand large sizes and a finished appearance is required, it is usually more economical to use laminated beams. They are manufactured in a wide range of sizes and finishes.

Solid timbers are available in:

- A range of standard cross sections.
- Lengths of 6' and longer.
- Longer lengths in multiples of 1 or 2 feet.

Surface finish is either rough sawn or planed. When beams are exposed, appearance becomes an important consideration. Figure 22-11 illustrates on-the-job treatment that may be applied to exposed beams.

Beam sizes must be based on the span (spacing between supports), deflection permitted, and the load they must carry. Design tables are available from lumber manufacturers which may be used to determine sizes for simple buildings. See Figure 22-12. Refer to Unit 8 for additional information on calculating beam loads.

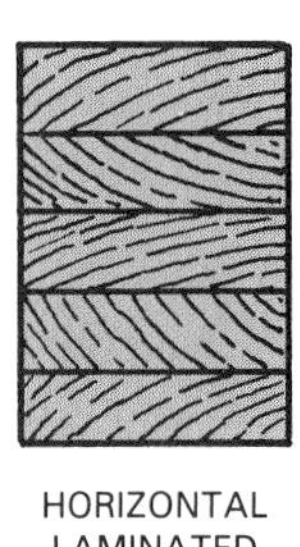

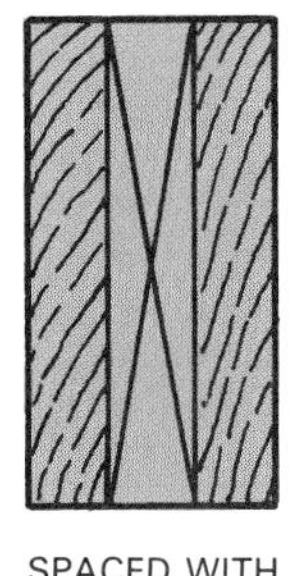

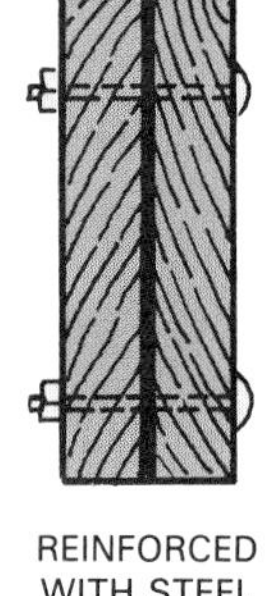

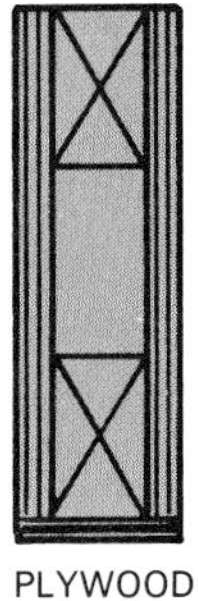

Figure 22-8. Beams can be solid or built-up from smaller dimension lumber.

Figure 22-9. Floor system with beams spaced 4' O.C. Decking is 1 1/8" plywood with edges tongue and grooved. (American Plywood Assoc.)

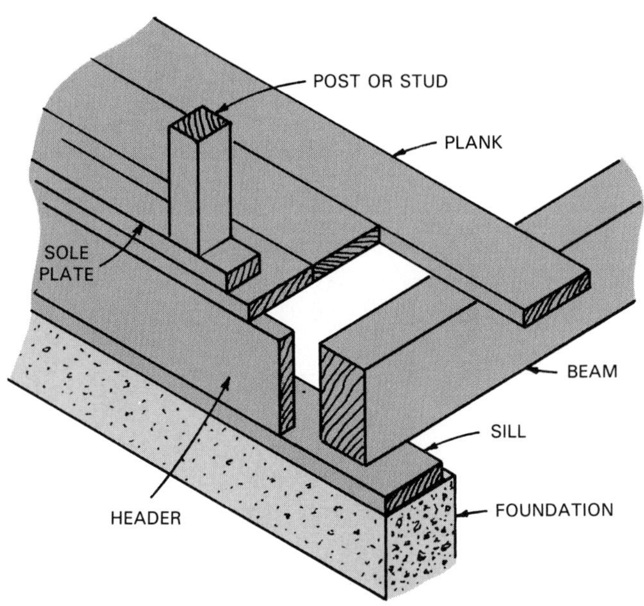

Figure 22-10. Typical sill construction for a post-and-beam frame. Add blocking under the post when it is not located over the beam.

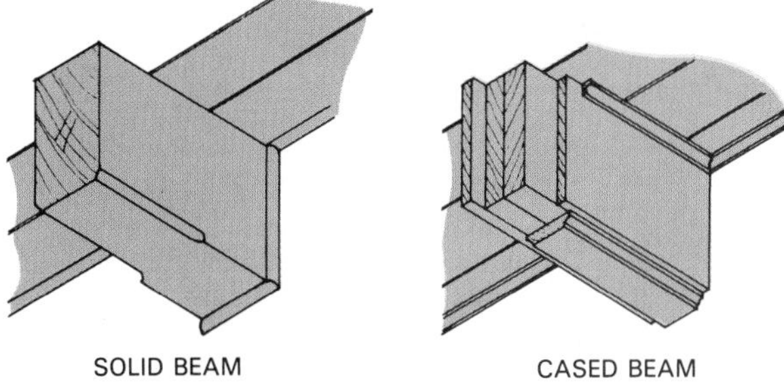

Figure 22-11. Either solid or built-up beams can be surface treated for better appearance when they are to be left exposed.

Roof Beams

Beam-supported roof systems are of two basic types, Figure 22-13:

- *Transverse beams,* which are similar to exposed rafters on wide spacings.
- *Longitudinal beams,* which run parallel to the supporting side walls and ridge beam. Their purpose is to support roof planking or panels.

Longitudinal roof beams, also called *purlin beams*, are usually larger in cross section than transverse beams because they have greater spans and carry heavier loads. The longitudinal beam permits many variations in end-wall design. Extensive use of glass and extended roof overhangs are special features.

Either type of beam must be adequately supported either on posts or stud walls that incorporate a heavy top plate. When supported on posts, the connection can be reinforced with a wide panel frame that extends to the top of the beam, Figure 22-14. A similar method of supporting a ridge beam is shown in Figure 22-15.

Size (Actual)	Wt. per Lineal Ft.	SIMPLE SPAN IN FEET													
		10	12	14	16	18	20	22	24	26	28	30	32	34	36
		Load Bearing Capacity—Lbs. per Lineal Ft.													
3"x5¼"	3.7 lbs.	151	85	—	—	—	—	—	—	—	—	—	—	—	—
3"x7¼"	4.9 lbs.	362	206	128	84	—	—	—	—	—	—	—	—	—	—
3"x9¼"	6.7 lbs.	566	448	300	199	137	99	—	—	—	—	—	—	—	—
3"x11¼"	8.0 lbs.	680	566	483	363	252	182	135	102	—	—	—	—	—	—
4½"x9¼"	9.8 lbs.	850	673	451	299	207	148	109	—	—	—	—	—	—	—
4½"x11¼"	12.0 lbs.	1,036	860	731	544	378	273	202	153	—	—	—	—	—	—
3¼"x13½"*	10.4 lbs.	1,100	916	784	685	479	347	258	197	152	120	—	—	—	—
3¼"x15"*	11.5 lbs.	1,145	1,015	870	759	650	473	352	267	206	163	128	104	—	—
5¼"x13½"*	16.7 lbs.	1,778	1,478	1,266	1,105	773	559	415	316	245	193	154	124	101	—
5¼"x15"*	18.6 lbs.	1,976	1,647	1,406	1,229	1,064	771	574	438	342	269	215	174	142	116
5¼"x16½"*	20.5 lbs.	2,180	1,810	1,550	1,352	1,155	933	768	586	457	362	290	236	183	160
5¼"x18"*	22.3 lbs.	2,378	1,978	1,688	1,478	1,308	1,113	918	766	598	474	382	311	254	204

*Horizontally Laminated Beams

TABLE 1

example: Clear Span = 18′ 0″
Beam Spacing = 8′ 0″
Dead Load = 8 lbs./sq. ft. (decking + roofing)
TABLE I Live Load = 20 lbs./sq. ft. (snow)
ROOF BEAM Total Load = (20 + 8) (8) = 224 lbs./lineal ft.
From Table I—Select 3″x11¼″ beam with capacity of 252 lbs./lin. ft.

Size (Actual)	Wt. per Lineal Ft.	SIMPLE SPAN IN FEET													
		10	12	14	16	18	20	22	24	26	28	30	32	34	36
		Load Bearing Capacity—Lbs. per Lineal Ft.													
3"x5¼"	3.7 lbs.	114	64	—	—	—	—	—	—	—	—	—	—	—	—
3"x7¼"	4.9 lbs.	275	156	84	55	—	—	—	—	—	—	—	—	—	—
3"x9¼"	6.7 lbs.	492	319	198	130	89	—	—	—	—	—	—	—	—	—
3"x11¼"	8.0 lbs.	590	491	361	239	165	119	—	—	—	—	—	—	—	—
4½"x9¼"	9.8 lbs.	738	479	298	196	134	96	—	—	—	—	—	—	—	—
4½"x11¼"	12.0 lbs.	900	748	541	359	248	178	131	92	—	—	—	—	—	—
3¼"x13½"*	10.4 lbs.	956	795	683	454	316	228	169	128	98	—	—	—	—	—
3¼"x15"*	11.5 lbs.	997	884	756	626	436	315	234	178	137	108	—	—	—	—
5¼"x13½"*	16.7 lbs.	1,541	1,283	1,095	732	509	367	271	205	158	123	96	—	—	—
5¼"x15"*	18.6 lbs.	1,713	1,423	1,219	1,009	703	508	376	286	221	173	137	109	—	—
5¼"x16½"*	20.5 lbs.	1,885	1,568	1,340	1,170	939	678	505	384	298	235	187	151	—	—
5¼"x18"*	22.3 lbs.	2,058	1,710	1,464	1,278	1,133	886	660	503	391	309	247	200	—	—

*Horizontally Laminated Beams

TABLE 2

Figure 22-12. Check tables to determine allowable spans for vertical glue-laminated beams. Table 1 is intended for roof beams where a deflection of 1/240 of span is permitted. Table 2 is for floor beams and is based on an allowable deflection of 1/360 or less. (Weyerhaeuser Co.)

Transverse beams are joined to the sides of the ridge beam or supported on top as illustrated in Figure 22-16. Metal tie plates, hangers, and straps are required to absorb the horizontal thrust.

Flat roof designs often consist of a plank-and-beam system. Details of construction are about the same as illustrated for low, sloping roofs.

Fasteners

A post-and-beam frame consists of a limited number of joints. Therefore, the loads and forces exerted upon the structure are concentrated at these points. Traditionally, mortise and tenon joints have been used with wood pegs as fasteners. If butt joints are used instead of mortise and

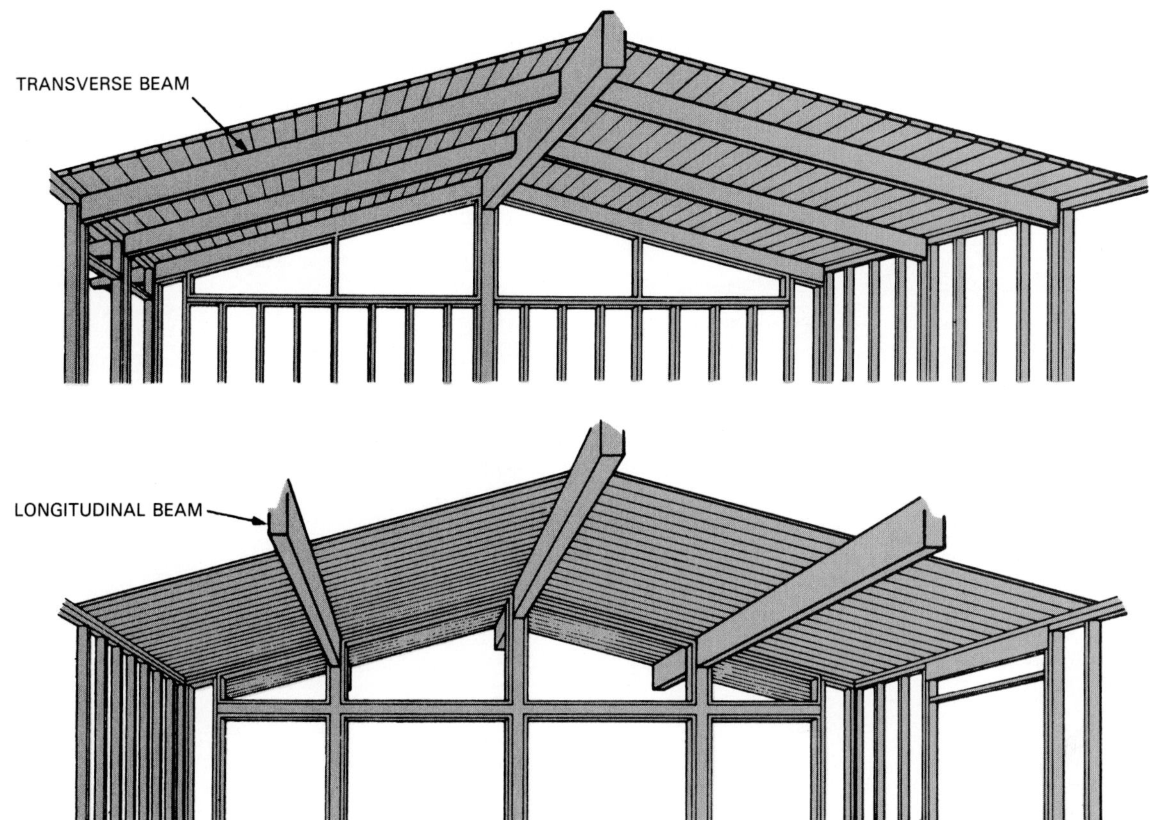

Figure 22-13. Plank-and-beam roof construction methods. Top—Transverse beams. Bottom—Longitudinal beams.

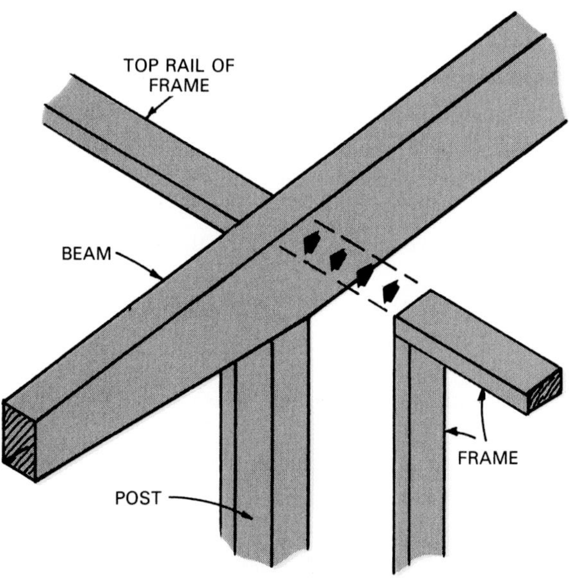

Figure 22-14. Transverse beam which bears on a post will need support against lateral (side-to-side) movement. Filler panel frame at right provides needed reinforcement.

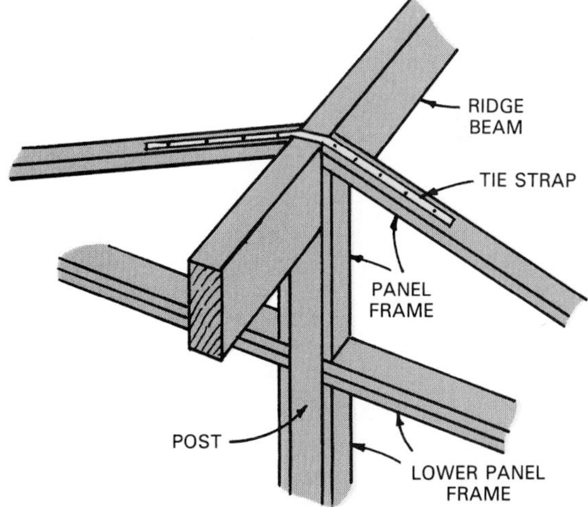

Figure 22-15. Ridge beam is held in place by panel frame and metal tie strip. Same detail of construction may be applied to transverse beams.

tenon, mechanical fasteners must be used. Regular nailing patterns used in conventional framing will usually not provide a satisfactory connection and the joints will need to be reinforced with metal connectors. See Figure 22-17. To increase the holding power of metal connectors, they should be attached with lag screws or bolts.

Since beam structures are usually exposed, some connectors will likely detract from the appearance. Concealed devices will need to be substituted. Steel or wood dowel pins of appropriate size can be used.

Notches and gains cut in the members may provide an interlocking effect or a recess for metal connectors. Figure 22-18 shows a heavy beam-and-truss system.

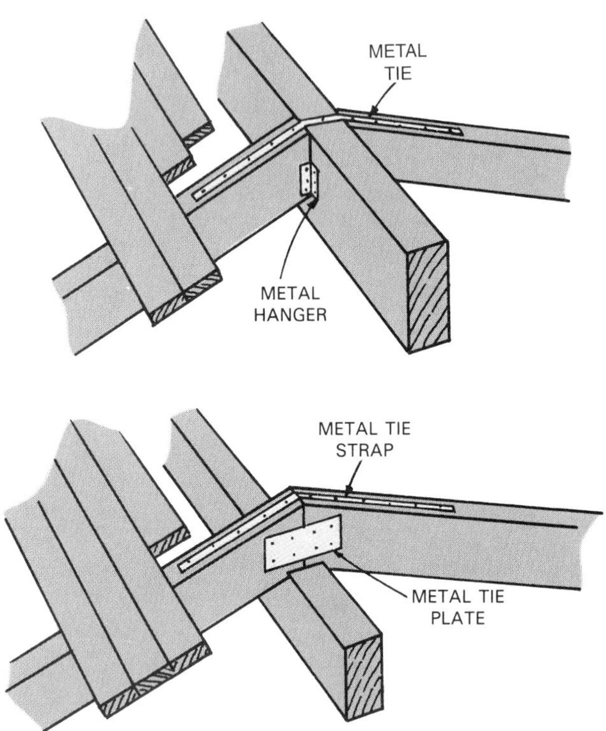

Figure 22-16. Construction details of transverse beam and ridge beam. Top—Beam attached to side of ridge. Below—Beam supported on top of ridge.

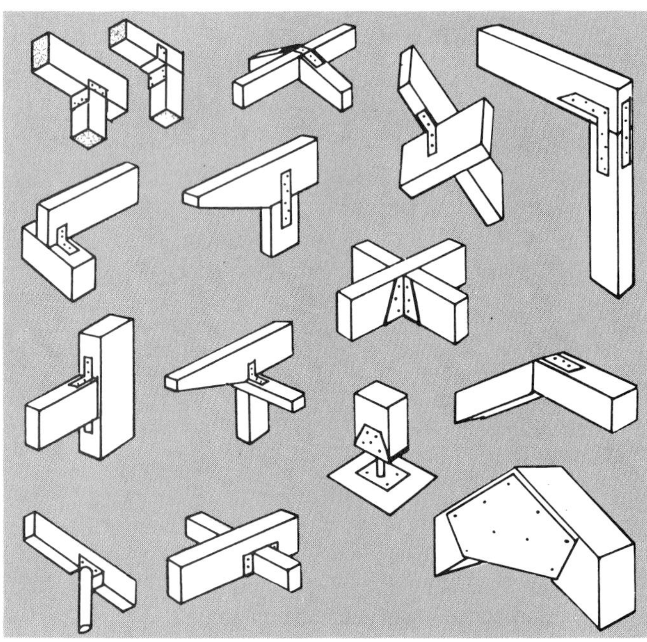

Figure 22-17. Many different kinds of metal fasteners are made for post-and-beam construction. (Western Wood Products Assoc. and Timber Engineering Co.)

Note how the metal plates and fasteners blend with the rough surface of the structural members to provide a special architectural effect.

Use extra care when assembling exposed posts and beams. Tool and hammer marks will detract from their final appearance.

Figure 22-18. Heavy beam-and-truss system supports roof beams at midpoint and ridge. All members are laminated and finished with a rough sawn surface that masks the laminate joints. Note the use of metal fasteners. (Boise-Cascade Corp.)

Partitions

Interior partitions are more difficult to construct under an exposed beam ceiling. Except for a load-bearing partition under a main ridge beam, it is usually best to make the installation after beams and planks are in place. Partitions running perpendicular to a sloping ceiling should have regular top plates with filler sections installed between the beams.

Partitions parallel to transverse beams will have a sloping top plate. Sometimes it is best to construct these partitions in two sections. First, build a conventional lower section the same height as the side walls. Then add a triangular section above.

When nonbearing partitions run at a right angle to a plank floor, no special framing is necessary. However, when nonbearing partitions run parallel, Figure 22-19, additional support must be provided. Replace the sole plate with a small beam or add the beam below the plank flooring.

Planks

Planks for floor and roof decking can be anywhere from 2" to 4" thick, depending on the span. Edges may be tongue and grooved or they may be grooved for a spline joint that can be assembled into a tight, strong surface.

Figure 22-20 illustrates standard designs. Identification numbers are those listed by the Western Wood Products Association. When planks are end-matched, Figure 22-21, the joints need not meet over beams.

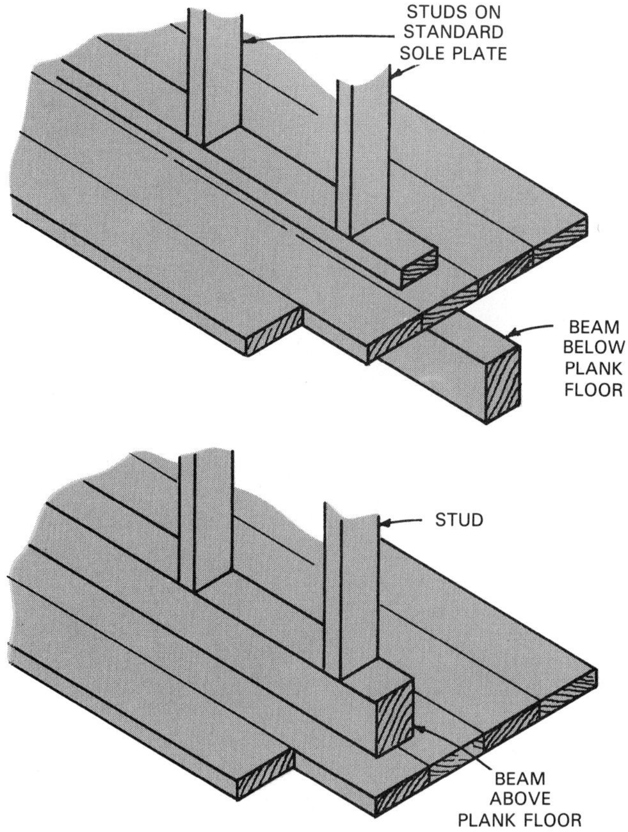

STUDS ON STANDARD SOLE PLATE

BEAM BELOW PLANK FLOOR

STUD

BEAM ABOVE PLANK FLOOR

Figure 22-19. Two methods of supporting nonbearing partitions that run parallel to flooring planks.

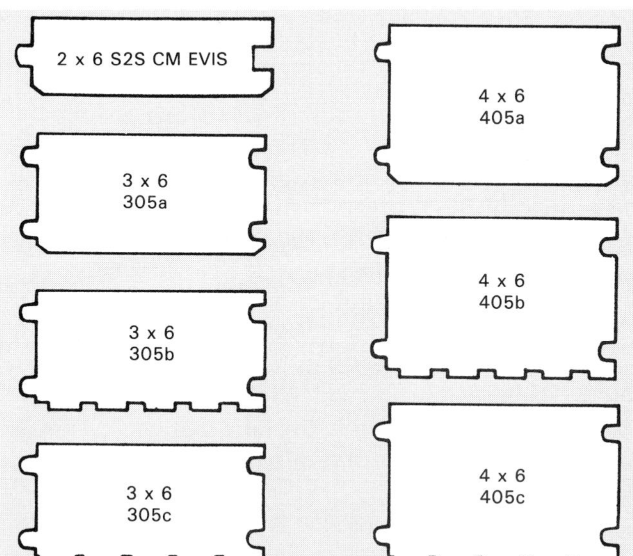

Figure 22-20. Standard plank patterns. Edges are tongue and groove. Faces are machined.

Planks can support greater loads if they continue over more than one span. This rule can also be applied to beams, plywood, and other support material. This is illustrated in Figure 22-22.

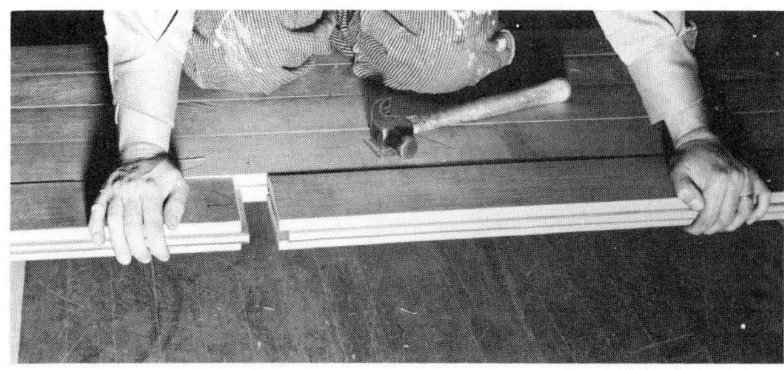

Figure 22-21. End-matched planks are being installed. (Weyerhaeuser Co.)

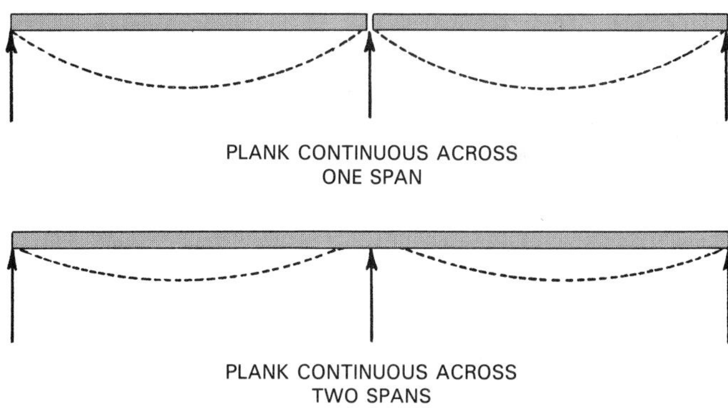

Figure 22-22. Stiffness of a plank is increased if it extends continuously over two or more spans.

Roof planks should be selected carefully, especially when the faces will be exposed. Figure 22-23 shows the application of planking with a V-joint along the edge. Solid materials should have a moisture content that will correspond closely to the Equilibrium Moisture Content (E.M.C.) of the interior of the structure when it has been placed in service. Because of the large cross-sectional size of posts, beams, and planks, special precautions should be observed in selecting material with proper M.C. levels. Otherwise, difficulties due to excessive swelling or shrinkage may be encountered.

In cold climates, plank roof structures directly over heated areas require insulation and a vapor barrier. Thickness of the insulation will depend on the climate. Refer to Unit 14. The insulation should be a rigid type that will support the finished roof surface and workers. An approved vapor barrier should be installed between the planks and the insulation as shown in Figure 22-24.

Several types of heavy structural composition board, 2" to 4" thick, are available for roof decks. The panel sizes are large, and the material is lightweight. See Figure 22-25. Edges usually have some type of interlocking joint that

Figure 22-23. Four inch double tongue-and-groove planks are being laid on glue-laminated beams. V-groove facing makes an attractive, durable ceiling.

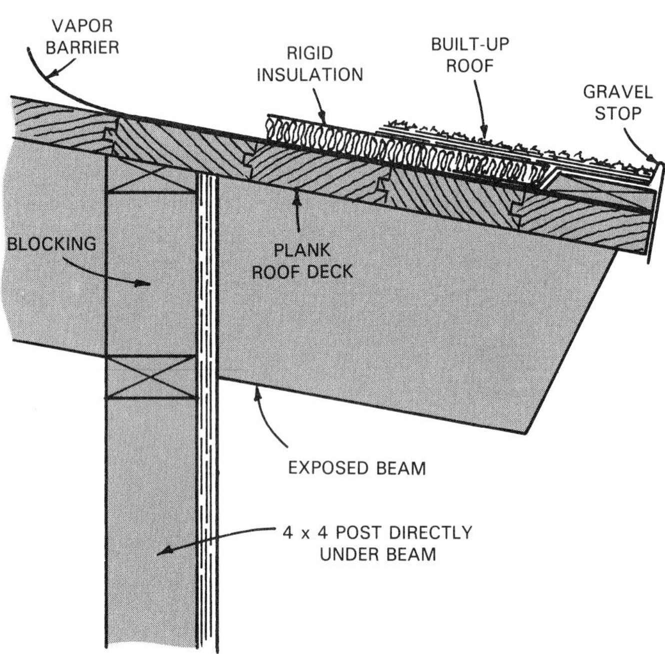

Figure 22-24. Cross section shows application of vapor barrier and insulation to a plank decking. This type of covering is necessary when the deck is located directly over heated space.

provides a tight, smooth deck. When the underside (ceiling side) is prefinished, no further decoration is usually necessary. Always follow the manufacturer's recommendations when selecting and installing these materials.

Figure 22-25. Application of special composition board decking. It must be strong enough to carry workers.

Stressed Skin Panels

Roof and floor decking, and also wall sections can be formed with plywood stressed skin panels. These can be designed to carry structural loads over wide spans. Figure 22-26 illustrates paneled deck sections being installed.

Stressed skin panels are made by gluing sheets of plywood (skins) to longitudinal framing members or other core materials. They form a structural unit with a supporting action similar to a series of built-up wooden I-beams. See Figure 22-27.

Panels are usually produced in factories where rigid specifications in design and construction can be maintained. Figure 22-28 provides general guidelines and constructional details. Note that insulation can easily be installed as a part of the manufacturing process.

Figure 22-26. Roof panels form the roof deck. Note how the panel rest on the purlins. (Yankee Barn Homes, Inc.)

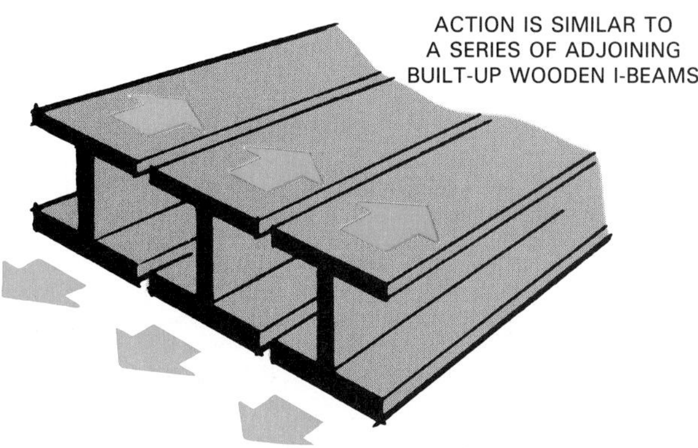

Figure 22-27. How a stressed skin panel provides support.

Sandwich panels with plywood skins and cores of such material as rigid polystyrene or paper honeycomb are similar to the stressed skin panel. They do not provide as much rigidity and strength, however. Such panels are used for curtain walls and various installations where the major support is carried by other components. Skins can be made from a wide variety of sheet materials including plywood, hardboard, plastic laminates, and aluminum.

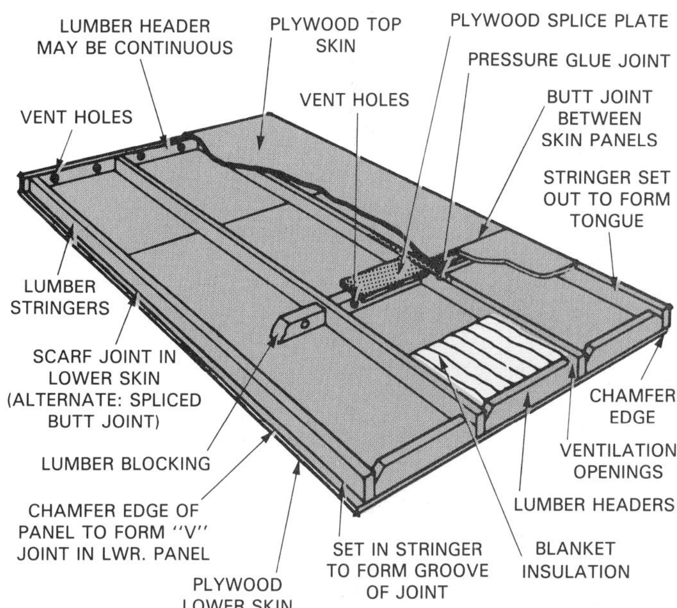

Figure 22-28. General construction details for stressed skin panels. Do you see how such panels conserve lumber and provide strength?

Box Beams

Modern *box beams* made of plywood webs offer a structural unit that can span distances up to 120'. The high strength-to-weight ratio offers a tremendous advantage in commercial structures where wide, unobstructed areas are required.

Basic design features of plywood box beams, illustrated in Figure 22-29, consist of one or more vertical plywood webs which are laminated to seasoned lumber flanges. The flanges are separated at regular intervals by vertical spacers (stiffeners) which help distribute the load between the upper and lower flange. Spacers also prevent buckling of the plywood webs. The strength of the unit depends, to a large extent, on the quality of the glue bond between the various members. Plywood box beams must be carefully designed and fabricated under controlled conditions.

Laminated Beams and Arches

Laminated wood beams and arches are available in many shapes and sizes. They have opened new dimensions of design in modern construction. In addition to its natural beauty, laminated wood offers strength, safety, economy, and permanence. See the graceful arches in Figure 22-30.

Most laminated structural members are made of softwoods. They are manufactured in industrial plants specializing in such production. When they arrive at the building site they are prefinished.

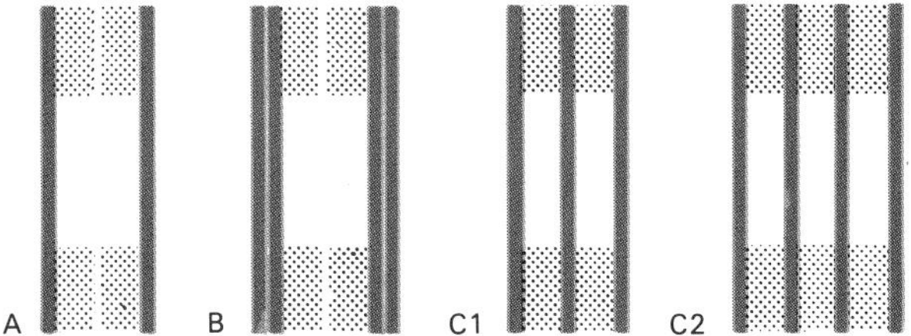

Figure 22-29. Box beams have great strength for their weight. These are the basic construction details.

Figure 22-30. Gracefully soaring parabolic arches of laminated wood are often used in church construction.

In residential work, beams are usually straight or tapered; however, in institutional and commercial buildings they are often formed into curves, arches, and other complicated shapes. Some of the standard forms are illustrated in Figure 22-31.

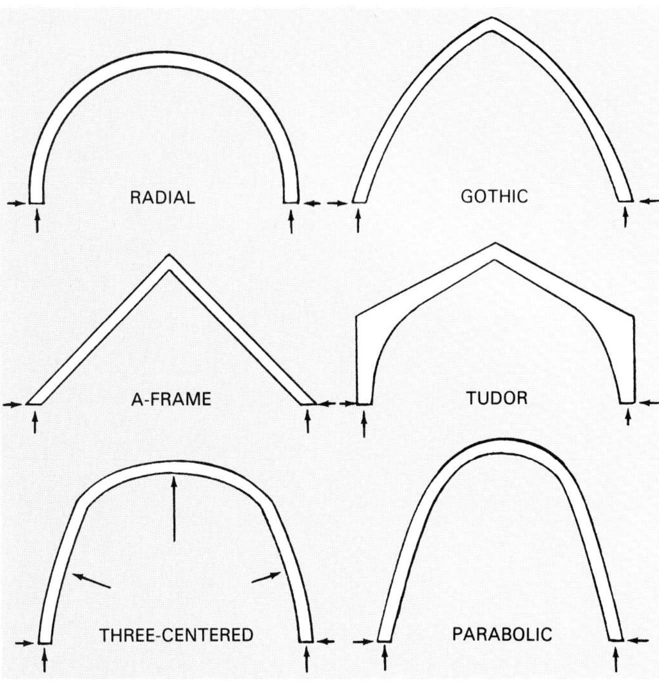

Figure 22-31. Types of laminated wood arches. Arrows indicate the support and lateral thrust that must be provided.

In the fabrication of beams and arches, lumber is carefully selected and machined to size. To get the required length, pieces must often be end-joined. Since end grain is hard to join, a special finger joint may be used. A number of these joints, Figure 22-32, may be required in each ply. The joints are staggered at least 2' from a similar joint in an adjacent layer. Figure 22-33 shows an installation view of laminated beams that will support the roof of a large commercial building.

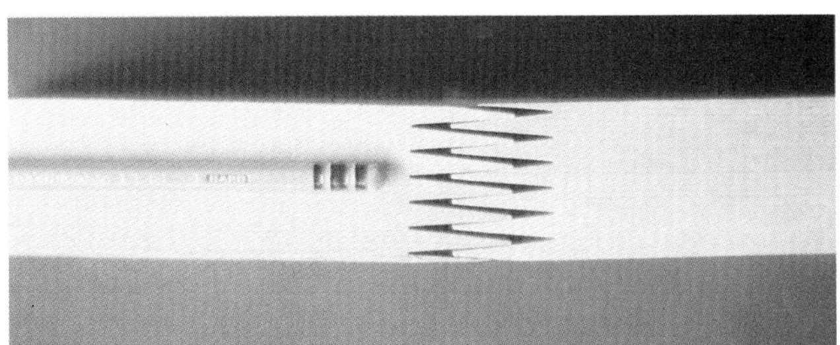

Figure 22-32. Finger joint used to join ends of laminations. Joint is formed with a special cutter head. Individual laminations should not exceed 2" in net thickness. (American Institute of Timber Construction)

Figure 22-33. Laminated beams provide a clear span of 48'. Purlins, spaced at 4' will support 4' x 8' prefabricated roof panels. Note the metal hangers used to fasten the purlins to the beams. (Boise-Cascade Corp.)

Prefabricating Post-and-Beam Structures

Prefabrication of post-and-beam homes is a major industry in certain regions of North America. As with other factory-built systems, the design process consists of customizing homes using any of several standard modules or structures and combining them to suit the wishes of the home owner. See Figure 22-34.

The client reviews standardized structures working with architects and engineers. They will work with the models shown in Figure 22-34 (top), combining them in ways that produce a plan that suits their wishes and lifestyle. They will also select the type of beams—planed, rough sawn, stained, etc., Figure 22-35.

Timbers are selected from the factory stock, measured, marked and cut. Huge production machines cut mortises and tenons, Figure 22-36.

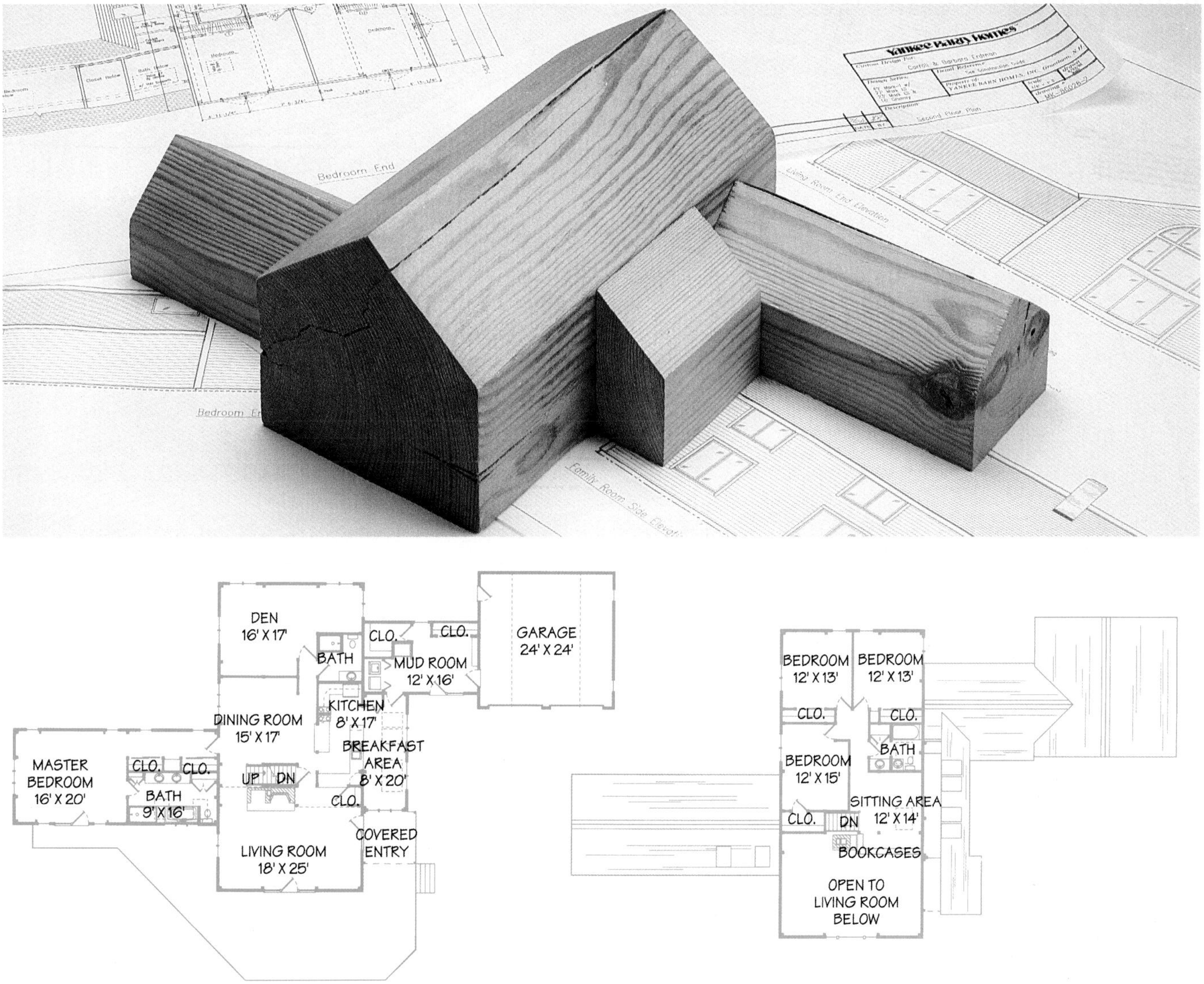

Figure 22-34. Top—Clients wanting a factory-built house use these 1/4" scale wood models to visualize their design. Bottom—Floor plan arranged from the block layout above. (Yankee Barn Homes, Inc.)

Figure 22-36. Huge circular saw cuts a mortise in a beam.

Figure 22-35. Working timbers in the factory. Top—Beams are planed if that is what the client wants. Bottom—Beams are stained to owner's specification. (Yankee Barn Homes, Inc.)

Wall and roof sections are fabricated into large panels on the factory floor. Rigid insulation is cut and installed in the sections along with windows, exterior siding, and interior wall coverings. See Figure 22-37. Completed panels are moved with overhead cranes, wrapped to protect them from weather, and fork-lifted onto trailers for transport to the building site as shown in Figure 22-38.

On site, builders erect the frame, Figure 22-39, and install wall and roof panels, Figure 22-40. Figure 22-41 shows a timber-framed addition to a New England farmhouse built in the 1800s.

Figure 22-37. Factory assembly. Top—Cutting and placing rigid insulation strips. Bottom—Panelized gable end is receiving exterior covering.

Figure 22-38. Transporting finished panelized home sections. Top—Huge overhead crane moves gable end to storage area. Middle—A fork lift places roof panels on the trailer that will move the home sections to a building site. Bottom—At the building site. (Yankee Barn Homes, Inc.)

Figure 22-39. First the frame goes up.

Figure 22-40. Top—Gable end is lifted into place with a crane. Bottom—Roof panels are in place. (Yankee Barn Homes, Inc.)

Figure 22-41. Post-and-beam structure is used to expand the living space in this old New England farmhouse.

Important Terms

Box beams
Longitudinal beams
Plank-and-beam construction
Planks
Post-and-beam construction
Purlin beams
Sandwich panels
Stressed skin panels
Transverse beams

Test Your Knowledge

1. The slenderness ratio of a post compares the total height in inches with the ____________ (*largest, smallest*) dimension of its cross section.
2. Two common types of roof beams include longitudinal beams and ____________ beams.
3. The live load on a residential floor is usually figured at ____________ lb. per sq. ft. and represents the weight of furniture, occupants, and equipment.
4. Planks for floor and roof decking are available in thicknesses of 2 to ____________ inches.
5. A fabricated building component, similar to a stressed skin unit with a core of foamed plastic or paper is called a ____________ panel.
6. In plywood box beam construction, the flanges are separated at regular intervals with spacer blocks which distribute the load and prevent buckling of the ____________.
7. Factory-built post-and-beam homes offer the owner little choice in customizing their home. True or False?

Outside Assignments

1. Visit a building supply center and obtain descriptive literature about floor and roof decking that can be used in post-and-beam construction. Include both solid and laminated planks and composition panels. Be sure to obtain prices. Study these materials thoroughly for qualities, characteristics, and installation procedures. Prepare a written or oral report.
2. Build a mock-up of a stress skin panel in which you can experiment with the design and size of the various parts. Use a scale of about 3" = 1'-0". Skins might be made of 1/8" plywood or 1/16" veneer.
3. Prepare a scale model of a factory-built post-and-beam house. Consult with your instructor regarding types of fasteners and dimensions of posts and beams.

23

Systems-Built Housing

The term *systems-built housing* refers to a house built of components assembled in a factory following precise design specifications. At one time, the term *prefabrication* was used to describe such construction. Either term indicates a process that cuts and assembles parts and sections of a home in factories. Then as now, the process included shipment to the building site for final assembly and erection. Once erected, a systems-built house cannot be told from a stick-built home, Figure 23-1.

Figure 23-2. A shopping center structure designed to house retail stores. It was constructed of superinsulated components built in a factory. (Enercept, Inc.)

Figure 23-1. A systems-built home cannot be told from a stick-built home. (Insulspan)

Systems- or factory-built also includes modules—three dimensional segments that are fully erected before leaving the manufacturing plant. Manufacturers of these modular structures also build commercial buildings such as banks, schools, office buildings, motels, and hotels. Figure 23-2 shows a shopping center structure assembled from factory-built components.

At one time, prefabrication or "factory-built" limited the buyer to a few styles and plans offered by the manufacturer. All that has changed, partly because computer-assisted drafting can quickly adapt architectural drawings to the buyer's needs and lifestyle.

Thus, architects and engineers employed in the systems-built housing industry can quickly produce any style of home desired by the buyer and will easily incorporate any custom designing required. Systems-built homes include geodesic domes and log homes as well as traditional and classic styles.

To make quality components, many tasks must be carried out under controlled assembly-floor conditions with special tools, Figure 23-3. Overhead cranes lift heavy assemblies and power tools are used in their assembly. Stressed-skin panels and plywood box beams require accurate glue applications, special presses, and handling equipment. Often, the simplest prefabrication process in modern plants is done on large production equipment. The machine reduces the amount of human energy required and increases production. Speed is increased both in the plant and on the building site.

Designs are completed with the assistance of computers and architectural software. During the manufacturing process, the components move from one workstation to another where all building trades activities are represented. A quality control process assures that all work is in conformity with state and local building codes.

Components

In house construction today, many factory-built components (parts) are used in place of job-site finish work and stick-built framing work. Some of these components are

Figure 23-3. View of a factory floor where modular structures are being built under roof. Skilled construction workers work with power tools as the modules move along an assembly line. (Cardinal Industries, Inc.)

windows, door units, soffit systems, stairs, and built-in cabinetwork.

One major framing component is the roof truss, Figure 23-4 (top). Manufacturing plants can usually furnish either simple Fink trusses (described in Unit 10) or matched units for a complex roof as shown in Figure 23-4 (bottom). Roof trusses must be carefully designed by structural engineers and built according to exact specifications. High production double-end saws cut truss members to length. Special presses and jig tables fasten the assembly with "gang-nail" connectors. Completed fabrications are shown being transported to the construction site in Figure 23-5.

The floor truss, Figure 23-6, is another example of a prefabricated component. It permits a wide unsupported span with a minimum of material. It also has openings for heating and air conditioning ducts, electrical conduit, and plumbing lines. Figure 23-7 shows an assembly floor where floor trusses are being built.

Two other types of panels are flat stressed-skin panels and flat sandwich panels. Both are described in Unit 22.

Wall and roof sections or panels are made in various shapes and sizes. Panels provide structural strength in addition to forming inside and outside surfaces. Some panel systems are constructed of conventional framing with 2" lumber, standard insulation, and coverings. Another type panel just mentioned is the "sandwich," consisting of a core of thick, rigid polystyrene insulation with facings of plywood, oriented strand board, or waferboard. These can be made to various heights and widths, depending on the design of the building. Voids, called channels or chases, can also be designed into the panels for running wires and pipes for mechanical systems. A roof panel may have channels simply to provide air circulation to prevent heat buildup. Figure 23-8 shows a wall panel assembly line where assemblers are fabricating a section that includes a large window with a half-round top. Figure 23-9 is a series of line drawings showing various types of panels designed for different applications in walls and roofs.

Sandwich panels, just discussed, made with a rigid insulation core provide enough strength for walls and partitions in some structures. See Figure 23-10. These lightweight panels may have an aluminum or waferboard outer skin and a hardboard or plywood inside surface. Special metal channels and angles are available which reduce labor at the building site.

Transporting Systems-Built Homes

A building in various stages of completion can be prepared in factories and then shipped as a "package" to the building site. Most such buildings are single-family homes.

Figure 23-4. Truss rafters are a major component of a building's structural system. Left—A crane lifts trusses into place where they are fastened by carpenters. (Wausau Homes, Inc.) Bottom—Prefabricated roof framing includes all the types of roof trusses used to form hips, valleys, overhangs, and gable ends. (Gang-Nail Systems, Inc.)

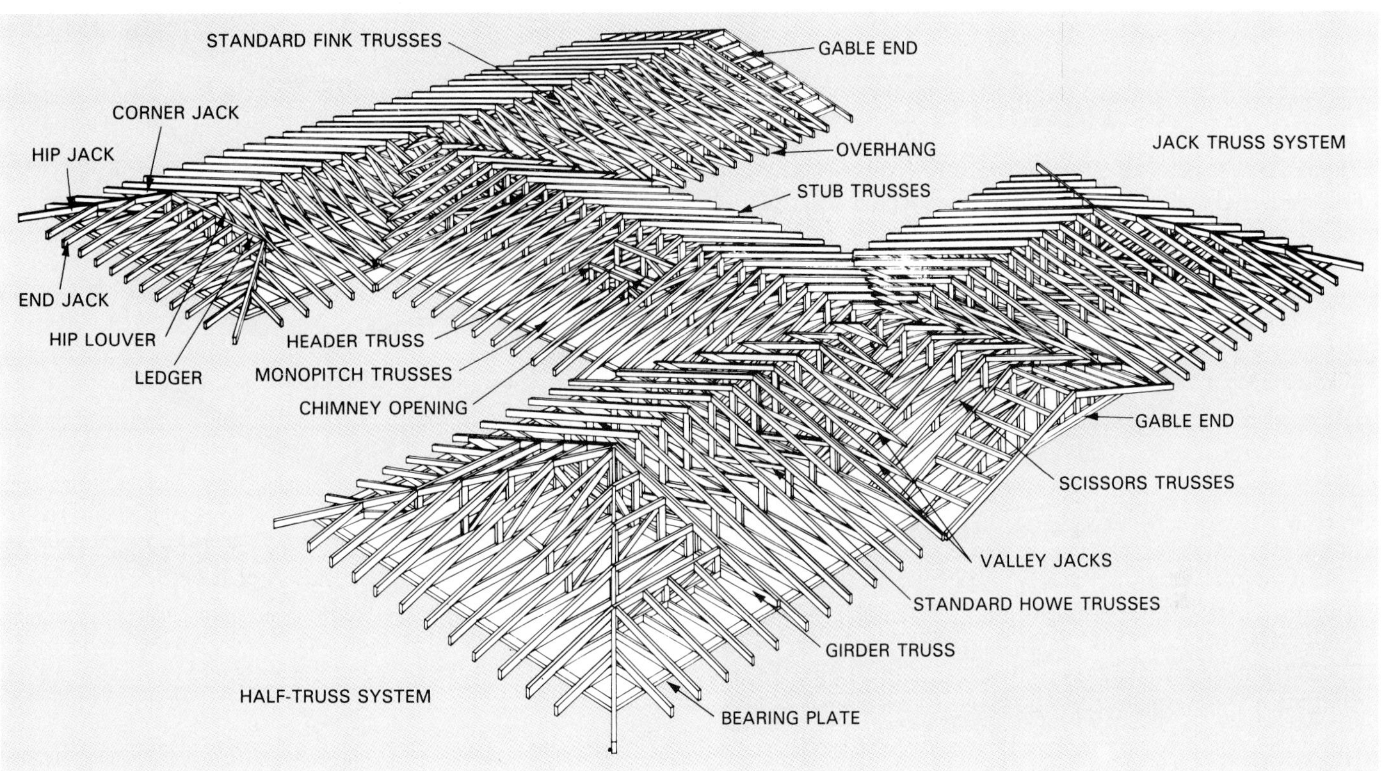

However, small commercial buildings, farm structures, and multi-family homes are also produced and moved this way.

Types of Factory-Built Homes

The Building Systems Council, which includes several categories of systems-built manufacturers, groups all such housing being manufactured today under three categories:

- Modular homes.
- Panelized homes.
- Log homes.

Other organizations include other categories besides these three. The North Carolina Manufactured Housing Institute adds two other categories:

- Precut homes.
- Manufactured homes (formerly called "mobile" homes).

Regardless of the type of housing chosen by the owner, all offer:

- Rapid construction.
- Quality materials.
- Lower building costs.
- Customizing.

Figure 23-5. Panels and other prefabricated components have been delivered to the home site. Materials are protected from the rain by plastic tarps. (Wausau Homes Inc.)

Figure 23-6. Floor truss system provides ample space for installation of plumbing lines. (Gang-Nail Systems, Inc.)

Figure 23-7. Building a metal web truss. Press (not shown) forces toothed metal webs into chords, top and bottom.

Figure 23-8. Workers attach a large half-round window to a tall wall panel. (Wausau Homes Inc.)

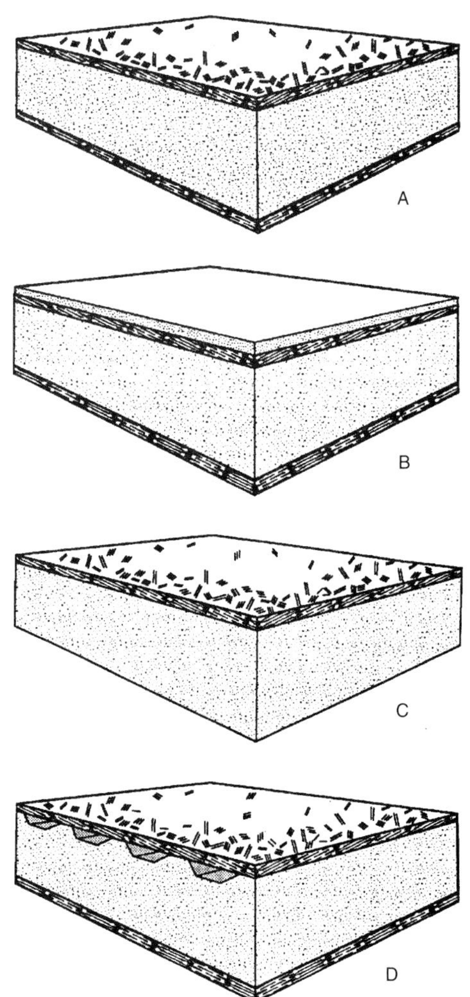

Figure 23-9. These four sandwich panels are typical of closed panels. *Closed* means that no additional material need be installed once the panels are erected. However, drywall and mechanical systems may be added later. A—A structural panel with facings of oriented strand board (OSB) and a polystyrene rigid insulation center that may be up to 11 3/8" thick. B—Drywall panel has a drywall interior facing over the OSB. C—Nailbase panel are designed to add a thick layer of insulation over a completed roof deck or wall. They sometimes have foil on the interior face. D—A vented panel designed to provide roof ventilation where it is desired. (Insulspan)

The Modular House

Modular homes are made up of two or more three-dimensional *modules*, also called mods. These modules are produced on a factory assembly line. One-, two-, and three-story homes are possible by stacking modules on each other, Figure 23-11. Modules may be smaller units such as a bathroom or a kitchen. See Figure 23-12.

Figure 23-10. This building is made of superinsulated floor and wall panels. Such components are basic to all prefabricated construction. (Enercept Inc.)

Each module is 12' or 14' wide and may be up to 60' long. Such units are nearly completed at the factory. Some even include cabinets and plumbing fixtures. At the building site, crews set them on a foundation with a crane, stacking them and fastening them with bolts. Many manufacturers of mods have their own erection crews.

Modular construction is efficient. A module can normally be built in 14 working days and erected in one or two days.

According to the Modular Council of the Buildings System Council, modular homes make up 10% of the new housing market. The number of modular homes sales grows by about 14% a year. In 1992, 84,000 units were sold.

In this type of prefabrication, entire sections (modules) of the structure are built and finished in manufacturing plants, Figure 23-13. The sections or modules are then hauled to the site where they are assembled. Widths of a module seldom exceed 14'. Trucks and roads cannot handle wider units.

Figure 23-11. A modular home being placed on an all-weather foundation. Sections are factory-built and completely finished inside and out. Sections will be bolted together by an erection crew. (American Plywood Assoc.)

Figure 23-12. Some modules are smaller, including only one room. Usually such modules include all plumbing, wiring, and fixtures. (Wausau Homes Inc.)

Figure 23-13. Final assembly line. Kitchen and living room modules are put together.

An advantage of modular construction is that nearly all of the detailed finish work can be done at the factory. Kitchen cabinets can be attached to the walls and other built-in features can be installed. Also wall, floor, and ceiling surfaces can be applied and finished.

Electrical wiring chases, heating and air conditioning ducts, plumbing lines, and even plumbing fixtures can be installed under close control in manufacturing plants using labor-saving tools. A section that has a group of plumbing and heating facilities is often included with other sections that consist mainly of panels. Sections of this nature, which include most of the utility hookups, are called *mechanical cores.*

Mechanical cores group the kitchen, bath, and utilities in one unit that requires only three connections at the site. A core with a bathroom and kitchen on opposite sides of the same wall is typical. Some units are designed to include heating and air conditioning equipment, as well as electrical and plumbing equipment.

Disadvantages of the sectionalized prefabrication are the problems involved with storage and transportation, and the need for large cranes to handle the units at the construction site, Figure 23-14.

Figure 23-14. Crane places 5 ton module on masonry foundation. Units are 95% complete when they arrive on site. (Cardinal Industries, Inc.)

Panelized Homes

While most homes today have some prebuilt parts, such as roof trusses and engineered lumber, *panelized homes* leave the factory as a series of wall panels. These panels are 8' high and may be from 4' to 40' long. Exterior walls are often "sandwich" panels of plywood, oriented strand board, or waferboard on either side of a core of rigid insulation. In other cases the panels may have 2 x 6 studs and an exterior sheath of plywood. Some panelizers cut in door and window openings, hang windows and predrill studs for electrical wiring and plumbing runs. Channels may be formed in rigid insulation for running electrical wiring. A few producers make panelized ceilings and floors the same way.

Panels arrive on the building site numbered according to their location in the house. Erections can begin during the unloading process.

In panelized prefabrication, flat sections of the structure are fabricated on assembly lines. Large woodworking machines cut framing members to length and angles, Figure 23-15. Parts are stored and delivered to the assembly stations as needed. Wall and floor frames are formed by placing the various members in positioning jigs on the production line. The parts are fastened with pneumatic nailers,

Figure 23-15. Double-end sawing machine cuts bottom chord members for roof trusses. Each sawing unit (left and right) has three blades that can be set at various angles. (Speed Cut, Inc.)

Figure 23-16. Electrical wiring or other mechanical facilities may be installed while the frame is being built.

As the completed frames move along conveyor lines, wall surface materials are placed in position and nailed. Nailing is done with powered gangnailers and/or staplers. See Figure 23-17. A special type of staple is being used to attach high R-value foamed plastic sheathing in Figure 23-18. Farther along the line, insulation is put in, Figure 23-19. Some of the wall sheathing steps may be repeated to close the panel. Then siding is attached.

Figure 23-16. An automatic machine, called a panel extruder, assembles 60 frames a day. The unit is fully computerized. (Cardinal Industries, Inc.)

In other areas of the plant, roof units are prepared. For post-and-beam structures, panels that form both the roof and ceiling surfaces are common. To avoid painting in high

Figure 23-17. A gang of pneumatic-powered nailers attach fiberboard sheathing to a wall frame in less than two minutes. (Duo-Fast Corp.)

Figure 23-18. Stapling foamed plastic sheathing to wall frame. Special steel strap is used for bracing. (Citation Homes)

Figure 23-20. Prefabricating stressed-skin floor panels. Full length joists and scarf-glued plywood sheets (foreground) are assembled with glue to form units as large as 8' x 24'. (Wausau Homes Inc.)

Figure 23-19. Installing blanket insulation in wall panel. Insulation is arranged around any electrical equipment. (Wausau Homes Inc.)

Figure 23-21. Two semi-trailer loads of building components are loaded at the plant and sent to a building site. (Wausau Homes Inc.)

places after erection, the ceiling side is painted at waist level.

Although most prefabricated houses of the panelized type use a first floor deck built by conventional methods, some manufacturers design and build floor panels. Full length joists are assembled with headers, Figure 23-20. Then long plywood sheets, formed with scarf joints, are glued in place. The resulting stressed-skin construction is rigid and strong. The panels prevent nail pops and squeaks.

As the panels near the end of the production lines, they receive a final inspection. After inspection, each panel is numbered for easy assembly and is then ready for storage or loading into a trailer. When all materials and millwork to complete the "package" are loaded, the truck is sent to the building site, Figure 23-21.

Log Homes

Log homes are essentially a type of precut home. Walls of log homes are built by stacking precut and machined logs on top of each other, rather than by framing with studs. This is the only difference between a log home and conventional stick-built homes or homes that are of modular or panelized construction.

Logs are combined in the factory with other modern building materials to produce any style of home for the buyer, customizing as the buyers wish. High-speed machines mill the logs to uniform shapes and lengths. At the same time, the logs receive the tongues, grooves, notches, and splines that hold them together.

Precut Homes

For a modern precut house, lumber is cut, shaped, and labeled to reduce labor and save time on the building site.

Manufacturers of this type of house include materials needed to form the outside and inside surfaces. Also shipped are such millwork items as windows, doors, stairs, and cabinets. Optional items include electrical, plumbing, and heating equipment.

Kitchen cabinets and other built-in units are usually made in plants specializing in these items. The units may be shipped either to the home fabricator or directly to the building site.

On-Site Erection

At the building site, the foundation should have been built. The various prefabricated units are loaded on the trailer so they can be removed in the proper order for matching the edges. The floor deck is built in the standard way or is assembled from panels. Then, mechanical core units are set in place, walls and partitions are joined, ceiling-roof units are installed, and roof panels close the structure.

Figure 23-22 shows in four views the construction of one modern systems-built home. The shell of a home can be finished in one day.

When prefabricated panels are finished on both the inside and outside surface, they are called *closed panels.* At one time, the terms "manufactured" and "industrialized" were used when referring to prefabricated housing.

One disadvantage of prefabrication is soon discovered at the building site: weather prevents work much of the time. The builder must choose a day without wind or rain. Rain will damage inside wall materials applied at the factory. Wind makes it hard to unload large panels.

Floor panels have been laid onto the foundation and a mechanical core for the kitchen is being lowered into place. Note bath module in place. Also note sun angle as work continues.

Front wall panel for living room is being installed. Outside surface has not been applied since this section of the house will be finished with brick veneer. Trailer in first photograph held panels in proper order.

Crane lowers ceiling-roof panels into place over bedroom area. Note how the panels are hinged together. The combined unit was closed for shipment. The unit is opened for fitting at the building site.

Installing final roof panel on garage. No ceiling panels were required. Note that carpenters have started shingle work over bedroom area. This factory-built home was enclosed in less than a day.

Figure 23-22. Sequence shows on-site erection of a modern prefabricated home. (Wausau Homes Inc.)

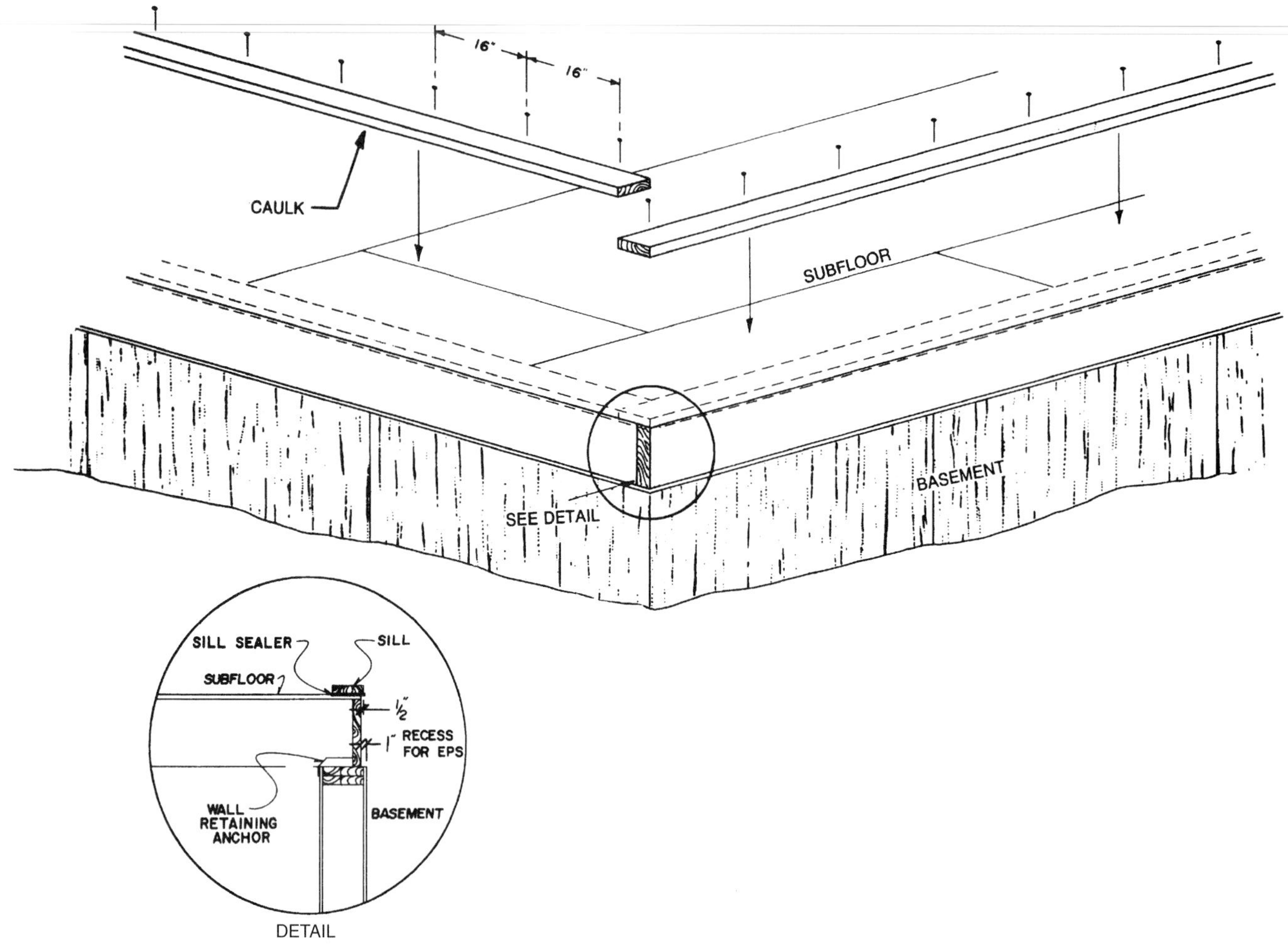

Figure 23-23. Exploded view shows sequence for attaching sill to subflooring prior to setting up wall panels. (Enercept Inc.)

There is also the problem of soft ground. It limits the size of the crane that can be used. Unless the reach of the crane is great, there is no way to safely bring large structures into wet areas.

In many ways, the weather is less important for systems-built than for conventional construction. A skilled crew can erect and enclose a manufactured single-family home in 5 to 20 hours. A few weeks are usually required to complete the wiring, heating, plumbing, and decorating. With factory-builts, the inside work begins on dry structures. Often the inside gets wet before conventional construction is closed in. Therefore, the finish work goes faster on a manufactured home. For this reason and for other reasons, prefabrication competes well with other building methods.

Assembling a Panelized Home

Typically, a panelized structure is assembled on site. Sometimes a block and tackle is used to lift and move the panels.

The following assembly procedures for sandwich panels of expanded polystyrene (EPS) are typical of this type of construction.

Assembling Sandwich Panels

1. Lay down a strip of polypropylene sill sealer along the outer edge of the subfloor. Then install 2 x 6 plate all around the perimeter with the outside edge 1/2" over the edge of the subfloor as shown in Figure 23-23.
2. Exterior wall panels are numbered in the order of their placement. Install them over the plate, Figure 23-24. Check each panel for vertical electrical wiring channels and drill a hole through the plate and subflooring in these locations.
3. Lay down a wavy line of caulk on the plate and vertical edges of each panel. Tip each panel into place carefully to avoid disturbing the caulk.
4. Plumb each panel and nail panels to the plate and to each other (on each side) with 8d nails spaced 6" apart.

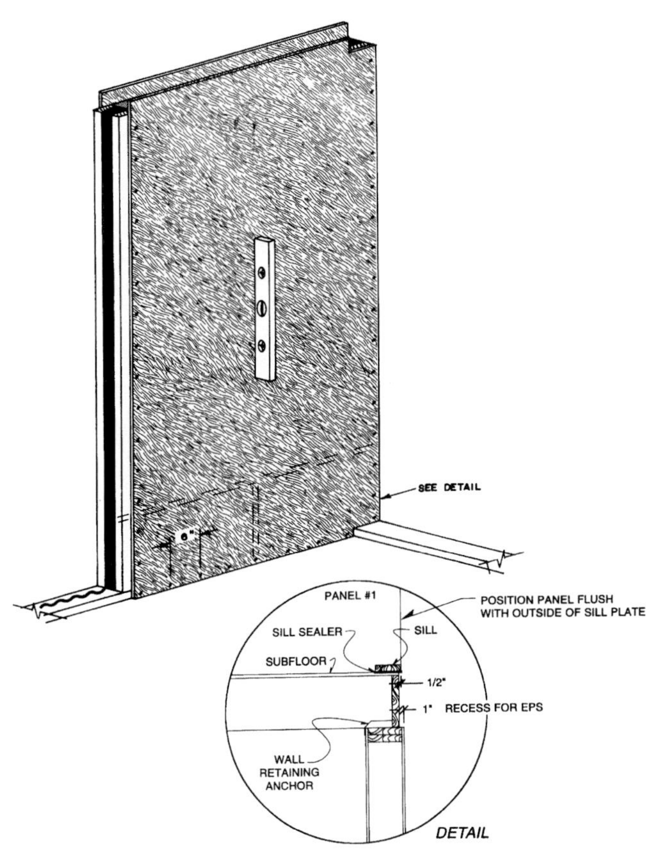

Figure 23-24. Setting the first panel in place. A bead of caulk is laid down and panel is plumbed before nailing it to the plate.

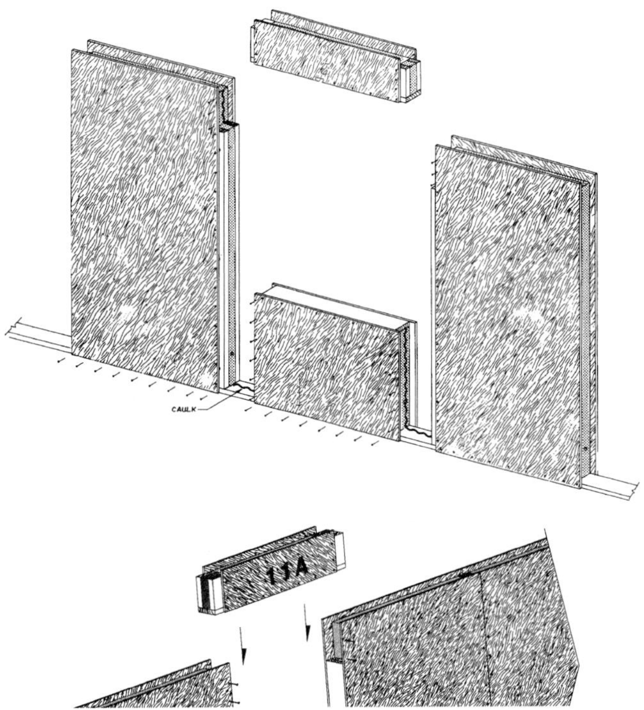

Figure 23-25. Top—Note sequence for installing a window panel. Bottom—Door header panel is dropped into place after wall panels are positioned. (Enercept Inc.)

5. Door and window panels are shown in Figure 23-25. Window rough openings will have a lower panel and a lintel (upper) panel. Door openings have only a lintel panel. Lintel panels are slid into position but not nailed immediately since height adjustments may have to be made.
6. Mating corner panels may have to have an inner facing trimmed. This should be marked and trimmed so panels can be plumbed and joined, Figure 23-26.
7. Gable ends have a "king" post. This is located and toenailed to the plate at the centerline of the building, Figure 23-27.
8. When the wall panels are installed and nailed, a double plate is installed. Before installing plates, mark the panels having vertical wiring cores. Place the mark on the interior of the panel. After the plates are stagger-nailed every 6", drill through both plates into the wiring cores.

Figure 23-28 shows details of how ceiling joists are attached to the top plate and end wall when the wall panels extend to the roof. Adhesive and 12d nails are used to attach the edge joist.

In multistory buildings, the construction method varies somewhat, Figure 23-29. The second-floor joists can

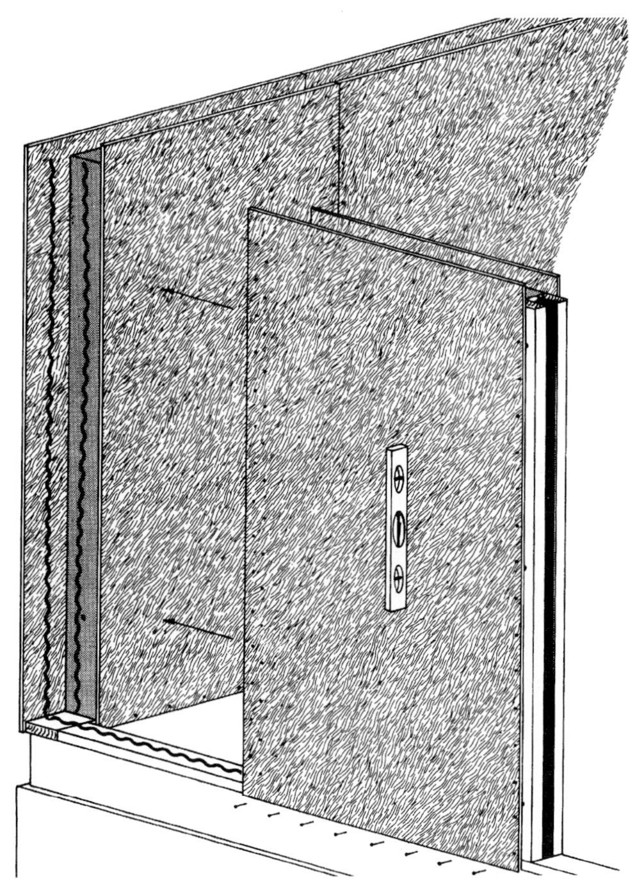

Figure 23-26. Inside face of corner panel may need to be marked and cut shorter. Note nailing pattern. Tilt the panel backward as it is moved into the corner.

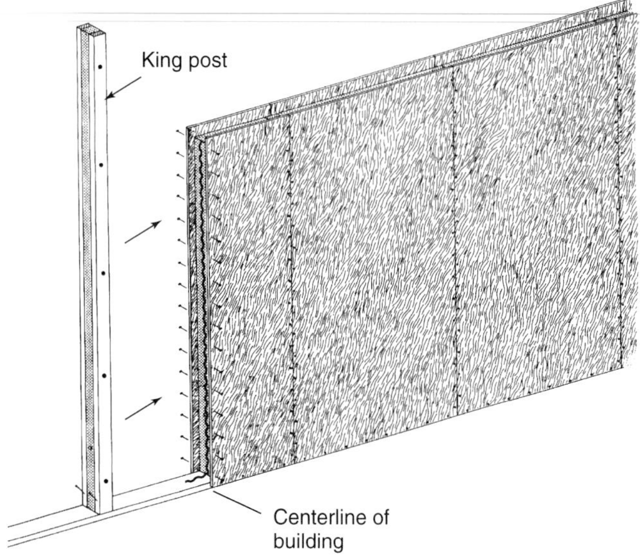

Figure 23-27. A king post is located at the center of the wall directly under the gable. It must align with the center of the building. It must be toenailed to the plate with 16d nails. (Enercept Inc.)

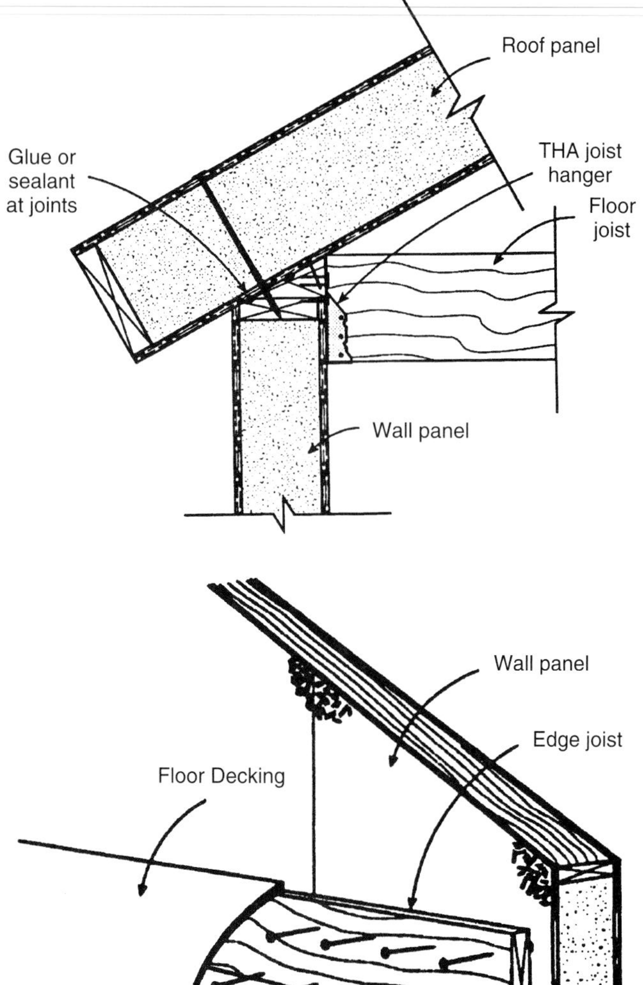

Figure 23-28. Top—Attaching a roof panel to the top of the wall plate. Note that the top plate must be angled to match the roof angle. Fasteners are threaded thin-shanked nails 1 1/2" to 2" longer than the panel thickness. Bottom—When gable end wall panels extend past the ceiling level, joists are glued and nailed to hold ceiling joists. (Insulspan)

be hung on joist hangers or they may rest on top of the first-floor wall.

Ceiling panels can be attached to walls and ridge beams in two ways. Figure 23-30 shows metal clips or ties being used as a means of securing the panels. Figure23-31 shows attachment with roof nails. These are thin-shanked, threaded nails 1 1/2" to 2" longer than the thickness of the panels.

Asphalt or fiberglass shingles can be applied directly over structural roof panels. When wood, tile, or slate shingles are used, it is customary to apply horizontal wood strapping first.

Manufactured Homes

Some 130 companies build manufactured homes in about 320 factories located throughout North America. There appears to be some distinctions that are important. At one time, homes that were trucked to a site on their own wheels were called "mobile" homes. However, a 1991 survey reported that 98% of such homes were never moved from their original site. Still HUD requires that they retain the original chassis on which they were delivered. The term *mobile* has been replaced with *manufactured.*

Manufactured homes are mounted on a chassis, Figure 23-32, and do not need a permanent foundation. They are defined by manufacturers as a trailer longer than 28' and heavier than 4500 lb.

The most common width is 12' to 14'. Lengths may be as great as 68'.

Designs sometimes include an "expandable" feature that permits wider living rooms. A slide out section is carried inside while the mobile home is being hauled. It is expanded to the side after the home is on the site.

Production methods used in building manufactured homes are like those used for standard factory-built housing. Some of the structures, however, are different. For example, the floor must be one rigid unit and usually consists of a wood frame attached to a welded steel chassis.

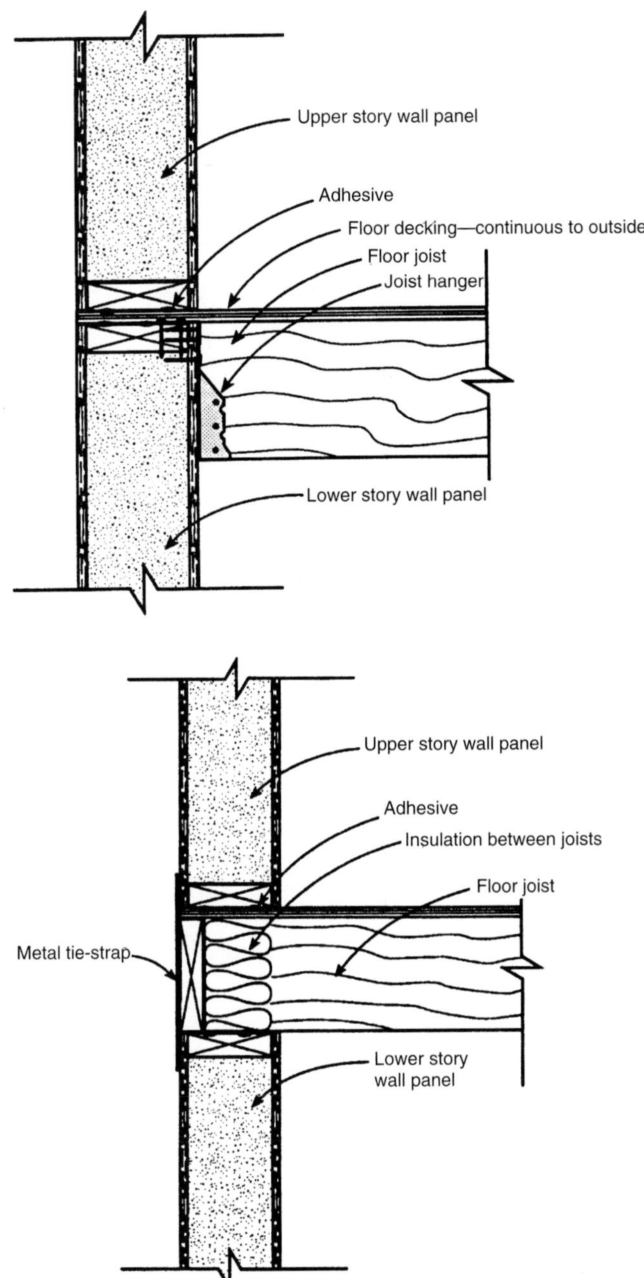

Figure 23-29. Methods vary for securing and supporting second floor joists. Top—Using joist hangers secured to top of plate. Bottom—Resting second-floor joists on top of the wall plate.

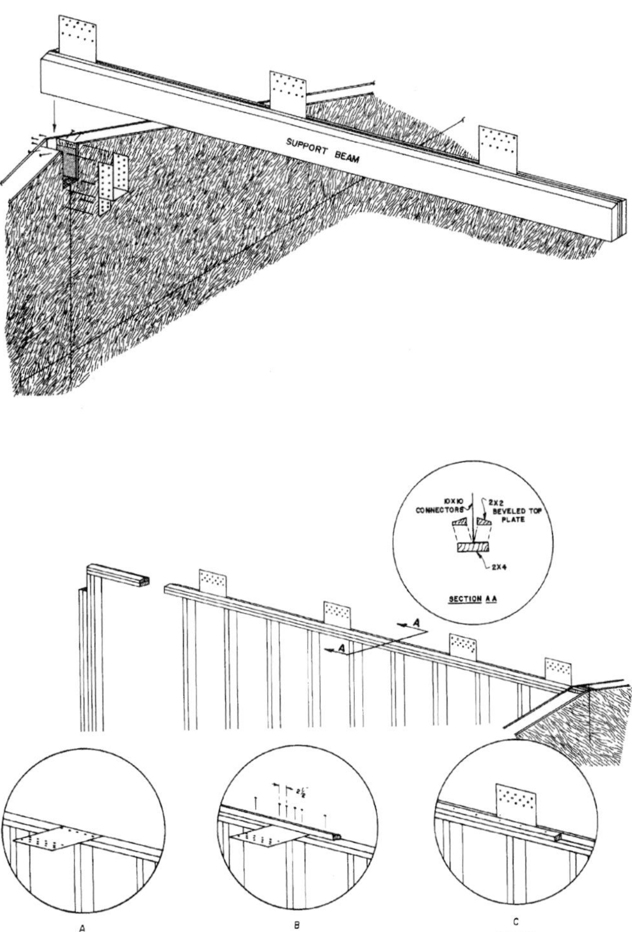

Figure 23-30. Alternate method for securing roof panels. Top—Steel connectors attached to the ridge beam will be used to secure tops of roof panels. Bottom—Clips attached to the wall plate will secure bottoms of roof panels. (Enercept Inc.)

A manufactured home is finished and fully equipped at the factory. It is then pulled to the dealer or directly to a trailer park. Utility hook-ups are ready to be made at the site. Interior furnishings include all major appliances, carpeting, drapes, furniture, and lamps. Figure 23-33 shows an inside view of a large mobile home. Figure 23-34 shows an old mobile home being removed from a retirement village to make room for a new unit.

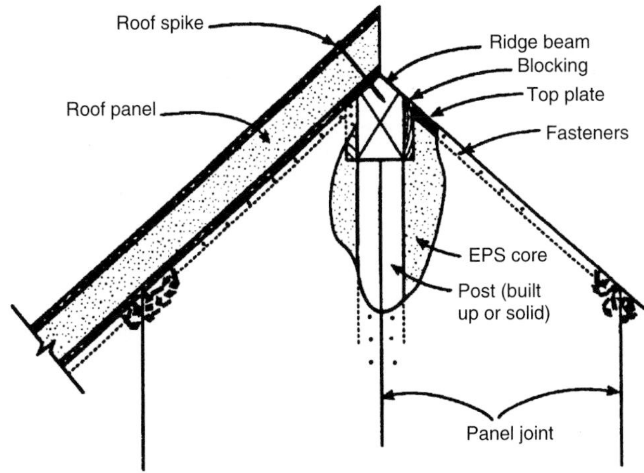

Figure 23-31. End view of a roof panel, ridge beam, and king (built-up) post. Beam must be solidly attached to the king post. (Insulspan)

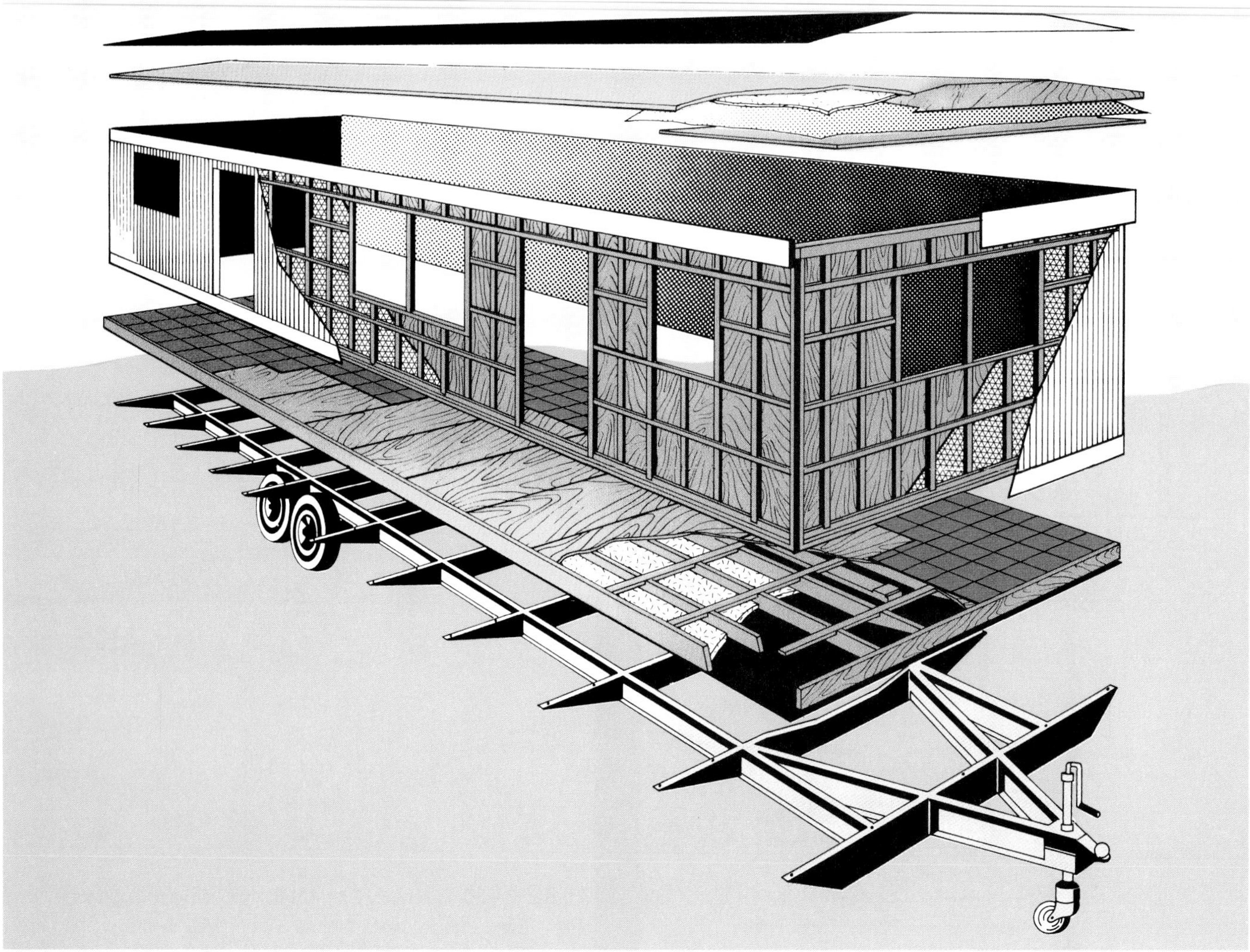

Figure 23-32. Exploded view of a mobile home. All structural members above the steel chassis are wood. Plywood deck is securely screwed to floor frame which is assembled with glue. Outside walls consist of 2 x 3 studs with 1 x 2 horizontal rails glued in place. Inside surface is 3/16" prefinished plywood. Prepainted aluminum panels cover outside. (Redman Industries Inc.)

Figure 23-33. This double-width manufactured home has wood framing under the floor to reduce heat flow and side stresses. The sections can be taken apart should the unit need to be moved to a new site. (Marshfield Homes, Inc.)

Figure 23-34. This double wide mobile home is being moved away from this site in a retirement subdivision to make way for a new unit. It will be refurbished in a factory and put back in service. Top—Sections have been unbolted. Bottom—Section is draped with mylar film to protect the interior while it is being moved.

Important Terms

Closed panels	Modular homes
Manufactured	Panelized homes
Mechanical cores	Precut homes
Mobile	Prefabrication
Modules (Mods)	Systems-built housing

Test Your Knowledge

1. What two terms are used in the industry to indicate that a home or its components were prefabricated?
2. What major factory-built framing component is used in all types of residential and light commercial construction?
3. Name two panel construction systems fabricated in factories.
4. ____________ systems-built structures are made up of two or more three-dimensional structural units.
5. A structural unit that contains plumbing, heating, and electrical systems is called a ____________ core.
6. A systems-built structure made up primarily of factory-built components that are assembled and then delivered to a site is called a ____________ structure.
7. A log home is a factory-built structure of the ____________ type.
8. Describe the building site at the point where the systems-built structure is delivered.
9. One disadvantage of a roof made up of sandwich panels is that asphalt or fiberglass shingles cannot be installed directly to it. True or False?
10. The outer structure of a panelized home can be finished in one ____________.
11. Mobile (manufactured) homes are constructed on a metal chassis and do not require a permanent ____________.
12. Due to speed of erection, a prefinished home will probably not be:
    A. Well-built.
    B. Wet inside.
    C. Checked by a Building Inspector.

Outside Assignments

1. Visit a local builder or the manager of a building supply center that represents a "prefab" home manufacturer in your region. Get descriptive literature and information concerning sizes, design, fabrication features, material quality, and erection procedures. Also check prices for standard models and time required for delivery. Organize the information carefully and make an oral presentation to your class. Supplement your oral descriptions with appropriate visual aids.
2. Prepare a written report on the development of factory building as applied to building construction in this country. Highlight early experiments and devote most of your study to progress since World War II. Find information in reference books, encyclopedias, and trade magazines.
3. Visit a mobile manufactured sales center and get information and descriptive literature concerning size, features, and price. Inspect models that are on display and give special attention to the quality of materials, work, and finish. Also note the quality of equipment and furnishings. Compare prices with costs of systems-built and conventionally built houses. Prepare carefully organized notes and make an oral report to your class on your findings.

Crane hoists huge two-story panel into place. (Wausau Homes Inc.)

This home was erected in a matter of weeks—an advantage of factory-built structures.

24 Passive Solar Construction

Because sunshine is free, it makes sense to capture it for heating buildings. Constructing buildings to take advantage of the sun requires some planning and an understanding of how to make the sun provide heat when it is most needed.

How Radiation and Heat Act

Solar radiation can travel through glass almost as well as it travels through the atmosphere. However, once transformed into heat energy (by striking a surface inside a glassed area) it cannot readily pass back through the glazing. This is known as the *greenhouse effect.* This effect is important to all solar construction. See Figure 24-1.

The very basis of solar construction is found in the way that heat moves from one place to another. Fortunately for us, heat always travels from a higher temperature to a lower temperature. It can travel by three different methods:

- Conduction.
- Convection.
- Radiation.

Conduction

In *conduction,* heat travels point-by-point through solid matter. One part of the solid body must be in touch with a heat source. Gradually the whole body becomes heated. For example, if the sun heats a block of concrete the entire block will eventually become hot. This is explained by the kinetic theory of matter. The molecules making up the concrete become more active as they become warmer, moving farther and faster and bumping into other molecules. This contact moves the heat from one molecule to another. See Figure 24-2.

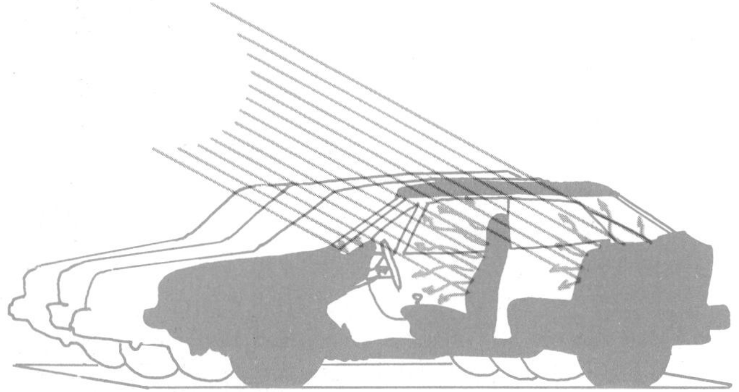

Figure 24-1. Solar radiation can pass through glass much easier than heat. This is known as the "greenhouse effect." It explains why an automobile with the windows closed will get so hot inside in direct sunlight. (Iowa Energy Policy Council)

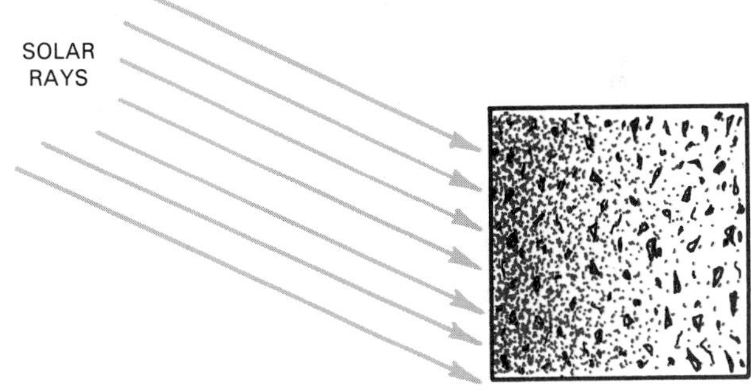

Figure 24-2. In conduction, heat travels when warm molecules of a solid bump into cooler ones. Eventually, all of them are heated.

Convection

Convection occurs only in fluids or gases. Heat causes them to expand and become lighter or less dense. Lighter elements always rise; cooler elements descend. Through this constant motion, heat moves through liquids or air, Figure 24-3. Older gravity furnaces used this principle. Heat rose through natural convection and spread through the living space. No mechanical means are needed to make the air move.

Radiation

Radiant energy is energy that moves through space in *waves.* When waves strike a solid object, Figure 24-4, the energy is absorbed by the solid matter. The internal energy of molecules in the matter increases and its temperature rises. An electric heater works this way. So do the sun's rays.

Figure 24-3. Convection is caused by heat contacting and entering liquids or gases. Heated gases and liquids expand, become lighter, and rise. Cooler gases and liquids are heavier so they settle. Left—Liquid action. Right—Convection in air.

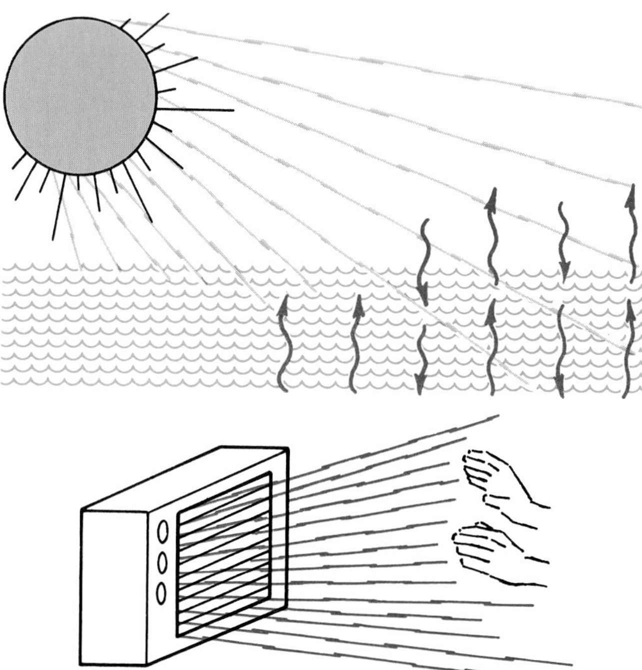

Figure 24-4. Two examples of radiant heating. In radiation, energy moves in waves from source to solid object. Top—Sun warms body of water as radiation strikes the surface. Bottom—Electric heater works the same way.

Thermosiphoning

Thermosiphoning is simply the result of a liquid or gas expanding and rising. This principle is put to work in both active and passive solar heating. Sunlight is captured in a closed space where a liquid such as water or air is heated. The system operates by the action of the expanded water or air rising. Cooler air or water comes into the lower space to take its place, Figure 24-5.

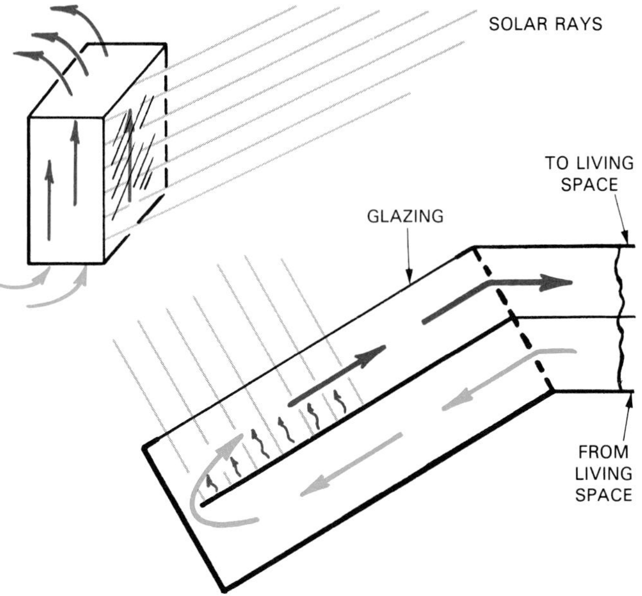

Figure 24-5. Simple example of thermosiphoning. Top—Sun shines into open-ended box. Heated air rises out the top, cooler air is pulled in the bottom. Bottom—When baffle is added, cold air moves in from lower side of box. Warm air rises along upper half.

Types of Solar Construction

There are two commonly used methods of incorporating solar energy into a building: *active solar* and/or *passive solar construction.* Active solar construction is actually a system of collecting and transporting solar energy apart from the structure of the house. It is a separate system that may be added to a dwelling during or after it is constructed. It has little, if any, effect on the way the building is built.

Solar collectors are, in simplest terms, boxes that trap solar energy. In some cases the solar collectors, Figure 24-6, are made a part of the roof. In other cases, the collectors are attached to the existing roof by a metal framework.

Active systems are so-called because electrically operated blowers or pumps are required to carry heated air or fluid to where it will be used or stored. Figure 24-7 is a simple sketch of an active solar system.

Passive solar construction, on the other hand, is so designed that energy is absorbed in the mass of the structure. It is not transported by any mechanical means. By its very nature, it requires a change in the structural makeup

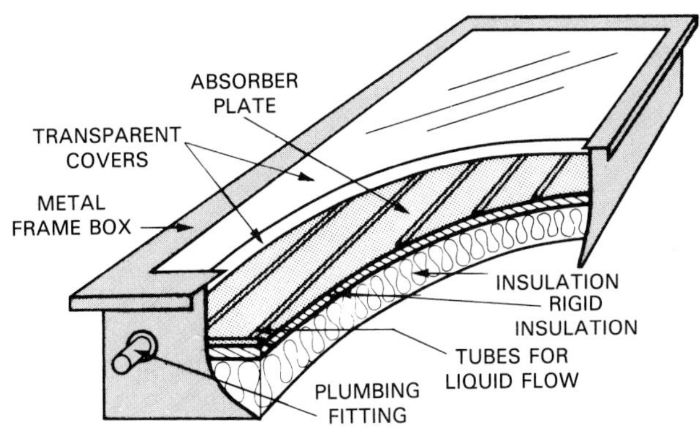

Figure 24-6. Cutaway of a flat plate solar collector designed to heat liquid. It is used in an active solar system. (National Solar Heating and Cooling Center)

of the building. It is called passive because it has few if any parts which must move or require power to make it work. Collection, storage, and transporting of heat is done naturally by the materials used in construction.

Types of Passive Solar Energy

Passive solar energy systems are named for the way in which they operate. The word "gain" refers to how the heat is picked up from solar radiation. There are three basic types of passive solar construction:

- Direct gain.
- Indirect gain.
- Isolated gain or sun space.

Depending upon the individual house design, more than one system may be used, Figure 24-8. Several approaches may be combined in the same house for greater efficiency, affordability, and beauty. No single floor plan is required and passive solar systems may be incorporated into many architectural styles. Passive designs work well with Cape Cods, colonials, contemporary, or ranch styles.

Direct Gain System

Direct gain means that the sun shines directly into the living space and heats it up. In the simplest of direct gain systems there may be no massive structures to store the heat produced. The furniture soaks up some heat and so does the air in the room. Usually, there is some means of storing the heat for use at night or on cloudy days. Thick masonry walls or floors, a masonry fireplace, a heavy stone planter, or containers of water are examples of storage systems that may be used. Suitable materials include stone, adobe, brick, or concrete.

At night, stored heat is distributed to the living space by radiation and convection. If the storage is adequate, it should give off heat throughout the night.

One major disadvantage of direct gain systems is the wide range of heat fluctuation (changes). During the hours of strongest sunlight, heat can build up to uncomfortable levels. Figure 24-9 is a simple drawing of a direct gain structure.

Indirect Gain System

In an *indirect gain system,* the rays of the sun enter through glazing and heat up thermal mass rather than the room itself. Often the mass is a thick masonry or concrete wall located directly behind large windows or glide-by glass doors. Sometimes the wall is vented at the top and bottom. This allows hot air to flow upward and enter the room quickly by way of convection currents, Figure 24-10.

Often called a *Trombe wall* for the French scientist, Felix Trombe, who designed it, the thermal storage wall is the most popular storage structure for indirect solar heating. The outer face is dark colored for greater absorption of heat. By conduction, the heat travels through the wall to its inner face. Then radiation and convection, together, distribute the heat throughout the space.

Thermal walls still perform their function even with windows cut in them. Inner masonry surfaces can be made to fit the interior design by covering them with stucco, gypsum plaster, ceramic tile, or other heat conducting materials. Pictures may be hung without greatly affecting the wall's performance.

Water Storage Wall

Water is an effective and inexpensive material for storage of solar heat. In fact, it can absorb about twice as much heat as rock. However, the materials for containing it are not usually cheap. Still, barrels and other empty containers that are rescued from salvage can be used.

The water wall is placed directly behind a south-facing window. As with masonry storage, the water absorbs the solar heat and transfers it by radiation and convection.

A more sophisticated water wall may use phase change materials. These are materials which, like ice, change from a solid to a liquid state as they absorb heat. They are able to absorb much more heat than materials which do not change their form. Usually, a type of salt solution is used. Common Glauber's salt is one of them. See Figure 24-11.

It is generally easier and more trouble free to use masonry with conventional structural features such as walls, floors, and interior fireplaces. Such materials and structures rarely need attention once they are in place.

Isolated Gain System

In an *isolated gain system,* the solar heat is collected and stored in an area remote (apart) from the living or working space. Its advantages:

- The living/working area is not directly exposed to the sun.

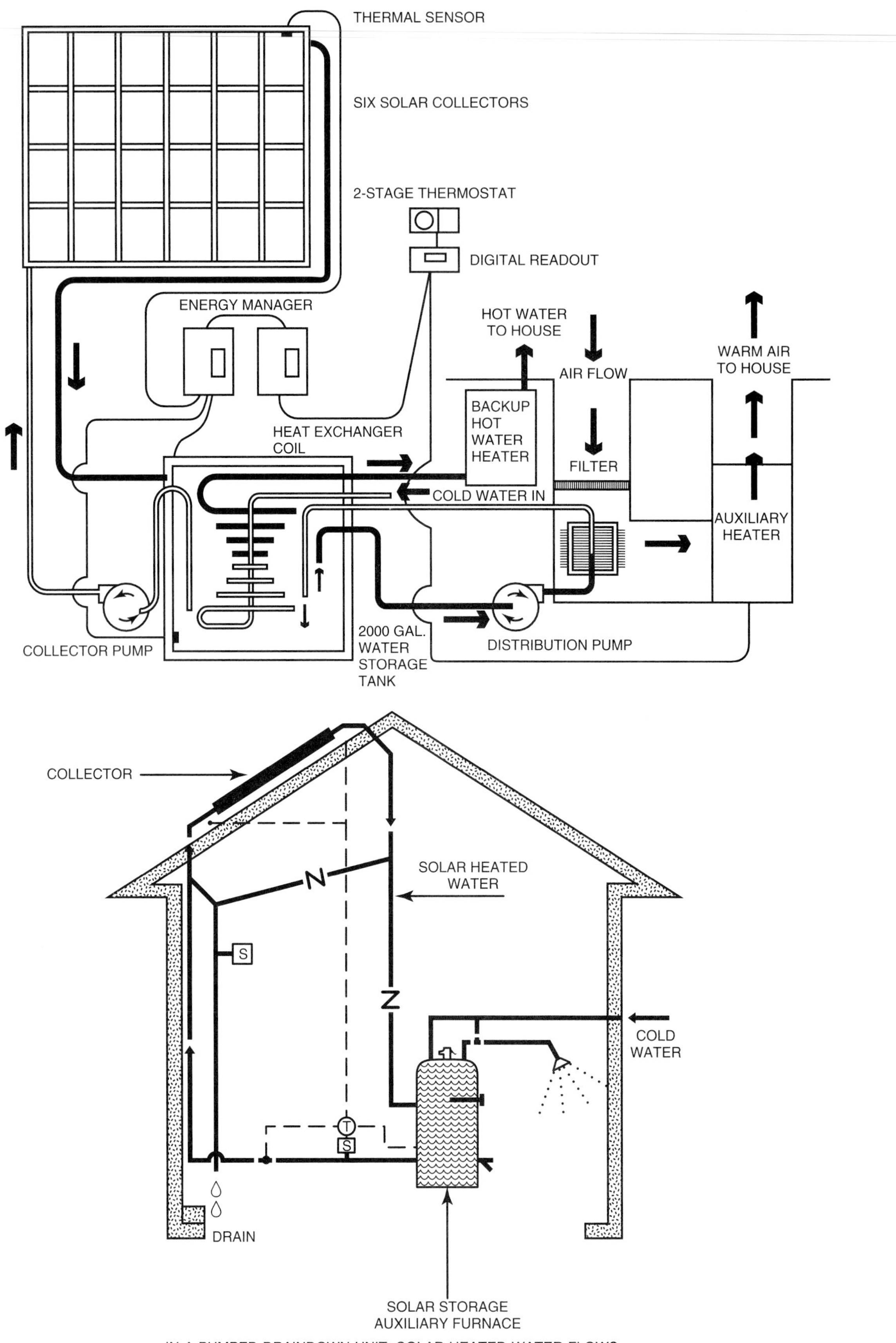

Figure 24-7. Two schematics of an active solar heating system using roof collector illustrated in Figure 24-6.

Figure 24-8. Home using both active and passive solar energy. Note solar panels on roof. Solar room at right is an example of isolated sun space illustrated in Figure 24-13. (Timberpeg)

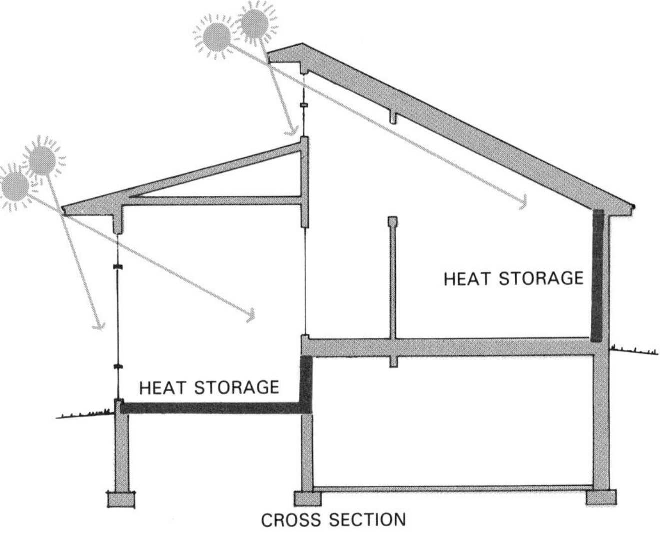

Figure 24-9. Basic design for a direct gain passive solar house. Floor and wall store direct sunlight.
(District 1 Technical Institute, Eau Claire, Wisconsin)

- The heat it collects is more easily controlled. Figure 24-12 shows a simple sketch of a thermosiphon isolated passive system. This is one type. Another is called a sun space.

The *sun space,* Figure 24-13, is a common method of providing solar heat. It is a separate room with large areas

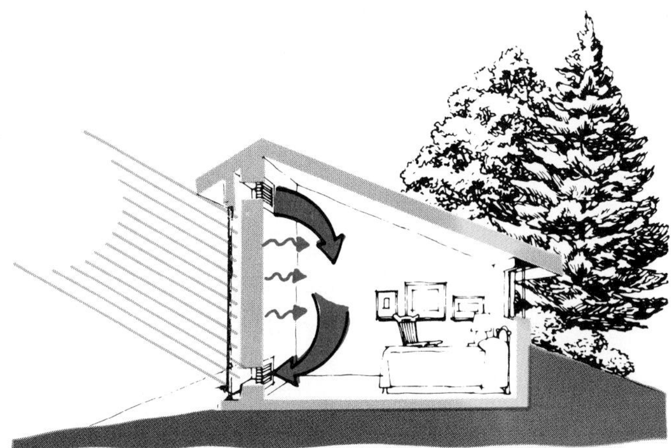

Figure 24-10. The Trombe wall is the most popular method of storing and distributing solar heat by the indirect gain method. (Iowa Energy Policy Council)

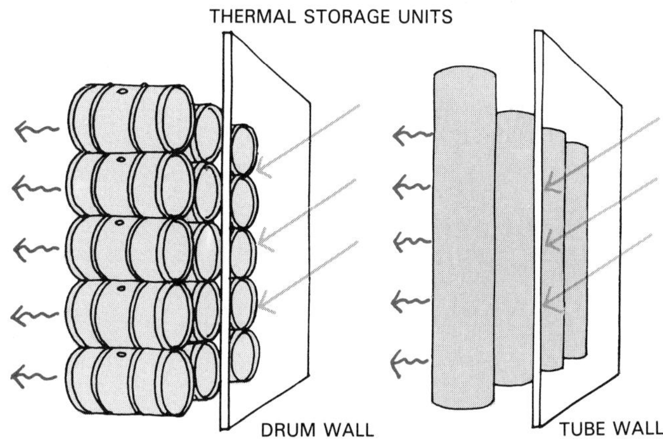

Figure 24-11. Water and phase change materials can also be used to store the sun's heat. (Pittsburgh Plate Glass Co.)

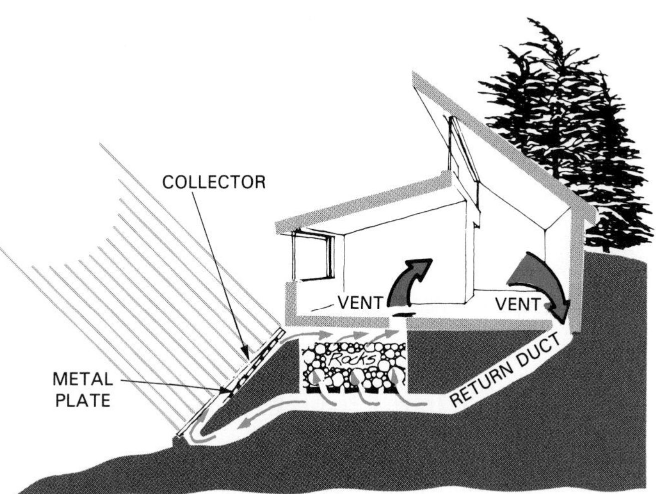

Figure 24-12. Thermosiphoning isolated passive system. Collector at ground level is the thermosiphon. Heated air flows on its own power to storage. Cool air is pulled into the collector from below.

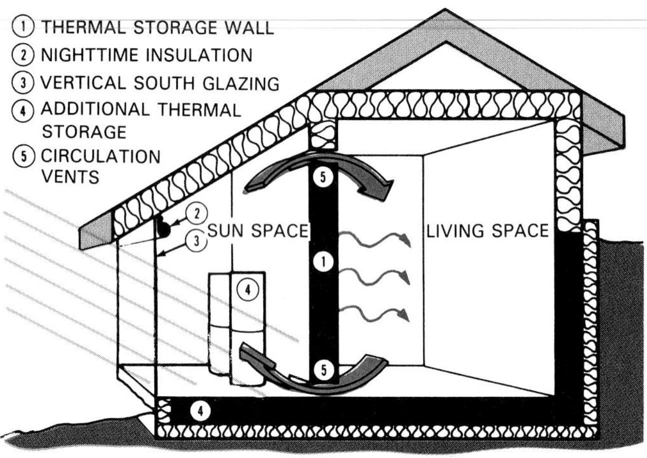

Figure 24-13. Another type of isolated gain heating is the sun space. Left—This is a room with plenty of storage mass to save solar generated heat. Note how insulation protects floor and north wall from colder surroundings. (Iowa Energy Policy Council) Right—Solar sunroom is an "add on" sun space. (Rolscreen Co.)

of its south wall glazed. Sometimes the room has thermal mass built into the inside wall or the floor. Additional thermal storage can be provided in water containers or bins of rock.

Passive Solar Advantages

Passive designs have several advantages over active solar heating and conventional heating. Some of these advantages include the following:

- Common building materials—glass, concrete, and masonry—can be used. Other manufactured materials are not needed although they are available.
- Conventional carpentry and masonry skills are sufficient. No additional skills need be learned.
- Passive components do not wear out and need little maintenance. Life expectancy is greater than active systems.
- Properly designed passive features can supply upwards of 50% of heat required.
- Passive solar is nonpolluting.
- Since heat is generated close to or in the space being heated, little heat is lost through transfer.

Passive Solar Disadvantages

Disadvantages of solar systems include:

- Control of the heat is not as responsive as either conventional heat systems or active solar systems. Building occupants are accustomed to temperature swings of 3°F to 5°F (2°C–3°C)Many passive systems have swings of 10°F to 15°F(6°C–8°C)
- It is not a simple matter to control heat and heat distribution in a passive system. Careful planning is required.

Solar Heat Control

Adequate control methods must be arranged when passive solar heating is being planned. There are times when solar heat is not wanted. In the summer, heat is not required. Then the same structural features that allow sunlight to enter must be sheltered from the sun. Shade can be provided either naturally or artificially.

Deciduous trees (those that lose their leaves in winter) should be located to protect south-facing windows in summer. It may be possible to locate a new building to take advantage of existing trees on the property. Fast-growing trees can also be planted to eventually provide this shade. They should be located to provide shade for west-facing walls as well.

Constructed protection includes outdoor roof structures including overhangs and shutters, Figures 24-14 and 24-15.

Overhangs

The width of the protective overhang must be determined by three variable factors. See Figure 24-16.

- The height of the window or collector.
- The height of the header above the window or collector.
- The latitude of the construction site.

As a rule of thumb, the overhang in southern states (roughly latitude 36° north) should be 25% of the combined height of header and window. In northern states (around 48° latitude) the percentage should be 50%.

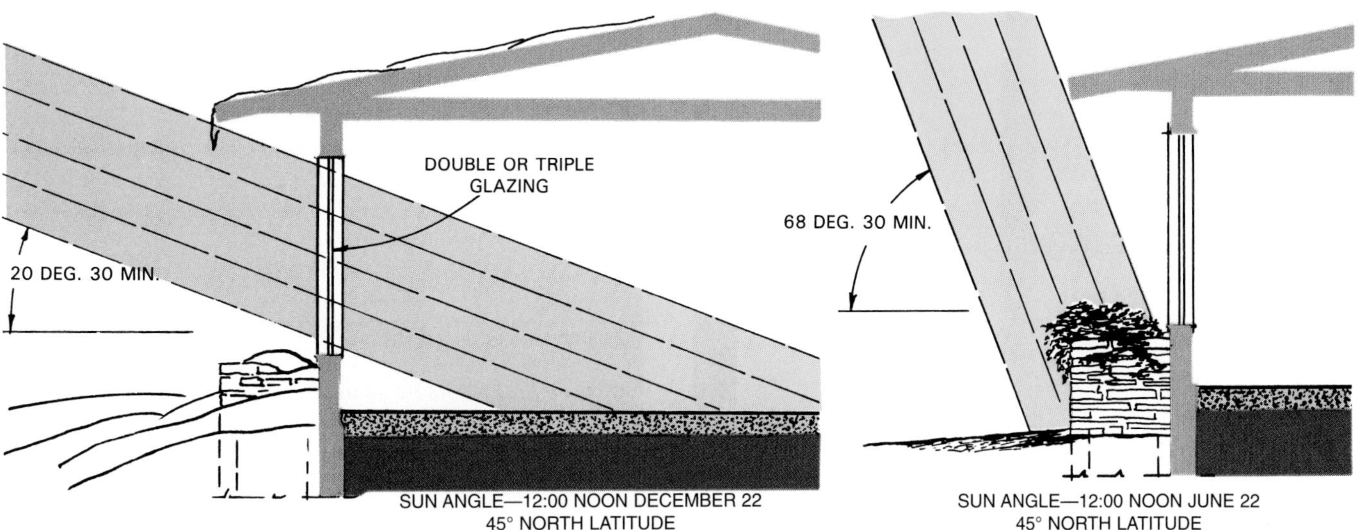

Figure 24-14. Roof extensions can be used to control solar rays in passive heating design. Latitude of 45° north is the general location of Portland, OR; Minneapolis, MN; and Montreal, Canada.

Figure 24-15. Exterior shading includes deciduous tree, overhang, and hinged shutters. (HUD)

A simple mathematical formula, used with the chart in Figure 24-17, will also provide a quick and quite accurate determination of how far the overhang should project:

$$\text{Projection} = \frac{\text{(window height + header height)}}{\text{F (factor from Figure 24-17)}}$$

Movable Insulation

While outdoor structures and natural shade are more effective protection against the hot summer sun, movable insulation is sometimes the only practical protection possible. It is also effectively used to reduce nighttime heat losses through glass.

Movable insulation is usually sheet or blanket materials that can be put over glazing temporarily. It is placed over the glass only when you do not want radiation or heat passing through.

Movable insulation solves a major problem of glazing—its poor insulating quality. The insulation cuts nighttime heat losses and will help keep the building cool during summer days.

Triple glazing of window area will also cut down heat losses. However, it also reduces the solar radiation transmission from 74% (for double glazing) to 64%. Removable insulation is preferred because it provides better heat retention and does not cut down on solar efficiency of the window.

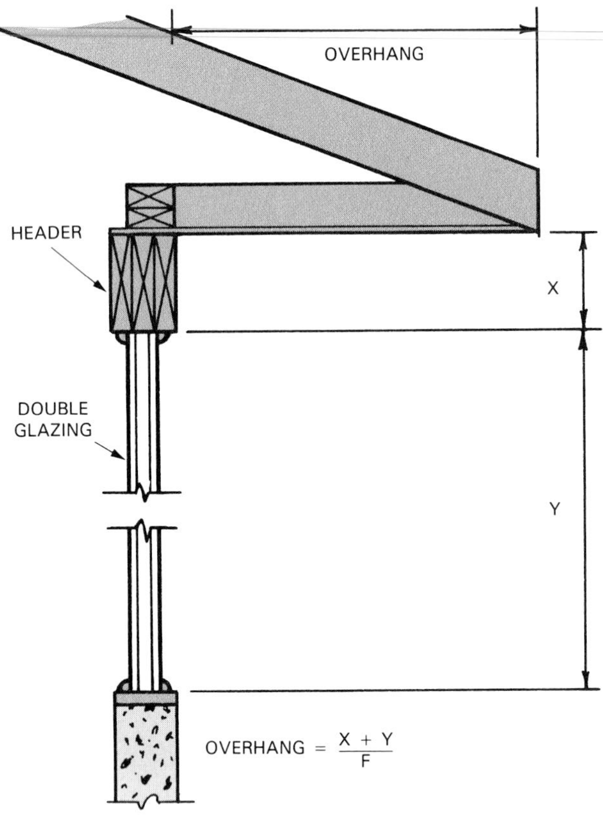

Figure 24-16. Correct amount of overhang is determined by dividing height of window and header by a factor based on latitude.

LATITUDE (in Degrees)	APPROXIMATE LOCATION (State)	FACTOR* (F)
28	Central Florida	5.6 to 11.1
32	Central Texas	4.0 to 6.3
36	Northern Oklahoma	3.0 to 4.5
40	Northern Missouri	2.5 to 3.4
44	Iowa-Minnesota Border	2.0 to 2.7
48	Northern Minnesota	1.7 to 2.2
52	Southern Canada	1.5 to 1.8
56	Central Canada	1.3 to 1.5

*Higher of two factors provides 100% shading at noon on June 21; lower factor until August 1.

Figure 24-17. Factors for finding width of overhang from south Texas to central Canada.

Movable insulation is available in a variety of types including:

- Sheets of rigid insulation.
- Framed and hinged insulation panels, Figure 24-18.
- Exterior mounted plastic or metal roller shades, Figure 24-19.
- Padded roller-mounted flexible cloth panels, Figure 24-20.
- Powered systems, such as "Beadwall," that depend upon blowers to fill air cavities between layers of double glazing with insulating beads.

Venting

Venting has an important role in controlling solar heat both in indirect and isolated gain systems. Trombe walls

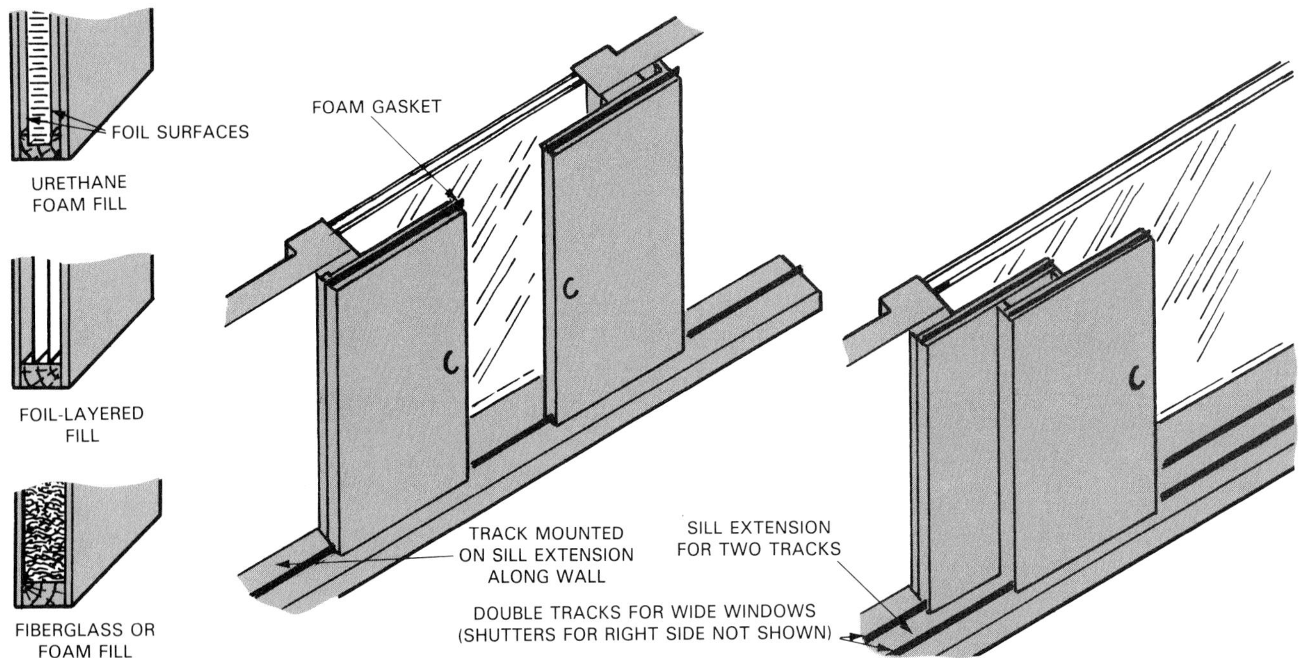

Figure 24-18. Sliding shutters are designed to prevent loss of heat through solar glazing at night.

Figure 24-19. Power-operated plastic or metal shutters shut out unwanted sunlight, provide security, and shut out cold at night. Left—Opened. Right—Closed to block out sunlight in warm weather. (Pease Industries, Inc.)

Figure 24-20. Roll-up fabric insulating shade.

sometimes have vents to provide daytime heating for living space. Refer once more to Figure 24-10. Opening the vents allows convection currents to carry heated air into the room through the vents. Meanwhile, cool air is drawn into the space between the glazing and the Trombe wall. Closing the vents, top and bottom, will prevent heat from entering the living space.

Isolated sun spaces are often vented to the outdoors to exhaust unwanted solar heat during hot weather. The vent acts like a chimney. The hot air rises through the roof. If there are vents at floor level, cooler air will be pulled into the sun space. This provides additional cooling.

Natural venting of buildings is also possible using the chimney effect. This is accomplished with a little sunlight and vents high up in the building. A window or skylight will let in sunlight to heat a "pocket" of inside air. The air rises and passes through the vent. Cooler air can then be brought into the building from another area of the living space. Figure 24-21 shows two methods of using venting for cooling.

Orientation

The first step in solar construction is to locate the building so one wall catches the sun's rays all day long. This is called *orientation,* Figure 24-22.

Most of the windows should be placed in the south wall so that the solar radiation can be collected. Few windows will be placed on the north side. Garages should be placed to the northwest, if possible, to block cold winter winds.

Energy Balance

Homes experience heat gains from three sources on a typical winter day:

- From conventional heat source, such as a stove or furnace.
- From solar radiation.
- From appliances and occupants. This is called *internal heat.*

Energy use is reduced when the heating system can operate at the lowest level that will provide comfort. Solar and internal heat gains can permit a cutback in operation of the heating plant if the house design can balance internal

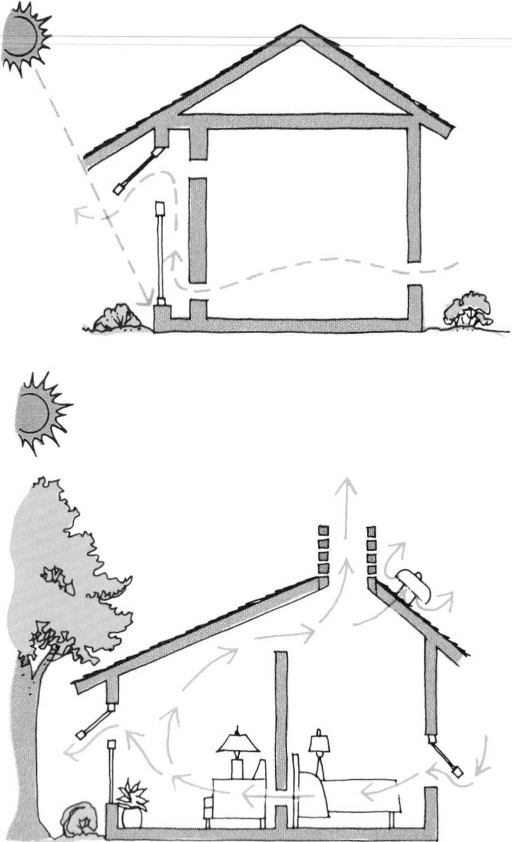

Figure 24-21. Two methods for using venting for warm weather cooling. Top—Vent in glazing draws air from cool, shaded side of dwelling across room, through Trombe wall, and out. Bottom—Thermal chimney effect. Warm air rises and exhausts through vents at top of house. Cool air enters from open window in shaded wall. (Pittsburgh Plate Glass Co.)

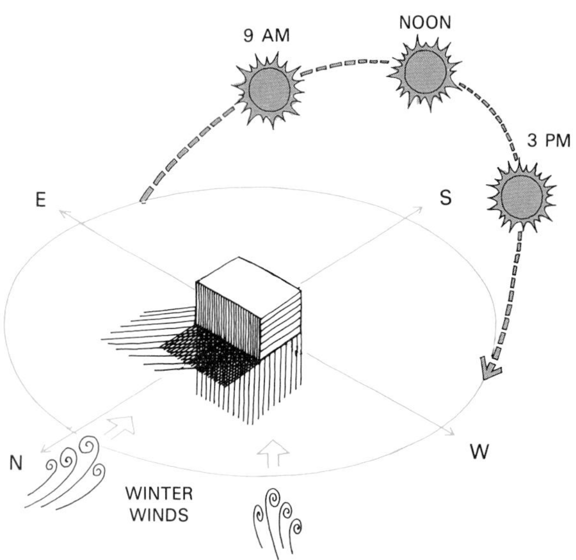

Figure 24-22. Home is properly placed with respect to the sun. South facing wall will catch all the sunlight it possibly can. Walls can be 10° to 20° off true south without losing much of their efficiency. (Pittsburgh Plate Glass Co.)

heat gain against solar gain. This means locating heat producing appliances on the opposite side of the house from space receiving solar gain. See Figure 24-23 for best use of internal and solar gains in the northern hemisphere.

Building Passive Solar Structures

Thermal storage walls used in passive solar construction need to store enough heat to even out the wild fluctuations of temperature which could make a building uncomfortable. All materials are capable of storing heat. However , some materials store heat more effectively than other materials. Some of the most commonly used materials include the folling:

- *Water*—Though cheap, it might be expensive to construct or purchase containers to store water, which has greater heat capacity than solid materials.
- *Concrete and concrete block*—Both have good heat storage capacity and will provide structural support for the building, as well. Cores of hollow block can be used as warm air ducts. However, the wall will hold more heat if they are filled with sand-mix concrete. Walls can be plastered or painted to conceal mass.
- *Brick*—Brick's properties are similar to those of concrete or block. Attractive in appearance, bricks are easier to blend with decor. Though more expensive than concrete, their appearance often makes them the preferred material.

These materials are good choices for thermal storage, because of their cost, appearance, and suitability. Figure 24-24 compares heat carrying capacity of these materials with other building materials.

Sizing Thermal Storage Systems

Sizes of thermal storage systems are determined by the square footage of the living space. For example, in climates where the average winter temperature is 20°F to 30°F (-4°C to -1°C), between 0.43 and 1 sq. ft. of masonry per sq. ft. of living area is adequate. Figure 24-25 lists adequate mass for different areas by degree days. Consult the chart, "Normal Number of Degree-Days per Year" in the Technical Information Section for degree-day information on your locality.

Wall Thickness

Storage wall thickness is important. A space will overheat if more energy is transmitted through the wall than is needed. This will happen if the wall is too narrow. Further, a wall that is too thin will not be able to transmit

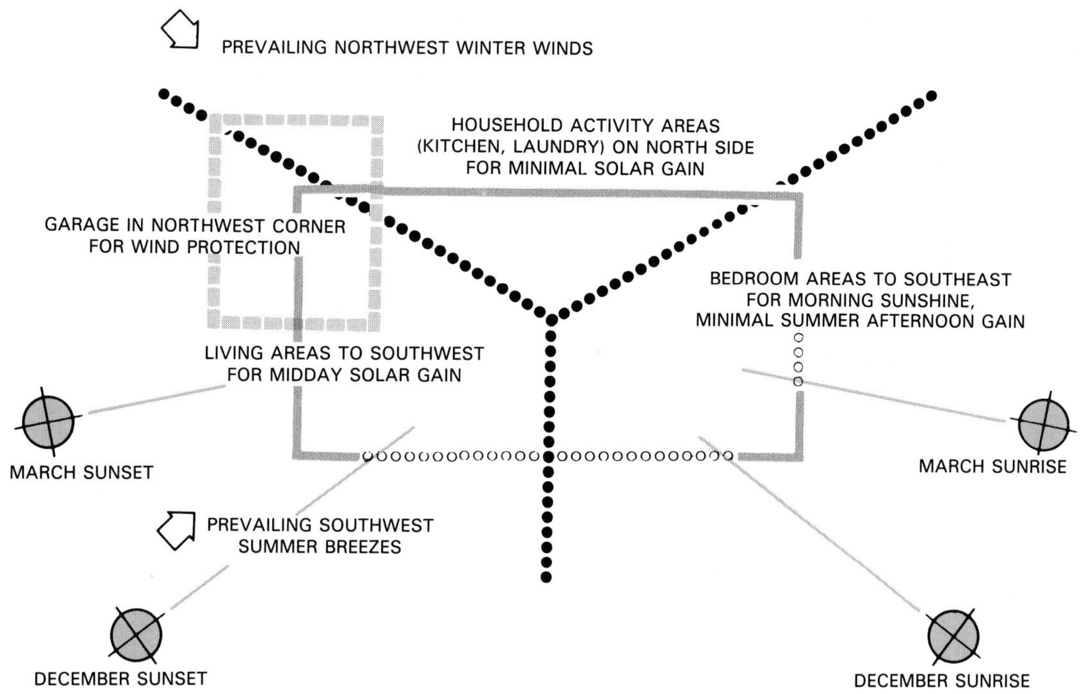

Figure 24-23. Locating living areas for best balance and use of solar and internal heat gains. Appliances represent internal gains. (Forest Products Laboratory)

Substance	Specific Heat (BTU/lb-°F)	Density (lbs/cu ft)	Heat Capacity (BTU/cu ft-°F)
*Water	1.00	62.4	62.4
Wood, oak	0.57	47.0	26.8
Fir, pine, and similar softwoods	0.33	32.0	10.6
Expanded polyurethane	0.38	1.5	0.57
Wool, fabric	0.32	6.9	2.2
Air	0.24	0.075	0.018
*Brick	0.20	123.0	25.0
*Concrete	0.156	144.0	22.0
Steel	0.12	489.0	59.0
Gypsum or plaster board	0.26	50.0	13.0
Plywood (Douglas Fir)	0.29	34.0	9.9

*Preferred materials.

Figure 24-24. Heat carrying capacities of water, concrete, and brick make them good choices for heat storage. (Iowa Energy Policy Council)

CLIMATE	WINTER TEMPERATURE For Coldest Months in Degree Days/Mo	SQUARE FOOTAGE FOR EACH SQUARE FOOT AREA	
		Masonry Wall	Water Wall
Cold	1500	0.72 to 1	0.55 to 1
	1350	0.62 to 1	0.45 to 0.85
	1200	0.51 to 0.93	0.38 to 0.70
	1050	0.43 to 0.78	0.31 to 0.55
Temperate	900	0.35 to 0.60	0.25 to 0.43
	750	0.28 to 0.46	0.20 to 0.34
	600	0.22 to 0.35	0.16 to 0.25

Figure 24-25. Solar storage mass must be based on average winter temperature and the size of the area to be heated.

heat throughout the night. It will have cooled before morning. Figure 24-26 suggests thicknesses for storage walls of various materials.

The more rapidly the material conducts heat the thicker it must be. Otherwise, it may deliver its heat load too soon. This makes the living space uncomfortably warm.

Sizing Direct Gain Storage

A rule of thumb for walls, slabs-on-grade, etc., used for storage in direct gain systems is to provide 150 lb. of

MATERIAL	THICKNESS (IN.)
Water	6 or more
Dense concrete and block	12 to 18
Brick	10 to 14
Adobe	8 to 12

Figure 24-26. General recommendations for thickness of storage walls when constructed of various materials. Thicker walls will store more heat as well as delay the time when the wall will conduct heat to the living space.

concrete or masonry for every square foot of glazing. This assumes that the storage is subjected to direct sunlight.

At the same time, there should be at least 150 lb. of concrete or masonry for every 2 sq. ft. of area to be heated. It should be about 4" to 6" thick. If thinner than 4", it will not be able to store enough heat per square foot. Floors used for thermal storage should not be carpeted wall to wall. However, scatter rugs, furniture, and other obstructions will have little effect on the performance of the floor.

Effect of Color on Collecting Surface

Since dark colors absorb heat more readily, the wall surface facing the glazing should be dark. Though black is most efficient as an absorber, other dark colors may be used with nearly the same efficiency. Dark blue, for example, works almost as well. Inside wall surfaces may be any color.

Wall Construction

Solid Trombe walls will store more heat than vented ones. The outside surface of the wall may reach 150°F (66°C) or more on a sunny day. The north-facing wall, however, will maintain a fairly uniform temperature throughout a 24-hour period. Vents will be necessary if the living space requires extra warmth during the day. Then the air space between the wall and the glazing sets up a thermosiphon action (natural convection).

Since the Trombe wall is a bearing wall, construct it based on the principles given in Unit 7, Footings and Foundations. Footings should be twice as wide as the wall's cross section and equal in thickness. Figure 24-27 shows a vertical cross section through a typical Trombe wall. Refer also to Figure 6-15 for a horizontal cross section.

Some provision for venting of unwanted heat in warm weather is advisable. Figure 24-28 is a detail drawing for venting the Trombe wall at the header. For aesthetic purposes, windows can be included in a Trombe wall. Adjustments should be made to the dimensions to assure sufficient area for proper heat collection and storage. Figure 24-29 shows several designs which incorporate windows.

Special Concerns

It is desirable to construct thermal glazing so that the glazing can be removed occasionally for cleaning the glass or to repaint the Trombe wall. Thus, it would be helpful if the glass panels are supported in their own sash. Panels should not be so large that the windows are difficult to handle.

Single panes of glazing are not as efficient as double glazing, and triple glazing is not generally recommended. It cuts down on the effectiveness of the solar radiation and does not appreciably reduce the heat loss over double glazing.

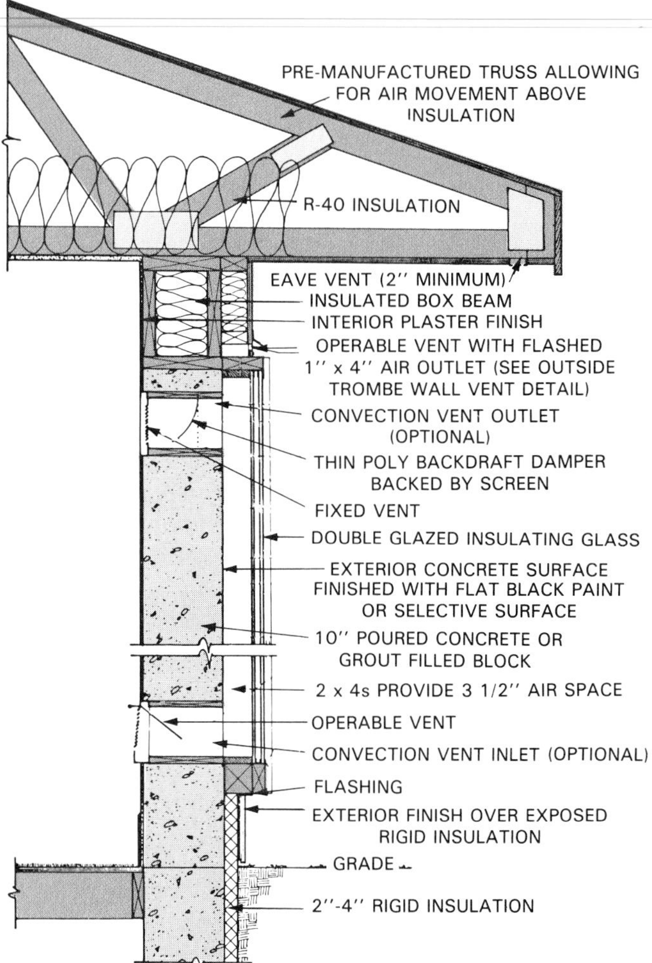

Figure 24-27. Typical section of a Trombe wall. Note insulation in hollow header. Air space between wall and glazing can vary from 2" to 5". (Iowa Energy Policy Council)

Designing the Isolated Gain System

Basically, the isolated gain system is a small room attached to the main structure of the building. Figure 24-30 shows, in simple sketches, a variety of designs for bringing the heat from the sun space to the living space. A greenhouse attached to a south exposure of a house is a good example of this type of structure.

The sun space is sometimes "embedded" in the house. That is, it is within the walls of the house and is separated from the house by only its interior partitions. One of these partitions can be a storage wall of concrete, masonry, or water.

Passive Thermosiphon System

Thermosiphoning space heating systems are not well researched or understood. Passive designers do suggest

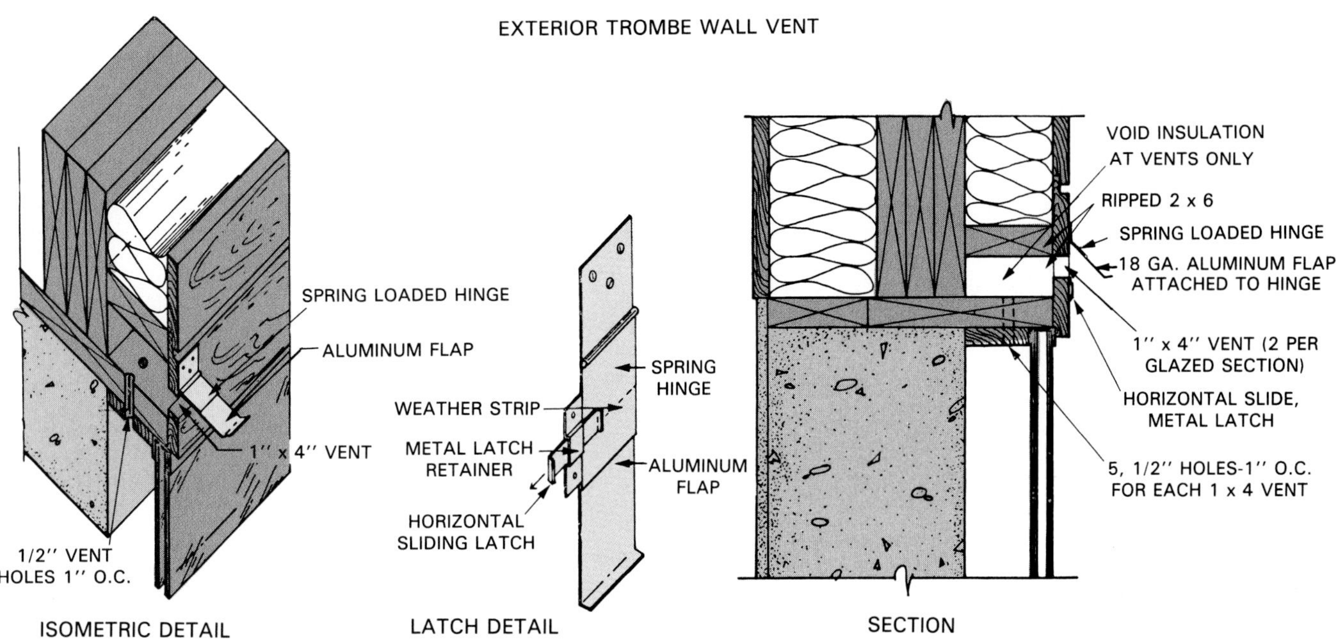

Figure 24-28. Method of venting Trombe wall to outside to disperse warm weather heat.

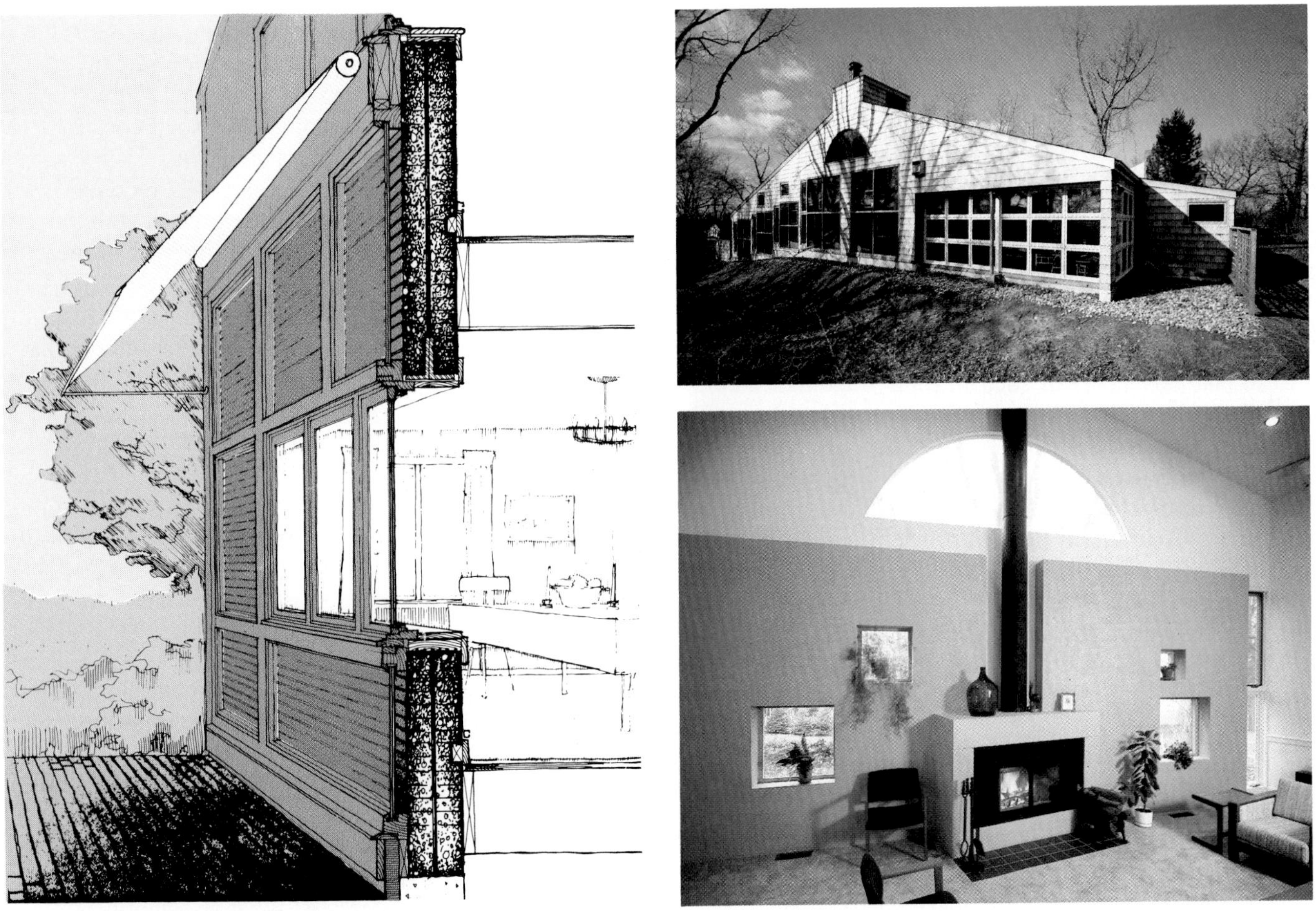

Figure 24-29. Two different designs for windows in Trombe walls. Left—Section of brick wall with casement windows. Storage serves two levels. Roll-up awning takes place of overhang. Right—Inside and outside view of design which uses small windows in free standing storage wall. Clerestory window, above, provides additional lighting. (Pittsburgh Plate Glass Co.)

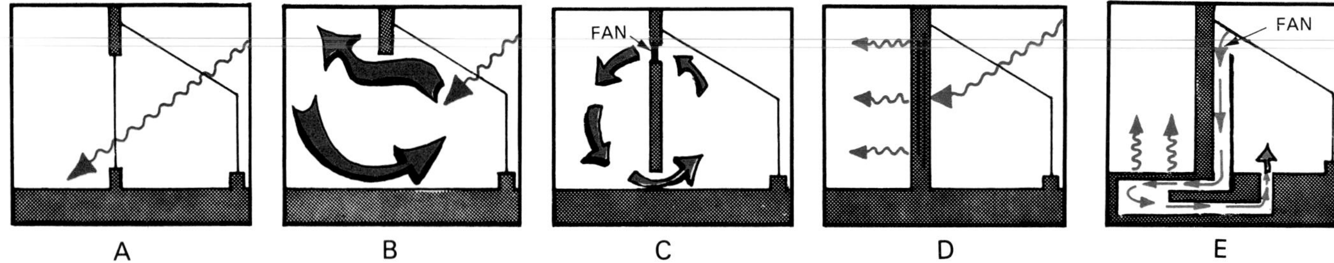

Figure 24-30. There are many methods of bringing heat from sun space into living area. A—Solar rays are transmitted through sun space into living space. B—Natural air exchange takes place. C—Fan pulls in heated air. D—Storage wall delivers heat through radiation. E—Gravel storage bin stores heat which radiates to building above.

that the rock bed needed for storage have a cross-sectional area that is about 50% to 75% of the area of the collectors. For example, if the area of the collector is 10' x 20' (200 sq. ft.), the cross section of the bed should be 100 or 150 sq. ft. Four-inch rock are recommended for easy circulation of the air around them. Depth of the bed should be 20 times the rock diameter, about 6'-6".

To find the proper volume of storage, multiply the collector area by a factor of 3.25 to 5. The 200 sq. ft. collector would require 650 to 1000 cu. ft. of storage space.

Rock bins need to be insulated against heat loss on their sides and bottoms. Dampers should be located in the ducts leading to the collector. These are closed every evening to prevent the loss of heat to the outside at night.

Insulating Passive Solar Buildings

To take advantage of solar heating, buildings should be well insulated. Refer to Unit 14, Thermal and Sound Insulation. Figure 24-31 is a cross section of a two-story house showing good insulating practice for passive solar construction.

When insulating, it pays to provide thermal barriers anywhere that heat is likely to leak out. There should be no place that air can flow between inside and outside without going through insulation. Seal up all areas where air could possibly leak through cracks. Refer to Figure 24-32 and Unit 6 for more information on passive solar structures.

Important Terms

Active solar construction
Conduction
Convection
Direct gain system
Greenhouse effect
Indirect gain system
Internal heat
Isolated gain system
Movable insulation
Passive solar construction
Radiant energy
Radiation
Solar collectors
Solar radiation
Sun space
Thermosiphoning
Trombe Wall

Test Your Knowledge

1. Solar radiation passes through glass readily but heat cannot pass back through the glass as readily. This phenomenon is called the ____________.
2. List the three different ways heat is transferred.
3. ____________ is simply the result of a liquid or gas expanding and rising.
4. The main characteristics of passive solar construction are (select all correct statements):
   A. It has few, if any, moving parts that require mechanical or electrical energy for operation.
   B. It is usually a stationary, structural part of the house.
   C. Maintenance of the system is more expensive.
   D. Collection, storage, and transporting of heat is done usually naturally by the materials used in construction.
   E. There is always ductwork to carry the heat to living space.
5. List two disadvantages of passive solar energy.
6. What factors govern the size of the overhang above solar glazing?
7. What is orientation?
8. Internal heat is the heat gained from ____________ and ____________.
9. As a general rule how much heat storage (in pounds of concrete) should be provided for every square foot of glazing?
10. Triple glazing is more effective than double glazing in passive solar construction. True or False?
11. A storage bin must be constructed to house the rock needed for a thermosiphon system. The glazed area of the collector is 8' by 20'. What amount of rock must the bin house?
    A. 160 to 180 cu. ft
    B. 520 to 800 cu. ft.
    C. 1600 to 1800 cu. ft.

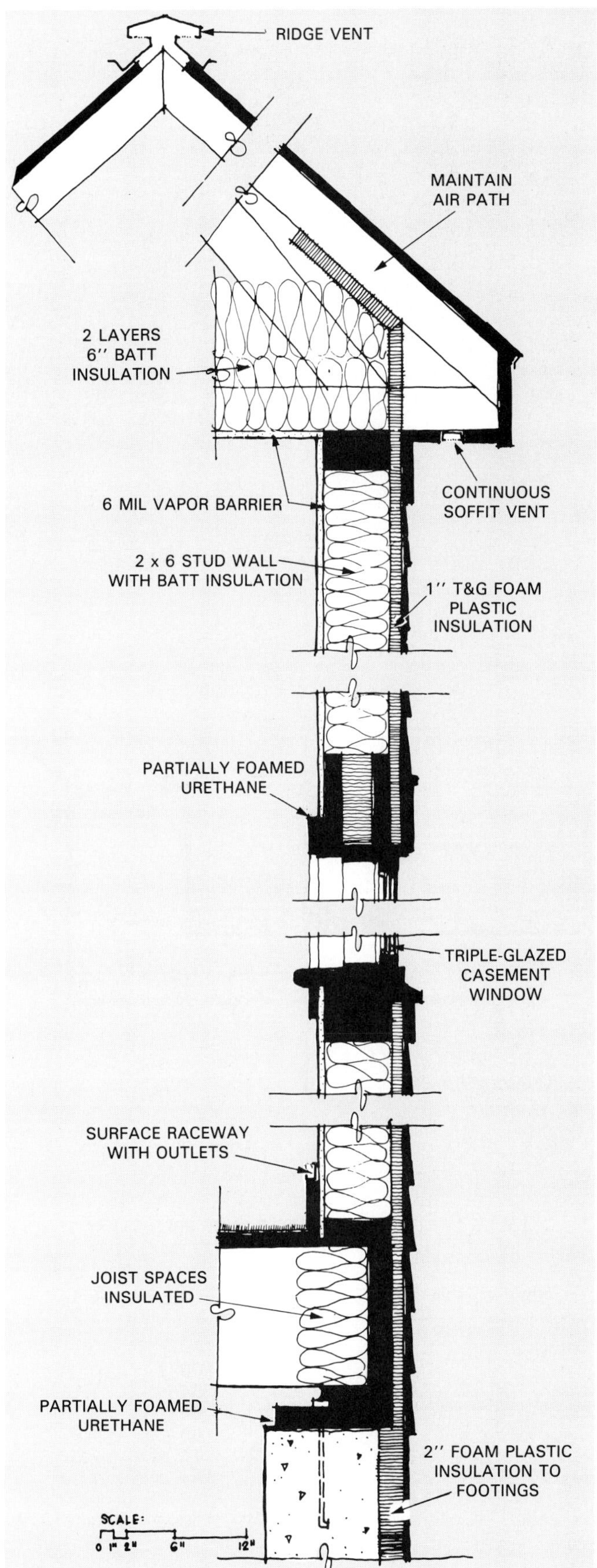

Figure 24-31. Cross section of well-insulated dwelling. Good insulation goes hand-in-hand with solar heating. (Iowa Energy Policy Council)

Outside Assignments

1. From your local library, select books on designing passive solar systems for housing. Report to your class on your findings.
2. Construct a small thermosiphoning solar heat unit and test it. Report to your class on its operation.
3. Write to companies offering passive solar structures. Study the literature and report to the class on items that seem to have practical advantages.
4. Using the charts and information in this chapter, design an indirect gain system (using a Trombe wall). Size the glazing, storage wall, and overhang according to the conventional principles.

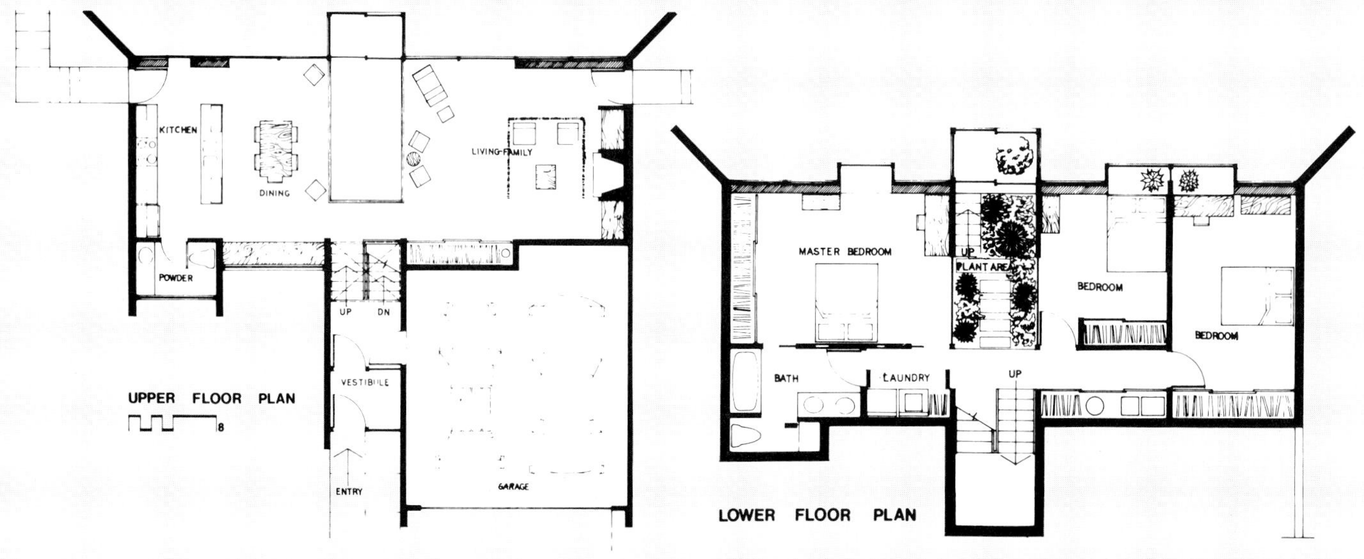

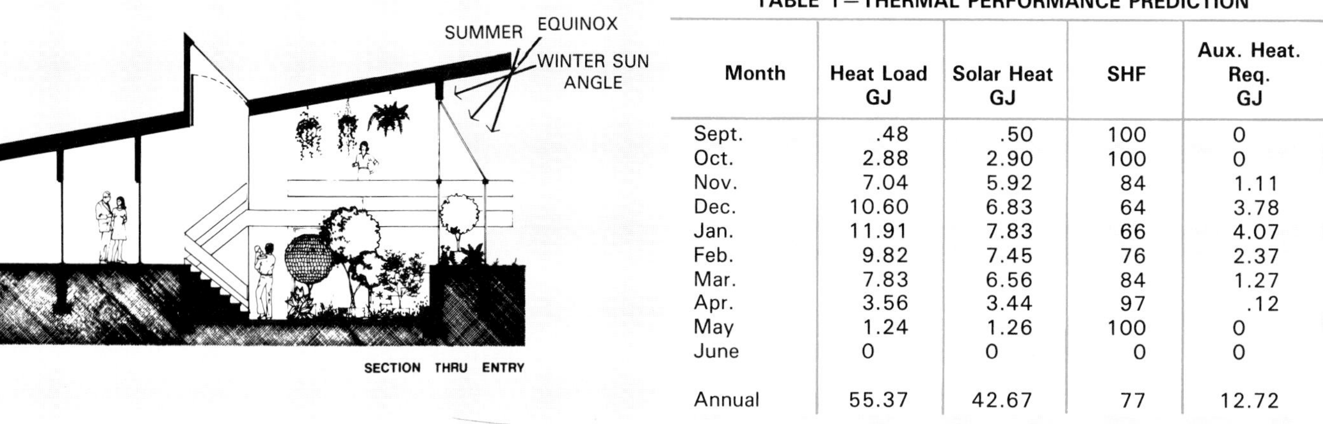

TABLE 1—THERMAL PERFORMANCE PREDICTION

Month	Heat Load GJ	Solar Heat GJ	SHF	Aux. Heat. Req. GJ
Sept.	.48	.50	100	0
Oct.	2.88	2.90	100	0
Nov.	7.04	5.92	84	1.11
Dec.	10.60	6.83	64	3.78
Jan.	11.91	7.83	66	4.07
Feb.	9.82	7.45	76	2.37
Mar.	7.83	6.56	84	1.27
Apr.	3.56	3.44	97	.12
May	1.24	1.26	100	0
June	0	0	0	0
Annual	55.37	42.67	77	12.72

Figure 24-32. Iowa passive solar project. Its heating load is half that of comparable homes. (Iowa Energy Policy Council)

25 Remodeling, Renovating, and Repairing

In some ways, remodeling and renovation work are more painstaking than new construction. It is more difficult to work with structures that may have defects such as sagging floors, rotting framework, and walls that are no longer plumb. Often, too, sections or components of the old building must be dismantled without destroying what is to be retained or what will be reused.

Frequently, the carpenter must be able to visualize how a structure was built so that no damage is done to it or any of its systems during demolition and removal of old walls.

Yet, remodeling and renovating are well worth the extra effort put into them. Remodeling brings greater comfort and usability for the owner, Figure 25-1. Further, either remodeling or repairing have other benefits such as enhancement of the value of the structure as well as forestalling its deterioration.

What Comes First?

In a long-neglected home, stopping further deterioration with temporary repairs may be the first order of business. Leaks cause further damage to the structure and their causes should be remedied. Replace missing shingles and seal leaks with roofing tar. If shingles and roof sheathing are beyond repair, replace the rotted sheathing and reshingle, unless structural changes are planned that would change roof lines. See Figure 25-2.

Figure 25-1. This fine old house stood in the way of a new commercial development. It was moved to this location, placed on a new foundation, and renovated.

The house should be examined to see if there are structural repairs that need to be made before attempting to repair or replace wall coverings, doors, and windows. Such items could be seriously affected by jacking or replacement of foundations, sills, studs or other frame members.

Figure 25-2. Roofing repairs are the most frequent renovation project. Top—Old shingles have been removed showing sheathing needing replacement. Middle—Quarter-inch waferwood is being installed over the old sheathing which was not rotted. Bottom—New drip edge is being applied.

Such repairs may be necessary for personal safety alone. Start with the foundation and sills and work your way up. Don't fix a structural problem with a roof and then jack up the frame. Everything will have shifted.

Sequence of Exterior Renovation

Renovating the exterior of the building should be the first order of business. Not all of the following steps will need to be taken; however, whatever steps are needed should be taken in the following order:

Renovating an Exterior

1. List what needs to be done.
2. Demolish what needs to be eliminated and clear away the debris.
3. Perform all structural work needed. Work from the bottom up. This step should include chimneys and masonry or concrete work. Take measures to protect open conditions from the weather.
4. Complete site work. Regrade, and provide drainage as needed.
5. Repair or replace the roof, including flashing, gutters, and roof vents.
6. Scrape or strip paint.
7. Repair masonry and tuck-point.
8. Repair or replace windows and outside doors, Figure 25-3.
9. Repair or replace siding, Figure 25-4.
10. Stain or prime wood siding and trim.
11. Caulk, glaze and putty.
12. Paint.

Figure 25-3. Windows and exterior door replacement. A—New energy efficient windows in this house are smaller than the old units. One is located where a door has been removed. B—Carpentry students check the operation of a newly installed replacement window. (Des Moines Iowa Public Schools) C—A window was removed and a door is being framed in this remodeling job. Siding was removed at left to get at a receptacle that has been relocated to the right of the door frame. D—New door being checked for clearance. Exterior trim will be added after the carpenter has made any needed adjustments to the door frame.

Figure 25-4. Replacing siding and soffits. Insulation with an R-value of 1 is installed under the new vinyl siding. Carpenter is installing soffit sections with a power stapler. (Monday Construction)

Figure 25-5. Demolition work. Top—Carpenter uses a hammer drill to remove a wall. (Des Moines Iowa Public Schools) Bottom—A dumpster is used to collect debris from demolition. Chute contains materials dumped from great heights.

Interior Renovation

Interior work can proceed while exterior work is underway. If the building is unoccupied, it is usually best to bring all phases of the work along at the same rate. All the demolition, all the mechanical systems, all of the drywalling, stripping and painting, etc.

If the house is occupied, consideration must be made for the comfort of the family. In this situation, there are two choices of operation:

- Completing a room or area to the point where the space is "livable" but without finishes or decoration.
- Zone by zone completion. In this approach a floor or a wing is completely finished before going on to the next floor or area.

Renovating an Interior

1. Demolish what is to be eliminated and remove the debris of demolition, Figure 25-5. This includes removal of structural elements such as studs.
2. Perform structural work such as alterations to bearing walls.
3. Insulating exterior walls and ceiling.
4. Changes or removal of nonbearing partitions Figure 25-6; installation or removal of soffits; pipe chases; subfloor repair; installation of nailers for built-ins, plumbing fixtures, light fixtures, etc.
5. Rough-in of plumbing and electrical wiring.
6. Installation of drywall, Figure 25-7, lath and plaster repairs, taping of drywall joints, applying skim coat of plaster.
7. Putting down underlayment for new flooring or tile.
8. Ceramic tile repair or installation.
9. Installation of fixtures for plumbing and heating. Set fixtures before any finishing to avoid damage to floors, walls, or trim.
10. Install light switches and receptacles.
11. Install unfinished floor coverings.
12. Do repair or installation of woodwork or trim. Refinish old woodwork and cabinetry, if in good repair, Figure 25-8.
13. Prime, paint, or wallpaper walls.
14. Apply finish to floors, as necessary, if unfinished.
15. Install prefinished flooring materials.
16. Touch up painted and clear-finished surfaces.
17. Install light fixtures and hardware such as cover plates, handrails, etc.
18. Clean finished areas and wash windows.

Figure 25-6. New steel studs are being plumbed and installed by this carpentry team. Steel is commonly used in commercial construction. (Des Moines Iowa Public Schools)

Figure 25-7. Drywall is installed after rough-in of electrical, plumbing, and HVAC systems. A drywall screw shooter drives screws into the steel studs.

Figure 25-8. Renovating an existing structure. Cabinets in good condition are often given a new coat of stain and varnish. (T.W. Lewis Construction Company)

Design of Old Structures

Many older homes were built using balloon framing. This method was popular in the United States from about 1850 to the 1930s. Its main feature is the long studs that run from the sill all the way to the plate on which the rafters rest. These are called "building height" studs. They may be spaced anywhere from 12" to 24" O.C.

Second floor joists rest on a horizontal member called a *ribbon*. This member, which is let into the stud, is usually a 1 x 6 board. Firestops, short pieces of blocking 2" thick, are installed between joists at each floor level. Their purpose is to prevent spread of fires from one part of the dwelling to another.

Figure 25-9 shows details of balloon type construction. Additional information on housing designs will be found in Units 8, 9, 10, and 22. Figure 25-10 shows details of post and beam construction.

Replacing Rotted Sills

Rotted sills and other structural members are not uncommon in older homes where leaks and other problems have not been corrected. Replacing a sill in a platform construction requires lifting and supporting the weight of the building while the old sill is removed and replaced with a new, pressure treated sill. See Figure 25-11.

Reinforcing rotted studs usually will not require temporary support unless the rot has caused the wall to settle and sag. Reinforcing is best done by *sistering*. This is accomplished by nailing another stud alongside the rotted member. Installing the sister will entail removal of interior wall covering or siding.

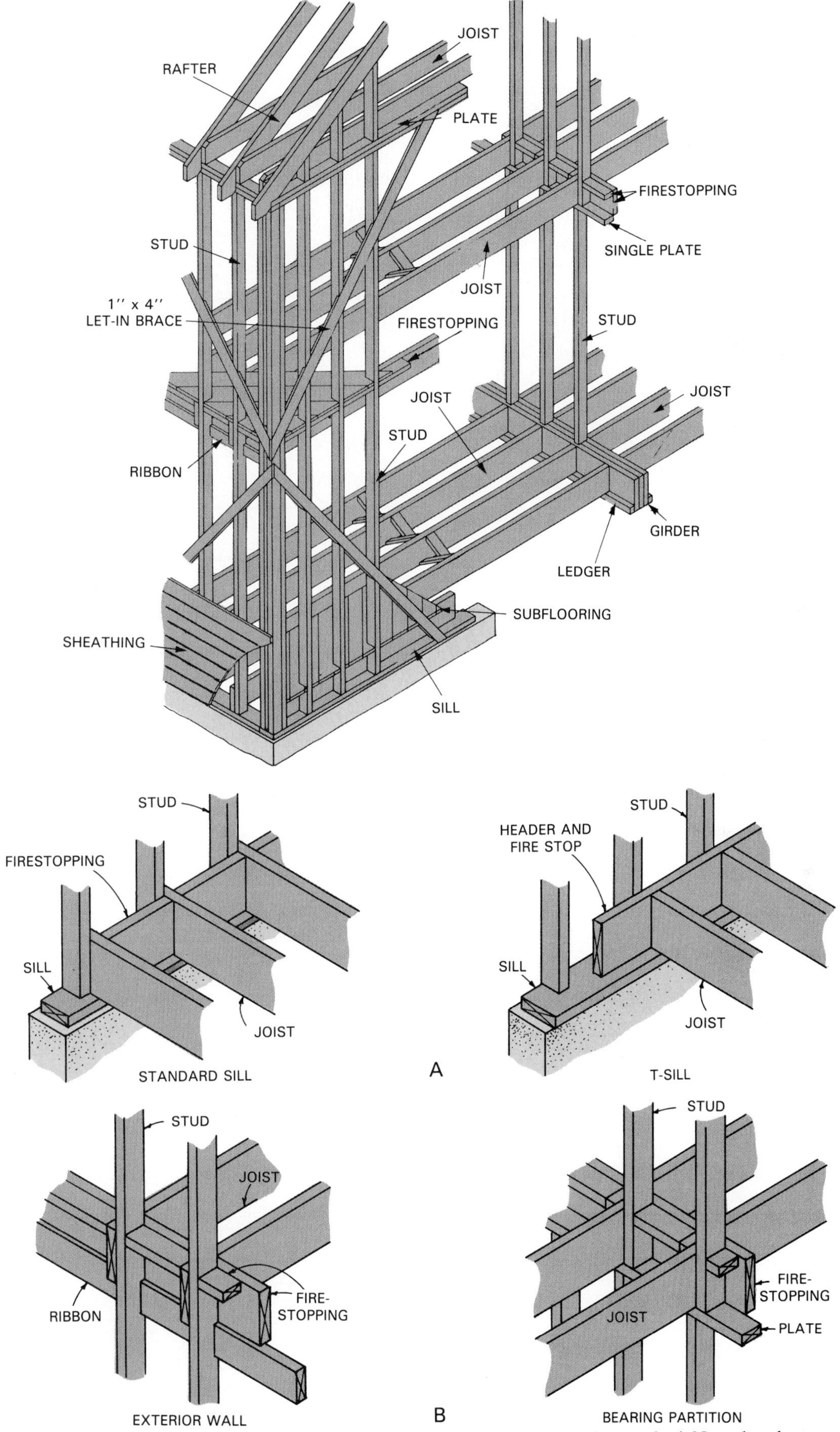

Figure 25-9. Top—A section of a balloon frame shows major features of this construction method. Note that there are no platforms with plates and band joists between floors. Bottom—Details of a balloon frame. A—First floor level, two methods of building the sill. B—Second floor level. Left, method of attaching second floor joists to studs, using a ribbon for support. Right, framing detail for bearing wall.

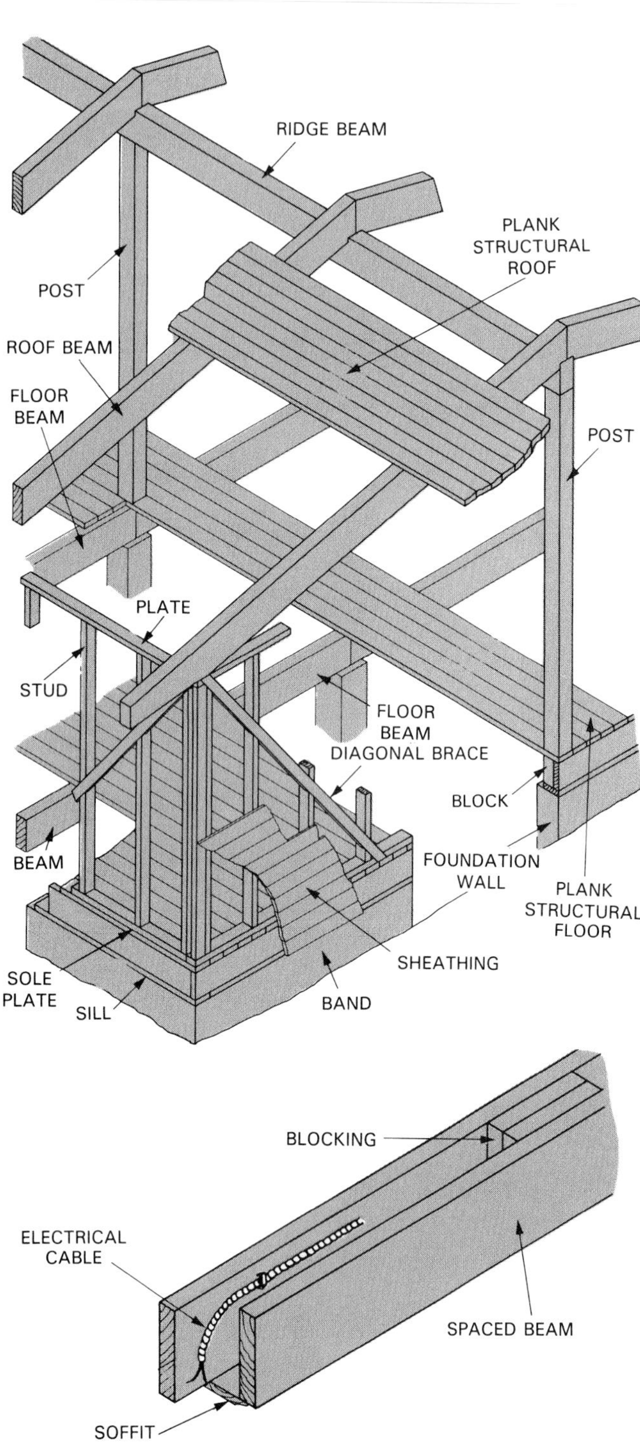

Figure 25-10. Basic post and beam construction. Top—General construction detail. Bottom—Often, pipes and electrical wiring are concealed in beams of this type construction.

Hidden Structural Details

Before demolition of any kind is attempted for inside changes, study the building to determine what type of construction was used. This will make removal of structural elements much easier and safer.

If the house has a basement, examine the area beneath upper level walls. Try to determine if there are gas or fuel lines, electric circuits, or plumbing service running through the walls or floors where you will be working. Locate and close valves to stop gas and fuel flow in service lines. As a precaution, cut power to any electrical circuit that might be disturbed. Pull the fuse or trip the circuit breaker. If there are other services on the same circuit that you do not want interrupted, an electrician should be called to do the work.

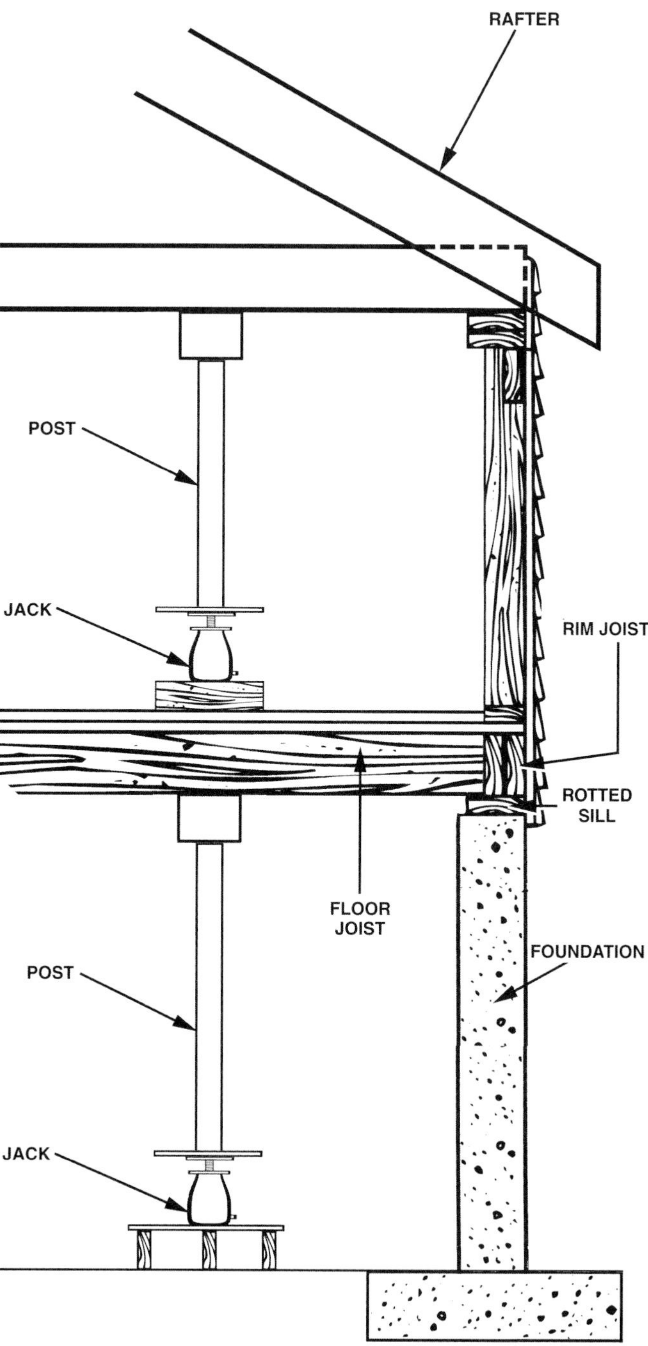

Figure 25-11. Replacing a rotted sill using shoring at two different levels. The second-level shoring may or may not use jacks.

Go outside and study the roof. Note where vents and chimneys are located. Greater care must be exercised in removing wall coverings where exhaust vents and plumbing stacks are located. See Figures 25-12 and 25-13.

If additions are to be built that require excavating for piers or foundations, Figure 25-14, be sure to locate any underground utility services. Pipes and cables could accidentally be cut or damaged by digging. Sometimes, inspection of sills and joists in a basement will reveal markings indicating where underground services are located. Checking with local utility companies and city departments for buried lines is also advisable.

Figure 25-13. Check the roof for other features such as vents which are not visible from the basement or inside of the house. (CertainTeed)

Removing Old Walls

The first step in opening up or removing a wall is to remove trim such as baseboards, cove strips at the ceiling, and trim around windows and doors. If these materials are to be reused, use care in removing them.

To salvage interior trim for reuse, do not attempt to remove the nails by driving them back through the front. Use a small wrecking bar or nippers, pull them through the back. This avoids splintering that would spoil the face of the trim.

Next, remove the wall covering to expose the wall frame. Most construction in the past 40 years has used drywall. In homes older than that, wood lath and plaster were used. Be sure to wear a mask to avoid breathing in dust and particles of construction materials.

Figure 25-14. Additions usually require footings or piers. This means holes or trenches must be dug. Care must be used in digging so as not to damage buried utilities. Local utility companies should be consulted about location of buried power cables, water mains, and gas mains. (Des Moines Public Schools)

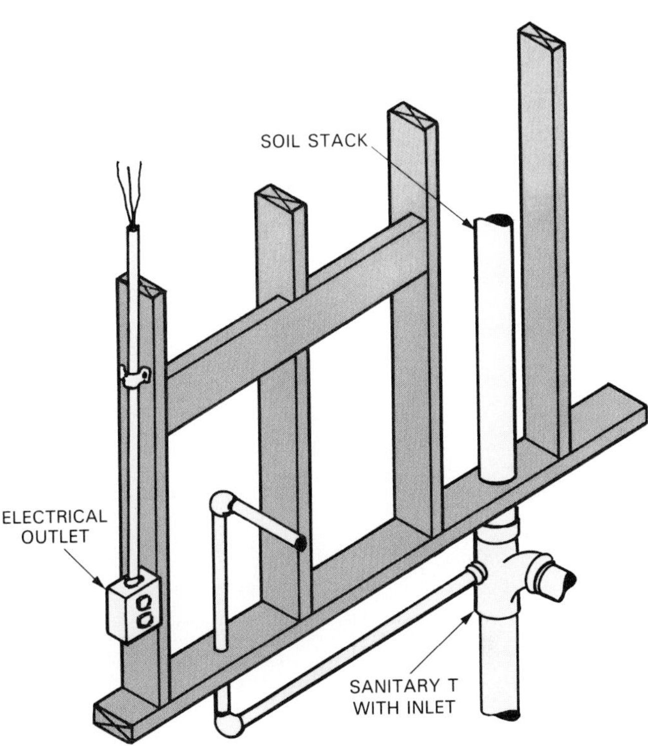

Figure 25-12. Wiring and plumbing may be concealed behind wall coverings.

Useful tools for demolition include a sledge hammer, Figure 25-15, a ripping or wrecking bar, a hammer or hatchet, and a rip chisel. A reciprocating saw is also useful for cutting framing. Most of these tools are shown in Unit 3. The rip chisel is essential if any of the trim is to be saved and reused.

Many carpenters use a flat-bladed garden spade to remove lath and drywall. This tool will also remove drywall and lathing nails as it is run over the studs.

Figure 25-15. A sledge is a useful tool for demolition of certain types of walls. (Des Moines Public Schools)

If the wall is a partition, removal of the covering will be all that is necessary to strip the framing. In outside walls, insulation and siding will also need to be removed to strip the framing.

For good housekeeping, it is often useful to rent a large trash container like the one shown in Figure 25-16. Then, debris can be cleared away immediately so that it does not clutter up the job site and present an eyesore as well as a safety hazard.

Recognizing Bearing Walls

Before removing a wall, or a portion of it, you will need to determine whether it is a bearing wall. Such walls will need *shoring* (temporary support) while the old wall is being removed. Permanent supports of some kind must be in place before the remodeling is completed.

Determining which are bearing walls is quite simple with outside walls. Interior partitions are another matter.

All outside walls support some weight of the structure. End walls of gabled, single story houses are the sole exception. They carry little weight except that of the wall section from the peak of the roof to the ceiling level. Outside walls running parallel to the ridge of the roof carry the most load since they provide most of the support for the roof.

Identifying which partitions (interior walls) support weight from structures above is not always simple. The following conditions are usually an indication of a bearing wall:

- The wall runs down the middle of the length of the house.
- Ceiling joists run perpendicular to the wall.
- Overhead joists are spliced over the wall, indicating they depend on the wall for support.
- The wall runs at right angles to overhead joists and breaks up a long span. The joist may not be spliced over the wall.) Check span tables for load-bearing ability of the joist.
- The wall is directly below a parallel wall on the upper level. If the walls are parallel to the run of the joists, the joists may have been doubled or tripled to carry the load of the upper wall. The only way to be certain is to break through the ceiling for a visual inspection.

Familiarize yourself with the framing principles used for the type of structure you are remodeling. Study the framing units in this book for details. Also see Figure 25-17.

Providing Shoring

Shoring is temporary support of a building's frame. It must be installed to support ceilings or upper floors when all or a portion of a bearing wall is being removed. Figure 25-18 shows one type. It is simply a short wall of 2 x 4 plates and studs.

Figure 25-16. Keep the job site clean. If there will be considerable debris, it is wise to rent a dumpster (trash container) like this that will be hauled away following the remodeling job. Left—Container on job site. Right—Typical scavenger service. (Homewood Scavenger Services, Inc.)

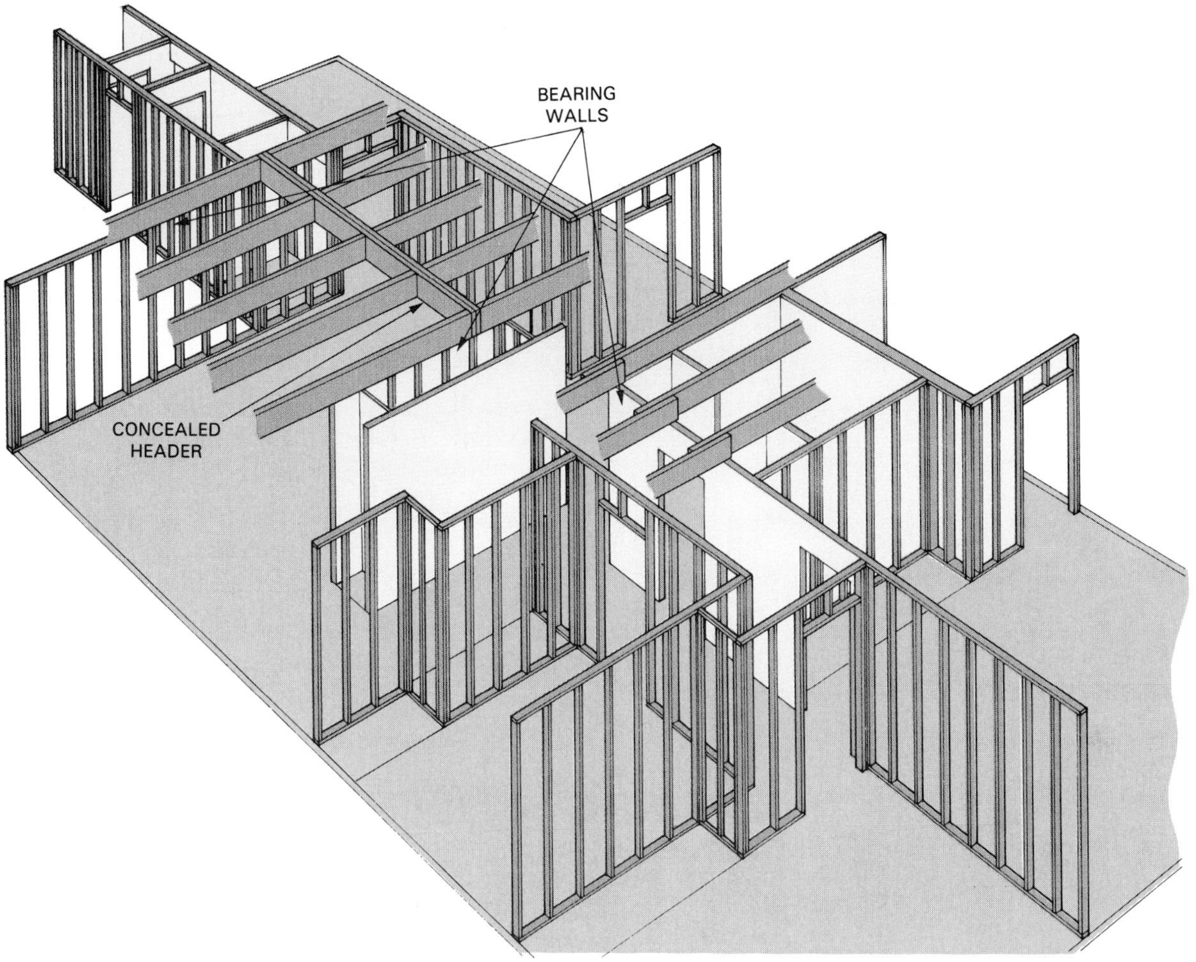

Figure 25-17. Looking for the bearing walls in a house. A view of an unfloored attic is the easiest way. Walls down the center usually support floor or ceiling joists.

The wall can be assembled on the floor and lifted into place. To strengthen it use double plates top and bottom.

Installing Shoring

1. Before cutting studs for the shoring, measure the distance from floor to ceiling. Cut studs shorter to allow 1/4" to 1/2" clearance.
2. Nail the studs to one top and one bottom plate.
3. Attach the second top plate. Leave the second bottom (or sole) plate off to allow clearance when raising the shoring.
4. Square up the shoring wall and attach a 1 x 4 diagonal brace.
5. Lift the shoring in place. If necessary, get a helper to do some of the lifting. (Keep the shoring far enough from the partition so you will have room to work.)
6. Slide the second plate under the bottom plate. Use wood shingle shims at regular intervals along the bottom of the wall to bring the temporary wall to full height.

In some cases it may be necessary to place a second shoring on the other side of the bearing wall. Usually, the joists are lapped and spiked well enough to provide stiffness to the joint so a second shoring is not necessary.

A second shoring method is used by many carpenters. Larger dimension lumber—4 x 6s, for example—are used for plates. Adjustable steel posts are used in place of the studs, Figure 25-19.

Be sure that the shoring is resting on adequate support. That is, it should rest across several joists. If shoring is running parallel to floor joists, lay down planking to distribute weight across several joists. Do the same at the ceiling.

When shoring is in place, you can remove studs in the old bearing wall. Remove studs first, being careful not to disturb plaster or drywall at the ceiling. Try to salvage the studs for future use. Use a reciprocating saw to cut the studs away from the sole plate if the nails are not accessible. If the studs are not being recycled, use a sledge at the bottom to loosen them from the sole plate.

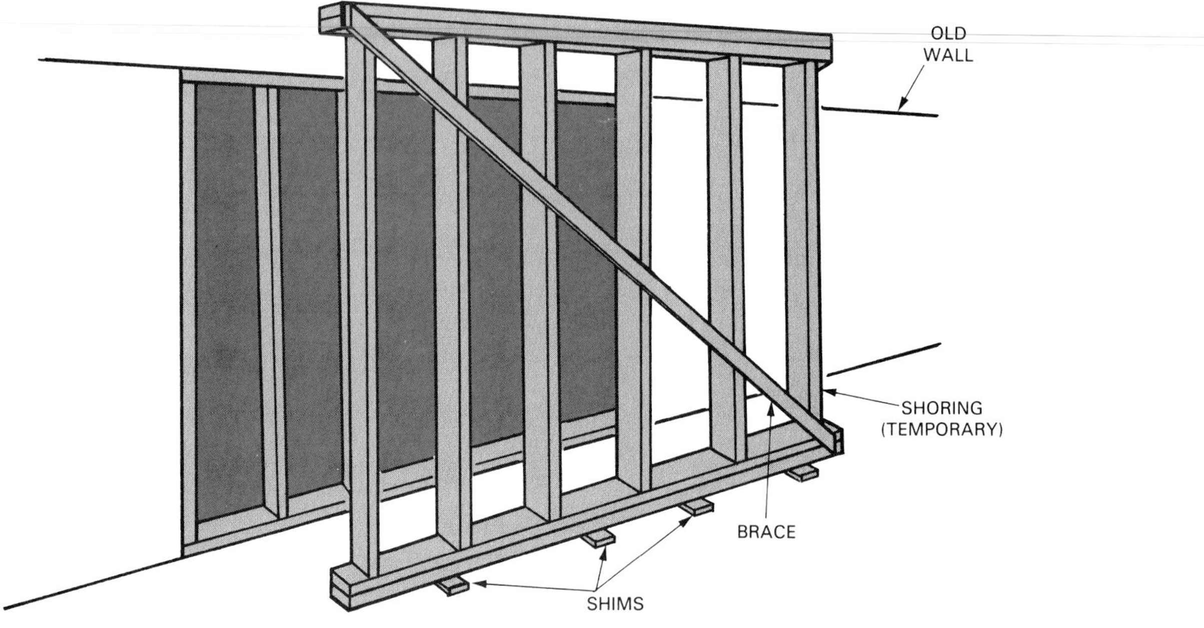

Figure 25-18. Temporary support is provided by the shoring in the foreground. Normally, one support is sufficient. However, sometimes it is necessary to place shoring on both sides of the wall.

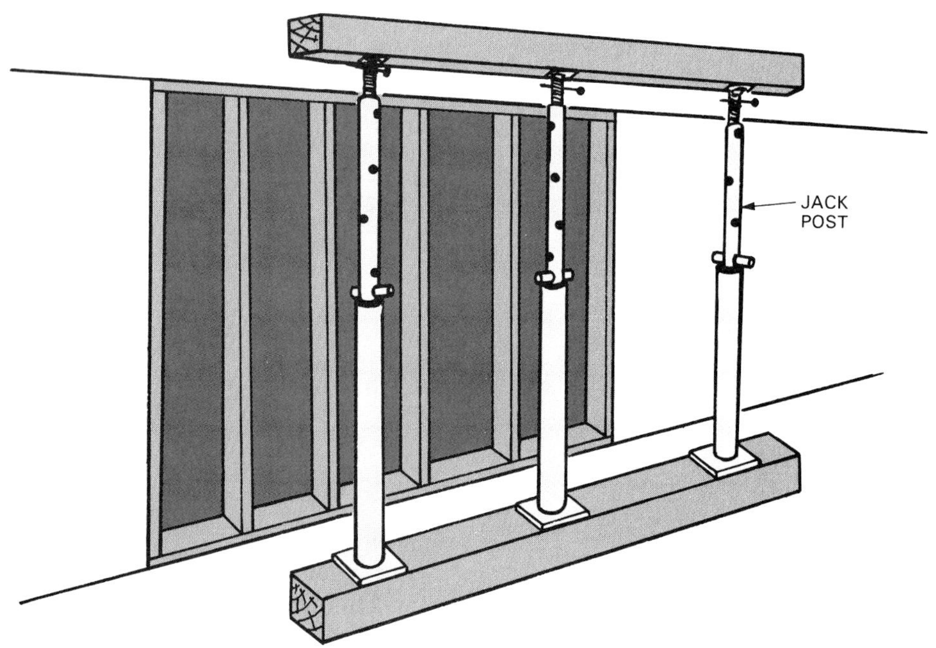

Figure 25-19. An alternate method of shoring up to support ceiling structure uses jack posts. These are adjustable steel pillars.

Leave the shoring in place until a permanent support has been framed in to take the place of the bearing wall. This may be a lintel or a concealed header.

Framing Openings in a Bearing Wall

When studs are removed to open up a wall for a passageway or room addition, a header (also called a lintel) must be installed to provide support. The header can be built up from 2" lumber. Shim between them with thin plywood to build out the width to that of the framed wall, Figure 25-20. On very long spans, the carpenter may prefer to construct or purchase a box beam. Refer to Unit 22, Figure 22-29, for construction information.

Supporting Headers

Headers are supported by resting them on the top of short studs called trimmers, or trimmer studs. Additional

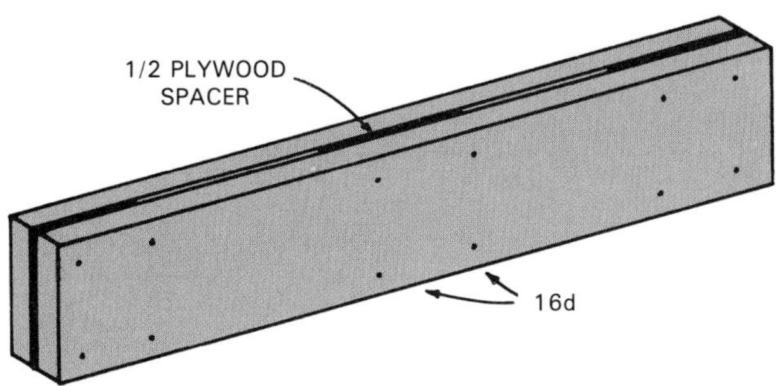

Figure 25-20. Basic header construction. Use plywood spacer to bring it to same width as the wall.

support is provided by nailing a full length stud alongside the trimmer at each end of the header. These are sometimes called *king studs*. Stagger nail the trimmers to the king studs. Toenail the header to the trimmer and end nail it to the king studs, Figure 25-21. On long spans, the header may be extended beyond the rough opening to the next full-length stud. This will help make the framed opening more rigid.

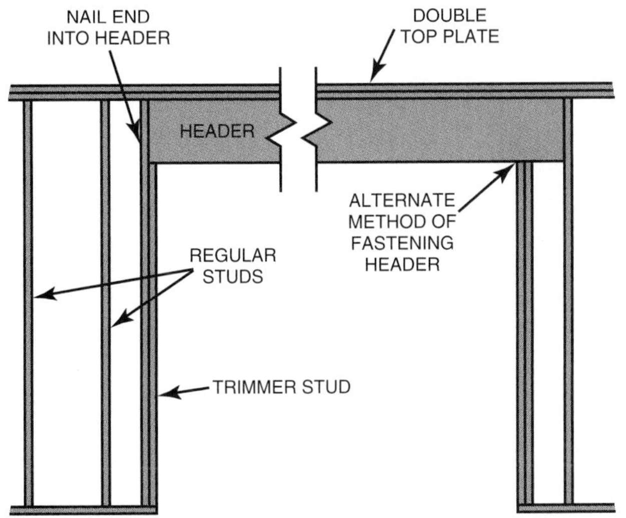

Figure 25-21. Method of installing a header and supporting it. Alternate method of fastening at right is used on longer spans. It gives the wall more rigidity and the doubled 2 x 4 post provides more support for the load.

Sizing Headers

Headers that replace sections of bearing walls should be sized according to recommended standards for load-bearing ability. The size will be specified in the architectural drawings for the remodeling job. The carpenter should also be aware of what the standards are and how the sizes are determined. Essential information is given in Figures 25-22 and 25-23. A careful builder will also check local building codes before fabricating the header.

To use the table in Figure 25-22, find the column that describes the load. Move down the column until you find the span corresponding with the opening you need. Then read to the left for the depth needed for the header.

The tables in Figure 25-23 will require some arithmetic. First compute the load using the table of average weights for different areas of the building.

Once you know the total load, locate the load in the column for the correct span in Figure 25-23. The size of the header needed will be at the top of the column.

If headroom is a concern, a steel-reinforced header will allow greater strength with less depth. A steel plate is sandwiched between lengths of 2" lumber. The plate is called a *flitch* and can be made up by a metal fabricator.

Computing the Load

To compute the load on a header, refer to Figure 25-24. First, you need to find the number of square feet. Multiply the length of the span you need by half the distance between load-bearing walls. (You can always assume that the header must support the load halfway to the next bearing wall.)

Thus, if the distance to the next bearing wall is 12', the square footage is 6' x the span. Add on any other loads which the header must support to find total load. On roof sections, load is figured the same as for the floor—use the horizontal distance covered by the roof.

	OUTSIDE WALLS			INSIDE WALLS			
Nominal depth of header (in inches)	Roof, with or without attic storage	Roof, with or without attic storage, + one floor	Roof, with or without attic storage, + two floors	Little attic storage	Full attic storage, or roof load, or little attic storage + one floor	Full attic storage + one floor, or roof load + one floor, or little attic storage + two floors	Full attic storage + two floors, or roof load + two floors
4	4′	2′	2′	4′	2′	No	No
6	6′	5′	4′	6′	3′	2′ 6″	2′
8	8′	7′	6′	8′	4′	3′	3′
10	10′	8′	7′	10′	5′	4′	3′ 6″
12	12′	9′	8′	12′ 6″	6′	5′	4′

Figure 25-22. Table of allowable spans for headers under different load conditions. This is a guide for sizing headers. Check local codes. Components of the structure, rather than load figures, are used.

Average weight of house by area	
Unit	Load/ft.²
Roof	40 lb.
Attic (low)	20 lb.
Attic (full)	30 lb.
Second floor	30 lb.
First floor	40 lb.
Wall	12 lb.

Built-up wood header (double or triple on two 4" x 4" posts)								
Span (in feet)	Weight (in pounds) safely supported by:							
	2-2x6	2-2x8	2-2x10	2-2x12	3-2x6	3-2x8	3-2x10	3-2x12
4	2250	4688	5000	5980	3780	5850	7410	8970
6	1680	3126	5000	5980	2520	4689	7410	8970
8		2657	3761	5511		3985	5641	8266
10		2125	3008	4409		3187	4512	6613
12			2507	3674			3760	5511
14				3149				4723

Steel plate header (on two 4" x 4" posts)									
Plate / Span (in feet)	Weight (in pounds) safely supported by wood sides and plate								
	2-2x8 + 7 1/2" by			2-2x10 + 9 1/2" by			2-2x12 + 11 1/2" by		
	3/8"	7/16"	1/2"	3/8"	7/16"	1/2"	3/8"	7/16"	1/2"
10	6754	7538	8242	10,973	12,199	13,418	15,933	17,729	19,604
12	5585	6216	6827	9095	10,131	11,106	13,224	14,517	16,265
14	4756	5293	5811	7751	8623	9463	11,295	12,561	13,876
16		4481	5036	6746	7494	8221	9815	10,953	12,086
18				5942	6606	7158	8675	9652	10,647
20						6466	7746	8618	9408

Figure 25-23. Another method of finding header sizes calculates the load. Top left—Basic load calculations for different parts of the house. Top right, Bottom—Recommendations for safely supporting different loads at different spans. Check local codes.

Concealed Headers and Saddle Beams

There are times when it is not desirable to have a header extending below the ceiling level. In such cases, the header must be at the same level or above the joists they support.

A *concealed header* is butted against the ends of the joists it is supporting. A *saddle beam* is positioned above the joists it supports. See Figure 25-25.

A concealed header is simple to add to an exterior wall in platform type constructions, Figure 25-26. Simply attach a second piece of 2" lumber to the band joist. Add studding at each end to provide support for the header.

Because they are placed above the joists, saddle beams can only be used in attic space. They should rest on top of joists supported by the remaining portions of the bearing wall.

It is recommended that loads be checked by an architect or engineer. He or she should also specify the header sizes.

Small Remodeling Jobs

Most of the units in this book deal with new construction. In general, these same construction steps can be used with some modification for small remodeling jobs. Figure 25-27 gives step-by-step instructions for installing a new nonbearing partition.

Replacing an Outside Door

An old, worn, ill-fitting exterior door, in addition to being an eyesore, wastes energy. It permits excessive air infiltration and heat loss. New designs, especially the prehung units, make replacement easier. Standard replacement units are made with an insulated steel or steel clad door mounted in a steel or wood frame. Figure 25-28 illustrates a unit with a steel frame and shows how it fits into the old door frame.

To install this unit, remove the old door, hinges, threshold, and trim. Then slide the new unit into the opening. Figure 25-29 shows the procedure step-by-step.

When making such an installation follow the directions provided by the manufacturer of the unit. Replacement doors are generally available to replace standard sizes: 2'-6" x 6'-8" and 3'-0" x 6'-8".

Replacing or Repairing Interior Doors

Remodeling may include replacing, moving, or adding interior doors and door jambs. The framing of doorways is explained in Unit 9. Installation of jambs and hanging of doors is explained in Unit 18. Figure 25-30 shows several steps in replacing a standard interior door.

Use of a special patented *jamb clip* presents another method of hanging an interior door, Figure 25-31. The clips, which eliminate the need to use wedging, are made of 19 gage cold-rolled steel. One size (4 1/2") fits the standard rough opening consisting of 2 x 4 plus 1/2" of drywall on each side. For other jamb sizes the clip snaps in two at

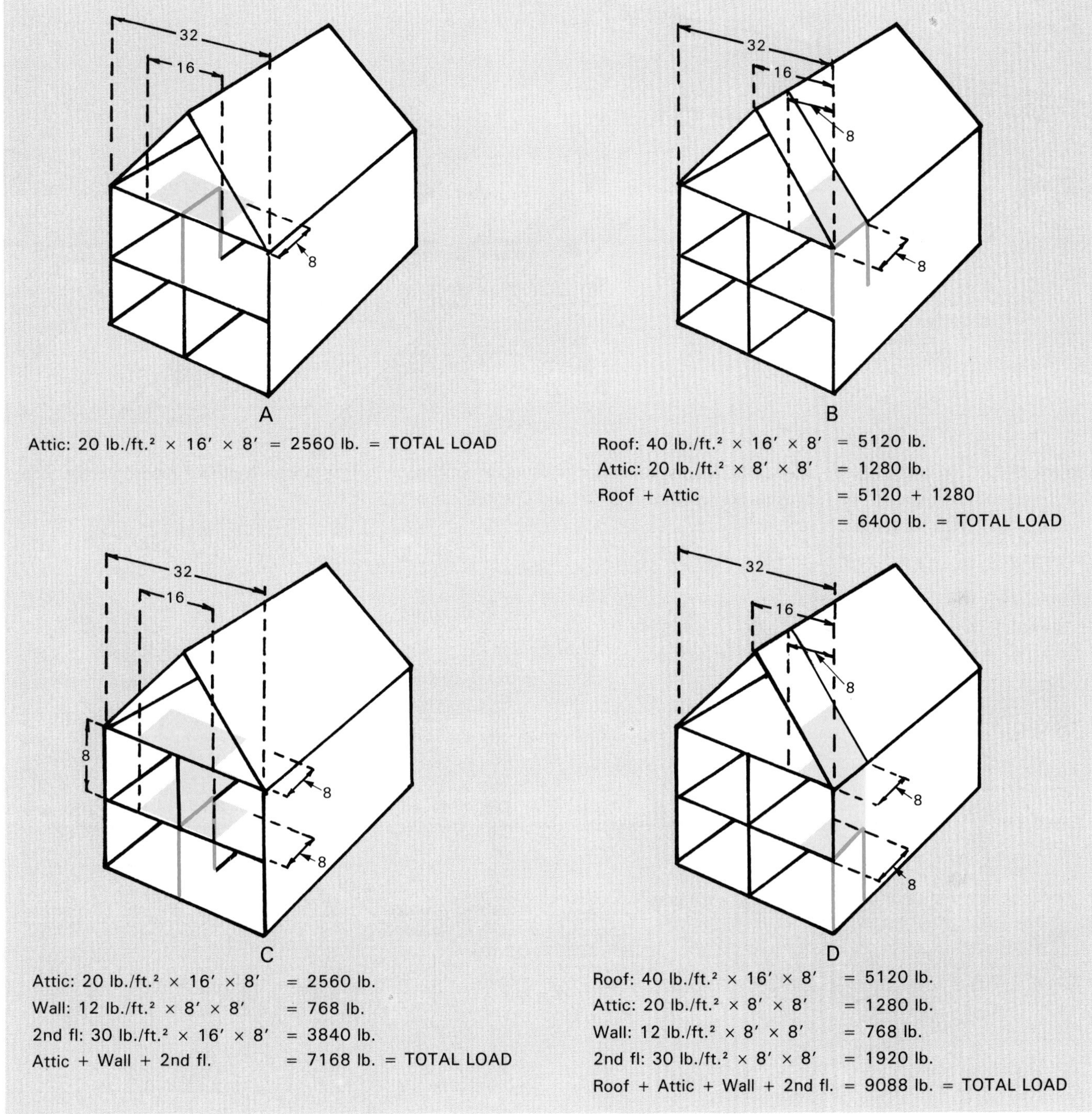

Figure 25-24. How to compute loads. Find horizontal distances between bearing walls. Determine what parts of the house the header will be supporting. Multiply load per sq. ft. by the length and width being supported. Add loads together.

a breakaway point. Eight clips should be used with hollow core doors, ten for solid core doors. Figure 25-32 is a cutaway of a jamb clip installed. Follow the installation instructions in Figure 25-33.

As a building ages, settling, day-to-day usage, and other factors cause problems with doors. Hinges wear and latches no longer work. Raising the floor level with new solid flooring material or carpeting necessitates trimming the door so it will clear the new flooring material. If the door itself is still in good condition, repairs are in order. Consider the following conditions and suggested repairs/remedies.

- If the door binds or will not close tightly, check for loose hinges. Reinstall the hinges either with longer screws or fill over-sized screw holes and reinstall screws. If door still binds, note where the door is binding with the frame and mark the spot with a pencil, Figure 25-34. Use a hand plane to remove small amount of material, being careful not to remove too much wood.

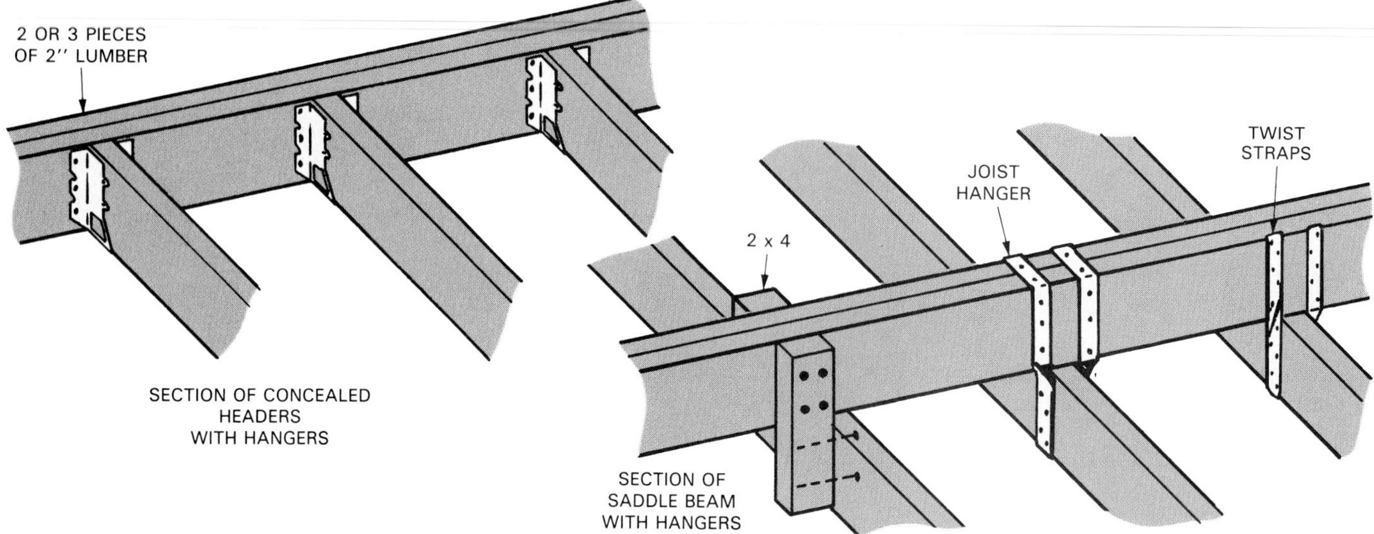

Figure 25-25. Concealed headers and saddle beams are used when support must be at ceiling height. Neither support member hangs lower than the ceiling joists. Joist hangers are recommended for these installations.

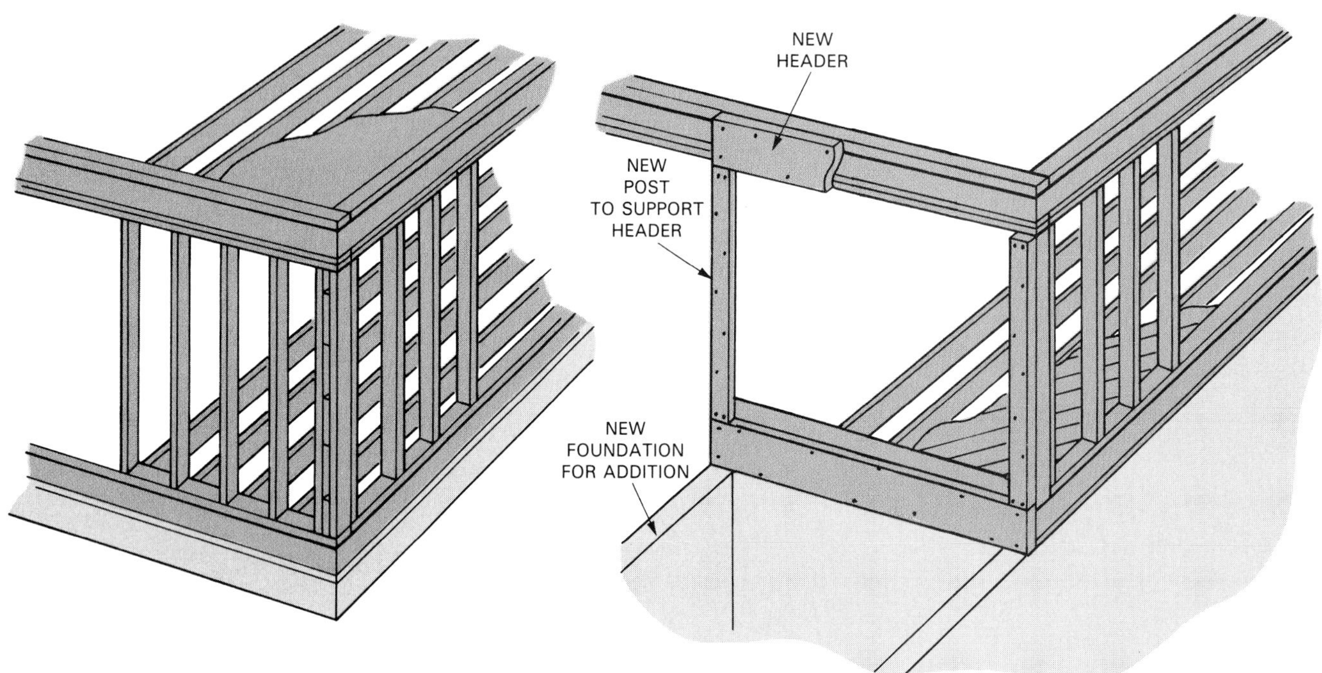

Figure 25-26. In remodeling a wall in platform construction, header can often be built by adding to a band joist on a second floor.

- Installation of new flooring material may make it necessary to saw 1/2" or more off the bottom of the door. Remove the door from its hinges and have a helper hold it in the door opening in its normal position with about 1/16" space at the top or with hinge leaves aligned. Mark the door bottom. Next, if the door has veneered facings, make a shallow cut on the line with a utility knife. Use a steel straight edge as a guide. Transfer the cut line to the opposite side and make a shallow cut there, as well. This will prevent splintering of the veneer when the excess is sawed or planed away.
- If the door binds on threshold or floor, check for a loose upper hinge or worn hinges, and replace if worn. Check for sagging door by measuring diagonals. Sagging is confirmed if diagonal from the top of the hinge side of door is longer. Replace the door or plane the bottom where the door is binding.
- Before replacing a seemingly defective door latch, check to see if it aligns with the striker. Reset the striker if alignment is the problem.

1. Carefully lay out the exact location of the partition. Snap a chalk line that will be used to locate one side of the sole plate. Lay out the top plate and sole plate.

2. Cut plates to length. Nail sole plate to the floor as shown. Space nails about 2 ft. O.C. and use a nail size that will extend through the subfloor.

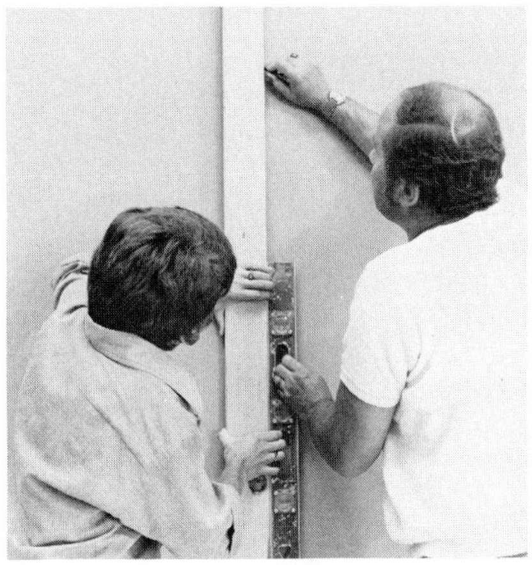

3. Nail top plate to the two end studs and then install this assembly on the sole plate. Use a level as shown. If top plate and studs cannot be nailed to structural members (studs and ceiling joists), use toggle bolts.

4. Install studs on 16 or 24 in. centers. Toenail them to the top plate and sole plate. Use two 8d box nails on each side of the stud.

5. Because floors and walls are seldom perfectly level, it is usually best to measure for each stud. This means cutting each stud so that it fits snugly between the sole and top plate.

6. Study wall for best arrangement of standard drywall sheets. Fasten sheets with 1 1/2 in. drywall nails. Cut panels as needed using utility knife and square or straightedge. Space nails 7 to 8 in. apart.

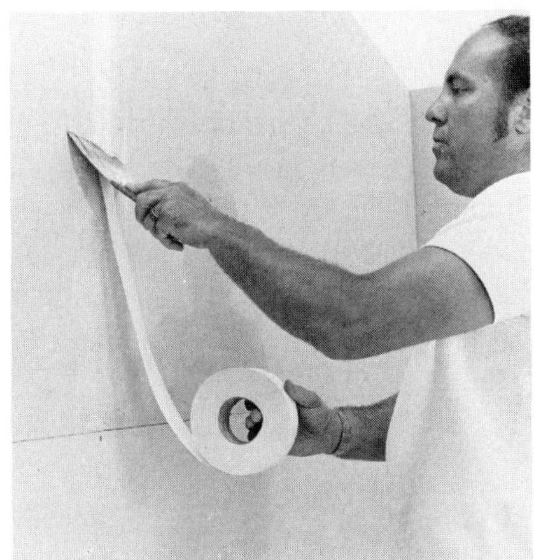

7. After the gypsum panels are installed, flat joints and inside corners are reinforced with tape bedded in joint compound. Outside corners are reinforced with a metal corner bead.

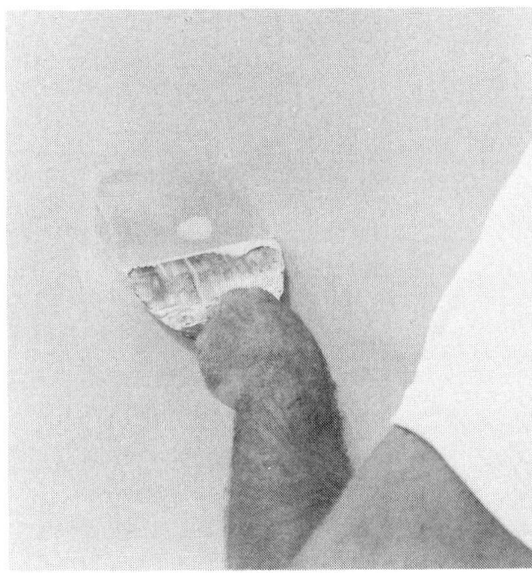

8. Cover nail heads and imperfections with compound as shown. Allow about 24 hours before applying a second coat.

Figure 25-27. Steps for constructing a nonbearing partition during remodeling. (Georgia-Pacific)

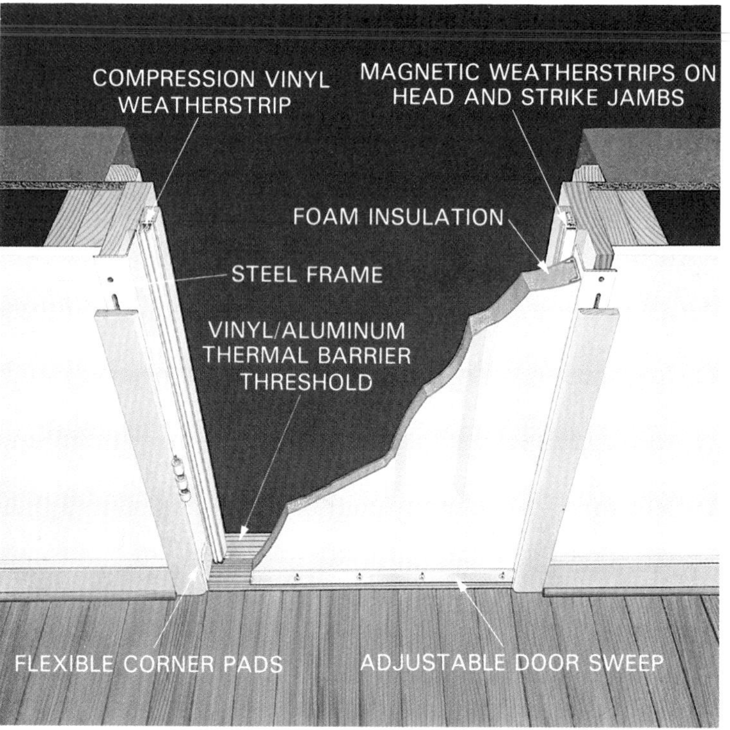

Figure 25-28. Top—Cutaway showing general details of modern replacement door unit. After alignment, frame is secured with long screws that extend into the wall studs. (General Products Co., Inc.) Bottom—A steel clad entrance door and its frame can be designed to fit inside an existing door frame.

Installing New Windows

Replacement windows are available in a variety of sizes and styles to fit rough openings of many old windows. Before ordering, take careful measurements. Measure again before installing to ensure proper fit. Check around the opening for signs of rot or leakage that a new window will not solve. Repair any problem before installing the new window.

Specific instructions for installation are provided by the window manufacturer. Read them carefully before beginning installation. Also, refer to Unit 12, Windows and Exterior Doors.

Repairing Wood Shingles

Wood shingles are used for both roofing and exterior wall coverings. Over time the shingles may sustain damage requiring repair or replacement of individual shingles.

Split shingles can be repaired. If the crack is small fill it with asphalt roofing cement. If the shingle appears loose, secure it with small galvanized nails. Cover the nail heads with asphalt roofing cement. If the crack is large or a piece of the shingle is missing, install a piece of flashing under the shingle. Make sure the flashing extends well above the top of the crack or that it extends under the next course. Secure the patch with galvanized nails. Use asphalt roofing cement over the patch as described for the smaller crack.

Badly damaged wood shingles can be replaced without disturbing surrounding shingles. Use a wood chisel or screwdriver to split the damaged shingle into narrow pieces. These can be pulled out with the fingers. Use an old hacksaw blade to cut off the nails. Slide in a new shingle of the same size but leave it about 1/4" below adjacent shingles. Secure with two galvanized nails driven at an angle against the upper course. Then drive the shingle upward until the nail heads are concealed.

Repairing Asphalt Shingles

Wind sometimes causes damage to asphalt shingles, flipping them up and sometimes tearing them. If not too badly damaged, apply a generous dab of asphalt cement to the underside. Press it down and secure it with broad-headed roofing nails. Apply asphalt cement to the nail heads.

Asphalt shingles usually are manufactured in two- or three-tab sections or strips. If badly damaged, they can be replaced without destroying adjacent shingles. Lift the upper shingle and remove the nails with a small, flat, ripping claw or cut them off with a hacksaw blade. Remove the damaged shingle; slip the new shingle into place; secure it by blind nailing. As an alternate method, secure the new shingle by nailing through the upper shingle. Cover the nail heads with asphalt cement.

1. Remove old door, hinges, strike plate, and threshold. Carefully pry off the interior trim. To secure a smooth separation between casing and paper or paint, pull a sharp knife or razor blade along the joint.

2. Nail wood shims in rabbet of old door frame to make a snug fit and plumb the opening. Be sure shims are located at each hinge and strike plate area. Apply a double-bead of caulking along base of opening. Set the new unit in place as shown. Check fit and nail through predrilled holes in front edge of frame.

3. Remove retainer bands and brackets, and open door carefully. Adjust and install screws on hinge side first. Use long screws that will go through the hinge, frame, and into the studs. Coating the screw threads with soap will make them easier to drive.

4. Adjust lock-side jamb and insert screws as specified. Install lockset according to manufacturer's recommendations. Steel replacement doors and frames will have openings that fit standard locksets. Chisel out wood behind strike plate as required.

5. Install weatherstrip. Use magnetic type for top and latch side, compression type for hinge side. Apply a 1/8 in. bead of caulk along the wood strips. With the door closed, position each strip against the door face and tack them into place. Check door operation and then complete the nailing as shown.

6. Adjust crown of threshold so there is smooth contact with door bottom weatherstrip. Apply a small bead of caulk along the outside edge of the threshold and bottom end of wood stops. Replace old interior trim or install new trim.

Figure 25-29. Basic procedures to follow when installing a modern replacement door unit. (Pease Industries Inc.)

Building Additions Onto Homes

When building additions to older homes, Figure 25-35, you need to check all dimensions of the existing construction carefully. In particular, check the following:

- *Foundations* — Modern masonry units may not match sizes of older units. You will need to adjust level of footings or alter mortar joint thickness.
- *Ceiling heights* — In balloon framing, ceiling heights may vary from house to house. Study local codes to see if your changes are acceptable.
- *Dimensions of framing members* — Standards of dimension lumber (2 x 4s, 2 x 6s, etc.) have changed with time. Check the actual dimensions of older lumber.

In recent years, regions that experience serious damage from earthquakes, high winds, tornadoes, and hurricanes have revised building codes to reduce such damage in the future. Figure 25-36 shows how and where to install ties that anchor a wood frame building from foundation to rafters with steel anchors and strapping.

Figure 25-30. Preparing to replace an interior door. A—Trim has been removed and a measurement is being taken before ordering a replacement. B—Old door is removed. C—Old hinge leaves are removed and discarded, if at all worn or damaged. D—Other hardware, such as the striker, is removed. (McDaniels Construction Co., Inc.)

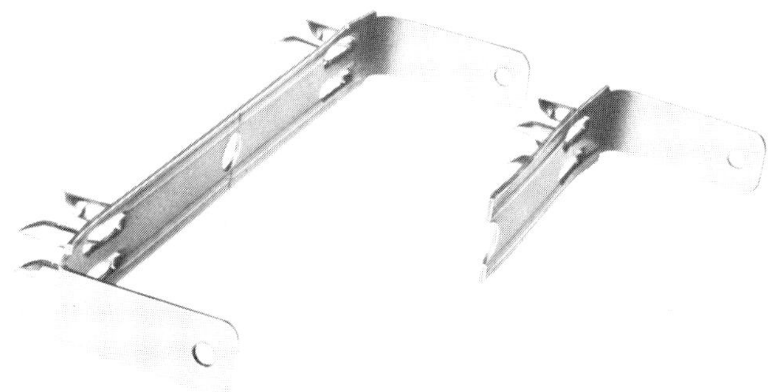

Figure 25-31. Patented jamb clips allow jamb installation without shims. Half clip at right, made by snapping full unit in two, is for jambs wider than standard. (Panel Clip Co.)

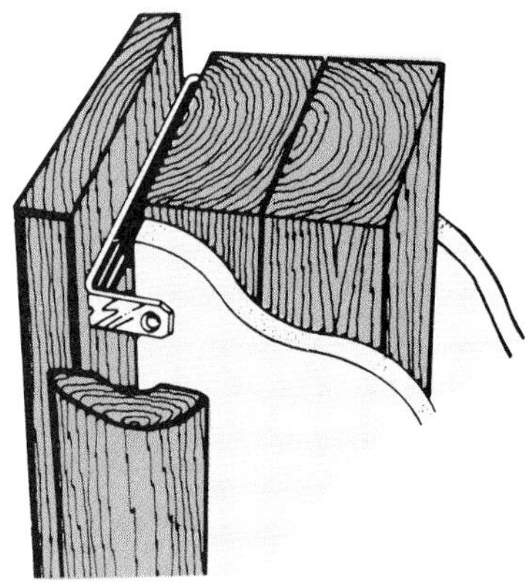

Figure 25-32. Cutaway of jamb and rough frame showing jamb clip attachment.

Figure 25-33. Steps for attaching jamb clips and installing door.

Figure 25-34. Marking a sticking door that needs planing to fit provides a guide for amount of material needing to be removed.

Figure 25-35. Additions to older homes should be carefully designed to complement their architectural style.

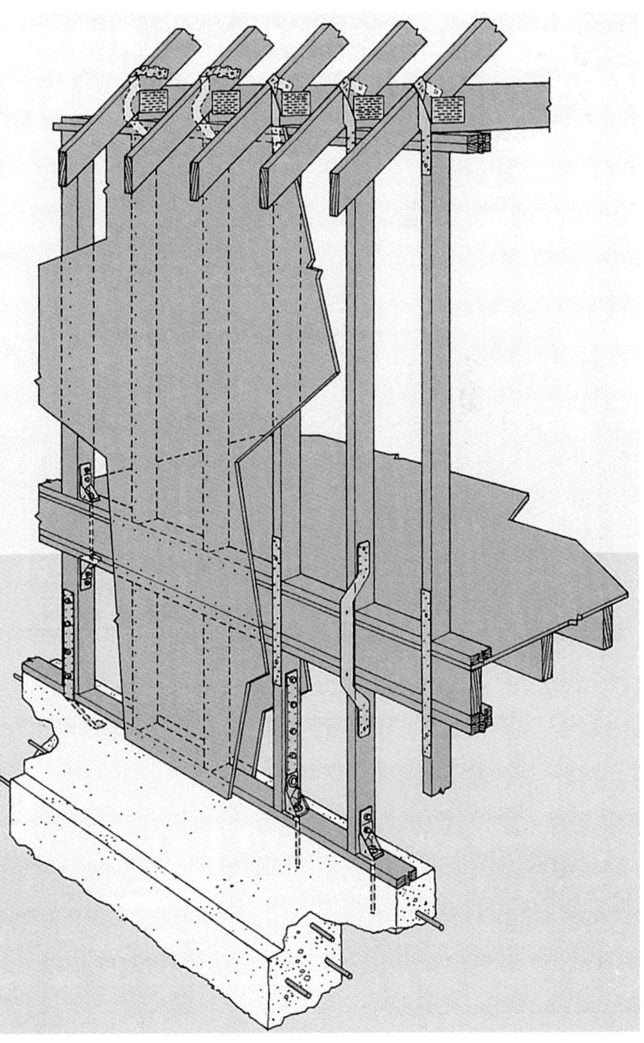

Figure 25-36. New code regulations for earthquake prone regions require special metal ties from foundation to rafters. (Simpson Strong-Tie Co., Inc.)

Solar Retrofitting

Older homes with a long wall to the south where the sunlight is not blocked by other structures are likely candidates for passive solar retrofit. A first consideration, however, is to investigate the insulation levels in the house. For insulation standards and methods, refer to Unit 14, Thermal and Sound Insulation.

Basic Solar Designs

The three basic passive solar designs are:

- *Direct gain* —South-facing double-glazed windows allow the sun direct access to living space. A masonry wall or floor is needed to act as storage for the solar heat.
- *Indirect gain* — In this design, a storage wall is located a few inches from the glazing. The wall soaks up the solar heat and radiates it to the living space.
- *Isolated gain (sun space)* — An example is shown in Figure 25-37. These are separate spaces—greenhouse or thermosiphon, for example—that have solar storage systems for collecting and distributing the heat to other parts of the house.

Figure 25-37. Adding solar space with a prefabricated greenhouse. Such units can be used to collect and store heat from solar energy.

Additional information on these systems and how to construct them is to be found in Unit 24, Passive Solar Construction. Direct gain and isolated gain systems are more practical as part of a remodeling program.

Adding a direct gain system involves removing a section of a south-facing wall and replacing it with windows. If glassed area is small, no storage is required. However, for large amounts of glass, a storage wall or floor may be needed. One way of getting storage is to install a fireplace on the opposite wall where sunlight could strike it. A brick planter can also serve as heat storage. These are not always practical because of weight.

The amount of thermal mass is important. For a large expanse of glass, about 150 lb. of masonry is needed for every square foot of south-facing glass. This could present serious structural problems for many older homes when you consider that 200 sq. ft. of glazing requires 15 ton of masonry.

In most situations, solar retrofit through adding sun space or a thermosiphon seems most practical.

Thermosiphon

Sometimes called a solar furnace, the *thermosiphon* is basically a glazed box that captures heat from sunlight. Natural convection (rising of heated air) moves the air into living space. See Figure 25-38.

Thermosiphons can be attached to window sills to provide solar heat for individual rooms. One experimental design involved glazing the entire south wall. At regular intervals, 6" wide boards were run edgewise vertically up the wall. Blocking was added to top and bottom enclosing the individual boxes. At first floor ceiling level, more blocking was added. Then vents were cut through the wall at floor and ceiling level on both floors. Finally, glazing was attached to the outer edge of the vertical boards.

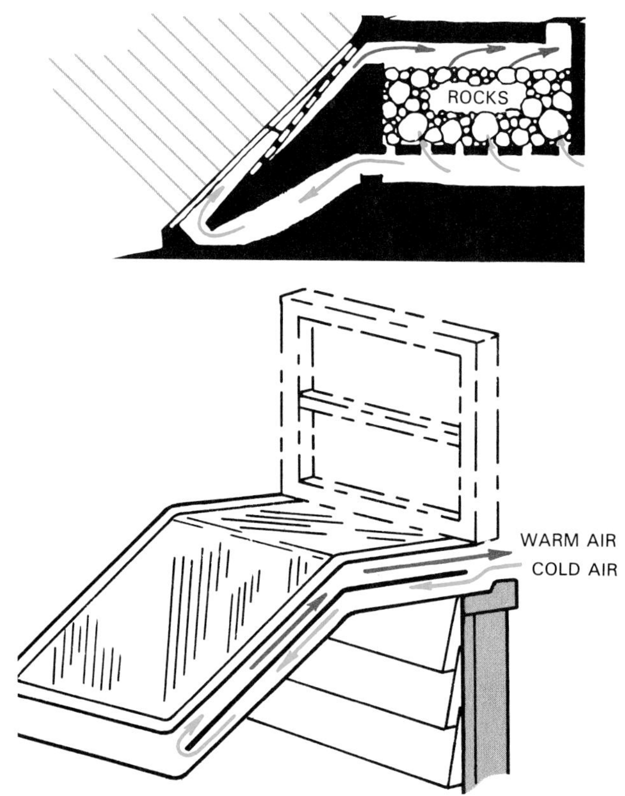

Figure 25-38. Two thermosiphon units. Top—Large unit designed to heat stored rock. Bottom—Small window unit.

As the sun shone through the glass, the air in the enclosed spaces heated up and set up a convection. Warm air rose entering the house at ceiling level. Cold air moved from the bottom. Figure 25-39 is a simple diagram of the system.

Figure 25-39. Thermosiphon system that occupies part of a south-facing wall. A series of boxes are formed by attaching 1" or 2" lumber on edge to the side of the house. Glazing covers it to trap heat. Vents allow circulation of the heated air into the home.

Responsible Renovation

A major remodeling of a building may seem to generate a great deal of waste. Even so, it is usually far less damaging to the environment than new construction. Inevitably, some waste ends up in a landfill. This waste can be kept to a minimum by careful salvage and attention to what can be recycled.

Window trim, doors, wood paneling, hardwood flooring, tile, brick, and framing lumber can be set aside and reused. If not used in the remodeling, someone else may use it. Careful removal of recyclable materials is more time-consuming than demolishing it, however. Many carpenters and contractors find it too costly. Salvage is most feasible for do-it-yourself projects.

Salvaged material should be carefully cleaned and stored out of the weather until it can be reused or sold. Some material will need to be given away. A "Free Material" sign on the lawn works well for giveaways. Disposal costs saved may even pay for the extra time spent in careful removal.

Hazardous material is a concern in every remodeling project. Painted material should be tested for lead content. Most houses built before 1945 could contain lead. Kits are available for testing lead content in paint. Lead contaminated materials should be professionally stripped before use. Some experts discourage reuse of such material and suggest it be disposed of in an approved landfill.

Other possible hazards include chlordane contaminated lumber, lead solder in old pipes, and asbestos in old flooring tile and exterior siding.

Inefficient fixtures such as flush toilets are water wasters and should not be recycled. Old windows are equally inefficient and waste energy. They should be used only for garages or sheds if they are not discarded.

Finally, in selecting building materials for remodeling, choose those made from recycled material wherever possible. By so doing you help create markets for recycled products. This is a key requirement to every recycling program.

For information on recycled products check a local library or bookstore. Another source of information is Environmental Building News, RR1, Box 161, Brattleboro VT 05301. For a cost of $2, they offer a bibliography of useful publications.

Safety: Fall Protection

OSHA has issued new fall protection rules that allow residential carpenters and contractors to write their own fall protection plans. The plan has to meet a set of guidelines supplied by OSHA. The new plan went into effect February 6, 1995.

The residential plan, which applies at heights of 6' or more, must be specific to the site, properly supervised, and can be applied only in situations where conventional fall protection procedures would cause a greater hazard than the work itself.

The OSHA regulations covering actions on scaffolding, stairs, ladders, roofs, and other high surfaces, such as top plates, are important to remodelers. Contact a local OSHA office or write Superintendent of Documents, U. S. Government Printing Office, Washington DC 20402. Ask for Code of Federal Regulations (CFR) Title 29, Part 1926.500. The title is "Safety Standards for Fall Protection in the Construction Industry; Final Rule."

The basic rules are:

- **Any height above 6' requires guardrails, safety nets, or a fall arrest system such as ropes and harnesses.**
- **Use of body belts as a fall arrest will be prohibited as of 1998 because the belts restrict breathing after a fall.**
- **Employers can choose among a variety of options, depending on circumstances on the job site. Often, a fall warning line is all that is needed.**

Important Terms

Concealed header
Fire stops
Flitch
Jamb clip
King studs
Ribbon
Saddle beam
Shoring
Sistering
Thermosiphon

Test Your Knowledge

1. The first concern when planning the renovation of a long-neglected house is (select best answer):
   A. Condition of the siding.
   B. Condition of the foundation.
   C. Assuring that mechanical systems are working.
   D. Repairing any leaks found.
2. What is the first step in a sequence of tasks for renovating the exterior?
3. The last step in renewing the exterior of a house is to _____________.
4. Why is floor finishing undertaken only after most other work is done.
5. Before rotted sills can be repaired :
   A. Foundation should be repaired.
   B. Building must be supported some way to remove load from the sills.
   C. Measurement should be taken of building's height.
   D. Carpenter must obtain a license to perform this complicated task.
6. What does the term "sistering" mean?
7. Before beginning a remodeling job, why is it wise to find out the type of construction used?
8. Before starting excavation for an addition you should determine what _____________ are _____________ beneath the ground.
9. List four indications that a wall is a bearing wall.
10. _____________ is a temporary wall installed to prevent the collapse or sagging of a structure when part of a bearing wall is removed.
11. How deep should a built-up wooden header be that must span 8' in a bearing partition and carry the load of a shallow attic above it?
12. When a header must be flush with the ceiling, one of two types of beam or header are used. Name them.
13. A home using direct gain solar heating must always include a method of storing heat gain. True or False?
14. A thermosiphon is a type of passive design called _____________ gain.
15. What does "solar refit" mean?
16. Give the main reason why use of recycled materials in a building project is important.

Outside Assignments

1. Write a report on the steps you would follow to add a 12' x 16' addition to the side of a ranch home built by the platform method. Include sketches for shoring and for final framing.
2. Compute the depth of the header needed to span an 8' opening in a bearing partition of a one-story house. The partition is in the center of a 28' span. Use the load tables in Figures 25-22 and 25-23. As necessary, consult Figure 25-24.
3. Discuss with your instructor the possibility of a field trip to an empty house in your community. Purpose of the visit: a visual inspection of the exterior to determine what renovation, if any, might be necessary. Each student should then prepare a written report on his or her recommendations.

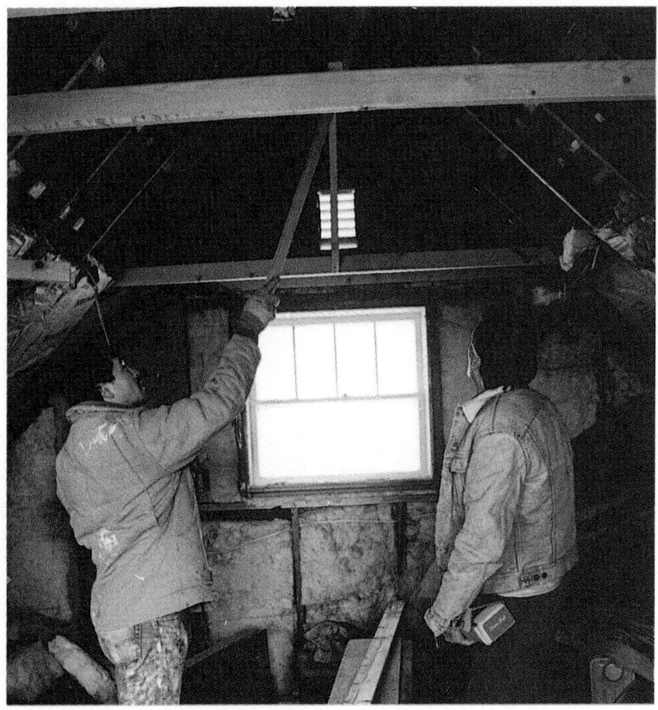

Architects can be important resources when remodeling is being considered. Carpenter, left, points out a framing detail to the architect. Space under eaves is being remodeled into an office. (Curran & La Pointe)

Addition of a deck is a simple home improvement project that adds appeal and value to an older home.

Mechanical Systems

Meter socket prior to meter installation. Electricity is fed to the building through the meter.

26

Electrical Wiring

Once a building is constructed and closed in (windows, exterior doors, and roof installed) it is ready for the installation of electrical wiring. This is one of three installations including plumbing, heating, and air conditioning, which are called mechanical systems.

Normally, these systems are the responsibility of other building trades craftspeople. These workers are generally hired by or work for subcontractors. Mechanical systems are installed before insulation is placed because they must be placed in the frame of the building.

Electrical wiring consists of wires, boxes, and a number of devices that control the distribution and use of electricity in the building. The system provides *current* (flow of electrons) that powers lights, heating units, and appliances such as refrigerators and washing machines.

Electricity is dangerous if it comes in contact with people or combustible materials, or metal that is not part of the electrical system. Current accidentally contacting a person can cause serious injury and even death from electrical shock. An electrical short may cause a fire. Therefore, electrical wiring must be installed by a qualified electrician according to methods prescribed by a national and local code.

The National Electrical Code is a collection of rules developed by experts. This code is organized and published by the National Fire Protection Association. It becomes law and thereby enforceable only when adopted by municipalities. Inspectors are hired to examine new installations and enforce the code locally when violations are found.

Tools, Equipment, and Materials

Certain tools, equipment, and materials are essential to install electrical wiring. Figure 26-1 shows a basic tool list. Electricians must be comfortable using all carpentry tools, Figure 26-2.

Hammers are used for driving fasteners and staples, attaching hangers and electrical boxes, and striking chisels. An electrician's hammer is preferred for attaching boxes because it has a longer neck than a carpenters hammer.

Saws are useful for making cuts in a building's frame. A handsaw can be used for notching or cutting studs. For heavier cuts, a power circular saw is better. A keyhole saw is better for working in tight quarters. A reciprocating saw will also work wherever a keyhole saw is used. It is faster

BASIC TOOL LIST

Striking Tools

- Claw hammer
- Lineman's or electrician's hammer

Drilling Tools

- Electric drill, 1/2" chuck
- Electric drill, 3/8" chuck
- Electric drill, 1/4" chuck
- Drill bits, various sizes
  - auger
  - wood twist
  - metal
  - masonry
  - expansive
  - bit extenders

Soldering & Wire-Joining Tools

- Soldering iron
- Soldering gun
- Propane torch
- Solder, rosin core
- Soldering paste
- Blow torch
- Crimping tool

Fastening Tools

- Standard screwdriver
- Phillips screwdriver
- Offset screwdriver
- Torque head screwdriver
- Adjustable wrench
- Allen wrenches
- Socket/ratchet wrenches
- Box end wrenches

Measuring Tools

- Folding ruler
- Carpenter's extension ruler
- Steel tape
- 12 " ruler or meter stick
- wire gage
- VOM

Cutting and Sawing Tools

- Files
- Crosscut saw
- Keyhole saw
- Hacksaw
- Circular saw, 7"
- Reciprocating saw
- Pocketknife or electrician's knife
- Cable cutters
- Chisel, wood
- Wire strippers
- Cable strippers

Pliers

- Slip joint pliers
- Lineman's pliers
- Side cutting pliers
- Diagonal pliers
- Long nose pliers
- End cutting pliers
- Curved jaw pliers

Special & Miscellaneous Tools

- Fish tape wire puller
- Wire pulling lubricant
- Conduit or pipe cutter
- Reamer
- Conduit bender (hickey)
- Fuse puller
- Tape and die set
- Flashlight
- Plumb bob
- Test light, continuity tester
- Level
- Conduit threader
- Trouble light
- Gas generator, about 1500 W
- Portable space heater
- Assorted wood or fiberglass ladders
- Wire grips
- Chalk line
- Tool pouch

Figure 26-1. An electrician's tool box should be stocked with this basic list.

Figure 26-2. Electricians use many of the tools that carpenters use. A—An electrician uses a cordless drill while working on a new home. B—In order to complete the electrical work, an electrician may need to remove brackets and members installed by carpenters. These items must be replaced.

and will handle thicker materials. A hacksaw will cut conduit and cable and may be used in place of a handsaw on wood.

Wood chisels are handy for trimming away small amounts of wood on framing members for mounting boxes and fixtures. They are preferred for notching studs, joists, plaster, flooring, and old-style lath.

A number of different tools can be used for wire and cable cutting. Multipurpose tools will cut, strip, and crimp. A cable stripper is used on larger conductors.

Pliers are good for holding, shaping, and cutting. The basic tools are the slip joint pliers and the lineman's pliers or side cutters. A versatile tool is the vice grips. It serves as pliers, lock wrench, open-end wrench, or a pipe wrench.

The standard straight blade screwdriver is the most-used fastening tool. It is needed primarily to tighten terminal screws, attach switches and receptacles, as well as fasten wires to devices. Phillips screwdrivers are used occasionally for some screw heads.

Drilling tools are used to bore holes prior to running conductors or conduit through studs and joists. Either brace and bit or a portable electric drill will do the job. For tight quarters, a right-angle drill is handy.

The electrician's tool kit will also contain solder tools, either propane torch or electric soldering iron. There is still some use for soldered connections, although the use of wire connectors have all but eliminated soldering.

A folding 6' rule and a steel tape are important for measuring. Typical rules are 6', 8', and 10'.

A *fish tape* is indispensable for pulling wires through conduit. Various other tools round out the electrician's tool box. These might include a wood or fiberglass ladder, trouble light, pipecutter, continuity tester, neon tester, conduit bender, conductor bender, and a set of sockets.

Ladders for electrical work should be wood or fiberglass, because neither is a good conductor of electricity. Metal ladders should never be used because of the danger in the ladder touching a current-carrying conductor.

Materials

An electrical system in a building is made up of:

- *Conductors*—These are the wires that carry current. See Figure 26-3.
- *Boxes and covers*—Various kinds are available, Figure 26-4. Their purpose is to enclose devices that control electric current or protect connections between two or more conductors. Covers conceal and protect devices, such as switches and receptacles, placed in boxes.
- *Fuses or circuit breakers*—These devices, Figure 26-5, protect conductors from overloads, shutting off power to the circuit in case of a fault in the circuit or a too-heavy current draw that would heat up the wire and cause it to fail.
- *Switches and receptacles*—Several types of switches are needed as shown in Figure 26-6. Switches control the current to lights and appliances. Receptacles allow convenient connection of appliances and lights to a circuit. They are sometimes called "outlets."
- *Conduit or raceways*—Figure 26-7. These are pipes or enclosed channels through which conductors are run.
- *Connectors*—These are secure connections between two or more conductors and conduit, Figure 26-8.

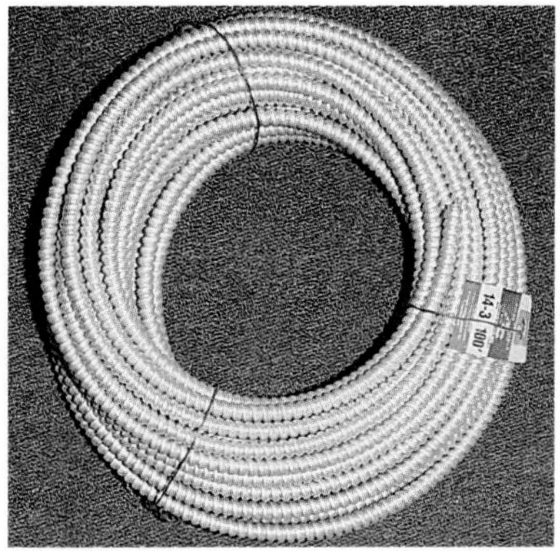

Figure 26-3. Electrical conductors are the copper or aluminum wires that provides a path for current in the building's circuits. The conductors are made up in different ways. Top—Armored cable, also called "BX," has a flexible metal cladding protecting several insulated conductors. Middle—Single, insulated conductors come in various gages (thicknesses). They must be located inside conduit for protection. Bottom—Nonmetallic (NM) cable is similar to armored cable but has a plastic protective covering. Codes vary on which types can be used.

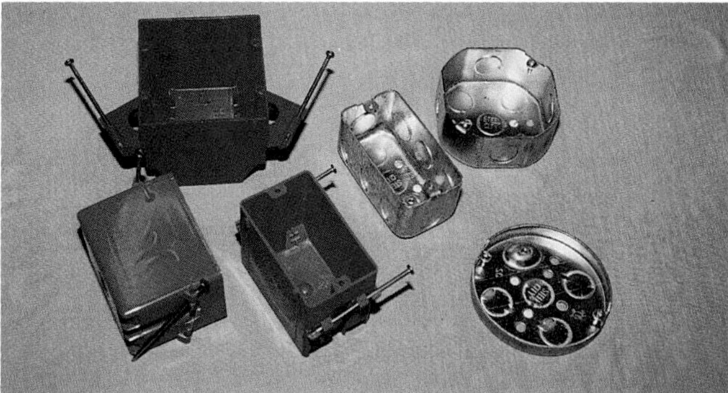

Figure 26-4. Electrical boxes house connections for conductors and support devices such as lights, switches, and receptacles. Covers in various configurations protect devices housed and conductor connections made in the boxes.

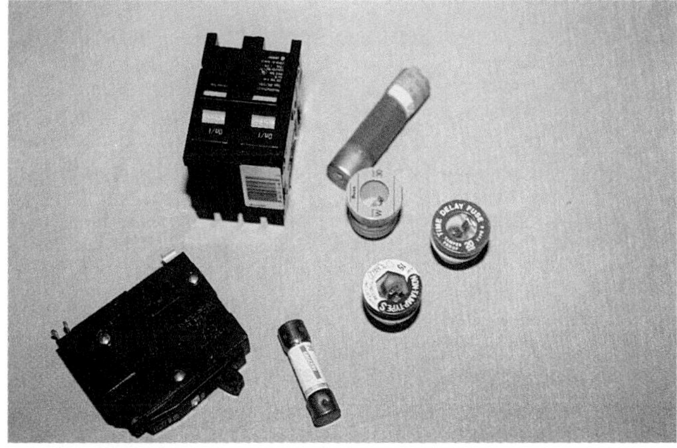

Figure 26-5. Circuit breakers and fuses protect circuits from overloads that could cause damage to the circuit or to the building.

Basic Electrical Wiring Theory

Electricity is generated when a conductor such as copper, silver, or aluminum is passed through a magnetic field. This causes *electrons* to move through the conductor in one direction. When the conductor (wire) passes through the magnetic field in the opposite direction the flow of electrons, known as *electric current,* flows in the opposite direction. This changing of direction is known as *alternating current.* This is the form of electric current found in homes today. Another form also used for various purposes is known as *direct current,* meaning that electrons flow in only one direction. This is the type of current supplied by batteries, for example.

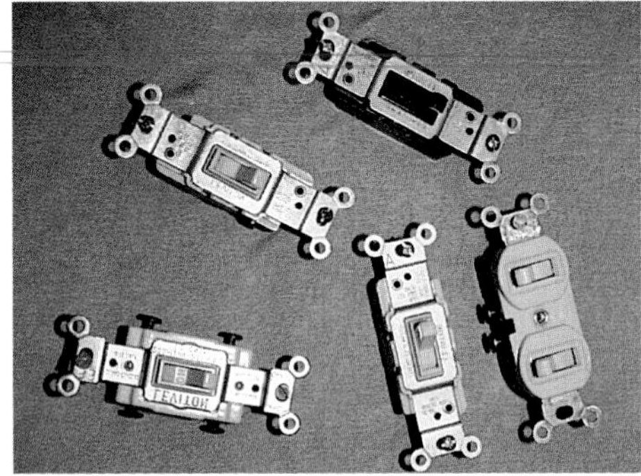

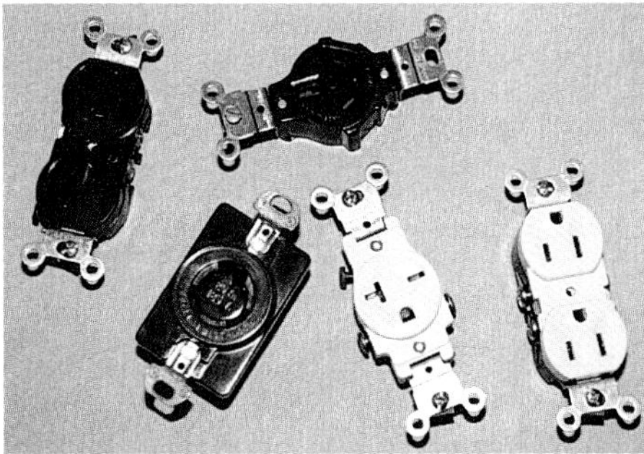

Figure 26-6. Top—Switches control the flow of electricity the same way as a faucet controls the flow of water. They interrupt the path of the current. Bottom—Receptacles allow quick temporary connection of appliances and lights to a circuit.

Figure 26-8. Fittings connect and hold conduit and cables.

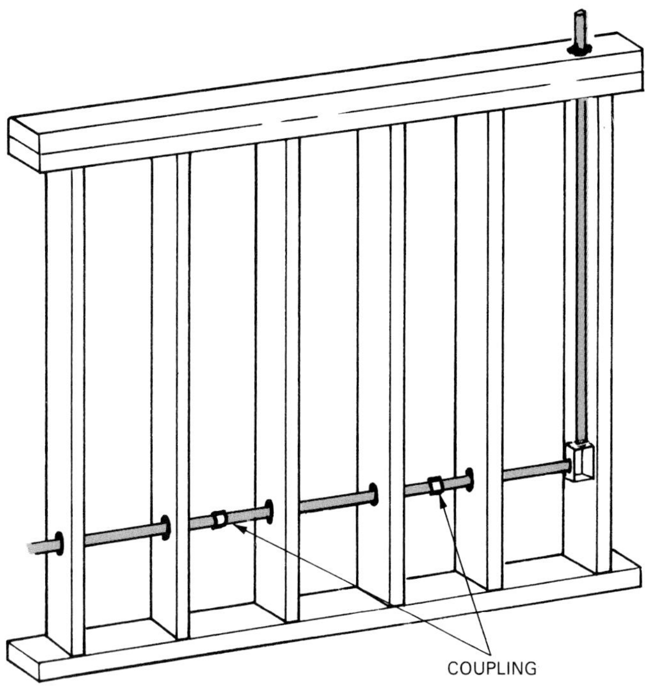

Figure 26-7. Conduit is used to protect electric wires.

All alternating electric power is generated at an electric power station and is sent out through conductors. *Transformers* along the way step up the voltage or reduce it as needed for power customers, Figure 26-9. A transformer is a device that can change voltage it receives to either a higher or lower voltage. It does this by way of a process called *induction.*

A transformer is made up of a core of iron around which are wrapped a number of turns of conducting wire. This is known as a *coil*. The transformer has two coils. One coil will have more turns of conducting wire than the other. This accounts for the change in voltage. There is no electrical connection between the two coils. When current enters one coil, a magnetic field is set up. This field reaches the other coil and sets up a current. This induced current is not of the same voltage because of the different number of turns between the two coils.

Transformers at the power station increase the voltage generated because it becomes easier to send high voltage current. This is done with a "step-up" transformer. Before current is delivered to a house, a "step-down" transformer reduces the line voltage to 120 or 240 volts before it is fed into a residence. Commercial and industrial customers may use much higher voltages.

Figure 26-10 shows three kinds of transformers for changing voltage. One is called a substation. It serves a whole area such as a town. The other serves one residence. From the transformer, conductors called a "service" bring electricity to the residence. The conductors may be strung from poles or buried in the ground. See Figure 26-11.

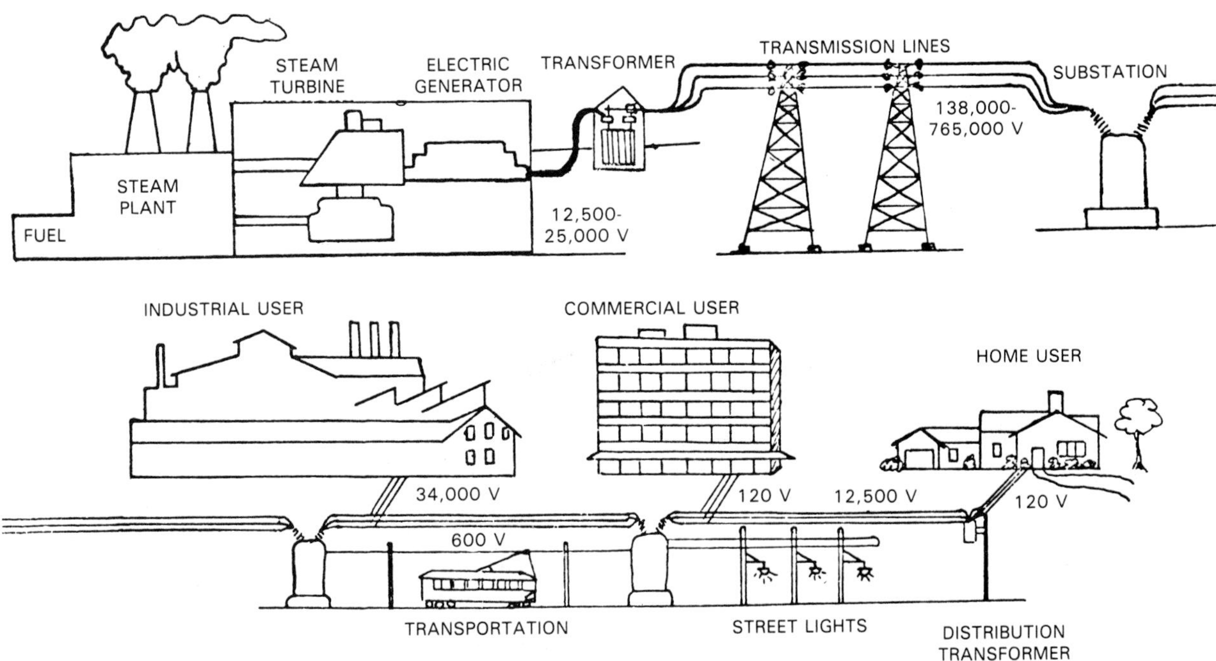

Figure 26-9. Diagram of an electrical power system delivering electric power from a generating plant to its customers.

Installing the Service

The *service* is the conductors that bring power from a transformer to a building. It extends only from the transformer, through a metering device, to a box in the dwelling called a distribution panel or entrance panel. Figure 26-12 is a line drawing of all the components of the service entrance at the dwelling. Installation of these components is the responsibility of the electrician employed by the homeowner. When the service is underground, the installation at the dwelling looks like the one shown in Figure 26-13.

Figure 26-10. Voltage is changed several times before being delivered to customers. A—A substation is a series of transformers that step down voltage before it is delivered to a large number of customers such as are in a community. B—Each building will have a step-down transformer to further reduce voltage. When underground service is provided to a building, a transformer is placed on the ground. C—Transformers can be very small, such as this 14 volt unit used to power doorbells and some other low-voltage energy user. (Ship and Shore General Store, Dauphin Island)

Figure 26-11. A—A trench must be dug to lay down underground service. An electric code will specify how deeply cables or conduit must be buried. B—Overhead service is strung from poles. Here a wireman for an electric power company completes a hook-up.

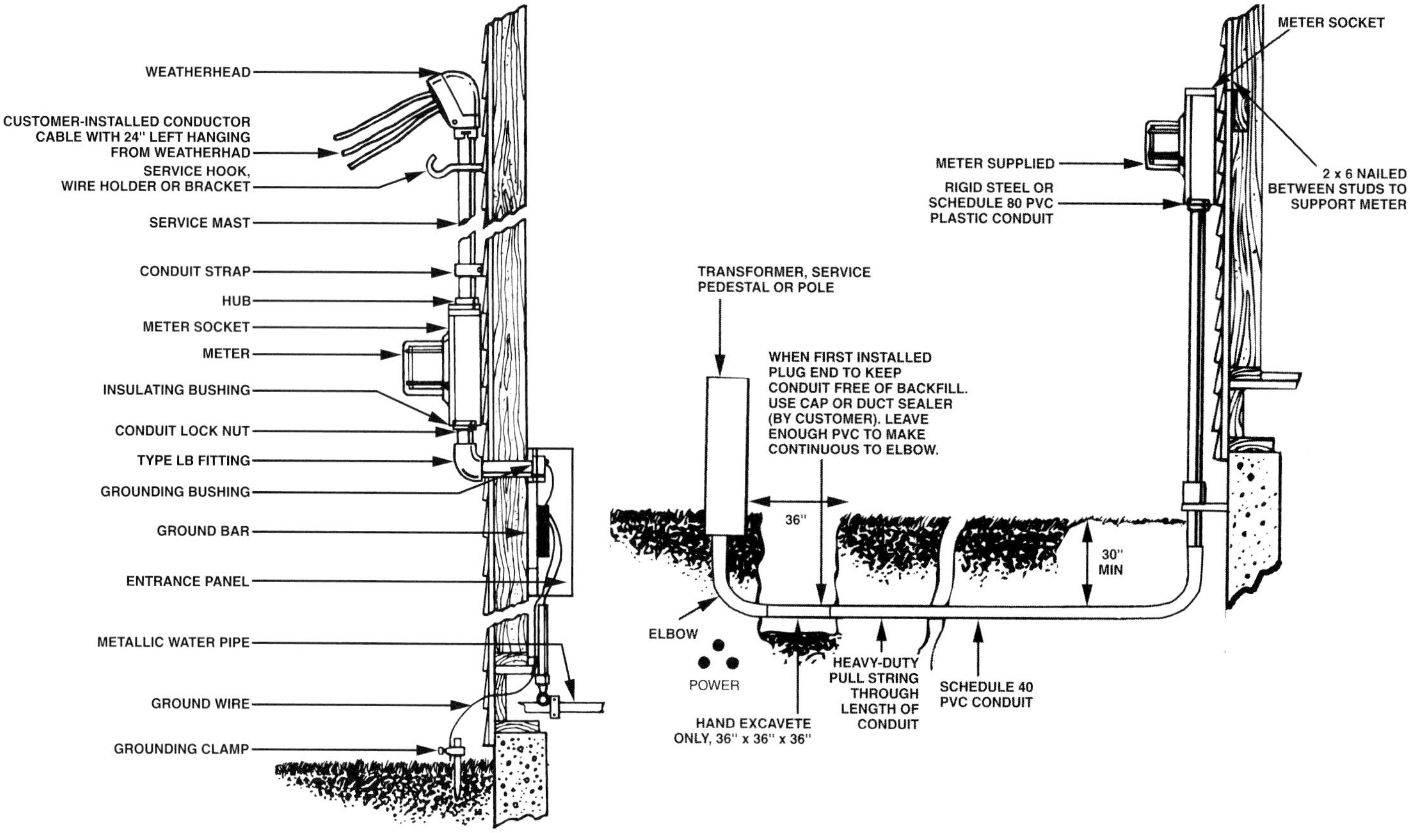

WEATHERHEAD
CUSTOMER-INSTALLED CONDUCTOR CABLE WITH 24" LEFT HANGING FROM WEATHERHAD
SERVICE HOOK, WIRE HOLDER OR BRACKET
SERVICE MAST
CONDUIT STRAP
HUB
METER SOCKET
METER
INSULATING BUSHING
CONDUIT LOCK NUT
TYPE LB FITTING
GROUNDING BUSHING
GROUND BAR
ENTRANCE PANEL
METALLIC WATER PIPE
GROUND WIRE
GROUNDING CLAMP

METER SOCKET
METER SUPPLIED
RIGID STEEL OR SCHEDULE 80 PVC PLASTIC CONDUIT
2 x 6 NAILED BETWEEN STUDS TO SUPPORT METER
TRANSFORMER, SERVICE PEDESTAL OR POLE
WHEN FIRST INSTALLED PLUG END TO KEEP CONDUIT FREE OF BACKFILL. USE CAP OR DUCT SEALER (BY CUSTOMER). LEAVE ENOUGH PVC TO MAKE CONTINUOUS TO ELBOW.
36"
30" MIN
ELBOW
POWER
HAND EXCAVETE ONLY, 36" x 36" x 36"
HEAVY-DUTY PULL STRING THROUGH LENGTH OF CONDUIT
SCHEDULE 40 PVC CONDUIT

Figure 26-12. Drawing of a three-wire service at a building. Left—An above-ground installation. Right—An underground installation. Note names of parts. (KCPL)

The service ends inside the dwelling at the entrance panel, Figure 26-14. The entrance panel houses a main breaker, circuit breakers, or fuses for all the home's circuits as well as the conductors for these circuits.

Figure 26-13. Exterior box is called a meter socket. A—Meter socket is shown before completion of service. B—When the wiring system of a building is completed, the meter is installed and the system is ready to be used.

Figure 26-14. The service entrance or distribution panel is located inside the building. Each white cable at the top represents a separate circuit. A—Each circuit is connected to a circuit breaker (black objects in the box). The heavy cables are the conductors that bring power from the transformer. The large black block just below the cables is the main breaker. B—When connections have been completed, a protective cover is placed over the installation.

Reading Blueprints

Wiring plans for the house are done by an architect. The plan, Figure 26-15, uses symbols to show the types of conductors and devices to install at each location. These symbols have been standardized so that there can be no misunderstanding, Figure 26-16.

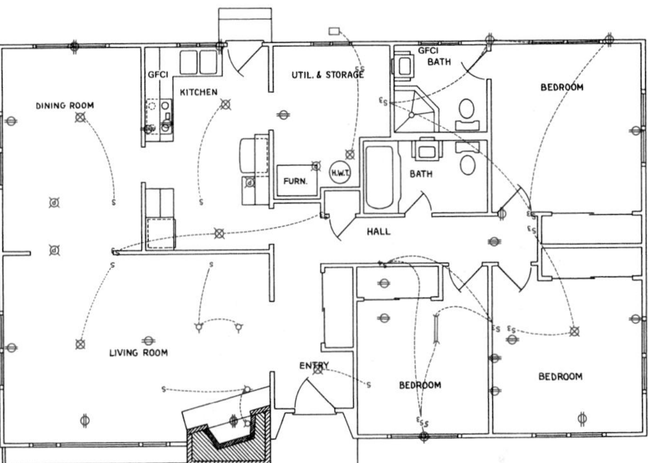

Figure 26-15. Wiring plans such as this are provided by the architect. It serves as a guide to the electrician when installing wiring, devices, and fixtures.

After studying blueprints, the electrician may draw a cable layout for each room. This layout will show where, in the room, electrical cable are to be run, the size of the conductors, and how many conductors will be in each run of cable. Refer to Figure 26-17 for an example of a cable layout.

Running Branch Circuits

Think of a circuit as a loop or continuous path from the source of electric power, through various current-using devices (such as lights or electric motors) and then back again to source. Figure 26-18 is a simple explanation of a simple circuit.

Branch circuits are all of the outlets on one run of conductors. The circuit begins at the distribution panel with the circuit being connected to a fuse or circuit breaker. The electrician will first install junction, switch, and receptacle boxes. Then, he or she will drill holes through the wood framing members for running cable or conduit, Figure 26-19. The cable or conduit is run through the building's frame, Figure 26-20, before insulation and drywall or plaster are installed. The conductors are run to each box and securely connected to them. See Figure 26-21.

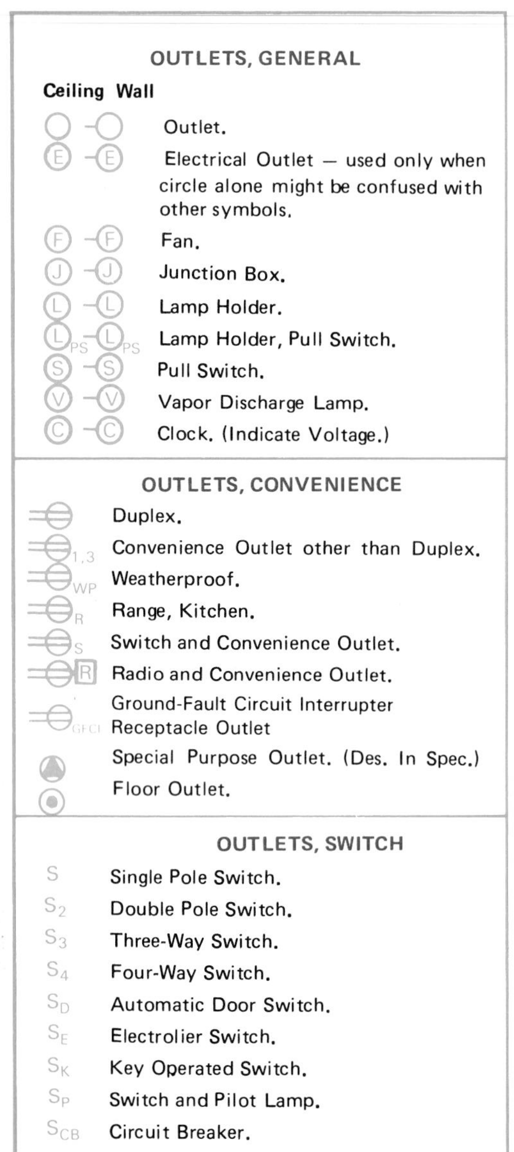

OUTLETS, GENERAL

Ceiling	Wall	
		Outlet.
E	E	Electrical Outlet — used only when circle alone might be confused with other symbols.
F	F	Fan.
J	J	Junction Box.
L	L	Lamp Holder.
L PS	L PS	Lamp Holder, Pull Switch.
S	S	Pull Switch.
V	V	Vapor Discharge Lamp.
C	C	Clock. (Indicate Voltage.)

OUTLETS, CONVENIENCE

	Duplex.
1,3	Convenience Outlet other than Duplex.
WP	Weatherproof.
R	Range, Kitchen.
S	Switch and Convenience Outlet.
R	Radio and Convenience Outlet.
GFCI	Ground-Fault Circuit Interrupter Receptacle Outlet
	Special Purpose Outlet. (Des. In Spec.)
	Floor Outlet.

OUTLETS, SWITCH

S	Single Pole Switch.
S_2	Double Pole Switch.
S_3	Three-Way Switch.
S_4	Four-Way Switch.
S_D	Automatic Door Switch.
S_E	Electrolier Switch.
S_K	Key Operated Switch.
S_P	Switch and Pilot Lamp.
S_{CB}	Circuit Breaker.

Figure 26-16. Symbols are a type of shorthand that tells the electrician what type of device or fixture is to be installed. Here is a sample of some symbols for electrical outlets.

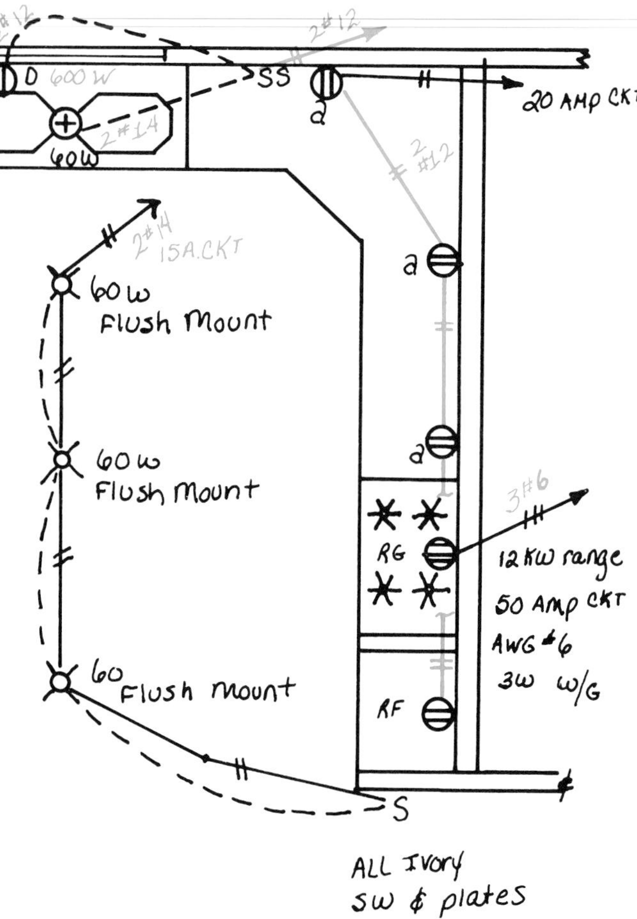

Figure 26-17. A cable drawing. Electricians sometimes produce such drawings as a reference while wiring a room.

Electrical cables must also be securely fastened to framing members, as shown in Figure 26-22. The *National Electrical Code* requires that fasteners must be installed at intervals not greater than 3' and within 12" of their entrance into an electrical box.

When doing any type of electrical work, it is necessary to be familiar with local building code requirements. The National Electrical Code (NEC) is the model code followed by most municipalities.

There is always a possibility of piercing the conductors with nails driven into the studs while installing drywall. Metal plates are available to protect the conductors. Figure 26-23 shows drawings of this plate and how they are attached to studs with a hammer.

Device Wiring

In order to complete a circuit, it is necessary to connect wires to one another. It is also necessary to connect wires to devices such as switches and receptacles. The Code tells how these connections are to be made. Before attempting to connect wires, several inches of insulation must be stripped away. The connection could be made by wrapping the bare wires together and then soldering this connection. The bare wires would then have to be wrapped. A more efficient way is to use a wire nut, Figure 26-24.

Terminals on switches and receptacles are usually screws that grip the conducting wires. Figure 26-25 shows the correct method of bending a conductor for attaching to a terminal.

Switches, Figure 26-26, control current through a circuit. The hot (current carrying) wire in the circuit is interrupted by the switch as shown in Figure 26-27.

Receptacles transfer electrical energy from the circuit conductors to lamps, toasters, and a variety of appliances. It has terminals on both sides. The hot wire is attached to the copper colored terminal; the neutral (white) wire is

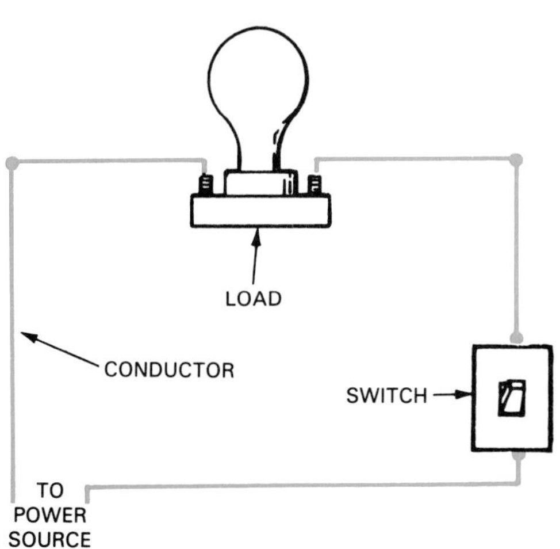

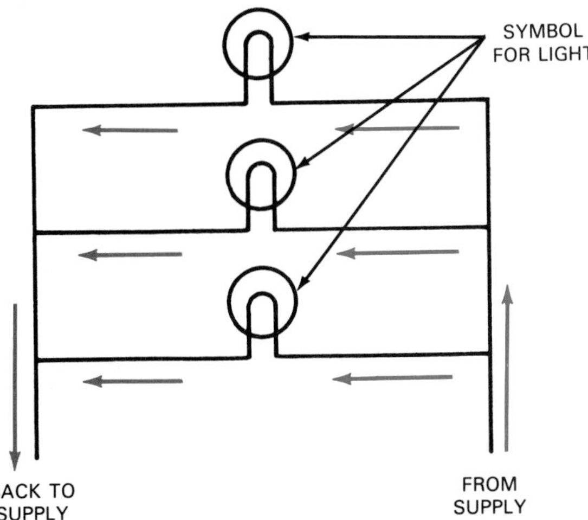

Figure 26-18. A circuit or branch circuit is simply a path from a source of power, through energy-using devices such as lights, back to the power source. Top—When there is only one path for current, it is called a series circuit. Bottom—When there are two or more paths for current, it is a parallel circuit.

Figure 26-19. Drilling holes for running electrical cable or conduit.

Figure 26-20. Running conduit through wall frames. Left—Running nonmetallic electrical cable through steel studs. The studs have prepunched holes for this purpose. The electrical codes requires plastic grommets (these are colored red) to protect the cable from abrasion. Right—Running nonmetallic cable through wood studs. Holes should be drilled approximately in the middle of the studs.

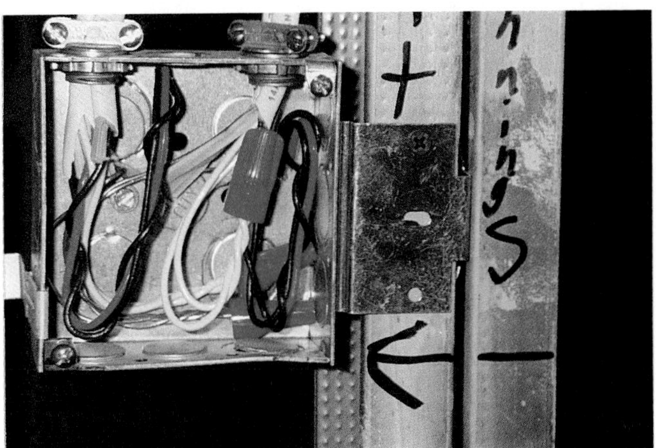

Figure 26-21. Connectors must be installed to hold cable or conduit securely to boxes.

attached to the lighter, silver colored terminal, Figure 26-28.

A single pole switch is simple to wire up, but three-way and four-way switches can be complicated unless the electrician fully understands what happens when a switch is activated. See Figure 26-29.

Electrical Troubleshooting

Electrical troubles show up as a wide range of symptoms. Usually, the problems involve faulty receptacles, switches, lighting fixtures, fuses, and breakers. Complex problems include open circuits, broken conductors, voltage fluctuation, and ground faults or current leakage.

Figure 26-22. Nonmetallic cable is shown properly fastened to framing members. A—Cable being attached at intervals to the side of a stud. B—Conductors fastened with a cable staple within 12" of entering a receptacle box.

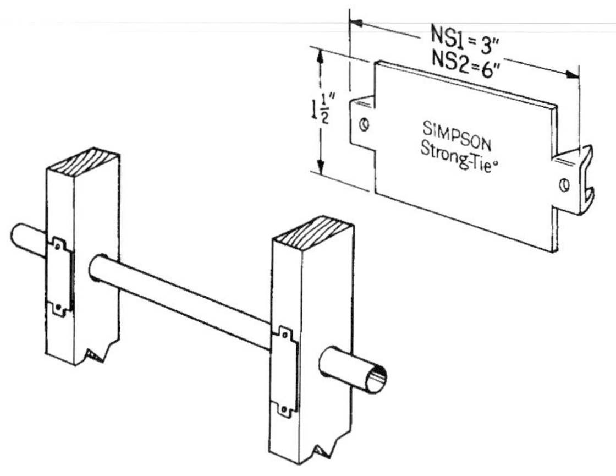

Figure 26-23. Protecting conduit and cable. Special metal plates are manufactured to be attached to wood studs so that nails cannot accidentally pierce the conductors and cause a dangerous short. (Simpson Strong-Tie Co., Inc. and Hadley-Hobley Construction)

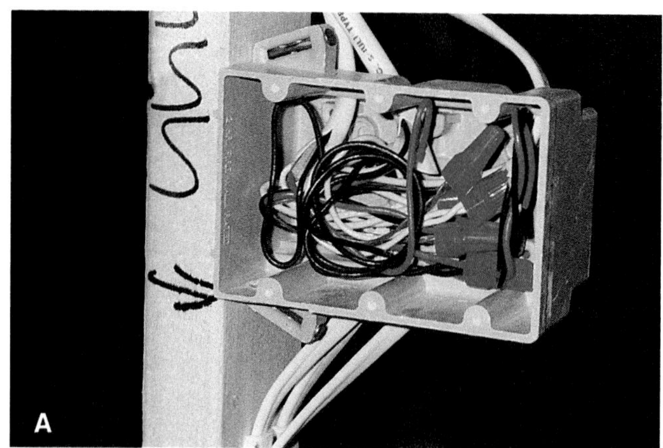

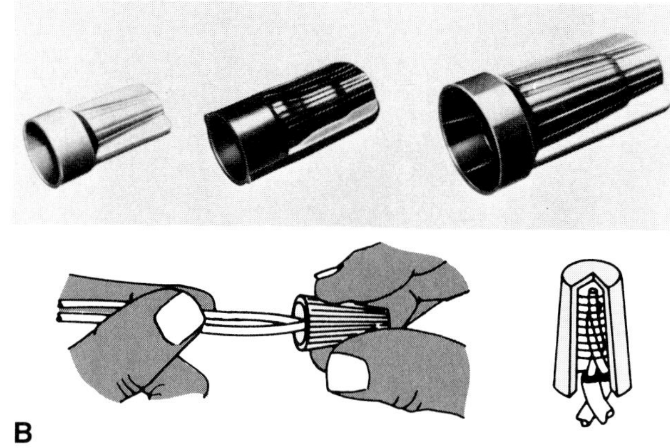

Figure 26-24. Proper use of a wire nut to connect bared conductors. A—Insert bare conductors into the wire nut and twist the nut clockwise until it is tight. It is usually best to twist the conductors together first, especially if there are three or more conductors being connected. B—Connected wires are insulated against accidental shorting by the wire nut.

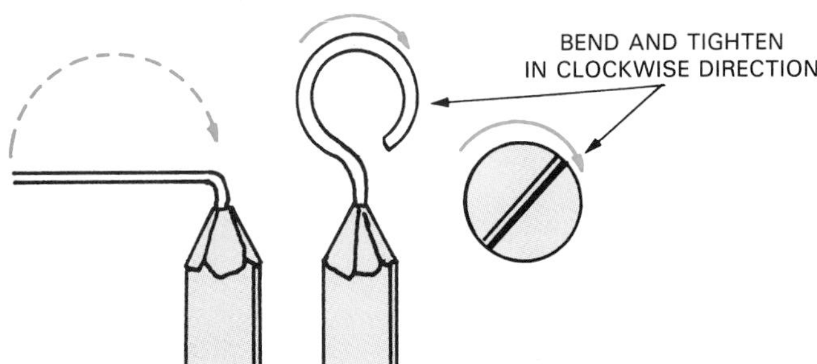

Figure 26-25. Correct method of preparing a conductor for attaching to a screw-type terminal. Use a needle-nosed pliers to form the open loop.

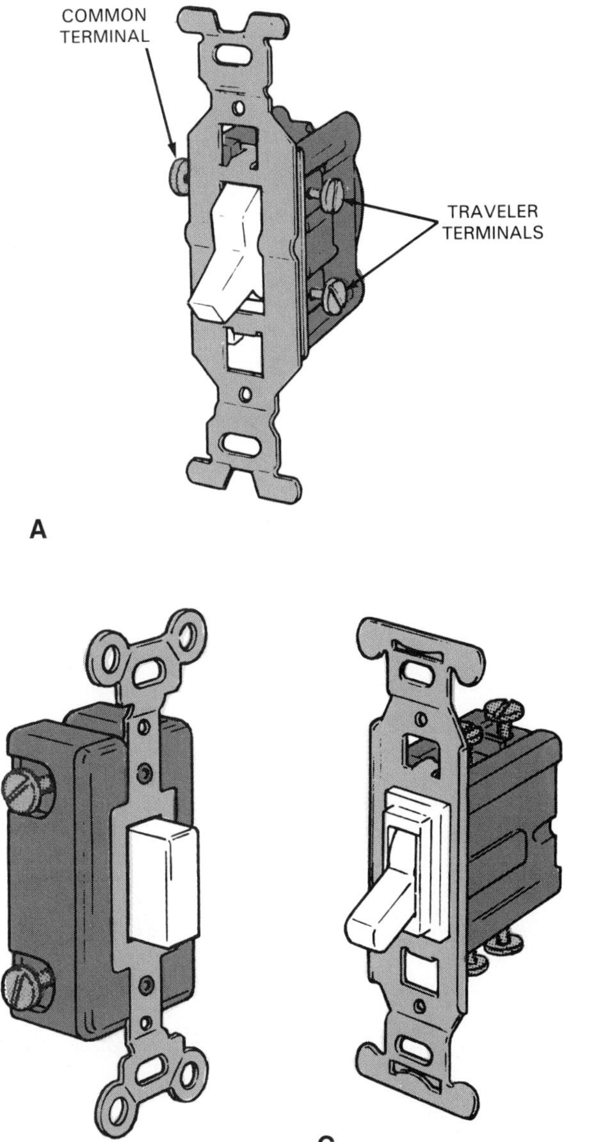

Figure 26-26. Three basic types of switches are available. A—Three-way. B—Single Pole. C—Four-way. Note that a single pole has only two terminals. A three-way has three terminals, and a four-way four terminals.

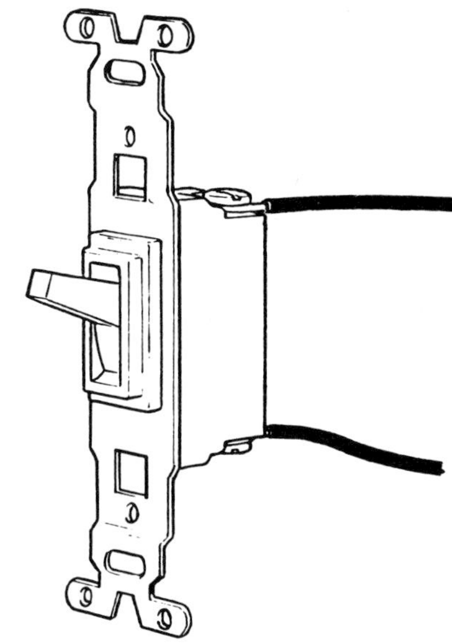

Figure 26-27. A switch interrupts a current-carrying wire and is always placed on a black wire or a red wire.

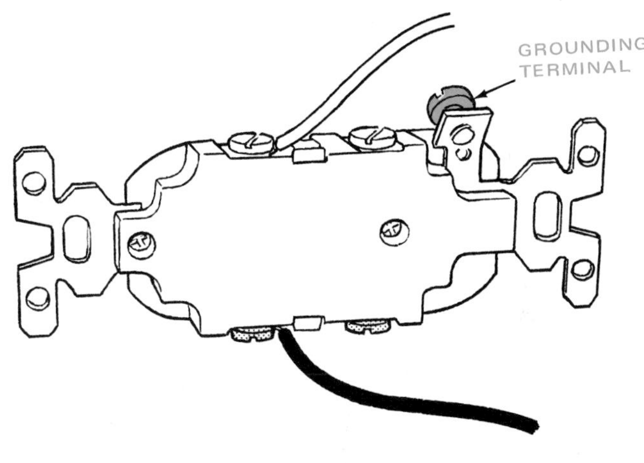

Figure 26-28. Wiring a receptacle. Top—Neutral (white) wire is connected to light colored terminals; black or red to darker terminal. Bottom—A receptacle properly wired and attached to a box.

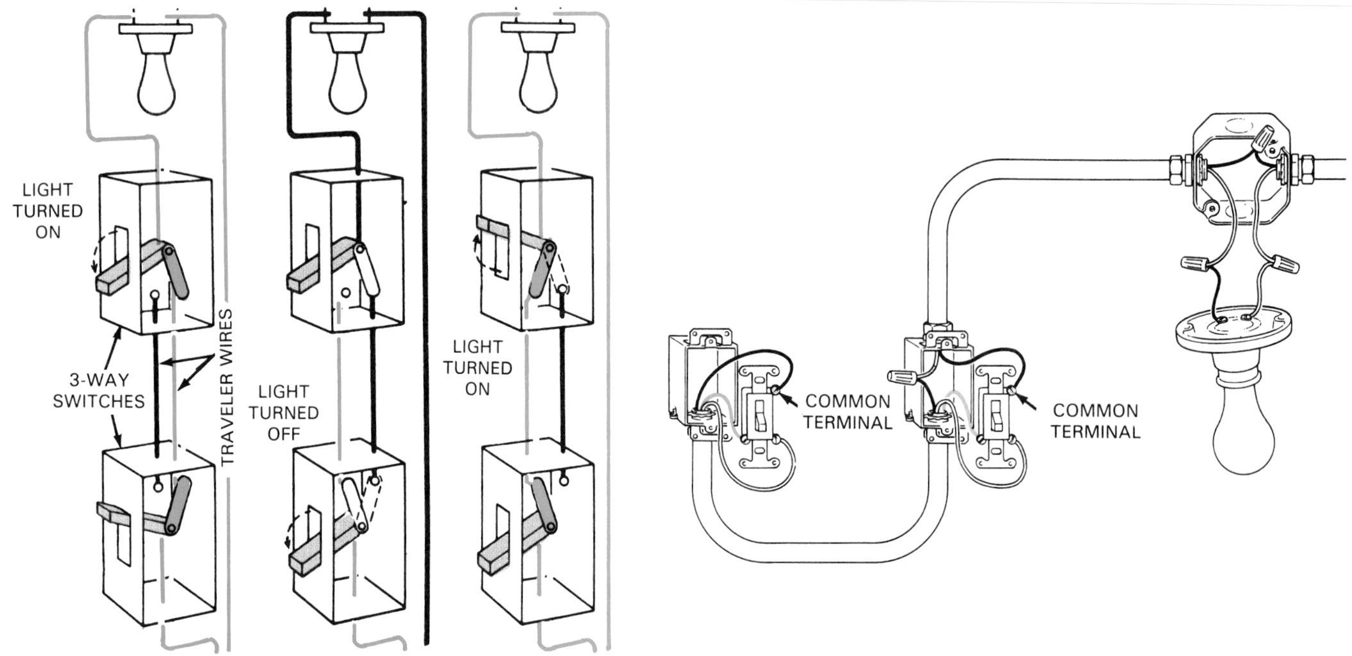

Figure 26-29. Explanation of how a three-way switch arrangement actually works. Illustration at right shows how an electrician might wire it up.

Testing Receptacles

When a receptacle is not functioning, either the receptacle or the circuit is at fault. Testing will determine which. If the neon tester does not light when the leads are pushed into the slots, Figure 26-30, check the line conductors. To do this, remove the wall plate and check for voltage at the terminals of the receptacle. See Figure 26-31. If the tester lights up, the receptacle is defective.

If the previous test does not cause the tester to light, the problem is in the circuit. Sometimes an open neutral is a problem. Test as shown in Figure 26-32 by placing one test probe in the ground slot and the other in the hot slot of the receptacle. If this indicates voltage, there is an "open" in the neutral conductor. Check other receptacles between the problem receptacle and the distribution panel. Check at the distribution panel, as well.

Figure 26-30. Checking for power at the outlet. Push probes of the neon tester in the slots. If tester glows, circuit is working. If there is no light, move on to another test to see if the problem is in the circuit or receptacle.

Testing Switches

To test switches, remove the cover plate and determine if there is power to the switch. Touch one probe of the neon tester to the metal box (or the ground wire if the box is plastic) while touching the other probe to the line side terminal. If the tester lights, the circuit is live. With the switch on, touch the probe to the load terminal of the switch and the other probe to ground. If test light glows, switch is good and the fixture or the wiring to it is defective. Refer to Figure 26-33.

Figure 26-31. Place probes on the two terminals having wires attached to them. If the tester lights up, replace the receptacle. If there is no voltage, the problem is in the circuit.

Figure 26-32. Place probes in the hot side slot and the ground. If the tester lights, there is an "open" in the neutral.

Testing Fixtures

Condition of fixtures can be checked by testing whether power reaches the supply conductors at the fixture outlet, Figure 26-34. If there is voltage, repair or replace the fixture.

Figure 26-33. Testing a switch. Top—With the switch off, tester should show voltage at the line side terminal screw if the circuit is "live." If there is no light, circuit is defective. Bottom—With the switch on, neon tester should indicate voltage on a live circuit. If there is no glow, the switch is faulty.

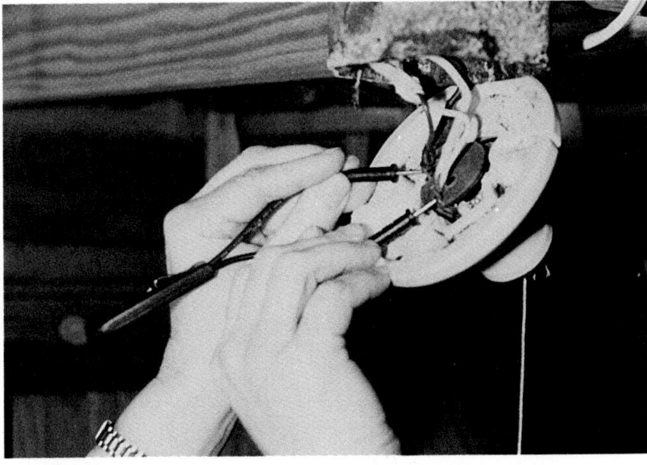

Figure 26-34. Test a fixture by checking if line voltage is reaching the terminals. Place one probe on each terminal. If the test causes the tester to glow, but the fixture is dead, replace or repair the fixture.

Important Terms

Alternating current
Branch circuit
Circuit breakers
Coil
Conductors
Conduit
Connectors
Current
Direct current
Fuses
Induction
National Electrical Code
Receptacles
Service
Switches
Transformers

Test Your Knowledge

1. What is included in an electrical system?
2. Define current.
3. The ____________ is a collection of rules for installing electrical systems.
4. Where are drilling tools used in electrical wiring?
5. Would an aluminum ladder be a suitable tool when working around electrical wires?
6. ____________ control current to lights and certain appliances.
7. Electrical current that goes one direction and then another is known as ____________ current.
8. Transformers are devices that change voltage by means of magnetism. True or False?
9. A wiring plan is the same as a cable layout. True or False?
10. Define an electrical circuit.
11. If in testing a receptacle, the neon tester does not light when the probes are placed on the terminals of the receptacle, where does the problem lie?
12. A fixture is not working. How do you test for the problem?

Outside Assignments

1. Wire a sample circuit that includes a plug, a convenience outlet, and a box-to-house outlet. Have your instructor inspect your circuit before testing it.
2. Using a junction box and two lengths of No. 14 nonmetallic cable (or metallic cable), show the proper method for making solderless connections between two conductors.
3. Prepare a drawing that shows the operation of two three-way switches wired to control a light from two locations. Discuss the drawing with your instructor and then wire it on a wall mockup.

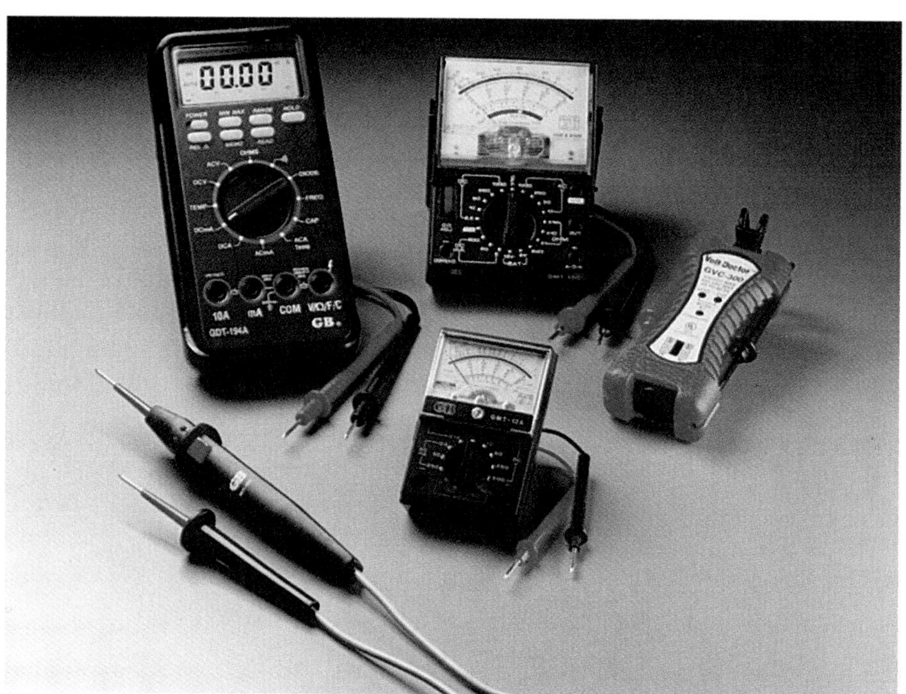

Voltage testers, continuity testers, battery testers, inductive voltage testers, and in-wall metal detectors are all required in the electrician's toolbox. (GB Electrical, Inc.)

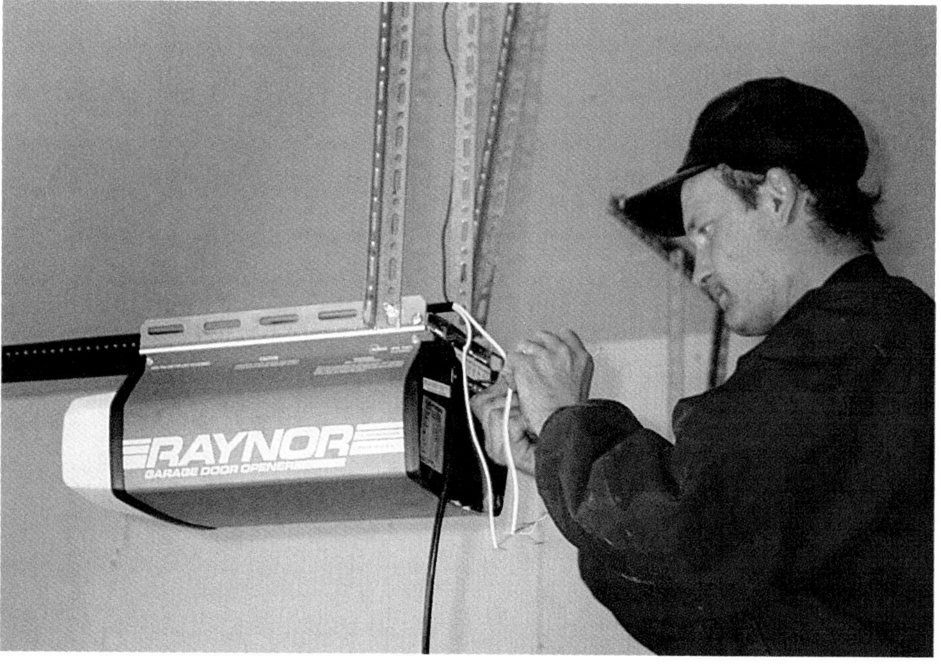

An electrician may be called upon to install and hook up the wiring for an overhead garage door opener.

27 Plumbing Systems

Plumbing includes all the piping and fixtures that provide water for drinking, cooking, bathing, and laundry, as well as a means of disposing of wastewater. See Figure 27-1.

Of necessity, plumbers must cut holes and notches in a building frame to install rough plumbing. It is important for the plumber to work closely with the carpenter to ensure that the holes will not weaken the joists and studs.

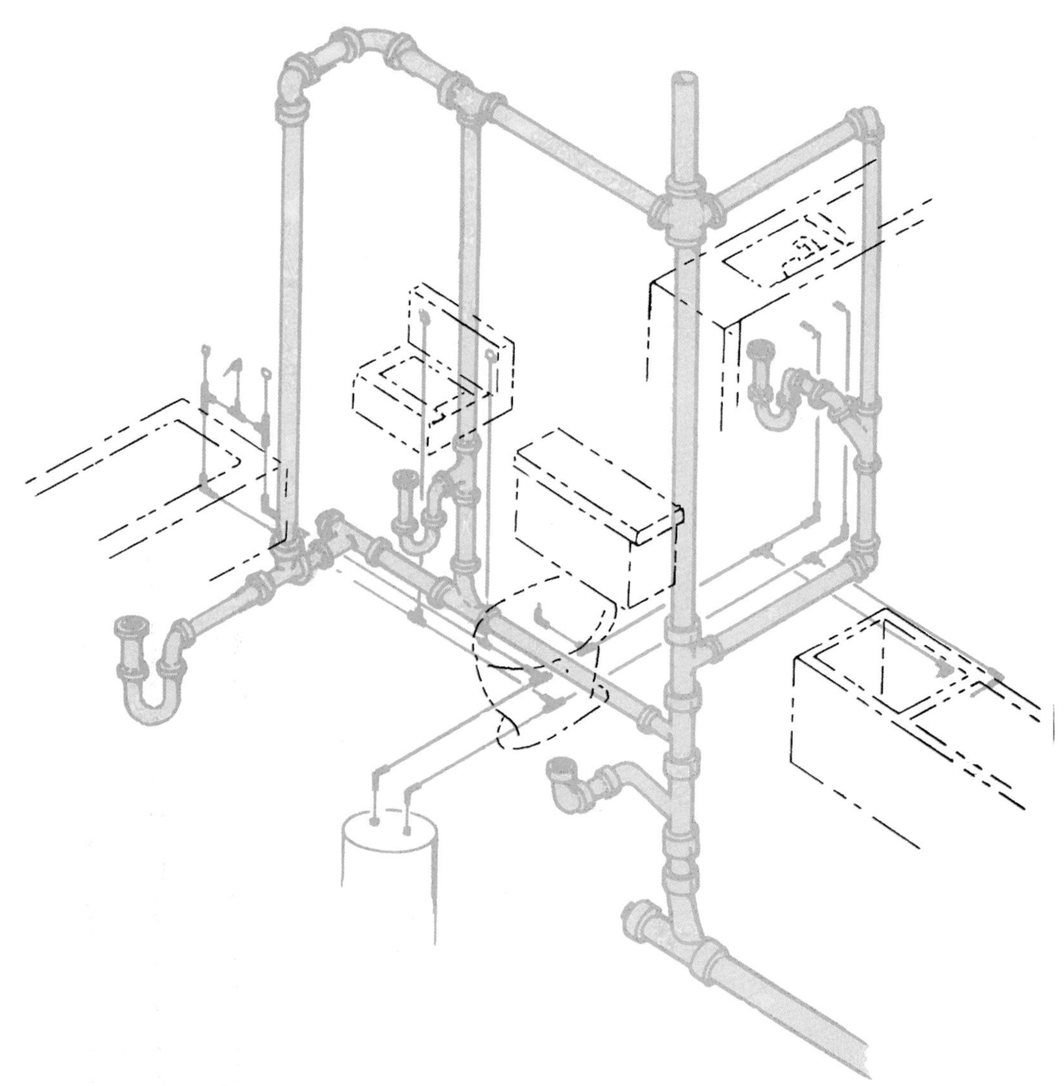

Figure 27-1. Pictorial view of a plumbing system. Large pipes are parts of the drainage, waste and venting (DWV) system. Upper portions of the DWV comprise the venting system. Small pipes are the water supply system.

At the same time the work should be done neatly. Holes should be smooth and cleanly made, only large enough to receive the pipe. Notches should be made square or rectangular. Once plumbing is installed, notched framing members should be reinforced with metal strapping. In certain situations, a joist may need to be partially removed. In such cases, a header should be installed to reinforce the floor at that point. Figure 27-2 illustrates how the plumber should handle these situations. Note the ease with which plumbing runs are made through truss joists.

Like installation of the electrical system, installation of the plumbing system is supervised by codes. It is important for plumbers to comply with this code for the following reasons:

- In communities covered by a plumbing code, an improperly installed system will not pass inspection and would have to be torn out and redone. This remedy is costly.
- A poorly designed system will not perform well. The owner will be dissatisfied and may insist on changes.
- Even worse, a poorly designed system could be unhealthy because it could allow wastewater to contaminate potable water.

Regulations for installing plumbing systems are covered by several model codes. Each is sponsored by a national, professional organization:

- The Uniform Plumbing Code (UPC) — International Association of Plumbing and Mechanical Officials.

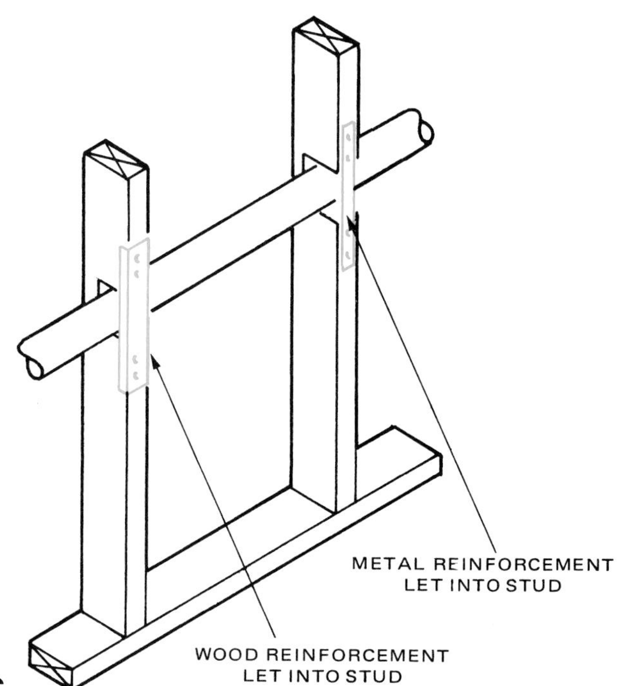

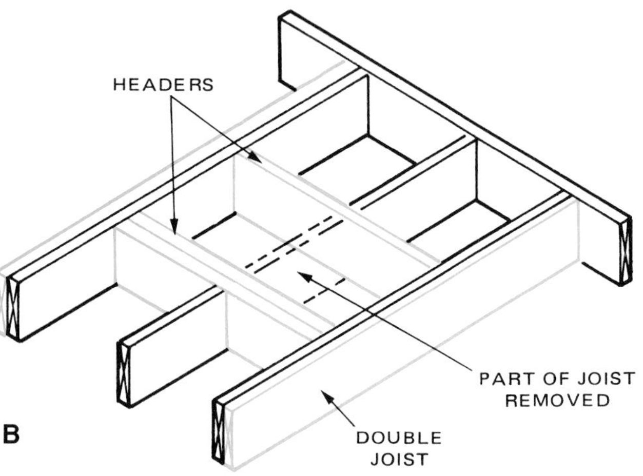

Figure 27-2. A plumber must be careful when altering framing to run plumbing pipes through joists and studs. A—Approved method of cutting holes in wood I-beams. B—Approved method of altering joists to get large openings for plumbing. Joists have been doubled and headers installed. C—Where studs must be notched, wood or metal reinforcing must be installed across the notch. D—When truss joists are used, no cutting is necessary. All mechanical systems can be run through the voids in the truss.

- Basic Building Code — Building Officials and Code Administrators International, Inc. (BOCA).
- ICBO Plumbing Code — International Conference of Building Officials.
- National Plumbing Code — American Standards Association.
- Standard Plumbing Code — Southern Building Code Congress International, Inc. (SBCCI).

Recently, BOCA has been working with ICBO and SBCCI to develop a model code acceptable to all three organizations.

Local codes usually adopt all or parts of the UPC or other plumbing codes. Once adopted by a municipality or other governmental entities, codes can be enforced as law. Since codes may vary in their requirements from one community to another, a plumber must always be familiar with the local code.

Plumbing requires a variety of skills. The plumber must use many different kinds of tools and equipment. Further, he or she must be skilled in woodworking, metalworking, pipe threading, welding, soldering, brazing, and caulking.

Plumbers need to make accurate measurements and calculations. They must add and subtract dimensions, figure pipe offsets, and determine the volume of tanks. As a precaution against costly errors, the plumber will check each measurement and each calculation twice before marking or cutting.

Two Separate Systems

Plumbing includes two subsystems that have important differences.

- The *water supply system* distributes water under pressure throughout the structure for drinking, cooking, bathing, and laundry. This is a two pipe system. One pipe carries cold water, and the other hot water.
- Drainage piping, commonly referred to as *drainage, waste, and venting*, or simply DWV, carries away wastewater and solid waste from bathrooms, kitchens, and laundries. This subsystem is not under pressure.

While both subsystems are watertight to prevent leakage, it is doubly important for the supply subsystem because it is under pressure and the smallest pinhole or fault in a connection will cause leakage.

It is also extremely important that the DWV be properly vented. Venting is that part of the drainage piping that permits air to circulate in the pipes. This prevents back pressure and siphoning of water from traps. Under certain conditions, it also prevents introduction of wastewater into the potable (drinkable) water supply.

Tools

A plumber's tool kit carries many of the same tools used in carpentry and electrical work. To perform simple plumbing tasks will require only a few tools. One of the plumber's most used tools is the wrench, Figure 27-3. Some other tools designed to perform tasks common only to plumbing are shown in Figure 27-4.

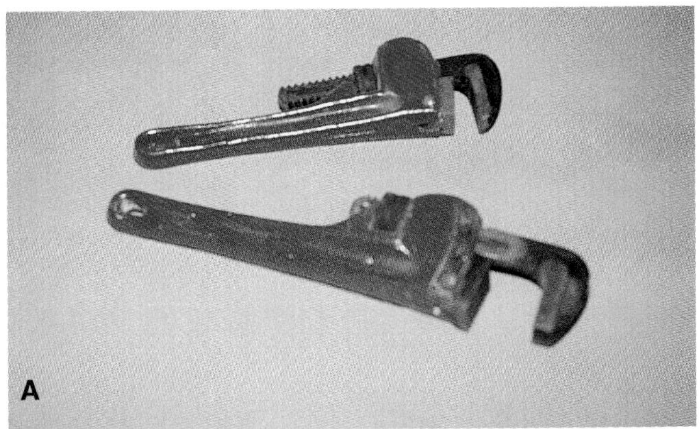

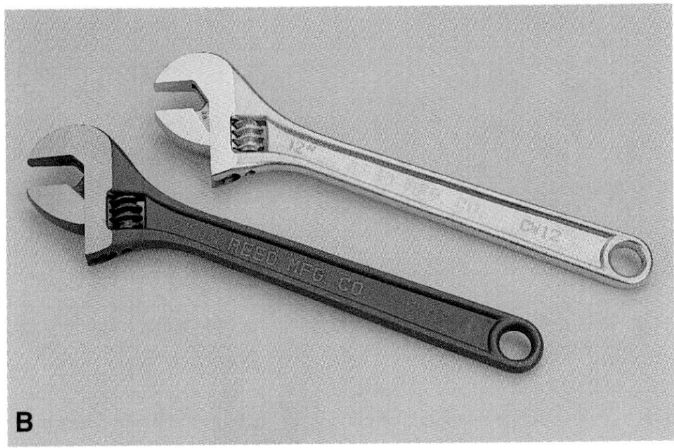

Figure 27-3. Wrenches are the plumber's most-used tool. A—Pipe wrench. B—Adjustable wrench. (Reed Manufacturing Co.)

Among the more important are measuring tools. These are needed to produce accurate lines, circles, and any other markings. Plumbing measurements must be accurate to fractions of an inch and the tools must be capable of this accuracy. Measuring tools include rules, tapes, squares, levels, transits, plumb bobs, chalk lines, compasses and dividers.

Alignment tools are also necessary for determining when pipes are level and plumb (exactly vertical). A level is needed to check both conditions. A good all-purpose tool, the level has at least three vials. A level is also used to determine the proper slope of a drainpipe. The pipe must slope neither too little nor too much.

A plumb bob transfers vertically locations of a pipe between floors. To measure accurately, the string on which the bob is suspended must come out of the center of the bob.

Marking tools such as pencils, chalk line, compasses, and dividers should also be part of the plumber's tool box.

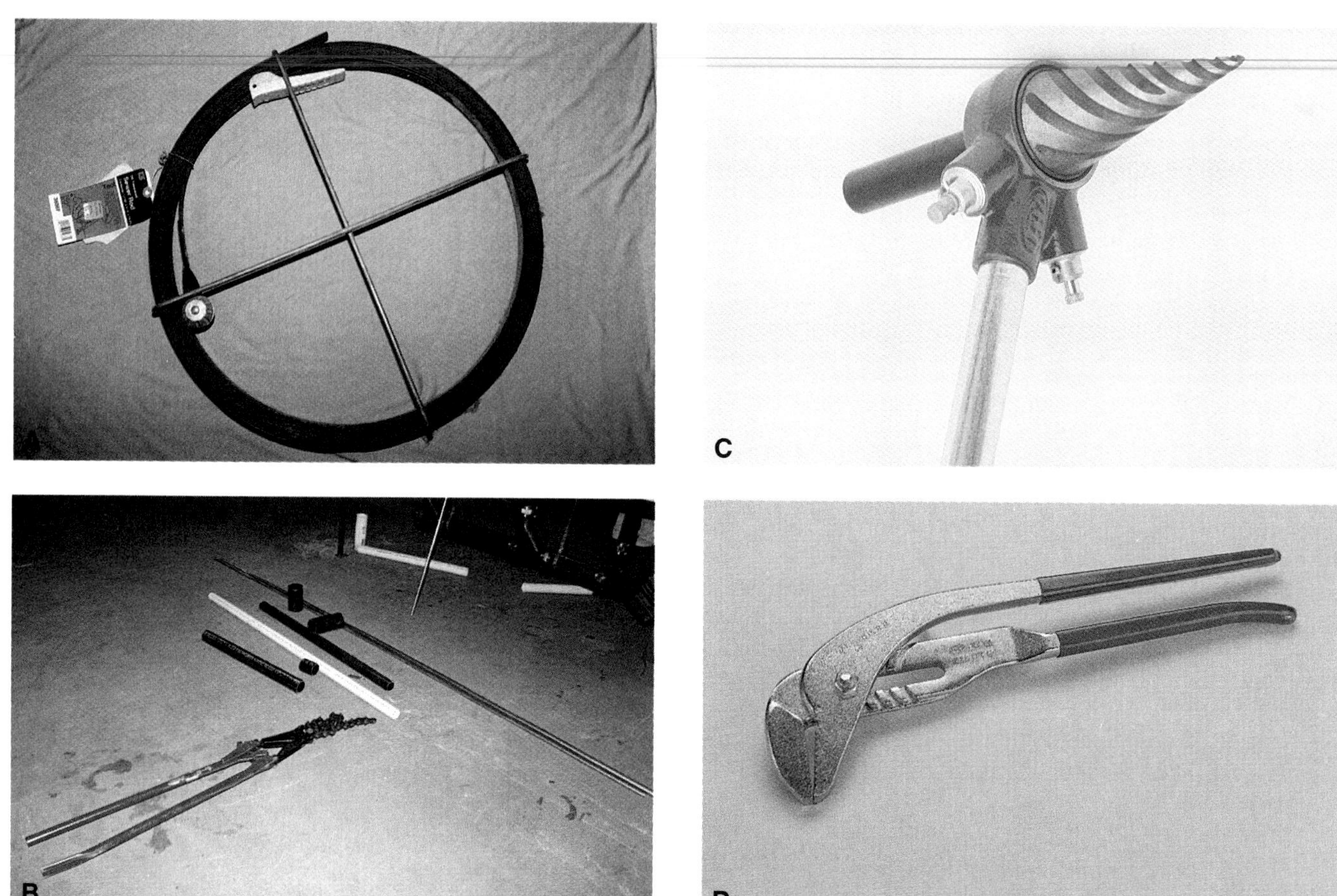

Figure 27-4. Certain tools are used only for plumbing. A—Snake is used to clear clogs from drains. B—A chain cutter is used for cutting large-diameter soil pipe. C—A reamer is used to remove burrs from any size pipe. D—Pipe wrench pliers. (Reed Manufacturing Co.)

Cutting tools needed include various saws, files, chisels, snips, and pipe cutters, Figure 27-5. Boring tools are used for making holes in wood framing members. Power tools save time and are less tiring than hand tools. Some cutting and boring tools should be metal-cutting since steel is often found today, even in residential construction.

Reaming and threading tools are used for work on metal pipe. Reaming the end of a pipe removes the burrs inside caused by cutting. Dies are used for cutting threads on galvanized pipe.

Various types of wrenches are designed for holding and turning pipe and pipe fittings. Pipe wrenches, pliers, vice grips, chain wrenches, strap wrenches, monkey wrenches, and adjustable wrenches should be included in the tool box.

Materials

Several different types of material are used for pipes and pipe fittings. Depending on codes, supply system pipe may be made of copper, galvanized steel, or plastic. DWV pipe may be of malleable iron, cast iron, copper, galvanized steel, or plastic.

Copper is available as rigid lengths or flexible coils, Figure 27-6. It can be cut with a pipe cutter or hacksaw. Copper tubing is joined with compression fittings that slide over the tubing. Another method is to *sweat solder* the fittings. Only lead-free solder is permitted, avoiding the hazard of lead getting into the water.

Plastic pipe is available in lengths that can be cut with a handsaw. Fittings are joined with a special adhesive. It is used for both supply and drainage systems. Plastic supply piping is made in two types:

- *Chlorinated polyvinyl chloride (CPVC)* is a buff-colored thermoplastic. It is light, easy to handle, and is resistant to cracking in case of freezing conditions. Connections may be threaded or solvent welded. It is suitable for piping hot water and has a rating of 180°F and 100 psi. See Figure 27-7.
- *Polybutylene (PB)* is flexible which allows easier installation. Connections are made mechanically with compression fittings.

Plastic piping has become popular for drainage systems, as well, Figure 27-7. It is used in most new construction. There are two types:

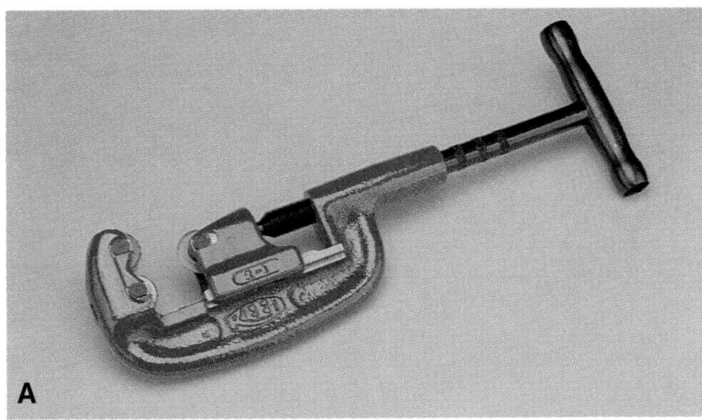

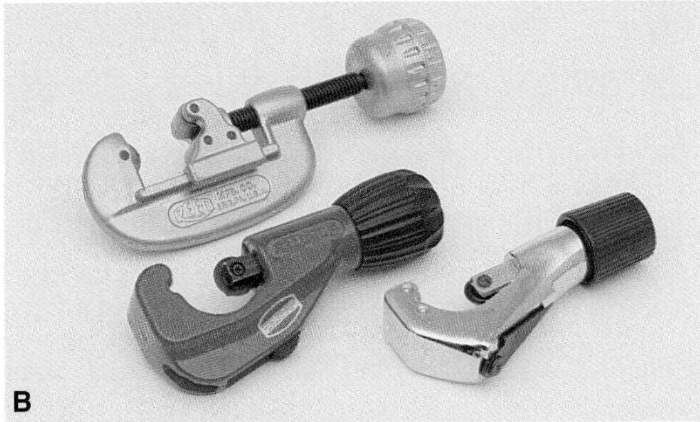

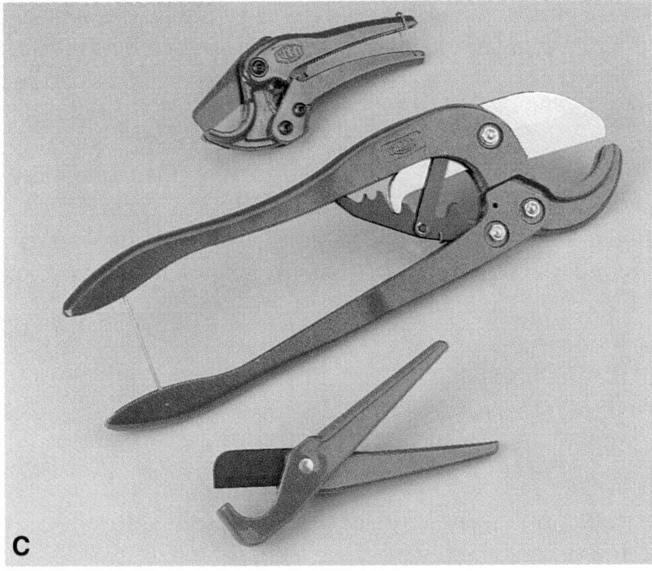

Figure 27-5. Pipe cutters. A—Standard steel pipe cutter. B—Tubing cutters. C—Plastic pipe shears. (Reed Manufacturing Co.)

- *Acrylonitrile butadiene styrene (ABS)* resists chemical attack and is inexpensive. It is usually joined with a one-step solvent.
- *Polyvinyl chloride (PVC)* has lower thermal expansion that makes long runs easier to control. It is joined with a two-step primer/solvent and suitable fittings.

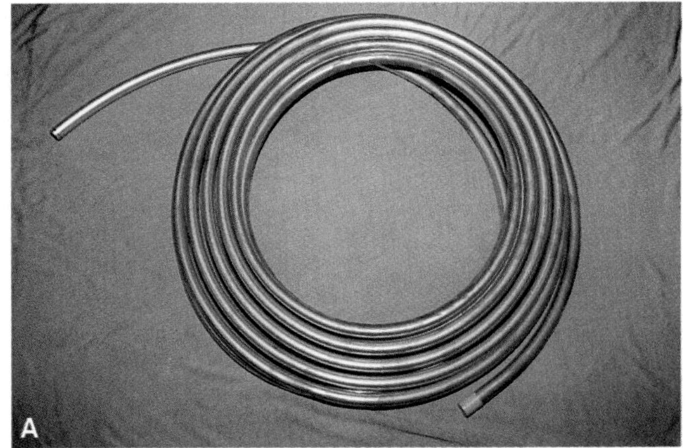

Figure 27-6. Typical copper materials. A—Copper tubing. B—Forged copper fittings. These must be sweat soldered to connect pipe.

Soft cast iron was once used extensively in plumbing and still is to some extent. Both pipe and fittings are cast from gray iron. Its strength and corrosion resistance stems from the formation of large graphite flakes during casting. Cast iron pipe will not leak or absorb water. It is often chosen because it is a quiet drainage system.

The Cast Iron Soil Pipe Institute specifies two grades of soil pipe: Service (SV) for above-grade use, and Extra Heavy (XH) for below-grade use.

Joining methods depend on which of two styles of cast iron soil pipe is used. The difference is in the shape of the ends. Both styles are shown in Figure 27-8.

There are two methods of sealing joints in hub and spigot soil pipe. In the first, molten lead is poured into the *hub* (bell-shaped end of the joint) which has been sealed with an oakum packing. *Oakum* is a rope-like material made up of tar-soaked fibers. The solidified lead is then tamped to form a seal against the bell. In the second method, the joint formed by a hub and spigot is sealed with a neoprene gasket. Joints of the second type of soil pipe, no-hub, are also sealed with neoprene held in place by a stainless steel clamp. Refer to Figure 27-9, which shows several methods of sealing joints.

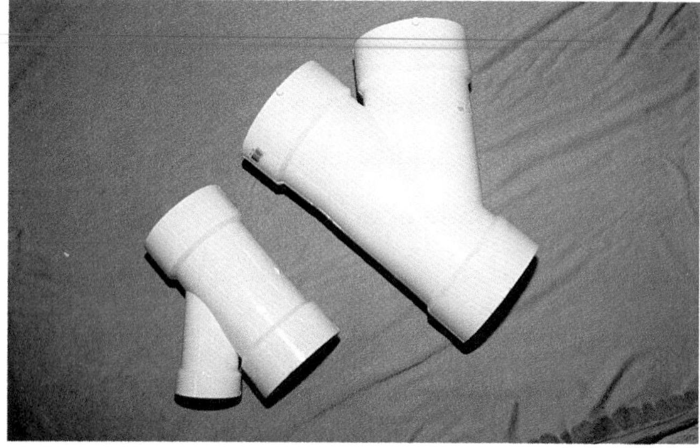

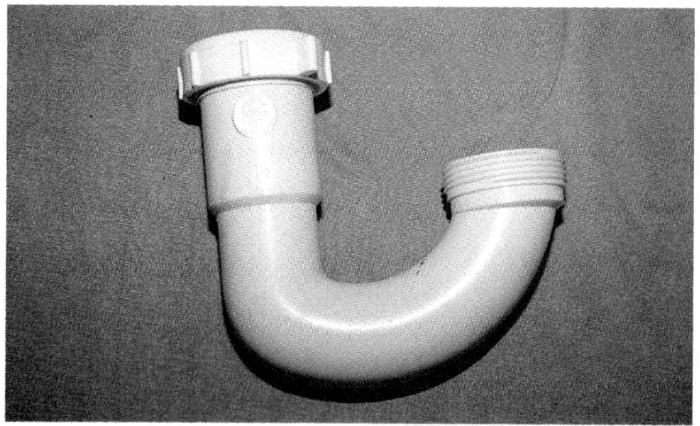

Figure 27-7. Plastic fittings like these are used in drainage systems. Top—Wye. Middle—Closet flange. Bottom—P-trap.

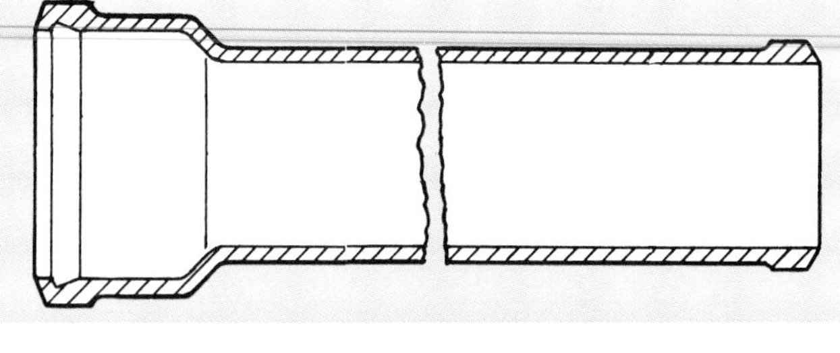

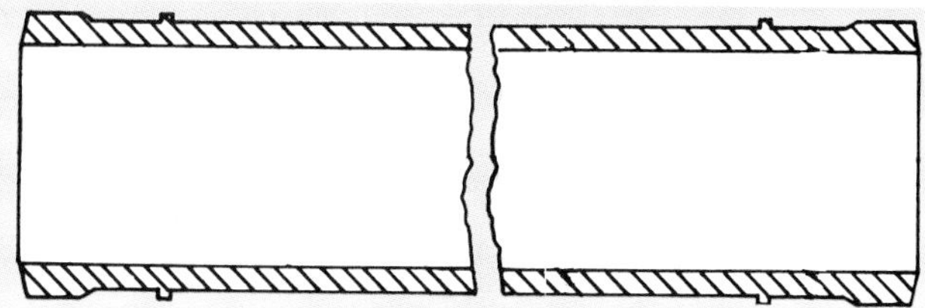

Figure 27-8. Two types of cast iron soil pipe. Top—Hub and spigot type. Bottom—No-hub type has no bell and must be connected with neoprene seals and stainless steel clamps.

Steel pipe used in plumbing is galvanized with a coating of zinc to retard rusting. It is manufactured in standard 21' lengths in diameters from 1/8" up to 2 1/2". The ends are given a tapered thread so that the joint will seal. Fittings are made of malleable iron to make them more resistant to stress. A putty-like compound or teflon tape is applied to threads before threaded joints are made. This helps seal the joints.

Valves are devices that control the flow of water in the water supply system. They are installed at certain places in the lines so that water can be shut off or its pressure reduced. They are made from a variety of materials including bronze, brass, malleable iron, cast iron, thermoset plastic, and thermoplastic. Threads may be on the inside or the outside.

A typical compression faucet and a cutaway are shown in Figure 27-10. Teflon tape is normally used to seal threads of plastic fittings, Figure 27-11. It is also sometimes used on galvanized threaded joints.

Faucets are valves that permit controlled amounts of water as needed for use in a building. They usually deliver the water to fixtures such as sinks, lavatories, showers, and bathtubs but may deliver it to an appliance, hose, or bucket. See Figure 27-12.

Fixtures

Fixtures, shown in Figure 27-13, are water-using devices, such as sinks, lavatories, bathtubs, bidets, urinals, stools, or showers. They are attached to plumbing systems. They receive water from the supply system and have a means for delivering wastewater to the DWV system.

Certain water-using appliances require attachment to plumbing systems. These include clothes washers, dish washers, and automatic icemakers in refrigerators. See Figure 27-14.

Printreading

To save time, certain drawing symbols have been standardized for the plumbing trade. These symbols are a type of shorthand. Figures 27-15 and 27-16 show generally accepted symbols for fixtures, appliances, mechanical equipment, pipes, and fittings.

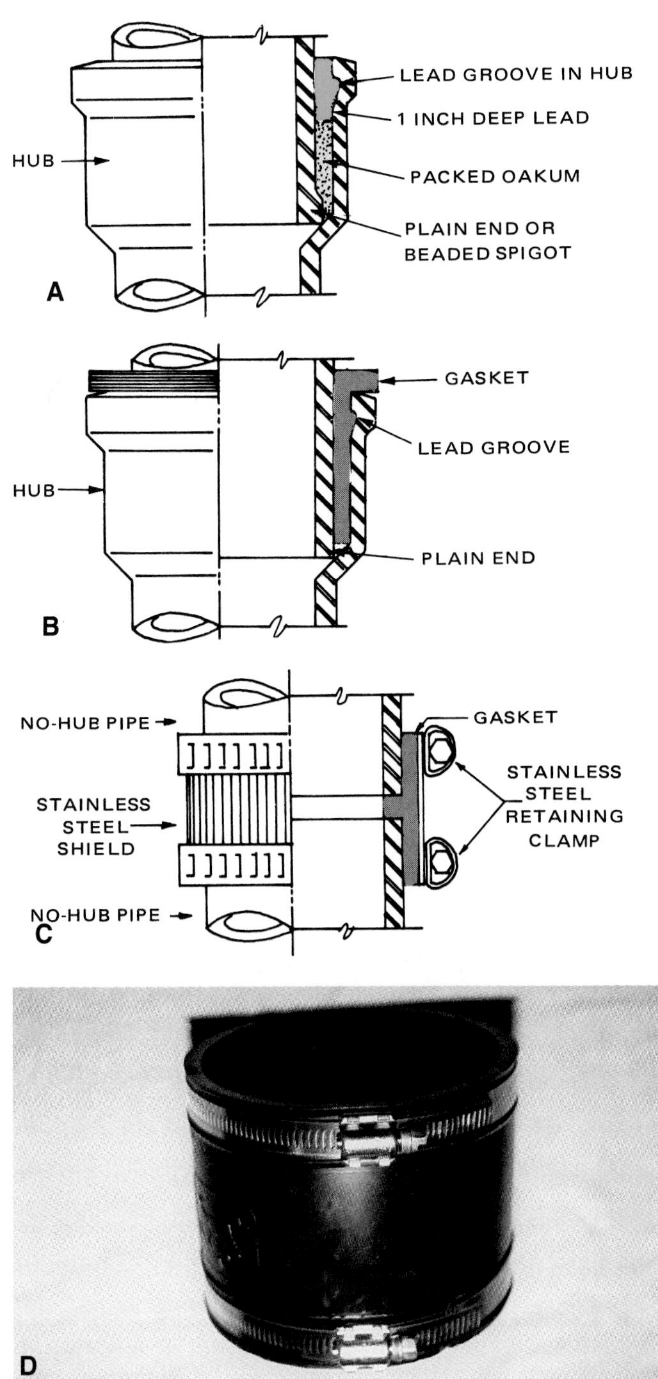

Figure 27-9. Methods of joining cast iron soil pipe. A—Lead and oakum are packed into the joint. B—Hub and spigot soil pipe sealed with a neoprene gasket. C—No-hub soil pipe joint sealed with a neoprene gasket and secured by a clamp. (E.I. duPont de Nemours & Co.) D—Photo of the fitting shown in C.

Since architectural drawings do not include pipe drawings, the plumber needs to develop some sketches as a guide to installation. This is especially needed if others are working on the same plumbing job. There are three types of drawings or sketches that will be made: riser diagrams, plan view sketches, and isometric sketches. These are shown in Figure 27-17.

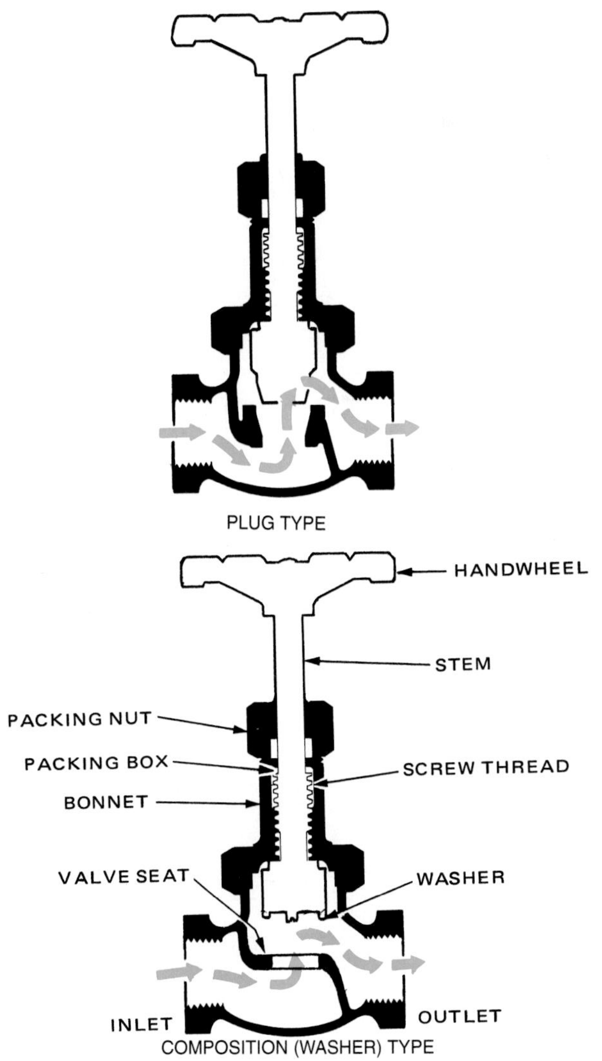

Figure 27-10. Typical valve is designed to be connected between two pipes to interrupt and control water flow. Cutaway shows two types of compression or globe valves. (William Powell Co.)

Figure 27-11. Teflon tape is used to seal threads.

Figure 27-13. Fixtures are water-using devices found in kitchens, bathrooms, and laundry areas. This whirlpool bathtub is just one of many types of fixtures. (Sterling)

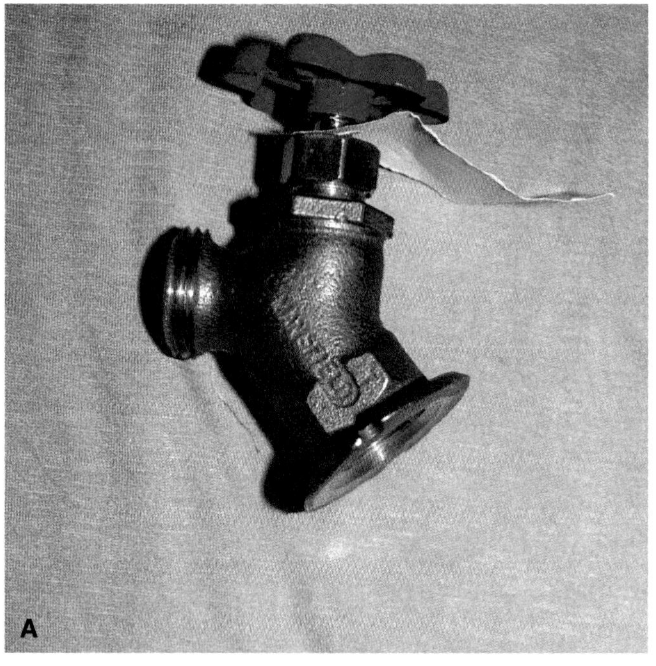

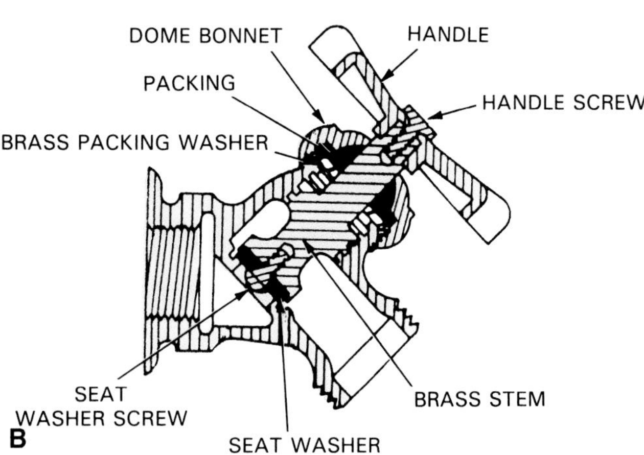

Figure 27-12. Faucets control water at a fixture. A—A hose bib (faucet) is used for outside water supply and is threaded to receive a hose. Flange allows it to fit flush to an outside wall. B—Cutaway of a hose bib. (Kunkle Valve Co., Inc.) C—A modern single control lavatory faucet. (American Standard) D—Single handle kitchen faucet with pull-out spout and integral vacuum breaker.

Figure 27-14. A dishwasher is an appliance that needs plumbing connections. (Whirlpool Corp.)

Installing Plumbing

Installation of plumbing is usually done by a subcontractor, although skilled homeowners may also do their own installation. However, it is not advisable to attempt installation without some professional training. Installation can begin as soon as the building is closed in. Mesurements are made to show exact location of pipes and fixtures. Then adjustments are made to the framing which must be cut or drilled to receive pipe or devices. Use a steel tape or plumber's rule to take careful measurements. CAUTION: Check all measurements carefully before making any cuts. It is a good idea to take the measurements twice and cut only once.

Cutters or saws are used to cut lengths of pipe or tubing. Be certain to ream pipes to remove interior ridges or ragged edges caused by the sawing or cutting.

Connecting Pipe

Sweat soldering copper plumbing is relatively easy but must be done carefully to avoid leaking joints. You will need a gas torch, lead-free solder, paste soldering flux, and a small pad of steel wool or a tube-cleaning brush.

Soldering Copper Pipe

1. Ream the cut end of the tubing to remove metal burrs. Clean the end of the copper tubing and the inside of the fitting with steel wool or a cleaning brush.
2. Using a brush, apply paste flux to the polished end of the tubing and to the inside of the fitting.
3. Slip the fitting over the tubing and heat the joint. It is best to heat the tubing first. Allow the inner cone of the flame to touch the metal. Heat the joint until the flux is smoking. Add solid core solder by touching the end of the solder wire to the joint area at the edge of the fitting. Feed the solder into the joint as the flame is moved to the center of the fitting. This movement will draw the solder evenly into the fitting, making a leakproof joint. Never try to melt the solder in the flame.
4. Once the solder is drawn into the joint, remove the heat.
5. Wipe excess solder off the joint with a piece of burlap or denim.

Compression or *flare fittings* require no soldering. The fittings are designed to fit tightly against the tubing and mating parts.

Making Compression Joints

1. Slide the flare nut over the tubing, tapered end facing away from the end of the tubing.
2. Select the correct size hole of the flaring tool and slip the tube into it. (The hole should be slightly smaller than the outside diameter of the tubing.)
3. With the tubing sticking out 1/32" to 1/16" above the bar of the tool, tighten the wing nut so the tubing is held tightly in the bar.
4. Slide the ram over the tube and screw the ram into the end of the tubing. Tighten the ram until a 45° angle is made in the tube.
5. Press the tapered end of the other part of the fitting into the flared end of the tubing.
6. Slide the nut to the tapered fitting and hand tighten.
7. Using two tubing wrenches or open end wrenches, tighten the assembly until a solid feel is apparent.
8. Test fitting under pressure of the water supply system.

Connections for plastic plumbing pipe are made using an adhesive that forms a permanent bond between the plastic parts.

Making Plastic Pipe Connections

1. Clean and ream the pipe end.
2. Apply the adhesive with a dauber (applicator usually an integral part of the cap) to outside of pipe and the inside of the fitting.
3. Quickly make the connection and make a quarter turn to spread the adhesive evenly.
4. Align the parts to the desired angle. The adhesive will bond within seconds; speed is essential or the joint will be spoiled.

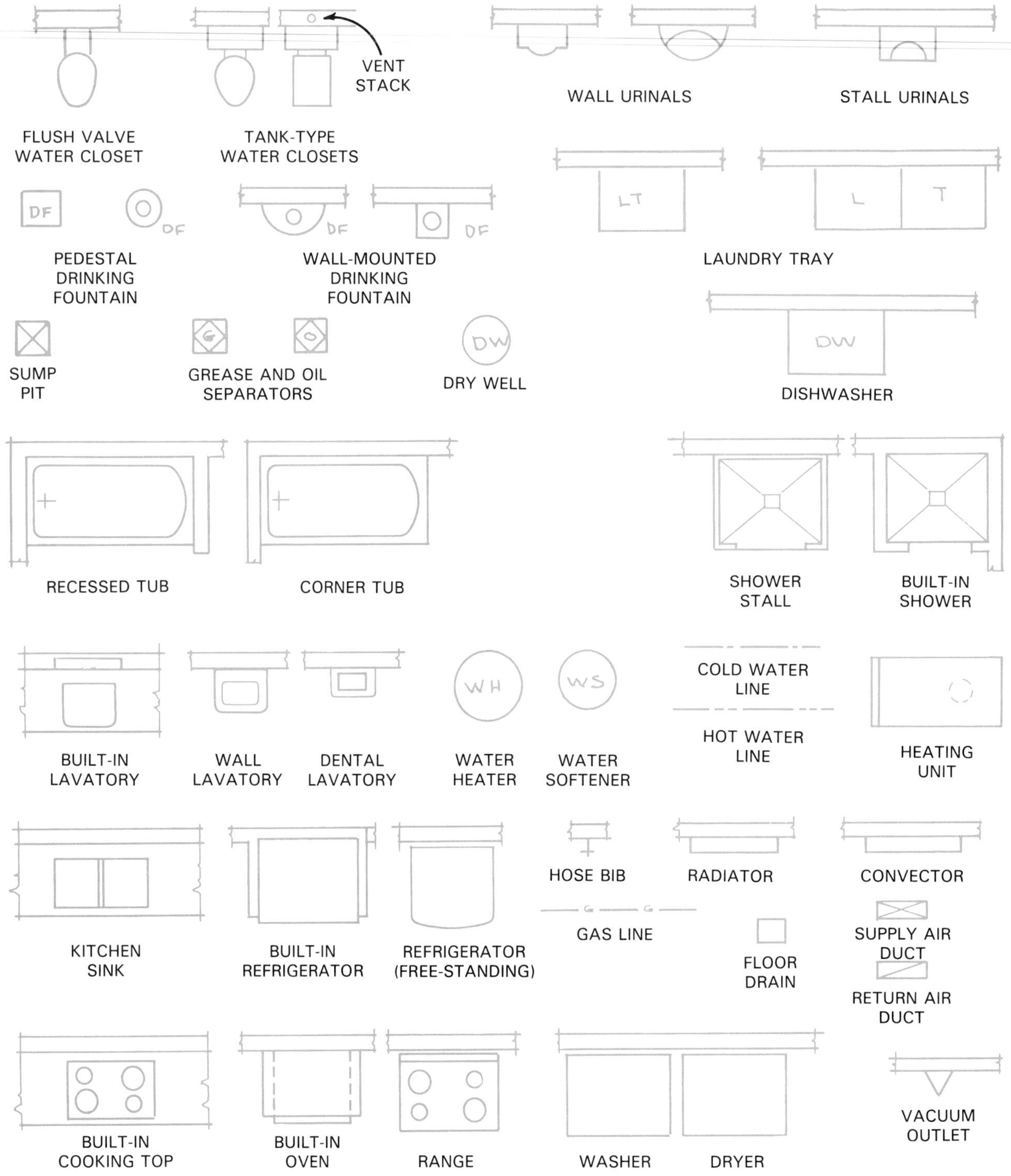

VENT STACK
WALL URINALS
STALL URINALS
FLUSH VALVE WATER CLOSET
TANK-TYPE WATER CLOSETS
DF
DF
DF
DF
LT
L
T
PEDESTAL DRINKING FOUNTAIN
WALL-MOUNTED DRINKING FOUNTAIN
LAUNDRY TRAY
DW
DW
SUMP PIT
GREASE AND OIL SEPARATORS
DRY WELL
DISHWASHER
RECESSED TUB
CORNER TUB
SHOWER STALL
BUILT-IN SHOWER
WH
WS
COLD WATER LINE
HOT WATER LINE
BUILT-IN LAVATORY
WALL LAVATORY
DENTAL LAVATORY
WATER HEATER
WATER SOFTENER
HEATING UNIT
HOSE BIB
RADIATOR
CONVECTOR
G
G
GAS LINE
SUPPLY AIR DUCT
RETURN AIR DUCT
FLOOR DRAIN
KITCHEN SINK
BUILT-IN REFRIGERATOR
REFRIGERATOR (FREE-STANDING)
BUILT-IN COOKING TOP
BUILT-IN OVEN
RANGE
WASHER
DRYER
VACUUM OUTLET

Figure 27-15. These symbols are used for plumbing fixtures, appliances, and mechanical equipment.

Galvanized pipe connections are probably the easiest of all to make.

Making Galvanized Pipe Connections

1. Check lengths to ensure accuracy. A "dry" assembly can be made to determine this.
2. Make sure threads are clean. Use a wire brush to remove any bits of metal or other debris.
3. Apply pipe compound or teflon tape. This will ensure leak proof joints.
4. Screw on the fittings finger tight.
5. Final tightening should be made with pipe wrenches. In long runs, tightening of all joints can be made by working on the last joint in the run. As each joint is tightened, the force is transferred to the next joint until all are tightened.

FITTING OR VALVE	TYPE OF CONNECTION		
	SCREWED	BELL AND SPIGOT	SOLDERED OR CEMENTED
ELBOW—90 DEGREES			
ELBOW—45 DEGREES			
ELBOW—TURNED UP			
ELBOW—TURNED DOWN			
ELBOW—LONG RADIUS			
ELBOW WITH SIDE INLET—OUTLET DOWN			
ELBOW WITH SIDE INLET—OUTLET UP			
REDUCING ELBOW			
SANITARY T			
T			
T—OUTLET UP			

FITTING OR VALVE	TYPE OF CONNECTION		
	SCREWED	BELL AND SPIGOT	SOLDERED OR CEMENTED
T—OUTLET DOWN			
CROSS			
REDUCER—CONCENTRIC			
REDUCER—OFFSET			
CONNECTOR			
Y OR WYE			
VALVE—GATE			
VALVE—GLOBE			
UNION			
BUSHING			
INCREASER			

Figure 27-16. Pipe and fitting symbols used by plumbers and which appear in prints.

An advantage of threaded joints is that they can be taken apart without destroying parts.

Bending and Unrolling Copper Tubing

Copper tubing, being flexible, can be easily bent to change its direction during an installation. To avoid crimping, which may lead to a leak, slide a tubing spring over the section to be bent. Bend the tube and spring slightly beyond the desired angle, then back to the correct angle. See Figure 27-18. Remove the spring by turning the flared end counterclockwise.

Copper tubing comes in coils. Be careful when uncoiling. A procedure that works well is to place a foot lightly on the end while holding the coil upright. Slowly roll the coil away from the secured end. Move the holding foot along until the proper length is unrolled.

Other Considerations

As noted in the opening of this unit, plumbing, like electrical wiring, must be concealed within the building frame. The single exception to this, is that plumbing runs are usually run below first-floor joists rather than through them. However, use of engineered lumber—specifically truss joists—allows mechanical systems to be run through the open spaces of the truss braces. Refer to Figure 27-19.

Large fixtures, such as bathtubs and showers, are too large to fit through doorways of framed-up partitions. They must be moved in prior to installation of studs for partitions, Figure 27-20.

The first phase of plumbing installed in a building is known as a *rough-in*. The endings of pipes that extend out from the walls and above the floors are known as *stub-ins*. All runs are placed and tested with pressurized air or water to ensure that no leaks exist. The venting pipes are connected to a large riser that extends through the roof. This pipe is known as a *vent stack*. It serves as an exhaust for any odors from the waste system and also allows the system to take in air and thus avoid siphoning problems with traps. Photographs in Figure 27-21 shows various stub-ins where fixtures will eventually be connected.

Wells and Pumps

Many buildings beyond the lines of municipal sewer and water systems have their own wells. Wells require pumps and supply tanks to maintain a constant water supply.

Wells are of different types, as shown in Figure 27-22. With modern technology, most wells today are either drilled or bored.

Drilled wells are used where great depths are needed to reach a suitable water table. There are two drilling methods. The percussion method uses a chisel-shaped bit. The bit is raised and lowered by a cable. This creates a pounding action that breaks up the subsurface material. The rotary method requires a tall mast and more complicated drilling equipment along with water. The water is the cutting medium. The slurry created is pumped to the surface and discarded. This method is used when the water table is far below the surface.

Bored wells are made with an earth auger. The auger cuts and lifts subsurface material in much the same way as a wood bit cuts and lifts wood particles.

The diameter of the well is large enough to accept a casing. The *casing* is a heavy metal tube that will keep contaminants out of the well water. Galvanized or plastic pipe is installed into the well and a pump is connected to this piping. This pump may be installed in the well (submersible pump), above ground near the well, or in the building.

Also located near the well or the pump is a supply tank. Its function is to maintain an adequate supply of water to meet the continuing needs of the buildings occupants. The tank's water is under constant pressure so that it

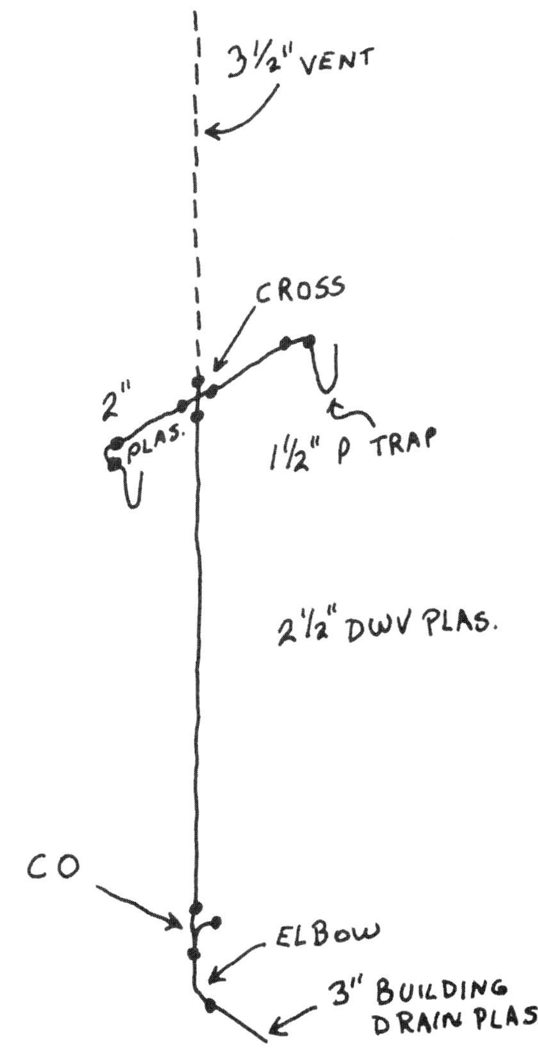

Figure 27-17. Three types of sketches. A—Riser diagram. B—Plan view. C—Isometric sketch.

can be delivered anywhere in the building at sufficient force to operate showers and water-using appliances. See Figure 27-23.

After extended service, a well water system requires maintenance and possibly repairs. Seals harden and leak; check valves wear out; pumps wear and lose efficiency; pipes rust and develop leaks. Problems within the well, the pump, or the pressure tank are usually the concern of the well driller. Problems with the supply pipes are referred to a plumber. Figure 27-24 shows a well driller and his assistant replacing seals, check valves, and leaking pipe.

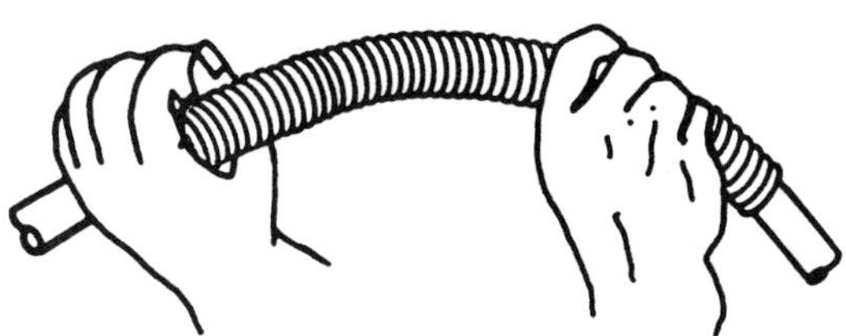

Figure 27-18. Bending tubing using a tube spring to avoid kinking the tubing.

Unclogging Drains

Special tools are used by the homeowner or plumber to clear clogged drains. For slow drains, a chemical cleaner may be employed. Plungers, snakes, and closet augers are used for more serious clogs. Plungers apply pressure to the clog, while snakes and augers reach and dislodge or retrieve the matter causing the obstruction. Figure 27-25 shows a plumber unclogging a toilet.

Leaking or malfunctioning toilet tanks or stools are common service problems, Figure 27-26. Service may involve replacing the ball cock assembly or a flush valve, adjusting the float level or renewing the linkage between the flush lever and the rod controlling the flush ball. A toilet leaking at its base must be removed to have the wax or rubber seal replaced.

Figure 27-19. Plumbing pipes can be run below joists where they are located in a basement or crawl space, as shown in A and B. C—Plumbing located in a 6" steel partition. Note closet flange ready to receive a toilet. D—Running plumbing through truss joists.

Figure 27-20. Large fixtures should be moved in before interior partitions are installed. A—Bathtub awaits installation. B—Two views of a tub and shower with surround. This unit was moved in before steel studs were erected.

Another common repair problem is the replacement of leaking P-traps, Figure 27-27. The slip nuts are unscrewed, releasing the trap. A new trap is slipped in place and secured with new slip washers and, sometimes, new slip nuts.

Safety

Accidents are usually the result of unsafe acts or failure to wear protective clothing or safety equipment. Eye protection, such as safety goggles, are necessary when doing hazardous tasks. These tasks include sawing, drilling, chipping, spraying, sand blasting, welding, operating powder-actuated tools, and using compressed air for cleaning.

OSHA regulations also require filter lenses of the proper shade and number for soldering, brazing, and welding operations. Laser safety goggles should be worn when using laser equipment for leveling and plumbing operations.

Figure 27-21. Stub-ins are shown ready for installation of fixtures.

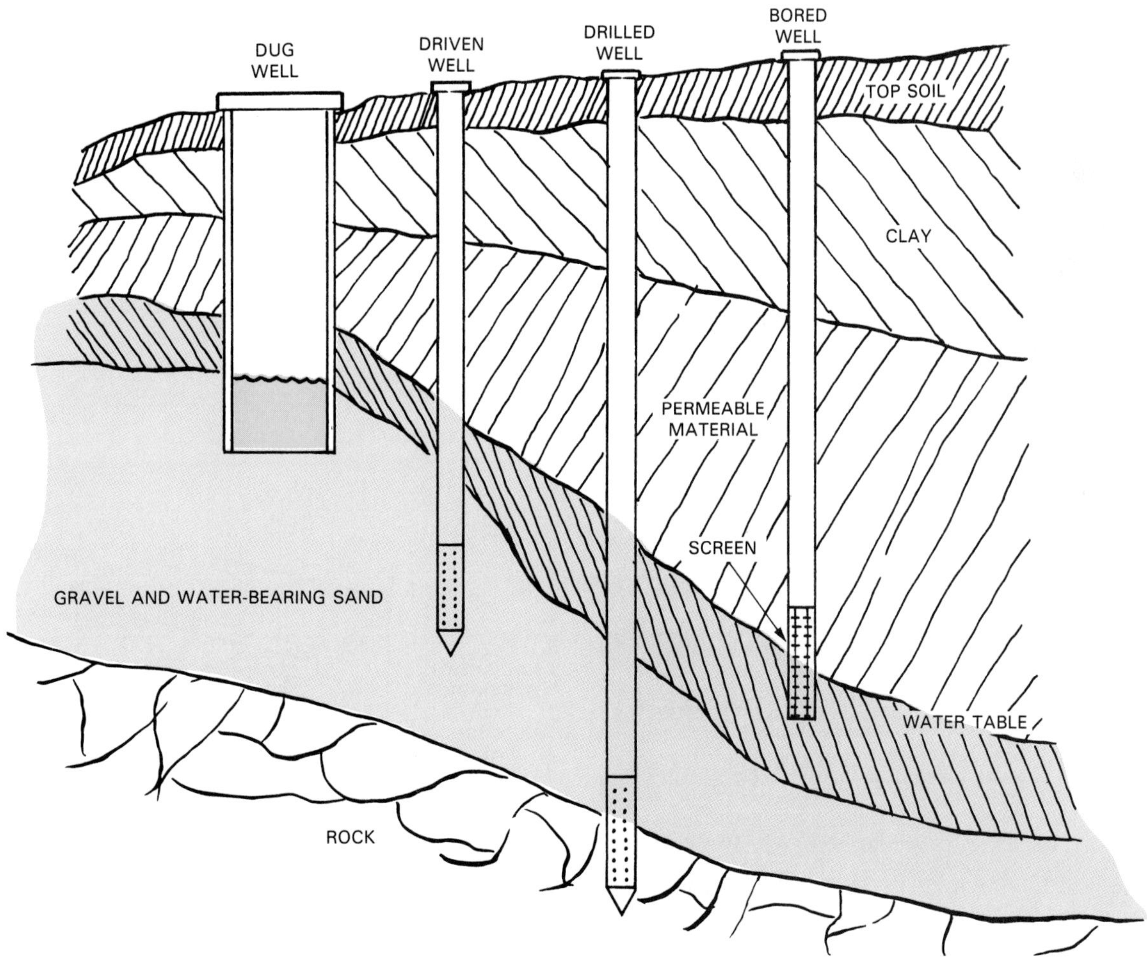

Figure 27-22. Types of wells. Most are drilled or bored.

An approved hard hat should be worn when there is danger from falling objects. Synthetic clothing should be carefully checked. Some of these fabrics are highly inflammable. Welding sparks or other high-temperature heat could set them afire. Wear gloves to protect hands when handling plumbing materials; they should NOT be worn when operating tools. Safety shoes protect the feet from falling objects as well as from puncture wounds.

Jewelry should always be removed. Rings are especially dangerous since they may catch on moving parts.

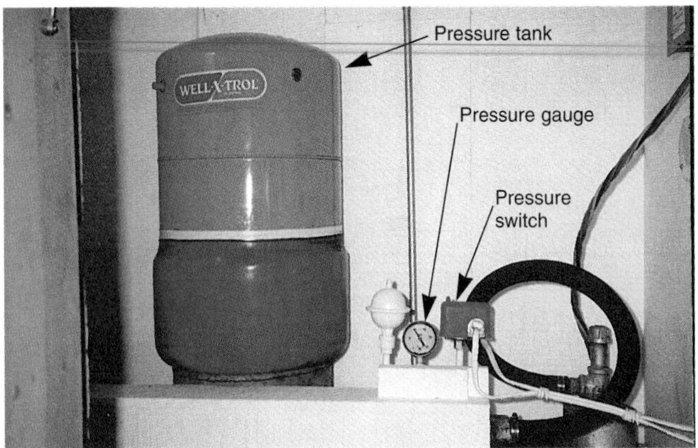

Figure 27-23. A typical setup for a water system designed for a well. A submersible pump is located in the well. Well casing is visible at lower right.

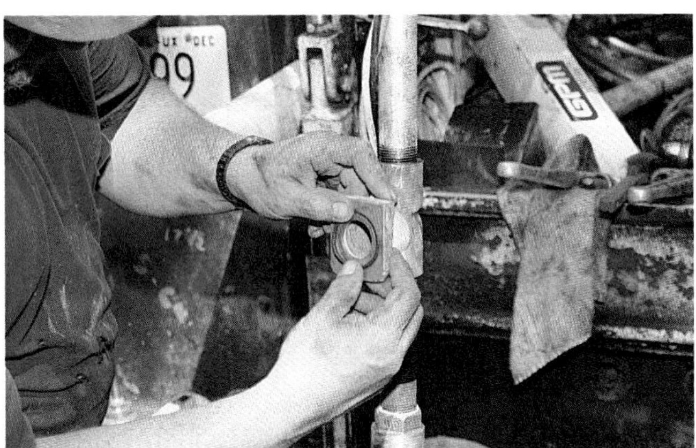

Figure 27-24. Wells require service from time to time. Top—Bad seal is being replaced. Bottom—Lengths of leaking pipe are being replaced. (Mellin Well Service)

Figure 27-25. Using a toilet auger to remove a clog from a toilet.

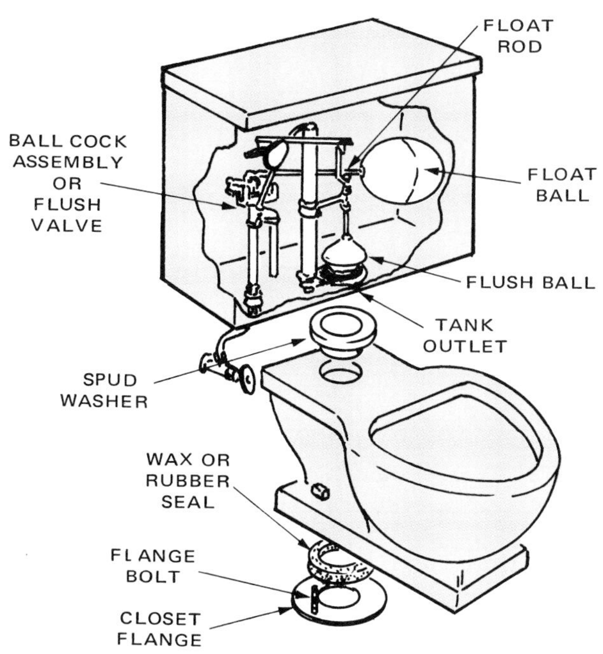

Figure 27-26. Mechanisms in a toilet frequently need adjustment or repairs.

Lifting and carrying heavy materials may cause back injuries if done improperly:

- **Make sure the area of travel is free of objects or liquids that might cause stumbling or slipping.**
- **Make certain that other workers are not near enough to be struck by long objects you might be carrying.**
- **Get help in lifting heavy or bulky objects.**
- **Use hand slings to carry large-diameter pipes.**
- **Keep the back straight when lifting heavy objects; let the legs do the bending and straightening.**
- **Avoid touching electrical wires and devices with metal objects.**
- **Use only grounded electrical tools. For added safety, use an extension cord with a ground fault circuit interrupter (GFCI). A GFCI can also be installed in a distribution panel.**

Danger of fires is always present on construction sites. Combustible construction debris should be cleared away promptly. Compressed gas cylinders must be protected from damage. Highly flammable materials such as liquid fuel, certain adhesives, and solvents should never be stored in large quantities on the job site. Fire extinguishers should be kept on hand. Study the charts in Figures 27-28 and 27-29 for classes of fires and the type of fire extinguishers required for each class.

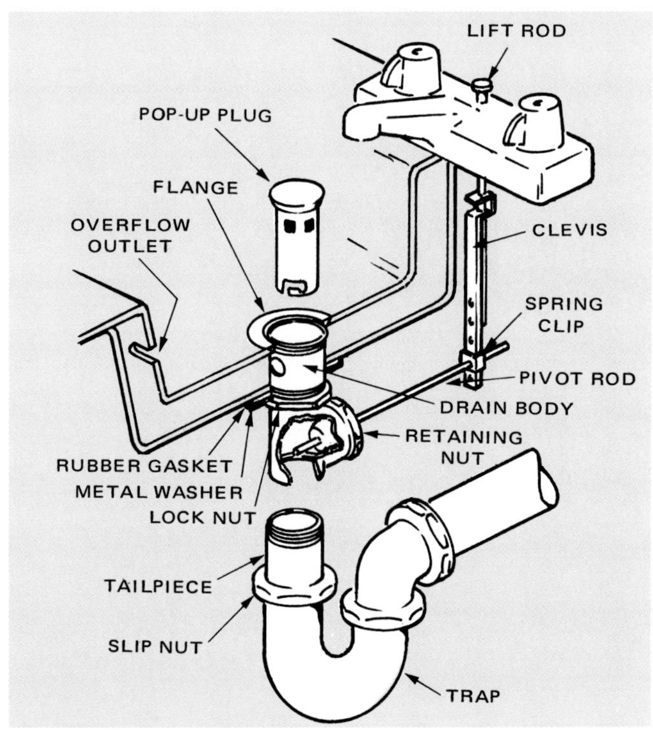

Figure 27-27. P-trap assembly. These can be replaced by loosening the slip nuts and replacing the U-shaped portion.

Class	Description
A	Fires in common combustible materials such as paper, wood, cloth, rubber, and many plastics. The cooling effects of water or solutions of water and chemicals will extinguish the fire. Also, these fires can be controlled by the coating effects of selected dry chemicals that retard combustion.
B	Fires in flammable liquids such as gasoline, grease, and oil. Preventing oxygen (air) from mixing with the vapors from the flammable liquid results in smothering the fire.
C	Fires in "hot" electrical equipment are especially dangerous because only non-conductive extinguishing agents can be safely used. Note that once the electricity is turned off, Class A or Class B extinguishers may be safe to use.
D	Fires in combustible metals such as sodium, magnesium, and titanium. These fires require the use of extinguishing agents that will not react with the burning metals. Also, the extinguishing agent needs to be heat absorbing.

Figure 27-28. Fires can be classified according to the combustible materials involved.

Ordinary A Combustibles	1. Extinguishers suitable for Class A fires should be identified by a triangle containing the letter "A." If colored, the triangle is colored green.
Flammable B Liquids	2. Extinguishers suitable for Class B fires should be identified by a square containing the letter "B." If colored, the square is colored red.
Electrical C Equipment	3. Extinguishers suitable for Class C fires should be identified by a circle containing the letter "B." If colored, the circle is colored blue.
Combustible D Metals	4. Extinguishers suitable for fires involving metals should be identified by a five-pointed star containing the letter "D." If colored, the star is colored yellow.

Figure 27-29. There are four classifications of fire extinguisher. Each class has its own distinguishing color in its symbol.

Important Terms

Casing
Compression fittings
Drainage, waste, venting (DWV)
Faucets
Fixtures
Flare fittings
Hub
Oakum
Rough-in
Stub-in
Sweat solder
Valves
Venting
Vent stack
Water supply system

Test Your Knowledge

1. Why is the cutting and notching of framing members an important skill?
2. List three reasons why it is important for plumbers to comply with the Uniform Plumbing Code.
3. The _____________ distributes pressurized water throughout the building.
4. What does DWV stand for and what does it do?
5. A level might be used to determine when pipes are _____________ and _____________.
6. What kinds of material may plumbing pipe be made of?
7. What is the full name of CPVC and what is it suited for?
8. Define a fixture and give three examples.
9. Where is sweat soldering used?
10. How are fittings attached to plastic pipe?
11. The first phase of installing plumbing in a house is called a _____________.
12. List the major components of a water system utilizing a well.
13. What materials or tools might be used to unclog a drain?
14. Explain how to replace a P-trap.

Outside Assignments

1. Make a sweat soldered joint in copper pipe.
2. Prepare a piping diagram for a bathroom having a stool, basin, and shower stall.
3. Using a blueprint supplied by your instructor, install a bathroom DWV in a mockup wall and floor. Use any type of DWV piping and appropriate fastening system.

28

Heating, Ventilation, and Air Conditioning

Heating, Ventilation, and Air Conditioning (HVAC) is an important building trade. Without it, buildings would not be comfortable places. They would be either too cold or too hot. Even prehistoric humans, with limited technology at hand, used open fires to heat their dwellings.

Industry has invested heavily in developing heating and cooling systems. Today's technology allows automatic control of systems so that our buildings are kept comfortable the entire year regardless of how hot or cool the climate might be.

Conservation Measures

Costs of fuel have been increasing greatly. Before discussing space conditioning mechanisms, it is necessary to consider how to increase their efficiency through conservation measures.

New homes are being designed to reduce the amount of energy needed to heat and cool them. Unit 14 of this text, Thermal and Sound Insulation, discusses many of these measures.

Savings in heating and cooling come about in several ways:

- Sealing of cracks at joints with caulking and housewrap cuts down loss of conditioned air through infiltration of outside air that is either too hot or too cool for comfort, Figure 28-1.
- Increasing the amount of insulation in floors, walls, and ceiling significantly reduces loss of conditioned air through radiation.
- Using an air exchanger that wrings heat out of the room air before it is exhausted to the outside reduces the heating load of the furnace.
- Modern heating and cooling appliances with high efficiency ratings get more heating or cooling from the energy they use.

It is not unusual to find new homes with R-ratings double those of 20 years ago. Windows also have much higher R-ratings. New technology in insulation makes it more efficient.

Superinsulated, panelized or modular systems-built (built in a factory) homes are now common in the home market. Panelized systems have wall and roof panels consisting of a 5 1/2" thick sheet of polystyrene sandwiched between two sheets of plywood, oriented strand board (OSB), or waferwood. Heating and cooling costs have been significantly reduced. Refer to Figure 28-2.

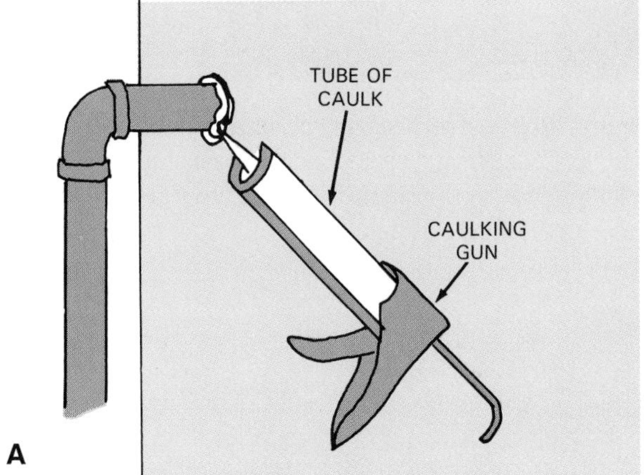

Figure 28-1. Cutting down on air infiltration makes a building easier to heat. A—Caulking up joints. B—Installing seals behind switch and receptacle plates.

Furnaces and air conditioners are a major user of energy in a home. Assume that the efficiency of an older unit is around 50 to 60%. For every 100 Btu of fuel energy consumed, the furnace delivers 50 to 60 Btu of heat energy.

Figure 28-2. A superinsulated panelized dwelling under construction. Sandwich panels of polystyrene and oriented strand board provide well-insulated walls and roof for a home. (Enercept Building Systems)

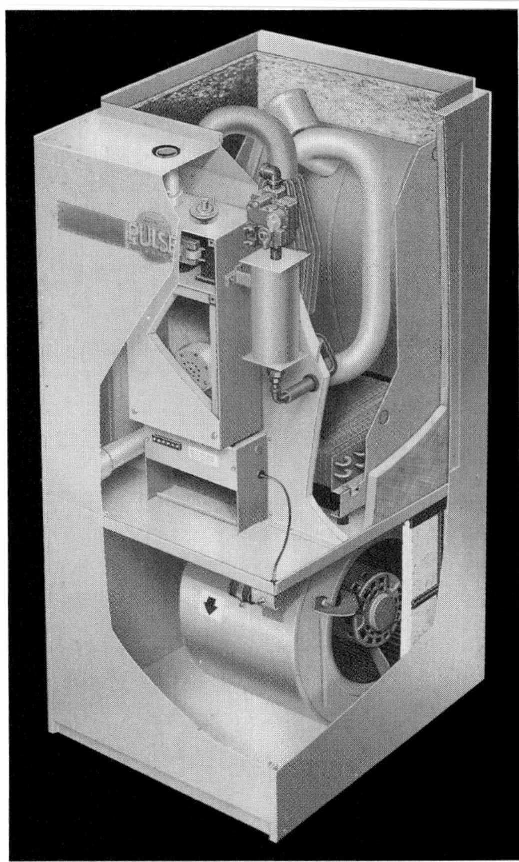

Figure 28-3. New furnaces have high Annual Fuel Utilization Efficiency (AFUE) ratings. (Lennox Industries, Inc.)

Much of this lost heat energy escapes up the chimney. Temperatures in a chimney must usually be in the 300°F (150°C) range for the chimney to "draw" (pull combustion gases to the atmosphere). When the furnace is not running, warm room air continues to exhaust through the chimney.

With current technology, modern furnaces, Figure 28-3, produce more heat energy out of fuel. Some of these systems are rated 91 to 96% efficient by the Department of Energy. Heat exchangers in the units are designed to extract more of the heat of combustion. Since spent gases leave the furnace at 100°F (38°C), the furnace does not require a chimney. The gases are exhausted through a small plastic pipe. Another energy-saving side effect is that the small pipe allows little loss of home heat when compared with the chimney's losses.

Following the energy crisis of 1973, the National Bureau of Standards set up an appliance labeling section. This section sponsors a testing program of efficiency ratings for all appliances. Since 1976, manufacturers have attached *EER (energy efficiency rating)* labels to all of their appliances. Covered by this rating program are furnaces, central and room air conditioners, and other appliances. EER labels contain the following information:

- The EER of the appliance. The higher the rating, the more efficient the appliance in its use of electricity, gas, or fuel oil.
- The efficiency range or energy consumption of comparable products on the market.
- An estimate of how much the appliance will cost to operate during the year at various levels of usage. See Figure 28-4.

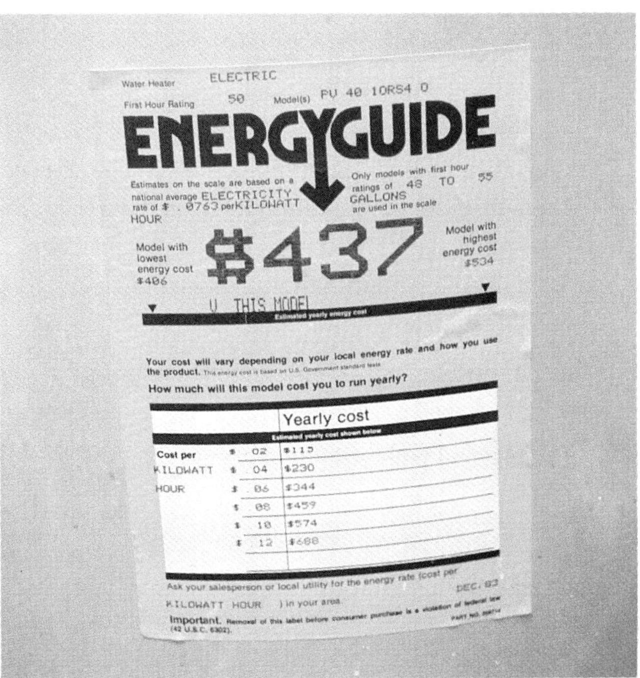

Figure 28-4. This is a sample of an EER label that all energy-using appliances now carry upon purchase.

Heating Systems

Central heating has all but replaced space heaters. In central heating, a single large heating plant produces heat at a central location. The heat is then circulated throughout the structure by perimeter heating or radiant heating.

There are several types of heating systems:

- Forced-air perimeter heating.
- Hydronic perimeter heating.
- Hydronic radiant heating.
- Electric resistance radiant heating in ceiling or walls.

Forced-Air Systems

In a forced-air system, Figure 28-5, air is circulated through large ducts that direct warmed air to every room in the building. In the furnace, the burner delivers high-temperature combustion gases to the furnace's heat exchanger. This is a series of tubes that absorb the heat of the gases and transfers it to a large sheet metal chamber called a *plenum.* A blower forces the heated air in the plenum to pass through ducts into the spaces being heated. At the same time, the blower draws cooled air back to the furnace through a series of ducts called the air return. An air filter, located in the cold air return, filters out dust particles picked up from the heated spaces.

Figure 28-6 shows several views of gas furnaces. One is a cutaway illustration of a modern forced-air furnace. Another view shows a high-efficiency furnace that exhausts spent combustion gases through plastic pipe.

Furnaces may be oil-fired or gas-fired. Oil-fired units require a storage tank for the fuel that is usually located either underground outside the building or in the basement near the heating unit. A filler pipe is usually installed that extends through the wall to the outside for easy access during fuel delivery. Gas-fired units may get their fuel supply directly from a gas main supplied by a gas utility or from a storage tank located on the property.

Registers are grill work that are installed over duct openings. They usually have shut-offs for controlling airflow. They are attached where the ducts open into the rooms. Some registers have adjustable vanes that control the direction of airflow as it exits the duct. Figure 28-7 shows a typical register located at floor level. Since warm air rises, being lighter, warm air registers are more efficient if located at floor level. Since cold air sinks, cold air returns should also be located at floor level.

Registers should be placed so that air is not discharged directly on occupants. Some people place them high on the sidewall—about 7' from the floor—so as not to interfere with furniture arrangement. Registers may also be located in the ceiling, floor, baseboard, or low in the sidewall.

Installing and Maintaining Forced-Air Systems.

Installation of forced-air furnaces usually follows installation of ductwork, Figure 28-8. The installer is often trained on the job but has probably also had some formal instruction in a vocational program, either in high school or in a posthigh program. She or he must be able to make mechanical connections and make electrical hookups. The same person may also install the ductwork. This requires a knowledge of certain metal trades as well as a thorough grounding in electrical principles and wiring practice. See Figure 28-9.

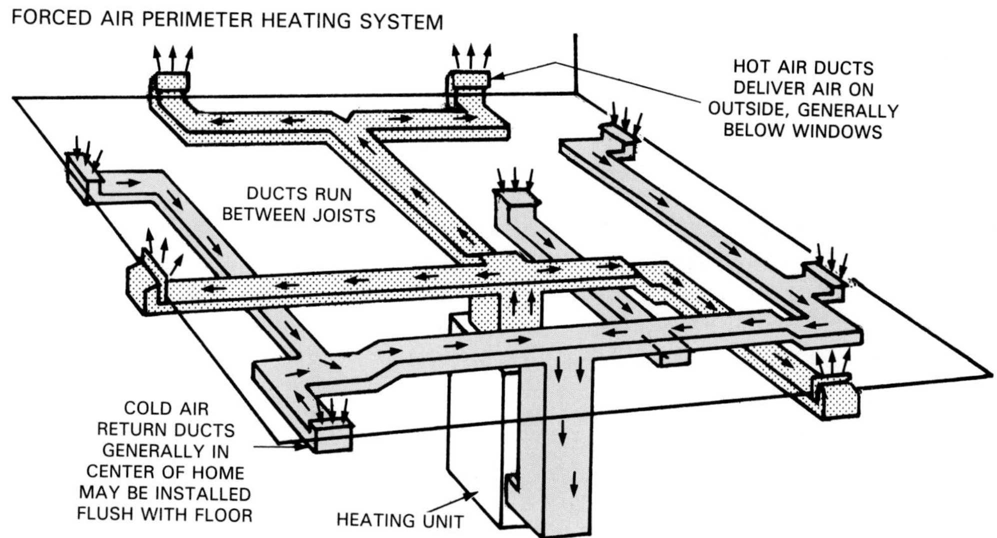

Figure 28-5. Basic, simplified sketch of a forced warm-air heating system. Warm-air ducts deliver heated air to rooms under a slight pressure. Cold air duct, called a "cold air return", brings cold air into the furnace each time the circulating blower runs.

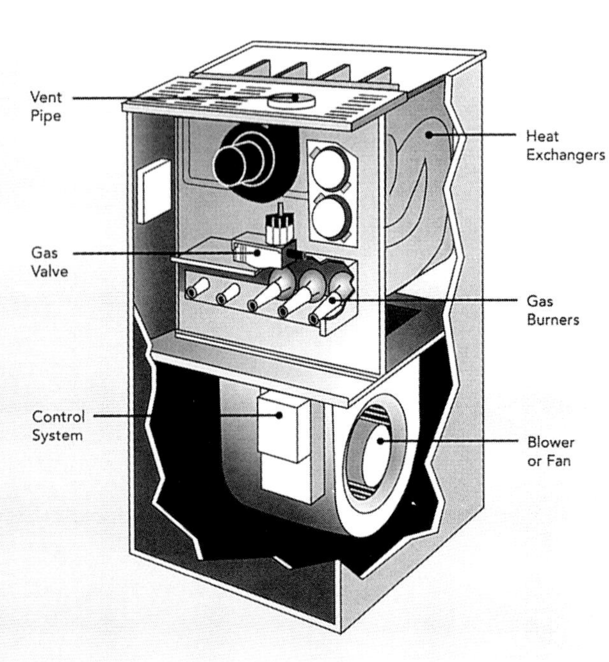

Figure 28-6. Examples of forced warm-air furnaces. Top—All gas furnaces have four main components: burners where fuel is delivered and burned; heat exchangers that transfer heat to the duct system; a blower that moves heated air through the ducts; a vent pipe or flue that exhausts gaseous by-products to the outdoors. (Trane) Bottom—Cutaway of a compact gas furnace illustrates main parts. (Luxaire Heating and Air Conditioning)

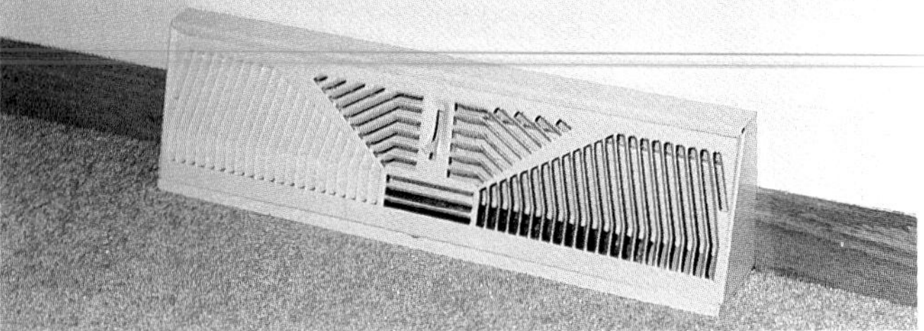

Figure 28-7. Hot air registers usually have vanes to direct flow of heated air and a "door" which can be positioned to reduce or even stop airflow. Hot air registers are usually placed on outside walls.

Figure 28-8. Sheet metal skills are involved in forming and installing ductwork. Normally this work is done and ducts installed before furnace installation.

Once placed in service, motor and blower bearings of the furnace must be oiled periodically—usually every three months unless the bearings are lubricated for life. Blower belts should be inspected before the beginning of every heating season. Several times during the heating season filters should be cleaned or replaced. Electronic air filters, if used, should be inspected several times during the heating season and cleaned as necessary. An annual furnace inspection and servicing is advisable to avoid malfunctioning.

If there is a humidifier in the system, it should be inspected at least once at the beginning of a heating season. Inspect panels and replace them when caked with minerals from the water supply. Check for proper flow of water, as well.

Warm-Air Perimeter Systems

This type of system is usually intended for a basement area or for a building without a basement. In all instances, the ductwork is installed before the slab-on-grade or the basement floor is poured, Figure 28-10.

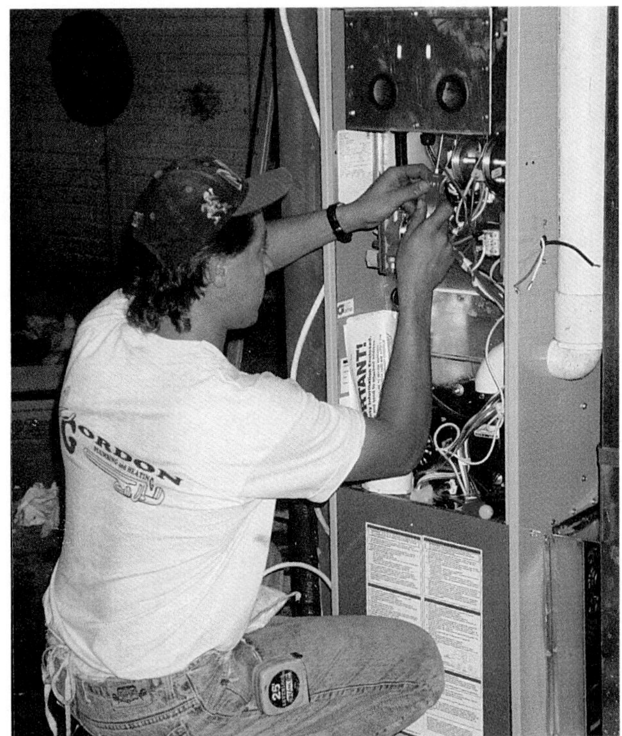

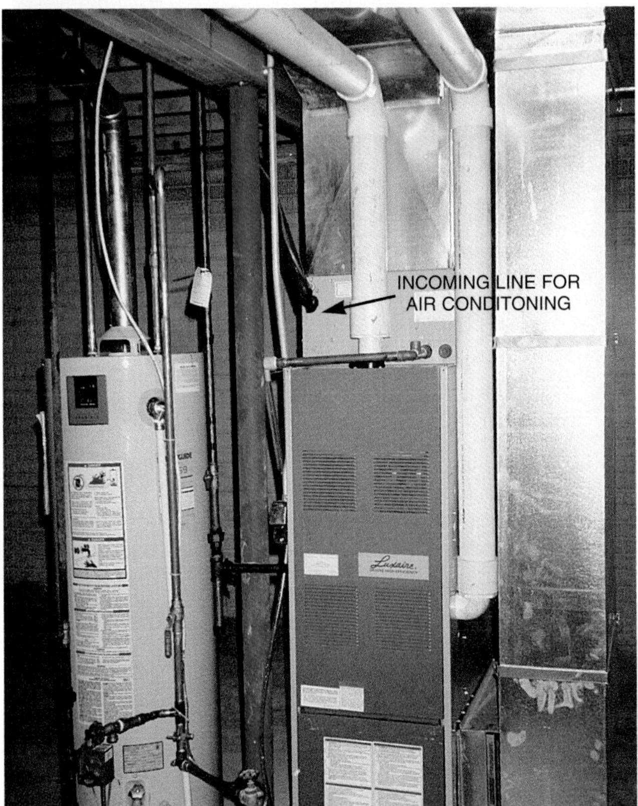

Figure 28-9. Top—Furnace is being installed and installer is making electrical connections. Bottom—A completed installation for a high-efficiency furnace. White plastic piping brings in combustion air from outdoors and carries away gaseous by-products. No chimney is required since combustion gases are cooled in an efficient heat exchanger before discharge. Note the air filter installed at lower right. Arrow shows hookup for air conditioning.

Warm air from the furnace circulates through the duct system embedded in the concrete. This duct system encircles the concrete slab at its outer edges. It connects to the furnace through a plenum and feeder ducts located beneath the furnace. As in the traditional *forced-air system*, the warm air is discharged to the rooms through registers. Floor or low sidewall registers are placed along the perimeter of the space. Air returns to the furnace through return-air intakes at high locations on sidewalls or on the sides of the furnace itself. Ducts may be of sheet metal, vitrified tile, concrete pipe, plastic pipe or other precast forms. If damp conditions are likely to exist, metal ducts are to be avoided.

Hydronic Perimeter Heating System

A *hydronic system,* Figure 28-11, is one that uses water as a medium of moving the heat from the heating unit. The heating unit, Figure 28-12, is called a boiler for the simple reason that it boils water much like a tea kettle sitting on a burner.

Many hydronic systems include a *manifold,* Figure 28-13, that balances the delivery of heated water to different *zones* (areas) of the building. The manifold has an adjusting valve for each zone. When the system is installed, the installer will place thermometers in each room and then increase or decrease heated water flow to each zone until the temperatures are suitable for the activities of each.

Hydronic systems vary somewhat. There are two types of piping layout:

- *One-pipe system*—This type was shown in Figure 28-11. A single pipe called a main supplies the heated water to room heating units. It also returns the cooled water from the units to the boiler.
- *Two-pipe system*—In this system, two mains are used. One, the supply main, supplies the heated water to the room units. The other, known as the return main, returns the cooled water to the boiler. See Figure 28-14.

In the single-pipe system, the heated water is pumped through the pipe from radiator to radiator, finally returning to the boiler where it is reheated.

An *expansion tank* acts as a reservoir for the heated water's increase in volume. A *relief valve* is also part of the system and acts to release excess water pressure that might otherwise damage the system. A pressure-reducing valve limits the pressure of the incoming freshwater supply. Air vents purge any air that has been introduced into the system. These vents are placed at high points in the piping. Refer to Figure 28-15.

Double-piped systems provide more uniform heating than a single-pipe system. Water enters each radiator at similar temperatures. This even temperature is the result of heated water not having to pass through so many radiators before it returns for reheating.

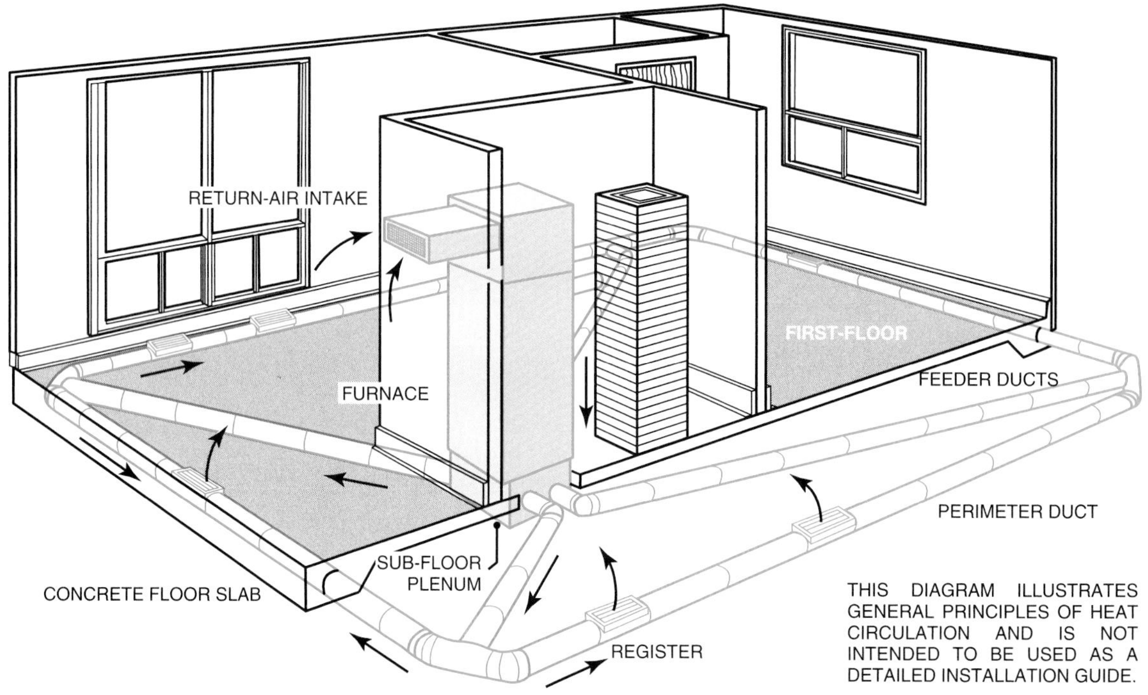

Figure 28-10. A warm-air perimeter system is being installed to heat a basement in a cold climate. A—Ducts are laid on top of rigid polystyrene insulation. B—Housing installed for an in-slab register. C—Sketch of a typical installation.

Either system can be designed for greater efficiency by adding zone valves and zone pumps. These allow areas of a building to be heated independently of temperatures in other spaces.

Hydronic Radiant Heat

In a hydronic radiant heat system, the piping is installed in a structural part of the building, Figure 28-16. Such systems require advanced planning and preparation. Piping is usually copper or flexible plastic laid out in a serpentine pattern that covers the area of installation from side to side. Floor radiant heating systems can be encased in concrete without further preparation. Wood or other types of flooring can be laid over the piping system after provisions are made to prevent damage or crushing of the pipe or tubing. When installing wood or other types of flooring, installers must be careful not to drive fasteners through the heating pipes.

Ceiling radiant heating should be encased in plaster. This provides a medium through which the heat can radiate.

Air Cooling Systems

Cooling systems are basically refrigeration systems designed to remove heat from a building. In many cases, the cooling mechanisms employed use the same air distribution system as for heating systems. See Figure 28-17. A system of tubes called a *cooling coil* is located in the furnace where return air from the rooms can pass over it.

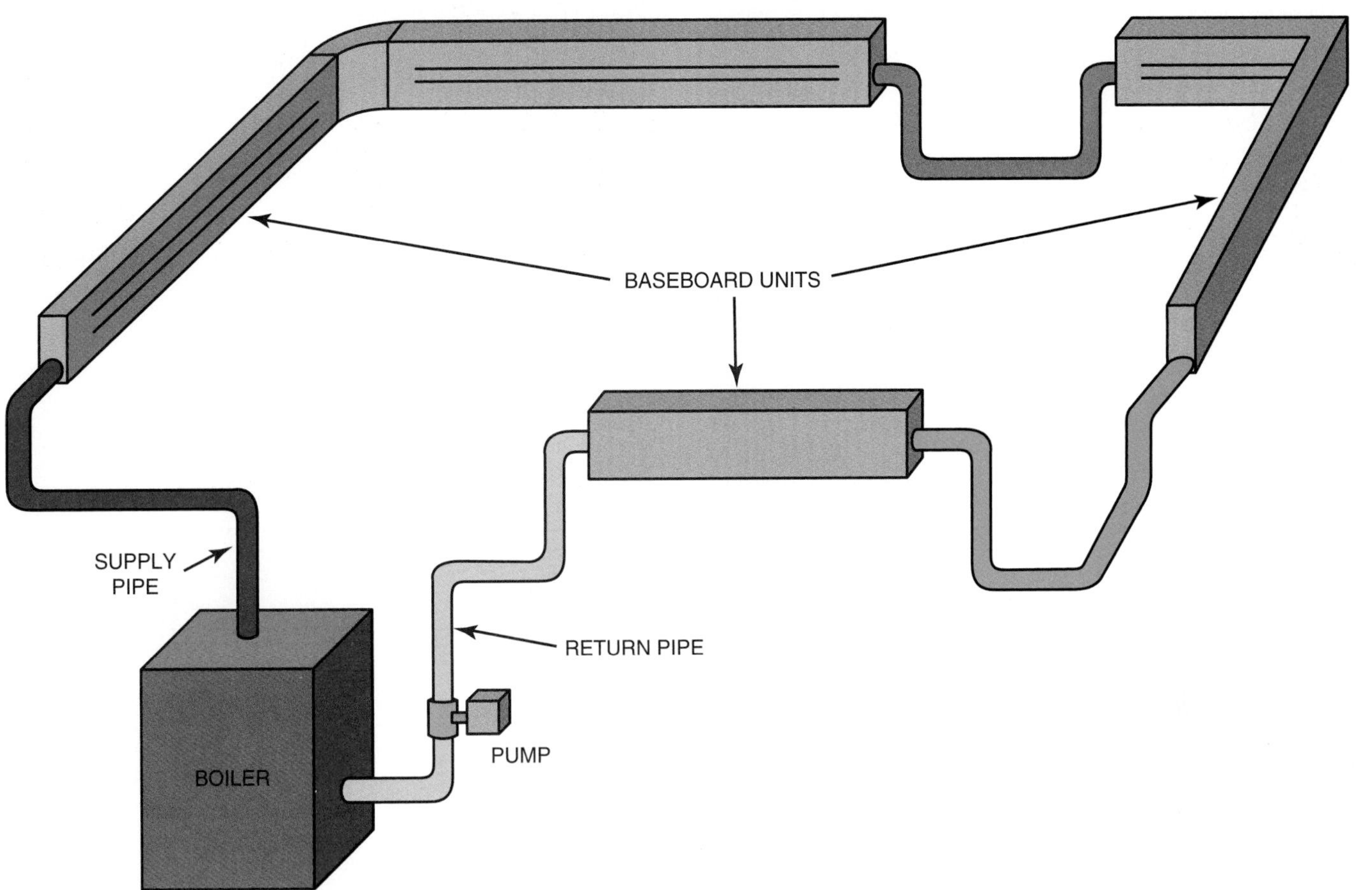

Figure 28-11. Drawing of a basic hydronic system. Since water is involved, hydronic systems are closed loop.

When the central unit is operating, a refrigerant is pumped through the cooling coil. Being colder than the room air passing over it, the refrigerant absorbs heat from the air. Then the refrigerant, warmed by the room air, is drawn back to the outside unit. Here, it passes through another series of coils and fins called a condenser. The *condenser* acts like the radiator of a car and allows the heated refrigerant to pass its heat to the atmosphere.

From the condenser, the refrigerant is pumped to a high pressure and sent once more to the cooling coil in the furnace. This completes one cycle of the air conditioner. This cycle is repeated over and over again as the building's air is passed over the cooling coil. Figure 28-18 shows an electric air-conditioning unit located outside of a home. It contains the compressor, a condensing coil, and related valves. The compressor is operated by an electric motor.

Ducts

Ducts can be constructed of various materials: metal, rigid plastic, or flexible plastic. Ducts should be as unobtrusive as possible, Figure 28-19. Most ducts are prefabricated and then fitted and cut on site. Metal duct sections are joined by the installer on site, Figure 28-20.

Controls

Modern systems for heating and cooling need devices that can sense the temperature and then send an electrical signal that activates the air-conditioning system. Such a device is called a *thermostat,* Figure 28-21.

Thermostats have a switch that is controlled by a bimetal strip, a thin coil of metal, or some other device that reacts to temperature changes. The thermostat can be set to activate and send a signal when the ambient (surrounding) temperature falls below or moves above the selected setting.

The temperature setting that causes the contacts in the thermostat to close is called the *cut-in point.* The temperature setting causing the contacts to open is called the *cut-out point.* The difference between these two points is called the *differential.*

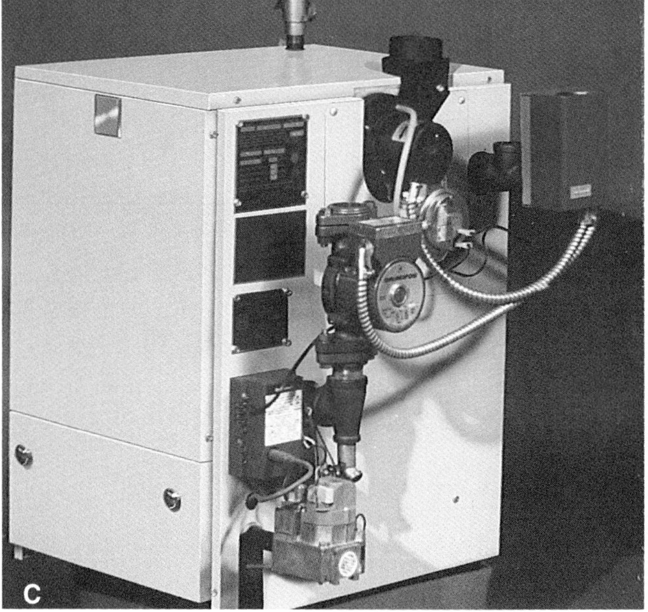

Figure 28-12. Boilers "then and now." A—This unit was installed 38 years ago. It is a single-pipe system connected to serpentine radiant heat in a concrete slab and a plastered ceiling. B—Modern boiler teamed with a hot water storage unit. Boiler delivers heated water to the tank until it is used by the radiant space heating system. Domestic hot water is also drawn from the tank. C—Rear view of a boiler shows circulator and burner controls. (Technology Systems, Carrier)

Figure 28-13. A manifold delivers heated water to five different pipes that go to five different zones of a house.

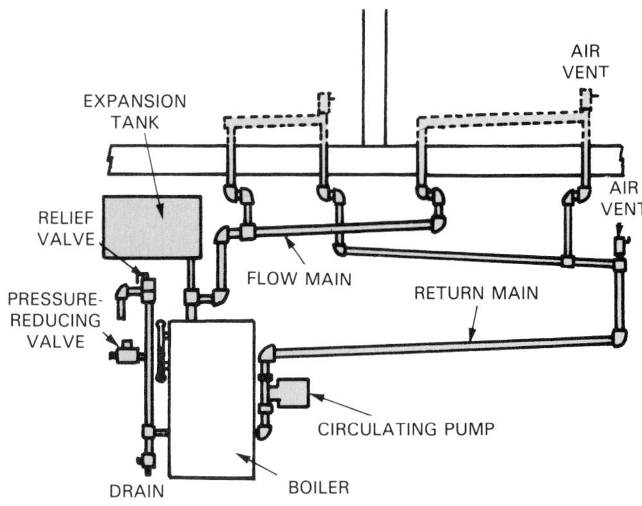

Figure 28-14. A double-pipe hydronic system delivers heat more efficiently than a single-pipe system.

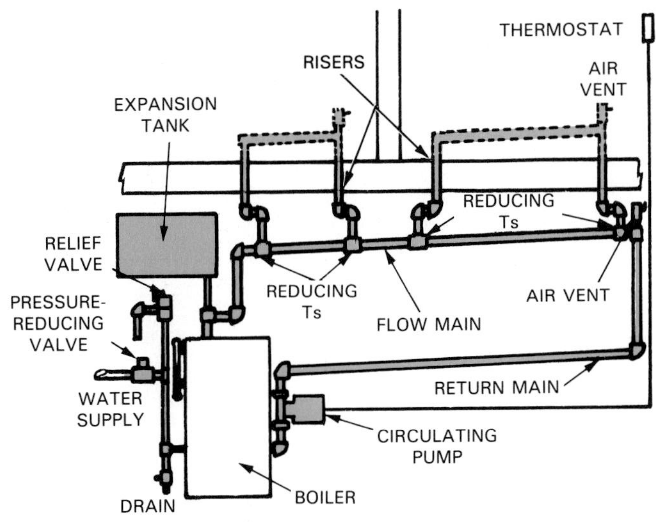

Figure 28-15. Various components have been added to this single-pipe system, making it more efficient.

For single-phase domestic use, the thermostat is connected in series to one of the wires supplying power to the relay. This permits the thermostat switch to disconnect the relay from the power supply.

Air Exchangers

Modern homes, being relatively airtight, may need a method of replacing indoor air polluted by gases and fumes such as radon, formaldehyde, cooking and tobacco smoke, as well as carbon dioxide. While it would seem that an open window and a fan might be an inexpensive remedy, it is not all that simple. While fresh air might be welcomed, the loss of heat in winter or of cool air in hot weather is not. Enter the *air exchanger,* Figure 28-22.

This device is designed to draw in fresh outside air, temper it with the heat of warm exhaust air, and then exhaust the polluted air to the atmosphere. At the same time, the exchanger can filter out most airborne particles,

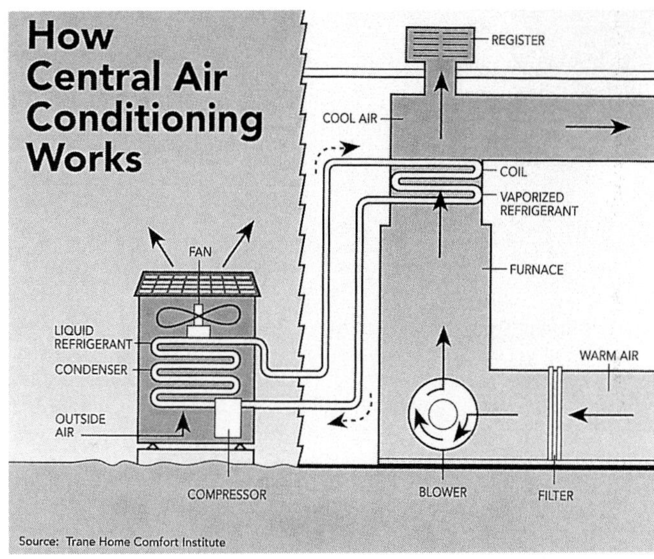

Figure 28-17. A central air-conditioning unit can be combined with a forced-air distribution system. A coil located in the furnace has cold refrigerant circulating through it. Air forced through the coils is delivered to rooms by way of the duct system. (Trane Home Comfort Institute)

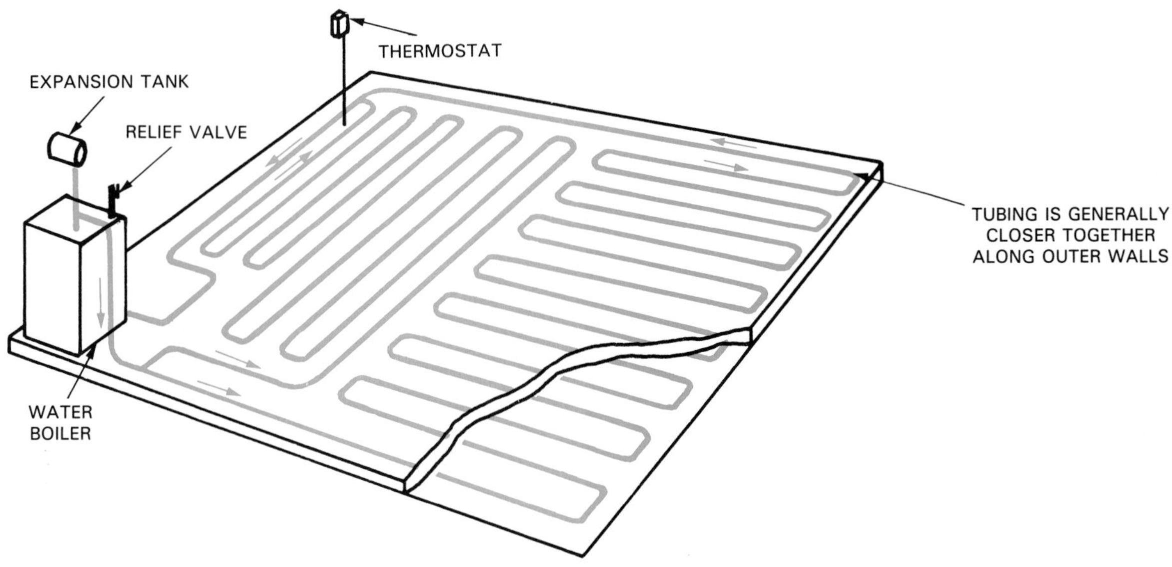

Figure 28-16. This simplified drawing shows a hydronic radiant heating system as it would be installed in a slab floor. Note location of thermostat.

Figure 28-18. A typical central air conditioner is located outside the building. The condenser totally surrounds the compressor and its related components. This arrangement increases the efficiency of the unit. Note the intake and exhaust pipes for the furnace.

Figure 28-19. Ductwork should be concealed wherever possible. A—Round ducts are located between floor joists. B—Ducts located in an attic where they are completely out of sight. C—Ducts run through openings in truss joists.

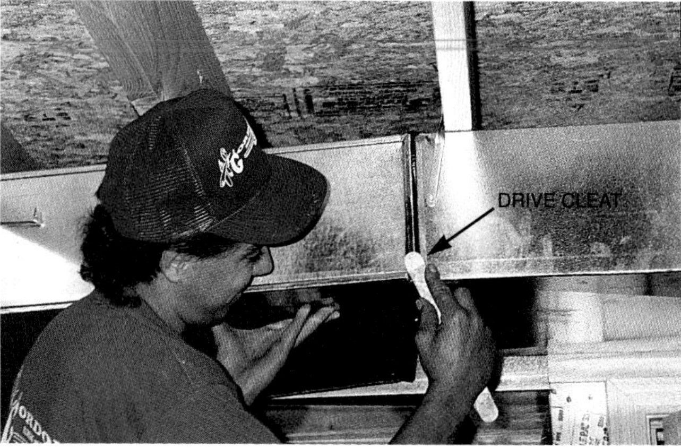

Figure 28-20. Using a metal strap-type clip, an installer connects two sections of sheet metal duct.

Figure 28-21. Thermostats control furnaces and boilers. A signal from the thermostat starts the burner when temperature goes below a set amount. It closes down the burner and then the blower when temperature reaches desired level.

even those microscopic in size. Some units will adjust incoming air for proper humidity in winter or summer.

Testing has indicated that indoor air should be exchanged with fresh outdoor air every two to three hours. In older homes, fresh air typically leaks in through cracks around doors, windows, and foundations.

Heat Pumps

A heat pump combines both heating and cooling in one unit. The basic system includes the following:

- A pump or compressor for compressing the refrigerant.
- Two chambers, like an air conditioner, one designed to collect heat and the other to give off heat.
- A reversing valve to switch the functions of the two chambers.
- Pipes or tubing connecting the chambers with the compressor.

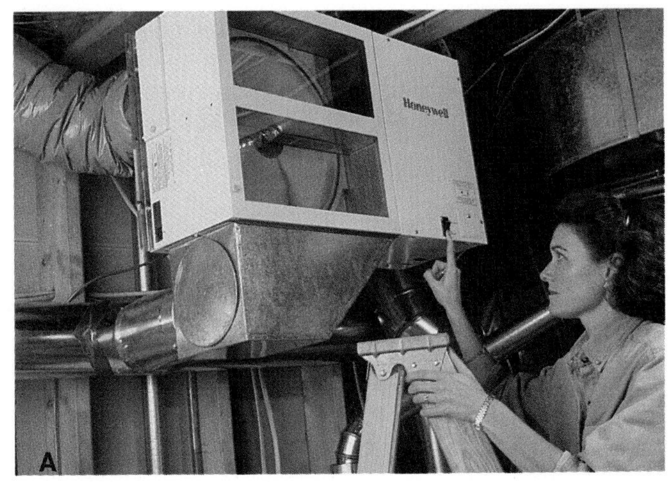

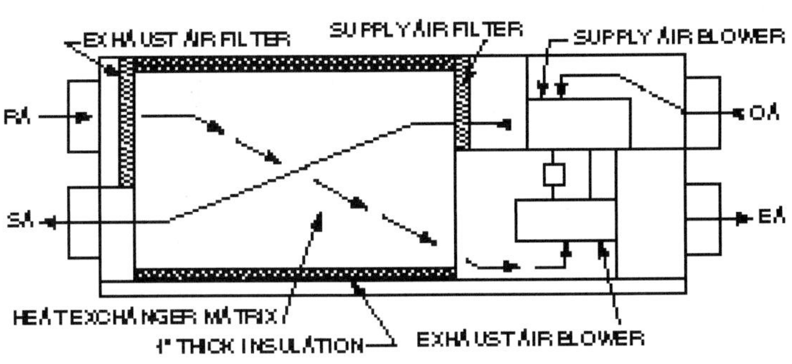

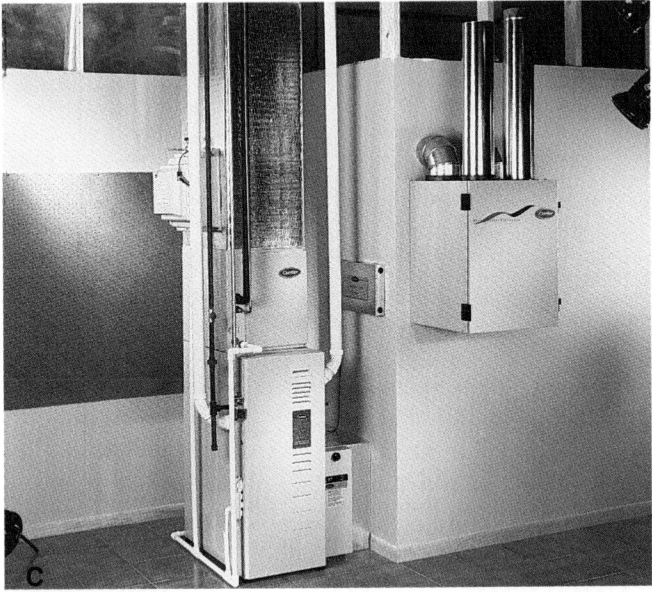

Figure 28-22. A—An air exchanger is a unit that brings fresh outside air into a building, filters it, and warms it with warmer air being exhausted. (Honeywell Inc.) B—Line drawing shows how air exchanger works. (Des Champs Laboratories Inc.) C—A furnace set up with an exchanger. Note the humidifier attached to the furnace at the left. (United Technologies Carrier)

Like an air conditioner, a heat pump uses a refrigerant to collect heat from one place and deliver it to another location. In winter, it can be used to collect heat from the outside and deliver it inside a building. In summer it can collect heat from inside and deliver it to the atmosphere.

The secret of its operation is in the reversing valve, Figures 28-23 and 28-24. By switching the valve, the two chambers switch functions. Thus, in winter, the inside coil becomes a condenser, giving off heat; in the summer, it becomes an evaporator (cooling coil).

Efficiency of the heat pump drops as outside temperatures dip below 20°F (-7°C). The system either must have an auxiliary heating system or the outside coil must be buried in the ground where winter temperatures are higher, or, as an alternative, located down a well where water temperatures may be around 40°F (4°C).

Direct Heating Systems

Direct heating systems are those whose heat source uses no ductwork to distribute the heat. This category includes many basic types of heating, such as stoves and fireplaces. It also includes later developments, such as electric resistance baseboard heating and radiant electric heat.

Baseboard electric resistance heating is extremely simple. It consists of a heating element wrapped in a tubular casing. Metal fins spaced along its length help radiate the heat to the room air.

Electric baseboard radiators come in various lengths—from 1' to 12' and are rated from 100 to 400 W/ft. Like most hydronic systems, the radiators are located around the room's outside wall and especially under windows to counteract drafts.

This system has some advantages:

- Easy to install.
- Affordable.
- Easy to zone heat. Some models have thermostatic controls so that rooms can be heated individually.
- Self-contained. No chimneys, no need for piping or storage of fuel.
- Quiet operating.

The downside, of course, is the operating cost. In some areas it is simply too expensive to consider.

Consequently, this type of system is usually used only as a backup system. In some parts of North America, it is used in combination with wood-burning stoves or gas-fired furnaces. Figure 28-25 shows a typical installation.

Another type of electric heat is known as *electric radiant.* Such systems are designed to directly heat objects rather than use radiators and convection currents to heat a room. Heating elements are placed in floors and ceilings in a manner similar to radiant hydronic systems. As the resistance elements heat up their heat radiates to concrete or wood floors and to gypsum board (drywall) or to tile when the elements are installed in ceilings.

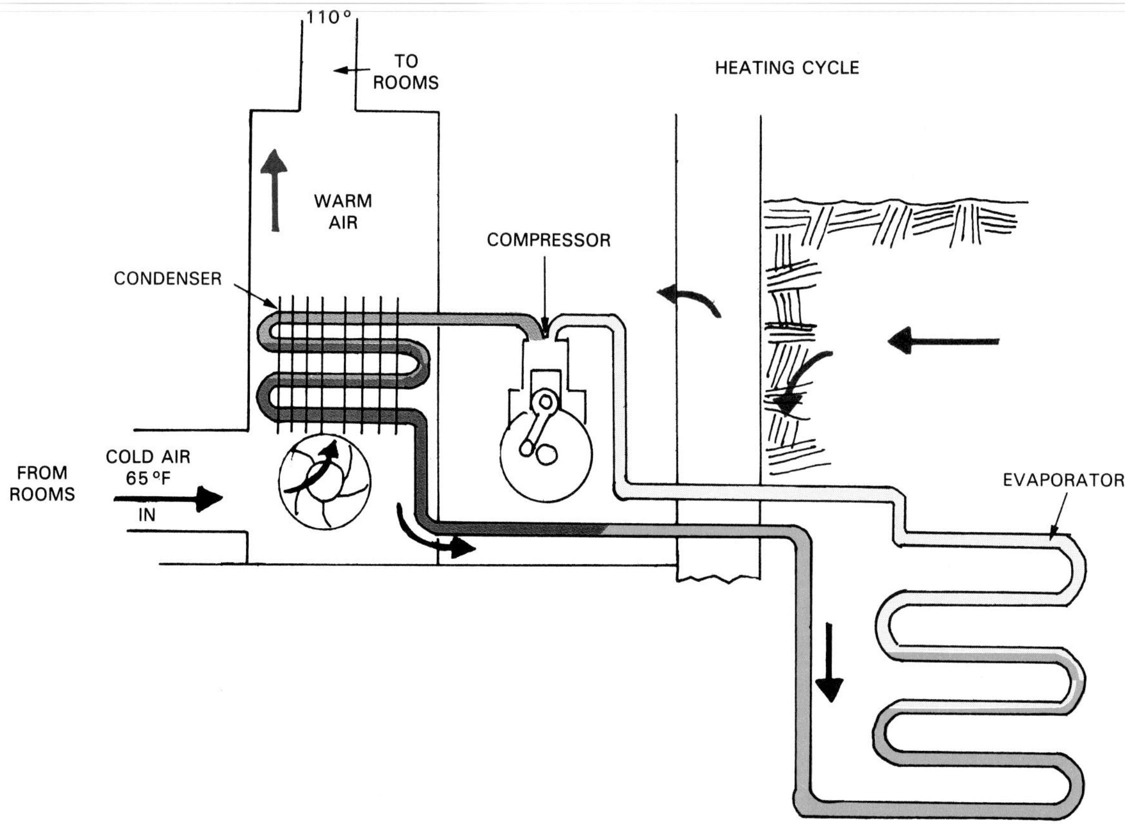

Figure 28-23. A heat pump is shown in the heating mode.

Another approach to radiant heating is to install a grid of specially sheathed electrical cable in the floor, Figure 28-26. The cables can be embedded directly in concrete or masonry when the floor is poured or laid.

In radiant installations it is important to install generous amounts of insulation on the cold side of the heating elements to reduce heat loss.

Safety

A forced-air furnace is frequently the source of leaks and should be carefully inspected. This is best done by a professional who has the appropriate equipment. An inspection should include the following:

- **Measuring concentration of carbon monoxide in flue gases.**
- **Check of furnace connections to flue pipes and venting systems to the outside of the building.**
- **Check of furnace filters for dirt and blockage.**
- **Check of forced-air fans for proper installation and to ensure correct airflow of flue gases.**
- **Examination of the combustion chamber and heat exchanger for cracks, holes, metal fatigue, or corrosion.**
- **Check of burners and ignition system. A flame that is mostly yellow in natural gas-fired furnaces is often a sign that the fuel in not burning completely and higher levels of carbon monoxide are being released. Oil furnaces with similar problems can give off an "oily" odor. CARBON MONOXIDE CANNOT BE SMELLED.**
- **Checking chimneys and other outside venting for cracks, corrosion, holes, debris, and blockages.**
- **Checking all other appliances that use flammable fuels such as natural gas, oil, propane, wood, and kerosene.**
- **Making certain that space heaters are properly vented.**
- **Check of fireplaces for closed, blocked, or bent flues, soot and debris.**
- **Checking gas clothes dryer venting outside the house for clogging by lint.**

Important Terms

Air exchanger
Air return
Air vents
Boiler
Cooling coil
Condenser
Cut-in point
Cut-out point
Differential
Direct heating systems
Ducts
Electric radiant heat

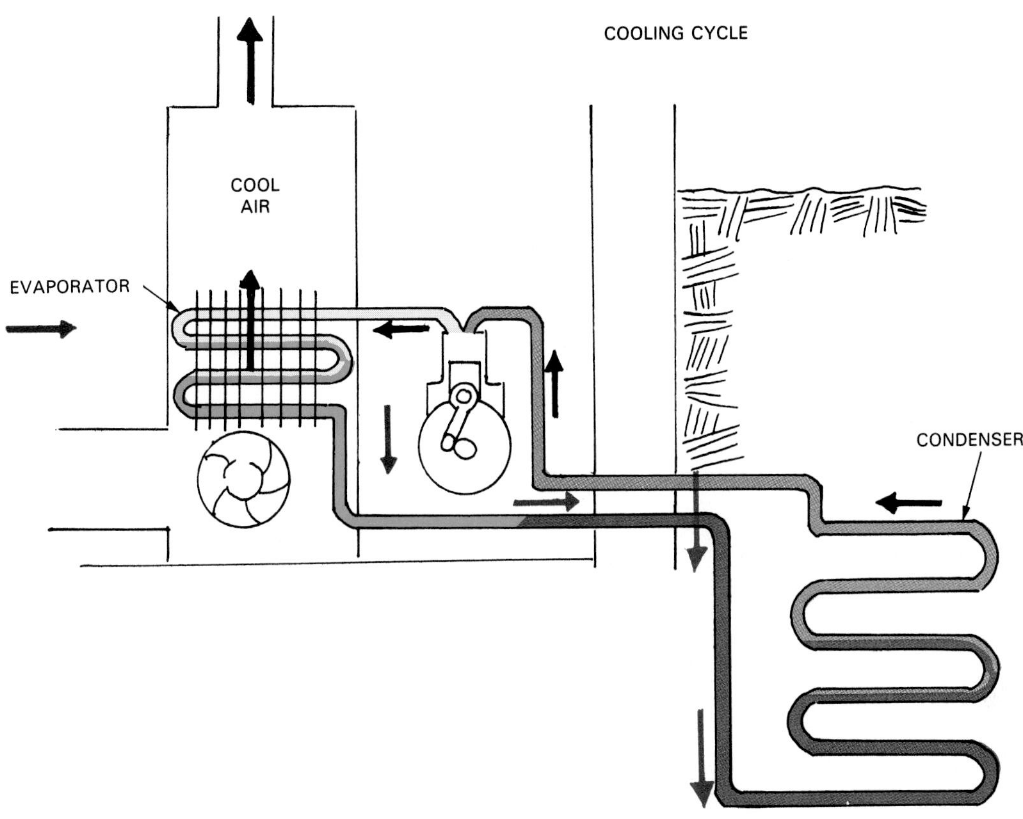

Figure 28-24. A heat pump, through use of the reversing valve, is now in the cooling mode. Valve makes condenser and evaporator change roles.

Energy efficiency rating
Expansion tank
Forced-air system
Heat exchanger
Hydronic systems
Main
Plenum
Register
Relief valve
Return main
Supply main
Thermostat
Zones

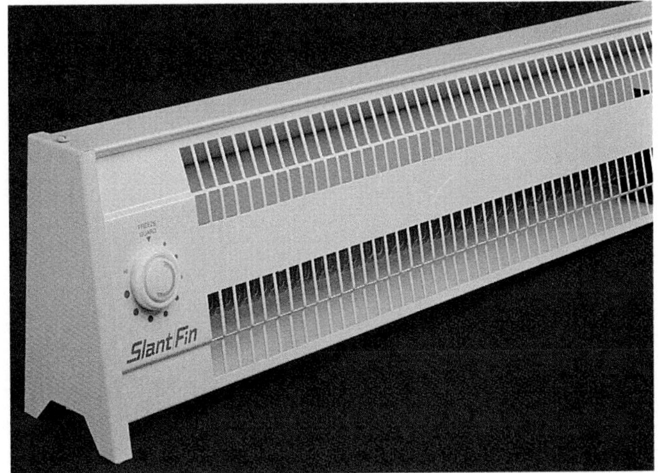

Figure 28-25. An electric baseboard unit is sometimes used as auxiliary heat. (Slant/Fin Corp.)

Test Your Knowledge

1. List four measures that would help save energy when heating or cooling a residence.
2. Chimneys are not a major source of energy waste. True or false?
3. What is the difference between a heat exchanger and an air exchanger?
4. What is an EER and what does it cover?
5. List four types of central heating systems.
6. What is done during maintenance of a humidifier?
7. A ____________ system is always installed in a slab-on-grade foundation.
8. What is the main difference between a forced-air heating system and a hydronic system?
9. Explain how a hydronic system accommodates different zones for heating a house.
10. A ____________ hydronic heating system has a supply main and a return main.
11. Give the function of an expansion tank in a hydronic heating system.
12. In a hydronic ____________ heating system, the piping is installed in a structural part of the building.

Figure 28-26. Electric radiant heat delivers its heat directly to solid material such as concrete. A—Heavy insulation is laid to the cold side of heat system being placed in new construction. Polystyrene sheets 2" thick are laid over several inches of gravel. A vapor barrier is laid over this. B—A grid of electrical resistance wire is laid on top of rebar and attached to the rebar with plastic ties before concrete is poured.
(Kasten-Weiler Construction)

13. In an air-conditioning system, the cooling coil is always located where it can deliver cooled air to the inside of a home. True or false?
14. A _____________ automatically signals a heating system to deliver more heat.
15. A _____________ is a combination heating and cooling system.
16. An air exchanger may also filter dust from incoming air. True or false?

Outside Assignments

1. Obtain a set of instructions for installation of a thermostat and explain to the class how to hook it up to a furnace.
2. Obtain a book on air-conditioning systems and study the function of various components of the system. Present a report to the rest of the class.
3. Examine EER rating sheets from two different furnace models and determine what energy savings are possible by choosing one over the other. (Note: Btu ratings should be the same for each in order to have a valid comparison.)

A gas-fired heating unit as installed in a garage area.

Snips and crimping tools are required for installation of ductwork. (Wiss)

Scaffolds and Careers

A personnel lift is far more versatile than regular scaffolding. The telescoping basket can support workers at various positions throughout the day. This devise is best-suited for small jobs that would normally require a substantial scaffold to be built.

Apprentice carpenters learn while working. (Johnson–Manley Lumber)

29

Scaffolds and Ladders

Carpenters use *scaffolding* (also called *staging*) to reach work areas that are out of reach while standing on the ground or floor deck. *Scaffolds*, Figure 29-1, are temporary and/or movable raised platforms that support workers, tools, and materials with a high degree of safety. Their design must help the worker avoid stooping and reaching.

The type of scaffold required depends on how many workers will be on it at one time, the distance above the ground, and whether it must support building materials as well.

Types of Scaffolding

Scaffolding includes a great variety of designs. In addition to those that are manufactured, there are wood scaffolds that can be built on site by the carpenter crew. Some manufactured types are meant to be assembled and disassembled on site and reused at other building sites. A third type is completely mobile and can be pulled from site to site. Typical scaffolds of various types are shown throughout this chapter.

Figure 29-1. Two types of scaffolding in use today. Left—Scaffold system for siding work consists of aluminum "pump-jack" units attached to aluminum poles 6', 12', 18', or 24' long. Poles can be spliced when additional height is needed. Units have brackets for platform, workbench, safety net, and guardrails. Foot pedal action raises the platform up the poles. For safety, all parts of the system must be used together. (Alum-A-Pole Corp.) Above—Another type of manufactured scaffolding. Standards and braces are assembled and stacked in many different sizes and shapes to fit the task at hand. Units have leveling devices for use on sloping ground. Wood planking is used for platforms.

Manufactured Scaffolding

In modern construction, many builders use sectional steel or aluminum scaffolding. The units are quickly and easily assembled from prefabricated frames. The type shown in Figure 29-2 has light aluminum uprights that are attached to the building with metal braces. The platform is adjusted to variable heights by a jacking arrangement that raises and lowers the brackets supporting the platform. The platform and bench are also manufactured of aluminum.

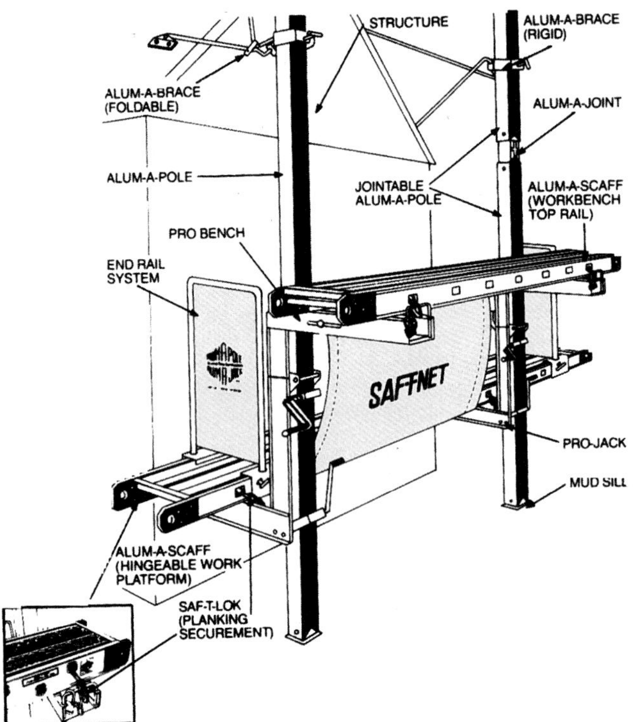

Figure 29-2. Details of a pole system for scaffolding. Sections can be added horizontally and vertically and are OSHA approved up to 50'. (Alum-A-Pole Corp.)

Figures 29-3 through 29-5 show another manufactured type. It consists of box-shaped (sectional) units that can be assembled horizontally and vertically to build a scaffold of nearly any safe height or length.

There are many styles of sectional steel scaffolding. Some types have adjustable legs that are attached to the ground-level sections. Frame sizes range from 2' to 5' wide and from 3' to 10' high. Various lengths of bracing provide frame spacings of 5' to 10'. The basic units are set up and joined vertically and horizontally. Figure 29-3 illustrates the erection details of a sectional scaffold. The frames can be equipped with casters when a rolling scaffold is desired.

Mobile Scaffolding

Movable or mobile scaffolding comes assembled and uses either a mechanical or hydraulic system to adjust its height. Usually mounted on a trailer arrangement, it is easily moved from one site to another. Figure 29-6 shows a mobile scaffolding raised by a scissors jack that is activated by a hydraulic cylinder. It is set up to be towed behind a truck or other construction vehicle.

Site-Constructed Wood Scaffolding

Where it is not possible or practical to use manufactured scaffolding, carpenters must build scaffolds from available lumber. In constructing wooden scaffolds, the uprights should be made of clear, straight-grained 2 x 4s. The lower ends should be placed on planks to prevent settling into the ground. See Figures 29-7 and 29-8.

Bearers (sometimes called cross ledgers) consist of 2 x 6s about 4' long. Use at least three 16d nails at each end of the bearers to fasten them to the uprights. For the single-pole scaffold, one end of the bearer should be fastened to a 2 x 6 block securely nailed to the wall. Braces may be made of 1 x 6 lumber, fastened to uprights with 10d nails, and with 8d nails where they cross. For the platform, 2 x 10 planks without large knots should be used. It is good practice to spike the planks to bearers to prevent slipping. Whether lumber is used for planks or uprights, select lengths with parallel grain. Refer once more to Figure 29-4.

Another scaffold type is the swinging scaffold. Swinging scaffolds are suspended from the roof or other overhead structures. These are used mainly by painters and should only be used with light equipment and materials.

Brackets, Jacks, and Trestles

Metal wall brackets, Figure 29-9, are used in residential construction because they can be quickly attached to a wall, easily moved from one construction site to another, and installed above overhangs or ledges.

Great care is needed when fastening the brackets to a wall. For light work at low levels, connections with nails may provide enough safety. Use at least four 16d or 20d nails. Be sure the nails penetrate sound framing lumber. Most carpenters prefer the greater safety provided by brackets that hook around studding. Holes in sheathing to accommodate such brackets must be repaired before exterior finish is applied.

Some metal wall brackets have posts for holding guardrails and toeboards, Figure 29-10. A guardrail protects a worker on the scaffold and a *toeboard* protects workers beneath from falling tools. A wire mesh screen on open sides is sometimes added.

The scaffold shown in Figure 29-10 is fastened to the wall with a bolt. Figure 29-11 is a more detailed cutaway view of the fastening system.

Wall brackets that support a wood platform should not be spaced more than 8' apart. Metal or reinforced platforms can extend more than 8' between supports. Figure 29-12 shows a metal platform on metal wall brackets.

Roofing brackets, Figure 29-13, are easy to use and provide safety when working on steep slopes. One type

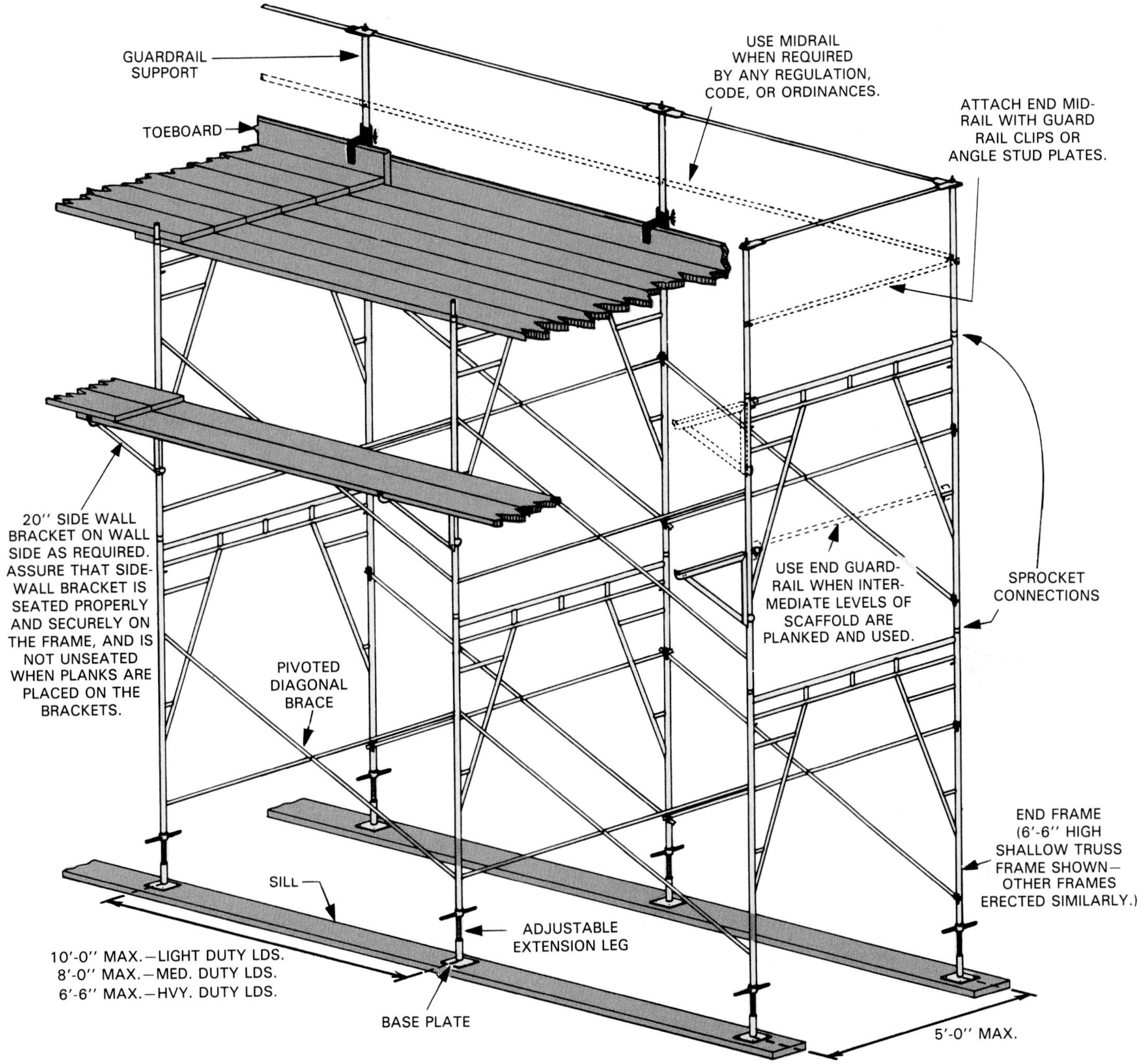

GUARDRAIL SUPPORT
TOEBOARD
USE MIDRAIL WHEN REQUIRED BY ANY REGULATION, CODE, OR ORDINANCES.
ATTACH END MID-RAIL WITH GUARD RAIL CLIPS OR ANGLE STUD PLATES.
SPROCKET CONNECTIONS
20″ SIDE WALL BRACKET ON WALL SIDE AS REQUIRED. ASSURE THAT SIDE-WALL BRACKET IS SEATED PROPERLY AND SECURELY ON THE FRAME, AND IS NOT UNSEATED WHEN PLANKS ARE PLACED ON THE BRACKETS.
USE END GUARD-RAIL WHEN INTER-MEDIATE LEVELS OF SCAFFOLD ARE PLANKED AND USED.
PIVOTED DIAGONAL BRACE
END FRAME (6′-6″ HIGH SHALLOW TRUSS FRAME SHOWN—OTHER FRAMES ERECTED SIMILARLY.)
SILL
ADJUSTABLE EXTENSION LEG
10′-0″ MAX.—LIGHT DUTY LDS.
8′-0″ MAX.—MED. DUTY LDS.
6′-6″ MAX.—HVY. DUTY LDS.
BASE PLATE
5′-0″ MAX.

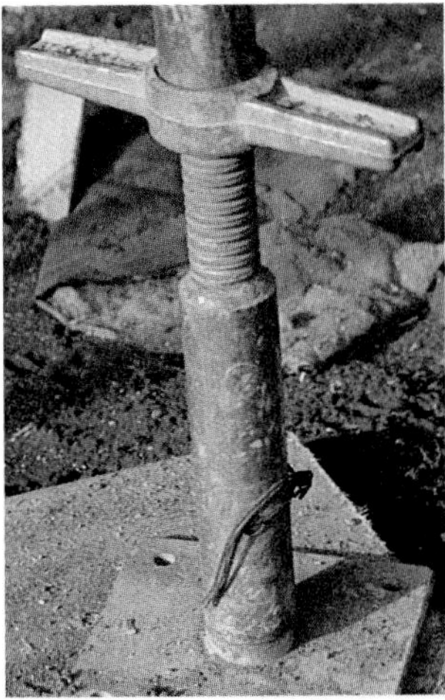

Figure 29-3. Top—A scaffold assembly made from sections joined vertically and horizontally. (Patent Scaffolding Co.) Left—Close-up of adjustable extension leg that is attached to the ground-level sections.

EDGE GRAIN WHICH IS PARALLEL TO THE SURFACE OF THE SCAFFOLD IS STRONG
STEEP ANGLED EDGE GRAIN INDICATES A WEAK BOARD

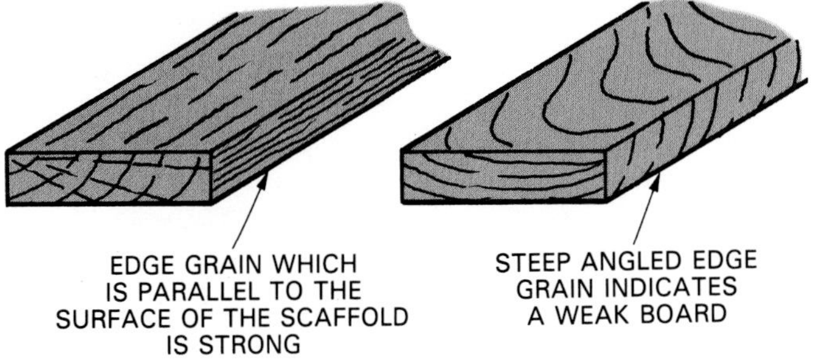

Figure 29-4. To build a safe platform, choose straight-grain lumber. Pick lumber for uprights the same way.

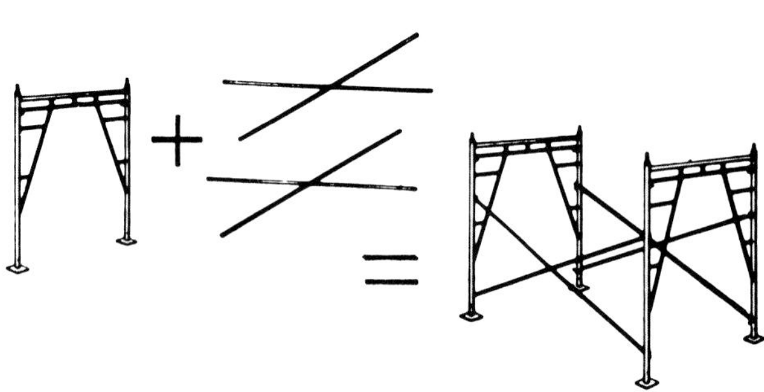

Figure 29-5. Metal scaffold sections can be rapidly assembled from prefabricated trussed frames and diagonal braces.

Figure 29-6. Mobile scaffolding has the great advantage of being quickly raised and lowered and moved to another site with no disassembly.

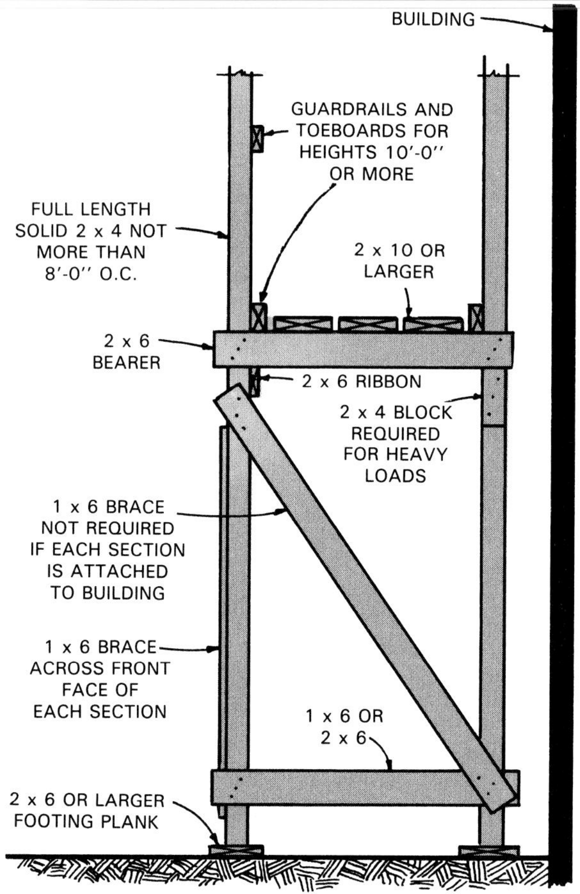

Figure 29-7. A typical design for a double-pole scaffold. Structure can be extended upward to form several platforms. Maximum height should be limited to 18'.

has an adjustable arm to support the plank. A level platform can be set up on nearly any angle of roof surface. Another type of roofing bracket is fixed at one angle.

Ladder jacks are used to support simple scaffolds for repair jobs. Setup requires two sturdy ladders of the same size and a strong plank. Figure 29-14 shows a jack that hangs below the ladder and hooks to the side rails. Another type of jack, Figure 29-15, is easily adjusted upward along the ladder.

Trestle jacks support low platforms for interior work. They are assembled as shown in Figure 29-16. Follow the manufacturer's guideline for the proper size and weight of jack. Be sure the material used for the ledger (horizontal board held by the jacks) is sound and large enough to support the load.

Safety Rules for Scaffolding

- **All scaffolds should be built under the direction of an experienced craftsperson.**

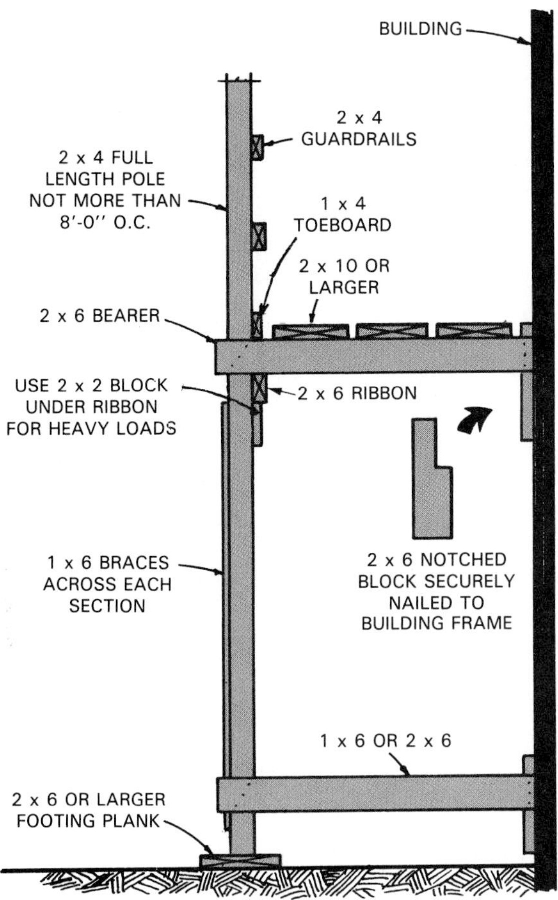

Figure 29-8. A typical design for a single-pole scaffold. Horizontal distance between bearer sections should never exceed 8'.

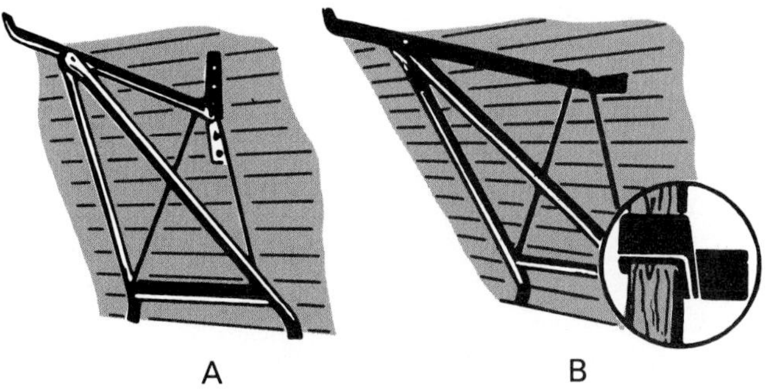

Figure 29-9. Kinds of metal wall brackets. A—Attached with nails securely set in building frame. B—Hooked directly to studding.

- **Follow design specifications as listed in local and state codes. Inspect scaffolds daily before use.**
- **Provide adequate pads or sills under scaffold posts.**
- **Plumb and level scaffold members as each is set.**

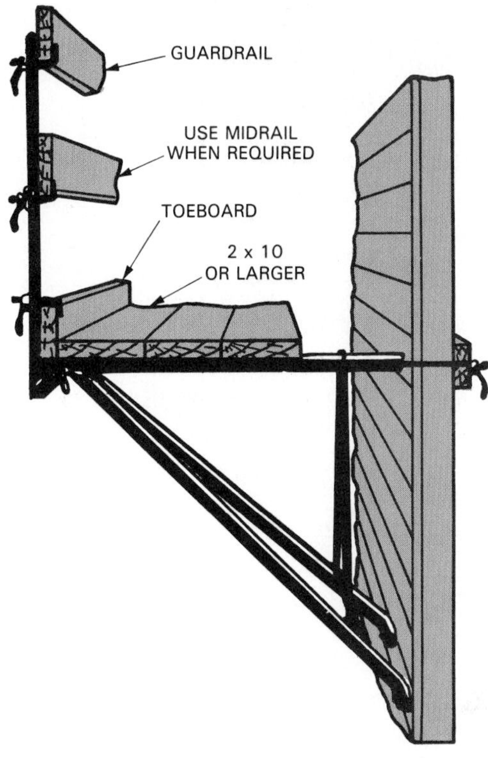

Figure 29-10. This metal wall bracket is approved for work at heights of 10' or more. A high scaffold must have a guardrail and toeboard. However, rails and other parts of the scaffold can be detached if not needed.

- **Equip planked areas with proper guardrails, toe-boards, and screens when required.**
- **Power lines near scaffolds are dangerous. Consult power company for advice and procedure.**
- **Do not use ladders or makeshift devices on top of scaffold platforms to increase the height.**
- **Be certain the planking is heavy enough to carry the load with a safe span length.**
- **Planking should be lapped at least 12" and extend 6" beyond all supports.**
- **Do not permit planking to extend an unsafe distance beyond supports.**
- **Remove all materials and equipment from the platform of rolling scaffolds before moving.**
- **The height of the platform on rolling scaffolds should not exceed four times the smallest base dimension.**

Ladders

Types of wood and aluminum ladders commonly used by the carpenter are illustrated in Figure 29-17. Stepladders range in size from 4 to 20'. A one piece ladder, the single straight ladder, is available in sizes of 8 to 26'. Extension ladders provide lengths up to 60'.

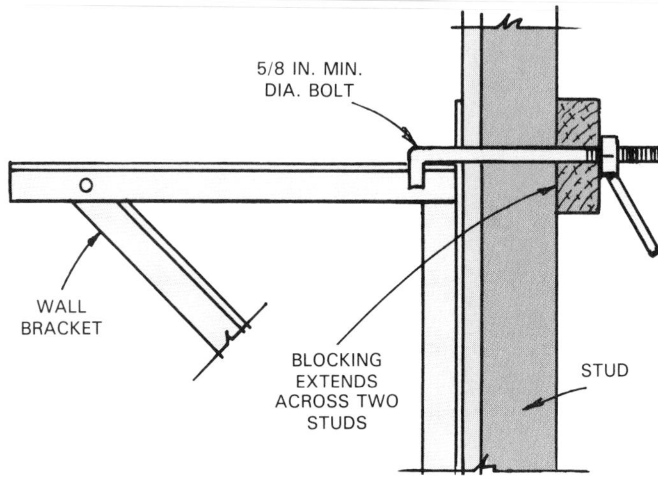

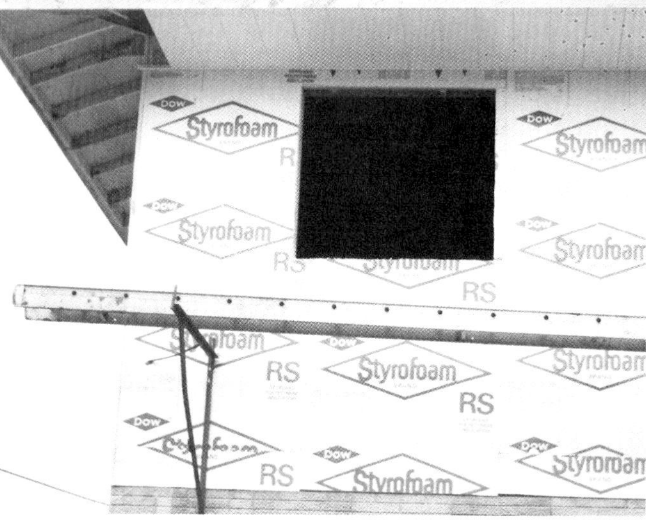

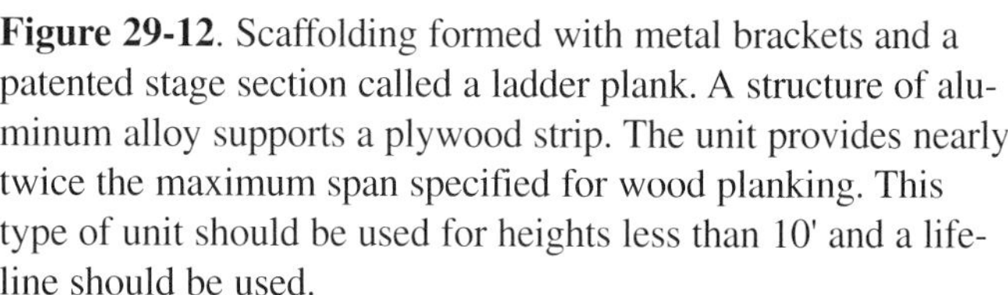

Figure 29-12. Scaffolding formed with metal brackets and a patented stage section called a ladder plank. A structure of aluminum alloy supports a plywood strip. The unit provides nearly twice the maximum span specified for wood planking. This type of unit should be used for heights less than 10' and a lifeline should be used.

Figure 29-11. Top—A strong metal wall bracket is secured to wall with bolt and nut assembly. Bolt passes through wall alongside the stud. Bracket is offset to bear against the sheathing directly over the stud. Bottom—A similar type bracket in use.

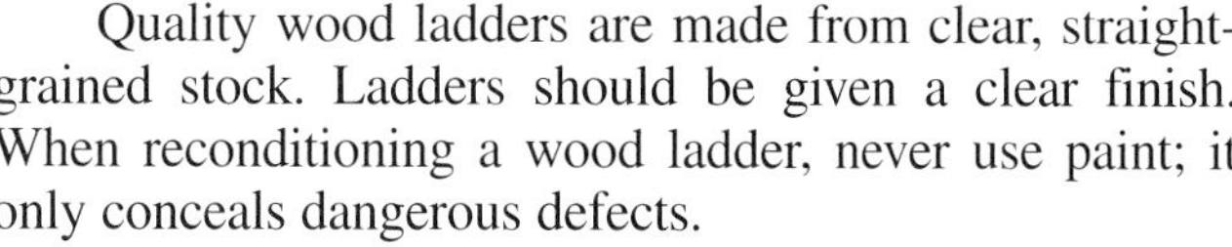

Quality wood ladders are made from clear, straight-grained stock. Ladders should be given a clear finish. When reconditioning a wood ladder, never use paint; it only conceals dangerous defects.

Basic care and handling of ladders is illustrated in Figure 29-18. Always keep ladders clean. Do not let grease, oil, or paint collect on the rails or rungs. On extension ladders, keep all fittings tight. Lubricate the locks and pulleys. Replace any frayed or worn rope.

To erect a ladder, place the lower end against a solid base so it cannot slide. Raise the top end. Walk toward the bottom end, grasping and raising the ladder rung by rung as you proceed. When vertical, lean it against the structure at the proper angle. Make sure bottom ends of both rails rest on a firm base.

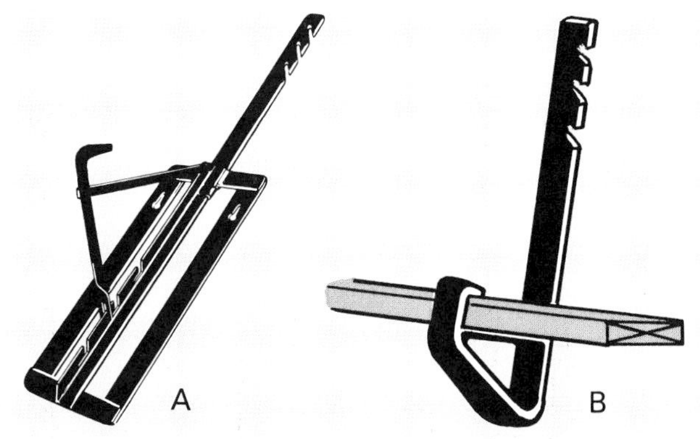

Figure 29-13. Roofing brackets. A—Slotted end will hook on nails driven into the roof deck and frame. Arm that carries planking is adjustable. Overall length is 30". B—Bracket is used on steep roofs. All brackets require 16d common nails driven into rafter. C—Roof brackets are being used on a reroofing job.

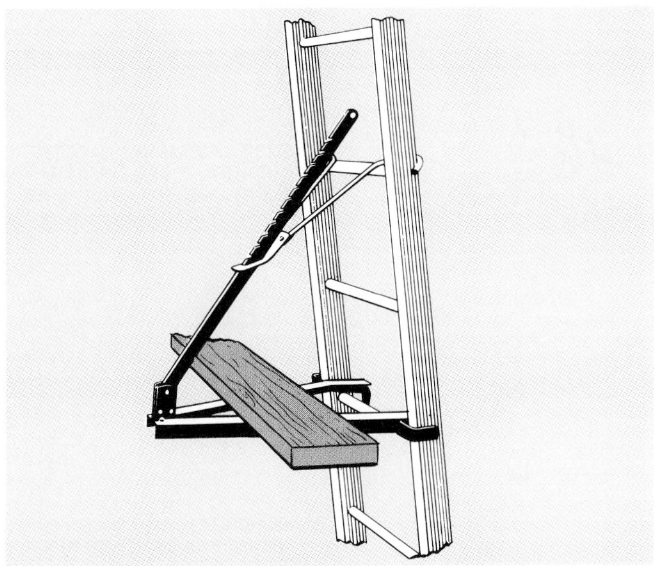

Figure 29-14. A ladder jack. This type is held in place with hooks that fit around side rails. (Patent Scaffolding Co.)

Figure 29-15. Ladder jacks that project on the front side. Top, left—Method of raising jack while plank is supported on arm and shoulder. Use lifeline when performing this operation and do not raise the scaffold more than one rung at a time. Use below 10'. Top, right—Lifeline is recommended when seated. Bottom—Ladder jacks used with a metal platform. (Trudeau Construction Co.)

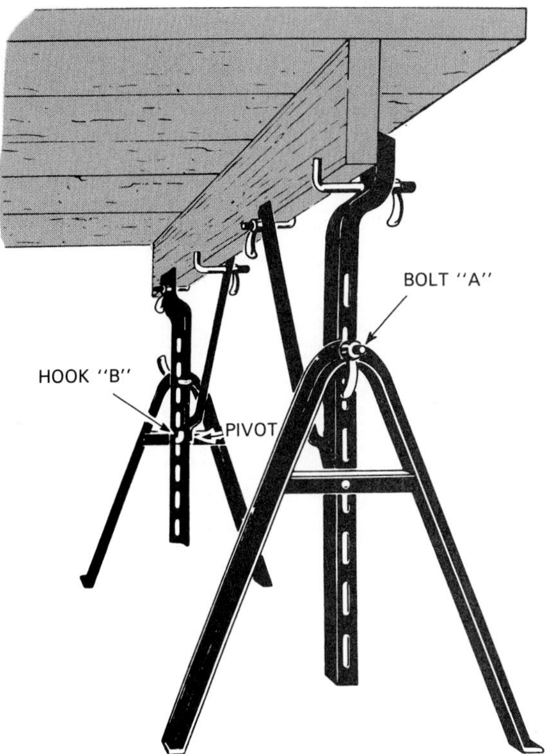

Figure 29-16. Trestle jacks. To adjust height, first loosen bolt "A" and then change hook "B" to another slot. (Patent Scaffolding Co.)

Safety Rules for Ladders

- **Always inspect a ladder before using it.**
- **Place ladder so the horizontal distance from its lower end to the vertical wall is one-fourth the length of the ladder, Figure 29-19.**
- **Before climbing the ladder, be sure both rails rest on solid footing.**
- **Equip the rails with safety shoes, Figure 29-20, when the ladder is used on surfaces where the bottom might slip.**
- **Never place a ladder in front of a doorway where the door can be opened toward the ladder.**
- **Never place ladders on boxes or any unstable base to get more height.**
- **Never splice together two short wood ladders to make a longer ladder.**
- **Always face a ladder when climbing up or down.**
- **Place ladders so work can be done without leaning beyond either side rail.**
- **Be sure extension ladders have sufficient lap between sections. A 36' length should lap at least 3'; a 48', at least 4'.**
- **When a ladder is used to get onto a roof, it should extend above the roof at least 3'.**

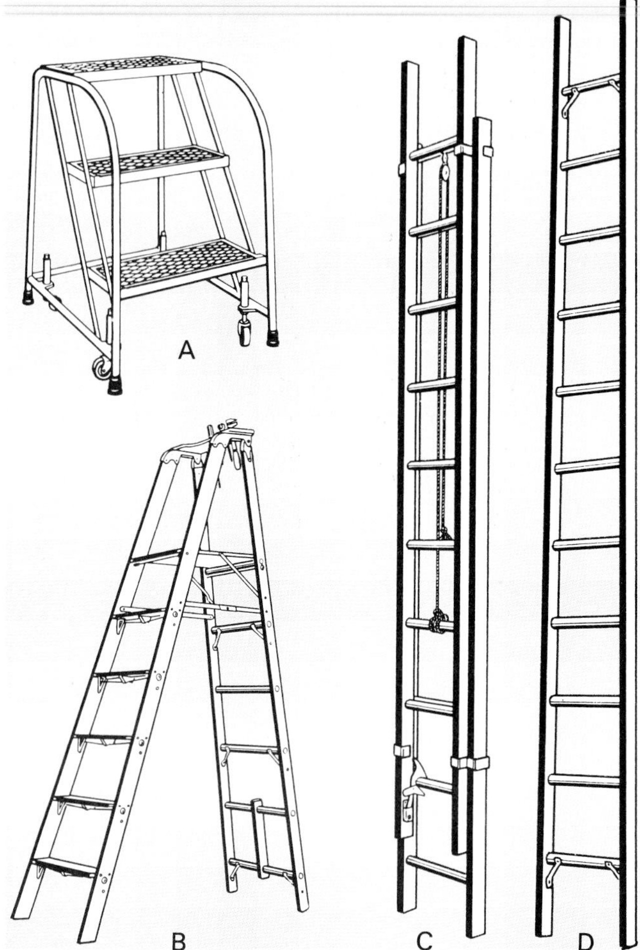

Figure 29-17. Ladder types used by the carpenter. A—Safety rolling ladder drops firmly to floor when weight applied. It is used for inside trim work. B—Stepladder with platform and toolholder. C—Extension ladder. D—Single straight ladder. (Patent Scaffolding Co.)

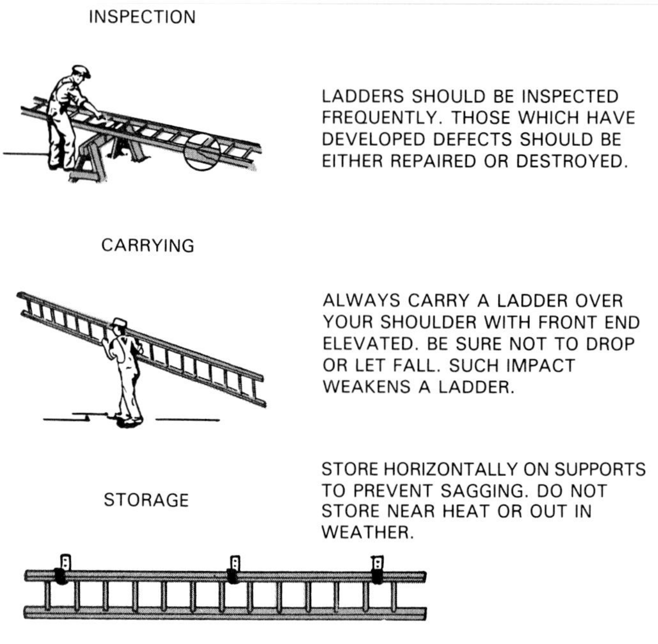

Figure 29-18. Ladder care and handling.

Figure 29-19. Proper placement of a ladder. Horizontal distance from base of the ladder is about one-quarter the length of the ladder. (Greco Painting)

- **Both hands should be free to grasp when climbing a ladder. Use a hand line to raise or lower tools and materials.**
- **Before mounting a stepladder, be sure it is fully open and locked, and all four legs are firmly supported.**
- **Do not stand on either of the two top steps of a stepladder.**
- **Do not leave tools on the top of a stepladder unless the ladder has a special holder.**
- **Never use metal ladders where contact with electric current is possible.**

Figure 29-20. Safety shoes are required for all ladders used on smooth surfaces. (Tilley Ladder Co.)

Important Terms

- Bearers
- Ladder jacks
- Roofing brackets
- Scaffolds
- Staging
- Toeboards
- Trestle jack

Test Your Knowledge

1. How are most steel sectional scaffolds made level?
2. What is the major advantage, if any, of mobile scaffolding?
3. Bearers that support the planking of wood scaffolds should be made of material with a nominal cross section of ____________.
4. In wooden scaffolds, horizontal spacing of support sections should not exceed ____________ feet.
5. Metal wall brackets that support a wood platform should not be spaced more than ____________ feet apart.
6. The height of the platform of a rolling scaffold unit should not exceed ____________ times the smallest dimension of the base.
7. Single straight ladders are available in sizes ____________ feet to ____________ feet.
8. The sides of a ladder are called ____________.
9. When a ladder is used to climb onto a roof, it should extend at least ____________ feet above the roof edge.

Outside Assignments

1. Get descriptive literature from a company that manufactures steel scaffolding. Check with your local building supply dealer or write directly to the company. From a study of this information, learn about the kinds and sizes of tubing that are used. Also learn how the various connecting devices operate. Try to find out approximate costs of this type of scaffolding. Prepare carefully organized notes and make an oral report to your class.

2. Make a study of the types of ladders used in construction and maintenance work and prepare a written report. Include information about the kind and quality of material used and how the ladders are assembled and finished. Also include commonly available sizes and list prices. Finally, develop a list of reasons a homeowner may have for buying or renting a ladder. Are these reasons the same for the builder?

"Pump-jack" type scaffolding is being used on condominium construction on a barrier island. Note piling foundation that raises the floor of the building about 10' above the ground.

Scaffolding formed with ladder jacks. A lifeline should be used when working at the height shown.

This simple type of scaffolding can be used near the ground. For greater safety, guardrails could be added.

30

Carpentry—A Career Path

Carpentry is a rewarding career. It is ideal for the person who has an interest in and an aptitude for working with tools and materials, Figure 30-1. The trade requires the development of manual skills. These skills involve both thinking and doing. Carpentry also requires a thorough knowledge of materials and methods used in construction work.

Figure 30-1. Potential carpenters enjoy working with tools and on machines.
(Montachusetts Regional Vocational-Technical School)

Industry	Percent change in employment, 1992-2005[1]
SERVICES	40
CONSTRUCTION	26
RETAIL TRADE	23
TOTAL, ALL INDUSTRIES	23
FINANCES, INSURANCE, AND REAL ESTATE	21
WHOLESALE TRADE	19
TRANSPORTAION AND PUBLIC UTILITIES	14
AGRICULTURE, FORESTRY, AND FISHING	14
GOVERNMENT	10
MANUFACTURING	-3
MINING	-11

SERVICE PRODUCING
GOODS PRODUCING

-20 -10 0 10 20 30 40

PERCENT CHANGE IN EMPLOYMENT, 1992-2005[1]

[1]ALL FIGURES ARE FOR WAGE AND SALARY EMPLOYMENT ONLY, EXCEPT FOR AGRICULTURE, FORESTRY, AND FISHING, WHICH INCLUDES SELF-EMPLOYED AND UNPAID FAMILY WORKERS.

Figure 30-2. Next to services, construction is the fastest growing occupation in the United States.

Economic Outlook for Construction

The population of the United States is expected to increase from about 192 million to 219 million from 1992 to the year 2005. This will increase demand for housing and construction workers who will build them. The Bureau of Census in 1992 projected population growth by regions. Fastest growing regions are the West (24%) and South (16%).

Construction employment is expected to grow by 24% (from 4.5 million to 5.6 million jobs). This represents growth beyond any other major industry except services, where growth is 40%. See the chart in Figure 30-2.

Most of this growth will come from the need to improve the nation's infrastructure—roads, bridges, and tunnels, Figure 30-3. Building construction will not grow as fast because there is currently an excess of commercial real estate. Residential construction continues to increase, but at a lesser rate because population growth is slowing.

Sharing with carpenters in the growth in construction are other building trades workers such as surveyors, electricians, bricklayers, plumbers, cabinetmakers, landscapers, and operating engineers (operators of heavy construction equipment). Figure 30-4 shows a variety of construction-related careers available to anyone who has the interest and aptitude to pursue them.

Employment Outlook

Carpenters are the largest group of trade workers, with 990,000 in 1992. Jobs in carpentry are expected to be plentiful through 2005 because of the need to replace those

Figure 30-3. Road repairs will be a major source of construction employment in the next 10 years.

who transfer into other jobs or retire. Well over 100,000 jobs are expected to be available every year. This demand is the largest for any craft group.

On the negative side is the prospect of cyclic economic turndowns that affect the construction industry. At such times, many people skilled in construction find it hard to get or hold a job. Those who have been in such occupations for a long time have come to expect both growth and decline in the industry. Also, working conditions can be uncomfortable in cold winters and hot summers.

Job Opportunities

Job opportunities for you as a carpenter depend somewhat on your choice of work. There are many areas in which you can find a job:

- *Building construction*—Three-fourths of all carpenters work in this area.
- *Remodeling, maintenance, and renovations*—In this category are many self-employed people, while others work for manufacturers and building management firms.
- *Prefabrication of buildings and building components*—While this area employs many workers with fewer skills than carpenters, the latter may hold higher level jobs in this type of fabrication.
- *Related occupations and specialization*—It is relatively easy for a carpenter to move into related fields such as terrazzo work, heavy equipment operation, bricklaying, electrician, or into specialties within the carpentry field. For example, a general building contractor may subcontract various phases of construction. Carpenters, as a result, may specialize in: foundation work, framing, drywalling, insulating, siding, and roofing. See Figure 30-5 for two specialties.

Four of every ten carpenters are self-employed, Figure 30-6. Self-employment has an advantage. Often profits are higher due to lower overhead.

The income earned by the carpenter is close to that received by other trades, and provides a good living. In addition, the carpenter has pride and satisfaction from doing quality work.

After gaining experience, the carpenter may want to undertake a small construction contract. This may be a first step toward the general contracting business. The experienced carpenter is usually able to handle the work of the estimator who figures labor and material costs.

The carpenter who prefers the sales and service aspects of construction work can often find a position with a lumber yard or building supply center as is shown in Figure 30-7.

The carpenter can also join the customer service department of a company making prefabricated structures. The customer may need advice on installation or choice of prefabricated units. Perhaps the carpenter will win repeat business for such a company.

Training

A high school education is highly desirable for a successful career in carpentry and other areas of building construction. This is as true for the aspiring apprentice as for the technician or engineer. Take as many woodworking and building construction courses as possible. Other courses are also valuable. Drafting is especially useful, as are courses in mechanical drawing, print reading and career education (formerly known as "industrial arts").

Students interested in the building trades sometimes avoid science and math. This is unfortunate because these studies are essential if you are to understand the technical aspects of modern methods and materials used in construction work. Include social studies and English because everyone should be prepared to improve our society and be competent in reading, writing, and speaking.

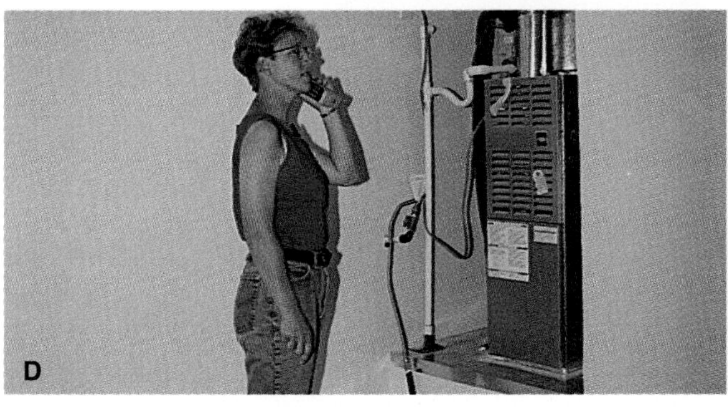

Figure 30-4. Carpentry is related to a wide variety of skilled occupations. A—Surveying. B—Landscaping. C—Electrician. D—Construction expediter. E—Cabinetmaker. F—Computer operator using construction software. G—Bricklayer. H—Operating engineer.

Figure 30-5. Often, carpenters will specialize in one particular phase of building. Top—This crew specializes in putting in foundations. This involves excavating, building forms, and laying down vapor barrier and reinforcing mat.
Bottom—This crew does only framing.

Figure 30-6. Self-employed carpenters have less overhead and no employee problems.

Figure 30-7. Work at a building supply center requires many carpentry skills. The carpenter has knowledge and experience, which assure that the customer receives quality work and advice.

After graduation from high school, you will probably have the skills to enter directly into an apprenticeship training program, Figure 30-8. If circumstances permit, you may wish to enroll in a vocational-technical school in your area. There you can take advanced courses in carpentry and related areas. Some high schools offer these vocational courses as a part of the normal programs before graduation. If possible, take classes in concrete work, bricklaying, plumbing, sheet-metal, and electric wiring. The carpenter usually works closely with tradespeople in these areas and a basic understanding of the methods and procedures will be very valuable. This is especially true if you want to become a supervisor.

Apprenticeship

Many carpenters learn their trade with informal training by their employer. However, many learn the trade through *apprenticeship* or through formal training in vocational training centers or high school programs. (Apprenticeship is a program that allows a student to both work and go to school to learn a trade—in this instance, carpentry.)

Our modern apprenticeship training program had its beginning in early times. It started as an arrangement between teenage children and their parents. This system

Figure 30-8. An apprenticeship program provides training on the job. Here a carpenter explains fastening techniques to an apprentice.
(United Brotherhood of Carpenters and Joiners of America)

passed the knowledge and skill of a trade on to each succeeding generation. As society and the economic structure became more complex, the trainee often left home and was placed under the guidance and direction of another master of the trade in the community. The student was called an apprentice and learned the trade of this master.

During the period of apprenticeship (sometimes it was as long as seven years) the apprentice often lived in the master's house. There were no wages, but board, room, and clothing were provided.

When the training was complete, the apprentice was granted the status of journeyman and could then work for wages. The word, journeyman, was derived from the fact that the apprentice who had completed the training period was then free to "journey" to other places in search of employment. The term journeyman is still used to denote a tradesperson who is fully qualified.

To gain more experience, the journeyman worked with other masters as an equal. Then this mature worker would often start a separate business.

This form of apprenticeship declined rapidly with the advent of the industrial revolution. A new kind of system developed. In it the apprentice lived at home and received wages for work done. Under this system, the apprentice was often exploited. Some became only low-paid workers who received little training. Such practices continued in varying degrees until federal legislation was enacted which established standards and specific requirements for apprenticeship training programs.

Today, apprenticeship programs are carefully set up and supervised. Local committees consisting of labor and management provide direct control. There is help from schools, state and federal organizations and agencies, Figure 30-9.

Apprenticeship Stages

The apprentice works under a signed agreement with an employer. The agreement includes the approval of local and state committees on apprenticeship training.

Applicants must be at least 17 years of age and shall satisfy the local committee that they have the ability or aptitude to master the trade. Then they are placed on a waiting list. The waiting period can last from one to five years depending on local demand.

The term of apprenticeship for the field of carpentry is normally four years. This may be adjusted for applicants with significant experience or those who may have completed certain advanced courses in vocational-technical schools.

In addition to the instruction and skills learned through regular work on the job, an apprentice attends construction classes in subjects related to the trade. These classes are usually held in the evening and total at least 144 hours per year. They cover technical information about tools, machines, methods, and processes. They provide practice in mathematical calculations, blueprint reading, sketching, layout work, estimating, and similar activities. A great deal of study is required to master the technical knowledge needed in carpentry work. See Figure 30-10.

The apprentice works and is paid while learning. The wage scale is determined by the local committee. It usually starts at about 50% of the amount received by a journeyman carpenter. This scale is advanced regularly and may approach 90% of a journeyman's pay during the last year.

When the training period is complete and the apprentice has passed a final examination, he or she becomes a journeyman carpenter. A certificate which affirms this status is issued. This document is recognized throughout the country.

Personal Qualifications

To become a successful carpenter you must:

- Be physically able to do the work.
- Display *manual dexterity*. This means being skilled with one's hands and having a talent for working with the tools and materials of the trade.

You also need sincere interest and enthusiasm that will intensify your efforts as you study and practice the skills and "know-how" required.

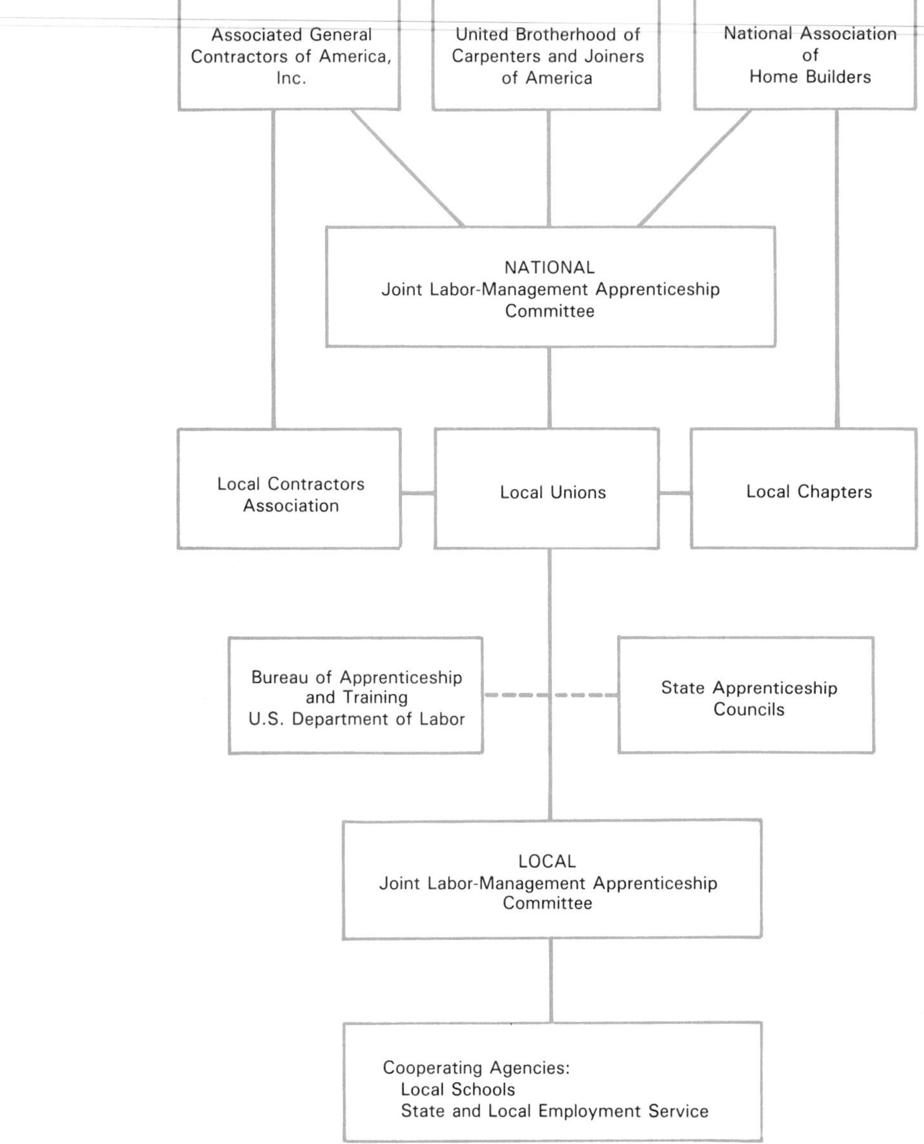

Figure 30-9. The apprenticeship and training system for the carpentry trade is well organized. Additional information may be obtained from the bulletin National Carpentry Apprenticeship and Training Standards, prepared by the U.S. Department of Labor.

In addition, you must possess certain character traits. Honesty in all your dealings is very important, especially in the quality and quantity of work performed. You must show courtesy, respect, and loyalty to those with whom you work. Punctuality and reliability reflect your general attitude and are important not only during your training program but later when you enter regular employment.

The ability to cooperate with others—students, coworker-workers, supervisors, and employers—is essential to success. Most people who fail in the carpentry trade do so because of a deficiency in personal characteristics, not because of their skill level.

The hazards associated with carpentry require that you develop a good attitude toward safety. This means that you must be willing to spend time learning the safest way to do your work. You must be willing to follow safety rules and regulations at all times.

Even after you complete your training program you must continue to perfect your skills and adjust to new methods and techniques, Figure 30-11. Each day brings new materials and improved procedures to the construction industry. This presents a special challenge to those in the carpentry trade. You should read and study new books and manufacturer's literature, along with trade journals and

Figure 30-10. Apprentice begins by working on simpler carpentry tasks. Top—Apprentice nailing waferboard on a wall frame. Bottom—Before any work is done, the carpenter or another person will consult the building plan. Some training in blueprint reading is required for all carpenters. (Orem Research)

magazines in the building construction field. Much information can be obtained at association meetings and conventions where new products are exhibited. To be a successful carpenter, you will also need to keep informed on code changes, new zoning ordinances, safety regulations, and other aspects of construction work that apply to the local community.

Figure 30-11. Worker installs acoustical ceiling tile in a new shopping mall. System provides easy access to mechanical systems. (USG Acoustical Products Co.)

For those who have ability and are willing to work, the field of carpentry is a satisfying, fulfilling, and lucrative occupation. Advancement is limited only by your willingness to try new skills and to seize opportunities as they present themselves.

Entrepreneurship

Earlier in this chapter you read that four out of every ten carpenters are self-employed. In other words, they are *entrepreneurs*. (This means starting and operating a business of one's own.)

There are advantages and disadvantages to operating your own construction company. First, you may find that you gain more satisfaction in certain tasks. As your own boss you can choose to specialize in these areas and thus offer your services to the community as a subcontractor. You are free to concentrate in the area of your specialty and develop superior skills and competence that allow you to compete with other firms.

Of course, there are disadvantages to owning the company. Failure is possible and you can spend years repaying loans and debts incurred. Your income could be uncertain, depending on how well the business goes or how healthy the economy is. You will have to work longer hours and many of the tasks—paperwork, maintenance, customer relations, setting prices, and organizing your day—are difficult.

Characteristics of Entrepreneurs

A successful entrepreneur, Figure 30-12, must possess certain characteristics to be successful. The list following summarizes the major ones:

- *Good health*—long hours and much physical labor make heavy demands on the carpenter or construction contractor.
- *Knowledgeable*—to make a profit, he or she must know all aspects of the trade or trade specialty. In addition, the person understands the industry and the products being used in the trade.
- *Good planner*—running a successful business means that nothing is left to chance. She or he must be able to foresee difficulties as well as plan how to take advantage of opportunities.
- *Willing to take calculated risks*—once a plan has been conceived that takes into account events likely to occur, the person must have the courage to risk money and future on making the plan work.
- *Innovative*—successful in finding ways to improve and find ways to produce better work and thus gain the confidence of customers.
- *Responsible*—willing to accept the consequences of decision whether good or bad. This includes paying debts, keeping promises, and accepting the responsibility for mistakes of his or her employees.
- *Goal oriented*—likes to set goals and works hard to achieve them. It has been predicted that by the year 2000, there will be a large increase in persons 30 to 60 years old. The population in this group will demand many services. Housing is one of the areas predicted to be in great demand. Thus, a business designed to offer services such as home construction or rehabbing of homes would likely do well.

Figure 30-12. Entrepreneurs are willing to take risks. This woman left another career to become a general contractor. Here she inspects a remodeling job with one of her subcontractors. (RECON —Reconstruction Unlimited, Inc.)

Teaching as a Construction Career

It is not unusual for successful and skilled carpenters to become vocational instructors. For some, this is a part-time occupation. Others may leave the construction field to teach their skills to others. This may involve returning to school to take courses leading to accreditation or getting a degree in education. There are also some who, as students, take construction courses to become proficient in it to teach shop courses.

Instructors of carpentry not only work in high schools, vocational schools, and universities, but may be employed by tool and equipment manufacturers. In such cases, they may visit educational institutions to instruct other teachers and trainees on how to use the products manufactured by their employers. See Figure 30-13.

Figure 30-13. Teaching as a profession. Top—Teaching construction to beginners is a rewarding profession. Bottom—Some construction people become expert in the use of new construction technology. A representative of a tool company demonstrates the proper use of a laser level. (St. Paul Technical College)

Organizations Promoting Construction Training

There are organizations that promote development of excellence in construction. One of them, Vocational Industrial Clubs of America (VICA) holds state and national competitions to encourage students in a variety of occupations, including building trades, Figure 30-14.

Figure 30-14. Carpentry trainees from all states compete at national VICA skills contest.
(Vocational Industrial Clubs of America)

In the construction trade, associations actively promote and support training in construction fields. Two of the most influential are the Associated General Contractors (AGC) and the National Association of Home Builders (NAHB). The major emphasis of AGC is to represent contractors who do heavy construction work. The NAHB concentrates its efforts on residential and light commercial construction.

The United Brotherhood of Carpenters and Joiners of America is a union that looks after the interests of construction workers. While most members are carpenters a wide spectrum of construction workers also belong. Membership includes cabinetworkers, pile drivers, millwrights, floor covering workers, and a host of industrial workers employed by factories turning out plywood, lumber, and other construction products. Among their membership are 27,000 women from many construction occupations.

Important Terms

Apprenticeship	Operating engineers
Entrepreneur	Manual dexterity

Test Your Knowledge

1. Opportunities for jobs in construction will be greatest in the next ten years in the ____________ and ____________ regions of the country.
2. Current employment in carpentry is ____________.
   A. 90,000
   B. 190,000
   C. 690,000
   D. 990,000
3. What are the main disadvantages of all carpentry occupations?
4. Four out of every ____________ carpenters work for themselves.
5. List three reasons why you are or may be planning a career in carpentry.
6. An entrepreneur is someone who ____________.

Outside Assignments

1. Find out how to enter a carpentry apprenticeship program in your locality. Contact a local carpentry contractor, a local of a union, and a state apprenticeship agency. What skills do you have now? What skills must you learn? Report your findings to your class.
2. Interview a carpenter about his or her attitude about what he or she does for a living. Prepare a list of possible questions to ask.
3. Obtain literature from the main associations in construction and report what they say about opportunities in construction.

Carpentry careers are open for anyone willing and able to acquire the necessary skills. Here, a student at Santa Rita High School, Tucson, Arizona helps build playhouses in the school's carpentry program. Students in the program have helped build houses in the "Habitat for Humanity" program.

A carpenter prepares to frame up a ceiling for indirect lighting in a building being rehabbed for an architect's studio. (Loren LaPointe)

Appendix A—Carpentry Math Review

A carpenter or carpenter apprentice must be able to take measurements and make calculations. These skills require an understanding of certain mathematical concepts. This appendix will review these basic concepts as well as give examples of their application to carpentry. Most will require the ability to make accurate measurements of distance and volume.

These basic math skills are needed for two reasons:

- In order to make the measurements necessary to cut, fit, and locate members in a structure.
- In order to calculate the quantities of building materials required for a project and the cost.

Using Rules, Tapes, and Squares

One of the first skills a carpenter must master is the accurate reading of rules, tapes, and squares. Figure A-1 shows a portion of a rule showing the various divisions. Note that inch lines go all the way across the rule while each progressively smaller unit has progressively shorter marks. Each of these marks represents a fraction of an inch. The carpenter must be able to recognize the fractional inch by the length of the division mark. In the illustration, the smallest mark represents 1/16". The next larger marks represent 1/8" spaces. Larger marks represent 1/4" and 1/2" spaces.

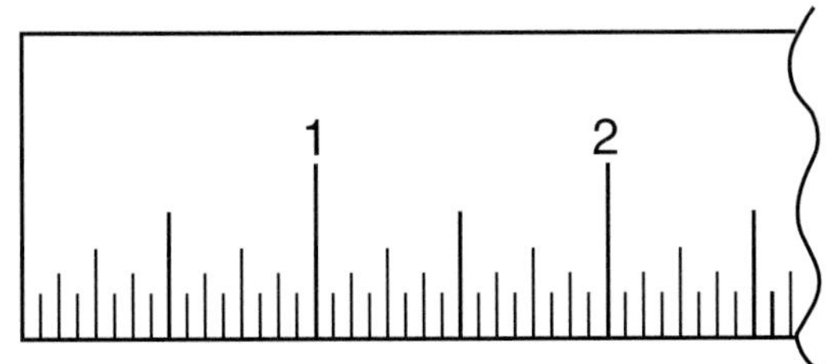

Figure A-1. A carpenter must be able to read a ruler accurately when taking measurements.

Working with Fractions and Decimals

A whole number is an undivided unit of anything—a number without a fractional or decimal portion. It can refer to measurements or quantities of material. Measurements deal with values such as length, area, or volume. There are several units of conventional measure. Inches, feet, and yards are units of length; square inches, square feet, and square yards are units of area; cubic inches, cubic feet, cubic yards, board feet, and gallons are units of volume.

A fraction or a decimal is part of a whole number. For example, 3/4 is three parts of a whole that has been separated into four parts; .1 is one part of a whole divided into 10 parts.

Fractions

Fractions are always written with one number above the other with a line separating them. The top number of a fraction is called the *numerator*; the bottom number is called the *denominator;* the line separating them is called a *fraction bar.*

The denominator tells how many parts the whole is divided into; the numerator tells how many parts are in the fraction concerned. See Figure A-2.

An *improper fraction* is a fraction where the numerator is a larger number than the denominator, such as 11/8. An improper fraction is always equal to a value greater than one.

Another way of expressing a value greater than one is as a *mixed number*—a number with a whole number part and a fractional part, such as 1 3/8. To convert a mixed number to an improper fraction, do the following:

- To find the numerator of the improper fraction, multiply the denominator by the whole number and then add the numerator.
- The denominator of the improper fraction is the same denominator as the fractional part of the mixed number.

Example:

Convert 2 5/8 to an improper fraction.

Numerator = (8 x 2) + 5 = 16 + 5 = 21
Denominator = 8
2 5/8 = 21/8

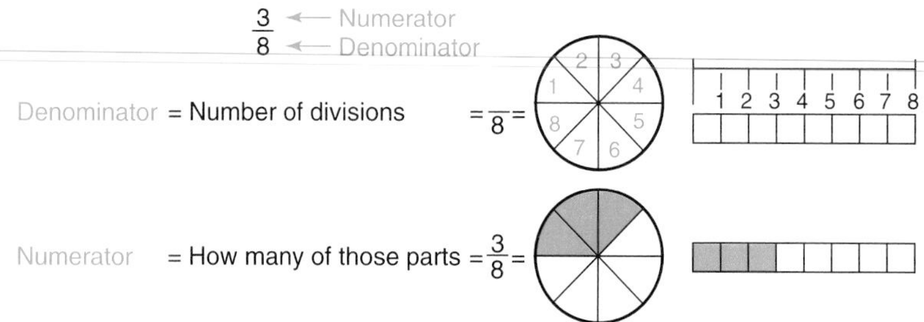

Figure A-2. The denominator tells how many pieces a whole is divided into. The numerator tells how many of those pieces are counted.

It is easier to visualize mixed numbers than improper fractions. For example, if someone told you that they drove exactly 15 3/16 miles to work, you can easily understand. But if they said they drove 243/16 miles, you wouldn't have an immediate idea of how far they traveled.

However, improper fractions are far easier to work with in calculations. Therefore, when you have a problem and a mixed number is given, convert it to an improper fraction. Then do the calculation and convert any improper fractions in your final answer to mixed numbers so it is easier to understand.

A whole number can also be expressed as a fraction. The number becomes the numerator and "1" is the denominator. For example, the whole number 7 is the same as the fraction 7/1.

Multiplying Fractions

To multiply fractions, change mixed numbers and whole numbers to improper fractions. Then multiply the numerators and then the denominators. The result is called a product. Reduce the product to its lowest form.

Examples:

Multiply 3/8 x 1/4
- Numerator: 3 x 1 = 3
- Denominator: 8 x 4 = 32
- 3/8 x 1/4 = 3/32

Multiply 3 5/8 x 1/2
- Convert mixed number to improper fraction:
- 3 5/8 = 29/8
- Numerator: 29 x 1 = 29
- Denominator: 8 x 2 = 16
- 3 5/8 x 1/2 = 29/16
- = 1 13/16

Dividing Fractions

There are three components to a division equation: the divisor, the dividend, and the quotient. These components are shown in Figure A-3, along with different ways of writing division equations. As you can see, a fraction is actually a division operation—the numerator is the dividend and the denominator is the divisor.

Dividend → 12 ÷ 3 = 4 ← Quotient (Divisor: 3)

Divisor → 3)12 ← Dividend; 4 ← Quotient

Dividend → 12 / Divisor → 3 = 4 ← Quotient

Dividend → $\frac{3}{8} \div \frac{4}{7} = \frac{21}{32}$ ← Quotient (Divisor: 4/7)

Figure A-3. There are several ways to write a division problem. The dividend divided by the divisor equals the quotient.

The process of switching the numerator and denominator of a fraction is called inversion. In order to divide fractions, simply invert the divisor (the number you are dividing by) and then multiply the two fractions.

Example:

Divide 2/3 by 7/4 (2/3/7/4)
- Invert the divisor: 7/4 → 4/7
- Multiply the dividend by the inverted divisor:
- 2/3 x 4/7 = 8/21
- 2/3 ÷ 7/4 = 8/21

Changing Denominators

Any fraction can be written in many ways. For example, 1/2 is the same as 3/6 and 6/12. There are situations where it is preferable to change the denominator of the fraction. In order to convert between equivalent (equal) fractions, you simply multiply both the numerator and denominator by the same number.

Example:

Find the equivalent fraction for 1/3 that has 12 as its denominator. Because you multiply 3 by 4 to get 12, multiply the numerator (1) and denominator (3) both by 4 to find the equivalent fraction:

Numerator: 1 x 4 = 4
Denominator: 3 x 4 = 12
1/3 = 4/12

Adding Fractions

Fractions must have identical denominators in order to add them together. They must have a *common denominator.* When fractions have a *common denominator,* they are added in the following way: the new numerator is obtained by adding the numerators together, the new denominator is the common denominator.

Example:

Add 3/8 + 1/8

Denominators are the same, so add numerators to get new numerator:
1 + 3 = 4
Keep the same denominator (8)
3/8 + 1/8 = 4/8
= 1/2

When fractions being added have different denomina tors, one or both of the fractions need to be converted to an equivalent fraction with a common denominator.

Example:

Add 3/4 + 3/16

The denominators are different, so one must be changed. Convert 3/4 to an equivalent fraction with 16 as a denominator:
3/4 x 4/4 = 12/16
Add: 12/16 + 3/16 = 15/16

Subtracting Fractions

Subtracting and adding fractions are very similar operations. As with addition, fractions must have common denominators to subtract. The only difference is that you subtract the numerators, rather than adding them.

Example:

Subtract: 3/4–1/4

The denominators are the same, so subtract numertors:
3/4 1/4 = 2/4
= 1/2

Example:

Subtract: 4/5–3/15

Multiply 4/5 by 3/3 to obtain a common denominator (15): 4/5 x 3/3 = 12/15

Subtract numerators:
12/15–3/15 = 9/15
= 3/5

Decimals

A decimal number has three parts—a whole number part, a decimal point, and a decimal part, Figure A-4. The decimal part is like the numerator in a fraction. The denominator in this fraction is "1" followed by the same number of zeros as there are numbers after the decimal point. A decimal number is another way of expressing a mixed number.

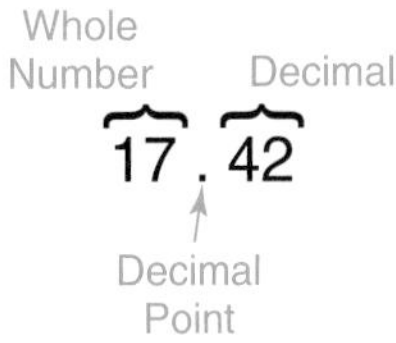

Figure A-4. A non-whole number can be expressed as a decimal.

Example:

Convert 17.31 to a mixed number.

There are two digits to the right of the decimal point, so there will be two zeros following the "1" in the denominator, with 31 in the numerator.
17.31 = 17 31/100

Example:

Convert 2.004 to a mixed number.

There are three digits to the right of the decimal point, so there will be three zeros following the "1" in the denominator.
2.004 = 2 4/1000 = 2 1/250

Figure A-5 is a place value chart. It shows the value of numbers to the left and right of a decimal point. Notice that as numbers are added to the left their values increase; as numbers are added further to the right, their values decrease.

Decimal Places

Ten-Thousand's Place	Thousand's Place	Hundred's Place	Ten's Place	One's Place	Tenth's Place	Hundredth's Place	Thousandth's Place	Ten-Thousandth's Place
1	2,	7	2	6.	6	5	7	9
10,000	1,000	100	10	1	1/10	1/100	1/1000	1/10,000

Figure A-5. Number place chart. Values of each places increases in multiples of ten as you move to the left of the chart.

While zeros have no value of themselves, it is very important to the value of the complete number. The zero is a "place holder." It is used to keep the other numbers in their place. For example, the decimal number 0.56 is larger than 0.056, the first being 56/100 and the second 56/1000; the zero makes the second number smaller than the first.

Adding and Subtracting Decimal Numbers

To add and subtract decimals, write the numbers in a column with their decimal points aligned. Then add or subtract as if they were whole numbers. Carry the decimal point down into the answer.

Example:

Add 2.6 + 10.5 + 8.803 + 5 + 0.0155

Arrange in a column aligning decimal points

```
 2.6000
10.5000
 8.8030
 5.0000
 0.0155
-------
26.9185
```

The number of zeros at the far right end of the decimal portion does not matter, as long as they are at the very end of the number. For example, 45.87 and 45.870000 are the same number; 23.002 and 23.0000002 are different numbers because there is a non-zero digit to the right of the zeros.

Multiplying Decimal Numbers

Multiplying decimal numbers is a two-step process. First, multiply the numbers together, ignoring the decimal points. Then count the total number of decimal places in the numbers being multiplied and put the same number of decimal places in the product. If there are more decimal places than numbers in the product, add zeros to the left to get enough places.

Example:

Multiply 3.511 x 2.2

Ignoring decimal points, multiply the numbers:
3511 x 22 = 77242

There are four decimal places in the numbers being multiplied, so there must be four decimal places in the product: 7.7242

3.511 x 2.2 = 7.7242

Whenever doing calculations of any kind, be sure to check and see if your answer is reasonable. In the last example, you could round the two factors to the nearest whole numbers (4 and 2) and multiply these together (8) to see if your answer makes sense.

Dividing Decimal Numbers

Using decimals allows you to divide numbers that will not divide evenly. The division remainder is a fractional part. It can be written as a fraction or the division can be continued for as many decimal places as desired.

To divide using decimal places:

- If neither divisor nor dividend contain decimal points and the division does not come out exactly (See 17÷8 in Figure A-6):
  - A. Place a decimal point to the right of the dividend and after the last number in the answer.
  - B. Add zeros to the dividend.
  - C. Continue dividing to as many decimal places as desired.

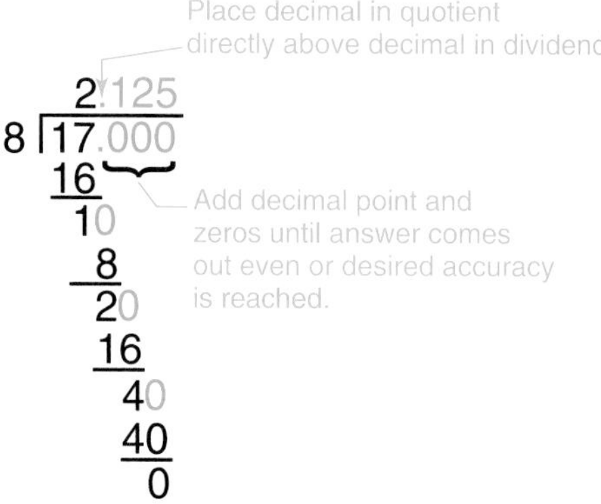

Figure A-6. Division problem in which both divisor and dividend are whole numbers.

- If the dividend contains a decimal point and the divisor does not (See 17.25÷8 in Figure A-7):
  - A. Divide as usual up to the decimal point.
  - B. Place a decimal point to the right of the last num ber in the answer.
  - C. Continue dividing the rest of the dividend. Add zeros to the dividend until the answer is carried out to the number of decimal places desired.
- If the divisor contains a decimal point (See 8.925÷2.1 in Figure A-8):
  - A. Move the decimal points to the right end of the divisor.
  - B. Move the decimal point in the dividend to the right the same number of places.
  - C. Divide, being sure to put the decimal point in the quotient directly above the new location of the decimal point in the dividend.

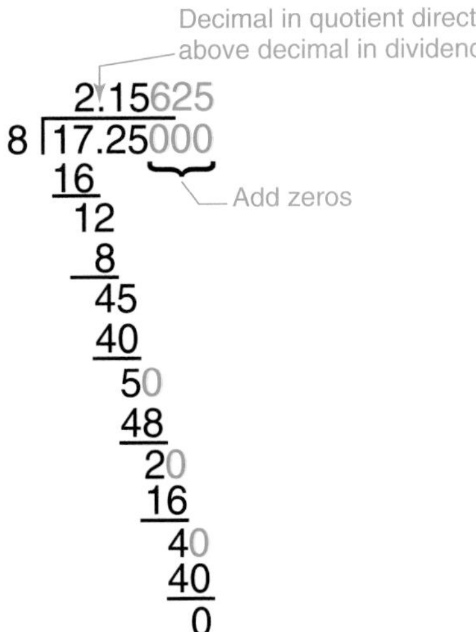

Figure A-7. Division problem in which the divisor is a whole number and the dividend is a decimal.

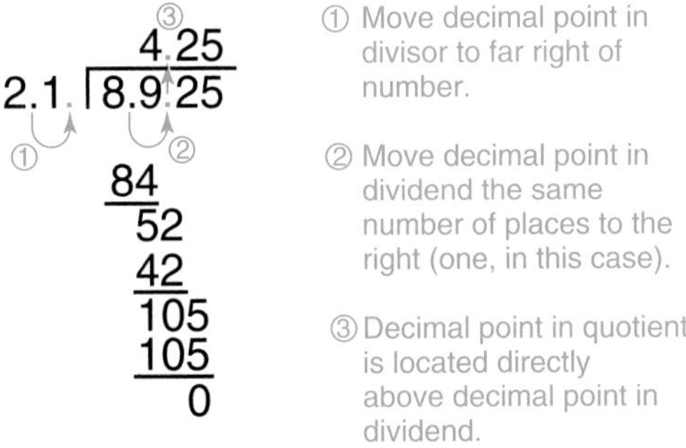

Figure A-8. Division problem in which the divisor is a decimal.

Converting Between Decimals and Fractions

The system for units of measure in the United States uses both fractions and decimals to indicate parts of a whole. While rulers are marked in inches and fractions, micrometers are marked in decimals. Sometimes it is necessary to convert from one to the other.

Converting Decimals to Fractions

1. Let the number or numbers to the right of the decimal point be the numerator.
2. Count off the places to the right of the decimal (one place = tenths; two places = hundredths; three places = thousandths, and so on).
3. Let this number be the denominator.
4. Reduce the resulting fraction to its lowest terms.

Examples:

0.5 = 5/10 = 1/2
0.50 = 50/100 = 5/10 = 1/2
0.175 = 175/1000 = 7/40

Converting Fractions to Decimals

1. Divide the numerator by the denominator.
2. If it does not divide exactly, round off to as many decimal places desired.
   See Figure A-9 for an example.

3 ÷ 4

```
   .75
4 )3.00
   28
    20
    20
     0
```

Figure A-9. Converting a fraction (3/4) to a decimal (0.75).

Knowing and Using Formulas

Often carpenters must use formulas. Given a certain problem to solve, the carpenter must know what formula to use and how to work the formula.

Area

The area of square or rectangular shapes can be determined by multiplying the length of two adjoining sides. A carpenter needs to find the area of rectangles in order to determine amounts of material needed to cover walls, floors, ceilings, and roofs. The product is given in "square" units, such as square feet or square yards.

Often, the two sides of a rectangle are referred to as length and width (L and W, respectively). Then the formula for area becomes: L x W=Area

Example:

What is the area of a wall 48' long and 8' high?

48' x 8' = 384 sq. ft.

When calculating for areas, or using any other formula, attention must be paid to the units of measurements. When multiplying two measurements with units of feet together, the answer will be in square feet. When the measurements are taken in inches, the answer will have units of square inches. Measurements with different units cannot be combined—they must be converted to a single unit.

Example:

What is the area of sidewalk that is 30" wide and 15' long?

The measurements are in different units, so they cannot be combined. Change 30" to 2.5' and multiply:

2.5' x 15'= 37.5 sq. ft.

Example:

How many sheets of 4' x 8 ' plywood are needed to cover a wall 8', high by 48' long?

Area of one sheet of plywood: 4' x 8'	= 32 sq. ft.
Area of wall: 8' x 48'	= 384 sq. ft.
Number of sheets: 384÷32	= 12 sheets required

Volume

Volume is a measure of three dimensions, length, width, and height. The product (of dry measure) is given in cubic inches, cubic feet, or cubic yards.

To find the volume of a cube, box, or rectangular cylinder, multiply the three dimensions together.

Example:

An excavation is to be dug for the basement of a build ing. The excavation is to be 78' long by 54' wide by 8' deep. How much dirt will be moved?

Earthwork and concrete quantities are usually mea sured by cubic yards (1 cubic yard = 27 cubic feet).

Volume = 78' x 54 ' x 8' = 33,696 cu. ft.

Volume = 33,696 ft. = 1248 cu. yd.

(33696 divided by 27)

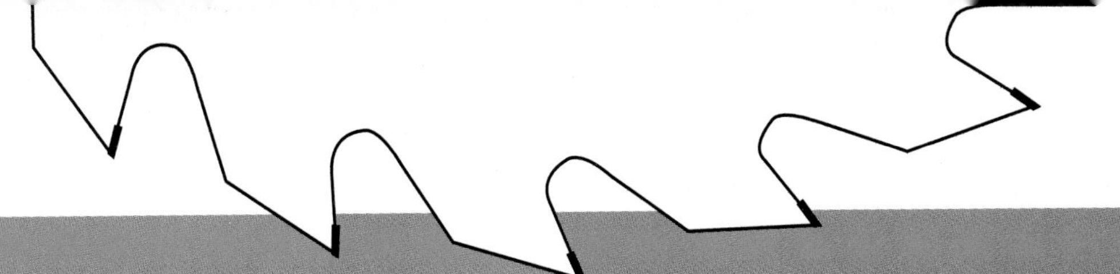

Appendix B—Technical Information

Contents

Standard Abbreviations for Use on Drawings

Above Finished Flooring	AFF
Acoustical Tile	AT or ACT
Aggregate	AGGR
Air Conditioning	AIR COND
Air Dried	AD
Alternate	ALT
Alternating Current	AC
Aluminum	AL
American Institute of Architects	AIA
American Institute of Electrial Engineers	AIEE
American Society for Testing and Materials	ASTM
American Standards Association, Inc.	ASA
Approximate	APPROX
Architectural	ARCH
Asbestos	ASB
Asphalt Roof Shingles	ASPHRS
Basement	BSMT
Batter	BAT
Beam	BM
Better	BTR
Beveled	BEV
Blocking	BLKG
Board	BD
Board Foot	BD FT
Brick	BRK
British Thermal Unit	BTU
Building	BLDG
Bundle	BDL
Cabinet	CAB
Carpenter	CPNTR
Casing	CSG
Cement	CEM
Cement Floor	CEM FL
Cement Motar	CEM MORT
Center Matched	CM
Chimney	CHM
Circuit	CKT
Closet	Cl or CLO
Column	COL
Common	COM
Concrete	CONC
Concrete Block	CONC B
Conduit	CND
Construction	CONST
Counter	CTR
Coupling	CPLG
Cubic Foot	CU FT
Cubic Yard	CU YD
Cutoff Valve	COV
Diagram	DIAG
Diameter	DIA or DIAM
Dimension	DIM
Direct Current	DC
Ditto	DO
Door	DR
Dormer	DRM
Double Strength Glass	DSG
Double-Hung Windows	DHW
Drain	DR
Drawing	DWG
Dressed and Matched	D&M
Each	EA
Edge	EDG
Edge Grain	EG
Electric Panel	EP
Elevation	EL
Entrance	ENT
Excavate	EXC
Exhaust Vent	
EXHV	
Exterior	EXT
Face Brick	FB
Federal Housing Authority	FHA
Finish	FIN
Finished Floor	FNSHFL or FF
Fixture	FIX
Flashing	FL
Flat Grain	FG
Flooring	FLG
Fluorescent	FLUOR
Flush	FL
Foot or Feet	FT
Footing	FTG
Foundation	FDN
Furring	FUR
Fuse	FU
Gage	GA
Gallon	GAL
Galvanize	GALV
Galvanized Iron	GI
Glass	GL
Grade	GR
Gypsum	GYP
Hardboard	HBD
Hardwood	HDWD
Header	HDR
Heat Exchanger	HE
Herringbone	HGBN
Horsepower	HP
Hose Bib	HB
Hot Water	HW
Hundred	C
Insulation	INS
Interior	INT
Iron Pipe	IP
Jamb	JMB
Joist	J
Keyway	KWY
Kiln-dried	KD
Kitchen	K
Knee Brace	KB
Knife Switch	KNSW
Laminate	LAM
Lath	LTH
Lattice	LTC
Lavatory	LAV
Length	LGTH
Level	LVL
Light	LT
Light Switch	LTSW
Linen Closet	L CL
Linoleum	LINO
Lintel	LNTL
Living Room	LR
Low Voltage	LV
Masonry Opening	MO
Mastic	MSTC
Material	MATL
Maximum	MAX
Medicine Cabinet	MC
Minimum	MIN
Miscellaneous	MISC
Modular	MOD
Molding	MLDG
Mortar	MOR
Nominal	NOM
Nosing	NOS
On Center	OC
Open Web Joint	OWJ
Opening	OPNG
Paint	PNT
Pair	PR
Partition	PTN
Perpendicular	PERP
Pilaster	P
Piping	PP
Plank	PLK
Plaster	PLAS
Plate	PL
Plate Glass	PL GL
Plumbing	PLBG
Power	PWR
Precast	PRCST
Prefabricated	PREFAB
Quart	QT
Random	RDM
Receptacle	RCPT
Recess	REC
Reference	REF
Refrigerator	REF
Reinforcing	REINF

Revision	REV
Roll Roofing	RR
Roof	RF
Rough	
RGH	
Rough Opening	RO
Saddle	SDL
Schedule	SCH
Screen	SCR
Select	SEL
Service	SERV
Sewer	SEW
Sheathing	SHTHG
Shelving	SHELV
Shiplap	S/LAP
Siding	SDG
Specifications	SPEC
Square	SQ
Square Feet	SQ FT
Stairway	STWY
Standard	STD
Steel	ST or STL
Stringer	STGR
Structural	STR
Surfaced Four Sides	S4S
Surfaced One Side	S1S
Surfaced One Side and Two Edges	S1S2E
Surfaced Two Sides	S2S
Switch	SW or S
Temperature	TEMP
Thermostat	THERMO
Thick	THK
Thousand	M
Timber	TMBR
Tongue and Groove	T&G
Transformer	XFMR
Truss	TR
Tubing	TBG
Typical	TYP
Union	UN
Valley	VAL
Vent Stack	VS
Ventilation	VENT
Volt	V
Voltmeter	VM
Water Closet	WC
Water Heater	WH
Weather Stripping	WS
Weep Hole	WH
Weight	WT
Welded Wire Fabric	WWF
Wide Flange	WF or W
With	W/
Without	W/O
Wood	WD

Metrics in Construction

Efficient building construction begins with standard sizes for construction parts. Designs geared for mass production are based on standard modules. Layouts (horizontal and vertical) use one specific size. Whole multiples of the size make up all larger measurements.

The U.S. customary system uses a basic module of 4". The layout grid is further divided into spaces of 16", 24", and 48". Modular design saves material and time.

As the United States converts to the metric system, the 4" module will be replaced by a 100 mm module. Smaller sizes (submultiples) would include 25 mm, 50 mm, and 75 mm. Large modules would measure 400 mm, 600 mm, and 1200 mm, as shown in the drawing below.

The 100 mm module is slightly smaller than the 4" module and the conversion can hardly be detected in small measurements. In a length of 48", however, the difference would be about 3/4" and a standard 4' x 8' plywood panel would be about 1 1/2" longer than the similar metric size (1200 mm x 2400 mm).

Manufacturers will need to change the size of many basic building materials. Also, some changes will be required in building components such as windows, doors, and cabinetwork.

Although the conversion to the metric system will cause some difficulties, there will be many advantages. International trade of building materials and products will be easier, especially for such items as plywood and hardwoods. Leaders in building construction believe the change will provide an unusual opportunity to discard obsolete sizing and grading patterns and develop new and more efficient systems.

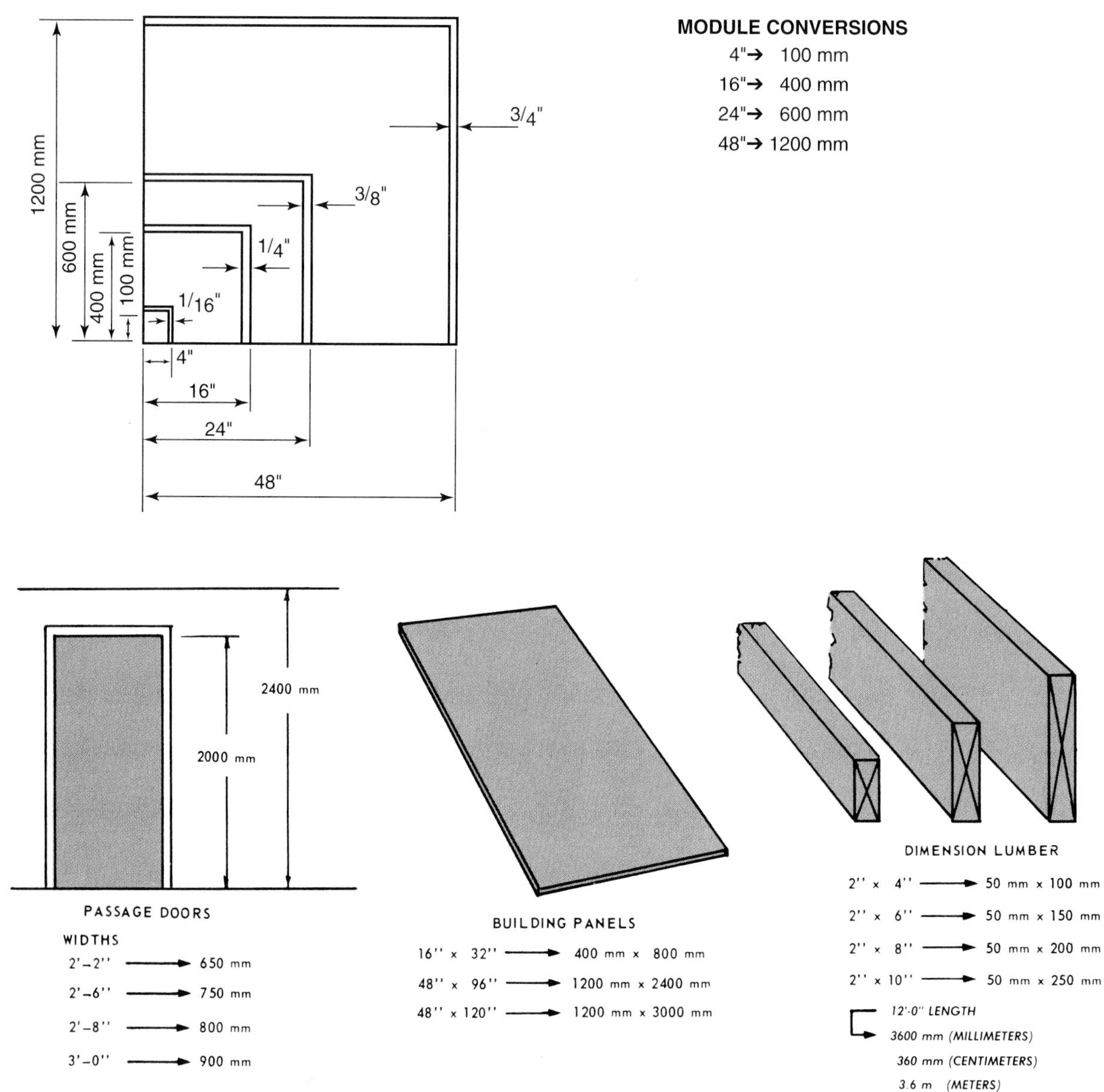

Metric standardization in building construction.

Millimeter-Inch Equivalents

INCHES		MILLI-METERS
FRACTIONS	DECIMALS	
	.00394	.1
	.00787	.2
	.01181	.3
1/64	.015625	.3969
	.01575	.4
	.01969	.5
	.02362	.6
	.02756	.7
1/32	.03125	.7938
	.0315	.8
	.03543	.9
	.03937	1.00
3/64	.046875	1.1906
1/16	.0625	1.5875
5/64	.078125	1.9844
	.07874	2.00
3/32	.09375	2.3813
7/64	.109375	2.7781
	.11811	3.00
1/8	.125	3.175
9/64	.140625	3.5719
5/32	.15625	3.9688
	.15748	4.00
11/64	.171875	4.3656
3/16	.1875	4.7625
	.19685	5.00
13/64	.203125	5.1594
7/32	.21875	5.5563
15/64	.234375	5.9531
	.23622	6.00
1/4	.2500	6.35
17/64	.265625	6.7469
	.27559	7.00
9/32	.28125	7.1438
19/64	.296875	7.5406
5/16	.3125	7.9375
	.31496	8.00
21/64	.328125	8.3344
11/32	.34375	8.7313
	.35433	9.00
23/64	.359375	9.1281
3/8	.375	9.525
25/64	.390625	9.9219
	.3937	10.00
13/32	.40625	10.3188
27/64	.421875	10.7156
	.43307	11.00
7/16	.4375	11.1125
29/64	.453125	11.5094
15/32	.46875	11.9063
	.47244	12.00
31/64	.484375	12.3031
1/2	.5000	12.70
	.51181	13.00
33/64	.515625	13.0969
17/32	.53125	13.4938
35/64	.546875	13.8907
	.55118	14.00
9/16	.5625	14.2875
37/64	.578125	14.6844
	.59055	15.00
19/32	.59375	15.0813
39/64	.609375	15.4782
5/8	.625	15.875
	.62992	16.00
41/64	.640625	16.2719
21/32	.65625	16.6688
	.66929	17.00
43/64	.671875	17.0657
11/16	.6875	17.4625
45/64	.703125	17.8594
	.70866	18.00
23/32	.71875	18.2563
47/64	.734375	18.6532
	.74803	19.00
3/4	.7500	19.05
49/64	.765625	19.4469
25/32	.78125	19.8438
	.7874	20.00
51/64	.796875	20.2407
13/16	.8125	20.6375
	.82677	21.00
53/64	.828125	21.0344
27/32	.84375	21.4313
55/64	.859375	21.8282
	.86614	22.00
7/8	.875	22.225
57/64	.890625	22.6219
	.90551	23.00
29/32	.90625	23.0188
59/64	.921875	23.4157
15/16	.9375	23.8125
	.94488	24.00
61/64	.953125	24.2094
31/32	.96875	24.6063
	.98425	25.00
63/64	.984375	25.0032
1	1.0000	25.4001

METRIC CONVERSIONS			
	To convert from	To	Multiply by
Length	inch	mm	25.4000
	mm	inch	0.0394
	foot	meter	0.3048
	meter	foot	3.2808
	mile	kilometer	1.6093
	kilometer	mile	0.6214
Area	in^2	mm^2	645.1600
	mm^2	in^2	0.0016
	ft^2	m^2	0.0929
	acre	hectare	0.4047
	hectare	acre	2.4104
Volume	ft^3	m^3	0.0283
	m^3	ft^3	35.3147
	in^3	mm^3	16,387.0640
	gallon	litter	3.7854
	liter	gallon	0.2642
	ft^3	litter	2.8316
	liter	ft^3	0.3532

Standard Sizes, Counts, and Weights

Nominal size is the size definition used by the trade, but it is not always the actual size. Sometimes the actual thickness of hardwood flooring is 1/32" less than the so-called nominal size.

Actual size is the mill size for thicknesses and face width. *Counted size* determines the board feet in a shipment. Pieces less than 1" in thickness are considered to be 1".

Nominal	Actual	Counted	Weights M Ft.
TONGUE AND GROOVE-END MATCHED			
** 3/4x3 1/4 in.	3/4x3 1/4 in.	1x4 in.	2210 lbs.
3/4x2 1/4 in.	3/4x2 1/4 in.	1x3 in.	2020 lbs.
3/4x2 in.	3/4x2 in.	1x2 3/4 in.	1920 lbs.
3/4x1 1/2 in.	3/4x1 1/2 in.	1x2 1/4 in.	1820 lbs.
** 3/8x2 in.	11/32x2 in.	1x2 1/2 in.	1000 lbs.
** 3/8x1 1/2 in.	11/32x1 1/2 in.	1x2 in.	1000 lbs.
** 1/2x2 in.	15/32x2 in.	1x2 1/2 in.	1350 lbs.
** 1/2x1 1/2 in.	15/32x1 1/2 in.	1x2 in.	1300 lbs.
SQUARE EDGE			
** 5/16x2 in.	5/16x2 in.	face count	1200 lbs.
** 5/16x1 1/2 in.	5/16x1 1/2 in.	face count	1200 lbs.

Nominal	Actual	Counted	Weights M Ft.
SPECIAL THICKNESSES (T and G, End Matched)			
** 33/32x3 1/4 in.	33/32x3 1/4 in.	5/4x4 in.	2400 lbs.
** 33/32x2 1/4 in.	33/32x2 1/4 in.	5/4x3 in.	2250 lbs.
** 33/32x2 in.	33/32x2 in.	5/4x2 3/4 in.	2250 lbs.
JOINTED FLOORING — i.e., SQUARE EDGE			
** 3/4x2 1/2 in.	3/4x2 1/2 in.	1x3 1/4 in.	2160 lbs.
** 3/4x3 1/4 in.	3/4x3 1/4 in.	1x4 in.	2300 lbs.
** 3/4x3 1/2 in.	3/4x3 1/2 in.	1x4 1/4 in.	2400 lbs.
** 33/32x2 1/2 in.	33/32x2 1/2 in.	5/4x3 1/4 in.	2500 lbs.
** 33/32x3 1/2 in.	33/32x3 1/2 in.	5/4x4 1/4 in.	2600 lbs.

**Special Order Only

NAIL SCHEDULE

Tongue And Groove Flooring Must Be Blind Nailed		
3/4x1 1/2, 2 1/4 & 3 1/4 in.	2 in. machine driven fasteners, 7d or 8d screw or cut nail.	10-12 in. apart*
3/4x3 in. to 8 in.** Plank	2 in. machine driven fasteners, 7d or 8d screw or cut nail.	8" apart into and between joists.

*If subfloor is 1/2 inch plywood, fasten into each joist, with additional fastening between.
**Plank Flooring over 4" wide must be installed over a subfloor.

Following Flooring Must Be Laid On A Subfloor		
1/2x1 1/2 & 2 in.	1 1/2 in. machine driven fastener, 5d screw, cut steel or wire casing nail.	10 in. apart
3/8x1 1/2 & 2 in.	1 1/4 in. machine driven fastener, or 4d bright wire casing nail.	8 in. apart
Square-Edge Flooring As Follows, Face-Nailed—Through Top Face		
5/16x1 1/2 & 2 in.	1 inch 15 gauge fully barbed flooring brad. 2 nails every 7 inches.	
5/16x1 1/3 in.	1 inch 15 gauge fully barbed flooring brad. 1 nail every 5 inches on alternate sides of strip.	

Tables showing hardwood flooring grades, sizes, counts, and weights make flooring selection easier. Follow the recommended nailing schedule for best results. (National Oak Flooring Manufacturer's Assoc.)

Guide to Hardwood Flooring Grades

Flooring is bundled by averaging the lengths. A bundle may include pieces from 6" under to 6" over the nominal length of the bundle. No piece is shorter than 9". Quantity with length under 4' held to stated percentage of total footage.

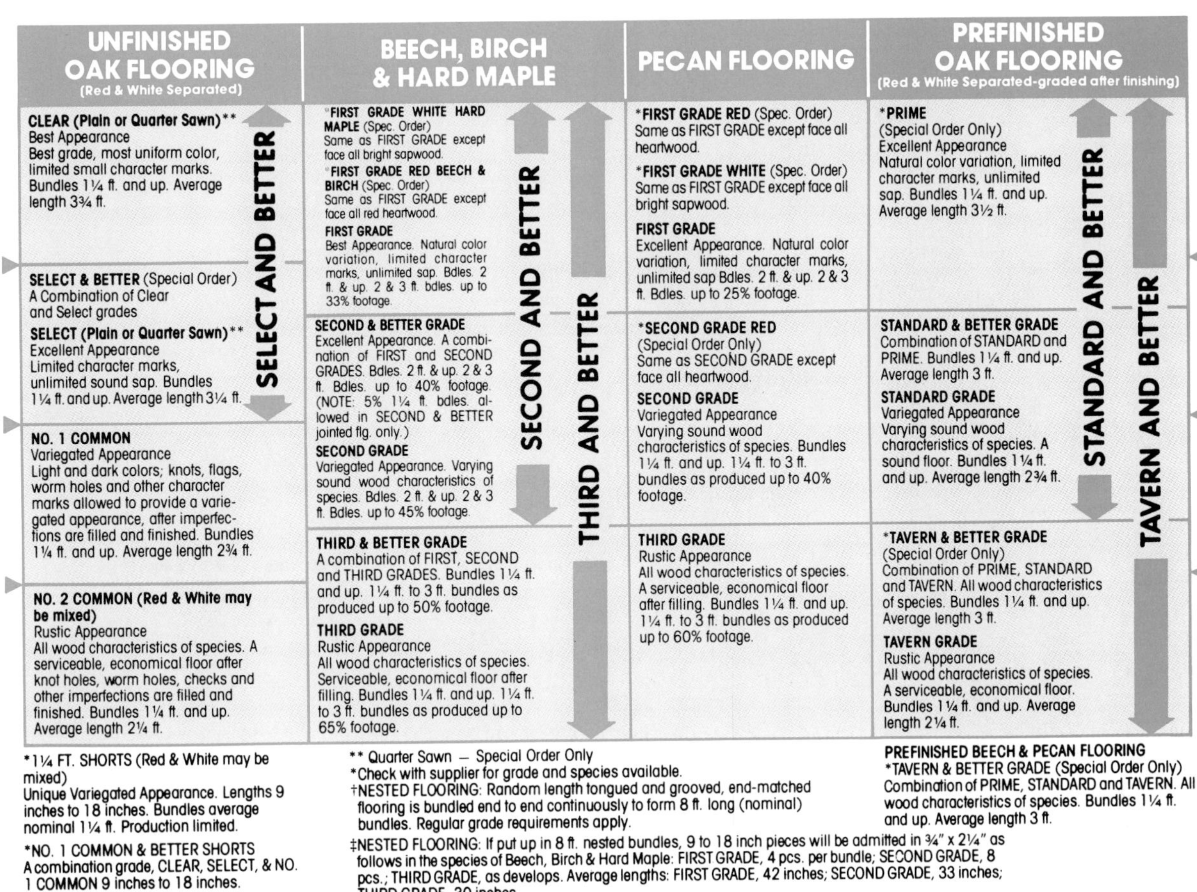

UNFINISHED OAK FLOORING (Red & White Separated)	BEECH, BIRCH & HARD MAPLE	PECAN FLOORING	PREFINISHED OAK FLOORING (Red & White Separated-graded after finishing)
CLEAR (Plain or Quarter Sawn)** Best Appearance Best grade, most uniform color, limited small character marks. Bundles 1¼ ft. and up. Average length 3¾ ft.	***FIRST GRADE WHITE HARD MAPLE** (Spec. Order) Same as FIRST GRADE except face all bright sapwood. ***FIRST GRADE RED BEECH & BIRCH** (Spec. Order) Same as FIRST GRADE except face all red heartwood. **FIRST GRADE** Best Appearance. Natural color variation, limited character marks, unlimited sap. Bdles. 2 ft. & up. 2 & 3 ft. bdles. up to 33% footage.	***FIRST GRADE RED** (Spec. Order) Same as FIRST GRADE except face all heartwood. ***FIRST GRADE WHITE** (Spec. Order) Same as FIRST GRADE except face all bright sapwood. **FIRST GRADE** Excellent Appearance. Natural color variation, limited character marks, unlimited sap Bdles. 2 ft. & up. 2 & 3 ft. Bdles. up to 25% footage.	***PRIME** (Special Order Only) Excellent Appearance Natural color variation, limited character marks, unlimited sap. Bundles 1¼ ft. and up. Average length 3½ ft.
SELECT & BETTER (Special Order) A Combination of Clear and Select grades **SELECT (Plain or Quarter Sawn)**** Excellent Appearance Limited character marks, unlimited sound sap. Bundles 1¼ ft. and up. Average length 3¼ ft.	**SECOND & BETTER GRADE** Excellent Appearance. A combination of FIRST and SECOND GRADES. Bdles. 2 ft. & up. 2 & 3 ft. Bdles. up to 40% footage. (NOTE: 5% 1¼ ft. bdles. allowed in SECOND & BETTER jointed flg. only.) **SECOND GRADE** Variegated Appearance. Varying sound wood characteristics of species. Bdles. 2 ft. & up. 2 & 3 ft. Bdles. up to 45% footage.	***SECOND GRADE RED** (Special Order Only) Same as SECOND GRADE except face all heartwood. **SECOND GRADE** Variegated Appearance Varying sound wood characteristics of species. Bundles 1¼ ft. and up. 1¼ ft. to 3 ft. bundles as produced up to 40% footage.	**STANDARD & BETTER GRADE** Combination of STANDARD and PRIME. Bundles 1¼ ft. and up. Average length 3 ft. **STANDARD GRADE** Variegated Appearance Varying sound wood characteristics of species. A sound floor. Bundles 1¼ ft. and up. Average length 2¾ ft.
NO. 1 COMMON Variegated Appearance Light and dark colors; knots, flags, worm holes and other character marks allowed to provide a variegated appearance, after imperfections are filled and finished. Bundles 1¼ ft. and up. Average length 2¾ ft.	**THIRD & BETTER GRADE** A combination of FIRST, SECOND and THIRD GRADES. Bundles 1¼ ft. and up. 1¼ ft. to 3 ft. bundles as produced up to 50% footage. **THIRD GRADE** Rustic Appearance All wood characteristics of species. Serviceable, economical floor after filling. Bundles 1¼ ft. and up. 1¼ ft. to 3 ft. bundles as produced up to 65% footage.	**THIRD GRADE** Rustic Appearance All wood characteristics of species. A serviceable, economical floor after filling. Bundles 1¼ ft. and up. 1¼ ft. to 3 ft. bundles as produced up to 60% footage.	***TAVERN & BETTER GRADE** (Special Order Only) Combination of PRIME, STANDARD and TAVERN. All wood characteristics of species. Bundles 1¼ ft. and up. Average length 3 ft. **TAVERN GRADE** Rustic Appearance All wood characteristics of species. A serviceable, economical floor. Bundles 1¼ ft. and up. Average length 2¼ ft.
NO. 2 COMMON (Red & White may be mixed) Rustic Appearance All wood characteristics of species. A serviceable, economical floor after knot holes, worm holes, checks and other imperfections are filled and finished. Bundles 1¼ ft. and up. Average length 2¼ ft.			

*1¼ FT. SHORTS (Red & White may be mixed)
Unique Variegated Appearance. Lengths 9 inches to 18 inches. Bundles average nominal 1¼ ft. Production limited.

*NO. 1 COMMON & BETTER SHORTS
A combination grade, CLEAR, SELECT, & NO. 1 COMMON 9 inches to 18 inches.

*NO. 2 COMMON SHORTS
Same as No. 2 COMMON, except length 9 inches to 18 inches.

** Quarter Sawn – Special Order Only
*Check with supplier for grade and species available.
†NESTED FLOORING: Random length tongued and grooved, end-matched flooring is bundled end to end continuously to form 8 ft. long (nominal) bundles. Regular grade requirements apply.
‡NESTED FLOORING: If put up in 8 ft. nested bundles, 9 to 18 inch pieces will be admitted in ¾" x 2¼" as follows in the species of Beech, Birch & Hard Maple: FIRST GRADE, 4 pcs. per bundle; SECOND GRADE, 8 pcs.; THIRD GRADE, as develops. Average lengths: FIRST GRADE, 42 inches; SECOND GRADE, 33 inches; THIRD GRADE, 30 inches.

PREFINISHED BEECH & PECAN FLOORING
*TAVERN & BETTER GRADE (Special Order Only)
Combination of PRIME, STANDARD and TAVERN. All wood characteristics of species. Bundles 1¼ ft. and up. Average length 3 ft.

NAILS

COMMON APPLICATIONS

Joining	Size & Type	Placement
Wall Framing		
Top plate	8d common 16d common	
Header	8d common 16d common	
Header to joist	16d common	
Studs	8d common 16d common	
Wall Sheathing		
Boards	8d common	6" o.c.
Plywood 5/16", 3/8", 1/2"	6d common	6" o.c.
Plywood (5/8", 3/4")	8d common	6" o.c.
Fiberboard	1 3/4" galv. roofing nail 8d galv. common nail	6" o.c. 6" o.c.
Foamboard	Cap nail, length sufficient for penetration of 1/2" into framing	12" o.c.
Gypsum	1 3/4" galv. roofing nail 8d galv. common nail	6" o.c. 6" o.c.
Subflooring	8d common	10"-12" o.c.
Underlayment	(1 1/4" x 14 ga. annular underlayment nail)	6" o.c. edges 12" o.c. face
Roof Framing		
Rafters, beveled or notched	12d common	
Rafter to joist	16d common	
Joist to rafter and stud	10d common	
Ridge beam	8d & 16d common	
Roof Sheathing		
Boards	8d common	
Plywood (5/16", 3/8", 1/2")	6d common	12" o.c. and 6" o.c. edges
Plywood (5/8", 3/4")	8d common	12" o.c. and 6" o.c. edges

*Aluminum nails are recommended for maximum protection from staining.

Joining	Size & Type	Placement
Roofing, Asphalt		
New construction shingles and felt	7/8" through 1 1/2" galv. roofing	4 per shingle
Re-roofing application shingles and felt	1 3/4" or 2" galv. roofing	4 per shingle
Roof deck/ Insulation	Thickness of insulation plus 1" insulation roof deck nail	
Roofing, Wood Shingles		
New construction	3d-4d galv. shingle	2-3 per shingle
Re-roofing application	5d-6d galv. shingle	2-3 per shingle
Soffit	6d-8d galv common	12" o.c. max.
Siding*		
Bevel and lap Drop and shiplap Plywood	Aluminum nails are recommended for optimum performance	Consult siding manufacturer's application instructions
Hardboard	Galvanized hardboard siding nail Galvanized box nail	Consult siding manufacturer's application instructions
Doors, Windows, Mouldings, Furring		
Wood strip to masonry Wood strip to stud or joist	Nail length is determined by thickness of siding and sheathing. Nails should penetrate at least 1 1/2" into solid wood framing.	
Paneling		
Wood	4d-8d casing-finishing	24" o.c.
Hardboard	2" x 16 ga. annular	8" o.c.
Plywood	3d casing-finishing	8" o.c.
Gypsum	1 1/4" annular drywall	6" o.c.
Lathing	4d common blued	4" o.c.
Exterior Projects:		
Decks, patios, etc.	8d-16d hot dipped galvanized common	

NOTE: Usage may vary somewhat due to regional differences and preferences.

(Georgia-Pacific)

DRYWALL SCREWS

Description	No.	Length	Applications
Bugle Phillips	1E 2E 3E 4E 5R 6R 7R 8R	6x1 6x1 1/8 6x1 1/4 6x1 5/8 6x2 6x2 1/4 8x2 1/2 8x3	For attaching drywall to metal studs from 25 ga. through 20 ga.
Coarse Thread	1C 2C 3C 4C 5C 6C	6x1 6x1 1/8 6x1 1/4 6x1 5/8 6x2 6x2 1/4	For attaching drywall to 25 ga. metal studs, and attaching drywall to wood studs
Pan Framing	19	6x7/16	For attaching stud to track up to 20 ga.
HWH Framing	21 22 35	6x7/16 8x9/16 10x3/4	For attaching stud to track up to 20 ga. where hex head is desired
K-Lath	28	8x9/16	For attaching wire lath, K-lath to 20 ga. studs
Laminating	8	10x1 1/2	Type G laminating screw for attaching gypsum to gypsum, a temporary fastener
Trim Head	9 10	6x1 5/8 6x2 1/4	Trim head screw for attaching wood trim and base to 25 ga. studs

(Compass International)

RECOMMENDED STYLES OF WELDED WIRE FABRIC REINFORCEMENT FOR CONCRETE

TYPE OF CONSTRUCTION	RECOMMENDED STYLE	REMARKS
Barbecue Foundation Slab	6x6-8/8 to 4x4-6/6	Use heavier style fabric for heavy, massive fireplaces or barbecue pits.
Basement Floors	6x6-10/10, 6x6-8/8 or 6x6-6/6	For small areas (15-foot maximum side dimension) use 6x6-10/10. As a rule of thumb, the larger the area or the poorer the sub-soil, the heavier the gauge.
Driveways	6x6-6/6	Continuous reinforcement between 25- to 30-foot contraction joints.
Foundation Slabs (Residential only)	6x6-10/10	Use heavier gauge over poorly drained sub-soil, or when maximum dimension is greater than 15 feet.
Garage Floors	6x6-6/6	Position at midpoint of 5- or 6-inch thick slab.
Patios and Terraces	6x6-10/10	Use 6x6-8/8 if sub-soil is poorly drained.
Porch Floor a. 6-inch thick slab up to 6-foot span b. 6-inch thick slab up to 8-foot span	 6x6-6/6 4x4-4/4	Position 1 inch from bottom form to resist tensile stresses.
Sidewalks	6x6-10/10 6x6-8/8	Use heavier gauge over poorly drained sub-soil. Construct 25- to 30-foot slabs as for driveways.
Steps (Free span)	6x6-6/6	Use heavier style if more than five risers. Position fabric 1 inch from bottom form.
Steps (On ground)	6x6-8/8	Use 6x6-6/6 for unstable sub-soil.

GYPSUM WALLBOARD APPLICATION DATA

THICKNESS	APPROX. WEIGHT LBS./SQ. FT.	SIZE	LOCATION	APPLICATION METHOD	MAX. SPACING OF FRAMING MEMBERS
¼″	1.1	4′ x 8′ to 12′	Over Existing Walls & Ceilings	Horizontal or Vertical	
⅜″	1.5	4′ x 8′ to 14′	Ceilings	Horizontal	16″
⅜″	1.5	4′ x 8′ to 14′	Sidewalls	Horizontal or Vertical	16″
½″	2.0	4′ x 8′ to 14′	Ceilings	Vertical Horizontal	16″ 24″
½″	2.0	4′ x 8′ to 14′	Sidewalls	Horizontal or Vertical	24″
⅝″	2.5	4′ x 8′ to 14′	Ceilings	Vertical Horizontal	16″ 24″
⅝″	2.5	4′ x 8′ to 14′	Sidewalls	Horizontal or Vertical	24″
1″	4.0	2′ x 8′ to 12′		For Laminated Partitions	

FLOOR JOIST SPAN DATA

Floor Joist Span Data

30 psi Live Load, 10 psi Dead Load, Def. <360										
Species or Group	Grade	2 x 8			2 x 10			2 x 12		
		12" oc	16" oc	24" oc	12" oc	16" oc	24" oc	12" oc	16" oc	24" oc
Douglas Fir and Larch	Sel. Struc.	16'-6"	15'-0"	13'-1"	21'-0"	19'-1"	16'-8"	25'-7"	23'-3"	20'-3"
	No. 1 & Btr.	16'-2"	14'-8"	12'-10"	20'-8"	18'-9"	16'-1"	25'-1"	22'-10"	18'-8"
	No. 1	15'-10"	14'-5"	12'-4"	20'-3"	18'-5"	15'-0"	24'-8"	21'-4"	17'-5"
	No. 2	15'-7"	14'-1"	11'-6"	19'-10"	17'-2"	14'-1"	23'-0"	19'-11"	16'-3"
	No. 3	12'-4"	10'-8"	8'-8"	15'-0"	13'-0"	10'-7"	17'-5"	15'-1"	12'-4"

40 psi Live Load, 10 psi Dead Load, Def. <360										
Species or Group	Grade	2 x 8			2 x 10			2 x 12		
		12" oc	16" oc	24" oc	12" oc	16" oc	24" oc	12" oc	16" oc	24" oc
Douglas Fir and Larch	Sel. Struc.	15'-0"	13'-7"	11'-11"	19'-1"	17'-4"	15'-2"	23'-3"	21'-1"	18'-5"
	No. 1 & Btr.	14'-8"	13'-4"	11'-8"	18'-9"	17'-0"	14'-5"	22'-10"	20'-5"	16'-8"
	No. 1	14'-5"	13'-1"	11'-0"	18'-5"	16'-5"	13'-5"	22'-0"	19'-1"	15'-7"
	No. 2	14'-2"	12'-7"	10'-3"	17'-9"	15'-5"	12'-7"	20'-7"	17'-10"	14'-7"
	No. 3	11'-0"	9'-6"	7'-9"	13'-5"	11'-8"	9'-6"	15'-7"	13'-6"	11'-0"

30 psi Live Load, 10 psi Dead Load, Def. <360										
Species or Group	Grade	2 x 8			2 x 10			2 x 12		
		12" oc	16" oc	24" oc	12" oc	16" oc	24" oc	12" oc	16" oc	24" oc
Southern Pine	Sel. Struc.	16'-2"	14'-8"	12'-10"	20'-8"	18'-9"	16'-5"	25'-1"	22'-10"	19'-11"
	No. 1	15'-10"	14'-5"	12'-7"	20'-3"	18'-5"	16'-1"	24'-8"	22'-5"	19'-6"
	No. 2	15'-7"	14'-2"	12'-4"	19'-10"	18'-0"	14'-8"	24'-2"	21'-1"	17'-2"
	No. 3	13'-3"	11'-6"	9'-5"	15'-8"	13'-7"	11'-1"	18'-8"	16'-2"	13'-2"

40 psi Live Load, 10 psi Dead Load, Def. <360										
Species or Group	Grade	2 x 8			2 x 10			2 x 12		
		12" oc	16" oc	24" oc	12" oc	16" oc	24" oc	12" oc	16" oc	24" oc
Southern Pine	Sel. Struc.	14'-8"	13'-4"	11'-8"	18'-9"	17'-0"	14'-11"	22'-10"	20'-9"	18'-1"
	No. 1	14'-5"	13'-1"	11'-5"	18'-5"	16'-9"	14'-7"	22'-5"	20'-4"	17'-5"
	No. 2	14'-2"	12'-10"	11'-0"	18'-0"	16'-1"	13'-2"	21'-9"	18'-10"	15'-4"
	No. 3	11'-11"	10'-3"	8'-5"	14'-0"	12'-2"	9'-11"	16'-8"	14'-5"	11'-10"

40 psi Live Load, 10 psi Dead Load, Def. <240										
Species or Group	Grade	2 x 6			2 x 8			2 x 10		
		12" oc	16" oc	24" oc	12" oc	16" oc	24" oc	12" oc	16" oc	24" oc
Red-wood	Cl. All Heart		7'-3"	6'-0"		10'-9"	8'-9"		13'-6"	11'-0"
	Const. Heart		7'-3"	6'-0"		10'-9"	8'-9"		13'-6"	11'-0"
	Const. Common		7'-3"	6'-0"		10'-9"	8'-9"		13'-6"	11'-0"

Span data are in feet and inches for floor joists of Douglas Fir/Larch, Southern Yellow Pine, and California Redwood. Spans are calculated on the basis of dry sizes with a moisture content equal to or less than 19%. Floor joist spans are for a single span.

CEILING JOIST AND RAFTER SPAN DATA

Ceiling Joist Span Data

20 psi Live Load, 10 psi Dead Load, Def. <240

Drywall ceiling, No future room development, Limited attic storage available

Species or Group	Grade	2 x 4			2 x 6			2 x 8			2 x 10		
		12"oc	16"oc	24"oc	12"oc	16"oc	24"oc	12"oc	16"oc	24"oc	12"oc	16"oc	24"oc
Douglas Fir and Larch	Sel. Struc.	10-5	9-6	8-3	16-4	14-11	13-0	21-7	19-7	17-1	27-6	25-0	20-11
	No. 1 & Btr.	10-3	9-4	8-1	16-1	14-7	12-0	21-2	18-8	15-3	26-4	22-9	18-7
	No. 1	10-0	9-1	7-8	15-9	13-9	11-2	20-1	17-5	14-2	24-6	21-3	17-4
	No. 2	9-10	8-9	7-2	14-10	12-10	10-6	18-9	16-3	13-3	22-11	19-10	16-3
	No. 3	7-8	6-8	5-5	11-2	9-8	7-11	14-2	12-4	10-0	17-4	15-0	12-3

20 psi Live Load, 10 psi Dead Load, Def. <240

Drywall ceiling, No future room development, Limited attic storage available

Species or Group	Grade	2 x 4			2 x 6			2 x 8			2 x 10		
		12"oc	16"oc	24"oc	12"oc	16"oc	24"oc	12"oc	16"oc	24"oc	12"oc	16"oc	24"oc
Southern Pine	Sel. Struc.	10-3	9-4	8-1	16-1	14-7	12-9	21-2	19-3	16-10	26-0	24-7	21-6
	No. 1	10-0	9-1	8-0	15-9	14-4	12-6	20-10	18-11	15-11	26-0	23-2	18-11
	No. 2	9-10	8-11	7-8	15-6	13-6	11-0	20-1	17-5	14-2	24-0	20-9	17-0
	No. 3	8-2	7-1	5-9	12-1	10-5	8-6	15-4	13-3	10-10	18-1	15-8	12-10

Roof Rafter Span Data

20 psi Live Load, 10 psi Dead Load, Def. <240

Roof slope 3:12 or less, Light roof covering, No ceiling finish

Species or Group	Grade	2 x 6			2 x 8			2 x 10			2 x 12		
		12"oc	16"oc	24"oc	12"oc	16"oc	24"oc	12"oc	16"oc	24"oc	12"oc	16"oc	24"oc
Douglas Fir and Larch	Sel. Struc.	16-4	14-11	13-0	21-7	19-7	17-2	27-6	25-0	21-10	33-6	30-5	26-7
	No. 1 & Btr.	16-1	14-7	12-9	21-2	19-3	16-10	27-1	24-7	20-9	32-11	29-6	24-1
	No. 1	15-9	14-4	12-6	20-10	18-11	15-10	26-6	23-9	19-5	31-10	27-6	22-6
	No. 2	15-6	14-1	11-9	20-5	18-2	14-10	25-8	22-3	18-2	29-9	25-9	21-0
	No. 3	12-6	10-10	8-10	15-10	13-9	11-3	19-5	16-9	13-8	22-6	19-6	15-11

20 psi Live Load, 15 psi Dead Load, Def. <240

Roof slope greater than 3:12, Light roof covering, Drywall ceiling, No snow load

Species or Group	Grade	2 x 6			2 x 8			2 x 10			2 x 12		
		12"oc	16"oc	24"oc	12"oc	16"oc	24"oc	12"oc	16"oc	24"oc	12"oc	16"oc	24"oc
Douglas Fir and Larch	Sel. Struc.	16-4	14-11	13-0	21-7	19-7	17-2	27-6	25-0	21-7	33-6	30-5	25-1
	No. 1 & Btr.	16-1	14-7	12-5	21-2	19-3	15-9	27-1	23-7	19-3	31-7	27-4	22-4
	No. 1	15-9	14-3	11-7	20-9	18-0	14-8	25-5	22-0	17-11	29-5	25-6	20-10
	No. 2	15-4	13-3	10-10	19-5	16-10	13-9	23-9	20-7	16-9	27-6	23-10	19-6
	No. 3	11-7	10-1	8-2	14-8	12-9	10-5	17-11	15-7	12-8	20-10	18-0	14-9

20 psi Live Load, 10 psi Dead Load, Def. <240

Drywall ceiling, Light roofing, Snow load

Species or Group	Grade	2 x 6			2 x 8			2 x 10			2 x 12		
		12"oc	16"oc	24"oc	12"oc	16"oc	24"oc	12"oc	16"oc	24"oc	12"oc	16"oc	24"oc
Southern Pine	Sel. Struc.	16-1	14-7	12-9	21-2	19-3	16-10	26-0	24-7	21-6	26-0	26-0	26-0
	No. 1	15-9	14-4	12-6	20-10	18-11	16-6	26-0	24-1	20-3	26-0	26-0	24-1
	No. 2	15-6	14-1	11-9	20-5	18-6	15-3	25-8	22-3	18-2	26-0	26-0	21-4
	No. 3	12-11	11-2	9-1	16-5	14-3	11-7	19-5	16-10	13-9	23-1	20-0	16-4

30 psi Live Load, 15 psi Dead Load, Def. <240

Drywall ceiling, Medium roofing, Snow load

Species or Group	Grade	2 x 6			2 x 8			2 x 10			2 x 12		
		12"oc	16"oc	24"oc	12"oc	16"oc	24"oc	12"oc	16"oc	24"oc	12"oc	16"oc	24"oc
Southern Pine	Sel. Struc.	14-1	12-9	11-2	18-6	16-10	14-8	23-8	21-6	18-9	26-0	26-0	22-10
	No. 1	13-9	12-6	10-11	18-2	16-6	13-11	23-2	20-3	16-6	26-0	24-1	19-8
	No. 2	13-6	11-9	9-7	17-7	15-3	12-5	21-0	18-2	14-10	24-7	21-4	17-5
	No. 3	10-6	9-1	7-5	13-5	11-7	9-6	15-10	13-9	11-3	18-10	16-4	13-4

These spans are based on the 1993 AFPA (formerly NFPA) Span Tables for Joists and Rafters.
These grades are the most commonly available.

Ceiling joist and rafter span data are in feet and inches for Douglas Fir/Larch and Southern Yellow Pine. Spans are based on dry lumber size with moisture content equal to or less than 19%.

SPAN DATA FOR ROOF DECKING
WITH A MAXIMUM DEFLECTION OF 1/240TH OF THE SPAN
LIVE LOAD = 20 LBS./SQ. FT.

THICKNESS IN INCHES (NOMINAL)	LUMBER GRADE	SIMPLE SPANS	
		DOUGLAS FIR, LARCH, SOUTHERN YELLOW PINE	WESTERN RED CEDAR
		SPAN	SPAN
2	CONSTRUCTION	9'- 5"	8'- 1"
2	STANDARD	9'- 5"	6'- 9"
3	SELECT DEX.	15'- 3"	13'- 0"
3	COMPL. DEX.	15'- 3"	13'- 0"
4	SELECT DEX.	20'- 3"	17'- 3"
4	COMPL. DEX.	20'- 3"	17'- 3"

THICKNESS IN INCHES (NOMINAL)	LUMBER GRADE	RANDOM LENGTHS	
		DOUGLAS FIR, LARCH, SOUTHERN YELLOW PINE	WESTERN RED CEDAR
		SPAN	SPAN
2	CONSTRUCTION	10'- 3"	8'- 10"
2	STANDARD	10'- 3"	6'- 9"
3	SELECT DEX.	16'- 9"	14'- 3"
3	COMPL. DEX.	16'- 9"	13'- 6"
4	SELECT DEX.	22'- 0"	19'- 0"
4	COMPL. DEX.	22'- 0"	18'- 0"

THICKNESS IN INCHES (NOMINAL)	LUMBER GRADE	COMB. SIMPLE AND TWO-SPAN CONTINUOUS	
		DOUGLAS FIR, LARCH, SOUTHERN YELLOW PINE	WESTERN RED CEDAR
		SPAN	SPAN
2	CONSTRUCTION	10'- 7"	8'- 9"
2	STANDARD	10'- 7"	6'- 9"
3	SELECT DEX.	17'- 3"	14'- 9"
3	COMPL. DEX.	17'- 3"	13'- 6"
4	SELECT DEX.	22'- 9"	19'- 6"
4	COMPL. DEX.	22'- 9"	18'- 0"

MANUFACTURED 2"x 4" WOOD FLOOR TRUSSES

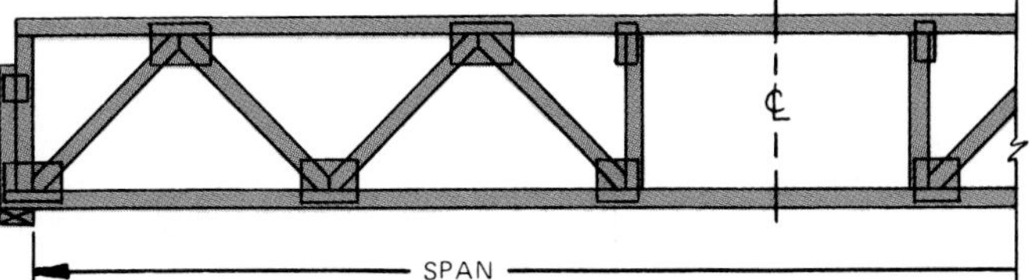

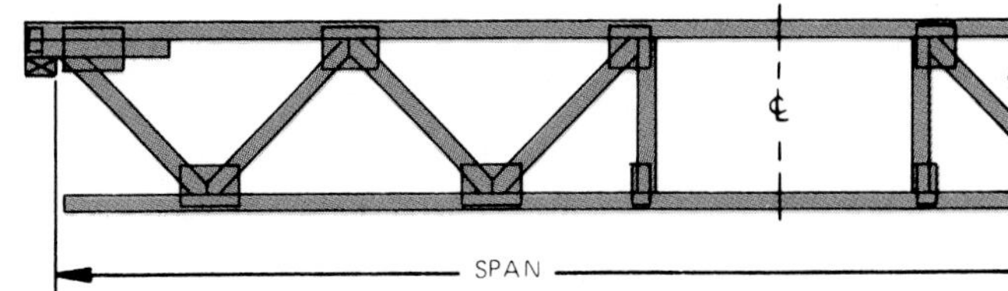

Bottom Chord Bearing Type			
DEPTH	CLEAR SPANS	# DIAGONAL WEBS	CAMBER
12"	7'-2"	4	0.063"
	9'-8"	6	0.063"
	12'-2"	8	0.063"
	14'-8"	10	0.134"
	17'-2"	12	0.237"
	19'-8"	14	0.365"
	21'-4"	16	0.507"
14"	9'-8"	6	0.063"
	12'-2"	8	0.063"
	14'-8"	10	0.095"
	17'-2"	12	0.178"
	19'-8"	14	0.288"
	22'-7"	16	0.449"
	24'-0"	18	0.569"
16"	12'-2"	8	0.065"
	14'-8"	10	0.070"
	17'-2"	12	0.132"
	19'-8"	14	0.228"
	22'-2"	16	0.346"
	25'-1"	18	0.505"
	26'-1"	20	0.596"
18"	14'-8"	10	0.065"
	17'-2"	12	0.120"
	19'-8"	14	0.176"
	22'-2"	16	0.268"
	24'-8"	18	0.367"
	27'-6"	20	0.600"
	27'-10"	22	0.630"
20"	14'-8"	10	0.063"
	17'-2"	12	0.081"
	19'-8"	14	0.140"
	22'-2"	16	0.226"
	24'-8"	18	0.327"
	27'-6"	20	0.451"
	29'-6"	22	0.630"
22"	17'-2"	10	0.066"
	19'-8"	12	0.114"
	22'-2"	14	0.184"
	24'-8"	16	0.266"
	27'-6"	18	0.367"
	30'-0"	20	0.520"
	31'-1"	22	0.630"
24"	17'-2"	12	0.063"
	19'-8"	14	0.095"
	22'-2"	16	0.153"
	24'-8"	18	0.235"
	27'-2"	20	0.325"
	30'-0"	22	0.431"
	32'-6"	24	0.630"

Top Chord Bearing Type			
DEPTH	CLEAR SPANS	# DIAGONAL WEBS	CAMBER
12"	6'-10"	4	0.063"
	9'-4"	6	0.063"
	11'-10"	8	0.063"
	14'-4"	10	0.122"
	16'-10"	12	0.233"
	19'-10"	14	0.376"
	21'-4"	16	0.507"
14"	9'-5"	6	0.063"
	11'-11"	8	0.063"
	14'-5"	10	0.088"
	16'-11"	12	0.167"
	19'-5"	14	0.273"
	21'-4"	16	0.429"
	24'-0"	18	0.569"
16"	12'-0"	8	0.063"
	14'-6"	10	0.067"
	17'-0"	12	0.126"
	19'-6"	14	0.219"
	22'-4"	16	0.337"
	24'-10"	18	0.489"
	26'-1"	20	0.596"
18"	14'-6"	10	0.063"
	17'-0"	12	0.098"
	19'-6"	14	0.170"
	22'-0"	16	0.260"
	24'-10"	18	0.378"
	27'-8"	20	0.617"
	27'-10"	22	0.630"
20"	14'-6"	10	0.063"
	17'-0"	12	0.079"
	19'-6"	14	0.136"
	22'-0"	16	0.221"
	24'-10"	18	0.337"
	27'-4"	20	0.442"
	29'-6"	22	0.630"
22"	17'-1"	12	0.065"
	19'-7"	14	0.112"
	22'-1"	16	0.181"
	24'-10"	18	0.275"
	27'-4"	20	0.381"
	30'-2"	22	0.534"
	31'-1"	24	0.630"
24"	17'-1"	12	0.063"
	19'-7"	14	0.093"
	22'-1"	16	0.150"
	24'-7"	18	0.231"
	27'-5"	20	0.335"
	30'-2"	22	0.443"
	32'-6"	24	0.630"

Wood floor trusses are typically manufactured from #3 Southern Yellow Pine. Pieces are joined together with 18 and 20 gauge galvanized steel plates applied to both faces of the truss at each joint. Where no sheathing is applied directly to top chords, they should be braced at intervals not to exceed 3'-0". Where no rigid ceiling is applied directly to bottom chords, they should be braced at intervals not to exceed 10'-0".

Manufactured wood floor trusses are generally spaced 24" o.c. and are designed to support various loads. Typical trusses shown here were designed to support 55 psf (live load - 40 psf, dead load - 10 psf, ceiling dead load - 5 psf). A slight bow (camber) is built into each joist so that it will produce a level floor when loaded. Allowable deflection is 1/360 of the span.

Some of the longer trusses require one or more double diagonal webs at both ends. Wood floor trusses are a manufactured product which must be engineered and produced with a high degree of accuracy to attain the desired performance. See your local manufacturer or lumber company for trusses available in your area.

LENGTHS OF COMMON RAFTERS

FEET OF RUN	2 in 12	2½ in 12	3 in 12	3½ in 12	4 in 12	4½ in 12
	Inclination (set saw at) 9°-28'	Inclination (set saw at) 11°-46'	Inclination (set saw at) 14°-2'	Inclination (set saw at) 16°-16'	Inclination (set saw at) 18°-26'	Inclination (set saw at) 20°-33'
	12.17 in. per ft. of run	12.26 in. per ft. of run	12.37 in. per ft. of run	12.5 in. per ft. of run	12.65 in. per ft. of run	12.82 in. per ft. of run
4'	4' 0-11/16"	4' 1-1/32"	4' 1-15/32"	4' 2"	4'-2-19/32"	4' 3-9/32"
5	5' 0-27/32"	5' 1-5/16"	5' 1-27/32"	5' 2-1/2"	5' 3-1/4"	5' 4-3/32"
6	6' 1-1/32"	6' 1-9/16"	6' 2-7/32"	6' 3"	6' 3-29/32"	6' 4-15/16"
7	7' 1-3/16"	7' 1-13/16"	7' 2-19/32"	7' 3-1/2"	7' 4-9/16"	7' 5-3/4"
8	8' 1-3/8"	8' 2-3/32"	8' 2-31/32"	8' 4"	8' 5-3/32"	8' 6-9/16"
9	9' 1-17/32"	9' 2-7/16"	9' 3-11/32"	9' 4-1/2"	9' 5-27/32"	9' 7-3/8"
10	10' 1-23/32"	10' 2-19/32"	10' 3-23/32"	10' 5"	10' 6-1/2"	10' 8-7/32"
11	11' 1-7/8"	11' 2-7/8"	11' 4-1/16"	11' 5-1/2"	11' 7-5/32"	11' 9-1/32"
12	12' 2-1/32"	12' 3-1/8"	12' 4-7/16"	12' 6"	12' 7-13/16"	12' 9-27/32"
13	13' 2-7/32"	13' 3-3/8"	13' 4-13/16"	13' 6-1/2"	13' 8-15/32"	13' 10-21/32"
14	14' 2-3/8"	14' 3-21/32"	14' 5-3/16"	14' 7"	14' 9-3/32"	14' 11-1/2"
15	15' 2-9/16"	15' 3-29/32"	15' 5-9/16"	15' 7-1/2	15' 9-3/4"	16' 0-5/16"
16	16' 2-23/32"	16' 4-5/32"	16' 5-15/16"	16' 8"	16' 10-13/32"	17' 1-1/8"
INCHES OF RUN	These lengths are to be added to those shown above when run involves inches					
1/4"	1/4"	1/4"	1/4"	1/4"	1/4"	9/32"
1/2"	1/2"	1/2"	1/2"	17/32"	17/32"	17/32"
1"	1"	1"	1-1/32"	1-1/32"	1-1/16"	1-1/16"
2"	2-1/32"	2-1/32"	2-1/16"	2-3/32"	2-3/32"	2-1/8"
3"	3-1/16"	3-1/16"	3-3/32"	3-1/8"	3-5/32"	3-7/32"
4"	4-1/16"	4-3/32"	4-1/8"	4-5/32"	4-7/32"	4-9/32"
5"	5-1/16"	5-1/8"	5-5/32"	5-7/32"	5-9/32"	5-11/32"
6"	6-3/32"	6-1/8"	6-3/16"	6-1/4"	6-5/16"	6-13/32"
7"	7-3/32"	7-5/32"	7-7/32"	7-9/32"	7-3/8"	7-15/32"
8"	8-1/8"	8-3/16"	8-1/4"	8-11/32"	8-7/16"	8-17/32"
9"	9-1/8"	9-3/16"	9-1/4"	9-3/8"	9-15/32"	9-5/8"
10"	10-5/32"	10-7/32"	10-5/16"	10-7/16"	10-17/32"	10-11/16"
11"	11-5/32"	11-1/4"	11-11/32"	11-15/32"	11-19/32"	11-3/4"

FEET OF RUN	5 in 12	5½ in 12	6 in 12	6½ in 12	7 in 12	7½ in 12	8 in 12
	Inclination (set saw at) 22°-37'	Inclination (set saw at) 24°-37'	Inclination (set saw at) 26°-34'	Inclination (set saw at) 28°-27'	Inclination (set saw at) 30°-15'	Inclination (set saw at) 32°-0'	Inclination (set saw at) 33°-41'
	13.00 in. per ft. of run	13.20 in. per ft. of run	13.42 in. per ft. of run	13.65 in. per ft. of run	13.89 in. per ft. of run	14.15 in. per ft. of run	14.42 in. per ft. of run
4	4' 4"	4' 4-13/16"	4' 5-11/16"	4' 6-19/32"	4' 7-9/16"	4' 8-19/32"	4' 9-11/16"
5	5' 5"	5'-6"	5' 7-1/8"	5' 8-1/4"	5' 9-15/32"	5' 10-3/4"	6' 0-1/8"
6	6' 6"	6' 7-3/32"	6' 8-1/2"	6' 9-29/32"	6' 11-11/32"	7' 0-29/32"	7' 2-17/32"
7	7' 7"	7' 8-13/32"	7' 9-15/16"	7' 11-9/16"	8' 1-7/32"	8' 3-1/16"	8' 4-15/16"
8	8' 8"	8' 9-19/32"	8' 11-3/8"	9' 1-7/32"	9' 3-1/8"	9' 5-7/32"	9' 7-3/8"
9	9' 9"	9' 10-13/16"	10' 0-25/32"	10' 2-27/32"	10' 5"	10' 7-11/32"	10' 9-25/32"
10	10' 10"	11' 0"	11' 2-7/32"	11' 4-1/2"	11' 6-29/32"	11' 9-1/2"	12' 0-7/32"
11	11' 11"	12' 1-3/32"	12' 3-5/8"	12' 6-5/32"	12' 8-13/16"	12' 11-21/32"	13' 2-5/8"
12	13' 0"	13' 2-13/32"	13' 5-1/32"	13' 7-13/16"	13' 10-11/16"	14' 1-13/16"	14' 5-1/32"
13	14' 1"	14' 3-19/32"	14' 6-15/32"	14' 9-15/32"	15' 0-9/16"	15' 3-31/32"	15' 7-15/32"
14	15' 2"	15' 4-13/16"	15' 7-7/8"	15' 11-1/8"	16' 2-15/32"	16' 6-1/8"	16' 9-7/8"
15	16' 3"	16' 6"	16' 9-5/16"	17' 0-3/4"	17' 4-11/32"	17' 8-1/4"	18' 0-5/16"
16	17' 4"	17' 7-7/32"	17' 10-23/32"	18' 2-13/32"	18' 6-1/4"	18' 10-13/32"	19' 2-23/32"
INCHES OF RUN	These lengths are to be added to those shown above when run involves inches						
1/4"	9/32"	9/32"	9/32"	9/32"	9/32"	5/16"	5/16"
1/2"	17/32"	9/16"	9/16"	9/16"	9/16"	1-9/32"	5/8"
1"	1-3/32"	1-3/32"	1-1/8"	1-1/8"	1-5/32"	1-3/16"	1-7/32"
2"	2-5/32"	2-7/32"	2-1/4"	2-9/32"	2-5/16"	2-11/32"	2-13/32"
3"	3-1/4"	3-5/16"	3-3/8"	3-13/32"	3-15/32"	3-17/32"	3-19/32"
4"	4-11/32"	4-13/32"	4-15/32"	4-9/16"	4-5/8"	4-23/32"	4-13/16"
5"	5-13/32"	5-1/2"	5-19/32"	5-11/16"	5-25/32"	5-29/32"	6"
6"	6-1/2"	6-19/32"	6-23/32"	6-13/16"	6-15/16"	7-1/32"	7-7/32"
7"	7-9/16"	7-23/32"	7-13/16"	7-31/32"	8-1/8"	8-1/4"	8-13/32"
8"	8-21/32"	8-13/16"	8-15/16"	9-1/8"	9-1/4"	9-7/16"	9-5/8"
9"	9-3/4"	9-29/32"	10-5/32"	10-1/4"	10-13/32"	10-19/32"	10-13/16"
10"	10-27/32"	11"	[illegible]	11-3/8"	11-9/16"	11-25/32"	1'0"
11"	11-29/32"	12-3/32"	1' 0-5/16"	1' 0-1/2"	1' 0-23/32"	1' 0-31/32"	1' 1-7/32"

Common rafter table. Subtract one-half ridge board thickness. (Building Supply News)

SPAN L	DIMENSIONS A	B	C	DESIGN STRESSES U_1	U_2	L_1	L_2	V_1	D_1
20'-0"	5'-5"	5'-3⅝"	2'-7⅝"	1756	1450	1614	1472	430	430
22'-0"	5'-11½"	5'-10 1/16"	2'-10 13/16"	1932	1595	1775	1619	473	473
24'-0"	6'-6"	6'-4 7/16"	3'-2"	2108	1740	1936	1766	516	516
26'-0"	7'-0½"	6'-10⅞"	3'-5¼"	2282	1885	2097	1913	559	559
28'-0"	7'-7"	7'-5¼"	3'-8 7/16"	2459	2030	2258	2060	602	602
30'-0"	8'-1½"	7'-11 11/16"	3'-11⅝"	2634	2175	2420	2207	645	645
32'-0"	8'-8"	8'-6 1/16"	4'-2 13/16"	2810	2320	2581	2354	688	688

Span	LUMBER 2" x 6" No.	2" x 6" Length	2" x 4" No.	2" x 4" Length	Total F.B.M.
20'-0"	2	12'-0"	2 2	12'-0" 10'-0"	53
22'-0"	2	14'-0"	2 2	14'-0" 10'-0"	60
24'-0"	2	14'-0"	2 2	14'-0" 12'-0"	63
26'-0"	2	16'-0"	2 2	16'-0" 12'-0"	70
28'-0"	2	16'-0"	2 2	16'-0" 14'-0"	73
30'-0"	2	18'-0"	2 2	18'-0" 14'-0"	79
32'-0"	2	20'-0"	2 2	18'-0" 14'-0"	83

HARDWARE No.	Item	Size
11	Split Rings	2½" Diam.
2	Trip-L-Grip	Type A
1	Bolt	½" x 7½"
2	Bolts	½" x 6"
4	Bolts	½" x 4"
14	Washers	½"

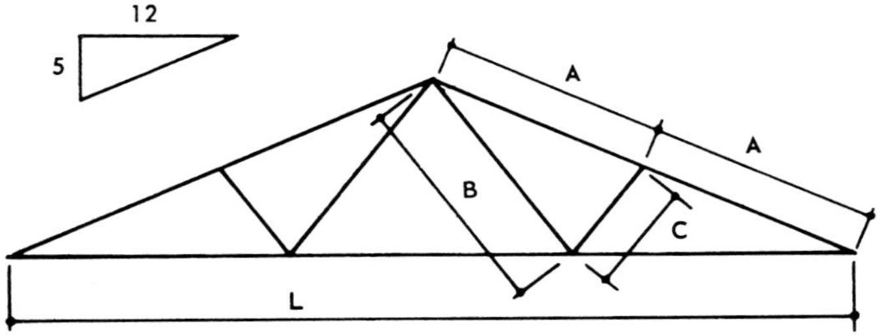

LUMBER--Lumber shall be of a good grade of sufficient quality to permit the following allowable unit stresses:
c = 900#/□" Compression parallel to grain.
f = 900#/□" Extreme fiber in bending.
E = 1,600,000#/□" Modulus of elasticity.

CONNECTORS--Timber connectors shall be 2-1/2" diameter split rings and Trip-L-Grip framing anchors.

BOLTS--Bolts shall be 1/2" diameter machine bolts with 2" x 2" x 1/8" plate washers, 2-1/8" diameter cast or malleable iron washers, or ordinary cut washers.

DIMENSIONS--Dimensions shown will provide approximately 1/2" camber at bottom chord panel points. Utilize full uncut length of bottom chord pieces by increasing the spacing of connectors in the splice.

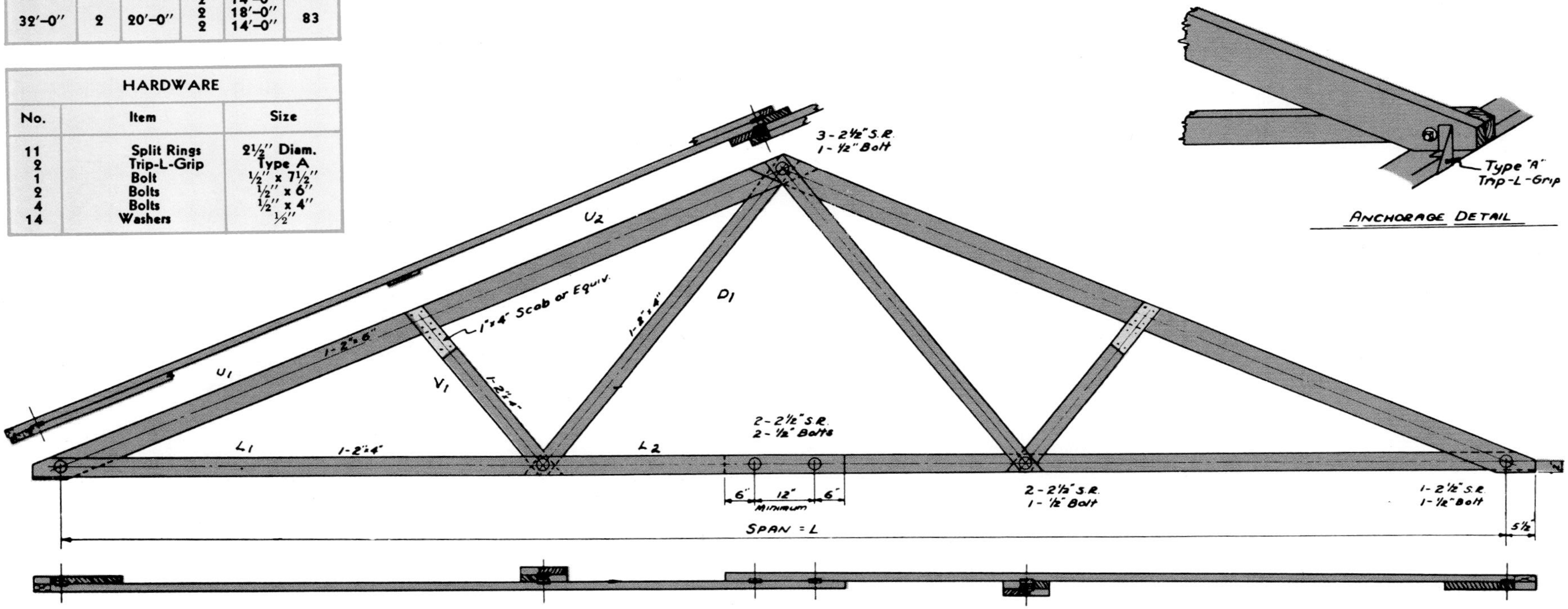

Detail sheet of a Fink truss designed for a roof slope of 5 in 12. (Timber Engineering Co.)

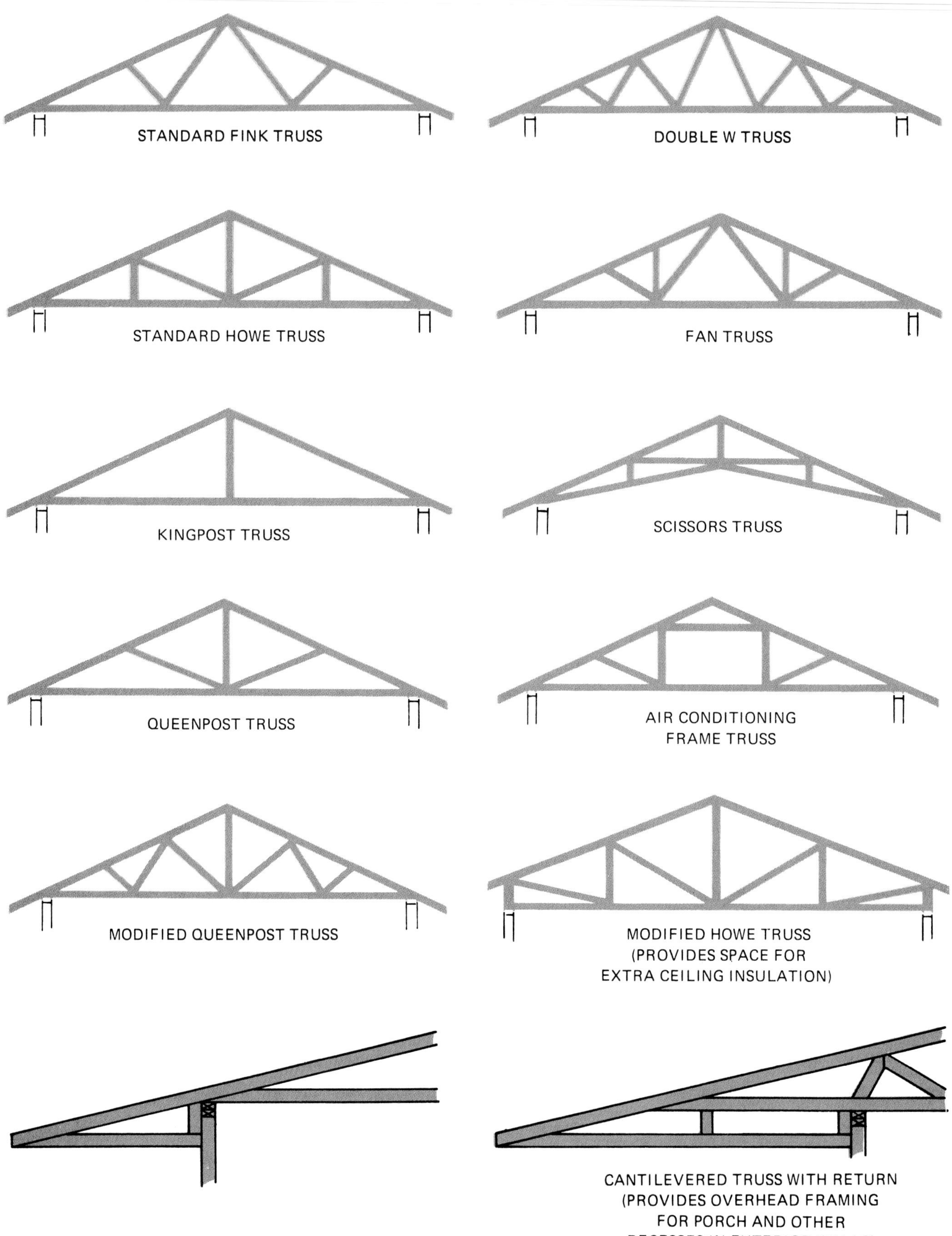

Types of roof trusses and overhangs. The Fink truss is also called a W truss. Any type of truss should be constructed according to designs developed from engineering data.

Better construction methods and materials used by builders today produce houses that are tighter, more draft-free than those built a half century ago.

In such snug shelters, moisture created inside the house often saturates insulation and causes paint to peel off exterior walls.

The use of a vapor barrier on the warm side of walls and ceilings will minimize this condensation, but adequate ventilation is needed to remove the moisture from the house.

FHA requires that attics have a total net free ventilating area not less than 1/150 of the square foot area, except that a ratio of 1/300 may be provided if:

(a) a vapor barrier having a transmission rate not exceeding one perm is installed on the warm side of the ceiling, or

(b) at least 50% of the required vent area is provided by ventilators located in the upper portion of the space to be ventilated, with the balance of the required ventilation provided by eave or cornice vents.

There are many types of screened ventilators available, including the handy miniature vents illustrated below, which can be used for problem areas or to supplement larger vents.

INSTALLING MINIATURE VENTS

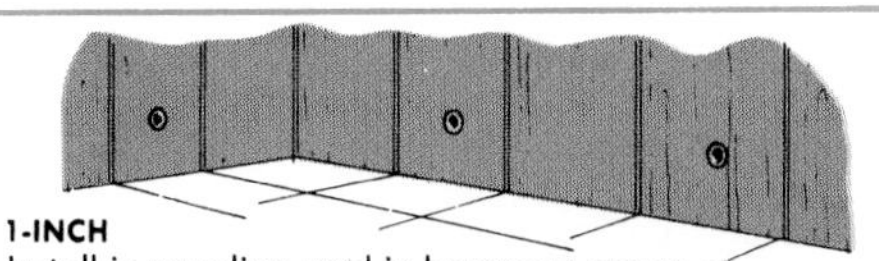

1-INCH
Install in paneling used in basement rooms.
Can be painted over to match panel finish.

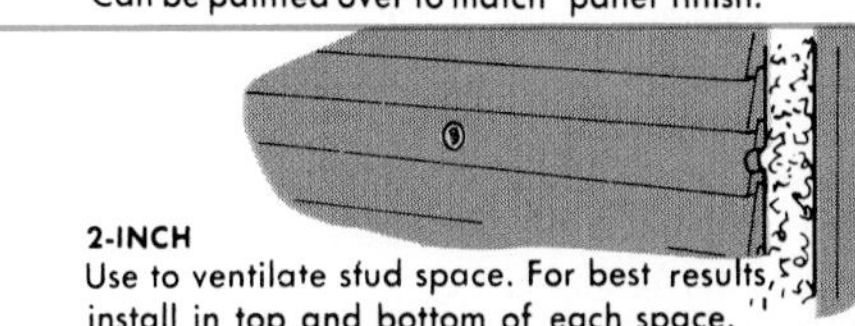

2-INCH
Use to ventilate stud space. For best results, install in top and bottom of each space.

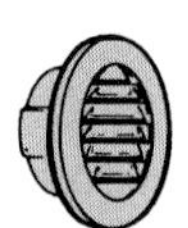

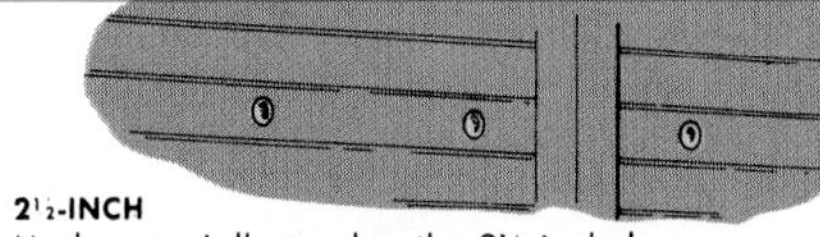

2½-INCH
Made especially to plug the 2½-in. hole cut to blow insulation between studs.

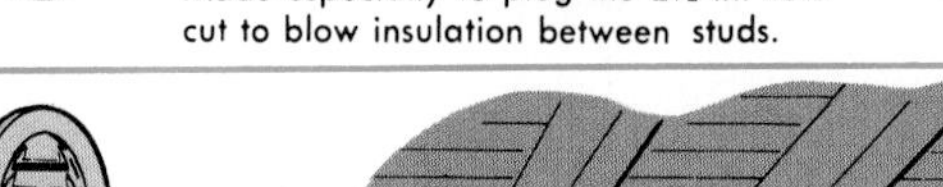

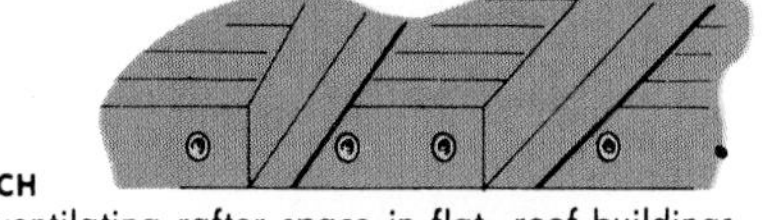

3-INCH
For ventilating rafter space in flat- roof buildings, or other jobs requiring fairly large free area.

4-INCH
For hard-to-reach spots needing large ventilating area. Large enough for venting soffits.

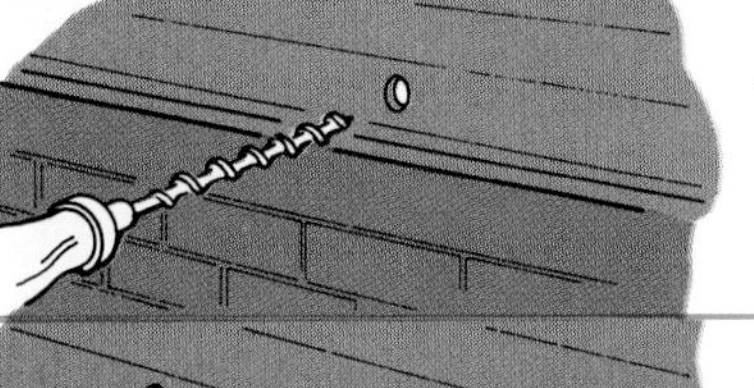

1. Drill or cut a hole the same size as the ventilator.

2. Insert the ventilator. Tension ridges hold ventilator in place . . . no nails or screws are needed.

3. Tap into place with a hammer, using a wood block to protect the margin. The louvers are recessed so there's no danger of damage during installation.

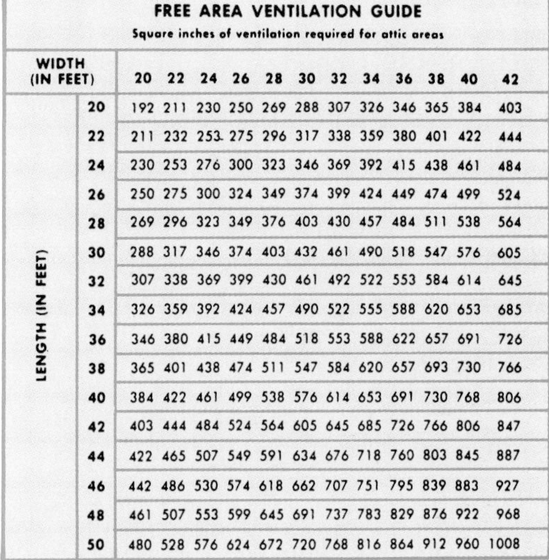

FREE AREA VENTILATION GUIDE
Square inches of ventilation required for attic areas

LENGTH (IN FEET) \ WIDTH (IN FEET)	20	22	24	26	28	30	32	34	36	38	40	42
20	192	211	230	250	269	288	307	326	346	365	384	403
22	211	232	253	275	296	317	338	359	380	401	422	444
24	230	253	276	300	323	346	369	392	415	438	461	484
26	250	275	300	324	349	374	399	424	449	474	499	524
28	269	296	323	349	376	403	430	457	484	511	538	564
30	288	317	346	374	403	432	461	490	518	547	576	605
32	307	338	369	399	430	461	492	522	553	584	614	645
34	326	359	392	424	457	490	522	555	588	620	653	685
36	346	380	415	449	484	518	553	588	622	657	691	726
38	365	401	438	474	511	547	584	620	657	693	730	766
40	384	422	461	499	538	576	614	653	691	730	768	806
42	403	444	484	524	564	605	645	685	726	766	806	847
44	422	465	507	549	591	634	676	718	760	803	845	887
46	442	486	530	574	618	662	707	751	795	839	883	927
48	461	507	553	599	645	691	737	783	829	876	922	968
50	480	528	576	624	672	720	768	816	864	912	960	1008

Using length and width dimensions of each rectangular or square attic space, find one dimension on vertical column, the other dimension on horizontal column. These will intersect at the number of square inches of ventilation required to provide 1/300th.

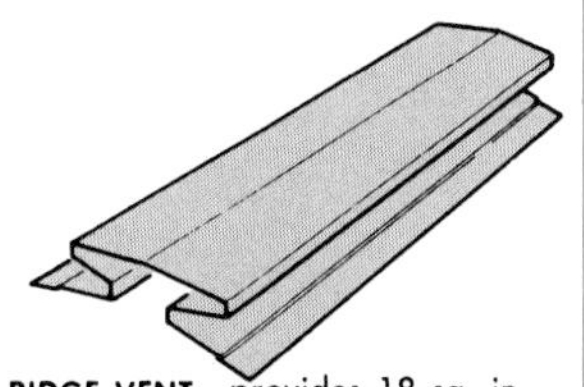

RIDGE VENT provides 18 sq. in. of net free area per lineal foot. Installed quickly over a 1½″ gap in the sheathing at the ridge.

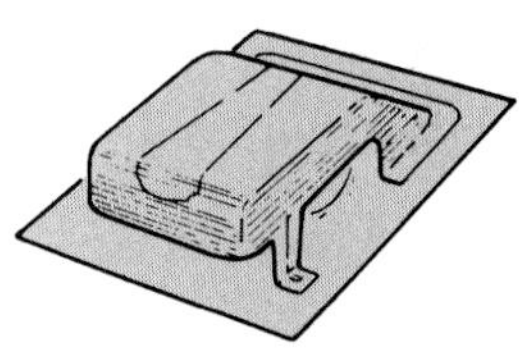

ROOF VENTS fit over openings cut between rafters to pull hot air out of attic.

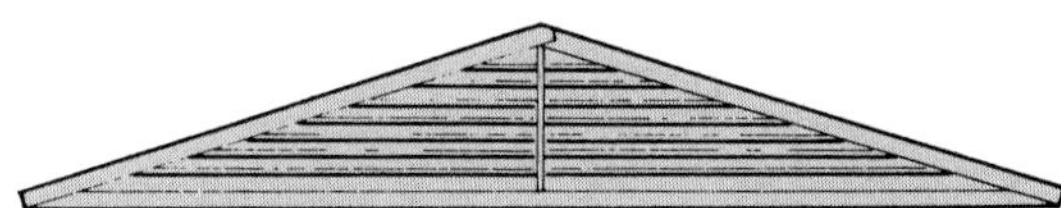

TRIANGLE VENT fits snugly under the roof gable to provide large vent areas at the highest point of the gable end.

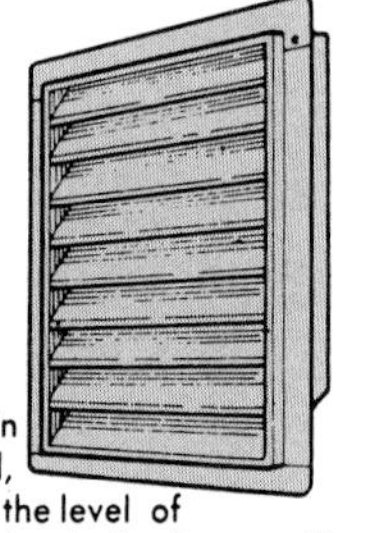

ATTIC VENTS are installed in the gable end, usually above the level of the probable level of a future ceiling should the attic be finished later.

SOFFIT VENT replaces a portion of the soffit material to provide continuous ventilation along its entire length.

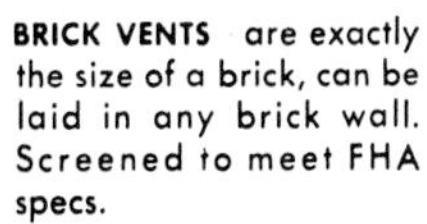

BRICK VENTS are exactly the size of a brick, can be laid in any brick wall. Screened to meet FHA specs.

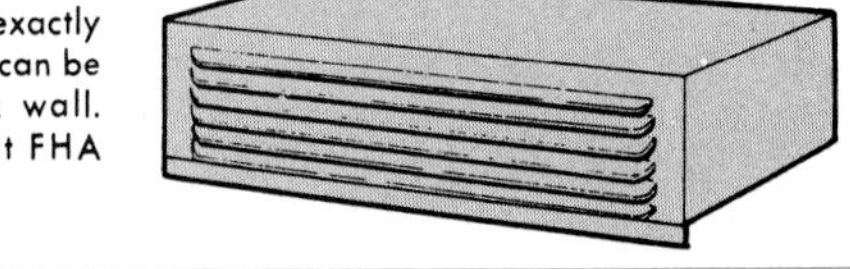

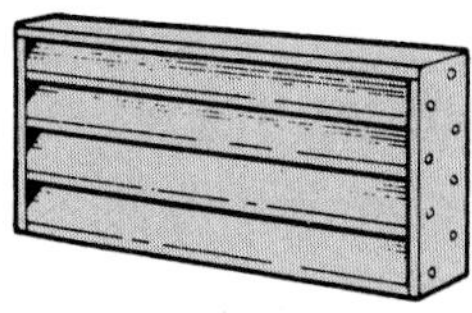

CEMENT BLOCK VENTS are designed to be mortared into the same space as an 8x16″ cement block. At least four should be used to vent crawl space.

Modern ventilators and vent applications.

EXTERIOR MATERIALS

Wood bevel siding, ½ x 8, lapped........ R-0.81
Wood bevel siding, ¾ x 10, lapped........ R-1.05
Wood siding shingles, 16," 7½" exposure ... R-0.87
Aluminum or Steel, over sheathing, hollow-backed........ R-0.61
Stucco, per inch........ R-0.20
Building paper........ R-0.06
½" nail-base insulating board sheathing.... R-1.14
½" insulating board sheathing, regular density........ R-1.32
25⁄32" insul. board sheathing, regular density . R-2.04
Insulating-board backed nominal ⅜"....... R-1.82
Insulating-board backed nominal ⅜" foil backed........ R-2.96
Plywood ¼"........ R-0.31
Plywood ⅜"........ R-0.47
Plywood ½"........ R-0.62
Plywood ⅝"........ R-0.78
Hardboard ¼"........ R-0.18
Hardboard, medium density siding 7⁄16"..... R-0.67
Softwood board, fir pine and similar softwoods
¾"........ R-0.94
1½"........ R-1.89
2½"........ R-3.12
3½"........ R-4.35
Gypsumboard ½"........ R-0.45
Gypsumboard ⅝"........ R-0.56

MASONRY MATERIALS

Concrete blocks, three oval cores
Cinder aggregate, 4" thick........ R-1.11
Cinder aggregate, 12" thick........ R-1.89
Cinder aggregate, 8" thick........ R-1.72
Sand and gravel aggregate, 8" thick..... R-1.11
Sand and gravel aggregate, 12" thick ... R-1.28
Lightweight aggregate (expanded clay, shale, slag, pumice, etc.), 8" thick..... R-2.00
Concrete blocks, two rectangular cores
Sand and gravel aggregate, 8" thick..... R-1.04
Lightweight aggregate, 8" thick........ R-2.18
Common brick, per inch........ R-0.20
Face brick, per inch........ R-0.11
Sand-and-gravel concrete, per inch....... R-0.08

GLASS

U-VALUES	Glass Only (Winter)
Single-pane glass	1.16
Double-pane ⅝" insulating glass (¼" air space)	.58
Double-pane xi insulating glass	.55
Double-pane 1" insulating glass (½" air space)	.49
Double-pane xi insulating glass with combination (2" air space)	.35

Glass U-Values obtained from PPG and Cardinal Glass Company.

INSULATION

Fiberglass 2" thick........ R-7.00
Fiberglass 3½" thick........ R-11.00
Fiberglass 6" thick........ R-19.00
Fiberglass 12" thick........ R-38.00
Styrofoam Board ¾" thick........ R-4.05
Styrofoam Board 1" tongue & groove...... R-5.40

ROOFING

Asphalt shingles........ R-0.44
Wood shingles, plain & plastic film faced ... R-0.94

SURFACE AIR FILMS

Inside, still air
Heat flow UP (through horizontal surface)
Non-reflective........ R-0.61
Reflective........ R-1.32
Heat flow DOWN (through horizontal surface)
Non-reflective........ R-0.92
Reflective........ R-4.55
Heat flow HORIZONTAL (through vertical surface)
Non-reflective........ R-0.68

Outside
Heat flow any direction, surface any position
15 mph wind (winter)........ R-0.17
7.5 mph wind (summer)........ R-0.25

EXAMPLE CALCULATIONS
(to determine the U value of an exterior wall)

Wall Construction	Insulated Wall Resistance
Outside surface (film), 15 mph wind.....	0.17
Wood bevel siding, lapped........	0.81
½" ins. bd. sheathing, reg. density......	1.32
3½" air space........	
R-11 insulation........	11.00
½" gypsumboard........	0.45
Inside surface (film)........	0.68
Totals........	14.43

For insulated wall, $U = \frac{1}{R} = \frac{1}{14.3} = 0.07$

TEMPERATURE CORRECTION FACTOR

Correction Factor is an ASHRAE standard to be applied for varying outdoor design temperatures. As follows:

If design temperature is:	−20	−10	0	+10	+20
Then correction factor is:	0.778	0.875	1.0	1.167	1.40

*Additional resistance values can be obtained from ASHRAE Handbook of Fundamentals published by the American Society of Heating, Refrigerating and Air-Conditioning Engineers.

Insulation values for common materials. The method used to calculate U- and R-values for a wall can also be used for ceilings and floors. Temperature correction factors are used by designers of heating and cooling systems. (Andersen Corp.)

DESIGN TEMPERATURES AND DEGREE DAYS
(Heating Season)

State	City	Outside Design Temperature (°F)	Degree Days (°F-Days)
Alabama	Birmingham	19	2,600
Alaska	Anchorage	-25	10,800
Arizona	Phoenix	31	1,800
Arkansas	Little Rock	19	3,200
California	San Francisco	35	3,000
California	Los Angeles	41	2,000
Colorado	Denver	-2	6,200
Connecticut	Hartford	1	6,200
Florida	Tampa	36	600
Georgia	Atlanta	18	3,000
Idaho	Boise	4	5,800
Illinois	Chicago	-3	6,600
Indiana	Indianapolis	0	5,600
Iowa	Des Moines	-7	6,600
Kansas	Wichita	5	4,600
Kentucky	Louisville	8	4,600
Louisiana	New Orleans	32	1,400
Maryland	Baltimore	12	4,600
Massachusetts	Boston	6	5,600
Michigan	Detroit	4	6,200
Minnesota	Minneapolis	-14	8,400
Mississippi	Jackson	21	2,200
Missouri	St. Louis	4	5,000
Montana	Helena	-17	8,200
Nebraska	Lincoln	-4	5,800
Nevada	Reno	2	6,400
New Hampshire	Concord	-11	7,400
New Mexico	Albuquerque	14	4,400
New York	Buffalo	3	7,000
New York	New York	12	5,000
North Carolina	Raleigh	16	3,400
North Dakota	Bismark	-24	8,800
Ohio	Columbus	2	5,600
Oklahoma	Tulsa	12	3,800
Oregon	Portland	21	4,600
Pennsylvania	Philadelphia	11	4,400
Pennsylvania	Pittsburg	5	6,000
Rhode Island	Providence	6	6,000
South Carolina	Charleston	23	2,000
South Dakota	Sioux Falls	-14	7,800
Tennessee	Chattanooga	15	3,200
Texas	Dallas	19	2,400
Texas	San Antonio	25	1,600
Utah	Salt Lake City	5	6,000
Vermont	Burlington	-12	8,200
Virginia	Richmond	14	3,800
Washington	Seattle	28	5,200
West Virginia	Charleston	9	4,400
Wisconsin	Madison	-9	7,800
Wyoming	Cheyenne	-6	7,400

A more complete listing of monthly and yearly degree days and outside design temperatures can be found in the ASHRAE Guide and Data Book.

This list of U.S. cities, with their outside design temperatures and degree days, is a useful resource for computer aided energy analysis.

PARTITIONS

STC-45

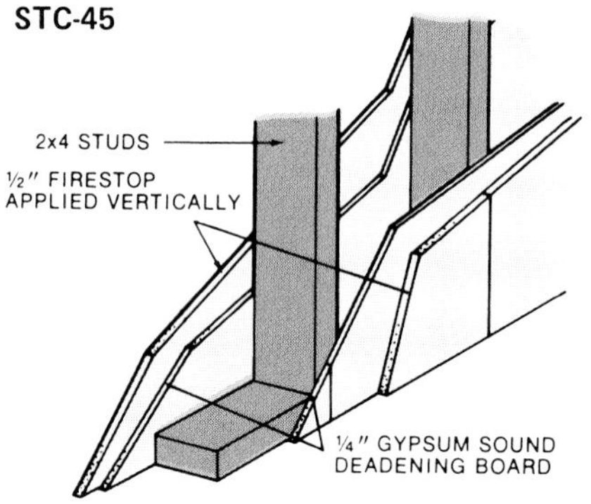

2x4 STUDS
½" FIRESTOP APPLIED VERTICALLY
¼" GYPSUM SOUND DEADENING BOARD

STC-47

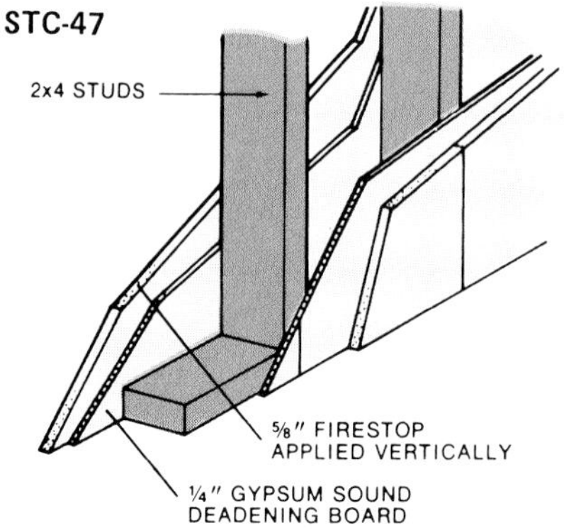

2x4 STUDS
⅝" FIRESTOP APPLIED VERTICALLY
¼" GYPSUM SOUND DEADENING BOARD

STC-50

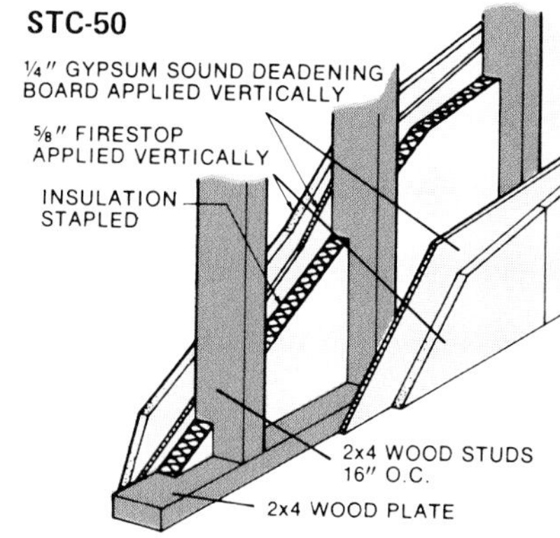

¼" GYPSUM SOUND DEADENING BOARD APPLIED VERTICALLY
⅝" FIRESTOP APPLIED VERTICALLY
INSULATION STAPLED
2x4 WOOD STUDS 16" O.C.
2x4 WOOD PLATE

STC-52

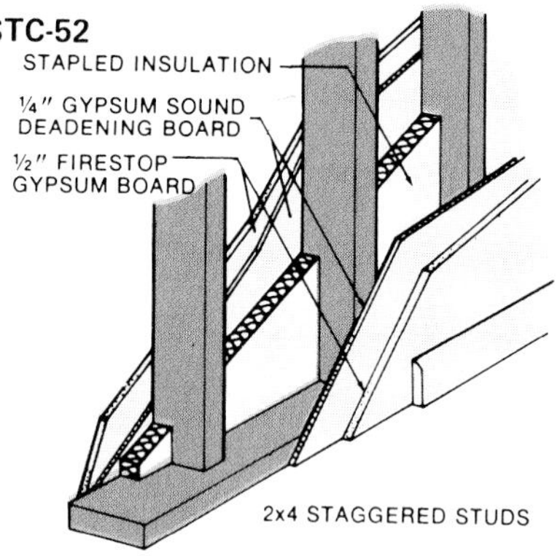

STAPLED INSULATION
¼" GYPSUM SOUND DEADENING BOARD
½" FIRESTOP GYPSUM BOARD
2x4 STAGGERED STUDS

STC-55

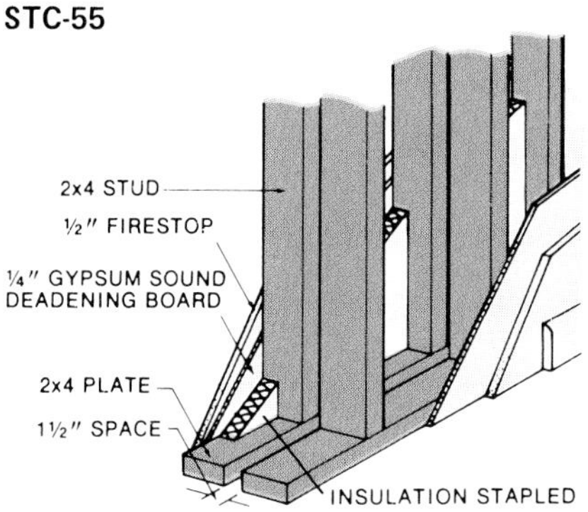

2x4 STUD
½" FIRESTOP
¼" GYPSUM SOUND DEADENING BOARD
2x4 PLATE
1½" SPACE
INSULATION STAPLED

STC-50

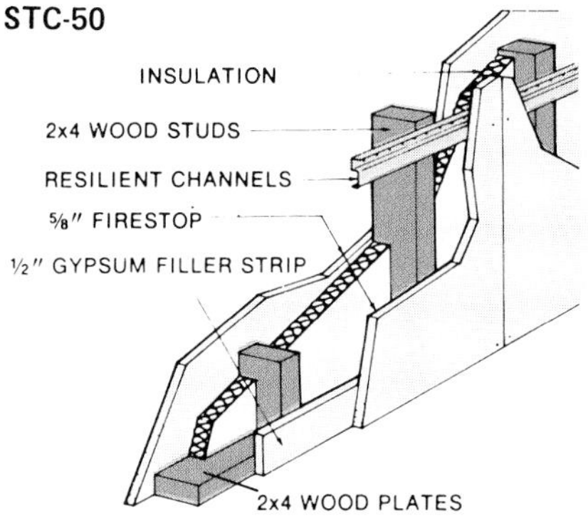

INSULATION
2x4 WOOD STUDS
RESILIENT CHANNELS
⅝" FIRESTOP
½" GYPSUM FILLER STRIP
2x4 WOOD PLATES

STC-41

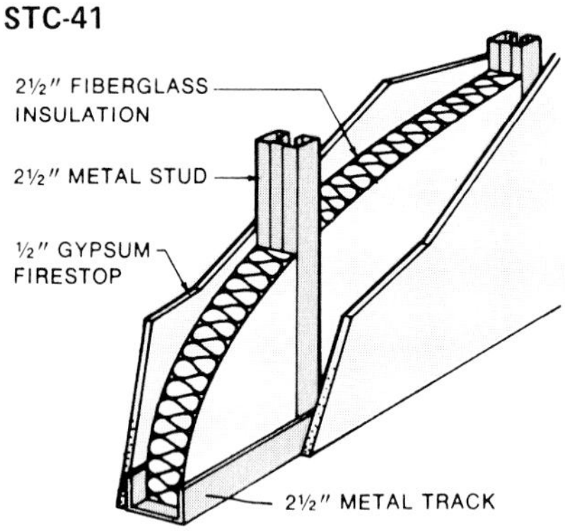

2½" FIBERGLASS INSULATION
2½" METAL STUD
½" GYPSUM FIRESTOP
2½" METAL TRACK

STC-50

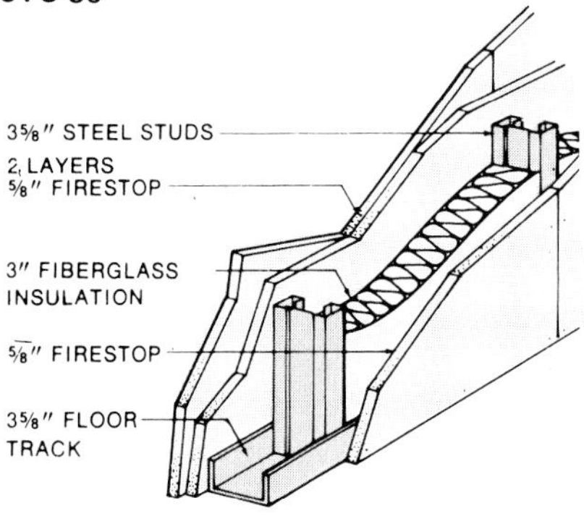

3⅝" STEEL STUDS
2 LAYERS ⅝" FIRESTOP
3" FIBERGLASS INSULATION
⅝" FIRESTOP
3⅝" FLOOR TRACK

STC-54

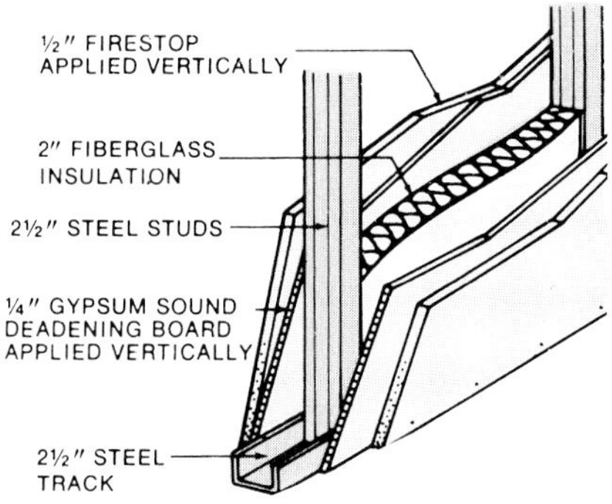

½" FIRESTOP APPLIED VERTICALLY
2" FIBERGLASS INSULATION
2½" STEEL STUDS
¼" GYPSUM SOUND DEADENING BOARD APPLIED VERTICALLY
2½" STEEL TRACK

FLOORS

STC-41

CARPET & PAD
½" PARTICLEBOARD UNDERLAYMENT
⅝" PLYWOOD SUB FLOOR
2x10 JOISTS
2x4s
¼" GYPSUM SOUND DEADENING BOARD
½" FIRESTOP GYPSUM BOARD

STC-47

1"x4" TONGUE & GROOVE FINISHED FLOORING
RESIN BUILDING PAPER
1"x6" TONGUE & GROOVE SUBFLOORING
½" FIRESTOP GYPSUM BOARD
2"x10" WOOD JOISTS
RESILIENT METAL FURRING CHANNELS

STC-53

CARPET
PAD
⅜" PARTICLEBOARD
⅝" PLYWOOD SUBFLOOR
3½" INSULATION
2x10 WOOD JOISTS
RESILIENT CHANNEL
½" FIRESTOP

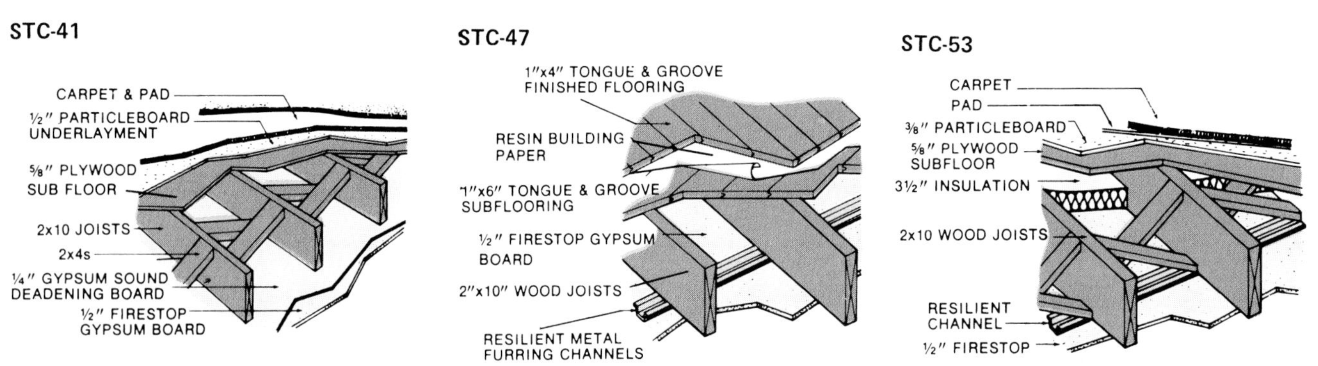

How to build partition and floor structures with high STC (sound transmission class) ratings. Rating are based on sound tests conducted according to ASTM-E90. (Georgia-Pacific)

FIRE RESISTANT CONSTRUCTION

– One-hour assembly – resilient channel ceiling system

5/8" plywood DFPA underlayment T & G

Building paper

1/2" Standard grade plywood with Exterior glue.

Joists 16" (2 x 10's min.)

1/2" galvanized metal resilient channels at 24" o.c.*

1/2" fire resistive Special Type X gypsum board—fasten to channels with self tapping screws 12" o.c.

*Channels may be suspended below joists.

– One-hour assembly – T-bar grid ceiling system

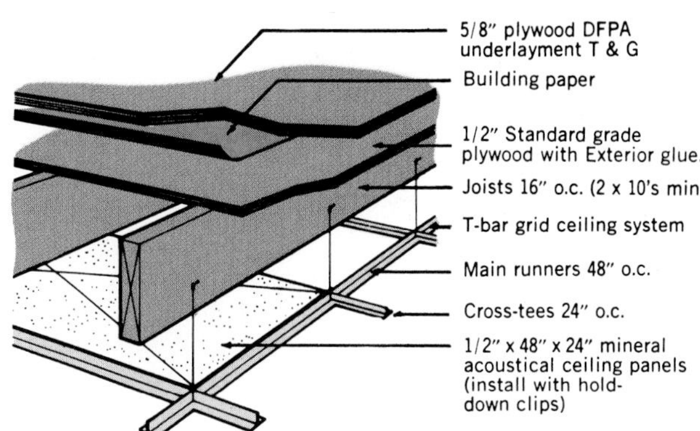

– One-hour interior shear wall construction

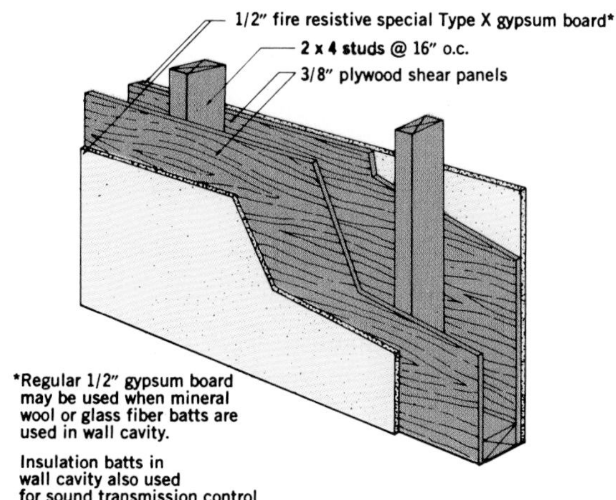

*Regular 1/2" gypsum board may be used when mineral wool or glass fiber batts are used in wall cavity.

Insulation batts in wall cavity also used for sound transmission control

– One-hour exterior wall construction

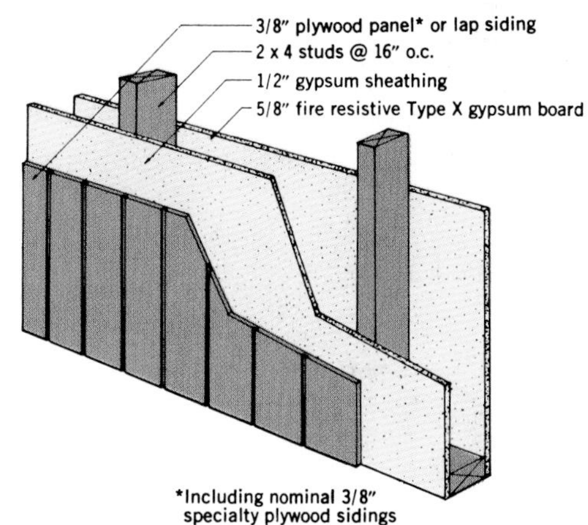

*Including nominal 3/8" specialty plywood sidings

– Treated stressed skin panel construction

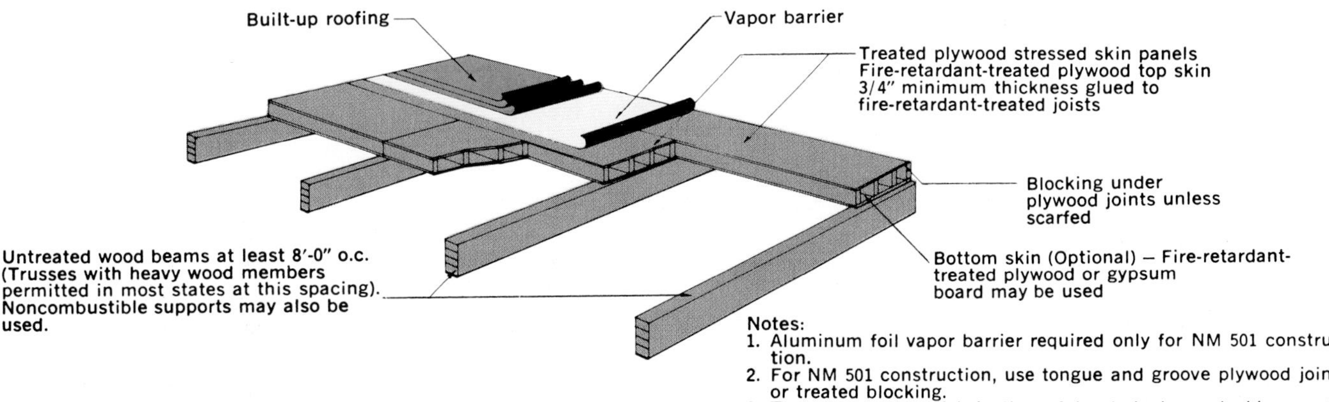

Notes:
1. Aluminum foil vapor barrier required only for NM 501 construction.
2. For NM 501 construction, use tongue and groove plywood joints or treated blocking.
3. To assure proper fabrication of treated stressed skin panels, components bearing the trademark of the Plywood Fabricator Service, Inc. are recommended.

Fire resistant construction. All assemblies shown provide a one-hour rating. (American Plywood Assoc.)

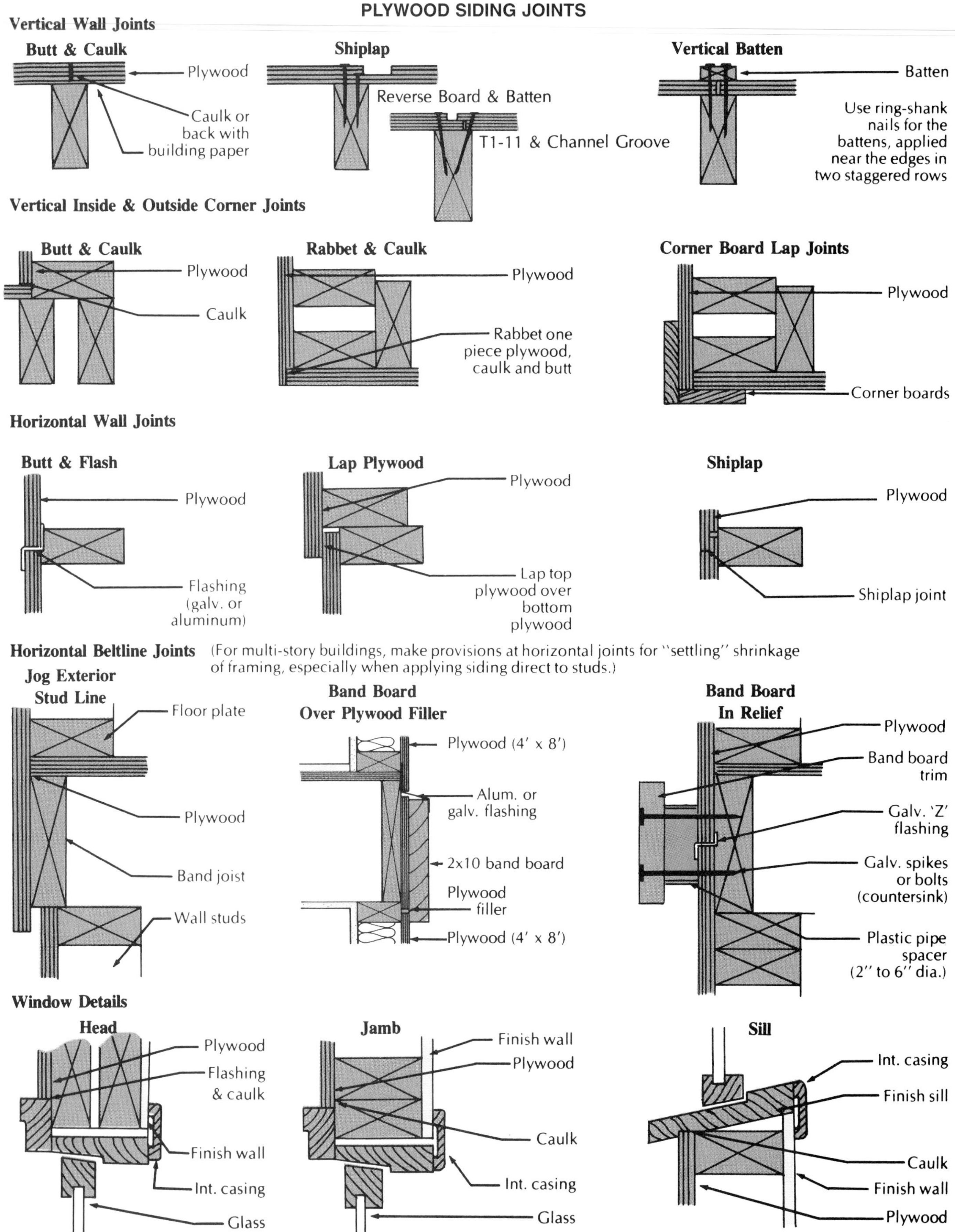

Approved joint details for plywood siding. (American Plywood Assoc.)

Technical Information
DECK CONSTRUCTION

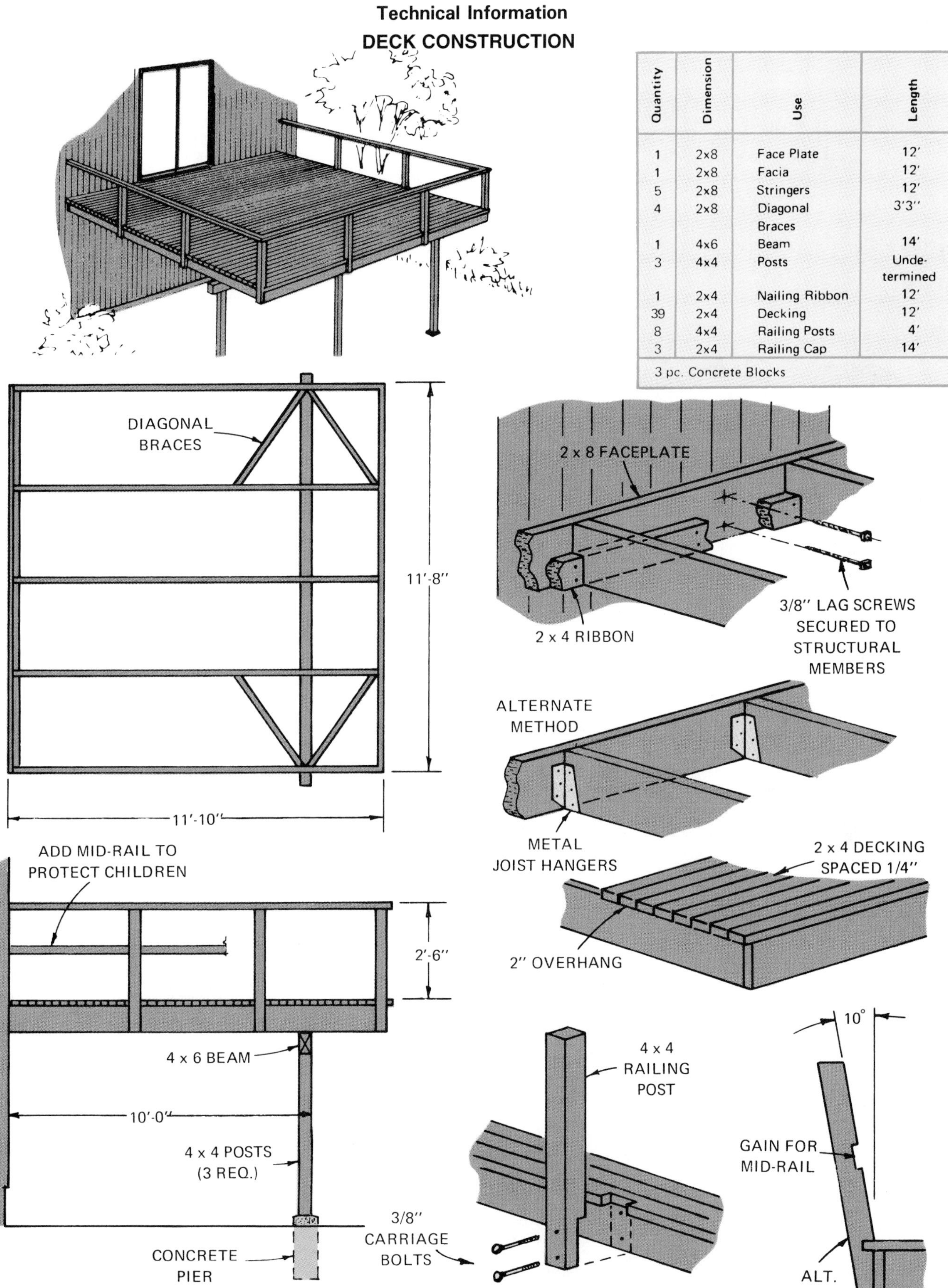

Quantity	Dimension	Use	Length
1	2x8	Face Plate	12′
1	2x8	Facia	12′
5	2x8	Stringers	12′
4	2x8	Diagonal Braces	3′3″
1	4x6	Beam	14′
3	4x4	Posts	Undetermined
1	2x4	Nailing Ribbon	12′
39	2x4	Decking	12′
8	4x4	Railing Posts	4′
3	2x4	Railing Cap	14′
3 pc. Concrete Blocks			

General construction details for sun deck. Use weatherproof metal fasteners and pressure treated wood. The American Wood Preservers Bureau (AWPB) provides standards and technical requirements for pressure treating.

BRANCH PIPE SIZES

FIXTURE	PIPE SIZE (INCH)	FIXTURE	PIPE SIZE (INCH)
BATHTUB	1/2	SHOWER	1/2
DISHWASHER	1/2	URINAL—FLUSH VALVE	3/4
DRINKING FOUNTAIN	3/8	—FLUSH TANK	1/2
HOSE BIB	1/2	WASHING MACHINE	1/2
KITCHEN SINK	1/2	WATER CLOSET—FLUSH VALVE	1
LAUNDRY TRAY	1/2	—FLUSH TANK	3/8
LAVATORY	1/2	WATER HEATER	1/2

PLASTIC PIPES

TYPE PLASTIC	USES	SIZES/GRADES AVAILABLE	JOINING TECHNIQUES	RIGID/FLEXIBLE
ABS	DRAIN/WASTE/VENT	1 1/4—6 SCHEDULE 40 (SCH. 40)	SOLVENT CEMENT	RIGID
	SEWER LINES	3—8 STANDARD DIMENSION RATIO (SDR)		
	WATER SUPPLY	1/2—2 SDR		
PB	HOT AND COLD WATER DISTRIBUTION	1/8—1'' CTS	INSERT FITTINGS	FLEXIBLE
		3/4—2'' IPS	COMPRESSION FITTINGS	
			INSTANT CONNECT FITTINGS	
PVC	WATER DISTRIBUTION	1/4—12 SCH. 40 & 80 AND SDR	SOLVENT WELDING	RIGID
	DWV, SEWERS, & PROCESS PIPING		SCH. 80—THREADED & FLANGES	
CPVC	HOT & COLD WATER DISTRIBUTION	1/2—3/4	SOLVENT WELDING	RIGID
	PROCESS PIPING	1/2—8 SCH. 40 & 80	THREADING & FLANGES	
PE	WATER DISTRIBUTION	1/2—2 IPS	COMPRESSION & FLANGE FITTINGS	FLEXIBLE
	NATURAL GAS DISTRIBUTION OIL FIELD PIPING	1/2—6 IPS		
	WATER TRANSMISSION, SEWERS	3—48 SDR	FUSION WELDING	
SR	AGRICULTURE FIELD DRAINS STORM DRAINS	3—8 SDR	SOLVENT WELDING	RIGID

COPPER PIPES

TYPE	COLOR CODE	APPLICATION	STRAIGHT LENGTHS	COILS (SOFT TEMPER ONLY)
K	GREEN	UNDERGROUND AND INTERIOR SERVICE	20 FT. IN DIAMETERS INCLUDING 8 IN. HARD AND SOFT TEMPER	60 FT. AND 100 FT. FOR DIAMETERS INCLUDING 1 IN.
L	BLUE	ABOVE GROUND SERVICE	20 FT. IN DIAMETERS INCLUDING 10 IN. HARD AND SOFT TEMPER	(SAME AS K)
M	RED	ABOVE GROUND WATER SUPPLY DRAINAGE, WASTE, AND VENT	20 FT. IN ALL DIAMETERS HARD TEMPER ONLY	NOT AVAILABLE
DWV	YELLOW	ABOVE GROUND DRAIN, WASTE, AND VENT PIPING	20 FT. IN ALL DIAMETERS 1 1/4 IN. AND GREATER HARD TEMPER ONLY	NOT AVAILABLE

INSULATED WIRE COVERINGS

Covering Types	Letter Designation
Rubber	RH, RHH, RHW,
Thermoplastic Compound	TW, THW, TBS, THHN
Thermoplastic and Asbestos	TA
Silicone and Asbestos	SA
Asbestos	A
Varnished Cambric	V
Asbestos and Varnished Cambric	AVA, AVL, AVB

Wire Type	Apply Where	Temperature Maximum °C	(°F)
RH	Dry/Damp	75	(167)
RHH	Dry/Damp	90	(194)
RHW	Dry/Wet	75	(167)
TW	Dry/Wet	60	(140)
THHN	Dry/Damp	90	(194)
THW	Dry/Wet	90	(194) Under Special Conditions
THWN	Dry/Wet	75	(167)
XHHW	Dry/Damp Wet	90 75	(194) (167)
MI	Dry Wet	90 250	(194) Under Special (482) Conditions
SA	Dry/Damp	90 125	(194) For Special (257) Applications
FEP	Dry/Damp	90	(194)
V	Dry	85	(185)
AVA	Dry	110	(230)
AVL	Dry/Wet	110	(230)
AVB	Dry	90	(194)

CONDUCTOR COLOR

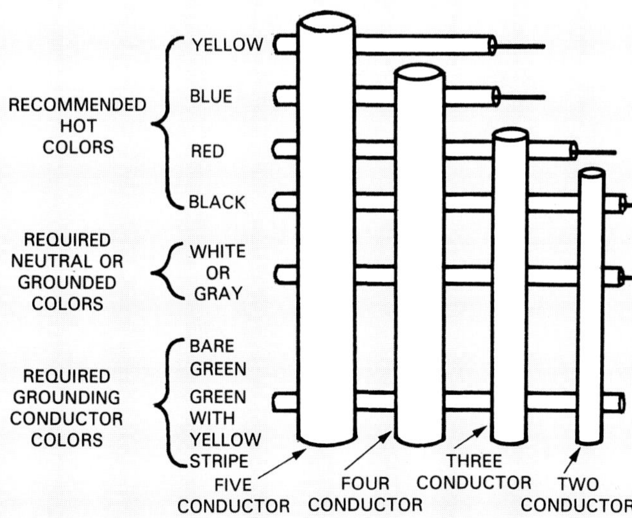

NOTE: THROUGHOUT THE TEXT WE OFTEN REFER TO THE HOT, UNGROUNDED CONDUCTOR AS THE BLACK OR RED WIRE. KEEP IN MIND, HOWEVER, THAT THE UNGROUNDED CONDUCTOR MAY BE ANY COLOR OTHER THAN THOSE SPECIFIED AND REQUIRED FOR THE NEUTRAL OR GROUND WIRE.

ELECTRICAL SWITCHES

Switches	Purposes
Single-pole, single-throw switch	Controls light or outlet from single location.
Three-way switch	Controls light or outlet from two locations.
Four-way switch	Controls light or outlet from other locations in between pair of three-way switches.
Dimmer switch	Like single-pole, single-throw switch, but also contains rheostat or voltage regulating device which allows all or only portion of electrical energy to outlet or light fixture.
Pilot lighted switches	Used to control light or outlet which is not in sight. It indicates, by use of pilot light, whether power to device is on or off.
Time-delay switch	Used where delayed shut-off is desirable.
Others: Toggle switch, Pushbotton switch, Pull Chain switch, Photoelectric, Knife switch, Tap switch, Plate switch, Locking switch.	

Glossary of Technical Terms

A

Acoustical materials: Types of tile, plaster, and other materials that absorb sound waves. Generally applied to interior wall surfaces to reduce reverberation or reflection of sound waves.
Active solar construction: A system that uses mechanical means to transfer a medium heated by solar energy.
Adhesive: A substance capable of holding material together by surface attachment. A general term that includes glue, cement, mastic, and paste.
Admixtures: Chemicals added to concrete to change the characteristics of the mix.
Aggregate: Materials such as sand, rock, and gravel used to make concrete.
Air conditioning: Control of temperature, humidity, movement, and purity of air in buildings.
Air dried: Wood seasoned by exposure to the atmosphere, either covered or uncovered, without artificial heat.
Air-entraining agents: Chemical additives that trap tiny air bubbles in concrete, improving the workability and freeze-thaw durability.
Air exchanger: A device that exhausts air from a building and also draws fresh air in. The fresh air is warmed by the heat of the exhausted air.
Air return: A series of ducts in a heating system that return cooled air to the furnace.
Air vents: Devices that purge air that has entered a hydronic heating system.
Alkyd: A paint in which the vehicle (binder) is an alkyd resin (a type of synthetic resin).
Alternating Current (AC): An electric current that regularly reverses direction, this type of electric current is common in home wiring.
Anchor bolts: Bolts embedded in concrete used to hold structural members in place.
Anchor straps: Strap fasteners that are embedded in concrete or masonry walls to hold sills in place.
Annual rings: Rings or layers of wood that represent one growth period of a tree. In cross section, the rings may indicate the age of the tree.
Anodizing: A method of coating metal objects for protection or decoration.
Apprenticeship: A formal method of learning a certain trade, such as carpentry, that involves instruction as well as working and learning on the job. Upon mastering prescribed tasks over a specified time period, the apprentice is certified as a journeyman and can work without supervision.
Apron: A piece of horizontal trim applied against the wall immediately below the stool of a window. Conceals rough edge of plaster.
Asphalt: A residue produced from evaporated petroleum. It is insoluble in water but is soluble in gasoline and melts when heated. Used for waterproofing roof coverings, exterior wall coverings, and flooring tile.

B

Backfill: The replacement of soil around foundations after excavating.
Backing board: In a two-layer drywall system, the base panel of gypsum drywall. It uses gray liner paper as facing and is not suitable as a top surface. Also referred to as backer board.
Balloon framing: A type of building construction with upright studs that extend from the foundation sill to the rafter plate. Its use is decreasing in favor of platform framing and other construction styles. Also called Western framing.
Baluster: Vertical stair member that supports the stair rail.
Balustrade: A railing consisting of a series of balusters resting on a base, usually the treads, that supports a continuous stair or hand rail.
Basement: The part of a house that is partly or completely below grade.
Base shoe: Small narrow molding used around the perimeter of a room where the baseboard meets the finish floor.
Batten: A strip of wood placed across a surface to cover joints.
Batter: The slope, or inclination from the vertical, of a wall or other structure or portion of a structure.
Batter board: A temporary framework used to assist in locating corners when laying out a foundation.
Bay: One of the intervals or spaces into which a building plan is divided by columns, piers, or division walls.
Bay window: A window or group of windows usually supported on a foundation extending beyond the main wall of a building.
Beam: A horizontal structural member used between posts, columns, or walls.
Bearing partition: A partition which supports a vertical load in addition to its own weight.
Bearers: Horizontal members (usually 2 x 6s) used to connect scaffold uprights. Also called cross ledgers.
Bearing wall: A wall which supports a vertical load in addition to its own weight.
Bedding: A layer of mortar, putty, or other substance used to secure a firm bearing.
Bed molding: A molding applied where two surfaces come together at an angle. Commonly used in cornice trim especially between the plancier and frieze.
Bench mark: A mark on a permanent object fixed to the ground from which land measurements and elevations are taken.
Bevel: To cut to an angle other than a right angle, such as the edge of a board or door.
Bevel siding: Used as finish covering on the exterior of a structure. It is usually manufactured by sawing boards diagonally to produce two wedge-shaped pieces.

Bid: An offer to supply, at a specified price, materials, supplies, equipment, an entire structure, or sections of a structure.
Bill of materials: A list of all materials, corresponding to a drawing or set of drawings for a project.
Bird's mouth: A notch cut on the underside of a rafter to fit the top plate. Not a full notch if rafter ends flush with top plate.
Blemish: Any defect, scar, or mark that tends to detract from the appearance of wood.
Blind nailed: Nailing concealed by installation of another strip of wood; used in securing door frames to framing and tongue-and-grooved flooring.
Blind stop: A member applied to the exterior edge of the side and head jamb of a window to serve as a stop for the top sash and to form a rabbet for storm sash, screens, blinds, and shutters.
Block flooring: Wood flooring cut in square blocks.
Blocking: Solid bridging.
Blue stain: A stain caused by a fungus growth in unseasoned lumber—especially pine. It does not affect the strength of the wood.
Board: Lumber less than 2" thick.
Board foot: The equivalent of a board 1' square and 1" thick.
Boiler: A heating unit or heat source for a hydronic heating system.
Box beam: A beam made of one or more plywood webs connected to lumber flanges.
Boxed cornice: A construction that encloses the rafters at the eaves.
Boxes: Enclosures in an electrical system that house devices and electrical connections.
Bracket: A projecting support for a shelf or other structure.
Branch circuit: A single circuit with a fuse or circuit breaker, running from a distribution panel.
Brick construction: A type of construction in which the exterior walls are bearing walls made of brick.
Brick molding: A molding for window and exterior door frames. Serves as the boundary molding for brick or other siding material and forms a rabbet for the screens, storm sash, or combination door.
Brick veneer construction: A type of construction in which a wood-frame construction has an exterior surface of single brick.
Bridging: Pieces fitted in pairs from the bottom of one floor joist to the top of adjacent joists used to distribute the floor load. Sometimes pieces of solid stock are used.
Brown coat: A second layer of plaster that is applied on top of the scratch coat to form a base for the finish coat.
Building code: A collection of rules and regulations for construction established by organizations and based on experience and experiment.
Built-in: A piece of furniture (normally cabinets, cupboards, and other large pieces) that is attached to the building frame and, therefore, can not be moved practically.
Built-up roof: A roofing composed of several layers of rag felt or jute saturated with coal tar, pitch, or asphalt. The top is finished with crushed slag or gravel. Generally used on flat or low-pitched roofs.
Built-up stringer: A stringer to which blocking has been added to form a base for adding treads and risers.
Butt: Type of door hinge. One leaf is fitted into space routed into the door frame jamb and the other into the edge of the door.

C

Cabinet: Case or box-like assembly consisting of shelves, doors, and drawers, used primarily for storage.
Cabinet drawer guide: A wood strip used to guide the drawer as it slides in and out of its opening.
Cabinet drawer kicker: Wood cabinet member placed immediately above and generally at the center of a drawer to prevent tilting down when drawer is pulled out.
Cable layout: A print to which a plumber or architect has added drawings and symbols to show conductor sizes, number of conductors in a run, and path of conductors as a guide to the electrician or electricians making the installation.
Camber: A slight arch in a beam or other horizontal member which prevents it from bending into a downward or concave shape due to its weight or load.
Cantilevered: Extending horizontally beyond a supporting surface.
Cant strip: A triangular shaped strip of wood used under shingles at gable ends or under the edges of roofing on flat decks.
Cap flashing: See *counter flashing*.
Carriage: See *stringer*.
Cased opening: An interior opening without a door that is finished with jambs or trim.
Casein glue: An adhesive of casein and hydrated lime suitable for gluing oily woods and for laminating wood which has a high moisture content.
Casement: A window in which the sash swings on its vertical edge, so it may be swung in or out.
Casing: The trimming around a door or window, either outside or inside, or the finished lumber around a post or beam.
Caulk: To seal and waterproof cracks and joints, especially around window and exterior door frames. Also *calk*.
Cement: Binding material that, when combined with water and aggregate, forms concrete.
Cement board: Fireproof, moisture-resistant fiber-reinforced cement panels used as a base for finishing materials on walls, floors, and countertops.
Center guides: A strip of wood centered between front and rear rails of a cabinet frame to keep a drawer aligned.
Chair: Small metal fixture used to hold reinforcing bars away from the ground prior to the casting of concrete.
Chair rail: An interior molding applied along the wall of a room to prevent the chair from marring the wall.
Chamfer: Corner of a board beveled at a 45° angle. Two boards butt-jointed with chamfered edges form a V-joint.
Chase: A wood frame jutting from an outside wall which supports a prefabricated chimney. A prefabricated fireplace is often enclosed.
Check rails: Meeting rails of a double-hung window which are made thicker to fill the opening between the top and bottom sash. They are usually beveled. Also called a *meeting rail.*
Chevron paneling: Strip paneling installed diagonally in a chevron pattern.
Chimney: A vertical hollow shaft that exhausts smoke and gases from heating units and incinerators.
Circuit: A path for electrical power provided by wires (conductors).
Circuit breaker: Device in a circuit that opens the circuit, preventing the flow of electricity, so that it will not be damaged by an overload or short.
Cleat: A strip of wood fastened across a door to add strength. Also a strip fastened to a wall to support a shelf, fixture, or other objects.
Closed panels: Factory-built housing panels that are finished on both sides.
Closet pole: A round wooden shaft installed in clothes closets to accommodate clothes hangers.
Coil: A core wrapped in conducting wires that, when current is run through, emits a magnetic field.
Collar beam: A tie beam connecting rafters considerably above the wall plate. Also called *rafter tie*.
Column: Vertical supporting member.
Commercial standard: A voluntary standard that establishes quality, methods of testing, certification, rating, and labeling of manufactured items. It provides a uniform base for fair competition.

Common rafter: A rafter connected to both the ridge and the wall plate.
Complimentary colors: Hues opposite each other on the color wheel.
Composite board: Boards consisting of a core of wood between veneered surfaces.
Concrete: Building material formed by combining cement, water, and aggregate.
Condenser: Part of an air conditioning unit that receives heated coolant and releases its heat to the atmosphere.
Conduction: Travel of heat in solid matter.
Conductor: A material, such as a wire, that carries an electric current.
Conduit, electrical: A pipe or tube in which wiring is installed.
Connectors: Fasteners that connect two or more conductors so that current may pass from one to the other.
Contact cement: Neoprene rubber-based adhesive that bonds instantly upon contact of parts being fastened.
Convection: Movement of heat through liquids and gases as a result of heated liquids and gases rising and being replaced by cooler liquids and gases.
Convenience outlet: Electrical outlet into which may be plugged portable equipment such as lamps.
Cooling coil: System of coils or tubes in an air conditioning unit that receives the high-pressure refrigerant and allows it to expand, causing the refrigerant to absorb heat.
Cope: To cut or shape the end of a molded wood member so it will cover and fit the contour of an adjoining piece of molding.
Corbel: To extend outward from the surface of a masonry wall one or more courses to form a supporting ledge.
Corner bead: Molding used to protect corners. Also a metal reinforcement placed on corners before plastering.
Corner braces: Diagonal braces let into studs to reinforce corners of frame structures.
Cornice: Exterior trim of a structure at the meeting of the roof and wall; usually consists of panels, boards, and moldings.
Counter flashing: Flashing used on chimneys at the roof-line to cover base flashing and prevent moisture entry. Also called cap flashing.
Cove molding: Molding with a concave profile used primarily where two members meet at a right angle.
Cricket: See *saddle*.
Cripple jack (also, cripple rafter): A rafter that intersects neither the wall plate nor the ridge and is terminated at each end by hip and valley rafters.
Cripple stud: A stud used above or below a wall opening. Extends from the header to the top plate or from sole plate to rough sill.
Cripple wall: See *underpinning*.
Current: Flow of electrons through a conductor.
Curtain wall: A wall, usually nonbearing, between piers or columns.
Cut-in point: The temperature setting on a thermostat at which the thermostat will signal the furnace to turn on.
Cut-out point: Temperature setting on a thermostat at which the thermostat will signal the furnace to turn off.
Cut-out stringer: A stair stringer into which the rise and run are cut.

D

Dado: A rectangular groove cut in wood across the grain.
Damper: A venting device in fireplaces used to control combustion, prevent heat loss, and redirect downdrafts.
Dead bolt: Special door security consisting of a hardened steel bolt and a lock. Lock is operated by a key on the outside and by either a key or handle on the inside.
Dead load: The weight of permanent, stationary construction and equipment included in a building.
Decay: Disintegration of wood substance due to action of wood-destroying fungi.
Degree day: Method of measuring the harshness of climate for insulation and heating purposes. A degree day is the product of one day and the number of degrees the mean temperature is below 65°F.
Dew point: Temperature at which air is sufficiently cooled for water vapor to condense out of it.
Diagonal paneling: Strip paneling installed at a 45° angle.
Differential: The difference between cut-in point and the cut-out point of a thermostat.
Dimensional stability: The ability of a material to resist changes in its dimensions due to temperature, moisture, and physical stress.
Dimension lumber: Lumber 2" to 5" thick, and up to 12" wide.
Direct current: An electric current flowing in one direction, such as that provided by a battery.
Direct gain system: Passive solar construction in which the sun shines directly into living space to heat it.
Direct heating system: A heating system that gives off heat directly to objects without benefit of ducts or air system.
Door frame: An assembly of wood parts that form an enclosure and support for a door. Door frames are classified as exterior and interior.
Door stop: A molding nailed to the faces of the door frame jambs to prevent the door from swinging through.
Dormer: A projecting structure built out from a sloping roof. Usually includes one or more windows.
Downdraft: A flow of air down a chimney.
Drawer guides: Wood strips or metal devices that support drawers at the sides or lower corners.
Drip cap: A molding that directs water away from a structure to prevent seepage under the exterior facing material. Applied mainly over window and exterior door frames.
Drip groove: Semicircular groove on the underside of a drip cap or the lip of a window sill that prevents water from running back under the member.
Drop siding: Siding, usually 3/4" thick and machined into various patterns. Drop siding has tongue and groove or shiplap joints.
Dry rot: A term loosely applied to many types of decay but especially to that which, when in an advanced stage, permits the wood to be easily crushed to a dry powder.
Drywall: Sheet material consisting of a uniform layer of gypsum sandwiched between facings of paper; used as an interior wall covering that can be painted or covered with decorative paper or paneling.
DWV: Stands for "drainage, waste, and venting"; the nonpressured part of a plumbing system which carries away waste water and solid waste.

E

Eased edge: Corner rounded or shaped to a slight radius.
Eaves: The lower part of a roof that projects over an exterior wall. Also called the overhang.
Electrical wiring: The wires, boxes, and a number of devices that control the distribution and use of electrical current in a building.
Electric moisture meter: Meter used to determine the moisture content of wood. Action is based on electrical resistance or capacitance which varies with change in moisture content.
Electric radiant heat: A system that radiates heat, without benefit of air, by moving through solid matter.
Elevation: The height of an object above grade level. Also means a type of drawing which shows the front, rear, and sides of a building.
Emissivity: The ability of a material to emit heat by radiation.
Entrepreneur: One who starts and operates a new business.
Equilibrium moisture content (EMC): The moisture content at which wood neither gains nor loses moisture when surrounded by air at a given relative humidity and temperature.

Expansion joint: A bituminous fiber strip used to separate blocks or units of concrete to prevent cracking due to thermal expansion.
Expansion tank: A reservoir in a hydronic heating system that accommodates increased water volume caused by expansion due to heating.

Facade: Main or front elevation of a building.
Face frame: See *face plate.*
Face frame construction: A framed cabinet where frame is covered by light-weight panels.
Face nail: A nail driven perpendicular to the surface of a piece.
Face plate: An assembly of solid lumber that is attached to the front edges of cabinets to give a finished appearance and provide support for doors. Also called a *face frame.*
Factory and shop lumber: Lumber intended to be cut up for use in further manufacture. It is graded on the basis of the percentage of the area which will produce a limited number of cuttings of a specified size and quality.
Fascia: A wood member nailed to the ends of the rafters and lookouts used for the outer face of a box cornice.
Faucets: Devices that deliver water, usually to a fixture, hose, appliance, or bucket.
Fiberboard: A broad term used to describe sheet material of widely varying densities; manufactured from wood, cane, or other vegetable fibers.
Fiber saturation point: The stage in the drying or wetting of wood at which the cell walls are saturated and the cell cavities are free from water. It is assumed to be 30% moisture content, based on oven-dry weight, and is the point below which shrinkage occurs.
Field: The middle area of a sheet of wallboard.
Finish flooring: The final floor covering.
Fire stop: A block or stop used in a building wall between studs to prevent the spread of fire and smoke through air space.
Fire wall: A wall which subdivides a building to restrict the spread of fire.
Fixtures: Devices that receive water from the water supply system and have a means for discharge into the DWV system.
Flashing: Sheet metal or other material used in roof and wall construction (especially around chimneys and vents) to prevent rain or other water from entering.
Flat roof: A roof that is either level or pitched only enough to provide for drainage.
Flight of stairs: Steps going in the same direction between landings or floors.
Flitch: A steel plate placed between two pieces of 2" lumber to form a reinforced header.
Floating: Smoothing the surface of wet concrete using a large flat tool (float).
Flue: The passage in a chimney through which smoke, gas, and fumes rise.
Flush: Adjacent surfaces even, or in same plane (with reference to two structural pieces).
Flush door: In cabinetry, a door that is perfectly flat; rails, if any, are covered.
Flush drawer: A drawer whose front fits flush in its opening.
Footing: The spreading course at the base of a foundation wall, pier, or column.
Foundation: The supporting portion of a structure located below the structure and supported only by soil or rock.
Frame construction: Cabinets where solid lumber provides a frame that is not covered with panels.
Frameless construction: Cabinetry where heavier panels provide support usually given by a framework of narrow pieces of solid wood.
Framing: The timber structure of a building which gives it shape and strength; including interior and exterior walls, floor, roof, and ceilings.
Frieze: A boxed cornice wood trim member attached to the structure where the soffit (plancier) and wall meet.
Furring: Narrow strips of wood spaced to form a nailing base for another surface. Furring is used to level, to form an air space between the two surfaces, and to give a thicker appearance to the base surface.
Fuses: Protective devices that shut off electrical power when an overload occurs.

Gable: That portion of a wall contained between the slopes of a double-sloped roof or that portion contained between the slope of a single-sloped roof and a line projected horizontally through the lowest elevation of the roof construction.
Gable roof: A roof consisting of a single ridge, made entirely of common rafters.
Gain: Recess or mortise cut to receive the end of another structural member, hinge, or other hardware.
Gambrel roof: A roof slope formed as if the top of a gable (triangular) roof were cut off and replaced with a less steeply sloped cap. This cap still has a peaked ridge in the center.
Girder: A principal beam used to support other beams.
Glazing: The process of installing glass into sash and doors. Also refers to glass panes inserted in various types of frames.
Glazing compound: A plastic substance of such consistency that it tends to remain soft and rubbery when used in glazing sash and doors.
Glue block: A wood block, triangular or rectangular in shape, which is glued into place to reinforce a right angle butt joint. Sometimes used at the intersection of the tread and riser in a stair.
Grade beam: Thickened and reinforced section of a slab foundation designed to rest on supporting piling.
Greenhouse effect: Heating effect of solar energy which can pass through glass easily, creating heat which cannot easily pass back through the glass.
Ground fault interrupter: An electrical safety device which can be installed either in an electrical circuit or at an outlet. It is able to detect a short circuit and shut off power automatically. Used as a protection against electrical shock.
Grounding: A system used for electrical safety. An electrical wire runs from the exposed metal of a power tool to a third prong on the power plug. When used with a grounded receptacle, this wire directs harmful currents away from the operator.
Grounds: Strips of wood installed as guides at the floor line and at openings in a wall to strike off plaster.
Grout: A thin mortar used in masonry work.
Gusset: A panel or bracket of either wood or metal attached to the corners or intersections of a frame to add strength and stiffness.
Gutter: Wood, metal, or plastic trough attached to the edge of a roof to collect and conduct water from rain or melting snow.
Gypsum wallboard: Wall covering panels consisting of a gypsum core with facing and backing of paper.

H

Half story: That part of a building situated wholly or partially within the roof frame, finished for occupancy.
Handrail: A pole above and parallel to stair steps to act as a support for persons using the stairs.
Hanger: Connecting hardware used to attach beam ends to other members.
Hardboard: A board material manufactured of wood fiber, formed into a panel having a density of approximately 50 to 80 lb. per cu. ft.

Header: Horizontal structural member that supports the load over an opening, such as a window or door. Also called a lintel.
Headroom: The clear space between floor line and ceiling, as in a stairway.
Hearth: That part of a fireplace that holds the fuel and contains the fire.
Heartwood: The wood extending from the pith or center of the tree to the sapwood, the cells of which no longer participate in the life processes of the tree.
Heat exchanger: A series of tubes in a furnace that absorb the heat of combustion and transfer it to the plenum.
Heat pump: An energy unit that can provide either heat or cooling.
Heat transmission coefficient: Hourly rate of heat transfer for one square foot of surface when there is a temperature difference of one degree Fahrenheit between the air on the two sides of the surface.
Herringbone paneling: Paneling installed on a wall in alternating diagonal pattern.
Hip jack rafter: A short rafter connecting to the wall plate and a hip rafter.
Hip roof: A roof which rises from all four sides of a building.
Hollow-back: Removal of a portion of the wood on the unexposed face of a wood member to more properly fit any irregularity in bearing surface.
Hollow core door: Flush door with a core assembly of strips or other units which support the outer faces.
Horn: The extension of a stile, jamb, or sill.
Hose bib: A water faucet that is threaded so a hose connection can be attached.
Housed stringer: A stair stringer where the edges of the steps are covered with a board.
House wrap: Plastic sheets used to seal exterior walls against air infiltration.
Hub: Bell-shaped end of a cast-iron soil pipe.
Hydration: Heat-producing chemical reaction occurring between cement and water to form concrete.
Hydronic heating system: A heating system that uses water to transfer heat from a heat source to space to be heated.

I

I beam: A beam made of a single, thin, vertical web connecting two horizontal flanges.
Incinerator: A device that consumes household waste by burning.
Indirect gain system: Passive solar construction in which solar heat is stored in structures of masonry, water, or other medium and then passed along to living space by radiation, conduction, or convection.
Induction: A process by which an electric current is produced by a magnetic field.
Insulation: (thermal) Any material high in resistance to heat transmission that is placed in structures to reduce the rate of heat flow; also, nonconducting covering on electrical conductors.
Interior trim: General term for all the molding, casing, baseboard, and other trim items applied within the building by finish carpenters.
Intermediate colors: A color that is a combination of a primary color and secondary color containing that primary color. Red-violet, blue-violet, blue-green, yellow-orange, red-orange.
Isolated gain system: Passive solar construction in which generated heat is stored in a separate sun space. It is transported to living space by mechanical means.

J

Jack rafter: A short rafter framing between the wall plate and a hip rafter; or a hip or valley rafter and ridge board.
Jalousie: A series of small horizontal overlapping glass slats, held together by an end metal frame attached to the faces of window frame side jambs or door stiles and rails. The slats or louvers rotate simultaneously like a Venetian blind.
Jamb: The top and two sides of a door or window frame which contact the door or sash; top jamb and side jambs.
Jig: A device used to position material for accurate cutting or assembly.
Joinery: A term used by woodworkers when referring to the various types of joints used in a structure.
Joist: One of a series of parallel framing members used to support floor and ceiling loads, and supported in turn by larger beams, girders, or bearing walls.

K

Kerfing: Longitudinal saw cuts or grooves of varying depths (dependent on the thickness of the wood member) made on the unexposed faces of millwork members to relieve stress and prevent warping; members are also kerfed to facilitate bending.
Kicker: A strip of wood centered between the rails above a drawer to keep it level as it is opened.
Kiln-dried: Wood seasoned in a kiln by means of artificial heat, controlled humidity, and air circulation.
Knocked down: Unassembled; refers to structural units requiring assembly after being delivered to the job.
Knot: Branch or limb embedded in the tree and cut through during lumber manufacture.
Kraft paper: A brown building paper which resists puncturing. Kraft paper is used to face some blanket insulation materials.

L

Lacquer: Clear finish used to protect and enhance the appearance of wood.
Lally column: A cylindrical shaped steel member used to support beams and girders. Sometimes filled with concrete.
Landing: A platform between flights of stairs.
Lath: A building material of wood, metal, gypsum, or insulating board, fastened to frame of building to act as a plaster base.
Lazy Susan: A circular revolving cabinet shelf used in corner kitchen cabinet unit.
Leader: A vertical pipe that carries rainwater from the gutter to the ground or a drain. Also called downspout.
Ledger: A strip attached to vertical framing or structural members to support joists or other horizontal framing. Similar to a ribbon strip.
Let in: Refers to any kind of notch in a stud, joist, block, or other piece which holds another piece. Somewhat like log cabin construction. The item which is supported in the notch is said to be "let in."
Level-transit: A surveying instrument used to check the plumb of walls in new structures. The telescope tube can swing in a vertical arc for comparing forward and backward readings.
Light construction: Construction generally restricted to conventional wood stud walls, floor and ceiling joists, and rafters. Primarily residential in nature although it does include small commercial buildings.
Lignin: Substance in wood that binds cell walls together.
Lineal foot: Having length only, pertaining to a distance of one foot long as distinguished from a square foot or cubic foot. Also called linear foot.
Lintel: A horizontal structural member which supports the load over an opening such as a door or window.
Lip drawer: Drawer front that has lip covering the joint between the drawer and a cabinet's face plate.
Lipped door: A cabinet door partially recessed in its opening but with a lip concealing the opening.
Live load: The total of all moving and variable loads that may be placed upon a building.

Lock block: A block of wood which is joined to the inside edge of the stile of a hollow core door and to which the lock is fitted. Flush doors have a lock block on each stile.
Lookout: Structural member running between the lower end of a rafter and the outside wall. Used to carry the underside of the overhang; plancier or soffit.
Lug: Part of a doorjamb extending upward beyond the head jamb.
Lugged tile: Tile with projections at all edges to maintain proper spacing between individual tiles.
Lumber: Wood that has been sawed to a workable size and planed. Some matching of ends and edges may be included.

M

Major module: A unit of measure for modular construction. In the conventional system of units, 48" is the length of a major module. In the metric system, a major module is 1200 mm long.
Main: In a one-pipe hydronic heating system, the section of the pipe that moves hot water from the boiler to the rooms being heated.
Mansard roof: A type of curb roof in which the pitch of the upper portion of a sloping side is slight and that of the lower portion is steep. The lower portion is usually interrupted by dormer windows.
Manual dexterity: The ability to accomplish tasks requiring use of the hands.
Marquetry: In carpentry, creating patterns in flooring materials by the use of different colors of wood.
Masonry: Stone, brick, hollow tile, concrete block, tile, or other similar materials bonded together with mortar to form a wall, pier, buttress, etc.
Matched lumber: Lumber that is edge dressed and shaped to make a close tongue-and-groove joint at the edges or ends. Also generally includes lumber with rabbeted edges.
Mechanical cores: Prefabricated building modules that contain one or more of the following utilities: electrical, plumbing, heating, ventilating, and air conditioning. Floor, ceiling, and wall framing are fully formed at the factory. Modules are joined at the building site.
Mechanical equipment: In architectural and engineering practice, all equipment included under the general heading of plumbing, heating, air conditioning, gas fitting, and electrical work.
Mechanical systems: Installations in a building including plumbing, electrical wiring, heating, ventilating, and air conditioning.
Medallion: A raised decorative piece, sometimes used on flush doors.
Meeting rail: The bottom rail of the upper sash, and the top rail of the lower sash of a double-hung window. Also called a *check rail.*
Millwork: The term used to describe products which are primarily manufactured from lumber in a planing mill or woodworking plant; including moldings, door frames and entrances, blinds and shutters, sash and window units, doors, stairwork, kitchen cabinets, mantels, cabinets, and porch work.
Minor module: A unit of measure for modular construction. In the conventional system of units, 24" is the length of a minor module. In the metric system, a minor module is 600 mm long.
Modular homes: Homes having two or more three-dimensional, factory-built units that are assembled on the building site.
Modules (Mods): Three-dimensional assembled housing units built in a factory and transported to a building site to be assembled with other modules.
Moisture content: The amount of water contained in wood expressed as a percentage of the weight of oven-dry wood.
Molder: A woodworking machine designed to run moldings and other wood members with regular or irregular profiles. Also called a *sticker.*
Molding: A relatively narrow strip of wood, usually shaped to a curved profile throughout its length, used to accent and emphasize the ornamentation of a structure and to conceal surface or angle joints.
Monolithic: Term used for concrete construction poured and cast in one unit, without joints.
Mortar: A combination of cement, sand, lime, and water used to bind masonry blocks.
Mortise: Recessed cavity in a piece of wood used to receive hardware or another piece of wood.
Mosaic tile: Small pieces of ceramic tile of different colors laid so as to form a pattern or picture.
Movable insulation: Usually sheet or blanket insulation that can be placed over windows temporarily to insulate against solar energy.
MR (moisture resistant) wallboard: A type of gypsum wallboard processed to resist the effects of moisture and high humidity. It is used as a base under ceramic tile and other nonabsorbent finishes used in showers and tub alcoves.
Mullion: A slender bar or pier forming a division between units of windows, screens, or similar generally nonstructural frames.
Muntin: Vertical member between two panels of the same piece of panel work. The vertical and horizontal sashbars separating the different panes of glass in a window.

N

National Electrical Code: Collection of rules developed to control and recommend methods of installing electrical systems.
Net floor area: The gross floor area, less the area of the partitions, columns, stairs, and other floor openings.
Newel: The main post at the start of a stair and the stiffening post at the landing.
Nominal size: As applied to timber or lumber, the ordinary commercial size by which lumber is known and sold, normally slightly larger than the actual size.
Nonbearing partition: A partition extending from floor to ceiling that supports no load other than its own weight.
Nosing: The part of a stair tread which projects beyond the riser, or any similar projection; a term applied to the rounded edge of a board.

Oakum: Asphalt-saturated ropelike material used as a packing in cast-iron soil pipe joints prior to pouring molten lead.
On center (O.C.): A method of indicating the spacing of framing members by stating the measurement from the center of one member to the center of the succeeding one.
Open grain wood: Woods with large pores, such as oak, ash, chestnut, and walnut.
Operating engineers: Those who operate heavy equipment such as cranes, bulldozers, and backhoes.
Oriel window: A window that projects from the main line of an enclosing wall of a building and is carried on brackets, corbels, or a cantilever.
Oriented strand board: A formed panel consisting of layers of compressed strand-like particles arranged at right angles to each other.
Overlay door: A slab door in cabinetry that covers the cabinet frame.

P

Panel door: Style of door having a frame and separate panels of plywood, hardboard, or solid steel set into the frame.
Panelized homes: Factory-built homes that have prebuilt parts such as wall sections and roof sections.
Parapet: A low wall or railing along the edge of a roof, balcony, or bridge. The part of a wall that extends above the roof line.
Parquet flooring: Squares of flooring made up of wood pieces.
Particleboard: A formed panel consisting of particles of wood flakes, shavings, slivers, etc., bonded together with a synthetic resin or other added binder.

Partition: A wall that subdivides space within any story of a building.
Passive solar construction: Designing a building to use solar energy as heat with no mechanical means to transport the resulting heat.
Paver tile: Concrete-based masonry units used as finish flooring and for walkways.
Penny: Term used to indicate nail length; abbreviated by the letter "d". Applies to common, box, casing, and finishing nails.
Photosynthesis: The chemical process plants use to store the sun's energy. Trees use sunlight to convert carbon dioxide and water into leaves and wood.
Pier: A column of masonry, usually rectangular in horizontal cross section, used to support other structural members.
Pilaster: A part of a wall that projects not more than one-half of its own width beyond the outside or inside face of a wall. Chief purpose is to add strength but may also be decorative.
Pile: A heavy timber, or pillar of metal or concrete, forced into the earth or cast in place to form a foundation member.
Pitch: Inclination or slope, as of roofs or stairs. Rise divided by the span.
Plan: A drawing representing any one of the floors or horizontal cross sections of a building, or the horizontal plane of any other object or area.
Plancier: The underside of an eave or cornice, usually horizontal.
Plaster: Mixture of gypsum and water that can be troweled wet onto interior walls and ceilings.
Plat: A map, plan, or chart of a city, town, section, or subdivision indicating the location and boundaries of individual properties.
Plate: A horizontal structural member placed on a wall or supported on posts, studs, or corbels to carry the trusses of a roof or to carry the rafters directly. Also a sole or base member of a partition or other frame.
Platform: A horizontal section between two flights of stairs.
Platform framing: A system of framing a building where the floor joists of each story rest on the top plates of the story below (or on the foundation wall for the first story) and the bearing walls and partitions rest on the subfloor of each story.
Plenum: A sheet metal chamber in a furnace that collects heat in preparation for its transfer to rooms.
Plinth block: Decorative, carved block set at left and right top corners of window and door trim; it eliminates need to make mitered corners.
Plot plan: A drawing of the view from above a building site. The plan shows distances from a structure to property lines. Sometimes called a *site plan.*
Plumb: Exactly perpendicular or vertical; at right angles to the horizon or floor.
Plumbing stack: A general term for the vertical main of a system of soil, waste, or vent piping.
Plumbing: The work or business of installing pipes, fixtures, and other apparatus for bringing in the water supply and removing liquid and water-borne wastes. This term is used also to denote the installed fixtures and piping of a building.
Plunge cutting: A cutting method used to make a starting hole for a saber saw. The saw is held with the blade teeth almost flush with the wood surface. A cut is made completely through as the saw is tilted to a normal position.
Polystyrene panels: Rigid insulation manufactured from expanded beads of plastic.
Polyvinyl resin emulsion glue: Wood adhesive intended for interiors. Made from polyvinyl acetates which are thermoplastic and not suited for temperatures over 165°F (74°C). Also called *white glue.*
Portico: A porch or covered walk consisting of a roof supported by columns. A porch with a continuous row of columns.
Post-and-beam framing: Framing method in which loads are carried by a frame comprised of posts connected with beams, eliminating the need for load-bearing walls.
Prefabricated construction: Type of construction with a minimum of assembly at the site, usually comprising a series of large units manufactured in a plant.
Preservative: Substance that will prevent the development of wood-destroying fungi, borers of various kinds, and other harmful insects that deteriorate wood.
Purlins: Horizontal roof members used to support rafters between the plate and ridge board.
Push stick: A pole or strip used to push a workpiece when cutting with power saws, jointers, and other power tools. Pushing a board by hand is usually unsafe with power equipment.

Quarry tile: Ceramic tile used in construction as finish flooring and counter tops.
Quarter round: Molding with a cross section of one-fourth of a circle.
Quarter-sawed: Lumber cut at about a 90° angle to the annular growth rings.

R

R value: A number related to the efficiency of an insulating material.
Rabbet: A rectangular shape consisting of two surfaces cut along the edge or end of a board.
Rafter: One of a series of structural members of a roof designed to support roof loads. The rafters of a flat roof are sometimes called roof joists.
Rail: Cross or horizontal members of the framework of a sash, door, blind, or other assembly.
Rake: The trim members that run parallel to the roof slope and form the finish between the roof and wall at a gable end.
Ramp: Inclined plane connecting separate levels.
Receptacle: See *convenience outlet.*
Registers: Grillwork installed over the ends of heating or air conditioning ducts for purpose of directing the heated air.
Reinforced concrete construction: A type of construction in which the principal structural members, such as floors, columns, and beams, are made of concrete poured around steel bars or steel meshwork in such a manner that the two materials act together to resist force.
Relative humidity: Ratio of amount of water vapor in air in terms of percentage to total amount it could hold at the same temperature.
Relief valve: Valve that opens an outlet when a set pressure is reached; used to protect pipes and equipment from being damaged by overpressure.
Resilient: The ability of a material to withstand temporary deformation and assume its original shape when the stresses are removed.
Retaining wall: Any wall subjected to lateral pressure other than wind pressures; for example, a wall built to support a bank of earth.
Return main: The part of a double-pipe hydronic heating system that routes cooled water back to the boiler.
Ribbon: A narrow board attached to studding or other vertical members of a frame that adds support to joists or other horizontal members.
Ridge: The horizontal line at the junction of the top edges of two roof surfaces.
Riser: The vertical stair member between two consecutive stair treads. In plumbing, a vertical pipe.
Roll roofing: Mineral granules on asphalt saturated felt or fiberglass. Roll roofing is the uncut form of mineral surfaced shingle material.
Roofing: The materials applied to the structural parts of a roof to make it waterproof.

Rotary cut veneer: Veneer cut on a lathe which rotates the log against a broad cutting knife. The veneer is cut in a continuous sheet much the same as paper is unwound from a roll.
Roughing-in: The work of installing all pipes in the drainage system and all water pipes to the point where connections are made with the plumbing fixtures. Also applies to partially completed electrical wiring and other mechanical aspects of the structure.
Rough lumber: Lumber that has been cut to rough size with saws but which has not been dressed or surfaced.
Rough opening: An opening formed by framing members.

S

Saddle: A small gable roof placed in back of a chimney on a sloping roof to shed water and debris.
Sapwood: The layers of wood next to the bark, usually lighter in color than the heartwood, that are actively involved in the life processes of the tree. More susceptible to decay than heartwood, sapwood is not necessarily weaker or stronger than heartwood of the same species.
Sash: The framework that holds the glass in a window.
Scaffold: A temporary structure or platform used to support workers and materials during building construction. Also called *staging*.
Scale: A term that specifies the size of a reduced size drawing. For example, a plan is drawn to 1/4" scale if every 1/4" represents 1' on the real structure.
Scantling: Lumber with a cross section ranging from 2x4 to 4x4.
Scarfing: Joining the ends of stock together with a sloping lap-joint so they appear to be a single piece.
Scotia: A concave molding consisting of an irregular curve. Used under the nosing of stair treads and for cornice trim.
Scratch coat: First layer of plaster that has its surface roughened to provide tooth for succeeding layers.
Screed: A tool used in concrete work to level and smooth a horizontal surface. Consists of a 3' to 5' wood or metal strip attached to a pole. Also the process of leveling off concrete slabs or plastering on interior walls.
Scuttle: An opening in a ceiling which provides access to the attic.
Seasoning: Removing moisture from green wood to improve its serviceability.
Secondary colors: Colors produced when primary colors are combined: violet, green, and orange.
Second growth: Timber that has grown after the removal of a large portion of the previous stand.
Section drawing: A type of drawing which shows how a part of a structure looks as if cut by a plane.
Selvage: The part of the width of roll roofing which is smooth. For example, a 36" width has a granular surfaced area 17" wide and a 19" wide selvage area.
Semihoused stringer: A stringer that is partially open.
Septic tank: A sewage settling tank intended to retain the sludge in immediate contact with the sewage flowing through the tank for a sufficient period to secure satisfactory decomposition of organic sludge solids by bacterial action.
Setting block: A wood block placed in the glass groove or rabbet of the bottom rail of an insulating glass sash to form a base or bed for the glass.
Setup time: Time it takes for an adhesive or concrete to stiffen.
Shakes: Handsplit shingles.
Sheathing: Boards or prefabricated panels that are attached to the exterior studding or rafters of a structure.
Sheathing paper: A building material used in wall, floor, and roof construction to resist the passage of air.
Shim: A thin strip of wood, sometimes wedge-shaped, for plumbing or leveling wood members. Especially helpful when setting door and window frames.
Shiplap: Lumber with edges that have been rabbeted to form a lap joint between adjacent pieces.
Shoring: Bracing used to provide temporary support.
Shutter: A wood assembly of stiles and rails to form a frame that encloses panels used in conjunction with door and window frames. Also may consist of vertical boards cleated together.
Siding: The finish covering of the outside wall of a frame building. Many different types are available.
Sill: The lowest member of the frame of a structure, usually horizontal, resting on the foundation and supporting the uprights of the frame. Also the lowest member of a window or outside door frame.
Skylight: Glazing framed into a roof.
Sleeper: A timber laid on or near the ground to support floor joists and other structures above. Also wood strips laid over or embedded in a concrete floor to which finish flooring is attached.
Smoke chamber: Fireplace chamber located between the smoke shelf and the entrance to the flue.
Smoke shelf: The horizontal shelf located adjacent to the damper in a fireplace.
Soffit: The underside of the members of a building, such as staircases, cornices, beams, and arches. Relatively minor in size as compared with ceilings. Also called drop ceiling and furred-down ceiling.
Softwoods: The botanical group of trees that have needle or scalelike leaves and are evergreen for the most part, cypress, larch, and tamarack being exceptions. The term has no reference to the actual hardness of the wood. Softwoods are often referred to as conifers, and botanically they are called gymnosperms.
Soil stack: A general term for the vertical main of a system of soil, waste, or vent piping.
Solar collectors: Boxlike structures with clear glazing to trap solar energy.
Solar furnace: Another name for a thermosiphon.
Solar orientation: Placement with respect to exposure to sunlight of a structure on a building site.
Sole plate: The lowest horizontal strip on wall and partition framing. The sole plate for a partition is supported by a wood subfloor, concrete slab, or another closed surface.
Span: The distance between structural supports.
Specification: A written document stipulating the type, quality, and, sometimes, the quantity of materials and work required for a construction job.
Specific gravity: The ratio of the weight of a body to the weight of an equal volume of water.
Splash block: A small masonry block laid with the top near the ground to receive roof drainage and carry it away from the building.
Splay: The rear face of the hearth that slopes toward the front.
Spline: A thin strip of wood that fits into mortises or grooves machined into boards that are to be joined.
Square: Unit of measure—100 square feet—usually applied to roofing material and to some types of siding.
Staging: See *scaffold*.
Stairwell: The framed opening which receives the stairs.
Station mark: The point where a level-transit is located. It is a reference point such as a stake or paint mark directly below the center of the instrument.
Steel-frame construction: A type of construction in which the structural members are steel or are dependent on a steel frame for support.
Stepped footing: A footing that changes grade levels at intervals to accommodate a sloping site.
Stickers: Strips of wood used to separate the layers in a pile of lumber so air can circulate.

Stile: The upright or vertical outside pieces of a sash, door, blind, screen, or face frame.
Stool: A molded interior trim member serving as a sash or window frame sill cap. Stools may be beveled or rabbeted to receive the window frame sill.
Stoop: A small porch, veranda, platform, or stairway outside an entrance to a building.
Story: A single floor of a structure.
Story pole: A strip of wood used to lay out and transfer measurements for door and window openings, siding and shingle courses, and stairways. Also called a *rod.*
Straightedge: A straight strip of wood or metal used to lay out or check the accuracy of work.
Straight run: A stairway that does not change direction.
Stressed skin: Two facings, one glued to one side and the other to the opposite side of an inner structural framework to form a panel. Facings may be of plywood or other suitable material.
Strike plate: A metal piece mortised into or fastened to the face of a door frame side jamb to receive the latch or bolt when the door is closed.
Stringer: A sloping member that supports the risers and treads of stairs. Also called *carriage.*
Strongback: L-shaped wooden support attached to tops of ceiling joists to strengthen them, maintain spacing, and bring them to the same level.
Stub in: Termination point of plumbing pipes that are to be connected to fixtures.
Stud: One of a series of vertical structural members in walls and partitions. Plural—studs or studding.
Subfloor: Boards or panels laid directly on floor joists over which a finished floor is laid.
Supply main: The part of a double-pipe hydronic heating system that routes heated water to rooms.
Surfaced lumber: Lumber that is dressed or finished by running it through a planer.
Switches: Device that controls current to lights and appliances.
Systems-built housing: Housing built of components designed, fabricated, and assembled in a factory.

T

Tail beam: A relatively short beam or joist supported by a wall on one end and a header on the other.
Termite shield: A shield, usually made of sheet metal, placed in or on a foundation wall or around pipes to keep termites out of the structure.
Terrazzo flooring: A floor produced by embedding small chips of marble or colored stone in concrete and then grinding and polishing the surface.
Thermosiphon: A solar collector consisting, in its simplest form, of a flat box. It is glazed on one side. A baffle, parallel to the glazing, divides the box in half. As sun heats the upper half of the box, the heated air moves out one end. Cool air moves in from the lower half of the box. The box will be vented either into living space or into ductwork leading to storage.
Thermosiphoning: Movement of heat through a fluid by the action of heated material rising in a narrowed space.
Thermostat: An instrument that automatically controls the operation of heating or cooling devices by responding to changes in temperature.
Three-way switch: A switch designed to operate in conjunction with a similar switch to control one outlet or light from two points.
Threshold: A wood member, beveled or tapered on each side, used to close the space between the bottom of a door and the sill or floor underneath. Sometimes called a *saddle.*
Throat: In a fireplace, the narrowed passage above the hearth and below the damper.
Tie beam (collar beam): A beam so situated that it ties the principal rafters of a roof together and prevents them from thrusting the plate out of line.
Timbers: Lumber 5" or larger in least dimension.
Toeboard: A board fastened horizontally slightly above planking to keep tools and materials from falling on workers below. Board should be at least 4" wide.
Toenailing: To drive a nail at a slant with the initial surface in order to permit it to penetrate into a second member.
Toe space: A recessed space at the floor line of a base kitchen cabinet or other built-in unit. Permits one to stand close without striking the vertical space with the toes.
Total rise: Vertical distance from one floor to another.
Total run: In a roof, half the span (width) of the building. In stairs, horizontal distance occupied by the stairs; measured from the foot of the stairs to a point directly beneath where the stairs rest on a floor or landing above.
Transformer: A device for transforming the voltage characteristics of an electric current.
Transom: A small opening above a door separated by a horizontal member (transom bar). Usually contains a sash or a louver panel hinged to the transom bar.
Trap: A plumbing fitting designed to provide a liquid trap seal which will prevent the sewer gases from passing through and entering a building.
Tread: Horizontal walking surface on a stair.
Trim: The finish materials in a building, such as moldings applied around openings (window trim, door trim) or at the floor and ceiling of rooms (baseboard, cornice, picture molding).
Trimmer: The beam or floor joist into which a header is framed. Adds strength to the side of the opening.
Trimmer stud: A stud that supports the header for a wall opening. The stud extends from the sole plate to the bottom of the header. It is parallel to and in contact with a full-length stud, which extends from sole plate to top plate.
Triple wall: A type of chimney flue made with three metal pipes, each inside another. The concentric arrangement provides safety from fire while its light weight makes installation easy.
Tripod: A support for a builder's level consisting of three legs.
Trombe wall: Thick wall of masonry placed next to exterior glazing to store solar energy in passive solar construction. Named for French physicist Felix Trombe.
Truss: A structural unit consisting of such members as beams, bars, and ties; usually arranged to form triangles. Provides rigid support over wide spans with a minimum amount of material.

U

Undercut: Cutting a wood member so that the back of the member is slightly shorter than the front surface.
Underpinning: A short wall section between the foundation sill and first floor framing. Also called *cripple wall.*
Unicom system: A term taken from "uniform manufacture of components." The system uses modules with sizes that are multiples of a standard size.
Unit rise: The number of inches a common rafter will rise for every foot of run; in stair building, the riser height calculated by dividing the total rise by the total number of risers.
Unit run: In roof framing, one unit of horizontal distance, based on 12'.
Unprotected-metal construction: A type of construction in which the structural parts are metal without fireproofing.

Urea-formaldehyde resin glue: Moisture resistant glue that hardens through chemical action when water is added to the powdered resin.

Valley: The internal angle formed by the two slopes of a roof.
Valley rafter: A rafter that forms the intersection of an internal roof angle.
Valves: Devices that control flow of water in a water supply system.
Vapor barrier: A watertight material used to prevent the passage of moisture or water vapor.
Varnish: Clear penetrating coating applied to wood to enhance appearance and provide protection.
Veneered wall: A frame building wall with a masonry facing (example—single brick). A veneered wall is nonload-bearing.
Veneer plaster: Interior wall covering consisting of a gypsum lath base and a surface of 1/8" gypsum plaster.
Vent: A pipe installed to provide a flow of air to or from a drainage system or to provide a circulation of air within such system to protect trap seals from siphonage and back pressure.
Ventilation: The process of supplying and removing air by natural or mechanical means. Such air may or may not have been conditioned.
Vent stack: That part of a DWV system that is open to the atmosphere for the purpose of allowing air into the system to prevent siphoning of traps.
Vermiculite: Mineral closely related to mica, with the faculty of expanding on heating to form lightweight material with insulating qualities. Used as bulk insulation, also as aggregate in insulating and acoustical plaster, and in insulating concrete floors.

Waferboard: Construction panels made up of long, thin chips of wood. Wafers are coated with a waterproof resin and wax and then bonded with heat and pressure. Also called *waferwood.*
Wainscot: A lower interior wall surface (usually 3' to 4' above the floor) that contrasts with the wall surface above. May consist of solid wood or plywood.
Wale: A horizontal wood or metal strip used on the outside of forms for concrete. Wales are used to keep the form walls from bending outward under the weight of poured concrete.
Wallboard: Wood pulp, gypsum, or other materials made into large rigid sheets that may be fastened to the frame of a building to provide a surface finish.
Wall tie: Metal strip or wire used to bind tiers of masonry in cavity wall construction, or to bind brick veneer to a wood frame wall.
Warp: Any variation from a true or plane surface. Warp includes any combination of bow, crook, cup, and twist.
Water repellent: A solution, primarily paraffin wax and resin in mineral spirits, which, upon penetrating wood, retards changes in the wood's moisture content.
Water supply system: System of smaller pipes that distribute water under pressure to kitchens, bathrooms, and laundry areas.
Water table: A ledge or slight projection at the bottom of a structure which carries the water away from the foundation. Also, the level of groundwater underground.
Weathering: The mechanical or chemical disintegration and discoloration of the surface of wood. It can be caused by exposure to light, the action of dust and sand carried by winds, and the alternate shrinking and swelling of the surface fibers that comes with the continual variation in moisture content brought by changes in the weather. Weathering does not include decay.
Weatherstrip: Narrow strips of metal, vinyl plastic, or other material that retards the passage of air, water, moisture, or dust around doors or windows.
Western framing: See *balloon framing*.
Weephole: A small hole, as in a retaining wall, to drain water to the outside. Commonly used at the lower edges of masonry cavity walls.
Wet wall: An interior wall finish surface usually consisting of 3/8" gypsum plaster lath and 1/2" gypsum plaster.
Whaler: A horizontal member used in concrete form construction to stiffen and support the walls of the form. Also called *waler.*
Wind: A term used to describe the warp in a board when twisted (winding). It will rest upon two diagonally opposite corners, if laid upon a perfectly flat surface.
Winders: The curved stringers in winding stairs.
Winding: Curving, as in a stairway that changes direction gradually.
Window unit: Consists of a combination of the frame, window, weather-stripping, and sash activation device. May also include screens and/or storm sash. All parts are assembled as a complete operating unit.
Wing: A lateral extension of a building from the main portion; one of two or more coordinate portions of a building that extend from a common junction.
Wire glass: Glass having a layer of meshed wire incorporated in the center of the sheet.

Z

Zero clearance: A quality of a prefabricated fireplace that, through double-walling, allows the fireplace to be placed adjacent to combustible material such as wood framing.
Zones: Areas of a building that can be heated separately and controlled by individual thermostats.

Index

C

E

F

G

H

I

N

O

P

Q

R

S

U

V

W

Z